fourteenth edition

FINITE MATHEMATICS

for Business, Economics, Life Sciences, and Social Sciences

RAYMOND A. BARNETT Merritt College

MICHAEL R. ZIEGLER Marquette University

KARL E. BYLEEN Marquette University

CHRISTOPHER J. STOCKER Marquette University

Pearson

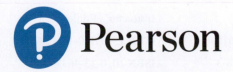

Director, Portfolio Management: Deirdre Lynch
Executive Editor: Jeff Weidenaar
Editorial Assistant: Jennifer Snyder
Content Producers: Sherry Berg, Ron Hampton
Managing Producer: Karen Wernholm
Senior Producer: Stephanie Green
Manager, Courseware QA: Mary Durnwald
Manager, Content Development: Kristina Evans
Product Marketing Manager: Emily Ockay
Field Marketing Manager: Evan St. Cyr
Marketing Assistants: Erin Rush, Shannon McCormack
Senior Author Support/Technology Specialist: Joe Vetere
Manager, Rights and Permissions: Gina Cheselka
Manufacturing Buyer: Carol Melville, LSC Communications
Art Director: Barbara Atkinson
Production Coordination, Composition, and Illustrations: Integra

Cover Image: Tai11/Shutterstock; Digital Storm/Shutterstock

Photo Credits: Page 1: Durk Talsma/Shuttertock; Page 41: Vlad61/Shutterstock; Page 126: Alan Sheldon/Shuttertock; Page 174: Arena Creative/Shutterstock; Page 257: Sspopov/123RF; Page 287: Kustov/Shutterstock; Page 351: PR Image Factory/Shutterstock; Page 394: S Duffett/Shutterstock; Page 458: Taaee/123RF; Page 500: Franck Boston/Shutterstock; G1: Zeljko Radojko/Shutterstock

Text Credits: Page 24: biggreenegg.com; Page 28: www.tradeshop.com; Page 29: www.tradeshop.com; Page 30: National Oceanic and Atmospheric Administration; Page 31: Jack Haggerty Forest, Lakehead University, Canada; Page 33: Energy Information Administration; Page 34: Bureau of TransportaLion Statistics; Page 34: Bureau of Transportation Statistics; Page 34: Centers for Disease Control; Page 34: Walmart Stores, Inc.; Page 35: Kagen Research; Page 35: Lakehead University; Page 35: T. Labuza, University of Minnesota; Page 36: Infoplease.com; Page 36: National Center for Education Statistics; Page 36: National Oceanic and Atmospheric Administration; Page 37: www.usda.gov/nass/pubslhistdata.htm; Page 40: Ashland Inc.; Page 40: U.S. Bureau of Labor Statistics; Page 499: American Lung Association; Page 501: Federal Aviation Administration; Page 501: U.S. Treasury; Page 502: Institute of Education Sciences; Page 502: UN Population Division; Page 509: American Petroleum Institute; Page 509: Fortune; Page 509: U. S . Postal Service; Page 509: U.S.Bureau of Economic Analysis; Page 509: U.S.Geological Survey; Page 509: U.S.Treasury; Page 520: American Petroleum Institute; Page 520: The Institute for College Access & Success; Page 520: World Tourism Organization; Page 521: U.S. Census Bureau; Page 549: U.S. Elections Project

Library of Congress Cataloging-in-Publication Data

Names: Barnett, Raymond A., author. | Ziegler, Michael R., author. | Byleen, Karl E., author. | Stocker, Christopher J., author.
Title: Finite mathematics for business, economics, life sciences, and social sciences.
Description: Fourteenth edition / Raymond A. Barnett, Michael R. Ziegler, Karl E. Byleen, Christopher J. Stocker. | Boston: Pearson, [2019] | "Designed for a one-term course in finite mathematics for students who have had one to two years of high school algebra or the equivalent"–Preface.
Identifiers: LCCN 2017037294 | ISBN 9780134675985 | ISBN 0134675983
Subjects: LCSH: Mathematics–Textbooks. | Social sciences–Mathematics–Textbooks. | Biomathematics–Textbooks.
Classification: LCC QA39.3 .B37 2019 | DDC 511/.1–dc23
LC record available at https://lccn.loc.gov/2017037294

1 17

Pearson

Student Edition
ISBN 13: 978-0-13-467598-5
ISBN 10: 0-13-467598-3

Instructor's Edition
ISBN 13: 978-0-13-467793-4
ISBN 10: 0-13-467793-5

CONTENTS

PREFACE

The fourteenth edition of *Finite Mathematics for Business, Economics, Life Sciences, and Social Sciences* is designed for a one-term course in finite mathematics for students who have had one to two years of high school algebra or the equivalent. The book's overall approach, refined by the authors' experience with large sections of college freshmen, addresses the challenges of teaching and learning when prerequisite knowledge varies greatly from student to student.

Note that Chapters 1–9 of this text also appear in *College Mathematics for Business, Economics, Life Sciences, and Social Sciences,* by the same author team. The College Mathematics text also contains coverage of applied calculus topics.

The authors had three main goals in writing this text:

1. To write a text that students can easily comprehend
2. To make connections between what students are learning and how they may apply that knowledge
3. To give flexibility to instructors to tailor a course to the needs of their students.

Many elements play a role in determining a book's effectiveness for students. Not only is it critical that the text be accurate and readable, but also, in order for a book to be effective, aspects such as the page design, the interactive nature of the presentation, and the ability to support and challenge all students have an incredible impact on how easily students comprehend the material. Here are some of the ways this text addresses the needs of students at all levels:

- Page layout is clean and free of potentially distracting elements.
- Matched Problems that accompany each of the completely worked examples help students gain solid knowledge of the basic topics and assess their own level of understanding before moving on.
- Review material (Appendix A and Chapters 1 and 2) can be used judiciously to help remedy gaps in prerequisite knowledge.
- A Diagnostic Prerequisite Test prior to Chapter 1 helps students assess their skills, while the Basic Algebra Review in Appendix A provides students with the content they need to remediate those skills.
- Explore and Discuss problems lead the discussion into new concepts or build upon a current topic. They help students of all levels gain better insight into the mathematical concepts through thought-provoking questions that are effective in both small and large classroom settings.
- Instructors are able to easily craft homework assignments that best meet the needs of their students by taking advantage of the variety of types and difficulty levels of the exercises. Exercise sets at the end of each section consist of a Skills Warm-up (four to eight problems that review prerequisite knowledge specific to that section) followed by problems divided into categories A, B, and C by level of difficulty, with level-C exercises being the most challenging.
- The MyLab Math course for this text is designed to help students help themselves and provide instructors with actionable information about their progress. The immediate feedback students receive when doing homework and practice in MyLab Math is invaluable, and the easily accessible eBook enhances student learning in a way that the printed page sometimes cannot.
- Most important, all students get substantial experience in modeling and solving real-world problems through application examples and exercises chosen from

business and economics, life sciences, and social sciences. Great care has been taken to write a book that is mathematically correct, with its emphasis on computational skills, ideas, and problem solving rather than mathematical theory.

- Finally, the choice and independence of topics make the text readily adaptable to a variety of courses.

New to This Edition

Fundamental to a book's effectiveness is classroom use and feedback. Now in its fourteenth edition, this text has had the benefit of a substantial amount of both. Improvements in this edition evolved out of the generous response from a large number of users of the last and previous editions as well as survey results from instructors. Additionally, we made the following improvements in this edition:

- Redesigned the text in full color to help students better use it and to help motivate students as they put in the hard work to learn the mathematics (because let's face it—a more modern looking book has more appeal).
- Updated graphing calculator screens to TI-84 Plus CE (color) format.
- Added *Reminder* features in the side margin to either remind students of a concept that is needed at that point in the book or direct the student back to the section in which it was covered earlier.
- Updated data in examples and exercises. Many modern and student-centered applications have been added to help students see the relevance of the content.
- Analyzed aggregated student performance data and assignment frequency data from MyLab Math for the previous edition of this text. The results of this analysis helped improve the quality and quantity of exercises that matter the most to instructors and students.
- Rewrote and simplified the treatment of cost, revenue, and profit in Section 2.1.
- Added 611 new exercises throughout the text.
- Moved the seldom-used chapter "Games and Decisions" online to `goo.gl/6VBjkQ`. Note that all of the resources that formerly accompanied this chapter are still available. They are housed within MyLab Math.

New to MyLab Math

Many improvements have been made to the overall functionality of MyLab Math since the previous edition. However, beyond that, we have also increased and improved the content specific to this text.

- Instructors now have **more exercises** than ever to choose from in assigning homework. Most new questions are application-oriented. There are approximately 3,340 assignable exercises in MyLab Math for this text. New exercise types include:
  - **Additional Conceptual Questions** provide support for assessing concepts and vocabulary. Many of these questions are application-oriented.
  - **Setup & Solve** exercises require students to show how they set up a problem as well as the solution, better mirroring what is required of students on tests.
- The **Guide to Video-Based Assignments** shows which MyLab Math exercises can be assigned for each video. (All videos are also assignable.) This resource is handy for online or flipped classes.
- The ***Note-Taking Guide*** provides support for students as they take notes in class. The Guide includes definitions, theorems, and statements of examples but has blank space for students to write solutions to examples and sample problems. The Note-Taking Guide corresponds to the Lecture PowerPoints that accompany the text. The Guide can be downloaded in PDF or Word format from within MyLab Math.

- A full suite of **Interactive Figures** has been added to support teaching and learning. The figures illustrate key concepts and allow manipulation. They have been designed to be used in lecture as well as by students independently.
- An **Integrated Review** version of the MyLab Math course contains premade quizzes to assess the prerequisite skills needed for each chapter, plus personalized remediation for any gaps in skills that are identified.
- **Study Skills Modules** help students with the life skills that can make the difference between passing and failing.
- **MathTalk and StatTalk videos** highlight applications of the content of the course to business. The videos are supported by assignable exercises.
- The ***Graphing Calculator Manual*** and ***Excel Spreadsheet Manual***, both specific to this course, have been updated to support the TI-84 Plus CE (color edition) and Excel 2016, respectively. Both manuals also contain additional topics to support the course. These manuals are within the Tools for Success tab.
- We heard from users that the Annotated Instructor's Edition for the previous edition required too much flipping of pages to find answers, so MyLab Math now contains a downloadable **Instructor's Answers document**—*with all answers in one place*. (This augments the downloadable *Instructor's Solutions Manual*, which contains even-numbered solutions.)

Trusted Features

- **Emphasis and Style**—As was stated earlier, this text is written for student comprehension. To that end, the focus has been on making the book both mathematically correct and accessible to students. Most derivations and proofs are omitted, except where their inclusion adds significant insight into a particular concept as the emphasis is on computational skills, ideas, and problem solving rather than mathematical theory. General concepts and results are typically presented only after particular cases have been discussed.
- **Design**—One of the hallmark features of this text is the clean, straightforward design of its pages. Navigation is made simple with an obvious hierarchy of key topics and a judicious use of call-outs and pedagogical features. A functional use of color improves the clarity of many illustrations, graphs, and explanations, and guides students through critical steps (see pages 59 and 60).
- **Examples**—More than 300 completely worked examples are used to introduce concepts and to demonstrate problem-solving techniques. Many examples have multiple parts, significantly increasing the total number of worked examples. The examples are annotated using blue text to the right of each step, and the problem-solving steps are clearly identified. To give students extra help in working through examples, dashed boxes are used to enclose steps that are usually performed mentally and rarely mentioned in other books (see Example 7 on page 7). Though some students may not need these additional steps, many will appreciate the fact that the authors do not assume too much in the way of prior knowledge.
- **Matched Problems**—Each example is followed by a similar Matched Problem for the student to work while reading the material. This actively involves the student in the learning process. The answers to these matched problems are included at the end of each section for easy reference.
- **Explore and Discuss**—Most every section contains Explore and Discuss problems at appropriate places to encourage students to think about a relationship or process before a result is stated or to investigate additional consequences of a development in the text (see pages 13 and 17). This serves to foster critical thinking and communication skills. The Explore and Discuss material can be

used for in-class discussions or out-of-class group activities and is effective in both small and large class settings.

- **Exercise Sets**—The book contains over 4,200 carefully selected and graded exercises. Many problems have multiple parts, significantly increasing the total number of exercises. Writing exercises, indicated by the icon, provide students with an opportunity to express their understanding of the topic in writing. Answers to all odd-numbered problems are in the back of the book. Exercises are paired so that consecutive odd- and even-numbered exercises are of the same type and difficulty level. Exercise sets are structured to facilitate crafting just the right assignment for students:
  - **Skills Warm-up** exercises, indicated by W, review key prerequisite knowledge.
  - **Graded exercises**: Levels A (routine, easy mechanics), B (more difficult mechanics), and C (difficult mechanics and some theory) make it easy for instructors to create assignments that are appropriate for their classes.
  - **Applications** conclude almost every exercise set. These exercises are labeled with the type of application to make it easy for instructors to select the right exercises for their audience.
- **Applications**—A major objective of this book is to give the student substantial experience in modeling and solving real-world problems. Enough applications are included to convince even the most skeptical student that mathematics is really useful (see the Index of Applications at the back of the book). Almost every exercise set contains application problems, including applications from business and economics, life sciences, and social sciences. An instructor with students from all three disciplines can let them choose applications from their own field of interest; if most students are from one of the three areas, then special emphasis can be placed there. Most of the applications are simplified versions of actual real-world problems inspired by professional journals and books. No specialized experience is required to solve any of the application problems.
- **Graphing Calculator and Spreadsheets**—Although access to a graphing calculator or spreadsheets is not assumed, it is likely that many students will want to make use of this technology. To assist these students, optional graphing calculator and spreadsheet activities are included in appropriate places. These include brief discussions in the text, examples or portions of examples solved on a graphing calculator or spreadsheet, and exercises for the students to solve. For example, linear regression is introduced in Section 1.3, and regression techniques on a graphing calculator are used at appropriate points to illustrate mathematical modeling with real data. All the optional graphing calculator material is clearly identified with the icon and can be omitted without loss of continuity, if desired. Graphing calculator screens displayed in the text are actual output from the TI-84 Plus CE (color version) graphing calculator.

Additional Pedagogical Features

The following features, while helpful to any student, are particularly helpful to students enrolled in a large classroom setting where access to the instructor is more challenging or just less frequent. These features provide much-needed guidance for students as they tackle difficult concepts.

- **Call-out boxes** highlight important definitions, results, and step-by-step processes (see pages 18, 62, and 69).
- **Caution** statements appear throughout the text where student errors often occur (see pages 50 and 115).

- **Conceptual Insights**, appearing in nearly every section, often make explicit connections to previous knowledge but sometimes encourage students to think beyond the particular skill they are working on and attain a more enlightened view of the concepts at hand (see pages 58 and 69).
- **Diagnostic Prerequisite Test**, located on page xvii, provides students with a tool to assess their prerequisite skills prior to taking the course. The **Basic Algebra Review**, in Appendix A, provides students with seven sections of content to help them remediate in specific areas of need. Answers to the Diagnostic Prerequisite Test are at the back of the book and reference specific sections in the Basic Algebra Review or Chapters 1 and 2 for students to use for remediation.
- **Chapter Reviews**—Often it is during the preparation for a chapter exam that concepts gel for students, making the chapter review material particularly important. The chapter review sections in this text include a comprehensive summary of important terms, symbols, and concepts, keyed to completely worked examples, followed by a comprehensive set of Review Exercises. Answers to Review Exercises are included at the back of the book; each answer contains a reference to the section in which that type of problem is discussed so students can remediate any deficiencies in their skills on their own.

Content

The text begins with the development of a library of elementary functions in Chapters 1 and 2, including their properties and applications. Many students will be familiar with most, if not all, of the material in these introductory chapters. Depending on students' preparation and the course syllabus, an instructor has several options for using the first two chapters, including the following:

(i) Skip Chapters 1 and 2 and refer to them only as necessary later in the course;
(ii) Cover Chapter 1 quickly in the first week of the course, emphasizing price–demand equations, price–supply equations, and linear regression, but skip Chapter 2;
(iii) Cover Chapters 1 and 2 systematically before moving on to other chapters.

The material in Part Two (Finite Mathematics) can be thought of as four units:

1. Mathematics of finance (Chapter 3)
2. Linear algebra, including matrices, linear systems, and linear programming (Chapters 4, 5, and 6)
3. Probability and statistics (Chapters 7, 8, and 10)
4. Applications of linear algebra and probability to Markov chains and game theory (Chapters 9 and 11)

The first three units are independent of each other, while the fourth unit is dependent on some of the earlier chapters. (Markov chains requires Chapters 4 and 8; game theory requires Chapters 4–6 and 8).

Chapter 3 presents a thorough treatment of simple and compound interest and present and future value of ordinary annuities. Appendix B.1 addresses arithmetic and geometric sequences and can be covered in conjunction with this chapter, if desired.

Chapter 4 covers linear systems and matrices with an emphasis on using row operations and Gauss–Jordan elimination to solve systems and to find matrix inverses. This chapter also contains numerous applications of mathematical modeling using systems and matrices. To assist students in formulating solutions, all answers at the back of the book for application exercises in Sections 4.3, 4.5, and the chapter Review Exercises contain both the mathematical model and its solution. The row operations discussed in Sections 4.2 and 4.3 are required for the simplex method in Chapter 6. Matrix multiplication, matrix inverses, and systems of equations are required for Markov chains in Chapter 9.

Chapters 5 and 6 provide a broad and flexible coverage of linear programming. Chapter 5 covers two-variable graphing techniques. Instructors who wish to emphasize linear programming techniques can cover the basic simplex method in Sections 6.1 and 6.2 and then discuss either or both of the following: the dual method (Section 6.3) and the big M method (Section 6.4). Those who want to emphasize modeling can discuss the formation of the mathematical model for any of the application examples in Sections 6.2–6.4, and either omit the solution or use software to find the solution. To facilitate this approach, all answers at the back of the book for application exercises in Sections 6.2–6.4 and the chapter Review Exercises contain both the mathematical model and its solution. The simplex and dual solution methods are required for portions of Chapter 11.

Chapter 7 provides a foundation for probability with a treatment of logic, sets, and counting techniques.

Chapter 8 covers basic probability, including Bayes' formula and random variables.

Chapter 10 deals with basic descriptive statistics and more advanced probability distributions, including the important normal distribution. Appendix B.3 contains a short discussion of the binomial theorem that can be used in conjunction with the development of the binomial distribution in Section 11.4.

Chapters 9 and 11 tie together concepts developed in earlier chapters and apply them to interesting topics. A study of Markov chains (Chapter 9) or game theory (Chapter 11) provides an excellent unifying conclusion to a finite mathematics course.

Appendix A contains a concise review of basic algebra that may be covered as part of the course or referenced as needed. As mentioned previously, Appendix B (online at `goo.gl/mjbXrG`) contains additional topics that can be covered in conjunction with certain sections in the text, if desired.

Accuracy Check—Because of the careful checking and proofing by a number of mathematics instructors (acting independently), the authors and publisher believe this book to be substantially error free. If an error should be found, the authors would be grateful if notification were sent to Karl E. Byleen, 9322 W. Garden Court, Hales Corners, WI 53130; or by e-mail to kbyleen@wi.rr.com.

Acknowledgments

In addition to the authors, many others are involved in the successful publication of a book. We wish to thank the following reviewers:

Ben Anilus, *Roger Williams University*
Paula F. Bailey, *Roger Williams University*
Mary Ann Barber, *University of North Texas*
Sandra Belcher, *Midwestern State University*
Chris Bendixen, *Lake Michigan College*
Brian C. Bright, *Ivy Tech Community College*
Ronnie L. Brown, *American University*
Dee Cochran, *Indiana University, Southeast*
Lauren Fern, *University of Montana*
Sonya Geisel, *Arizona State University*
Jeffrey Hakim, *American University*
Patricia Ann Hussey, *Triton College*
Chuck Jackson, *Lone Star College, Kingwood*
Sheila C. Luke, *Colorado Technical University*
Mari M. Menard, *Lone Star College, Kingwood*
Faramarz Mokhtari, *Montgomery College, Rockville*
Jimmy Mopecha, *Delaware Valley University*
Luis Molina, *Lone Star College, CyFair*

Michael Puente, *Richland College*
Rick Pugsley, *Ivy Tech Community College*
Douglas Ray, *Texas State University*
Frances Smith, *Richland College*
John C. Treacy, *Ivy Tech Community College*

The following faculty members provided direction on the development of the MyLab Math course for this edition:

Mary Ann Barber, *University of North Texas*
Ronnie L. Brown, *University of Baltimore*
David Busekist, *Southeastern Louisiana University*
Kristen A. Ceballos, *Southern Illinois University, Carbondale*
Brandi L. Cline, *Lone Star College, Tomball*
James Colton, *Shaker High School*
Natalia Gavryshova, *College of San Mateo*
Liz Hawk, *Eastbrook High School*
Natalie Herndon, *Southwest Tennessee State Univerity*
Alan Kunkle, *Ivy Tech Community College, Kokomo*
Steve Lam, *Guam Community College*
Mari M. Menard, *Lone Star College, Kingwood*
Faramarz Mokhtari, *Montgomery College, Rockville*
Jimmy Mopecha, *Delaware Valley University*
Mike Panahi, *El Centro College*
Terri Seiver, *San Jacinto College*
Jack Wagner, *Fordham University*
Dennis Wolf, *Indiana University, South Bend*
Dinesh S. Yadav, *Richland College*

We also express our thanks to John Samons and Damon Demas for providing a careful and thorough accuracy check of the text, problems, and answers. Our thanks to Garret Etgen, John Samons, Salvatore Sciandra, Victoria Baker, Ben Rushing, and Stela Pudar-Hozo for developing the supplemental materials so important to the success of a text. And finally, thanks to all the people at Pearson and Integra who contributed their efforts to the production of this book.

MyLab Math Online Course for *Finite Mathematics for Business, Economics, Life Sciences, and Social Sciences*, 14e (access code required)

MyLab™ Math is available to accompany Pearson's market-leading text offerings. To give students a consistent tone, voice, and teaching method, each text's flavor and approach are tightly integrated throughout the accompanying MyLab Math course, making learning the material as seamless as possible.

Preparedness

One of the biggest challenges in applied math courses is making sure students are adequately prepared with the prerequisite skills needed to successfully complete their course work. MyLab Math supports students with just-in-time remediation and key-concept review.

NEW! Integrated Review Course

An Integrated Review version of the MyLab Math course contains premade, assignable quizzes to assess the prerequisite skills needed for each chapter, plus personalized remediation for any gaps in skills that are identified. Each student, therefore, receives just the help that he or she needs—no more, no less.

NEW! Study Skills Modules

Study skills modules help students with the life skills that can make the difference between passing and failing.

Developing Deeper Understanding

MyLab Math provides content and tools that help students build a deeper understanding of course content than would otherwise be possible.

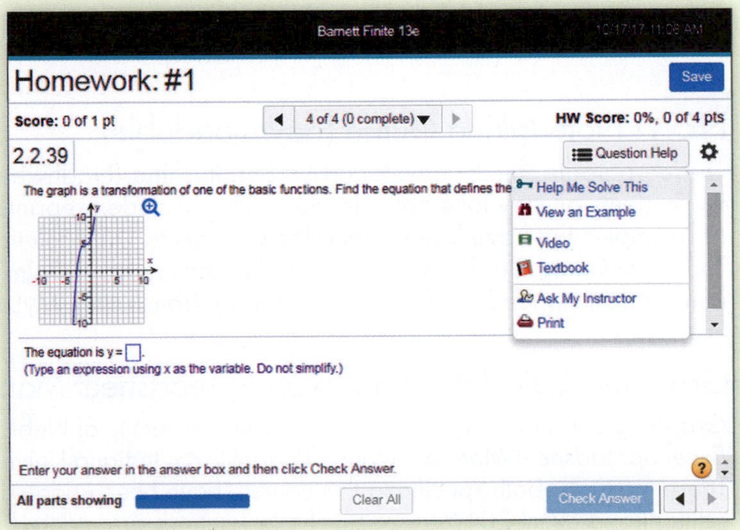

Exercises with Immediate Feedback

Homework and practice exercises for this text regenerate algorithmically to give students unlimited opportunity for practice and mastery. MyLab Math provides helpful feedback when students enter incorrect answers and includes the optional learning aids Help Me Solve This, View an Example, videos, and/or the eText.

NEW! Additional Conceptual Questions

Additional Conceptual Questions provide support for assessing concepts and vocabulary. Many of these questions are application-oriented. They are clearly labeled "Conceptual" in the Assignment Manager.

NEW! Setup & Solve Exercises

These exercises require students to show how they set up a problem as well as the solution, better mirroring what is required on tests.

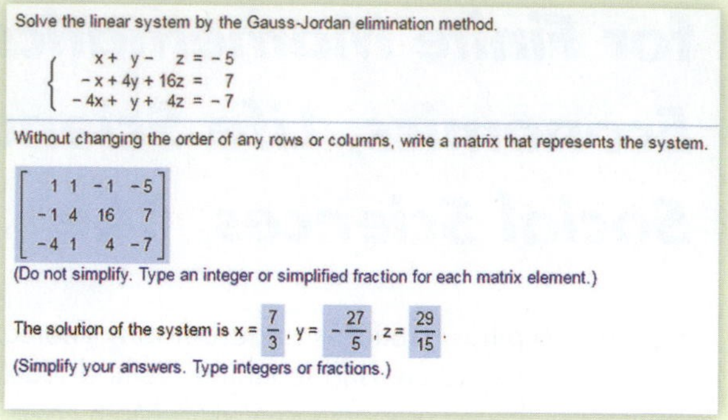

Solve the linear system by the Gauss-Jordan elimination method.
x + y − z = −5
−x + 4y + 16z = 7
−4x + y + 4z = −7
Without changing the order of any rows or columns, write a matrix that represents the system.
1 1 −1 −5
−1 4 16 7
−4 1 4 −7
(Do not simplify. Type an integer or simplified fraction for each matrix element.)
The solution of the system is x = 7/3, y = −27/5, z = 29/15.
(Simplify your answers. Type integers or fractions.)

NEW! Interactive Figures

A full suite of Interactive Figures has been added to support teaching and learning. The figures illustrate key concepts and allow manipulation. They are designed to be used in lecture as well as by students independently.

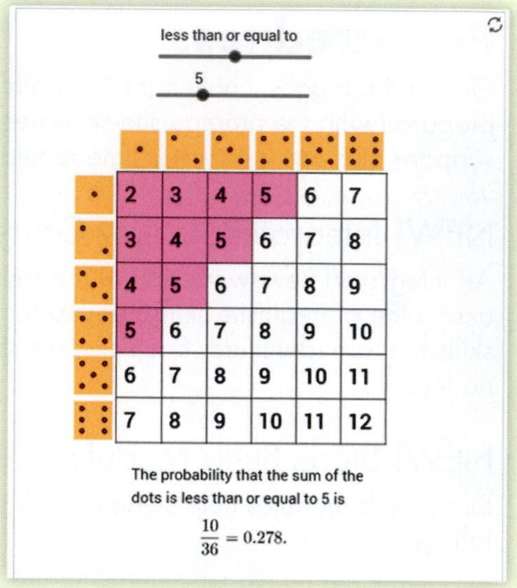

Instructional Videos

Every example in the text has an instructional video tied to it that can be used as a learning aid or for self-study. MathTalk videos were added to highlight business applications to the course content, and a Guide to Video-Based Assignments shows which MyLab Math exercises can be assigned for each video.

NEW! Note-Taking Guide (downloadable)

These printable sheets, developed by Ben Rushing (Northwestern State University) provide support for students as they take notes in class. They include preprinted definitions, theorems, and statements of examples but have blank space for students to write solutions to examples and sample problems. The *Note-Taking Guide* corresponds to the Lecture PowerPoints that accompany the text. The *Guide* can be downloaded in PDF or Word format from within MyLab Math from the Tools for Success tab.

Graphing Calculator and Excel Spreadsheet Manuals (downloadable)

Graphing Calculator Manual by Chris True, University of Nebraska
Excel Spreadsheet Manual by Stela Pudar-Hozo, Indiana University–Northwest
These manuals, both specific to this course, have been updated to support the TI-84 Plus CE (color edition) and Excel 2016, respectively. Instructions are ordered by mathematical topic. The files can be downloaded from within MyLab Math from the Tools for Success tab.

Student's Solutions Manual (softcover and downloadable)

ISBN: 0-13-467741-2 • 978-0-13-467741-5

Written by John Samons (Florida State College), the *Student's Solutions Manual* contains worked-out solutions to all the odd-numbered exercises. This manual is available in print and can be downloaded from within MyLab Math within the Chapter Contents tab.

A Complete eText

Students get unlimited access to the eText within any MyLab Math course using that edition of the textbook. The Pearson eText app allows existing subscribers to access their titles on an iPad or Android tablet for either online or offline viewing.

Supporting Instruction

MyLab Math comes from an experienced partner with educational expertise and an eye on the future. It provides resources to help you assess and improve students' results at every turn and unparalleled flexibility to create a course tailored to you and your students.

Learning Catalytics™

Now included in all MyLab Math courses, this student response tool uses students' smartphones, tablets, or laptops to engage them in more interactive tasks and thinking during lecture. Learning Catalytics™ fosters student engagement and peer-to-peer learning with real-time analytics. Access pre-built exercises created specifically for this course.

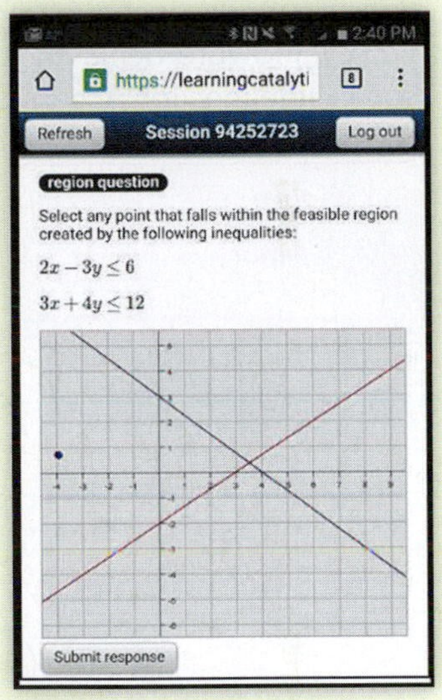

PowerPoint® Lecture Slides (downloadable)

Classroom presentation slides feature key concepts, examples, and definitions from this text. They are designed to be used in conjunction with the Note-Taking Guide that accompanies the text. They can be downloaded from within MyLab Math or from Pearson's online catalog, **www.pearson.com**.

Learning Worksheets

Written by Salvatore Sciandra (Niagara County Community College), these worksheets include key chapter definitions and formulas, followed by exercises for students to practice in class, for homework, or for independent study. They are downloadable as PDFs or Word documents from within MyLab Math.

Comprehensive Gradebook

The gradebook includes enhanced reporting functionality such as item analysis and a reporting dashboard to allow you to efficiently manage your course. Student performance data is presented at the class, section, and program levels in an accessible, visual manner so you'll have the information you need to keep your students on track.

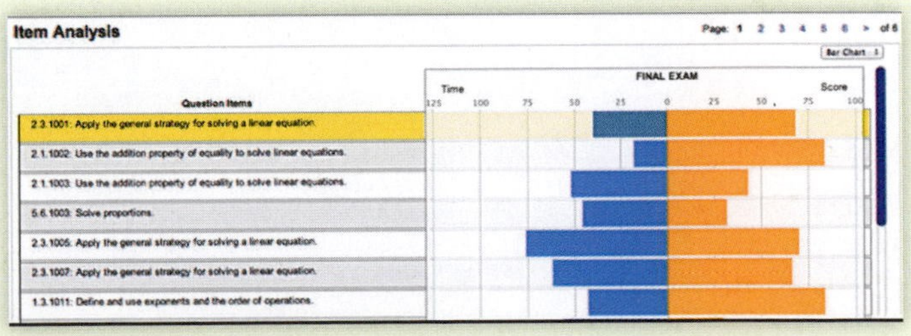

TestGen®

TestGen® (**www.pearson.com/testgen**) enables instructors to build, edit, print, and administer tests using a computerized bank of questions developed to cover all the objectives of the text. TestGen is algorithmically based, allowing instructors to create multiple but equivalent versions of the same question or test with the click of a button. Instructors can also modify test bank questions or add new questions. The software and test bank are available for download from Pearson's online catalog, **www.pearson.com**. The questions are also assignable in MyLab Math.

Instructor's Solutions Manual (downloadable)

Written by Garret J. Etgen (University of Houston) and John Samons (Florida State College), the *Instructor's Solutions Manual* contains worked-out solutions to all the even-numbered exercises. It can be downloaded from within MyLab Math or from Pearson's online catalog, **www.pearson.com**.

Accessibility

Pearson works continuously to ensure our products are as accessible as possible to all students. We are working toward achieving WCAG 2.0 Level AA and Section 508 standards, as expressed in the Pearson Guidelines for Accessible Educational Web Media, **www.pearson.com/mylab/math/accessibility**.

Diagnostic Prerequisite Test

Work all of the problems in this self-test without using a calculator. Then check your work by consulting the answers in the back of the book. Where weaknesses show up, use the reference that follows each answer to find the section in the text that provides the necessary review.

1. Replace each question mark with an appropriate expression that will illustrate the use of the indicated real number property:

(A) Commutative $(\cdot)$: $x(y + z) = ?$

(B) Associative $(+)$: $2 + (x + y) = ?$

(C) Distributive: $(2 + 3)x = ?$

Problems 2–6 refer to the following polynomials:

(A) $3x - 4$ (B) $x + 2$

(C) $2 - 3x^2$ (D) $x^3 + 8$

2. Add all four.

3. Subtract the sum of (A) and (C) from the sum of (B) and (D).

4. Multiply (C) and (D).

5. What is the degree of each polynomial?

6. What is the leading coefficient of each polynomial?

In Problems 7 and 8, perform the indicated operations and simplify.

7. $5x^2 - 3x[4 - 3(x - 2)]$

8. $(2x + y)(3x - 4y)$

In Problems 9 and 10, factor completely.

9. $x^2 + 7x + 10$ **10.** $x^3 - 2x^2 - 15x$

11. Write 0.35 as a fraction reduced to lowest terms.

12. Write $\frac{7}{8}$ in decimal form.

13. Write in scientific notation:

(A) 4,065,000,000,000 (B) 0.0073

14. Write in standard decimal form:

(A) 2.55×10^8 (B) 4.06×10^{-4}

15. Indicate true (T) or false (F):

(A) A natural number is a rational number.

(B) A number with a repeating decimal expansion is an irrational number.

16. Give an example of an integer that is not a natural number.

In Problems 17–24, simplify and write answers using positive exponents only. All variables represent positive real numbers.

17. $6(xy^3)^5$ **18.** $\frac{9u^8v^6}{3u^4v^8}$

19. $(2 \times 10^5)(3 \times 10^{-3})$ **20.** $(x^{-3}y^2)^{-2}$

21. $u^{5/3}u^{2/3}$ **22.** $(9a^4b^{-2})^{1/2}$

23. $\frac{5^0}{3^2} + \frac{3^{-2}}{2^{-2}}$ **24.** $(x^{1/2} + y^{1/2})^2$

In Problems 25–30, perform the indicated operation and write the answer as a simple fraction reduced to lowest terms. All variables represent positive real numbers.

25. $\frac{a}{b} + \frac{b}{a}$ **26.** $\frac{a}{bc} - \frac{c}{ab}$

27. $\frac{x^2}{y} \cdot \frac{y^6}{x^3}$ **28.** $\frac{x}{y^3} \div \frac{x^2}{y}$

29. $\dfrac{\frac{1}{7 + h} - \frac{1}{7}}{h}$ **30.** $\dfrac{x^{-1} + y^{-1}}{x^{-2} - y^{-2}}$

31. Each statement illustrates the use of one of the following real number properties or definitions. Indicate which one.

Commutative $(+, \cdot)$	Associative $(+, \cdot)$	Distributive
Identity $(+, \cdot)$	Inverse $(+, \cdot)$	Subtraction
Division	Negatives	Zero

(A) $(-7) - (-5) = (-7) + [-(-5)]$

(B) $5u + (3v + 2) = (3v + 2) + 5u$

(C) $(5m - 2)(2m + 3) = (5m - 2)2m + (5m - 2)3$

(D) $9 \cdot (4y) = (9 \cdot 4)y$

(E) $\frac{u}{-(v - w)} = \frac{u}{w - v}$

(F) $(x - y) + 0 = (x - y)$

32. Round to the nearest integer:

(A) $\frac{17}{3}$ (B) $-\frac{5}{19}$

33. Multiplying a number x by 4 gives the same result as subtracting 4 from x. Express as an equation, and solve for x.

34. Find the slope of the line that contains the points $(3, -5)$ and $(-4, 10)$.

35. Find the x and y coordinates of the point at which the graph of $y = 7x - 4$ intersects the x axis.

36. Find the x and y coordinates of the point at which the graph of $y = 7x - 4$ intersects the y axis.

In Problems 37–40, solve for x.

37. $x^2 = 5x$

38. $3x^2 - 21 = 0$

39. $x^2 - x - 20 = 0$

40. $-6x^2 + 7x - 1 = 0$

1 Linear Equations and Graphs

Introduction

How far will a glacier advance or retreat in the next ten years? The key to answering such a question, and to making other climate-related predictions, is mathematical modeling. In Chapter 1, we study one of the simplest mathematical models, a linear equation. We introduce a technique called linear regression to construct mathematical models from numerical data. We use mathematical models to predict average annual temperature and average annual precipitation (see Problems 23 and 24 in Section 1.3), the atmospheric concentration of carbon dioxide, the consumption of fossil fuels, and many other quantities in business, economics, life sciences, and social sciences.

1.1 Linear Equations and Inequalities

- Linear Equations
- Linear Inequalities
- Applications

The equation

$$3 - 2(x + 3) = \frac{x}{3} - 5$$

and the inequality

$$\frac{x}{2} + 2(3x - 1) \geq 5$$

are both first degree in one variable. In general, a **first-degree**, or **linear**, **equation** in one variable is any equation that can be written in the form

$$\textbf{Standard form:} \quad ax + b = 0 \qquad a \neq 0 \tag{1}$$

If the equality symbol, $=$, in (1) is replaced by $<$, $>$, $\leq$, or $\geq$, the resulting expression is called a **first-degree**, or **linear**, **inequality**.

A **solution** of an equation (or inequality) involving a single variable is a number that, when substituted for the variable, makes the equation (or inequality) true. The set of all solutions is called the **solution set**. To **solve an equation** (or inequality) means to find its solution set.

Knowing what is meant by the solution set is one thing; finding it is another. We start by recalling the idea of equivalent equations and equivalent inequalities. If we perform an operation on an equation (or inequality) that produces another equation (or inequality) with the same solution set, then the two equations (or inequalities) are said to be **equivalent**. The basic idea in solving equations or inequalities is to perform operations that produce simpler equivalent equations or inequalities and to continue the process until we obtain an equation or inequality with an obvious solution.

Linear Equations

Linear equations are generally solved using the following equality properties.

THEOREM 1 Equality Properties

An equivalent equation will result if

1. The same quantity is added to or subtracted from each side of a given equation.
2. Each side of a given equation is multiplied by or divided by the same nonzero quantity.

EXAMPLE 1 **Solving a Linear Equation** Solve and check:

$$8x - 3(x - 4) = 3(x - 4) + 6$$

SOLUTION

$8x - 3(x - 4) = 3(x - 4) + 6$	Use the distributive property.
$8x - 3x + 12 = 3x - 12 + 6$	Combine like terms.
$5x + 12 = 3x - 6$	Subtract $3x$ from both sides.
$2x + 12 = -6$	Subtract 12 from both sides.
$2x = -18$	Divide both sides by 2.
$x = -9$	

CHECK

$$8x - 3(x - 4) = 3(x - 4) + 6$$
$$8(-9) - 3[(-9) - 4] \stackrel{?}{=} 3[(-9) - 4] + 6$$
$$-72 - 3(-13) \stackrel{?}{=} 3(-13) + 6$$
$$-33 \stackrel{\checkmark}{=} -33$$

Matched Problem 1 Solve and check: $3x - 2(2x - 5) = 2(x + 3) - 8$

Explore and Discuss 1

According to equality property 2, multiplying both sides of an equation by a nonzero number always produces an equivalent equation. What is the smallest positive number that you could use to multiply both sides of the following equation to produce an equivalent equation without fractions?

$$\frac{x + 1}{3} - \frac{x}{4} = \frac{1}{2}$$

EXAMPLE 2 **Solving a Linear Equation** Solve and check: $\frac{x + 2}{2} - \frac{x}{3} = 5$

SOLUTION What operations can we perform on

$$\frac{x + 2}{2} - \frac{x}{3} = 5$$

to eliminate the denominators? If we can find a number that is exactly divisible by each denominator, we can use the multiplication property of equality to clear the denominators. The LCD (least common denominator) of the fractions, 6, is exactly what we are looking for! Actually, any common denominator will do, but the LCD results in a simpler equivalent equation. So we multiply both sides of the equation by 6:

Reminder

Dashed boxes are used throughout the book to denote steps that are usually performed mentally.

$$6\left(\frac{x + 2}{2} - \frac{x}{3}\right) = 6 \cdot 5$$
$$\overset{3}{\cancel{6}} \cdot \frac{(x + 2)}{\underset{1}{\cancel{2}}} - \overset{2}{\cancel{6}} \cdot \frac{x}{\underset{1}{\cancel{3}}} = 30$$

$$3(x + 2) - 2x = 30 \quad \text{Use the distributive property.}$$
$$3x + 6 - 2x = 30 \quad \text{Combine like terms.}$$
$$x + 6 = 30 \quad \text{Subtract 6 from both sides.}$$
$$x = 24$$

CHECK

$$\frac{x + 2}{2} - \frac{x}{3} = 5$$
$$\frac{24 + 2}{2} - \frac{24}{3} \stackrel{?}{=} 5$$
$$13 - 8 \stackrel{?}{=} 5$$
$$5 \stackrel{\checkmark}{=} 5$$

Matched Problem 2 Solve and check: $\frac{x + 1}{3} - \frac{x}{4} = \frac{1}{2}$

In many applications of algebra, formulas or equations must be changed to alternative equivalent forms. The following example is typical.

EXAMPLE 3 **Solving a Formula for a Particular Variable** If you deposit a principal P in an account that earns simple interest at an annual rate r, then the amount A in the account after t years is given by $A = P + Prt$. Solve for

(A) r in terms of A, P, and t

(B) P in terms of A, r, and t

SOLUTION (A)

$$A = P + Prt \quad \text{Reverse equation.}$$
$$P + Prt = A \quad \text{Subtract } P \text{ from both sides.}$$
$$Prt = A - P \quad \text{Divide both members by } Pt.$$
$$r = \frac{A - P}{Pt}$$

(B)

$$A = P + Prt \quad \text{Reverse equation.}$$
$$P + Prt = A \quad \text{Factor out } P \text{ (note the use of the distributive property).}$$
$$P(1 + rt) = A \quad \text{Divide by } (1 + rt).$$
$$P = \frac{A}{1 + rt}$$

Matched Problem 3 If a cardboard box has length L, width W, and height H, then its surface area is given by the formula $S = 2LW + 2LH + 2WH$. Solve the formula for

(A) L in terms of S, W, and H (B) H in terms of S, L, and W

Linear Inequalities

Before we start solving linear inequalities, let us recall what we mean by < (less than) and > (greater than). If a and b are real numbers, we write

$$a < b \quad a \text{ is less than } b$$

if there exists a positive number p such that $a + p = b$. Certainly, we would expect that if a positive number was added to any real number, the sum would be larger than the original. That is essentially what the definition states. If $a < b$, we may also write

$$b > a \quad b \text{ is greater than } a.$$

EXAMPLE 4 **Inequalities** Replace each question mark with either < or >.

(A) 3 ? 5 (B) −6 ? −2 (C) 0 ? −10

SOLUTION

(A) $3 < 5$ because $3 + 2 = 5$.

(B) $-6 < -2$ because $-6 + 4 = -2$.

(C) $0 > -10$ because $-10 < 0$ (because $-10 + 10 = 0$).

Matched Problem 4 Replace each question mark with either < or >.

(A) 2 ? 8 (B) −20 ? 0 (C) −3 ? −30

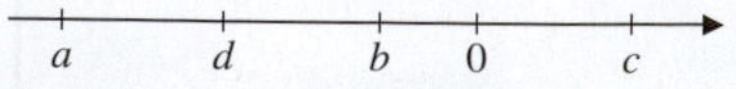

Figure 1 $a < b$, $c > d$

The inequality symbols have a very clear geometric interpretation on the real number line. If $a < b$, then a is to the left of b on the number line; if $c > d$, then c is to the right of d on the number line (Fig. 1). Check this geometric property with the inequalities in Example 4.

Explore and Discuss 2

Replace ? with < or > in each of the following:

(A) $-1 \ ? \ 3$ and $2(-1) \ ? \ 2(3)$

(B) $-1 \ ? \ 3$ and $-2(-1) \ ? \ -2(3)$

(C) $12 \ ? \ -8$ and $\frac{12}{4} \ ? \ \frac{-8}{4}$

(D) $12 \ ? \ -8$ and $\frac{12}{-4} \ ? \ \frac{-8}{-4}$

Based on these examples, describe the effect of multiplying both sides of an inequality by a number.

The procedures used to solve linear inequalities in one variable are almost the same as those used to solve linear equations in one variable, but with one important exception, as noted in item 3 of Theorem 2.

THEOREM 2 Inequality Properties

An equivalent inequality will result, and the **sense or direction will remain the same**, if each side of the original inequality

1. has the same real number added to or subtracted from it.
2. is multiplied or divided by the same *positive* number.

An equivalent inequality will result, and the **sense or direction will reverse**, if each side of the original inequality

3. is multiplied or divided by the same *negative* number.

Note: Multiplication by 0 and division by 0 are not permitted.

Therefore, we can perform essentially the same operations on inequalities that we perform on equations, with the exception that **the sense of the inequality reverses if we multiply or divide both sides by a negative number**. Otherwise, the sense of the inequality does not change. For example, if we start with the true statement

$$-3 > -7$$

and multiply both sides by 2, we obtain

$$-6 > -14$$

and the sense of the inequality stays the same. But if we multiply both sides of $-3 > -7$ by -2, the left side becomes 6 and the right side becomes 14, so we must write

$$6 < 14$$

to have a true statement. The sense of the inequality reverses.

If $a < b$, the **double inequality** $a < x < b$ means that $\boldsymbol{a < x}$ **and** $\boldsymbol{x < b}$; that is, x is between a and b. **Interval notation** is also used to describe sets defined by inequalities, as shown in Table 1.

The numbers a and b in Table 1 are called the **endpoints** of the interval. An interval is **closed** if it contains all its endpoints and **open** if it does not contain any of its endpoints. The intervals $[a, b]$, $(-\infty, a]$, and $[b, \infty)$ are closed, and the intervals (a, b), $(-\infty, a)$, and (b, ∞) are open. Note that the symbol ∞ (read infinity) is not a number. When we write $[b, \infty)$, we are simply referring to the interval that starts at b and continues indefinitely to the right. We never refer to ∞ as an endpoint, and we never write $[b, \infty]$. The interval $(-\infty, \infty)$ is the entire real number line.

Table 1 Interval Notation

Interval Notation	Inequality Notation	Line Graph
$[a, b]$	$a \le x \le b$	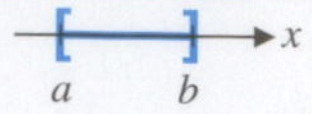
$[a, b)$	$a \le x < b$	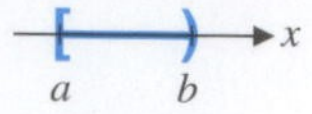
$(a, b]$	$a < x \le b$	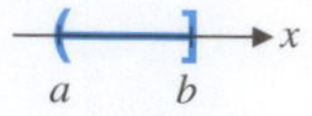
(a, b)	$a < x < b$	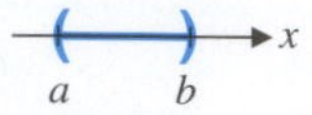
$(-\infty, a]$	$x \le a$	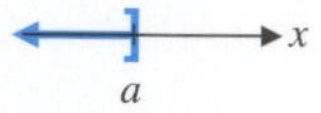
$(-\infty, a)$	$x < a$	
$[b, \infty)$	$x \ge b$	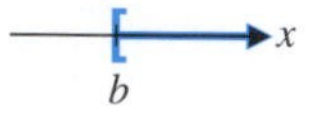
(b, ∞)	$x > b$	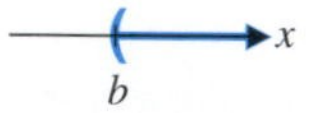

Note that an endpoint of a line graph in Table 1 has a square bracket through it if the endpoint is included in the interval; a parenthesis through an endpoint indicates that it is not included.

CONCEPTUAL INSIGHT

The notation $(2, 7)$ has two common mathematical interpretations: the ordered pair with first coordinate 2 and second coordinate 7, and the open interval consisting of all real numbers between 2 and 7. The choice of interpretation is usually determined by the context in which the notation is used. The notation $(2, -7)$ could be interpreted as an ordered pair but not as an interval. In interval notation, the left endpoint is always written first. So, $(-7, 2)$ is correct interval notation, but $(2, -7)$ is not.

EXAMPLE 5 **Interval and Inequality Notation, and Line Graphs**

(A) Write $[-2, 3)$ as a double inequality and graph.

(B) Write $x \ge -5$ in interval notation and graph.

SOLUTION

(A) $[-2, 3)$ is equivalent to $-2 \le x < 3$.

(B) $x \ge -5$ is equivalent to $[-5, \infty)$.

Matched Problem 5

(A) Write $(-7, 4]$ as a double inequality and graph.

(B) Write $x < 3$ in interval notation and graph.

Explore and Discuss 3

The solution to Example 5B shows the graph of the inequality $x \ge -5$. What is the graph of $x < -5$? What is the corresponding interval? Describe the relationship between these sets.

EXAMPLE 6 **Solving a Linear Inequality** Solve and graph:

$$2(2x + 3) < 6(x - 2) + 10$$

SOLUTION

$$2(2x + 3) < 6(x - 2) + 10 \quad \text{Remove parentheses.}$$
$$4x + 6 < 6x - 12 + 10 \quad \text{Combine like terms.}$$
$$4x + 6 < 6x - 2 \quad \text{Subtract } 6x \text{ from both sides.}$$
$$-2x + 6 < -2 \quad \text{Subtract 6 from both sides.}$$
$$-2x < -8 \quad \text{Divide both sides by } -2 \text{ and reverse the sense of the inequality.}$$
$$x > 4 \quad \text{or} \quad (4, \infty)$$

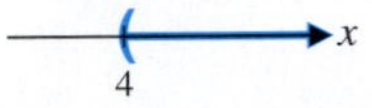

Notice that in the graph of $x > 4$, we use a parenthesis through 4, since the point 4 is not included in the graph.

Matched Problem 6 Solve and graph: $3(x - 1) \le 5(x + 2) - 5$

EXAMPLE 7 **Solving a Double Inequality** Solve and graph: $-3 < 2x + 3 \le 9$

SOLUTION We are looking for all numbers x such that $2x + 3$ is between -3 and 9, including 9 but not -3. We proceed as before except that we try to isolate x in the middle:

$$-3 < 2x + 3 \le 9$$
$$-3 - 3 < 2x + 3 - 3 \le 9 - 3$$
$$-6 < 2x \le 6$$
$$\frac{-6}{2} < \frac{2x}{2} \le \frac{6}{2}$$
$$-3 < x \le 3 \quad \text{or} \quad (-3, 3]$$

Matched Problem 7 Solve and graph: $-8 \le 3x - 5 < 7$

Note that a linear equation usually has exactly one solution, while a linear inequality usually has infinitely many solutions.

Applications

To realize the full potential of algebra, we must be able to translate real-world problems into mathematics. In short, we must be able to do word problems.

Here are some suggestions that will help you get started:

PROCEDURE For Solving Word Problems

1. Read the problem carefully and introduce a variable to represent an unknown quantity in the problem. Often the question asked in a problem will indicate the unknown quantity that should be represented by a variable.
2. Identify other quantities in the problem (known or unknown), and whenever possible, express unknown quantities in terms of the variable you introduced in Step 1.
3. Write a verbal statement using the conditions stated in the problem and then write an equivalent mathematical statement (equation or inequality).
4. Solve the equation or inequality and answer the questions posed in the problem.
5. Check the solution(s) in the original problem.

EXAMPLE 8 **Purchase Price** Alex purchases a big screen TV, pays 7% state sales tax, and is charged \$65 for delivery. If Alex's total cost is \$1,668.93, what was the purchase price of the TV?

SOLUTION

Step 1 **Introduce a variable for the unknown quantity.** After reading the problem, we decide to let x represent the purchase price of the TV.

Step 2 **Identify quantities in the problem.**

Delivery charge: \$65

Sales tax: $0.07x$

Total cost: \$1,668.93

Step 3 **Write a verbal statement and an equation.**

$$\text{Price} + \text{Delivery Charge} + \text{Sales Tax} = \text{Total Cost}$$
$$x + 65 + 0.07x = 1{,}668.93$$

Step 4 **Solve the equation and answer the question.**

$$x + 65 + 0.07x = 1{,}668.93 \quad \text{Combine like terms.}$$
$$1.07x + 65 = 1{,}668.93 \quad \text{Subtract 65 from both sides.}$$
$$1.07x = 1{,}603.93 \quad \text{Divide both sides by 1.07.}$$
$$x = 1{,}499$$

The price of the TV is \$1,499.

Step 5 **Check the answer in the original problem.**

$$\text{Price} = \$1{,}499.00$$
$$\text{Delivery charge} = \$\ \ 65.00$$
$$\text{Tax} = 0.07 \cdot 1{,}499 = \$\ 104.93$$
$$\text{Total} = \$1{,}668.93$$

Matched Problem 8 Mary paid 8.5% sales tax and a \$190 title and license fee when she bought a new car for a total of \$28,400. What is the purchase price of the car?

Any manufacturing company has **costs**, C, which include **fixed costs** such as plant overhead, product design, setup, and promotion; and **variable costs** that depend on the number of items produced. The **revenue**, R, is the amount of money received from the sale of its product. The company **breaks even** if the revenue is equal to the costs, that is, if $R = C$. Example 9 provides an introduction to cost, revenue, and break-even analysis.

EXAMPLE 9 **Break-Even Analysis** A manufacturing company makes bike computers. Fixed costs are \$48,000, and variable costs are \$12.40 per computer. If the computers are sold at a price of \$17.40 each, how many bike computers must be manufactured and sold in order for the company to break even?

SOLUTION

Step 1 Let x = number of bike computers manufactured and sold.

Step 2 $C = \text{Fixed costs} + \text{Variable costs}$

$= \$48{,}000 + \$12.40x$

$R = \$17.40x$

Step 3 The company breaks even if $R = C$; that is, if

$$\$17.40x = \$48,000 + \$12.40x$$

Step 4 $17.4x = 48,000 + 12.4x$ Subtract $12.4x$ from both sides.

$5x = 48,000$ Divide both sides by 5.

$x = 9,600$

The company must make and sell 9,600 bike computers to break even.

Step 5 Check:

Costs	Revenue
$48,000 + 12.4(9,600)$	$17.4(9,600)$
$= \$167,040$	$= \$167,040$

Matched Problem 9 How many bike computers would a company have to make and sell to break even if the fixed costs are \$36,000, variable costs are \$10.40 per computer, and the computers are sold to retailers for \$15.20 each?

EXAMPLE 10 **Consumer Price Index** The Consumer Price Index (CPI) is a measure of the average change in prices over time from a designated reference period, which equals 100. The index is based on prices of basic consumer goods and services. Table 2 lists the CPI for several years from 1960 to 2016. What net annual salary in 2016 would have the same purchasing power as a net annual salary of \$13,000 in 1960? Compute the answer to the nearest dollar. (*Source:* U.S. Bureau of Labor Statistics)

Table 2 CPI (1982–1984 = 100)

Year	Index
1960	29.6
1973	44.4
1986	109.6
1999	156.9
2016	241.7

SOLUTION

Step 1 Let $x =$ the purchasing power of an annual salary in 2016.

Step 2 Annual salary in 1960 $= \$13,000$

CPI in 1960 $= 29.6$

CPI in 2016 $= 241.7$

Step 3 The ratio of a salary in 2016 to a salary in 1960 is the same as the ratio of the CPI in 2016 to the CPI in 1960.

$$\frac{x}{13,000} = \frac{241.7}{29.6}$$ Multiply both sides by 13,000.

Step 4 $$x = 13,000 \cdot \frac{241.7}{29.6}$$

$$= \$106,152 \text{ per year}$$

Step 5 To check the answer, we confirm that the salary ratio agrees with the CPI ratio:

Salary Ratio	CPI Ratio
$\frac{106,152}{13,000} = 8.166$	$\frac{241.7}{29.6} = 8.166$

Matched Problem 10 What net annual salary in 1973 would have had the same purchasing power as a net annual salary of \$100,000 in 2016? Compute the answer to the nearest dollar.

Exercises 1.1

A *In Problems 1–6, solve for x.*

1. $5x + 3 = x + 23$

2. $7x - 6 = 5x - 24$

3. $9(4 - x) = 2(x + 7)$

4. $3(x + 6) = 5 - 2(x + 1)$

5. $\frac{x+1}{4} = \frac{x}{2} + 5$

6. $\frac{2x+1}{3} - \frac{5x}{2} = 4$

In Problems 7–12, write the interval as an inequality or double inequality.

7. $[4, 13)$

8. $(-3, 5]$

9. $(-2, 7)$

10. $[-6, -1]$

11. $(-\infty, 4]$

12. $[9, \infty)$

In Problems 13–18, write the solution set using interval notation.

13. $-8 < x \le 2$

14. $-1 \le x < 5$

15. $2x < 18$

16. $3x \ge 12$

17. $15 \le -3x < 21$

18. $-8 < -4x \le 12$

B *In Problems 19–32, find the solution set.*

19. $\frac{x}{4} + \frac{1}{2} = \frac{1}{8}$

20. $\frac{m}{3} - 4 = \frac{2}{3}$

21. $\frac{y}{-5} > \frac{3}{2}$

22. $\frac{x}{-4} < \frac{5}{6}$

23. $2u + 4 = 5u + 1 - 7u$

24. $-3y + 9 + y = 13 - 8y$

25. $10x + 25(x - 3) = 275$

26. $-3(4 - x) = 5 - (x + 1)$

27. $3 - y \le 4(y - 3)$

28. $x - 2 \ge 2(x - 5)$

29. $\frac{x}{5} - \frac{x}{6} = \frac{6}{5}$

30. $\frac{y}{4} - \frac{y}{3} = \frac{1}{2}$

31. $\frac{m}{5} - 3 < \frac{3}{5} - \frac{m}{2}$

32. $\frac{u}{2} - \frac{2}{3} < \frac{u}{3} + 2$

In Problems 33–36, solve and graph.

33. $2 \le 3x - 7 < 14$

34. $-4 \le 5x + 6 < 21$

35. $-4 \le \frac{9}{5}C + 32 \le 68$

36. $-1 \le \frac{2}{3}t + 5 \le 11$

In Problems 37–42, solve for the indicated variable.

37. $3x - 4y = 12$; for y

38. $y = -\frac{2}{3}x + 8$; for x

39. $Ax + By = C$; for y $(B \ne 0)$

40. $y = mx + b$; for m

41. $F = \frac{9}{5}C + 32$; for C

42. $C = \frac{5}{9}(F - 32)$; for F

C *In Problems 43 and 44, solve and graph.*

43. $-3 \le 4 - 7x < 18$

44. $-10 \le 8 - 3u \le -6$

45. If both a and b are positive numbers and b/a is greater than 1, then is $a - b$ positive or negative?

46. If both a and b are negative numbers and b/a is greater than 1, then is $a - b$ positive or negative?

Applications

47. Ticket sales. A rock concert brought in $432,500 on the sale of 9,500 tickets. If the tickets sold for $35 and $55 each, how many of each type of ticket were sold?

48. Parking meter coins. An all-day parking meter takes only dimes and quarters. If it contains 100 coins with a total value of $14.50, how many of each type of coin are in the meter?

49. IRA. You have $500,000 in an IRA (Individual Retirement Account) at the time you retire. You have the option of investing this money in two funds: Fund A pays 5.2% annually and Fund B pays 7.7% annually. How should you divide your money between Fund A and Fund B to produce an annual interest income of $34,000?

50. IRA. Refer to Problem 49. How should you divide your money between Fund A and Fund B to produce an annual interest income of $30,000?

51. Car prices. If the price change of cars parallels the change in the CPI (see Table 2 in Example 10), what would a car sell for (to the nearest dollar) in 2016 if a comparable model sold for $10,000 in 1999?

52. Home values. If the price change in houses parallels the CPI (see Table 2 in Example 10), what would a house valued at $200,000 in 2016 be valued at (to the nearest dollar) in 1960?

53. Retail and wholesale prices. Retail prices in a department store are obtained by marking up the wholesale price by 40%. That is, the retail price is obtained by adding 40% of the wholesale price to the wholesale price.

(A) What is the retail price of a suit if the wholesale price is $300?

(B) What is the wholesale price of a pair of jeans if the retail price is $77?

54. Retail and sale prices. Sale prices in a department store are obtained by marking down the retail price by 15%. That is, the sale price is obtained by subtracting 15% of the retail price from the retail price.

(A) What is the sale price of a hat that has a retail price of $60?

(B) What is the retail price of a dress that has a sale price of $136?

55. Equipment rental. A golf course charges \$52 for a round of golf using a set of their clubs, and \$44 if you have your own clubs. If you buy a set of clubs for \$270, how many rounds must you play to recover the cost of the clubs?

56. Equipment rental. The local supermarket rents carpet cleaners for \$20 a day. These cleaners use shampoo in a special cartridge that sells for \$16 and is available only from the supermarket. A home carpet cleaner can be purchased for \$300. Shampoo for the home cleaner is readily available for \$9 a bottle. Past experience has shown that it takes two shampoo cartridges to clean the 10-foot-by-12-foot carpet in your living room with the rented cleaner. Cleaning the same area with the home cleaner will consume three bottles of shampoo. If you buy the home cleaner, how many times must you clean the living-room carpet to make buying cheaper than renting?

57. Sales commissions. One employee of a computer store is paid a base salary of \$2,000 a month plus an 8% commission on all sales over \$7,000 during the month. How much must the employee sell in one month to earn a total of \$4,000 for the month?

58. Sales commissions. A second employee of the computer store in Problem 57 is paid a base salary of \$3,000 a month plus a 5% commission on all sales during the month.

(A) How much must this employee sell in one month to earn a total of \$4,000 for the month?

(B) Determine the sales level at which both employees receive the same monthly income.

(C) If employees can select either of these payment methods, how would you advise an employee to make this selection?

59. Break-even analysis. A publisher for a promising new novel figures fixed costs (overhead, advances, promotion, copy editing, typesetting) at \$55,000, and variable costs (printing, paper, binding, shipping) at \$1.60 for each book produced. If the book is sold to distributors for \$11 each, how many must be produced and sold for the publisher to break even?

60. Break-even analysis. The publisher of a new book figures fixed costs at \$92,000 and variable costs at \$2.10 for each book produced. If the book is sold to distributors for \$15 each, how many must be sold for the publisher to break even?

61. Break-even analysis. The publisher in Problem 59 finds that rising prices for paper increase the variable costs to \$2.10 per book.

(A) Discuss possible strategies the company might use to deal with this increase in costs.

(B) If the company continues to sell the books for \$11, how many books must they sell now to make a profit?

(C) If the company wants to start making a profit at the same production level as before the cost increase, how much should they sell the book for now?

62. Break-even analysis. The publisher in Problem 60 finds that rising prices for paper increase the variable costs to \$2.70 per book.

(A) Discuss possible strategies the company might use to deal with this increase in costs.

(B) If the company continues to sell the books for \$15, how many books must they sell now to make a profit?

(C) If the company wants to start making a profit at the same production level as before the cost increase, how much should they sell the book for now?

63. Wildlife management. A naturalist estimated the total number of rainbow trout in a certain lake using the capture–mark–recapture technique. He netted, marked, and released 200 rainbow trout. A week later, allowing for thorough mixing, he again netted 200 trout, and found 8 marked ones among them. Assuming that the proportion of marked fish in the second sample was the same as the proportion of all marked fish in the total population, estimate the number of rainbow trout in the lake.

64. Temperature conversion. If the temperature for a 24-hour period at an Antarctic station ranged between $-49°F$ and $14°F$ (that is, $-49 \le F \le 14$), what was the range in degrees Celsius? [*Note:* $F = \frac{9}{5}C + 32$.]

65. Psychology. The IQ (intelligence quotient) is found by dividing the mental age (MA), as indicated on standard tests, by the chronological age (CA) and multiplying by 100. For example, if a child has a mental age of 12 and a chronological age of 8, the calculated IQ is 150. If a 9-year-old girl has an IQ of 140, compute her mental age.

66. Psychology. Refer to Problem 65. If the IQ of a group of 12-year-old children varies between 80 and 140, what is the range of their mental ages?

Answers to Matched Problems

1. $x = 4$

2. $x = 2$

3. (A) $L = \dfrac{S - 2WH}{2W + 2H}$ (B) $H = \dfrac{S - 2LW}{2L + 2W}$

4. (A) $<$ (B) $<$ (C) $>$

5. (A) $-7 < x \le 4$;

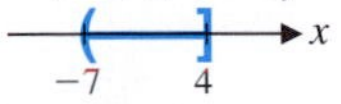

(B) $(-\infty, 3)$

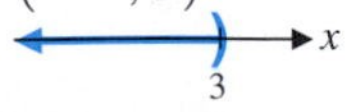

6. $x \ge -4$ or $[-4, \infty)$

7. $-1 \le x < 4$ or $[-1, 4)$

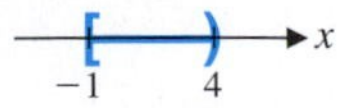

8. \$26,000

9. 7,500 bike computers

10. \$18,370

1.2 Graphs and Lines

- Cartesian Coordinate System
- Graphs of $Ax + By = C$
- Slope of a Line
- Equations of Lines: Special Forms
- Applications

In this section, we will consider one of the most basic geometric figures—a line. When we use the term *line* in this book, we mean *straight line*. We will learn how to recognize and graph a line and how to use information concerning a line to find its equation. Examining the graph of any equation often results in additional insight into the nature of the equation's solutions.

Cartesian Coordinate System

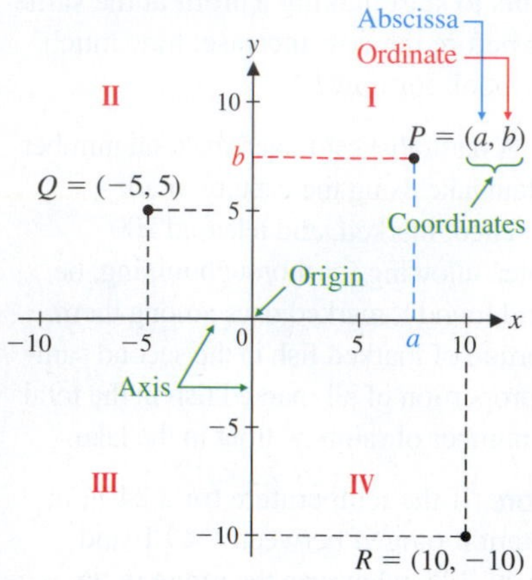

Figure 1 The Cartesian (rectangular) coordinate system

Recall that to form a **Cartesian** or **rectangular coordinate system**, we select two real number lines—one horizontal and one vertical—and let them cross through their origins as indicated in Figure 1. Up and to the right are the usual choices for the positive directions. These two number lines are called the **horizontal axis** and the **vertical axis**, or, together, the **coordinate axes**. The horizontal axis is usually referred to as the ***x* axis** and the vertical axis as the ***y* axis**, and each is labeled accordingly. The coordinate axes divide the plane into four parts called **quadrants**, which are numbered counterclockwise from I to IV (see Fig. 1).

Now we want to assign *coordinates* to each point in the plane. Given an arbitrary point P in the plane, pass horizontal and vertical lines through the point (Fig. 1). The vertical line will intersect the horizontal axis at a point with coordinate a, and the horizontal line will intersect the vertical axis at a point with coordinate b. These two numbers, written as the **ordered pair** (a, b), form the **coordinates** of the point P. The first coordinate, a, is called the **abscissa** of P; the second coordinate, b, is called the **ordinate** of P. The abscissa of Q in Figure 1 is -5, and the ordinate of Q is 5. The coordinates of a point can also be referenced in terms of the axis labels. The ***x* coordinate** of R in Figure 1 is 10, and the ***y* coordinate** of R is -10. The point with coordinates $(0, 0)$ is called the **origin**.

The procedure we have just described assigns to each point P in the plane a unique pair of real numbers (a, b). Conversely, if we are given an ordered pair of real numbers (a, b), then, reversing this procedure, we can determine a unique point P in the plane. Thus,

There is a one-to-one correspondence between the points in a plane and the elements in the set of all ordered pairs of real numbers.

This is often referred to as the **fundamental theorem of analytic geometry**.

Graphs of $Ax + By = C$

In Section 1.1, we called an equation of the form $ax + b = 0$ $(a \neq 0)$ a linear equation in one variable. Now we want to consider linear equations in two variables:

DEFINITION Linear Equations in Two Variables

A **linear equation in two variables** is an equation that can be written in the **standard form**

$$Ax + By = C$$

where A, B, and C are constants (A and B not both 0), and x and y are variables.

A **solution** of an equation in two variables is an ordered pair of real numbers that satisfies the equation. For example, $(4, 3)$ is a solution of $3x - 2y = 6$. The **solution set** of an equation in two variables is the set of all solutions of the equation. The **graph** of an equation is the graph of its solution set.

Explore and Discuss 1

(A) As noted earlier, $(4, 3)$ is a solution of the equation

$$3x - 2y = 6$$

Find three more solutions of this equation. Plot these solutions in a Cartesian coordinate system. What familiar geometric shape could be used to describe the solution set of this equation?

(B) Repeat part (A) for the equation $x = 2$.

(C) Repeat part (A) for the equation $y = -3$.

In Explore and Discuss 1, you may have recognized that the graph of each equation is a (straight) line. Theorem 1 confirms this fact.

THEOREM 1 Graph of a Linear Equation in Two Variables

The graph of any equation of the form

$$Ax + By = C \qquad (A \text{ and } B \text{ not both } 0) \tag{1}$$

is a line, and any line in a Cartesian coordinate system is the graph of an equation of this form.

If $A \neq 0$ and $B \neq 0$, then equation (1) can be written as

$$y = -\frac{A}{B}x + \frac{C}{B} = mx + b,\ m \neq 0$$

If $A = 0$ and $B \neq 0$, then equation (1) can be written as

$$y = \frac{C}{B}$$

and its graph is a **horizontal line**. If $A \neq 0$ and $B = 0$, then equation (1) can be written as

$$x = \frac{C}{A}$$

and its graph is a **vertical line**. To graph equation (1), or any of its special cases, plot any two points in the solution set and use a straightedge to draw the line through these two points. The points where the line crosses the axes are often the easiest to find. The ***y* intercept** is the y coordinate of the point where the graph crosses the y axis, and the ***x* intercept** is the x coordinate of the point where the graph crosses the x axis. To find the y intercept, let $x = 0$ and solve for y. To find the x intercept, let $y = 0$ and solve for x. It is a good idea to find a third point as a check point.

Reminder

If the x intercept is a and the y intercept is b, then the graph of the line passes through the points $(a, 0)$ and $(0, b)$. It is common practice to refer to both the numbers a and b and the points $(a, 0)$ and $(0, b)$ as the x and y intercepts of the line.

EXAMPLE 1 **Using Intercepts to Graph a Line** Graph: $3x - 4y = 12$

SOLUTION Find and plot the y intercept, the x intercept, and a check point (Fig. 2).

x	y	
0	−3	y intercept
4	0	x intercept
8	3	Check point

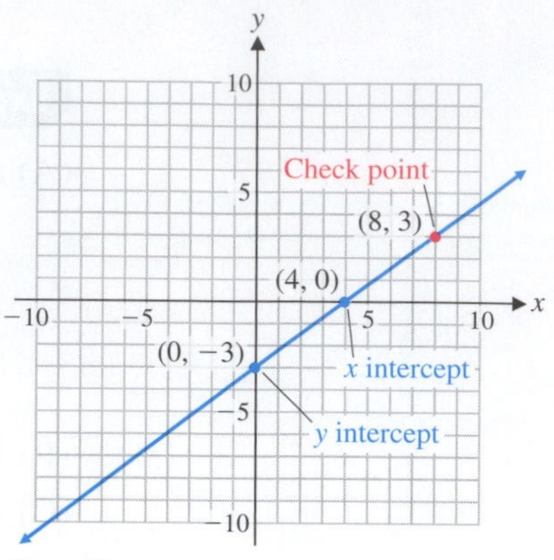

Figure 2

Matched Problem 1 Graph: $4x - 3y = 12$

The icon in the margin is used throughout this book to identify optional graphing calculator activities that are intended to give you additional insight into the concepts under discussion. You may have to consult the manual for your calculator for the details necessary to carry out these activities.

EXAMPLE 2 **Using a Graphing Calculator** Graph $3x - 4y = 12$ on a graphing calculator and find the intercepts.

SOLUTION First, we solve $3x - 4y = 12$ for y.

$$3x - 4y = 12 \qquad \text{Add } -3x \text{ to both sides.}$$

$$-4y = -3x + 12 \qquad \text{Divide both sides by } -4.$$

$$y = \frac{-3x + 12}{-4} \qquad \text{Simplify.}$$

$$y = \frac{3}{4}x - 3 \qquad (2)$$

Now we enter the right side of equation (2) in a calculator (Fig. 3A), enter values for the window variables (Fig. 3B), and graph the line (Fig. 3C). (The numerals to the left and right of the screen in Figure 3C are Xmin and Xmax, respectively. Similarly, the numerals below and above the screen are Ymin and Ymax.)

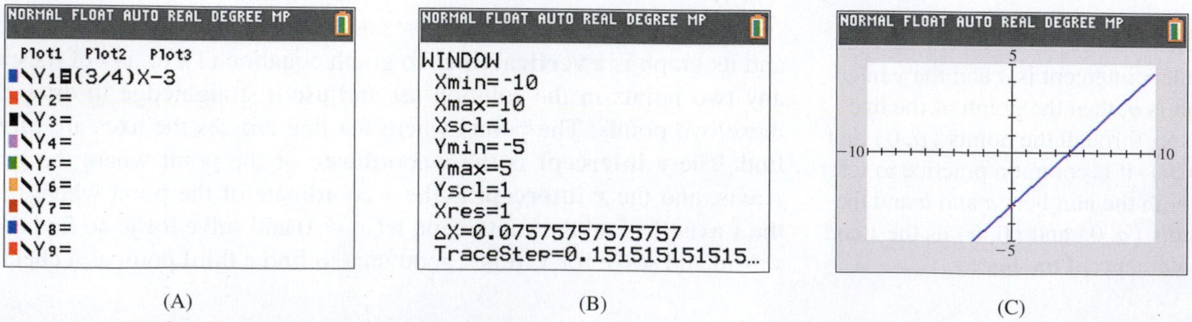

(A) (B) (C)

Figure 3 Graphing a line on a graphing calculator

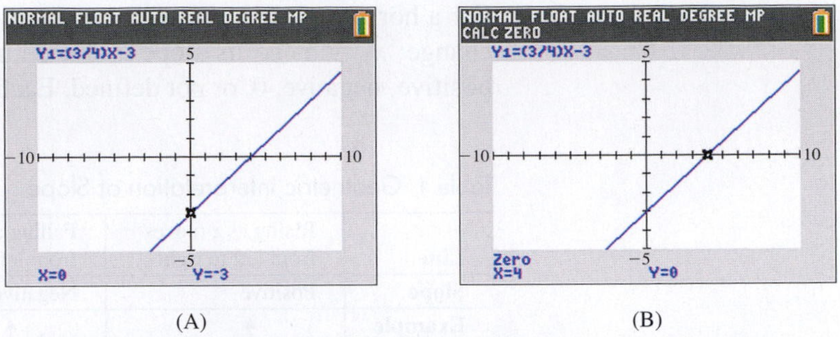

Figure 4 Using TRACE and ZERO on a graphing calculator

Next we use two calculator commands to find the intercepts: TRACE (Fig. 4A) and ZERO (Fig. 4B). The y intercept is -3 (Fig. 4A), and the x intercept is 4 (Fig. 4B).

Matched Problem 2 Graph $4x - 3y = 12$ on a graphing calculator and find the intercepts.

EXAMPLE 3 **Horizontal and Vertical Lines**

(A) Graph $x = -4$ and $y = 6$ simultaneously in the same rectangular coordinate system.

(B) Write the equations of the vertical and horizontal lines that pass through the point $(7, -5)$.

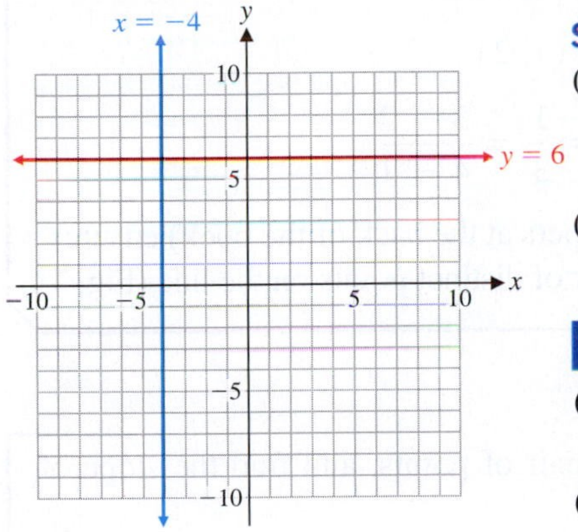

Figure 5

SOLUTION

(A) The line $x = -4$ consists of all points with x coordinate -4. To graph it, draw the vertical line through $(-4, 0)$. The line $y = 6$ consists of all points with y coordinate 6. To graph it, draw the horizontal line through $(0, 6)$. See Figure 5.

(B) Horizontal line through $(7, -5)$: $y = -5$
Vertical line through $(7, -5)$: $x = 7$

Matched Problem 3

(A) Graph $x = 5$ and $y = -3$ simultaneously in the same rectangular coordinate system.

(B) Write the equations of the vertical and horizontal lines that pass through the point $(-8, 2)$.

Slope of a Line

If we take two points, (x_1, y_1) and (x_2, y_2), on a line, then the ratio of the change in y to the change in x is called the **slope** of the line. In a sense, slope provides a measure of the "steepness" of a line relative to the x axis. The change in x is often called the **run**, and the change in y is the **rise**.

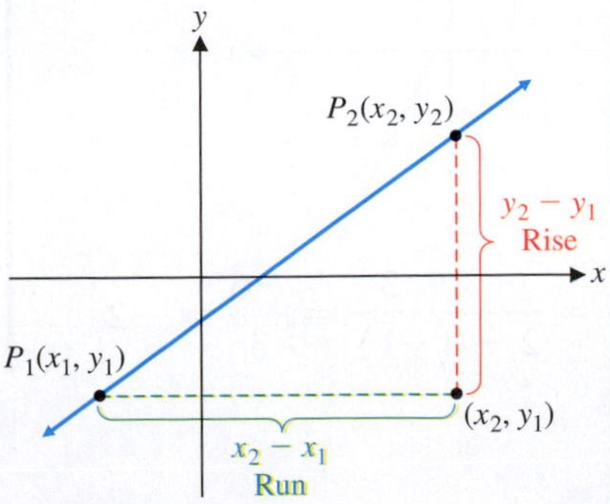

Figure 6

DEFINITION **Slope of a Line**

If a line passes through two distinct points, (x_1, y_1) and (x_2, y_2) (see Fig. 6), then its slope is given by the formula

$$m = \frac{y_2 - y_1}{x_2 - x_1} \qquad x_1 \neq x_2$$

$$= \frac{\text{vertical change (rise)}}{\text{horizontal change (run)}}$$

For a horizontal line, y does not change; its slope is 0. For a vertical line, x does not change; $x_1 = x_2$ so its slope is not defined. In general, the slope of a line may be positive, negative, 0, or not defined. Each case is illustrated geometrically in Table 1.

Table 1 Geometric Interpretation of Slope

Line	Rising as x moves from left to right	Falling as x moves from left to right	Horizontal	Vertical
Slope	Positive	Negative	0	Not defined
Example	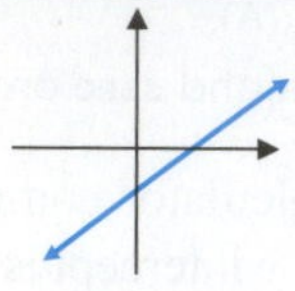	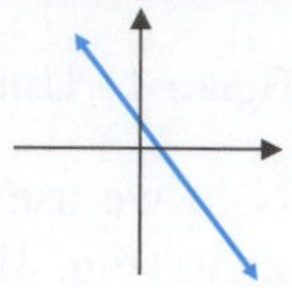	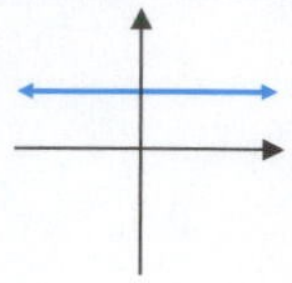	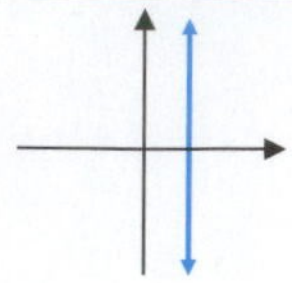

CONCEPTUAL INSIGHT

One property of real numbers discussed in Appendix A, Section A.1, is

$$\frac{-a}{-b} = -\frac{-a}{b} = -\frac{a}{-b} = \frac{a}{b}, \quad b \neq 0$$

This property implies that the slope of the line through A and B is equal to the slope of the line through B and A. For example, if $A = (4, 3)$ and $B = (1, 2)$, then

$$B = (1, 2) \qquad A = (4, 3)$$
$$A = (4, 3) \qquad B = (1, 2)$$

$$m = \frac{2-3}{1-4} = \frac{-1}{-3} = \frac{1}{3} = \frac{3-2}{4-1}$$

A property of similar triangles (see the endpapers at the back of the book) ensures that the slope of a line is the same for any pair of distinct points on the line (Fig. 7).

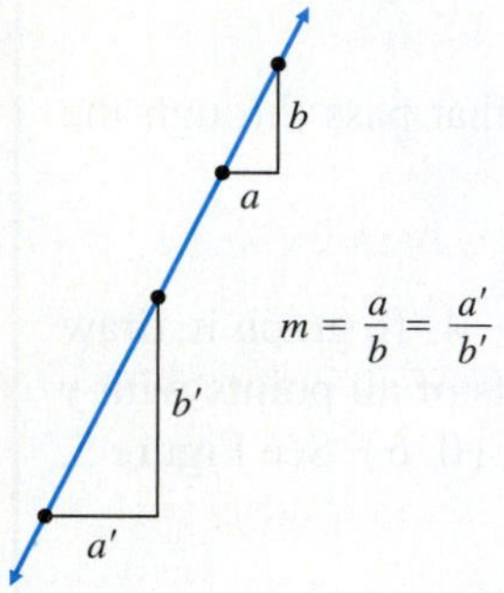

Figure 7

EXAMPLE 4 **Finding Slopes** Sketch a line through each pair of points, and find the slope of each line.

(A) $(-3, -2), (3, 4)$ (B) $(-1, 3), (2, -3)$

(C) $(-2, -3), (3, -3)$ (D) $(-2, 4), (-2, -2)$

SOLUTION

(A)

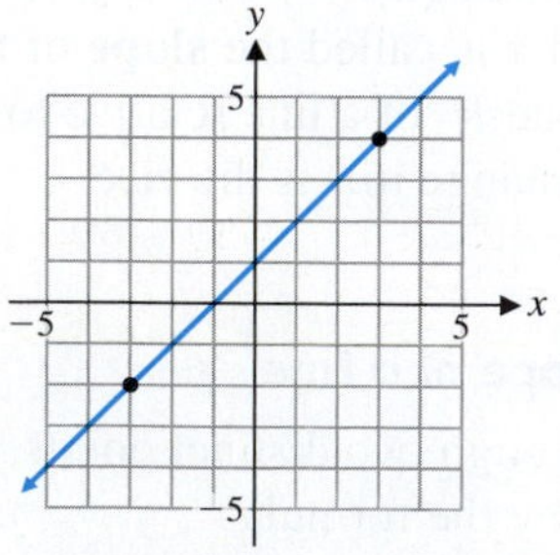

$$m = \frac{4 - (-2)}{3 - (-3)} = \frac{6}{6} = 1$$

(B)

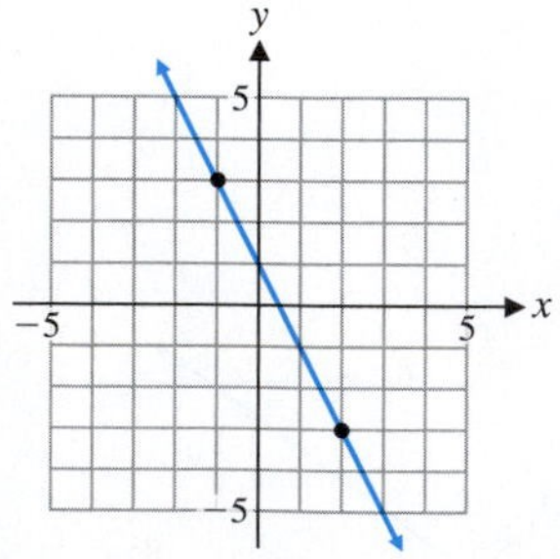

$$m = \frac{-3 - 3}{2 - (-1)} = \frac{-6}{3} = -2$$

(C)

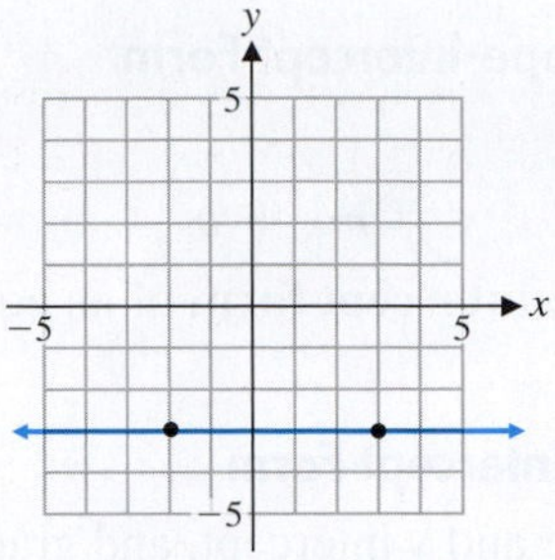

$$m = \frac{-3 - (-3)}{3 - (-2)} = \frac{0}{5} = 0$$

(D)

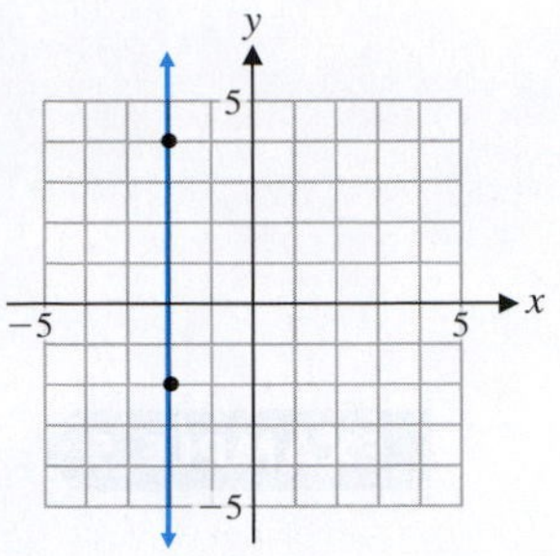

$$m = \frac{-2 - 4}{-2 - (-2)} = \frac{-6}{0}$$

Slope is not defined.

Matched Problem 4 Find the slope of the line through each pair of points.

(A) $(-2, 4), (3, 4)$ (B) $(-2, 4), (0, -4)$

(C) $(-1, 5), (-1, -2)$ (D) $(-1, -2), (2, 1)$

Equations of Lines: Special Forms

Let us start by investigating why $y = mx + b$ is called the *slope-intercept form* for a line.

Explore and Discuss 2

(A) Graph $y = x + b$ for $b = -5, -3, 0, 3$, and 5 simultaneously in the same coordinate system. Verbally describe the geometric significance of b.

(B) Graph $y = mx - 1$ for $m = -2, -1, 0, 1$, and 2 simultaneously in the same coordinate system. Verbally describe the geometric significance of m.

(C) Using a graphing calculator, explore the graph of $y = mx + b$ for different values of m and b.

As you may have deduced from Explore and Discuss 2, constants m and b in $y = mx + b$ have the following geometric interpretations.

If we let $x = 0$, then $y = b$. So the graph of $y = mx + b$ crosses the y axis at $(0, b)$. The constant b is the *y intercept*. For example, the y intercept of the graph of $y = -4x - 1$ is -1.

To determine the geometric significance of m, we proceed as follows: If $y = mx + b$, then by setting $x = 0$ and $x = 1$, we conclude that $(0, b)$ and $(1, m + b)$ lie on its graph (Fig. 8). The slope of this line is given by:

$$\text{Slope} = \frac{y_2 - y_1}{x_2 - x_1} = \frac{(m + b) - b}{1 - 0} = m$$

So m is the slope of the line given by $y = mx + b$.

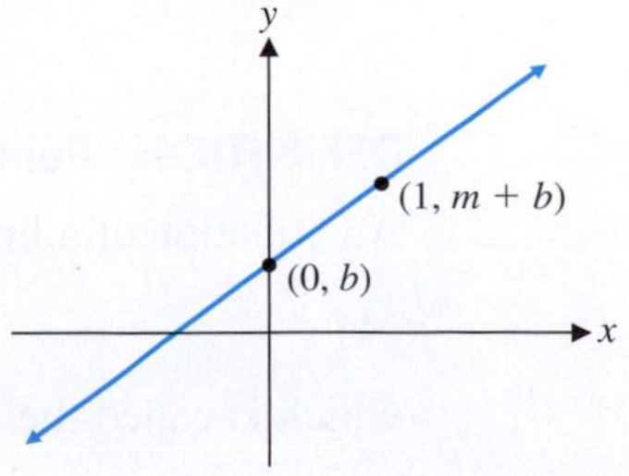

Figure 8

DEFINITION Slope-Intercept Form

The equation

$$y = mx + b \qquad m = \text{slope}, b = y \text{ intercept} \tag{3}$$

is called the **slope-intercept form** of an equation of a line.

EXAMPLE 5 **Using the Slope-Intercept Form**

(A) Find the slope and y intercept, and graph $y = -\frac{2}{3}x - 3$.

(B) Write the equation of the line with slope $\frac{2}{3}$ and y intercept -2.

SOLUTION

(A) Slope $= m = -\frac{2}{3}$; y intercept $= b = -3$.

To graph the line, first plot the y intercept $(0, -3)$. Then, since the slope is $-\frac{2}{3}$, locate a second point on the line by moving 3 units in the x direction (run) and -2 units in the y direction (rise). Draw the line through these two points (see Fig. 9).

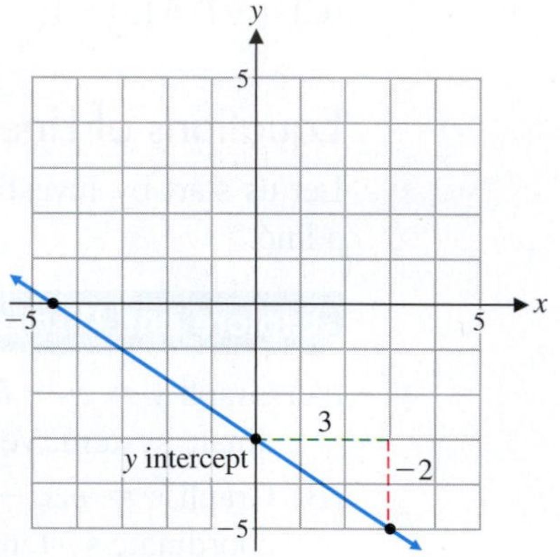

Figure 9

(B) $m = \frac{2}{3}$ and $b = -2$; so, $y = \frac{2}{3}x - 2$

Matched Problem 5 Write the equation of the line with slope $\frac{1}{2}$ and y intercept -1. Graph.

Suppose that a line has slope m and passes through a fixed point (x_1, y_1). If the point (x, y) is any other point on the line (Fig. 10), then

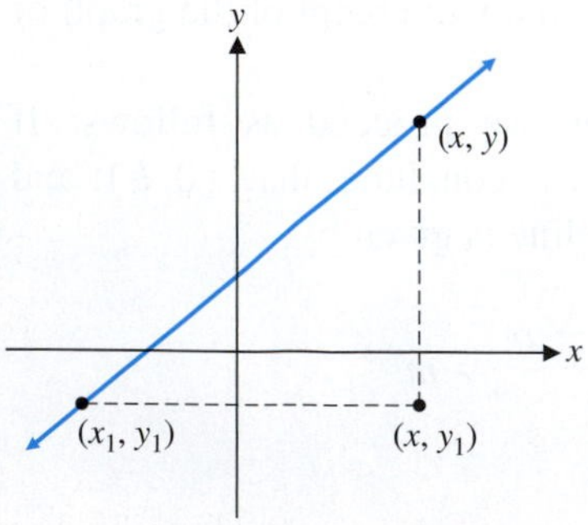

Figure 10

$$\frac{y - y_1}{x - x_1} = m$$

That is,

$$y - y_1 = m(x - x_1) \tag{4}$$

We now observe that (x_1, y_1) also satisfies equation (4) and conclude that equation (4) is an equation of a line with slope m that passes through (x_1, y_1).

DEFINITION Point-Slope Form

An equation of a line with slope m that passes through (x_1, y_1) is

$$y - y_1 = m(x - x_1) \tag{4}$$

which is called the **point-slope form** of an equation of a line.

The point-slope form is extremely useful, since it enables us to find an equation for a line if we know its slope and the coordinates of a point on the line or if we know the coordinates of two points on the line.

EXAMPLE 6 **Using the Point-Slope Form**

(A) Find an equation for the line that has slope $\frac{1}{2}$ and passes through $(-4, 3)$. Write the final answer in the form $Ax + By = C$.

(B) Find an equation for the line that passes through the points $(-3, 2)$ and $(-4, 5)$. Write the resulting equation in the form $y = mx + b$.

SOLUTION

(A) Use $y - y_1 = m(x - x_1)$. Let $m = \frac{1}{2}$ and $(x_1, y_1) = (-4, 3)$. Then

$$y - 3 = \tfrac{1}{2}[x - (-4)]$$

$$y - 3 = \tfrac{1}{2}(x + 4) \qquad \text{Multiply both sides by 2.}$$

$$2y - 6 = x + 4$$

$$-x + 2y = 10 \quad \text{or} \quad x - 2y = -10$$

(B) First, find the slope of the line by using the slope formula:

$$m = \frac{y_2 - y_1}{x_2 - x_1} = \frac{5 - 2}{-4 - (-3)} = \frac{3}{-1} = -3$$

Now use $y - y_1 = m(x - x_1)$ with $m = -3$ and $(x_1, y_1) = (-3, 2)$:

$$y - 2 = -3[x - (-3)]$$

$$y - 2 = -3(x + 3)$$

$$y - 2 = -3x - 9$$

$$y = -3x - 7$$

Matched Problem 6

(A) Find an equation for the line that has slope $\frac{2}{3}$ and passes through $(6, -2)$. Write the resulting equation in the form $Ax + By = C, A > 0$.

(B) Find an equation for the line that passes through $(2, -3)$ and $(4, 3)$. Write the resulting equation in the form $y = mx + b$.

The various forms of the equation of a line that we have discussed are summarized in Table 2 for quick reference.

Table 2 Equations of a Line

Standard form	$Ax + By = C$	A and B not both 0
Slope-intercept form	$y = mx + b$	Slope: m; y intercept: b
Point-slope form	$y - y_1 = m(x - x_1)$	Slope: m; point: (x_1, y_1)
Horizontal line	$y = b$	Slope: 0
Vertical line	$x = a$	Slope: undefined

Applications

We will now see how equations of lines occur in certain applications.

EXAMPLE 7 **Cost Equation** The management of a company that manufactures skateboards has fixed costs (costs at 0 output) of \$300 per day and total costs of \$4,300 per day at an output of 100 skateboards per day. Assume that cost C is linearly related to output x.

(A) Find the slope of the line joining the points associated with outputs of 0 and 100, that is, the line passing through $(0, 300)$ and $(100, 4{,}300)$.

(B) Find an equation of the line relating output to cost. Write the final answer in the form $C = mx + b$.

(C) Graph the cost equation from part (B) for $0 \le x \le 200$.

SOLUTION

(A)
$$m = \frac{y_2 - y_1}{x_2 - x_1} = \frac{4{,}300 - 300}{100 - 0} = \frac{4{,}000}{100} = 40$$

(B) We must find an equation of the line that passes through $(0, 300)$ with slope 40. We use the slope-intercept form:

$$C = mx + b$$
$$C = 40x + 300$$

(C) To graph $C = 40x + 300$ for $0 \le x \le 200$, we first calculate $C(200)$:

$$C(200) = 40(200) + 300 = 8{,}300$$

Plot the points $(200, 8{,}300)$ and $(0, 300)$ and draw the line segment joining them (see Fig. 11).

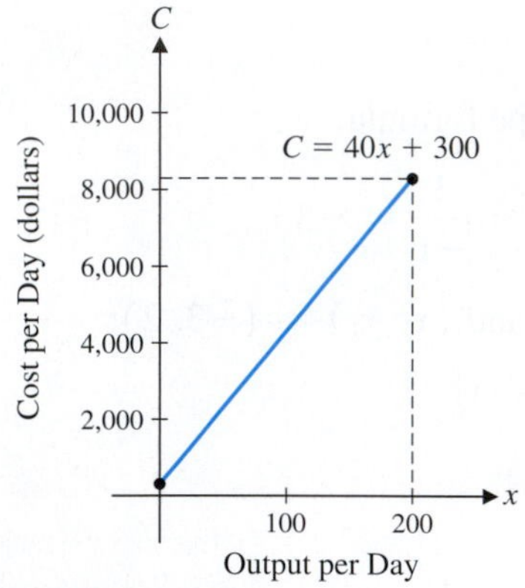

Figure 11

In Example 7, the *fixed cost* of \$300 per day covers plant cost, insurance, and so on. This cost is incurred whether or not there is any production. The *variable cost* is $40x$, which depends on the day's output. Since increasing production from x to $x + 1$ will increase the cost by \$40 (from $40x + 300$ to $40x + 340$), the slope 40 can be interpreted as the **rate of change** of the cost function with respect to production x.

Matched Problem 7 Answer parts (A) and (B) in Example 7 for fixed costs of \$250 per day and total costs of \$3,450 per day at an output of 80 skateboards per day.

In a free competitive market, the price of a product is determined by the relationship between supply and demand. If there is a surplus—that is, the supply is greater than the demand—the price tends to come down. If there is a shortage—that is, the demand is greater than the supply—the price tends to go up. The price tends to move toward an equilibrium price at which the supply and demand are equal. Example 8 introduces the basic concepts.

EXAMPLE 8 **Supply and Demand** At a price of \$9.00 per box of oranges, the supply is 320,000 boxes and the demand is 200,000 boxes. At a price of \$8.50 per box, the supply is 270,000 boxes and the demand is 300,000 boxes.

(A) Find a price–supply equation of the form $p = mx + b$, where p is the price in dollars and x is the corresponding supply in thousands of boxes.

(B) Find a price–demand equation of the form $p = mx + b$, where p is the price in dollars and x is the corresponding demand in thousands of boxes.

(C) Graph the price–supply and price–demand equations in the same coordinate system and find their point of intersection.

SOLUTION

(A) To find a price–supply equation of the form $p = mx + b$, we must find two points of the form (x, p) that are on the supply line. From the given supply data, $(320, 9)$ and $(270, 8.5)$ are two such points. First, find the slope of the line:

$$m = \frac{9 - 8.5}{320 - 270} = \frac{0.5}{50} = 0.01$$

Now use the point-slope form to find the equation of the line:

$$\begin{aligned} p - p_1 &= m(x - x_1) && (x_1, p_1) = (320, 9) \\ p - 9 &= 0.01(x - 320) \\ p - 9 &= 0.01x - 3.2 \\ p &= 0.01x + 5.8 && \text{Price–supply equation} \end{aligned}$$

(B) From the given demand data, $(200, 9)$ and $(300, 8.5)$ are two points on the demand line.

$$m = \frac{8.5 - 9}{300 - 200} = \frac{-0.5}{100} = -0.005$$

$$\begin{aligned} p - p_1 &= m(x - x_1) && \text{Substitute } x_1 = 200,\ p_1 = 9. \\ p - 9 &= -0.005(x - 200) \\ p - 9 &= -0.005x + 1 \\ p &= -0.005x + 10 && \text{Price–demand equation} \end{aligned}$$

(C) From part (A), we plot the points $(320, 9)$ and $(270, 8.5)$ and then draw the line through them. We do the same with the points $(200, 9)$ and $(300, 8.5)$ from part (B) (Fig. 12). (Note that we restricted the axes to intervals that contain these data points.) To find the intersection point of the two lines, we equate the right-hand sides of the price–supply and price–demand equations and solve for x:

Price–supply Price–demand

$$\begin{aligned} 0.01x + 5.8 &= -0.005x + 10 \\ 0.015x &= 4.2 \\ x &= 280 \end{aligned}$$

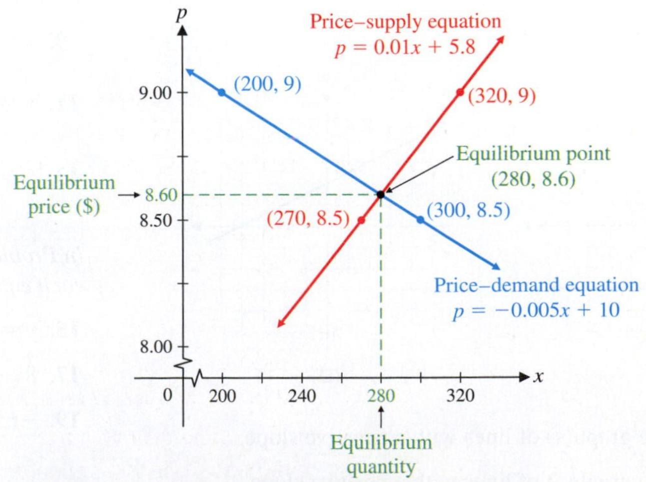

Figure 12 Graphs of price–supply and price–demand equations

Now use the price–supply equation to find p when $x = 280$:

$$p = 0.01x + 5.8$$

$$p = 0.01(280) + 5.8 = 8.6$$

As a check, we use the price–demand equation to find p when $x = 280$:

$$p = -0.005x + 10$$

$$p = -0.005(280) + 10 = 8.6$$

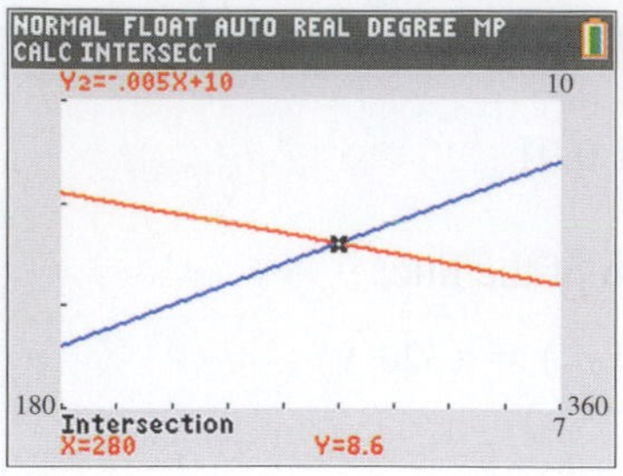

Figure 13 Finding an intersection point

The lines intersect at $(280, 8.6)$. The intersection point of the price–supply and price–demand equations is called the **equilibrium point**, and its coordinates are the **equilibrium quantity** (280) and the **equilibrium price** ($8.60). These terms are illustrated in Figure 12. The intersection point can also be found by using the INTERSECT command on a graphing calculator (Fig. 13). To summarize, the price of a box of oranges tends toward the equilibrium price of $8.60, at which the supply and demand are both equal to 280,000 boxes.

Matched Problem 8 At a price of $12.59 per box of grapefruit, the supply is 595,000 boxes and the demand is 650,000 boxes. At a price of $13.19 per box, the supply is 695,000 boxes and the demand is 590,000 boxes. Assume that the relationship between price and supply is linear and that the relationship between price and demand is linear.

(A) Find a price–supply equation of the form $p = mx + b$.

(B) Find a price–demand equation of the form $p = mx + b$.

(C) Find the equilibrium point.

Exercises 1.2

A *Problems 1–4 refer to graphs (A)–(D).*

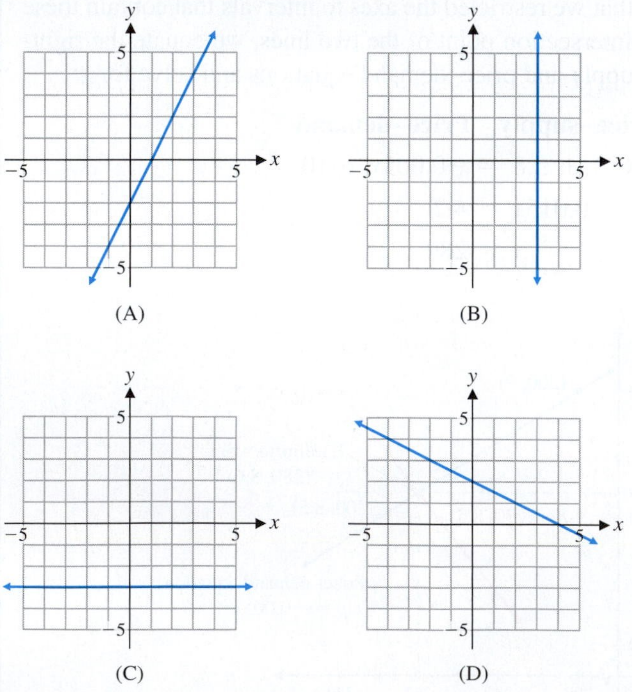

1. Identify the graph(s) of lines with a negative slope.

2. Identify the graph(s) of lines with a positive slope.

3. Identify the graph(s) of any lines with slope zero.

4. Identify the graph(s) of any lines with undefined slope.

In Problems 5–8, sketch a graph of each equation in a rectangular coordinate system.

5. $y = 2x - 3$

6. $y = \frac{x}{2} + 1$

7. $2x + 3y = 12$

8. $8x - 3y = 24$

In Problems 9–14, find the slope and y intercept of the graph of each equation.

9. $y = 5x - 7$

10. $y = 3x + 2$

11. $y = -\frac{5}{2}x - 9$

12. $y = -\frac{10}{3}x + 4$

13. $y = \frac{x}{4} + \frac{2}{3}$

14. $y = \frac{x}{5} - \frac{1}{2}$

In Problems 15–20, find the slope and x intercept of the graph of each equation.

15. $y = 2x + 10$

16. $y = -4x + 12$

17. $8x - y = 40$

18. $3x + y = 6$

19. $-6x + 7y = 42$

20. $9x + 2y = 4$

In Problems 21–24, write an equation of the line with the indicated slope and y intercept.

21. Slope = 2
y intercept = 1

22. Slope = 1
y intercept = 5

23. Slope $= -\frac{1}{3}$
y intercept = 6

24. Slope $= \frac{6}{7}$
y intercept $= -\frac{9}{2}$

B *In Problems 25–28, use the graph of each line to find the x intercept, y intercept, and slope. Write the slope-intercept form of the equation of the line.*

25.

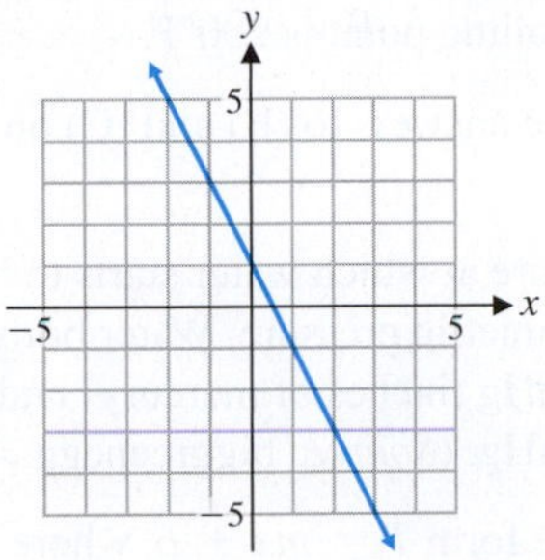

26.

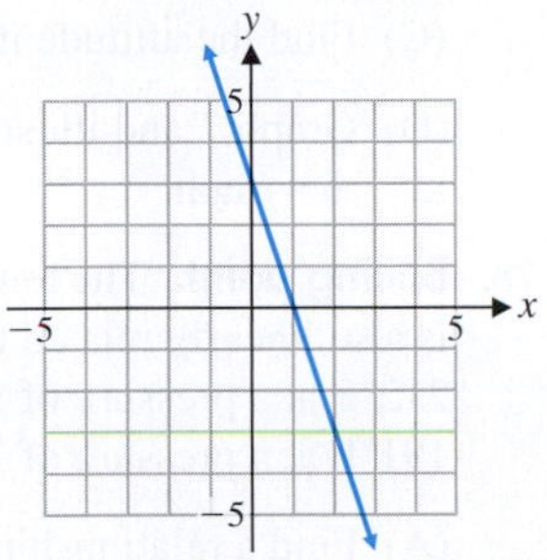

27.

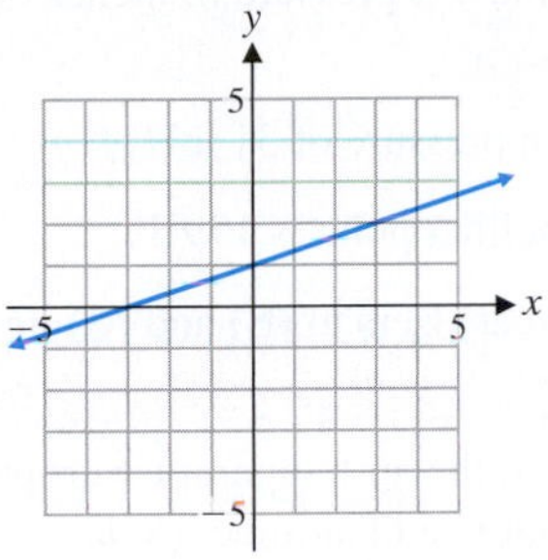

28.

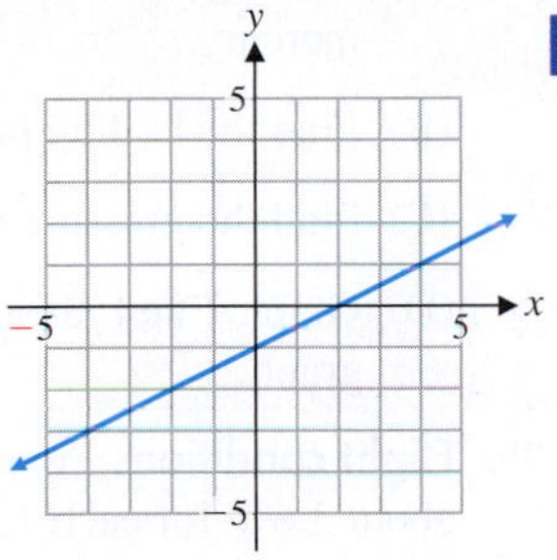

In Problems 29–34, sketch a graph of each equation or pair of equations in a rectangular coordinate system.

29. $y = -\frac{2}{3}x - 2$

30. $y = -\frac{3}{2}x + 1$

31. $3x - 2y = 10$

32. $5x - 6y = 15$

33. $x = 3; y = -2$

34. $x = -3; y = 2$

In Problems 35–40, find the slope of the graph of each equation.

35. $4x + y = 3$

36. $5x - y = -2$

37. $3x + 5y = 15$

38. $2x - 3y = 18$

39. $-4x + 2y = 9$

40. $-x + 8y = 4$

41. Given $Ax + By = 12$, graph each of the following three cases in the same coordinate system.

(A) $A = 2$ and $B = 0$

(B) $A = 0$ and $B = 3$

(C) $A = 3$ and $B = 4$

42. Given $Ax + By = 24$, graph each of the following three cases in the same coordinate system.

(A) $A = 6$ and $B = 0$

(B) $A = 0$ and $B = 8$

(C) $A = 2$ and $B = 3$

43. Graph $y = 25x + 200, x \ge 0$.

44. Graph $y = 40x + 160, x \ge 0$.

45. (A) Graph $y = 1.2x - 4.2$ in a rectangular coordinate system.

(B) Find the x and y intercepts algebraically to one decimal place.

(C) Graph $y = 1.2x - 4.2$ in a graphing calculator.

(D) Find the x and y intercepts to one decimal place using TRACE and the ZERO command.

46. (A) Graph $y = -0.8x + 5.2$ in a rectangular coordinate system.

(B) Find the x and y intercepts algebraically to one decimal place.

(C) Graph $y = -0.8x + 5.2$ in a graphing calculator.

(D) Find the x and y intercepts to one decimal place using TRACE and the ZERO command.

In Problems 47–50, write the equations of the vertical and horizontal lines through each point.

47. $(4, -3)$

48. $(-5, 6)$

49. $(-1.5, -3.5)$

50. $(2.6, 3.8)$

C *In Problems 51–58, write the slope-intercept form of the equation of the line with the indicated slope that goes through the given point.*

51. $m = 5; (3, 0)$

52. $m = 4; (0, 6)$

53. $m = -2; (-1, 9)$

54. $m = -10; (2, -5)$

55. $m = \frac{1}{3}; (-4, -8)$

56. $m = \frac{2}{7}; (7, 1)$

57. $m = -3.2; (5.8, 12.3)$

58. $m = 0.9; (2.3, 6.7)$

In Problems 59–66,

(A) *Find the slope of the line that passes through the given points.*

(B) *Find the standard form of the equation of the line.*

(C) *Find the slope-intercept form of the equation of the line.*

59. $(2, 5)$ and $(5, 7)$

60. $(1, 2)$ and $(3, 5)$

61. $(-2, -1)$ and $(2, -6)$

62. $(2, 3)$ and $(-3, 7)$

63. $(5, 3)$ and $(5, -3)$

64. $(1, 4)$ and $(0, 4)$

65. $(-2, 5)$ and $(3, 5)$

66. $(2, 0)$ and $(2, -3)$

67. Discuss the relationship among the graphs of the lines with equation $y = mx + 2$, where m is any real number.

68. Discuss the relationship among the graphs of the lines with equation $y = -0.5x + b$, where b is any real number.

Applications

69. Cost analysis. A donut shop has a fixed cost of \$124 per day and a variable cost of \$0.12 per donut. Find the total daily cost of producing x donuts. How many donuts can be produced for a total daily cost of \$250?

70. Cost analysis. A small company manufactures picnic tables. The weekly fixed cost is \$1,200 and the variable cost is \$45 per table. Find the total weekly cost of producing x picnic tables. How many picnic tables can be produced for a total weekly cost of \$4,800?

71. Cost analysis. A plant can manufacture 80 golf clubs per day for a total daily cost of \$7,647 and 100 golf clubs per day for a total daily cost of \$9,147.

(A) Assuming that daily cost and production are linearly related, find the total daily cost of producing x golf clubs.

(B) Graph the total daily cost for $0 \le x \le 200$.

(C) Interpret the slope and y intercept of this cost equation.

72. Cost analysis. A plant can manufacture 50 tennis rackets per day for a total daily cost of \$3,855 and 60 tennis rackets per day for a total daily cost of \$4,245.

(A) Assuming that daily cost and production are linearly related, find the total daily cost of producing x tennis rackets.

(B) Graph the total daily cost for $0 \le x \le 100$.

(C) Interpret the slope and y intercept of this cost equation.

73. Business—Markup policy. A drugstore sells a drug costing \$85 for \$112 and a drug costing \$175 for \$238.

(A) If the markup policy of the drugstore is assumed to be linear, write an equation that expresses retail price R in terms of cost C (wholesale price).

(B) What does a store pay (to the nearest dollar) for a drug that retails for \$185?

74. Business—Markup policy. A clothing store sells a shirt costing \$20 for \$33 and a jacket costing \$60 for \$93.

(A) If the markup policy of the store is assumed to be linear, write an equation that expresses retail price R in terms of cost C (wholesale price).

(B) What does a store pay for a suit that retails for \$240?

75. Business—Depreciation. A farmer buys a new tractor for \$157,000 and assumes that it will have a trade-in value of \$82,000 after 10 years. The farmer uses a constant rate of depreciation (commonly called **straight-line depreciation**—one of several methods permitted by the IRS) to determine the annual value of the tractor.

(A) Find a linear model for the depreciated value V of the tractor t years after it was purchased.

(B) What is the depreciated value of the tractor after 6 years?

(C) When will the depreciated value fall below \$70,000?

(D) Graph V for $0 \le t \le 20$ and illustrate the answers from parts (B) and (C) on the graph.

76. Business—Depreciation. A charter fishing company buys a new boat for \$224,000 and assumes that it will have a trade-in value of \$115,200 after 16 years.

(A) Find a linear model for the depreciated value V of the boat t years after it was purchased.

(B) What is the depreciated value of the boat after 10 years?

(C) When will the depreciated value fall below \$100,000?

(D) Graph V for $0 \le t \le 30$ and illustrate the answers from (B) and (C) on the graph.

77. Boiling point. The temperature at which water starts to boil is called its **boiling point** and is linearly related to the altitude. Water boils at 212°F at sea level and at 193.6°F at an altitude of 10,000 feet. (*Source:* biggreenegg.com)

(A) Find a relationship of the form $T = mx + b$ where T is degrees Fahrenheit and x is altitude in thousands of feet.

(B) Find the boiling point at an altitude of 3,500 feet.

(C) Find the altitude if the boiling point is 200°F.

(D) Graph T and illustrate the answers to (B) and (C) on the graph.

78. Boiling point. The temperature at which water starts to boil is also linearly related to barometric pressure. Water boils at 212°F at a pressure of 29.9 inHg (inches of mercury) and at 191°F at a pressure of 28.4 inHg. (*Source:* biggreenegg.com)

(A) Find a relationship of the form $T = mx + b$, where T is degrees Fahrenheit and x is pressure in inches of mercury.

(B) Find the boiling point at a pressure of 31 inHg.

(C) Find the pressure if the boiling point is 199°F.

(D) Graph T and illustrate the answers to (B) and (C) on the graph.

79. Flight conditions. In stable air, the air temperature drops about 3.6°F for each 1,000-foot rise in altitude. (*Source:* Federal Aviation Administration)

(A) If the temperature at sea level is 70°F, write a linear equation that expresses temperature T in terms of altitude A in thousands of feet.

(B) At what altitude is the temperature 34°F?

80. Flight navigation. The airspeed indicator on some aircraft is affected by the changes in atmospheric pressure at different altitudes. A pilot can estimate the true airspeed by observing the indicated airspeed and adding to it about 1.6% for every 1,000 feet of altitude. (*Source:* Megginson Technologies Ltd.)

(A) A pilot maintains a constant reading of 200 miles per hour on the airspeed indicator as the aircraft climbs from sea level to an altitude of 10,000 feet. Write a linear equation that expresses true airspeed T (in miles per hour) in terms of altitude A (in thousands of feet).

(B) What would be the true airspeed of the aircraft at 6,500 feet?

81. Demographics. The average number of persons per household in the United States has been shrinking steadily for as long as statistics have been kept and is approximately linear with respect to time. In 1980 there were about 2.76 persons per household, and in 2015 about 2.54. (*Source:* U.S. Census Bureau)

(A) If N represents the average number of persons per household and t represents the number of years since 1980, write a linear equation that expresses N in terms of t.

(B) Use this equation to estimate household size in the year 2030.

82. Demographics. The **median** household income divides the households into two groups: the half whose income is less than or equal to the median, and the half whose income is greater than the median. The median household income in the United States grew from about \$30,000 in 1990 to about \$55,775 in 2015. (*Source:* U.S. Census Bureau)

(A) If I represents the median household income and t represents the number of years since 1990, write a linear equation that expresses I in terms of t.

(B) Use this equation to estimate median household income in the year 2030.

83. Cigarette smoking. The percentage of female cigarette smokers in the United States declined from 21.0% in 2000 to 13.6% in 2015. (*Source:* Centers for Disease Control)

(A) Find a linear equation relating percentage of female smokers (f) to years since 2000 (t).

(B) Use this equation to predict the year in which the percentage of female smokers falls below 7%.

84. Cigarette smoking. The percentage of male cigarette smokers in the United States declined from 25.7% in 2000 to 16.7% in 2015. (*Source:* Centers for Disease Control)

(A) Find a linear equation relating percentage of male smokers (m) to years since 2000 (t).

(B) Use this equation to predict the year in which the percentage of male smokers falls below 7%.

85. Supply and demand. At a price of \$9.00 per bushel, the supply of soybeans is 3,600 million bushels and the demand is 4,000 million bushels. At a price of \$9.50 per bushel, the supply is 4,100 million bushels and the demand is 3,500 million bushels.

(A) Find a price–supply equation of the form $p = mx + b$.

(B) Find a price–demand equation of the form $p = mx + b$.

(C) Find the equilibrium point.

(D) Graph the price–supply equation, price–demand equation, and equilibrium point in the same coordinate system.

86. Supply and demand. At a price of \$3.20 per bushel, the supply of corn is 9,800 million bushels and the demand is 9,200 million bushels. At a price of \$2.95 per bushel, the supply is 9,300 million bushels and the demand is 9,700 million bushels.

(A) Find a price–supply equation of the form $p = mx + b$.

(B) Find a price–demand equation of the form $p = mx + b$.

(C) Find the equilibrium point.

(D) Graph the price–supply equation, price–demand equation, and equilibrium point in the same coordinate system.

87. Physics. Hooke's law states that the relationship between the stretch s of a spring and the weight w causing the stretch is linear. For a particular spring, a 5-pound weight causes a stretch of 2 inches, while with no weight, the stretch of the spring is 0.

(A) Find a linear equation that expresses s in terms of w.

(B) What is the stretch for a weight of 20 pounds?

(C) What weight will cause a stretch of 3.6 inches?

88. Physics. The distance d between a fixed spring and the floor is a linear function of the weight w attached to the bottom of the spring. The bottom of the spring is 18 inches from the floor when the weight is 3 pounds, and 10 inches from the floor when the weight is 5 pounds.

(A) Find a linear equation that expresses d in terms of w.

(B) Find the distance from the bottom of the spring to the floor if no weight is attached.

(C) Find the smallest weight that will make the bottom of the spring touch the floor. (Ignore the height of the weight.)

Answers to Matched Problems

1.

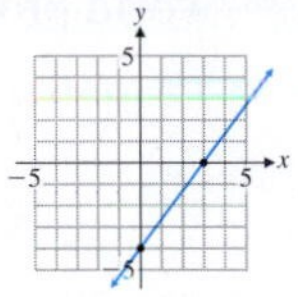

2. y intercept $= -4$, x intercept $= 3$

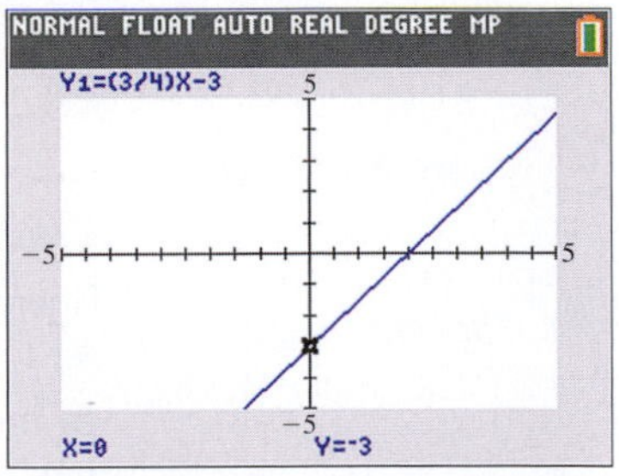

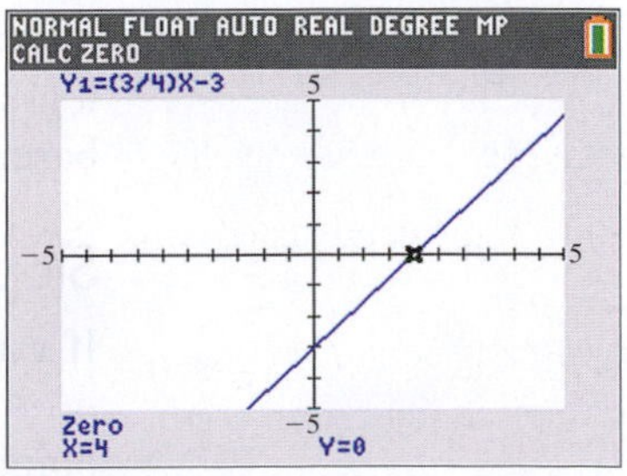

3. (A)

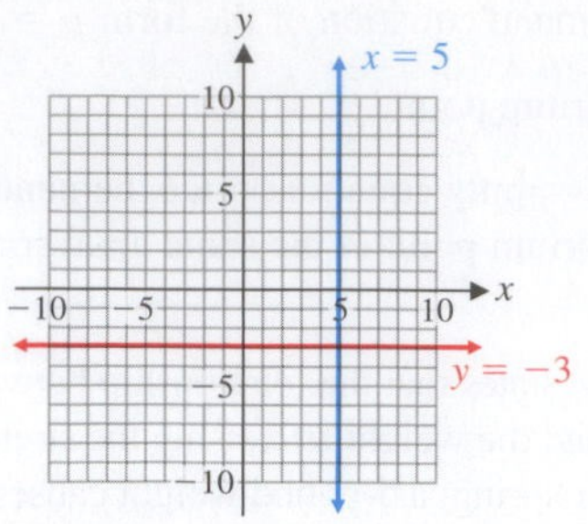

(B) Horizontal line: $y = 2$; vertical line: $x = -8$

4. (A) 0 (B) -4

(C) Not defined (D) 1

5. $y = \frac{1}{2}x - 1$

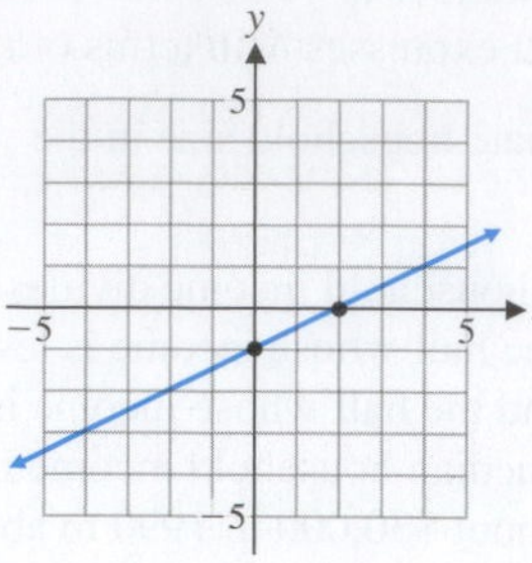

6. (A) $2x - 3y = 18$ (B) $y = 3x - 9$

7. (A) $m = 40$ (B) $C = 40x + 250$

8. (A) $p = 0.006x + 9.02$ (B) $p = -0.01x + 19.09$

(C) $(629, 12.80)$

1.3 Linear Regression

- Slope as a Rate of Change
- Linear Regression

Mathematical modeling is the process of using mathematics to solve real-world problems. This process can be broken down into three steps (Fig. 1):

Step 1 *Construct* the **mathematical model** (that is, a mathematics problem that, when solved, will provide information about the real-world problem).

Step 2 *Solve* the mathematical model.

Step 3 *Interpret* the solution to the mathematical model in terms of the original real-world problem.

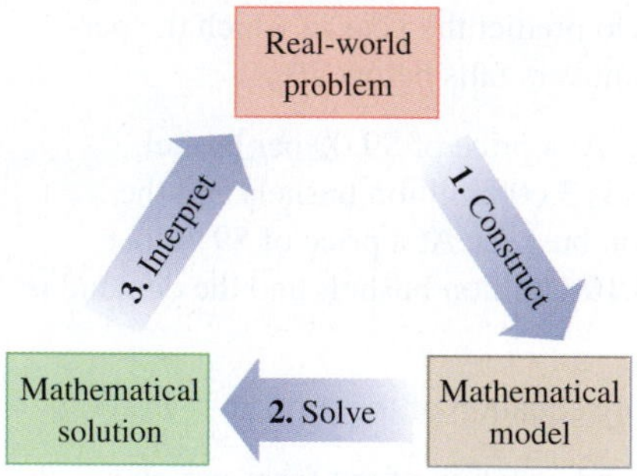

Figure 1

In more complex problems, this cycle may have to be repeated several times to obtain the required information about the real-world problem. In this section, we will discuss one of the simplest mathematical models, a linear equation. With the aid of a graphing calculator or computer, we also will learn how to analyze a linear model based on real-world data.

Slope as a Rate of Change

If x and y are related by the equation $y = mx + b$, where m and b are constants with $m \neq 0$, then x and y are **linearly related**. If (x_1, y_1) and (x_2, y_2) are two distinct points on this line, then the slope of the line is

$$m = \frac{y_2 - y_1}{x_2 - x_1} = \frac{\text{Change in } y}{\text{Change in } x} \tag{1}$$

In applications, ratio (1) is called the **rate of change** of y with respect to x. Since the slope of a line is unique, **the rate of change of two linearly related variables is constant**. Here are some examples of familiar rates of change: miles per hour, revolutions per minute, price per pound, passengers per plane, and so on. If the relationship between x and y is not linear, ratio (1) is called the **average rate of change** of y with respect to x.

EXAMPLE 1 **Estimating Body Surface Area** Appropriate doses of medicine for both animals and humans are often based on body surface area (BSA). Since weight is much easier to determine than BSA, veterinarians use the weight of an animal to estimate BSA. The following linear equation expresses BSA for canines in terms of weight:

$$a = 16.12w + 375.6$$

where a is BSA in square inches and w is weight in pounds. (*Source:* Veterinary Oncology Consultants, PTY LTD)

(A) Interpret the slope of the BSA equation.

(B) What is the effect of a one-pound increase in weight?

SOLUTION

(A) The rate-of-change of BSA with respect to weight is 16.12 square inches per pound.

(B) Since slope is the ratio of rise to run, increasing w by 1 pound (run) increases a by 16.12 square inches (rise).

Matched Problem 1 The equation $a = 28.55w + 118.7$ expresses BSA for felines in terms of weight, where a is BSA in square inches and w is weight in pounds.

(A) Interpret the slope of the BSA equation.

(B) What is the effect of a one-pound increase in weight?

Explore and Discuss 1

As illustrated in Example 1A, the slope m of a line with equation $y = mx + b$ has two interpretations:

1. m is the rate of change of y with respect to x.
2. Increasing x by one unit will change y by m units.

How are these two interpretations related?

Parachutes are used to deliver cargo to areas that cannot be reached by other means. The **rate of descent** of the cargo is the rate of change of altitude with respect to time. The absolute value of the rate of descent is called the **speed** of the cargo. At low altitudes, the altitude of the cargo and the time in the air are linearly related. The appropriate rate of descent varies widely with the item. Bulk food (rice, flour, beans, etc.) and clothing can tolerate nearly any rate of descent under 40 ft/sec. Machinery and electronics (pumps, generators, radios, etc.) should generally be dropped at 15 ft/sec or less. Butler Tactical Parachute Systems in Roanoke, Virginia, manufactures a variety of canopies for dropping cargo. The following example uses information taken from the company's brochures.

EXAMPLE 2 **Finding the Rate of Descent** A 100-pound cargo of delicate electronic equipment is dropped from an altitude of 2,880 feet and lands 200 seconds later. (*Source:* Butler Tactical Parachute Systems)

(A) Find a linear model relating altitude a (in feet) and time in the air t (in seconds).

(B) How fast is the cargo moving when it lands?

SOLUTION

(A) If $a = mt + b$ is the linear equation relating altitude a and time in air t, then the graph of this equation must pass through the following points:

$$(t_1, a_1) = (0, 2{,}880) \quad \text{Cargo is dropped from plane.}$$

$$(t_2, a_2) = (200, 0) \quad \text{Cargo lands.}$$

The slope of this line is

$$m = \frac{a_2 - a_1}{t_2 - t_1} = \frac{0 - 2{,}880}{200 - 0} = -14.4$$

and the equation of this line is

$$a - 0 = -14.4(t - 200)$$

$$a = -14.4t + 2{,}880$$

(B) The rate of descent is the slope $m = -14.4$, so the speed of the cargo at landing is $|-14.4| = 14.4$ ft/sec.

Matched Problem 2 A 400-pound load of grain is dropped from an altitude of 2,880 feet and lands 80 seconds later.

(A) Find a linear model relating altitude a (in feet) and time in the air t (in seconds).

(B) How fast is the cargo moving when it lands?

Linear Regression

In real-world applications, we often encounter numerical data in the form of a table. **Regression analysis** is a process for finding a function that provides a useful model for a set of data points. Graphs of equations are often called **curves**, and regression analysis is also referred to as **curve fitting**. In the next example, we use a linear model obtained by using **linear regression** on a graphing calculator.

EXAMPLE 3 **Diamond Prices** Prices for round-shaped diamonds taken from an online trader are given in Table 1.

(A) A linear model for the data in Table 1 is given by

$$p = 6{,}140c - 480 \qquad (2)$$

where p is the price of a diamond weighing c carats. (We will discuss the source of models like this later in this section.) Plot the points in Table 1 on a Cartesian coordinate system, producing a *scatter plot*, and graph the model on the same axes.

(B) Interpret the slope of the model in (2).

(C) Use the model to estimate the cost of a 0.85-carat diamond and the cost of a 1.2-carat diamond. Round answers to the nearest dollar.

(D) Use the model to estimate the weight of a diamond (to two decimal places) that sells for $4,000.

Table 1 Round-Shaped Diamond Prices

Weight (carats)	Price
0.5	$2,790
0.6	$3,191
0.7	$3,694
0.8	$4,154
0.9	$5,018
1.0	$5,898

Source: www.tradeshop.com

SOLUTION

(A) A **scatter plot** is simply a graph of the points in Table 1 (Fig. 2A). To add the graph of the model to the scatter plot, we find any two points that satisfy equation (2) [we choose (0.4, 1,976) and (1.1, 6,274)]. Plotting these points and drawing a line through them gives us Figure 2B.

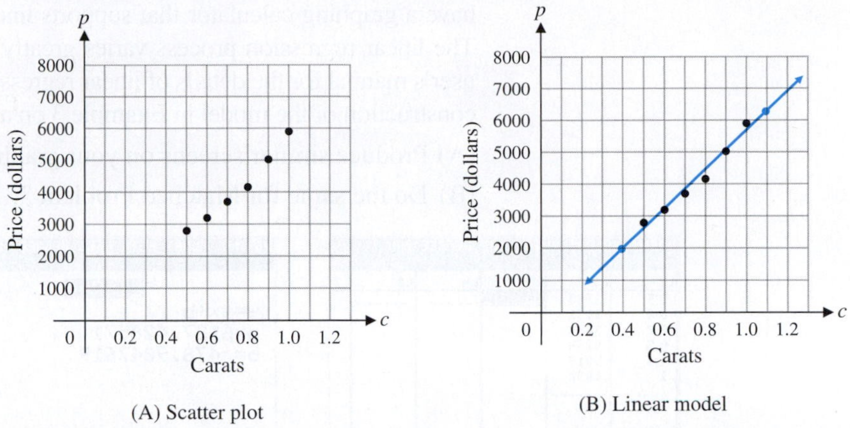

Figure 2

(B) The rate of change of the price of a diamond with respect to its weight is 6,140. Increasing the weight by one carat will increase the price by about \$6,140.

(C) The graph of the model (Fig. 2B) does not pass through any of the points in the scatter plot, but it comes close to all of them. [Verify this by evaluating equation (2) at $c = 0.5, 0.6, \ldots, 1$.] So we can use equation (2) to approximate points not in Table 1.

$$c = 0.85 \qquad\qquad c = 1.2$$
$$p \approx 6{,}140(0.85) - 480 \qquad p \approx 6{,}140(1.2) - 480$$
$$= \$4{,}739 \qquad\qquad = \$6{,}888$$

A 0.85-carat diamond will cost about \$4,739, and a 1.2-carat diamond will cost about \$6,888.

(D) To find the weight of a \$4,000 diamond, we solve the following equation for c:

$$6{,}140c - 480 = 4{,}000 \qquad \text{Add 480 to both sides.}$$
$$6{,}140c = 4{,}480 \qquad \text{Divide both sides by 6,140.}$$
$$c = \frac{4{,}480}{6{,}140} \approx 0.73 \qquad \text{Rounded to two decimal places.}$$

A \$4,000 diamond will weigh about 0.73 carat.

Table 2 Emerald-Shaped Diamond Prices

Weight (carats)	Price
0.5	\$1,677
0.6	\$2,353
0.7	\$2,718
0.8	\$3,218
0.9	\$3,982
1.0	\$4,510

Source: www.tradeshop.com

Matched Problem 3 Prices for emerald-shaped diamonds from an online trader are given in Table 2. Repeat Example 3 for this data with the linear model

$$p = 5{,}600c - 1{,}100$$

where p is the price of an emerald-shaped diamond weighing c carats.

The model we used in Example 3 was obtained using a technique called **linear regression**, and the model is called the **regression line**. This technique produces a line that is the **best fit** for a given data set. (The line of best fit is the line that minimizes the sum of the squares of the vertical distances from the data points to the line.) Although you can find a linear regression line by hand, we prefer to leave the calculations to a graphing calculator or a computer. Don't be concerned if you don't have either of these electronic devices. We will supply the regression model in most of the applications we discuss, as we did in Example 3.

Explore and Discuss 2

As stated previously, we used linear regression to produce the model in Example 3. If you have a graphing calculator that supports linear regression, then you can find this model. The linear regression process varies greatly from one calculator to another. Consult the user's manual for the details of linear regression. The screens in Figure 3 are related to the construction of the model in Example 3 on a Texas Instruments TI-84 Plus CE.

(A) Produce similar screens on your graphing calculator.

(B) Do the same for Matched Problem 3.

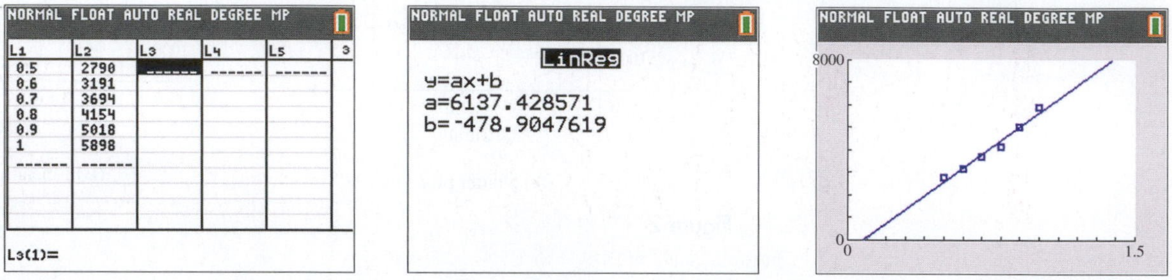

(A) Entering the data (B) Finding the model (C) Graphing the data and the model

Figure 3 Linear regression on a graphing calculator

In Example 3, we used the regression model to approximate points that were not given in Table 1 but would fit between points in the table. This process is called **interpolation**. In the next example, we use a regression model to approximate points outside the given data set. This process is called **extrapolation**, and the approximations are often referred to as **predictions**.

EXAMPLE 4 **Atmospheric Concentration of Carbon Dioxide** Table 3 contains information about the concentration of carbon dioxide (CO_2) in the atmosphere. The linear regression model for the data is

$$C = 360 + 2.04t$$

where C is the concentration (in parts per million) of carbon dioxide and t is the time in years with $t = 0$ corresponding to the year 1995.

(A) Interpret the slope of the regression line as a rate of change.

(B) Use the regression model to predict the concentration of CO_2 in the atmosphere in 2025.

Table 3 Atmospheric Concentration of CO_2 (parts per million)

1995	2000	2005	2010	2015
361	370	380	390	402

Source: National Oceanic and Atmospheric Administration

SOLUTION

(A) The slope $m = 2.04$ is the rate of change of concentration of CO_2 with respect to time. Since the slope is positive, the concentration of CO_2 is increasing at a rate of 2.04 parts per million per year.

(B) If $t = 30$, then

$$C = 360 + 2.04(30) \approx 421$$

So the model predicts that the atmospheric concentration of CO_2 will be approximately 421 parts per million in 2025.

Matched Problem 4 Using the model of Example 4, estimate the concentration of carbon dioxide in the atmosphere in the year 1990.

Forest managers estimate growth, volume, yield, and forest potential. One common measure is the diameter of a tree at breast height (Dbh), which is defined as the diameter of the tree at a point 4.5 feet above the ground on the uphill side of the tree. Example 5 uses Dbh to estimate the height of balsam fir trees.

EXAMPLE 5 **Forestry** A linear regression model for the height of balsam fir trees is

$$h = 3.8d + 18.73$$

where d is Dbh in inches and h is the height in feet.

(A) Interpret the slope of this model.

(B) What is the effect of a 1-inch increase in Dbh?

(C) Estimate the height of a balsam fir with a Dbh of 8 inches. Round your answer to the nearest foot.

(D) Estimate the Dbh of a balsam fir that is 30 feet tall. Round your answer to the nearest inch.

SOLUTION

(A) The rate of change of height with respect to breast height diameter is 3.8 feet per inch.

(B) Height increases by 3.8 feet.

(C) We must find h when $d = 8$:

$$h = 3.8d + 18.73 \qquad \text{Substitute } d = 8.$$

$$h = 3.8(8) + 18.73 \qquad \text{Evaluate.}$$

$$h = 49.13 \approx 49 \text{ ft}$$

(D) We must find d when $h = 30$:

$$h = 3.8d + 18.73 \qquad \text{Substitute } h = 30.$$

$$30 = 3.8d + 18.73 \qquad \text{Subtract 18.73 from both sides.}$$

$$11.27 = 3.8d \qquad \text{Divide both sides by 3.8.}$$

$$d = \frac{11.27}{3.8} \approx 3 \text{ in.}$$

The data used to produce the regression model in Example 5 are from the Jack Haggerty Forest at Lakehead University in Canada (Table 4). We used the popular

Table 4 Height and Diameter of the Balsam Fir

Dbh (in.)	Height (ft)	Dbh (in.)	Height (ft)	Dbh (in.)	Height (ft)	Dbh (in.)	Height (ft)
6.5	51.8	6.4	44.0	3.1	19.7	4.6	26.6
8.6	50.9	4.4	46.9	7.1	55.8	4.8	33.1
5.7	49.2	6.5	52.2	6.3	32.8	3.1	28.5
4.9	46.3	4.1	46.9	2.4	26.2	3.2	29.2
6.4	44.3	8.8	51.2	2.5	29.5	5.0	34.1
4.1	46.9	5.0	36.7	6.9	45.9	3.0	28.2
1.7	13.1	4.9	34.1	2.4	32.8	4.8	33.8
1.8	19.0	3.8	32.2	4.3	39.4	4.4	35.4
3.2	20.0	5.5	49.2	7.3	36.7	11.3	55.4
5.1	46.6	6.3	39.4	10.9	51.5	3.7	32.2

Source: Jack Haggerty Forest, Lakehead University, Canada

spreadsheet Excel to produce a scatter plot of the data in Table 4 and to find the regression model (Fig. 4).

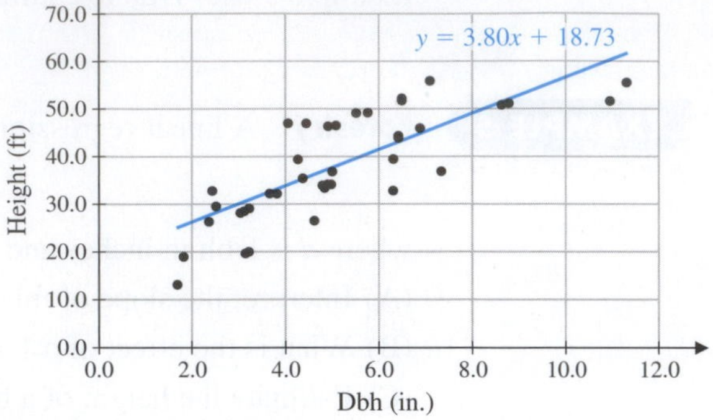

Figure 4 Linear regression with a spreadsheet

Matched Problem 5 Figure 5 shows the scatter plot for white spruce trees in the Jack Haggerty Forest at Lakehead University in Canada. A regression model produced by a spreadsheet (Fig. 5), after rounding, is

$$h = 1.8d + 34$$

where d is Dbh in inches and h is the height in feet.

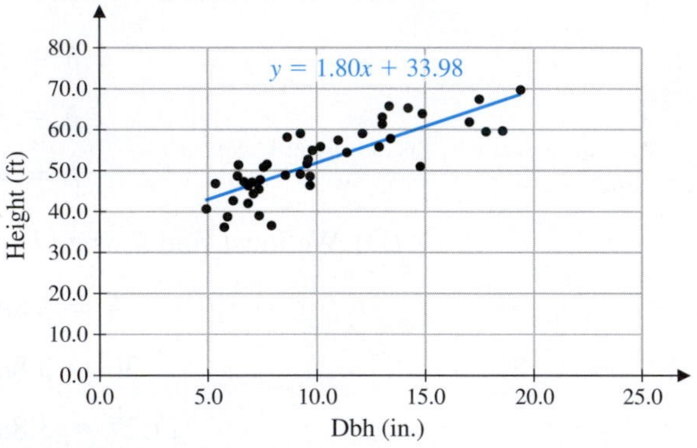

Figure 5 Linear regression for white spruce trees

(A) Interpret the slope of this model.

(B) What is the effect of a 1-inch increase in Dbh?

(C) Estimate the height of a white spruce with a Dbh of 10 inches. Round your answer to the nearest foot.

(D) Estimate the Dbh of a white spruce that is 65 feet tall. Round your answer to the nearest inch.

Exercises 1.3

Applications

1. **Ideal weight.** Dr. J. D. Robinson published the following estimate of the ideal body weight of a woman:

$$49 \text{ kg} + 1.7 \text{ kg for each inch over 5 ft}$$

(A) Find a linear model for Robinson's estimate of the ideal weight of a woman using w for ideal body weight (in kilograms) and h for height over 5 ft (in inches).

(B) Interpret the slope of the model.

(C) If a woman is 5′4″ tall, what does the model predict her weight to be?

(D) If a woman weighs 60 kg, what does the model predict her height to be?

2. **Ideal weight.** Dr. J. D. Robinson also published the following estimate of the ideal body weight of a man:

$$52 \text{ kg} + 1.9 \text{ kg for each inch over 5 ft}$$

(A) Find a linear model for Robinson's estimate of the ideal weight of a man using w for ideal body weight (in kilograms) and h for height over 5 ft (in inches).

(B) Interpret the slope of the model.

(C) If a man is 5′8″ tall, what does the model predict his weight to be?

(D) If a man weighs 70 kg, what does the model predict his height to be?

3. **Underwater pressure.** At sea level, the weight of the atmosphere exerts a pressure of 14.7 pounds per square inch, commonly referred to as 1 **atmosphere of pressure**. As an object descends in water, pressure P and depth d are linearly related. In salt water, the pressure at a depth of 33 ft is 2 atms, or 29.4 pounds per square inch.

(A) Find a linear model that relates pressure P (in pounds per square inch) to depth d (in feet).

(B) Interpret the slope of the model.

(C) Find the pressure at a depth of 50 ft.

(D) Find the depth at which the pressure is 4 atms.

4. **Underwater pressure.** Refer to Problem 3. In fresh water, the pressure at a depth of 34 ft is 2 atms, or 29.4 pounds per square inch.

(A) Find a linear model that relates pressure P (in pounds per square inch) to depth d (in feet).

(B) Interpret the slope of the model.

(C) Find the pressure at a depth of 50 ft.

(D) Find the depth at which the pressure is 4 atms.

5. **Rate of descent—Parachutes.** At low altitudes, the altitude of a parachutist and time in the air are linearly related. A jump at 2,880 ft using the U.S. Army's T-10 parachute system lasts 120 secs.

(A) Find a linear model relating altitude a (in feet) and time in the air t (in seconds).

(B) Find the rate of descent for a T-10 system.

(C) Find the speed of the parachutist at landing.

6. **Rate of descent—Parachutes.** The U.S Army is considering a new parachute, the Advanced Tactical Parachute System (ATPS). A jump at 2,880 ft using the ATPS system lasts 180 secs.

(A) Find a linear model relating altitude a (in feet) and time in the air t (in seconds).

(B) Find the rate of descent for an ATPS system parachute.

(C) Find the speed of the parachutist at landing.

7. **Speed of sound.** The speed of sound through air is linearly related to the temperature of the air. If sound travels at 331 m/sec at 0°C and at 343 m/sec at 20°C, construct a linear model relating the speed of sound (s) and the air temperature (t). Interpret the slope of this model. (*Source:* Engineering Toolbox)

8. **Speed of sound.** The speed of sound through sea water is linearly related to the temperature of the water. If sound travels at 1,403 m/sec at 0°C and at 1,481 m/sec at 20°C, construct a linear model relating the speed of sound (s) and the air temperature (t). Interpret the slope of this model. (*Source:* Engineering Toolbox)

9. **Energy production.** Table 5 lists U.S. fossil fuel production as a percentage of total energy production for selected years. A linear regression model for this data is

$$y = -0.19x + 83.75$$

where x represents years since 1985 and y represents the corresponding percentage of total energy production.

Table 5 U.S. Fossil Fuel Production

Year	Production (%)
1985	85
1990	83
1995	81
2000	80
2005	79
2010	78
2015	80

Source: Energy Information Administration

(A) Draw a scatter plot of the data and a graph of the model on the same axes.

(B) Interpret the slope of the model.

(C) Use the model to predict fossil fuel production in 2025.

(D) Use the model to estimate the first year for which fossil fuel production is less than 70% of total energy production.

10. **Energy consumption.** Table 6 lists U.S. fossil fuel consumption as a percentage of total energy consumption for selected years. A linear regression model for this data is

$$y = -0.14x + 86.18$$

where x represents years since 1985 and y represents the corresponding percentage of fossil fuel consumption.

Table 6 U.S. Fossil Fuel Consumption

Year	Consumption (%)
1985	86
1990	85
1995	85
2000	84
2005	85
2010	83
2015	81

Source: Energy Information Administration

(A) Draw a scatter plot of the data and a graph of the model on the same axes.

(B) Interpret the slope of the model.

(C) Use the model to predict fossil fuel consumption in 2025.

(D) Use the model to estimate the first year for which fossil fuel consumption is less than 80% of total energy consumption.

11. Cigarette smoking. The data in Table 7 shows that the percentage of female cigarette smokers in the United States declined from 22.1% in 1997 to 13.6% in 2015.

Table 7 Percentage of Smoking Prevalence among U.S. Adults

Year	Males (%)	Females (%)
1997	27.6	22.1
2000	25.7	21.0
2003	24.1	19.2
2006	23.9	18.0
2010	21.5	17.3
2015	16.7	13.6

Source: Centers for Disease Control

(A) Applying linear regression to the data for females in Table 7 produces the model

$$f = -0.45t + 22.20$$

where f is percentage of female smokers and t is time in years since 1997. Draw a scatter plot of the female smoker data and a graph of the regression model on the same axes.

(B) Estimate the first year in which the percentage of female smokers is less than 10%.

12. Cigarette smoking. The data in Table 7 shows that the percentage of male cigarette smokers in the United States declined from 27.6% in 1997 to 16.7% in 2015.

(A) Applying linear regression to the data for males in Table 7 produces the model

$$m = -0.56t + 27.82$$

where m is percentage of male smokers and t is time in years since 1997. Draw a scatter plot of the male smoker data and a graph of the regression model.

(B) Estimate the first year in which the percentage of male smokers is less than 10%.

13. Licensed drivers. Table 8 contains the state population and the number of licensed drivers in the state (both in millions) for the states with population under 1 million in 2014. The regression model for this data is

$$y = 0.75x$$

where x is the state population (in millions) and y is the number of licensed drivers (in millions) in the state.

(A) Draw a scatter plot of the data and a graph of the model on the same axes.

Table 8 Licensed Drivers in 2014

State	Population	Licensed Drivers
Alaska	0.74	0.53
Delaware	0.94	0.73
Montana	1.00	0.77
North Dakota	0.74	0.53
South Dakota	0.85	0.61
Vermont	0.63	0.55
Wyoming	0.58	0.42

Source: Bureau of Transportation Statistics

(B) If the population of Hawaii in 2014 was about 1.4 million, use the model to estimate the number of licensed drivers in Hawaii in 2014 to the nearest thousand.

(C) If the number of licensed drivers in Maine in 2014 was about 1,019,000, use the model to estimate the population of Maine in 2014 to the nearest thousand.

14. Licensed drivers. Table 9 contains the state population and the number of licensed drivers in the state (both in millions) for the most populous states in 2014. The regression model for this data is

$$y = 0.62x + 0.29$$

where x is the state population (in millions) and y is the number of licensed drivers (in millions) in the state.

Table 9 Licensed Drivers in 2014

State	Population	Licensed Drivers
California	39	25
Florida	20	14
Illinois	13	8
New York	20	11
Ohio	12	8
Pennsylvania	13	9
Texas	27	16

Source: Bureau of Transportation Statistics

(A) Draw a scatter plot of the data and a graph of the model on the same axes.

(B) If the population of Michigan in 2014 was about 9.9 million, use the model to estimate the number of licensed drivers in Michigan in 2014 to the nearest thousand.

(C) If the number of licensed drivers in Georgia in 2014 was about 6.7 million, use the model to estimate the population of Georgia in 2014 to the nearest thousand.

15. Net sales. A linear regression model for the net sales data in Table 10 is

$$S = 15.85t + 250.1$$

where S is net sales and t is time since 2000 in years.

Table 10 Walmart Stores, Inc.

Billions of U.S. Dollars	2008	2009	2010	2011	2012	2015
Net sales	374	401	405	419	444	488
Operating income	21.9	22.8	24.0	25.5	26.6	27.3

Source: Walmart Stores, Inc.

(A) Draw a scatter plot of the data and a graph of the model on the same axes.

(B) Predict Walmart's net sales for 2026.

16. Operating income. A linear regression model for the operating income data in Table 10 is

$$I = 0.82t + 15.84$$

where I is operating income and t is time since 2000 in years.

(A) Draw a scatter plot of the data and a graph of the model on the same axes.

(B) Predict Walmart's annual operating income for 2026.

17. Freezing temperature. Ethylene glycol and propylene glycol are liquids used in antifreeze and deicing solutions. Ethylene glycol is listed as a hazardous chemical by the Environmental Protection Agency, while propylene glycol is generally regarded as safe. Table 11 lists the freezing temperature for various concentrations (as a percentage of total weight) of each chemical in a solution used to deice airplanes. A linear regression model for the ethylene glycol data in Table 11 is

$$E = -0.55T + 31$$

where E is the percentage of ethylene glycol in the deicing solution and T is the temperature at which the solution freezes.

Table 11 Freezing Temperatures

Freezing Temperature (°F)	Ethylene Glycol (% Wt.)	Propylene Glycol (% Wt.)
−50	56	58
−40	53	55
−30	49	52
−20	45	48
−10	40	43
0	33	36
10	25	29
20	16	19

Source: T. Labuza, University of Minnesota

(A) Draw a scatter plot of the data and a graph of the model on the same axes.

(B) Use the model to estimate the freezing temperature to the nearest degree of a solution that is 30% ethylene glycol.

(C) Use the model to estimate the percentage of ethylene glycol in a solution that freezes at 15°F.

18. Freezing temperature. A linear regression model for the propylene glycol data in Table 11 is

$$P = -0.54T + 34$$

where P is the percentage of propylene glycol in the deicing solution and T is the temperature at which the solution freezes.

(A) Draw a scatter plot of the data and a graph of the model on the same axes.

(B) Use the model to estimate the freezing temperature to the nearest degree of a solution that is 30% propylene glycol.

(C) Use the model to estimate the percentage of propylene glycol in a solution that freezes at 15°F.

19. Forestry. The figure contains a scatter plot of 100 data points for black spruce trees and the linear regression model for this data.

(A) Interpret the slope of the model.

(B) What is the effect of a 1-in. increase in Dbh?

(C) Estimate the height of a black spruce with a Dbh of 15 in. Round your answer to the nearest foot.

(D) Estimate the Dbh of a black spruce that is 25 ft tall. Round your answer to the nearest inch.

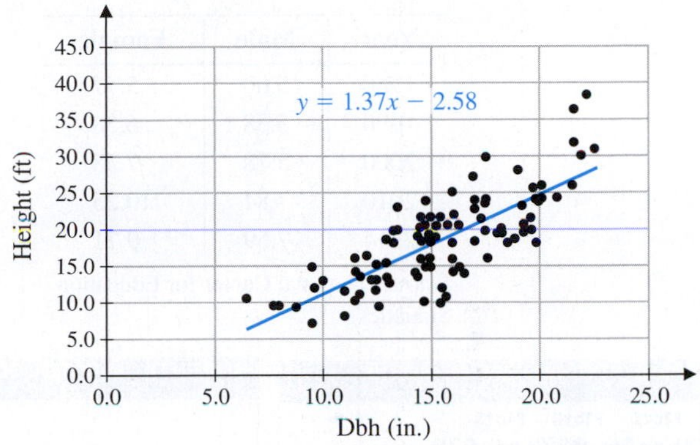

black spruce
Source: Lakehead University

20. Forestry. The figure contains a scatter plot of 100 data points for black walnut trees and the linear regression model for this data.

(A) Interpret the slope of the model.

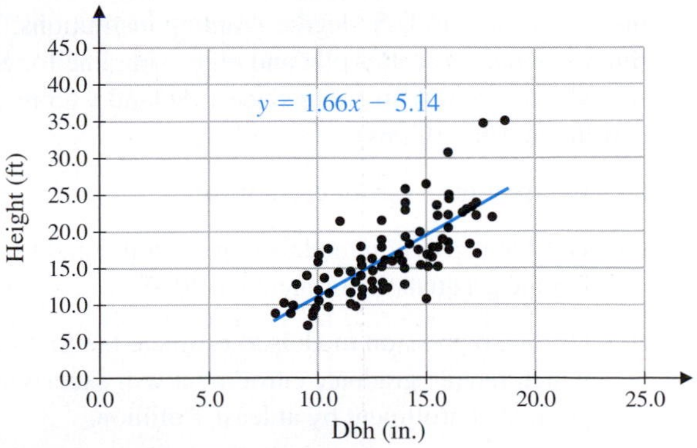

black walnut
Source: Kagen Research

(B) What is the effect of a 1-in. increase in Dbh?

(C) Estimate the height of a black walnut with a Dbh of 12 in. Round your answer to the nearest foot.

(D) Estimate the Dbh of a black walnut that is 25 ft tall. Round your answer to the nearest inch.

21. Undergraduate enrollment. Table 12 lists fall undergraduate enrollment by gender in U.S. degree-granting institutions. The figure contains a scatter plot and regression line for each data set, where x represents years since 1980 and y represents enrollment (in millions).

(A) Interpret the slope of each model.

(B) Use the regression models to predict the male and female undergraduate enrollments in 2025.

(C) Use the regression models to estimate the first year in which female undergraduate enrollment will exceed male undergraduate enrollment by at least 3 million.

Table 12 Fall Undergraduate Enrollment (in millions)

Year	Male	Female
1980	5.00	5.47
1990	5.38	6.58
2000	5.78	7.38
2010	7.84	10.25
2014	7.59	9.71

Source: National Center for Education Statistics

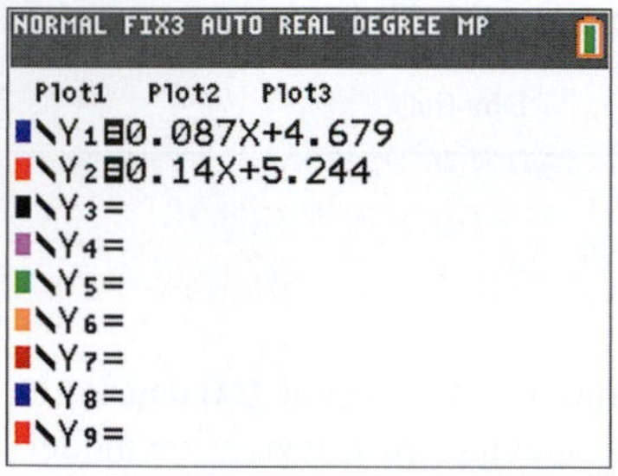

(A)

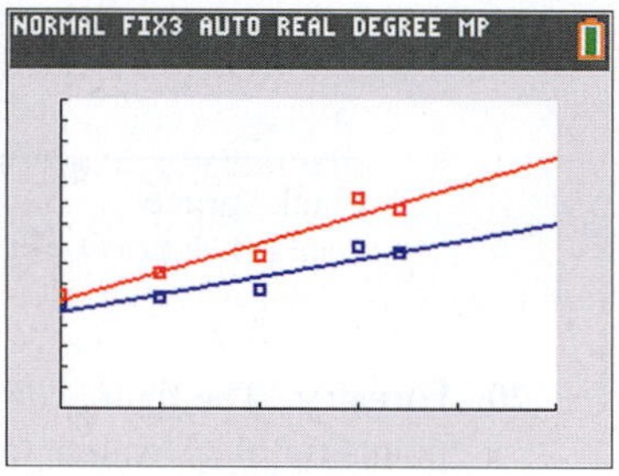

(B)

Figure for 21

22. Graduate enrollment. Table 13 lists fall graduate enrollment by gender in U.S. degree-granting institutions. The figure contains a scatter plot and regression line for each data set, where x represents years since 1980 and y represents enrollment (in millions).

(A) Interpret the slope of each model.

(B) Use the regression models to predict the male and female graduate enrollments in 2025.

(C) Use the regression models to estimate the first year in which female graduate enrollment will exceed male graduate enrollment by at least 1 million.

Table 13 Fall Graduate Enrollment (in millions)

Year	Male	Female
1980	0.87	0.75
1990	0.90	0.96
2000	0.94	1.21
2010	1.21	1.73
2014	1.21	1.70

Source: National Center for Education Statistics

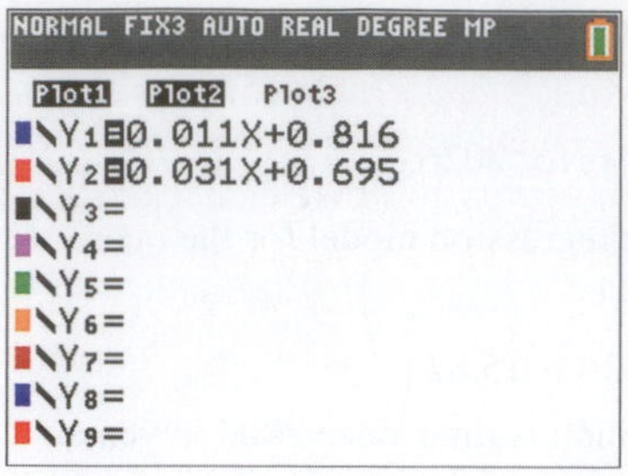

(A)

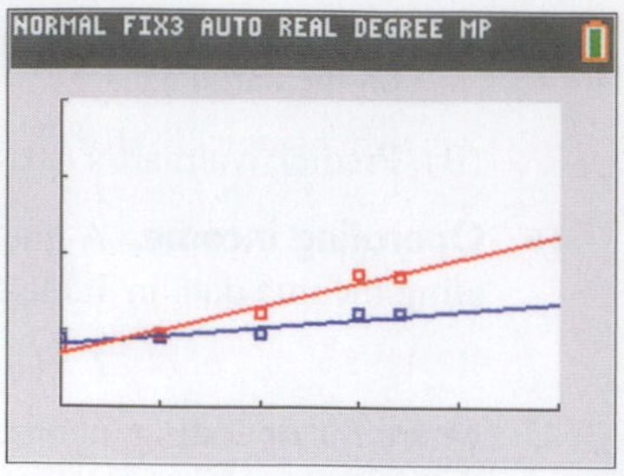

(B)

Figure for 22

Problems 23–28 require a graphing calculator or a computer that can calculate the linear regression line for a given data set.

23. Climate. Find a linear regression model for the data on average annual temperature in Table 14, where x is years since 1960 and y is temperature (in °F). (Round regression coefficients to three decimal places). Use the model to estimate the average annual temperature in the contiguous United States in 2025.

Table 14 Climate Data for the Contiguous United States

Year	Average Annual Temperature (°F)	Average Annual Precipitation (in)
1965	51.69	29.80
1975	51.50	33.03
1985	51.30	29.97
1995	52.65	32.69
2005	53.64	30.08
2015	54.40	34.59

Source: National Oceanic and Atmospheric Administration

24. Climate. Find a linear regression model for the data on average annual precipitation in Table 14, where x is years since 1960 and y is precipitation (in inches). (Round regression coefficients to three decimal places). Use the model to estimate the average annual precipitation in the contiguous United States in 2025.

25. Olympic Games. Find a linear regression model for the men's 100-meter freestyle data given in Table 15, where x is years since 1990 and y is winning time (in seconds). Do the same for the women's 100-meter freestyle data. (Round regression coefficients to three decimal places.) Do these models indicate that the women will eventually catch up with the men?

Table 15 Winning Times in Olympic Swimming Events

	100-Meter Freestyle		200-Meter Backstroke	
	Men	Women	Men	Women
1992	49.02	54.65	1:58.47	2:07.06
1996	48.74	54.50	1.58.54	2:07.83
2000	48.30	53.83	1:56.76	2:08.16
2004	48.17	53.84	1:54.76	2:09.16
2008	47.21	53.12	1:53.94	2:05.24
2012	47.52	53.00	1:53.41	2:04.06
2016	47.58	52.70	1:53:62	2:05.99

Source: www.infoplease.com

26. Olympic Games. Find a linear regression model for the men's 200-meter backstroke data given in Table 15, where x is years since 1990 and y is winning time (in seconds). Do the same for the women's 200-meter backstroke data. (Round regression coefficients to three decimal places.) Do these models indicate that the women will eventually catch up with the men?

27. Supply and demand. Table 16 contains price–supply data and price–demand data for corn. Find a linear regression model for the price–supply data where x is supply (in billions of bushels) and y is price (in dollars). Do the same for the price–demand data. (Round regression coefficients to two decimal places.) Find the equilibrium price for corn.

Table 16 Supply and Demand for U.S. Corn

Price ($/bu)	Supply (billion bu)	Price ($/bu)	Demand (billion bu)
2.15	6.29	2.07	9.78
2.29	7.27	2.15	9.35
2.36	7.53	2.22	8.47
2.48	7.93	2.34	8.12
2.47	8.12	2.39	7.76
2.55	8.24	2.47	6.98

Source: www.usda.gov/nass/pubs/histdata.htm

28. Supply and demand. Table 17 contains price–supply data and price–demand data for soybeans. Find a linear regression model for the price–supply data where x is supply (in billions of bushels) and y is price (in dollars). Do the same for the price–demand data. (Round regression coefficients to two decimal places.) Find the equilibrium price for soybeans.

Table 17 Supply and Demand for U.S. Soybeans

Price ($/bu)	Supply (billion bu)	Price ($/bu)	Demand (billion bu)
5.15	1.55	4.93	2.60
5.79	1.86	5.48	2.40
5.88	1.94	5.71	2.18
6.07	2.08	6.07	2.05
6.15	2.15	6.40	1.95
6.25	2.27	6.66	1.85

Source: www.usda.gov/nass/pubs/histdata.htm

Answers to Matched Problems

1. (A) The rate of change of BSA with respect to weight is 28.55 square inches per pound.

(B) Increasing w by 1 pound increases a by 28.55 square inches.

2. (A) $a = -36t + 2{,}880$

(B) 36 ft/sec

3. (A)

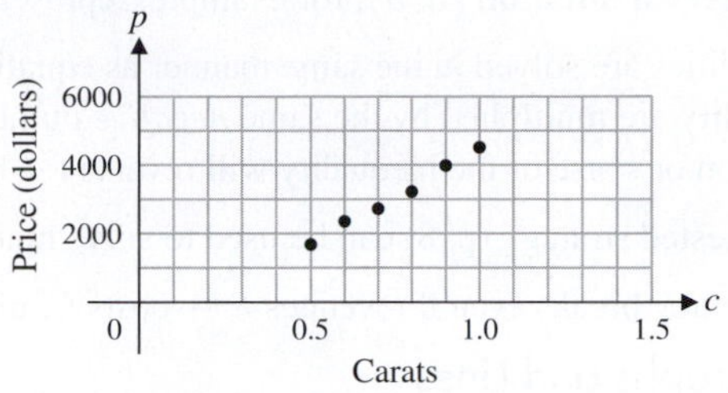

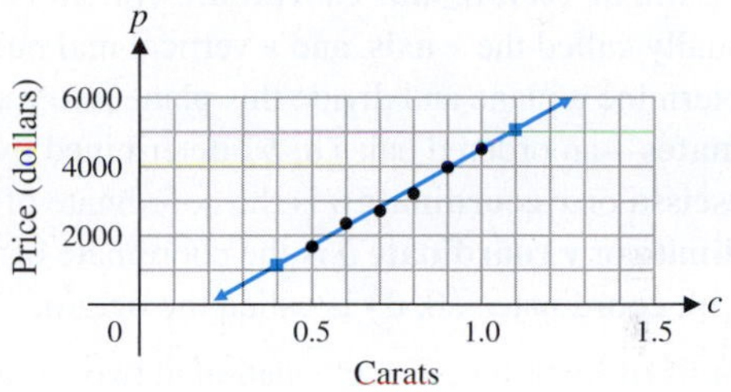

(B) The rate of change of the price of a diamond with respect to its weight is \$5,600. Increasing the weight by one carat will increase the price by about \$5,600.

(C) \$3,660; \$5,620

(D) 0.91 carat

4. Approximately 350 parts per million.

5. (A) The slope is 1.8, so the rate of change of height with respect to breast height diameter is 1.8 feet per inch.

(B) Height increases by 1.8 feet.

(C) 52 ft

(D) 17 in.

Chapter 1 Summary and Review

Important Terms, Symbols, and Concepts

1.1 Linear Equations and Inequalities

EXAMPLES

- A **first-degree, or linear, equation** in one variable is any equation that can be written in the form

Standard form: $ax + b = 0 \quad a \neq 0$

If the equality sign in the standard form is replaced by $<$, $>$, $\leq$, or $\geq$, the resulting expression is called a **first-degree, or linear, inequality**.

1.1 Linear Equations and Inequalities (*Continued*)

- A **solution** of an equation (or inequality) involving a single variable is a number that, when substituted for the variable, makes the equation (inequality) true. The set of all solutions is called the **solution set**.
- If we perform an operation on an equation (or inequality) that produces another equation (or inequality) with the same solution set, then the two equations (or inequalities) are **equivalent**. Equations are solved by adding or subtracting the same quantity to both sides, or by multiplying both sides by the same *nonzero* quantity until an equation with an obvious solution is obtained. Ex. 1, p. 2 Ex. 2, p. 3 Ex. 3, p. 4 Ex. 10, p. 9
- The **interval notation** $[a, b)$, for example, represents the solution of the **double inequality** $a \le x < b$. Ex. 5, p. 6
- Inequalities are solved in the same manner as equations with one important exception. If both sides of an inequality are multiplied by the same *negative* number or divided by the same *negative* number, then the direction or sense of the inequality will reverse ($<$ becomes $>$, $\ge$ becomes $\le$, and so on). Ex. 4, p. 4 Ex. 6, p. 7 Ex. 7, p. 7
- A suggested strategy (p. 8) can be used to solve many word problems. Ex. 8, p. 8
- A company breaks even if revenues $R =$ costs C, makes a profit if $R > C$, and incurs a loss if $R < C$. Ex. 9, p. 8

1.2 Graphs and Lines

- A **Cartesian or rectangular coordinate system** is formed by the intersection of a horizontal real number line, usually called the ***x* axis**, and a vertical real number line, usually called the ***y* axis**, at their origins. The axes determine a plane and divide this plane into four **quadrants**. Each point in the plane corresponds to its **coordinates**—an ordered pair (a, b) determined by passing horizontal and vertical lines through the point. The **abscissa** or ***x* coordinate** a is the coordinate of the intersection of the vertical line and the x axis, and the **ordinate** or ***y* coordinate** b is the coordinate of the intersection of the horizontal line and the y axis. The point with coordinates $(0, 0)$ is called the **origin**. Fig 1, p. 12
- The **standard form** for a linear equation in two variables is $Ax + By = C$, with A and B not both zero. The graph of this equation is a line, and every line in a Cartesian coordinate system is the graph of a linear equation. Ex. 1, p. 14 Ex. 2, p. 14
- The graph of the equation $x = a$ is a **vertical line** and the graph of $y = b$ is a **horizontal line**. Ex. 3, p. 15
- If (x_1, y_1) and (x_2, y_2) are two distinct points on a line, then $m = (y_2 - y_1)/(x_2 - x_1)$ is the **slope** of the line. Ex. 4, p. 16
- The equation $y = mx + b$ is the **slope-intercept form** of the equation of the line with slope m and y intercept b. Ex. 5, p. 18 Ex. 7, p. 20
- The **point-slope form** of the equation of the line with slope m that passes through (x_1, y_1) is $y - y_1 = m(x - x_1)$. Ex. 6, p. 19
- In a competitive market, the intersection of the supply equation and the demand equation is called the **equilibrium point**, the corresponding price is called the **equilibrium price**, and the common value of supply and demand is called the **equilibrium quantity**. Ex. 8, p. 20

1.3 Linear Regression

- A **mathematical model** is a mathematics problem that, when solved, will provide information about a real-world problem.
- If the variables x and y are related by the equation $y = mx + b$, then x and y are **linearly related** and the slope m is the **rate of change** of y with respect to x. Ex. 1, p. 27 Ex. 2, p. 28
- A graph of the points in a data set is called a **scatter plot**. **Linear regression** is used to find the line that is the **best fit** for a data set. A regression model can be used to **interpolate** between points in a data set or to **extrapolate** or predict points outside the data set. Ex. 3, p. 28 Ex. 4, p. 30 Ex. 5, p. 31

Review Exercises

Work through all the problems in this chapter review and check answers in the back of the book. Following each answer you will find a number in italics indicating the section where that type of problem is discussed. Where weaknesses show up, review appropriate sections in the text.

A

1. Solve $2x + 3 = 7x - 11$.

2. Solve $\frac{x}{12} - \frac{x - 3}{3} = \frac{1}{2}$.

3. Solve $2x + 5y = 9$ for y.

4. Solve $3x - 4y = 7$ for x.

Solve Problems 5–7 and graph on a real number line.

5. $4y - 3 < 10$

6. $-1 < -2x + 5 \le 3$

7. $1 - \frac{x - 3}{3} \le \frac{1}{2}$

8. Sketch a graph of $3x + 2y = 9$.

9. Write an equation of a line with x intercept 6 and y intercept 4. Write the final answer in the form $Ax + By = C$.

10. Sketch a graph of $2x - 3y = 18$. What are the intercepts and slope of the line?

11. Write an equation in the form $y = mx + b$ for a line with slope $-\frac{2}{3}$ and y intercept 6.

12. Write the equations of the vertical line and the horizontal line that pass through $(-6, 5)$.

13. Write the equation of a line through each indicated point with the indicated slope. Write the final answer in the form $y = mx + b$.

(A) $m = -\frac{2}{3}$; $(-3, 2)$ (B) $m = 0$; $(3, 3)$

14. Write the equation of the line through the two indicated points. Write the final answer in the form $Ax + By = C$.

(A) $(-3, 5), (1, -1)$ (B) $(-1, 5), (4, 5)$

(C) $(-2, 7), (-2, -2)$

B *Solve Problems 15–19.*

15. $3x + 25 = 5x$

16. $\frac{u}{5} = \frac{u}{6} + \frac{6}{5}$

17. $\frac{5x}{3} - \frac{4 + x}{2} = \frac{x - 2}{4} + 1$

18. $0.05x + 0.25(30 - x) = 3.3$

19. $0.2(x - 3) + 0.05x = 0.4$

Solve Problems 20–24 and graph on a real number line.

20. $2(x + 4) > 5x - 4$

21. $3(2 - x) - 2 \le 2x - 1$

22. $\frac{x + 3}{8} - \frac{4 + x}{2} > 5 - \frac{2 - x}{3}$

23. $-5 \le 3 - 2x < 1$

24. $-1.5 \le 2 - 4x \le 0.5$

25. Given $Ax + By = 30$, graph each of the following cases on the same coordinate axes.

(A) $A = 5$ and $B = 0$ (B) $A = 0$ and $B = 6$

(C) $A = 6$ and $B = 5$

26. Describe the graphs of $x = -3$ and $y = 2$. Graph both simultaneously in the same coordinate system.

27. Describe the lines defined by the following equations:

(A) $3x + 4y = 0$ (B) $3x + 4 = 0$

(C) $4y = 0$ (D) $3x + 4y - 36 = 0$

C *Solve Problems 28 and 29 for the indicated variable.*

28. $A = \frac{1}{2}(a + b)h$; for a $(h \ne 0)$

29. $S = \frac{P}{1 - dt}$; for d $(dt \ne 1)$

30. For what values of a and b is the inequality $a + b < b - a$ true?

31. If a and b are negative numbers and $a > b$, then is a/b greater than 1 or less than 1?

32. Graph $y = mx + b$ and $y = -\frac{1}{m}x + b$ simultaneously in the same coordinate system for b fixed and several different values of m, $m \ne 0$. Describe the apparent relationship between the graphs of the two equations.

Applications

33. Investing. An investor has \$300,000 to invest. If part is invested at 5% and the rest at 9%, how much should be invested at 5% to yield 8% on the total amount?

34. Break-even analysis. A producer of educational DVDs is producing an instructional DVD. She estimates that it will cost \$90,000 to record the DVD and \$5.10 per unit to copy and distribute the DVD. If the wholesale price of the DVD is \$14.70, how many DVDs must be sold for the producer to break even?

35. Sports medicine. A simple rule of thumb for determining your maximum safe heart rate (in beats per minute) is to subtract your age from 220. While exercising, you should maintain a heart rate between 60% and 85% of your maximum safe rate.

(A) Find a linear model for the minimum heart rate m that a person of age x years should maintain while exercising.

(B) Find a linear model for the maximum heart rate M that a person of age x years should maintain while exercising.

(C) What range of heartbeats should you maintain while exercising if you are 20 years old?

(D) What range of heartbeats should you maintain while exercising if you are 50 years old?

36. Linear depreciation. A bulldozer was purchased by a construction company for \$224,000 and has a depreciated value of \$100,000 after 8 years. If the value is depreciated linearly from \$224,000 to \$100,000,

(A) Find the linear equation that relates value V (in dollars) to time t (in years).

(B) What would be the depreciated value after 12 years?

37. Business—Pricing. A sporting goods store sells tennis rackets that cost \$130 for \$208 and court shoes that cost \$50 for \$80.

(A) If the markup policy of the store for items that cost over \$10 is linear and is reflected in the pricing of these two items, write an equation that expresses retail price R in terms of cost C.

(B) What would be the retail price of a pair of in-line skates that cost \$120?

(C) What would be the cost of a pair of cross-country skis that had a retail price of \$176?

(D) What is the slope of the graph of the equation found in part (A)? Interpret the slope relative to the problem.

38. Income. A salesperson receives a base salary of \$400 per week and a commission of 10% on all sales over \$6,000 during the week. Find the weekly earnings for weekly sales of \$4,000 and for weekly sales of \$10,000.

39. Price–demand. The weekly demand for mouthwash in a chain of drug stores is 1,160 bottles at a price of \$3.79 each. If the price is lowered to \$3.59, the weekly demand increases to 1,320 bottles. Assuming that the relationship between the weekly demand x and price per bottle p is linear, express p in terms of x. How many bottles would the stores sell each week if the price were lowered to \$3.29?

40. Freezing temperature. Methanol, also known as wood alcohol, can be used as a fuel for suitably equipped vehicles. Table 1 lists the freezing temperature for various concentrations (as a percentage of total weight) of methanol in water. A linear regression model for the data in Table 1 is

$$T = 40 - 2M$$

where M is the percentage of methanol in the solution and T is the temperature at which the solution freezes.

(A) Draw a scatter plot of the data and a graph of the model on the same axes.

(B) Use the model to estimate the freezing temperature to the nearest degree of a solution that is 35% methanol.

(C) Use the model to estimate the percentage of methanol in a solution that freezes at −50°F.

Table 1

Methanol (%Wt)	Freezing temperature (°F)
0	32
10	20
20	0
30	−15
40	−40
50	−65
60	−95

Source: Ashland Inc.

41. High school dropout rates. Table 2 gives U.S. high school dropout rates as percentages for selected years since 1990. A linear regression model for the data is

$$r = -0.308t + 13.9$$

where t represents years since 1990 and r is the dropout rate.

(A) Interpret the slope of the model.

Table 2 High School Dropout Rates (%)

1996	2002	2006	2010	2014
11.8	10.5	9.3	7.4	6.5

(B) Draw a scatter plot of the data and the model in the same coordinate system.

(C) Use the model to predict the first year for which the dropout rate is less than 3%.

42. Consumer Price Index. The U.S. Consumer Price Index (CPI) in recent years is given in Table 3. A scatter plot of the data and linear regression line are shown in the figure, where x represents years since 1990.

(A) Interpret the slope of the model.

(B) Predict the CPI in 2024.

Table 3 Consumer Price Index (1982–1984 = 100)

Year	CPI
1990	130.7
1995	152.4
2000	172.2
2005	195.3
2010	218.1
2015	237.0

Source: U.S. Bureau of Labor Statistics

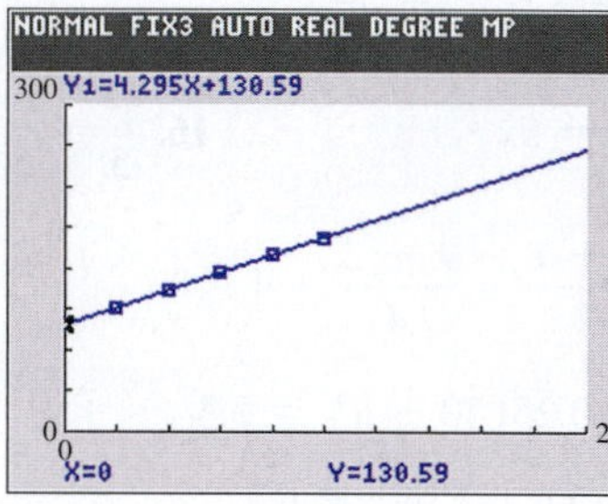

43. Forestry. The figure contains a scatter plot of 20 data points for white pine trees and the linear regression model for this data.

(A) Interpret the slope of the model.

(B) What is the effect of a 1-in. increase in Dbh?

(C) Estimate the height of a white pine tree with a Dbh of 25 in. Round your answer to the nearest foot.

(D) Estimate the Dbh of a white pine tree that is 15 ft tall. Round your answer to the nearest inch.

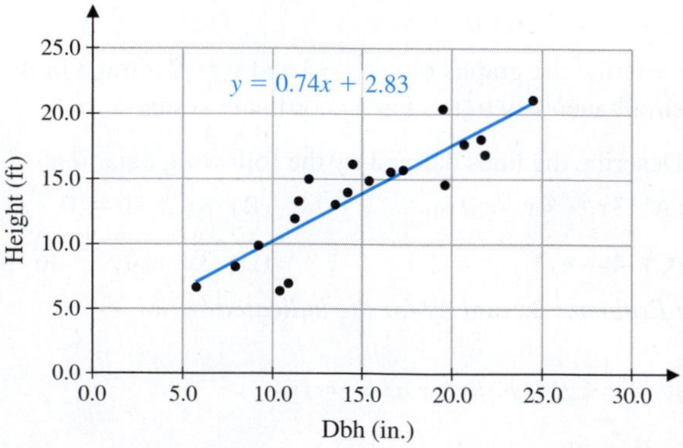

2 Functions and Graphs

Introduction

Many marine species are dependent on light from the sun. They will be found near the surface of the ocean, because light intensity decreases dramatically with depth. We use the function concept, one of the most important ideas in mathematics, to express the precise relationship between light intensity and ocean depth (see Problems 63 and 64 in Section 2.5).

The study of mathematics beyond the elementary level requires a firm understanding of a basic list of elementary functions (see the endpapers at the back of the book). In Chapter 2, we introduce the elementary functions and study their properties, graphs, and many applications.

2.1 Functions

- Equations in Two Variables
- Definition of a Function
- Functions Specified by Equations
- Function Notation
- Applications

We introduce the general notion of a *function* as a correspondence between two sets. Then we restrict attention to functions for which the two sets are both sets of real numbers. The most useful are those functions that are specified by equations in two variables. We discuss the terminology and notation associated with functions, graphs of functions, and applications.

Equations in Two Variables

In Chapter 1, we found that the graph of an equation of the form $Ax + By = C$, where A and B are not both zero, is a line. Because a line is determined by any two of its points, such an equation is easy to graph: Just plot *any* two points in its solution set and sketch the unique line through them (Fig. 1).

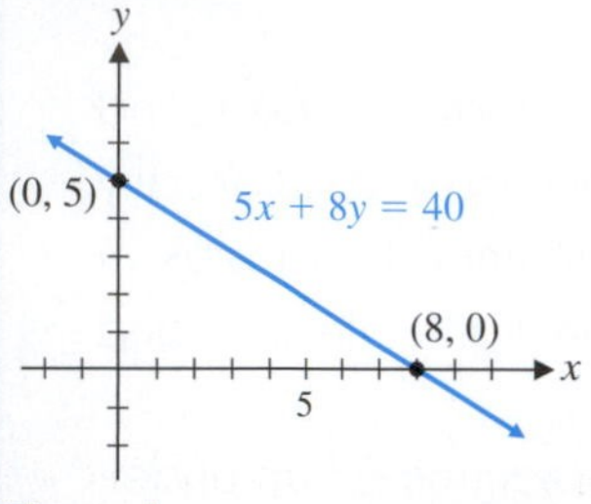

Figure 1

More complicated equations in two variables, such as $y = 9 - x^2$ or $x^2 = y^4$, are more difficult to graph. To **sketch the graph** of an equation, we plot enough points from its solution set in a rectangular coordinate system so that the total graph is apparent, and then we connect these points with a smooth curve. This process is called **point-by-point plotting**.

EXAMPLE 1 **Point-by-Point Plotting** Sketch the graph of each equation.

(A) $y = 9 - x^2$ (B) $x^2 = y^4$

SOLUTION

(A) Make up a table of solutions—that is, ordered pairs of real numbers that satisfy the given equation. For easy mental calculation, choose integer values for x.

x	-4	-3	-2	-1	0	1	2	3	4
y	-7	0	5	8	9	8	5	0	-7

After plotting these solutions, if there are any portions of the graph that are unclear, plot additional points until the shape of the graph is apparent. Then join all the plotted points with a smooth curve (Fig. 2). Arrowheads are used to indicate that the graph continues beyond the portion shown here with no significant changes in shape.

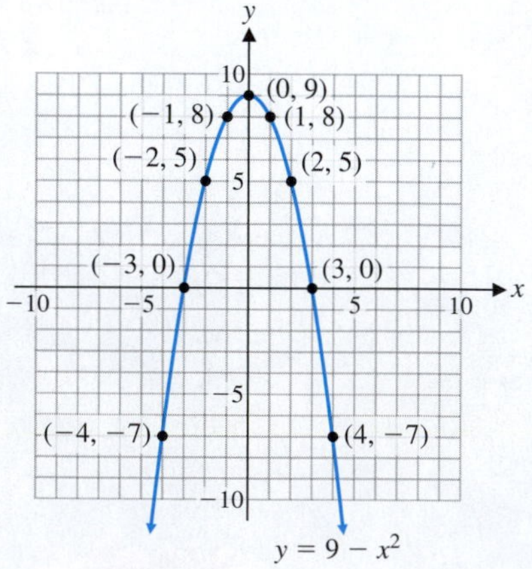

Figure 2 $y = 9 - x^2$

(B) Again we make a table of solutions—here it may be easier to choose integer values for y and calculate values for x. Note, for example, that if $y = 2$, then $x = \pm 4$; that is, the ordered pairs $(4, 2)$ and $(-4, 2)$ are both in the solution set.

x	± 9	± 4	± 1	0	± 1	± 4	± 9
y	-3	-2	-1	0	1	2	3

We plot these points and join them with a smooth curve (Fig. 3).

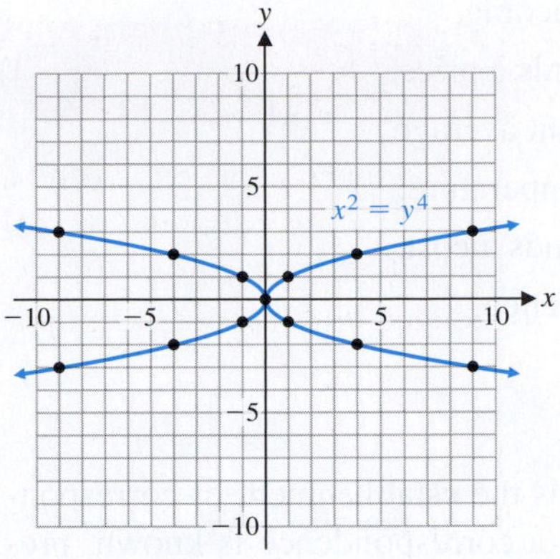

Figure 3 $x^2 = y^4$

Matched Problem 1 Sketch the graph of each equation.

(A) $y = x^2 - 4$ (B) $y^2 = \dfrac{100}{x^2 + 1}$

Explore and Discuss 1

To graph the equation $y = -x^3 + 3x$, we use point-by-point plotting to obtain the graph in Figure 4.

x	y
-1	-2
0	0
1	2

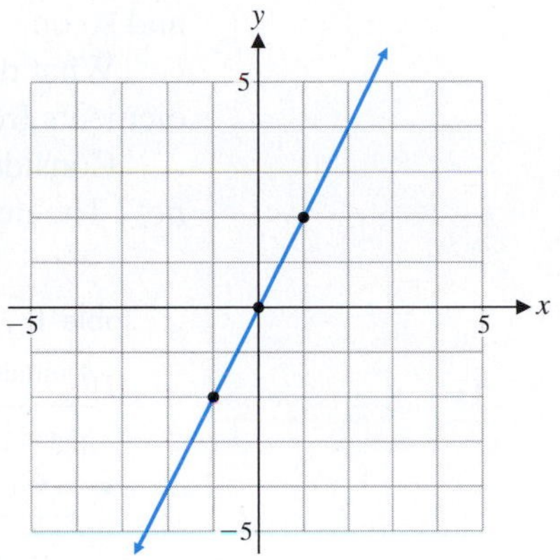

Figure 4

(A) Do you think this is the correct graph of the equation? Why or why not?

(B) Add points on the graph for $x = -2, -1.5, -0.5, 0.5, 1.5$, and 2.

(C) Now, what do you think the graph looks like? Sketch your version of the graph, adding more points as necessary.

(D) Graph this equation on a graphing calculator and compare it with your graph from part (C).

The icon in the margin is used throughout this book to identify optional graphing calculator activities that are intended to give you additional insight into the concepts under discussion. You may have to consult the manual for your graphing calculator for the details necessary to carry out these activities. For example, to graph the equation in Explore and Discuss 1 on most graphing calculators, you must enter the equation (Fig. 5A) and the window variables (Fig. 5B).

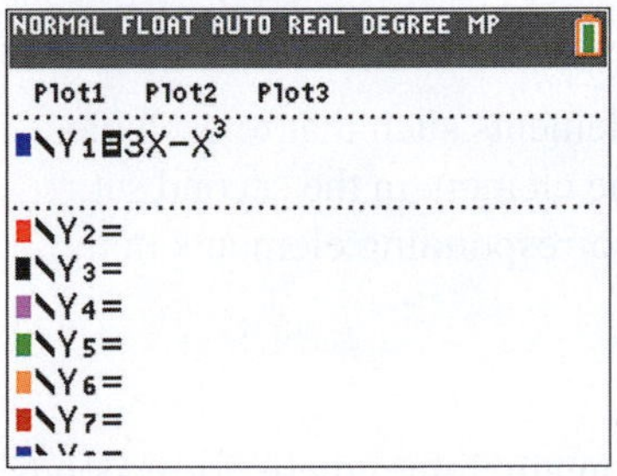

(A)

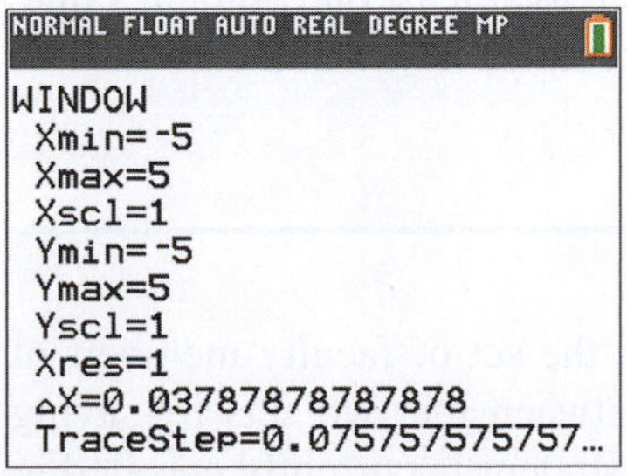

(B)

Figure 5

As Explore and Discuss 1 illustrates, the shape of a graph may not be apparent from your first choice of points. Using point-by-point plotting, it may be difficult to find points in the solution set of the equation, and it may be difficult to determine when you have found enough points to understand the shape of the graph. We will supplement the technique of point-by-point plotting with a detailed analysis of several basic equations, giving you the ability to sketch graphs with accuracy and confidence.

Definition of a Function

Central to the concept of function is correspondence. You are familiar with correspondences in daily life. For example,

To each person, there corresponds an annual income.

To each item in a supermarket, there corresponds a price.

To each student, there corresponds a grade-point average.

To each day, there corresponds a maximum temperature.

For the manufacture of x items, there corresponds a cost.

For the sale of x items, there corresponds a revenue.

To each square, there corresponds an area.

To each number, there corresponds its cube.

One of the most important aspects of any science is the establishment of correspondences among various types of phenomena. Once a correspondence is known, predictions can be made. A cost analyst would like to predict costs for various levels of output in a manufacturing process, a medical researcher would like to know the correspondence between heart disease and obesity, a psychologist would like to predict the level of performance after a subject has repeated a task a given number of times, and so on.

What do all of these examples have in common? Each describes the matching of elements from one set with the elements in a second set.

Consider Tables 1–3. Tables 1 and 2 specify functions, but Table 3 does not. Why not? The definition of the term *function* will explain.

Table 1

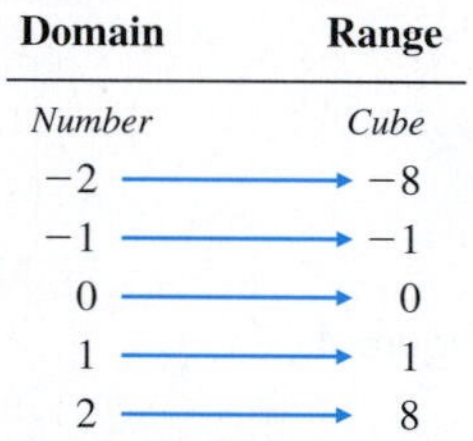
Domain
Range
Number
Cube
−2
−8
−1
−1
0
0
1
1
2
8

Table 2

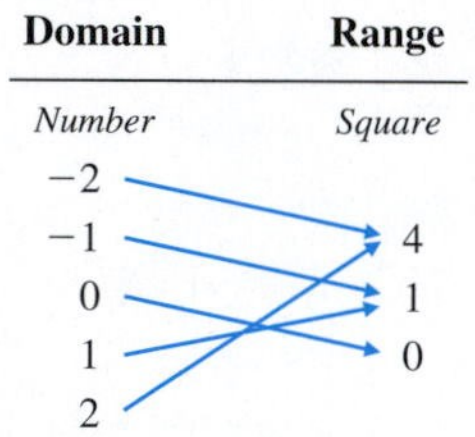
Domain
Range
Number
Square
−2
−1
0
1
2
4
1
0

Table 3

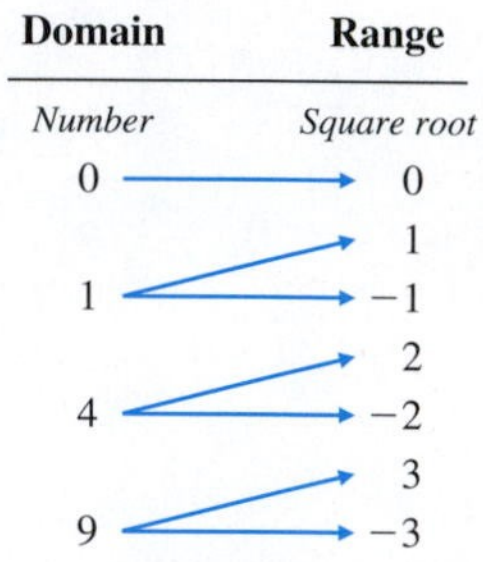
Domain
Range
Number
Square root
0
1
4
9
0
1
−1
2
−2
3
−3

DEFINITION Function

A **function** is a correspondence between two sets of elements such that to each element in the first set, there corresponds one and only one element in the second set.

The first set is called the **domain**, and the set of corresponding elements in the second set is called the **range**.

Tables 1 and 2 specify functions since to each domain value, there corresponds exactly one range value (for example, the cube of -2 is -8 and no other number). On the other hand, Table 3 does not specify a function since to at least one domain value, there corresponds more than one range value (for example, to the domain value 9, there corresponds -3 and 3, both square roots of 9).

Explore and Discuss 2

Consider the set of students enrolled in a college and the set of faculty members at that college. Suppose we define a correspondence between the two sets by saying that a student corresponds to a faculty member if the student is currently enrolled in a course taught by that faculty member. Is this correspondence a function? Discuss.

Functions Specified by Equations

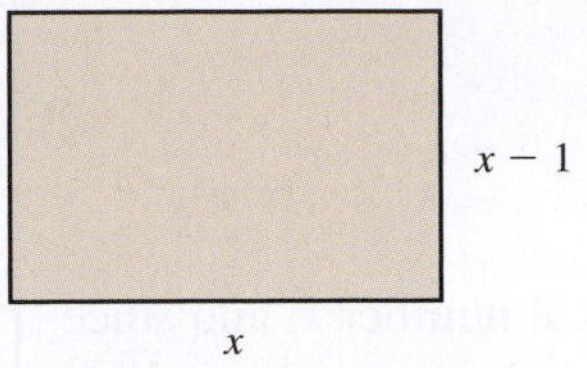

Figure 6

Most of the functions in this book will have domains and ranges that are (infinite) sets of real numbers. The **graph** of such a function is the set of all points (x, y) in the Cartesian plane such that x is an element of the domain and y is the corresponding element in the range. The correspondence between domain and range elements is often specified by an equation in two variables. Consider, for example, the equation for the area of a rectangle with width 1 inch less than its length (Fig. 6). If x is the length, then the area y is given by

$$y = x(x - 1) \qquad x \geq 1$$

For each **input** x (length), we obtain an **output** y (area). For example,

$$\text{If} \quad x = 5, \quad \text{then} \quad y = 5(5 - 1) = 5 \cdot 4 = 20.$$
$$\text{If} \quad x = 1, \quad \text{then} \quad y = 1(1 - 1) = 1 \cdot 0 = 0.$$
$$\text{If} \quad x = \sqrt{5}, \quad \text{then} \quad y = \sqrt{5}(\sqrt{5} - 1) = 5 - \sqrt{5}$$
$$\approx 2.7639.$$

The input values are domain values, and the output values are range values. The equation assigns each domain value x a range value y. The variable x is called an *independent variable* (since values can be "independently" assigned to x from the domain), and y is called a *dependent variable* (since the value of y "depends" on the value assigned to x). In general, any variable used as a placeholder for domain values is called an **independent variable**; any variable that is used as a placeholder for range values is called a **dependent variable**.

When does an equation specify a function?

DEFINITION Functions Specified by Equations

If in an equation in two variables, we get exactly one output (value for the dependent variable) for each input (value for the independent variable), then the equation specifies a function. The graph of such a function is just the graph of the specifying equation.

If we get more than one output for a given input, the equation does not specify a function.

EXAMPLE 2 **Functions and Equations** Determine which of the following equations specify functions with independent variable x.

(A) $4y - 3x = 8$, x a real number (B) $y^2 - x^2 = 9$, x a real number

SOLUTION

(A) Solving for the dependent variable y, we have

$$4y - 3x = 8$$
$$4y = 8 + 3x \qquad (1)$$
$$y = 2 + \frac{3}{4}x$$

Since each input value x corresponds to exactly one output value $(y = 2 + \frac{3}{4}x)$, we see that equation (1) specifies a function.

(B) Solving for the dependent variable y, we have

$$y^2 - x^2 = 9$$

$$y^2 = 9 + x^2 \qquad (2)$$

$$y = \pm\sqrt{9 + x^2}$$

Reminder

Each positive real number u has two square roots: $\sqrt{u}$, the principal square root, and $-\sqrt{u}$, the negative of the principal square root (see Appendix A, Section A.6).

Since $9 + x^2$ is always a positive real number for any real number x, and since each positive real number has two square roots, then to each input value x there corresponds two output values ($y = -\sqrt{9 + x^2}$ and $y = \sqrt{9 + x^2}$). For example, if $x = 4$, then equation (2) is satisfied for $y = 5$ and for $y = -5$. So equation (2) does not specify a function.

Matched Problem 2 Determine which of the following equations specify functions with independent variable x.

(A) $y^2 - x^4 = 9$, x a real number (B) $3y - 2x = 3$, x a real number

Since the graph of an equation is the graph of all the ordered pairs that satisfy the equation, it is very easy to determine whether an equation specifies a function by examining its graph. The graphs of the two equations we considered in Example 2 are shown in Figure 7.

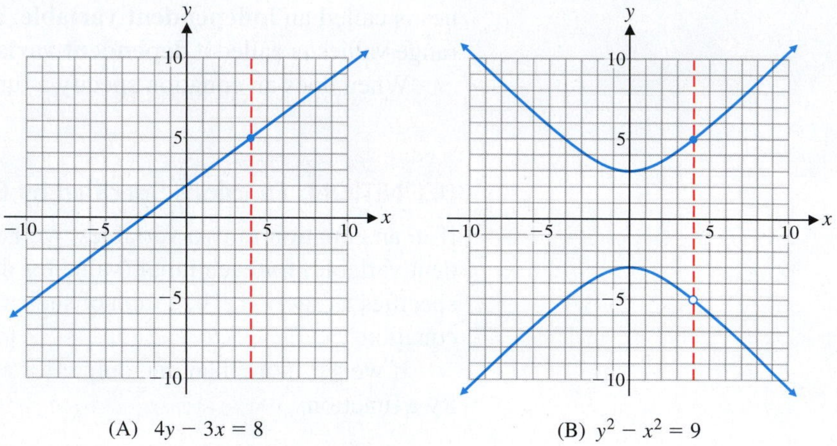

Figure 7

In Figure 7A, notice that any vertical line will intersect the graph of the equation $4y - 3x = 8$ in exactly one point. This shows that to each x value, there corresponds exactly one y value, confirming our conclusion that this equation specifies a function. On the other hand, Figure 7B shows that there exist vertical lines that intersect the graph of $y^2 - x^2 = 9$ in two points. This indicates that there exist x values to which there correspond two different y values and verifies our conclusion that this equation does not specify a function. These observations are generalized in Theorem 1.

THEOREM 1 Vertical-Line Test for a Function

An equation specifies a function if each vertical line in the coordinate system passes through, at most, one point on the graph of the equation.

If any vertical line passes through two or more points on the graph of an equation, then the equation does not specify a function.

The function graphed in Figure 7A is an example of a *linear function.* The vertical-line test implies that equations of the form $y = mx + b$, where $m \neq 0$, specify functions; they are called **linear functions**. Similarly, equations of the form $y = b$ specify functions; they are called **constant functions**, and their graphs are horizontal lines. The vertical-line test implies that equations of the form $x = a$ do not specify functions; note that the graph of $x = a$ is a vertical line.

In Example 2, the domains were explicitly stated along with the given equations. In many cases, this will not be done. Unless stated to the contrary, we shall adhere to the following convention regarding domains and ranges for functions specified by equations:

If a function is specified by an equation and the domain is not indicated, then we assume that the domain is the set of all real-number replacements of the independent variable (inputs) that produce real values for the dependent variable (outputs). The range is the set of all outputs corresponding to input values.

EXAMPLE 3 **Finding a Domain** Find the domain of the function specified by the equation $y = \sqrt{4 - x}$, assuming that x is the independent variable.

SOLUTION For y to be real, $4 - x$ must be greater than or equal to 0; that is,

$$4 - x \geq 0$$
$$-x \geq -4$$
$$x \leq 4 \qquad \text{Sense of inequality reverses when both sides are divided by } -1.$$

Domain: $x \leq 4$ (inequality notation) or $(-\infty, 4]$ (interval notation)

Matched Problem 3 Find the domain of the function specified by the equation $y = \sqrt{x - 2}$, assuming x is the independent variable.

Function Notation

We have seen that a function involves two sets, a domain and a range, and a correspondence that assigns to each element in the domain exactly one element in the range. Just as we use letters as names for numbers, now we will use letters as names for functions. For example, f and g may be used to name the functions specified by the equations $y = 2x + 1$ and $y = x^2 + 2x - 3$:

$$f: \quad y = 2x + 1$$
$$g: \quad y = x^2 + 2x - 3 \qquad (3)$$

If x represents an element in the domain of a function f, then we frequently use the symbol

$$f(x)$$

in place of y to designate the number in the range of the function f to which x is paired (Fig. 8). This symbol does *not* represent the product of f and x. The symbol $f(x)$ is read as "f of x," "f at x," or "the value of f at x." Whenever we write $y = f(x)$, we assume that the variable x is an independent variable and that both y and $f(x)$ are dependent variables.

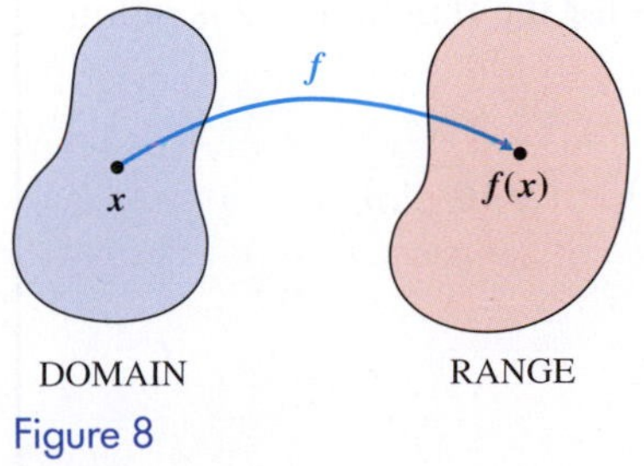

Figure 8

Using function notation, we can now write functions f and g in equation (3) as

$$f(x) = 2x + 1 \qquad \text{and} \qquad g(x) = x^2 + 2x - 3$$

Let us find $f(3)$ and $g(-5)$. To find $f(3)$, we replace x with 3 wherever x occurs in $f(x) = 2x + 1$ and evaluate the right side:

$$f(x) = 2x + 1$$
$$f(3) = 2 \cdot 3 + 1$$
$$= 6 + 1 = 7 \qquad \text{For input 3, the output is 7.}$$

Therefore,

$$f(3) = 7 \qquad \text{The function } f \text{ assigns the range value 7 to the domain value 3.}$$

To find $g(-5)$, we replace each x by -5 in $g(x) = x^2 + 2x - 3$ and evaluate the right side:

$$g(x) = x^2 + 2x - 3$$
$$g(-5) = (-5)^2 + 2(-5) - 3$$
$$= 25 - 10 - 3 = 12 \qquad \text{For input } -5\text{, the output is 12.}$$

Therefore,

$$g(-5) = 12 \qquad \text{The function } g \text{ assigns the range value 12 to the domain value } -5.$$

It is very important to understand and remember the definition of $f(x)$:

For any element x in the domain of the function f, the symbol $f(x)$ represents the element in the range of f corresponding to x in the domain of f. If x is an input value, then $f(x)$ is the corresponding output value. If x is an element that is not in the domain of f, then f *is not defined at* x and $f(x)$ *does not exist*.

EXAMPLE 4 **Function Evaluation** For $f(x) = 12/(x - 2)$, $g(x) = 1 - x^2$, and $h(x) = \sqrt{x - 1}$, evaluate:

(A) $f(6)$ (B) $g(-2)$ (C) $h(-2)$ (D) $f(0) + g(1) - h(10)$

SOLUTION

(A) $f(6) = \dfrac{12}{6 - 2} = \dfrac{12}{4} = 3$

(B) $g(-2) = 1 - (-2)^2 = 1 - 4 = -3$

(C) $h(-2) = \sqrt{-2 - 1} = \sqrt{-3}$

Reminder

Dashed boxes are used throughout the book to represent steps that are usually performed mentally.

But $\sqrt{-3}$ is not a real number. Since we have agreed to restrict the domain of a function to values of x that produce real values for the function, -2 is not in the domain of h, and $h(-2)$ does not exist.

(D) $$f(0) + g(1) - h(10) = \frac{12}{0 - 2} + (1 - 1^2) - \sqrt{10 - 1}$$
$$= \frac{12}{-2} + 0 - \sqrt{9}$$
$$= -6 - 3 = -9$$

Matched Problem 4 Use the functions in Example 4 to find

(A) $f(-2)$ (B) $g(-1)$ (C) $h(-8)$ (D) $\dfrac{f(3)}{h(5)}$

EXAMPLE 5 **Finding Domains** Find the domains of functions f, g, and h:

$$f(x) = \frac{12}{x - 2} \qquad g(x) = 1 - x^2 \qquad h(x) = \sqrt{x - 1}$$

SOLUTION *Domain of f:* $12/(x - 2)$ represents a real number for all replacements of x by real numbers except for $x = 2$ (division by 0 is not defined). Thus, $f(2)$ does not exist, and the domain of f is the set of all real numbers except 2. We often indicate this by writing

$$f(x) = \frac{12}{x - 2} \qquad x \neq 2$$

Domain of g: The domain is R, the set of all real numbers, since $1 - x^2$ represents a real number for all replacements of x by real numbers.

Domain of h: The domain is the set of all real numbers x such that $\sqrt{x - 1}$ is a real number; so

$$x - 1 \geq 0$$

$$x \geq 1 \quad \text{or, in interval notation,} \quad [1, \infty)$$

Matched Problem 5 Find the domains of functions F, G, and H:

$$F(x) = x^2 - 3x + 1 \qquad G(x) = \frac{5}{x + 3} \qquad H(x) = \sqrt{2 - x}$$

In addition to evaluating functions at specific numbers, it is important to be able to evaluate functions at expressions that involve one or more variables. For example, the **difference quotient**

$$\frac{f(x + h) - f(x)}{h} \qquad x \text{ and } x + h \text{ in the domain of } f,\ h \neq 0$$

is studied extensively in calculus.

CONCEPTUAL INSIGHT

In algebra, you learned to use parentheses for grouping variables. For example,

$$2(x + h) = 2x + 2h$$

Now we are using parentheses in the function symbol $f(x)$. For example, if $f(x) = x^2$, then

$$f(x + h) = (x + h)^2 = x^2 + 2xh + h^2$$

Note that $f(x) + f(h) = x^2 + h^2 \neq f(x + h)$. That is, the function name f does not distribute across the grouped variables $(x + h)$, as the "2" does in $2(x + h)$ (see Appendix A, Section A.2).

EXAMPLE 6 **Using Function Notation** For $f(x) = x^2 - 2x + 7$, find

(A) $f(a)$

(B) $f(a + h)$

(C) $f(a + h) - f(a)$

(D) $\dfrac{f(a + h) - f(a)}{h}, \quad h \neq 0$

SOLUTION

(A) $f(a) = a^2 - 2a + 7$

(B) $f(a + h) = (a + h)^2 - 2(a + h) + 7 = a^2 + 2ah + h^2 - 2a - 2h + 7$

(C) $f(a + h) - f(a) = (a^2 + 2ah + h^2 - 2a - 2h + 7) - (a^2 - 2a + 7)$
$= 2ah + h^2 - 2h$

(D) $\dfrac{f(a+h) - f(a)}{h} = \dfrac{2ah + h^2 - 2h}{h} = \dfrac{h(2a + h - 2)}{h}$ Because $h \neq 0$, $\dfrac{h}{h} = 1$.
$= 2a + h - 2$

Matched Problem 6 Repeat Example 6 for $f(x) = x^2 - 4x + 9$.

Applications

If we reduce the price of a product, will we generate more revenue? If we increase production, will our profits rise? **Profit–loss analysis** is a method for answering such questions in order to make sound business decisions.

Here are the basic concepts of profit–loss analysis: A manufacturing company has **costs**, C, which include **fixed costs** such as plant overhead, product design, setup, and promotion and **variable costs** that depend on the number of items produced. The **revenue**, R, is the amount of money received from the sale of its product. The company takes a **loss** if $R < C$, **breaks even** if $R = C$, and has a **profit** if $R > C$. The **profit** P is equal to revenue minus cost; that is, $P = R - C$. (So the company takes a loss if $P < 0$, breaks even if $P = 0$, and has a profit if $P > 0$.) To predict its revenue, a company uses a **price–demand** function, $p(x)$, determined using historical data or sampling techniques, that specifies the relationship between the demand x and the price p. A point (x, p) is on the graph of the price–demand function if x items can be sold at a price of \$$p$ per item. (Normally, a reduction in the price p will increase the demand x, so the graph of the price–demand function is expected to go downhill as you move from left to right.) The revenue R is equal to the number of items sold multiplied by the price per item; that is, $R = xp$.

Cost, revenue, and profit can be written as functions $C(x)$, $R(x)$, and $P(x)$ of the independent variable x, the number of items manufactured and sold. The functions $C(x)$, $R(x)$, $P(x)$, and $p(x)$ often have the following forms, where a, b, m, and n are positive constants determined from the context of a particular problem:

Cost function

$C(x) = a + bx$ C = fixed costs + variable costs

Price–demand function

$p(x) = m - nx$ x is the number of items that can be sold at \$$p$ per item

Revenue function

$R(x) = xp$ R = number of items sold × price per item
$= x(m - nx)$

Profit function

$P(x) = R(x) - C(x)$
$= x(m - nx) - (a + bx)$

⚠ CAUTION Do not confuse the price–demand function $p(x)$ with the profit function $P(x)$. Price is always denoted by the lowercase p. Profit is always denoted by the uppercase P. Note that the revenue and profit functions, $R(x)$ and $P(x)$, depend on the price–demand function $p(x)$, but $C(x)$ does not. ▲

Example 7 and Matched Problem 7 provide an introduction to profit–loss analysis.

EXAMPLE 7 **Price–Demand and Revenue** A manufacturer of a popular digital camera wholesales the camera to retail outlets throughout the United States. Using statistical methods, the financial department in the company produced the price–demand data in Table 4, where p is the wholesale price per camera at which x million cameras are sold. Notice that as the price goes down, the number sold goes up.

Table 4 Price–Demand

x (millions)	p(\$)
2	87
5	68
8	53
12	37

Using special analytical techniques (regression analysis), an analyst obtained the following price–demand function to model the Table 4 data:

$$p(x) = 94.8 - 5x \qquad 1 \le x \le 15 \tag{4}$$

Table 5 Revenue

x (millions)	$R(x)$ (million \$)
1	90
3	
6	
9	
12	
15	

(A) Plot the data in Table 4. Then sketch a graph of the price–demand function in the same coordinate system.

(B) What is the company's revenue function for this camera, and what is its domain?

(C) Complete Table 5, computing revenues to the nearest million dollars.

(D) Plot the data in Table 5. Then sketch a graph of the revenue function using these points.

(E) Graph the revenue function on a graphing calculator.

SOLUTION

(A) The four data points are plotted in Figure 9. Note that $p(1) = 89.8$ and $p(15) = 19.8$. So the graph of the price–demand function is the line through (1, 89.8) and (15, 19.8) (see Fig. 9).

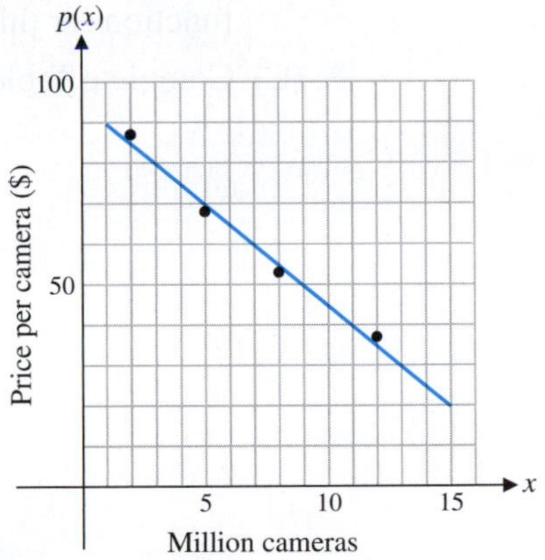

Figure 9 Price–demand

In Figure 9, notice that the model approximates the actual data in Table 4, and it is assumed that it gives realistic and useful results for all other values of x between 1 million and 15 million.

(B) $R(x) = xp(x) = x(94.8 - 5x)$ million dollars

Domain: $1 \le x \le 15$

[Same domain as the price–demand function, equation (4).]

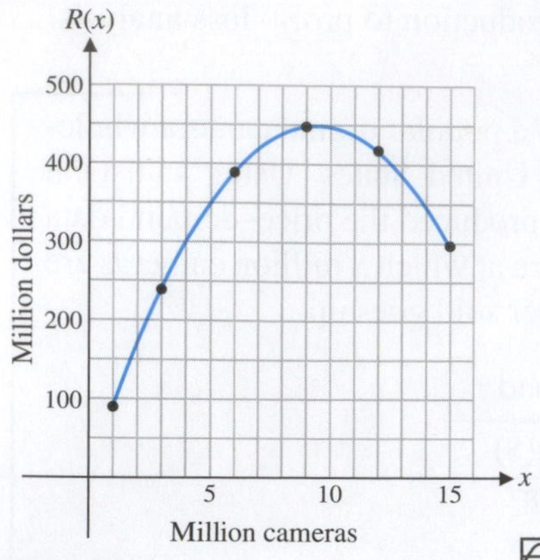

Figure 10

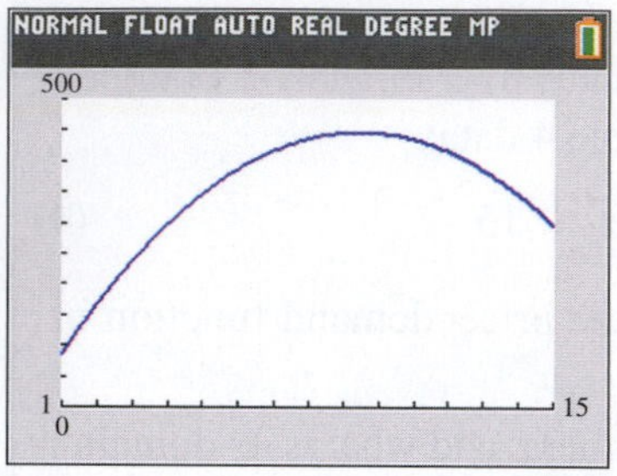

Figure 11

(C)

Table 5 Revenue

x (millions)	$R(x)$ (million \$)
1	90
3	239
6	389
9	448
12	418
15	297

(D) The six points from Table 5 are plotted in Figure 10. The graph of the revenue function is the smooth curve drawn through those six points.

(E) Figure 11 shows the graph of $R(x) = x(94.8 - 5x)$ on a graphing calculator.

Matched Problem 7 The financial department in Example 7, using statistical techniques, produced the data in Table 6, where $C(x)$ is the cost in millions of dollars for manufacturing and selling x million cameras.

Table 6 Cost Data

x (millions)	$C(x)$ (million \$)
1	175
5	260
8	305
12	395

Using special analytical techniques (regression analysis), an analyst produced the following cost function to model the Table 6 data:

$$C(x) = 156 + 19.7x \qquad 1 \le x \le 15 \tag{5}$$

(A) Plot the data in Table 6. Then sketch a graph of equation (5) in the same coordinate system.

(B) Using the revenue function from Example 7(B), what is the company's profit function for this camera, and what is its domain?

(C) Complete Table 7, computing profits to the nearest million dollars.

Table 7 Profit

x (millions)	$P(x)$ (million \$)
1	−86
3	
6	
9	
12	
15	

(D) Plot the data in Table 7. Then sketch a graph of the profit function using these points.

(E) Graph the profit function on a graphing calculator.

Exercises 2.1

A *In Problems 1–8, use point-by-point plotting to sketch the graph of each equation.*

1. $y = x + 1$ **2.** $x = y + 1$

3. $x = y^2$ **4.** $y = x^2$

5. $y = x^3$ **6.** $x = y^3$

7. $xy = -6$ **8.** $xy = 12$

Indicate whether each table in Problems 9–14 specifies a function.

9.

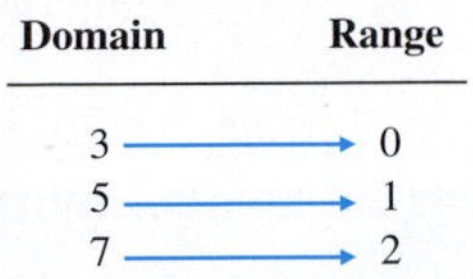

10.

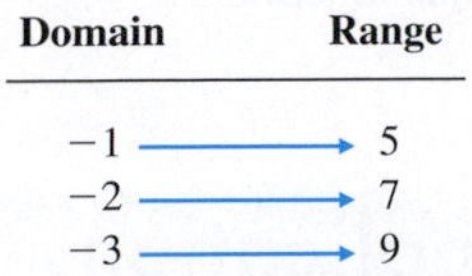

11.

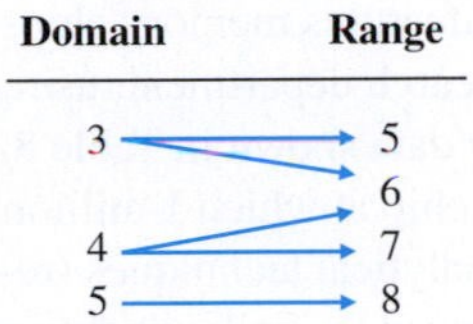

12.

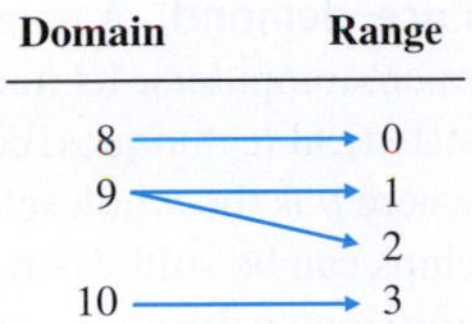

13.

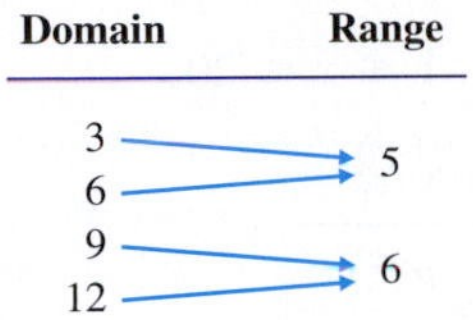

14.

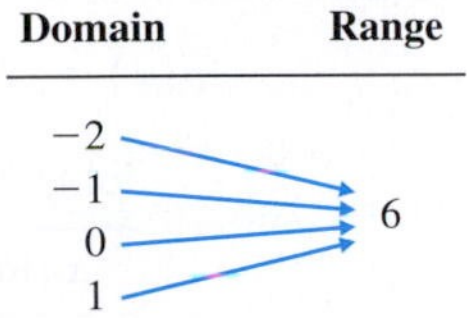

Indicate whether each graph in Problems 15–20 specifies a function.

15.

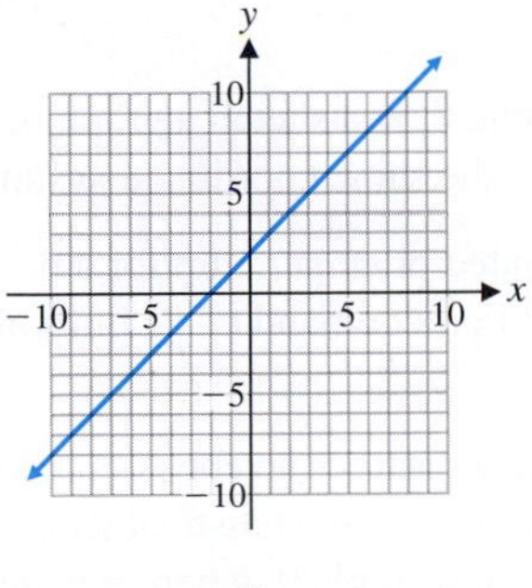

16.

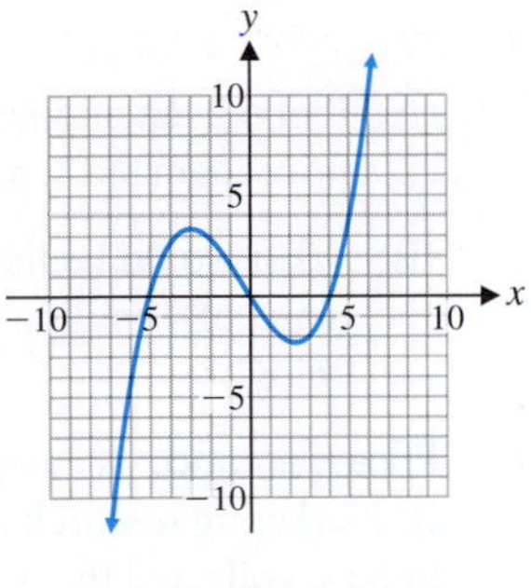

17.

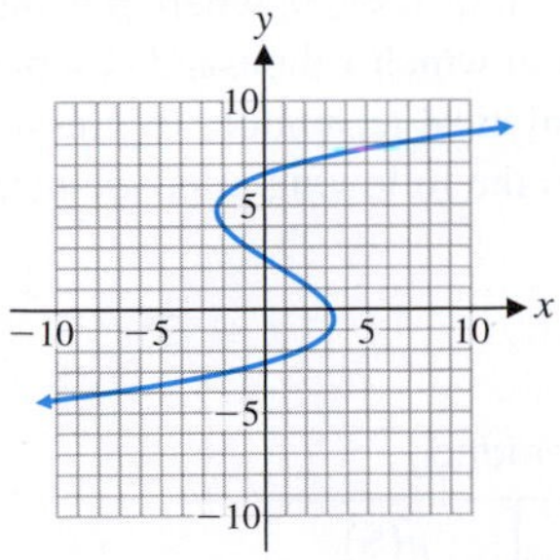

18.

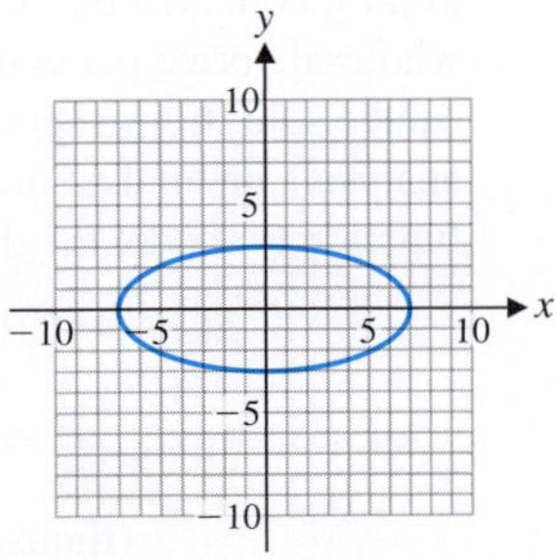

19.

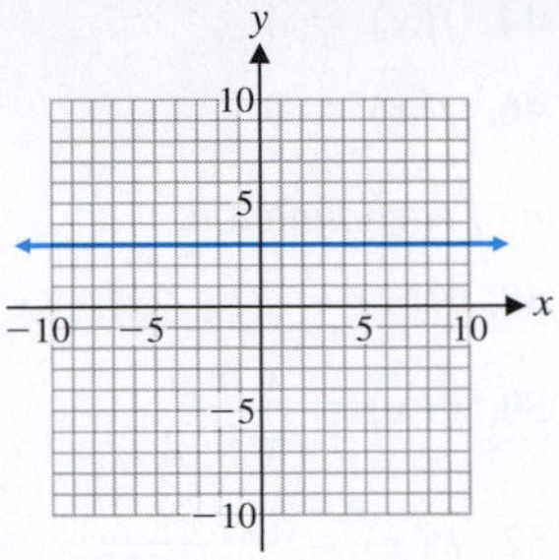

20.

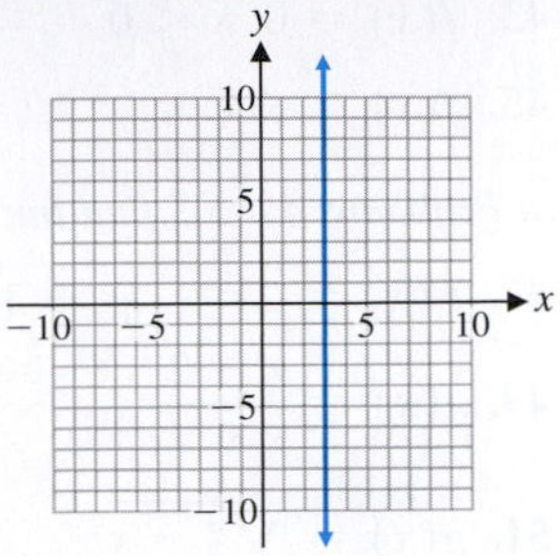

In Problems 21–28, each equation specifies a function with independent variable x. Determine whether the function is linear, constant, or neither.

21. $y = -3x + \frac{1}{8}$ **22.** $y = 4x + \frac{1}{x}$

23. $y + 5x^2 = 7$ **24.** $2x - 4y - 6 = 0$

25. $x = 8y + 9$ **26.** $x + xy + 1 = 0$

27. $y - x^2 + 2 = 10 - x^2$ **28.** $\frac{y - x}{2} + \frac{3 + 2x}{4} = 1$

In Problems 29–36, use point-by-point plotting to sketch the graph of each function.

29. $f(x) = 1 - x$ **30.** $f(x) = \frac{x}{2} - 3$

31. $f(x) = x^2 - 1$ **32.** $f(x) = 3 - x^2$

33. $f(x) = 4 - x^3$ **34.** $f(x) = x^3 - 2$

35. $f(x) = \frac{8}{x}$ **36.** $f(x) = \frac{-6}{x}$

In Problems 37 and 38, the three points in the table are on the graph of the indicated function f. Do these three points provide sufficient information for you to sketch the graph of $y = f(x)$? Add more points to the table until you are satisfied that your sketch is a good representation of the graph of $y = f(x)$ for $-5 \le x \le 5$.

37.

x	−1	0	1
$f(x)$	−1	0	1

$f(x) = \frac{2x}{x^2 + 1}$

38.

x	0	1	2
$f(x)$	0	1	2

$f(x) = \frac{3x^2}{x^2 + 2}$

In Problems 39–46, use the following graph of a function f to determine x or y to the nearest integer, as indicated. Some problems may have more than one answer.

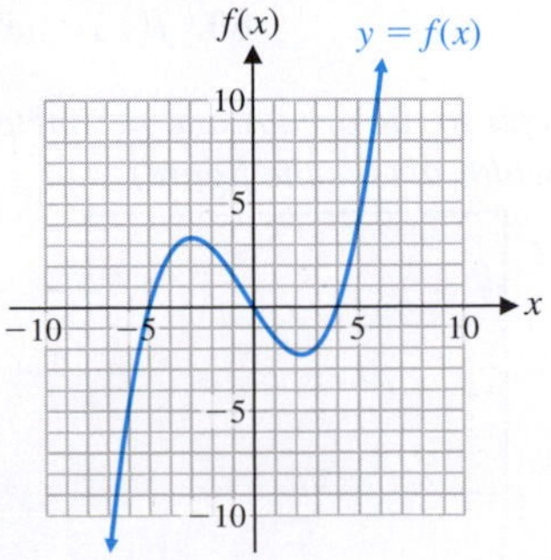

39. $y = f(-5)$

40. $y = f(4)$

41. $y = f(5)$

42. $y = f(-2)$

43. $f(x) = 0, x < 0$

44. $f(x) = 4$

45. $f(x) = -5$

46. $f(x) = 0$

B *In Problems 47–52, find the domain of each function.*

47. $F(x) = 2x^3 - x^2 + 3$

48. $H(x) = 7 - 2x^2 - x^4$

49. $f(x) = \dfrac{x - 2}{x + 4}$

50. $g(x) = \dfrac{x + 1}{x - 2}$

51. $g(x) = \sqrt{7 - x}$

52. $F(x) = \dfrac{1}{\sqrt{5 + x}}$

In Problems 53–60, does the equation specify a function with independent variable x? If so, find the domain of the function. If not, find a value of x to which there corresponds more than one value of y.

53. $2x + 5y = 10$

54. $6x - 7y = 21$

55. $y(x + y) = 4$

56. $x(x + y) = 4$

57. $x^{-3} + y^3 = 27$

58. $x^2 + y^2 = 9$

59. $x^3 - y^2 = 0$

60. $\sqrt{x} - y^3 = 0$

In Problems 61–74, find and simplify the expression if $f(x) = x^2 - 4$.

61. $f(5x)$

62. $f(-3x)$

63. $f(x + 2)$

64. $f(x - 1)$

65. $f(x^2)$

66. $f(x^3)$

67. $f(\sqrt{x})$

68. $f(\sqrt[4]{x})$

69. $f(2) + f(h)$

70. $f(-3) + f(h)$

71. $f(2 + h)$

72. $f(-3 + h)$

73. $f(2 + h) - f(2)$

74. $f(-3 + h) - f(-3)$

C *In Problems 75–80, find and simplify each of the following, assuming $h \neq 0$ in (C).*

(A) $f(x + h)$

(B) $f(x + h) - f(x)$

(C) $\dfrac{f(x + h) - f(x)}{h}$

75. $f(x) = 4x - 3$

76. $f(x) = -3x + 9$

77. $f(x) = 4x^2 - 7x + 6$

78. $f(x) = 3x^2 + 5x - 8$

79. $f(x) = x(20 - x)$

80. $f(x) = x(x + 40)$

Problems 81–84 refer to the area A and perimeter P of a rectangle with length l and width w (see the figure).

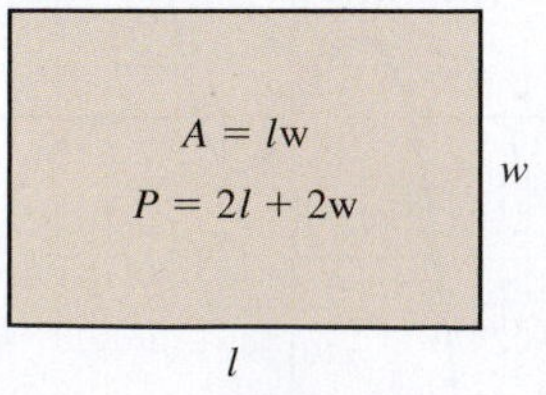

81. The area of a rectangle is 25 sq in. Express the perimeter $P(w)$ as a function of the width w, and state the domain of this function.

82. The area of a rectangle is 81 sq in. Express the perimeter $P(l)$ as a function of the length l, and state the domain of this function.

83. The perimeter of a rectangle is 100 m. Express the area $A(l)$ as a function of the length l, and state the domain of this function.

84. The perimeter of a rectangle is 160 m. Express the area $A(w)$ as a function of the width w, and state the domain of this function.

Applications

85. Price–demand. A company manufactures memory chips for microcomputers. Its marketing research department, using statistical techniques, collected the data shown in Table 8, where p is the wholesale price per chip at which x million chips can be sold. Using special analytical techniques (regression analysis), an analyst produced the following price–demand function to model the data:

$$p(x) = 75 - 3x \qquad 1 \le x \le 20$$

Table 8 Price–Demand

x (millions)	p($)
1	72
4	63
9	48
14	33
20	15

(A) Plot the data points in Table 8, and sketch a graph of the price–demand function in the same coordinate system.

(B) What would be the estimated price per chip for a demand of 7 million chips? For a demand of 11 million chips?

86. Price–demand. A company manufactures notebook computers. Its marketing research department, using statistical techniques, collected the data shown in Table 9, where p is the wholesale price per computer at which x thousand computers can be sold. Using special analytical techniques (regression analysis), an analyst produced the following price–demand function to model the data:

$$p(x) = 2{,}000 - 60x \qquad 1 \le x \le 25$$

Table 9 Price–Demand

x (thousands)	p($)
1	1,940
8	1,520
16	1,040
21	740
25	500

(A) Plot the data points in Table 9, and sketch a graph of the price–demand function in the same coordinate system.

(B) What would be the estimated price per computer for a demand of 11,000 computers? For a demand of 18,000 computers?

87. Revenue.

(A) Using the price–demand function

$$p(x) = 75 - 3x \qquad 1 \le x \le 20$$

from Problem 85, write the company's revenue function and indicate its domain.

(B) Complete Table 10, computing revenues to the nearest million dollars.

Table 10 Revenue

x (millions)	$R(x)$ (million \$)
1	72
4	
8	
12	
16	
20	

(C) Plot the points from part (B) and sketch a graph of the revenue function using these points. Choose millions for the units on the horizontal and vertical axes.

88. Revenue.

(A) Using the price–demand function

$$p(x) = 2{,}000 - 60x \qquad 1 \le x \le 25$$

from Problem 86, write the company's revenue function and indicate its domain.

(B) Complete Table 11, computing revenues to the nearest thousand dollars.

Table 11 Revenue

x (thousands)	$R(x)$ (thousand \$)
1	1,940
5	
10	
15	
20	
25	

(C) Plot the points from part (B) and sketch a graph of the revenue function using these points. Choose thousands for the units on the horizontal and vertical axes.

89. Profit. The financial department for the company in Problems 85 and 87 established the following cost function for producing and selling x million memory chips:

$$C(x) = 125 + 16x \text{ million dollars}$$

(A) Write a profit function for producing and selling x million memory chips and indicate its domain.

(B) Complete Table 12, computing profits to the nearest million dollars.

Table 12 Profit

x (millions)	$P(x)$ (million \$)
1	−69
4	
8	
12	
16	
20	

(C) Plot the points in part (B) and sketch a graph of the profit function using these points.

90. Profit. The financial department for the company in Problems 86 and 88 established the following cost function for producing and selling x thousand notebook computers:

$$C(x) = 4{,}000 + 500x \text{ thousand dollars}$$

(A) Write a profit function for producing and selling x thousand notebook computers and indicate its domain.

(B) Complete Table 13, computing profits to the nearest thousand dollars.

Table 13 Profit

x (thousands)	$P(x)$ (thousand \$)
1	−2,560
5	
10	
15	
20	
25	

(C) Plot the points in part (B) and sketch a graph of the profit function using these points.

91. Muscle contraction. In a study of the speed of muscle contraction in frogs under various loads, British biophysicist A. W. Hill determined that the weight w (in grams) placed on the muscle and the speed of contraction v (in centimeters per second) are approximately related by an equation of the form

$$(w + a)(v + b) = c$$

where a, b, and c are constants. Suppose that for a certain muscle, $a = 15$, $b = 1$, and $c = 90$. Express v as a function of w. Find the speed of contraction if a weight of 16 g is placed on the muscle.

92. Politics. The percentage s of seats in the House of Representatives won by Democrats and the percentage v of votes cast for Democrats (when expressed as decimal fractions) are related by the equation

$$5v - 2s = 1.4 \qquad 0 < s < 1, \quad 0.28 < v < 0.68$$

(A) Express v as a function of s and find the percentage of votes required for the Democrats to win 51% of the seats.

(B) Express s as a function of v and find the percentage of seats won if Democrats receive 51% of the votes.

Answers to Matched Problems

1. (A)

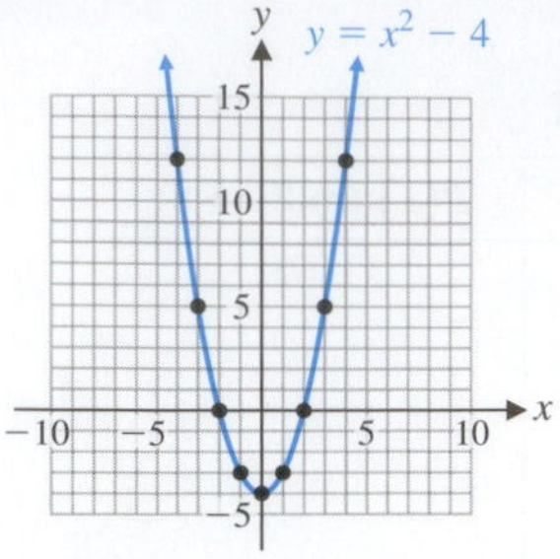

(B)

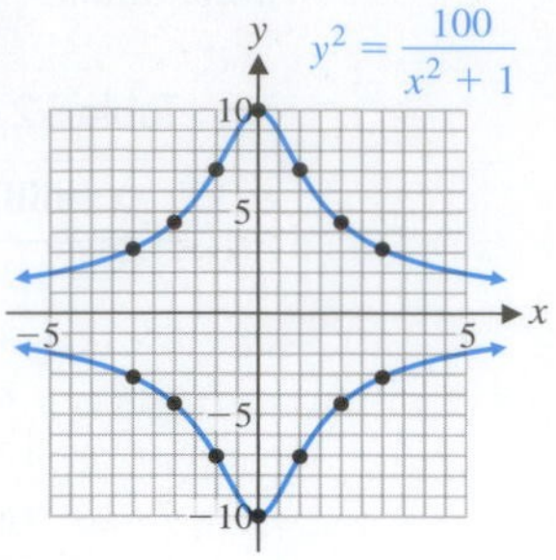

2. (A) Does not specify a function

(B) Specifies a function

3. $x \geq 2$ (inequality notation) or $[2, \infty)$ (interval notation)

4. (A) -3 (B) 0

(C) Does not exist (D) 6

5. Domain of F: R; domain of G: all real numbers except -3; domain of H: $x \leq 2$ (inequality notation) or $(-\infty, 2]$ (interval notation)

6. (A) $a^2 - 4a + 9$

(B) $a^2 + 2ah + h^2 - 4a - 4h + 9$

(C) $2ah + h^2 - 4h$

(D) $2a + h - 4$

7. (A)

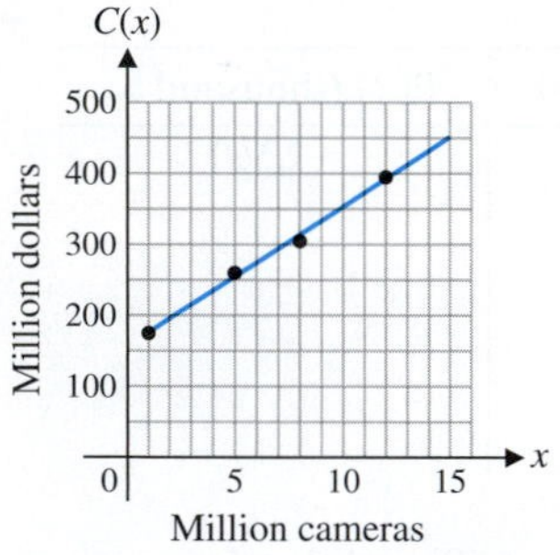

(B) $P(x) = R(x) - C(x)$
$= x(94.8 - 5x) - (156 + 19.7x)$;
domain: $1 \leq x \leq 15$

(C) Table 7 Profit

x (millions)	$P(x)$ (million \$)
1	-86
3	24
6	115
9	115
12	25
15	-155

(D)

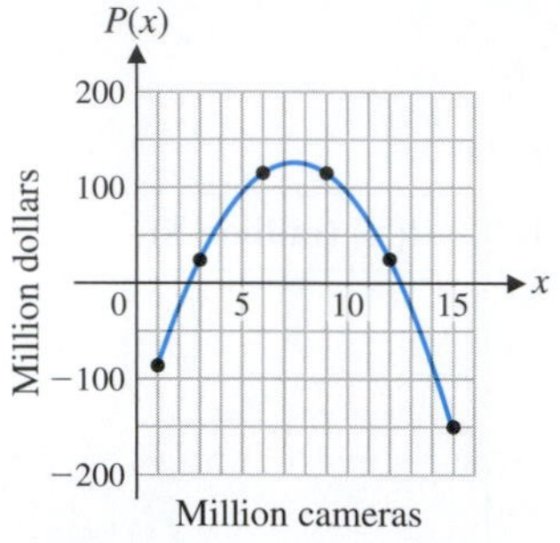

(E)

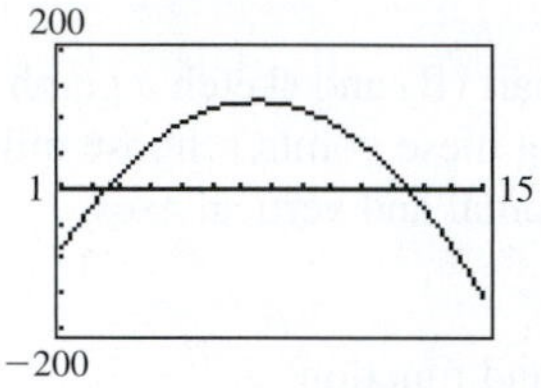

2.2 Elementary Functions: Graphs and Transformations

- A Beginning Library of Elementary Functions
- Vertical and Horizontal Shifts
- Reflections, Stretches, and Shrinks
- Piecewise-Defined Functions

Each of the functions

$$g(x) = x^2 - 4 \qquad h(x) = (x - 4)^2 \qquad k(x) = -4x^2$$

can be expressed in terms of the function $f(x) = x^2$:

$$g(x) = f(x) - 4 \qquad h(x) = f(x - 4) \qquad k(x) = -4f(x)$$

In this section, we will see that the graphs of functions g, h, and k are closely related to the graph of function f. Insight gained by understanding these relationships will help us analyze and interpret the graphs of many different functions.

A Beginning Library of Elementary Functions

As you progress through this book, you will repeatedly encounter a small number of elementary functions. We will identify these functions, study their basic properties, and include them in a library of elementary functions (see the endpapers at the back of the book). This library will become an important addition to your mathematical toolbox and can be used in any course or activity where mathematics is applied.

We begin by placing six basic functions in our library.

DEFINITION Basic Elementary Functions

$f(x) = x$	Identity function
$h(x) = x^2$	Square function
$m(x) = x^3$	Cube function
$n(x) = \sqrt{x}$	Square root function
$p(x) = \sqrt[3]{x}$	Cube root function
$g(x) = \lvert x \rvert$	Absolute value function

These elementary functions can be evaluated by hand for certain values of x and with a calculator for all values of x for which they are defined.

EXAMPLE 1 **Evaluating Basic Elementary Functions** Evaluate each basic elementary function at

(A) $x = 64$ (B) $x = -12.75$

Round any approximate values to four decimal places.

SOLUTION

(A) $f(64) = 64$

$h(64) = 64^2 = 4{,}096$ Use a calculator.

$m(64) = 64^3 = 262{,}144$ Use a calculator.

$n(64) = \sqrt{64} = 8$

$p(64) = \sqrt[3]{64} = 4$

$g(64) = \lvert 64 \rvert = 64$

(B) $f(-12.75) = -12.75$

$h(-12.75) = (-12.75)^2 = 162.5625$ Use a calculator.

$m(-12.75) = (-12.75)^3 \approx -2{,}072.6719$ Use a calculator.

$n(-12.75) = \sqrt{-12.75}$ Not a real number.

$p(-12.75) = \sqrt[3]{-12.75} \approx -2.3362$ Use a calculator.

$g(-12.75) = \lvert -12.75 \rvert = 12.75$

Matched Problem 1 Evaluate each basic elementary function at

(A) $x = 729$ (B) $x = -5.25$

Round any approximate values to four decimal places.

Remark Most computers and graphing calculators use ABS(x) to represent the absolute value function. The following representation can also be useful:

$$\lvert x \rvert = \sqrt{x^2}$$

Figure 1 shows the graph, range, and domain of each of the basic elementary functions.

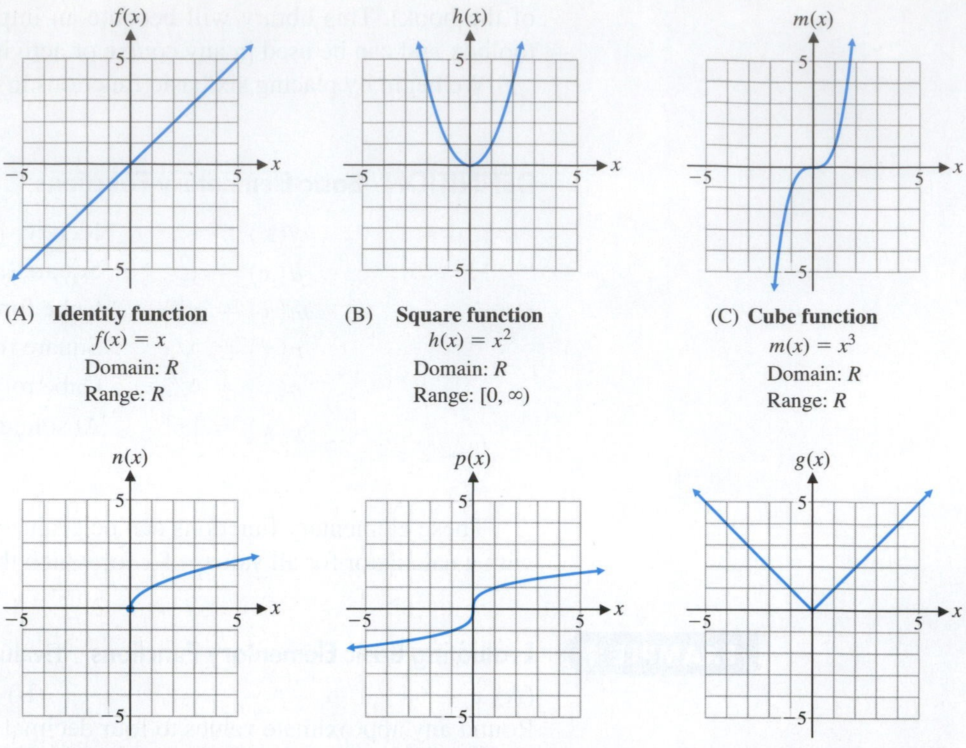

(A) **Identity function**
$f(x) = x$
Domain: R
Range: R

(B) **Square function**
$h(x) = x^2$
Domain: R
Range: $[0, \infty)$

(C) **Cube function**
$m(x) = x^3$
Domain: R
Range: R

(D) **Square root function**
$n(x) = \sqrt{x}$
Domain: $[0, \infty)$
Range: $[0, \infty)$

(E) **Cube root function**
$p(x) = \sqrt[3]{x}$
Domain: R
Range: R

(F) **Absolute value function**
$g(x) = |x|$
Domain: R
Range: $[0, \infty)$

Figure 1 Some basic functions and their graphs

Reminder

Letters used to designate these functions may vary from context to context; R is the set of all real numbers.

CONCEPTUAL INSIGHT

Absolute Value In beginning algebra, absolute value is often interpreted as distance from the origin on a real number line (see Appendix A, Section A.1).

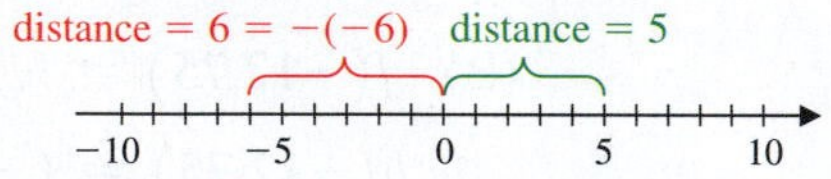

If $x < 0$, then $-x$ is the *positive* distance from the origin to x, and if $x > 0$, then x is the positive distance from the origin to x. Thus,

$$|x| = \begin{cases} -x & \text{if } x < 0 \\ x & \text{if } x \geq 0 \end{cases}$$

Vertical and Horizontal Shifts

If a new function is formed by performing an operation on a given function, then the graph of the new function is called a **transformation** of the graph of the original function. For example, graphs of $y = f(x) + k$ and $y = f(x + h)$ are transformations of the graph of $y = f(x)$.

Explore and Discuss 1

Let $f(x) = x^2$.

(A) Graph $y = f(x) + k$ for $k = -4$, 0, and 2 simultaneously in the same coordinate system. Describe the relationship between the graph of $y = f(x)$ and the graph of $y = f(x) + k$ for any real number k.

(B) Graph $y = f(x + h)$ for $h = -4$, 0, and 2 simultaneously in the same coordinate system. Describe the relationship between the graph of $y = f(x)$ and the graph of $y = f(x + h)$ for any real number h.

EXAMPLE 2

Vertical and Horizontal Shifts

(A) How are the graphs of $y = |x| + 4$ and $y = |x| - 5$ related to the graph of $y = |x|$? Confirm your answer by graphing all three functions simultaneously in the same coordinate system.

(B) How are the graphs of $y = |x + 4|$ and $y = |x - 5|$ related to the graph of $y = |x|$? Confirm your answer by graphing all three functions simultaneously in the same coordinate system.

SOLUTION

(A) The graph of $y = |x| + 4$ is the same as the graph of $y = |x|$ shifted upward 4 units, and the graph of $y = |x| - 5$ is the same as the graph of $y = |x|$ shifted downward 5 units. Figure 2 confirms these conclusions. [It appears that the graph of $y = f(x) + k$ is the graph of $y = f(x)$ shifted up if k is positive and down if k is negative.]

(B) The graph of $y = |x + 4|$ is the same as the graph of $y = |x|$ shifted to the left 4 units, and the graph of $y = |x - 5|$ is the same as the graph of $y = |x|$ shifted to the right 5 units. Figure 3 confirms these conclusions. [It appears that the graph of $y = f(x + h)$ is the graph of $y = f(x)$ shifted right if h is negative and left if h is positive—the opposite of what you might expect.]

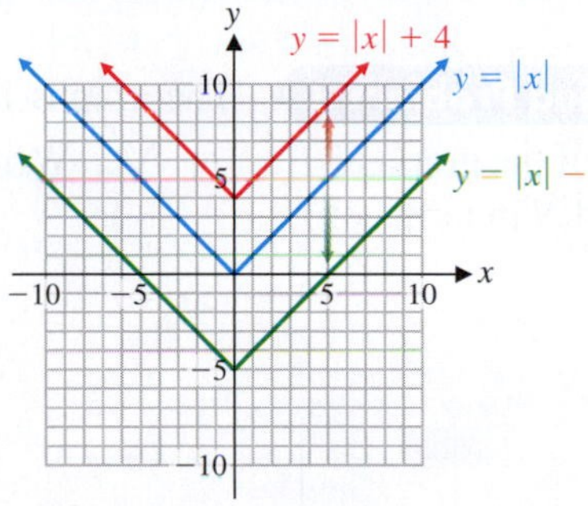

Figure 2 Vertical shifts

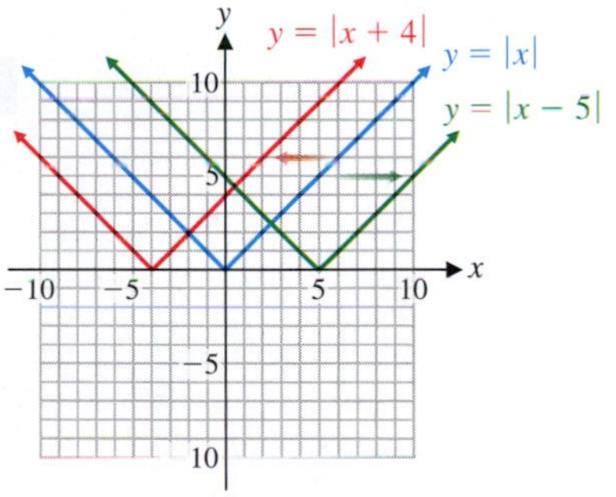

Figure 3 Horizontal shifts

Matched Problem 2

(A) How are the graphs of $y = \sqrt{x} + 5$ and $y = \sqrt{x} - 4$ related to the graph of $y = \sqrt{x}$? Confirm your answer by graphing all three functions simultaneously in the same coordinate system.

(B) How are the graphs of $y = \sqrt{x + 5}$ and $y = \sqrt{x - 4}$ related to the graph of $y = \sqrt{x}$? Confirm your answer by graphing all three functions simultaneously in the same coordinate system.

Comparing the graphs of $y = f(x) + k$ with the graph of $y = f(x)$, we see that the graph of $y = f(x) + k$ can be obtained from the graph of $y = f(x)$ by **vertically translating** (shifting) the graph of the latter upward k units if k is positive and downward $|k|$ units if k is negative. Comparing the graphs of $y = f(x + h)$ with the graph of $y = f(x)$, we see that the graph of $y = f(x + h)$ can be obtained from the graph of $y = f(x)$ by **horizontally translating** (shifting) the graph of the latter h units to the left if h is positive and $|h|$ units to the right if h is negative.

EXAMPLE 3 **Vertical and Horizontal Translations (Shifts)** The graphs in Figure 4 are either horizontal or vertical shifts of the graph of $f(x) = x^2$. Write appropriate equations for functions H, G, M, and N in terms of f.

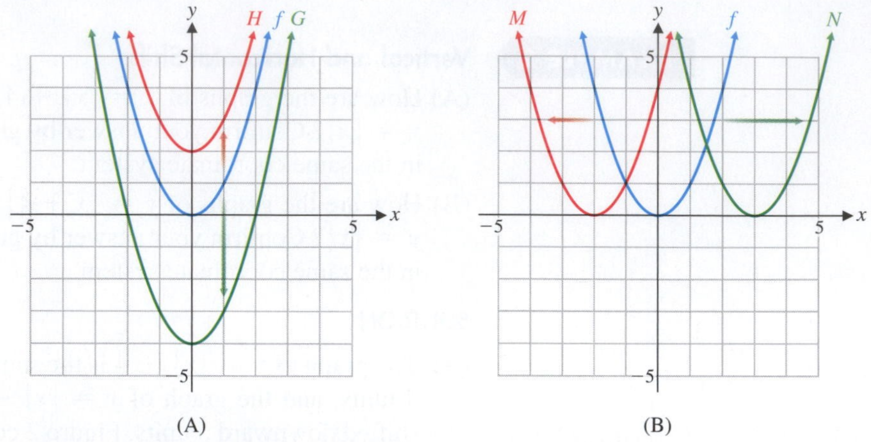

Figure 4 Vertical and horizontal shifts

SOLUTION Functions H and G are vertical shifts given by

$$H(x) = x^2 + 2 \qquad G(x) = x^2 - 4$$

Functions M and N are horizontal shifts given by

$$M(x) = (x + 2)^2 \qquad N(x) = (x - 3)^2$$

Matched Problem 3 The graphs in Figure 5 are either horizontal or vertical shifts of the graph of $f(x) = \sqrt[3]{x}$. Write appropriate equations for functions H, G, M, and N in terms of f.

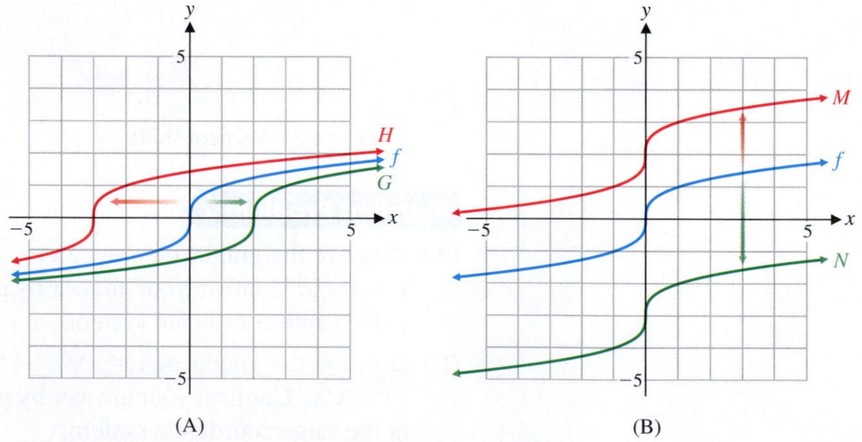

Figure 5 Vertical and horizontal shifts

Reflections, Stretches, and Shrinks

We now investigate how the graph of $y = Af(x)$ is related to the graph of $y = f(x)$ for different real numbers A.

Explore and Discuss 2

(A) Graph $y = Ax^2$ for $A = 1, 4,$ and $\frac{1}{4}$ simultaneously in the same coordinate system.

(B) Graph $y = Ax^2$ for $A = -1, -4,$ and $-\frac{1}{4}$ simultaneously in the same coordinate system.

(C) Describe the relationship between the graph of $h(x) = x^2$ and the graph of $G(x) = Ax^2$ for any real number A.

Comparing $y = Af(x)$ to $y = f(x)$, we see that the graph of $y = Af(x)$ can be obtained from the graph of $y = f(x)$ by multiplying each ordinate value of the latter by A. The result is a **vertical stretch** of the graph of $y = f(x)$ if $A > 1$, a **vertical shrink** of the graph of $y = f(x)$ if $0 < A < 1$, and a **reflection in the x axis** if $A = -1$. If A is a negative number other than -1, then the result is a combination of a reflection in the x axis and either a vertical stretch or a vertical shrink.

EXAMPLE 4 **Reflections, Stretches, and Shrinks**

(A) How are the graphs of $y = 2|x|$ and $y = 0.5|x|$ related to the graph of $y = |x|$? Confirm your answer by graphing all three functions simultaneously in the same coordinate system.

(B) How is the graph of $y = -2|x|$ related to the graph of $y = |x|$? Confirm your answer by graphing both functions simultaneously in the same coordinate system.

SOLUTION

(A) The graph of $y = 2|x|$ is a vertical stretch of the graph of $y = |x|$ by a factor of 2, and the graph of $y = 0.5|x|$ is a vertical shrink of the graph of $y = |x|$ by a factor of 0.5. Figure 6 confirms this conclusion.

(B) The graph of $y = -2|x|$ is a reflection in the x axis and a vertical stretch of the graph of $y = |x|$. Figure 7 confirms this conclusion.

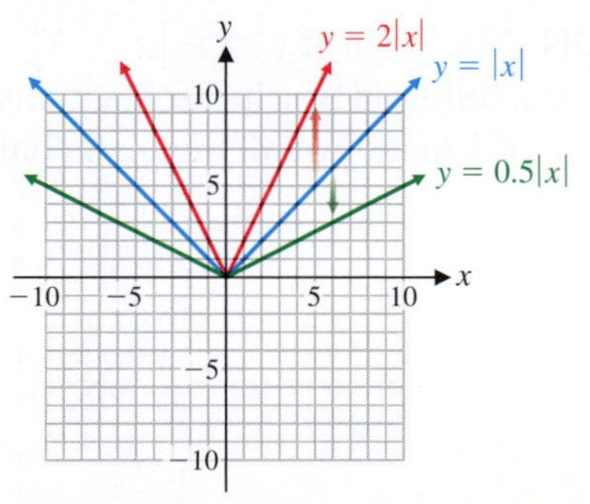

Figure 6 Vertical stretch and shrink

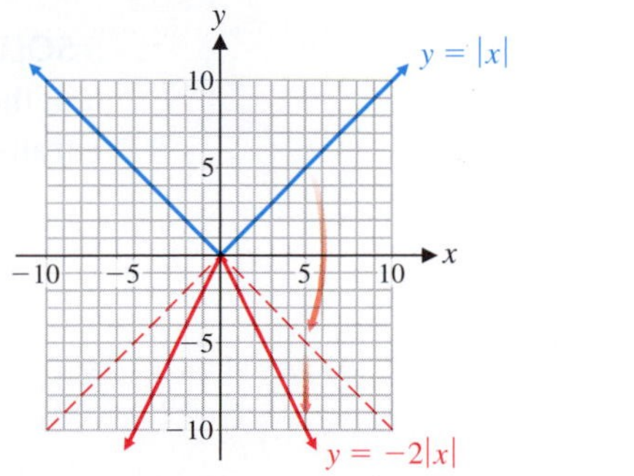

Figure 7 Reflection and vertical stretch

Matched Problem 4

(A) How are the graphs of $y = 2x$ and $y = 0.5x$ related to the graph of $y = x$? Confirm your answer by graphing all three functions simultaneously in the same coordinate system.

(B) How is the graph of $y = -0.5x$ related to the graph of $y = x$? Confirm your answer by graphing both functions in the same coordinate system.

The various transformations considered above are summarized in the following box for easy reference:

SUMMARY Graph Transformations

Vertical Translation:

$$y = f(x) + k \quad \begin{cases} k > 0 & \text{Shift graph of } y = f(x) \text{ up } k \text{ units.} \\ k < 0 & \text{Shift graph of } y = f(x) \text{ down } |k| \text{ units.} \end{cases}$$

Horizontal Translation:

$$y = f(x + h) \quad \begin{cases} h > 0 & \text{Shift graph of } y = f(x) \text{ left } h \text{ units.} \\ h < 0 & \text{Shift graph of } y = f(x) \text{ right } |h| \text{ units.} \end{cases}$$

Reflection:

$y = -f(x)$ Reflect the graph of $y = f(x)$ in the x axis.

Vertical Stretch and Shrink:

$$y = Af(x) \quad \begin{cases} A > 1 & \text{Stretch graph of } y = f(x) \text{ vertically by multiplying each ordinate value by } A. \\ 0 < A < 1 & \text{Shrink graph of } y = f(x) \text{ vertically by multiplying each ordinate value by } A. \end{cases}$$

Explore and Discuss 3

Explain why applying any of the graph transformations in the summary box to a linear function produces another linear function.

EXAMPLE 5 **Combining Graph Transformations** Discuss the relationship between the graphs of $y = -|x - 3| + 1$ and $y = |x|$. Confirm your answer by graphing both functions simultaneously in the same coordinate system.

SOLUTION The graph of $y = -|x - 3| + 1$ is a reflection of the graph of $y = |x|$ in the x axis, followed by a horizontal translation of 3 units to the right and a vertical translation of 1 unit upward. Figure 8 confirms this description.

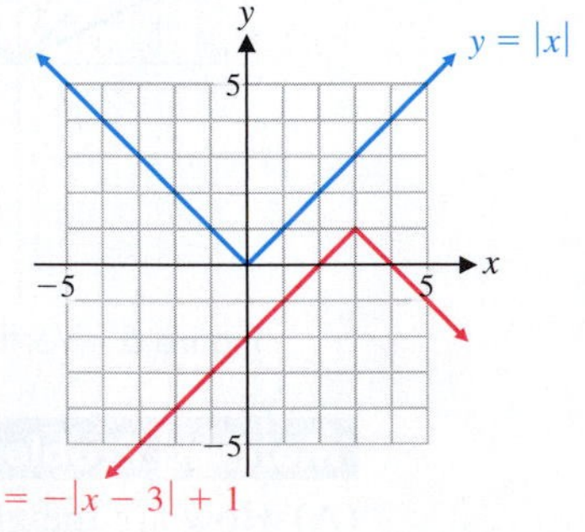

Figure 8 Combined transformations

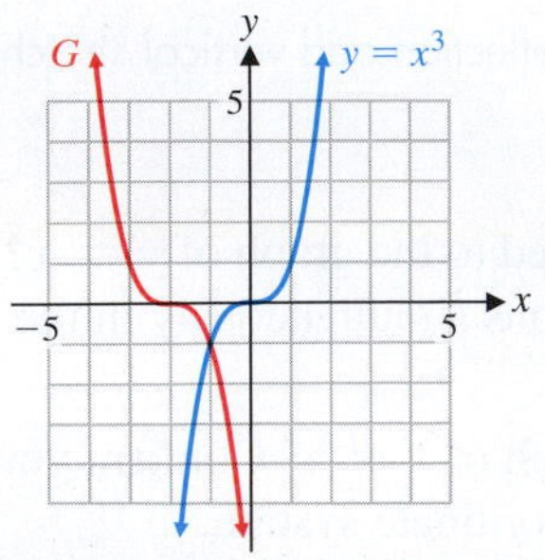

Figure 9 Combined transformations

Matched Problem 5 The graph of $y = G(x)$ in Figure 9 involves a reflection and a translation of the graph of $y = x^3$. Describe how the graph of function G is related to the graph of $y = x^3$ and find an equation of the function G.

Piecewise-Defined Functions

Earlier we noted that the absolute value of a real number x can be defined as

$$|x| = \begin{cases} -x & \text{if } x < 0 \\ x & \text{if } x \geq 0 \end{cases}$$

Notice that this function is defined by different rules for different parts of its domain. Functions whose definitions involve more than one rule are called **piecewise-defined functions**. Graphing one of these functions involves graphing each rule over the appropriate portion of the domain (Fig. 10). In Figure 10C, notice that an open dot is used to show that the point $(0, -2)$ is not part of the graph and a solid dot is used to show that (0, 2) is part of the graph.

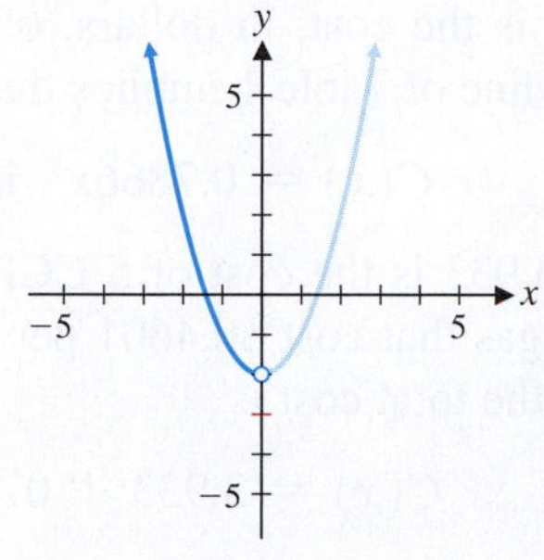

(A) $y = x^2 - 2$

$x < 0$

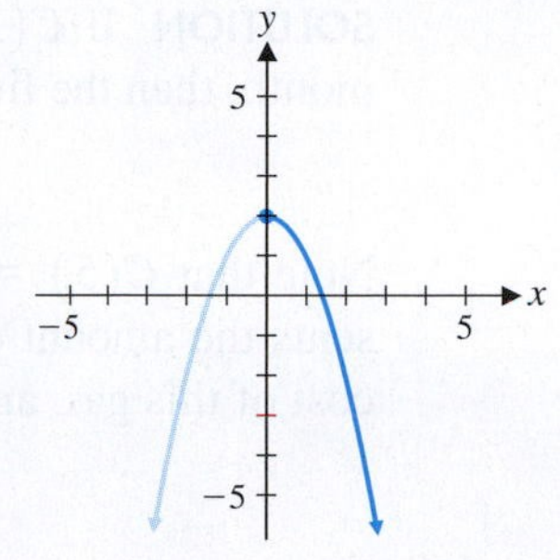

(B) $y = 2 - x^2$

$x \geq 0$

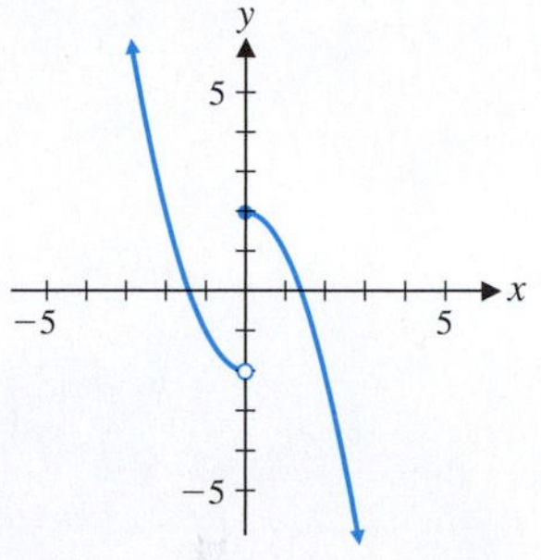

(C) $y = \begin{cases} x^2 - 2 & \text{if } x < 0 \\ 2 - x^2 & \text{if } x \geq 0 \end{cases}$

Figure 10 Graphing a piecewise-defined function

EXAMPLE 6 **Graphing Piecewise-Defined Functions** Graph the piecewise-defined function

$$g(x) = \begin{cases} x + 1 & \text{if } 0 \leq x < 2 \\ 0.5x & \text{if } x \geq 2 \end{cases}$$

SOLUTION If $0 \leq x < 2$, then the first rule applies and the graph of g lies on the line $y = x + 1$ (a vertical shift of the identity function $y = x$). If $x = 0$, then (0, 1) lies on $y = x + 1$; we plot (0, 1) with a solid dot (Fig. 11) because $g(0) = 1$. If $x = 2$, then (2, 3) lies on $y = x + 1$; we plot (2, 3) with an open dot because $g(2) \neq 3$. The line segment from (0, 1) to (2, 3) is the graph of g for $0 \leq x < 2$. If $x \geq 2$, then the second rule applies and the graph of g lies on the line $y = 0.5x$ (a vertical shrink of the identity function $y = x$). If $x = 2$, then (2, 1) lies on the line $y = 0.5x$; we plot (2, 1) with a solid dot because $g(2) = 1$. The portion of $y = 0.5x$ that starts at (2, 1) and extends to the right is the graph of g for $x \geq 2$.

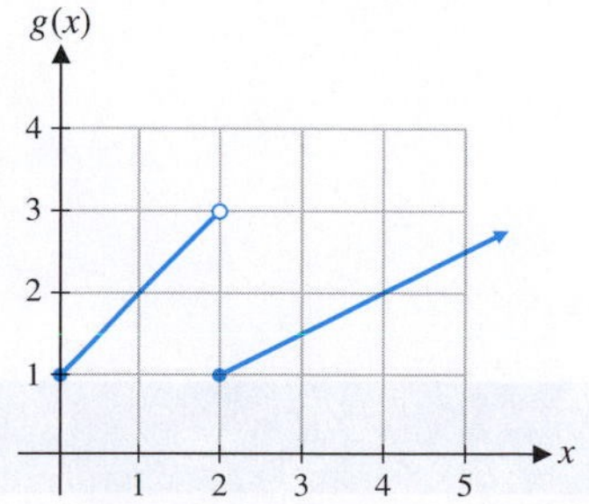

Figure 11

Matched Problem 6 Graph the piecewise-defined function

$$h(x) = \begin{cases} -2x + 4 & \text{if } 0 \leq x \leq 2 \\ x - 1 & \text{if } x > 2 \end{cases}$$

As the next example illustrates, piecewise-defined functions occur naturally in many applications.

EXAMPLE 7 **Natural Gas Rates** Easton Utilities uses the rates shown in Table 1 to compute the monthly cost of natural gas for each customer. Write a piecewise definition for the cost of consuming x CCF (cubic hundred feet) of natural gas and graph the function.

Table 1 Charges per Month

\$0.7866 per CCF for the first 5 CCF
\$0.4601 per CCF for the next 35 CCF
\$0.2508 per CCF for all over 40 CCF

SOLUTION If $C(x)$ is the cost, in dollars, of using x CCF of natural gas in one month, then the first line of Table 1 implies that

$$C(x) = 0.7866x \quad \text{if } 0 \le x \le 5$$

Note that $C(5) = 3.933$ is the cost of 5 CCF. If $5 < x \le 40$, then $x - 5$ represents the amount of gas that cost \$0.4601 per CCF, $0.4601(x - 5)$ represents the cost of this gas, and the total cost is

$$C(x) = 3.933 + 0.4601(x - 5)$$

If $x > 40$, then

$$C(x) = 20.0365 + 0.2508(x - 40)$$

where $20.0365 = C(40)$, the cost of the first 40 CCF. Combining all these equations, we have the following piecewise definition for $C(x)$:

$$C(x) = \begin{cases} 0.7866x & \text{if } 0 \le x \le 5 \\ 3.933 + 0.4601(x - 5) & \text{if } 5 < x \le 40 \\ 20.0365 + 0.2508(x - 40) & \text{if } x > 40 \end{cases}$$

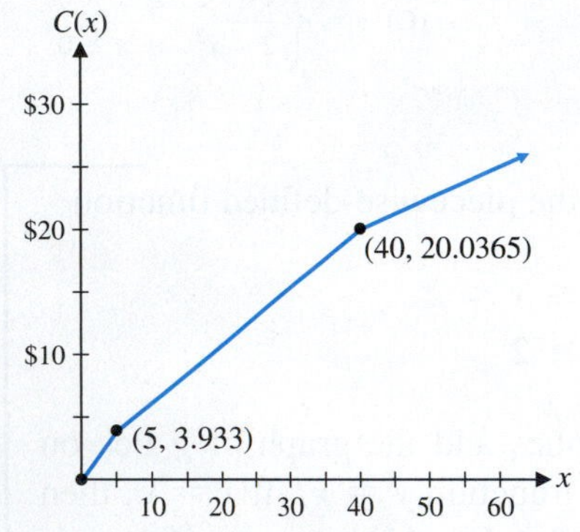

Figure 12 Cost of purchasing x CCF of natural gas

To graph C, first note that each rule in the definition of C represents a transformation of the identity function $f(x) = x$. Graphing each transformation over the indicated interval produces the graph of C shown in Figure 12.

Matched Problem 7 Trussville Utilities uses the rates shown in Table 2 to compute the monthly cost of natural gas for residential customers. Write a piecewise definition for the cost of consuming x CCF of natural gas and graph the function.

Table 2 Charges per Month

\$0.7675 per CCF for the first 50 CCF
\$0.6400 per CCF for the next 150 CCF
\$0.6130 per CCF for all over 200 CCF

Exercises 2.2

A *In Problems 1–10, find the domain and range of each function.*

1. $f(x) = x^2 - 4$

2. $f(x) = 1 + \sqrt{x}$

3. $f(x) = 7 - 2x$

4. $f(x) = x^2 + 10$

5. $f(x) = 8 - \sqrt{x}$

6. $f(x) = 5x + 3$

7. $f(x) = 27 + \sqrt[3]{x}$

8. $f(x) = 15 - 20|x|$

9. $f(x) = 6|x| + 9$

10. $f(x) = -8 + \sqrt[3]{x}$

In Problems 11–26, graph each of the functions using the graphs of functions f and g below.

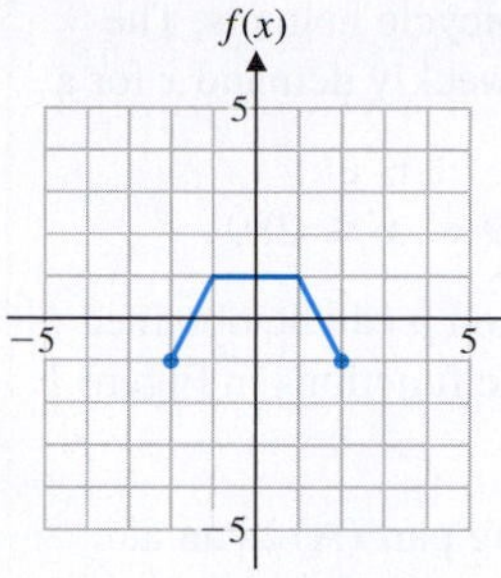

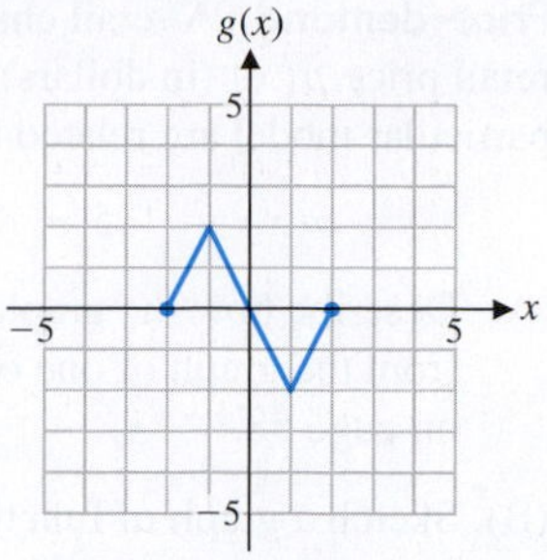

11. $y = f(x) + 2$ **12.** $y = g(x) - 1$

13. $y = f(x + 2)$ **14.** $y = g(x - 1)$

15. $y = g(x - 3)$ **16.** $y = f(x + 3)$

17. $y = g(x) - 3$ **18.** $y = f(x) + 3$

19. $y = -f(x)$ **20.** $y = -g(x)$

21. $y = 0.5g(x)$ **22.** $y = 2f(x)$

23. $y = 2f(x) + 1$ **24.** $y = -0.5g(x) + 3$

25. $y = 2(f(x) + 1)$ **26.** $y = -(0.5g(x) + 3)$

B *In Problems 27–34, describe how the graph of each function is related to the graph of one of the six basic functions in Figure 1 on page 58. Sketch a graph of each function.*

27. $g(x) = -|x + 3|$ **28.** $h(x) = -|x - 5|$

29. $f(x) = (x - 4)^2 - 3$ **30.** $m(x) = (x + 3)^2 + 4$

31. $f(x) = 7 - \sqrt{x}$ **32.** $g(x) = -6 + \sqrt[3]{x}$

33. $h(x) = -3|x|$ **34.** $m(x) = -0.4x^2$

Each graph in Problems 35–42 is the result of applying a sequence of transformations to the graph of one of the six basic functions in Figure 1 on page 58. Identify the basic function and describe the transformation verbally. Write an equation for the given graph.

35.

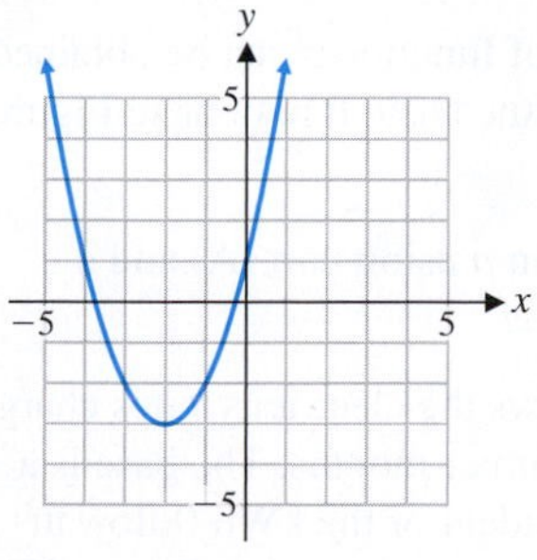

36.

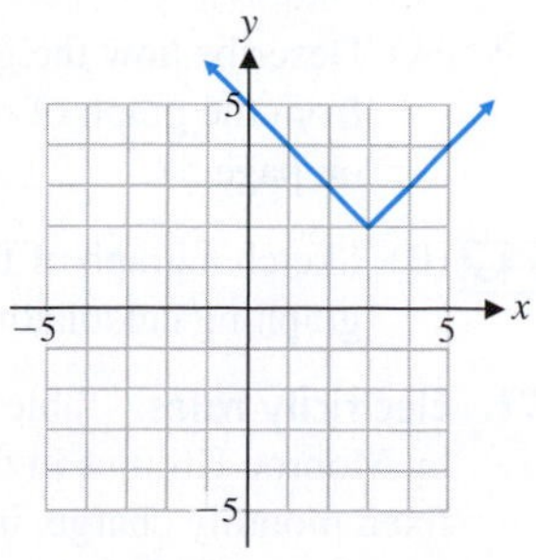

37.

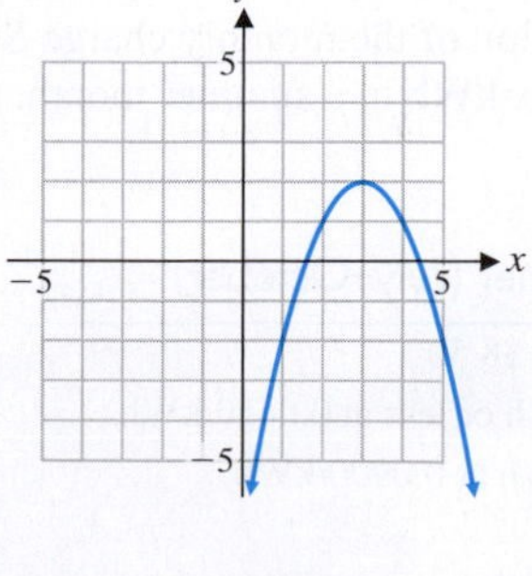

38.

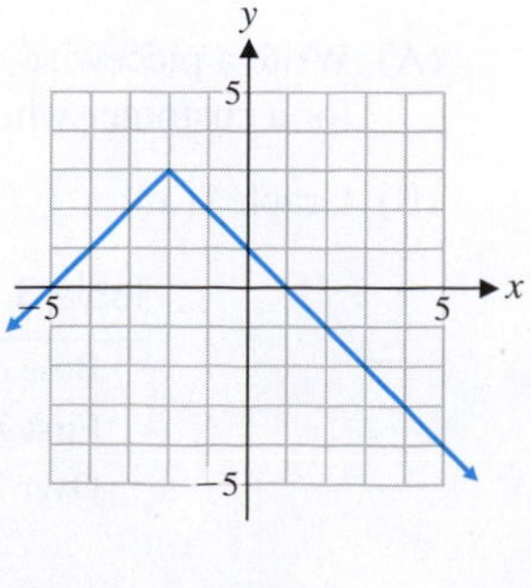

39.

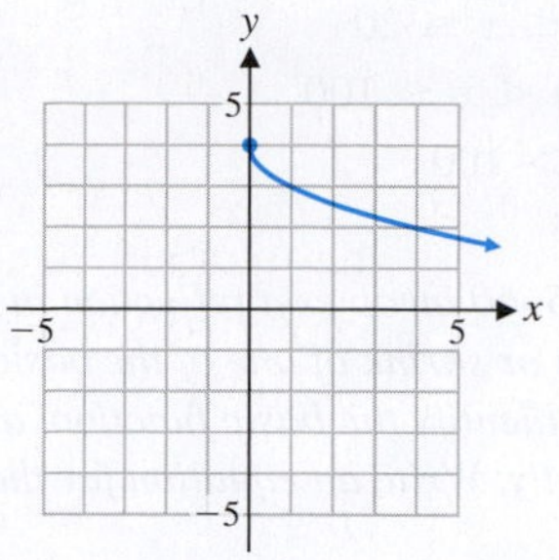

40.

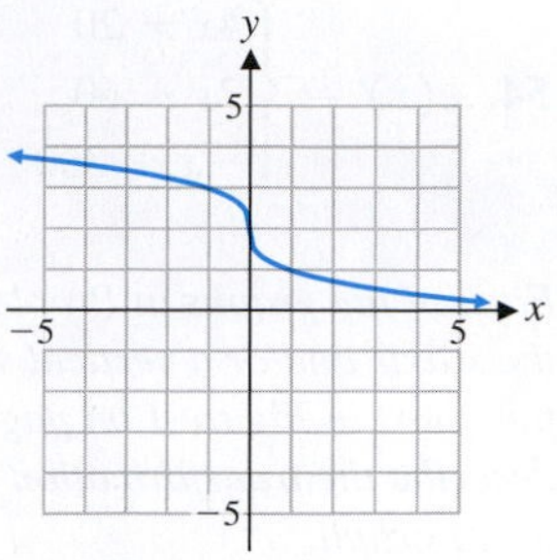

41.

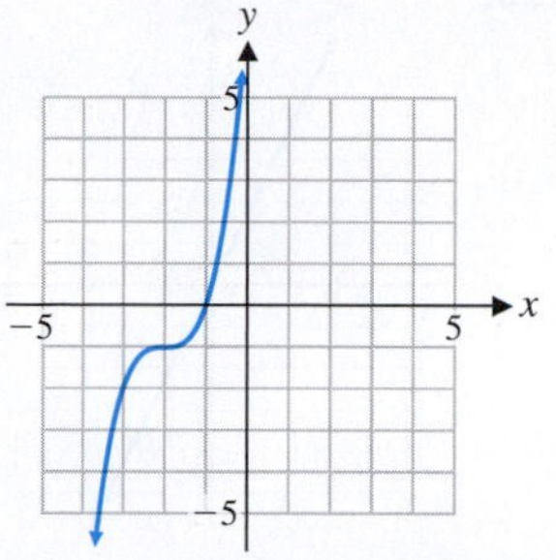

42.

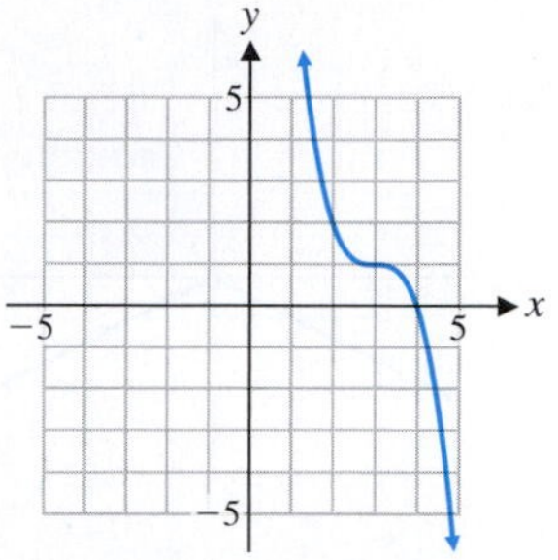

In Problems 43–48, the graph of the function g is formed by applying the indicated sequence of transformations to the given function f. Find an equation for the function g and graph g using $-5 \le x \le 5$ and $-5 \le y \le 5$.

43. The graph of $f(x) = \sqrt{x}$ is shifted 2 units to the right and 3 units down.

44. The graph of $f(x) = \sqrt[3]{x}$ is shifted 3 units to the left and 2 units up.

45. The graph of $f(x) = |x|$ is reflected in the x axis and shifted to the left 3 units.

46. The graph of $f(x) = |x|$ is reflected in the x axis and shifted to the right 1 unit.

47. The graph of $f(x) = x^3$ is reflected in the x axis and shifted 2 units to the right and down 1 unit.

48. The graph of $f(x) = x^2$ is reflected in the x axis and shifted to the left 2 units and up 4 units.

Graph each function in Problems 49–54.

49. $f(x) = \begin{cases} 2 - 2x & \text{if } x < 2 \\ x - 2 & \text{if } x \ge 2 \end{cases}$

50. $g(x) = \begin{cases} x + 1 & \text{if } x < -1 \\ 2 + 2x & \text{if } x \ge -1 \end{cases}$

51. $h(x) = \begin{cases} 5 + 0.5x & \text{if } 0 \le x \le 10 \\ -10 + 2x & \text{if } x > 10 \end{cases}$

52. $h(x) = \begin{cases} 10 + 2x & \text{if } 0 \le x \le 20 \\ 40 + 0.5x & \text{if } x > 20 \end{cases}$

53. $h(x) = \begin{cases} 2x & \text{if } 0 \le x \le 20 \\ x + 20 & \text{if } 20 < x \le 40 \\ 0.5x + 40 & \text{if } x > 40 \end{cases}$

54. $h(x) = \begin{cases} 4x + 20 & \text{if } 0 \le x \le 20 \\ 2x + 60 & \text{if } 20 < x \le 100 \\ -x + 360 & \text{if } x > 100 \end{cases}$

C *Each of the graphs in Problems 55–60 involves a reflection in the x axis and/or a vertical stretch or shrink of one of the basic functions in Figure 1 on page 58. Identify the basic function, and describe the transformation verbally. Write an equation for the given graph.*

55.

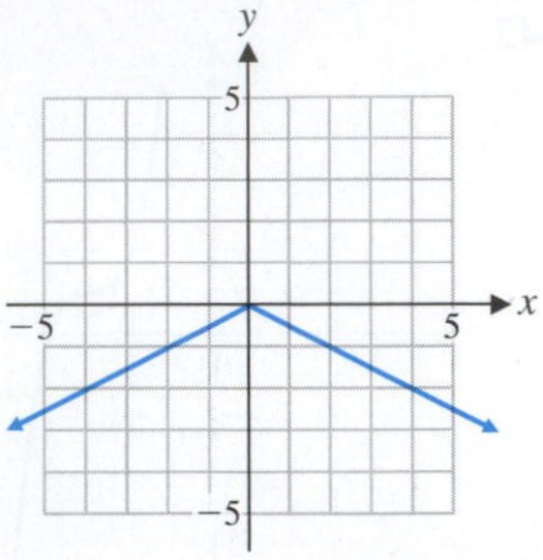

56.

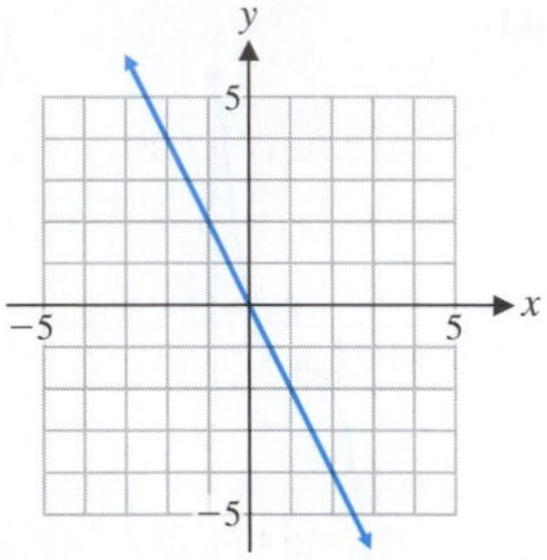

57.

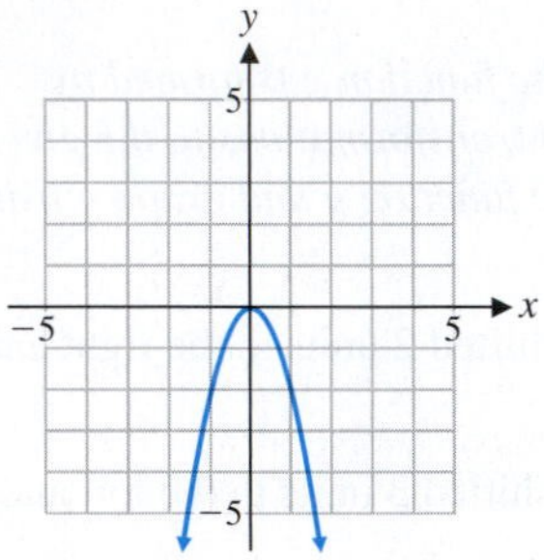

58.

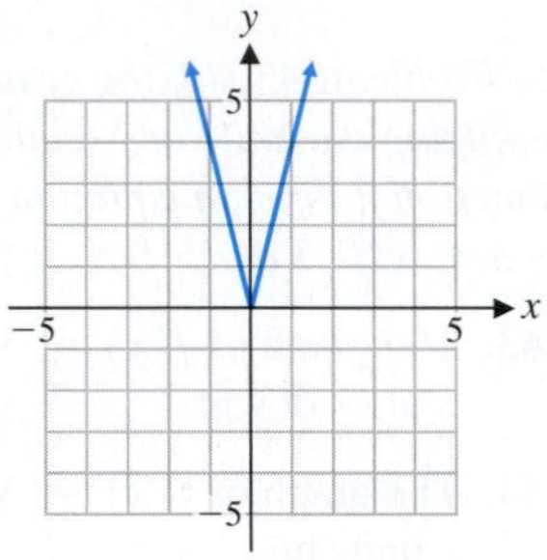

59.

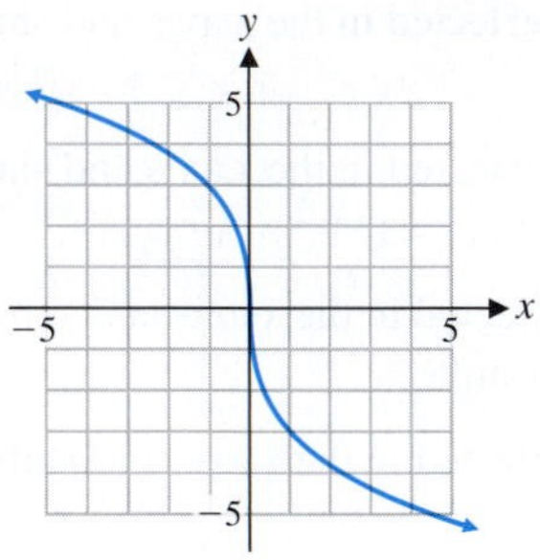

60.

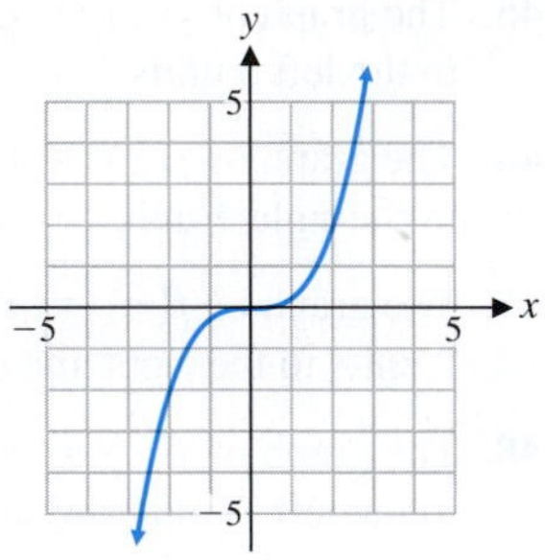

Changing the order in a sequence of transformations may change the final result. Investigate each pair of transformations in Problems 61–66 to determine if reversing their order can produce a different result. Support your conclusions with specific examples and/or mathematical arguments. (The graph of $y = f(-x)$ is the reflection of $y = f(x)$ in the y axis.)

61. Vertical shift; horizontal shift

62. Vertical shift; reflection in y axis

63. Vertical shift; reflection in x axis

64. Vertical shift; vertical stretch

65. Horizontal shift; reflection in y axis

66. Horizontal shift; vertical shrink

Applications

67. Price–demand. A retail chain sells bicycle helmets. The retail price $p(x)$ (in dollars) and the weekly demand x for a particular model are related by

$$p(x) = 115 - 4\sqrt{x} \qquad 9 \le x \le 289$$

(A) Describe how the graph of function p can be obtained from the graph of one of the basic functions in Figure 1 on page 58.

(B) Sketch a graph of function p using part (A) as an aid.

68. Price–supply. The manufacturer of the bicycle helmets in Problem 67 is willing to supply x helmets at a price of $p(x)$ as given by the equation

$$p(x) = 4\sqrt{x} \qquad 9 \le x \le 289$$

(A) Describe how the graph of function p can be obtained from the graph of one of the basic functions in Figure 1 on page 58.

(B) Sketch a graph of function p using part (A) as an aid.

69. Hospital costs. Using statistical methods, the financial department of a hospital arrived at the cost equation

$$C(x) = 0.00048(x - 500)^3 + 60{,}000 \qquad 100 \le x \le 1{,}000$$

where $C(x)$ is the cost in dollars for handling x cases per month.

(A) Describe how the graph of function C can be obtained from the graph of one of the basic functions in Figure 1 on page 58.

(B) Sketch a graph of function C using part (A) and a graphing calculator as aids.

70. Price–demand. A company manufactures and sells in-line skates. Its financial department has established the price–demand function

$$p(x) = 190 - 0.013(x - 10)^2 \qquad 10 \le x \le 100$$

where $p(x)$ is the price at which x thousand pairs of in-line skates can be sold.

(A) Describe how the graph of function p can be obtained from the graph of one of the basic functions in Figure 1 on page 58.

(B) Sketch a graph of function p using part (A) and a graphing calculator as aids.

71. Electricity rates. Table 3 shows the electricity rates charged by Monroe Utilities in the summer months. The base is a fixed monthly charge, independent of the kWh (kilowatt-hours) used during the month.

(A) Write a piecewise definition of the monthly charge $S(x)$ for a customer who uses x kWh in a summer month.

(B) Graph $S(x)$.

Table 3 Summer (July–October)

Base charge, \$8.50
First 700 kWh or less at 0.0650/kWh
Over 700 kWh at 0.0900/kWh

72. Electricity rates. Table 4 shows the electricity rates charged by Monroe Utilities in the winter months.

(A) Write a piecewise definition of the monthly charge $W(x)$ for a customer who uses x kWh in a winter month.

(B) Graph $W(x)$.

Table 4 Winter (November–June)

Base charge, \$8.50
First 700 kWh or less at 0.0650/kWh
Over 700 kWh at 0.0530/kWh

73. State income tax. Table 5 shows state income tax rates for married couples filing a joint return in Louisiana.

(A) Write a piecewise definition for $T(x)$, the tax due on a taxable income of x dollars.

(B) Graph $T(x)$.

(C) Find the tax due on a taxable income of \$55,000. Of \$110,000.

Table 5 Louisiana State Income Tax

Married filing jointly or qualified surviving spouse:		
Over	But not over	Tax due is
\$0	\$25,000	2% of taxable income
\$25,000	\$100,000	\$500 plus 4% of excess over \$25,000
\$100,000		\$3,500 plus 6% of excess over \$100,000

74. State income tax. Table 6 shows state income tax rates for individuals filing a return in Louisiana.

(A) Write a piecewise definition for $T(x)$, the tax due on a taxable income of x dollars.

(B) Graph $T(x)$.

(C) Find the tax due on a taxable income of \$32,000. Of \$64,000.

(D) Would it be better for a married couple in Louisiana with two equal incomes to file jointly or separately? Discuss.

Table 6 Louisiana State Income Tax

Single, married filing separately, or head of household:		
Over	But not over	Tax due is
\$0	\$12,500	2% of taxable income
\$12,500	\$50,000	\$250 plus 4% of excess over \$12,500
\$50,000		\$1,750 plus 6% of excess over \$50,000

75. Human weight. A good approximation of the normal weight of a person 60 inches or taller but not taller than 80 inches is given by $w(x) = 5.5x - 220$, where x is height in inches and $w(x)$ is weight in pounds.

(A) Describe how the graph of function w can be obtained from the graph of one of the basic functions in Figure 1, page 58.

(B) Sketch a graph of function w using part (A) as an aid.

76. Herpetology. The average weight of a particular species of snake is given by $w(x) = 463x^3$, $0.2 \le x \le 0.8$, where x is length in meters and $w(x)$ is weight in grams.

(A) Describe how the graph of function w can be obtained from the graph of one of the basic functions in Figure 1, page 58.

(B) Sketch a graph of function w using part (A) as an aid.

77. Safety research. Under ideal conditions, if a person driving a vehicle slams on the brakes and skids to a stop, the speed of the vehicle $v(x)$ (in miles per hour) is given approximately by $v(x) = C\sqrt{x}$, where x is the length of skid marks (in feet) and C is a constant that depends on the road conditions and the weight of the vehicle. For a particular vehicle, $v(x) = 7.08\sqrt{x}$ and $4 \le x \le 144$.

(A) Describe how the graph of function v can be obtained from the graph of one of the basic functions in Figure 1, page 58.

(B) Sketch a graph of function v using part (A) as an aid.

78. Learning. A production analyst has found that on average it takes a new person $T(x)$ minutes to perform a particular assembly operation after x performances of the operation, where $T(x) = 10 - \sqrt[3]{x}$, $0 \le x \le 125$.

(A) Describe how the graph of function T can be obtained from the graph of one of the basic functions in Figure 1, page 58.

(B) Sketch a graph of function T using part (A) as an aid.

Answers to Matched Problems

1. (A) $f(729) = 729$, $h(729) = 531{,}441$, $m(729) = 387{,}420{,}489$, $n(729) = 27$, $p(729) = 9$, $g(729) = 729$

(B) $f(-5.25) = -5.25$, $h(-5.25) = 27.5625$, $m(-5.25) = -144.7031$, $n(-5.25)$ is not a real number, $p(-5.25) = -1.7380$, $g(-5.25) = 5.25$

2. (A) The graph of $y = \sqrt{x} + 5$ is the same as the graph of $y = \sqrt{x}$ shifted upward 5 units, and the graph of $y = \sqrt{x} - 4$ is the same as the graph of $y = \sqrt{x}$ shifted downward 4 units. The figure confirms these conclusions.

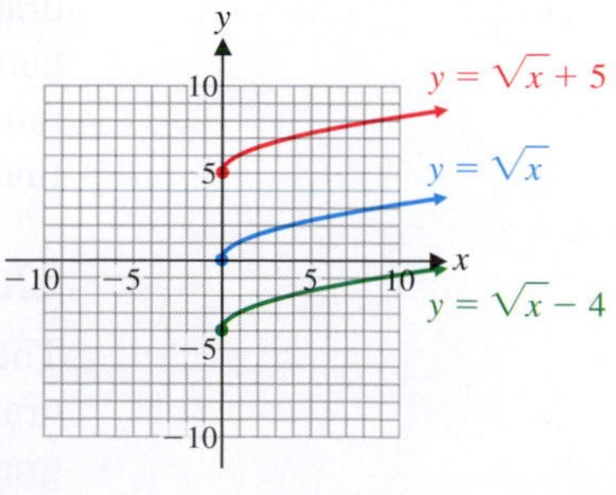

(B) The graph of $y = \sqrt{x + 5}$ is the same as the graph of $y = \sqrt{x}$ shifted to the left 5 units, and the graph of $y = \sqrt{x - 4}$ is the same as the graph of $y = \sqrt{x}$ shifted to the right 4 units. The figure confirms these conclusions.

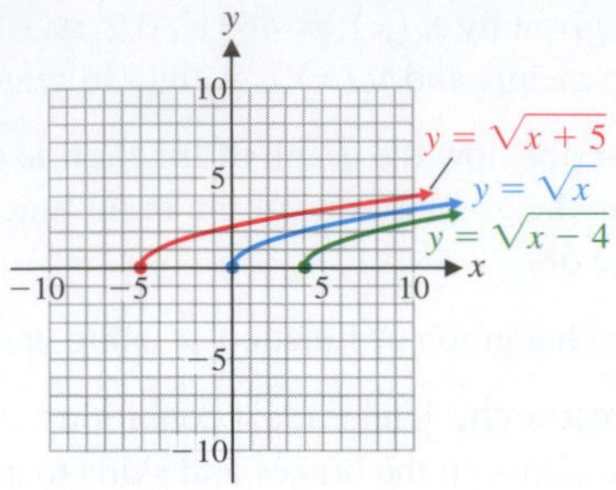

3. $H(x) = \sqrt[3]{x} + 3,\ G(x) = \sqrt[3]{x} - 2,\ M(x) = \sqrt[3]{x + 2},$
$N(x) = \sqrt[3]{x - 3}$

4. (A) The graph of $y = 2x$ is a vertical stretch of the graph of $y = x$, and the graph of $y = 0.5x$ is a vertical shrink of the graph of $y = x$. The figure confirms these conclusions.

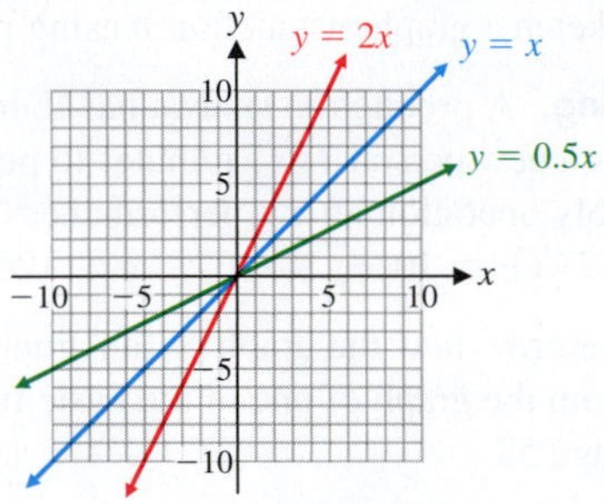

(B) The graph of $y = -0.5x$ is a vertical shrink and a reflection in the x axis of the graph of $y = x$. The figure confirms this conclusion.

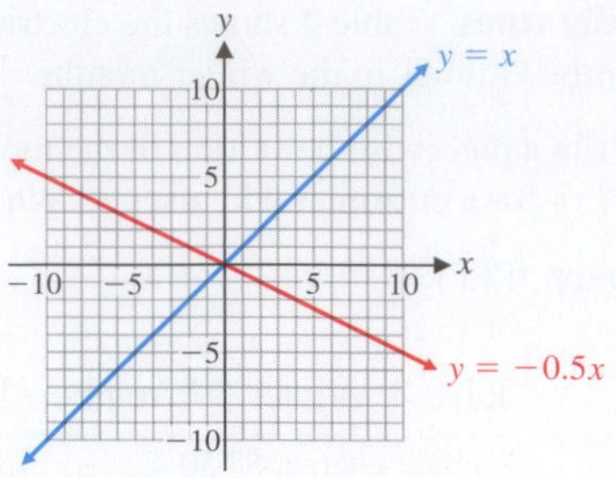

5. The graph of function G is a reflection in the x axis and a horizontal translation of 2 units to the left of the graph of $y = x^3$. An equation for G is $G(x) = -(x + 2)^3$.

6.

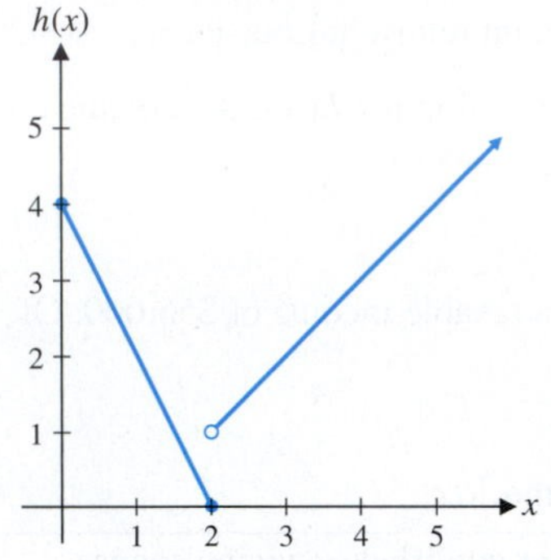

7. $C(x) = \begin{cases} 0.7675x & \text{if } 0 \le x \le 50 \\ 38.375 + 0.64\,(x - 50) & \text{if } 50 < x \le 200 \\ 134.375 + 0.613\,(x - 200) & \text{if } 200 < x \end{cases}$

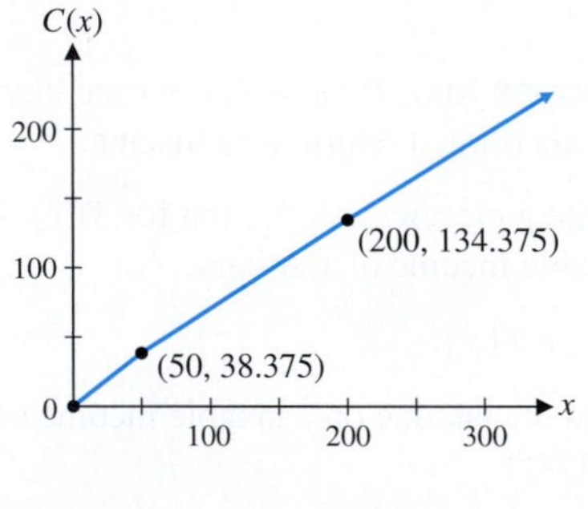

2.3 Quadratic Functions

- Quadratic Functions, Equations, and Inequalities
- Properties of Quadratic Functions and Their Graphs
- Applications

If the degree of a linear function is increased by one, we obtain a *second-degree function,* usually called a *quadratic function,* another basic function that we will need in our library of elementary functions. We will investigate relationships between quadratic functions and the solutions to quadratic equations and inequalities. Other important properties of quadratic functions will also be investigated, including maxima and minima. We will then be in a position to solve important practical problems such as finding production levels that will generate maximum revenue or maximum profit.

Quadratic Functions, Equations, and Inequalities

The graph of the square function $h(x) = x^2$ is shown in Figure 1. Notice that the graph is symmetric with respect to the y axis and that $(0, 0)$ is the lowest point on the graph. Let's explore the effect of applying graph transformations to the graph of h.

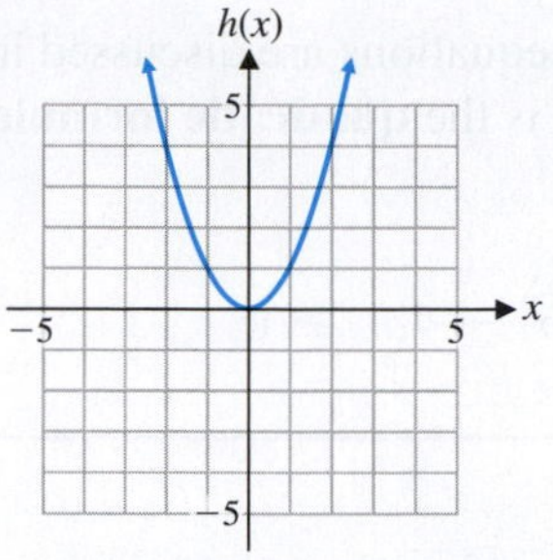

Figure 1 Square function $h(x) = x^2$

Explore and Discuss 1

Indicate how the graph of each function is related to the graph of the function $h(x) = x^2$. Find the highest or lowest point, whichever exists, on each graph.

(A) $f(x) = (x - 3)^2 - 7 = x^2 - 6x + 2$

(B) $g(x) = 0.5(x + 2)^2 + 3 = 0.5x^2 + 2x + 5$

(C) $m(x) = -(x - 4)^2 + 8 = -x^2 + 8x - 8$

(D) $n(x) = -3(x + 1)^2 - 1 = -3x^2 - 6x - 4$

The graphs of the functions in Explore and Discuss 1 are similar in shape to the graph of the square function in Figure 1. All are *parabolas*. The arc of a basketball shot is a parabola. Reflecting telescopes, solar furnaces, and automobile headlights are some of the many applications of parabolas. Each of the functions in Explore and Discuss 1 is a *quadratic function*.

DEFINITION Quadratic Functions

If a, b, and c are real numbers with $a \neq 0$, then the function

$$f(x) = ax^2 + bx + c \qquad \text{Standard form}$$

is a **quadratic function** and its graph is a **parabola**.

CONCEPTUAL INSIGHT

If x is any real number, then $ax^2 + bx + c$ is also a real number. According to the agreement on domain and range in Section 2.1, the domain of a quadratic function is R, the set of real numbers.

We will discuss methods for determining the range of a quadratic function later in this section. Typical graphs of quadratic functions are illustrated in Figure 2.

Reminder

The *union* of two sets A and B, denoted $A \cup B$, is the set of all elements that belong to A or B (or both). So the set of all real numbers x such that $x^2 - 4 \geq 0$ (see Fig. 2A) is $(-\infty, -2] \cup [2, \infty)$.

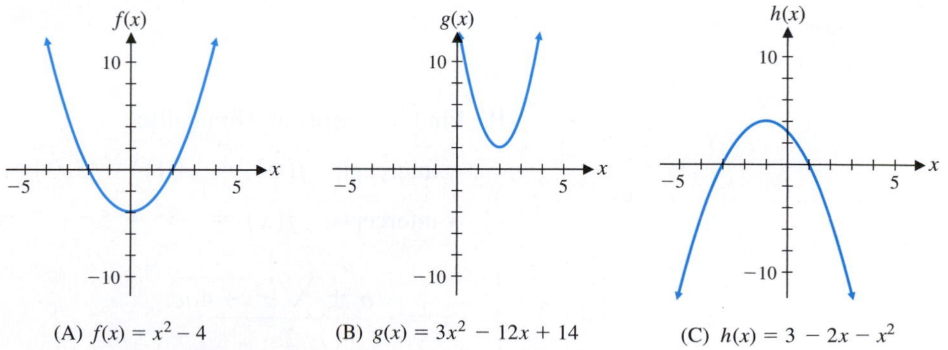

Figure 2 Graphs of quadratic functions

CONCEPTUAL INSIGHT

An x intercept of a function is also called a **zero** of the function. The x intercept of a linear function can be found by solving the linear equation $y = mx + b = 0$ for x, $m \neq 0$ (see Section 1.2). Similarly, the x intercepts of a quadratic function can be found by solving the quadratic equation $y = ax^2 + bx + c = 0$

for x, $a \neq 0$. Several methods for solving quadratic equations are discussed in Appendix A, Section A.7. The most popular of these is the **quadratic formula**.

If $ax^2 + bx + c = 0$, $a \neq 0$, then

$$x = \frac{-b \pm \sqrt{b^2 - 4ac}}{2a}, \text{ provided } b^2 - 4ac \geq 0$$

EXAMPLE 1 **Intercepts, Equations, and Inequalities**

(A) Sketch a graph of $f(x) = -x^2 + 5x + 3$ in a rectangular coordinate system.

(B) Find x and y intercepts algebraically to four decimal places.

(C) Graph $f(x) = -x^2 + 5x + 3$ in a standard viewing window.

(D) Find the x and y intercepts to four decimal places using TRACE and ZERO on your graphing calculator.

(E) Solve the quadratic inequality $-x^2 + 5x + 3 \geq 0$ graphically to four decimal places using the results of parts (A) and (B) or (C) and (D).

(F) Solve the equation $-x^2 + 5x + 3 = 4$ graphically to four decimal places using INTERSECT on your graphing calculator.

SOLUTION

(A) Hand-sketch a graph of f by drawing a smooth curve through the plotted points (Fig. 3).

x	y
−1	−3
0	3
1	7
2	9
3	9
4	7
5	3
6	−3

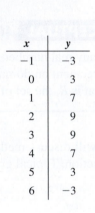

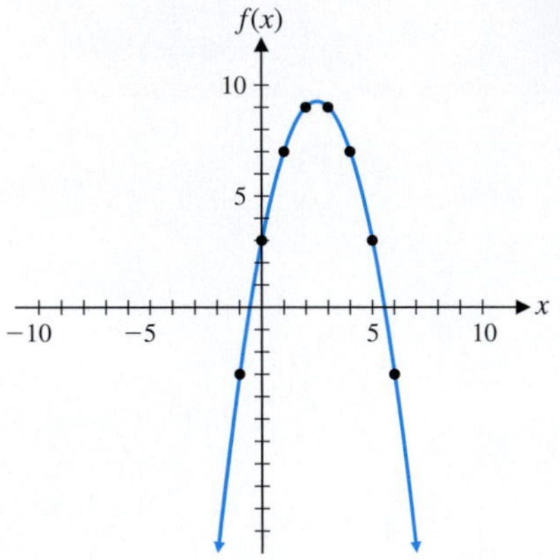

Figure 3

(B) Find intercepts algebraically:

y intercept: $f(0) = -(0)^2 + 5(0) + 3 = 3$

x intercepts: $f(x) = -x^2 + 5x + 3 = 0$ — Use the quadratic formula.

$$x = \frac{-b \pm \sqrt{b^2 - 4ac}}{2a}$$ Substitute $a = -1, b = 5, c = 3$.

$$x = \frac{-(5) \pm \sqrt{5^2 - 4(-1)(3)}}{2(-1)}$$ Simplify.

$$= \frac{-5 \pm \sqrt{37}}{-2} = -0.5414 \quad \text{or} \quad 5.5414$$

(C) Use in a graphing calculator (Fig. 4).

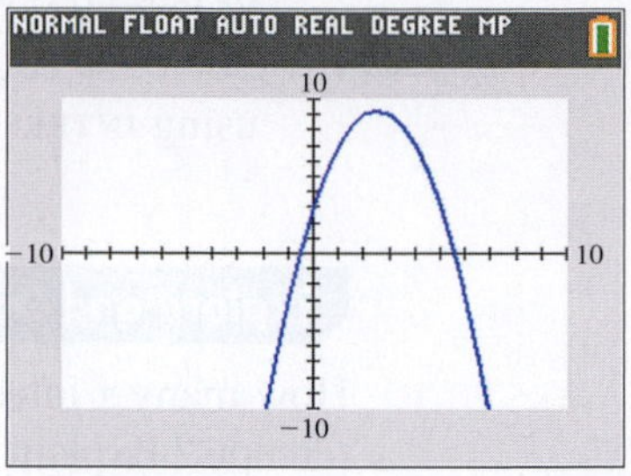

Figure 4

(D) Find intercepts using a graphing calculator (Fig. 5).

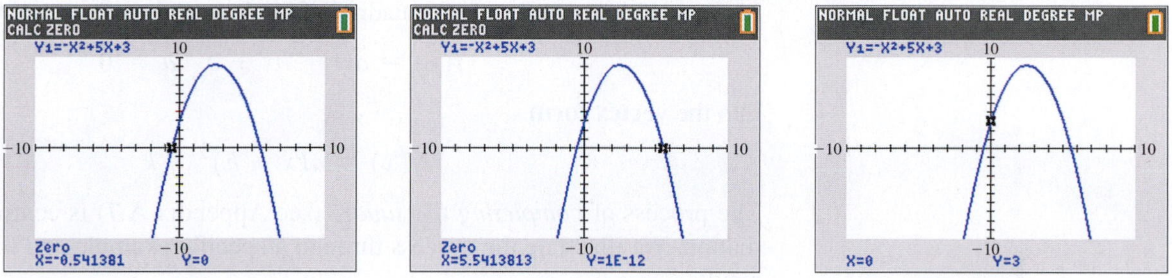

(A) x intercept: -0.5414 (B) x intercept: 5.5414 (C) y intercept: 3

Figure 5

(E) Solve $-x^2 + 5x + 3 \geq 0$ graphically: The quadratic inequality

$$-x^2 + 5x + 3 \geq 0$$

holds for those values of x for which the graph of $f(x) = -x^2 + 5x + 3$ in the figures in parts (A) and (C) is at or above the x axis. This happens for x between the two x intercepts [found in part (B) or (D)], including the two x intercepts. The solution set for the quadratic inequality is $-0.5414 \leq x \leq 5.5414$ or $[-0.5414, 5.5414]$.

(F) Solve the equation $-x^2 + 5x + 3 = 4$ using a graphing calculator (Fig. 6).

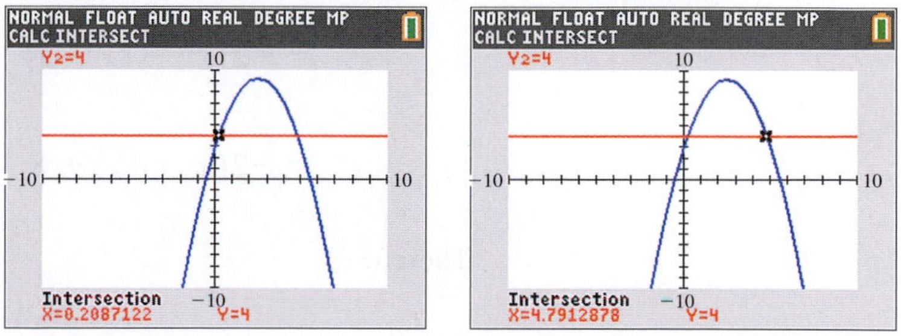

(A) $-x^2 + 5x + 3 = 4$ at $x = 0.2087$ (B) $-x^2 + 5x + 3 = 4$ at $x = 4.7913$

Figure 6

Matched Problem 1

(A) Sketch a graph of $g(x) = 2x^2 - 5x - 5$ in a rectangular coordinate system.

(B) Find x and y intercepts algebraically to four decimal places.

(C) Graph $g(x) = 2x^2 - 5x - 5$ in a standard viewing window.

(D) Find the x and y intercepts to four decimal places using TRACE and the ZERO command on your graphing calculator.

(E) Solve $2x^2 - 5x - 5 \geq 0$ graphically to four decimal places using the results of parts (A) and (B) or (C) and (D).

(F) Solve the equation $2x^2 - 5x - 5 = -3$ graphically to four decimal places using INTERSECT on your graphing calculator.

Explore and Discuss 2

How many x intercepts can the graph of a quadratic function have? How many y intercepts? Explain your reasoning.

Properties of Quadratic Functions and Their Graphs

Many useful properties of the quadratic function can be uncovered by transforming

$$f(x) = ax^2 + bx + c \quad a \neq 0$$

into the **vertex form**

$$f(x) = a(x - h)^2 + k$$

The process of *completing the square* (see Appendix A.7) is central to the transformation. We illustrate the process through a specific example and then generalize the results.

Consider the quadratic function given by

$$f(x) = -2x^2 + 16x - 24 \tag{1}$$

We use completing the square to transform this function into vertex form:

$$f(x) = -2x^2 + 16x - 24$$

$$= -2(x^2 - 8x) - 24$$ Factor the coefficient of x^2 out of the first two terms.

$$= -2(x^2 - 8x + ?) - 24$$ Add 16 to complete the square inside the parentheses. Because of the -2 outside the parentheses, we have actually added -32, so we must add 32 to the outside.

$$= -2(x^2 - 8x + 16) - 24 + 32$$ Factor, simplify

$$= -2(x - 4)^2 + 8$$ The transformation is complete and can be checked by multiplying out.

Therefore,

$$f(x) = -2(x - 4)^2 + 8 \tag{2}$$

If $x = 4$, then $-2(x - 4)^2 = 0$ and $f(4) = 8$. For any other value of x, the negative number $-2(x - 4)^2$ is added to 8, making it smaller. Therefore,

$$f(4) = 8$$

is the *maximum value* of $f(x)$ for all x. Furthermore, if we choose any two x values that are the same distance from 4, we will obtain the same function value. For example, $x = 3$ and $x = 5$ are each one unit from $x = 4$ and their function values are

$$f(3) = -2(3 - 4)^2 + 8 = 6$$

$$f(5) = -2(5 - 4)^2 + 8 = 6$$

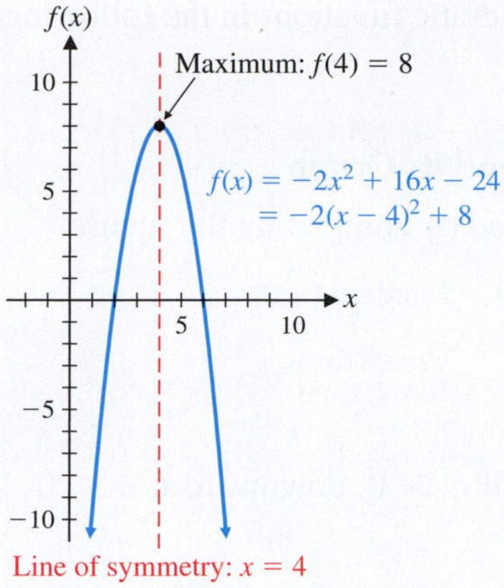

Figure 7 Graph of a quadratic function

Therefore, the vertical line $x = 4$ is a line of symmetry. That is, if the graph of equation (1) is drawn on a piece of paper and the paper is folded along the line $x = 4$, then the two sides of the parabola will match exactly. All these results are illustrated by graphing equations (1) and (2) and the line $x = 4$ simultaneously in the same coordinate system (Fig. 7).

From the preceding discussion, we see that as x moves from left to right, $f(x)$ is increasing on $(-\infty, 4]$, and decreasing on $[4, \infty)$, and that $f(x)$ can assume no value greater than 8. Thus,

$$\text{Range of } f\text{:} \quad y \le 8 \quad \text{or} \quad (-\infty, 8]$$

In general, the graph of a quadratic function is a parabola with line of symmetry parallel to the vertical axis. The lowest or highest point on the parabola, whichever exists, is called the **vertex**. The maximum or minimum value of a quadratic function always occurs at the vertex of the parabola. The line of symmetry through the vertex is called the **axis** of the parabola. In the example above, $x = 4$ is the axis of the parabola and $(4, 8)$ is its vertex.

CONCEPTUAL INSIGHT

Applying the graph transformation properties discussed in Section 2.2 to the transformed equation,

$$\begin{aligned} f(x) &= -2x^2 + 16x - 24 \\ &= -2(x - 4)^2 + 8 \end{aligned}$$

we see that the graph of $f(x) = -2x^2 + 16x - 24$ is the graph of $g(x) = x^2$ vertically stretched by a factor of 2, reflected in the x axis, and shifted to the right 4 units and up 8 units, as shown in Figure 8.

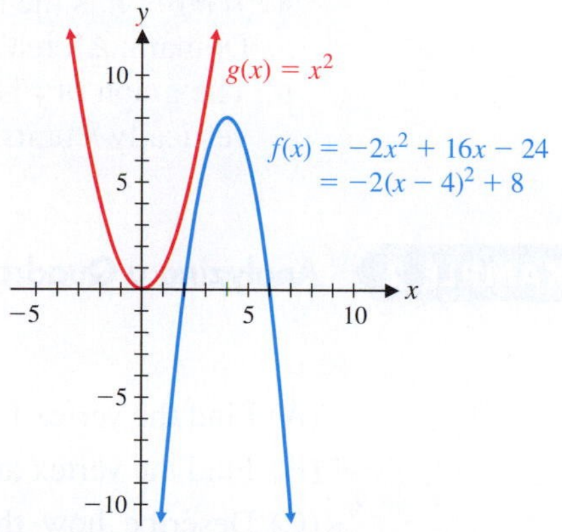

Figure 8 Graph of **f** is the graph of **g** transformed

Note the important results we have obtained from the vertex form of the quadratic function f:

- The vertex of the parabola
- The axis of the parabola
- The maximum value of $f(x)$
- The range of the function f
- The relationship between the graph of $g(x) = x^2$ and the graph of $f(x) = -2x^2 + 16x - 24$

The preceding discussion is generalized to all quadratic functions in the following summary:

SUMMARY Properties of a Quadratic Function and Its Graph

Given a quadratic function and the vertex form obtained by completing the square

$$f(x) = ax^2 + bx + c \qquad a \neq 0 \qquad \text{Standard form}$$
$$= a(x - h)^2 + k \qquad \text{Vertex form}$$

we summarize general properties as follows:

1. The graph of f is a parabola that opens upward if $a > 0$, downward if $a < 0$ (Fig. 9).

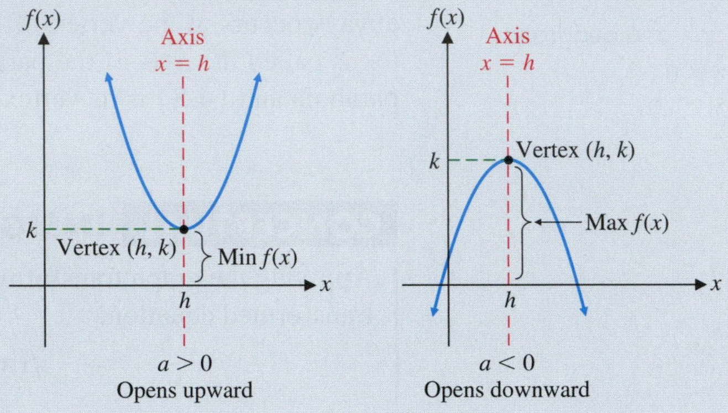

Figure 9

2. Vertex: (h, k) (parabola increases on one side of the vertex and decreases on the other)
3. Axis (of symmetry): $x = h$ (parallel to y axis)
4. $f(h) = k$ is the minimum if $a > 0$ and the maximum if $a < 0$
5. Domain: All real numbers. Range: $(-\infty, k]$ if $a < 0$ or $[k, \infty)$ if $a > 0$
6. The graph of f is the graph of $g(x) = ax^2$ translated horizontally h units and vertically k units.

EXAMPLE 2 **Analyzing a Quadratic Function** Given the quadratic function

$$f(x) = 0.5x^2 - 6x + 21$$

(A) Find the vertex form for f.

(B) Find the vertex and the maximum or minimum. State the range of f.

(C) Describe how the graph of function f can be obtained from the graph of $g(x) = x^2$ using transformations.

(D) Sketch a graph of function f in a rectangular coordinate system.

(E) Graph function f using a suitable viewing window.

(F) Find the vertex and the maximum or minimum using the appropriate graphing calculator command.

SOLUTION

(A) Complete the square to find the vertex form:

$$\begin{aligned} f(x) &= 0.5x^2 - 6x + 21 \\ &= 0.5(x^2 - 12x + ?) + 21 \\ &= 0.5(x^2 - 12x + 36) + 21 - 18 \\ &= 0.5(x - 6)^2 + 3 \end{aligned}$$

(B) From the vertex form, we see that $h = 6$ and $k = 3$. Thus, vertex: $(6,3)$; minimum: $f(6) = 3$; range: $y \geq 3$ or $[3, \infty)$.

(C) The graph of $f(x) = 0.5(x - 6)^2 + 3$ is the same as the graph of $g(x) = x^2$ vertically shrunk by a factor of 0.5, and shifted to the right 6 units and up 3 units.

(D) Graph in a rectangular coordinate system (Fig. 10).

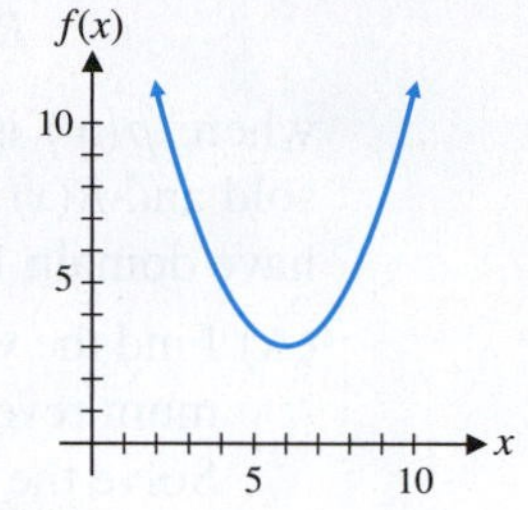

Figure 10

(E) Use a graphing calculator (Fig. 11).

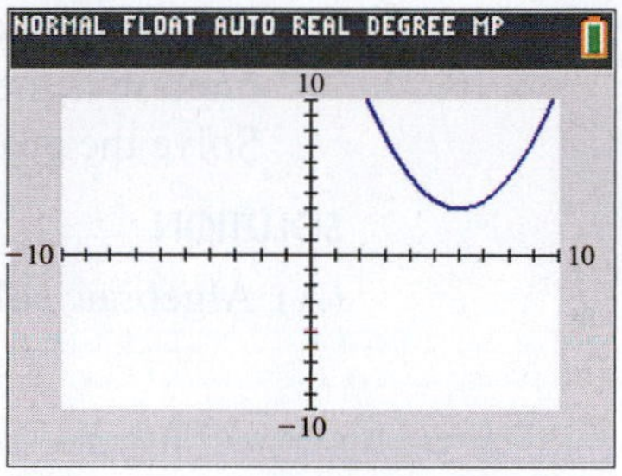

Figure 11

(F) Find the vertex and minimum using the minimum command (Fig. 12).

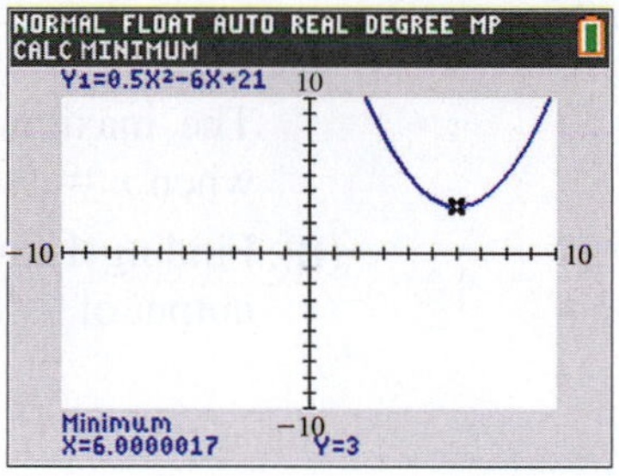

Figure 12

Vertex: $(6,3)$; minimum: $f(6) = 3$

Matched Problem 2 Given the quadratic function $f(x) = -0.25x^2 - 2x + 2$

(A) Find the vertex form for f.

(B) Find the vertex and the maximum or minimum. State the range of f.

(C) Describe how the graph of function f can be obtained from the graph of $g(x) = x^2$ using transformations.

(D) Sketch a graph of function f in a rectangular coordinate system.

(E) Graph function f using a suitable viewing window.

(F) Find the vertex and the maximum or minimum using the appropriate graphing calculator command.

Applications

EXAMPLE 3 **Maximum Revenue** This is a continuation of Example 7 in Section 2.1. Recall that the financial department in the company that produces a digital camera arrived at the following price–demand function and the corresponding revenue function:

$$p(x) = 94.8 - 5x \qquad \text{Price–demand function}$$

$$R(x) = xp(x) = x(94.8 - 5x) \qquad \text{Revenue function}$$

where $p(x)$ is the wholesale price per camera at which x million cameras can be sold and $R(x)$ is the corresponding revenue (in millions of dollars). Both functions have domain $1 \le x \le 15$.

(A) Find the value of x to the nearest thousand cameras that will generate the maximum revenue. What is the maximum revenue to the nearest thousand dollars? Solve the problem algebraically by completing the square.

(B) What is the wholesale price per camera (to the nearest dollar) that generates the maximum revenue?

(C) Graph the revenue function using an appropriate viewing window.

(D) Find the value of x to the nearest thousand cameras that will generate the maximum revenue. What is the maximum revenue to the nearest thousand dollars? Solve the problem graphically using the maximum command.

SOLUTION

(A) Algebraic solution:

$$\begin{aligned} R(x) &= x(94.8 - 5x) \\ &= -5x^2 + 94.8x \\ &= -5(x^2 - 18.96x + ?) \\ &= -5(x^2 - 18.96x + 89.8704) + 449.352 \\ &= -5(x - 9.48)^2 + 449.352 \end{aligned}$$

The maximum revenue of 449.352 million dollars (\$449,352,000) occurs when $x = 9.480$ million cameras (9,480,000 cameras).

(B) Finding the wholesale price per camera: Use the price–demand function for an output of 9.480 million cameras:

$$\begin{aligned} p(x) &= 94.8 - 5x \\ p(9.480) &= 94.8 - 5(9.480) \\ &= \$47 \text{ per camera} \end{aligned}$$

(C) Use a graphing calculator (Fig. 13).

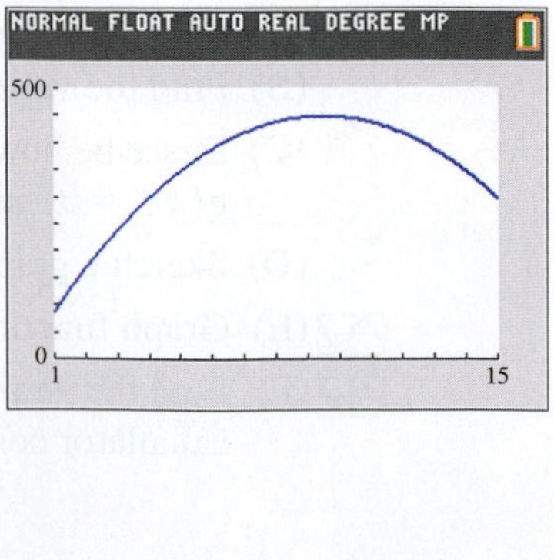

Figure 13

(D) Use the maximum command on a graphing calculator (Fig. 14).

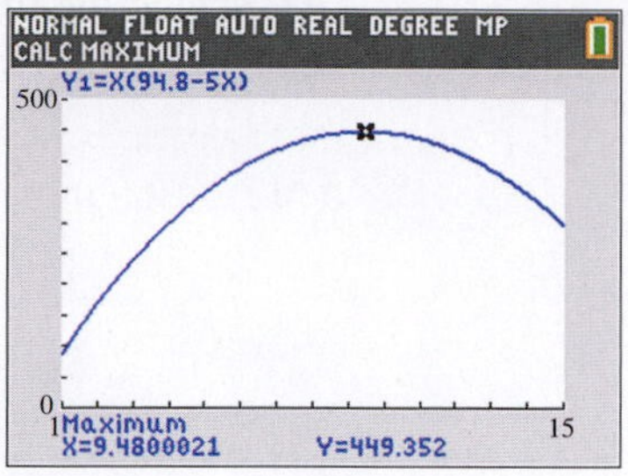

Figure 14

The manufacture and sale of 9.480 million cameras (9,480,000 cameras) will generate a maximum revenue of 449.352 million dollars ($449,352,000).

Matched Problem 3 The financial department in Example 3, using statistical and analytical techniques (see Matched Problem 7 in Section 2.1), arrived at the cost function

$$C(x) = 156 + 19.7x \quad \text{Cost function}$$

where $C(x)$ is the cost (in millions of dollars) for manufacturing and selling x million cameras.

(A) Using the revenue function from Example 3 and the preceding cost function, write an equation for the profit function.

(B) Find the value of x to the nearest thousand cameras that will generate the maximum profit. What is the maximum profit to the nearest thousand dollars? Solve the problem algebraically by completing the square.

(C) What is the wholesale price per camera (to the nearest dollar) that generates the maximum profit?

(D) Graph the profit function using an appropriate viewing window.

(E) Find the output to the nearest thousand cameras that will generate the maximum profit. What is the maximum profit to the nearest thousand dollars? Solve the problem graphically using the maximum command.

EXAMPLE 4 **Break-Even Analysis** Use the revenue function from Example 3 and the cost function from Matched Problem 3:

$$R(x) = x(94.8 - 5x) \quad \text{Revenue function}$$

$$C(x) = 156 + 19.7x \quad \text{Cost function}$$

Both have domain $1 \le x \le 15$.

(A) Sketch the graphs of both functions in the same coordinate system.

(B) **Break-even points** are the production levels at which $R(x) = C(x)$. Find the break-even points algebraically to the nearest thousand cameras.

(C) Plot both functions simultaneously in the same viewing window.

(D) Use INTERSECT to find the break-even points graphically to the nearest thousand cameras.

(E) Recall that a loss occurs if $R(x) < C(x)$ and a profit occurs if $R(x) > C(x)$. For what values of x (to the nearest thousand cameras) will a loss occur? A profit?

SOLUTION

(A) Sketch the functions (Fig. 15).

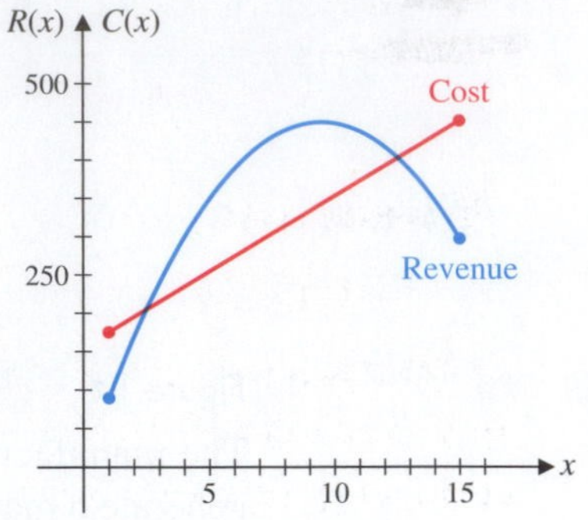

Figure 15

(B) Find x such that $R(x) = C(x)$:

$$x(94.8 - 5x) = 156 + 19.7x \quad \text{Simplify.}$$

$$-5x^2 + 75.1x - 156 = 0 \quad \text{Use the quadratic formula.}$$

$$x = \frac{-75.1 \pm \sqrt{75.1^2 - 4(-5)(-156)}}{2(-5)}$$

$$= \frac{-75.1 \pm \sqrt{2{,}520.01}}{-10}$$

$$x = 2.490 \quad \text{and} \quad 12.530$$

The company breaks even at $x = 2.490$ million cameras (2,490,000 cameras) and at $x = 12.530$ million cameras (12,530,000 cameras).

(C) Use a graphing calculator (Fig. 16).

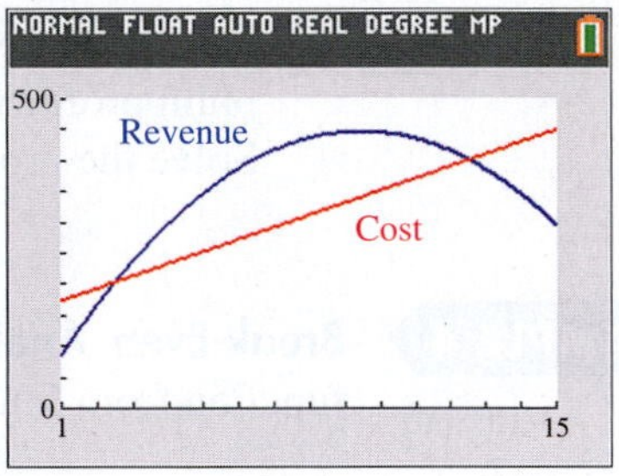

Figure 16

(D) Use INTERSECT on a graphing calculator (Fig. 17).

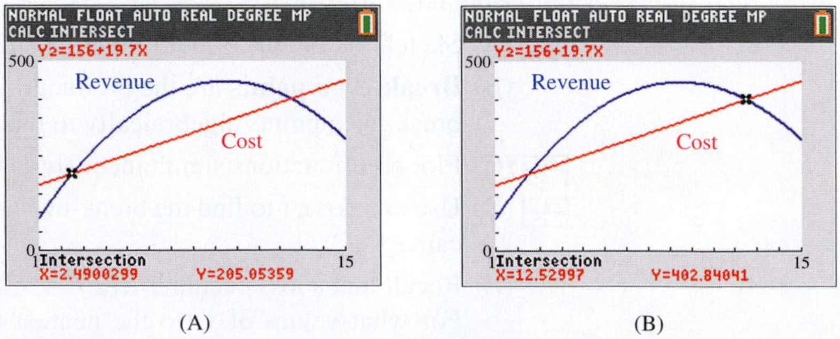

Figure 17

The company breaks even at $x = 2.490$ million cameras (2,490,000 cameras) and at $x = 12.530$ million cameras (12,530,000 cameras).

(E) Use the results from parts (A) and (B) or (C) and (D):

$$\text{Loss: } 1 \le x < 2.490 \quad \text{or} \quad 12.530 < x \le 15$$

$$\text{Profit: } 2.490 < x < 12.530$$

Matched Problem 4 Use the profit equation from Matched Problem 3:

$$P(x) = R(x) - C(x)$$
$$= -5x^2 + 75.1x - 156 \qquad \text{Profit function}$$
$$\text{Domain: } 1 \le x \le 15$$

(A) Sketch a graph of the profit function in a rectangular coordinate system.

(B) Break-even points occur when $P(x) = 0$. Find the break-even points algebraically to the nearest thousand cameras.

(C) Plot the profit function in an appropriate viewing window.

(D) Find the break-even points graphically to the nearest thousand cameras.

(E) A loss occurs if $P(x) < 0$, and a profit occurs if $P(x) > 0$. For what values of x (to the nearest thousand cameras) will a loss occur? A profit?

A visual inspection of the plot of a data set might indicate that a parabola would be a better model of the data than a straight line. In that case, rather than using linear regression to fit a linear model to the data, we would use **quadratic regression** on a graphing calculator to find the function of the form $y = ax^2 + bx + c$ that best fits the data.

EXAMPLE 5 **Outboard Motors** Table 1 gives performance data for a boat powered by an Evinrude outboard motor. Use quadratic regression to find the best model of the form $y = ax^2 + bx + c$ for fuel consumption y (in miles per gallon) as a function of speed x (in miles per hour). Estimate the fuel consumption (to one decimal place) at a speed of 12 miles per hour.

Table 1

rpm	mph	mpg
2,500	10.3	4.1
3,000	18.3	5.6
3,500	24.6	6.6
4,000	29.1	6.4
4,500	33.0	6.1
5,000	36.0	5.4
5,400	38.9	4.9

SOLUTION Enter the data in a graphing calculator (Fig. 18A) and find the quadratic regression equation (Fig. 18B). The data set and the regression equation are graphed in Figure 18C. Using trace, we see that the estimated fuel consumption at a speed of 12 mph is 4.5 mpg.

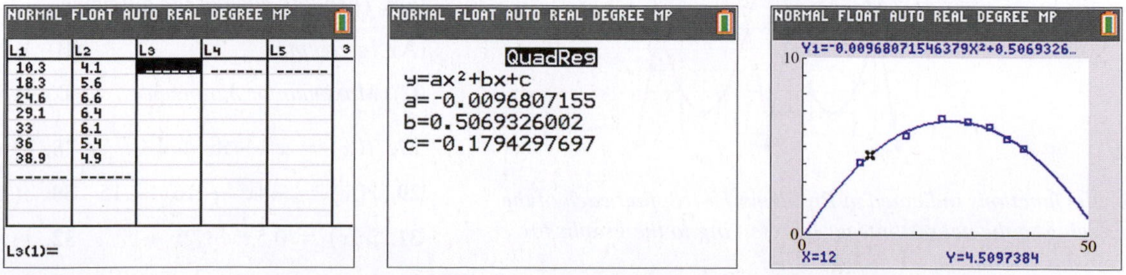

(A) (B) (C)

Figure 18

Matched Problem 5 Refer to Table 1. Use quadratic regression to find the best model of the form $y = ax^2 + bx + c$ for boat speed y (in miles per hour) as a function of engine speed x (in revolutions per minute). Estimate the boat speed (in miles per hour, to one decimal place) at an engine speed of 3,400 rpm.

Exercises 2.3

A *In Problems 1–8, find the vertex form of each quadratic function by completing the square.*

1. $f(x) = x^2 - 10x$

2. $f(x) = x^2 + 16x$

3. $f(x) = x^2 + 20x + 50$

4. $f(x) = x^2 - 12x - 8$

5. $f(x) = -2x^2 + 4x - 5$

6. $f(x) = 3x^2 + 18x + 21$

7. $f(x) = 2x^2 + 2x + 1$

8. $f(x) = -5x^2 + 15x - 11$

In Problems 9–12, write a brief verbal description of the relationship between the graph of the indicated function and the graph of $y = x^2$.

9. $f(x) = x^2 - 4x + 3$

10. $g(x) = x^2 - 2x - 5$

11. $m(x) = -x^2 + 6x - 4$

12. $n(x) = -x^2 + 8x - 9$

13. Match each equation with a graph of one of the functions f, g, m, or n in the figure.

(A) $y = -(x + 2)^2 + 1$

(B) $y = (x - 2)^2 - 1$

(C) $y = (x + 2)^2 - 1$

(D) $y = -(x - 2)^2 + 1$

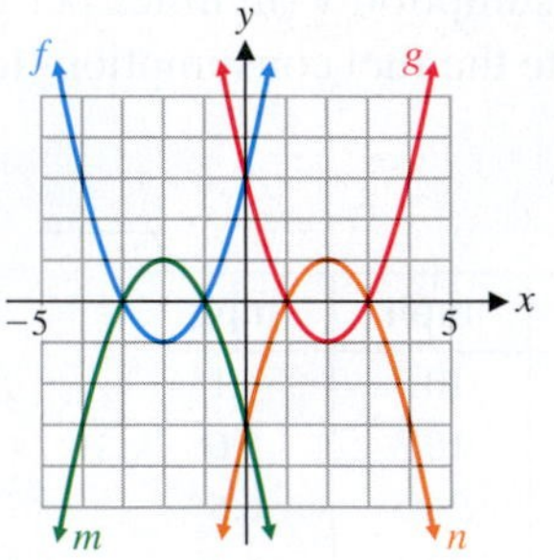

14. Match each equation with a graph of one of the functions f, g, m, or n in the figure.

(A) $y = (x - 3)^2 - 4$

(B) $y = -(x + 3)^2 + 4$

(C) $y = -(x - 3)^2 + 4$

(D) $y = (x + 3)^2 - 4$

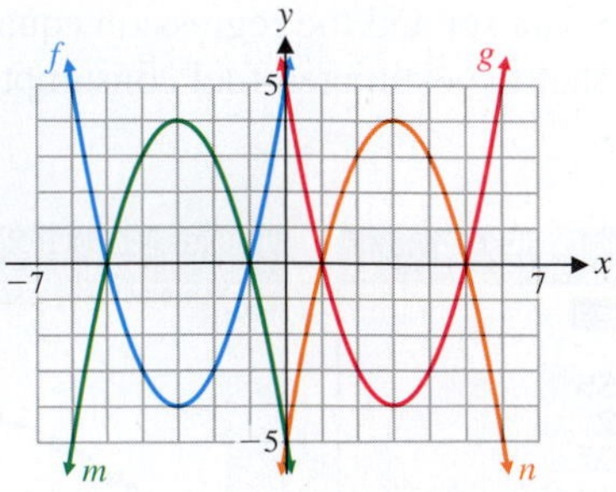

For the functions indicated in Problems 15–18, find each of the following to the nearest integer by referring to the graphs for Problems 13 and 14.

(A) *Intercepts*

(B) *Vertex*

(C) *Maximum or minimum*

(D) *Range*

15. Function n in the figure for Problem 13

16. Function m in the figure for Problem 14

17. Function f in the figure for Problem 13

18. Function g in the figure for Problem 14

In Problems 19–22, find each of the following:

(A) *Intercepts*

(B) *Vertex*

(C) *Maximum or minimum*

(D) *Range*

19. $f(x) = -(x - 3)^2 + 2$

20. $g(x) = -(x + 2)^2 + 3$

21. $m(x) = (x + 1)^2 - 2$

22. $n(x) = (x - 4)^2 - 3$

B *In Problems 23–26, write an equation for each graph in the form $y = a(x - h)^2 + k$, where a is either 1 or −1 and h and k are integers.*

23.

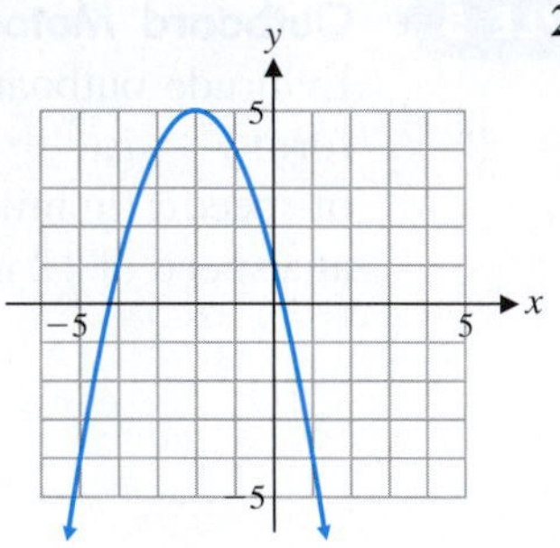

24.

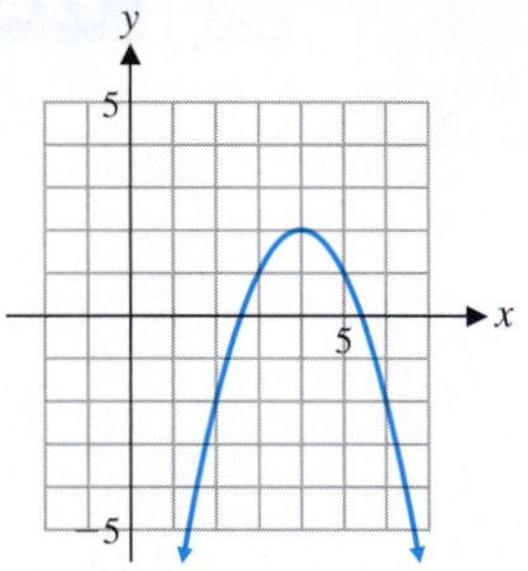

25.

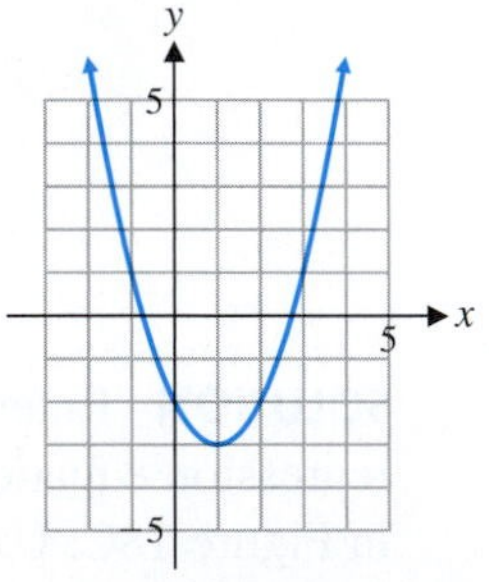

26.

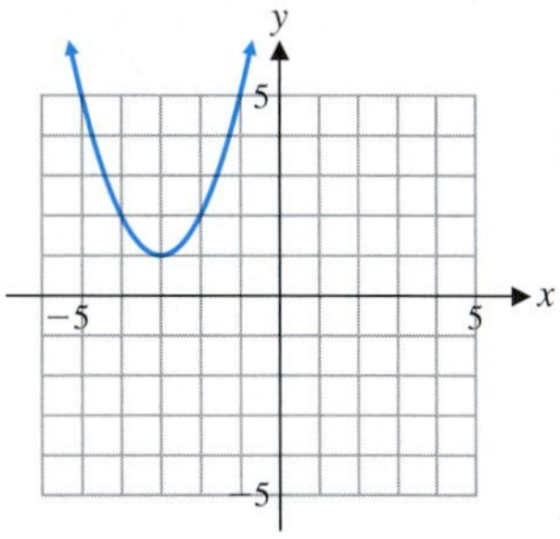

In Problems 27–32, find the vertex form for each quadratic function. Then find each of the following:

(A) *Intercepts*

(B) *Vertex*

(C) *Maximum or minimum*

(D) *Range*

27. $f(x) = x^2 - 8x + 12$

28. $g(x) = x^2 - 6x + 5$

29. $r(x) = -4x^2 + 16x - 15$

30. $s(x) = -4x^2 - 8x - 3$

31. $u(x) = 0.5x^2 - 2x + 5$

32. $v(x) = 0.5x^2 + 4x + 10$

33. Let $f(x) = 0.3x^2 - x - 8$. Solve each equation graphically to two decimal places.

(A) $f(x) = 4$

(B) $f(x) = -1$

(C) $f(x) = -9$

34. Let $g(x) = -0.6x^2 + 3x + 4$. Solve each equation graphically to two decimal places.

(A) $g(x) = -2$ (B) $g(x) = 5$ (C) $g(x) = 8$

35. Let $f(x) = 125x - 6x^2$. Find the maximum value of f to four decimal places graphically.

36. Let $f(x) = 100x - 7x^2 - 10$. Find the maximum value of f to four decimal places graphically.

C *In Problems 37–40, first write each function in vertex form; then find each of the following (to two decimal places):*

(A) *Intercepts* (B) *Vertex*

(C) *Maximum or minimum* (D) *Range*

37. $g(x) = 0.25x^2 - 1.5x - 7$

38. $m(x) = 0.20x^2 - 1.6x - 1$

39. $f(x) = -0.12x^2 + 0.96x + 1.2$

40. $n(x) = -0.15x^2 - 0.90x + 3.3$

In Problems 41–44, use interval notation to write the solution set of the inequality.

41. $(x - 3)(x + 5) > 0$ **42.** $(x + 6)(x - 3) < 0$

43. $x^2 + x - 6 \le 0$ **44.** $x^2 + 7x + 12 \ge 0$

Solve Problems 45–50 graphically to two decimal places using a graphing calculator.

45. $2 - 5x - x^2 = 0$ **46.** $7 + 3x - 2x^2 = 0$

47. $1.9x^2 - 1.5x - 5.6 < 0$ **48.** $3.4 + 2.9x - 1.1x^2 \ge 0$

49. $2.8 + 3.1x - 0.9x^2 \le 0$ **50.** $1.8x^2 - 3.1x - 4.9 > 0$

51. Given that f is a quadratic function with minimum $f(x) = f(2) = 4$, find the axis, vertex, range, and x intercepts.

52. Given that f is a quadratic function with maximum $f(x) = f(-3) = -5$, find the axis, vertex, range, and x intercepts.

In Problems 53–56,

(A) *Graph f and g in the same coordinate system.*

(B) *Solve $f(x) = g(x)$ algebraically to two decimal places.*

(C) *Solve $f(x) > g(x)$ using parts (A) and (B).*

(D) *Solve $f(x) < g(x)$ using parts (A) and (B).*

53. $f(x) = -0.4x(x - 10)$

$g(x) = 0.3x + 5$

$0 \le x \le 10$

54. $f(x) = -0.7x(x - 7)$

$g(x) = 0.5x + 3.5$

$0 \le x \le 7$

55. $f(x) = -0.9x^2 + 7.2x$

$g(x) = 1.2x + 5.5$

$0 \le x \le 8$

56. $f(x) = -0.7x^2 + 6.3x$

$g(x) = 1.1x + 4.8$

$0 \le x \le 9$

57. How can you tell from the graph of a quadratic function whether it has exactly one real zero?

58. How can you tell from the graph of a quadratic function whether it has no real zeros?

59. How can you tell from the standard form $y = ax^2 + bx + c$ whether a quadratic function has two real zeros?

60. How can you tell from the standard form $y = ax^2 + bx + c$ whether a quadratic function has exactly one real zero?

61. How can you tell from the vertex form $y = a(x - h)^2 + k$ whether a quadratic function has no real zeros?

62. How can you tell from the vertex form $y = a(x - h)^2 + k$ whether a quadratic function has two real zeros?

In Problems 63 and 64, assume that a, b, c, h, and k are constants with $a \ne 0$ such that

$$ax^2 + bx + c = a(x - h)^2 + k$$

for all real numbers x.

63. Show that $h = -\dfrac{b}{2a}$. **64.** Show that $k = \dfrac{4ac - b^2}{4a}$.

Applications

65. Tire mileage. An automobile tire manufacturer collected the data in the table relating tire pressure x (in pounds per square inch) and mileage (in thousands of miles):

x	Mileage
28	45
30	52
32	55
34	51
36	47

A mathematical model for the data is given by

$$f(x) = -0.518x^2 + 33.3x - 481$$

(A) Complete the following table. Round values of $f(x)$ to one decimal place.

x	Mileage	f(x)
28	45	
30	52	
32	55	
34	51	
36	47	

(B) Sketch the graph of f and the mileage data in the same coordinate system.

(C) Use the modeling function $f(x)$ to estimate the mileage for a tire pressure of 31 lbs/sq in. and for 35 lbs/sq in. Round answers to two decimal places.

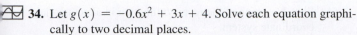

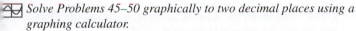

(D) Write a brief description of the relationship between tire pressure and mileage.

66. Automobile production. The table shows the retail market share of passenger cars from Ford Motor Company as a percentage of the U.S. market.

Year	Market Share
1985	18.8%
1990	20.0%
1995	20.7%
2000	20.2%
2005	17.4%
2010	16.4%
2015	15.3%

A mathematical model for this data is given by

$$f(x) = -0.0117x^2 + 0.32x + 17.9$$

where $x = 0$ corresponds to 1980.

(A) Complete the following table. Round values of $f(x)$ to one decimal place.

x	Market Share	$f(x)$
5	18.8	
10	20.0	
15	20.7	
20	20.2	
25	17.4	
30	16.4	
35	15.3	

(B) Sketch the graph of f and the market share data in the same coordinate system.

(C) Use values of the modeling function f to estimate Ford's market share in 2025 and in 2028.

(D) Write a brief verbal description of Ford's market share from 1985 to 2015.

67. Tire mileage. Using quadratic regression on a graphing calculator, show that the quadratic function that best fits the data on tire mileage in Problem 65 is

$$f(x) = -0.518x^2 + 33.3x - 481$$

68. Automobile production. Using quadratic regression on a graphing calculator, show that the quadratic function that best fits the data on market share in Problem 66 is

$$f(x) = -0.0117x^2 + 0.32x + 17.9$$

69. Revenue. The marketing research department for a company that manufactures and sells memory chips for microcomputers established the following price–demand and revenue functions:

$$p(x) = 75 - 3x \qquad \text{Price–demand function}$$
$$R(x) = xp(x) = x(75 - 3x) \qquad \text{Revenue function}$$

where $p(x)$ is the wholesale price in dollars at which x million chips can be sold, and $R(x)$ is in millions of dollars. Both functions have domain $1 \le x \le 20$.

(A) Sketch a graph of the revenue function in a rectangular coordinate system.

(B) Find the value of x that will produce the maximum revenue. What is the maximum revenue?

(C) What is the wholesale price per chip that produces the maximum revenue?

70. Revenue. The marketing research department for a company that manufactures and sells notebook computers established the following price–demand and revenue functions:

$$p(x) = 2{,}000 - 60x \qquad \text{Price–demand function}$$
$$R(x) = xp(x) \qquad \text{Revenue function}$$
$$= x(2{,}000 - 60x)$$

where $p(x)$ is the wholesale price in dollars at which x thousand computers can be sold, and $R(x)$ is in thousands of dollars. Both functions have domain $1 \le x \le 25$.

(A) Sketch a graph of the revenue function in a rectangular coordinate system.

(B) Find the value of x that will produce the maximum revenue. What is the maximum revenue to the nearest thousand dollars?

(C) What is the wholesale price per computer (to the nearest dollar) that produces the maximum revenue?

71. Break-even analysis. Use the revenue function from Problem 69 and the given cost function:

$$R(x) = x(75 - 3x) \qquad \text{Revenue function}$$
$$C(x) = 125 + 16x \qquad \text{Cost function}$$

where x is in millions of chips, and $R(x)$ and $C(x)$ are in millions of dollars. Both functions have domain $1 \le x \le 20$.

(A) Sketch a graph of both functions in the same rectangular coordinate system.

(B) Find the break-even points to the nearest thousand chips.

(C) For what values of x will a loss occur? A profit?

72. Break-even analysis. Use the revenue function from Problem 70, and the given cost function:

$$R(x) = x(2{,}000 - 60x) \qquad \text{Revenue function}$$
$$C(x) = 4{,}000 + 500x \qquad \text{Cost function}$$

where x is thousands of computers, and $C(x)$ and $R(x)$ are in thousands of dollars. Both functions have domain $1 \le x \le 25$.

(A) Sketch a graph of both functions in the same rectangular coordinate system.

(B) Find the break-even points.

(C) For what values of x will a loss occur? A profit?

73. Profit-loss analysis. Use the revenue and cost functions from Problem 71:

$$R(x) = x(75 - 3x) \qquad \text{Revenue function}$$
$$C(x) = 125 + 16x \qquad \text{Cost function}$$

where x is in millions of chips, and $R(x)$ and $C(x)$ are in millions of dollars. Both functions have domain $1 \le x \le 20$.

(A) Form a profit function P, and graph R, C, and P in the same rectangular coordinate system.

(B) Discuss the relationship between the intersection points of the graphs of R and C and the x intercepts of P.

(C) Find the x intercepts of P and the break-even points to the nearest thousand chips.

(D) Find the value of x (to the nearest thousand chips) that produces the maximum profit. Find the maximum profit (to the nearest thousand dollars), and compare with Problem 69B.

74. Profit-loss analysis. Use the revenue function from Problem 70 and the given cost function:

$$R(x) = x(2{,}000 - 60x) \quad \text{Revenue function}$$
$$C(x) = 4{,}000 + 500x \quad \text{Cost function}$$

where x is thousands of computers, and $R(x)$ and $C(x)$ are in thousands of dollars. Both functions have domain $1 \le x \le 25$.

(A) Form a profit function P, and graph R, C, and P in the same rectangular coordinate system.

(B) Discuss the relationship between the intersection points of the graphs of R and C and the x intercepts of P.

(C) Find the x intercepts of P and the break-even points.

(D) Find the value of x that produces the maximum profit. Find the maximum profit and compare with Problem 70B.

75. Medicine. The French physician Poiseuille was the first to discover that blood flows faster near the center of an artery than near the edge. Experimental evidence has shown that the rate of flow v (in centimeters per second) at a point x centimeters from the center of an artery (see the figure) is given by

$$v = f(x) = 1{,}000(0.04 - x^2) \quad 0 \le x \le 0.2$$

Find the distance from the center that the rate of flow is 20 centimeters per second. Round answer to two decimal places.

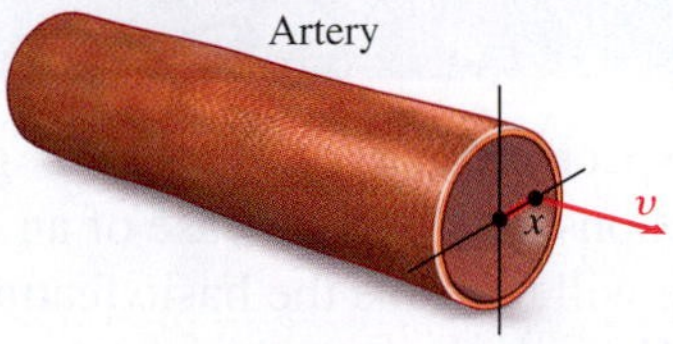

Figure for 75 and 76

76. Medicine. Refer to Problem 75. Find the distance from the center that the rate of flow is 30 centimeters per second. Round answer to two decimal places.

77. Outboard motors. The table gives performance data for a boat powered by an Evinrude outboard motor. Find a quadratic regression model ($y = ax^2 + bx + c$) for boat speed y (in miles per hour) as a function of engine speed (in revolutions per minute). Estimate the boat speed at an engine speed of 3,100 revolutions per minute.

Table for 77 and 78

rpm	mph	mpg
1,500	4.5	8.2
2,000	5.7	6.9
2,500	7.8	4.8
3,000	9.6	4.1
3,500	13.4	3.7

78. Outboard motors. The table gives performance data for a boat powered by an Evinrude outboard motor. Find a quadratic regression model ($y = ax^2 + bx + c$) for fuel consumption y (in miles per gallon) as a function of engine speed (in revolutions per minute). Estimate the fuel consumption at an engine speed of 2,300 revolutions per minute.

Answers to Matched Problems

1. (A)

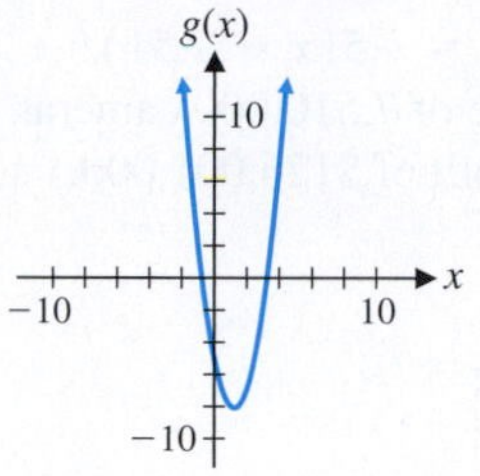

(B) x intercepts: $-0.7656, 3.2656$; y intercept: -5

(C)

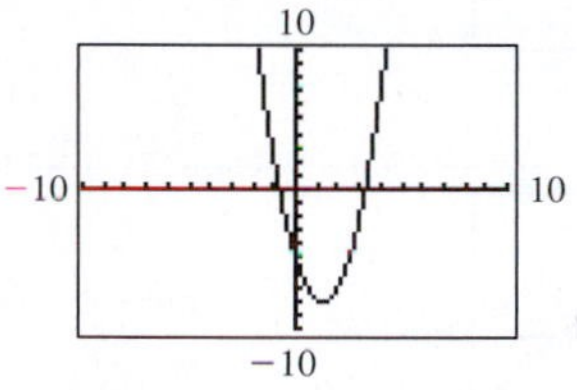

(D) x intercepts: $-0.7656, 3.2656$; y intercept: -5

(E) $x \le -0.7656$ or $x \ge 3.2656$; or $(-\infty, -0.7656]$ or $[3.2656, \infty)$

(F) $x = -0.3508, 2.8508$

2. (A) $f(x) = -0.25(x + 4)^2 + 6.$

(B) Vertex: $(-4, 6)$; maximum: $f(-4) = 6$; range: $y \le 6$ or $(-\infty, 6]$

(C) The graph of $f(x) = -0.25(x + 4)^2 + 6$ is the same as the graph of $g(x) = x^2$ vertically shrunk by a factor of 0.25, reflected in the x axis, and shifted 4 units to the left and 6 units up.

(D)

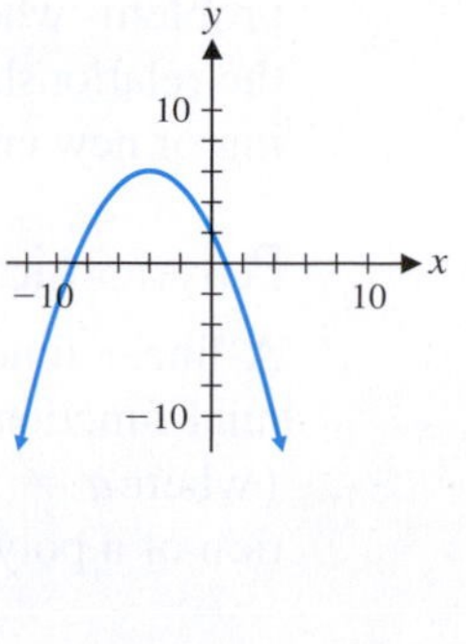

(E)

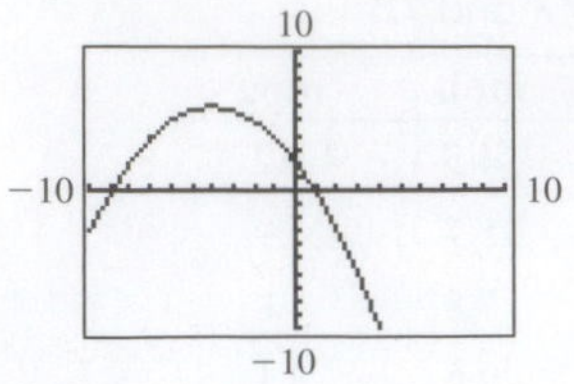

(F) Vertex: $(-4, 6)$; maximum: $f(-4) = 6$

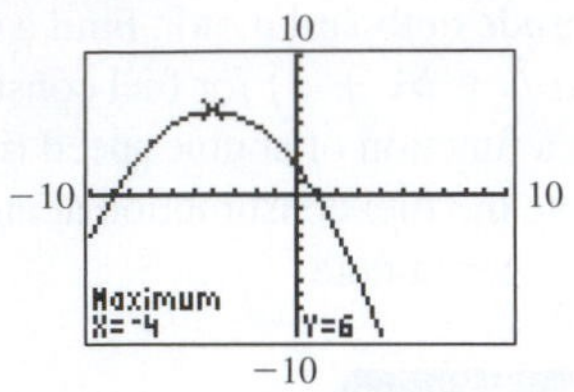

3. (A) $P(x) = R(x) - C(x) = -5x^2 + 75.1x - 156$

(B) $P(x) = R(x) - C(x) = -5(x - 7.51)^2 + 126.0005$; the manufacture and sale of 7,510,000 cameras will produce a maximum profit of \$126,001,000.

(C) $p(7.510) = \$57$

(D)

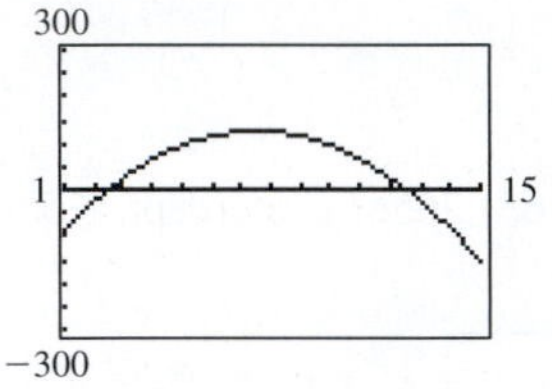

(E)

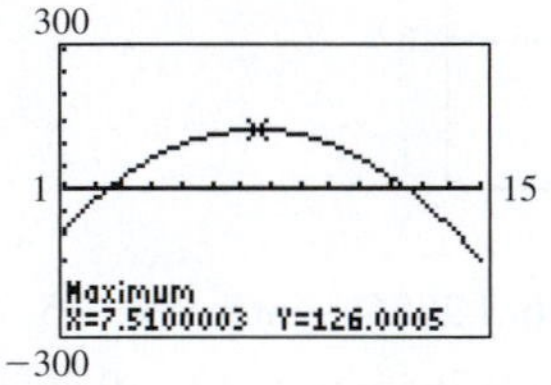

The manufacture and sale of 7,510,000 cameras will produce a maximum profit of \$126,001,000. (Notice that maximum profit does not occur at the same value of x where maximum revenue occurs.)

4. (A)

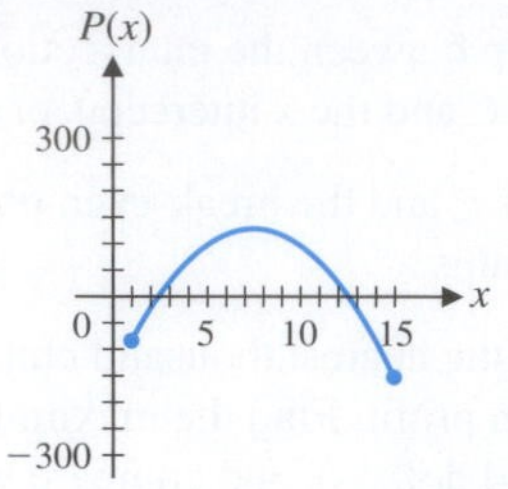

(B) $x = 2.490$ million cameras (2,490,000 cameras) and $x = 12.530$ million cameras (12,530,000 cameras)

(C)

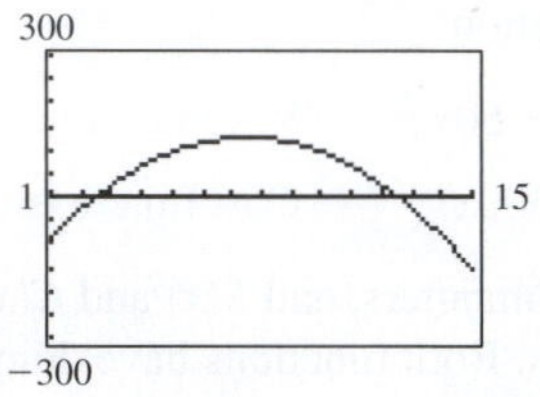

(D) $x = 2.490$ million cameras (2,490,000 cameras) and $x = 12.530$ million cameras (12,530,000 cameras)

(E) Loss: $1 \le x < 2.490$ or $12.530 < x \le 15$; profit: $2.490 < x < 12.530$

5.

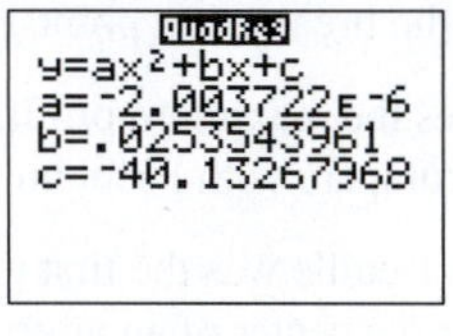

22.9 mph

2.4 Polynomial and Rational Functions

- Polynomial Functions
- Regression Polynomials
- Rational Functions
- Applications

Linear and quadratic functions are special cases of the more general class of *polynomial functions*. Polynomial functions are a special case of an even larger class of functions, the *rational functions*. We will describe the basic features of the graphs of polynomial and rational functions. We will use these functions to solve real-world problems where linear or quadratic models are inadequate, for example, to determine the relationship between length and weight of a species of fish, or to model the training of new employees.

Polynomial Functions

A linear function has the form $f(x) = mx + b$ (where $m \ne 0$) and is a polynomial function of degree 1. A quadratic function has the form $f(x) = ax^2 + bx + c$ (where $a \ne 0$) and is a polynomial function of degree 2. Here is the general definition of a polynomial function.

DEFINITION Polynomial Function

A **polynomial function** is a function that can be written in the form

$$f(x) = a_n x^n + a_{n-1}x^{n-1} + \cdots + a_1 x + a_0$$

for n a nonnegative integer, called the **degree** of the polynomial. The coefficients $a_0, a_1, \ldots, a_n$ are real numbers with $a_n \neq 0$. The **domain** of a polynomial function is the set of all real numbers.

Figure 1 shows graphs of representative polynomial functions of degrees 1 through 6. The figure, which also appears on the inside back cover, suggests some general properties of graphs of polynomial functions.

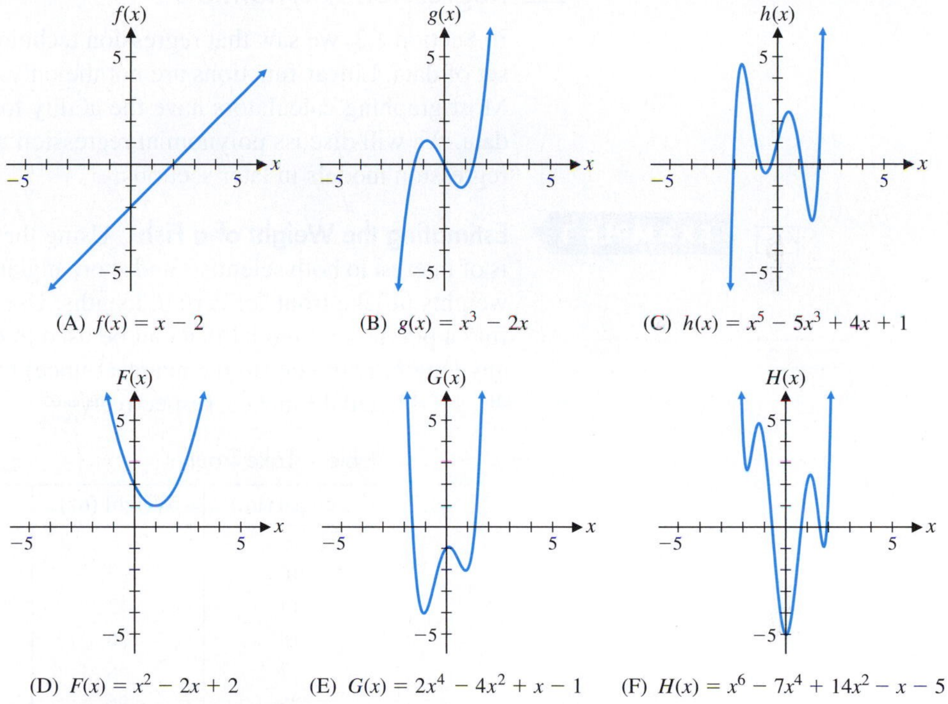

Figure 1 Graphs of polynomial functions

Notice that the odd-degree polynomial graphs start negative, end positive, and cross the x axis at least once. The even-degree polynomial graphs start positive, end positive, and may not cross the x axis at all. In all cases in Figure 1, the **leading coefficient**—that is, the coefficient of the highest-degree term—was chosen positive. If any leading coefficient had been chosen negative, then we would have a similar graph but reflected in the x axis.

A polynomial of degree n can have, at most, n linear factors. Therefore, the graph of a polynomial function of positive degree n can intersect the x axis at most n times. Note from Figure 1 that a polynomial of degree n may intersect the x axis fewer than n times. An x intercept of a function is also called a **zero** or **root** of the function.

The graph of a polynomial function is **continuous**, with no holes or breaks. That is, the graph can be drawn without removing a pen from the paper. Also, the graph of a polynomial has no sharp corners. Figure 2 shows the graphs of two functions—one that is not continuous, and the other that is continuous but with a sharp corner. Neither function is a polynomial.

Reminder

Only real numbers can be x intercepts. Functions may have complex zeros that are not real numbers, but such zeros, which are not x intercepts, will not be discussed in this book.

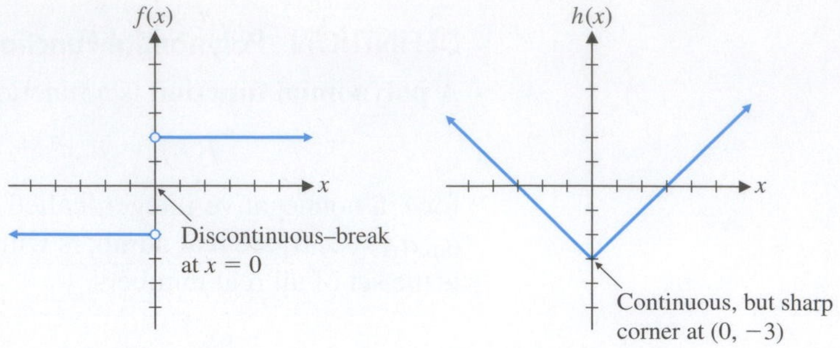

(A) $f(x) = \frac{2|x|}{x}$ (B) $h(x) = |x| - 3$

Figure 2 Discontinuous and sharp-corner functions

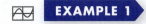

Regression Polynomials

In Section 1.3, we saw that regression techniques can be used to fit a straight line to a set of data. Linear functions are not the only ones that can be applied in this manner. Most graphing calculators have the ability to fit a variety of curves to a given set of data. We will discuss polynomial regression models in this section and other types of regression models in later sections.

EXAMPLE 1 **Estimating the Weight of a Fish** Using the length of a fish to estimate its weight is of interest to both scientists and sport anglers. The data in Table 1 give the average weights of lake trout for certain lengths. Use the data and regression techniques to find a polynomial model that can be used to estimate the weight of a lake trout for any length. Estimate (to the nearest ounce) the weights of lake trout of lengths 39, 40, 41, 42, and 43 inches, respectively.

Table 1 Lake Trout

Length (in.)	Weight (oz)	Length (in.)	Weight (oz)
x	y	x	y
10	5	30	152
14	12	34	226
18	26	38	326
22	56	44	536
26	96		

SOLUTION The graph of the data in Table 1 (Fig. 3A) indicates that a linear regression model would not be appropriate in this case. And, in fact, we would not expect a linear relationship between length and weight. Instead, it is more likely that the weight would be related to the cube of the length. We use a cubic regression polynomial to model the data (Fig. 3B). Figure 3C adds the graph of the polynomial model to the graph of the data. The graph in Figure 3C shows that this cubic polynomial

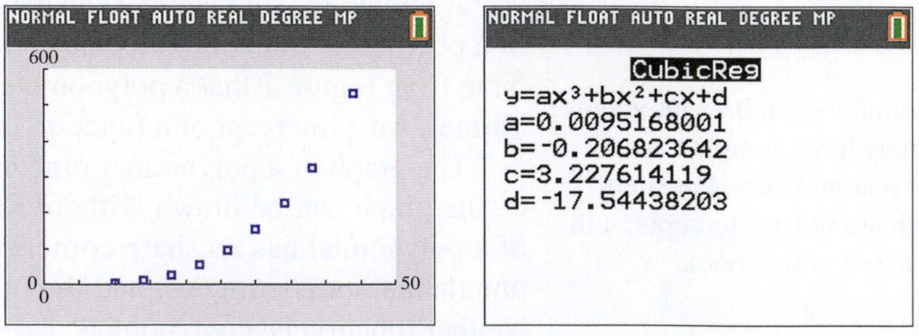

(A) (B)

Figure 3

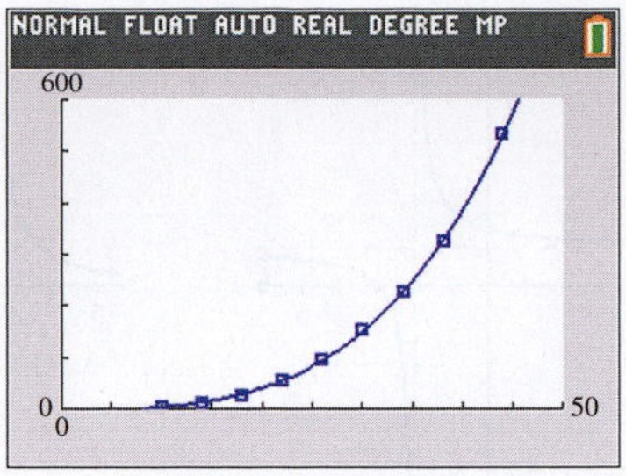

(C)

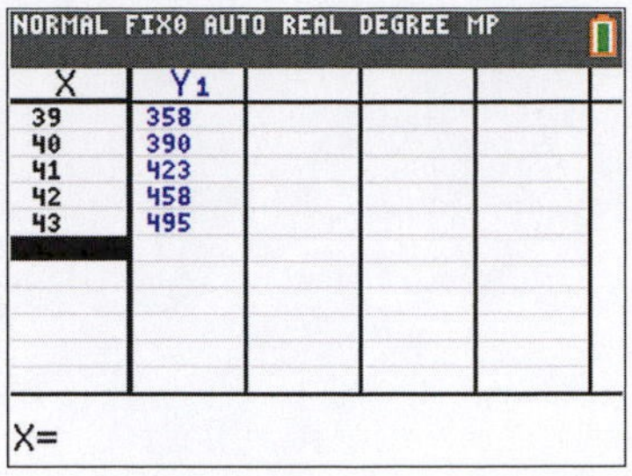

(D)

does provide a good fit for the data. (We will have more to say about the choice of functions and the accuracy of the fit provided by regression analysis later in the book.) Figure 3D shows the estimated weights for the lengths requested.

Matched Problem 1 The data in Table 2 give the average weights of pike for certain lengths. Use a cubic regression polynomial to model the data. Estimate (to the nearest ounce) the weights of pike of lengths 39, 40, 41, 42, and 43 inches, respectively.

Table 2 Pike

Length (in.)	Weight (oz)	Length (in.)	Weight (oz)
x	y	x	y
10	5	30	108
14	12	34	154
18	26	38	210
22	44	44	326
26	72	52	522

Rational Functions

Just as rational numbers are defined in terms of quotients of integers, *rational functions* are defined in terms of quotients of polynomials. The following equations specify rational functions:

$$f(x) = \frac{1}{x} \qquad g(x) = \frac{x - 2}{x^2 - x - 6} \qquad h(x) = \frac{x^3 - 8}{x}$$

$$p(x) = 3x^2 - 5x \qquad q(x) = 7 \qquad r(x) = 0$$

DEFINITION Rational Function

A **rational function** is any function that can be written in the form

$$f(x) = \frac{n(x)}{d(x)} \qquad d(x) \neq 0$$

where $n(x)$ and $d(x)$ are polynomials. The **domain** is the set of all real numbers such that $d(x) \neq 0$.

Figure 4 shows the graphs of representative rational functions. Note, for example, that in Figure 4A the line $x = 2$ is a *vertical asymptote* for the function. The graph of f gets closer to this line as x gets closer to 2. The line $y = 1$ in Figure 4A is a *horizontal asymptote* for the function. The graph of f gets closer to this line as x increases or decreases without bound.

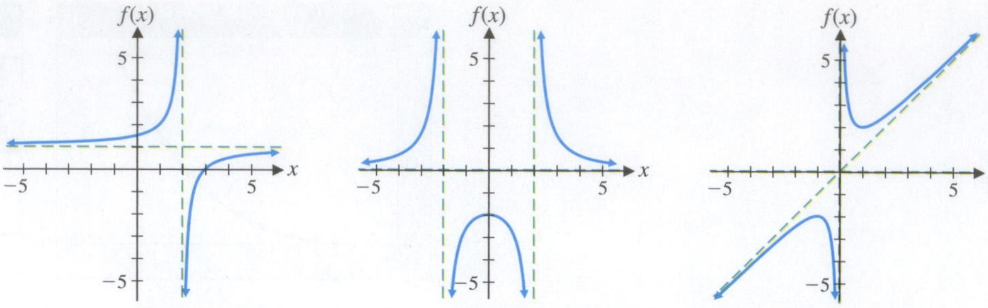

(A) $f(x) = \frac{x - 3}{x - 2}$ (B) $f(x) = \frac{8}{x^2 - 4}$ (C) $f(x) = x + \frac{1}{x}$

Figure 4 Graphs of rational functions

The number of vertical asymptotes of a rational function $f(x) = n(x)/d(x)$ is at most equal to the degree of $d(x)$. A rational function has at most one horizontal asymptote (note that the graph in Fig. 4C does not have a horizontal asymptote). Moreover, the graph of a rational function approaches the horizontal asymptote (when one exists) both as x increases and decreases without bound.

EXAMPLE 2 **Graphing Rational Functions** Given the rational function

$$f(x) = \frac{3x}{x^2 - 4}$$

(A) Find the domain.

(B) Find the x and y intercepts.

(C) Find the equations of all vertical asymptotes.

(D) If there is a horizontal asymptote, find its equation.

(E) Using the information from (A)–(D) and additional points as necessary, sketch a graph of f.

SOLUTION

(A) $x^2 - 4 = (x - 2)(x + 2)$, so the denominator is 0 if $x = -2$ or $x = 2$. Therefore the domain is the set of all real numbers except -2 and 2.

(B) *x intercepts:* $f(x) = 0$ only if $3x = 0$, or $x = 0$. So the only x intercept is 0.
y intercept:

$$f(0) = \frac{3 \cdot 0}{0^2 - 4} = \frac{0}{-4} = 0$$

So the y intercept is 0.

(C) Consider individually the values of x for which the denominator is 0, namely, 2 and -2, found in part (A).

(i) If $x = 2$, the numerator is 6, and the denominator is 0, so $f(2)$ is undefined. But for numbers just to the right of 2 (like 2.1, 2.01, 2.001), the numerator is close to 6, and the denominator is a positive number close to 0, so the fraction $f(x)$ is large and positive. For numbers just to the left of 2 (like 1.9, 1.99, 1.999), the numerator is close to 6, and the denominator is a negative number close to 0, so the fraction $f(x)$ is large (in absolute value) and negative. Therefore, the line $x = 2$ is a vertical asymptote, and $f(x)$ is positive to the right of the asymptote, and negative to the left.

(ii) If $x = -2$, the numerator is -6, and the denominator is 0, so $f(2)$ is undefined. But for numbers just to the right of -2 (like -1.9, -1.99, -1.999), the numerator is close to -6, and the denominator is a negative number close to 0, so the fraction $f(x)$ is large and positive. For numbers just to the left of -2 (like $-2.1, -2.01, -2.001$), the numerator is close to -6, and the denominator is a positive number close to 0, so the fraction $f(x)$ is large (in absolute value) and negative. Therefore, the line $x = -2$ is a vertical asymptote, and $f(x)$ is positive to the right of the asymptote and negative to the left.

(D) Rewrite $f(x)$ by dividing each term in the numerator and denominator by the highest power of x in $f(x)$.

$$f(x) = \frac{3x}{x^2 - 4} = \frac{\frac{3x}{x^2}}{\frac{x^2}{x^2} - \frac{4}{x^2}} = \frac{\frac{3}{x}}{1 - \frac{4}{x^2}}$$

As x increases or decreases without bound, the numerator tends to 0 and the denominator tends to 1; so, $f(x)$ tends to 0. The line $y = 0$ is a horizontal asymptote.

(E) Use the information from parts (A)–(D) and plot additional points as necessary to complete the graph, as shown in Figure 5.

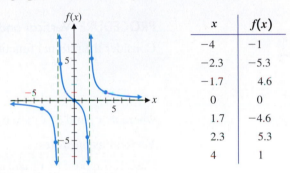

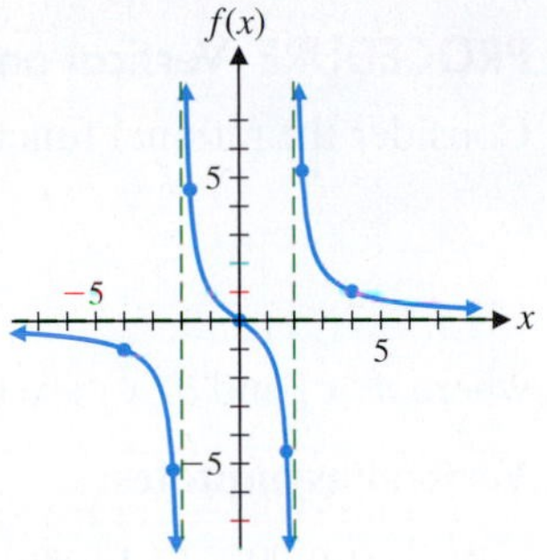

x	$f(x)$
−4	−1
−2.3	−5.3
−1.7	4.6
0	0
1.7	−4.6
2.3	5.3
4	1

Figure 5

Matched Problem 2 Given the rational function $g(x) = \dfrac{3x + 3}{x^2 - 9}$

(A) Find the domain.

(B) Find the x and y intercepts.

(C) Find the equations of all vertical asymptotes.

(D) If there is a horizontal asymptote, find its equation.

(E) Using the information from parts (A)–(D) and additional points as necessary, sketch a graph of g.

CONCEPTUAL INSIGHT

Consider the rational function

$$g(x) = \frac{3x^2 - 12x}{x^3 - 4x^2 - 4x + 16} = \frac{3x(x - 4)}{(x^2 - 4)(x - 4)}$$

The numerator and denominator of g have a common zero, $x = 4$. If $x \neq 4$, then we can cancel the factor $x - 4$ from the numerator and denominator, leaving the function $f(x)$ of Example 2. So the graph of g (Fig. 6) is identical to the graph of f (Fig. 5), except that the graph of g has an open dot at $(4, 1)$, indicating that 4 is not in the domain of g. In particular, f and g have the same asymptotes. Note that the line $x = 4$ is *not* a vertical asymptote of g, even though 4 is a zero of its denominator.

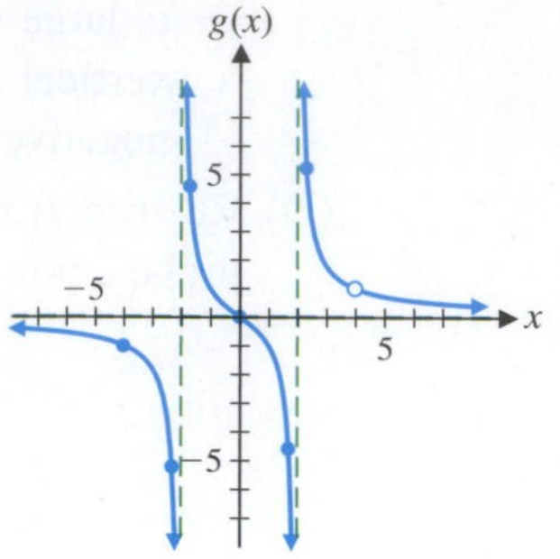

Figure 6

Graphing rational functions is aided by locating vertical and horizontal asymptotes first, if they exist. The following general procedure is suggested by Example 2 and the Conceptual Insight above.

PROCEDURE Vertical and Horizontal Asymptotes of Rational Functions

Consider the rational function

$$f(x) = \frac{n(x)}{d(x)}$$

where $n(x)$ and $d(x)$ are polynomials.

Vertical asymptotes:

Case 1. Suppose $n(x)$ and $d(x)$ have no real zero in common. If c is a real number such that $d(c) = 0$, then the line $x = c$ is a vertical asymptote of the graph of f.

Case 2. If $n(x)$ and $d(x)$ have one or more real zeros in common, cancel common linear factors, and apply Case 1 to the reduced function. (The reduced function has the same asymptotes as f.)

Horizontal asymptote:

Case 1. If degree $n(x) <$ degree $d(x)$, then $y = 0$ is the horizontal asymptote.

Case 2. If degree $n(x) =$ degree $d(x)$, then $y = a/b$ is the horizontal asymptote, where a is the leading coefficient of $n(x)$, and b is the leading coefficient of $d(x)$.

Case 3. If degree $n(x) >$ degree $d(x)$, there is no horizontal asymptote.

Example 2 illustrates Case 1 of the procedure for horizontal asymptotes. Cases 2 and 3 are illustrated in Example 3 and Matched Problem 3.

EXAMPLE 3 **Finding Asymptotes** Find the vertical and horizontal asymptotes of the rational function

$$f(x) = \frac{3x^2 + 3x - 6}{2x^2 - 2}$$

SOLUTION Vertical asymptotes We factor the numerator $n(x)$ and the denominator $d(x)$:

$$n(x) = 3(x^2 + x - 2) = 3(x - 1)(x + 2)$$
$$d(x) = 2(x^2 - 1) = 2(x - 1)(x + 1)$$

The reduced function is

$$\frac{3(x + 2)}{2(x + 1)}$$

which, by the procedure, has the vertical asymptote $x = -1$. Therefore, $x = -1$ is the only vertical asymptote of f.

Horizontal asymptote Both $n(x)$ and $d(x)$ have degree 2 (Case 2 of the procedure for horizontal asymptotes). The leading coefficient of the numerator $n(x)$ is 3, and the leading coefficient of the denominator $d(x)$ is 2. So $y = 3/2$ is the horizontal asymptote.

Matched Problem 3 Find the vertical and horizontal asymptotes of the rational function

$$f(x) = \frac{x^3 - 4x}{x^2 + 5x}$$

Explore and Discuss 1

A function f is **bounded** if the entire graph of f lies between two horizontal lines. The only polynomials that are bounded are the constant functions, but there are many rational functions that are bounded. Give an example of a bounded rational function, with domain the set of all real numbers, that is not a constant function.

Applications

Rational functions occur naturally in many types of applications.

EXAMPLE 4 **Employee Training** A company that manufactures computers has established that, on the average, a new employee can assemble $N(t)$ components per day after t days of on-the-job training, as given by

$$N(t) = \frac{50t}{t + 4} \qquad t \geq 0$$

Sketch a graph of N, $0 \leq t \leq 100$, including any vertical or horizontal asymptotes. What does $N(t)$ approach as t increases without bound?

SOLUTION Vertical asymptotes None for $t \geq 0$

Horizontal asymptote

$$N(t) = \frac{50t}{t + 4} = \frac{50}{1 + \frac{4}{t}}$$

$N(t)$ approaches 50 (the leading coefficient of $50t$ divided by the leading coefficient of $t + 4$) as t increases without bound. So $y = 50$ is a horizontal asymptote.

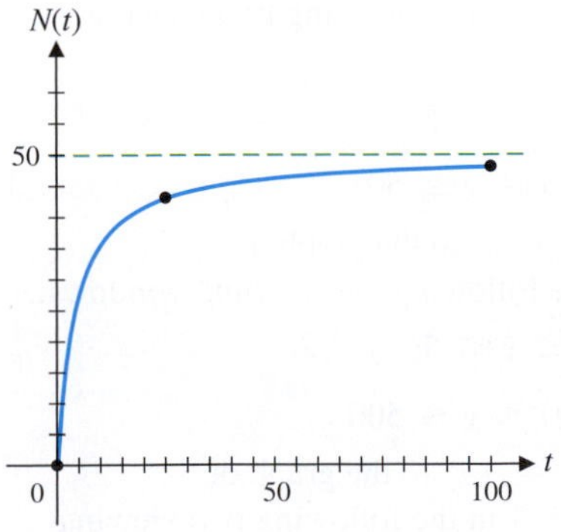

Figure 7

Sketch of graph Note that $N(0) = 0$, $N(25) \approx 43$, and $N(100) \approx 48$. We draw a smooth curve through (0, 0), (25, 43) and (100, 48) (Fig. 7).

$N(t)$ approaches 50 as t increases without bound. It appears that 50 components per day would be the upper limit that an employee would be expected to assemble.

Matched Problem 4 Repeat Example 4 for $N(t) = \dfrac{25t + 5}{t + 5} \quad t \geq 0.$

Exercises 2.4

A *In Problems 1–10, for each polynomial function find the following:*

(A) *Degree of the polynomial*

(B) *All x intercepts*

(C) *The y intercept*

1. $f(x) = 7x + 21$

2. $f(x) = x^2 - 5x + 6$

3. $f(x) = x^2 + 9x + 20$

4. $f(x) = 30 - 3x$

5. $f(x) = x^2 + 2x^6 + 3x^4 + 15$

6. $f(x) = 5x^6 + x^4 + 4x^8 + 10$

7. $f(x) = x^2(x + 6)^3$

8. $f(x) = (x - 5)^2(x + 7)^2$

9. $f(x) = (x^2 - 25)(x^3 + 8)^3$

10. $f(x) = (2x - 5)^2(x^2 - 9)^4$

Each graph in Problems 11–18 is the graph of a polynomial function. Answer the following questions for each graph:

(A) *What is the minimum degree of a polynomial function that could have the graph?*

(B) *Is the leading coefficient of the polynomial negative or positive?*

11.

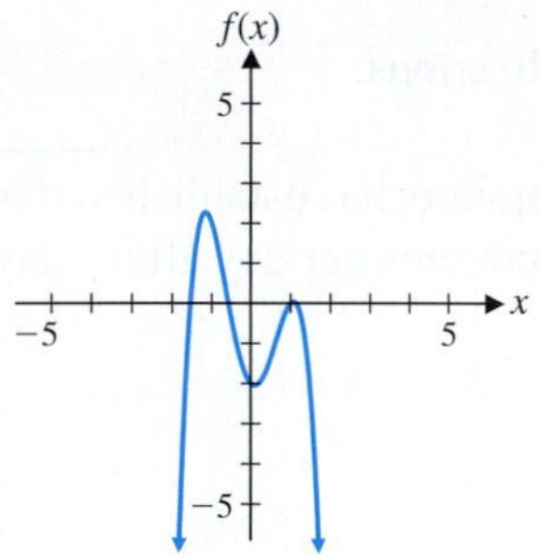

12.

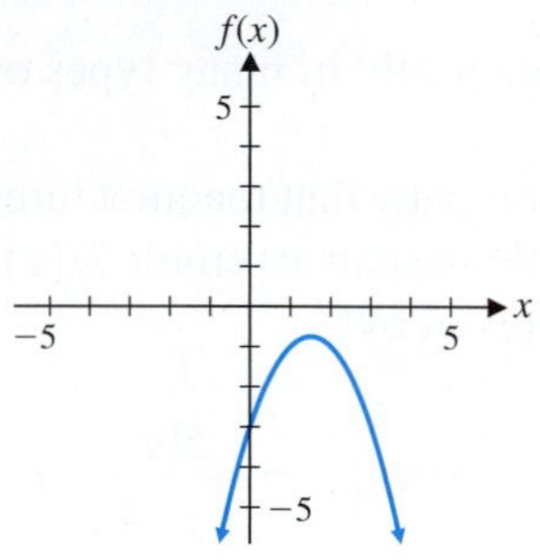

13.

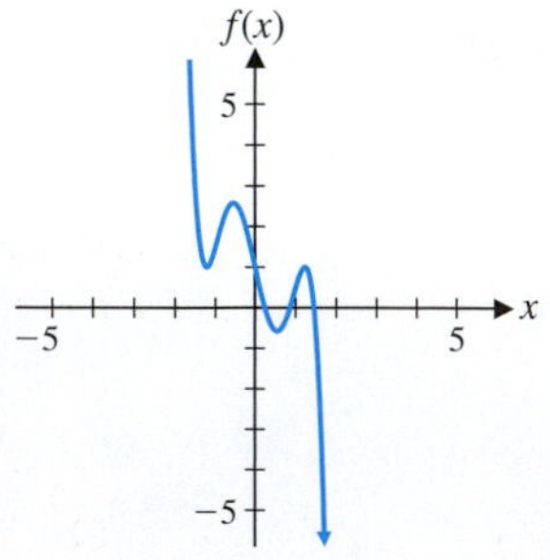

14.

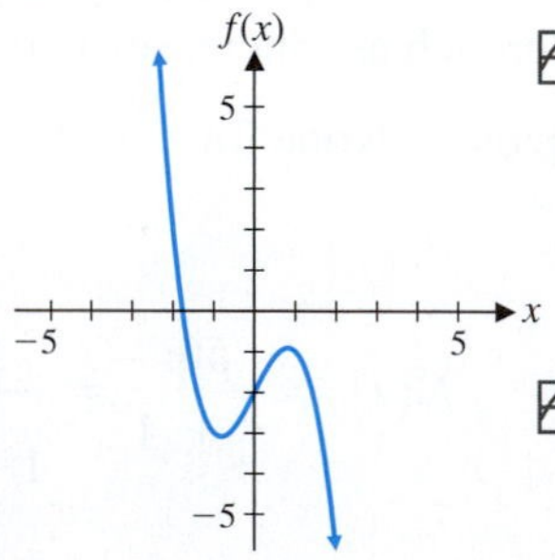

15.

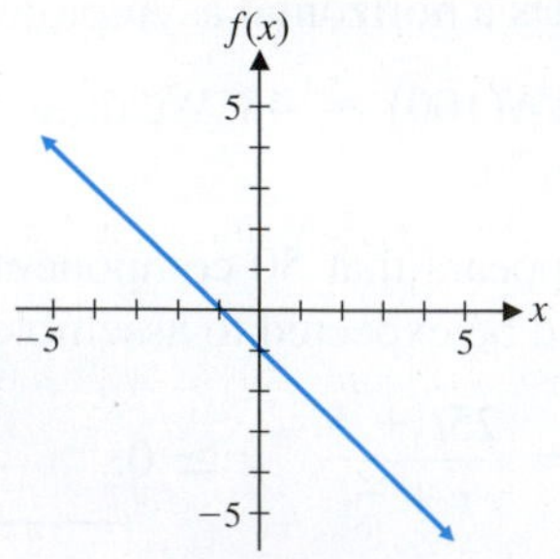

16.

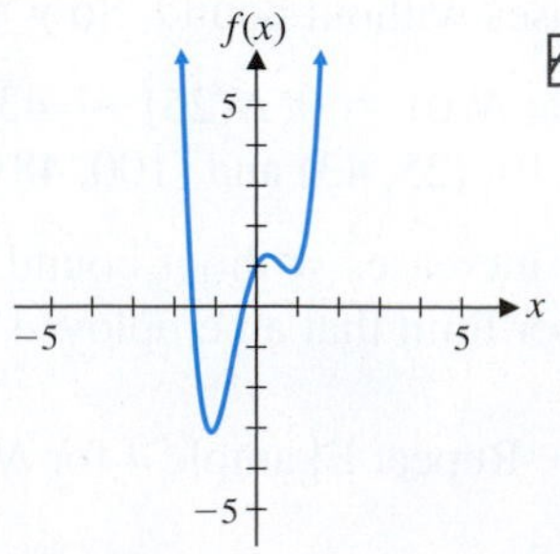

17.

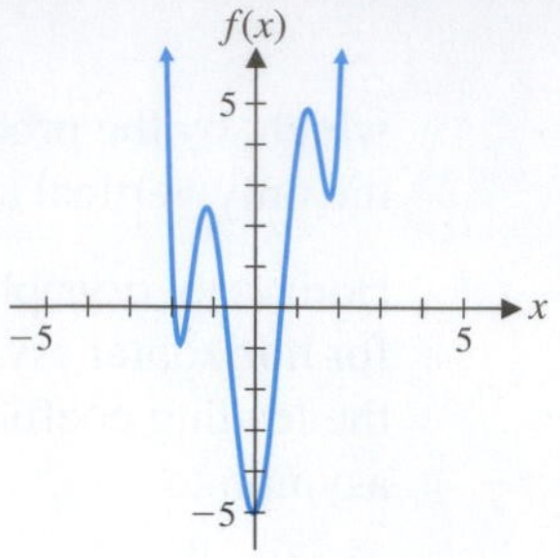

18.

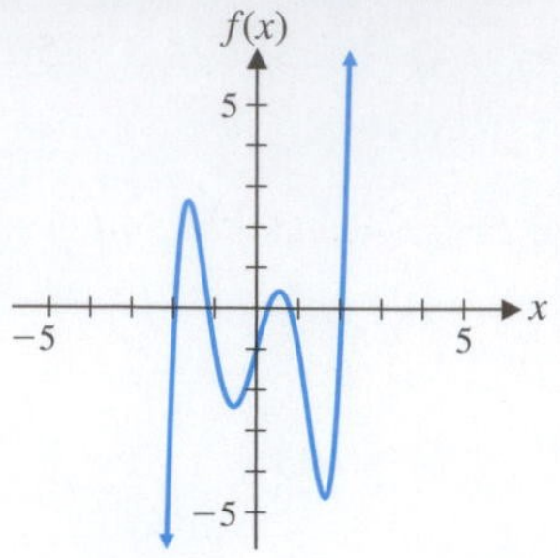

19. What is the maximum number of x intercepts that a polynomial of degree 10 can have?

20. What is the maximum number of x intercepts that a polynomial of degree 7 can have?

21. What is the minimum number of x intercepts that a polynomial of degree 9 can have? Explain.

22. What is the minimum number of x intercepts that a polynomial of degree 6 can have? Explain.

B *For each rational function in Problems 23–28,*

(A) *Find the intercepts for the graph.*

(B) *Determine the domain.*

(C) *Find any vertical or horizontal asymptotes for the graph.*

(D) *Sketch any asymptotes as dashed lines. Then sketch a graph of $y = f(x)$.*

23. $f(x) = \dfrac{x + 2}{x - 2}$

24. $f(x) = \dfrac{x - 3}{x + 3}$

25. $f(x) = \dfrac{3x}{x + 2}$

26. $f(x) = \dfrac{2x}{x - 3}$

27. $f(x) = \dfrac{4 - 2x}{x - 4}$

28. $f(x) = \dfrac{3 - 3x}{x - 2}$

29. Compare the graph of $y = 2x^4$ to the graph of $y = 2x^4 - 5x^2 + x + 2$ in the following two viewing windows:

(A) $-5 \le x \le 5, -5 \le y \le 5$

(B) $-5 \le x \le 5, -500 \le y \le 500$

30. Compare the graph of $y = x^3$ to the graph of $y = x^3 - 2x + 2$ in the following two viewing windows:

(A) $-5 \le x \le 5, -5 \le y \le 5$

(B) $-5 \le x \le 5, -500 \le y \le 500$

31. Compare the graph of $y = -x^5$ to the graph of $y = -x^5 + 4x^3 - 4x + 1$ in the following two viewing windows:

(A) $-5 \le x \le 5, -5 \le y \le 5$

(B) $-5 \le x \le 5, -500 \le y \le 500$

32. Compare the graph of $y = -x^5$ to the graph of $y = -x^5 + 5x^3 - 5x + 2$ in the following two viewing windows:

(A) $-5 \le x \le 5, -5 \le y \le 5$

(B) $-5 \le x \le 5, -500 \le y \le 500$

In Problems 33–40, find the equation of any horizontal asymptote.

33. $f(x) = \dfrac{5x^3 + 2x - 3}{6x^3 - 7x + 1}$ **34.** $f(x) = \dfrac{6x^4 - x^3 + 2}{4x^4 + 10x + 5}$

35. $f(x) = \dfrac{1 - 5x + x^2}{2 + 3x + 4x^2}$ **36.** $f(x) = \dfrac{8 - x^3}{1 + 2x^3}$

37. $f(x) = \dfrac{x^4 + 2x^2 + 1}{1 - x^5}$ **38.** $f(x) = \dfrac{3 + 5x}{x^2 + x + 3}$

39. $f(x) = \dfrac{x^2 + 6x + 1}{x - 5}$ **40.** $f(x) = \dfrac{x^2 + x^4 + 1}{x^3 + 2x - 4}$

In Problems 41–46, find the equations of any vertical asymptotes.

41. $f(x) = \dfrac{x^2 + 1}{(x^2 - 1)(x^2 - 9)}$ **42.** $f(x) = \dfrac{2x + 5}{(x^2 - 4)(x^2 - 16)}$

43. $f(x) = \dfrac{x^2 - x - 6}{x^2 - 3x - 10}$ **44.** $f(x) = \dfrac{x^2 - 8x + 7}{x^2 + 7x - 8}$

45. $f(x) = \dfrac{x^2 + 3x}{x^3 - 36x}$ **46.** $f(x) = \dfrac{x^2 + x - 2}{x^3 - 3x^2 + 2x}$

C *For each rational function in Problems 47–52,*

(A) *Find any intercepts for the graph.*

(B) *Find any vertical and horizontal asymptotes for the graph.*

(C) *Sketch any asymptotes as dashed lines. Then sketch a graph of f.*

(D) *Graph the function in a standard viewing window using a graphing calculator.*

47. $f(x) = \dfrac{2x^2}{x^2 - x - 6}$ **48.** $f(x) = \dfrac{3x^2}{x^2 + x - 6}$

49. $f(x) = \dfrac{6 - 2x^2}{x^2 - 9}$ **50.** $f(x) = \dfrac{3 - 3x^2}{x^2 - 4}$

51. $f(x) = \dfrac{-4x + 24}{x^2 + x - 6}$ **52.** $f(x) = \dfrac{5x - 10}{x^2 + x - 12}$

53. Write an equation for the lowest-degree polynomial function with the graph and intercepts shown in the figure.

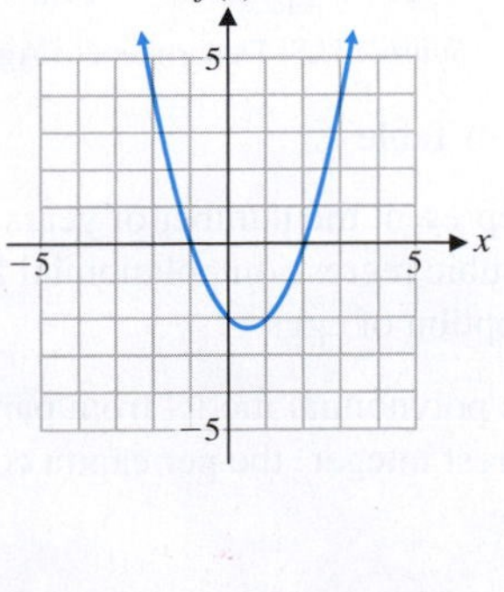

54. Write an equation for the lowest-degree polynomial function with the graph and intercepts shown in the figure.

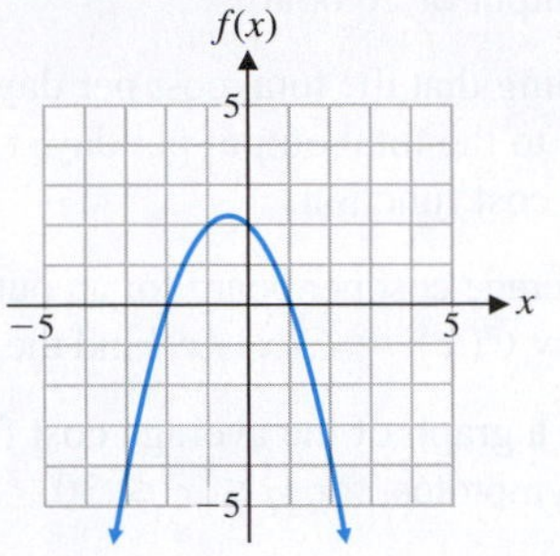

55. Write an equation for the lowest-degree polynomial function with the graph and intercepts shown in the figure.

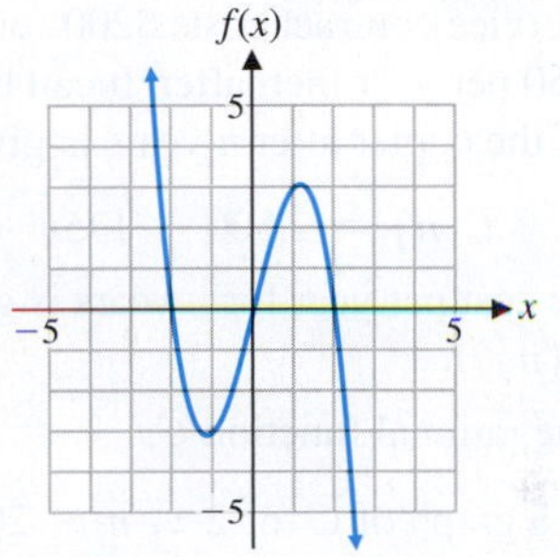

56. Write an equation for the lowest-degree polynomial function with the graph and intercepts shown in the figure.

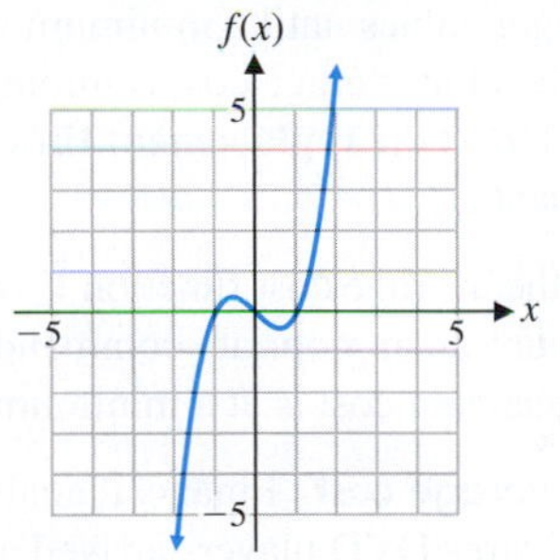

Applications

57. Average cost. A company manufacturing snowboards has fixed costs of \$200 per day and total costs of \$3,800 per day at a daily output of 20 boards.

(A) Assuming that the total cost per day, $C(x)$, is linearly related to the total output per day, x, write an equation for the cost function.

(B) The average cost per board for an output of x boards is given by $\overline{C}(x) = C(x)/x$. Find the average cost function.

(C) Sketch a graph of the average cost function, including any asymptotes, for $1 \le x \le 30$.

(D) What does the average cost per board tend to as production increases?

58. Average cost. A company manufacturing surfboards has fixed costs of \$300 per day and total costs of \$5,100 per day at a daily output of 20 boards.

(A) Assuming that the total cost per day, $C(x)$, is linearly related to the total output per day, x, write an equation for the cost function.

(B) The average cost per board for an output of x boards is given by $\overline{C}(x) = C(x)/x$. Find the average cost function.

(C) Sketch a graph of the average cost function, including any asymptotes, for $1 \le x \le 30$.

(D) What does the average cost per board tend to as production increases?

59. Replacement time. An office copier has an initial price of \$2,500. A service contract costs \$200 for the first year and increases \$50 per year thereafter. It can be shown that the total cost of the copier after n years is given by

$$C(n) = 2{,}500 + 175n + 25n^2$$

The average cost per year for n years is given by $\overline{C}(n) = C(n)/n$.

(A) Find the rational function $\overline{C}$.

(B) Sketch a graph of $\overline{C}$ for $2 \le n \le 20$.

(C) When is the average cost per year at a minimum, and what is the minimum average annual cost? [*Hint:* Refer to the sketch in part (B) and evaluate $\overline{C}(n)$ at appropriate integer values until a minimum value is found.] The time when the average cost is minimum is frequently referred to as the **replacement time** for the piece of equipment.

(D) Graph the average cost function $\overline{C}$ on a graphing calculator and use an appropriate command to find when the average annual cost is at a minimum.

60. Minimum average cost. Financial analysts in a company that manufactures DVD players arrived at the following daily cost equation for manufacturing x DVD players per day:

$$C(x) = x^2 + 2x + 2{,}000$$

The average cost per unit at a production level of x players per day is $\overline{C}(x) = C(x)/x$.

(A) Find the rational function $\overline{C}$.

(B) Sketch a graph of $\overline{C}$ for $5 \le x \le 150$.

(C) For what daily production level (to the nearest integer) is the average cost per unit at a minimum, and what is the minimum average cost per player (to the nearest cent)? [*Hint:* Refer to the sketch in part (B) and evaluate $\overline{C}(x)$ at appropriate integer values until a minimum value is found.]

(D) Graph the average cost function $\overline{C}$ on a graphing calculator and use an appropriate command to find the daily production level (to the nearest integer) at which the average cost per player is at a minimum. What is the minimum average cost to the nearest cent?

61. Minimum average cost. A consulting firm, using statistical methods, provided a veterinary clinic with the cost equation

$$C(x) = 0.00048(x - 500)^3 + 60{,}000$$
$$100 \le x \le 1{,}000$$

where $C(x)$ is the cost in dollars for handling x cases per month. The average cost per case is given by $\overline{C}(x) = C(x)/x$.

(A) Write the equation for the average cost function $\overline{C}$.

(B) Graph $\overline{C}$ on a graphing calculator.

(C) Use an appropriate command to find the monthly caseload for the minimum average cost per case. What is the minimum average cost per case?

62. Minimum average cost. The financial department of a hospital, using statistical methods, arrived at the cost equation

$$C(x) = 20x^3 - 360x^2 + 2{,}300x - 1{,}000$$
$$1 \le x \le 12$$

where $C(x)$ is the cost in thousands of dollars for handling x thousand cases per month. The average cost per case is given by $\overline{C}(x) = C(x)/x$.

(A) Write the equation for the average cost function $\overline{C}$.

(B) Graph $\overline{C}$ on a graphing calculator.

(C) Use an appropriate command to find the monthly caseload for the minimum average cost per case. What is the minimum average cost per case to the nearest dollar?

63. Diet. Table 3 shows the per capita consumption of ice cream in the United States for selected years since 1987.

(A) Let x represent the number of years since 1980 and find a cubic regression polynomial for the per capita consumption of ice cream.

(B) Use the polynomial model from part (A) to estimate (to the nearest tenth of a pound) the per capita consumption of ice cream in 2023.

Table 3 Per Capita Consumption of Ice Cream

Year	Ice Cream (pounds)
1987	18.0
1992	15.8
1997	15.7
2002	16.4
2007	14.9
2012	13.4
2014	12.8

Source: U.S. Department of Agriculture

64. Diet. Refer to Table 4.

(A) Let x represent the number of years since 2000 and find a cubic regression polynomial for the per capita consumption of eggs.

(B) Use the polynomial model from part (A) to estimate (to the nearest integer) the per capita consumption of eggs in 2023.

Table 4 Per Capita Consumption of Eggs

Year	Number of Eggs
2002	255
2004	257
2006	258
2008	247
2010	243
2012	254
2014	263

Source: U.S. Department of Agriculture

Table 5 Marriages and Divorces (per 1,000 population)

Date	Marriages	Divorces
1960	8.5	2.2
1970	10.6	3.5
1980	10.6	5.2
1990	9.8	4.7
2000	8.5	4.1
2010	6.8	3.6

Source: National Center for Health Statistics

65. Physiology. In a study on the speed of muscle contraction in frogs under various loads, researchers W. O. Fems and J. Marsh found that the speed of contraction decreases with increasing loads. In particular, they found that the relationship between speed of contraction v (in centimeters per second) and load x (in grams) is given approximately by

$$v(x) = \frac{26 + 0.06x}{x} \quad x \geq 5$$

(A) What does $v(x)$ approach as x increases?

(B) Sketch a graph of function v.

66. Learning theory. In 1917, L. L. Thurstone, a pioneer in quantitative learning theory, proposed the rational function

$$f(x) = \frac{a(x + c)}{(x + c) + b}$$

to model the number of successful acts per unit time that a person could accomplish after x practice sessions. Suppose that for a particular person enrolled in a typing class,

$$f(x) = \frac{55(x + 1)}{(x + 8)} \quad x \geq 0$$

where $f(x)$ is the number of words per minute the person is able to type after x weeks of lessons.

(A) What does $f(x)$ approach as x increases?

(B) Sketch a graph of function f, including any vertical or horizontal asymptotes.

67. Marriage. Table 5 shows the marriage and divorce rates per 1,000 population for selected years since 1960.

(A) Let x represent the number of years since 1960 and find a cubic regression polynomial for the marriage rate.

(B) Use the polynomial model from part (A) to estimate the marriage rate (to one decimal place) for 2025.

68. Divorce. Refer to Table 5.

(A) Let x represent the number of years since 1960 and find a cubic regression polynomial for the divorce rate.

(B) Use the polynomial model from part (A) to estimate the divorce rate (to one decimal place) for 2025.

Answers to Matched Problems

1.

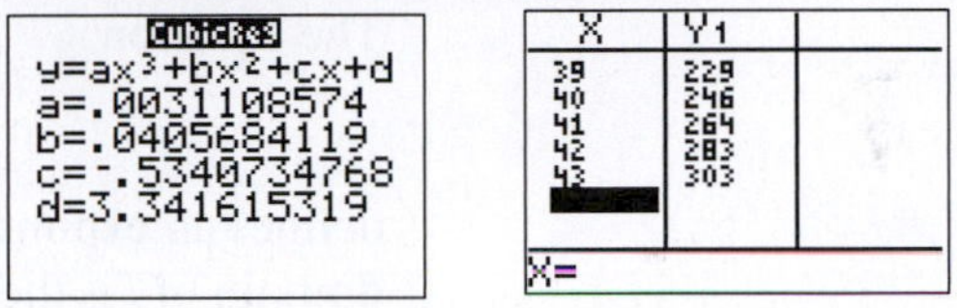

2. (A) Domain: all real numbers except -3 and 3

(B) x intercept: -1; y intercept: $-\frac{1}{3}$

(C) Vertical asymptotes: $x = -3$ and $x = 3$;

(D) Horizontal asymptote: $y = 0$

(E)

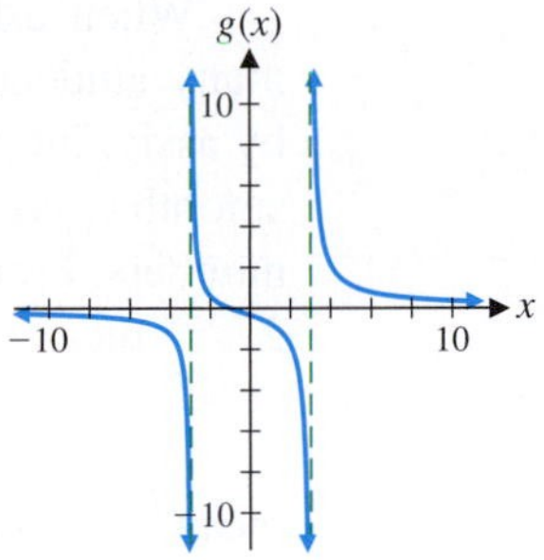

3. Vertical asymptote: $x = -5$
Horizontal asymptote: none

4. No vertical asymptotes for $t \geq 0$; $y = 25$ is a horizontal asymptote. $N(t)$ approaches 25 as t increases without bound. It appears that 25 components per day would be the upper limit that an employee would be expected to assemble.

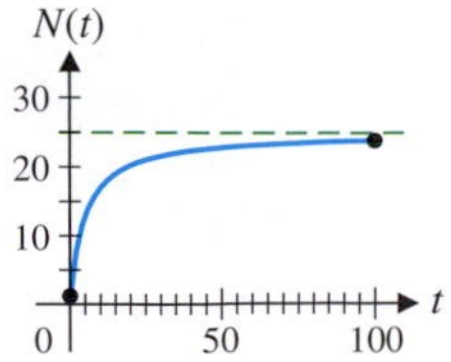

2.5 Exponential Functions

- Exponential Functions
- Base *e* Exponential Function
- Growth and Decay Applications
- Compound Interest

This section introduces an important class of functions called *exponential functions.* These functions are used extensively in modeling and solving a wide variety of real-world problems, including growth of money at compound interest, growth of populations, radioactive decay, and learning associated with the mastery of such devices as a new computer or an assembly process in a manufacturing plant.

Exponential Functions

We start by noting that

$$f(x) = 2^x \quad \text{and} \quad g(x) = x^2$$

are not the same function. Whether a variable appears as an exponent with a constant base or as a base with a constant exponent makes a big difference. The function g is a quadratic function, which we have already discussed. The function f is a new type of function called an *exponential function.* In general,

DEFINITION Exponential Function

The equation

$$f(x) = b^x \quad b > 0, b \neq 1$$

defines an **exponential function** for each different constant b, called the **base**. The **domain** of f is the set of all real numbers, and the **range** of f is the set of all positive real numbers.

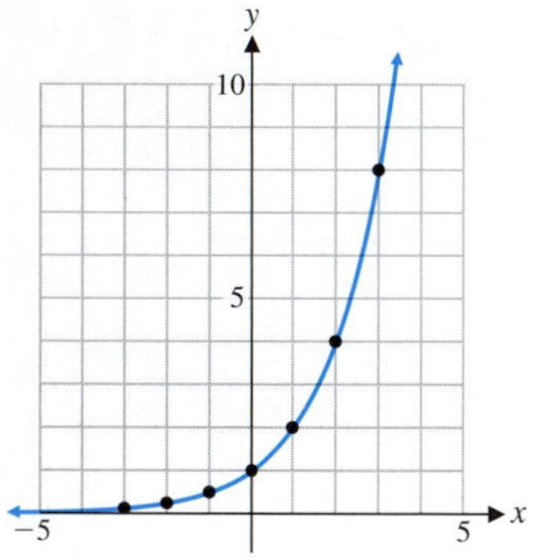

Figure 1 $y = 2^x$

We require the base b to be positive to avoid imaginary numbers such as $(-2)^{1/2} = \sqrt{-2} = i\sqrt{2}$. We exclude $b = 1$ as a base, since $f(x) = 1^x = 1$ is a constant function, which we have already considered.

When asked to hand-sketch graphs of equations such as $y = 2^x$ or $y = 2^{-x}$, many students do not hesitate. [*Note:* $2^{-x} = 1/2^x = (1/2)^x$.] They make tables by assigning integers to x, plot the resulting points, and then join these points with a smooth curve as in Figure 1. The only catch is that we have not defined 2^x for all real numbers. From Appendix A, Section A.6, we know what 2^5, 2^{-3}, $2^{2/3}$, $2^{-3/5}$, $2^{1.4}$, and $2^{-3.14}$ mean (that is, 2^p, where p is a rational number), but what does

$$2^{\sqrt{2}}$$

mean? The question is not easy to answer at this time. In fact, a precise definition of $2^{\sqrt{2}}$ must wait for more advanced courses, where it is shown that

$$2^x$$

names a positive real number for x any real number, and that the graph of $y = 2^x$ is as indicated in Figure 1.

It is useful to compare the graphs of $y = 2^x$ and $y = 2^{-x}$ by plotting both on the same set of coordinate axes, as shown in Figure 2A. The graph of

$$f(x) = b^x \quad b > 1 \text{ (Fig. 2B)}$$

looks very much like the graph of $y = 2^x$, and the graph of

$$f(x) = b^x \quad 0 < b < 1 \text{ (Fig. 2B)}$$

looks very much like the graph of $y = 2^{-x}$. Note that in both cases the x axis is a horizontal asymptote for the graphs.

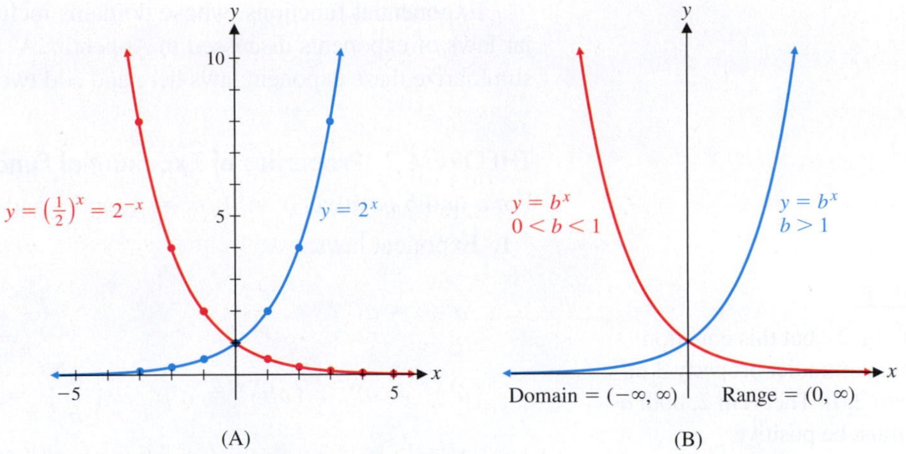

Figure 2 Exponential functions

The graphs in Figure 2 suggest the following general properties of exponential functions, which we state without proof:

THEOREM 1 Basic Properties of the Graph of $f(x) = b^x$, $b > 0$, $b \neq 1$

1. All graphs will pass through the point $(0, 1)$.
2. All graphs are continuous curves, with no holes or jumps.
3. The x axis is a horizontal asymptote.
4. If $b > 1$, then b^x increases as x increases.
5. If $0 < b < 1$, then b^x decreases as x increases.

$b^0 = 1$ for any permissible base b.

CONCEPTUAL INSIGHT

Recall that the graph of a rational function has at most one horizontal asymptote and that it approaches the horizontal asymptote (if one exists) both as $x \to \infty$ *and* as $x \to -\infty$ (see Section 2.4). The graph of an exponential function, on the other hand, approaches its horizontal asymptote as $x \to \infty$ *or* as $x \to -\infty$, but not both. In particular, there is no rational function that has the same graph as an exponential function.

The use of a calculator with the key y^x, or its equivalent, makes the graphing of exponential functions almost routine. Example 1 illustrates the process.

EXAMPLE 1 **Graphing Exponential Functions** Sketch a graph of $y = \left(\frac{1}{2}\right)4^x$, $-2 \leq x \leq 2$.

SOLUTION Use a calculator to create the table of values shown. Plot these points, and then join them with a smooth curve as in Figure 3.

x	y
−2	0.031
−1	0.125
0	0.50
1	2.00
2	8.00

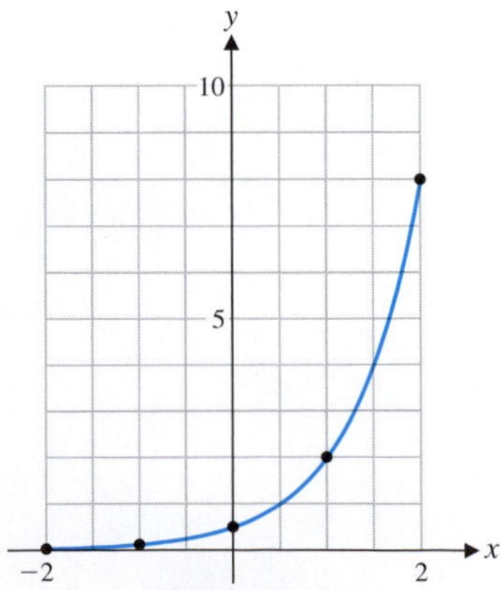

Figure 3 Graph of $y = \left(\frac{1}{2}\right)4^x$

Matched Problem 1 Sketch a graph of $y = \left(\frac{1}{2}\right)4^{-x}$, $-2 \leq x \leq 2$.

Exponential functions, whose domains include irrational numbers, obey the familiar laws of exponents discussed in Appendix A, Section A.6 for rational exponents. We summarize these exponent laws here and add two other important and useful properties.

THEOREM 2 Properties of Exponential Functions

For a and b positive, $a \neq 1, b \neq 1$, and x and y real,

1. Exponent laws:

$$a^x a^y = a^{x+y} \qquad \frac{a^x}{a^y} = a^{x-y} \qquad \frac{4^{2y}}{4^{5y}} = 4^{2y-5y} = 4^{-3y}$$

$$(a^x)^y = a^{xy} \qquad (ab)^x = a^x b^x \qquad \left(\frac{a}{b}\right)^x = \frac{a^x}{b^x}$$

2. $a^x = a^y$ if and only if $x = y$ — If $7^{5t+1} = 7^{3t-3}$, then $5t + 1 = 3t - 3$, and $t = -2$.

3. For $x \neq 0$, $a^x = b^x$ if and only if $a = b$ — If $a^5 = 2^5$, then $a = 2$.

Reminder

$(-2)^2 = 2^2$, but this equation does not contradict property 3 of Theorem 2. In Theorem 2, both a and b must be positive.

Base e Exponential Function

Of all the possible bases b we can use for the exponential function $y = b^x$, which ones are the most useful? If you look at the keys on a calculator, you will probably see 10^x and e^x. It is clear why base 10 would be important, because our number system is a base 10 system. But what is e, and why is it included as a base? It turns out that base e is used more frequently than all other bases combined. The reason for this is that certain formulas and the results of certain processes found in calculus and more advanced mathematics take on their simplest form if this base is used. This is why you will see e used extensively in expressions and formulas that model real-world phenomena. In fact, its use is so prevalent that you will often hear people refer to $y = e^x$ as *the* exponential function.

The base e is an irrational number and, like π, it cannot be represented exactly by any finite decimal or fraction. However, e can be approximated as closely as we like by evaluating the expression

$$\left(1 + \frac{1}{x}\right)^x \tag{1}$$

for sufficiently large values of x. What happens to the value of expression (1) as x increases without bound? Think about this for a moment before proceeding. Maybe you guessed that the value approaches 1, because

$$1 + \frac{1}{x}$$

approaches 1, and 1 raised to any power is 1. Let us see if this reasoning is correct by actually calculating the value of the expression for larger and larger values of x. Table 1 summarizes the results.

Table 1

x	$\left(1 + \frac{1}{x}\right)^x$
1	2
10	2.593 74...
100	2.704 81...
1,000	2.716 92...
10,000	2.718 14...
100,000	2.718 26...
1,000,000	2.718 28...

Interestingly, the value of expression (1) is never close to 1 but seems to be approaching a number close to 2.7183. In fact, as x increases without bound, the value of expression (1) approaches an irrational number that we call e. The irrational number e to 12 decimal places is

$$e = 2.718281828459$$

Compare this value of e with the value of e^1 from a calculator.

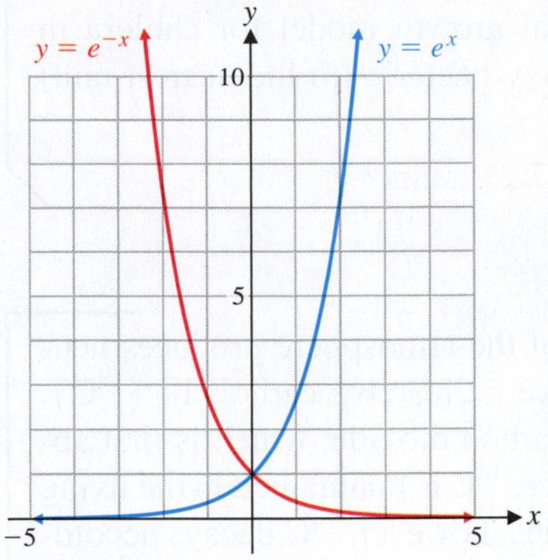

Figure 4

DEFINITION Exponential Functions with Base *e* and Base $1/e$

The exponential functions with base e and base $1/e$, respectively, are defined by

$$y = e^x \quad \text{and} \quad y = e^{-x}$$

Domain: $(-\infty, \infty)$

Range: $(0, \infty)$ (see Fig. 4)

Explore and Discuss 1

Graph the functions $f(x) = e^x$, $g(x) = 2^x$, and $h(x) = 3^x$ on the same set of coordinate axes. At which values of x do the graphs intersect? For positive values of x, which of the three graphs lies above the other two? Below the other two? How does your answer change for negative values of x?

Growth and Decay Applications

Functions of the form $y = ce^{kt}$, where c and k are constants and the independent variable t represents time, are often used to model population growth and radioactive decay. Note that if $t = 0$, then $y = c$. So the constant c represents the initial population (or initial amount). The constant k is called the **relative growth rate** and has the following interpretation: Suppose that $y = ce^{kt}$ models the population of a country, where y is the number of persons and t is time in years. If the relative growth rate is $k = 0.02$, then at any time t, the population is growing at a rate of $0.02y$ persons (that is, 2% of the population) per year.

We say that **population is growing continuously at relative growth rate *k*** to mean that the population y is given by the model $y = ce^{kt}$.

EXAMPLE 2 **Exponential Growth** Cholera, an intestinal disease, is caused by a cholera bacterium that multiplies exponentially. The number of bacteria grows continuously at relative growth rate 1.386, that is,

$$N = N_0 e^{1.386t}$$

where N is the number of bacteria present after t hours and N_0 is the number of bacteria present at the start $(t = 0)$. If we start with 25 bacteria, how many bacteria (to the nearest unit) will be present

(A) In 0.6 hour? (B) In 3.5 hours?

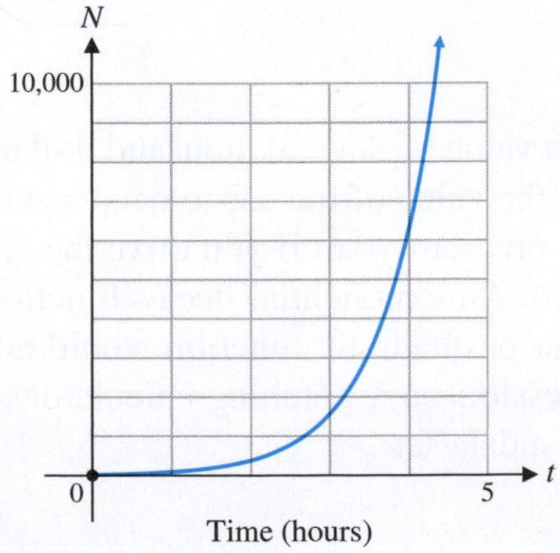

Figure 5

SOLUTION Substituting $N_0 = 25$ into the preceding equation, we obtain

$$N = 25e^{1.386t}$$ The graph is shown in Figure 5.

(A) Solve for N when $t = 0.6$:

$$N = 25e^{1.386(0.6)}$$ Use a calculator.

$$= 57 \text{ bacteria}$$

(B) Solve for N when $t = 3.5$:

$$N = 25e^{1.386(3.5)} \quad \text{Use a calculator.}$$
$$= 3{,}197 \text{ bacteria}$$

Matched Problem 2 Refer to the exponential growth model for cholera in Example 2. If we start with 55 bacteria, how many bacteria (to the nearest unit) will be present

(A) In 0.85 hour? (B) In 7.25 hours?

EXAMPLE 3 **Exponential Decay** Cosmic-ray bombardment of the atmosphere produces neutrons, which in turn react with nitrogen to produce radioactive carbon-14 (^{14}C). Radioactive ^{14}C enters all living tissues through carbon dioxide, which is first absorbed by plants. As long as a plant or animal is alive, ^{14}C is maintained in the living organism at a constant level. Once the organism dies, however, ^{14}C decays according to the equation

$$A = A_0e^{-0.000124t}$$

where A is the amount present after t years and A_0 is the amount present at time $t = 0$.

(A) If 500 milligrams of ^{14}C is present in a sample from a skull at the time of death, how many milligrams will be present in the sample in 15,000 years? Compute the answer to two decimal places.

(B) The **half-life** of ^{14}C is the time t at which the amount present is one-half the amount at time $t = 0$. Use Figure 6 to estimate the half-life of ^{14}C.

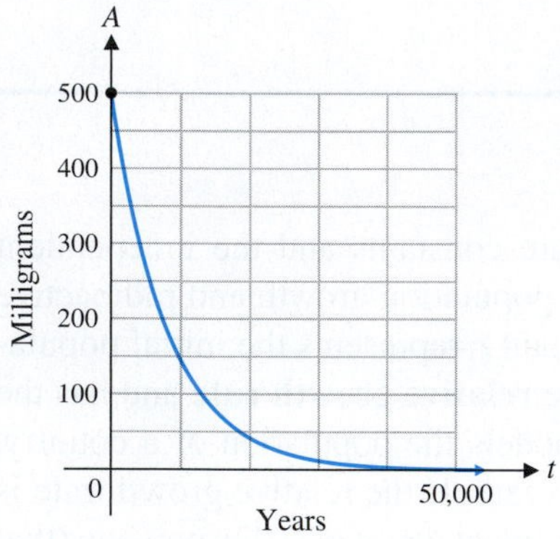

Figure 6

SOLUTION Substituting $A_0 = 500$ in the decay equation, we have

$$A = 500e^{-0.000124t} \quad \text{See the graph in Figure 6.}$$

(A) Solve for A when $t = 15{,}000$:

$$A = 500e^{-0.000124(15{,}000)} \quad \text{Use a calculator.}$$
$$= 77.84 \text{ milligrams}$$

(B) Refer to Figure 6, and estimate the time t at which the amount A has fallen to 250 milligrams: $t \approx 6{,}000$ years. (Finding the intersection of $y_1 = 500e^{-0.000124x}$ and $y_2 = 250$ on a graphing calculator gives a better estimate: $t \approx 5{,}590$ years.)

Matched Problem 3 Refer to the exponential decay model in Example 3. How many milligrams of ^{14}C would have to be present at the beginning in order to have 25 milligrams present after 18,000 years? Compute the answer to the nearest milligram.

If you buy a new car, it is likely to depreciate in value by several thousand dollars during the first year you own it. You would expect the value of the car to decrease in each subsequent year, but not by as much as in the previous year. If you drive the car long enough, its resale value will get close to zero. An exponential decay function will often be a good model of depreciation; a linear or quadratic function would not be suitable (why?). We can use **exponential regression** on a graphing calculator to find the function of the form $y = ab^x$ that best fits a data set.

EXAMPLE 4 **Depreciation** Table 2 gives the market value of a hybrid sedan (in dollars) x years after its purchase. Find an exponential regression model of the form $y = ab^x$ for this data set. Estimate the purchase price of the hybrid. Estimate the value of the hybrid 10 years after its purchase. Round answers to the nearest dollar.

Table 2

x	Value ($)
1	12,575
2	9,455
3	8,115
4	6,845
5	5,225
6	4,485

SOLUTION Enter the data into a graphing calculator (Fig. 7A) and find the exponential regression equation (Fig. 7B). The estimated purchase price is $y_1(0) = \$14,910$. The data set and the regression equation are graphed in Figure 7C. Using trace, we see that the estimated value after 10 years is $1,959.

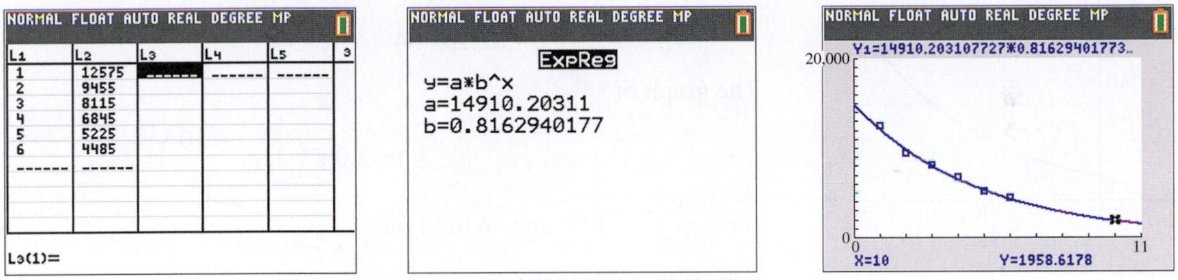

(A) (B) (C)

Figure 7

Matched Problem 4 Table 3 gives the market value of a midsize sedan (in dollars) x years after its purchase. Find an exponential regression model of the form $y = ab^x$ for this data set. Estimate the purchase price of the sedan. Estimate the value of the sedan 10 years after its purchase. Round answers to the nearest dollar.

Table 3

x	Value ($)
1	23,125
2	19,050
3	15,625
4	11,875
5	9,450
6	7,125

Compound Interest

The fee paid to use another's money is called **interest**. It is usually computed as a percent (called **interest rate**) of the principal over a given period of time. If, at the end of a payment period, the interest due is reinvested at the same rate, then the interest earned as well as the principal will earn interest during the next payment period. Interest paid on interest reinvested is called **compound interest** and may be calculated using the following compound interest formula:

If a **principal P (present value)** is invested at an annual **rate r** (expressed as a decimal) compounded m times a year, then the **amount A (future value)** in the account at the end of t years is given by

$$A = P\left(1 + \frac{r}{m}\right)^{mt} \qquad \text{Compound interest formula}$$

For given r and m, the amount A is equal to the principal P multiplied by the exponential function b^t, where $b = (1 + r/m)^m$.

EXAMPLE 5 **Compound Growth** If \$1,000 is invested in an account paying 10% compounded monthly, how much will be in the account at the end of 10 years? Compute the answer to the nearest cent.

SOLUTION We use the compound interest formula as follows:

$$A = P\left(1 + \frac{r}{m}\right)^{mt}$$

$$= 1{,}000\left(1 + \frac{0.10}{12}\right)^{(12)(10)} \qquad \text{Use a calculator.}$$

$$= \$2{,}707.04$$

The graph of

$$A = 1{,}000\left(1 + \frac{0.10}{12}\right)^{12t}$$

for $0 \le t \le 20$ is shown in Figure 8.

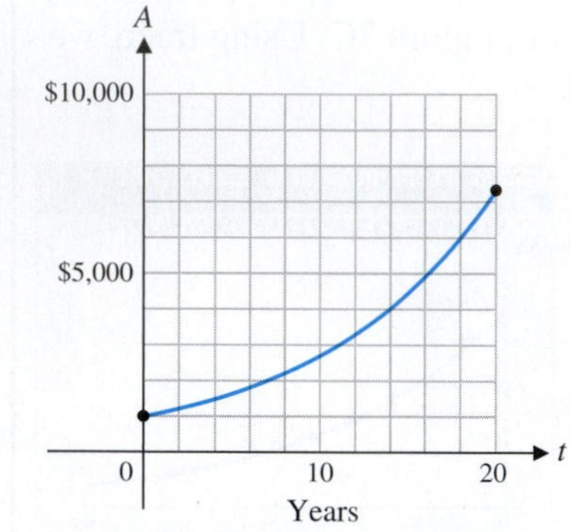

Figure 8

Matched Problem 5 If you deposit \$5,000 in an account paying 9% compounded daily, how much will you have in the account in 5 years? Compute the answer to the nearest cent.

Explore and Discuss 2

Suppose that \$1,000 is deposited in a savings account at an annual rate of 5%. Guess the amount in the account at the end of 1 year if interest is compounded (1) quarterly, (2) monthly, (3) daily, (4) hourly. Use the compound interest formula to compute the amounts at the end of 1 year to the nearest cent. Discuss the accuracy of your initial guesses.

Explore and Discuss 2 suggests that if \$1,000 were deposited in a savings account at an annual interest rate of 5%, then the amount at the end of 1 year would be less than \$1,051.28, even if interest were compounded every minute or every second. The limiting value, approximately \$1,051.271 096, is said to be the amount in the account if interest were compounded continuously.

If a principal, P, is invested at an annual rate, r, and compounded continuously, then the amount in the account at the end of t years is given by

$$A = Pe^{rt} \qquad \text{Continuous compound interest formula}$$

where the constant $e \approx 2.718\,28$ is the base of the exponential function.

EXAMPLE 6 **Continuous Compound Interest** If $1,000 is invested in an account paying 10% compounded continuously, how much will be in the account at the end of 10 years? Compute the answer to the nearest cent.

SOLUTION We use the continuous compound interest formula:

$$A = Pe^{rt} = 1000e^{0.10(10)} = 1000e = \$2{,}718.28$$

Compare with the answer to Example 5.

Matched Problem 6 If you deposit $5,000 in an account paying 9% compounded continuously, how much will you have in the account in 5 years? Compute the answer to the nearest cent.

The formulas for compound interest and continuous compound interest are summarized below for convenient reference.

SUMMARY

Compound Interest: $A = P\left(1 + \frac{r}{m}\right)^{mt}$

Continuous Compound Interest: $A = Pe^{rt}$

where A = amount (future value) at the end of t years
P = principal (present value)
r = annual rate (expressed as a decimal)
m = number of compounding periods per year
t = time in years

Exercises 2.5

A **1.** Match each equation with the graph of f, g, h, or k in the figure.

(A) $y = 2^x$ (B) $y = (0.2)^x$
(C) $y = 4^x$ (D) $y = \left(\frac{1}{3}\right)^x$

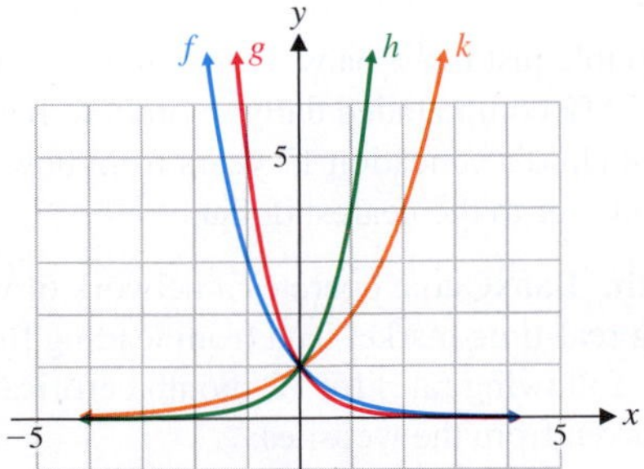

2. Match each equation with the graph of f, g, h, or k in the figure.

(A) $y = \left(\frac{1}{4}\right)^x$ (B) $y = (0.5)^x$
(C) $y = 5^x$ (D) $y = 3^x$

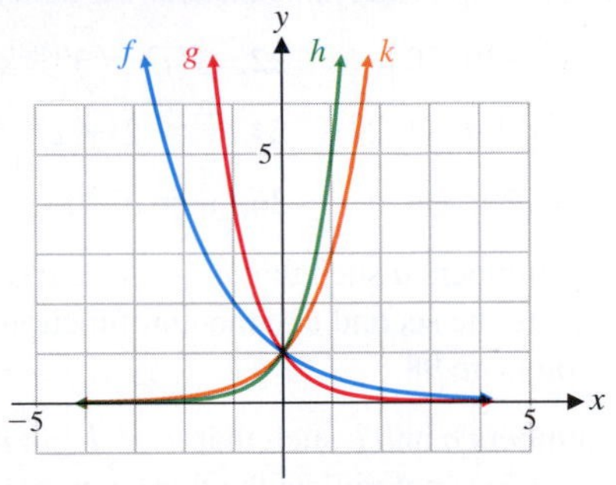

Graph each function in Problems 3–10 over the indicated interval.

3. $y = 5^x$; $[-2, 2]$

4. $y = 3^x$; $[-3, 3]$

5. $y = \left(\frac{1}{5}\right)^x = 5^{-x}$; $[-2, 2]$

6. $y = \left(\frac{1}{3}\right)^x = 3^{-x}$; $[-3, 3]$

7. $f(x) = -5^x$; $[-2, 2]$

8. $g(x) = -3^{-x}$; $[-3, 3]$

9. $y = -e^{-x}$; $[-3, 3]$

10. $y = -e^x$; $[-3, 3]$

B *In Problems 11–18, describe verbally the transformations that can be used to obtain the graph of g from the graph of f (see Section 2.2).*

11. $g(x) = -2^x$; $f(x) = 2^x$

12. $g(x) = 2^{x-2}$; $f(x) = 2^x$

13. $g(x) = 3^{x+1}$; $f(x) = 3^x$

14. $g(x) = -3^x$; $f(x) = 3^x$

15. $g(x) = e^x + 1$; $f(x) = e^x$

16. $g(x) = e^x - 2$; $f(x) = e^x$

17. $g(x) = 2e^{-(x+2)}$; $f(x) = e^{-x}$

18. $g(x) = 0.5e^{-(x-1)}$; $f(x) = e^{-x}$

19. Use the graph of *f* shown in the figure to sketch the graph of each of the following.

(A) $y = f(x) - 1$ (B) $y = f(x + 2)$

(C) $y = 3f(x) - 2$ (D) $y = 2 - f(x - 3)$

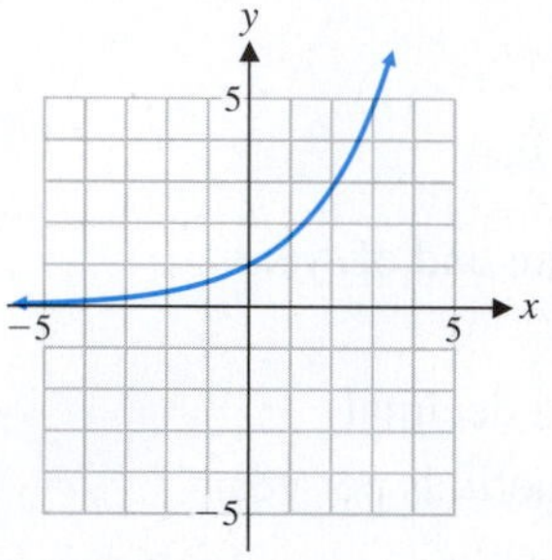

Figure for 19 and 20

20. Use the graph of *f* shown in the figure to sketch the graph of each of the following.

(A) $y = f(x) + 2$ (B) $y = f(x - 3)$

(C) $y = 2f(x) - 4$ (D) $y = 4 - f(x + 2)$

In Problems 21–26, graph each function over the indicated interval.

21. $f(t) = 2^{t/10}$; $[-30, 30]$

22. $G(t) = 3^{t/100}$; $[-200, 200]$

23. $y = -3 + e^{1+x}$; $[-4, 2]$

24. $y = 2 + e^{x-2}$; $[-1, 5]$

25. $y = e^{|x|}$; $[-3, 3]$

26. $y = e^{-|x|}$; $[-3, 3]$

27. Find all real numbers a such that $a^2 = a^{-2}$. Explain why this does not violate the second exponential function property in Theorem 2 on page 98.

28. Find real numbers a and b such that $a \neq b$ but $a^4 = b^4$. Explain why this does not violate the third exponential function property in Theorem 2 on page 98.

In Problems 29–38, solve each equation for x.

29. $2^{2x+5} = 2^{101}$

30. $3^{x+4} = 3^{2x-5}$

31. $7^{x^2} = 7^{3x+10}$

32. $5^{x^2-x} = 5^{42}$

33. $(3x + 9)^5 = 32x^5$

34. $(3x + 4)^3 = 52^3$

35. $(x + 5)^2 = (2x - 14)^2$

36. $(2x + 1)^2 = (3x - 1)^2$

37. $(5x + 18)^4 = (x + 6)^4$

38. $(4x + 1)^4 = (5x - 10)^4$

C *In Problems 39–46, solve each equation for x. (Remember: $e^x \neq 0$ and $e^{-x} \neq 0$ for all values of x).*

39. $xe^{-x} + 7e^{-x} = 0$

40. $10xe^x - 5e^x = 0$

41. $2x^2e^x - 8e^x = 0$

42. $x^2e^{-x} - 9e^{-x} = 0$

43. $e^{4x} - e = 0$

44. $e^{4x} + e = 0$

45. $e^{3x-1} + e = 0$

46. $e^{3x-1} - e = 0$

Graph each function in Problems 47–50 over the indicated interval.

47. $h(x) = x(2^x)$; $[-5, 0]$

48. $m(x) = x(3^{-x})$; $[0, 3]$

49. $N = \dfrac{100}{1 + e^{-t}}$; $[0, 5]$

50. $N = \dfrac{200}{1 + 3e^{-t}}$; $[0, 5]$

Applications

In all problems involving days, a 365-day year is assumed.

51. Continuous compound interest. Find the value of an investment of \$10,000 in 12 years if it earns an annual rate of 3.95% compounded continuously.

52. Continuous compound interest. Find the value of an investment of \$24,000 in 7 years if it earns an annual rate of 4.35% compounded continuously.

53. Compound growth. Suppose that \$2,500 is invested at 7% compounded quarterly. How much money will be in the account in

(A) $\frac{3}{4}$ year? (B) 15 years?

Compute answers to the nearest cent.

54. Compound growth. Suppose that \$4,000 is invested at 6% compounded weekly. How much money will be in the account in

(A) $\frac{1}{2}$ year? (B) 10 years?

Compute answers to the nearest cent.

55. Finance. A person wishes to have \$15,000 cash for a new car 5 years from now. How much should be placed in an account now, if the account pays 6.75% compounded weekly? Compute the answer to the nearest dollar.

56. Finance. A couple just had a baby. How much should they invest now at 5.5% compounded daily in order to have \$40,000 for the child's education 17 years from now? Compute the answer to the nearest dollar.

57. Money growth. BanxQuote operates a network of websites providing real-time market data from leading financial providers. The following rates for 12-month certificates of deposit were taken from the websites:

(A) Stonebridge Bank, 0.95% compounded monthly

(B) DeepGreen Bank, 0.80% compounded daily

(C) Provident Bank, 0.85% compounded quarterly

Compute the value of \$10,000 invested in each account at the end of 1 year.

58. Money growth. Refer to Problem 57. The following rates for 60-month certificates of deposit were also taken from BanxQuote websites:

(A) Oriental Bank & Trust, 1.35% compounded quarterly

(B) BMW Bank of North America, 1.30% compounded monthly

(C) BankFirst Corporation, 1.25% compounded daily

Compute the value of $10,000 invested in each account at the end of 5 years.

59. Advertising. A company is trying to introduce a new product to as many people as possible through television advertising in a large metropolitan area with 2 million possible viewers. A model for the number of people N (in millions) who are aware of the product after t days of advertising was found to be

$$N = 2\left(1 - e^{-0.037t}\right)$$

Graph this function for $0 \le t \le 50$. What value does N approach as t increases without bound?

60. Learning curve. People assigned to assemble circuit boards for a computer manufacturing company undergo on-the-job training. From past experience, the learning curve for the average employee is given by

$$N = 40(1 - e^{-0.12t})$$

where N is the number of boards assembled per day after t days of training. Graph this function for $0 \le t \le 30$. What is the maximum number of boards an average employee can be expected to produce in 1 day?

61. Internet users. Table 4 shows the number of individuals worldwide who could access the internet from home for selected years since 2000.

(A) Let x represent the number of years since 2000 and find an exponential regression model ($y = ab^x$) for the number of internet users.

(B) Use the model to estimate the number of internet users in 2024.

Table 4 Internet Users (billions)

Year	Users
2000	0.41
2004	0.91
2008	1.58
2012	2.02
2016	3.42

Source: Internet Stats Live

62. Mobile data traffic. Table 5 shows estimates of mobile data traffic, in exabytes (10^{18} bytes) per month, for years from 2015 to 2020.

(A) Let x represent the number of years since 2015 and find an exponential regression model ($y = ab^x$) for mobile data traffic.

(B) Use the model to estimate the mobile data traffic in 2025.

Table 5 Mobile Data Traffic (exabytes per month)

Year	Traffic
2015	3.7
2016	6.2
2017	9.9
2018	14.9
2019	21.7
2020	30.6

Source: Cisco Systems Inc.

63. Marine biology. Marine life depends on the microscopic plant life that exists in the photic zone, a zone that goes to a depth where only 1% of surface light remains. In some waters with a great deal of sediment, the photic zone may go down only 15 to 20 feet. In some murky harbors, the intensity of light d feet below the surface is given approximately by

$$I = I_0 e^{-0.23d}$$

where I_0 is the intensity of light at the surface. What percentage of the surface light will reach a depth of

(A) 10 feet? (B) 20 feet?

64. Marine biology. Refer to Problem 63. Light intensity I relative to depth d (in feet) for one of the clearest bodies of water in the world, the Sargasso Sea, can be approximated by

$$I = I_0 e^{-0.00942d}$$

where I_0 is the intensity of light at the surface. What percentage of the surface light will reach a depth of

(A) 50 feet? (B) 100 feet?

65. Population growth. In 2015, the estimated population of South Sudan was 12 million with a relative growth rate of 4.02%.

(A) Write an equation that models the population growth in South Sudan, letting 2015 be year 0.

(B) Based on the model, what is the expected population of South Sudan in 2025?

66. Population growth. In 2015, the estimated population of Brazil was 204 million with a relative growth rate of 0.77%.

(A) Write an equation that models the population growth in Brazil, letting 2015 be year 0.

(B) Based on the model, what is the expected population of Brazil in 2030?

67. Population growth. In 2015, the estimated population of Japan was 127 million with a relative growth rate of -0.16%.

(A) Write an equation that models the population growth in Japan, letting 2015 be year 0.

(B) Based on the model, what is the expected population in Japan in 2030?

68. World population growth. From the dawn of humanity to 1830, world population grew to one billion people. In 100 more years (by 1930) it grew to two billion, and 3 billion more were added in only 60 years (by 1990). In 2016, the estimated world population was 7.4 billion with a relative growth rate of 1.13%.

(A) Write an equation that models the world population growth, letting 2016 be year 0.

(B) Based on the model, what is the expected world population (to the nearest hundred million) in 2025? In 2033?

Answers to Matched Problems

1.

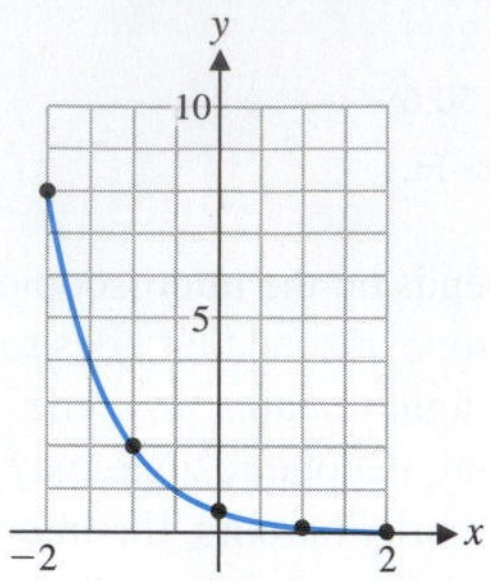

2. (A) 179 bacteria (B) 1,271,659 bacteria

3. 233 mg

4. Purchase price: \$30,363; value after 10 yr: \$2,864

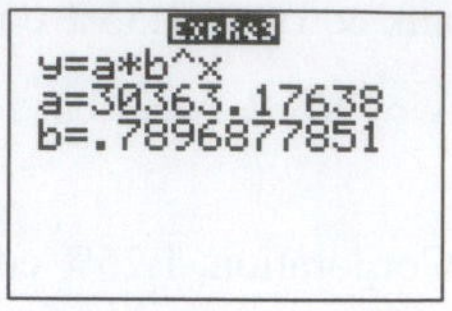

5. \$7,841.13

6. \$7,841.56

2.6 Logarithmic Functions

- Inverse Functions
- Logarithmic Functions
- Properties of Logarithmic Functions
- Calculator Evaluation of Logarithms
- Applications

Find the exponential function keys 10^x and e^x on your calculator. Close to these keys you will find the LOG and LN keys. The latter two keys represent *logarithmic functions,* and each is closely related to its nearby exponential function. In fact, the exponential function and the corresponding logarithmic function are said to be *inverses* of each other. In this section we will develop the concept of inverse functions and use it to define a logarithmic function as the inverse of an exponential function. We will then investigate basic properties of logarithmic functions, use a calculator to evaluate them for particular values of x, and apply them to real-world problems.

Logarithmic functions are used in modeling and solving many types of problems. For example, the decibel scale is a logarithmic scale used to measure sound intensity, and the Richter scale is a logarithmic scale used to measure the force of an earthquake. An important business application has to do with finding the time it takes money to double if it is invested at a certain rate compounded a given number of times a year or compounded continuously. This requires the solution of an exponential equation, and logarithms play a central role in the process.

Inverse Functions

Look at the graphs of $f(x) = \dfrac{x}{2}$ and $g(x) = \dfrac{|x|}{2}$ in Figure 1:

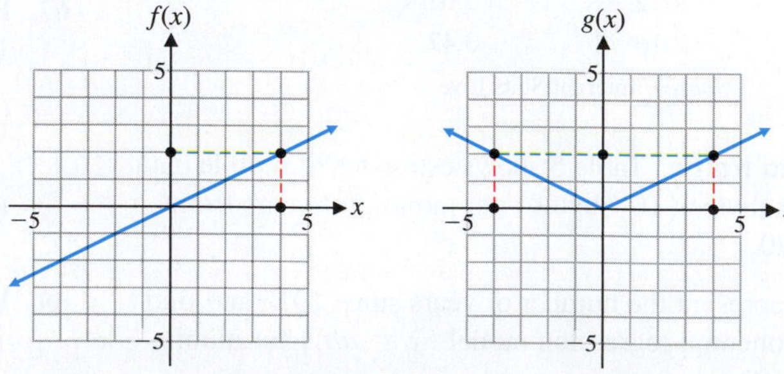

(A) $f(x) = \dfrac{x}{2}$ (B) $g(x) = \dfrac{|x|}{2}$

Figure 1

Because both f and g are functions, each domain value corresponds to exactly one range value. For which function does each range value correspond to exactly one domain value? This is the case only for function f. Note that for function f, the range value 2 corresponds to the domain value 4. For function g the range value 2 corresponds to both -4 and 4. Function f is said to be *one-to-one*.

DEFINITION One-to-One Functions

A function f is said to be **one-to-one** if each range value corresponds to exactly one domain value.

Reminder

We say that the function f is **increasing** on an interval (a, b) if $f(x_2) > f(x_1)$ whenever $a < x_1 < x_2 < b$ and f is **decreasing** on (a, b) if $f(x_2) < f(x_1)$ whenever $a < x_1 < x_2 < b$.

It can be shown that any continuous function that is either increasing or decreasing for all domain values is one-to-one. If a continuous function increases for some domain values and decreases for others, then it cannot be one-to-one. Figure 1 shows an example of each case.

Explore and Discuss 1

Graph $f(x) = 2^x$ and $g(x) = x^2$. For a range value of 4, what are the corresponding domain values for each function? Which of the two functions is one-to-one? Explain why.

Starting with a one-to-one function f, we can obtain a new function called the *inverse* of f.

DEFINITION Inverse of a Function

If f is a one-to-one function, then the **inverse** of f is the function formed by interchanging the independent and dependent variables for f. Thus, if (a, b) is a point on the graph of f, then (b, a) is a point on the graph of the inverse of f.

Note: If f is not one-to-one, then f **does not have an inverse**.

In this course, we are interested in the inverses of exponential functions, called *logarithmic functions*.

Logarithmic Functions

If we start with the exponential function f defined by

$$y = 2^x \tag{1}$$

and interchange the variables, we obtain the inverse of f:

$$x = 2^y \tag{2}$$

We call the inverse the **logarithmic function with base 2**, and write

$$y = \log_2 x \quad \text{if and only if} \quad x = 2^y$$

We can graph $y = \log_2 x$ by graphing $x = 2^y$ since they are equivalent. Any ordered pair of numbers on the graph of the exponential function will be on the graph of the logarithmic function if we interchange the order of the components. For example, $(3, 8)$ satisfies equation (1) and $(8, 3)$ satisfies equation (2). The graphs of $y = 2^x$ and $y = \log_2 x$ are shown in Figure 2. Note that if we fold the paper along the dashed line $y = x$ in Figure 2, the two graphs match exactly. The line $y = x$ is a line of symmetry for the two graphs.

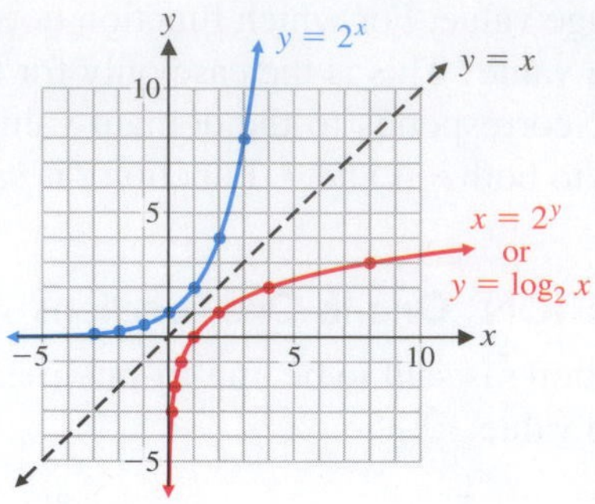

Figure 2

Exponential Function		Logarithmic Function	
x	$y = 2^x$	$x = 2^y$	y
-3	$\frac{1}{8}$	$\frac{1}{8}$	-3
-2	$\frac{1}{4}$	$\frac{1}{4}$	-2
-1	$\frac{1}{2}$	$\frac{1}{2}$	-1
0	1	1	0
1	2	2	1
2	4	4	2
3	8	8	3

Ordered pairs reversed

In general, since the graphs of all exponential functions of the form $f(x) = b^x$, $b \neq 1$, $b > 0$, are either increasing or decreasing, exponential functions have inverses.

DEFINITION Logarithmic Functions

The inverse of an exponential function is called a **logarithmic function**. For $b > 0$ and $b \neq 1$,

Logarithmic form		Exponential form
$y = \log_b x$	is equivalent to	$x = b^y$

The **log to the base *b* of *x*** is the exponent to which b must be raised to obtain x. [*Remember:* A logarithm is an exponent.] The **domain** of the logarithmic function is the set of all positive real numbers, which is also the range of the corresponding exponential function, and the **range** of the logarithmic function is the set of all real numbers, which is also the domain of the corresponding exponential function. Typical graphs of an exponential function and its inverse, a logarithmic function, are shown in Figure 3.

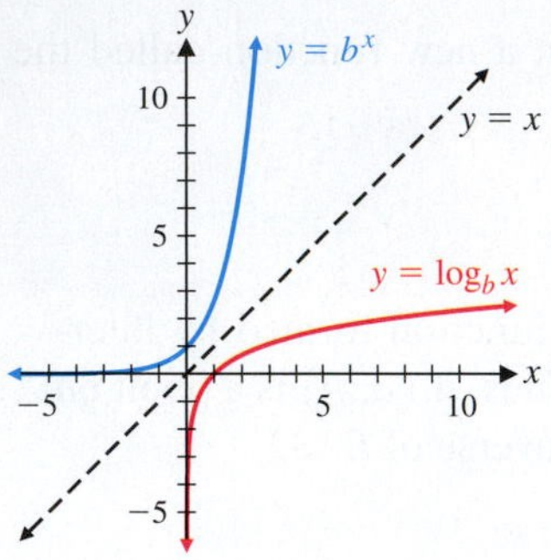

Figure 3

CONCEPTUAL INSIGHT

Because the domain of a logarithmic function consists of the positive real numbers, the entire graph of a logarithmic function lies to the right of the y axis. In contrast, the graphs of polynomial and exponential functions intersect every vertical line, and the graphs of rational functions intersect all but a finite number of vertical lines.

The following examples involve converting logarithmic forms to equivalent exponential forms, and vice versa.

EXAMPLE 1 **Logarithmic–Exponential Conversions** Change each logarithmic form to an equivalent exponential form:

(A) $\log_5 25 = 2$ (B) $\log_9 3 = \frac{1}{2}$ (C) $\log_2\left(\frac{1}{4}\right) = -2$

SOLUTION

(A) $\log_5 25 = 2$ is equivalent to $25 = 5^2$

(B) $\log_9 3 = \frac{1}{2}$ is equivalent to $3 = 9^{1/2}$

(C) $\log_2\left(\frac{1}{4}\right) = -2$ is equivalent to $\frac{1}{4} = 2^{-2}$

Matched Problem 1 Change each logarithmic form to an equivalent exponential form:

(A) $\log_3 9 = 2$ (B) $\log_4 2 = \frac{1}{2}$ (C) $\log_3\left(\frac{1}{9}\right) = -2$

EXAMPLE 2 **Exponential–Logarithmic Conversions** Change each exponential form to an equivalent logarithmic form:

(A) $64 = 4^3$ (B) $6 = \sqrt{36}$ (C) $\frac{1}{8} = 2^{-3}$

SOLUTION

(A) $64 = 4^3$ is equivalent to $\log_4 64 = 3$

(B) $6 = \sqrt{36}$ is equivalent to $\log_{36} 6 = \frac{1}{2}$

(C) $\frac{1}{8} = 2^{-3}$ is equivalent to $\log_2\left(\frac{1}{8}\right) = -3$

Matched Problem 2 Change each exponential form to an equivalent logarithmic form:

(A) $49 = 7^2$ (B) $3 = \sqrt{9}$ (C) $\frac{1}{3} = 3^{-1}$

To gain a deeper understanding of logarithmic functions and their relationship to exponential functions, we consider a few problems where we want to find x, b, or y in $y = \log_b x$, given the other two values. All values are chosen so that the problems can be solved exactly without a calculator.

EXAMPLE 3 **Solutions of the Equation $y = \log_b x$** Find y, b, or x, as indicated.

(A) Find y: $y = \log_4 16$ (B) Find x: $\log_2 x = -3$

(C) Find b: $\log_b 100 = 2$

SOLUTION

(A) $y = \log_4 16$ is equivalent to $16 = 4^y$. So,

$$y = 2$$

(B) $\log_2 x = -3$ is equivalent to $x = 2^{-3}$. So,

$$x = \frac{1}{2^3} = \frac{1}{8}$$

(C) $\log_b 100 = 2$ is equivalent to $100 = b^2$. So,

$$b = 10$$ Recall that b cannot be negative.

Matched Problem 3 Find y, b, or x, as indicated.

(A) Find y: $y = \log_9 27$ (B) Find x: $\log_3 x = -1$

(C) Find b: $\log_b 1{,}000 = 3$

Properties of Logarithmic Functions

The properties of exponential functions (Section 2.5) lead to properties of logarithmic functions. For example, consider the exponential property $b^x b^y = b^{x+y}$. Let $M = b^x$, $N = b^y$. Then

$$\log_b MN = \log_b(b^x b^y) = \log_b b^{x+y} = x + y = \log_b M + \log_b N$$

So $\log_b MN = \log_b M + \log_b N$, that is, the logarithm of a product is the sum of the logarithms. Similarly, the logarithm of a quotient is the difference of the logarithms. These properties are among the eight useful properties of logarithms that are listed in Theorem 1.

THEOREM 1 Properties of Logarithmic Functions

If b, M, and N are positive real numbers, $b \neq 1$, and p and x are real numbers, then

1. $\log_b 1 = 0$
2. $\log_b b = 1$
3. $\log_b b^x = x$
4. $b^{\log_b x} = x, \quad x > 0$
5. $\log_b MN = \log_b M + \log_b N$
6. $\log_b \frac{M}{N} = \log_b M - \log_b N$
7. $\log_b M^p = p \log_b M$
8. $\log_b M = \log_b N$ if and only if $M = N$

EXAMPLE 4 **Using Logarithmic Properties** Use logarithmic properties to write in simpler form:

(A) $\log_b \frac{wx}{yz}$ (B) $\log_b (wx)^{3/5}$ (C) $e^{x \log_e b}$ (D) $\frac{\log_e x}{\log_e b}$

SOLUTION

(A)
$$\begin{aligned}\log_b \frac{wx}{yz} &= \log_b wx - \log_b yz \\ &= \log_b w + \log_b x - (\log_b y + \log_b z) \\ &= \log_b w + \log_b x - \log_b y - \log_b z\end{aligned}$$

(B) $\log_b (wx)^{3/5} = \frac{3}{5}\log_b wx = \frac{3}{5}(\log_b w + \log_b x)$

(C) $e^{x \log_e b} = e^{\log_e b^x} = b^x$

(D) $\frac{\log_e x}{\log_e b} = \frac{\log_e (b^{\log_b x})}{\log_e b} = \frac{(\log_b x)(\log_e b)}{\log_e b} = \log_b x$

Matched Problem 4 Write in simpler form, as in Example 4.

(A) $\log_b \frac{R}{ST}$ (B) $\log_b \left(\frac{R}{S}\right)^{2/3}$ (C) $2^{u \log_2 b}$ (D) $\frac{\log_2 x}{\log_2 b}$

The following examples and problems will give you additional practice in using basic logarithmic properties.

EXAMPLE 5 **Solving Logarithmic Equations** Find x so that

$$\frac{3}{2}\log_b 4 - \frac{2}{3}\log_b 8 + \log_b 2 = \log_b x$$

SOLUTION

$\frac{3}{2}\log_b 4 - \frac{2}{3}\log_b 8 + \log_b 2 = \log_b x$	Use property 7.
$\log_b 4^{3/2} - \log_b 8^{2/3} + \log_b 2 = \log_b x$	Simplify.
$\log_b 8 - \log_b 4 + \log_b 2 = \log_b x$	Use properties 5 and 6.
$\log_b \frac{8 \cdot 2}{4} = \log_b x$	Simplify.
$\log_b 4 = \log_b x$	Use property 8.
$x = 4$	

Matched Problem 5 Find x so that $3 \log_b 2 + \frac{1}{2}\log_b 25 - \log_b 20 = \log_b x$.

EXAMPLE 6 **Solving Logarithmic Equations** Solve: $\log_{10} x + \log_{10}(x + 1) = \log_{10} 6$.

SOLUTION

$$\log_{10} x + \log_{10}(x + 1) = \log_{10} 6 \qquad \text{Use property 5.}$$
$$\log_{10}[x(x + 1)] = \log_{10} 6 \qquad \text{Use property 8.}$$
$$x(x + 1) = 6 \qquad \text{Expand.}$$
$$x^2 + x - 6 = 0 \qquad \text{Solve by factoring.}$$
$$(x + 3)(x - 2) = 0$$
$$x = -3, 2$$

We must exclude $x = -3$, since the domain of the function $\log_{10} x$ is $(0, \infty)$; so $x = 2$ is the only solution.

Matched Problem 6 Solve: $\log_3 x + \log_3(x - 3) = \log_3 10$.

Calculator Evaluation of Logarithms

Of all possible logarithmic bases, e and 10 are used almost exclusively. Before we can use logarithms in certain practical problems, we need to be able to approximate the logarithm of any positive number either to base 10 or to base e. And conversely, if we are given the logarithm of a number to base 10 or base e, we need to be able to approximate the number. Historically, tables were used for this purpose, but now calculators make computations faster and far more accurate.

Common logarithms are logarithms with base 10. **Natural logarithms** are logarithms with base e. Most calculators have a key labeled "log" (or "LOG") and a key labeled "ln" (or "LN"). The former represents a common (base 10) logarithm and the latter a natural (base e) logarithm. In fact, "log" and "ln" are both used extensively in mathematical literature, and whenever you see either used in this book without a base indicated, they will be interpreted as follows:

Common logarithm: $\log x$ **means** $\log_{10} x$
Natural logarithm: $\ln x$ **means** $\log_e x$

Finding the common or natural logarithm using a calculator is very easy. On some calculators, you simply enter a number from the domain of the function and press LOG or LN. On other calculators, you press either LOG or LN, enter a number from the domain, and then press ENTER. Check the user's manual for your calculator.

EXAMPLE 7 **Calculator Evaluation of Logarithms** Use a calculator to evaluate each to six decimal places:

(A) log 3,184 (B) ln 0.000 349 (C) log (−3.24)

SOLUTION

(A) $\log 3{,}184 = 3.502\,973$
(B) $\ln 0.000\,349 = -7.960\,439$
(C) $\log(-3.24) = \text{Error}$ $\quad$ −3.24 is not in the domain of the log function.

Matched Problem 7 Use a calculator to evaluate each to six decimal places:

(A) log 0.013 529 (B) ln 28.693 28 (C) ln (−0.438)

Given the logarithm of a number, how do you find the number? We make direct use of the logarithmic-exponential relationships, which follow from the definition of logarithmic function given at the beginning of this section.

$$\log x = y \quad \textbf{is equivalent to} \quad x = 10^y$$

$$\ln x = y \quad \textbf{is equivalent to} \quad x = e^y$$

EXAMPLE 8 **Solving $\log_b x = y$ for x** Find x to four decimal places, given the indicated logarithm:

(A) $\log x = -2.315$ (B) $\ln x = 2.386$

SOLUTION

(A) $\log x = -2.315$ Change to equivalent exponential form.

$x = 10^{-2.315}$ Evaluate with a calculator.

$= 0.0048$

(B) $\ln x = 2.386$ Change to equivalent exponential form.

$x = e^{2.386}$ Evaluate with a calculator.

$= 10.8699$

Matched Problem 8 Find x to four decimal places, given the indicated logarithm:

(A) $\ln x = -5.062$ (B) $\log x = 2.0821$

We can use logarithms to solve exponential equations.

EXAMPLE 9 **Solving Exponential Equations** Solve for x to four decimal places:

(A) $10^x = 2$ (B) $e^x = 3$ (C) $3^x = 4$

SOLUTION

(A) $10^x = 2$ Take common logarithms of both sides.

$\log 10^x = \log 2$ Use property 3.

$x = \log 2$ Use a calculator.

$= 0.3010$ To four decimal places

(B) $e^x = 3$ Take natural logarithms of both sides.

$\ln e^x = \ln 3$ Use property 3.

$x = \ln 3$ Use a calculator.

$= 1.0986$ To four decimal places

(C) $3^x = 4$ Take either natural or common logarithms of both sides. (We choose common logarithms.)

$\log 3^x = \log 4$ Use property 7.

$x \log 3 = \log 4$ Solve for x.

$x = \dfrac{\log 4}{\log 3}$ Use a calculator.

$= 1.2619$ To four decimal places

Matched Problem 9 Solve for x to four decimal places:

(A) $10^x = 7$ (B) $e^x = 6$ (C) $4^x = 5$

Exponential equations can also be solved graphically by graphing both sides of an equation and finding the points of intersection. Figure 4 illustrates this approach for the equations in Example 9.

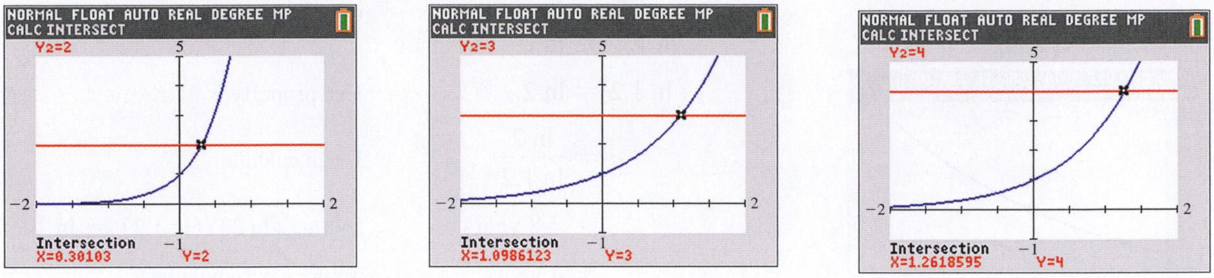

(A) $y_1 = 10^x$
$y_2 = 2$

(B) $y_1 = e^x$
$y_2 = 3$

(C) $y_1 = 3^x$
$y_2 = 4$

Figure 4 Graphical solution of exponential equations

Explore and Discuss 2

Discuss how you could find $y = \log_5 38.25$ using either natural or common logarithms on a calculator. [*Hint:* Start by rewriting the equation in exponential form.]

Remark In the usual notation for natural logarithms, the simplifications of Example 4, parts (C) and (D) on page 110, become

$$e^{x \ln b} = b^x \qquad \text{and} \qquad \frac{\ln x}{\ln b} = \log_b x$$

With these formulas, we can change an exponential function with base b, or a logarithmic function with base b, to expressions involving exponential or logarithmic functions, respectively, to the base e. Such **change-of-base formulas** are useful in calculus.

Applications

A convenient and easily understood way of comparing different investments is to use their **doubling times**—the length of time it takes the value of an investment to double. Logarithm properties, as you will see in Example 10, provide us with just the right tool for solving some doubling-time problems.

EXAMPLE 10 **Doubling Time for an Investment** How long (to the next whole year) will it take money to double if it is invested at 20% compounded annually?

SOLUTION We use the compound interest formula discussed in Section 2.5:

$$A = P\left(1 + \frac{r}{m}\right)^{mt} \qquad \text{Compound interest}$$

The problem is to find t, given $r = 0.20$, $m = 1$, and $A = 2P$; that is,

$$2P = P(1 + 0.2)^t$$

$$2 = 1.2^t$$

Solve for t by taking the natural or common logarithm of both sides (we choose the natural logarithm).

$$1.2^t = 2$$

$$\ln 1.2^t = \ln 2$$

$$t \ln 1.2 = \ln 2 \quad \text{Use property 7.}$$

$$t = \frac{\ln 2}{\ln 1.2} \quad \text{Use a calculator.}$$

$$\approx 3.8 \text{ years} \quad [\textit{Note: } (\ln 2)/(\ln 1.2) \neq \ln 2 - \ln 1.2]$$

$$\approx 4 \text{ years} \quad \text{To the next whole year}$$

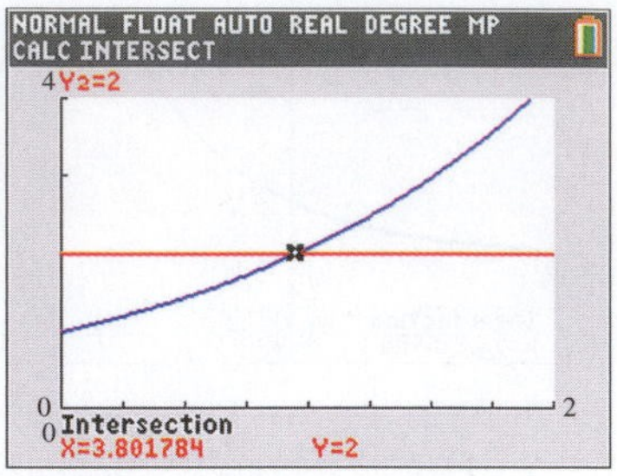

Figure 5 $y_1 = 1.2^x$, $y_2 = 2$

When interest is paid at the end of 3 years, the money will not be doubled; when paid at the end of 4 years, the money will be slightly more than doubled.

Example 10 can also be solved graphically by graphing both sides of the equation $2 = 1.2^t$, and finding the intersection point (Fig. 5).

Matched Problem 10 How long (to the next whole year) will it take money to triple if it is invested at 13% compounded annually?

It is interesting and instructive to graph the doubling times for various rates compounded annually. We proceed as follows:

$$A = P(1 + r)^t$$

$$2P = P(1 + r)^t$$

$$2 = (1 + r)^t$$

$$(1 + r)^t = 2$$

$$\ln (1 + r)^t = \ln 2$$

$$t \ln (1 + r) = \ln 2$$

$$t = \frac{\ln 2}{\ln(1 + r)}$$

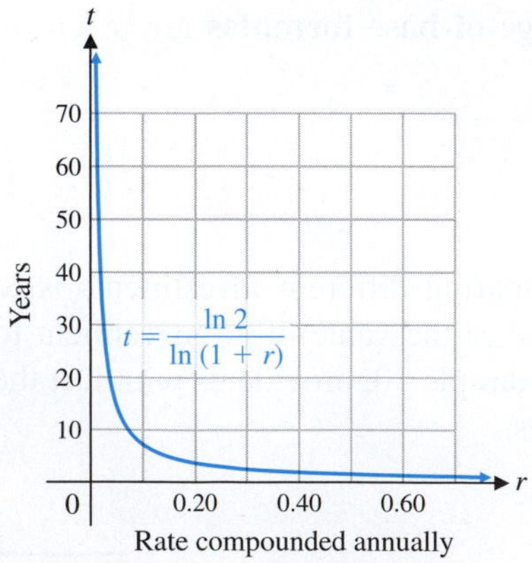

Figure 6

Figure 6 shows the graph of this equation (doubling time in years) for interest rates compounded annually from 1 to 70% (expressed as decimals). Note the dramatic change in doubling time as rates change from 1 to 20% (from 0.01 to 0.20).

Among increasing functions, the logarithmic functions (with bases $b > 1$) increase much more slowly for large values of x than either exponential or polynomial functions. When a visual inspection of the plot of a data set indicates a slowly increasing function, a logarithmic function often provides a good model. We use **logarithmic regression** on a graphing calculator to find the function of the form $y = a + b \ln x$ that best fits the data.

EXAMPLE 11 **Home Ownership Rates** The U.S. Census Bureau published the data in Table 1 on home ownership rates. Let x represent time in years with $x = 0$ representing 1900. Use logarithmic regression to find the best model of the form $y = a + b \ln x$ for the home ownership rate y as a function of time x. Use the model to predict the home ownership rate in the United States in 2025 (to the nearest tenth of a percent).

Table 1 Home Ownership Rates

Year	Rate (%)
1950	55.0
1960	61.9
1970	62.9
1980	64.4
1990	64.2
2000	67.4
2010	66.9

SOLUTION Enter the data in a graphing calculator (Fig. 7A) and find the logarithmic regression equation (Fig. 7B). The data set and the regression equation are graphed in Figure 7C. Using trace, we predict that the home ownership rate in 2025 would be 69.8%.

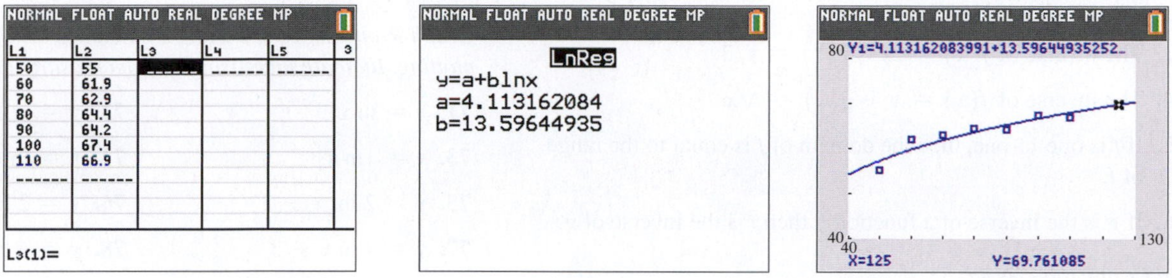

(A) (B) (C)

Figure 7

Matched Problem 11 Refer to Example 11. Use the model to predict the home ownership rate in the United States in 2030 (to the nearest tenth of a percent).

⚠ CAUTION Note that in Example 11 we let $x = 0$ represent 1900. If we let $x = 0$ represent 1940, for example, we would obtain a different logarithmic regression equation. We would *not* let $x = 0$ represent 1950 (the first year in Table 1) or any later year, because logarithmic functions are undefined at 0. ▲

Exercises 2.6

A *For Problems 1–6, rewrite in equivalent exponential form.*

1. $\log_3 27 = 3$

2. $\log_2 32 = 5$

3. $\log_{10} 1 = 0$

4. $\log_e 1 = 0$

5. $\log_4 8 = \frac{3}{2}$

6. $\log_9 27 = \frac{3}{2}$

For Problems 7–12, rewrite in equivalent logarithmic form.

7. $49 = 7^2$

8. $36 = 6^2$

9. $8 = 4^{3/2}$

10. $9 = 27^{2/3}$

11. $A = b^u$

12. $M = b^x$

In Problems 13–22, evaluate the expression without using a calculator.

13. $\log_{10} 1{,}000{,}000$

14. $\log_{10} \frac{1}{1{,}000}$

15. $\log_{10} \frac{1}{100{,}000}$

16. $\log_{10} 10{,}000$

17. $\log_2 128$

18. $\log_2 \frac{1}{64}$

19. $\ln e^{-3}$

20. $e^{\ln(-1)}$

21. $e^{\ln(-3)}$

22. $\ln e^{-1}$

For Problems 23–28, write in simpler form, as in Example 4.

23. $\log_b \frac{P}{Q}$

24. $\log_b FG$

25. $\log_b L^5$

26. $\log_b w^{15}$

27. $3^{p \log_3 q}$

28. $\frac{\log_3 P}{\log_3 R}$

B *For Problems 29–38, find x, y, or b without using a calculator.*

29. $\log_{10} x = -1$

30. $\log_{10} x = 1$

31. $\log_b 64 = 3$

32. $\log_b \frac{1}{25} = 2$

33. $\log_2 \frac{1}{8} = y$

34. $\log_{49} 7 = y$

35. $\log_b 81 = -4$

36. $\log_b 10{,}000 = 2$

37. $\log_4 x = \frac{3}{2}$

38. $\log_8 x = \frac{5}{3}$

In Problems 39–46, discuss the validity of each statement. If the statement is always true, explain why. If not, give a counterexample.

39. Every polynomial function is one-to-one.

40. Every polynomial function of odd degree is one-to-one.

41. If g is the inverse of a function f, then g is one-to-one.

42. The graph of a one-to-one function intersects each vertical line exactly once.

43. The inverse of $f(x) = 2x$ is $g(x) = x/2$.

44. The inverse of $f(x) = x^2$ is $g(x) = \sqrt{x}$.

45. If f is one-to-one, then the domain of f is equal to the range of f.

46. If g is the inverse of a function f, then f is the inverse of g.

C *Find x in Problems 47–54.*

47. $\log_b x = \frac{2}{3}\log_b 8 + \frac{1}{2}\log_b 9 - \log_b 6$

48. $\log_b x = \frac{2}{3}\log_b 27 + 2\log_b 2 - \log_b 3$

49. $\log_b x = \frac{3}{2}\log_b 4 - \frac{2}{3}\log_b 8 + 2\log_b 2$

50. $\log_b x = 3\log_b 2 + \frac{1}{2}\log_b 25 - \log_b 20$

51. $\log_b x + \log_b (x - 4) = \log_b 21$

52. $\log_b(x + 2) + \log_b x = \log_b 24$

53. $\log_{10}(x - 1) - \log_{10}(x + 1) = 1$

54. $\log_{10}(x + 6) - \log_{10}(x - 3) = 1$

Graph Problems 55 and 56 by converting to exponential form first.

55. $y = \log_2 (x - 2)$

56. $y = \log_3 (x + 2)$

57. Explain how the graph of the equation in Problem 55 can be obtained from the graph of $y = \log_2 x$ using a simple transformation (see Section 2.2).

58. Explain how the graph of the equation in Problem 56 can be obtained from the graph of $y = \log_3 x$ using a simple transformation (see Section 2.2).

59. What are the domain and range of the function defined by $y = 1 + \ln(x + 1)$?

60. What are the domain and range of the function defined by $y = \log(x - 1) - 1$?

For Problems 61 and 62, evaluate to five decimal places using a calculator.

61. (A) $\log 3{,}527.2$ (B) $\log 0.006\,913\,2$
(C) $\ln 277.63$ (D) $\ln 0.040\,883$

62. (A) $\log 72.604$ (B) $\log 0.033\,041$
(C) $\ln 40{,}257$ (D) $\ln 0.005\,926\,3$

For Problems 63 and 64, find x to four decimal places.

63. (A) $\log x = 1.1285$ (B) $\log x = -2.0497$
(C) $\ln x = 2.7763$ (D) $\ln x = -1.8879$

64. (A) $\log x = 2.0832$ (B) $\log x = -1.1577$
(C) $\ln x = 3.1336$ (D) $\ln x = -4.3281$

For Problems 65–70, solve each equation to four decimal places.

65. $10^x = 12$

66. $10^x = 153$

67. $e^x = 4.304$

68. $e^x = 0.3059$

69. $1.005^{12t} = 3$

70. $1.02^{4t} = 2$

Graph Problems 71–78 using a calculator and point-by-point plotting. Indicate increasing and decreasing intervals.

71. $y = \ln x$

72. $y = -\ln x$

73. $y = |\ln x|$

74. $y = \ln|x|$

75. $y = 2\ln(x + 2)$

76. $y = 2\ln x + 2$

77. $y = 4\ln x - 3$

78. $y = 4\ln(x - 3)$

79. Explain why the logarithm of 1 for any permissible base is 0.

80. Explain why 1 is not a suitable logarithmic base.

81. Let $p(x) = \ln x$, $q(x) = \sqrt{x}$, and $r(x) = x$. Use a graphing calculator to draw graphs of all three functions in the same viewing window for $1 \le x \le 16$. Discuss what it means for one function to be larger than another on an interval, and then order the three functions from largest to smallest for $1 < x \le 16$.

82. Let $p(x) = \log x$, $q(x) = \sqrt[3]{x}$, and $r(x) = x$. Use a graphing calculator to draw graphs of all three functions in the same viewing window for $1 \le x \le 16$. Discuss what it means for one function to be smaller than another on an interval, and then order the three functions from smallest to largest for $1 < x \le 16$.

Applications

83. Doubling time. In its first 10 years the Gabelli Growth Fund produced an average annual return of 21.36%. Assume that money invested in this fund continues to earn 21.36% compounded annually. How long (to the nearest year) will it take money invested in this fund to double?

84. Doubling time. In its first 10 years the Janus Flexible Income Fund produced an average annual return of 9.58%. Assume that money invested in this fund continues to earn 9.58% compounded annually. How long (to the nearest year) will it take money invested in this fund to double?

85. Investing. How many years (to two decimal places) will it take \$1,000 to grow to \$1,800 if it is invested at 6% compounded quarterly? Compounded daily?

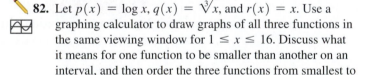

86. Investing. How many years (to two decimal places) will it take \$5,000 to grow to \$7,500 if it is invested at 8% compounded semiannually? Compounded monthly?

87. Continuous compound interest. How many years (to two decimal places) will it take an investment of \$35,000 to grow to \$50,000 if it is invested at 4.75% compounded continuously?

88. Continuous compound interest. How many years (to two decimal places) will it take an investment of \$17,000 to grow to \$41,000 if it is invested at 2.95% compounded continuously?

89. Supply and demand. A cordless screwdriver is sold through a national chain of discount stores. A marketing company established price–demand and price–supply tables (Tables 2 and 3), where x is the number of screwdrivers people are willing to buy and the store is willing to sell each month at a price of p dollars per screwdriver.

(A) Find a logarithmic regression model $(y = a + b \ln x)$ for the data in Table 2. Estimate the demand (to the nearest unit) at a price level of \$50.

Table 2 Price–Demand

x	$p = D(x)$ (\$)
1,000	91
2,000	73
3,000	64
4,000	56
5,000	53

(B) Find a logarithmic regression model $(y = a + b \ln x)$ for the data in Table 3. Estimate the supply (to the nearest unit) at a price level of \$50.

Table 3 Price–Supply

x	$p = S(x)$ (\$)
1,000	9
2,000	26
3,000	34
4,000	38
5,000	41

(C) Does a price level of \$50 represent a stable condition, or is the price likely to increase or decrease? Explain.

90. Equilibrium point. Use the models constructed in Problem 89 to find the equilibrium point. Write the equilibrium price to the nearest cent and the equilibrium quantity to the nearest unit.

91. Sound intensity: decibels. Because of the extraordinary range of sensitivity of the human ear (a range of over 1,000 million millions to 1), it is helpful to use a logarithmic scale, rather than an absolute scale, to measure sound intensity over this range. The unit of measure is called the *decibel,* after the inventor of the telephone, Alexander Graham Bell. If we let N be the number of decibels, I the power of the sound in question (in watts per square centimeter), and I_0 the power of sound just below the threshold of hearing (approximately 10^{-16} watt per square centimeter), then

$$I = I_0 10^{N/10}$$

Show that this formula can be written in the form

$$N = 10 \log \frac{I}{I_0}$$

92. Sound intensity: decibels. Use the formula in Problem 91 (with $I_0 = 10^{-16}$ W/cm^2) to find the decibel ratings of the following sounds:

(A) Whisper: 10^{-13} W/cm^2

(B) Normal conversation: 3.16×10^{-10} W/cm^2

(C) Heavy traffic: 10^{-8} W/cm^2

(D) Jet plane with afterburner: 10^{-1} W/cm^2

93. Agriculture. Table 4 shows the yield (in bushels per acre) and the total production (in millions of bushels) for corn in the United States for selected years since 1950. Let x represent years since 1900. Find a logarithmic regression model $(y = a + b \ln x)$ for the yield. Estimate (to the nearest bushel per acre) the yield in 2024.

Table 4 United States Corn Production

Year	x	Yield (bushels per acre)	Total Production (million bushels)
1950	50	38	2,782
1960	60	56	3,479
1970	70	81	4,802
1980	80	98	6,867
1990	90	116	7,802
2000	100	140	10,192
2010	110	153	12,447

94. Agriculture. Refer to Table 4. Find a logarithmic regression model $(y = a + b \ln x)$ for the total production. Estimate (to the nearest million) the production in 2024.

95. World population. If the world population is now 7.4 billion people and if it continues to grow at an annual rate of 1.1% compounded continuously, how long (to the nearest year) would it take before there is only 1 square yard of land per person? (The Earth contains approximately 1.68×10^{14} square yards of land.)

96. Archaeology: carbon-14 dating. The radioactive carbon-14 $\left({}^{14}\text{C}\right)$ in an organism at the time of its death decays according to the equation

$$A = A_0 e^{-0.000124t}$$

where t is time in years and A_0 is the amount of ^{14}C present at time $t = 0$. (See Example 3 in Section 2.5.) Estimate the age of a skull uncovered in an archaeological site if 10% of the original amount of ^{14}C is still present. [*Hint:* Find t such that $A = 0.1A_0$.]

Answers to Matched Problems

1. (A) $9 = 3^2$ (B) $2 = 4^{1/2}$ (C) $\frac{1}{9} = 3^{-2}$
2. (A) $\log_7 49 = 2$ (B) $\log_9 3 = \frac{1}{2}$ (C) $\log_3\left(\frac{1}{3}\right) = -1$
3. (A) $y = \frac{3}{2}$ (B) $x = \frac{1}{3}$ (C) $b = 10$
4. (A) $\log_b R - \log_b S - \log_b T$ (B) $\frac{2}{3}(\log_b R - \log_b S)$ (C) b^u (D) $\log_b x$
5. $x = 2$
6. $x = 5$
7. (A) -1.868734 (B) 3.356 663 (C) Not defined
8. (A) 0.0063 (B) 120.8092
9. (A) 0.8451 (B) 1.7918 (C) 1.1610
10. 9 yr
11. 70.3%

Chapter 2 Summary and Review

Important Terms, Symbols, and Concepts

2.1 Functions

EXAMPLES

- **Point-by-point plotting** may be used to **sketch the graph** of an equation in two variables: Plot enough points from its solution set in a rectangular coordinate system so that the total graph is apparent and then connect these points with a smooth curve. Ex. 1, p. 42
- A **function** is a correspondence between two sets of elements such that to each element in the first set there corresponds one and only one element in the second set. The first set is called the **domain** and the set of corresponding elements in the second set is called the **range**.
- If x is a placeholder for the elements in the domain of a function, then x is called the **independent variable** or the **input**. If y is a placeholder for the elements in the range, then y is called the **dependent variable** or the **output**.
- If in an equation in two variables we get exactly one output for each input, then the equation specifies a function. The graph of such a function is just the graph of the specifying equation. If we get more than one output for a given input, then the equation does not specify a function. Ex. 2, p. 45
- The **vertical-line test** can be used to determine whether or not an equation in two variables specifies a function (Theorem 1, p. 46).
- The functions specified by equations of the form $y = mx + b$, where $m \neq 0$, are called **linear functions**. Functions specified by equations of the form $y = b$ are called **constant functions**.
- If a function is specified by an equation and the domain is not indicated, we agree to assume that the domain is the set of all inputs that produce outputs that are real numbers. Ex. 3, p. 47; Ex. 5, p. 49
- The symbol $f(x)$ represents the element in the range of f that corresponds to the element x of the domain. Ex. 4, p. 48; Ex. 6, p. 10
- **Break-even** and **profit–loss** analysis use a cost function C and a revenue function R to determine when a company will have a loss $(R < C)$, will break even $(R = C)$, or will have a profit $(R > C)$. Typical **cost, revenue, profit,** and **price–demand functions** are given on page 50. Ex. 7, p. 51

2.2 Elementary Functions: Graphs and Transformations

- The graphs of **six basic elementary functions** (the identity function, the square and cube functions, the square root and cube root functions, and the absolute value function) are shown on page 58. Ex. 1, p. 57
- Performing an operation on a function produces a **transformation** of the graph of the function. The basic graph transformations, **vertical and horizontal translations** (shifts), **reflection in the x axis**, and **vertical stretches and shrinks**, are summarized on page 62. Ex. 2, p. 59; Ex. 3, p. 60; Ex. 4, p. 61; Ex. 5, p. 62
- A **piecewise-defined function** is a function whose definition involves more than one rule. Ex. 6, p. 63; Ex. 7, p. 64

2.3 Quadratic Functions

- If a, b, and c are real numbers with $a \neq 0$, then the function Ex. 1, p. 70

$$f(x) = ax^2 + bx + c \qquad \text{Standard form}$$

is a **quadratic function** in **standard form** and its graph is a **parabola**.

- The quadratic formula

$$x = \frac{-b \pm \sqrt{b^2 - 4ac}}{2a} \qquad b^2 - 4ac \geq 0$$

can be used to find the x intercepts of a quadratic function.

- Completing the square in the standard form of a quadratic function produces the **vertex form**

$$f(x) = a(x - h)^2 + k \qquad \text{Vertex form}$$

- From the vertex form of a quadratic function, we can read off the vertex, axis of symmetry, maximum or minimum, and range, and can easily sketch the graph (page 74). Ex. 2, p. 74 Ex. 3, p. 76
- If a revenue function $R(x)$ and a cost function $C(x)$ intersect at a point (x_0, y_0), then both this point and its x coordinate x_0 are referred to as **break-even points**. Ex. 4, p. 77
- **Quadratic regression** on a graphing calculator produces the function of the form $y = ax^2 + bx + c$ that best fits a data set. Ex. 5, p. 79

2.4 Polynomial and Rational Functions

- A **polynomial function** is a function that can be written in the form

$$f(x) = a_n x^n + a_{n-1}x^{n-1} + \cdots + a_1 x + a_0$$

for n a nonnegative integer called the **degree** of the polynomial. The coefficients $a_0, a_1, \ldots, a_n$ are real numbers with **leading coefficient** $a_n \neq 0$. The **domain** of a polynomial function is the set of all real numbers. Graphs of representative polynomial functions are shown on page 85 and the endpapers at the back of the book.

- The graph of a polynomial function of degree n can intersect the x axis at most n times. An x intercept is also called a **zero** or **root**.
- The graph of a polynomial function has no sharp corners and is **continuous**; that is; it has no holes or breaks.
- **Polynomial regression** produces a polynomial of specified degree that best fits a data set. Ex. 1, p. 86
- A **rational function** is any function that can be written in the form

$$f(x) = \frac{n(x)}{d(x)} \qquad d(x) \neq 0$$

where $n(x)$ and $d(x)$ are polynomials. The **domain** is the set of all real numbers such that $d(x) \neq 0$. Graphs of representative rational functions are shown on page 88 and the endpapers at the back of the book.

- Unlike polynomial functions, a rational function can have vertical asymptotes [but not more than the degree of the denominator $d(x)$] and at most one horizontal asymptote. Ex. 2, p. 88
- A procedure for finding the vertical and horizontal asymptotes of a rational function is given on page 90. Ex. 3, p. 90 Ex. 4, p. 91

2.5 Exponential Functions

- An **exponential function** is a function of the form

$$f(x) = b^x$$

where $b \neq 1$ is a positive constant called the **base**. The **domain** of f is the set of all real numbers, and the **range** is the set of positive real numbers.

- The graph of an exponential function is continuous, passes through $(0, 1)$, and has the x axis as a horizontal asymptote. If $b > 1$, then b^x increases as x increases; if $0 < b < 1$, then b^x decreases as x increases (Theorem 1, p. 97). Ex. 1, p. 97
- Exponential functions obey the familiar laws of exponents and satisfy additional properties (Theorem 2, p. 98).
- The base that is used most frequently in mathematics is the irrational number $e \approx 2.7183$.

2.5 Exponential Functions (*Continued*)

- Exponential functions can be used to model population growth and radioactive decay.
- **Exponential regression** on a graphing calculator produces the function of the form $y = ab^x$ that best fits a data set.

Ex. 2, p. 99
Ex. 3, p. 100
Ex. 4, p. 101

- Exponential functions are used in computations of **compound interest** and **continuous compound interest**:

Ex. 5, p. 102
Ex. 6, p. 103

$$A = P\left(1 + \frac{r}{m}\right)^{mt} \qquad \text{Compound interest}$$

$$A = Pe^{rt} \qquad \text{Continuous compound interest}$$

(see summary on page 103).

2.6 Logarithmic Functions

- A function is said to be **one-to-one** if each range value corresponds to exactly one domain value.
- The **inverse** of a one-to-one function f is the function formed by interchanging the independent and dependent variables of f. That is, (a, b) is a point on the graph of f if and only if (b, a) is a point on the graph of the inverse of f. A function that is not one-to-one does not have an inverse.
- The inverse of the exponential function with base b is called the **logarithmic function with base** $\boldsymbol{b}$, denoted $y = \log_b x$. The **domain** of $\log_b x$ is the set of all positive real numbers (which is the range of b^x), and the **range** of $\log_b x$ is the set of all real numbers (which is the domain of b^x).
- Because $\log_b x$ is the inverse of the function b^x,

Logarithmic form		Exponential form
$y = \log_b x$	is equivalent to	$x = b^y$

Ex. 1, p. 108
Ex. 2, p. 109
Ex. 3, p. 109

- Properties of logarithmic functions can be obtained from corresponding properties of exponential functions (Theorem 1, p. 110).

Ex. 4, p. 110
Ex. 5, p. 110
Ex. 6, p. 111

- Logarithms to the base 10 are called **common logarithms**, often denoted simply by $\log x$. Logarithms to the base e are called **natural logarithms**, often denoted by $\ln x$.
- Logarithms can be used to find an investment's **doubling time**—the length of time it takes for the value of an investment to double.

Ex. 7, p. 111
Ex. 8, p. 112
Ex. 9, p. 112
Ex. 10, p. 113

- **Logarithmic regression** on a graphing calculator produces the function of the form $y = a + b \ln x$ that best fits a data set.

Ex. 11, p. 114

Review Exercises

Work through all the problems in this chapter review and check your answers in the back of the book. Answers to all review problems are there along with section numbers in italics to indicate where each type of problem is discussed. Where weaknesses show up, review appropriate sections in the text.

A *In Problems 1–3, use point-by-point plotting to sketch the graph of each equation.*

1. $y = 5 - x^2$

2. $x^2 = y^2$

3. $y^2 = 4x^2$

4. Indicate whether each graph specifies a function:

(A)

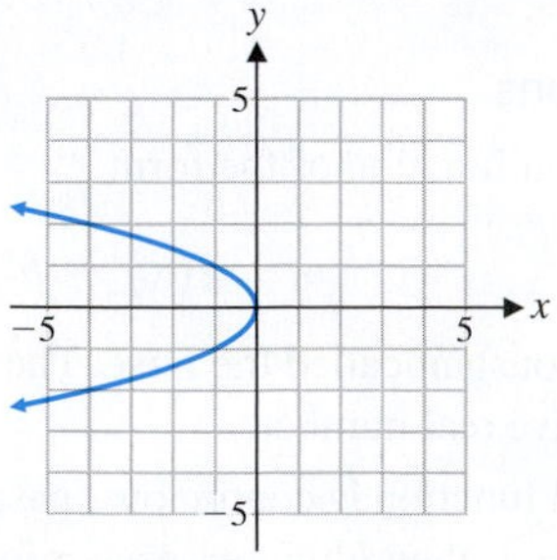

(B)

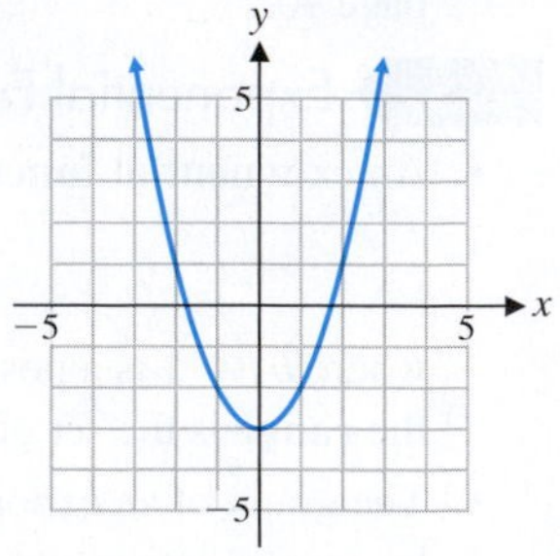

(C)

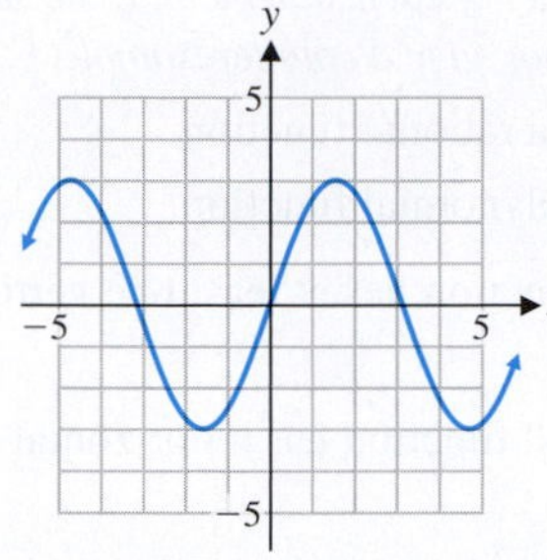

(D)

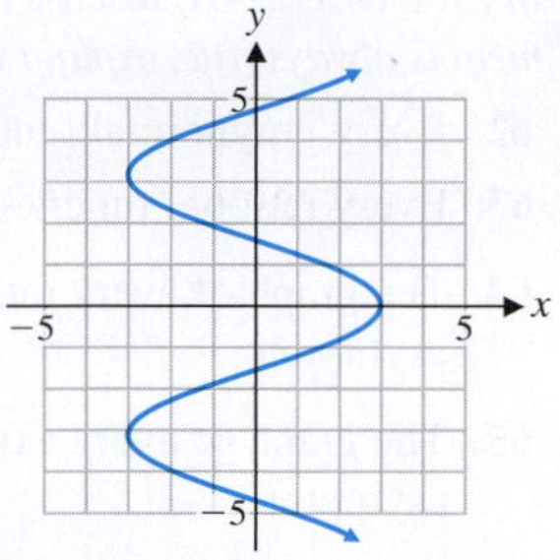

5. For $f(x) = 2x - 1$ and $g(x) = x^2 - 2x$, find:

(A) $f(-2) + g(-1)$ (B) $f(0) \cdot g(4)$

(C) $\dfrac{g(2)}{f(3)}$ (D) $\dfrac{f(3)}{g(2)}$

6. Write in logarithmic form using base e: $u = e^v$.

7. Write in logarithmic form using base 10: $x = 10^y$.

8. Write in exponential form using base e: In $M = N$.

9. Write in exponential form using base 10: $\log u = v$.

Solve Problems 10–12 for x exactly without using a calculator.

10. $\log_3 x = 2$ **11.** $\log_x 36 = 2$

12. $\log_2 16 = x$

Solve Problems 13–16 for x to three decimal places.

13. $10^x = 143.7$ **14.** $e^x = 503{,}000$

15. $\log x = 3.105$ **16.** In $x = -1.147$

17. Use the graph of function f in the figure to determine (to the nearest integer) x or y as indicated.

(A) $y = f(0)$ (B) $4 = f(x)$

(C) $y = f(3)$ (D) $3 = f(x)$

(E) $y = f(-6)$ (F) $-1 = f(x)$

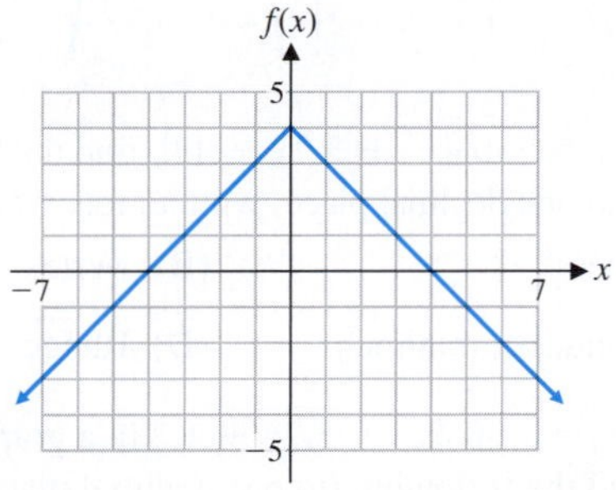

B **18.** Sketch a graph of each of the functions in parts (A)–(D) using the graph of function f in the figure below.

(A) $y = -f(x)$ (B) $y = f(x) + 4$

(C) $y = f(x - 2)$ (D) $y = -f(x + 3) - 3$

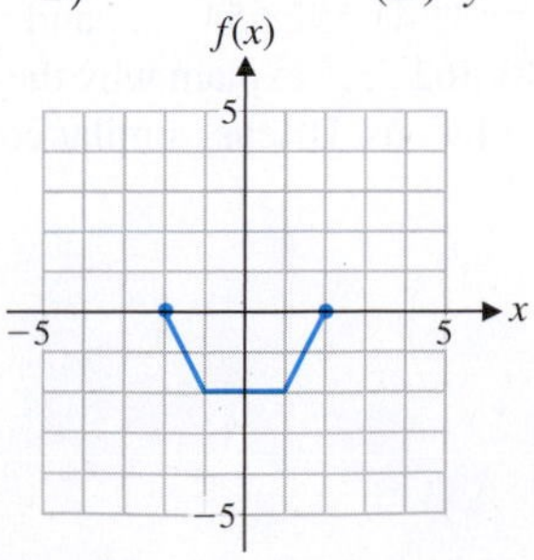

19. Complete the square and find the standard form for the quadratic function

$$f(x) = -x^2 + 4x$$

Then write a brief verbal description of the relationship between the graph of f and the graph of $y = x^2$.

20. Match each equation with a graph of one of the functions f, g, m, or n in the figure.

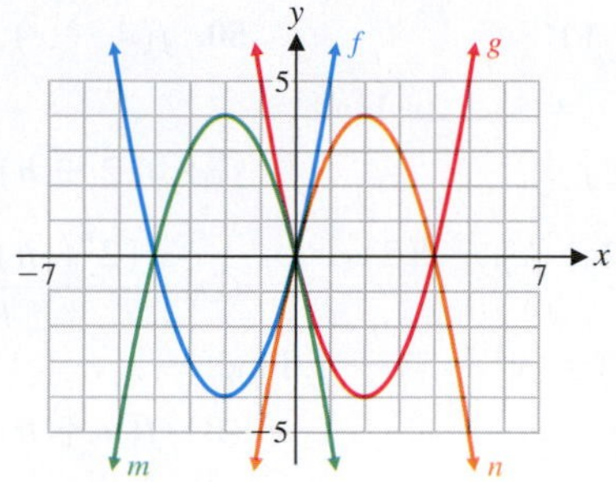

(A) $y = (x - 2)^2 - 4$ (B) $y = -(x + 2)^2 + 4$

(C) $y = -(x - 2)^2 + 4$ (D) $y = (x + 2)^2 - 4$

21. Referring to the graph of function f in the figure for Problem 20 and using known properties of quadratic functions, find each of the following to the nearest integer:

(A) Intercepts (B) Vertex

(C) Maximum or minimum (D) Range

In Problems 22–25, each equation specifies a function. Determine whether the function is linear, quadratic, constant, or none of these.

22. $y = 4 - x + 3x^2$ **23.** $y = \dfrac{1 + 5x}{6}$

24. $y = \dfrac{7 - 4x}{2x}$ **25.** $y = 8x + 2(10 - 4x)$

Solve Problems 26–33 for x exactly without using a calculator.

26. $\log(x + 5) = \log(2x - 3)$ **27.** $2\,\text{In}(x - 1) = \text{In}(x^2 - 5)$

28. $9^{x-1} = 3^{1+x}$ **29.** $e^{2x} = e^{x^2-3}$

30. $2x^2e^x = 3xe^x$ **31.** $\log_{1/3} 9 = x$

32. $\log_x 8 = -3$ **33.** $\log_9 x = \frac{3}{2}$

Solve Problems 34–41 for x to four decimal places.

34. $x = 3(e^{1.49})$ **35.** $x = 230(10^{-0.161})$

36. $\log x = -2.0144$ **37.** In $x = 0.3618$

38. $35 = 7(3^x)$ **39.** $0.01 = e^{-0.05x}$

40. $8{,}000 = 4{,}000(1.08^x)$ **41.** $5^{2x-3} = 7.08$

42. Find the domain of each function:

(A) $f(x) = \dfrac{2x - 5}{x^2 - x - 6}$ (B) $g(x) = \dfrac{3x}{\sqrt{5 - x}}$

43. Find the vertex form for $f(x) = 4x^2 + 4x - 3$ and then find the intercepts, the vertex, the maximum or minimum, and the range.

44. Let $f(x) = e^x - 1$ and $g(x) = \text{In}(x + 2)$. Find all points of intersection for the graphs of f and g. Round answers to two decimal places.

In Problems 45 and 46, use point-by-point plotting to sketch the graph of each function.

45. $f(x) = \dfrac{50}{x^2 + 1}$ **46.** $f(x) = \dfrac{-66}{2 + x^2}$

If $f(x) = 5x + 1$, find and simplify each of the following in Problems 47–50.

47. $f(f(0))$ **48.** $f(f(-1))$

49. $f(2x - 1)$ **50.** $f(4 - x)$

51. Let $f(x) = 3 - 2x$. Find

(A) $f(2)$ (B) $f(2 + h)$

(C) $f(2 + h) - f(2)$ (D) $\dfrac{f(2 + h) - f(2)}{h}, h \neq 0$

52. Let $f(x) = x^2 - 3x + 1$. Find

(A) $f(a)$ (B) $f(a + h)$

(C) $f(a + h) - f(a)$ (D) $\dfrac{f(a + h) - f(a)}{h}, h \neq 0$

53. Explain how the graph of $m(x) = -|x - 4|$ is related to the graph of $y = |x|$.

54. Explain how the graph of $g(x) = 0.3x^3 + 3$ is related to the graph of $y = x^3$.

55. The following graph is the result of applying a sequence of transformations to the graph of $y = x^2$. Describe the transformations verbally and write an equation for the graph.

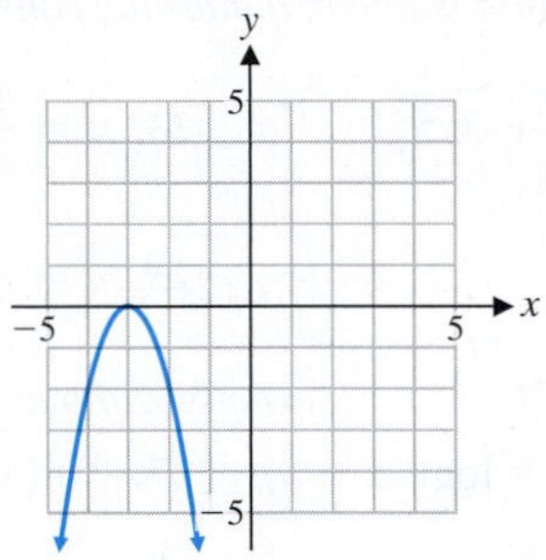

56. The graph of a function f is formed by vertically stretching the graph of $y = \sqrt{x}$ by a factor of 2, and shifting it to the left 3 units and down 1 unit. Find an equation for function f and graph it for $-5 \leq x \leq 5$ and $-5 \leq y \leq 5$.

In Problems 57–59, find the equation of any horizontal asymptote.

57. $f(x) = \dfrac{5x + 4}{x^2 - 3x + 1}$ **58.** $f(x) = \dfrac{3x^2 + 2x - 1}{4x^2 - 5x + 3}$

59. $f(x) = \dfrac{x^2 + 4}{100x + 1}$

In Problems 60 and 61, find the equations of any vertical asymptotes.

60. $f(x) = \dfrac{x^2 + 100}{x^2 - 100}$ **61.** $f(x) = \dfrac{x^2 + 3x}{x^2 + 2x}$

In Problems 62–67, discuss the validity of each statement. If the statement is always true, explain why. If not, give a counterexample.

62. Every polynomial function is a rational function.

63. Every rational function is a polynomial function.

64. The graph of every rational function has at least one vertical asymptote.

65. The graph of every exponential function has a horizontal asymptote.

66. The graph of every logarithmic function has a vertical asymptote.

67. There exists a rational function that has both a vertical and horizontal asymptote.

68. Sketch the graph of f for $x \geq 0$.

$$f(x) = \begin{cases} 9 + 0.3x & \text{if } 0 \leq x \leq 20 \\ 5 + 0.2x & \text{if } x > 20 \end{cases}$$

69. Sketch the graph of g for $x \geq 0$.

$$f(x) = \begin{cases} 0.5x + 5 & \text{if } 0 \leq x \leq 10 \\ 1.2x - 2 & \text{if } 10 < x \leq 30 \\ 2x - 26 & \text{if } x > 30 \end{cases}$$

70. Write an equation for the graph shown in the form $y = a(x - h)^2 + k$, where a is either -1 or $+1$ and h and k are integers.

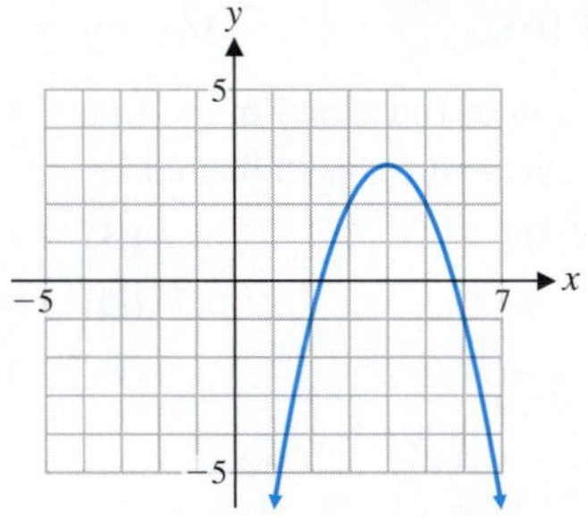

71. Given $f(x) = -0.4x^2 + 3.2x + 1.2$, find the following algebraically (to one decimal place) without referring to a graph:

(A) Intercepts (B) Vertex

(C) Maximum or minimum (D) Range

72. Graph $f(x) = -0.4x^2 + 3.2x + 1.2$ in a graphing calculator and find the following (to one decimal place) using TRACE and appropriate commands:

(A) Intercepts (B) Vertex

(C) Maximum or minimum (D) Range

73. Noting that $\pi = 3.141\,592\,654\ldots$ and $\sqrt{2} = 1.414\,213\,562\ldots$ explain why the calculator results shown here are obvious. Discuss similar connections between the natural logarithmic function and the exponential function with base e.

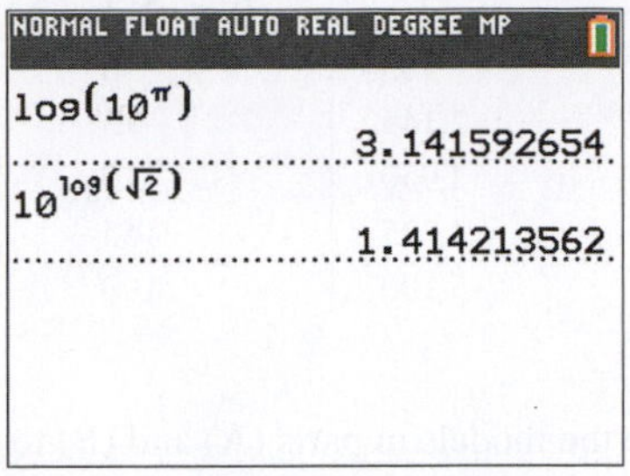

Solve Problems 74–77 exactly without using a calculator.

74. $\log x - \log 3 = \log 4 - \log(x + 4)$

75. $\ln(2x - 2) - \ln(x - 1) = \ln x$

76. $\ln(x + 3) - \ln x = 2 \ln 2$

77. $\log 3x^2 = 2 + \log 9x$

78. Write $\ln y = -5t + \ln c$ in an exponential form free of logarithms. Then solve for y in terms of the remaining variables.

79. Explain why 1 cannot be used as a logarithmic base.

80. The following graph is the result of applying a sequence of transformations to the graph of $y = \sqrt[3]{x}$. Describe the transformations verbally, and write an equation for the graph.

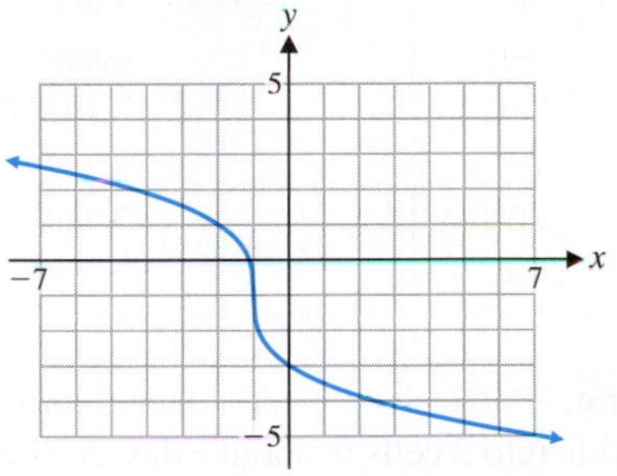

81. Given $G(x) = 0.3x^2 + 1.2x - 6.9$, find the following algebraically (to one decimal place) without the use of a graph:

(A) Intercepts (B) Vertex

(C) Maximum or minimum (D) Range

82. Graph $G(x) = 0.3x^2 + 1.2x - 6.9$ in a standard viewing window. Then find each of the following (to one decimal place) using appropriate commands.

(A) Intercepts (B) Vertex

(C) Maximum or minimum (D) Range

Applications

In all problems involving days, a 365-day year is assumed.

83. Electricity rates. The table shows the electricity rates charged by Easton Utilities in the summer months.

(A) Write a piecewise definition of the monthly charge $S(x)$ (in dollars) for a customer who uses x kWh in a summer month.

(B) Graph $S(x)$.

Energy Charge (June–September)
\$3.00 for the first 20 kWh or less
5.70¢ per kWh for the next 180 kWh
3.46¢ per kWh for the next 800 kWh
2.17¢ per kWh for all over 1,000 kWh

84. Money growth. Provident Bank of Cincinnati, Ohio, offered a certificate of deposit that paid 1.25% compounded quarterly. If a \$5,000 CD earns this rate for 5 years, how much will it be worth?

85. Money growth. Capital One Bank of Glen Allen, Virginia, offered a certificate of deposit that paid 1.05% compounded daily. If a \$5,000 CD earns this rate for 5 years, how much will it be worth?

86. Money growth. How long will it take for money invested at 6.59% compounded monthly to triple?

87. Money growth. How long will it take for money invested at 7.39% compounded continuously to double?

88. Break-even analysis. The research department in a company that manufactures AM/FM clock radios established the following price-demand, cost, and revenue functions:

$$p(x) = 50 - 1.25x \quad \text{Price–demand function}$$
$$C(x) = 160 + 10x \quad \text{Cost function}$$
$$R(x) = xp(x) = x(50 - 1.25x) \quad \text{Revenue function}$$

where x is in thousands of units, and $C(x)$ and $R(x)$ are in thousands of dollars. All three functions have domain $1 \le x \le 40$.

(A) Graph the cost function and the revenue function simutaneously in the same coordinate system.

(B) Determine algebraically when $R = C$. Then, with the aid of part (A), determine when $R < C$ and $R > C$ to the nearest unit.

(C) Determine algebraically the maximum revenue (to the nearest thousand dollars) and the output (to the nearest unit) that produces the maximum revenue. What is the wholesale price of the radio (to the nearest dollar) at this output?

89. Profit–loss analysis. Use the cost and revenue functions from Problem 88.

(A) Write a profit function and graph it in a graphing calculator.

(B) Determine graphically when $P = 0$, $P < 0$, and $P > 0$ to the nearest unit.

(C) Determine graphically the maximum profit (to the nearest thousand dollars) and the output (to the nearest unit) that produces the maximum profit. What is the wholesale price of the radio (to the nearest dollar) at this output? [Compare with Problem 88C.]

90. Construction. A construction company has 840 feet of chain-link fence that is used to enclose storage areas for equipment and materials at construction sites. The supervisor wants to set up two identical rectangular storage areas sharing a common fence (see the figure).

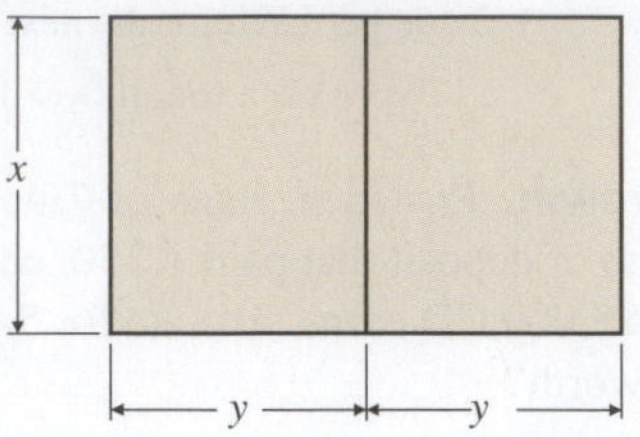

Assuming that all fencing is used,

(A) Express the total area $A(x)$ enclosed by both pens as a function of x.

(B) From physical considerations, what is the domain of the function A?

(C) Graph function A in a rectangular coordinate system.

(D) Use the graph to discuss the number and approximate locations of values of x that would produce storage areas with a combined area of 25,000 square feet.

(E) Approximate graphically (to the nearest foot) the values of x that would produce storage areas with a combined area of 25,000 square feet.

(F) Determine algebraically the dimensions of the storage areas that have the maximum total combined area. What is the maximum area?

91. Equilibrium point. A company is planning to introduce a 10-piece set of nonstick cookware. A marketing company established price–demand and price–supply tables for selected prices (Tables 1 and 2), where x is the number of cookware sets people are willing to buy and the company is willing to sell each month at a price of p dollars per set.

(A) Find a quadratic regression model for the data in Table 1. Estimate the demand at a price level of \$180.

(B) Find a linear regression model for the data in Table 2. Estimate the supply at a price level of \$180.

(C) Does a price level of \$180 represent a stable condition, or is the price likely to increase or decrease? Explain.

Table 1 Price–Demand

x	$p = D(x)(\$)$
985	330
2,145	225
2,950	170
4,225	105
5,100	50

Table 2 Price–Supply

x	$p = S(x)(\$)$
985	30
2,145	75
2,950	110
4,225	155
5,100	190

(D) Use the models in parts (A) and (B) to find the equilibrium point. Write the equilibrium price to the nearest cent and the equilibrium quantity to the nearest unit.

92. Crime statistics. According to data published by the FBI, the crime index in the United States has shown a downward trend since the early 1990s (Table 3).

(A) Find a cubic regression model for the crime index if $x = 0$ represents 1987.

(B) Use the cubic regression model to predict the crime index in 2025.

Table 3 Crime Index

Year	Crimes per 100,000 Inhabitants
1987	5,550
1992	5,660
1997	4,930
2002	4,125
2007	3,749
2010	3,350
2013	3,099

93. Medicine. One leukemic cell injected into a healthy mouse will divide into 2 cells in about $\frac{1}{2}$ day. At the end of the day these 2 cells will divide into 4. This doubling continues until 1 billion cells are formed; then the animal dies with leukemic cells in every part of the body.

(A) Write an equation that will give the number N of leukemic cells at the end of t days.

(B) When, to the nearest day, will the mouse die?

94. Marine biology. The intensity of light entering water is reduced according to the exponential equation

$$I = I_0 e^{-kd}$$

where I is the intensity d feet below the surface, I_0 is the intensity at the surface, and k is the coefficient of extinction. Measurements in the Sargasso Sea have indicated that half of the surface light reaches a depth of 73.6 feet. Find k (to five decimal places), and find the depth (to the nearest foot) at which 1% of the surface light remains.

95. Agriculture. The number of dairy cows on farms in the United States is shown in Table 4 for selected years since 1950. Let 1940 be year 0.

Table 4 Dairy Cows on Farms in the United States

Year	Dairy Cows (thousands)
1950	23,853
1960	19,527
1970	12,091
1980	10,758
1990	10,015
2000	9,190
2010	9,117

(A) Find a logarithmic regression model $(y = a + b \ln x)$ for the data. Estimate (to the nearest thousand) the number of dairy cows in 2023.

(B) Explain why it is not a good idea to let 1950 be year 0.

96. Population growth. The population of some countries has a relative growth rate of 3% (or more) per year. At this rate, how many years (to the nearest tenth of a year) will it take a population to double?

97. Medicare. The annual expenditures for Medicare (in billions of dollars) by the U.S. government for selected years since 1980 are shown in Table 5. Let x represent years since 1980.

(A) Find an exponential regression model $(y = ab^x)$ for the data. Estimate (to the nearest billion) the annual expenditures in 2025.

(B) When will the annual expenditures exceed two trillion dollars?

Table 5 Medicare Expenditures

Year	Billion $
1980	37
1985	72
1990	111
1995	181
2000	197
2005	299
2010	452
2015	546

3 Mathematics of Finance

Introduction

How do I choose the right loan for college? Would it be better to take the dealer's financing or the rebate for my new car? How much should my parents offer for the new home they want to buy? To make wise decisions in such matters, you need a basic understanding of the mathematics of finance.

In Chapter 3 we study the mathematics of simple and compound interest, ordinary annuities, auto loans, and home mortage loans (see Problems 47–48 in Section 3.4). You will need a calculator with logarithmic and exponential keys. A graphing calculator would be even better: It can help you visualize the rate at which an investment grows or the rate at which principal on a loan is amortized.

You may wish to review arithmetic and geometric sequences, discussed in Appendix B.2, before beginning this chapter.

Finally, to avoid repeating the following reminder many times, we emphasize it here: Throughout the chapter, interest rates are to be converted to decimal form before they are used in a formula.

3.1 Simple Interest

- The Simple Interest Formula
- Simple Interest and Investments

The Simple Interest Formula

Simple interest is used on short-term notes—often of duration less than 1 year. The concept of simple interest, however, forms the basis of much of the rest of the material developed in this chapter, for which time periods may be much longer than a year.

If you deposit a sum of money P in a savings account or if you borrow a sum of money P from a lender, then P is referred to as the **principal**. When money is borrowed—whether it is a savings institution borrowing from you when you deposit money in your account, or you borrowing from a lender—a fee is charged for the money borrowed. This fee is rent paid for the use of another's money, just as rent is paid for the use of another's house. The fee is called **interest**. It is usually computed as a percentage (called the **interest rate**) of the principal over a given period of time. The interest rate, unless otherwise stated, is an annual rate. **Simple interest** is given by the following formula:

Reminder

To write a percentage as a decimal, divide by 100 and delete the % sign. To write a decimal as a percentage, multiply by 100 and add the % sign. So 12% and 0.12 are equivalent expressions.

DEFINITION Simple Interest

$$I = Prt \tag{1}$$

where
I = interest
P = principal
r = annual simple interest rate (written as a decimal)
t = time in years

For example, the interest on a loan of \$100 at 12% for 9 months would be

$$\begin{aligned} I &= Prt \\ &= (100)(0.12)(0.75) \\ &= \$9 \end{aligned}$$

Convert 12% to a decimal (0.12) and 9 months to years ($\frac{9}{12} = 0.75$).

At the end of 9 months, the borrower would repay the principal (\$100) plus the interest (\$9), or a total of \$109.

In general, if a principal P is borrowed at a rate r, then after t years, the borrower will owe the lender an amount A that will include the principal P plus the interest I. Since P is the amount that is borrowed now and A is the amount that must be paid back in the future, P is often referred to as the **present value** and A as the **future value**. The formula relating A and P follows:

THEOREM 1 Simple Interest

$$\begin{aligned} A &= P + Prt \\ &= P(1 + rt) \end{aligned} \tag{2}$$

where
A = amount, or future value
P = principal, or present value
r = annual simple interest rate (written as a decimal)
t = time in years

Given any three of the four variables A, P, r, and t in (2), we can solve for the fourth. The following examples illustrate several types of common problems that can be solved by using formula (2).

EXAMPLE 1 **Total Amount Due on a Loan** Find the total amount due on a loan of \$800 at 9% simple interest at the end of 4 months.

SOLUTION To find the amount A (future value) due in 4 months, we use formula (2) with $P = 800$, $r = 0.09$, and $t = \frac{4}{12} = \frac{1}{3}$ year. Thus,

$$\begin{aligned} A &= P(1 + rt) \\ &= 800\left[1 + 0.09\left(\tfrac{1}{3}\right)\right] \\ &= 800(1.03) \\ &= \$824 \end{aligned}$$

Matched Problem 1 Find the total amount due on a loan of \$500 at 12% simple interest at the end of 30 months.

Explore and Discuss 1

(A) Your sister has loaned you \$1,000 with the understanding that you will repay the principal plus 4% simple interest when you can. How much would you owe her if you repaid the loan after 1 year? After 2 years? After 5 years? After 10 years?

(B) How is the interest after 10 years related to the interest after 1 year? After 2 years? After 5 years?

(C) Explain why your answers are consistent with the fact that for simple interest, the graph of future value as a function of time is a straight line (Fig. 1).

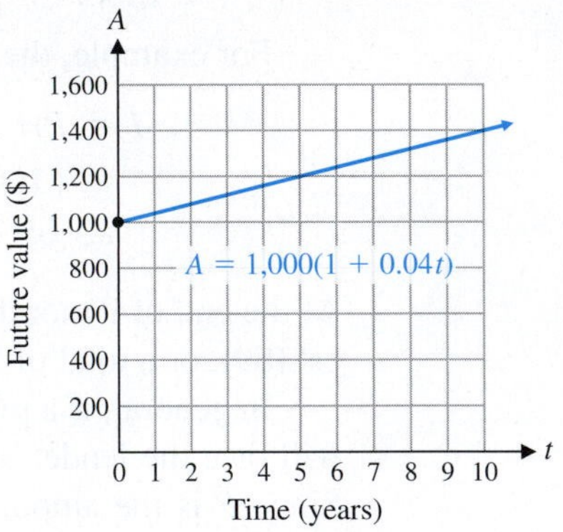

Figure 1

EXAMPLE 2 **Present Value of an Investment** If you want to earn an annual rate of 10% on your investments, how much (to the nearest cent) should you pay for a note that will be worth \$5,000 in 9 months?

SOLUTION We again use formula (2), but now we are interested in finding the principal P (present value), given $A = \$5{,}000$, $r = 0.1$, and $t = \frac{9}{12} = 0.75$ year. Thus,

$$\begin{aligned} A &= P(1 + rt) \\ 5{,}000 &= P[1 + 0.1(0.75)] \\ 5{,}000 &= (1.075)P \\ P &= \$4{,}651.16 \end{aligned}$$

Replace A, r, and t with the given values, and solve for P.

Matched Problem 2 Repeat Example 2 with a time period of 6 months.

CONCEPTUAL INSIGHT

If we consider future value A as a function of time t with the present value P and the annual rate r being fixed, then $A = P + Prt$ is a linear function of t with y intercept P and slope Pr. For example, if $P = 1{,}000$ and $r = 0.04$ (Fig. 1), then

$$A = 1{,}000(1 + 0.04t) = 1{,}000 + 40t$$

is a linear function with y intercept 1,000 and slope 40.

Simple Interest and Investments

Because simple interest is used on short-term notes, the time period is often given in days rather than months or years. How should a time period given in days be converted to years? In this section, we will divide by 360, assuming a 360-day year called a *banker's year*. In other sections, we will assume a 365-day year. The choice will always be clearly stated. This assumption does not affect calculations in which the time period is not given in days. If the time period is given in months or quarters, for example, we would divide by 12 or 4, respectively, to convert the time period to years.

EXAMPLE 3 **Interest Rate Earned on a Note** T-bills (Treasury bills) are one of the instruments that the U.S. Treasury Department uses to finance the public debt. If you buy a 180-day T-bill with a maturity value of \$10,000 for \$9,893.78, what annual simple interest rate will you earn? (Express answer as a percentage, correct to three decimal places.)

SOLUTION Again we use formula (2), but this time we are interested in finding r, given $P = \$9{,}893.78$, $A = \$10{,}000$, and $t = 180/360 = 0.5$ year.

Replace P, A, and t with the given values, and solve for r.

$$\begin{aligned} A &= P(1 + rt) \\ 10{,}000 &= 9{,}893.78(1 + 0.5r) \\ 10{,}000 &= 9{,}893.78 + 4{,}946.89r \\ 106.22 &= 4{,}946.89r \\ r &= \frac{106.22}{4{,}946.89} \approx 0.02147 \quad \text{or} \quad 2.147\% \end{aligned}$$

Matched Problem 3 Repeat Example 3, assuming that you pay \$9,828.74 for the T-bill.

EXAMPLE 4 **Interest Rate Earned on an Investment** Suppose that after buying a new car you decide to sell your old car to a friend. You accept a 270-day note for \$3,500 at 10% simple interest as payment. (Both principal and interest are paid at the end of 270 days.) Sixty days later you find that you need the money and sell the note to a third party for \$3,550. What annual interest rate will the third party receive for the investment? Express the answer as a percentage, correct to three decimal places.

SOLUTION

Step 1 Find the amount that will be paid at the end of 270 days to the holder of the note.

$$\begin{aligned} A &= P(1 + rt) \\ &= \$3{,}500\left[1 + (0.1)\left(\tfrac{270}{360}\right)\right] \\ &= \$3{,}762.50 \end{aligned}$$

Step 2 For the third party, we are to find the annual rate of interest r required to make \$3,550 grow to \$3,762.50 in 210 days $(270 - 60)$; that is, we are to find r (which is to be converted to $100r\%$), given $A = \$3{,}762.50$, $P = \$3{,}550$, and $t = \frac{210}{360}$.

$$A = P + Prt \qquad \text{Solve for } r.$$

$$r = \frac{A - P}{Pt}$$

$$r = \frac{3{,}762.50 - 3{,}550}{(3{,}550)\left(\frac{210}{360}\right)} \approx 0.10262 \quad \text{or} \quad 10.262\%$$

Matched Problem 4 Repeat Example 4 assuming that 90 days after it was initially signed, the note was sold to a third party for \$3,500.

Some online discount brokerage firms offer flat rates for trading stock, but many still charge commissions based on the transaction amount (principal). Table 1 shows the commission schedule for one of these firms.

Table 1 Commission Schedule

Principal	Commission
\$0–\$2,499	\$29 + 1.6% of principal
\$2,500–\$9,999	\$49 + 0.8% of principal
\$10,000+	\$99 + 0.3% of principal

EXAMPLE 5 **Interest on an Investment** An investor purchases 50 shares of a stock at \$47.52 per share. After 200 days, the investor sells the stock for \$52.19 per share. Using Table 1, find the annual rate of interest earned by this investment. Express the answer as a percentage, correct to three decimal places.

SOLUTION The principal referred to in Table 1 is the value of the stock. The total cost for the investor is the cost of the stock plus the commission:

$$47.52(50) = \$2{,}376 \qquad \text{Principal}$$

$$29 + 0.016(2{,}376) = \$67.02 \qquad \text{Commission, using line 1 of Table 1}$$

$$2{,}376 + 67.02 = \$2{,}443.02 \qquad \text{Total investment}$$

When the stock is sold, the commission is subtracted from the proceeds of the sale and the remainder is returned to the investor:

$$52.19(50) = \$2{,}609.50 \qquad \text{Principal}$$

$$49 + 0.008(2{,}609.50) = \$69.88 \qquad \text{Commission, using line 2 of Table 1}$$

$$2{,}609.50 - 69.88 = \$2{,}539.62 \qquad \text{Total return}$$

Now using formula (2) with $A = 2{,}539.62$, $P = 2{,}443.02$, and $t = \frac{200}{360} = \frac{5}{9}$, we have

$$A = P(1 + rt)$$

$$2{,}539.62 = 2{,}443.02\left(1 + \frac{5}{9}r\right)$$

$$= 2{,}443.02 + 1{,}357.23r$$

$$96.60 = 1{,}357.23r$$

$$r = \frac{96.60}{1{,}357.23} \approx 0.07117 \quad \text{or} \quad 7.117\%$$

Matched Problem 5 Repeat Example 5 if 500 shares of stock were purchased for \$17.64 per share and sold 270 days later for \$22.36 per share.

CONCEPTUAL INSIGHT

The commission schedule in Table 1 specifies a piecewise-defined function C with independent variable p, the principal (see Section 2.2).

$$C = \begin{cases} 29 + 0.016p & \text{if } 0 \le p < 2{,}500 \\ 49 + 0.008p & \text{if } 2{,}500 \le p < 10{,}000 \\ 99 + 0.003p & \text{if } 10{,}000 \le p \end{cases}$$

Two credit card accounts may differ in a number of ways, including annual interest rates, credit limits, minimum payments, annual fees, billing cycles, and even the methods for calculating interest. A common method for calculating the interest owed on a credit card is the **average daily balance method**. In this method, a balance is calculated at the end of each day, incorporating any purchases, credits, or payments that were made that day. Interest is calculated at the end of the billing cycle on the average of those daily balances. The average daily balance method is considered in Example 6.

EXAMPLE 6 **Credit Card Accounts** A credit card has an annual interest rate of 21.99%, and interest is calculated by the average daily balance method. In a 30-day billing cycle, purchases of \$56.75, \$184.36, and \$49.19 were made on days 12, 19, and 24, respectively, and a payment of \$100.00 was credited to the account on day 10. If the unpaid balance at the start of the billing cycle was \$842.67, how much interest will be charged at the end of the billing cycle? What will the unpaid balance be at the start of the next billing cycle?

SOLUTION First calculate the unpaid balance on each day of the billing cycle:

$$\begin{aligned} &\text{Days } 1\text{–}9: && \$842.67 \\ &\text{Days } 10\text{–}11: && \$842.67 - \$100.00 = \$742.67 \\ &\text{Days } 12\text{–}18: && \$742.67 + \$56.75 = \$799.42 \\ &\text{Days } 19\text{–}23: && \$799.42 + \$184.36 = \$983.78 \\ &\text{Days } 24\text{–}30: && \$983.78 + \$49.19 = \$1{,}032.97 \end{aligned}$$

So the unpaid balance was \$842.67 for the first 9 days of the billing cycle, \$742.67 for the next 2 days, \$799.42 for the next 7 days, and so on. To calculate the average daily balance, we find the sum of the 30 daily balances, and then divide by 30:

$$\begin{aligned} \text{Sum: } & 9(\$842.67) + 2(\$742.67) + 7(\$799.42) + 5(\$983.78) \\ & + 7(\$1032.97) = \$26{,}815.00 \end{aligned}$$

$$\text{Average daily balance: } \$26{,}815.00/30 = \$893.83$$

To calculate the interest, use the formula $I = Prt$ with $P = \$893.83$, $r = 0.2199$, and $t = 30/360$:

$$I = Prt = \$893.83(0.2199)(30/360) = \$16.38$$

Therefore, the interest charged at the end of the billing cycle is \$16.38, and the unpaid balance at the start of the next cycle is \$1,032.97 + \$16.38 = \$1,049.35.

Matched Problem 6 A credit card has an annual interest rate of 16.99%, and interest is calculated by the average daily balance method. In a 30-day billing cycle, purchases of \$345.86 and \$246.71 were made on days 9 and 16, respectively, and a payment of \$500.00 was credited to the account on day 15. If the unpaid balance at the start of the billing cycle was \$1,792.19, how much interest will be charged at the end of the billing cycle? What will the unpaid balance be at the start of the next billing cycle?

Exercises 3.1

W Skills Warm-up Exercises

In Problems 1–4, if necessary, review Section A.1.

1. If your state sales tax rate is 5.65%, how much tax will you pay on a bicycle that sells for \$449.99?

2. If your state sales tax rate is 8.25%, what is the total cost of a motor scooter that sells for \$1,349.95?

3. A baseball team had a 103–58 win–loss record. Find its winning percentage to the nearest percentage point.

4. A basketball team played 21 games with a winning percentage of 81%. How many games did it lose?

In Problems 5–8, give the slope and y intercept of each line. (If necessary, review Section 1.2.)

5. $y = 12{,}000 + 120x$

6. $y = 15{,}000 + 300x$

7. $y = 2{,}000(1 + 0.025x)$

8. $y = 5{,}000(1 + 0.035x)$

A *In Problems 9–16, convert the given interest rate to decimal form if it is given as a percentage, and to a percentage if it is given in decimal form.*

9. 6.2%

10. 0.085

11. 0.137

12. 4.35%

13. 0.25%

14. 0.0019

15. 0.0084

16. 0.36%

In Problems 17–24, convert the given time period to years, in reduced fraction form, assuming a 360-day year [this assumption does not affect the number of quarters (4), months (12), or weeks (52) in a year].

17. 180 days

18. 9 months

19. 4 months

20. 90 days

21. 5 quarters

22. 6 weeks

23. 40 weeks

24. 7 quarters

In Problems 25–32, use formula (1) for simple interest to find each of the indicated quantities.

25. $P = \$300$; $r = 7\%$; $t = 2$ years; $I = ?$

26. $P = \$950$; $r = 9\%$; $t = 1$ year; $I = ?$

27. $I = \$36$; $r = 4\%$; $t = 6$ months; $P = ?$

28. $I = \$15$; $r = 8\%$; $t = 3$ quarters; $P = ?$

29. $I = \$48$; $P = \$600$; $t = 240$ days; $r = ?$

30. $I = \$28$; $P = \$700$; $t = 13$ weeks; $r = ?$

31. $I = \$60$; $P = \$2{,}400$; $r = 5\%$; $t = ?$

32. $I = \$96$; $P = \$3{,}200$; $r = 4\%$; $t = ?$

B *In Problems 33–40, use formula (2) for the amount to find each of the indicated quantities.*

33. $P = \$4{,}500$; $r = 10\%$; $t = 1$ quarter; $A = ?$

34. $P = \$3{,}000$; $r = 4.5\%$; $t = 30$ days; $A = ?$

35. $A = \$910$; $r = 16\%$; $t = 13$ weeks; $P = ?$

36. $A = \$6{,}608$; $r = 24\%$; $t = 3$ quarters; $P = ?$

37. $A = \$14{,}560$; $P = \$13{,}000$; $t = 4$ months; $r = ?$

38. $A = \$22{,}135$; $P = \$19{,}000$; $t = 39$ weeks; $r = ?$

39. $A = \$736$; $P = \$640$; $r = 15\%$; $t = ?$

40. $A = \$410$; $P = \$400$; $r = 10\%$; $t = ?$

C *In Problems 41–46, solve each formula for the indicated variable.*

41. $I = Prt$; for r

42. $I = Prt$; for P

43. $A = P + Prt$; for P

44. $A = P + Prt$; for r

45. $A = P(1 + rt)$; for t

46. $I = Prt$; for t

47. Discuss the similarities and differences in the graphs of future value A as a function of time t if \$1,000 is invested at simple interest at rates of 4%, 8%, and 12%, respectively (see the figure).

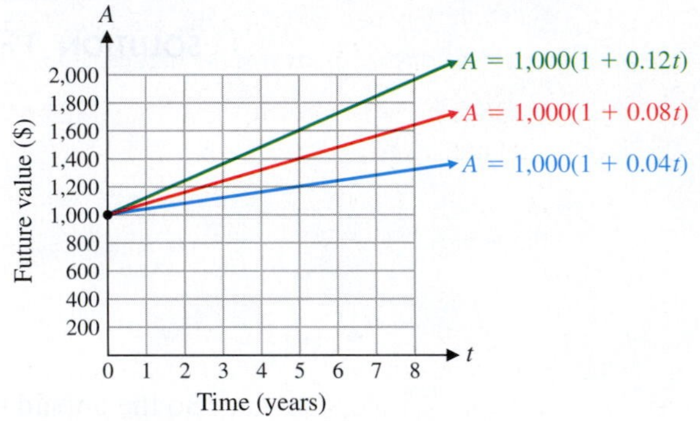

48. Discuss the similarities and differences in the graphs of future value A as a function of time t for loans of \$400, \$800, and \$1,200, respectively, each at 7.5% simple interest (see the figure).

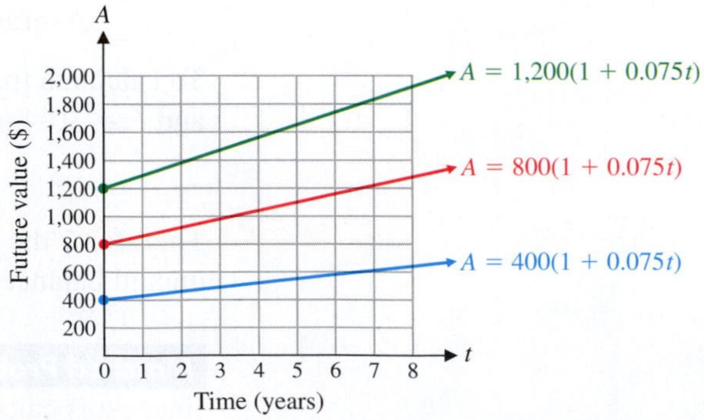

Applications*

In all problems involving days, a 360-day year is assumed. When annual rates are requested as an answer, express the rate as a percentage, correct to three decimal places, unless directed otherwise. Round dollar amounts to the nearest cent.

49. If $3,000 is loaned for 4 months at a 4.5% annual rate, how much interest is earned?

50. If $5,000 is loaned for 9 months at a 6.2% annual rate, how much interest is earned?

51. How much interest will you have to pay for a 60-day loan of $500, if a 36% annual rate is charged?

52. If a 50% annual rate is charged, how much interest will be owed on a loan of $1,000 for 30 days?

53. A loan of $7,260 was repaid at the end of 8 months. What size repayment check (principal and interest) was written, if an 8% annual rate of interest was charged?

54. A loan of $10,000 was repaid at the end of 6 months. What amount (principal and interest) was repaid, if a 6.5% annual rate of interest was charged?

55. A loan of $4,000 was repaid at the end of 10 months with a check for $4,270. What annual rate of interest was charged?

56. A check for $3,097.50 was used to retire a 5-month $3,000 loan. What annual rate of interest was charged?

57. If you paid $30 to a loan company for the use of $1,000 for 60 days, what annual rate of interest did they charge?

58. If you paid $120 to a loan company for the use of $2,000 for 90 days, what annual rate of interest did they charge?

59. A radio commercial for a loan company states: "You only pay 29¢ a day for each $500 borrowed." If you borrow $1,500 for 120 days, what amount will you repay, and what annual interest rate is the company charging?

60. George finds a company that charges 59¢ per day for each $1,000 borrowed. If he borrows $3,000 for 60 days, what amount will he repay, and what annual interest rate will he pay the company?

61. What annual interest rate is earned by a 13-week T-bill with a maturity value of $1,000 that sells for $989.37?

62. What annual interest rate is earned by a 33-day T-bill with a maturity value of $1,000 that sells for $996.16?

63. What is the purchase price of a 50-day T-bill with a maturity value of $1,000 that earns an annual interest rate of 5.53%?

64. What is the purchase price of a 26-week T-bill with a maturity value of $1,000 that earns an annual interest rate of 4.903%?

In Problems 65 and 66, assume that the minimum payment on a credit card is the greater of $20 or 2% of the unpaid balance.

65. Find the minimum payment on an unpaid balance of $1,215.45.

66. Find the minimum payment on an unpaid balance of $936.24.

In Problems 67 and 68, assume that the minimum payment on a credit card is the greater of $27 or 3% of the unpaid balance.

67. Find the minimum payment on an unpaid balance of $815.69.

68. Find the minimum payment on an unpaid balance of $927.38.

69. For services rendered, an attorney accepts a 90-day note for $5,500 at 8% simple interest from a client. (Both interest and principal are repaid at the end of 90 days.) Wishing to use her money sooner, the attorney sells the note to a third party for $5,560 after 30 days. What annual interest rate will the third party receive for the investment?

70. To complete the sale of a house, the seller accepts a 180-day note for $10,000 at 7% simple interest. (Both interest and principal are repaid at the end of 180 days.) Wishing to use the money sooner for the purchase of another house, the seller sells the note to a third party for $10,124 after 60 days. What annual interest rate will the third party receive for the investment?

Use the commission schedule from Company A shown in Table 2 to find the annual rate of interest earned by each investment in Problems 71 and 72.

Table 2 Company A

Principal	Commission
Under $3,000	$25 + 1.8% of principal
$3,000–$10,000	$37 + 1.4% of principal
Over $10,000	$107 + 0.7% of principal

71. An investor purchases 200 shares at $14.20 a share, holds the stock for 39 weeks, and then sells the stock for $15.75 a share.

72. An investor purchases 450 shares at $21.40 a share, holds the stock for 26 weeks, and then sells the stock for $24.60 a share.

Use the commission schedule from Company B shown in Table 3 to find the annual rate of interest earned by each investment in Problems 73 and 74.

Table 3 Company B

Principal	Commission
Under $3,000	$32 + 1.8% of principal
$3,000–$10,000	$56 + 1% of principal
Over $10,000	$106 + 0.5% of principal

73. An investor purchases 215 shares at $45.75 a share, holds the stock for 300 days, and then sells the stock for $51.90 a share.

74. An investor purchases 75 shares at $37.90 a share, holds the stock for 150 days, and then sells the stock for $41.20 a share.

*The authors wish to thank Professor Roy Luke of Pierce College and Professor Dennis Pence of Western Michigan University for their many useful suggestions of applications for this chapter.

Many tax preparation firms offer their clients a refund anticipation loan (RAL). For a fee, the firm will give a client his refund when the return is filed. The loan is repaid when the IRS refund is sent to the firm. The RAL fee is equivalent to the interest charge for a loan. The schedule in Table 4 is from a major RAL lender. Use this schedule to find the annual rate of interest for the RALs in Problems 75–78.

Table 4

RAL Amount	RAL Fee
\$0–\$500	\$29.00
\$501–\$1,000	\$39.00
\$1,001–\$1,500	\$49.00
\$1,501–\$2,000	\$69.00
\$2,001–\$5,000	\$89.00

75. A client receives a \$475 RAL, which is paid back in 20 days.

76. A client receives a \$1,100 RAL, which is paid back in 30 days.

77. A client receives a \$1,900 RAL, which is paid back in 15 days.

78. A client receives a \$3,000 RAL, which is paid back in 25 days.

In Problems 79–82, assume that the annual interest rate on a credit card is 25.74% and interest is calculated by the average daily balance method.

79. The unpaid balance at the start of a 28-day billing cycle was \$955.13. A \$5,000 purchase was made on the first day of the billing cycle and a \$50 payment was credited to the account on day 21. How much interest will be charged at the end of the billing cycle?

80. The unpaid balance at the start of a 28-day billing cycle was \$955.13. A \$50 payment was credited to the account on day 21 of the billing cycle and a \$5,000 purchase was made on the last day of the billing cycle. How much interest will be charged at the end of the billing cycle?

81. The unpaid balance at the start of a 28-day billing cycle was \$1,472.35. Purchases of \$154.15 and \$38.76 were made on days 5 and 12, respectively, and a payment of \$250 was credited to the account on day 18. Find the unpaid balance at the end of the billing cycle.

82. The unpaid balance at the start of a 28-day billing cycle was \$1,837.23. Purchases of \$126.54 and \$52.89 were made on days 21 and 27, respectively, and a payment of \$100 was credited to the account on day 20. Find the unpaid balance at the end of the billing cycle.

In Problems 83–86, assume that the annual interest rate on a credit card is 19.99% and interest is calculated by the average daily balance method.

83. The unpaid balance at the start of a 30-day billing cycle was \$654.71. No purchases were made during the billing cycle and a payment of \$654.71 was credited to the account on day 21. Find the unpaid balance at the end of the billing cycle.

84. The unpaid balance at the start of a 30-day billing cycle was \$1,583.44. No purchases were made during the billing cycle and a payment of \$1,583.44 was credited to the account on day 21. Find the unpaid balance at the end of the billing cycle.

85. The unpaid balance at the start of a 30-day billing cycle was \$725.38. A purchase of \$49.82 was made on day 15. No payment was made during the billing cycle and a late fee of \$37 was charged to the account on day 25. Find the unpaid balance at the end of the billing cycle.

86. The unpaid balance at the start of a 30-day billing cycle was \$475.17. A purchase of \$125.93 was made on day 3. No payment was made during the billing cycle and a late fee of \$37 was charged to the account on day 25. Find the unpaid balance at the end of the billing cycle.

*A **payday loan** is a short-term loan that is repaid on the next payday, often by giving the lender electronic access to a personal checking account. Some states have statutes that regulate the fees that may be charged for payday loans. In Problems 87–90, express the annual interest rate as a percentage, rounded to the nearest integer.*

87. In Alabama, finance charges on a payday loan may not exceed 17.5% of the amount advanced. Find the annual interest rate if \$500 is borrowed for 10 days at the maximum allowable charge.

88. In Illinois, charges on a payday loan may not exceed \$15.50 per \$100 borrowed. Find the annual interest rate if \$400 is borrowed for 13 days at the maximum allowable charge.

89. In Kansas, charges on a payday loan may not exceed 15% of the amount advanced. Find the annual interest rate if \$450 is borrowed for 7 days at the maximum allowable charge.

90. In Louisiana, charges on a payday loan may not exceed 16.75% of the amount advanced. Find the annual interest rate if \$350 is borrowed for 14 days at the maximum allowable charge.

Answers to Matched Problems

1. \$650 **2.** \$4,761.90 **3.** 3.485%
4. 15.0% **5.** 31.439% **6.** \$26.94; \$1,911.70

3.2 Compound and Continuous Compound Interest

- Compound Interest
- Continuous Compound Interest
- Growth and Time
- Annual Percentage Yield

Compound Interest

If at the end of a payment period the interest due is reinvested at the same rate, then the interest as well as the original principal will earn interest during the next payment period. Interest paid on interest reinvested is called **compound interest**.

For example, suppose you deposit \$1,000 in a bank that pays 8% compounded quarterly. How much will the bank owe you at the end of a year? *Compounding quarterly* means that earned interest is paid to your account at the end of each 3-month period and that interest as well as the principal earns interest for the next quarter. Using the simple interest formula (2) from the preceding section, we compute the amount in the account at the end of the first quarter after interest has been paid:

$$\begin{aligned} A &= P(1 + rt) \\ &= 1{,}000[1 + 0.08(\tfrac{1}{4})] \\ &= 1{,}000(1.02) = \$1{,}020 \end{aligned}$$

Now, $1,020 is your new principal for the second quarter. At the end of the second quarter, after interest is paid, the account will have

$$\begin{aligned} A &= \$1{,}020[1 + 0.08(\tfrac{1}{4})] \\ &= \$1{,}020(1.02) = \$1{,}040.40 \end{aligned}$$

Similarly, at the end of the third quarter, you will have

$$\begin{aligned} A &= \$1{,}040.40[1 + 0.08(\tfrac{1}{4})] \\ &= \$1{,}040.40(1.02) = \$1{,}061.21 \end{aligned}$$

Finally, at the end of the fourth quarter, the account will have

$$\begin{aligned} A &= \$1{,}061.21[1 + 0.08(\tfrac{1}{4})] \\ &= \$1{,}061.21(1.02) = \$1{,}082.43 \end{aligned}$$

How does this compounded amount compare with simple interest? The amount with simple interest would be

$$\begin{aligned} A &= P(1 + rt) \\ &= \$1{,}000[1 + 0.08(1)] \\ &= \$1{,}000(1.08) = \$1{,}080 \end{aligned}$$

We see that compounding quarterly yields $2.43 more than simple interest would provide.

Let's look over the calculations for compound interest above to see if we can uncover a pattern that might lead to a general formula for computing compound interest:

$$\begin{aligned} A &= 1{,}000(1.02) && \text{End of first quarter} \\ A &= [1{,}000(1.02)](1.02) = 1{,}000(1.02)^2 && \text{End of second quarter} \\ A &= [1{,}000(1.02)^2](1.02) = 1{,}000(1.02)^3 && \text{End of third quarter} \\ A &= [1{,}000(1.02)^3](1.02) = 1{,}000(1.02)^4 && \text{End of fourth quarter} \end{aligned}$$

It appears that at the end of n quarters, we would have

$$A = 1{,}000(1.02)^n \qquad \text{End of } n\text{th quarter}$$

or

$$\begin{aligned} A &= 1{,}000[1 + 0.08(\tfrac{1}{4})]^n \\ &= 1{,}000[1 + \tfrac{0.08}{4}]^n \end{aligned}$$

where $\frac{0.08}{4} = 0.02$ is the interest rate per quarter. Since interest rates are generally quoted as *annual nominal rates*, the **rate per compounding period** is found by dividing the annual nominal rate by the number of compounding periods per year.

In general, if P is the principal earning interest compounded m times a year at an annual rate of r, then (by repeated use of the simple interest formula, using $i = r/m$, the rate per period) the amount A at the end of each period is

$$A = P(1 + i) \quad \text{End of first period}$$
$$A = [P(1 + i)](1 + i) = P(1 + i)^2 \quad \text{End of second period}$$
$$A = [P(1 + i)^2](1 + i) = P(1 + i)^3 \quad \text{End of third period}$$
$$\vdots$$
$$A = [P(1 + i)^{n-1}](1 + i) = P(1 + i)^n \quad \text{End of } n\text{th period}$$

We summarize this important result in Theorem 1:

THEOREM 1 Compound Interest

$$A = P(1 + i)^n \tag{1}$$

where $i = r/m$ and A = amount (future value) at the end of n periods

P = principal (present value)

r = annual nominal rate*

m = number of compounding periods per year

i = rate per compounding period

n = total number of compounding periods

* This is often shortened to "annual rate" or just "rate."

CONCEPTUAL INSIGHT

Formula (1) of Theorem 1 is equivalent to the formula

$$A = P\left(1 + \frac{r}{m}\right)^{mt} \tag{2}$$

where t is the time, in years, that the principal is invested. For a compound interest calculation, formula (2) may seem more natural to use than (1), if r (the annual interest rate) and t (time in years) are given. On the other hand, if i (the interest rate per period), and n (the number of compounding periods) are given, formula (1) may seem easier to use. It is not necessary to memorize both formulas, but it is important to understand how they are related.

EXAMPLE 1 **Comparing Interest for Various Compounding Periods** If \$1,000 is invested at 8% compounded

(A) annually, (B) semiannually,

(C) quarterly, (D) monthly,

what is the amount after 5 years? Write answers to the nearest cent.

SOLUTION

(A) Compounding annually means that there is one interest payment period per year. So, $n = 5$ and $i = r = 0.08$.

$$A = P(1 + i)^n$$
$$= 1{,}000(1 + 0.08)^5 \quad \text{Use a calculator.}$$
$$= 1{,}000(1.469\,328)$$
$$= \$1{,}469.33 \quad \text{Interest earned} = A - p = \$469.33.$$

(B) Compounding semiannually means that there are two interest payment periods per year. The number of payment periods in 5 years is $n = 2(5) = 10$, and the interest rate per period is

$$i = \frac{r}{m} = \frac{0.08}{2} = 0.04$$

$$\begin{aligned} A &= P(1+i)^n \\ &= 1{,}000(1+0.04)^{10} && \text{Use a calculator.} \\ &= 1{,}000(1.480\,244) \\ &= \$1{,}480.24 && \text{Interest earned} = A - P = \$480.24. \end{aligned}$$

(C) Compounding quarterly means that there are four interest payments per year. So, $n = 4(5) = 20$ and $i = \frac{0.08}{4} = 0.02$.

$$\begin{aligned} A &= P(1+i)^n \\ &= 1{,}000(1+0.02)^{20} && \text{Use a calculator.} \\ &= 1{,}000(1.485\,947) \\ &= \$1{,}485.95 && \text{Interest earned} = A - P = \$485.95. \end{aligned}$$

(D) Compounding monthly means that there are twelve interest payments per year. So, $n = 12(5) = 60$ and $i = \frac{0.08}{12} = 0.006\,66\overline{6}$ (see the Reminder).

$$\begin{aligned} A &= P(1+i)^n \\ &= 1{,}000\left(1+\frac{0.08}{12}\right)^{60} && \text{Use a calculator.} \\ &= 1{,}000(1.489\,846) \\ &= \$1{,}489.85 && \text{Interest earned} = A - P = \$489.85. \end{aligned}$$

Reminder

The bar over the 6 in $i = 0.006\,66\overline{6}$ indicates a repeating decimal expansion. Rounding i to a small number of decimal places, such as 0.007 or 0.0067, can result in round-off errors. To avoid this, use as many decimal places for i as your calculator is capable of displaying.

Matched Problem 1 Repeat Example 1 with an annual interest rate of 6% over an 8-year period.

Continuous Compound Interest

In Example 1, we considered an investment of $1,000 at an annual rate of 8%. We calculated the amount after 5 years for interest compounded annually, semiannually, quarterly, and monthly. What would happen to the amount if interest were compounded daily, or every minute, or every second?

Although the difference in amounts in Example 1 between compounding semiannually and annually is $1,480.24 − $1,469.33 = $10.91, the difference between compounding monthly and quarterly is only $1,489.85 − $1,485.95 = $3.90. This suggests that as the number m of compounding periods per year increases without bound, the amount will approach some limiting value. To see that this is indeed the case, we rewrite the amount A as follows:

$$\begin{aligned} A &= P(1+i)^n && \text{Substitute } i = \frac{r}{m},\ n = mt. \\ &= P\left(1+\frac{r}{m}\right)^{mt} && \text{Multiply the exponent by } \frac{r}{r}\ (=1). \\ &= P\left(1+\frac{r}{m}\right)^{[m/r]rt} && \text{Let } x = \frac{m}{r}\text{; then } \frac{1}{x} = \frac{r}{m}. \\ &= P\left(1+\frac{1}{x}\right)^{xrt} && \text{Use a law of exponents: } a^{xy} = (a^x)^y. \\ &= P\left[\left(1+\frac{1}{x}\right)^{x}\right]^{rt} \end{aligned}$$

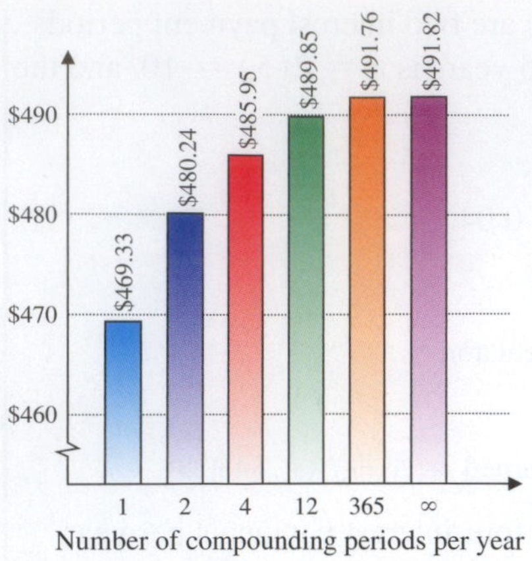

Figure 1 Interest on $1,000 for 5 years at 8% with various compounding periods

As the number m of compounding periods increases without bound, so does x. So the expression in square brackets gets close to the irrational number $e \approx 2.7183$ (see Table 1 in Section 2.5), and the amount approaches the limiting value

$$A = Pe^{rt} = 1{,}000e^{0.08(5)} \approx \$1{,}491.8247$$

In other words, no matter how often interest is compounded, the amount in the account after 5 years will never equal or exceed $1,491.83. Therefore, the interest $I(= A - P)$ will never equal or exceed $491.83 (Fig. 1).

CONCEPTUAL INSIGHT

One column in Figure 1 is labeled with the symbol ∞, read as "infinity." This symbol does not represent a real number. We use ∞ to denote the process of allowing m, the number of compounding periods per year, to get larger and larger with no upper limit on its size.

The formula we have obtained, $A = Pe^{rt}$, is known as the **continuous compound interest formula**. It is used when interest is **compounded continuously**, that is, when the number of compounding periods per year increases without bound.

THEOREM 2 Continuous Compound Interest Formula

If a principal P is invested at an annual rate r (expressed as a decimal) compounded continuously, then the amount A in the account at the end of t years is given by

$$A = Pe^{rt} \tag{3}$$

EXAMPLE 2 **Compounding Daily and Continuously** What amount will an account have after 2 years if $5,000 is invested at an annual rate of 8%

(A) compounded daily? (B) compounded continuously?

Compute answers to the nearest cent.

SOLUTION

(A) Use the compound interest formula

$$A = P\left(1 + \frac{r}{m}\right)^{mt}$$

with $P = 5{,}000$, $r = 0.08$, $m = 365$, and $t = 2$:

$$A = 5{,}000\left(1 + \frac{0.08}{365}\right)^{(365)(2)} \quad \text{Use a calculator.}$$
$$= \$5{,}867.45$$

(B) Use the continuous compound interest formula

$$A = Pe^{rt}$$

with $P = 5{,}000$, $r = 0.08$, and $t = 2$:

$$A = 5{,}000e^{(0.08)(2)} \quad \text{Use a calculator.}$$
$$= \$5{,}867.55$$

CAUTION In Example 2B, do not use the approximation 2.7183 for e; it is not accurate enough to compute the correct amount to the nearest cent. Instead, use your calculator's built-in e. Avoid any rounding off until the end of the calculation, when you round the amount to the nearest cent. ▲

Matched Problem 2 What amount will an account have after 1.5 years if \$8,000 is invested at an annual rate of 9%

(A) compounded weekly? (B) compounded continuously?

Compute answers to the nearest cent.

CONCEPTUAL INSIGHT

The continuous compound interest formula $A = Pe^{rt}$ is identical, except for the names of the variables, to the equation $y = ce^{kt}$ that we used to model population growth in Section 2.5. Like the growth of an investment that earns continuous compound interest, we usually consider the population growth of a country to be continuous: Births and deaths occur all the time, not just at the end of a month or quarter.

Growth and Time

How much should you invest now to have a given amount at a future date? What annual rate of return have your investments earned? How long will it take your investment to double in value? The formulas for compound interest and continuous compound interest can be used to answer such questions. If the values of all but one of the variables in the formula are known, then we can solve for the remaining variable.

EXAMPLE 3 **Finding Present Value** How much should you invest now at 10% to have \$8,000 toward the purchase of a car in 5 years if interest is

(A) compounded quarterly? (B) compounded continuously?

SOLUTION

(A) We are given a future value $A = \$8{,}000$ for a compound interest investment, and we need to find the present value P given $i = \frac{0.10}{4} = 0.025$ and $n = 4(5) = 20$.

$$A = P(1 + i)^n$$

$$8{,}000 = P(1 + 0.025)^{20}$$

$$P = \frac{8{,}000}{(1 + 0.025)^{20}} \qquad \text{Use a calculator.}$$

$$= \frac{8{,}000}{1.638\,616}$$

$$= \$4{,}882.17$$

Your initial investment of \$4,882.17 will grow to \$8,000 in 5 years.

(B) We are given $A = \$8{,}000$ for an investment at continuous compound interest, and we need to find the present value P given $r = 0.10$ and $t = 5$.

$$A = Pe^{rt}$$

$$8{,}000 = Pe^{0.10(5)}$$

$$P = \frac{8{,}000}{e^{0.10(5)}} \qquad \text{Use a calculator.}$$

$$P = \$4{,}852.25$$

Your initial investment of \$4,852.25 will grow to \$8,000 in 5 years.

Matched Problem 3 How much should new parents invest at 8% to have \$80,000 toward their child's college education in 17 years if interest is

(A) compounded semiannually? (B) compounded continuously?

A graphing calculator is a useful tool for studying compound interest. In Figure 2, we use a spreadsheet to illustrate the growth of the investment in Example 3A both numerically and graphically. Similar results can be obtained from most graphing calculators.

	A	B	C
1	Period	Interest	Amount
2	0		$4,882.17
3	1	$122.05	$5,004.22
4	2	$125.11	$5,129.33
5	3	$128.23	$5,257.56
6	4	$131.44	$5,389.00
7	5	$134.73	$5,523.73
8	6	$138.09	$5,661.82
9	7	$141.55	$5,803.37
10	8	$145.08	$5,948.45
11	9	$148.71	$6,097.16
12	10	$152.43	$6,249.59
13	11	$156.24	$6,405.83
14	12	$160.15	$6,565.98
15	13	$164.15	$6,730.13
16	14	$168.25	$6,898.38
17	15	$172.46	$7,070.84
18	16	$176.77	$7,247.61
19	17	$181.19	$7,428.80
20	18	$185.72	$7,614.52
21	19	$190.36	$7,804.88
22	20	$195.12	$8,000.00

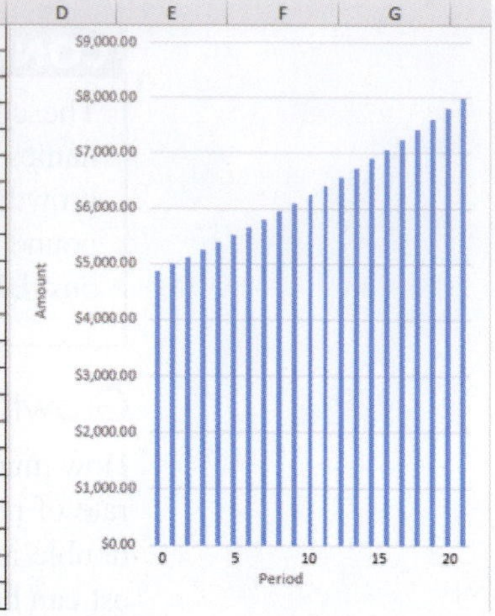

Figure 2 Growth of $4,882.17 at 10% compounded quarterly for 5 years

Solving the compound interest formula or the continuous compound interest formula for r enables us to determine the rate of growth of an investment.

EXAMPLE 4 **Computing Growth Rate** Figure 3 shows that a $10,000 investment in a growth-oriented mutual fund over a 10-year period would have grown to $126,000. What annual nominal rate would produce the same growth if interest was:

(A) compounded annually? (B) compounded continuously?

Express answers as percentages, rounded to three decimal places.

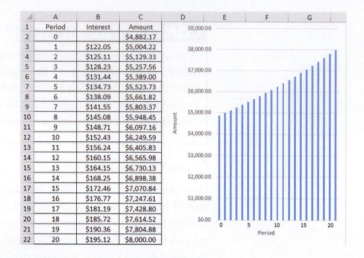

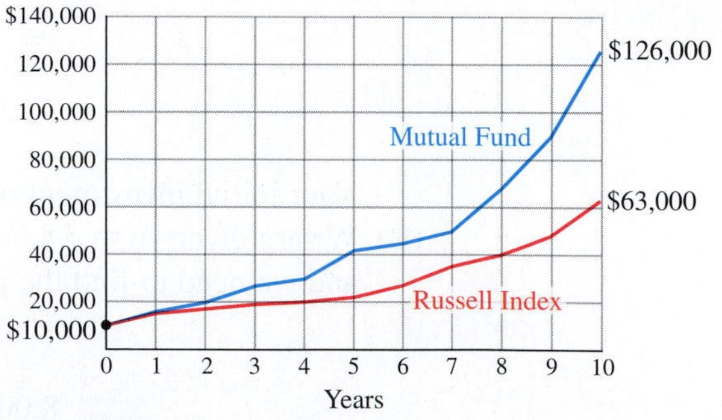

Figure 3 Growth of a $10,000 investment

SOLUTION

(A) $126{,}000 = 10{,}000(1 + r)^{10}$

$$12.6 = (1 + r)^{10}$$

$$\sqrt[10]{12.6} = 1 + r$$

$$r = \sqrt[10]{12.6} - 1 = 0.28836 \quad \text{or} \quad 28.836\%$$

(B) $126{,}000 = 10{,}000e^{r(10)}$

$$12.6 = e^{10r}$$ Take ln of both sides.

$$\ln 12.6 = 10r$$

$$r = \frac{\ln 12.6}{10} = 0.25337 \quad \text{or} \quad 25.337\%$$

Matched Problem 4 The Russell Index tracks the average performance of various groups of stocks. Figure 3 shows that, on average, a $10,000 investment in midcap growth funds over a 10-year period would have grown to $63,000. What annual nominal rate would produce the same growth if interest were

(A) compounded annually? (B) compounded continuously?

Express answers as percentages, rounded to three decimal places.

CONCEPTUAL INSIGHT

We can solve $A = P(1 + i)^n$ for n using a property of logarithms:

$$\log_b M^p = p \log_b M$$

Theoretically, any base can be used for the logarithm, but most calculators only evaluate logarithms with base 10 (denoted log) or base e (denoted ln).

Finally, if we solve the compound interest formula for n (or the continuous compound interest formula for t), we can determine the **growth time** of an investment—the time it takes a given principal to grow to a particular value (the shorter the time, the greater the return on the investment).

Example 5 illustrates three methods for solving for growth time.

EXAMPLE 5 **Computing Growth Time** How long will it take $10,000 to grow to $12,000 if it is invested at 9% compounded monthly?

SOLUTION

Method 1. Use logarithms and a calculator:

$$A = P(1 + i)^n$$

$$12{,}000 = 10{,}000\left(1 + \frac{0.09}{12}\right)^n$$

$$1.2 = 1.0075^n$$

Now, solve for n by taking logarithms of both sides:

$$\ln 1.2 = \ln 1.0075^n$$ We choose the natural logarithm (base e)

$$\ln 1.2 = n \ln 1.0075$$ and use the property $\ln M^p = p \ln M$.

$$n = \frac{\ln 1.2}{\ln 1.0075}$$

$$\approx 24.40 \approx 25 \text{ months} \quad \text{or} \quad 2 \text{ years and } 1 \text{ month}$$

Note: 24.40 is rounded up to 25 to guarantee reaching $12,000 since interest is paid at the end of each month.

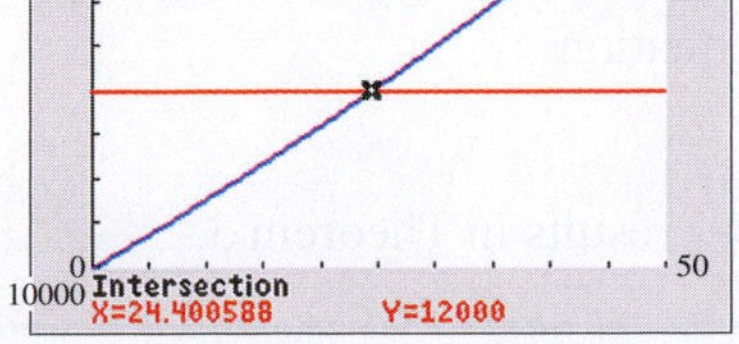

$y_1 = 10{,}000(1.0075)^x$
$y_2 = 12{,}000$

Figure 4

Method 2. Use a graphing calculator: To solve this problem using graphical approximation techniques, we graph both sides of the equation $12{,}000 = 10{,}000(1.0075)^n$ and find that the graphs intersect at $x = n = 24.40$ months (Fig. 4). So the growth time is 25 months.

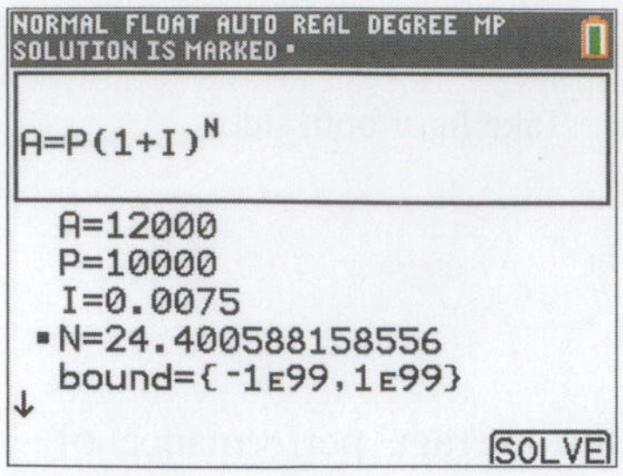

Figure 5 TI-84 Plus CE equation solver

Method 3. Most graphing calculators have an approximation process that is referred to as an **equation solver**. Figure 5 shows the equation solver on a TI-84 Plus CE. After entering values for three of the four variables, the solver will approximate the value of the remaining variable. Once again, we see that the growth time is 25 months (Fig. 5).

Matched Problem 5 How long will it take \$10,000 to grow to \$25,000 if it is invested at 8% compounded quarterly?

Annual Percentage Yield

Table 1 lists the rate and compounding period for certificates of deposit (CDs) offered by four banks. How can we tell which of these CDs has the best return?

Table 1 Certificates of Deposit (CDs)

Bank	Rate	Compounded
Advanta	4.93%	monthly
DeepGreen	4.95%	daily
Charter One	4.97%	quarterly
Liberty	4.94%	continuously

Explore and Discuss 1

Determine the value after 1 year of a \$1,000 CD purchased from each of the banks in Table 1. Which CD offers the greatest return? Which offers the least return?

If a principal P is invested at an annual rate r compounded m times a year, then the amount after 1 year is

$$A = P\left(1 + \frac{r}{m}\right)^m$$

The simple interest rate that will produce the same amount A in 1 year is called the **annual percentage yield** (APY). To find the APY, we proceed as follows:

$$\begin{pmatrix}\text{amount at}\\ \text{simple interest}\\ \text{after 1 year}\end{pmatrix} = \begin{pmatrix}\text{amount at}\\ \text{compound interest}\\ \text{after 1 year}\end{pmatrix}$$

$$P(1 + \text{APY}) = P\left(1 + \frac{r}{m}\right)^m \qquad \text{Divide both sides by } P.$$

$$1 + \text{APY} = \left(1 + \frac{r}{m}\right)^m \qquad \text{Isolate APY on the left side.}$$

$$\text{APY} = \left(1 + \frac{r}{m}\right)^m - 1$$

If interest is compounded continuously, then the amount after 1 year is $A = Pe^r$. So to find the annual percentage yield, we solve the equation

$$P(1 + \text{APY}) = Pe^r$$

for APY, obtaining $\text{APY} = e^r - 1$. We summarize our results in Theorem 3.

THEOREM 3 Annual Percentage Yield

If a principal is invested at the annual (nominal) rate r compounded m times a year, then the annual percentage yield is

$$\text{APY} = \left(1 + \frac{r}{m}\right)^m - 1$$

If a principal is invested at the annual (nominal) rate r compounded continuously, then the annual percentage yield is

$$\text{APY} = e^r - 1$$

The annual percentage yield is also referred to as the **effective rate** or **true interest rate**.

Compound rates with different compounding periods cannot be compared directly (see Explore and Discuss 1). But since the annual percentage yield is a simple interest rate, the annual percentage yields for two different compound rates can be compared.

EXAMPLE 6 **Using APY to Compare Investments** Find the APYs (expressed as a percentage, correct to three decimal places) for each of the banks in Table 1 and compare these CDs.

SOLUTION

$$\text{Advanta: APY} = \left(1 + \frac{0.0493}{12}\right)^{12} - 1 = 0.05043 \quad \text{or} \quad 5.043\%$$

$$\text{DeepGreen: APY} = \left(1 + \frac{0.0495}{365}\right)^{365} - 1 = 0.05074 \quad \text{or} \quad 5.074\%$$

$$\text{Charter One: APY} = \left(1 + \frac{0.0497}{4}\right)^{4} - 1 = 0.05063 \quad \text{or} \quad 5.063\%$$

$$\text{Liberty: APY} = e^{0.0494} - 1 = 0.05064 \quad \text{or} \quad 5.064\%$$

Comparing these APYs, we conclude that the DeepGreen CD will have the largest return and the Advanta CD will have the smallest.

Matched Problem 6 Southern Pacific Bank offered a 1-year CD that paid 4.8% compounded daily and Washington Savings Bank offered one that paid 4.85% compounded quarterly. Find the APY (expressed as a percentage, correct to three decimal places) for each CD. Which has the higher return?

EXAMPLE 7 **Computing the Annual Nominal Rate Given the APY** A savings and loan wants to offer a CD with a monthly compounding rate that has an APY of 7.5%. What annual nominal rate compounded monthly should it use? Check with a graphing calculator.

SOLUTION

$$\text{APY} = \left(1 + \frac{r}{m}\right)^m - 1$$

$$0.075 = \left(1 + \frac{r}{12}\right)^{12} - 1$$

$$1.075 = \left(1 + \frac{r}{12}\right)^{12}$$

$$\sqrt[12]{1.075} = 1 + \frac{r}{12}$$

$$\sqrt[12]{1.075} - 1 = \frac{r}{12}$$

$$r = 12(\sqrt[12]{1.075} - 1) \qquad \text{Use a calculator.}$$

$$= 0.072\,539 \quad \text{or} \quad 7.254\%$$

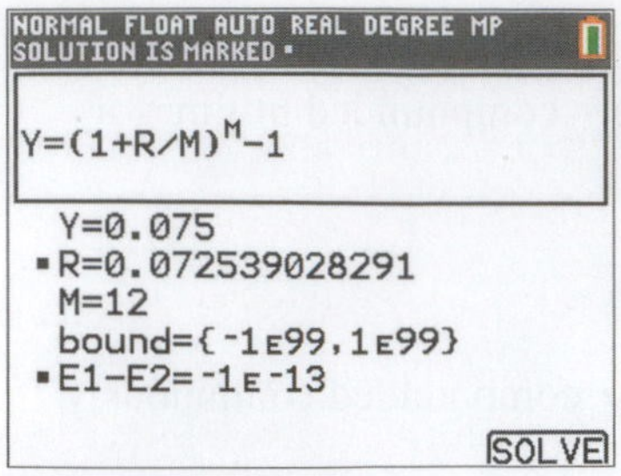

Figure 6 TI-84 Plus CE equation solver

So an annual nominal rate of 7.254% compounded monthly is equivalent to an APY of 7.5%.

CHECK We use an equation solver on a graphing calculator to check this result (Fig. 6).

Matched Problem 7 What is the annual nominal rate compounded quarterly for a bond that has an APY of 5.8%?

CAUTION Each compound interest problem involves two interest rates. Referring to Example 5, $r = 0.09$ or 9% is the annual nominal compounding rate, and $i = r/12 = 0.0075$ or 0.75% is the interest rate per month. Do not confuse these two rates by using r in place of i in the compound interest formula. If interest is compounded annually, then $i = r/1 = r$. In all other cases, r and i are not the same. ▲

Explore and Discuss 2

(A) Which would be the better way to invest $1,000: at 9% simple interest for 10 years, or at 7% compounded monthly for 10 years?

(B) Explain why the graph of future value as a function of time is a straight line for simple interest, but for compound interest the graph curves upward (see Fig. 7).

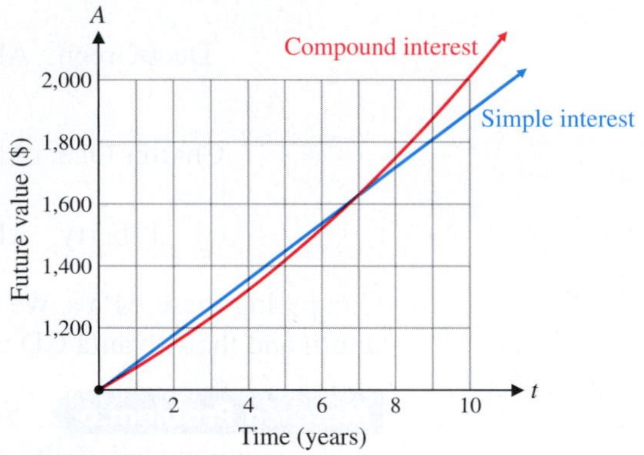

Figure 7

CONCEPTUAL INSIGHT

The two curves in Figure 7 intersect at $t = 0$ and again near $t = 7$. The t coordinate of each intersection point is a solution of the equation

$$1{,}000(1 + 0.09t) = 1{,}000(1 + 0.07/12)^{12t}$$

Don't try to use algebra to solve this equation. It can't be done. But the solutions are easily approximated on a graphing calculator (Fig. 8).

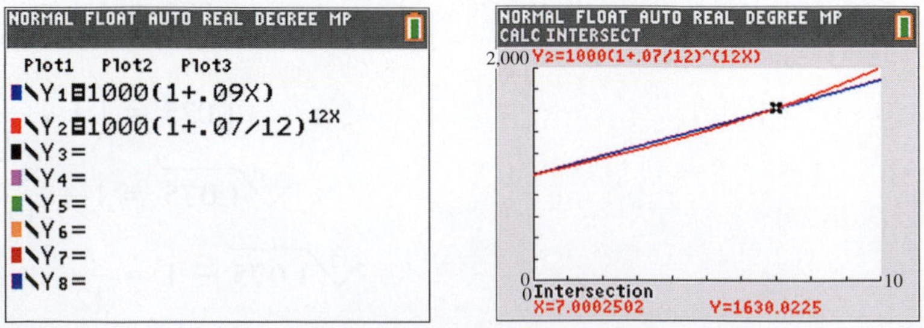

Figure 8

Exercises 3.2

Find all dollar amounts to the nearest cent. When an interest rate is requested as an answer, express the rate as a percentage correct to two decimal places, unless directed otherwise. In all problems involving days, use a 365-day year.

W Skills Warm-up Exercises

In Problems 1–8, solve the equation for the unknown quantity. (If necessary, review sections A.7, 2.5, and 2.6.)

1. $1{,}641.6 = P(1.2)^3$ **2.** $2{,}652.25 = P(1.03)^2$

3. $12x^3 = 58{,}956$ **4.** $100x^4 = 15{,}006.25$

5. $6.75 = 3(1 + i)^2$ **6.** $13.72 = 5(1 + i)^3$

7. $14{,}641 = 10{,}000(1.1)^n$ **8.** $2{,}488.32 = 1{,}000(1.2)^n$

A *In Problems 9–12, use compound interest formula (1) to find each of the indicated values.*

9. $P = \$5{,}000;\ i = 0.005;\ n = 36;\ A = ?$

10. $P = \$2{,}800;\ i = 0.003;\ n = 24;\ A = ?$

11. $A = \$8{,}000;\ i = 0.02;\ n = 32;\ P = ?$

12. $A = \$15{,}000;\ i = 0.01;\ n = 28;\ P = ?$

In Problems 13–20, use the continuous compound interest formula (3) to find each of the indicated values.

13. $P = \$2{,}450;\ r = 8.12\%;\ t = 3$ years; $A = ?$

14. $P = \$995;\ r = 22\%;\ t = 2$ years; $A = ?$

15. $A = \$6{,}300;\ r = 9.45\%;\ t = 8$ years; $P = ?$

16. $A = \$19{,}000;\ r = 7.69\%;\ t = 5$ years; $P = ?$

17. $A = \$88{,}000;\ P = \$71{,}153;\ r = 8.5\%;\ t = ?$

18. $A = \$32{,}982;\ P = \$27{,}200;\ r = 5.93\%;\ t = ?$

19. $A = \$15{,}875;\ P = \$12{,}100;\ t = 48$ months; $r = ?$

20. $A = \$23{,}600;\ P = \$19{,}150;\ t = 60$ months; $r = ?$

In Problems 21–28, use the given annual interest rate r and the compounding period to find i, the interest rate per compounding period.

21. 6.6% compounded quarterly

22. 3.84% compounded monthly

23. 5.52% compounded monthly

24. 2.94% compounded semiannually

25. 7.3% compounded daily

26. 5.44% compounded quarterly

27. 4.86% compounded semiannually

28. 10.95% compounded daily

In Problems 29–36, use the given interest rate i per compounding period to find r, the annual rate.

29. 1.73% per half-year

30. 1.57% per quarter

31. 0.53% per month

32. 0.012% per day

33. 2.19% per quarter

34. 3.69% per half-year

35. 0.008% per day

36. 0.47% per month

B **37.** If \$100 is invested at 6% compounded

(A) annually (B) quarterly (C) monthly

what is the amount after 4 years? How much interest is earned?

38. If \$2,000 is invested at 7% compounded

(A) annually (B) quarterly (C) monthly

what is the amount after 5 years? How much interest is earned?

39. If \$5,000 is invested at 5% compounded monthly, what is the amount after

(A) 2 years? (B) 4 years?

40. If \$20,000 is invested at 4% compounded monthly, what is the amount after

(A) 5 years? (B) 8 years?

41. If \$8,000 is invested at 7% compounded continuously, what is the amount after 6 years?

42. If \$23,000 is invested at 13.5% compounded continuously, what is the amount after 15 years?

43. Discuss the similarities and the differences in the graphs of future value A as a function of time t if \$1,000 is invested for 8 years and interest is compounded monthly at annual rates of 4%, 8%, and 12%, respectively (see the figure).

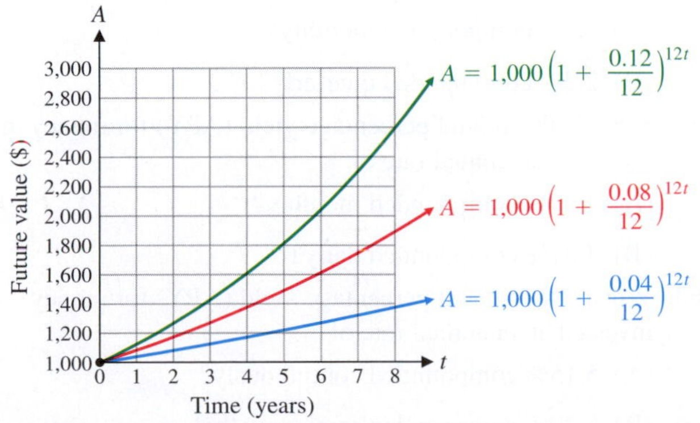

44. Discuss the similarities and differences in the graphs of future value A as a function of time t for loans of \$4,000, \$8,000, and \$12,000, respectively, each at 7.5% compounded monthly for 8 years (see the figure).

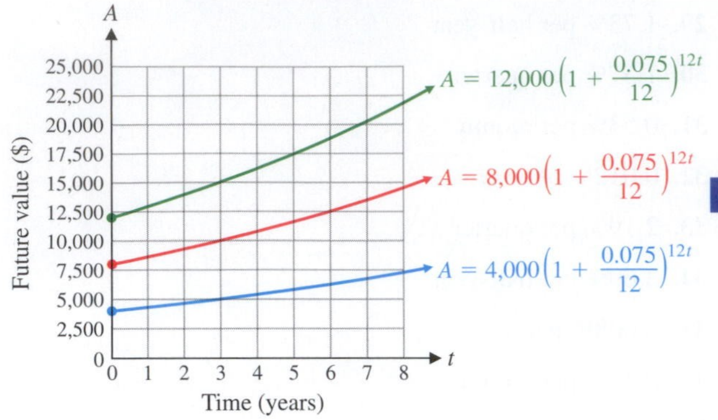

45. If \$1,000 is invested in an account that earns 9.75% compounded annually for 6 years, find the interest earned during each year and the amount in the account at the end of each year. Organize your results in a table.

46. If \$2,000 is invested in an account that earns 8.25% compounded annually for 5 years, find the interest earned during each year and the amount in the account at the end of each year. Organize your results in a table.

47. If an investment company pays 6% compounded semiannually, how much should you deposit now to have \$10,000

(A) 5 years from now? (B) 10 years from now?

48. If an investment company pays 8% compounded quarterly, how much should you deposit now to have \$6,000

(A) 3 years from now? (B) 6 years from now?

49. If an investment earns 9% compounded continuously, how much should you deposit now to have \$25,000

(A) 36 months from now? (B) 9 years from now?

50. If an investment earns 12% compounded continuously, how much should you deposit now to have \$4,800

(A) 48 months from now? (B) 7 years from now?

51. What is the annual percentage yield (APY) for money invested at an annual rate of

(A) 3.9% compounded monthly?

(B) 2.3% compounded quarterly?

52. What is the annual percentage yield (APY) for money invested at an annual rate of

(A) 4.32% compounded monthly?

(B) 4.31% compounded daily?

53. What is the annual percentage yield (APY) for money invested at an annual rate of

(A) 5.15% compounded continuously?

(B) 5.20% compounded semiannually?

54. What is the annual percentage yield (APY) for money invested at an annual rate of

(A) 3.05% compounded quarterly?

(B) 2.95% compounded continuously?

55. How long will it take \$4,000 to grow to \$9,000 if it is invested at 7% compounded monthly?

56. How long will it take \$5,000 to grow to \$7,000 if it is invested at 6% compounded quarterly?

57. How long will it take \$6,000 to grow to \$8,600 if it is invested at 9.6% compounded continuously?

58. How long will it take \$42,000 to grow to \$60,276 if it is invested at 4.25% compounded continuously?

C *In Problems 59 and 60, use compound interest formula (1) to find n to the nearest larger integer value.*

59. $A = 2P;\ i = 0.06;\ n = ?$

60. $A = 2P;\ i = 0.05;\ n = ?$

61. How long will it take money to double if it is invested at

(A) 10% compounded quarterly?

(B) 12% compounded quarterly?

62. How long will it take money to double if it is invested at

(A) 8% compounded semiannually?

(B) 7% compounded semiannually?

63. How long will it take money to double if it is invested at

(A) 9% compounded continuously?

(B) 11% compounded continuously?

64. How long will it take money to double if it is invested at

(A) 21% compounded continuously?

(B) 33% compounded continuously?

Applications

65. A newborn child receives a \$20,000 gift toward college from her grandparents. How much will the \$20,000 be worth in 17 years if it is invested at 7% compounded quarterly?

66. A person with \$14,000 is trying to decide whether to purchase a car now, or to invest the money at 6.5% compounded semiannually and then buy a more expensive car. How much will be available for the purchase of a car at the end of 3 years?

67. What will a \$210,000 house cost 10 years from now if the inflation rate over that period averages 3% compounded annually?

68. If the inflation rate averages 4% per year compounded annually for the next 5 years, what will a car that costs \$17,000 now cost 5 years from now?

69. Rental costs for office space have been going up at 4.8% per year compounded annually for the past 5 years. If office space rent is now \$25 per square foot per month, what were the rental rates 5 years ago?

70. In a suburb, housing costs have been increasing at 5.2% per year compounded annually for the past 8 years. A house worth \$260,000 now would have had what value 8 years ago?

71. (A) If an investment of \$100 were made in 1776, and if it earned 3% compounded quarterly, how much would it be worth in 2026?

(B) Discuss the effect of compounding interest monthly, daily, and continuously (rather than quarterly) on the \$100 investment.

(C) Use a graphing calculator to graph the growth of the investment of part (A).

72. (A) Starting with formula (1), derive each of the following formulas:

$$P = \frac{A}{(1+i)^n}, \quad i = \left(\frac{A}{P}\right)^{1/n} - 1, \quad n = \frac{\ln A - \ln P}{\ln(1+i)}$$

(B) Explain why it is unnecessary to memorize the formulas above for *P*, *i*, and *n* if you know formula (1).

73. A promissory note will pay \$50,000 at maturity 6 years from now. If you pay \$28,000 for the note now, what rate compounded continuously would you earn?

74. If you deposit \$10,000 in a savings account now, what rate compounded continuously would be required for you to withdraw \$12,500 at the end of 4 years?

75. You have saved \$7,000 toward the purchase of a car costing \$9,000. How long will the \$7,000 have to be invested at 9% compounded monthly to grow to \$9,000? (Round up to the next-higher month if not exact.)

76. A married couple has \$15,000 toward the purchase of a house. For the house that the couple wants to buy, a down payment of \$20,000 is required. How long will the money have to be invested at 7% compounded quarterly to grow to \$20,000? (Round up to the next-higher quarter if not exact.)

77. An Individual Retirement Account (IRA) has \$20,000 in it, and the owner decides not to add any more money to the account other than interest earned at 6% compounded daily. How much will be in the account 35 years from now when the owner reaches retirement age?

78. If \$1 had been placed in a bank account in the year 1066 and forgotten until now, how much would be in the account at the end of 2026 if the money earned 2% interest compounded annually? 2% simple interest? (Now you can see the power of compounding and why inactive accounts are closed after a relatively short period of time.)

79. How long will it take money to double if it is invested at 7% compounded daily? 8.2% compounded continuously?

80. How long will it take money to triple if it is invested at 5% compounded daily? 6% compounded continuously?

81. In a conversation with a friend, you note that you have two real estate investments, one that has doubled in value in the past 9 years and another that has doubled in value in the past 12 years. Your friend says that the first investment has been growing at approximately 8% compounded annually and the second at 6% compounded annually. How did your friend make these estimates? The **rule of 72** states that the annual compound rate of growth *r* of an investment that doubles in *n* years can be approximated by $r = 72/n$. Construct a table comparing the exact rate of growth and the approximate rate provided by the rule of 72 for doubling times of $n = 6, 7, \ldots, 12$ years. Round both rates to one decimal place.

82. Refer to Problem 81. Show that the exact annual compound rate of growth of an investment that doubles in *n* years is given by $r = 100(2^{1/n} - 1)$. Graph this equation and the rule of 72 on a graphing calculator for $5 \le n \le 20$.

Solve Problems 83–86 using graphical approximation techniques on a graphing calculator.

83. How long does it take for a \$2,400 investment at 13% compounded quarterly to be worth more than a \$3,000 investment at 6% compounded quarterly?

84. How long does it take for a \$4,800 investment at 8% compounded monthly to be worth more than a \$5,000 investment at 5% compounded monthly?

85. One investment pays 10% simple interest and another pays 7% compounded annually. Which investment would you choose? Why?

86. One investment pays 9% simple interest and another pays 6% compounded monthly. Which investment would you choose? Why?

87. What is the annual nominal rate compounded daily for a bond that has an annual percentage yield of 3.39%?

88. What is the annual nominal rate compounded monthly for a bond that has an annual percentage yield of 2.95%?

89. What annual nominal rate compounded monthly has the same annual percentage yield as 7% compounded continuously?

90. What annual nominal rate compounded continuously has the same annual percentage yield as 6% compounded monthly?

*Problems 91–94 refer to zero coupon bonds. A **zero coupon bond** is a bond that is sold now at a discount and will pay its **face value** at some time in the future when it matures—no interest payments are made.*

91. A zero coupon bond with a face value of \$30,000 matures in 15 years. What should the bond be sold for now if its rate of return is to be 4.348% compounded annually?

92. A zero coupon bond with a face value of \$20,000 matures in 10 years. What should the bond be sold for now if its rate of return is to be 4.194% compounded annually?

93. If you pay \$4,126 for a 20-year zero coupon bond with a face value of \$10,000, what is your annual compound rate of return?

94. If you pay \$32,000 for a 5-year zero coupon bond with a face value of \$40,000, what is your annual compound rate of return?

The buying and selling commission schedule shown in the table is from an online discount brokerage firm. Taking into consideration the buying and selling commissions in this schedule, find the annual compound rate of interest earned by each investment in Problems 95–98.

Transaction Size	Commission Rate
\$0–\$1,500	\$29 + 2.5% of principal
\$1,501–\$6,000	\$57 + 0.6% of principal
\$6,001–\$22,000	\$75 + 0.30% of principal
\$22,001–\$50,000	\$97 + 0.20% of principal
\$50,001–\$500,000	\$147 + 0.10% of principal
\$500,001+	\$247 + 0.08% of principal

95. An investor purchases 100 shares of stock at \$65 per share, holds the stock for 5 years, and then sells the stock for \$125 a share.

96. An investor purchases 300 shares of stock at \$95 per share, holds the stock for 3 years, and then sells the stock for \$156 a share.

97. An investor purchases 200 shares of stock at \$28 per share, holds the stock for 4 years, and then sells the stock for \$55 a share.

98. An investor purchases 400 shares of stock at \$48 per share, holds the stock for 6 years, and then sells the stock for \$147 a share.

Answers to Matched Problems

1. (A) \$1,593.85 (B) \$1,604.71
(C) \$1,610.32 (D) \$1,614.14

2. (A) \$9,155.23 (B) \$9,156.29

3. (A) \$21,084.17 (B) \$20,532.86

4. (A) 20.208% (B) 18.405%

5. 47 quarters, or 11 years and 3 quarters

6. Southern Pacific Bank: 4.917%
Washington Savings Bank: 4.939%
Washington Savings Bank has the higher return.

7. 5.678%

3.3 Future Value of an Annuity; Sinking Funds

- Future Value of an Annuity
- Sinking Funds
- Approximating Interest Rates

Future Value of an Annuity

An **annuity** is any sequence of equal periodic payments. If payments are made at the end of each time interval, then the annuity is called an **ordinary annuity**. We consider only ordinary annuities in this book. The amount, or **future value**, of an annuity is the sum of all payments plus all interest earned.

Suppose you decide to deposit \$100 every 6 months into an account that pays 6% compounded semiannually. If you make six deposits, one at the end of each interest payment period, over 3 years, how much money will be in the account after the last deposit is made? To solve this problem, let's look at it in terms of a time line. Using the compound amount formula $A = P(1 + i)^n$, we can find the value of each deposit after it has earned compound interest up through the sixth deposit, as shown in Figure 1.

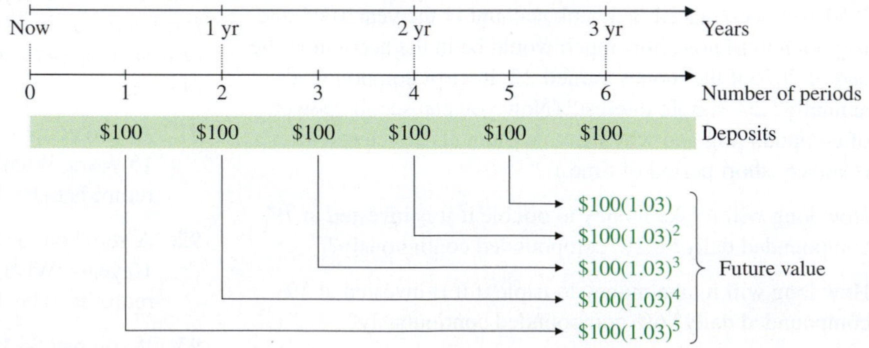

Figure 1

We could, of course, evaluate each of the future values in Figure 1 using a calculator and then add the results to find the amount in the account at the time of the sixth deposit—a tedious project at best. Instead, we take another approach, which leads directly to a formula that will produce the same result in a few steps (even when the number of deposits is very large). We start by writing the total amount in the account after the sixth deposit in the form

$$S = 100 + 100(1.03) + 100(1.03)^2 + 100(1.03)^3 + 100(1.03)^4 + 100(1.03)^5 \quad (1)$$

We would like a simple way to sum these terms. Let us multiply each side of (1) by 1.03 to obtain

$$1.03S = 100(1.03) + 100(1.03)^2 + 100(1.03)^3 + 100(1.03)^4 + 100(1.03)^5 + 100(1.03)^6 \quad (2)$$

Subtracting equation (1) from equation (2), left side from left side and right side from right side, we obtain

$$1.03S - S = 100(1.03)^6 - 100 \qquad \text{Factor each side.}$$

$$0.03S = 100[(1.03)^6 - 1] \qquad \text{Divide by 0.03.}$$

$$S = 100\frac{(1 + 0.03)^6 - 1}{0.03} \qquad \text{We write } S \text{ in this form to observe a general pattern.} \quad (3)$$

In general, if R is the periodic deposit, i the rate per period, and n the number of periods, then the future value is given by

$$S = R + R(1 + i) + R(1 + i)^2 + \cdots + R(1 + i)^{n-1} \qquad \text{Note how this compares to (1).} \quad (4)$$

and proceeding as in the above example, we obtain the general formula for the future value of an ordinary annuity:

$$S = R\frac{(1 + i)^n - 1}{i} \qquad \text{Note how this compares to (3).} \quad (5)$$

Returning to the example above, we use a calculator to complete the problem:

$$S = 100\frac{(1.03)^6 - 1}{0.03} \qquad \text{For improved accuracy, keep all values in the calculator until the end; round to the required number of decimal places.}$$

$$= \$646.84$$

CONCEPTUAL INSIGHT

In general, an expression of the form

$$a + ar + ar^2 + \cdots + ar^{n-1}$$

is called a finite geometric series (each term is obtained from the preceding term by multiplying by r). The sum of the terms of a finite geometric series is (see Section B.2)

$$a + ar + ar^2 + \cdots + ar^{n-1} = a\frac{r^n - 1}{r - 1}$$

If $a = R$ and $r = 1 + i$, then equation (4) is the sum of the terms of a finite geometric series and, using the preceding formula, we have

$$S = R + R(1 + i) + R(1 + i)^2 + \cdots + R(1 + i)^{n-1}$$

$$= R\frac{(1 + i)^n - 1}{1 + i - 1} \qquad a = R,\ r = 1 + i$$

$$= R\frac{(1 + i)^n - 1}{i} \quad (5)$$

So formula (5) is a direct consequence of the sum formula for a finite geometric series.

It is common to use FV (future value) for S and PMT (payment) for R in formula (5). Making these changes, we have the formula in Theorem 1.

THEOREM 1 Future Value of an Ordinary Annuity

$$FV = PMT\frac{(1+i)^n - 1}{i} \tag{6}$$

where

FV = future value (amount)

PMT = periodic payment

i = rate per period

n = number of payments (periods)

Note: Payments are made at the end of each period.

EXAMPLE 1 **Future Value of an Ordinary Annuity** What is the value of an annuity at the end of 20 years if \$2,000 is deposited each year into an account earning 8.5% compounded annually? How much of this value is interest?

SOLUTION To find the value of the annuity, use formula (6) with $PMT = \$2,000$, $i = r = 0.085$, and $n = 20$.

$$FV = PMT\frac{(1+i)^n - 1}{i}$$

$$= 2{,}000\frac{(1.085)^{20} - 1}{0.085} = \$96{,}754.03 \qquad \text{Use a calculator.}$$

To find the amount of interest earned, subtract the total amount deposited in the annuity (20 payments of \$2,000) from the total value of the annuity after the 20th payment.

$$\text{Deposits} = 20(2{,}000) = \$40{,}000$$

$$\text{Interest} = \text{value} - \text{deposits} = 96{,}754.03 - 40{,}000 = \$56{,}754.03$$

Figure 2, which was generated using a spreadsheet, illustrates the growth of this account over 20 years.

	A	B	C	D
1	Period	Payment	Interest	Balance
2	1	$2,000.00	$0.00	$2,000.00
3	2	$2,000.00	$170.00	$4,170.00
4	3	$2,000.00	$354.45	$6,524.45
5	4	$2,000.00	$554.58	$9,079.03
6	5	$2,000.00	$771.72	$11,850.75
7	6	$2,000.00	$1,007.31	$14,858.06
8	7	$2,000.00	$1,262.94	$18,120.99
9	8	$2,000.00	$1,540.28	$21,661.28
10	9	$2,000.00	$1,841.21	$25,502.49
11	10	$2,000.00	$2,167.71	$29,670.20
12	11	$2,000.00	$2,521.97	$34,192.17
13	12	$2,000.00	$2,906.33	$39,098.50
14	13	$2,000.00	$3,323.37	$44,421.87
15	14	$2,000.00	$3,775.86	$50,197.73
16	15	$2,000.00	$4,266.81	$56,464.54
17	16	$2,000.00	$4,799.49	$63,264.02
18	17	$2,000.00	$5,377.44	$70,641.47
19	18	$2,000.00	$6,004.52	$78,645.99
20	19	$2,000.00	$6,684.91	$87,330.90
21	20	$2,000.00	$7,423.13	$96,754.03
22	Totals	$40,000.00	$56,754.03	

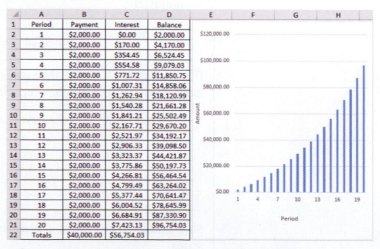

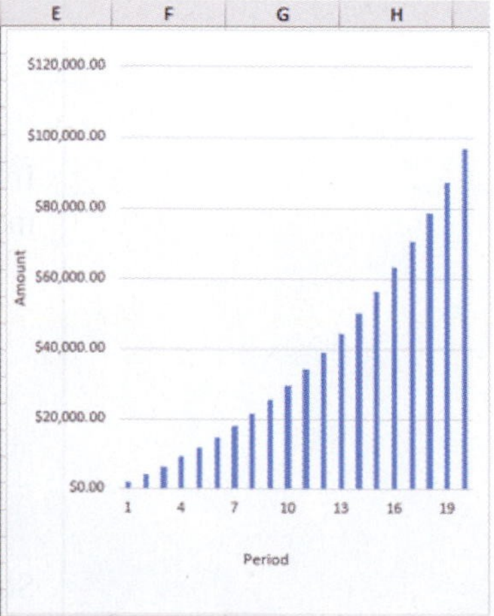

Figure 2 Ordinary annuity at 8.5% compounded annually for 20 years

Matched Problem 1 What is the value of an annuity at the end of 10 years if \$1,000 is deposited every 6 months into an account earning 8% compounded semiannually? How much of this value is interest?

The table in Figure 2 is called a **balance sheet**. Let's take a closer look at the construction of this table. The first line is a special case because the payment is made at the end of the period and no interest is earned. Each subsequent line of the table is computed as follows:

payment	+ interest	+ old balance	= new balance	
2,000	+ 0.085(2,000)	+ 2,000	= 4,170	Period 2
2,000	+ 0.085(4,170)	+ 4,170	= 6,524.45	Period 3

And so on. The amounts at the bottom of each column in the balance sheet agree with the results we obtained by using formula (6), as you would expect. Although balance sheets are appropriate for certain situations, we will concentrate on applications of formula (6). There are many important problems that can be solved only by using this formula.

Explore and Discuss 1

(A) Discuss the similarities and differences in the graphs of future value FV as a function of time t for ordinary annuities in which \$100 is deposited each month for 8 years and interest is compounded monthly at annual rates of 4%, 8%, and 12%, respectively (Fig. 3).

(B) Discuss the connections between the graph of the equation $y = 100t$, where t is time in months, and the graphs of part (A).

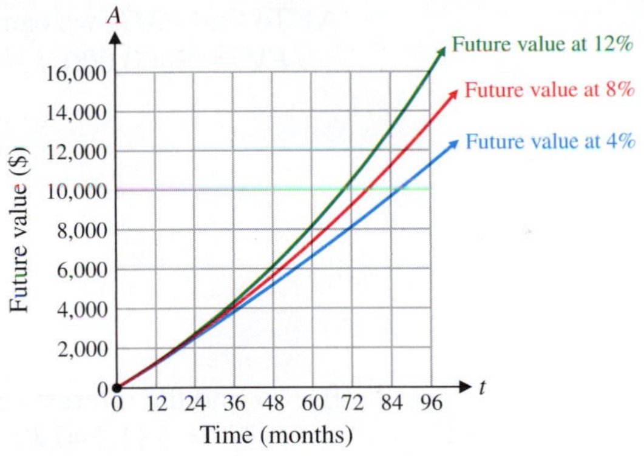

Figure 3

Sinking Funds

The formula for the future value of an ordinary annuity has another important application. Suppose the parents of a newborn child decide that on each of the child's birthdays up to the 17th year, they will deposit \$$PMT$ in an account that pays 6% compounded annually. The money is to be used for college expenses. What should the annual deposit (\$$PMT$) be in order for the amount in the account to be \$80,000 after the 17th deposit?

We are given FV, i, and n in formula (6), and we must find PMT:

$$FV = PMT\frac{(1 + i)^n - 1}{i}$$

$$80{,}000 = PMT\frac{(1.06)^{17} - 1}{0.06} \qquad \text{Solve for } PMT.$$

$$PMT = 80{,}000\frac{0.06}{(1.06)^{17} - 1} \qquad \text{Use a calculator.}$$

$$= \$2{,}835.58 \text{ per year}$$

An annuity of 17 annual deposits of $2,835.58 at 6% compounded annually will amount to $80,000 in 17 years.

This is an example of a *sinking fund problem*. In general, any account that is established for accumulating funds to meet future obligations or debts is called a **sinking fund**. If the payments are to be made in the form of an ordinary annuity, then we have only to solve formula (6) for the **sinking fund payment** *PMT*:

$$PMT = FV\frac{i}{(1 + i)^n - 1} \tag{7}$$

It is important to understand that formula (7), which is convenient to use, is simply a variation of formula (6). You can always find the sinking fund payment by first substituting the appropriate values into formula (6) and then solving for *PMT*, as we did in the college fund example discussed above. Or you can substitute directly into formula (7), as we do in the next example. Use whichever method is easier for you.

EXAMPLE 2 **Computing the Payment for a Sinking Fund** A company estimates that it will have to replace a piece of equipment at a cost of $800,000 in 5 years. To have this money available in 5 years, a sinking fund is established by making equal monthly payments into an account paying 6.6% compounded monthly.

(A) How much should each payment be?

(B) How much interest is earned during the last year?

SOLUTION

(A) To find *PMT*, we can use either formula (6) or (7). We choose formula (7) with $FV = \$800{,}000$, $i = \frac{0.066}{12} = 0.0055$, and $n = 12 \cdot 5 = 60$:

$$\begin{aligned} PMT &= FV\frac{i}{(1 + i)^n - 1} \\ &= 800{,}000\frac{0.0055}{(1.0055)^{60} - 1} \\ &= \$11{,}290.42 \text{ per month} \end{aligned}$$

(B) To find the interest earned during the fifth year, we first use formula (6) with $PMT = \$11{,}290.42$, $i = 0.0055$, and $n = 12 \cdot 4 = 48$ to find the amount in the account after 4 years:

$$\begin{aligned} FV &= PMT\frac{(1 + i)^n - 1}{i} \\ &= 11{,}290.42\frac{(1.0055)^{48} - 1}{0.0055} \\ &= \$618{,}277.04 \qquad \text{Amount after 4 years} \end{aligned}$$

During the 5th year, the amount in the account grew from $618,277.04 to $800,000. A portion of this growth was due to the 12 monthly payments of $11,290.42. The remainder of the growth was interest. Thus,

$$\begin{aligned} 800{,}000 - 618{,}277.04 &= 181{,}722.96 \qquad \text{Growth in the 5th year} \\ 12 \cdot 11{,}290.42 &= 135{,}485.04 \qquad \text{Payments during the 5th year} \\ 181{,}722.96 - 135{,}485.04 &= \$46{,}237.92 \qquad \text{Interest during the 5th year} \end{aligned}$$

Matched Problem 2 A bond issue is approved for building a marina in a city. The city is required to make regular payments every 3 months into a sinking fund paying 5.4% compounded quarterly. At the end of 10 years, the bond obligation will be retired with a cost of $5,000,000.

(A) What should each payment be?

(B) How much interest is earned during the 10th year?

EXAMPLE 3 **Growth in an IRA** Jane deposits $2,000 annually into a Roth IRA that earns 6.85% compounded annually. (The interest earned by a Roth IRA is tax free.) Due to a change in employment, these deposits stop after 10 years, but the account continues to earn interest until Jane retires 25 years after the last deposit was made. How much is in the account when Jane retires?

SOLUTION First, we use the future value formula with $PMT = \$2{,}000$, $i = 0.0685$, and $n = 10$ to find the amount in the account after 10 years:

$$\begin{aligned} FV &= PMT\frac{(1+i)^n - 1}{i} \\ &= 2{,}000\frac{(1.0685)^{10} - 1}{0.0685} \\ &= \$27{,}437.89 \end{aligned}$$

Now we use the compound interest formula from Section 3.2 with $P = \$27{,}437.89$, $i = 0.0685$, and $n = 25$ to find the amount in the account when Jane retires:

$$\begin{aligned} A &= P(1+i)^n \\ &= 27{,}437.89(1.0685)^{25} \\ &= \$143{,}785.10 \end{aligned}$$

Matched Problem 3 Refer to Example 3. Mary starts a Roth IRA earning the same rate of interest at the time Jane stops making payments into her IRA. How much must Mary deposit each year for the next 25 years in order to have the same amount at retirement as Jane?

Explore and Discuss 2

Refer to Example 3 and Matched Problem 3. What was the total amount Jane deposited in order to have $143,785.10 at retirement? What was the total amount Mary deposited in order to have the same amount at retirement? Do you think it is advisable to start saving for retirement as early as possible?

Approximating Interest Rates

Algebra can be used to solve the future value formula (6) for PMT or n but not for i. However, graphical techniques or equation solvers can be used to approximate i to as many decimal places as desired.

EXAMPLE 4 **Approximating an Interest Rate** A person makes monthly deposits of $100 into an ordinary annuity. After 30 years, the annuity is worth $160,000. What annual rate compounded monthly has this annuity earned during this 30-year period? Express the answer as a percentage, correct to two decimal places.

SOLUTION Substituting $FV = \$160{,}000$, $PMT = \$100$, and $n = 30(12) = 360$ in (6) produces the following equation:

$$160{,}000 = 100\frac{(1+i)^{360}-1}{i}$$

We can approximate the solution to this equation by using graphical techniques (Figs. 4A, 4B) or an equation solver (Fig. 4C). From Figure 4B or 4C, we see that $i = 0.006\,956\,7$ and $12(i) = 0.083\,480\,4$. So the annual rate (to two decimal places) is $r = 8.35\%$.

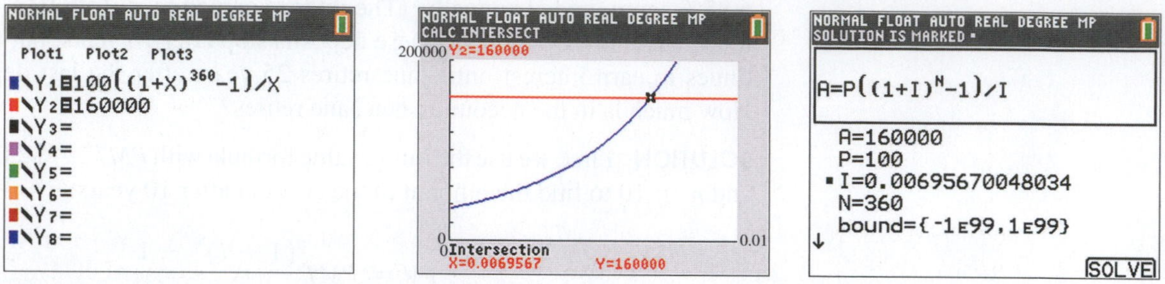

(A) (B) (C)

Figure 4

Matched Problem 4 A person makes annual deposits of \$1,000 into an ordinary annuity. After 20 years, the annuity is worth \$55,000. What annual compound rate has this annuity earned during this 20-year period? Express the answer as a percentage, correct to two decimal places.

Exercises 3.3

W **Skills Warm-up Exercises**

In Problems 1–8, find the sum of the finite geometric series $a + ar + ar^2 + \cdots + ar^{n-1}$. (If necessary, review Section B.2.)

1. $1 + 2 + 4 + 8 + \cdots + 2^9$

2. $1 + 5 + 25 + 125 + \cdots + 5^8$

3. $a = 30, r = 1, n = 100$

4. $a = 25, r = -1, n = 81$

5. $a = 10, r = 3, n = 15$

6. $a = 4, r = 10, n = 6$

A *In Problems 7–14, find i (the rate per period) and n (the number of periods) for each annuity.*

7. Quarterly deposits of \$500 are made for 20 years into an annuity that pays 8% compounded quarterly.

8. Monthly deposits of \$350 are made for 6 years into an annuity that pays 6% compounded monthly.

9. Semiannual deposits of \$900 are made for 12 years into an annuity that pays 7.5% compounded semiannually.

10. Annual deposits of \$2,500 are made for 15 years into an annuity that pays 6.25% compounded annually.

11. Monthly deposits of \$235 are made for 4 years into an annuity that pays 9% compounded monthly.

12. Semiannual deposits of \$1,900 are made for 7 years into an annuity that pays 8.5% compounded semiannually.

13. Annual deposits of \$3,100 are made for 12 years into an annuity that pays 5.95% compounded annually.

14. Quarterly deposits of \$1,200 are made for 18 years into an annuity that pays 7.6% compounded quarterly.

B *In Problems 15–22, use the future value formula (6) to find each of the indicated values.*

15. $n = 20$; $i = 0.03$; $PMT = \$500$; $FV = ?$

16. $n = 25$; $i = 0.04$; $PMT = \$100$; $FV = ?$

17. $FV = \$5{,}000$; $n = 15$; $i = 0.01$; $PMT = ?$

18. $FV = \$2{,}500$; $n = 10$; $i = 0.08$; $PMT = ?$

19. $FV = \$4{,}000$; $i = 0.02$; $PMT = 200$; $n = ?$

20. $FV = \$8{,}000$; $i = 0.04$; $PMT = 500$; $n = ?$

21. $FV = \$7{,}600$; $PMT = \$500$; $n = 10$; $i = ?$ (Round answer to two decimal places.)

22. $FV = \$4{,}100$; $PMT = \$100$; $n = 20$; $i = ?$ (Round answer to two decimal places.)

C

23. Explain what is meant by an ordinary annuity.

24. Explain why no interest is credited to an ordinary annuity at the end of the first period.

25. Solve the future value formula (6) for n.

26. Solve the future value formula (6) for i if $n = 2$.

Applications

27. Guaranty Income Life offered an annuity that pays 6.65% compounded monthly. If \$500 is deposited into this annuity every month, how much is in the account after 10 years? How much of this is interest?

28. USG Annuity and Life offered an annuity that pays 7.25% compounded monthly. If \$1,000 is deposited into this annuity every month, how much is in the account after 15 years? How much of this is interest?

29. In order to accumulate enough money for a down payment on a house, a couple deposits \$300 per month into an account paying 6% compounded monthly. If payments are made at the end of each period, how much money will be in the account in 5 years?

30. A self-employed person has a Keogh retirement plan. (This type of plan is free of taxes until money is withdrawn.) If deposits of \$7,500 are made each year into an account paying 8% compounded annually, how much will be in the account after 20 years?

31. Sun America offered an annuity that pays 6.35% compounded monthly. What equal monthly deposit should be made into this annuity in order to have \$200,000 in 15 years?

32. The Hartford offered an annuity that pays 5.5% compounded monthly. What equal monthly deposit should be made into this annuity in order to have \$100,000 in 10 years?

33. A company estimates that it will need \$100,000 in 8 years to replace a computer. If it establishes a sinking fund by making fixed monthly payments into an account paying 7.5% compounded monthly, how much should each payment be?

34. Parents have set up a sinking fund in order to have \$120,000 in 15 years for their children's college education. How much should be paid semiannually into an account paying 6.8% compounded semiannually?

35. If \$1,000 is deposited at the end of each year for 5 years into an ordinary annuity earning 8.32% compounded annually, construct a balance sheet showing the interest earned during each year and the balance at the end of each year.

36. If \$2,000 is deposited at the end of each quarter for 2 years into an ordinary annuity earning 7.9% compounded quarterly, construct a balance sheet showing the interest earned during each quarter and the balance at the end of each quarter.

37. Beginning in January, a person plans to deposit \$100 at the end of each month into an account earning 6% compounded monthly. Each year taxes must be paid on the interest earned during that year. Find the interest earned during each year for the first 3 years.

38. If \$500 is deposited each quarter into an account paying 8% compounded quarterly for 3 years, find the interest earned during each of the 3 years.

39. Bob makes his first \$1,000 deposit into an IRA earning 6.4% compounded annually on his 24th birthday and his last \$1,000 deposit on his 35th birthday (12 equal deposits in all). With no additional deposits, the money in the IRA continues to earn 6.4% interest compounded annually until Bob retires on his 65th birthday. How much is in the IRA when Bob retires?

40. Refer to Problem 39. John procrastinates and does not make his first \$1,000 deposit into an IRA until he is 36, but then he continues to deposit \$1,000 each year until he is 65 (30 deposits in all). If John's IRA also earns 6.4% compounded annually, how much is in his IRA when he makes his last deposit on his 65th birthday?

41. Refer to Problems 39 and 40. How much would John have to deposit each year in order to have the same amount at retirement as Bob has?

42. Refer to Problems 39 and 40. Suppose that Bob decides to continue to make \$1,000 deposits into his IRA every year until his 65th birthday. If John still waits until he is 36 to start his IRA, how much must he deposit each year in order to have the same amount at age 65 as Bob has?

43. Compubank, an online banking service, offered a money market account with an APY of 1.551%.

(A) If interest is compounded monthly, what is the equivalent annual nominal rate?

(B) If you wish to have \$10,000 in this account after 4 years, what equal deposit should you make each month?

44. American Express's online banking division offered a money market account with an APY of 2.243%.

(A) If interest is compounded monthly, what is the equivalent annual nominal rate?

(B) If a company wishes to have \$1,000,000 in this account after 8 years, what equal deposit should be made each month?

45. You can afford monthly deposits of \$200 into an account that pays 5.7% compounded monthly. How long will it be until you have \$7,000? (Round to the next-higher month if not exact.)

46. A company establishes a sinking fund for upgrading office equipment with monthly payments of \$2,000 into an account paying 6.6% compounded monthly. How long will it be before the account has \$100,000? (Round up to the next-higher month if not exact.)

In Problems 47–50, use graphical approximation techniques or an equation solver to approximate the desired interest rate. Express each answer as a percentage, correct to two decimal places.

47. A person makes annual payments of \$1,000 into an ordinary annuity. At the end of 5 years, the amount in the annuity is \$5,840. What annual nominal compounding rate has this annuity earned?

48. A person invests \$2,000 annually in an IRA. At the end of 6 years, the amount in the fund is \$14,000. What annual nominal compounding rate has this fund earned?

49. An employee opens a credit union account and deposits \$120 at the end of each month. After one year, the account contains \$1,444.96. What annual nominal rate compounded monthly has the account earned?

50. An employee opens a credit union account and deposits \$90 at the end of each month. After two years, the account contains \$2,177.48. What annual nominal rate compounded monthly has the account earned?

In Problems 51 and 52, use graphical approximation techniques to answer the questions.

51. When would an ordinary annuity consisting of quarterly payments of \$500 at 6% compounded quarterly be worth more than a principal of \$5,000 invested at 4% simple interest?

52. When would an ordinary annuity consisting of monthly payments of \$200 at 5% compounded monthly be worth more than a principal of \$10,000 invested at 7.5% compounded monthly?

Answers to Matched Problems

1. Value: \$29,778.08; interest: \$9,778.08
2. (A) \$95,094.67 (B) \$248,628.88
3. \$2,322.73 **4.** 9.64%

3.4 Present Value of an Annuity; Amortization

- Present Value of an Annuity
- Amortization
- Amortization Schedules
- General Problem-Solving Strategy

Present Value of an Annuity

How much should you deposit in an account paying 6% compounded semiannually in order to be able to withdraw \$1,000 every 6 months for the next 3 years? (After the last payment is made, no money is to be left in the account.)

Actually, we are interested in finding the present value of each \$1,000 that is paid out during the 3 years. We can do this by solving for P in the compound interest formula:

$$A = P(1 + i)^n$$

$$P = \frac{A}{(1 + i)^n} = A(1 + i)^{-n}$$

The rate per period is $i = \frac{0.06}{2} = 0.03$. The present value P of the first payment is $1{,}000(1.03)^{-1}$, the present value of the second payment is $1{,}000(1.03)^{-2}$, and so on. Figure 1 shows this in terms of a time line.

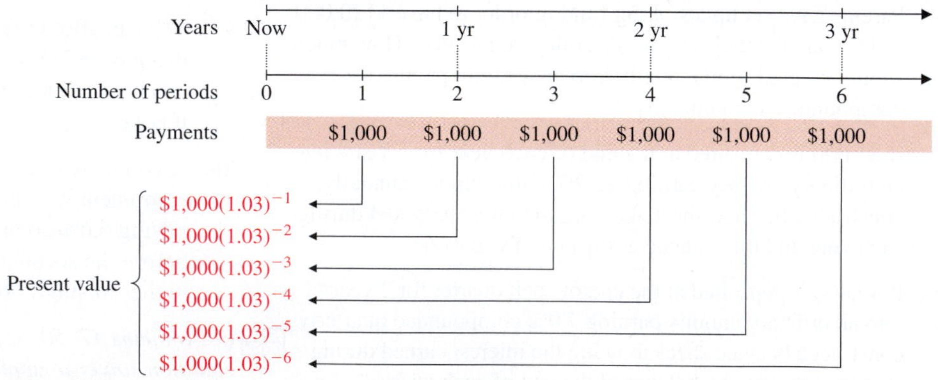

Figure 1

We could evaluate each of the present values in Figure 1 using a calculator and add the results to find the total present values of all the payments (which will be the amount needed now to buy the annuity). Since this is a tedious process, particularly when the number of payments is large, we will use the same device we used in the preceding section to produce a formula that will accomplish the same result in a couple of steps. We start by writing the sum of the present values in the form

$$P = 1{,}000(1.03)^{-1} + 1{,}000(1.03)^{-2} + \cdots + 1{,}000(1.03)^{-6} \qquad (1)$$

Multiplying both sides of equation (1) by 1.03, we obtain

$$1.03P = 1{,}000 + 1{,}000(1.03)^{-1} + \cdots + 1{,}000(1.03)^{-5} \qquad (2)$$

Now subtract equation (1) from equation (2):

$$1.03P - P = 1{,}000 - 1{,}000(1.03)^{-6}$$ Factor each side.

$$0.03P = 1{,}000[1 - (1 + 0.03)^{-6}]$$ Divide by 0.03.

$$P = 1{,}000\frac{1 - (1 + 0.03)^{-6}}{0.03} \qquad (3)$$ We write P in this form to observe a general pattern.

In general, if R is the periodic payment, i the rate per period, and n the number of periods, then the present value of all payments is given by

$$P = R(1 + i)^{-1} + R(1 + i)^{-2} + \cdots + R(1 + i)^{-n}$$ Note how this compares to (1).

Proceeding as in the above example, we obtain the general formula for the present value of an ordinary annuity:

$$P = R\frac{1 - (1 + i)^{-n}}{i} \qquad (4)$$ Note how this compares to (3).

Returning to the preceding example, we use a calculator to complete the problem:

$$P = 1{,}000\frac{1 - (1.03)^{-6}}{0.03}$$

$$= \$5{,}417.19$$

CONCEPTUAL INSIGHT

Formulas (3) and (4) can also be established by using the sum formula for a finite geometric series (see Section B.2):

$$a + ar + ar^2 + \cdots + ar^{n-1} = a\frac{r^n - 1}{r - 1}$$

It is common to use PV (present value) for P and PMT (payment) for R in formula (4). Making these changes, we have the following:

THEOREM 1 Present Value of an Ordinary Annuity

$$PV = PMT\frac{1 - (1 + i)^{-n}}{i} \qquad (5)$$

where

PV = present value of all payments

PMT = periodic payment

i = rate per period

n = number of periods

Note: Payments are made at the end of each period.

EXAMPLE 1 **Present Value of an Annuity** What is the present value of an annuity that pays \$200 per month for 5 years if money is worth 6% compounded monthly?

SOLUTION To solve this problem, use formula (5) with $PMT = \$200$, $i = \frac{0.06}{12} = 0.005$, and $n = 12(5) = 60$:

$$\begin{aligned} PV &= PMT\frac{1-(1+i)^{-n}}{i} \\ &= 200\frac{1-(1.005)^{-60}}{0.005} \quad \text{Use a calculator.} \\ &= \$10{,}345.11 \end{aligned}$$

Matched Problem 1 How much should you deposit in an account paying 8% compounded quarterly in order to receive quarterly payments of \$1,000 for the next 4 years?

EXAMPLE 2 **Retirement Planning** Lincoln Benefit Life offered an ordinary annuity that earned 6.5% compounded annually. A person plans to make equal annual deposits into this account for 25 years and then make 20 equal annual withdrawals of \$25,000, reducing the balance in the account to zero. How much must be deposited annually to accumulate sufficient funds to provide for these payments? How much total interest is earned during this entire 45-year process?

SOLUTION This problem involves both future and present values. Figure 2 illustrates the flow of money into and out of the annuity.

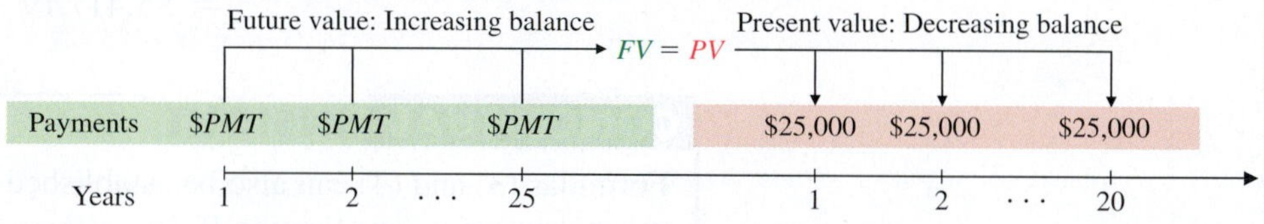

Figure 2

Since we are given the required withdrawals, we begin by finding the present value necessary to provide for these withdrawals. Using formula (5) with $PMT = \$25{,}000$, $i = 0.065$, and $n = 20$, we have

$$\begin{aligned} PV &= PMT\frac{1-(1+i)^{-n}}{i} \\ &= 25{,}000\frac{1-(1.065)^{-20}}{0.065} \quad \text{Use a calculator.} \\ &= \$275{,}462.68 \end{aligned}$$

Now we find the deposits that will produce a future value of \$275,462.68 in 25 years. Using formula (7) from Section 3.3 with $FV = \$275{,}462.68$, $i = 0.065$, and $n = 25$, we have

$$\begin{aligned} PMT &= FV\frac{i}{(1+i)^n - 1} \\ &= 275{,}462.68\frac{0.065}{(1.065)^{25} - 1} \quad \text{Use a calculator.} \\ &= \$4{,}677.76 \end{aligned}$$

Thus, depositing \$4,677.76 annually for 25 years will provide for 20 annual withdrawals of \$25,000. The interest earned during the entire 45-year process is

$$\begin{aligned} \text{interest} &= (\text{total withdrawals}) - (\text{total deposits}) \\ &= 20(25{,}000) - 25(\$4{,}677.76) \\ &= \$383{,}056 \end{aligned}$$

Matched Problem 2 Refer to Example 2. If \$2,000 is deposited annually for the first 25 years, how much can be withdrawn annually for the next 20 years?

Amortization

The present value formula for an ordinary annuity, formula (5), has another important use. Suppose that you borrow \$5,000 from a bank to buy a car and agree to repay the loan in 36 equal monthly payments, including all interest due. If the bank charges 1% per month on the unpaid balance (12% per year compounded monthly), how much should each payment be to retire the total debt, including interest, in 36 months?

Actually, the bank has bought an annuity from you. The question is: If the bank pays you \$5,000 (present value) for an annuity paying them \$*PMT* per month for 36 months at 12% interest compounded monthly, what are the monthly payments (*PMT*)? (Note that the value of the annuity at the end of 36 months is zero.) To find *PMT*, we have only to use formula (5) with $PV = \$5{,}000$, $i = 0.01$, and $n = 36$:

$$PV = PMT\frac{1 - (1 + i)^{-n}}{i}$$

$$5{,}000 = PMT\frac{1 - (1.01)^{-36}}{0.01} \qquad \text{Solve for } PMT \text{ and use a calculator.}$$

$$PMT = \$166.07 \text{ per month}$$

At \$166.07 per month, the car will be yours after 36 months. That is, you have *amortized* the debt in 36 equal monthly payments. (*Mort* means "death"; you have "killed" the loan in 36 months.) In general, **amortizing a debt** means that the debt is retired in a given length of time by equal periodic payments that include compound interest. We are interested in computing the equal periodic payments. Solving the present value formula (5) for *PMT* in terms of the other variables, we obtain the following **amortization formula**:

$$PMT = PV\frac{i}{1 - (1 + i)^{-n}} \tag{6}$$

Formula (6) is simply a variation of formula (5), and either formula can be used to find the periodic payment *PMT*.

EXAMPLE 3 **Monthly Payment and Total Interest on an Amortized Debt** Assume that you buy a TV for \$800 and agree to pay for it in 18 equal monthly payments at $1\frac{1}{2}\%$ interest per month on the unpaid balance.

(A) How much are your payments? (B) How much interest will you pay?

SOLUTION

(A) Use formula (5) or (6) with $PV = \$800$, $i = 0.015$, $n = 18$, and solve for *PMT*:

$$\begin{aligned} PMT &= PV\frac{i}{1 - (1 + i)^{-n}} \\ &= 800\frac{0.015}{1 - (1.015)^{-18}} \qquad \text{Use a calculator.} \\ &= \$51.04 \text{ per month} \end{aligned}$$

(B) Total interest paid = (amount of all payments) − (initial loan)

$$= 18(\$51.04) - \$800$$

$$= \$118.72$$

Matched Problem 3 If you sell your car to someone for $2,400 and agree to finance it at 1% per month on the unpaid balance, how much should you receive each month to amortize the loan in 24 months? How much interest will you receive?

Explore and Discuss 1

To purchase a home, a family plans to sign a mortgage of $70,000 at 8% on the unpaid balance. Discuss the advantages and disadvantages of a 20-year mortgage as opposed to a 30-year mortgage. Include a comparison of monthly payments and total interest paid.

Amortization Schedules

What happens if you are amortizing a debt with equal periodic payments and later decide to pay off the remainder of the debt in one lump-sum payment? This occurs each time a home with an outstanding mortgage is sold. In order to understand what happens in this situation, we must take a closer look at the amortization process. We begin with an example that allows us to examine the effect each payment has on the debt.

EXAMPLE 4 **Constructing an Amortization Schedule** If you borrow $500 that you agree to repay in six equal monthly payments at 1% interest per month on the unpaid balance, how much of each monthly payment is used for interest and how much is used to reduce the unpaid balance?

SOLUTION First, we compute the required monthly payment using formula (5) or (6). We choose formula (6) with $PV = \$500$, $i = 0.01$, and $n = 6$:

$$PMT = PV\frac{i}{1 - (1 + i)^{-n}}$$

$$= 500\frac{0.01}{1 - (1.01)^{-6}} \quad \text{Use a calculator.}$$

$$= \$86.27 \text{ per month}$$

At the end of the first month, the interest due is

$$\$500(0.01) = \$5.00$$

The amortization payment is divided into two parts, payment of the interest due and reduction of the unpaid balance (repayment of principal):

Monthly payment		Interest due		Unpaid balance reduction
$86.27	=	$5.00	+	$81.27

The unpaid balance for the next month is

Previous unpaid balance		Unpaid balance reduction		New unpaid balance
$500.00	−	$81.27	=	$418.73

At the end of the second month, the interest due on the unpaid balance of \$418.73 is

$$\$418.73(0.01) = \$4.19$$

Thus, at the end of the second month, the monthly payment of \$86.27 covers interest and unpaid balance reduction as follows:

$$\$86.27 = \$4.19 + \$82.08$$

and the unpaid balance for the third month is

$$\$418.73 - \$82.08 = \$336.65$$

This process continues until all payments have been made and the unpaid balance is reduced to zero. The calculations for each month are listed in Table 1, often referred to as an **amortization schedule**.

Table 1 Amortization Schedule

Payment Number	Payment	Interest	Unpaid Balance Reduction	Unpaid Balance
0				\$500.00
1	\$86.27	\$5.00	\$81.27	418.73
2	86.27	4.19	82.08	336.65
3	86.27	3.37	82.90	253.75
4	86.27	2.54	83.73	170.02
5	86.27	1.70	84.57	85.45
6	86.30	0.85	85.45	0.00
Totals	\$517.65	\$17.65	\$500.00	

Matched Problem 4 Construct the amortization schedule for a \$1,000 debt that is to be amortized in six equal monthly payments at 1.25% interest per month on the unpaid balance.

CONCEPTUAL INSIGHT

In Table 1, notice that the last payment had to be increased by \$0.03 in order to reduce the unpaid balance to zero. This small discrepancy is due to rounding the monthly payment and the entries in the interest column to two decimal places.

Suppose that a family is making monthly payments on a home mortgage loan. If the family decides to borrow money to make home improvements, they might take out a *home equity loan*. The amount of the loan will depend on the *equity* in their home, defined as the current net market value (the amount that would be received if the home were sold, after subtracting all costs involved in selling the house) minus the unpaid loan balance:

$$\textbf{Equity} = (\textbf{current net market value}) - (\textbf{unpaid loan balance}).$$

Similarly, if a family decides to sell the home they own and buy a more expensive home, the equity in their current home will be an important factor in determining the new home price that they can afford.

EXAMPLE 5 **Equity in a Home** A family purchased a home 10 years ago for \$80,000. The home was financed by paying 20% down and signing a 30-year mortgage at 9% on the unpaid balance. The net market value of the house is now \$120,000, and the family wishes to sell the house. How much equity (to the nearest dollar) does the family have in the house now after making 120 monthly payments?

SOLUTION How can we find the unpaid loan balance after 10 years or 120 monthly payments? One way to proceed would be to construct an amortization schedule, but this would require a table with 120 lines. Fortunately, there is an easier way. The unpaid balance after 120 payments is the amount of the loan that can be paid off with the remaining 240 monthly payments (20 remaining years on the loan). Since the lending institution views a loan as an annuity that they bought from the family, **the unpaid balance of a loan with *n* remaining payments is the present value of that annuity and can be computed by using formula (5)**. Since formula (5) requires knowledge of the monthly payment, we compute *PMT* first using formula (6).

Step 1 Find the monthly payment:

$$PMT = PV\frac{i}{1-(1+i)^{-n}} \qquad PV = (0.80)(\$80{,}000) = \$64{,}000 \quad i = \tfrac{0.09}{12} = 0.0075$$

$$= 64{,}000\frac{0.0075}{1-(1.0075)^{-360}} \qquad n = 12(30) = 360$$

$$= \$514.96 \text{ per month} \qquad \text{Use a calculator.}$$

Step 2 Find the present value of a \$514.96 per month, 20-year annuity:

$$PV = PMT\frac{1-(1+i)^{-n}}{i} \qquad PMT = \$514.96 \quad n = 12(20) = 240$$

$$= 514.96\frac{1-(1.0075)^{-240}}{0.0075} \qquad i = \tfrac{0.09}{12} = 0.0075 \quad \text{Use a calculator.}$$

$$= \$57{,}235 \qquad \text{Unpaid loan balance}$$

Step 3 Find the equity:

$$\text{equity} = (\text{current net market value}) - (\text{unpaid loan balance})$$
$$= \$120{,}000 - \$57{,}235$$
$$= \$62{,}765$$

So the equity in the home is \$62,765. In other words, if the family sells the house for \$120,000 net, the family will walk away with \$62,765 after paying off the unpaid loan balance of \$57,235.

Matched Problem 5 A couple purchased a home 20 years ago for \$65,000. The home was financed by paying 20% down and signing a 30-year mortgage at 8% on the unpaid balance. The net market value of the house is now \$130,000, and the couple wishes to sell the house. How much equity (to the nearest dollar) does the couple have in the house now after making 240 monthly payments?

The unpaid loan balance in Example 5 may seem a surprisingly large amount to owe after having made payments for 10 years, but long-term amortizations start out with very small reductions in the unpaid balance. For example, the interest due at the end of the very first period of the loan in Example 5 was

$$\$64{,}000(0.0075) = \$480.00$$

The first monthly payment was divided as follows:

$$\underset{\text{Monthly payment}}{\$514.96} - \underset{\text{Interest due}}{\$480.00} = \underset{\text{Unpaid balance reduction}}{\$34.96}$$

Only \$34.96 was applied to the unpaid balance.

Explore and Discuss 2

(A) A family has an \$85,000, 30-year mortgage at 9.6% compounded monthly. Show that the monthly payments are \$720.94.

(B) Explain why the equation

$$y = 720.94\frac{1 - (1.008)^{-12(30-x)}}{0.008}$$

gives the unpaid balance of the loan after x years.

(C) Find the unpaid balance after 5 years, after 10 years, and after 15 years.

(D) When does the unpaid balance drop below half of the original \$85,000?

(E) Solve part (D) using graphical approximation techniques on a graphing calculator (see Fig. 3).

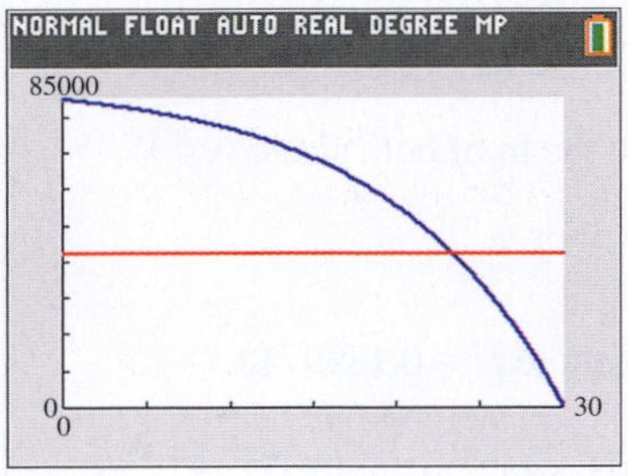

Figure 3

EXAMPLE 6 **Automobile Financing** You have negotiated a price of \$25,200 for a new Bison pickup truck. Now you must choose between 0% financing for 48 months or a \$3,000 rebate. If you choose the rebate, you can obtain a credit union loan for the balance at 4.5% compounded monthly for 48 months. Which option should you choose?

SOLUTION If you choose 0% financing, your monthly payment will be

$$PMT_1 = \frac{25{,}200}{48} = \$525$$

If you choose the \$3,000 rebate, and borrow \$22,200 at 4.5% compounded monthly for 48 months, the monthly payment is

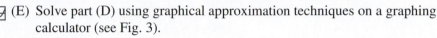

$$PMT_2 = PV\frac{i}{1 - (1 + i)^{-n}} \qquad PV = \$22{,}200$$

$$= 22{,}200\frac{0.00375}{1 - 1.00375^{-48}} \qquad i = \tfrac{.045}{12} = 0.00375$$

$$= \$506.24 \qquad n = 48$$

You should choose the \$3,000 rebate. You will save $525 - 506.24 = \$18.76$ monthly or $48(18.76) = \$900.48$ over the life of the loan.

Matched Problem 6 Which option should you choose if your credit union raises its loan rate to 7.5% compounded monthly and all other data remain the same?

EXAMPLE 7 **Credit Cards** The annual interest rate on a credit card is 18.99%. How long will it take to pay off an unpaid balance of \$847.29 if no new purchases are made and the minimum payment of \$20.00 is made each month?

SOLUTION It is necessary to make some simplifying assumptions because the lengths of the billing cycles, the days on which payments are credited, and the method for calculating interest are not specified. So we assume that there are 12 equal billing cycles per year, and that the \$20 payments are credited at the end of each cycle. With these assumptions, it is reasonable to use the present value formula with $PV = \$847.29$, $PMT = \$20.00$, and $i = 0.1899/12$ in order to solve for n, the number of payments:

$$
\begin{aligned}
PV &= PMT\frac{1-(1+i)^{-n}}{i} && \text{Multiply by } i \text{ and divide by } PMT.\\
i(PV/PMT) &= 1-(1+i)^{-n} && \text{Solve for } (1+i)^{-n}.\\
(1+i)^{-n} &= 1-i(PV/PMT) && \text{Take the ln of both sides.}\\
-n\ln(1+i) &= \ln(1-i(PV/PMT)) && \text{Solve for } n.\\
n &= -\frac{\ln(1-i(PV/PMT))}{\ln(1+i)} && \text{Substitute } i = 0.1899/12,\\
n &\approx 70.69 && PV = \$847.29, \text{ and } PMT = \$20.
\end{aligned}
$$

We conclude that the unpaid balance will be paid off in 71 months.

Matched Problem 7 The annual interest rate on a credit card is 24.99%. How long will it take to pay off an unpaid balance of $1,485.73 if no new purchases are made and a $50.00 payment is made each month?

General Problem-Solving Strategy

After working the problems in Exercises 3.4, it is important to work the problems in the Review Exercises. This will give you valuable experience in distinguishing among the various types of problems we have considered in this chapter. It is impossible to completely categorize all the problems you will encounter, but you may find the following guidelines helpful in determining which of the four basic formulas is involved in a particular problem. Be aware that some problems may involve more than one of these formulas and others may not involve any of them.

SUMMARY Strategy for Solving Mathematics of Finance Problems

Step 1 Determine whether the problem involves a single payment or a sequence of equal periodic payments. Simple and compound interest problems involve a single present value and a single future value. Ordinary annuities may be concerned with a present value or a future value but always involve a sequence of equal periodic payments.

Step 2 If a single payment is involved, determine whether simple or compound interest is used. Often simple interest is used for durations of a year or less and compound interest for longer periods.

Step 3 If a sequence of periodic payments is involved, determine whether the payments are being made into an account that is increasing in value—a future value problem—or the payments are being made out of an account that is decreasing in value—a present value problem. Remember that amortization problems always involve the present value of an ordinary annuity.

Steps 1–3 will help you choose the correct formula for a problem, as indicated in Figure 4. Then you must determine the values of the quantities in the formula that are given in the problem and those that must be computed, and solve the problem.

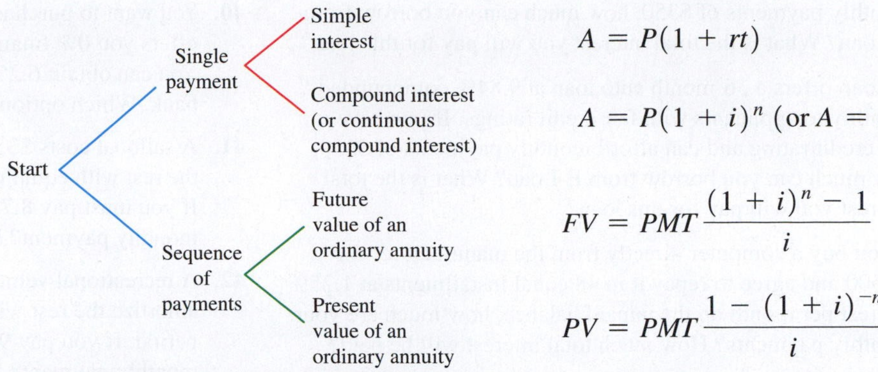

Figure 4 Selecting the correct formula for a problem

Exercises 3.4

W Skills Warm-up Exercises

In Problems 1–6, find the sum of the finite geometric series $a + ar + ar^2 + \cdots + ar^{n-1}$. Write the answer as a quotient of integers. (If necessary, review Section B.2).

1. $1 + \frac{1}{2} + \frac{1}{4} + \frac{1}{8} + \cdots + \frac{1}{2^8}$

2. $1 + \frac{1}{5} + \frac{1}{25} + \frac{1}{125} + \cdots + \frac{1}{5^7}$

3. $30 + 3 + \frac{3}{10} + \frac{3}{100} + \cdots + \frac{3}{1{,}000{,}000}$

4. $10{,}000 + 1{,}000 + 100 + 10 + \cdots + \frac{1}{10{,}000}$

5. $1 - \frac{1}{2} + \frac{1}{4} - \frac{1}{8} + \cdots + \frac{1}{2^8}$

6. $1 - \frac{1}{10} + \frac{1}{100} - \frac{1}{1{,}000} + \frac{1}{10{,}000} - \frac{1}{100{,}000}$

A *In Problems 7–14, find i (the rate per period) and n (the number of periods) for each loan at the given annual rate.*

7. Monthly payments of \$245.65 are made for 4 years to repay a loan at 7.2% compounded monthly.

8. Semiannual payments of \$3,200 are made for 12 years to repay a loan at 9.9% compounded semiannually.

9. Quarterly payments of \$975 are made for 10 years to repay a loan at 9.9% compounded quarterly.

10. Annual payments of \$1,045 are made for 5 years to repay a loan at 4.75% compounded annually.

11. Semiannual payments of \$4,500 are made for 16 years to repay a loan at 5.05% compounded semiannually.

12. Quarterly payments of \$610 are made for 6 years to repay loan at 8.24% compounded quarterly.

13. Annual payments of \$5,195 are made for 9 years to repay a loan at 5.48% compounded annually.

14. Monthly payments of \$433 are made for 3 years to repay a loan at 10.8% compounded monthly.

B *In Problems 15–22, use formula (5) or (6) to solve each problem.*

15. $n = 30$; $i = 0.04$; $PMT = \$200$; $PV = ?$

16. $n = 40$; $i = 0.01$; $PMT = \$400$; $PV = ?$

17. $PV = \$40{,}000$; $n = 96$; $i = 0.0075$; $PMT = ?$

18. $PV = \$14{,}000$; $n = 72$; $i = 0.005$; $PMT = ?$

19. $PV = \$5{,}000$; $i = 0.01$; $PMT = \$200$; $n = ?$

20. $PV = \$20{,}000$; $i = 0.0175$; $PMT = \$500$; $n = ?$

21. $PV = \$9{,}000$; $PMT = \$600$; $n = 20$; $i = ?$ (Round answer to three decimal places.)

22. $PV = \$12{,}000$; $PMT = \$400$; $n = 40$; $i = ?$ (Round answer to three decimal places.)

C

23. Explain what is meant by the present value of an ordinary annuity.

24. Solve the present value formula (5) for n.

25. Explain how an ordinary annuity is involved when you take out an auto loan from a bank.

26. Explain why the last payment in an amortization schedule might differ from the other payments.

Applications

27. American General offers a 10-year ordinary annuity with a guaranteed rate of 6.65% compounded annually. How much should you pay for one of these annuities if you want to receive payments of \$5,000 annually over the 10-year period?

28. American General offers a 7-year ordinary annuity with a guaranteed rate of 6.35% compounded annually. How much should you pay for one of these annuities if you want to receive payments of \$10,000 annually over the 7-year period?

29. E-Loan, an online lending service, offers a 36-month auto loan at 7.56% compounded monthly to applicants with good credit ratings. If you have a good credit rating and can afford monthly payments of $350, how much can you borrow from E-Loan? What is the total interest you will pay for this loan?

30. E-Loan offers a 36-month auto loan at 9.84% compounded monthly to applicants with fair credit ratings. If you have a fair credit rating and can afford monthly payments of $350, how much can you borrow from E-Loan? What is the total interest you will pay for this loan?

31. If you buy a computer directly from the manufacturer for $2,500 and agree to repay it in 48 equal installments at 1.25% interest per month on the unpaid balance, how much are your monthly payments? How much total interest will be paid?

32. If you buy a computer directly from the manufacturer for $3,500 and agree to repay it in 60 equal installments at 1.75% interest per month on the unpaid balance, how much are your monthly payments? How much total interest will be paid?

In Problems 33–36, assume that no new purchases are made with the credit card.

33. The annual interest rate on a credit card is 16.99%. If a payment of $100.00 is made each month, how long will it take to pay off an unpaid balance of $2,487.56?

34. The annual interest rate on a credit card is 24.99%. If a payment of $100.00 is made each month, how long will it take to pay off an unpaid balance of $2,487.56?

35. The annual interest rate on a credit card is 14.99%. If the minimum payment of $20 is made each month, how long will it take to pay off an unpaid balance of $937.14?

36. The annual interest rate on a credit card is 22.99%. If the minimum payment of $25 is made each month, how long will it take to pay off an unpaid balance of $860.22?

Problems 37 and 38 refer to the following ads.

37. The ad for a Bison sedan claims that a monthly payment of $299 constitutes 0% financing. Explain why that is false. Find the annual interest rate compounded monthly that is actually being charged for financing $17,485 with 72 monthly payments of $299.

2020 BISON SEDAN
Zero down – 0% financing
$299 per month*
Buy for $17,485.
* Bison sedan, 0% down, 0% for 72 months

2020 BISON SUV
Zero down – 0% financing
$399 per month*
Buy for $23,997.
* Bison SUV, 0% down, 0% for 72 months

38. The ad for a Bison SUV claims that a monthly payment of $399 constitutes 0% financing. Explain why that is false. Find the annual interest rate compounded monthly that is actually being charged for financing $23,997 with 72 monthly payments of $399.

39. You want to purchase an automobile for $27,300. The dealer offers you 0% financing for 60 months or a $5,000 rebate. You can obtain 6.3% financing for 60 months at the local bank. Which option should you choose? Explain.

40. You want to purchase an automobile for $28,500. The dealer offers you 0% financing for 60 months or a $6,000 rebate. You can obtain 6.2% financing for 60 months at the local bank. Which option should you choose? Explain.

41. A sailboat costs $35,000. You pay 20% down and amortize the rest with equal monthly payments over a 12-year period. If you must pay 8.75% compounded monthly, what is your monthly payment? How much interest will you pay?

42. A recreational vehicle costs $80,000. You pay 10% down and amortize the rest with equal monthly payments over a 7-year period. If you pay 9.25% compounded monthly, what is your monthly payment? How much interest will you pay?

43. Construct the amortization schedule for a $5,000 debt that is to be amortized in eight equal quarterly payments at 2.8% interest per quarter on the unpaid balance.

44. Construct the amortization schedule for a $10,000 debt that is to be amortized in six equal quarterly payments at 2.6% interest per quarter on the unpaid balance.

45. A woman borrows $6,000 at 9% compounded monthly, which is to be amortized over 3 years in equal monthly payments. For tax purposes, she needs to know the amount of interest paid during each year of the loan. Find the interest paid during the first year, the second year, and the third year of the loan. [*Hint:* Find the unpaid balance after 12 payments and after 24 payments.]

46. A man establishes an annuity for retirement by depositing $50,000 into an account that pays 7.2% compounded monthly. Equal monthly withdrawals will be made each month for 5 years, at which time the account will have a zero balance. Each year taxes must be paid on the interest earned by the account during that year. How much interest was earned during the first year? [*Hint:* The amount in the account at the end of the first year is the present value of a 4-year annuity.]

47. Some friends tell you that they paid $25,000 down on a new house and are to pay $525 per month for 30 years. If interest is 7.8% compounded monthly, what was the selling price of the house? How much interest will they pay in 30 years?

48. A family is thinking about buying a new house costing $120,000. The family must pay 20% down, and the rest is to be amortized over 30 years in equal monthly payments. If money costs 7.5% compounded monthly, what will the monthly payment be? How much total interest will be paid over 30 years?

49. A student receives a federally backed student loan of $6,000 at 3.5% interest compounded monthly. After finishing college in 2 years, the student must amortize the loan in the next 4 years by making equal monthly payments. What will the payments be and what total interest will the student pay? [*Hint:* This is a two-part problem. First, find the amount of the debt at the end of the first 2 years; then amortize this amount over the next 4 years.]

50. A person establishes a sinking fund for retirement by contributing $7,500 per year at the end of each year for 20 years. For the next 20 years, equal yearly payments are withdrawn, at the end of which time the account will have a zero balance. If money is worth 9% compounded annually, what yearly payments will the person receive for the last 20 years?

51. A family has a $150,000, 30-year mortgage at 6.1% compounded monthly. Find the monthly payment. Also find the unpaid balance after

(A) 10 years (B) 20 years

(C) 25 years

52. A family has a $210,000, 20-year mortgage at 6.75% compounded monthly. Find the monthly payment. Also find the unpaid balance after

(A) 5 years (B) 10 years

(C) 15 years

53. A family has a $129,000, 20-year mortgage at 7.2% compounded monthly.

(A) Find the monthly payment and the total interest paid.

(B) Suppose the family decides to add an extra $102.41 to its mortgage payment each month starting with the very first payment. How long will it take the family to pay off the mortgage? How much interest will be saved?

54. At the time they retire, a couple has $200,000 in an account that pays 8.4% compounded monthly.

(A) If the couple decides to withdraw equal monthly payments for 10 years, at the end of which time the account will have a zero balance, how much should the couple withdraw each month?

(B) If the couple decides to withdraw $3,000 a month until the balance in the account is zero, how many withdrawals can the couple make?

55. An ordinary annuity that earns 7.5% compounded monthly has a current balance of $500,000. The owner of the account is about to retire and has to decide how much to withdraw from the account each month. Find the number of withdrawals under each of the following options:

(A) $5,000 monthly (B) $4,000 monthly

(C) $3,000 monthly

56. Refer to Problem 55. If the account owner decides to withdraw $3,000 monthly, how much is in the account after 10 years? After 20 years? After 30 years?

57. An ordinary annuity pays 7.44% compounded monthly.

(A) A person deposits $100 monthly for 30 years and then makes equal monthly withdrawals for the next 15 years, reducing the balance to zero. What are the monthly withdrawals? How much interest is earned during the entire 45-year process?

(B) If the person wants to make withdrawals of $2,000 per month for the last 15 years, how much must be deposited monthly for the first 30 years?

58. An ordinary annuity pays 6.48% compounded monthly.

(A) A person wants to make equal monthly deposits into the account for 15 years in order to then make equal monthly withdrawals of $1,500 for the next 20 years, reducing the balance to zero. How much should be deposited each month for the first 15 years? What is the total interest earned during this 35-year process?

(B) If the person makes monthly deposits of $1,000 for the first 15 years, how much can be withdrawn monthly for the next 20 years?

59. A couple wishes to borrow money using the equity in their home for collateral. A loan company will loan the couple up to 70% of their equity. The couple purchased the home 12 years ago for $179,000. The home was financed by paying 20% down and signing a 30-year mortgage at 8.4% on the unpaid balance. Equal monthly payments were made to amortize the loan over the 30-year period. The net market value of the house is now $215,000. After making the 144th payment, the couple applied to the loan company for the maximum loan. How much (to the nearest dollar) will the couple receive?

60. A person purchased a house 10 years ago for $160,000. The house was financed by paying 20% down and signing a 30-year mortgage at 7.75% on the unpaid balance. Equal monthly payments were made to amortize the loan over a 30-year period. The owner now (after the 120th payment) wishes to refinance the house due to a need for additional cash. If the loan company agrees to a new 30-year mortgage of 80% of the new appraised value of the house, which is $225,000, how much cash (to the nearest dollar) will the owner receive after repaying the balance of the original mortgage?

61. A person purchased a $145,000 home 10 years ago by paying 20% down and signing a 30-year mortgage at 7.9% compounded monthly. Interest rates have dropped and the owner wants to refinance the unpaid balance by signing a new 20-year mortgage at 5.5% compounded monthly. How much interest will refinancing save?

62. A person purchased a $200,000 home 20 years ago by paying 20% down and signing a 30-year mortgage at 13.2% compounded monthly. Interest rates have dropped and the owner wants to refinance the unpaid balance by signing a new 10-year mortgage at 8.2% compounded monthly. How much interest will refinancing save?

63. Discuss the similarities and differences in the graphs of unpaid balance as a function of time for 30-year mortgages of $50,000, $75,000, and $100,000, respectively, each at 9% compounded monthly (see the figure). Include computations of the monthly payment and total interest paid in each case.

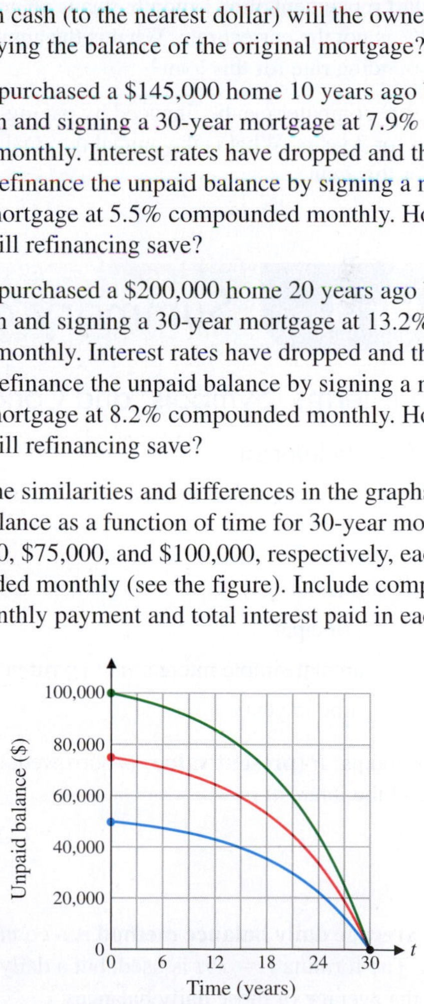

Figure for 63

64. Discuss the similarities and differences in the graphs of unpaid balance as a function of time for 30-year mortgages of \$60,000 at rates of 7%, 10%, and 13%, respectively (see the figure). Include computations of the monthly payment and total interest paid in each case.

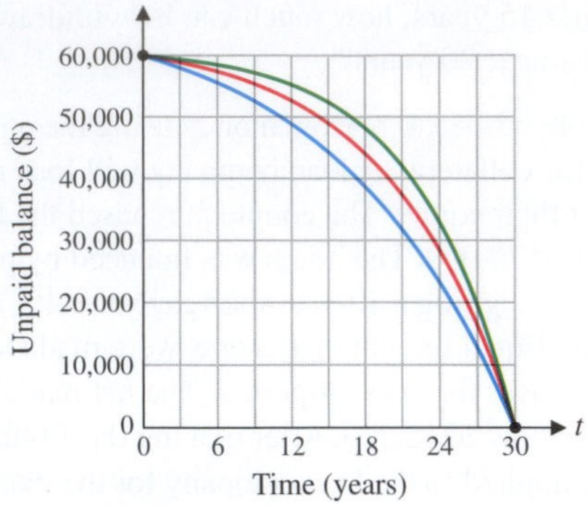

In Problems 65–68, use graphical approximation techniques or an equation solver to approximate the desired interest rate. Express each answer as a percentage, correct to two decimal places.

65. A discount electronics store offers to let you pay for a \$1,000 stereo in 12 equal \$90 installments. The store claims that since you repay \$1,080 in 1 year, the \$80 finance charge represents an 8% annual rate. This would be true if you repaid the loan in a single payment at the end of the year. But since you start repayment after 1 month, this is an amortized loan, and 8% is not the correct rate. What is the annual nominal compounding rate for this loan?

66. A \$2,000 computer can be financed by paying \$100 per month for 2 years. What is the annual nominal compounding rate for this loan?

67. The owner of a small business has received two offers of purchase. The first prospective buyer offers to pay the owner \$100,000 in cash now. The second offers to pay the owner \$10,000 now and monthly payments of \$1,200 for 10 years. In effect, the second buyer is asking the owner for a \$90,000 loan. If the owner accepts the second offer, what annual nominal compounding rate will the owner receive for financing this purchase?

68. At the time they retire, a couple has \$200,000 invested in an annuity. The couple can take the entire amount in a single payment, or receive monthly payments of \$2,000 for 15 years. If the couple elects to receive the monthly payments, what annual nominal compounding rate will the couple earn on the money invested in the annuity?

Answers to Matched Problems

1. \$13,577.71 **2.** \$10,688.87

3. $PMT = \$112.98/\text{mo}$; total interest = \$311.52

4.

Payment Number	Payment	Interest	Unpaid Balance Reduction	Unpaid Balance
0				\$1,000.00
1	\$174.03	\$12.50	\$161.53	838.47
2	174.03	10.48	163.55	674.92
3	174.03	8.44	165.59	509.33
4	174.03	6.37	167.66	341.67
5	174.03	4.27	169.76	171.91
6	174.06	2.15	171.91	0.00
Totals	\$1,044.21	\$44.21	\$1,000.00	

5. \$98,551 **6.** Choose the 0% financing. **7.** 47 months

Chapter 3 Summary and Review

Important Terms, Symbols, and Concepts

3.1 Simple Interest

EXAMPLES

- **Interest** is the fee paid for the use of a sum of money P, called the **principal**. **Simple interest** is given by

$$I = Prt$$

where I = interest
P = principal
r = annual simple interest rate (written as a decimal)
t = time in years

- If a principal P **(present value)** is borrowed, then the **amount** A **(future value)** is the total of the principal and the interest:

$$A = P + Prt = P(1 + rt)$$

- The **average daily balance method** is a common method for calculating the interest owed on a credit card. The formula $I = Prt$ is used, but a daily balance is calculated for each day of the billing cycle, and P is the average of those daily balances.

Ex. 1, p. 128
Ex. 2, p. 128
Ex. 3, p. 129
Ex. 4, p. 129
Ex. 5, p. 130
Ex. 6, p. 131

3.2 Compound and Continuous Compound Interest

- **Compound interest** is interest paid on the principal plus reinvested interest. The future and present values are related by Ex. 1, p. 136

$$A = P(1 + i)^n$$

where $i = r/m$ and
A = amount or future value
P = principal or present value
r = annual nominal rate (or just rate)
m = number of compounding periods per year
i = rate per compounding period
n = total number of compounding periods

- If a principal P is invested at an annual rate r earning **continuous compound interest**, then the amount A after t years is given by Ex. 2, p. 138 Ex. 3, p. 139 Ex. 4, p. 140

$$A = Pe^{rt}$$

- The **growth time** of an investment is the time it takes for a given principal to grow to a particular amount. Three methods for finding the growth time are as follows: Ex. 5, p. 141
  1. Use logarithms and a calculator.
  2. Use graphical approximation on a graphing calculator.
  3. Use an **equation solver** on a graphing calculator or a computer.
- The **annual percentage yield** (APY; also called the **effective rate** or **true interest rate**) is the simple interest rate that would earn the same amount as a given annual rate for which interest is compounded.
- If a principal is invested at the annual rate r compounded m times a year, then the annual percentage yield is given by Ex. 6, p. 143

$$\text{APY} = \left(1 + \frac{r}{m}\right)^m - 1$$

- If a principal is invested at the annual rate r compounded continuously, then the annual percentage yield is given by Ex. 7, p. 143

$$\text{APY} = e^r - 1$$

- A **zero coupon bond** is a bond that is sold now at a discount and will pay its **face value** at some time in the future when it matures.

3.3 Future Value of an Annuity; Sinking Funds

- An **annuity** is any sequence of equal periodic payments. If payments are made at the end of each time interval, then the annuity is called an **ordinary annuity**. The amount, or **future value**, of an annuity is the sum of all payments plus all interest earned and is given by Ex. 1, p. 150 Ex. 3, p. 153 Ex. 4, p. 153

$$FV = PMT\frac{(1 + i)^n - 1}{i}$$

where FV = future value (amount)
PMT = periodic payment
i = rate per period
n = number of payments (periods)

- A **balance sheet** is a table that shows the interest and balance for each payment of an annuity.
- An account that is established to accumulate funds to meet future obligations or debts is called a **sinking fund**. The **sinking fund payment** can be found by solving the future value formula for PMT: Ex. 2, p. 152

$$PMT = FV\frac{i}{(1 + i)^n - 1}$$

3.4 Present Value of an Annuity; Amortization

- If equal payments are made from an account until the amount in the account is 0, the payment and the **present value** are related by the following formula: Ex. 1, p. 158 Ex. 2, p. 158

$$PV = PMT\frac{1 - (1 + i)^{-n}}{i}$$

where PV = present value of all payments
PMT = periodic payment
i = rate per period
n = number of periods

- **Amortizing** a debt means that the debt is retired in a given length of time by equal periodic payments that include compound interest. Solving the present value formula for the payment gives us the **amortization formula**: Ex. 3, p. 159 Ex. 6, p. 163 Ex. 7, p. 163

$$PMT = PV\frac{i}{1 - (1 + i)^{-n}}$$

- An **amortization schedule** is a table that shows the interest due and the balance reduction for each payment of a loan. Ex. 4, p. 160
- The **equity** in a property is the difference between the current net market value and the unpaid loan balance. The unpaid balance of a loan with n **remaining payments** is given by the present value formula. Ex. 5, p. 161
- A strategy for solving problems in the mathematics of finance is presented on page 164.

Review Exercises

Work through all the problems in this chapter review and check your answers in the back of the book. Answers to all review problems are there along with section numbers in italics to indicate where each type of problem is discussed. Where weaknesses show up, review appropriate sections in the text.

A *In Problems 1–4, find the indicated quantity, given* $A = P(1 + rt)$.

1. $A = ?; P = \$100; r = 9\%; t = 6$ months

2. $A = \$808; P = ?; r = 12\%; t = 1$ month

3. $A = \$212; P = \$200; r = 8\%; t = ?$

4. $A = \$4{,}120; P = \$4{,}000; r = ?; t = 6$ months

In Problems 5 and 6, find the indicated quantity, given $A = P(1 + i)^n$.

5. $A = ?; P = \$1{,}200; i = 0.005; n = 30$

6. $A = \$5{,}000; P = ?; i = 0.0075; n = 60$

In Problems 7 and 8, find the indicated quantity, given $A = Pe^{rt}$.

7. $A = ?; P = \$4{,}750; r = 6.8\%; t = 3$ years

8. $A = \$36{,}000; P = ?; r = 9.3\%; t = 60$ months

B *In Problems 9 and 10, find the indicated quantity, given*

$$FV = PMT\frac{(1 + i)^n - 1}{i}$$

9. $FV = ?; PMT = \$1{,}000; i = 0.005; n = 60$

10. $FV = \$8{,}000; PMT = ?; i = 0.015; n = 48$

In Problems 11 and 12, find the indicated quantity, given

$$PV = PMT\frac{1 - (1 + i)^{-n}}{i}$$

11. $PV = ?; PMT = \$2{,}500; i = 0.02; n = 16$

12. $PV = \$8{,}000; PMT = ?; i = 0.0075; n = 60$

C **13.** Solve the equation $2{,}500 = 1{,}000(1.06)^n$ for n to the nearest integer using:

(A) Logarithms

(B) Graphical approximation techniques or an equation solver on a graphing calculator

14. Solve the equation

$$5{,}000 = 100\,\frac{(1.01)^n - 1}{0.01}$$

for n to the nearest integer using:

(A) Logarithms

(B) Graphical approximation techniques or an equation solver on a graphing calculator.

Applications

Find all dollar amounts correct to the nearest cent. When an interest rate is requested as an answer, express the rate as a percentage, correct to two decimal places.

15. If you borrow \$3,000 at 14% simple interest for 10 months, how much will you owe in 10 months? How much interest will you pay?

16. Grandparents deposited \$6,000 into a grandchild's account toward a college education. How much money (to the nearest dollar) will be in the account 17 years from now if the account earns 7% compounded monthly?

17. How much should you pay for a corporate bond paying 6.6% compounded monthly in order to have \$25,000 in 10 years?

18. An investment account pays 5.4% compounded annually. Construct a balance sheet showing the interest earned during each year and the balance at the end of each year for 4 years if

(A) A single deposit of \$400 is made at the beginning of the first year.

(B) Four deposits of \$100 are made at the end of each year.

19. One investment pays 13% simple interest and another 9% compounded annually. Which investment would you choose? Why?

20. A \$10,000 retirement account is left to earn interest at 7% compounded daily. How much money will be in the account 40 years from now when the owner reaches 65? (Use a 365-day year and round answer to the nearest dollar.)

21. A couple wishes to have \$40,000 in 6 years for the down payment on a house. At what rate of interest compounded continuously must \$25,000 be invested now to accomplish this goal?

22. Which is the better investment and why: 9% compounded quarterly or 9.25% compounded annually?

23. What is the value of an ordinary annuity at the end of 8 years if \$200 per month is deposited into an account earning 7.2% compounded monthly? How much of this value is interest?

24. A payday lender charges \$60 for a loan of \$500 for 15 days. Find the annual interest rate. (Use a 360-day year.)

25. The annual interest rate on a credit card is 25.74% and interest is calculated by the average daily balance method. The unpaid balance at the start of a 30-day billing cycle was \$1,672.18. A purchase of \$265.12 was made on day 8 and a payment of \$250 was credited to the account on day 20. Find the unpaid balance at the end of the billing cycle. (Use a 360-day year.)

26. What will a \$23,000 car cost (to the nearest dollar) 5 years from now if the inflation rate over that period averages 5% compounded annually?

27. What would the \$23,000 car in Problem 25 have cost (to the nearest dollar) 5 years ago if the inflation rate over that period had averaged 5% compounded annually?

28. A loan of \$2,500 was repaid at the end of 10 months with a check for \$2,812.50. What annual rate of interest was charged?

29. You want to purchase an automobile for \$21,600. The dealer offers you 0% financing for 48 months or a \$3,000 rebate. You can obtain 4.8% financing for 48 months at the local bank. Which option should you choose? Explain.

30. Find the annual percentage yield on a bond earning 6.25% if interest is compounded

(A) monthly.

(B) continuously.

31. You have \$5,000 toward the purchase of a boat that will cost \$6,000. How long will it take the \$5,000 to grow to \$6,000 if it is invested at 9% compounded quarterly? (Round up to the next-higher quarter if not exact.)

32. How long will it take money to double if it is invested at 6% compounded monthly? 9% compounded monthly? (Round up to the next-higher month if not exact.)

33. Starting on his 21st birthday, and continuing on every birthday up to and including his 65th, John deposits \$2,000 a year into an IRA. How much (to the nearest dollar) will be in the account on John's 65th birthday, if the account earns:

(A) 7% compounded annually?

(B) 11% compounded annually?

34. If you just sold a stock for \$17,388.17 (net) that cost you \$12,903.28 (net) 3 years ago, what annual compound rate of return did you make on your investment?

35. The table shows the fees for refund anticipation loans (RALs) offered by an online tax preparation firm. Find the annual rate of interest for each of the following loans. Assume a 360-day year.

(A) A \$400 RAL paid back in 15 days

(B) A \$1,800 RAL paid back in 21 days

RAL Amount	RAL Fee
\$10–\$500	\$29.00
\$501–\$1,000	\$39.00
\$1,001–\$1,500	\$49.00
\$1,501–\$2,000	\$69.00
\$2,001–\$5,000	\$82.00

36. Lincoln Benefit Life offered an annuity that pays 5.5% compounded monthly. What equal monthly deposit should be made into this annuity in order to have \$50,000 in 5 years?

37. A person wants to establish an annuity for retirement purposes. He wants to make quarterly deposits for 20 years so that he can then make quarterly withdrawals of \$5,000 for 10 years. The annuity earns 7.32% interest compounded quarterly.

(A) How much will have to be in the account at the time he retires?

(B) How much should be deposited each quarter for 20 years in order to accumulate the required amount?

(C) What is the total amount of interest earned during the 30-year period?

38. If you borrow \$4,000 from an online lending firm for the purchase of a computer and agree to repay it in 48 equal installments at 0.9% interest per month on the unpaid balance, how much are your monthly payments? How much total interest will be paid?

39. A company decides to establish a sinking fund to replace a piece of equipment in 6 years at an estimated cost of \$50,000. To accomplish this, they decide to make fixed monthly payments into an account that pays 6.12% compounded monthly. How much should each payment be?

40. How long will it take money to double if it is invested at 7.5% compounded daily? 7.5% compounded annually?

41. A student receives a student loan for \$8,000 at 5.5% interest compounded monthly to help her finish the last 1.5 years of college. Starting 1 year after finishing college, the student must amortize the loan in the next 5 years by making equal monthly payments. What will the payments be and what total interest will the student pay?

42. If you invest \$5,650 in an account paying 8.65% compounded continuously, how much money will be in the account at the end of 10 years?

43. A company makes a payment of \$1,200 each month into a sinking fund that earns 6% compounded monthly. Use graphical approximation techniques on a graphing calculator to determine when the fund will be worth \$100,000.

44. A couple has a \$50,000, 20-year mortgage at 9% compounded monthly. Use graphical approximation techniques on a graphing calculator to determine when the unpaid balance will drop below \$10,000.

45. A loan company advertises in the paper that you will pay only 8¢ a day for each \$100 borrowed. What annual rate of interest are they charging? (Use a 360-day year.)

46. Construct the amortization schedule for a \$1,000 debt that is to be amortized in four equal quarterly payments at 2.5% interest per quarter on the unpaid balance.

47. You can afford monthly deposits of only \$300 into an account that pays 7.98% compounded monthly. How long will it be until you will have \$9,000 to purchase a used car? (Round to the next-higher month if not exact.)

48. A company establishes a sinking fund for plant retooling in 6 years at an estimated cost of \$850,000. How much should be invested semiannually into an account paying 8.76% compounded semiannually? How much interest will the account earn in the 6 years?

49. What is the annual nominal rate compounded monthly for a CD that has an annual percentage yield of 2.50%?

50. If you buy a 13-week T-bill with a maturity value of \$5,000 for \$4,922.15 from the U.S. Treasury Department, what annual interest rate will you earn?

51. In order to save enough money for the down payment on a condominium, a young couple deposits \$200 each month into an account that pays 7.02% interest compounded monthly. If the couple needs \$10,000 for a down payment, how many deposits will the couple have to make?

52. A business borrows \$80,000 at 9.42% interest compounded monthly for 8 years.

(A) What is the monthly payment?

(B) What is the unpaid balance at the end of the first year?

(C) How much interest was paid during the first year?

53. You unexpectedly inherit \$10,000 just after you have made the 72nd monthly payment on a 30-year mortgage of \$60,000 at 8.2% compounded monthly. Discuss the relative merits of using the inheritance to reduce the principal of the loan or to buy a certificate of deposit paying 7% compounded monthly.

54. Your parents are considering a \$75,000, 30-year mortgage to purchase a new home. The bank at which they have done business for many years offers a rate of 7.54% compounded monthly. A competitor is offering 6.87% compounded monthly. Would it be worthwhile for your parents to switch banks? Explain.

55. How much should a \$5,000 face value zero coupon bond, maturing in 5 years, be sold for now, if its rate of return is to be 5.6% compounded annually?

56. If you pay \$5,695 for a \$10,000 face value zero coupon bond that matures in 10 years, what is your annual compound rate of return?

57. If an investor wants to earn an annual interest rate of 6.4% on a 26-week T-bill with a maturity value of \$5,000, how much should the investor pay for the T-bill?

58. Two years ago you borrowed \$10,000 at 12% interest compounded monthly, which was to be amortized over 5 years. Now you have acquired some additional funds and decide that you want to pay off this loan. What is the unpaid balance after making equal monthly payments for 2 years?

59. What annual nominal rate compounded monthly has the same annual percentage yield as 7.28% compounded quarterly?

60. (A) A man deposits \$2,000 in an IRA on his 21st birthday and on each subsequent birthday up to, and including, his 29th (nine deposits in all). The account earns 8% compounded annually. If he leaves the money in the account without making any more deposits, how much will he have on his 65th birthday, assuming the account continues to earn the same rate of interest?

(B) How much would be in the account (to the nearest dollar) on his 65th birthday if he had started the deposits on his 30th birthday and continued making deposits on each birthday until (and including) his 65th birthday?

61. A promissory note will pay $27,000 at maturity 10 years from now. How much money should you be willing to pay now if money is worth 5.5% compounded continuously?

62. In a new housing development, the houses are selling for $100,000 and require a 20% down payment. The buyer is given a choice of 30-year or 15-year financing, both at 7.68% compounded monthly.

(A) What is the monthly payment for the 30-year choice? For the 15-year choice?

(B) What is the unpaid balance after 10 years for the 30-year choice? For the 15-year choice?

63. A loan company will loan up to 60% of the equity in a home. A family purchased their home 8 years ago for $83,000. The home was financed by paying 20% down and signing a 30-year mortgage at 8.4% for the balance. Equal monthly payments were made to amortize the loan over the 30-year period. The market value of the house is now $95,000. After making the 96th payment, the family applied to the loan company for the maximum loan. How much (to the nearest dollar) will the family receive?

64. A $600 stereo is financed for 6 months by making monthly payments of $110. What is the annual nominal compounding rate for this loan?

65. A person deposits $2,000 each year for 25 years into an IRA. When she retires immediately after making the 25th deposit, the IRA is worth $220,000.

(A) Find the interest rate earned by the IRA over the 25-year period leading up to retirement.

(B) Assume that the IRA continues to earn the interest rate found in part (A). How long can the retiree withdraw $30,000 per year? How long can she withdraw $24,000 per year?

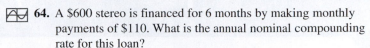

4 Systems of Linear Equations; Matrices

Introduction

Traffic congestion in urban areas is a major problem. Traffic control by on-ramp metering, traffic signals, and message signs is one approach to reducing congestion. Traffic control by vehicle-to-vehicle communication shows promise for the future. The mathematical prerequisite for solving problems in traffic flow is the ability to solve systems of linear equations (see Problems 97 and 98 in Section 4.3).

In Chapter 4 we discuss methods for solving systems of two linear equations by hand calculation. We also introduce matrix methods, including Gauss–Jordan elimination, that can be used to solve systems with many equations on a calculator or computer. We consider a number of important applications, including resource allocation, production scheduling, and economic planning.

4.1 Review: Systems of Linear Equations in Two Variables

- Systems of Linear Equations in Two Variables
- Graphing
- Substitution
- Elimination by Addition
- Applications

Systems of Linear Equations in Two Variables

To establish basic concepts, let's consider the following simple example: If 2 adult tickets and 1 child ticket cost \$32, and if 1 adult ticket and 3 child tickets cost \$36, what is the price of each?

$$\begin{aligned} \text{Let:}\quad & x = \text{price of adult ticket} \\ & y = \text{price of child ticket} \\ \text{Then:}\quad & 2x + y = 32 \\ & x + 3y = 36 \end{aligned}$$

Now we have a system of two linear equations in two variables. It is easy to find ordered pairs (x, y) that satisfy one or the other of these equations. For example, the ordered pair $(16, 0)$ satisfies the first equation but not the second, and the ordered pair $(24, 4)$ satisfies the second but not the first. To solve this system, we must find all ordered pairs of real numbers that satisfy both equations at the same time. In general, we have the following definition:

DEFINITION Systems of Two Linear Equations in Two Variables

Given the **linear system**

$$ax + by = h$$
$$cx + dy = k$$

where a, b, c, d, h, and k are real constants, a pair of numbers $x = x_0$ and $y = y_0$ [also written as an ordered pair (x_0, y_0)] is a **solution** of this system if each equation is satisfied by the pair. The set of all such ordered pairs is called the **solution set** for the system. To **solve** a system is to find its solution set.

We will consider three methods of solving such systems: *graphing, substitution,* and *elimination by addition*. Each method has its advantages, depending on the situation.

Graphing

Recall that the graph of a line is a graph of all the ordered pairs that satisfy the equation of the line. To solve the ticket problem by graphing, we graph both equations in the same coordinate system. The coordinates of any points that the graphs have in common must be solutions to the system since they satisfy both equations.

EXAMPLE 1 **Solving a System by Graphing** Solve the ticket problem by graphing:

$$2x + y = 32$$
$$x + 3y = 36$$

SOLUTION An easy way to find two distinct points on the first line is to find the x and y intercepts. Substitute $y = 0$ to find the x intercept ($2x = 32$, so $x = 16$), and substitute $x = 0$ to find the y intercept ($y = 32$). Then draw the line through

Reminder

Recall that we may graph a line by finding its intercepts (Section 1.2). For the equation $2x + y = 32$:

If $x = 0$, then $y = 32$.

If $y = 0$, then $x = 16$.

This gives the points (0, 32) and (16, 0), which uniquely define the line.

(16, 0) and (0, 32). After graphing both lines in the same coordinate system (Fig. 1), estimate the coordinates of the intersection point:

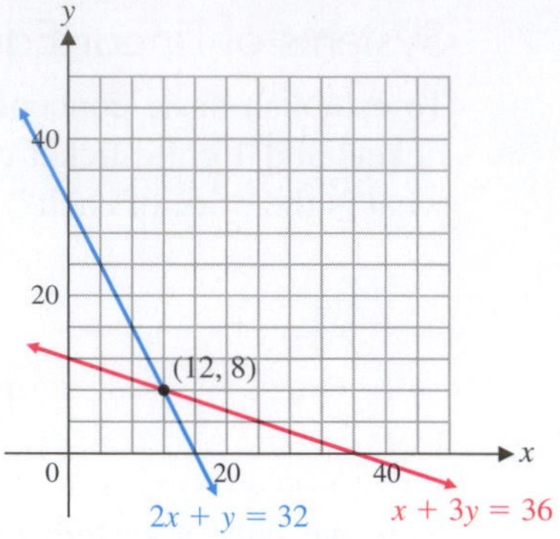

Figure 1

$x = \$12$ Adult ticket

$y = \$8$ Child ticket

$$2x + y = 32 \qquad x + 3y = 36$$

CHECK $2(12) + 8 \stackrel{?}{=} 32 \qquad 12 + 3(8) \stackrel{?}{=} 36$

$$32 \stackrel{\checkmark}{=} 32 \qquad 36 \stackrel{\checkmark}{=} 36$$

Check that (12, 8) satisfies each of the original equations.

Matched Problem 1 Solve by graphing and check:

$$2x - y = -3$$
$$x + 2y = -4$$

It is clear that Example 1 has exactly one solution since the lines have exactly one point in common. In general, lines in a rectangular coordinate system are related to each other in one of the three ways illustrated in the next example.

EXAMPLE 2 **Solving a System by Graphing** Solve each of the following systems by graphing:

(A) $x - 2y = 2$, $x + y = 5$ (B) $x + 2y = -4$, $2x + 4y = 8$ (C) $2x + 4y = 8$, $x + 2y = 4$

SOLUTION

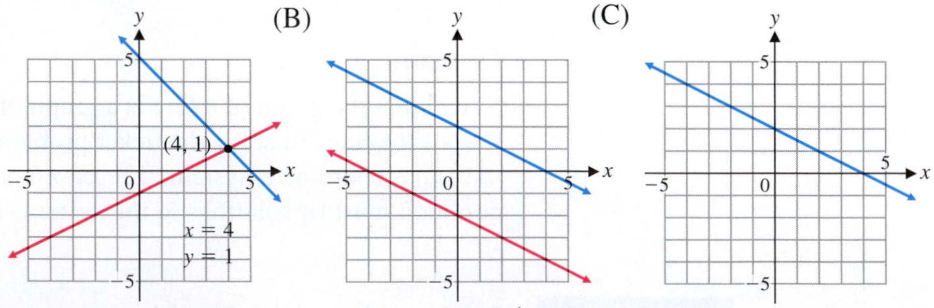

(A) Intersection at one point only—exactly one solution

(B) Lines are parallel (each has slope $-\frac{1}{2}$)—no solutions

(C) Lines coincide—infinite number of solutions

Matched Problem 2 Solve each of the following systems by graphing:

(A) $x + y = 4$, $2x - y = 2$ (B) $6x - 3y = 9$, $2x - y = 3$ (C) $2x - y = 4$, $6x - 3y = -18$

We introduce some terms that describe the different types of solutions to systems of equations.

DEFINITION Systems of Linear Equations: Basic Terms

A system of linear equations is **consistent** if it has one or more solutions and **inconsistent** if no solutions exist. Furthermore, a consistent system is said to be **independent** if it has exactly one solution (often referred to as the **unique solution**) and **dependent** if it has more than one solution. Two systems of equations are **equivalent** if they have the same solution set.

Referring to the three systems in Example 2, the system in part (A) is consistent and independent with the unique solution $x = 4$, $y = 1$. The system in part (B) is inconsistent. And the system in part (C) is consistent and dependent with an infinite number of solutions (all points on the two coinciding lines).

CAUTION Given a system of equations, do not confuse the *number of variables* with the *number of solutions*. The systems of Example 2 involve two variables, x and y. A solution to such a system is a *pair* of numbers, one for x and one for y. So the system in Example 2A has two variables, but exactly one solution, namely, $x = 4$, $y = 1$.

Explore and Discuss 1

Can a consistent and dependent system have exactly two solutions? Exactly three solutions? Explain.

By graphing a system of two linear equations in two variables, we gain useful information about the solution set of the system. In general, any two lines in a coordinate plane must intersect in exactly one point, be parallel, or coincide (have identical graphs). So the systems in Example 2 illustrate the only three possible types of solutions for systems of two linear equations in two variables. These ideas are summarized in Theorem 1.

THEOREM 1 Possible Solutions to a Linear System

The linear system

$$ax + by = h$$
$$cx + dy = k$$

must have

(A) Exactly one solution — Consistent and independent

or

(B) No solution — Inconsistent

or

(C) Infinitely many solutions — Consistent and dependent

There are no other possibilities.

In the past, one drawback to solving systems by graphing was the inaccuracy of hand-drawn graphs. Graphing calculators have changed that. Graphical solutions on a graphing calculator provide an accurate approximation of the solution to a system of linear equations in two variables. Example 3 demonstrates this.

EXAMPLE 3 **Solving a System Using a Graphing Calculator** Solve to two decimal places using graphical approximation techniques on a graphing calculator:

$$5x + 2y = 15$$
$$2x - 3y = 16$$

SOLUTION First, solve each equation for y:

$$\begin{aligned} 5x + 2y &= 15 \\ 2y &= -5x + 15 \\ y &= -2.5x + 7.5 \end{aligned} \qquad \begin{aligned} 2x - 3y &= 16 \\ -3y &= -2x + 16 \\ y &= \frac{2}{3}x - \frac{16}{3} \end{aligned}$$

Next, enter each equation in the graphing calculator (Fig. 2A), graph in an appropriate viewing window, and approximate the intersection point (Fig. 2B).

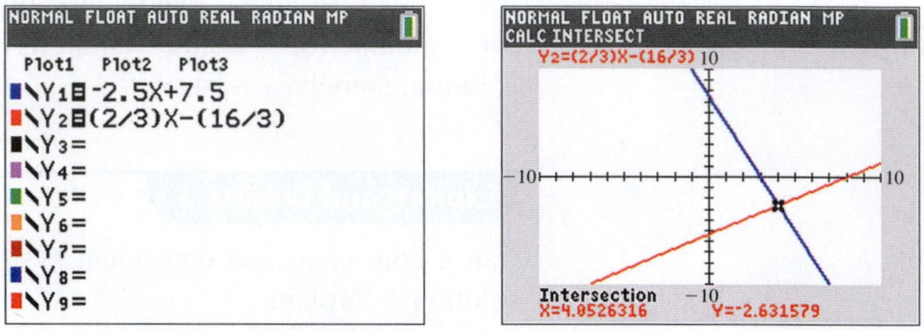

(A) Equation definitions (B) Intersection points

Figure 2

Rounding the values in Figure 2B to two decimal places, we see that the solution is $x = 4.05$ and $y = -2.63$, or $(4.05, -2.63)$.

CHECK

$$\begin{aligned} 5x + 2y &= 15 \\ 5(4.05) + 2(-2.63) &\overset{?}{=} 15 \\ 14.99 &\overset{\checkmark}{\approx} 15 \end{aligned} \qquad \begin{aligned} 2x - 3y &= 16 \\ 2(4.05) - 3(-2.63) &\overset{?}{=} 16 \\ 15.99 &\overset{\checkmark}{\approx} 16 \end{aligned}$$

The checks are sufficiently close but, due to rounding, not exact.

Matched Problem 3 Solve to two decimal places using graphical approximation techniques on a graphing calculator:

$$2x - 5y = -25$$
$$4x + 3y = 5$$

Graphical methods help us to visualize a system and its solutions, reveal relationships that might otherwise be hidden, and, with the assistance of a graphing calculator, provide accurate approximations to solutions.

Substitution

Now we review an algebraic method that is easy to use and provides exact solutions to a system of two equations in two variables, provided that solutions exist. In this method, first we choose one of two equations in a system and solve for one variable in terms of the other. (We make a choice that avoids fractions, if possible.) Then we **substitute** the result into the other equation and solve the resulting linear equation in one variable. Finally, we substitute this result back into the results of the first step to find the second variable.

EXAMPLE 4 **Solving a System by Substitution** Solve by substitution:

$$5x + y = 4$$
$$2x - 3y = 5$$

SOLUTION Solve either equation for one variable in terms of the other; then substitute into the remaining equation. In this problem, we avoid fractions by choosing the first equation and solving for y in terms of x:

$$5x + y = 4 \qquad \text{Solve the first equation for } y \text{ in terms of } x.$$
$$y = 4 - 5x \qquad \text{Substitute into the second equation.}$$

$$2x - 3y = 5 \qquad \text{Second equation}$$
$$2x - 3(4 - 5x) = 5 \qquad \text{Solve for } x.$$
$$2x - 12 + 15x = 5$$
$$17x = 17$$
$$x = 1$$

Now, replace x with 1 in $y = 4 - 5x$ to find y:

$$y = 4 - 5x$$
$$y = 4 - 5(1)$$
$$y = -1$$

The solution is $x = 1, y = -1$ or $(1, -1)$.

CHECK

$$5x + y = 4 \qquad 2x - 3y = 5$$
$$5(1) + (-1) \stackrel{?}{=} 4 \qquad 2(1) - 3(-1) \stackrel{?}{=} 5$$
$$4 \stackrel{\checkmark}{=} 4 \qquad 5 \stackrel{\checkmark}{=} 5$$

Matched Problem 4 Solve by substitution:

$$3x + 2y = -2$$
$$2x - y = -6$$

Explore and Discuss 2

Return to Example 2 and solve each system by substitution. Based on your results, describe how you can recognize a dependent system or an inconsistent system when using substitution.

Elimination by Addition

The methods of graphing and substitution both work well for systems involving two variables. However, neither is easily extended to larger systems. Now we turn to **elimination by addition**. This is probably the most important method of solution. It readily generalizes to larger systems and forms the basis for computer-based solution methods.

To solve an equation such as $2x - 5 = 3$, we perform operations on the equation until we reach an equivalent equation whose solution is obvious (see Appendix A, Section A.7).

$$2x - 5 = 3 \qquad \text{Add 5 to both sides.}$$
$$2x = 8 \qquad \text{Divide both sides by 2.}$$
$$x = 4$$

Theorem 2 indicates that we can solve systems of linear equations in a similar manner.

THEOREM 2 Operations That Produce Equivalent Systems

A system of linear equations is transformed into an equivalent system if

(A) Two equations are interchanged.
(B) An equation is multiplied by a nonzero constant.
(C) A constant multiple of one equation is added to another equation.

Any one of the three operations in Theorem 2 can be used to produce an equivalent system, but the operations in parts (B) and (C) will be of most use to us now. Part (A) becomes useful when we apply the theorem to larger systems. The use of Theorem 2 is best illustrated by examples.

EXAMPLE 5 **Solving a System Using Elimination by Addition** Solve the following system using elimination by addition:

$$\begin{aligned} 3x - 2y &= 8 \\ 2x + 5y &= -1 \end{aligned}$$

SOLUTION We use Theorem 2 to eliminate one of the variables, obtaining a system with an obvious solution:

$$\begin{aligned} 3x - 2y &= 8 \\ 2x + 5y &= -1 \end{aligned}$$

Multiply the top equation by 5 and the bottom equation by 2 (Theorem 2B).

$$\begin{aligned} 5(3x - 2y) &= 5(8) \\ 2(2x + 5y) &= 2(-1) \end{aligned}$$

$$\begin{aligned} 15x - 10y &= 40 \\ 4x + 10y &= -2 \\ \hline 19x \quad\quad &= 38 \end{aligned}$$

Add the top equation to the bottom equation (Theorem 2C), eliminating the y terms.

Divide both sides by 19, which is the same as multiplying the equation by $\frac{1}{19}$ (Theorem 2B).

$$x = 2$$

This equation paired with either of the two original equations produces a system equivalent to the original system.

Knowing that $x = 2$, we substitute this number back into either of the two original equations (we choose the second) to solve for y:

$$\begin{aligned} 2(2) + 5y &= -1 \\ 5y &= -5 \\ y &= -1 \end{aligned}$$

The solution is $x = 2$, $y = -1$ or $(2, -1)$.

CHECK

$$\begin{aligned} 3x - 2y &= 8 \\ 3(2) - 2(-1) &\stackrel{?}{=} 8 \\ 8 &\stackrel{\checkmark}{=} 8 \end{aligned} \qquad \begin{aligned} 2x + 5y &= -1 \\ 2(2) + 5(-1) &\stackrel{?}{=} -1 \\ -1 &\stackrel{\checkmark}{=} -1 \end{aligned}$$

Matched Problem 5 Solve the following system using elimination by addition:

$$\begin{aligned} 5x - 2y &= 12 \\ 2x + 3y &= 1 \end{aligned}$$

Let's see what happens in the elimination process when a system has either no solution or infinitely many solutions. Consider the following system:

$$\begin{aligned} 2x + 6y &= -3 \\ x + 3y &= 2 \end{aligned}$$

Multiplying the second equation by -2 and adding, we obtain

$$\begin{aligned} 2x + 6y &= -3 \\ -2x - 6y &= -4 \\ \hline 0 &= -7 \end{aligned} \quad \text{Not possible}$$

We have obtained a contradiction. The assumption that the original system has solutions must be false. So the system has no solutions, and its solution set is the empty set. The graphs of the equations are parallel lines, and the system is inconsistent.

Now consider the system

$$\begin{aligned} x - \tfrac{1}{2}y &= 4 \\ -2x + y &= -8 \end{aligned}$$

If we multiply the top equation by 2 and add the result to the bottom equation, we obtain

$$\begin{aligned} 2x - y &= 8 \\ -2x + y &= -8 \\ \hline 0 &= 0 \end{aligned}$$

Obtaining $0 = 0$ implies that the equations are equivalent; that is, their graphs coincide and the system is dependent. If we let $x = k$, where k is any real number, and solve either equation for y, we obtain $y = 2k - 8$. So $(k, 2k - 8)$ is a solution to this system for any real number k. The variable k is called a **parameter** and replacing k with a real number produces a **particular solution** to the system. For example, some particular solutions to this system are

$k = -1$	$k = 2$	$k = 5$	$k = 9.4$
$(-1, -10)$	$(2, -4)$	$(5, 2)$	$(9.4, 10.8)$

Applications

Many real-world problems are solved readily by constructing a mathematical model consisting of two linear equations in two variables and applying the solution methods that we have discussed. We shall examine two applications in detail.

EXAMPLE 6 **Diet** Jasmine wants to use milk and orange juice to increase the amount of calcium and vitamin A in her daily diet. An ounce of milk contains 37 milligrams of calcium and 57 micrograms* of vitamin A. An ounce of orange juice contains 5 milligrams of calcium and 65 micrograms of vitamin A. How many ounces of milk and orange juice should Jasmine drink each day to provide exactly 500 milligrams of calcium and 1,200 micrograms of vitamin A?

SOLUTION The first step in solving an application problem is to introduce the proper variables. Often, the question asked in the problem will guide you in this decision. Reading the last sentence in Example 6, we see that we must determine a certain number of ounces of milk and orange juice. So we introduce variables to represent these unknown quantities:

$$\begin{aligned} x &= \text{number of ounces of milk} \\ y &= \text{number of ounces of orange juice} \end{aligned}$$

*A microgram (μg) is one millionth (10^{-6}) of a gram.

Next, we summarize the given information using a table. It is convenient to organize the table so that the quantities represented by the variables correspond to columns in the table (rather than to rows) as shown.

	Milk	Orange Juice	Total Needed
Calcium	37 mg/oz	5 mg/oz	500 mg
Vitamin A	57 μg/oz	65 μg/oz	1,200 μg

Now we use the information in the table to form equations involving x and y:

$$\begin{pmatrix}\text{calcium in } x \text{ oz}\\ \text{of milk}\end{pmatrix} + \begin{pmatrix}\text{calcium in } y \text{ oz}\\ \text{of orange juice}\end{pmatrix} = \begin{pmatrix}\text{total calcium}\\ \text{needed (mg)}\end{pmatrix}$$

$$37x \quad + \quad 5y \quad = \quad 500$$

$$\begin{pmatrix}\text{vitamin A in } x \text{ oz}\\ \text{of milk}\end{pmatrix} + \begin{pmatrix}\text{vitamin A in } y \text{ oz}\\ \text{of orange juice}\end{pmatrix} = \begin{pmatrix}\text{total vitamin A}\\ \text{needed } (\mu\text{g})\end{pmatrix}$$

$$57x \quad + \quad 65y \quad = \quad 1{,}200$$

So we have the following model to solve:

$$\begin{aligned} 37x + 5y &= 500 \\ 57x + 65y &= 1{,}200 \end{aligned}$$

We can multiply the first equation by -13 and use elimination by addition:

$$\begin{aligned} -481x - 65y &= -6{,}500 \\ 57x + 65y &= 1{,}200 \\ \hline -424x \qquad &= -5{,}300 \\ x &= 12.5 \end{aligned} \qquad \begin{aligned} 37(12.5) + 5y &= 500 \\ 5y &= 37.5 \\ y &= 7.5 \end{aligned}$$

Drinking 12.5 ounces of milk and 7.5 ounces of orange juice each day will provide Jasmine with the required amounts of calcium and vitamin A.

CHECK

$$\begin{aligned} 37x + 5y &= 500 \\ 37(12.5) + 5(7.5) &\overset{?}{=} 500 \\ 500 &\overset{\checkmark}{=} 500 \end{aligned} \qquad \begin{aligned} 57x + 65y &= 1{,}200 \\ 57(12.5) + 65(7.5) &\overset{?}{=} 1{,}200 \\ 1{,}200 &\overset{\checkmark}{=} 1{,}200 \end{aligned}$$

Figure 3 illustrates a solution to Example 6 using graphical approximation techniques.

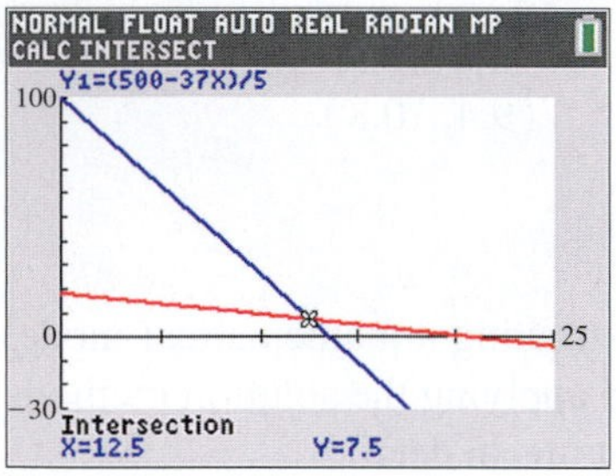

Figure 3

$y_1 = (500 - 37x)/5$

$y_2 = (1{,}200 - 57x)/65$

Matched Problem 6 Dennis wants to use cottage cheese and yogurt to increase the amount of protein and calcium in his daily diet. An ounce of cottage cheese contains 3 grams of protein and 15 milligrams of calcium. An ounce of yogurt contains 1 gram of protein and 41 milligrams of calcium. How many ounces of cottage cheese and yogurt should Dennis eat each day to provide exactly 62 grams of protein and 760 milligrams of calcium?

In a free market economy, the price of a product is determined by the relationship between supply and demand. Suppliers are more willing to supply a product at higher prices. So when the price is high, the supply is high. If the relationship between price and supply is linear, then the graph of the price–supply equation is a line with positive slope. On the other hand, consumers of a product are generally less willing to buy a product at higher prices. So when the price is high, demand is low. If the relationship between price and demand is linear, the graph of the price–demand equation is a line with negative slope. In a free competitive market, the price of a product tends to move toward an **equilibrium price**, in which the supply and demand are equal; that common value of the supply and demand is the **equilibrium quantity**. To find the equilibrium price, we solve the system consisting of the price–supply and price–demand equations.

EXAMPLE 7 **Supply and Demand** At a price of \$1.88 per pound, the supply for cherries in a large city is 16,000 pounds, and the demand is 10,600 pounds. When the price drops to \$1.46 per pound, the supply decreases to 10,000 pounds, and the demand increases to 12,700 pounds. Assume that the price–supply and price–demand equations are linear.

(A) Find the price–supply equation.

(B) Find the price–demand equation.

(C) Find the supply and demand at a price of \$2.09 per pound.

(D) Find the supply and demand at a price of \$1.32 per pound.

(E) Use the substitution method to find the equilibrium price and equilibrium demand.

SOLUTION

(A) Let p be the price per pound, and let x be the quantity in thousands of pounds. Then (16, 1.88) and (10, 1.46) are solutions of the price–supply equation. Use the point–slope form for the equation of a line, $y - y_1 = m(x - x_1)$, to obtain the price–supply equation:

$$p - 1.88 = \frac{1.46 - 1.88}{10 - 16}(x - 16) \qquad \text{Simplify.}$$

$$p - 1.88 = 0.07(x - 16) \qquad \text{Solve for } p.$$

$$p = 0.07x + 0.76 \qquad \text{Price–supply equation}$$

(B) Again, let p be the price per pound, and let x be the quantity in thousands of pounds. Then (10.6, 1.88) and (12.7, 1.46) are solutions of the price–demand equation.

$$p - 1.88 = \frac{1.46 - 1.88}{12.7 - 10.6}(x - 10.6) \qquad \text{Simplify.}$$

$$p - 1.88 = -0.2(x - 10.6) \qquad \text{Solve for } p.$$

$$p = -0.2x + 4 \qquad \text{Price–demand equation}$$

(C) Substitute $p = 2.09$ into the price–supply equation, and also into the price–demand equation, and solve for x:

Price–supply equation	Price–demand equation
$p = 0.07x + 0.76$	$p = -0.2x + 4$
$2.09 = 0.07x + 0.76$	$2.09 = -0.2x + 4$
$x = 19$	$x = 9.55$

At a price of \$2.09 per pound, the supply is 19,000 pounds of cherries and the demand is 9,550 pounds. (The supply is greater than the demand, so the price will tend to come down.)

(D) Substitute $p = 1.32$ in each equation and solve for x:

Price–supply equation	Price–demand equation
$p = 0.07x + 0.76$	$p = -0.2x + 4$
$1.32 = 0.07x + 0.76$	$1.32 = -0.2x + 4$
$x = 8$	$x = 13.4$

At a price of \$1.32 per pound, the supply is 8,000 pounds of cherries, and the demand is 13,400 pounds. (The demand is greater than the supply, so the price will tend to go up.)

(E) We solve the linear system

$$p = 0.07x + 0.76 \qquad \text{Price–supply equation}$$
$$p = -0.2x + 4 \qquad \text{Price–demand equation}$$

using substitution (substitute $p = -0.2x + 4$ in the first equation):

$$-0.2x + 4 = 0.07x + 0.76$$
$$-0.27x = -3.24 \qquad \text{Equilibrium quantity}$$
$$x = 12 \text{ thousand pounds}$$

Now substitute $x = 12$ into the price–demand equation:

$$p = -0.2(12) + 4$$
$$p = \$1.60 \text{ per pound} \qquad \text{Equilibrium price}$$

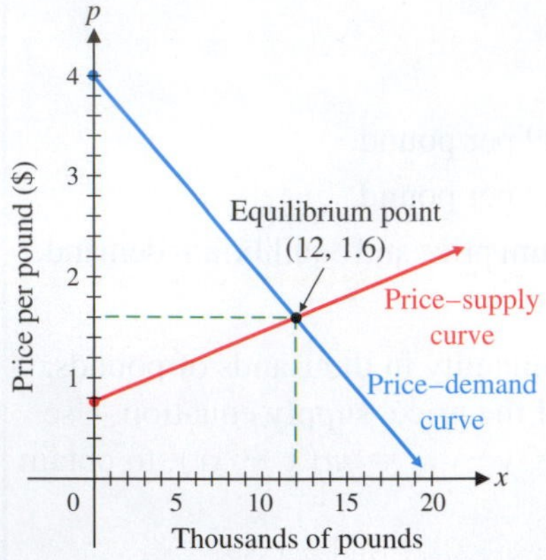

Figure 4

The results are interpreted graphically in Figure 4 (it is customary to refer to the graphs of price–supply and price–demand equations as "curves" even when they are lines). Note that if the price is above the equilibrium price of \$1.60 per pound, the supply will exceed the demand and the price will come down. If the price is below the equilibrium price of \$1.60 per pound, the demand will exceed the supply and the price will go up. So the price will stabilize at \$1.60 per pound. At this equilibrium price, suppliers will supply 12,000 pounds of cherries and consumers will purchase 12,000 pounds.

Matched Problem 7 Find the equilibrium quantity and equilibrium price, and graph the following price–supply and price–demand equations:

$$p = 0.08q + 0.66 \qquad \text{Price–supply equation}$$
$$p = -0.1q + 3 \qquad \text{Price–demand equation}$$

Exercises 4.1

Skills Warm-up Exercises

W *In Problems 1–6, find the x and y coordinates of the intersection of the given lines. (If necessary, review Section 1.2).*

1. $y = 5x + 7$ and the y axis

2. $y = 5x + 7$ and the x axis

3. $3x + 4y = 72$ and the x axis

4. $3x + 4y = 72$ and the y axis

5. $6x - 5y = 120$ and $x = 5$

6. $6x - 5y = 120$ and $y = 3$

In Problems 7 and 8, find an equation in point–slope form, $y - y_1 = m(x - x_1)$, of the line through the given points.

7. (2, 7) and (4, −5).

8. (3, 20) and (−5, 4).

A *Match each system in Problems 9–12 with one of the following graphs, and use the graph to solve the system.*

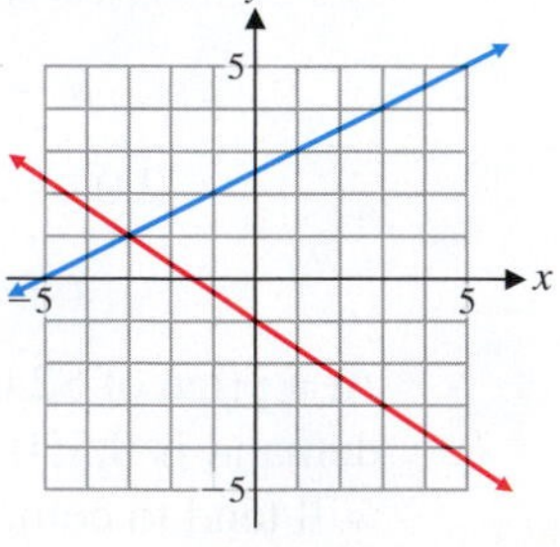

(A)

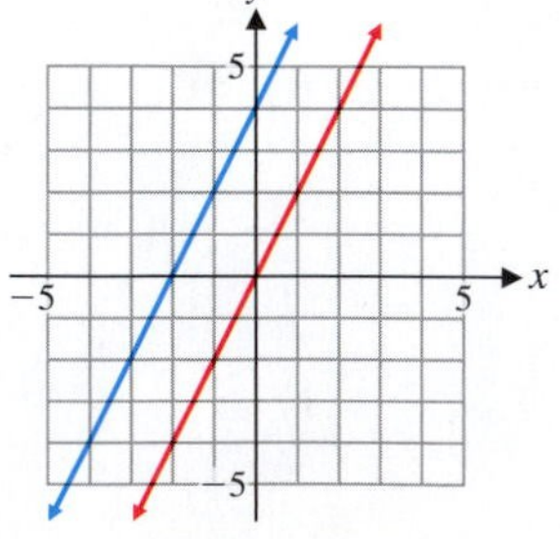

(B)

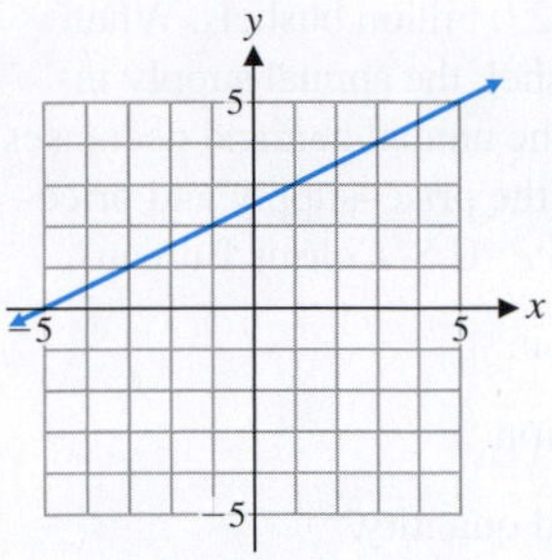

(C)

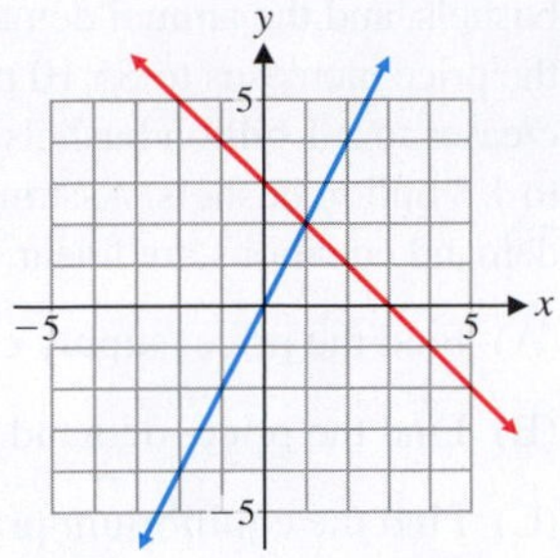

(D)

9. $-4x + 2y = 8$
$2x - y = 0$

10. $x + y = 3$
$2x - y = 0$

11. $-x + 2y = 5$
$2x + 3y = -3$

12. $2x - 4y = -10$
$-x + 2y = 5$

Solve Problems 13–16 by graphing.

13. $3x - y = 2$
$x + 2y = 10$

14. $3x - 2y = 12$
$7x + 2y = 8$

15. $m + 2n = 4$
$2m + 4n = -8$

16. $3u + 5v = 15$
$6u + 10v = -30$

Solve Problems 17–20 using substitution.

17. $y = 2x - 3$
$x + 2y = 14$

18. $y = x - 4$
$x + 3y = 12$

19. $2x + y = 6$
$x - y = -3$

20. $3x - y = 7$
$2x + 3y = 1$

Solve Problems 21–24 using elimination by addition.

21. $3u - 2v = 12$
$7u + 2v = 8$

22. $2x - 3y = -8$
$5x + 3y = 1$

23. $2m - n = 10$
$m - 2n = -4$

24. $2x + 3y = 1$
$3x - y = 7$

B *Solve Problems 25–34 using substitution or elimination by addition.*

25. $6x - 2y = 4$
$5x + 3y = 8$

26. $3x + 9y = 6$
$4x - 3y = 8$

27. $4x - 2y = 10$
$-6x + 3y = 15$

28. $-5x + 15y = 10$
$5x - 15y = 10$

29. $4x - 2y = 10$
$-6x + 3y = -15$

30. $-5x + 15y = 10$
$5x - 15y = -10$

31. $3m + 5n = 7$
$2m + 10n = 12$

32. $5m - 7n = 9$
$2m - 12n = 22$

33. $x + y = 1$
$0.3x + 0.5y = 0.7$

34. $x + y = 1$
$0.4x + 0.7y = 0.1$

In Problems 35–42, solve the system. Note that each solution can be found mentally, without the use of a calculator or pencil-and-paper calculation; try to visualize the graphs of both lines.

35. $x + 0y = 7$
$0x + y = 3$

36. $x + 0y = -4$
$0x + y = 9$

37. $5x + 0y = 4$
$0x + 3y = -2$

38. $6x + 0y = 7$
$0x + 4y = 9$

39. $x + y = 0$
$x - y = 0$

40. $-2x + y = 0$
$5x - y = 0$

41. $x - 2y = 4$
$0x + y = 5$

42. $x + 3y = 9$
$0x + y = -2$

43. In a free competitive market, if the supply of a good is greater than the demand, will the price tend to go up or come down?

44. In a free competitive market, if the demand for a good is greater than the supply, will the price tend to go up or come down?

Problems 45–48 are concerned with the linear system

$$y = mx + b$$
$$y = nx + c$$

where m, b, n, and c are nonzero constants.

45. If the system has a unique solution, discuss the relationships among the four constants.

46. If the system has no solution, discuss the relationships among the four constants.

47. If the system has an infinite number of solutions, discuss the relationships among the four constants.

48. If $m = 0$, how many solutions does the system have?

In Problems 49–56, use a graphing calculator to find the solution to each system. Round any approximate solutions to three decimal places.

49. $y = 9x - 10$
$y = -7x + 8$

50. $y = 5x - 13$
$y = -11x + 7$

51. $y = 0.2x + 0.7$
$y = 0.2x - 0.1$

52. $y = -1.7x + 2.3$
$y = -1.7x - 1.3$

53. $3x - 2y = 15$
$4x + 3y = 13$

54. $3x - 7y = -20$
$2x + 5y = 8$

55. $-2.4x + 3.5y = 0.1$
$-1.7x + 2.6y = -0.2$

56. $4.2x + 5.4y = -12.9$
$6.4x + 3.7y = -4.5$

C *In Problems 57–62, graph the equations in the same coordinate system. Find the coordinates of any points where two or more lines intersect. Is there a point that is a solution to all three equations?*

57. $x - 2y = -6$
$2x + y = 8$
$x + 2y = -2$

58. $x + y = 3$
$x + 3y = 15$
$3x - y = 5$

59. $x + y = 1$
$x - 2y = -8$
$3x + y = -3$

60. $x - y = 6$
$x - 2y = 8$
$x + 4y = -4$

61. $4x - 3y = -24$
$2x + 3y = 12$
$8x - 6y = 24$

62. $2x + 3y = 18$
$2x - 6y = -6$
$4x + 6y = -24$

63. The coefficients of the three systems given below are similar. One might guess that the solution sets to the three systems would be nearly identical. Develop evidence for or against this guess by considering graphs of the systems and solutions obtained using substitution or elimination by addition.

(A) $5x + 4y = 4$
$11x + 9y = 4$

(B) $5x + 4y = 4$
$11x + 8y = 4$

(C) $5x + 4y = 4$
$10x + 8y = 4$

64. Repeat Problem 63 for the following systems:

(A) $6x - 5y = 10$
$-13x + 11y = -20$

(B) $6x - 5y = 10$
$-13x + 10y = -20$

(C) $6x - 5y = 10$
$-12x + 10y = -20$

Applications

65. Supply and demand for T-shirts. Suppose that the supply and demand equations for printed T-shirts for a particular week are

$p = 0.7q + 3$ Price–supply equation

$p = -1.7q + 15$ Price–demand equation

where p is the price in dollars and q is the quantity in hundreds.

(A) Find the supply and demand (to the nearest unit) if T-shirts are \$4 each. Discuss the stability of the T-shirt market at this price level.

(B) Find the supply and demand (to the nearest unit) if T-shirts are \$9 each. Discuss the stability of the T-shirt market at this price level.

(C) Find the equilibrium price and quantity.

(D) Graph the two equations in the same coordinate system and identify the equilibrium point, supply curve, and demand curve.

66. Supply and demand for baseball caps. Suppose that the supply and demand for printed baseball caps for a particular week are

$p = 0.4q + 3.2$ Price–supply equation

$p = -1.9q + 17$ Price–demand equation

where p is the price in dollars and q is the quantity in hundreds.

(A) Find the supply and demand (to the nearest unit) if baseball caps are \$4 each. Discuss the stability of the baseball cap market at this price level.

(B) Find the supply and demand (to the nearest unit) if baseball caps are \$9 each. Discuss the stability of the baseball cap market at this price level.

(C) Find the equilibrium price and quantity.

(D) Graph the two equations in the same coordinate system and identify the equilibrium point, supply curve, and demand curve.

67. Supply and demand for soybeans. At \$4.80 per bushel, the annual supply for soybeans in the Midwest is 1.9 billion bushels, and the annual demand is 2.0 billion bushels. When the price increases to \$5.10 per bushel, the annual supply increases to 2.1 billion bushels, and the annual demand decreases to 1.8 billion bushels. Assume that the price–supply and price–demand equations are linear. (*Source:* U.S. Census Bureau)

(A) Find the price–supply equation.

(B) Find the price–demand equation.

(C) Find the equilibrium price and quantity.

(D) Graph the two equations in the same coordinate system and identify the equilibrium point, supply curve, and demand curve.

68. Supply and demand for corn. At \$2.13 per bushel, the annual supply for corn in the Midwest is 8.9 billion bushels and the annual demand is 6.5 billion bushels. When the price falls to \$1.50 per bushel, the annual supply decreases to 8.2 billion bushels and the annual demand increases to 7.4 billion bushels. Assume that the price–supply and price–demand equations are linear. (*Source:* U.S. Census Bureau)

(A) Find the price–supply equation.

(B) Find the price–demand equation.

(C) Find the equilibrium price and quantity.

(D) Graph the two equations in the same coordinate system and identify the equilibrium point, supply curve, and demand curve.

69. Break-even analysis. A small plant manufactures riding lawn mowers. The plant has fixed costs (leases, insurance, etc.) of \$48,000 per day and variable costs (labor, materials, etc.) of \$1,400 per unit produced. The mowers are sold for \$1,800 each. So the cost and revenue equations are

$y = 48{,}000 + 1{,}400x$ Cost equation

$y = 1{,}800x$ Revenue equation

where x is the total number of mowers produced and sold each day. The daily costs and revenue are in dollars.

(A) How many units must be manufactured and sold each day for the company to break even?

(B) Graph both equations in the same coordinate system and show the break-even point. Interpret the regions between the lines to the left and to the right of the break-even point.

70. Break-even analysis. Repeat Problem 69 with the cost and revenue equations

$y = 65{,}000 + 1{,}100x$ Cost equation

$y = 1{,}600x$ Revenue equation

71. Break-even analysis. A company markets exercise DVDs that sell for \$19.95, including shipping and handling. The monthly fixed costs (advertising, rent, etc.) are \$24,000 and the variable costs (materials, shipping, etc) are \$7.45 per DVD.

(A) Find the cost equation and the revenue equation.

(B) How many DVDs must be sold each month for the company to break even?

(C) Graph the cost and revenue equations in the same coordinate system and show the break-even point. Interpret the regions between the lines to the left and to the right of the break-even point.

72. Break-even analysis. Repeat Problem 71 if the monthly fixed costs increase to \$27,200, the variable costs increase to \$9.15, and the company raises the selling price of the DVDs to \$21.95.

73. Delivery charges. United Express, a national package delivery service, charges a base price for overnight delivery of packages weighing 1 pound or less and a surcharge for each additional pound (or fraction thereof). A customer is billed \$27.75 for shipping a 5-pound package and \$64.50 for a 20-pound package. Find the base price and the surcharge for each additional pound.

74. Delivery charges. Refer to Problem 73. Federated Shipping, a competing overnight delivery service, informs the customer in Problem 73 that they would ship the 5-pound package for \$29.95 and the 20-pound package for \$59.20.

(A) If Federated Shipping computes its cost in the same manner as United Express, find the base price and the surcharge for Federated Shipping.

(B) Devise a simple rule that the customer can use to choose the cheaper of the two services for each package shipped. Justify your answer.

75. Coffee blends. A coffee company uses Colombian and Brazilian coffee beans to produce two blends, robust and mild. A pound of the robust blend requires 12 ounces of Colombian beans and 4 ounces of Brazilian beans. A pound of the mild blend requires 6 ounces of Colombian beans and 10 ounces of Brazilian beans. Coffee is shipped in 132-pound burlap bags. The company has 50 bags of Colombian beans and 40 bags of Brazilian beans on hand. How many pounds of each blend should the company produce in order to use all the available beans?

76. Coffee blends. Refer to Problem 75.

(A) If the company decides to discontinue production of the robust blend and produce only the mild blend, how many pounds of the mild blend can the company produce? How many beans of each type will the company use? Are there any beans that are not used?

(B) Repeat part (A) if the company decides to discontinue production of the mild blend and produce only the robust blend.

77. Animal diet. Animals in an experiment are to be kept under a strict diet. Each animal should receive 20 grams of protein and 6 grams of fat. The laboratory technician is able to purchase two food mixes: Mix A has 10% protein and 6% fat; mix B has 20% protein and 2% fat. How many grams of each mix should be used to obtain the right diet for one animal?

78. Fertilizer. A fruit grower uses two types of fertilizer in an orange grove, brand A and brand B. Each bag of brand A contains 8 pounds of nitrogen and 4 pounds of phosphoric acid. Each bag of brand B contains 7 pounds of nitrogen and 6 pounds of phosphoric acid. Tests indicate that the grove needs 720 pounds of nitrogen and 500 pounds of phosphoric acid. How many bags of each brand should be used to provide the required amounts of nitrogen and phosphoric acid?

79. Electronics. A supplier for the electronics industry manufactures keyboards and screens for graphing calculators at plants in Mexico and Taiwan. The hourly production rates at each plant are given in the table. How many hours should each plant be operated to exactly fill an order for 4,000 keyboards and 4,000 screens?

Plant	Keyboards	Screens
Mexico	40	32
Taiwan	20	32

80. Sausage. A company produces Italian sausages and bratwursts at plants in Green Bay and Sheboygan. The hourly production rates at each plant are given in the table. How many hours should each plant operate to exactly fill an order for 62,250 Italian sausages and 76,500 bratwursts?

Plant	Italian Sausage	Bratwurst
Green Bay	800	800
Sheboygan	500	1,000

81. Physics. An object dropped off the top of a tall building falls vertically with constant acceleration. If s is the distance of the object above the ground (in feet) t seconds after its release, then s and t are related by an equation of the form $s = a + bt^2$, where a and b are constants. Suppose the object is 180 feet above the ground 1 second after its release and 132 feet above the ground 2 seconds after its release.

(A) Find the constants a and b.

(B) How tall is the building?

(C) How long does the object fall?

82. Physics. Repeat Problem 81 if the object is 240 feet above the ground after 1 second and 192 feet above the ground after 2 seconds.

83. Earthquakes. An earthquake emits a primary wave and a secondary wave. Near the surface of the Earth the primary wave travels at 5 miles per second and the secondary wave at 3 miles per second. From the time lag between the two waves arriving at a given receiving station, it is possible to estimate the distance to the quake. Suppose a station measured a time difference of 16 seconds between the arrival of the two waves. How long did each wave travel, and how far was the earthquake from the station?

84. Sound waves. A ship using sound-sensing devices above and below water recorded a surface explosion 6 seconds sooner by its underwater device than its above-water device. Sound travels in air at 1,100 feet per second and in seawater at 5,000 feet per second. How long did it take each sound wave to reach the ship? How far was the explosion from the ship?

85. Psychology. People approach certain situations with "mixed emotions." For example, public speaking often brings forth the positive response of recognition and the negative response of failure. Which dominates? J. S. Brown, in an experiment on approach and avoidance, trained rats by feeding them from a goal box. The rats received mild electric shocks from the same goal box. This established an approach–avoidance conflict relative to the goal box. Using an appropriate apparatus, Brown arrived at the following relationships:

$$p = -\tfrac{1}{5}d + 70 \quad \text{Approach equation}$$
$$p = -\tfrac{4}{3}d + 230 \quad \text{Avoidance equation}$$

where $30 \le d \le 172.5$. The approach equation gives the pull (in grams) toward the food goal box when the rat is placed d centimeters away from it. The avoidance equation gives the pull (in grams) away from the shock goal box when the rat is placed d centimeters from it.

(A) Graph the approach equation and the avoidance equation in the same coordinate system.

(B) Find the value of d for the point of intersection of these two equations.

(C) What do you think the rat would do when placed the distance d from the box found in part (B)?

(*Source: Journal of Comparative and Physiological Psychology*, 41:450–465.)

Answers to Matched Problems

1. $x = -2, y = -1$

$$2x - \quad y = -3$$
$$2(-2) - (-1) \stackrel{?}{=} -3$$
$$-3 \stackrel{\checkmark}{=} -3$$
$$x + \quad 2y = -4$$
$$(-2) + 2(-1) \stackrel{?}{=} -4$$
$$-4 \stackrel{\checkmark}{=} -4$$

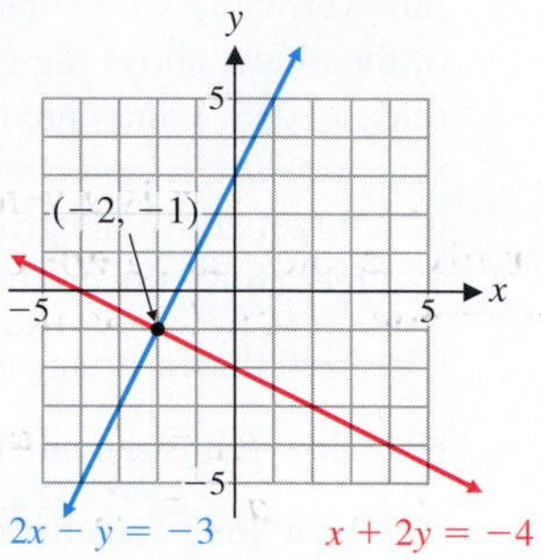

2. (A) $x = 2, y = 2$
(B) Infinitely many solutions
(C) No solution

3. $x = -1.92, y = 4.23$

4. $x = -2, y = 2$

5. $x = 2, y = -1$

6. 16.5 oz of cottage cheese, 12.5 oz of yogurt

7. Equilibrium quantity = 13 thousand pounds; equilibrium price = \$1.70 per pound

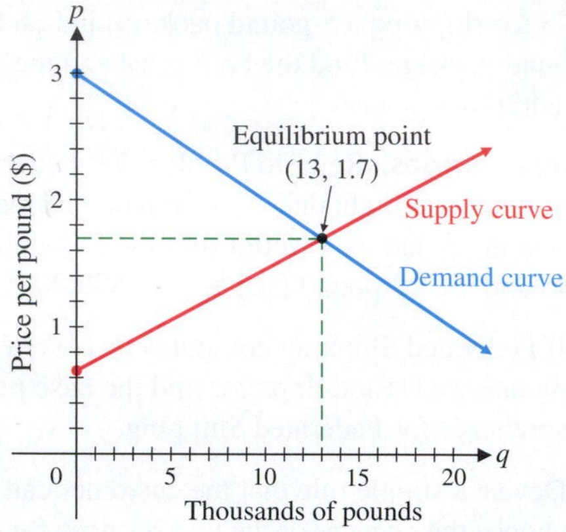

4.2 Systems of Linear Equations and Augmented Matrices

- Matrices
- Solving Linear Systems Using Augmented Matrices
- Summary

Most linear systems of any consequence involve large numbers of equations and variables. It is impractical to try to solve such systems by hand. In the past, these complex systems could be solved only on large computers. Now there are a wide array of approaches to solving linear systems, ranging from graphing calculators to software and spreadsheets. In the rest of this chapter, we develop several *matrix methods* for solving systems with the understanding that these methods are generally used with a graphing calculator. It is important to keep in mind that we are not presenting these techniques as efficient methods for solving linear systems by hand. Instead, we emphasize formulation of mathematical models and interpretation of the results—two activities that graphing calculators cannot perform for you.

Matrices

In solving systems of equations using elimination by addition, the coefficients of the variables and the constant terms played a central role. The process can be made more efficient for generalization and computer work by the introduction of a mathematical form called a *matrix*. A **matrix** is a rectangular array of numbers written within brackets. Two examples are

$$A = \begin{bmatrix} 1 & -4 & 5 \\ 7 & 0 & -2 \end{bmatrix} \qquad B = \begin{bmatrix} -4 & 5 & 12 \\ 0 & 1 & 8 \\ -3 & 10 & 9 \\ -6 & 0 & -1 \end{bmatrix} \qquad (1)$$

Each number in a matrix is called an **element** of the matrix. Matrix A has 6 elements arranged in 2 rows and 3 columns. Matrix B has 12 elements arranged in 4 rows and 3 columns. If a matrix has m rows and n columns, it is called an $\boldsymbol{m \times n}$ **matrix** (read "m by n matrix"). The expression $m \times n$ is called the **size** of the matrix, and the numbers m and n are called the **dimensions** of the matrix. It is important to note that the number of rows is always given first. Referring to equations (1), A is a 2×3 matrix and B is a 4×3 matrix. A matrix with n rows and n columns is called a **square matrix of order** $\boldsymbol{n}$. A matrix with only 1 column is called a **column matrix**, and a matrix with only 1 row is called a **row matrix**.

$$\underset{\text{Square matrix of order 3}}{\overset{3 \times 3}{\begin{bmatrix} 0.5 & 0.2 & 1.0 \\ 0.0 & 0.3 & 0.5 \\ 0.7 & 0.0 & 0.2 \end{bmatrix}}} \qquad \underset{\text{Column matrix}}{\overset{4 \times 1}{\begin{bmatrix} 3 \\ -2 \\ 1 \\ 0 \end{bmatrix}}} \qquad \underset{\text{Row matrix}}{\overset{1 \times 4}{\begin{bmatrix} 2 & \frac{1}{2} & 0 & -\frac{2}{3} \end{bmatrix}}}$$

The **position** of an element in a matrix is given by the row and column containing the element. This is usually denoted using **double subscript notation** a_{ij}, where i is the row and j is the column containing the element a_{ij}, as illustrated below:

$$A = \begin{bmatrix} 1 & -4 & 5 \\ 7 & 0 & -2 \end{bmatrix} \qquad \begin{matrix} a_{11} = 1, & a_{12} = -4, & a_{13} = 5 \\ a_{21} = 7, & a_{22} = 0, & a_{23} = -2 \end{matrix}$$

Note that a_{12} is read "a sub one two" (*not* "a sub twelve"). The elements $a_{11} = 1$ and $a_{22} = 0$ make up the *principal diagonal* of A. In general, the **principal diagonal** of a matrix A consists of the elements $a_{11}, a_{22}, a_{33}, \ldots$.

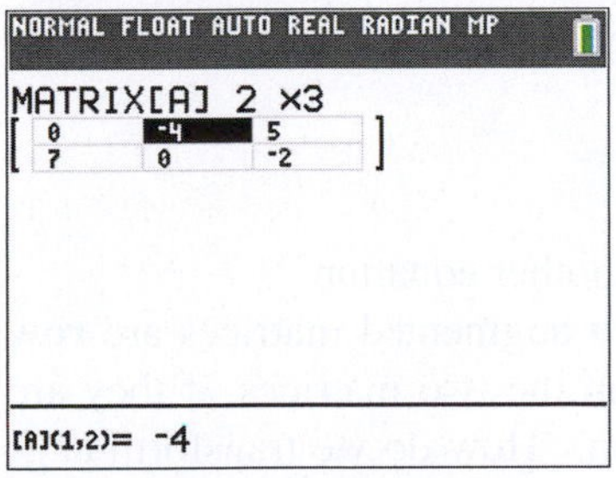

Figure 1 Matrix notation on a graphing calculator

Remark—Most graphing calculators are capable of storing and manipulating matrices. Figure 1 shows matrix A displayed in the editing screen of a graphing calculator. The size of the matrix is given at the top of the screen. The position and value of the currently selected element is given at the bottom. Note that a comma is used in the notation for the position. This is common practice on many graphing calculators but not in mathematical literature. In a spreadsheet, matrices are referred to by their location (upper left corner to lower right corner), using either row and column numbers (Fig. 2A) or row numbers and column letters (Fig. 2B).

	1	2	3
1	1	-4	5
2	7	0	-2

(A) Location of matrix A: R1C1:R2C3

	A	B	C	D	E	F
1						
2						
3						
4						
5				1	-4	5
6				7	0	-2

(B) Location of matrix A: D5:F6

Figure 2 Matrix notation in a spreadsheet

Matrices serve as a shorthand for solving systems of linear equations. Associated with the system

$$\begin{aligned} 2x - 3y &= 5 \\ x + 2y &= -3 \end{aligned} \tag{2}$$

are its **coefficient matrix**, **constant matrix**, and **augmented matrix**:

Coefficient matrix	Constant matrix	Augmented matrix
$\begin{bmatrix} 2 & -3 \\ 1 & 2 \end{bmatrix}$	$\begin{bmatrix} 5 \\ -3 \end{bmatrix}$	$\left[\begin{array}{cc\|c} 2 & -3 & 5 \\ 1 & 2 & -3 \end{array}\right]$

Note that the augmented matrix is just the coefficient matrix, augmented by the constant matrix. The vertical bar is included only as a visual aid to separate the coefficients from the constant terms. The augmented matrix contains all of the essential information about the linear system—everything but the names of the variables.

For ease of generalization to the larger systems in later sections, we will change the notation for the variables in system (2) to a subscript form. That is, in place of x and y, we use x_1 and x_2, respectively, and system (2) is rewritten as

$$\begin{aligned} 2x_1 - 3x_2 &= 5 \\ x_1 + 2x_2 &= -3 \end{aligned}$$

In general, associated with each linear system of the form

$$\begin{aligned} a_{11}x_1 + a_{12}x_2 &= k_1 \\ a_{21}x_1 + a_{22}x_2 &= k_2 \end{aligned} \tag{3}$$

where x_1 and x_2 are variables, is the *augmented matrix* of the system:

Column 1 (C_1)
Column 2 (C_2)
Column 3 (C_3)

$$\left[\begin{array}{cc|c} a_{11} & a_{12} & k_1 \\ a_{21} & a_{22} & k_2 \end{array}\right] \begin{array}{l} \leftarrow \text{Row 1 } (R_1) \\ \leftarrow \text{Row 2 } (R_2) \end{array}$$

This matrix contains the essential parts of system (3). Our objective is to learn how to manipulate augmented matrices in order to solve system (3), if a solution exists. The manipulative process is closely related to the elimination process discussed in Section 4.1.

Recall that two linear systems are said to be equivalent if they have the same solution set. In Theorem 2, Section 4.1, we used the operations listed below to transform linear systems into equivalent systems:

(A) Two equations are interchanged.
(B) An equation is multiplied by a nonzero constant.
(C) A constant multiple of one equation is added to another equation.

Paralleling the earlier discussion, we say that two augmented matrices are **row equivalent**, denoted by the symbol ~ placed between the two matrices, if they are augmented matrices of equivalent systems of equations. How do we transform augmented matrices into row-equivalent matrices? We use Theorem 1, which is a direct consequence of the operations listed in Section 4.1.

THEOREM 1 Operations That Produce Row-Equivalent Matrices

An augmented matrix is transformed into a row-equivalent matrix by performing any of the following **row operations**:

(A) Two rows are interchanged $(R_i \leftrightarrow R_j)$.
(B) A row is multiplied by a nonzero constant $(kR_i \rightarrow R_i)$.
(C) A constant multiple of one row is added to another row $(kR_j + R_i \rightarrow R_i)$.

Note: The arrow $\rightarrow$ means "replaces."

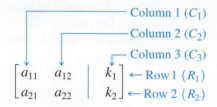

Solving Linear Systems Using Augmented Matrices

We illustrate the use of Theorem 1 by several examples.

EXAMPLE 1 **Solving a System Using Augmented Matrix Methods** Solve using augmented matrix methods:

$$\begin{aligned} 3x_1 + 4x_2 &= 1 \\ x_1 - 2x_2 &= 7 \end{aligned} \tag{4}$$

SOLUTION We start by writing the augmented matrix corresponding to system (4):

$$\left[\begin{array}{cc|c} 3 & 4 & 1 \\ 1 & -2 & 7 \end{array}\right] \tag{5}$$

Our objective is to use row operations from Theorem 1 to try to transform matrix (5) into the form

$$\left[\begin{array}{cc|c} 1 & 0 & m \\ 0 & 1 & n \end{array}\right] \tag{6}$$

where m and n are real numbers. Then the solution to system (4) will be obvious, since matrix (6) will be the augmented matrix of the following system (a row in an augmented matrix always corresponds to an equation in a linear system):

$$\begin{aligned} x_1 &= m \qquad & x_1 + 0x_2 &= m \\ x_2 &= n \qquad & 0x_1 + x_2 &= n \end{aligned}$$

Now we use row operations to transform matrix (5) into form (6).

Step 1 To get a 1 in the upper left corner, we interchange R_1 and R_2 (Theorem 1A):

$$\left[\begin{array}{cc|c} 3 & 4 & 1 \\ 1 & -2 & 7 \end{array}\right] \underset{\sim}{R_1 \leftrightarrow R_2} \left[\begin{array}{cc|c} 1 & -2 & 7 \\ 3 & 4 & 1 \end{array}\right]$$

Step 2 To get a 0 in the lower left corner, we multiply R_1 by (-3) and add to R_2 (Theorem 1C)—this changes R_2 but not R_1. Some people find it useful to write $(-3R_1)$ outside the matrix to help reduce errors in arithmetic, as shown:

$$\left[\begin{array}{cc|c} 1 & -2 & 7 \\ 3 & 4 & 1 \end{array}\right] \underset{\sim}{(-3)R_1 + R_2 \rightarrow R_2} \left[\begin{array}{cc|c} 1 & -2 & 7 \\ 0 & 10 & -20 \end{array}\right]$$

$$\begin{array}{ccc} -3 & 6 & -21 \end{array}$$

Step 3 To get a 1 in the second row, second column, we multiply R_2 by $\frac{1}{10}$ (Theorem 1B):

$$\left[\begin{array}{cc|c} 1 & -2 & 7 \\ 0 & 10 & -20 \end{array}\right] \underset{\sim}{\tfrac{1}{10}R_2 \rightarrow R_2} \left[\begin{array}{cc|c} 1 & -2 & 7 \\ 0 & 1 & -2 \end{array}\right]$$

Step 4 To get a 0 in the first row, second column, we multiply R_2 by 2 and add the result to R_1 (Theorem 1C)—this changes R_1 but not R_2:

$$\begin{array}{ccc} 0 & 2 & -4 \end{array}$$

$$\left[\begin{array}{cc|c} 1 & -2 & 7 \\ 0 & 1 & -2 \end{array}\right] \underset{\sim}{2R_2 + R1 \rightarrow R1} \left[\begin{array}{cc|c} 1 & 0 & 3 \\ 0 & 1 & -2 \end{array}\right]$$

We have accomplished our objective! The last matrix is the augmented matrix for the system

$$\begin{aligned} x_1 &= 3 \qquad & x_1 + 0x_2 &= 3 \\ x_2 &= -2 \qquad & 0x_1 + x_2 &= -2 \end{aligned} \tag{7}$$

Since system (7) is equivalent to system (4), our starting system, we have solved system (4); that is, $x_1 = 3$ and $x_2 = -2$.

CHECK

$$\begin{aligned} 3x_1 + 4x_2 &= 1 & \qquad x_1 - 2x_2 &= 7 \\ 3(3) + 4(-2) &\stackrel{?}{=} 1 & 3 - 2(-2) &\stackrel{?}{=} 7 \\ 1 &\stackrel{\checkmark}{=} 1 & 7 &\stackrel{\checkmark}{=} 7 \end{aligned}$$

The preceding process may be written more compactly as follows:

Step 1: Need a 1 here.

$$\left[\begin{array}{cc|c} 3 & 4 & 1 \\ 1 & -2 & 7 \end{array}\right] \quad R_1 \leftrightarrow R_2$$

Step 2: Need a 0 here.

$$\sim \left[\begin{array}{cc|c} 1 & -2 & 7 \\ 3 & 4 & 1 \end{array}\right] \quad (-3)R_1 + R_2 \rightarrow R_2$$

$$\begin{array}{ccc} -3 & 6 & -21 \end{array}$$

Step 3: Need a 1 here.

$$\sim \left[\begin{array}{cc|c} 1 & -2 & 7 \\ 0 & 10 & -20 \end{array}\right] \quad \tfrac{1}{10}R_2 \rightarrow R_2$$

$$\begin{array}{ccc} 0 & 2 & -4 \end{array}$$

Step 4: Need a 0 here.

$$\sim \left[\begin{array}{cc|c} 1 & -2 & 7 \\ 0 & 1 & -2 \end{array}\right] \quad 2R_2 + R_1 \rightarrow R_1$$

$$\sim \left[\begin{array}{cc|c} 1 & 0 & 3 \\ 0 & 1 & -2 \end{array}\right]$$

Therefore, $x_1 = 3$ and $x_2 = -2$.

Matched Problem 1 Solve using augmented matrix methods:

$$\begin{aligned} 2x_1 - x_2 &= -7 \\ x_1 + 2x_2 &= 4 \end{aligned}$$

Many graphing calculators can perform row operations. Figure 3 shows the results of performing the row operations used in the solution of Example 1. Consult your manual for the details of performing row operations on your graphing calculator.

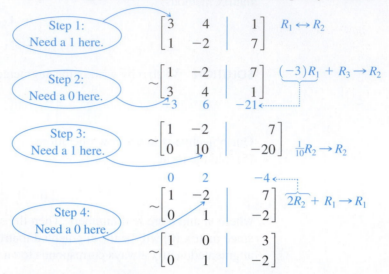

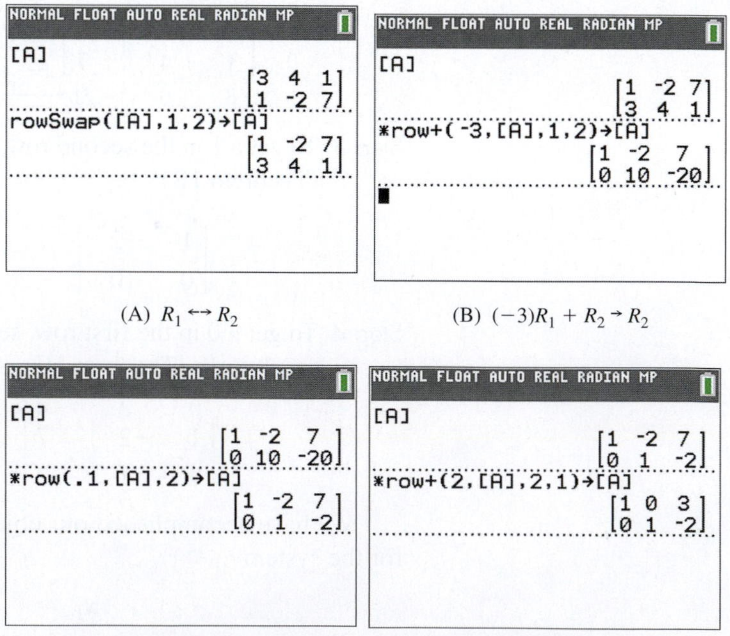

(A) $R_1 \leftrightarrow R_2$ (B) $(-3)R_1 + R_2 \rightarrow R_2$

(C) $\frac{1}{10}R_2 \rightarrow R_2$ (D) $2R_2 + R_1 \rightarrow R_1$

Figure 3 Row operations on a graphing calculator

Explore and Discuss 1

The summary following the solution of Example 1 shows five augmented matrices. Write the linear system that each matrix represents, solve each system graphically, and discuss the relationships among these solutions.

EXAMPLE 2 **Solving a System Using Augmented Matrix Methods** Solve using augmented matrix methods:

$$\begin{aligned} 2x_1 - 3x_2 &= 6 \\ 3x_1 + 4x_2 &= \frac{1}{2} \end{aligned}$$

SOLUTION

Step 1: Need a 1 here.

$$\left[\begin{array}{cc|c} 2 & -3 & 6 \\ 3 & 4 & \frac{1}{2} \end{array}\right] \qquad \tfrac{1}{2}R_1 \rightarrow R_1$$

Step 2: Need a 0 here.

$$\sim \left[\begin{array}{cc|c} 1 & -\frac{3}{2} & 3 \\ 3 & 4 & \frac{1}{2} \end{array}\right] \qquad (-3)R_1 + R_2 \rightarrow R_2$$

$$\begin{array}{ccc} -3 & \frac{9}{2} & -9 \end{array}$$

Step 3: Need a 1 here.

$$\sim \left[\begin{array}{cc|c} 1 & -\frac{3}{2} & 3 \\ 0 & \frac{17}{2} & -\frac{17}{2} \end{array}\right] \qquad \tfrac{2}{17}R_2 \rightarrow R_2$$

$$\begin{array}{ccc} 0 & \frac{3}{2} & -\frac{3}{2} \end{array}$$

Step 4: Need a 0 here.

$$\sim \left[\begin{array}{cc|c} 1 & -\frac{3}{2} & 3 \\ 0 & 1 & -1 \end{array}\right] \qquad \tfrac{3}{2}R_2 + R_1 \rightarrow R_1$$

$$\sim \left[\begin{array}{cc|c} 1 & 0 & \frac{3}{2} \\ 0 & 1 & -1 \end{array}\right]$$

So, $x_1 = \frac{3}{2}$ and $x_2 = -1$. The check is left for you.

Matched Problem 2 Solve using augmented matrix methods:

$$\begin{aligned} 5x_1 - 2x_2 &= 11 \\ 2x_1 + 3x_2 &= \frac{5}{2} \end{aligned}$$

EXAMPLE 3 **Solving a System Using Augmented Matrix Methods** Solve using augmented matrix methods:

$$\begin{aligned} 2x_1 - x_2 &= 4 \\ -6x_1 + 3x_2 &= -12 \end{aligned} \qquad (8)$$

SOLUTION

$$\left[\begin{array}{cc|c} 2 & -1 & 4 \\ -6 & 3 & -12 \end{array}\right] \qquad \begin{array}{l} \tfrac{1}{2}R_1 \rightarrow R_1 \text{ (to get a 1 in the upper left corner)} \\ \tfrac{1}{3}R_2 \rightarrow R_2 \text{ (this simplifies } R_2\text{)} \end{array}$$

$$\sim \left[\begin{array}{cc|c} 1 & -\frac{1}{2} & 2 \\ -2 & 1 & -4 \end{array}\right] \qquad 2R_1 + R_2 \rightarrow R_2 \text{ (to get a 0 in the lower left corner)}$$

$$\begin{array}{ccc} 2 & -1 & 4 \end{array}$$

$$\sim \left[\begin{array}{cc|c} 1 & -\frac{1}{2} & 2 \\ 0 & 0 & 0 \end{array}\right]$$

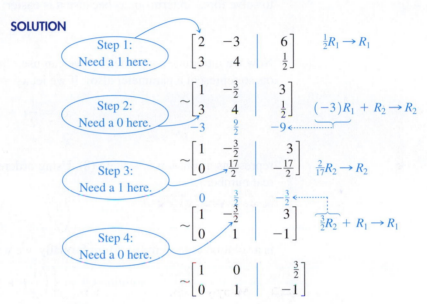

The last matrix corresponds to the system

$$\begin{aligned} x_1 - \frac{1}{2}x_2 &= 2 \qquad & x_1 - \frac{1}{2}x_2 &= 2 \\ 0 &= 0 \qquad & 0x_1 + 0x_2 &= 0 \end{aligned} \tag{9}$$

This system is equivalent to the original system. Geometrically, the graphs of the two original equations coincide, and there are infinitely many solutions. In general, if we end up with a row of zeros in an augmented matrix for a two-equation, two-variable system, the system is dependent, and there are infinitely many solutions.

We represent the infinitely many solutions using the same method that was used in Section 4.1, that is, by introducing a parameter. We start by solving $x_1 - \frac{1}{2}x_2 = 2$, the first equation in system (9), for either variable in terms of the other. We choose to solve for x_1 in terms of x_2 because it is easier:

$$x_1 = \frac{1}{2}x_2 + 2 \tag{10}$$

Now we introduce a parameter t (we can use other letters, such as k, s, p, q, and so on, to represent a parameter also). If we let $x_2 = t$, then for any real number t,

$$\begin{aligned} x_1 &= \frac{1}{2}t + 2 \\ x_2 &= t \end{aligned} \tag{11}$$

represents a solution of system (8). Using ordered-pair notation, we write: For any real number t,

$$\left(\frac{1}{2}t + 2, t\right) \tag{12}$$

is a solution of system (8). More formally, we write

$$\text{solution set} = \left\{\left(\frac{1}{2}t + 2, t\right) \middle| t \in R\right\} \tag{13}$$

Typically we use the less formal notations (11) or (12) to represent the solution set for problems of this type.

CHECK The following is a check that system (11) provides a solution to system (8) for any real number t:

$$\begin{aligned} 2x_1 - x_2 &= 4 \\ 2\left(\frac{1}{2}t + 2\right) - t &\stackrel{?}{=} 4 \\ t + 4 - t &\stackrel{?}{=} 4 \\ 4 &\stackrel{\checkmark}{=} 4 \end{aligned} \qquad \begin{aligned} -6x_1 + 3x_2 &= -12 \\ -6\left(\frac{1}{2}t + 2\right) + 3t &\stackrel{?}{=} -12 \\ -3t - 12 + 3t &\stackrel{?}{=} -12 \\ -12 &\stackrel{\checkmark}{=} -12 \end{aligned}$$

Matched Problem 3 Solve using augmented matrix methods:

$$\begin{aligned} -2x_1 + 6x_2 &= 6 \\ 3x_1 - 9x_2 &= -9 \end{aligned}$$

Explore and Discuss 2

The solution of Example 3 involved three augmented matrices. Write the linear system that each matrix represents, solve each system graphically, and discuss the relationships among these solutions.

EXAMPLE 4 **Solving a System Using Augmented Matrix Methods** Solve using augmented matrix methods:

$$\begin{aligned} 2x_1 + 6x_2 &= -3 \\ x_1 + 3x_2 &= 2 \end{aligned}$$

SOLUTION

$$\left[\begin{array}{cc|c} 2 & 6 & -3 \\ 1 & 3 & 2 \end{array}\right] \quad R_1 \leftrightarrow R_2$$

$$\sim \left[\begin{array}{cc|c} 1 & 3 & 2 \\ 2 & 6 & -3 \end{array}\right] \quad (-2)R_1 + R_2 \rightarrow R_2$$

$$\begin{array}{ccc} -2 & -6 & -4 \end{array}$$

$$\sim \left[\begin{array}{cc|c} 1 & 3 & 2 \\ 0 & 0 & -7 \end{array}\right] \quad R_2 \text{ implies the contradiction } 0 = -7.$$

This is the augmented matrix of the system

$$\begin{aligned} x_1 + 3x_2 &= 2 & \qquad x_1 + 3x_2 &= 2 \\ 0 &= -7 & \qquad 0x_1 + 0x_2 &= -7 \end{aligned}$$

The second equation is not satisfied by any ordered pair of real numbers. As we saw in Section 4.1, the original system is inconsistent and has no solution. If in a row of an augmented matrix, we obtain all zeros to the left of the vertical bar and a nonzero number to the right, the system is inconsistent and there are no solutions.

Matched Problem 4 Solve using augmented matrix methods:

$$\begin{aligned} 2x_1 - x_2 &= 3 \\ 4x_1 - 2x_2 &= -1 \end{aligned}$$

Summary

Examples 2, 3, and 4 illustrate the three possible solution types for a system of two linear equations in two variables, as discussed in Theorem 1, Section 4.1. Examining the final matrix form in each of these solutions leads to the following summary.

SUMMARY Possible Final Matrix Forms for a System of Two Linear Equations in Two Variables

Form 1: Exactly one solution (consistent and independent)

$$\left[\begin{array}{cc|c} 1 & 0 & m \\ 0 & 1 & n \end{array}\right]$$

Form 2: Infinitely many solutions (consistent and dependent)

$$\left[\begin{array}{cc|c} 1 & m & n \\ 0 & 0 & 0 \end{array}\right]$$

Form 3: No solution (inconsistent)

$$\left[\begin{array}{cc|c} 1 & m & n \\ 0 & 0 & p \end{array}\right]$$

m, n, p are real numbers; $p \neq 0$

The process of solving systems of equations described in this section is referred to as **Gauss–Jordan elimination**. We formalize this method in the next section so that it will apply to systems of any size, including systems where the number of equations and the number of variables are not the same.

Exercises 4.2

W Skills Warm-up Exercises

Problems 1–14 refer to the following matrices: (If necessary, review the terminology at the beginning of Section 4.2.)

$$A = \begin{bmatrix} 2 & -4 & 0 \\ 6 & 1 & -5 \end{bmatrix}$$

$$B = \begin{bmatrix} -1 & 9 & 0 \\ -4 & 8 & 7 \\ 2 & 4 & 0 \end{bmatrix}$$

$$C = \begin{bmatrix} 2 & -3 & 0 \end{bmatrix}$$

$$D = \begin{bmatrix} -5 \\ 8 \end{bmatrix}$$

1. How many elements are there in A? In C?

2. How many elements are there in B? In D?

3. What is the size of B? Of D?

4. What is the size of A? Of C?

5. Which of the matrices is a column matrix?

6. Which of the matrices is a row matrix?

7. Which of the matrices is a square matrix?

8. Which of the matrices does not contain the element 0?

A

9. List the elements on the principal diagonal of A.

10. List the elements on the principal diagonal of B.

11. For matrix B, list the elements b_{31}, b_{22}, b_{13}.

12. For matrix A, list the elements a_{21}, a_{12}.

13. For matrix C, find $c_{11} + c_{12} + c_{13}$.

14. For matrix D, find $d_{11} + d_{21}$.

In Problems 15–18, write the coefficient matrix and the augmented matrix of the given system of linear equations.

15. $\begin{aligned} 3x_1 + 5x_2 &= 8 \\ 2x_1 - 4x_2 &= -7 \end{aligned}$

16. $\begin{aligned} -8x_1 + 3x_2 &= 10 \\ 6x_1 + 5x_2 &= 13 \end{aligned}$

17. $\begin{aligned} x_1 + 4x_2 &= 15 \\ 6x_1 \quad &= 18 \end{aligned}$

18. $\begin{aligned} 5x_1 - x_2 &= 10 \\ 3x_2 &= 21 \end{aligned}$

In Problems 19–22, write the system of linear equations that is represented by the given augmented matrix. Assume that the variables are x_1 and x_2.

19. $\left[\begin{array}{cc|c} 2 & 5 & 7 \\ 1 & 4 & 9 \end{array}\right]$

20. $\left[\begin{array}{cc|c} 0 & 3 & 15 \\ -8 & 2 & 25 \end{array}\right]$

21. $\left[\begin{array}{cc|c} 4 & 0 & -10 \\ 0 & 8 & 40 \end{array}\right]$

22. $\left[\begin{array}{cc|c} 1 & -2 & 12 \\ 0 & 1 & 6 \end{array}\right]$

Perform the row operations indicated in Problems 23–34 on the following matrix:

$$\left[\begin{array}{cc|c} 2 & -4 & 6 \\ 1 & -3 & 5 \end{array}\right]$$

23. $R_1 \leftrightarrow R_2$

24. $R_2 \leftrightarrow R_1$

25. $2R_2 \rightarrow R_2$

26. $-2R_2 \rightarrow R_2$

27. $R_1 + R_2 \rightarrow R_1$

28. $R_1 + R_2 \rightarrow R_2$

29. $-\frac{1}{2}R_1 \rightarrow R_1$

30. $\frac{1}{2}R_1 \rightarrow R_1$

31. $(-1)R_2 + R_1 \rightarrow R_1$

32. $(-2)R_2 + R_1 \rightarrow R_1$

33. $(-1)R_1 + R_2 \rightarrow R_2$

34. $(-\frac{1}{2})R_1 + R_2 \rightarrow R_2$

Each of the matrices in Problems 35–42 is the result of performing a single row operation on the matrix A shown below. Identify the row operation.

$$A = \left[\begin{array}{cc|c} -1 & 2 & -3 \\ 6 & -3 & 12 \end{array}\right]$$

35. $\left[\begin{array}{cc|c} -1 & 2 & -3 \\ 2 & -1 & 4 \end{array}\right]$

36. $\left[\begin{array}{cc|c} -2 & 4 & -6 \\ 6 & -3 & 12 \end{array}\right]$

37. $\left[\begin{array}{cc|c} -1 & 2 & -3 \\ 0 & 9 & -6 \end{array}\right]$

38. $\left[\begin{array}{cc|c} 3 & 0 & 5 \\ 6 & -3 & 12 \end{array}\right]$

39. $\left[\begin{array}{cc|c} 1 & 1 & 1 \\ 6 & -3 & 12 \end{array}\right]$

40. $\left[\begin{array}{cc|c} -1 & 2 & -3 \\ 2 & 5 & 0 \end{array}\right]$

41. $\left[\begin{array}{cc|c} 6 & -3 & 12 \\ -1 & 2 & -3 \end{array}\right]$

42. $\left[\begin{array}{cc|c} -1 & 2 & -3 \\ 0 & 9 & -6 \end{array}\right]$

Solve Problems 43–46 using augmented matrix methods. Graph each solution set. Discuss the differences between the graph of an equation in the system and the graph of the system's solution set.

43. $\begin{aligned} 3x_1 - 2x_2 &= 6 \\ 4x_1 - 3x_2 &= 6 \end{aligned}$

44. $\begin{aligned} x_1 - 2x_2 &= 5 \\ -2x_1 + 4x_2 &= -10 \end{aligned}$

45. $\begin{aligned} 3x_1 - 2x_2 &= -3 \\ -6x_1 + 4x_2 &= 6 \end{aligned}$

46. $\begin{aligned} x_1 - 2x_2 &= 1 \\ -2x_1 + 5x_2 &= 2 \end{aligned}$

Solve Problems 47 and 48 using augmented matrix methods. Write the linear system represented by each augmented matrix in your solution, and solve each of these systems graphically. Discuss the relationships among the solutions of these systems.

47. $\begin{aligned} x_1 + x_2 &= 5 \\ x_1 - x_2 &= 1 \end{aligned}$

48. $\begin{aligned} x_1 - x_2 &= 2 \\ x_1 + x_2 &= 6 \end{aligned}$

Each of the matrices in Problems 49–54 is the final matrix form for a system of two linear equations in the variables x_1 and x_2. Write the solution of the system.

49. $\left[\begin{array}{cc|c} 1 & 0 & -4 \\ 0 & 1 & 6 \end{array}\right]$

50. $\left[\begin{array}{cc|c} 1 & 0 & 3 \\ 0 & 1 & -5 \end{array}\right]$

51. $\left[\begin{array}{cc|c} 1 & 3 & 2 \\ 0 & 0 & 4 \end{array}\right]$

52. $\left[\begin{array}{cc|c} 1 & -2 & 7 \\ 0 & 0 & -9 \end{array}\right]$

53. $\left[\begin{array}{cc|c} 1 & -2 & 15 \\ 0 & 0 & 0 \end{array}\right]$

54. $\left[\begin{array}{cc|c} 1 & 5 & 10 \\ 0 & 0 & 0 \end{array}\right]$

B *Solve Problems 55–74 using augmented matrix methods.*

55. $\begin{aligned} x_1 - 2x_2 &= 1 \\ 2x_1 - x_2 &= 5 \end{aligned}$

56. $\begin{aligned} x_1 + 3x_2 &= 1 \\ 3x_1 - 2x_2 &= 14 \end{aligned}$

57. $\begin{aligned} x_1 - 4x_2 &= -2 \\ -2x_1 + x_2 &= -3 \end{aligned}$

58. $\begin{aligned} x_1 - 3x_2 &= -5 \\ -3x_1 - x_2 &= 5 \end{aligned}$

59. $\begin{aligned} 3x_1 - x_2 &= 2 \\ x_1 + 2x_2 &= 10 \end{aligned}$

60. $\begin{aligned} 2x_1 + x_2 &= 0 \\ x_1 - 2x_2 &= -5 \end{aligned}$

61. $\begin{aligned} x_1 + 2x_2 &= 4 \\ 2x_1 + 4x_2 &= -8 \end{aligned}$

62. $\begin{aligned} 2x_1 - 3x_2 &= -2 \\ -4x_1 + 6x_2 &= 7 \end{aligned}$

63. $\begin{aligned} 2x_1 + x_2 &= 6 \\ x_1 - x_2 &= -3 \end{aligned}$

64. $\begin{aligned} 3x_1 - x_2 &= -5 \\ x_1 + 3x_2 &= 5 \end{aligned}$

65. $\begin{aligned} 3x_1 - 6x_2 &= -9 \\ -2x_1 + 4x_2 &= 6 \end{aligned}$

66. $\begin{aligned} 2x_1 - 4x_2 &= -2 \\ -3x_1 + 6x_2 &= 3 \end{aligned}$

67. $\begin{aligned} 4x_1 - 2x_2 &= 2 \\ -6x_1 + 3x_2 &= -3 \end{aligned}$

68. $\begin{aligned} -6x_1 + 2x_2 &= 4 \\ 3x_1 - x_2 &= -2 \end{aligned}$

69. $\begin{aligned} 2x_1 + x_2 &= 1 \\ 4x_1 - x_2 &= -7 \end{aligned}$

70. $\begin{aligned} 2x_1 - x_2 &= -8 \\ 2x_1 + x_2 &= 8 \end{aligned}$

71. $\begin{aligned} 4x_1 - 6x_2 &= 8 \\ -6x_1 + 9x_2 &= -10 \end{aligned}$

72. $\begin{aligned} 2x_1 - 4x_2 &= -4 \\ -3x_1 + 6x_2 &= 4 \end{aligned}$

73. $\begin{aligned} -4x_1 + 6x_2 &= -8 \\ 6x_1 - 9x_2 &= 12 \end{aligned}$

74. $\begin{aligned} -2x_1 + 4x_2 &= 4 \\ 3x_1 - 6x_2 &= -6 \end{aligned}$

C *Solve Problems 75–80 using augmented matrix methods.*

75. $\begin{aligned} 3x_1 - x_2 &= 7 \\ 2x_1 + 3x_2 &= 1 \end{aligned}$

76. $\begin{aligned} 2x_1 - 3x_2 &= -8 \\ 5x_1 + 3x_2 &= 1 \end{aligned}$

77. $\begin{aligned} 3x_1 + 2x_2 &= 4 \\ 2x_1 - x_2 &= 5 \end{aligned}$

78. $\begin{aligned} 4x_1 + 3x_2 &= 26 \\ 3x_1 - 11x_2 &= -7 \end{aligned}$

79. $\begin{aligned} 0.2x_1 - 0.5x_2 &= 0.07 \\ 0.8x_1 - 0.3x_2 &= 0.79 \end{aligned}$

80. $\begin{aligned} 0.3x_1 - 0.6x_2 &= 0.18 \\ 0.5x_1 - 0.2x_2 &= 0.54 \end{aligned}$

Solve Problems 81–84 using augmented matrix methods. Use a graphing calculator to perform the row operations.

81. $\begin{aligned} 0.8x_1 + 2.88x_2 &= 4 \\ 1.25x_1 + 4.34x_2 &= 5 \end{aligned}$

82. $\begin{aligned} 2.7x_1 - 15.12x_2 &= 27 \\ 3.25x_1 - 18.52x_2 &= 33 \end{aligned}$

83. $\begin{aligned} 4.8x_1 - 40.32x_2 &= 295.2 \\ -3.75x_1 + 28.7x_2 &= -211.2 \end{aligned}$

84. $\begin{aligned} 5.7x_1 - 8.55x_2 &= -35.91 \\ 4.5x_1 + 5.73x_2 &= 76.17 \end{aligned}$

Answers to Matched Problems

1. $x_1 = -2, x_2 = 3$
2. $x_1 = 2, x_2 = -\frac{1}{2}$
3. The system is dependent. For t any real number, a solution is $x_1 = 3t - 3, x_2 = t$.
4. Inconsistent—no solution

4.3 Gauss–Jordan Elimination

- Reduced Matrices
- Solving Systems by Gauss–Jordan Elimination
- Application

Now that you have had some experience with row operations on simple augmented matrices, we consider systems involving more than two variables. We will not require a system to have the same number of equations as variables. Just as for systems of two linear equations in two variables, any linear system, regardless of the number of equations or number of variables, has either

1. Exactly one solution (consistent and independent), or
2. Infinitely many solutions (consistent and dependent), or
3. No solution (inconsistent).

Reduced Matrices

In the preceding section we used row operations to transform the augmented matrix for a system of two equations in two variables,

$$\left[\begin{array}{cc|c} a_{11} & a_{12} & k_1 \\ a_{21} & a_{22} & k_2 \end{array}\right] \qquad \begin{aligned} a_{11}x_1 + a_{12}x_2 &= k_1 \\ a_{21}x_1 + a_{22}x_2 &= k_2 \end{aligned}$$

into one of the following simplified forms:

$$\overset{\text{Form 1}}{\left[\begin{array}{cc|c} 1 & 0 & m \\ 0 & 1 & n \end{array}\right]} \quad \overset{\text{Form 2}}{\left[\begin{array}{cc|c} 1 & m & n \\ 0 & 0 & 0 \end{array}\right]} \quad \overset{\text{Form 3}}{\left[\begin{array}{cc|c} 1 & m & n \\ 0 & 0 & p \end{array}\right]} \tag{1}$$

where m, n, and p are real numbers, $p \neq 0$. Each of these reduced forms represents a system that has a different type of solution set, and no two of these forms are row equivalent.

For large linear systems, it is not practical to list all such simplified forms; there are too many of them. Instead, we give a general definition of a simplified form called a **reduced matrix**, which can be applied to all matrices and systems, regardless of size.

DEFINITION Reduced Form

A matrix is said to be in **reduced row echelon form**, or, more simply, in **reduced form**, if

1. Each row consisting entirely of zeros is below any row having at least one nonzero element.
2. The leftmost nonzero element in each row is 1.
3. All other elements in the column containing the leftmost 1 of a given row are zeros.
4. The leftmost 1 in any row is to the right of the leftmost 1 in the row above.

The following matrices are in reduced form. Check each one carefully to convince yourself that the conditions in the definition are met.

Note that a row of a reduced matrix may have a leftmost 1 in the last column.

$$\left[\begin{array}{cc|c} 1 & 0 & 2 \\ 0 & 1 & -3 \end{array}\right] \quad \left[\begin{array}{ccc|c} 1 & 0 & 0 & 2 \\ 0 & 1 & 0 & -1 \\ 0 & 0 & 1 & 3 \end{array}\right] \quad \left[\begin{array}{cc|c} 1 & 0 & 3 \\ 0 & 1 & -1 \\ 0 & 0 & 0 \end{array}\right]$$

$$\left[\begin{array}{cccc|c} 1 & 4 & 0 & 0 & -3 \\ 0 & 0 & 1 & 0 & 2 \\ 0 & 0 & 0 & 1 & 6 \end{array}\right] \quad \left[\begin{array}{ccc|c} 1 & 0 & 4 & 0 \\ 0 & 1 & 3 & 0 \\ 0 & 0 & 0 & 1 \end{array}\right]$$

EXAMPLE 1 **Reduced Forms** The following matrices are not in reduced form. Indicate which condition in the definition is violated for each matrix. State the row operation(s) required to transform the matrix into reduced form, and find the reduced form.

(A) $\left[\begin{array}{cc|c} 0 & 1 & -2 \\ 1 & 0 & 3 \end{array}\right]$ (B) $\left[\begin{array}{ccc|c} 1 & 2 & -2 & 3 \\ 0 & 0 & 1 & -1 \end{array}\right]$

(C) $\left[\begin{array}{cc|c} 1 & 0 & -3 \\ 0 & 0 & 0 \\ 0 & 1 & -2 \end{array}\right]$ (D) $\left[\begin{array}{ccc|c} 1 & 0 & 0 & -1 \\ 0 & 2 & 0 & 3 \\ 0 & 0 & 1 & -5 \end{array}\right]$

SOLUTION

(A) Condition 4 is violated: The leftmost 1 in row 2 is not to the right of the leftmost 1 in row 1. Perform the row operation $R_1 \leftrightarrow R_2$ to obtain

$$\left[\begin{array}{cc|c} 1 & 0 & 3 \\ 0 & 1 & -2 \end{array}\right]$$

(B) Condition 3 is violated: The column containing the leftmost 1 in row 2 has a nonzero element above the 1. Perform the row operation $2R_2 + R_1 \rightarrow R_1$ to obtain

$$\left[\begin{array}{ccc|c} 1 & 2 & 0 & 1 \\ 0 & 0 & 1 & -1 \end{array}\right]$$

(C) Condition 1 is violated: The second row contains all zeros and is not below any row having at least one nonzero element. Perform the row operation $R_2 \leftrightarrow R_3$ to obtain

$$\left[\begin{array}{cc|c} 1 & 0 & -3 \\ 0 & 1 & -2 \\ 0 & 0 & 0 \end{array}\right]$$

(D) Condition 2 is violated: The leftmost nonzero element in row 2 is not a 1. Perform the row operation $\frac{1}{2}R_2 \rightarrow R_2$ to obtain

$$\left[\begin{array}{ccc|c} 1 & 0 & 0 & -1 \\ 0 & 1 & 0 & \frac{3}{2} \\ 0 & 0 & 1 & -5 \end{array}\right]$$

Matched Problem 1 The matrices below are not in reduced form. Indicate which condition in the definition is violated for each matrix. State the row operation(s) required to transform the matrix into reduced form, and find the reduced form.

(A) $\left[\begin{array}{cc|c} 1 & 0 & 2 \\ 0 & 3 & -6 \end{array}\right]$ (B) $\left[\begin{array}{ccc|c} 1 & 5 & 4 & 3 \\ 0 & 1 & 2 & -1 \\ 0 & 0 & 0 & 0 \end{array}\right]$

(C) $\left[\begin{array}{ccc|c} 0 & 1 & 0 & -3 \\ 1 & 0 & 0 & 0 \\ 0 & 0 & 1 & 2 \end{array}\right]$ (D) $\left[\begin{array}{ccc|c} 1 & 2 & 0 & 3 \\ 0 & 0 & 0 & 0 \\ 0 & 0 & 1 & 4 \end{array}\right]$

Solving Systems by Gauss–Jordan Elimination

We are now ready to outline the Gauss–Jordan method for solving systems of linear equations. The method systematically transforms an augmented matrix into a reduced form. The system corresponding to a reduced augmented matrix is called a **reduced system**. As we shall see, reduced systems are easy to solve.

The Gauss–Jordan elimination method is named after the German mathematician Carl Friedrich Gauss (1777–1885) and the German geodesist Wilhelm Jordan (1842–1899). Gauss, one of the greatest mathematicians of all time, used a method of solving systems of equations in his astronomical work that was later generalized by Jordan to solve problems in large-scale surveying.

EXAMPLE 2 **Solving a System Using Gauss–Jordan Elimination** Solve by Gauss–Jordan elimination:

$$\begin{aligned} 2x_1 - 2x_2 + x_3 &= 3 \\ 3x_1 + x_2 - x_3 &= 7 \\ x_1 - 3x_2 + 2x_3 &= 0 \end{aligned}$$

SOLUTION Write the augmented matrix and follow the steps indicated at the right.

Need a 1 here.

$$\left[\begin{array}{rrr|r} 2 & -2 & 1 & 3 \\ 3 & 1 & -1 & 7 \\ 1 & -3 & 2 & 0 \end{array}\right] \quad R_1 \leftrightarrow R_3$$

Step 1 Choose the leftmost nonzero column and get a 1 at the top.

Need 0's here.

$$\sim \left[\begin{array}{rrr|r} 1 & -3 & 2 & 0 \\ 3 & 1 & -1 & 7 \\ 2 & -2 & 1 & 3 \end{array}\right] \quad \begin{array}{l} (-3)R_1 + R_2 \rightarrow R_2 \\ (-2)R_1 + R_3 \rightarrow R_3 \end{array}$$

Step 2 Use multiples of the row containing the 1 from step 1 to get zeros in all remaining places in the column containing this 1.

Need a 1 here.

$$\sim \left[\begin{array}{rrr|r} 1 & -3 & 2 & 0 \\ 0 & 10 & -7 & 7 \\ 0 & 4 & -3 & 3 \end{array}\right] \quad 0.1R_2 \rightarrow R_2$$

Step 3 Repeat step 1 with the *submatrix* formed by (mentally) deleting the top row.

Need 0's here.

$$\sim \left[\begin{array}{rrr|r} 1 & -3 & 2 & 0 \\ 0 & 1 & -0.7 & 0.7 \\ 0 & 4 & -3 & 3 \end{array}\right] \quad \begin{array}{l} 3R_2 + R_1 \rightarrow R_1 \\ \\ (-4)R_2 + R_3 \rightarrow R_3 \end{array}$$

Step 4 Repeat step 2 with the *entire matrix*.

Need a 1 here.

$$\sim \left[\begin{array}{rrr|r} 1 & 0 & -0.1 & 2.1 \\ 0 & 1 & -0.7 & 0.7 \\ 0 & 0 & -0.2 & 0.2 \end{array}\right] \quad (-5)R_3 \rightarrow R_3$$

Step 3 Repeat step 1 with the *submatrix* formed by (mentally) deleting the top rows.

Need 0's here.

$$\sim \left[\begin{array}{rrr|r} 1 & 0 & -0.1 & 2.1 \\ 0 & 1 & -0.7 & 0.7 \\ 0 & 0 & 1 & -1 \end{array}\right] \quad \begin{array}{l} 0.1R_3 + R_1 \rightarrow R_1 \\ 0.7R_3 + R_2 \rightarrow R_2 \end{array}$$

Step 4 Repeat step 2 with the *entire matrix*.

$$\sim \left[\begin{array}{rrr|r} 1 & 0 & 0 & 2 \\ 0 & 1 & 0 & 0 \\ 0 & 0 & 1 & -1 \end{array}\right]$$

The matrix is now in reduced form, and we can solve the corresponding reduced system.

$$\begin{aligned} x_1 \qquad\qquad &= 2 \\ x_2 \qquad &= 0 \\ x_3 &= -1 \end{aligned}$$

The solution to this system is $x_1 = 2, x_2 = 0, x_3 = -1$. You should check this solution in the original system.

Matched Problem 2 Solve by Gauss–Jordan elimination:

$$\begin{aligned} 3x_1 + x_2 - 2x_3 &= 2 \\ x_1 - 2x_2 + x_3 &= 3 \\ 2x_1 - x_2 - 3x_3 &= 3 \end{aligned}$$

PROCEDURE Gauss–Jordan Elimination

Step 1 Choose the leftmost nonzero column and use appropriate row operations to get a 1 at the top.

Step 2 Use multiples of the row containing the 1 from step 1 to get zeros in all remaining places in the column containing this 1.

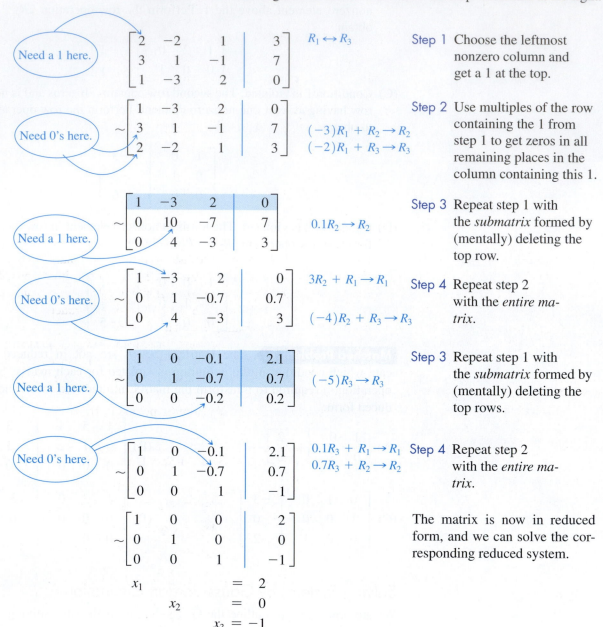

Step 3 Repeat step 1 with the **submatrix** formed by (mentally) deleting the row used in step 2 and all rows above this row.

Step 4 Repeat step 2 with the **entire matrix,** including the rows deleted mentally. Continue this process until the entire matrix is in reduced form.

Note: If at any point in this process we obtain a row with all zeros to the left of the vertical line and a nonzero number to the right, we can stop before we find the reduced form since we will have a contradiction: $0 = n, n \neq 0$. We can then conclude that the system has no solution.

Remarks

1. Even though each matrix has a unique reduced form, the sequence of steps presented here for transforming a matrix into a reduced form is not unique. For example, it is possible to use row operations in such a way that computations involving fractions are minimized. But we emphasize again that we are not interested in the most efficient hand methods for transforming small matrices into reduced forms. Our main interest is in giving you a little experience with a method that is suitable for solving large-scale systems on a graphing calculator or computer.

2. Most graphing calculators have the ability to find reduced forms. Figure 1 illustrates the solution of Example 2 on a TI-84 Plus CE graphing calculator using the rref command (rref is an acronym for reduced row echelon form). Notice that in row 2 and column 4 of the reduced form the graphing calculator has displayed the very small number –3.5E – 13, instead of the exact value 0. This is a common occurrence on a graphing calculator and causes no problems. Just replace any very small numbers displayed in scientific notation with 0.

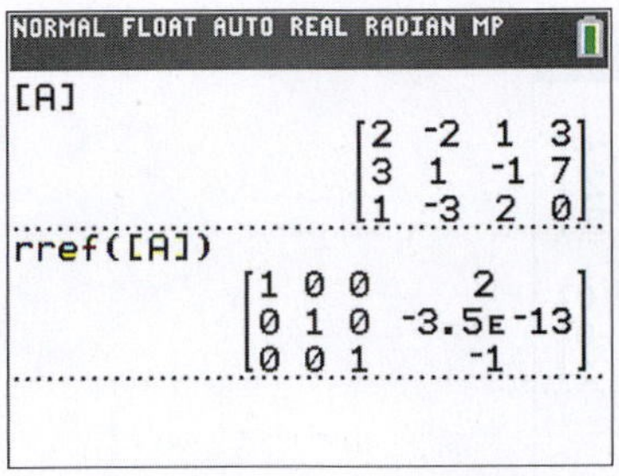

Figure 1 Gauss–Jordan elimination on a graphing calculator

EXAMPLE 3 **Solving a System Using Gauss–Jordan Elimination** Solve by Gauss–Jordan elimination:

$$\begin{aligned} 2x_1 - 4x_2 + x_3 &= -4 \\ 4x_1 - 8x_2 + 7x_3 &= 2 \\ -2x_1 + 4x_2 - 3x_3 &= 5 \end{aligned}$$

SOLUTION

$$\left[\begin{array}{ccc|c} 2 & -4 & 1 & -4 \\ 4 & -8 & 7 & 2 \\ -2 & 4 & -3 & 5 \end{array}\right] \quad 0.5R_1 \rightarrow R_1$$

$$\sim \left[\begin{array}{ccc|c} 1 & -2 & 0.5 & -2 \\ 4 & -8 & 7 & 2 \\ -2 & 4 & -3 & 5 \end{array}\right] \quad \begin{array}{l} (-4)R_1 + R_2 \rightarrow R_2 \\ 2R_1 + R_3 \rightarrow R_3 \end{array}$$

$$\sim \left[\begin{array}{ccc|c} 1 & -2 & 0.5 & -2 \\ 0 & 0 & 5 & 10 \\ 0 & 0 & -2 & 1 \end{array}\right] \quad 0.2R_2 \rightarrow R_2$$

Note that column 3 is the leftmost nonzero column in this submatrix.

$$\sim \left[\begin{array}{ccc|c} 1 & -2 & 0.5 & -2 \\ 0 & 0 & 1 & 2 \\ 0 & 0 & -2 & 1 \end{array}\right] \quad \begin{array}{l} (-0.5)R_2 + R_1 \rightarrow R_1 \\ \\ 2R_2 + R_3 \rightarrow R_3 \end{array}$$

$$\sim \left[\begin{array}{ccc|c} 1 & -2 & 0 & -3 \\ 0 & 0 & 1 & 2 \\ 0 & 0 & 0 & 5 \end{array}\right]$$

We stop the Gauss–Jordan elimination, even though the matrix is not in reduced form, since the last row produces a contradiction.

The system has no solution.

Matched Problem 3 Solve by Gauss–Jordan elimination:

$$\begin{aligned} 2x_1 - 4x_2 - x_3 &= -8 \\ 4x_1 - 8x_2 + 3x_3 &= 4 \\ -2x_1 + 4x_2 + x_3 &= 11 \end{aligned}$$

CAUTION Figure 2 shows the solution to Example 3 on a graphing calculator with a built-in reduced-form routine. Notice that the graphing calculator does not stop when a contradiction first occurs but continues on to find the reduced form. Nevertheless, the last row in the reduced form still produces a contradiction. Do not confuse this type of reduced form with one that represents a consistent system (see Fig. 1). ▲

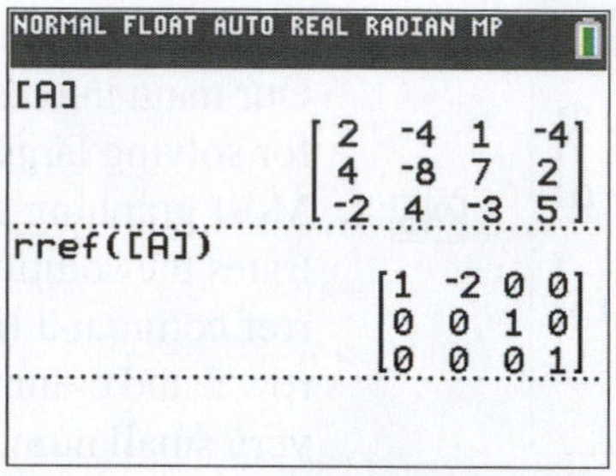

Figure 2 Recognizing contradictions on a graphing calculator

EXAMPLE 4 **Solving a System Using Gauss–Jordan Elimination** Solve by Gauss–Jordan elimination:

$$\begin{aligned} 3x_1 + 6x_2 - 9x_3 &= 15 \\ 2x_1 + 4x_2 - 6x_3 &= 10 \\ -2x_1 - 3x_2 + 4x_3 &= -6 \end{aligned}$$

SOLUTION

$$\left[\begin{array}{rrr|r} 3 & 6 & -9 & 15 \\ 2 & 4 & -6 & 10 \\ -2 & -3 & 4 & -6 \end{array}\right] \quad \tfrac{1}{3}R_1 \to R_1$$

$$\sim \left[\begin{array}{rrr|r} 1 & 2 & -3 & 5 \\ 2 & 4 & -6 & 10 \\ -2 & -3 & 4 & -6 \end{array}\right] \quad \begin{array}{l} (-2)R_1 + R_2 \to R_2 \\ 2R_1 + R_3 \to R_3 \end{array}$$

$$\sim \left[\begin{array}{rrr|r} 1 & 2 & -3 & 5 \\ 0 & 0 & 0 & 0 \\ 0 & 1 & -2 & 4 \end{array}\right] \quad R_2 \leftrightarrow R_3$$

Note that we must interchange rows 2 and 3 to obtain a nonzero entry at the top of the second column of this submatrix.

$$\sim \left[\begin{array}{rrr|r} 1 & 2 & -3 & 5 \\ 0 & 1 & -2 & 4 \\ 0 & 0 & 0 & 0 \end{array}\right] \quad (-2)R_2 + R_1 \to R_1$$

$$\sim \left[\begin{array}{rrr|r} 1 & 0 & 1 & -3 \\ 0 & 1 & -2 & 4 \\ 0 & 0 & 0 & 0 \end{array}\right]$$

The matrix is now in reduced form. Write the corresponding reduced system and solve.

CAUTION All-zero rows do not necessarily indicate that there are infinitely many solutions. These rows show that some of the information given by the equations was redundant.

$$\begin{aligned} x_1 + x_3 &= -3 \\ x_2 - 2x_3 &= 4 \end{aligned}$$

We discard the equation corresponding to the third (all zero) row in the reduced form, since it is satisfied by all values of x_1, x_2, and x_3.

Note that the leftmost variable in each equation appears in one and only one equation. We solve for the leftmost variables x_1 and x_2 in terms of the remaining variable, x_3:

$$\begin{aligned} x_1 &= -x_3 - 3 \\ x_2 &= 2x_3 + 4 \end{aligned}$$

If we let $x_3 = t$, then for any real number t,

$$\begin{aligned} x_1 &= -t - 3 \\ x_2 &= 2t + 4 \\ x_3 &= t \end{aligned}$$

You should check that $(-t - 3, 2t + 4, t)$ is a solution of the original system for any real number t. Some particular solutions are

$t = 0$	$t = -2$	$t = 3.5$
$(-3, 4, 0)$	$(-1, 0, -2)$	$(-6.5, 11, 3.5)$

More generally,

If the number of leftmost 1's in a reduced augmented coefficient matrix is less than the number of variables in the system and there are no contradictions, then the system is dependent and has infinitely many solutions.

Describing the solution set to such a dependent system is not difficult. In a reduced system without contradictions, the **leftmost variables** correspond to the leftmost 1's in the corresponding reduced augmented matrix. The definition of reduced form for an augmented matrix ensures that each leftmost variable in the corresponding reduced system appears in one and only one equation of the system. Solving for each leftmost variable in terms of the remaining variables and writing a general solution to the system is usually easy. Example 5 illustrates a slightly more involved case.

Matched Problem 4 Solve by Gauss–Jordan elimination:

$$\begin{aligned} 2x_1 - 2x_2 - 4x_3 &= -2 \\ 3x_1 - 3x_2 - 6x_3 &= -3 \\ -2x_1 + 3x_2 + x_3 &= 7 \end{aligned}$$

Explore and Discuss 1

Explain why the definition of reduced form ensures that each leftmost variable in a reduced system appears in one and only one equation and no equation contains more than one leftmost variable. Discuss methods for determining whether a consistent system is independent or dependent by examining the reduced form.

EXAMPLE 5 **Solving a System Using Gauss–Jordan Elimination** Solve by Gauss–Jordan elimination:

$$\begin{aligned} x_1 + 2x_2 + 4x_3 + x_4 - x_5 &= 1 \\ 2x_1 + 4x_2 + 8x_3 + 3x_4 - 4x_5 &= 2 \\ x_1 + 3x_2 + 7x_3 \qquad\quad + 3x_5 &= -2 \end{aligned}$$

SOLUTION

$$\left[\begin{array}{ccccc|c} 1 & 2 & 4 & 1 & -1 & 1 \\ 2 & 4 & 8 & 3 & -4 & 2 \\ 1 & 3 & 7 & 0 & 3 & -2 \end{array}\right] \quad \begin{array}{l} (-2)R_1 + R_2 \to R_2 \\ (-1)R_1 + R_3 \to R_3 \end{array}$$

$$\sim \left[\begin{array}{ccccc|c} 1 & 2 & 4 & 1 & -1 & 1 \\ 0 & 0 & 0 & 1 & -2 & 0 \\ 0 & 1 & 3 & -1 & 4 & -3 \end{array}\right] \quad R_2 \leftrightarrow R_3$$

$$\sim \left[\begin{array}{ccccc|c} 1 & 2 & 4 & 1 & -1 & 1 \\ 0 & 1 & 3 & -1 & 4 & -3 \\ 0 & 0 & 0 & 1 & -2 & 0 \end{array}\right] \quad (-2)R_2 + R_1 \to R_1$$

$$\sim \left[\begin{array}{ccccc|c} 1 & 0 & -2 & 3 & -9 & 7 \\ 0 & 1 & 3 & -1 & 4 & -3 \\ 0 & 0 & 0 & 1 & -2 & 0 \end{array}\right] \quad \begin{array}{l} (-3)R_3 + R_1 \to R_1 \\ R_3 + R_2 \to R_2 \end{array}$$

$$\sim \left[\begin{array}{ccccc|c} 1 & 0 & -2 & 0 & -3 & 7 \\ 0 & 1 & 3 & 0 & 2 & -3 \\ 0 & 0 & 0 & 1 & -2 & 0 \end{array}\right] \quad \text{Matrix is in reduced form.}$$

$$\begin{aligned} x_1 - 2x_3 - 3x_5 &= 7 \\ x_2 + 3x_3 + 2x_5 &= -3 \\ x_4 - 2x_5 &= 0 \end{aligned}$$

Solve for the leftmost variables x_1, x_2, and x_4 in terms of the remaining variables x_3 and x_5:

$$\begin{aligned} x_1 &= 2x_3 + 3x_5 + 7 \\ x_2 &= -3x_3 - 2x_5 - 3 \\ x_4 &= 2x_5 \end{aligned}$$

If we let $x_3 = s$ and $x_5 = t$, then for any real numbers s and t,

$$\begin{aligned} x_1 &= 2s + 3t + 7 \\ x_2 &= -3s - 2t - 3 \\ x_3 &= s \\ x_4 &= 2t \\ x_5 &= t \end{aligned}$$

is a solution. The check is left for you.

Matched Problem 5 Solve by Gauss–Jordan elimination:

$$\begin{aligned} x_1 - x_2 + 2x_3 \qquad - 2x_5 &= 3 \\ -2x_1 + 2x_2 - 4x_3 - x_4 + x_5 &= -5 \\ 3x_1 - 3x_2 + 7x_3 + x_4 - 4x_5 &= 6 \end{aligned}$$

Application

Dependent systems of linear equations provide an excellent opportunity to discuss mathematical modeling in more detail. The process of using mathematics to solve real-world problems can be broken down into three steps (Fig. 3):

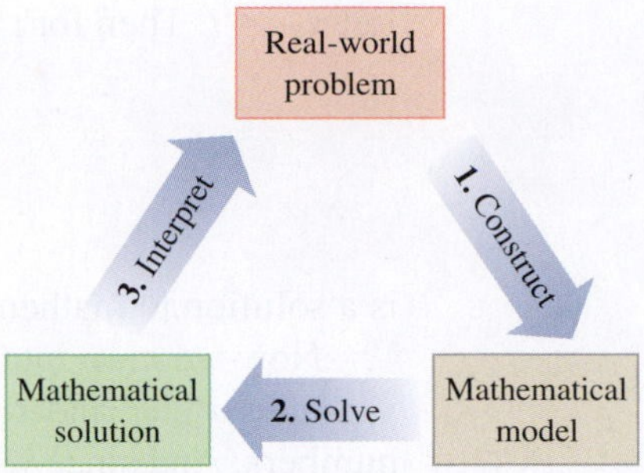

Figure 3

Step 1 *Construct* a mathematical model whose solution will provide information about the real-world problem.

Step 2 *Solve* the mathematical model.

Step 3 *Interpret* the solution to the mathematical model in terms of the original real-world problem.

In more complex problems, this cycle may have to be repeated several times to obtain the required information about the real-world problem.

EXAMPLE 6 **Purchasing** A company that rents small moving trucks wants to purchase 25 trucks with a combined capacity of 28,000 cubic feet. Three different types of trucks are available: a 10-foot truck with a capacity of 350 cubic feet, a 14-foot truck with a capacity of 700 cubic feet, and a 24-foot truck with a capacity of 1,400 cubic feet. How many of each type of truck should the company purchase?

SOLUTION The question in this example indicates that the relevant variables are the number of each type of truck:

$$x_1 = \text{number of 10-foot trucks}$$
$$x_2 = \text{number of 14-foot trucks}$$
$$x_3 = \text{number of 24-foot trucks}$$

Next we form the mathematical model:

$$\begin{aligned} x_1 + x_2 + x_3 &= 25 && \text{Total number of trucks} \\ 350x_1 + 700x_2 + 1{,}400x_3 &= 28{,}000 && \text{Total capacity} \end{aligned} \qquad (2)$$

Now we form the augmented matrix of the system and solve by Gauss–Jordan elimination:

$$\left[\begin{array}{ccc|c} 1 & 1 & 1 & 25 \\ 350 & 700 & 1{,}400 & 28{,}000 \end{array}\right] \quad \tfrac{1}{350}R_2 \to R_2$$

$$\sim \left[\begin{array}{ccc|c} 1 & 1 & 1 & 25 \\ 1 & 2 & 4 & 80 \end{array}\right] \quad -R_1 + R_2 \to R_2$$

$$\sim \left[\begin{array}{ccc|c} 1 & 1 & 1 & 25 \\ 0 & 1 & 3 & 55 \end{array}\right] \quad -R_2 + R_1 \to R_1$$

$$\sim \left[\begin{array}{ccc|c} 1 & 0 & -2 & -30 \\ 0 & 1 & 3 & 55 \end{array}\right] \quad \text{Matrix is in reduced form.}$$

$$x_1 - 2x_3 = -30 \quad \text{or} \quad x_1 = 2x_3 - 30$$

$$x_2 + 3x_3 = 55 \quad \text{or} \quad x_2 = -3x_3 + 55$$

Let $x_3 = t$. Then for t any real number,

$$\begin{aligned} x_1 &= 2t - 30 \\ x_2 &= -3t + 55 \\ x_3 &= t \end{aligned} \tag{3}$$

is a solution to mathematical model (2).

Now we must interpret this solution in terms of the original problem. Since the variables x_1, x_2, and x_3 represent numbers of trucks, they must be nonnegative real numbers. And since we can't purchase a fractional number of trucks, each must be a nonnegative whole number. Since $t = x_3$, it follows that t must also be a nonnegative whole number. The first and second equations in model (3) place additional restrictions on the values that t can assume:

$$x_1 = 2t - 30 \geq 0 \quad \text{implies that} \quad t \geq 15$$

$$x_2 = -3t + 55 \geq 0 \quad \text{implies that} \quad t \leq \frac{55}{3} = 18\tfrac{1}{3}$$

So the only possible values of t that will produce meaningful solutions to the original problem are 15, 16, 17, and 18. That is, the only combinations of 25 trucks that will result in a combined capacity of 28,000 cubic feet are $x_1 = 2t - 30$ 10-foot trucks, $x_2 = -3t + 55$ 14-foot trucks, and $x_3 = t$ 24-foot trucks, where $t = 15, 16, 17$, or 18. A table is a convenient way to display these solutions:

	10-Foot Truck	14-Foot Truck	24-Foot Truck
t	x_1	x_2	x_3
15	0	10	15
16	2	7	16
17	4	4	17
18	6	1	18

Matched Problem 6 A company that rents small moving trucks wants to purchase 16 trucks with a combined capacity of 19,200 cubic feet. Three different types of trucks are available: a cargo van with a capacity of 300 cubic feet, a 15-foot truck with a capacity of 900 cubic feet, and a 24-foot truck with a capacity of 1,500 cubic feet. How many of each type of truck should the company purchase?

Explore and Discuss 2

Refer to Example 6. The rental company charges \$19.95 per day for a 10-foot truck, \$29.95 per day for a 14-foot truck, and \$39.95 per day for a 24-foot truck. Which of the four possible choices in the table would produce the largest daily income from truck rentals?

Exercises 4.3

Skills Warm-up Exercises

W *In Problems 1–4, write the augmented matrix of the system of linear equations. (If necessary, review the terminology of Section 4.2.)*

1. $\begin{aligned} x_1 + 2x_2 + 3x_3 &= 12 \\ x_1 + 7x_2 - 5x_3 &= 15 \end{aligned}$

2. $\begin{aligned} 4x_1 + x_2 &= 8 \\ 3x_1 - 5x_2 &= 6 \\ x_1 + 9x_2 &= 4 \end{aligned}$

3. $\begin{aligned} x_1 + 6x_3 &= 2 \\ x_2 - x_3 &= 5 \\ x_1 + 3x_2 &= 7 \end{aligned}$

4. $\begin{aligned} 3x_1 + 4x_2 &= 10 \\ x_1 + 5x_3 &= 15 \\ - x_2 + x_3 &= 20 \end{aligned}$

In Problems 5–8, write the system of linear equations that is represented by the augmented matrix. Assume that the variables are $x_1, x_2, \ldots$.

5. $\left[\begin{array}{cc|c} 1 & -3 & 4 \\ 3 & 2 & 5 \\ -1 & 6 & 3 \end{array}\right]$

6. $\left[\begin{array}{ccc|c} -1 & 5 & 2 & 8 \\ 4 & 0 & -3 & 7 \end{array}\right]$

7. $\left[\begin{array}{cccc|c} 5 & -2 & 0 & 8 & 4 \end{array}\right]$

8. $\left[\begin{array}{ccc|c} 1 & 0 & -1 & 1 \\ -1 & 1 & 0 & 3 \\ 0 & 2 & 1 & -5 \end{array}\right]$

A *In Problems 9–18, if a matrix is in reduced form, say so. If not, explain why and indicate a row operation that completes the next step of Gauss–Jordan elimination.*

9. $\left[\begin{array}{cc|c} 1 & 0 & 3 \\ 0 & 1 & -2 \end{array}\right]$

10. $\left[\begin{array}{cc|c} 0 & 1 & 3 \\ 1 & 0 & -2 \end{array}\right]$

11. $\left[\begin{array}{cc|c} 0 & 1 & 5 \\ 1 & 0 & -1 \end{array}\right]$

12. $\left[\begin{array}{cc|c} 1 & 0 & 5 \\ 0 & 1 & -1 \end{array}\right]$

13. $\left[\begin{array}{ccc|c} 3 & 6 & -3 & 9 \\ 0 & 2 & -2 & 7 \\ 0 & 5 & -1 & 0 \end{array}\right]$

14. $\left[\begin{array}{ccc|c} 5 & 10 & -5 & -15 \\ 0 & 2 & -2 & 7 \\ 0 & 5 & -1 & 0 \end{array}\right]$

15. $\left[\begin{array}{ccc|c} 1 & 0 & 2 & 9 \\ 0 & 1 & -2 & 7 \\ 0 & 5 & -1 & 0 \end{array}\right]$

16. $\left[\begin{array}{ccc|c} 1 & 10 & -5 & -15 \\ 0 & 0 & -2 & 6 \\ 0 & 0 & 0 & 0 \end{array}\right]$

17. $\left[\begin{array}{ccc|c} 1 & 2 & 2 & 9 \\ 0 & 0 & 2 & 8 \\ 0 & 0 & 0 & 0 \end{array}\right]$

18. $\left[\begin{array}{ccc|c} 1 & 0 & -5 & -15 \\ 0 & 1 & -2 & 7 \\ 0 & 5 & -1 & 0 \end{array}\right]$

In Problems 19–28, write the solution of the linear system corresponding to each reduced augmented matrix.

19. $\left[\begin{array}{ccc|c} 1 & 0 & 0 & -2 \\ 0 & 1 & 0 & 3 \\ 0 & 0 & 1 & 0 \end{array}\right]$

20. $\left[\begin{array}{cccc|c} 1 & 0 & 0 & 0 & -2 \\ 0 & 1 & 0 & 0 & 0 \\ 0 & 0 & 1 & 0 & 1 \\ 0 & 0 & 0 & 1 & 3 \end{array}\right]$

21. $\left[\begin{array}{ccc|c} 1 & 0 & -2 & 3 \\ 0 & 1 & 1 & -5 \\ 0 & 0 & 0 & 0 \end{array}\right]$

22. $\left[\begin{array}{ccc|c} 1 & -2 & 0 & -3 \\ 0 & 0 & 1 & 5 \\ 0 & 0 & 0 & 0 \end{array}\right]$

23. $\left[\begin{array}{cc|c} 1 & 0 & 0 \\ 0 & 1 & 0 \\ 0 & 0 & 1 \end{array}\right]$

24. $\left[\begin{array}{cc|c} 1 & 0 & 5 \\ 0 & 1 & -3 \\ 0 & 0 & 0 \end{array}\right]$

25. $\left[\begin{array}{ccc|c} 1 & 0 & -3 & 5 \\ 0 & 1 & 2 & -7 \end{array}\right]$

26. $\left[\begin{array}{ccc|c} 1 & 0 & 1 & -4 \\ 0 & 1 & -1 & 6 \end{array}\right]$

27. $\left[\begin{array}{cccc|c} 1 & -2 & 0 & -3 & -5 \\ 0 & 0 & 1 & 3 & 2 \end{array}\right]$

28. $\left[\begin{array}{cccc|c} 1 & 0 & -2 & 3 & 4 \\ 0 & 1 & -1 & 2 & -1 \end{array}\right]$

29. In which of Problems 19, 21, 23, 25, and 27 is the number of leftmost ones equal to the number of variables?

30. In which of Problems 20, 22, 24, 26, and 28 is the number of leftmost ones equal to the number of variables?

31. In which of Problems 19, 21, 23, 25, and 27 is the number of leftmost ones less than the number of variables?

32. In which of Problems 20, 22, 24, 26, and 28 is the number of leftmost ones less than the number of variables?

In Problems 33–38, discuss the validity of each statement about linear systems. If the statement is always true, explain why. If not, give a counterexample.

33. If the number of leftmost ones is equal to the number of variables, then the system has exactly one solution.

34. If the number of leftmost ones is less than the number of variables, then the system has infinitely many solutions.

35. If the number of leftmost ones is equal to the number of variables and the system is consistent, then the system has exactly one solution.

36. If the number of leftmost ones is less than the number of variables and the system is consistent, then the system has infinitely many solutions.

37. If there is an all-zero row, then the system has infinitely many solutions.

38. If there are no all-zero rows, then the system has exactly one solution.

B *Use row operations to change each matrix in Problems 39–46 to reduced form.*

39. $\left[\begin{array}{cc|c} 1 & 2 & -1 \\ 0 & 1 & 3 \end{array}\right]$

40. $\left[\begin{array}{cc|c} 1 & 3 & 1 \\ 0 & 2 & -4 \end{array}\right]$

41. $\left[\begin{array}{ccc|c} 1 & 1 & 1 & 16 \\ 2 & 3 & 4 & 25 \end{array}\right]$

42. $\left[\begin{array}{ccc|c} 1 & 1 & 1 & 8 \\ 3 & 5 & 7 & 30 \end{array}\right]$

43. $\left[\begin{array}{ccc|c} 1 & 0 & -3 & 1 \\ 0 & 1 & 2 & 0 \\ 0 & 0 & 3 & -6 \end{array}\right]$

44. $\left[\begin{array}{ccc|c} 1 & 0 & 4 & 0 \\ 0 & 1 & -3 & -1 \\ 0 & 0 & -2 & 2 \end{array}\right]$

45. $\left[\begin{array}{ccc|c} 1 & 2 & -2 & -1 \\ 0 & 3 & -6 & -3 \\ 0 & -1 & 2 & 1 \end{array}\right]$

46. $\left[\begin{array}{ccc|c} 1 & -2 & 6 & -4 \\ 0 & -2 & 8 & 2 \\ 0 & -1 & 4 & 1 \end{array}\right]$

Solve Problems 47–62 using Gauss–Jordan elimination.

47. $\begin{aligned} 2x_1 + 4x_2 - 10x_3 &= -2 \\ 3x_1 + 9x_2 - 21x_3 &= 0 \\ x_1 + 5x_2 - 12x_3 &= 1 \end{aligned}$

48. $\begin{aligned} 3x_1 + 5x_2 - x_3 &= -7 \\ x_1 + x_2 + x_3 &= -1 \\ 2x_1 + 11x_3 &= 7 \end{aligned}$

49. $\begin{aligned} 3x_1 + 8x_2 - x_3 &= -18 \\ 2x_1 + x_2 + 5x_3 &= 8 \\ 2x_1 + 4x_2 + 2x_3 &= -4 \end{aligned}$

50. $\begin{aligned} 2x_1 + 6x_2 + 15x_3 &= -12 \\ 4x_1 + 7x_2 + 13x_3 &= -10 \\ 3x_1 + 6x_2 + 12x_3 &= -9 \end{aligned}$

51. $\begin{aligned} 2x_1 - x_2 - 3x_3 &= 8 \\ x_1 - 2x_2 &= 7 \end{aligned}$

52. $\begin{aligned} 2x_1 + 4x_2 - 6x_3 &= 10 \\ 3x_1 + 3x_2 - 3x_3 &= 6 \end{aligned}$

53. $\begin{aligned} 2x_1 - x_2 &= 0 \\ 3x_1 + 2x_2 &= 7 \\ x_1 - x_2 &= -1 \end{aligned}$

54. $\begin{aligned} 2x_1 - x_2 &= 0 \\ 3x_1 + 2x_2 &= 7 \\ x_1 - x_2 &= -2 \end{aligned}$

55. $\begin{aligned} 3x_1 - 4x_2 - x_3 &= 1 \\ 2x_1 - 3x_2 + x_3 &= 1 \\ x_1 - 2x_2 + 3x_3 &= 2 \end{aligned}$

56. $\begin{aligned} 3x_1 + 7x_2 - x_3 &= 11 \\ x_1 + 2x_2 - x_3 &= 3 \\ 2x_1 + 4x_2 - 2x_3 &= 10 \end{aligned}$

57. $3x_1 - 2x_2 + x_3 = -7$
$2x_1 + x_2 - 4x_3 = 0$
$x_1 + x_2 - 3x_3 = 1$

58. $2x_1 + 3x_2 + 5x_3 = 21$
$x_1 - x_2 - 5x_3 = -2$
$2x_1 + x_2 - x_3 = 11$

59. $2x_1 + 4x_2 - 2x_3 = 2$
$-3x_1 - 6x_2 + 3x_3 = -3$

60. $3x_1 - 9x_2 + 12x_3 = 6$
$-2x_1 + 6x_2 - 8x_3 = -4$

61. $4x_1 - x_2 + 2x_3 = 3$
$-4x_1 + x_2 - 3x_3 = -10$
$8x_1 - 2x_2 + 9x_3 = -1$

62. $4x_1 - 2x_2 + 2x_3 = 5$
$-6x_1 + 3x_2 - 3x_3 = -2$
$10x_1 - 5x_2 + 9x_3 = 4$

63. Consider a consistent system of three linear equations in three variables. Discuss the nature of the system and its solution set if the reduced form of the augmented coefficient matrix has

(A) One leftmost 1 (B) Two leftmost 1's

(C) Three leftmost 1's (D) Four leftmost 1's

64. Consider a system of three linear equations in three variables. Give examples of two reduced forms that are not row-equivalent if the system is

(A) Consistent and dependent

(B) Inconsistent

C *Solve Problems 65–70 using Gauss–Jordan elimination.*

65. $x_1 + 2x_2 - 4x_3 - x_4 = 7$
$2x_1 + 5x_2 - 9x_3 - 4x_4 = 16$
$x_1 + 5x_2 - 7x_3 - 7x_4 = 13$

66. $2x_1 + 4x_2 + 5x_3 + 4x_4 = 8$
$x_1 + 2x_2 + 2x_3 + x_4 = 3$

67. $x_1 - x_2 + 3x_3 - 2x_4 = 1$
$-2x_1 + 4x_2 - 3x_3 + x_4 = 0.5$
$3x_1 - x_2 + 10x_3 - 4x_4 = 2.9$
$4x_1 - 3x_2 + 8x_3 - 2x_4 = 0.6$

68. $x_1 + x_2 + 4x_3 + x_4 = 1.3$
$-x_1 + x_2 - x_3 = 1.1$
$2x_1 + x_3 + 3x_4 = -4.4$
$2x_1 + 5x_2 + 11x_3 + 3x_4 = 5.6$

69. $x_1 - 2x_2 + x_3 + x_4 + 2x_5 = 2$
$-2x_1 + 4x_2 + 2x_3 + 2x_4 - 2x_5 = 0$
$3x_1 - 6x_2 + x_3 + x_4 + 5x_5 = 4$
$-x_1 + 2x_2 + 3x_3 + x_4 + x_5 = 3$

70. $x_1 - 3x_2 + x_3 + x_4 + 2x_5 = 2$
$-x_1 + 5x_2 + 2x_3 + 2x_4 - 2x_5 = 0$
$2x_1 - 6x_2 + 2x_3 + 2x_4 + 4x_5 = 4$
$-x_1 + 3x_2 - x_3 + x_5 = -3$

71. Find a, b, and c so that the graph of the quadratic equation $y = ax^2 + bx + c$ passes through the points $(-2, 9)$, $(1, -9)$, and $(4, 9)$.

72. Find a, b, and c so that the graph of the quadratic equation $y = ax^2 + bx + c$ passes through the points $(-1, -5)$, $(2, 7)$, and $(5, 1)$.

Applications

Construct a mathematical model for each of the following problems. (The answers in the back of the book include both the mathematical model and the interpretation of its solution.) Use Gauss–Jordan elimination to solve the model and then interpret the solution.

73. Boat production. A small manufacturing plant makes three types of inflatable boats: one-person, two-person, and four-person models. Each boat requires the services of three departments, as listed in the table. The cutting, assembly, and packaging departments have available a maximum of 380, 330, and 120 labor-hours per week, respectively.

Department	One-Person Boat	Two-Person Boat	Four-Person Boat
Cutting	0.5 hr	1.0 hr	1.5 hr
Assembly	0.6 hr	0.9 hr	1.2 hr
Packaging	0.2 hr	0.3 hr	0.5 hr

(A) How many boats of each type must be produced each week for the plant to operate at full capacity?

(B) How is the production schedule in part (A) affected if the packaging department is no longer used?

(C) How is the production schedule in part (A) affected if the four-person boat is no longer produced?

74. Production scheduling. Repeat Problem 73 assuming that the cutting, assembly, and packaging departments have available a maximum of 350, 330, and 115 labor-hours per week, respectively.

75. Tank car leases. A chemical manufacturer wants to lease a fleet of 24 railroad tank cars with a combined carrying capacity of 520,000 gallons. Tank cars with three different carrying capacities are available: 8,000 gallons, 16,000 gallons, and 24,000 gallons. How many of each type of tank car should be leased?

76. Airplane leases. A corporation wants to lease a fleet of 12 airplanes with a combined carrying capacity of 220 passengers. The three available types of planes carry 10, 15, and 20 passengers, respectively. How many of each type of plane should be leased?

77. Tank car leases. Refer to Problem 75. The cost of leasing an 8,000-gallon tank car is \$450 per month, a 16,000-gallon tank car is \$650 per month, and a 24,000-gallon tank car is \$1,150 per month. Which of the solutions to Problem 75 would minimize the monthly leasing cost?

78. Airplane leases. Refer to Problem 76. The cost of leasing a 10-passenger airplane is \$8,000 per month, a 15-passenger airplane is \$14,000 per month, and a 20-passenger airplane is \$16,000 per month. Which of the solutions to Problem 76 would minimize the monthly leasing cost?

79. Income tax. A corporation has a taxable income of $7,650,000. At this income level, the federal income tax rate is 50%, the state tax rate is 20%, and the local tax rate is 10%. If each tax rate is applied to the total taxable income, the resulting tax liability for the corporation would be 80% of taxable income. However, it is customary to deduct taxes paid to one agency before computing taxes for the other agencies. Assume that the federal taxes are based on the income that remains after the state and local taxes are deducted, and that state and local taxes are computed in a similar manner. What is the tax liability of the corporation (as a percentage of taxable income) if these deductions are taken into consideration?

80. Income tax. Repeat Problem 79 if local taxes are not allowed as a deduction for federal and state taxes.

81. Taxable income. As a result of several mergers and acquisitions, stock in four companies has been distributed among the companies. Each row of the following table gives the percentage of stock in the four companies that a particular company owns and the annual net income of each company (in millions of dollars):

	Percentage of Stock Owned in Company				Annual Net Income
Company	A	B	C	D	Million $
A	71	8	3	7	3.2
B	12	81	11	13	2.6
C	11	9	72	8	3.8
D	6	2	14	72	4.4

So company *A* holds 71% of its own stock, 8% of the stock in company *B*, 3% of the stock in company *C*, etc. For the purpose of assessing a state tax on corporate income, the taxable income of each company is defined to be its share of its own annual net income plus its share of the taxable income of each of the other companies, as determined by the percentages in the table. What is the taxable income of each company (to the nearest thousand dollars)?

82. Taxable income. Repeat Problem 81 if tax law is changed so that the taxable income of a company is defined to be all of its own annual net income plus its share of the taxable income of each of the other companies.

83. Nutrition. A dietitian in a hospital is to arrange a special diet composed of three basic foods. The diet is to include exactly 340 units of calcium, 180 units of iron, and 220 units of vitamin A. The number of units per ounce of each special ingredient for each of the foods is indicated in the table.

	Units per Ounce		
	Food A	Food B	Food C
Calcium	30	10	20
Iron	10	10	20
Vitamin A	10	30	20

(A) How many ounces of each food must be used to meet the diet requirements?

(B) How is the diet in part (A) affected if food *C* is not used?

(C) How is the diet in part (A) affected if the vitamin A requirement is dropped?

84. Nutrition. Repeat Problem 83 if the diet is to include exactly 400 units of calcium, 160 units of iron, and 240 units of vitamin A.

85. Plant food. A farmer can buy four types of plant food. Each barrel of mix *A* contains 30 pounds of phosphoric acid, 50 pounds of nitrogen, and 30 pounds of potash; each barrel of mix *B* contains 30 pounds of phosphoric acid, 75 pounds of nitrogen, and 20 pounds of potash; each barrel of mix *C* contains 30 pounds of phosphoric acid, 25 pounds of nitrogen, and 20 pounds of potash; and each barrel of mix *D* contains 60 pounds of phosphoric acid, 25 pounds of nitrogen, and 50 pounds of potash. Soil tests indicate that a particular field needs 900 pounds of phosphoric acid, 750 pounds of nitrogen, and 700 pounds of potash. How many barrels of each type of food should the farmer mix together to supply the necessary nutrients for the field?

86. Animal feed. In a laboratory experiment, rats are to be fed 5 packets of food containing a total of 80 units of vitamin E. There are four different brands of food packets that can be used. A packet of brand *A* contains 5 units of vitamin E, a packet of brand *B* contains 10 units of vitamin E, a packet of brand *C* contains 15 units of vitamin E, and a packet of brand *D* contains 20 units of vitamin E. How many packets of each brand should be mixed and fed to the rats?

87. Plant food. Refer to Problem 85. The costs of the four mixes are Mix *A*, $46; Mix *B*, $72; Mix *C*, $57; and Mix *D*, $63. Which of the solutions to Problem 85 would minimize the cost of the plant food?

88. Animal feed. Refer to Problem 86. The costs of the four brands are Brand *A*, $1.50; Brand *B*, $3.00; Brand *C*, $3.75; and Brand *D*, $2.25. Which of the solutions to Problem 86 would minimize the cost of the rat food?

89. Population growth. The U.S. population was approximately 75 million in 1900, 150 million in 1950, and 275 million in 2000. Construct a model for this data by finding a quadratic equation whose graph passes through the points (0, 75), (50, 150), and (100, 275). Use this model to estimate the population in 2050.

90. Population growth. The population of California was approximately 30 million in 1990, 34 million in 2000, and 37 million in 2010. Construct a model for this data by finding a quadratic equation whose graph passes through the points (0, 30), (10, 34), and (20, 37). Use this model to estimate the population in 2030. Do you think the estimate is plausible? Explain. (Source: US Census Bureau)

91. Female life expectancy. The life expectancy for females born during 1980–1985 was approximately 77.6 years. This grew to 78 years during 1985–1990 and to 78.6 years during 1990–1995. Construct a model for this data by finding a quadratic equation whose graph passes through the points (0, 77.6), (5, 78), and (10, 78.6). Use this model to estimate the life expectancy for females born between 1995 and 2000 and for those born between 2000 and 2005.

92. Male life expectancy. The life expectancy for males born during 1980–1985 was approximately 70.7 years. This grew to 71.1 years during 1985–1990 and to 71.8 years during 1990–1995. Construct a model for this data by finding a quadratic equation whose graph passes through the points (0, 70.7), (5, 71.1), and (10, 71.8). Use this model to estimate the life expectancy for males born between 1995 and 2000 and for those born between 2000 and 2005.

93. Female life expectancy. Refer to Problem 91. Subsequent data indicated that life expectancy grew to 79.1 years for females born during 1995–2000 and to 79.7 years for females born during 2000–2005. Add the points (15, 79.1) and (20, 79.7) to the data set in Problem 91. Use a graphing calculator to find a quadratic regression model for all five data points. Graph the data and the model in the same viewing window.

94. Male life expectancy. Refer to Problem 92. Subsequent data indicated that life expectancy grew to 73.2 years for males born during 1995–2000 and to 74.3 years for males born during 2000–2005. Add the points (15, 73.2) and (20, 74.3) to the data set in Problem 92. Use a graphing calculator to find a quadratic regression model for all five data points. Graph the data and the model in the same viewing window.

95. Sociology. Two sociologists have grant money to study school busing in a particular city. They wish to conduct an opinion survey using 600 telephone contacts and 400 house contacts. Survey company *A* has personnel to do 30 telephone and 10 house contacts per hour; survey company *B* can handle 20 telephone and 20 house contacts per hour. How many hours should be scheduled for each firm to produce exactly the number of contacts needed?

96. Sociology. Repeat Problem 95 if 650 telephone contacts and 350 house contacts are needed.

97. Traffic flow. The rush-hour traffic flow for a network of four one-way streets in a city is shown in the figure. The numbers next to each street indicate the number of vehicles per hour that enter and leave the network on that street. The variables x_1, x_2, x_3, and x_4 represent the flow of traffic between the four intersections in the network.

(A) For a smooth traffic flow, the number of vehicles entering each intersection should always equal the number leaving. For example, since 1,500 vehicles enter the intersection of 5th Street and Washington Avenue each hour and $x_1 + x_4$ vehicles leave this intersection, we see that $x_1 + x_4 = 1{,}500$. Find the equations determined by the traffic flow at each of the other three intersections.

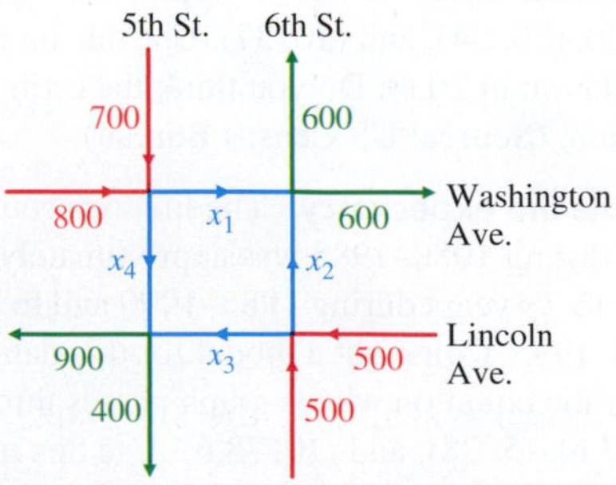

(B) Find the solution to the system in part (A).

(C) What is the maximum number of vehicles that can travel from Washington Avenue to Lincoln Avenue on 5th Street? What is the minimum number?

(D) If traffic lights are adjusted so that 1,000 vehicles per hour travel from Washington Avenue to Lincoln Avenue on 5th Street, determine the flow around the rest of the network.

98. Traffic flow. Refer to Problem 97. Closing Washington Avenue east of 6th Street for construction changes the traffic flow for the network as indicated in the figure. Repeat parts (A)–(D) of Problem 97 for this traffic flow.

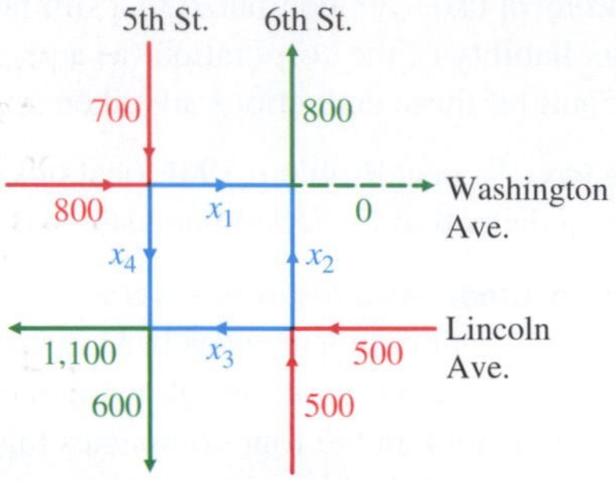

Answers to Matched Problems

1. (A) Condition 2 is violated: The 3 in row 2 and column 2 should be a 1. Perform the operation $\frac{1}{3}R_2 \rightarrow R_2$ to obtain

$$\left[\begin{array}{cc|c} 1 & 0 & 2 \\ 0 & 1 & -2 \end{array}\right]$$

(B) Condition 3 is violated: The 5 in row 1 and column 2 should be a 0. Perform the operation $(-5)R_2 + R_1 \rightarrow R_1$ to obtain

$$\left[\begin{array}{ccc|c} 1 & 0 & -6 & 8 \\ 0 & 1 & 2 & -1 \\ 0 & 0 & 0 & 0 \end{array}\right]$$

(C) Condition 4 is violated. The leftmost 1 in the second row is not to the right of the leftmost 1 in the first row. Perform the operation $R_1 \leftrightarrow R_2$ to obtain

$$\left[\begin{array}{ccc|c} 1 & 0 & 0 & 0 \\ 0 & 1 & 0 & -3 \\ 0 & 0 & 1 & 2 \end{array}\right]$$

(D) Condition 1 is violated: The all-zero second row should be at the bottom. Perform the operation $R_2 \leftrightarrow R_3$ to obtain

$$\left[\begin{array}{ccc|c} 1 & 2 & 0 & 3 \\ 0 & 0 & 1 & 4 \\ 0 & 0 & 0 & 0 \end{array}\right]$$

2. $x_1 = 1, x_2 = -1, x_3 = 0$
3. Inconsistent; no solution
4. $x_1 = 5t + 4, x_2 = 3t + 5, x_3 = t$, t any real number
5. $x_1 = s + 7, x_2 = s, x_3 = t - 2, x_4 = -3t - 1, x_5 = t$, s and t any real numbers
6. $t - 8$ cargo vans, $-2t + 24$ 15-foot trucks, and t 24-foot trucks, where $t = 8, 9, 10, 11$, or 12

4.4 Matrices: Basic Operations

- Addition and Subtraction
- Product of a Number k and a Matrix M
- Matrix Product

In the two preceding sections we introduced the important idea of matrices. In this and following sections, we develop this concept further. Matrices are both an ancient and a current mathematical concept. References to matrices and systems of equations can be found in Chinese manuscripts dating back to about 200 B.C. More recently, computers have made matrices a useful tool for a wide variety of applications. Most graphing calculators and computers are capable of performing calculations with matrices.

As we will see, matrix addition and multiplication are similar to real number addition and multiplication in many respects, but there are some important differences. A brief review of Appendix A, Section A.1, where real number operations are discussed, will help you understand the similarities and the differences.

Addition and Subtraction

Before we can discuss arithmetic operations for matrices, we have to define equality for matrices. Two matrices are **equal** if they have the same size and their corresponding elements are equal. For example,

$$\underset{2 \times 3}{\begin{bmatrix} a & b & c \\ d & e & f \end{bmatrix}} = \underset{2 \times 3}{\begin{bmatrix} u & v & w \\ x & y & z \end{bmatrix}} \quad \text{if and only if} \quad \begin{matrix} a = u & b = v & c = w \\ d = x & e = y & f = z \end{matrix}$$

The **sum of two matrices of the same size** is the matrix with elements that are the sum of the corresponding elements of the two given matrices. Addition is not defined for matrices of different sizes.

EXAMPLE 1 **Matrix Addition**

(A) $\begin{bmatrix} a & b \\ c & d \end{bmatrix} + \begin{bmatrix} w & x \\ y & z \end{bmatrix} = \begin{bmatrix} (a + w) & (b + x) \\ (c + y) & (d + z) \end{bmatrix}$

(B) $\begin{bmatrix} 2 & -3 & 0 \\ 1 & 2 & -5 \end{bmatrix} + \begin{bmatrix} 3 & 1 & 2 \\ -3 & 2 & 5 \end{bmatrix} = \begin{bmatrix} 5 & -2 & 2 \\ -2 & 4 & 0 \end{bmatrix}$

(C) $\begin{bmatrix} 5 & 0 & -2 \\ 1 & -3 & 8 \end{bmatrix} + \begin{bmatrix} -1 & 7 \\ 0 & 6 \\ -2 & 8 \end{bmatrix}$ Not defined

Matched Problem 1 Add: $\begin{bmatrix} 3 & 2 \\ -1 & -1 \\ 0 & 3 \end{bmatrix} + \begin{bmatrix} -2 & 3 \\ 1 & -1 \\ 2 & -2 \end{bmatrix}$

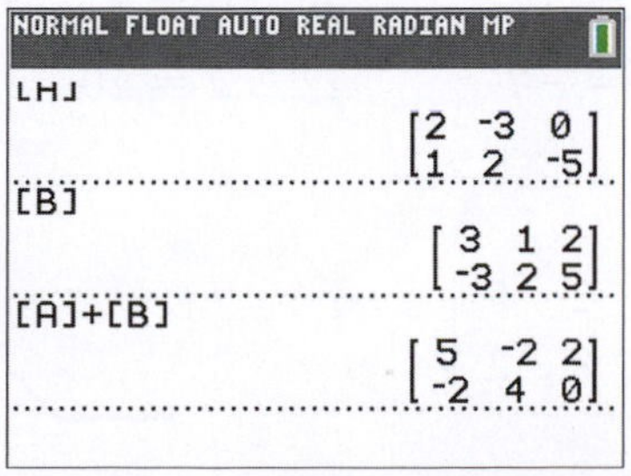

NORMAL FLOAT AUTO REAL RADIAN MP
[A]
[2 -3 0]
[1 2 -5]
[B]
[3 1 2]
[-3 2 5]
[A]+[B]
[5 -2 2]
[-2 4 0]

Figure 1 Addition on a graphing calculator

Graphing calculators can be used to solve problems involving matrix operations. Figure 1 illustrates the solution to Example 1B on a TI-84 Plus CE.

Because we add two matrices by adding their corresponding elements, it follows from the properties of real numbers that matrices of the same size are commutative and associative relative to addition. That is, if A, B, and C are matrices of the same size, then

Commutative: $A + B = B + A$

Associative: $(A + B) + C = A + (B + C)$

A matrix with elements that are all zeros is called a **zero matrix**. For example,

$$[0 \quad 0 \quad 0] \quad \begin{bmatrix} 0 & 0 \\ 0 & 0 \end{bmatrix} \quad \begin{bmatrix} 0 \\ 0 \\ 0 \\ 0 \end{bmatrix} \quad \begin{bmatrix} 0 & 0 & 0 & 0 \\ 0 & 0 & 0 & 0 \\ 0 & 0 & 0 & 0 \end{bmatrix}$$

are zero matrices of different sizes. [*Note:* The simpler notation "0" is often used to denote the zero matrix of an arbitrary size.] The **negative of a matrix** M, denoted by $-M$, is a matrix with elements that are the negatives of the elements in M. Thus, if

$$M = \begin{bmatrix} a & b \\ c & d \end{bmatrix} \quad \text{then} \quad -M = \begin{bmatrix} -a & -b \\ -c & -d \end{bmatrix}$$

Note that $M + (-M) = 0$ (a zero matrix).

If A and B are matrices of the same size, we define **subtraction** as follows:

$$\mathbf{A - B = A + (-B)}$$

So to subtract matrix B from matrix A, we simply add the negative of B to A.

EXAMPLE 2 **Matrix Subtraction**

$$\begin{bmatrix} 3 & -2 \\ 5 & 0 \end{bmatrix} - \begin{bmatrix} -2 & 2 \\ 3 & 4 \end{bmatrix} = \begin{bmatrix} 3 & -2 \\ 5 & 0 \end{bmatrix} + \begin{bmatrix} 2 & -2 \\ -3 & -4 \end{bmatrix} = \begin{bmatrix} 5 & -4 \\ 2 & -4 \end{bmatrix}$$

Matched Problem 2 Subtract: $[2 \quad -3 \quad 5] - [3 \quad -2 \quad 1]$

EXAMPLE 3 **Matrix Equations** Find a, b, c, and d so that

$$\begin{bmatrix} a & b \\ c & d \end{bmatrix} - \begin{bmatrix} 2 & -1 \\ -5 & 6 \end{bmatrix} = \begin{bmatrix} 4 & 3 \\ -2 & 4 \end{bmatrix}$$

SOLUTION

$$\begin{bmatrix} a & b \\ c & d \end{bmatrix} - \begin{bmatrix} 2 & -1 \\ -5 & 6 \end{bmatrix} = \begin{bmatrix} 4 & 3 \\ -2 & 4 \end{bmatrix}$$

Subtract the matrices on the left side.

$$\begin{bmatrix} a - 2 & b - (-1) \\ c - (-5) & d - 6 \end{bmatrix} = \begin{bmatrix} 4 & 3 \\ -2 & 4 \end{bmatrix}$$

Remove parentheses.

$$\begin{bmatrix} a - 2 & b + 1 \\ c + 5 & d - 6 \end{bmatrix} = \begin{bmatrix} 4 & 3 \\ -2 & 4 \end{bmatrix}$$

Use the definition of *equality* to change this matrix equation into four real number equations.

$$a - 2 = 4 \qquad b + 1 = 3 \qquad c + 5 = -2 \qquad d - 6 = 4$$

$$a = 6 \qquad b = 2 \qquad c = -7 \qquad d = 10$$

Matched Problem 3 Find a, b, c, and d so that

$$\begin{bmatrix} a & b \\ c & d \end{bmatrix} - \begin{bmatrix} -4 & 2 \\ 1 & -3 \end{bmatrix} = \begin{bmatrix} -2 & 5 \\ 8 & 2 \end{bmatrix}$$

Product of a Number k and a Matrix M

The **product of a number k and a matrix M**, denoted by kM, is a matrix formed by multiplying each element of M by k.

EXAMPLE 4 **Multiplication of a Matrix by a Number**

$$-2 \begin{bmatrix} 3 & -1 & 0 \\ -2 & 1 & 3 \\ 0 & -1 & -2 \end{bmatrix} = \begin{bmatrix} -6 & 2 & 0 \\ 4 & -2 & -6 \\ 0 & 2 & 4 \end{bmatrix}$$

Matched Problem 4 Find: $10 \begin{bmatrix} 1.3 \\ 0.2 \\ 3.5 \end{bmatrix}$

The next example illustrates the use of matrix operations in an applied setting.

EXAMPLE 5 **Sales Commissions** Ms. Smith and Mr. Jones are salespeople in a new-car agency that sells only two models. August was the last month for this year's models, and next year's models were introduced in September. Gross dollar sales for each month are given in the following matrices:

	August sales			September sales		
	Compact	Luxury		Compact	Luxury	
Ms. Smith	$54,000	$88,000	$= A$	$228,000	$368,000	$= B$
Mr. Jones	$126,000	0		$304,000	$322,000	

For example, Ms. Smith had $54,000 in compact sales in August, and Mr. Jones had $322,000 in luxury car sales in September.

(A) What were the combined dollar sales in August and September for each salesperson and each model?

(B) What was the increase in dollar sales from August to September?

(C) If both salespeople receive 5% commissions on gross dollar sales, compute the commission for each person for each model sold in September.

SOLUTION

(A) $A + B = \begin{bmatrix} \text{Compact} & \text{Luxury} \\ \$282{,}000 & \$456{,}000 \\ \$430{,}000 & \$322{,}000 \end{bmatrix} \begin{matrix} \text{Ms. Smith} \\ \text{Mr. Jones} \end{matrix}$

(B) $B - A = \begin{bmatrix} \text{Compact} & \text{Luxury} \\ \$174{,}000 & \$280{,}000 \\ \$178{,}000 & \$322{,}000 \end{bmatrix} \begin{matrix} \text{Ms. Smith} \\ \text{Mr. Jones} \end{matrix}$

(C) $0.05B = \begin{bmatrix} (0.05)(\$228{,}000) & (0.05)(\$368{,}000) \\ (0.05)(\$304{,}000) & (0.05)(\$322{,}000) \end{bmatrix}$

$= \begin{bmatrix} \$11{,}400 & \$18{,}400 \\ \$15{,}200 & \$16{,}100 \end{bmatrix} \begin{matrix} \text{Ms. Smith} \\ \text{Mr. Jones} \end{matrix}$

Matched Problem 5 Repeat Example 5 with

$$A = \begin{bmatrix} \$45{,}000 & \$77{,}000 \\ \$106{,}000 & \$22{,}000 \end{bmatrix} \quad \text{and} \quad B = \begin{bmatrix} \$190{,}000 & \$345{,}000 \\ \$266{,}000 & \$276{,}000 \end{bmatrix}$$

Figure 2 illustrates a solution for Example 5 on a spreadsheet.

	A	B	C	D	E	F	G	H	I
1		August Sales			September Sales			September Commissions	
2		Compact	Luxury		Compact	Luxury		Compact	Luxury
3	Smith	$ 54,000	$ 88,000		$228,000	$368,000		$ 11,400	$ 18,400
4	Jones	$126,000	$ 0		$304,000	$322,000		$ 15,200	$ 16,100
5									
6		Combined Sales			Sales Increase				
7	Smith	$282,000	$456,000		$174,000	$280,000			
8	Jones	$430,000	$322,000		$178,000	$322,000			

Figure 2

Matrix Product

Matrix multiplication was introduced by the English mathematician Arthur Cayley (1821–1895) in studies of systems of linear equations and linear transformations. Although this multiplication may seem strange at first, it is extremely useful in many practical problems.

We start by defining the product of two special matrices, a row matrix and a column matrix.

DEFINITION Product of a Row Matrix and a Column Matrix

The **product** of a $1 \times n$ row matrix and an $n \times 1$ column matrix is a 1×1 matrix given by

$$\underset{1 \times n}{[a_1\ a_2\ \dots\ a_n]}\ \underset{n \times 1}{\begin{bmatrix} b_1 \\ b_2 \\ \vdots \\ b_n \end{bmatrix}} = [a_1b_1 + a_2b_2 + \cdots + a_nb_n]$$

Note that the number of elements in the row matrix and in the column matrix must be the same for the product to be defined.

EXAMPLE 6 Product of a Row Matrix and a Column Matrix

$$[2 \quad -3 \quad 0]\begin{bmatrix} -5 \\ 2 \\ -2 \end{bmatrix} = [(2)(-5) + (-3)(2) + (0)(-2)]$$

$$= [-10 - 6 + 0] = [-16]$$

Matched Problem 6 $[-1 \quad 0 \quad 3 \quad 2]\begin{bmatrix} 2 \\ 3 \\ 4 \\ -1 \end{bmatrix} = ?$

Refer to Example 6. The distinction between the real number -16 and the 1×1 matrix $[-16]$ is a technical one, and it is common to see 1×1 matrices written as real numbers without brackets. In the work that follows, we will frequently refer to 1×1 matrices as real numbers and omit the brackets whenever it is convenient to do so.

EXAMPLE 7 **Labor Costs** A factory produces a slalom water ski that requires 3 labor-hours in the assembly department and 1 labor-hour in the finishing department. Assembly personnel receive \$9 per hour and finishing personnel receive \$6 per hour. Total labor cost per ski is given by the product:

$$\begin{bmatrix} 3 & 1 \end{bmatrix}\begin{bmatrix} 9 \\ 6 \end{bmatrix} = [(3)(9) + (1)(6)] = [27 + 6] = [33] \quad \text{or} \quad \$33 \text{ per ski}$$

Matched Problem 7 If the factory in Example 7 also produces a trick water ski that requires 5 labor-hours in the assembly department and 1.5 labor-hours in the finishing department, write a product between appropriate row and column matrices that will give the total labor cost for this ski. Compute the cost.

We now use the product of a $1 \times n$ row matrix and an $n \times 1$ column matrix to extend the definition of matrix product to more general matrices.

DEFINITION Matrix Product

If A is an $m \times p$ matrix and B is a $p \times n$ matrix, then the **matrix product** of A and B, denoted AB, is an $m \times n$ matrix whose element in the ith row and jth column is the real number obtained from the product of the ith row of A and the jth column of B. If the number of columns in A does not equal the number of rows in B, the matrix product AB is **not defined**.

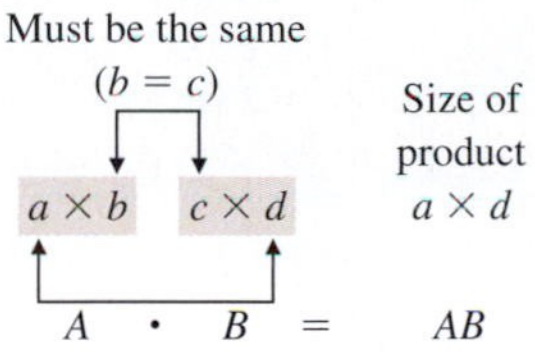

Figure 3

It is important to check sizes before starting the multiplication process. If A is an $a \times b$ matrix and B is a $c \times d$ matrix, then if $b = c$, the product AB will exist and will be an $a \times d$ matrix (see Fig. 3). If $b \neq c$, the product AB does not exist. The definition is not as complicated as it might first seem. An example should help clarify the process.

For

$$A = \begin{bmatrix} 2 & 3 & -1 \\ -2 & 1 & 2 \end{bmatrix} \quad \text{and} \quad B = \begin{bmatrix} 1 & 3 \\ 2 & 0 \\ -1 & 2 \end{bmatrix}$$

A is 2×3 and B is 3×2, so AB is 2×2. To find the first row of AB, we take the product of the first row of A with every column of B and write each result as a real number, not as a 1×1 matrix. The second row of AB is computed in the same manner. The four products of row and column matrices used to produce the four elements in AB are shown in the following dashed box. These products are usually calculated mentally or with the aid of a calculator, and need not be written out. The shaded portions highlight the steps involved in computing the element in the first row and second column of AB.

$$\underset{2 \times 3}{\begin{bmatrix} 2 & 3 & -1 \\ -2 & 1 & 2 \end{bmatrix}}\underset{3 \times 2}{\begin{bmatrix} 1 & 3 \\ 2 & 0 \\ -1 & 2 \end{bmatrix}} = \begin{bmatrix} \begin{bmatrix} 2 & 3 & -1 \end{bmatrix}\begin{bmatrix} 1 \\ 2 \\ -1 \end{bmatrix} & \begin{bmatrix} 2 & 3 & -1 \end{bmatrix}\begin{bmatrix} 3 \\ 0 \\ 2 \end{bmatrix} \\ \begin{bmatrix} -2 & 1 & 2 \end{bmatrix}\begin{bmatrix} 1 \\ 2 \\ -1 \end{bmatrix} & \begin{bmatrix} -2 & 1 & 2 \end{bmatrix}\begin{bmatrix} 3 \\ 0 \\ 2 \end{bmatrix} \end{bmatrix}$$

$$= \begin{bmatrix} (2)(1) + (3)(2) + (-1)(-1) & (2)(3) + (3)(0) + (-1)(2) \\ (-2)(1) + (1)(2) + (2)(-1) & (-2)(3) + (1)(0) + (2)(2) \end{bmatrix} = \underset{2 \times 2}{\begin{bmatrix} 9 & 4 \\ -2 & -2 \end{bmatrix}}$$

EXAMPLE 8 **Matrix Multiplication** Find the indicated matrix product, if it exists, where:

$$A = \begin{bmatrix} 2 & 1 \\ 1 & 0 \\ -1 & 2 \end{bmatrix} \quad B = \begin{bmatrix} 1 & -1 & 0 & 1 \\ 2 & 1 & 2 & 0 \end{bmatrix} \quad C = \begin{bmatrix} 2 & 6 \\ -1 & -3 \end{bmatrix}$$

$$D = \begin{bmatrix} 1 & 2 \\ 3 & 6 \end{bmatrix} \quad E = \begin{bmatrix} 2 & -3 & 0 \end{bmatrix} \quad F = \begin{bmatrix} -5 \\ 2 \\ -2 \end{bmatrix}$$

(A) $AB = \underset{3 \times 2}{\begin{bmatrix} 2 & 1 \\ 1 & 0 \\ -1 & 2 \end{bmatrix}} \underset{2 \times 4}{\begin{bmatrix} 1 & -1 & 0 & 1 \\ 2 & 1 & 2 & 0 \end{bmatrix}}$

$$= \begin{bmatrix} (2)(1)+(1)(2) & (2)(-1)+(1)(1) & (2)(0)+(1)(2) & (2)(1)+(1)(0) \\ (1)(1)+(0)(2) & (1)(-1)+(0)(1) & (1)(0)+(0)(2) & (1)(1)+(0)(0) \\ (-1)(1)+(2)(2) & (-1)(-1)+(2)(1) & (-1)(0)+(2)(2) & (-1)(1)+(2)(0) \end{bmatrix}$$

$$= \underset{3 \times 4}{\begin{bmatrix} 4 & -1 & 2 & 2 \\ 1 & -1 & 0 & 1 \\ 3 & 3 & 4 & -1 \end{bmatrix}}$$

(B) $BA = \underset{2 \times 4}{\begin{bmatrix} 1 & -1 & 0 & 1 \\ 2 & 1 & 2 & 0 \end{bmatrix}} \underset{3 \times 2}{\begin{bmatrix} 2 & 1 \\ 1 & 0 \\ -1 & 2 \end{bmatrix}}$ Not defined

(C) $CD = \begin{bmatrix} 2 & 6 \\ -1 & -3 \end{bmatrix} \begin{bmatrix} 1 & 2 \\ 3 & 6 \end{bmatrix} = \begin{bmatrix} (2)(1)+(6)(3) & (2)(2)+(6)(6) \\ (-1)(1)+(-3)(3) & (-1)(2)+(-3)(6) \end{bmatrix}$

$$= \begin{bmatrix} 20 & 40 \\ -10 & -20 \end{bmatrix}$$

(D) $DC = \begin{bmatrix} 1 & 2 \\ 3 & 6 \end{bmatrix} \begin{bmatrix} 2 & 6 \\ -1 & -3 \end{bmatrix} = \begin{bmatrix} (1)(2)+(2)(-1) & (1)(6)+(2)(-3) \\ (3)(2)+(6)(-1) & (3)(6)+(6)(-3) \end{bmatrix}$

$$= \begin{bmatrix} 0 & 0 \\ 0 & 0 \end{bmatrix}$$

(E) $EF = \begin{bmatrix} 2 & -3 & 0 \end{bmatrix} \begin{bmatrix} -5 \\ 2 \\ -2 \end{bmatrix} = [(2)(-5)+(-3)(2)+(0)(-2)] = [-16]$

(F) $FE = \begin{bmatrix} -5 \\ 2 \\ -2 \end{bmatrix} [2 \quad -3 \quad 0] = \begin{bmatrix} (-5)(2) & (-5)(-3) & (-5)(0) \\ (2)(2) & (2)(-3) & (2)(0) \\ (-2)(2) & (-2)(-3) & (-2)(0) \end{bmatrix}$

$= \begin{bmatrix} -10 & 15 & 0 \\ 4 & -6 & 0 \\ -4 & 6 & 0 \end{bmatrix}$

(G) $A^{2*} = AA = \underset{3 \times 2}{\begin{bmatrix} 2 & 1 \\ 1 & 0 \\ -1 & 2 \end{bmatrix}} \underset{3 \times 2}{\begin{bmatrix} 2 & 1 \\ 1 & 0 \\ -1 & 2 \end{bmatrix}}$ Not defined

(H) $C^2 = CC = \begin{bmatrix} 2 & 6 \\ -1 & -3 \end{bmatrix} \begin{bmatrix} 2 & 6 \\ -1 & -3 \end{bmatrix}$

$= \begin{bmatrix} (2)(2)+(6)(-1) & (2)(6)+(6)(-3) \\ (-1)(2)+(-3)(-1) & (-1)(6)+(-3)(-3) \end{bmatrix}$

$= \begin{bmatrix} -2 & -6 \\ 1 & 3 \end{bmatrix}$

Matched Problem 8 Find each product, if it is defined:

(A) $\begin{bmatrix} -1 & 0 & 3 & -2 \\ 1 & 2 & 2 & 0 \end{bmatrix} \begin{bmatrix} -1 & 1 \\ 2 & 3 \\ 1 & 0 \end{bmatrix}$

(B) $\begin{bmatrix} -1 & 1 \\ 2 & 3 \\ 1 & 0 \end{bmatrix} \begin{bmatrix} -1 & 0 & 3 & -2 \\ 1 & 2 & 2 & 0 \end{bmatrix}$

(C) $\begin{bmatrix} 1 & 2 \\ -1 & -2 \end{bmatrix} \begin{bmatrix} -2 & 4 \\ 1 & -2 \end{bmatrix}$

(D) $\begin{bmatrix} -2 & 4 \\ 1 & -2 \end{bmatrix} \begin{bmatrix} 1 & 2 \\ -1 & -2 \end{bmatrix}$

(E) $[3 \quad -2 \quad 1] \begin{bmatrix} 4 \\ 2 \\ 3 \end{bmatrix}$

(F) $\begin{bmatrix} 4 \\ 2 \\ 3 \end{bmatrix} [3 \quad -2 \quad 1]$

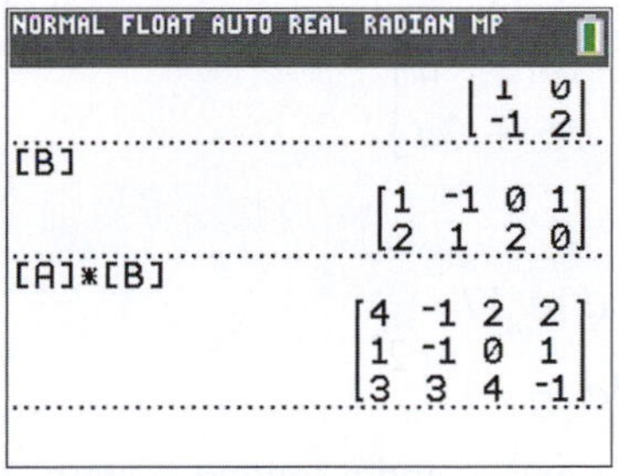

Figure 4 Multiplication on a graphing calculator

Figure 4 illustrates a graphing calculator solution to Example 8A. What would you expect to happen if you tried to solve Example 8B on a graphing calculator?

CONCEPTUAL INSIGHT

In the arithmetic of real numbers, it does not matter in which order we multiply. For example, $5 \times 7 = 7 \times 5$. In matrix multiplication, however, it does make a difference. That is, AB does not always equal BA, even if both multiplications are defined and both products are the same size (see Examples 8C and 8D).

Matrix multiplication is not commutative.

The zero property of real numbers states that if the product of two real numbers is 0, then one of the numbers must be 0 (see Appendix A, Section A.1). This property is very important when solving equations. For example,

(*Continued*)

*Following standard algebraic notation, we write $A^2 = AA$, $A^3 = AAA$, and so on.

$$x^2 - 4x + 3 = 0$$
$$(x-1)(x-3) = 0$$
$$x - 1 = 0 \quad \text{or} \quad x - 3 = 0$$
$$x = 1 \qquad\qquad x = 3$$

For matrices, it is possible to find nonzero matrices A and B such that AB is a zero matrix (see Example 8D).

The zero property does not hold for matrix multiplication.

Explore and Discuss 1

In addition to the commutative and zero properties, there are other significant differences between real number multiplication and matrix multiplication.

(A) In real number multiplication, the only real number whose square is 0 is the real number 0 $(0^2 = 0)$. Find at least one 2×2 matrix A with all elements nonzero such that $A^2 = 0$, where 0 is the 2×2 zero matrix.

(B) In real number multiplication, the only nonzero real number that is equal to its square is the real number 1 $(1^2 = 1)$. Find at least one 2×2 matrix B with all elements nonzero such that $B^2 = B$.

EXAMPLE 9 **Matrix Multiplication** Find a, b, c, and d so that

$$\begin{bmatrix} 2 & -1 \\ 5 & 3 \end{bmatrix}\begin{bmatrix} a & b \\ c & d \end{bmatrix} = \begin{bmatrix} -6 & 17 \\ 7 & 4 \end{bmatrix}$$

SOLUTION The product of the matrices on the left side of the equation is

$$\begin{bmatrix} 2 & -1 \\ 5 & 3 \end{bmatrix}\begin{bmatrix} a & b \\ c & d \end{bmatrix} = \begin{bmatrix} 2a - c & 2b - d \\ 5a + 3c & 5b + 3d \end{bmatrix}$$

Therefore,

$$2a - c = -6 \qquad 2b - d = 17$$
$$5a + 3c = 7 \qquad 5b + 3d = 4$$

This gives a system of two equations in the variables a and c, and a second system of two equations in the variables b and d. Each system can be solved by substitution, or elimination by addition, or Gauss–Jordan elimination (the details are omitted). The solution of the first system is $a = -1$, $c = 4$, and the solution of the second system is $b = 5$, $d = -7$.

Matched Problem 9 Find a, b, c, and d so that

$$\begin{bmatrix} 6 & -5 \\ 0 & 3 \end{bmatrix}\begin{bmatrix} a & b \\ c & d \end{bmatrix} = \begin{bmatrix} -16 & 64 \\ 24 & -6 \end{bmatrix}$$

Now we consider an application of matrix multiplication.

EXAMPLE 10 **Labor Costs** We can combine the time requirements for slalom and trick water skis discussed in Example 7 and Matched Problem 7 into one matrix:

Labor-hours per ski

$$\begin{matrix} & \text{Assembly department} & \text{Finishing department} \\ \text{Trick ski} & \\ \text{Slalom ski} & \end{matrix} \begin{bmatrix} 5\text{ hr} & 1.5\text{ hr} \\ 3\text{ hr} & 1\text{ hr} \end{bmatrix} = L$$

Now suppose that the company has two manufacturing plants, one in California and the other in Maryland, and that their hourly rates for each department are given in the following matrix:

Hourly wages

$$\begin{matrix} \text{Assembly department} \\ \text{Finishing department} \end{matrix} \begin{bmatrix} \$12 & \$13 \\ \$7 & \$8 \end{bmatrix} = H \quad (\text{columns: California, Maryland})$$

Since H and L are both 2×2 matrices, we can take the product of H and L in either order and the result will be a 2×2 matrix:

$$HL = \begin{bmatrix} 12 & 13 \\ 7 & 8 \end{bmatrix}\begin{bmatrix} 5 & 1.5 \\ 3 & 1 \end{bmatrix} = \begin{bmatrix} 99 & 31 \\ 59 & 18.5 \end{bmatrix}$$

$$LH = \begin{bmatrix} 5 & 1.5 \\ 3 & 1 \end{bmatrix}\begin{bmatrix} 12 & 13 \\ 7 & 8 \end{bmatrix} = \begin{bmatrix} 70.5 & 77 \\ 43 & 47 \end{bmatrix}$$

How can we interpret the elements in these products? Let's begin with the product HL. The element 99 in the first row and first column of HL is the product of the first row matrix of H and the first column matrix of L:

$$\overset{\text{CA}\quad\text{MD}}{[12 \quad 13]}\begin{bmatrix} 5 \\ 3 \end{bmatrix}\begin{matrix}\text{Trick} \\ \text{Slalom}\end{matrix} = 12(5) + 13(3) = 60 + 39 = 99$$

Notice that \$60 is the labor cost for assembling a trick ski at the California plant and \$39 is the labor cost for assembling a slalom ski at the Maryland plant. Although both numbers represent labor costs, it makes no sense to add them together. They do not pertain to the same type of ski or to the same plant. So even though the product HL happens to be defined mathematically, it has no useful interpretation in this problem.

Now let's consider the product LH. The element 70.5 in the first row and first column of LH is given by the following product:

$$\overset{\text{Assembly}\quad\text{Finishing}}{[5 \qquad 1.5]}\begin{bmatrix} 12 \\ 7 \end{bmatrix}\begin{matrix}\text{Assembly} \\ \text{Finishing}\end{matrix} = 5(12) + 1.5(7) = 60 + 10.5 = 70.5$$

This time, \$60 is the labor cost for assembling a trick ski at the California plant and \$10.50 is the labor cost for finishing a trick ski at the California plant. So the sum is the total labor cost for producing a trick ski at the California plant. The other elements in LH also represent total labor costs, as indicated by the row and column labels shown below:

Labor costs per ski

$$LH = \begin{bmatrix} \$70.50 & \$77 \\ \$43 & \$47 \end{bmatrix}\begin{matrix}\text{Trick} \\ \text{Slalom}\end{matrix} \quad (\text{columns: CA, MD})$$

Figure 5 shows a solution to Example 9 on a spreadsheet.

	A	B	C	D	E	F	G
1		Labor-hours per ski				Hourly wages	
2		Assembly	Finishing			California	Maryland
3	Trick ski	5	1.5		Assembly	$ 12.00	$ 13.00
4	Slalom ski	3	1		Finishing	$ 7.00	$ 8.00
5							
6		Labor costs per ski					
7		California	Maryland				
8	Trick ski	$ 75.50	$ 77.00				
9	Slalom ski	$ 43.00	$ 47.00				

Figure 5 Matrix multiplication in a spreadsheet: The command MMULT(B3:C4, F3:G4) produces the matrix in B8:C9

Matched Problem 10 Refer to Example 10. The company wants to know how many hours to schedule in each department in order to produce 2,000 trick skis and 1,000 slalom skis. These production requirements can be represented by either of the following matrices:

$$P = \overset{\text{Trick skis}\quad\text{Slalom skis}}{[2{,}000 \quad 1{,}000]} \qquad Q = \begin{bmatrix} 2{,}000 \\ 1{,}000 \end{bmatrix} \begin{matrix} \text{Trick skis} \\ \text{Slalom skis} \end{matrix}$$

Using the labor-hour matrix L from Example 10, find PL or LQ, whichever has a meaningful interpretation for this problem, and label the rows and columns accordingly.

CONCEPTUAL INSIGHT

Example 10 and Matched Problem 10 illustrate an important point about matrix multiplication. Even if you are using a graphing calculator to perform the calculations in a matrix product, it is still necessary for you to know the definition of matrix multiplication so that you can interpret the results correctly.

Exercises 4.4

Skills Warm-up Exercises

W *In Problems 1–14, perform the indicated operation, if possible. (If necessary, review the definitions at the beginning of Section 4.4.)*

1. $[1 \quad 5] + [3 \quad 10]$

2. $\begin{bmatrix} 3 \\ 2 \end{bmatrix} - \begin{bmatrix} 7 \\ -4 \end{bmatrix}$

3. $\begin{bmatrix} 2 & 0 \\ -3 & 6 \end{bmatrix} - \begin{bmatrix} 0 & -4 \\ -1 & 0 \end{bmatrix}$

4. $\begin{bmatrix} -9 & 2 \\ 8 & 0 \end{bmatrix} + \begin{bmatrix} 9 & 0 \\ 0 & 8 \end{bmatrix}$

5. $\begin{bmatrix} 3 \\ 6 \end{bmatrix} + \begin{bmatrix} -1 & 9 \end{bmatrix}$

6. $4\begin{bmatrix} 2 & -6 & 1 \\ 8 & 5 & -3 \end{bmatrix}$

7. $7\begin{bmatrix} 3 & -5 & 9 & 4 \end{bmatrix}$

8. $[10 \quad 12] + \begin{bmatrix} 4 \\ 3 \end{bmatrix}$

A **9.** $\begin{bmatrix} 3 & 4 \\ -1 & -2 \end{bmatrix}\begin{bmatrix} -1 \\ 2 \end{bmatrix}$

10. $\begin{bmatrix} -1 & 1 \\ 2 & -3 \end{bmatrix}\begin{bmatrix} 4 \\ -2 \end{bmatrix}$

11. $\begin{bmatrix} 2 & -3 \\ 1 & 2 \end{bmatrix}\begin{bmatrix} 1 & -1 \\ 0 & -2 \end{bmatrix}$

12. $\begin{bmatrix} -3 & 2 \\ 4 & -1 \end{bmatrix}\begin{bmatrix} -2 & 5 \\ -1 & 3 \end{bmatrix}$

13. $\begin{bmatrix} 1 & -1 \\ 0 & -2 \end{bmatrix}\begin{bmatrix} 2 & -3 \\ 1 & 2 \end{bmatrix}$

14. $\begin{bmatrix} -2 & 5 \\ -1 & 3 \end{bmatrix}\begin{bmatrix} -3 & 2 \\ 4 & -1 \end{bmatrix}$

In Problems 15–22, find the matrix product. Note that each product can be found mentally, without the use of a calculator or pencil-and-paper calculations.

15. $\begin{bmatrix} 5 & 0 \\ 0 & 5 \end{bmatrix}\begin{bmatrix} 1 & 2 \\ 3 & 4 \end{bmatrix}$

16. $\begin{bmatrix} 3 & 0 \\ 0 & 3 \end{bmatrix}\begin{bmatrix} 1 & 3 \\ 5 & 7 \end{bmatrix}$

17. $\begin{bmatrix} 1 & 2 \\ 3 & 4 \end{bmatrix}\begin{bmatrix} 5 & 0 \\ 0 & 5 \end{bmatrix}$

18. $\begin{bmatrix} 1 & 3 \\ 5 & 7 \end{bmatrix}\begin{bmatrix} 3 & 0 \\ 0 & 3 \end{bmatrix}$

19. $\begin{bmatrix} 0 & 1 \\ 0 & 0 \end{bmatrix}\begin{bmatrix} 3 & 5 \\ 7 & 9 \end{bmatrix}$

20. $\begin{bmatrix} 0 & 0 \\ 1 & 0 \end{bmatrix}\begin{bmatrix} 2 & 4 \\ 6 & 8 \end{bmatrix}$

21. $\begin{bmatrix} 3 & 5 \\ 7 & 9 \end{bmatrix}\begin{bmatrix} 0 & 1 \\ 0 & 0 \end{bmatrix}$

22. $\begin{bmatrix} 2 & 4 \\ 6 & 8 \end{bmatrix}\begin{bmatrix} 0 & 0 \\ 1 & 0 \end{bmatrix}$

B *Find the products in Problems 23–30.*

23. $[3 \quad -2]\begin{bmatrix} -5 \\ 4 \end{bmatrix}$

24. $[-3 \quad 5]\begin{bmatrix} 2 \\ -4 \end{bmatrix}$

25. $\begin{bmatrix} -5 \\ 4 \end{bmatrix}[3 \quad -2]$

26. $\begin{bmatrix} 2 \\ -4 \end{bmatrix}[-3 \quad 5]$

27. $[-1 \quad 0 \quad 1]\begin{bmatrix} 2 \\ 1 \\ 3 \end{bmatrix}$

28. $[1 \quad -1 \quad 3]\begin{bmatrix} -2 \\ 0 \\ 1 \end{bmatrix}$

29. $\begin{bmatrix} 2 \\ 1 \\ 3 \end{bmatrix}[-1 \quad 0 \quad 1]$

30. $\begin{bmatrix} -2 \\ 0 \\ 1 \end{bmatrix}[1 \quad -1 \quad 3]$

Problems 31–48 refer to the following matrices:

$$A = \begin{bmatrix} 2 & -1 & 3 \\ 0 & 4 & -2 \end{bmatrix} \qquad B = \begin{bmatrix} -3 & 1 \\ 2 & 5 \end{bmatrix}$$

$$C = \begin{bmatrix} -1 & 0 & 2 \\ 4 & -3 & 1 \\ -2 & 3 & 5 \end{bmatrix} \qquad D = \begin{bmatrix} 3 & -2 \\ 0 & -1 \\ 1 & 2 \end{bmatrix}$$

Perform the indicated operations, if possible.

31. AC

32. CA

33. AB

34. BA

35. B^2

36. C^2

37. $B + AD$

38. $C + DA$

39. $(0.1)DB$

40. $(0.2)CD$

41. $(3)BA + (4)AC$

42. $(2)DB + (5)CD$

43. $(-2)BA + (6)CD$

44. $(-1)AC + (3)DB$

45. ACD

46. CDA

47. DBA

48. BAD

49. If a and b are nonzero real numbers,

$$A = \begin{bmatrix} a & a \\ b & b \end{bmatrix}, \quad \text{and} \quad B = \begin{bmatrix} a & a \\ -a & -a \end{bmatrix}$$

find AB and BA.

50. If a and b are nonzero real numbers,

$$A = \begin{bmatrix} a & b \\ -a & -b \end{bmatrix}, \quad \text{and} \quad B = \begin{bmatrix} a & a \\ a & a \end{bmatrix}$$

find AB and BA.

51. If a and b are nonzero real numbers and

$$A = \begin{bmatrix} ab & b^2 \\ -a^2 & -ab \end{bmatrix}$$

find A^2.

52. If a and b are nonzero real numbers and

$$A = \begin{bmatrix} ab & b - ab^2 \\ a & 1 - ab \end{bmatrix}$$

find A^2.

In Problems 53 and 54, use a graphing calculator to calculate $B, B^2, B^3, \ldots$ and $AB, AB^2, AB^3, \ldots$. Describe any patterns you observe in each sequence of matrices.

53. $A = [0.3 \quad 0.7]$ and $B = \begin{bmatrix} 0.4 & 0.6 \\ 0.2 & 0.8 \end{bmatrix}$

54. $A = [0.4 \quad 0.6]$ and $B = \begin{bmatrix} 0.9 & 0.1 \\ 0.3 & 0.7 \end{bmatrix}$

C **55.** Find a, b, c, and d so that

$$\begin{bmatrix} a & b \\ c & d \end{bmatrix} + \begin{bmatrix} 2 & -3 \\ 0 & 1 \end{bmatrix} = \begin{bmatrix} 1 & -2 \\ 3 & -4 \end{bmatrix}$$

56. Find w, x, y, and z so that

$$\begin{bmatrix} 4 & -2 \\ -3 & 0 \end{bmatrix} + \begin{bmatrix} w & x \\ y & z \end{bmatrix} = \begin{bmatrix} 2 & -3 \\ 0 & 5 \end{bmatrix}$$

57. Find a, b, c, and d so that

$$\begin{bmatrix} 1 & -2 \\ 2 & -3 \end{bmatrix}\begin{bmatrix} a & b \\ c & d \end{bmatrix} = \begin{bmatrix} 1 & 0 \\ 3 & 2 \end{bmatrix}$$

58. Find a, b, c, and d so that

$$\begin{bmatrix} 1 & 3 \\ 1 & 4 \end{bmatrix}\begin{bmatrix} a & b \\ c & d \end{bmatrix} = \begin{bmatrix} 6 & -5 \\ 7 & -7 \end{bmatrix}$$

In Problems 59–62, determine whether the statement is true or false.

59. There exist two 1×1 matrices A and B such that $AB \neq BA$.

60. There exist two 2×2 matrices A and B such that $AB \neq BA$.

61. There exist two nonzero 2×2 matrices A and B such that AB is the 2×2 zero matrix.

62. There exist two nonzero 1×1 matrices A and B such that AB is the 1×1 zero matrix.

63. A square matrix is a **diagonal matrix** if all elements not on the principal diagonal are zero. So a 2×2 diagonal matrix has the form

$$A = \begin{bmatrix} a & 0 \\ 0 & d \end{bmatrix}$$

where a and d are real numbers. Discuss the validity of each of the following statements. If the statement is always true, explain why. If not, give examples.

(A) If A and B are 2×2 diagonal matrices, then $A + B$ is a 2×2 diagonal matrix.

(B) If A and B are 2×2 diagonal matrices, then AB is a 2×2 diagonal matrix.

(C) If A and B are 2×2 diagonal matrices, then $AB = BA$.

64. A square matrix is an **upper triangular matrix** if all elements below the principal diagonal are zero. So a 2×2 upper triangular matrix has the form

$$A = \begin{bmatrix} a & b \\ 0 & d \end{bmatrix}$$

where a, b, and d are real numbers. Discuss the validity of each of the following statements. If the statement is always true, explain why. If not, give examples.

(A) If A and B are 2×2 upper triangular matrices, then $A + B$ is a 2×2 upper triangular matrix.

(B) If A and B are 2×2 upper triangular matrices, then AB is a 2×2 upper triangular matrix.

(C) If A and B are 2×2 upper triangular matrices, then $AB = BA$.

Applications

65. Cost analysis. A company with two different plants manufactures guitars and banjos. Its production costs for each instrument are given in the following matrices:

$$\begin{array}{c}\text{Plant X}\\ \begin{array}{r} \\ \text{Materials}\\ \text{Labor}\end{array}\begin{array}{c}\text{Guitar}\quad\text{Banjo}\\ \begin{bmatrix}\$47 & \$39\\ \$90 & \$125\end{bmatrix}\end{array} = A\end{array} \qquad \begin{array}{c}\text{Plant Y}\\ \begin{array}{c}\text{Guitar}\quad\text{Banjo}\\ \begin{bmatrix}\$56 & \$42\\ \$84 & \$115\end{bmatrix}\end{array} = B\end{array}$$

Find $\frac{1}{2}(A + B)$, the average cost of production for the two plants.

66. Cost analysis. If both labor and materials at plant X in Problem 65 are increased by 20%, find $\frac{1}{2}(1.2A + B)$, the new average cost of production for the two plants.

67. Markup. An import car dealer sells three models of a car. The retail prices and the current dealer invoice prices (costs) for the basic models and options indicated are given in the following two matrices (where "Air" means air-conditioning):

Retail price

	Basic Car	Air	AM/FM radio	Cruise control
Model A	\$35,075	\$2,560	\$1,070	\$640
Model B	\$39,045	\$1,840	\$770	\$460
Model C	\$45,535	\$3,400	\$1,415	\$850

$= M$

Dealer invoice price

	Basic Car	Air	AM/FM radio	Cruise control
Model A	\$30,996	\$2,050	\$850	\$510
Model B	\$34,857	\$1,585	\$660	\$395
Model C	\$41,667	\$2,890	\$1,200	\$725

$= N$

We define the markup matrix to be $M - N$ (**markup** is the difference between the retail price and the dealer invoice price). Suppose that the value of the dollar has had a sharp decline and the dealer invoice price is to have an across-the-board 15% increase next year. To stay competitive with domestic cars, the dealer increases the retail prices 10%. Calculate a markup matrix for next year's models and the options indicated. (Compute results to the nearest dollar.)

68. Markup. Referring to Problem 67, what is the markup matrix resulting from a 20% increase in dealer invoice prices and an increase in retail prices of 15%? (Compute results to the nearest dollar.)

69. Labor costs. A company with manufacturing plants located in Massachusetts (MA) and Virginia (VA) has labor-hour and wage requirements for the manufacture of three types of inflatable boats as given in the following two matrices:

Labor-hours per boat

	Cutting department	Assembly department	Packaging department	
$M =$	0.6 hr	0.6 hr	0.2 hr	One-person boat
	1.0 hr	0.9 hr	0.3 hr	Two-person boat
	1.5 hr	1.2 hr	0.4 hr	Four-person boat

Hourly wages

	MA	VA	
$N =$	\$17.30	\$14.65	Cutting department
	\$12.22	\$10.29	Assembly department
	\$10.63	\$9.66	Packaging department

(A) Find the labor costs for a one-person boat manufactured at the Massachusetts plant.

(B) Find the labor costs for a four-person boat manufactured at the Virginia plant.

(C) Discuss possible interpretations of the elements in the matrix products MN and NM.

(D) If either of the products MN or NM has a meaningful interpretation, find the product and label its rows and columns.

70. Inventory value. A personal computer retail company sells five different computer models through three stores. The inventory of each model on hand in each store is summarized in matrix M. Wholesale (W) and retail (R) values of each model computer are summarized in matrix N.

Model

	A	B	C	D	E	
$M =$	4	2	3	7	1	Store 1
	2	3	5	0	6	Store 2
	10	4	3	4	3	Store 3

	W	R	
$N =$	\$700	\$840	A
	\$1,400	\$1,800	B
	\$1,800	\$2,400	C Model
	\$2,700	\$3,300	D
	\$3,500	\$4,900	E

(A) What is the retail value of the inventory at store 2?

(B) What is the wholesale value of the inventory at store 3?

(C) If either product MN or NM has a meaningful interpretation, find the product and label its rows and columns. What do the entries represent?

(D) Discuss methods of matrix multiplication that can be used to find the total inventory of each model on hand at all three stores. State the matrices that can be used and perform the necessary operations.

71. Cereal. A nutritionist for a cereal company blends two cereals in three different mixes. The amounts of protein, carbohydrate, and fat (in grams per ounce) in each cereal are given by matrix M. The amounts of each cereal used in the three mixes are given by matrix N.

$$M = \begin{array}{c} \text{Cereal A} \quad \text{Cereal B} \\ \begin{bmatrix} 4\text{ g/oz} & 2\text{ g/oz} \\ 20\text{ g/oz} & 16\text{ g/oz} \\ 3\text{ g/oz} & 1\text{ g/oz} \end{bmatrix} \end{array} \begin{array}{l} \\ \text{Protein} \\ \text{Carbohydrate} \\ \text{Fat} \end{array}$$

$$N = \begin{array}{c} \text{Mix X} \quad \text{Mix Y} \quad \text{Mix Z} \\ \begin{bmatrix} 15\text{ oz} & 10\text{ oz} & 5\text{ oz} \\ 5\text{ oz} & 10\text{ oz} & 15\text{ oz} \end{bmatrix} \end{array} \begin{array}{l} \\ \text{Cereal A} \\ \text{Cereal B} \end{array}$$

(A) Find the amount of protein in mix *X*.

(B) Find the amount of fat in mix *Z*.

(C) Discuss possible interpretations of the elements in the matrix products *MN* and *NM*.

(D) If either of the products *MN* or *NM* has a meaningful interpretation, find the product and label its rows and columns.

72. Heredity. Gregor Mendel (1822–1884) made discoveries that revolutionized the science of genetics. In one experiment, he crossed dihybrid yellow round peas (yellow and round are dominant characteristics; the peas also contained genes for the recessive characteristics green and wrinkled) and obtained peas of the types indicated in the matrix:

$$\begin{array}{l} \\ \text{Yellow} \\ \text{Green} \end{array} \begin{array}{c} \text{Round} \quad \text{Wrinkled} \\ \begin{bmatrix} 315 & 101 \\ 108 & 32 \end{bmatrix} \end{array} = M$$

Suppose he carried out a second experiment of the same type and obtained peas of the types indicated in this matrix:

$$\begin{array}{l} \\ \text{Yellow} \\ \text{Green} \end{array} \begin{array}{c} \text{Round} \quad \text{Wrinkled} \\ \begin{bmatrix} 370 & 128 \\ 110 & 36 \end{bmatrix} \end{array} = N$$

If the results of the two experiments are combined, discuss matrix multiplication methods that can be used to find the following quantities. State the matrices that can be used and perform the necessary operations.

(A) The total number of peas in each category

(B) The total number of peas in all four categories

(C) The percentage of peas in each category

73. Politics. In a local California election, a public relations firm promoted its candidate in three ways: telephone calls, house calls, and letters. The cost per contact is given in matrix *M*, and the number of contacts of each type made in two adjacent cities is given in matrix *N*.

$$M = \begin{array}{c} \text{Cost per contact} \\ \begin{bmatrix} \$1.20 \\ \$3.00 \\ \$1.45 \end{bmatrix} \end{array} \begin{array}{l} \\ \text{Telephone call} \\ \text{House call} \\ \text{Letter} \end{array}$$

$$N = \begin{array}{c} \text{Telephone call} \quad \text{House call} \quad \text{Letter} \\ \begin{bmatrix} 1{,}000 & 500 & 5{,}000 \\ 2{,}000 & 800 & 8{,}000 \end{bmatrix} \end{array} \begin{array}{l} \\ \text{Berkeley} \\ \text{Oakland} \end{array}$$

(A) Find the total amount spent in Berkeley.

(B) Find the total amount spent in Oakland.

(C) If either product *MN* or *NM* has a meaningful interpretation, find the product and label its rows and columns. What do the entries represent?

(D) Discuss methods of matrix multiplication that can be used to find the total number of telephone calls, house calls, and letters. State the matrices that can be used and perform the necessary operations.

74. Test averages. A teacher has given four tests to a class of five students and stored the results in the following matrix:

$$\begin{array}{l} \\ \text{Ann} \\ \text{Bob} \\ \text{Carol} \\ \text{Dan} \\ \text{Eric} \end{array} \begin{array}{c} \text{Tests} \\ 1 \quad 2 \quad 3 \quad 4 \\ \begin{bmatrix} 78 & 84 & 81 & 86 \\ 91 & 65 & 84 & 92 \\ 95 & 90 & 92 & 91 \\ 75 & 82 & 87 & 91 \\ 83 & 88 & 81 & 76 \end{bmatrix} \end{array} = M$$

Discuss methods of matrix multiplication that the teacher can use to obtain the information indicated below. In each case, state the matrices to be used and then perform the necessary operations.

(A) The average on all four tests for each student, assuming that all four tests are given equal weight

(B) The average on all four tests for each student, assuming that the first three tests are given equal weight and the fourth is given twice this weight

(C) The class average on each of the four tests

Answers to Matched Problems

1. $\begin{bmatrix} 1 & 5 \\ 0 & -2 \\ 2 & 1 \end{bmatrix}$ **2.** $[-1 \quad -1 \quad 4]$ **3.** $a = -6$, $b = 7$, $c = 9$, $d = -1$ **4.** $\begin{bmatrix} 13 \\ 2 \\ 35 \end{bmatrix}$

5. (A) $\begin{bmatrix} \$235{,}000 & \$422{,}000 \\ \$372{,}000 & \$298{,}000 \end{bmatrix}$

(B) $\begin{bmatrix} \$145{,}000 & \$268{,}000 \\ \$160{,}000 & \$254{,}000 \end{bmatrix}$

(C) $\begin{bmatrix} \$9{,}500 & \$17{,}250 \\ \$13{,}300 & \$13{,}800 \end{bmatrix}$

6. [8] **7.** $[5 \quad 1.5]\begin{bmatrix} 9 \\ 6 \end{bmatrix} = [54]$, or \$54

8. (A) Not defined

(B) $\begin{bmatrix} 2 & 2 & -1 & 2 \\ 1 & 6 & 12 & -4 \\ -1 & 0 & 3 & -2 \end{bmatrix}$ (C) $\begin{bmatrix} 0 & 0 \\ 0 & 0 \end{bmatrix}$

(D) $\begin{bmatrix} -6 & -12 \\ 3 & 6 \end{bmatrix}$ (E) [11]

(F) $\begin{bmatrix} 12 & -8 & 4 \\ 6 & -4 & 2 \\ 9 & -6 & 3 \end{bmatrix}$

9. $a = 4, c = 8, b = 9, d = -2$

10. $PL = \begin{array}{c} \text{Assembly} \quad \text{Finishing} \\ [13{,}000 \quad 4{,}000] \end{array}$ Labor-hours

4.5 Inverse of a Square Matrix

- Identity Matrix for Multiplication
- Inverse of a Square Matrix
- Application: Cryptography

Identity Matrix for Multiplication

Does the set of all matrices of a given size have an identity element for multiplication? That is, if M is an arbitrary $m \times n$ matrix, does there exist an identity element I such that $IM = MI = M$? The answer, in general, is no. However, the set of all **square matrices of order n** (matrices with n rows and n columns) does have an identity element.

DEFINITION Identity Matrix

The **identity element for multiplication** for the set of all square matrices of order n is the square matrix of order n, denoted by I, with 1's along the principal diagonal (from the upper left corner to the lower right) and 0's elsewhere.

For example,

$$\begin{bmatrix} 1 & 0 \\ 0 & 1 \end{bmatrix} \quad \text{and} \quad \begin{bmatrix} 1 & 0 & 0 \\ 0 & 1 & 0 \\ 0 & 0 & 1 \end{bmatrix}$$

are the identity matrices for all square matrices of order 2 and 3, respectively.

Most graphing calculators have a built-in command for generating the identity matrix of a given order (see Fig. 1).

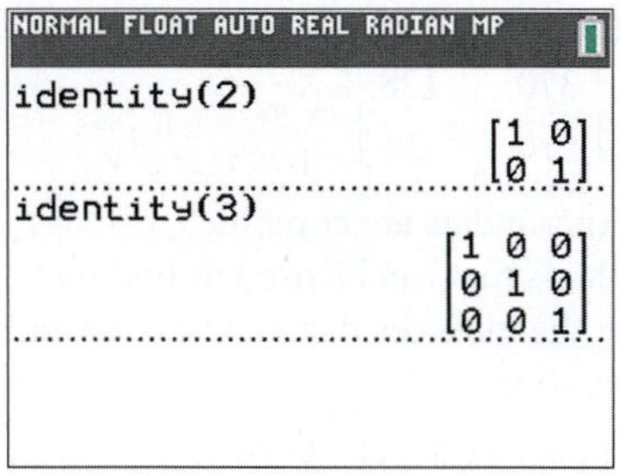

Figure 1 Identity matrices

EXAMPLE 1 Identity Matrix Multiplication

(A) $\begin{bmatrix} 1 & 0 & 0 \\ 0 & 1 & 0 \\ 0 & 0 & 1 \end{bmatrix} \begin{bmatrix} 3 & -2 & 5 \\ 0 & 2 & -3 \\ -1 & 4 & -2 \end{bmatrix} = \begin{bmatrix} 3 & -2 & 5 \\ 0 & 2 & -3 \\ -1 & 4 & -2 \end{bmatrix}$

(B) $\begin{bmatrix} 3 & -2 & 5 \\ 0 & 2 & -3 \\ -1 & 4 & -2 \end{bmatrix} \begin{bmatrix} 1 & 0 & 0 \\ 0 & 1 & 0 \\ 0 & 0 & 1 \end{bmatrix} = \begin{bmatrix} 3 & -2 & 5 \\ 0 & 2 & -3 \\ -1 & 4 & -2 \end{bmatrix}$

(C) $\begin{bmatrix} 1 & 0 \\ 0 & 1 \end{bmatrix} \begin{bmatrix} 2 & -1 & 3 \\ -2 & 0 & 4 \end{bmatrix} = \begin{bmatrix} 2 & -1 & 3 \\ -2 & 0 & 4 \end{bmatrix}$

(D) $\begin{bmatrix} 2 & -1 & 3 \\ -2 & 0 & 4 \end{bmatrix} \begin{bmatrix} 1 & 0 & 0 \\ 0 & 1 & 0 \\ 0 & 0 & 1 \end{bmatrix} = \begin{bmatrix} 2 & -1 & 3 \\ -2 & 0 & 4 \end{bmatrix}$

Matched Problem 1 Multiply:

(A) $\begin{bmatrix} 1 & 0 \\ 0 & 1 \end{bmatrix}\begin{bmatrix} 2 & -3 \\ 5 & 7 \end{bmatrix}$ and $\begin{bmatrix} 2 & -3 \\ 5 & 7 \end{bmatrix}\begin{bmatrix} 1 & 0 \\ 0 & 1 \end{bmatrix}$

(B) $\begin{bmatrix} 1 & 0 & 0 \\ 0 & 1 & 0 \\ 0 & 0 & 1 \end{bmatrix}\begin{bmatrix} 4 & 2 \\ 3 & -5 \\ 6 & 8 \end{bmatrix}$ and $\begin{bmatrix} 4 & 2 \\ 3 & -5 \\ 6 & 8 \end{bmatrix}\begin{bmatrix} 1 & 0 \\ 0 & 1 \end{bmatrix}$

In general, we can show that if M is a square matrix of order n and I is the identity matrix of order n, then

$$IM = MI = M$$

If M is an $m \times n$ matrix that is not square $(m \neq n)$, it is still possible to multiply M on the left and on the right by an identity matrix, but not with the same size identity matrix (see Example 1C and D). To avoid the complications involved with associating two different identity matrices with each nonsquare matrix, we restrict our attention in this section to square matrices.

Explore and Discuss 1

The only real number solutions to the equation $x^2 = 1$ are $x = 1$ and $x = -1$.

(A) Show that $A = \begin{bmatrix} 0 & 1 \\ 1 & 0 \end{bmatrix}$ satisfies $A^2 = I$, where I is the 2×2 identity.

(B) Show that $B = \begin{bmatrix} 0 & -1 \\ -1 & 0 \end{bmatrix}$ satisfies $B^2 = I$.

(C) Find a 2×2 matrix with all elements nonzero whose square is the 2×2 identity matrix.

Inverse of a Square Matrix

If r is an arbitary real number, then its **additive inverse** is the solution x to the equation $r + x = 0$. So the additive inverse of 3 is -3, and the additive inverse of -7 is 7. Similarly, if M is an arbitary $m \times n$ matrix, then M has an additive inverse $-M$, whose elements are just the additive inverses of the elements of M.

The situation is more complicated for *multiplicative* inverses. The **multiplicative inverse** of an arbitrary real number r is the solution x to the equation $r \cdot x = 1$. So the multiplicative inverse of 3 is $\frac{1}{3}$, and the multiplicative inverse of $\frac{-15}{4}$ is $\frac{-4}{15}$. Every real number has a multiplicative inverse except for 0. Because the equation $0 \cdot x = 1$ has no real solution, 0 does not have a multiplicative inverse.

Can we extend the multiplicative inverse concept to matrices? That is, given a matrix M, can we find another matrix N such that $MN = NM = I$, the matrix identity for multiplication? To begin, we consider the size of these matrices. Let M be an $n \times m$ matrix and N a $p \times q$ matrix. If both MN and NM are defined, then $m = p$ and $q = n$ (Fig. 2). If $MN = NM$, then $n = p$ and $q = m$ (Fig. 3). Thus, we have $m = p = n = q$. In other words, M and N must be square matrices of the same order. Later we will see that not all square matrices have inverses.

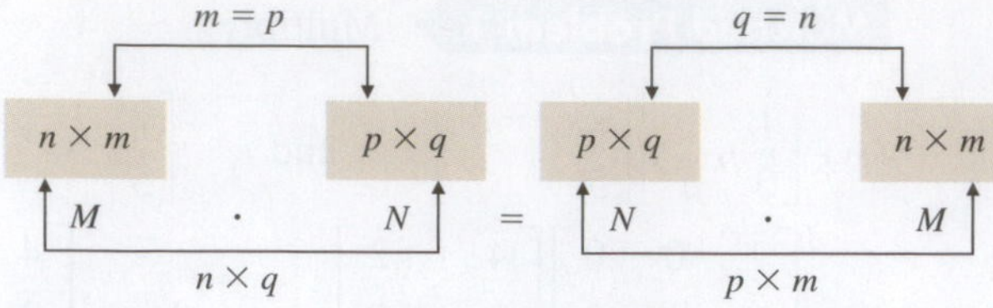

Figure 2

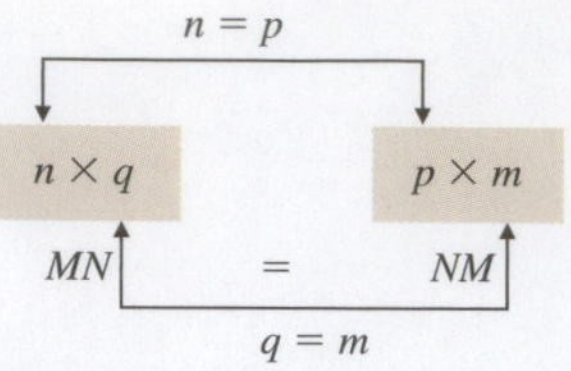

Figure 3

DEFINITION Inverse of a Square Matrix

Let M be a square matrix of order n and I be the identity matrix of order n. If there exists a matrix M^{-1} (read "M inverse") such that

$$M^{-1}M = MM^{-1} = I$$

then M^{-1} is called the **multiplicative inverse of M** or, more simply, the **inverse of M**. If no such matrix exists, then M is said to be a **singular matrix**.

Let us use the definition above to find M^{-1} for

$$M = \begin{bmatrix} 2 & 3 \\ 1 & 2 \end{bmatrix}$$

We are looking for

$$M^{-1} = \begin{bmatrix} a & c \\ b & d \end{bmatrix}$$

such that

$$MM^{-1} = M^{-1}M = I$$

So we write

$$\overset{M}{\begin{bmatrix} 2 & 3 \\ 1 & 2 \end{bmatrix}}\overset{M^{-1}}{\begin{bmatrix} a & c \\ b & d \end{bmatrix}} = \overset{I}{\begin{bmatrix} 1 & 0 \\ 0 & 1 \end{bmatrix}}$$

and try to find a, b, c, and d so that the product of M and M^{-1} is the identity matrix I. Multiplying M and M^{-1} on the left side, we obtain

$$\begin{bmatrix} (2a + 3b) & (2c + 3d) \\ (a + 2b) & (c + 2d) \end{bmatrix} = \begin{bmatrix} 1 & 0 \\ 0 & 1 \end{bmatrix}$$

which is true only if

Use Gauss–Jordan elimination to solve each system.

$$\begin{aligned} 2a + 3b &= 1 \\ a + 2b &= 0 \end{aligned} \qquad\qquad \begin{aligned} 2c + 3d &= 0 \\ c + 2d &= 1 \end{aligned}$$

$$\left[\begin{array}{cc|c} 2 & 3 & 1 \\ 1 & 2 & 0 \end{array}\right] \quad R_1 \leftrightarrow R_2 \qquad \left[\begin{array}{cc|c} 2 & 3 & 0 \\ 1 & 2 & 1 \end{array}\right] \quad R_1 \leftrightarrow R_2$$

$$\left[\begin{array}{cc|c} 1 & 2 & 0 \\ 2 & 3 & 1 \end{array}\right] \quad (-2)R_1 + R_2 \rightarrow R_2 \qquad \left[\begin{array}{cc|c} 1 & 2 & 1 \\ 2 & 3 & 0 \end{array}\right] \quad (-2)R_1 + R_2 \rightarrow R_2$$

$$\left[\begin{array}{cc|c} 1 & 2 & 0 \\ 0 & -1 & 1 \end{array}\right] \quad (-1)R_2 \rightarrow R_2 \qquad \left[\begin{array}{cc|c} 1 & 2 & 1 \\ 0 & -1 & -2 \end{array}\right] \quad (-1)R_2 \rightarrow R_2$$

$$\left[\begin{array}{cc|c} 1 & 2 & 0 \\ 0 & 1 & -1 \end{array}\right] \quad (-2)R_2 + R_1 \rightarrow R_1 \qquad \left[\begin{array}{cc|c} 1 & 2 & 1 \\ 0 & 1 & 2 \end{array}\right] \quad (-2)R_2 + R_1 \rightarrow R_1$$

$$\left[\begin{array}{cc|c} 1 & 0 & 2 \\ 0 & 1 & -1 \end{array}\right] \qquad \left[\begin{array}{cc|c} 1 & 0 & -3 \\ 0 & 1 & 2 \end{array}\right]$$

$a = 2, b = -1 \qquad c = -3, d = 2$

$$M^{-1} = \begin{bmatrix} a & c \\ b & d \end{bmatrix} = \begin{bmatrix} 2 & -3 \\ -1 & 2 \end{bmatrix}$$

CHECK

$$\overset{M}{\begin{bmatrix} 2 & 3 \\ 1 & 2 \end{bmatrix}} \overset{M^{-1}}{\begin{bmatrix} 2 & -3 \\ -1 & 2 \end{bmatrix}} = \overset{I}{\begin{bmatrix} 1 & 0 \\ 0 & 1 \end{bmatrix}} = \overset{M^{-1}}{\begin{bmatrix} 2 & -3 \\ -1 & 2 \end{bmatrix}} \overset{M}{\begin{bmatrix} 2 & 3 \\ 1 & 2 \end{bmatrix}}$$

Unlike nonzero real numbers, inverses do not always exist for square matrices. For example, if

$$N = \begin{bmatrix} 2 & 1 \\ 4 & 2 \end{bmatrix}$$

then, using the previous process, we are led to the systems

$$\begin{aligned} 2a + b &= 1 \\ 4a + 2b &= 0 \end{aligned} \qquad \begin{aligned} 2c + d &= 0 \\ 4c + 2d &= 1 \end{aligned}$$

Use Gauss–Jordan elimination to solve each system.

$$\left[\begin{array}{cc|c} 2 & 1 & 1 \\ 4 & 2 & 0 \end{array}\right] \; (-2)R_1 + R_2 \to R_2 \qquad \left[\begin{array}{cc|c} 2 & 1 & 0 \\ 4 & 2 & 1 \end{array}\right] \; (-2)R_1 + R_2 \to R_2$$

$$\left[\begin{array}{cc|c} 2 & 1 & 0 \\ 0 & 0 & -2 \end{array}\right] \qquad \left[\begin{array}{cc|c} 2 & 1 & 0 \\ 0 & 0 & 1 \end{array}\right]$$

The last row of each augmented matrix contains a contradiction. So each system is inconsistent and has no solution. We conclude that N^{-1} does not exist and N is a singular matrix.

Being able to find inverses, when they exist, leads to direct and simple solutions to many practical problems. In the next section, we show how inverses can be used to solve systems of linear equations.

The method outlined previously for finding M^{-1}, if it exists, gets very involved for matrices of order larger than 2. Now that we know what we are looking for, we can use augmented matrices (see Sections 4.2 and 4.3) to make the process more efficient.

EXAMPLE 2 **Finding the Inverse of a Matrix** Find the inverse, if it exists, of the matrix

$$M = \begin{bmatrix} 1 & -1 & 1 \\ 0 & 2 & -1 \\ 2 & 3 & 0 \end{bmatrix}$$

SOLUTION We start as before and write

$$\overset{M}{\begin{bmatrix} 1 & -1 & 1 \\ 0 & 2 & -1 \\ 2 & 3 & 0 \end{bmatrix}} \overset{M^{-1}}{\begin{bmatrix} a & d & g \\ b & e & h \\ c & f & i \end{bmatrix}} = \overset{I}{\begin{bmatrix} 1 & 0 & 0 \\ 0 & 1 & 0 \\ 0 & 0 & 1 \end{bmatrix}}$$

which is true only if

$$\begin{aligned} a - b + c &= 1 \\ 2b - c &= 0 \\ 2a + 3b &= 0 \end{aligned} \qquad \begin{aligned} d - e + f &= 0 \\ 2e - f &= 1 \\ 2d + 3e &= 0 \end{aligned} \qquad \begin{aligned} g - h + i &= 0 \\ 2h - i &= 0 \\ 2g + 3h &= 1 \end{aligned}$$

Now we write augmented matrices for each of the three systems:

$$\overset{\text{First}}{\left[\begin{array}{ccc|c} 1 & -1 & 1 & 1 \\ 0 & 2 & -1 & 0 \\ 2 & 3 & 0 & 0 \end{array}\right]} \quad \overset{\text{Second}}{\left[\begin{array}{ccc|c} 1 & -1 & 1 & 0 \\ 0 & 2 & -1 & 1 \\ 2 & 3 & 0 & 0 \end{array}\right]} \quad \overset{\text{Third}}{\left[\begin{array}{ccc|c} 1 & -1 & 1 & 0 \\ 0 & 2 & -1 & 0 \\ 2 & 3 & 0 & 1 \end{array}\right]}$$

Since each matrix to the left of the vertical bar is the same, exactly the same row operations can be used on each augmented matrix to transform it into a reduced form. We can speed up the process substantially by combining all three augmented matrices into the single augmented matrix form below:

$$\left[\begin{array}{ccc|ccc} 1 & -1 & 1 & 1 & 0 & 0 \\ 0 & 2 & -1 & 0 & 1 & 0 \\ 2 & 3 & 0 & 0 & 0 & 1 \end{array}\right] = [M \mid I] \tag{1}$$

We now try to perform row operations on matrix (1) until we obtain a row-equivalent matrix of the form

$$\left[\begin{array}{ccc|ccc} 1 & 0 & 0 & a & d & g \\ 0 & 1 & 0 & b & e & h \\ 0 & 0 & 1 & c & f & i \end{array}\right] = [I \mid B] \tag{2}$$

(with I labelling the left block and B the right block)

If this can be done, the new matrix B to the right of the vertical bar will be M^{-1}. Now let's try to transform matrix (1) into a form like matrix (2). We follow the same sequence of steps as we did in the solution of linear systems by Gauss–Jordan elimination (see Section 4.3).

$$\overset{M \qquad\qquad I}{\left[\begin{array}{ccc|ccc} 1 & -1 & 1 & 1 & 0 & 0 \\ 0 & 2 & -1 & 0 & 1 & 0 \\ 2 & 3 & 0 & 0 & 0 & 1 \end{array}\right]} \quad (-2)R_1 + R_3 \rightarrow R_3$$

$$\sim \left[\begin{array}{ccc|ccc} 1 & -1 & 1 & 1 & 0 & 0 \\ 0 & 2 & -1 & 0 & 1 & 0 \\ 0 & 5 & -2 & -2 & 0 & 1 \end{array}\right] \quad \tfrac{1}{2}R_2 \rightarrow R_2$$

$$\sim \left[\begin{array}{ccc|ccc} 1 & -1 & 1 & 1 & 0 & 0 \\ 0 & 1 & -\frac{1}{2} & 0 & \frac{1}{2} & 0 \\ 0 & 5 & -2 & -2 & 0 & 1 \end{array}\right] \quad \begin{array}{l} R_2 + R_1 \rightarrow R_1 \\ (-5)R_2 + R_3 \rightarrow R_3 \end{array}$$

$$\sim\left[\begin{array}{ccc|ccc} 1 & 0 & \frac{1}{2} & 1 & \frac{1}{2} & 0 \\ 0 & 1 & -\frac{1}{2} & 0 & \frac{1}{2} & 0 \\ 0 & 0 & \frac{1}{2} & -2 & -\frac{5}{2} & 1 \end{array}\right] \quad 2R_3 \rightarrow R_3$$

$$\sim\left[\begin{array}{ccc|ccc} 1 & 0 & \frac{1}{2} & 1 & \frac{1}{2} & 0 \\ 0 & 1 & -\frac{1}{2} & 0 & \frac{1}{2} & 0 \\ 0 & 0 & 1 & -4 & -5 & 2 \end{array}\right] \quad \begin{array}{l} (-\frac{1}{2})R_3 + R_1 \rightarrow R_1 \\ \frac{1}{2}R_3 + R_2 \rightarrow R_2 \end{array}$$

$$\sim\left[\begin{array}{ccc|ccc} 1 & 0 & 0 & 3 & 3 & -1 \\ 0 & 1 & 0 & -2 & -2 & 1 \\ 0 & 0 & 1 & -4 & -5 & 2 \end{array}\right] = [I \mid B]$$

Converting back to systems of equations equivalent to our three original systems, we have

$$\begin{array}{lll} a = 3 & d = 3 & g = -1 \\ b = -2 & e = -2 & h = 1 \\ c = -4 & f = -5 & i = 2 \end{array}$$

And these are just the elements of M^{-1} that we are looking for!

$$M^{-1} = \begin{bmatrix} 3 & 3 & -1 \\ -2 & -2 & 1 \\ -4 & -5 & 2 \end{bmatrix}$$

Note that this is the matrix to the right of the vertical line in the last augmented matrix. That is, $M^{-1} = B$.

Since the definition of matrix inverse requires that

$$M^{-1}M = I \quad \text{and} \quad MM^{-1} = I \tag{3}$$

it appears that we must compute both $M^{-1}M$ and MM^{-1} to check our work. However, it can be shown that if one of the equations in (3) is satisfied, the other is also satisfied. So to check our answer it is sufficient to compute either $M^{-1}M$ or MM^{-1}; we do not need to do both.

CHECK

$$M^{-1}M = \begin{bmatrix} 3 & 3 & -1 \\ -2 & -2 & 1 \\ -4 & -5 & 2 \end{bmatrix}\begin{bmatrix} 1 & -1 & 1 \\ 0 & 2 & -1 \\ 2 & 3 & 0 \end{bmatrix} = \begin{bmatrix} 1 & 0 & 0 \\ 0 & 1 & 0 \\ 0 & 0 & 1 \end{bmatrix} = I$$

Matched Problem 2 Let $M = \begin{bmatrix} 3 & -1 & 1 \\ -1 & 1 & 0 \\ 1 & 0 & 1 \end{bmatrix}$.

(A) Form the augmented matrix $[M \mid I]$.

(B) Use row operations to transform $[M \mid I]$ into $[I \mid B]$.

(C) Verify by multiplication that $B = M^{-1}$ (that is, show that $BM = I$).

The procedure shown in Example 2 can be used to find the inverse of any square matrix, if the inverse exists, and will also indicate when the inverse does not exist. These ideas are summarized in Theorem 1.

THEOREM 1 Inverse of a Square Matrix M

If $[M \mid I]$ is transformed by row operations into $[I \mid B]$, then the resulting matrix B is M^{-1}. However, if we obtain all 0's in one or more rows to the left of the vertical line, then M^{-1} does not exist.

Explore and Discuss 2

(A) Suppose that the square matrix M has a row of all zeros. Explain why M has no inverse.

(B) Suppose that the square matrix M has a column of all zeros. Explain why M has no inverse.

EXAMPLE 3 **Finding a Matrix Inverse** Find M^{-1}, given $M = \begin{bmatrix} 4 & -1 \\ -6 & 2 \end{bmatrix}$.

SOLUTION

$$\left[\begin{array}{cc|cc} 4 & -1 & 1 & 0 \\ -6 & 2 & 0 & 1 \end{array}\right] \qquad \tfrac{1}{4}R_1 \rightarrow R_1$$

$$\sim \left[\begin{array}{cc|cc} 1 & -\frac{1}{4} & \frac{1}{4} & 0 \\ -6 & 2 & 0 & 1 \end{array}\right] \qquad 6R_1 + R_2 \rightarrow R_2$$

$$\sim \left[\begin{array}{cc|cc} 1 & -\frac{1}{4} & \frac{1}{4} & 0 \\ 0 & \frac{1}{2} & \frac{3}{2} & 1 \end{array}\right] \qquad 2R_2 \rightarrow R_2$$

$$\sim \left[\begin{array}{cc|cc} 1 & -\frac{1}{4} & \frac{1}{4} & 0 \\ 0 & 1 & 3 & 2 \end{array}\right] \qquad \tfrac{1}{4}R_2 + R_1 \rightarrow R_1$$

$$\sim \left[\begin{array}{cc|cc} 1 & 0 & 1 & \frac{1}{2} \\ 0 & 1 & 3 & 2 \end{array}\right]$$

Therefore,

$$M^{-1} = \begin{bmatrix} 1 & \frac{1}{2} \\ 3 & 2 \end{bmatrix}$$

Check by showing that $M^{-1}M = I$.

Matched Problem 3 Find M^{-1}, given $M = \begin{bmatrix} 2 & -6 \\ 1 & -2 \end{bmatrix}$.

Most graphing calculators and spreadsheets can compute matrix inverses, as illustrated in Figure 4 for the solution to Example 3.

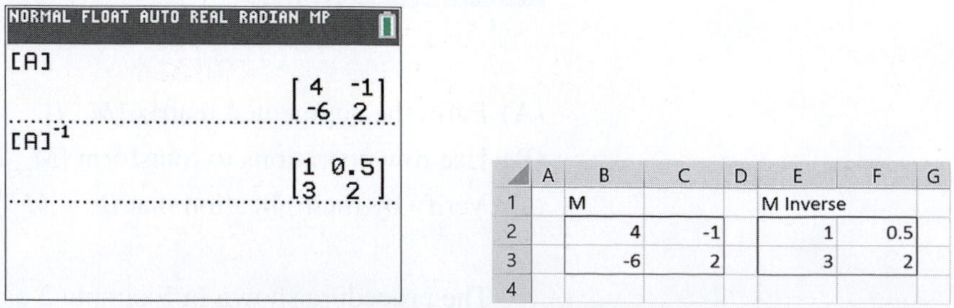

(A) The command $[A]^{-1}$ produces the inverse on this graphing calculator

(B) The command MINVERSE (B2:C3) produces the inverse in this spreadsheet

Figure 4 Finding a matrix inverse

Explore and Discuss 3

The inverse of

$$A = \begin{bmatrix} a & b \\ c & d \end{bmatrix}$$

is

$$A^{-1} = \begin{bmatrix} \dfrac{d}{ad - bc} & \dfrac{-b}{ad - bc} \\ \dfrac{-c}{ad - bc} & \dfrac{a}{ad - bc} \end{bmatrix} = \frac{1}{D}\begin{bmatrix} d & -b \\ -c & a \end{bmatrix} \qquad D = ad - bc$$

provided that $D \neq 0$.

(A) Use matrix multiplication to verify this formula. What can you conclude about A^{-1} if $D = 0$?

(B) Use this formula to find the inverse of matrix M in Example 3.

EXAMPLE 4 **Finding a Matrix Inverse** Find M^{-1}, given $M = \begin{bmatrix} 2 & -4 \\ -3 & 6 \end{bmatrix}$.

SOLUTION

$$\left[\begin{array}{rr|rr} 2 & -4 & 1 & 0 \\ -3 & 6 & 0 & 1 \end{array}\right] \quad \tfrac{1}{2}R_1 \to R_1$$

$$\sim \left[\begin{array}{rr|rr} 1 & -2 & \tfrac{1}{2} & 0 \\ -3 & 6 & 0 & 1 \end{array}\right] \quad 3R_1 + R_2 \to R_2$$

$$\sim \left[\begin{array}{rr|rr} 1 & -2 & \tfrac{1}{2} & 0 \\ 0 & 0 & \tfrac{3}{2} & 1 \end{array}\right]$$

We have all 0's in the second row to the left of the vertical bar; therefore, the inverse does not exist.

Matched Problem 4 Find N^{-1}, given $N = \begin{bmatrix} 3 & 1 \\ 6 & 2 \end{bmatrix}$.

Square matrices that do not have inverses are called singular matrices. Graphing calculators and spreadsheets recognize singular matrices and generally respond with some type of error message, as illustrated in Figure 5 for the solution to Example 4.

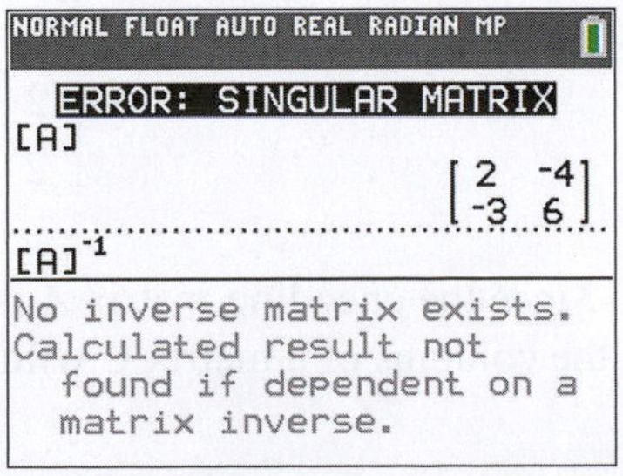

(A) A graphing calculator displays a clear error message

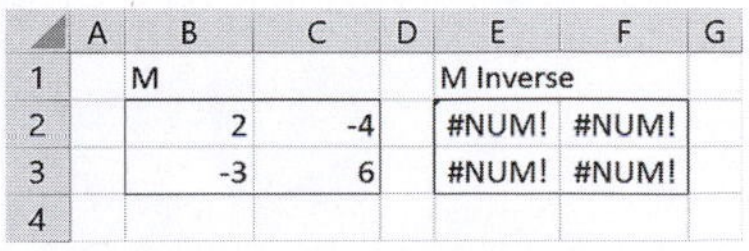

(B) A spreadsheet displays a more cryptic error message

Figure 5

Application: Cryptography

Matrix inverses can provide a simple and effective procedure for encoding and decoding messages. To begin, assign the numbers 1–26 to the letters in the alphabet, as shown below. Also assign the number 0 to a blank to provide for space between words. (A more sophisticated code could include both capital and lowercase letters and punctuation symbols.)

Blank	A	B	C	D	E	F	G	H	I	J	K	L	M	N	O	P	Q	R	S	T	U	V	W	X	Y	Z
0	1	2	3	4	5	6	7	8	9	10	11	12	13	14	15	16	17	18	19	20	21	22	23	24	25	26

The message "SECRET CODE" corresponds to the sequence

19 5 3 18 5 20 0 3 15 4 5

Any matrix whose elements are positive integers and whose inverse exists can be used as an **encoding matrix**. For example, to use the 2×2 matrix

$$A = \begin{bmatrix} 4 & 3 \\ 1 & 1 \end{bmatrix}$$

to encode the preceding message, first we divide the numbers in the sequence into groups of 2 and use these groups as the columns of a matrix B with 2 rows:

$$B = \begin{bmatrix} 19 & 3 & 5 & 0 & 15 & 5 \\ 5 & 18 & 20 & 3 & 4 & 0 \end{bmatrix}$$

Proceed down the columns, not across the rows.

Notice that we added an extra blank at the end of the message to make the columns come out even. Then we multiply this matrix on the left by A:

$$AB = \begin{bmatrix} 4 & 3 \\ 1 & 1 \end{bmatrix} \begin{bmatrix} 19 & 3 & 5 & 0 & 15 & 5 \\ 5 & 18 & 20 & 3 & 4 & 0 \end{bmatrix}$$

$$= \begin{bmatrix} 91 & 66 & 80 & 9 & 72 & 20 \\ 24 & 21 & 25 & 3 & 19 & 5 \end{bmatrix}$$

The coded message is

91 24 66 21 80 25 9 3 72 19 20 5

This message can be decoded simply by putting it back into matrix form and multiplying on the left by the **decoding matrix** A^{-1}. Since A^{-1} is easily determined if A is known, the encoding matrix A is the only key needed to decode messages that are encoded in this manner.

EXAMPLE 5 **Cryptography** The message

46 84 85 28 47 46 4 5 10 30 48 72 29 57 38 38 57 95

was encoded with the matrix A shown next. Decode this message.

$$A = \begin{bmatrix} 1 & 1 & 1 \\ 2 & 1 & 2 \\ 2 & 3 & 1 \end{bmatrix}$$

SOLUTION Since the encoding matrix A is 3×3, we begin by entering the coded message in the columns of a matrix C with three rows:

$$C = \begin{bmatrix} 46 & 28 & 4 & 30 & 29 & 38 \\ 84 & 47 & 5 & 48 & 57 & 57 \\ 85 & 46 & 10 & 72 & 38 & 95 \end{bmatrix}$$

If B is the matrix containing the uncoded message, then B and C are related by $C = AB$. To recover B, we find A^{-1} (details omitted) and multiply both sides of the equation $C = AB$ by A^{-1}:

$$B = A^{-1}C$$

$$= \begin{bmatrix} -5 & 2 & 1 \\ 2 & -1 & 0 \\ 4 & -1 & -1 \end{bmatrix} \begin{bmatrix} 46 & 28 & 4 & 30 & 29 & 38 \\ 84 & 47 & 5 & 48 & 57 & 57 \\ 85 & 46 & 10 & 72 & 38 & 95 \end{bmatrix}$$

$$= \begin{bmatrix} 23 & 0 & 0 & 18 & 7 & 19 \\ 8 & 9 & 3 & 12 & 1 & 19 \\ 15 & 19 & 1 & 0 & 21 & 0 \end{bmatrix}$$

Writing the numbers in the columns of this matrix in sequence and using the correspondence between numbers and letters noted earlier produces the decoded message:

23	8	15	0	9	19	0	3	1	18	12	0	7	1	21	19	19	0
W	H	O		I	S		C	A	R	L		G	A	U	S	S	

The answer to this question can be found earlier in this chapter.

Matched Problem 5 The message below was also encoded with the matrix A in Example 5. Decode this message:

46 84 85 28 47 46 32 41 78 25 42 53 25 37 63 43 71 83 19 37 25

Exercises 4.5

Skills Warm-up Exercises

In Problems 1–4, find the additive inverse and the multiplicative inverse, if defined, of each real number. (If necessary, review Section A.1).

1. (A) 4 (B) −3 (C) 0

2. (A) −7 (B) 2 (C) −1

3. (A) $\frac{2}{3}$ (B) $\frac{-1}{7}$ (C) 1.6

4. (A) $\frac{4}{5}$ (B) $\frac{12}{7}$ (C) −2.5

In Problems 5–8, does the given matrix have a multiplicative inverse? Explain your answer.

5. $\begin{bmatrix} 2 & 5 \end{bmatrix}$ **6.** $\begin{bmatrix} 4 \\ 8 \end{bmatrix}$ **7.** $\begin{bmatrix} 0 & 0 \\ 0 & 0 \end{bmatrix}$ **8.** $\begin{bmatrix} 1 & 0 \\ 0 & 1 \end{bmatrix}$

In Problems 9–18, find the matrix products. Note that each product can be found mentally, without the use of a calculator or pencil-and-paper calculations.

9. (A) $\begin{bmatrix} 1 & 0 \\ 0 & 0 \end{bmatrix}\begin{bmatrix} 2 & -3 \\ 4 & 5 \end{bmatrix}$ (B) $\begin{bmatrix} 2 & -3 \\ 4 & 5 \end{bmatrix}\begin{bmatrix} 1 & 0 \\ 0 & 0 \end{bmatrix}$

10. (A) $\begin{bmatrix} 1 & 0 \\ 0 & 0 \end{bmatrix}\begin{bmatrix} -1 & 6 \\ 5 & 2 \end{bmatrix}$ (B) $\begin{bmatrix} -1 & 6 \\ 5 & 2 \end{bmatrix}\begin{bmatrix} 1 & 0 \\ 0 & 0 \end{bmatrix}$

11. (A) $\begin{bmatrix} 0 & 0 \\ 0 & 1 \end{bmatrix}\begin{bmatrix} 2 & -3 \\ 4 & 5 \end{bmatrix}$ (B) $\begin{bmatrix} 2 & -3 \\ 4 & 5 \end{bmatrix}\begin{bmatrix} 0 & 0 \\ 0 & 1 \end{bmatrix}$

12. (A) $\begin{bmatrix} 0 & 0 \\ 0 & 1 \end{bmatrix}\begin{bmatrix} -1 & 6 \\ 5 & 2 \end{bmatrix}$ (B) $\begin{bmatrix} -1 & 6 \\ 5 & 2 \end{bmatrix}\begin{bmatrix} 0 & 0 \\ 0 & 1 \end{bmatrix}$

13. (A) $\begin{bmatrix} 1 & 0 \\ 0 & 1 \end{bmatrix}\begin{bmatrix} 2 & -3 \\ 4 & 5 \end{bmatrix}$ (B) $\begin{bmatrix} 2 & -3 \\ 4 & 5 \end{bmatrix}\begin{bmatrix} 1 & 0 \\ 0 & 1 \end{bmatrix}$

14. (A) $\begin{bmatrix} 1 & 0 \\ 0 & 1 \end{bmatrix}\begin{bmatrix} -1 & 6 \\ 5 & 2 \end{bmatrix}$ (B) $\begin{bmatrix} -1 & 6 \\ 5 & 2 \end{bmatrix}\begin{bmatrix} 1 & 0 \\ 0 & 1 \end{bmatrix}$

15. $\begin{bmatrix} 1 & 0 & 0 \\ 0 & 1 & 0 \\ 0 & 0 & 1 \end{bmatrix}\begin{bmatrix} -2 & 1 & 3 \\ 2 & 4 & -2 \\ 5 & 1 & 0 \end{bmatrix}$

16. $\begin{bmatrix} 1 & 0 & 0 \\ 0 & 1 & 0 \\ 0 & 0 & 1 \end{bmatrix}\begin{bmatrix} 3 & -4 & 0 \\ 1 & 2 & -5 \\ 6 & -3 & -1 \end{bmatrix}$

17. $\begin{bmatrix} -2 & 1 & 3 \\ 2 & 4 & -2 \\ 5 & 1 & 0 \end{bmatrix}\begin{bmatrix} 1 & 0 & 0 \\ 0 & 1 & 0 \\ 0 & 0 & 1 \end{bmatrix}$

18. $\begin{bmatrix} 3 & -4 & 0 \\ 1 & 2 & -5 \\ 6 & -3 & -1 \end{bmatrix}\begin{bmatrix} 1 & 0 & 0 \\ 0 & 1 & 0 \\ 0 & 0 & 1 \end{bmatrix}$

In Problems 19–28, examine the product of the two matrices to determine if each is the inverse of the other.

19. $\begin{bmatrix} 3 & -4 \\ -2 & 3 \end{bmatrix}; \begin{bmatrix} 3 & 4 \\ 2 & 3 \end{bmatrix}$

20. $\begin{bmatrix} -2 & -1 \\ -4 & 2 \end{bmatrix}; \begin{bmatrix} 1 & -1 \\ 2 & -2 \end{bmatrix}$

21. $\begin{bmatrix} 2 & 2 \\ -1 & -1 \end{bmatrix}; \begin{bmatrix} 1 & 1 \\ -1 & -1 \end{bmatrix}$

22. $\begin{bmatrix} 5 & -7 \\ -2 & 3 \end{bmatrix}; \begin{bmatrix} 3 & 7 \\ 2 & 5 \end{bmatrix}$

23. $\begin{bmatrix} -5 & 2 \\ -8 & 3 \end{bmatrix}; \begin{bmatrix} 3 & -2 \\ 8 & -5 \end{bmatrix}$

24. $\begin{bmatrix} 7 & 4 \\ -5 & -3 \end{bmatrix}; \begin{bmatrix} 3 & 4 \\ -5 & -7 \end{bmatrix}$

25. $\begin{bmatrix} 1 & 2 & 0 \\ 0 & 1 & 0 \\ -1 & -1 & 1 \end{bmatrix}; \begin{bmatrix} 1 & -2 & 0 \\ 0 & 1 & 0 \\ 1 & -1 & 0 \end{bmatrix}$

26. $\begin{bmatrix} 1 & 0 & 1 \\ -3 & 1 & -2 \\ 0 & 0 & 1 \end{bmatrix}; \begin{bmatrix} 1 & 0 & -1 \\ 3 & 1 & -1 \\ 0 & 0 & 1 \end{bmatrix}$

27. $\begin{bmatrix} 1 & -1 & 1 \\ 0 & 2 & -1 \\ 2 & 3 & 0 \end{bmatrix}; \begin{bmatrix} 3 & 3 & -1 \\ -2 & -2 & 1 \\ -4 & -5 & 2 \end{bmatrix}$

28. $\begin{bmatrix} 1 & 0 & -1 \\ 3 & 1 & -1 \\ 0 & 0 & 0 \end{bmatrix}; \begin{bmatrix} 1 & 0 & -1 \\ -3 & 1 & -2 \\ 0 & 0 & 1 \end{bmatrix}$

Without performing any row operations, explain why each of the matrices in Problems 29–38 does not have an inverse.

29. $\begin{bmatrix} 1 & 2 & 0 \\ -3 & 2 & -1 \end{bmatrix}$

30. $\begin{bmatrix} -2 & 3 & -1 \\ 4 & 0 & 1 \end{bmatrix}$

31. $\begin{bmatrix} 1 & -2 \\ 3 & 0 \\ 2 & -1 \end{bmatrix}$

32. $\begin{bmatrix} 0 & -1 \\ 2 & -2 \\ 1 & -3 \end{bmatrix}$

33. $\begin{bmatrix} 0 & 1 \\ 0 & -2 \end{bmatrix}$

34. $\begin{bmatrix} 0 & -1 \\ 0 & 2 \end{bmatrix}$

35. $\begin{bmatrix} 1 & -2 \\ 0 & 0 \end{bmatrix}$

36. $\begin{bmatrix} -1 & 2 \\ 0 & 0 \end{bmatrix}$

37. $\begin{bmatrix} 1 & -2 \\ 2 & -4 \end{bmatrix}$

38. $\begin{bmatrix} -1 & 2 \\ 1 & -2 \end{bmatrix}$

B *Given M in Problems 39–48, find M^{-1} and show that $M^{-1}M = I$.*

39. $\begin{bmatrix} -1 & 0 \\ -3 & 1 \end{bmatrix}$

40. $\begin{bmatrix} 1 & -5 \\ 0 & -1 \end{bmatrix}$

41. $\begin{bmatrix} 1 & 2 \\ 1 & 3 \end{bmatrix}$

42. $\begin{bmatrix} 2 & 1 \\ 5 & 3 \end{bmatrix}$

43. $\begin{bmatrix} 1 & 3 \\ 2 & 7 \end{bmatrix}$

44. $\begin{bmatrix} 2 & 1 \\ 1 & 1 \end{bmatrix}$

45. $\begin{bmatrix} 1 & -3 & 0 \\ 0 & 1 & 1 \\ 2 & -1 & 4 \end{bmatrix}$

46. $\begin{bmatrix} 2 & 3 & 0 \\ 1 & 2 & 3 \\ 0 & -1 & -5 \end{bmatrix}$

47. $\begin{bmatrix} 1 & 1 & 0 \\ 2 & 3 & -1 \\ 1 & 0 & 2 \end{bmatrix}$

48. $\begin{bmatrix} 1 & 0 & -1 \\ 2 & -1 & 0 \\ 1 & 1 & -2 \end{bmatrix}$

Find the inverse of each matrix in Problems 49–56, if it exists.

49. $\begin{bmatrix} 4 & 3 \\ -3 & -2 \end{bmatrix}$

50. $\begin{bmatrix} -4 & 3 \\ -5 & 4 \end{bmatrix}$

51. $\begin{bmatrix} 2 & 6 \\ 3 & 9 \end{bmatrix}$

52. $\begin{bmatrix} 2 & -4 \\ -3 & 6 \end{bmatrix}$

53. $\begin{bmatrix} 2 & 1 \\ 4 & 3 \end{bmatrix}$

54. $\begin{bmatrix} -5 & 3 \\ 2 & -2 \end{bmatrix}$

55. $\begin{bmatrix} 3 & -1 \\ -5 & 35 \end{bmatrix}$

56. $\begin{bmatrix} 5 & -10 \\ -2 & 24 \end{bmatrix}$

In Problems 57–60, find the inverse. Note that each inverse can be found mentally, without the use of a calculator or pencil-and-paper calculations.

57. $\begin{bmatrix} 3 & 0 \\ 0 & 3 \end{bmatrix}$

58. $\begin{bmatrix} 5 & 0 \\ 0 & 5 \end{bmatrix}$

59. $\begin{bmatrix} 2 & 0 \\ 0 & \frac{1}{2} \end{bmatrix}$

60. $\begin{bmatrix} -1 & 0 \\ 0 & \frac{1}{3} \end{bmatrix}$

C *Find the inverse of each matrix in Problems 61–68, if it exists.*

61. $\begin{bmatrix} -5 & -2 & -2 \\ 2 & 1 & 0 \\ 1 & 0 & 1 \end{bmatrix}$

62. $\begin{bmatrix} 2 & -2 & 4 \\ 1 & 1 & 1 \\ 1 & 0 & 1 \end{bmatrix}$

63. $\begin{bmatrix} 2 & 1 & 1 \\ 1 & 1 & 0 \\ -1 & -1 & 0 \end{bmatrix}$

64. $\begin{bmatrix} 1 & -1 & 0 \\ 2 & -1 & 1 \\ 0 & 1 & 1 \end{bmatrix}$

65. $\begin{bmatrix} -1 & -2 & 2 \\ 4 & 3 & 0 \\ 4 & 0 & 4 \end{bmatrix}$

66. $\begin{bmatrix} 4 & 2 & 2 \\ 4 & 2 & 0 \\ 5 & 0 & 5 \end{bmatrix}$

67. $\begin{bmatrix} 2 & -1 & -2 \\ -4 & 2 & 8 \\ 6 & -2 & -1 \end{bmatrix}$

68. $\begin{bmatrix} -1 & -1 & 4 \\ 3 & 3 & -22 \\ -2 & -1 & 19 \end{bmatrix}$

69. Show that $(A^{-1})^{-1} = A$ for: $A = \begin{bmatrix} 4 & 3 \\ 3 & 2 \end{bmatrix}$

70. Show that $(AB)^{-1} = B^{-1}A^{-1}$ for

$$A = \begin{bmatrix} 4 & 3 \\ 3 & 2 \end{bmatrix} \quad \text{and} \quad B = \begin{bmatrix} 2 & 5 \\ 3 & 7 \end{bmatrix}$$

71. Discuss the existence of M^{-1} for 2×2 diagonal matrices of the form

$$M = \begin{bmatrix} a & 0 \\ 0 & d \end{bmatrix}$$

Generalize your conclusions to $n \times n$ diagonal matrices.

72. Discuss the existence of M^{-1} for 2×2 upper triangular matrices of the form

$$M = \begin{bmatrix} a & b \\ 0 & d \end{bmatrix}$$

Generalize your conclusions to $n \times n$ upper triangular matrices.

In Problems 73–75, find A^{-1} and A^2.

73. $\begin{bmatrix} -1 & 3 \\ 0 & 1 \end{bmatrix}$ **74.** $\begin{bmatrix} -5 & 4 \\ -6 & 5 \end{bmatrix}$ **75.** $\begin{bmatrix} 5 & -3 \\ 8 & -5 \end{bmatrix}$

76. Based on your observations in Problems 73–75, if $A = A^{-1}$ for a square matrix A, what is A^2? Give a mathematical argument to support your conclusion.

Applications

Problems 77–80 refer to the encoding matrix

$$A = \begin{bmatrix} 1 & 2 \\ 1 & 3 \end{bmatrix}$$

77. Cryptography. Encode the message "WINGARDIUM LEVIOSA" using matrix A.

78. Cryptography. Encode the message "FINITE INCANTATEM" using matrix A.

79. Cryptography. The following message was encoded with matrix A. Decode this message:

52 70 17 21 5 5 29 43 4 4 52 70 25
35 29 33 15 18 5 5

80. Cryptography. The following message was encoded with matrix A. Decode this message:

36 44 5 5 38 56 55 75 18 23 56 75
22 33 37 55 27 40 53 79 59 81

Problems 81–84 require the use of a graphing calculator or computer. Use the 4×4 encoding matrix B given below. Form a matrix with 4 rows and as many columns as necessary to accommodate the message.

$$B = \begin{bmatrix} 2 & 2 & 1 & 3 \\ 1 & 2 & 2 & 1 \\ 1 & 1 & 0 & 1 \\ 2 & 3 & 2 & 3 \end{bmatrix}$$

81. Cryptography. Encode the message "DEPART ISTANBUL ORIENT EXPRESS" using matrix B.

82. Cryptography. Encode the message "SAIL FROM LISBON IN MORNING" using matrix B.

83. Cryptography. The following message was encoded with matrix B. Decode this message:

85 74 27 109 31 27 13 40 139 73 58 154
61 70 18 93 69 59 23 87 18 13 9 22

84. Cryptography. The following message was encoded with matrix B. Decode this message:

75 61 28 94 35 22 13 40 49 21 16 52
42 45 19 64 38 55 10 65 69 75 24 102
67 49 19 82 10 5 5 10

Problems 85–88 require the use of a graphing calculator or a computer. Use the 5×5 encoding matrix C given below. Form a matrix with 5 rows and as many columns as necessary to accommodate the message.

$$C = \begin{bmatrix} 1 & 0 & 1 & 0 & 1 \\ 0 & 1 & 1 & 0 & 3 \\ 2 & 1 & 1 & 1 & 1 \\ 0 & 0 & 1 & 0 & 2 \\ 1 & 1 & 1 & 2 & 1 \end{bmatrix}$$

85. Cryptography. Encode the message "THE EAGLE HAS LANDED" using matrix C.

86. Cryptography. Encode the message "ONE IF BY LAND AND TWO IF BY SEA" using matrix C.

87. Cryptography. The following message was encoded with matrix C. Decode this message:

37 72 58 45 56 30 67 50 46 60 27 77
41 45 39 28 24 52 14 37 32 58 70 36
76 22 38 70 12 67

88. Cryptography. The following message was encoded with matrix C. Decode this message:

25 75 55 35 50 43 83 54 60 53 25 13
59 9 53 15 35 40 15 45 33 60 60 36
51 15 7 37 0 22

Answers to Matched Problems

1. (A) $\begin{bmatrix} 2 & -3 \\ 5 & 7 \end{bmatrix}$ (B) $\begin{bmatrix} 4 & 2 \\ 3 & -5 \\ 6 & 8 \end{bmatrix}$

2. (A) $\left[\begin{array}{rrr|rrr} 3 & -1 & 1 & 1 & 0 & 0 \\ -1 & 1 & 0 & 0 & 1 & 0 \\ 1 & 0 & 1 & 0 & 0 & 1 \end{array}\right]$

(B) $\left[\begin{array}{rrr|rrr} 1 & 0 & 0 & 1 & 1 & -1 \\ 0 & 1 & 0 & 1 & 2 & -1 \\ 0 & 0 & 1 & -1 & -1 & 2 \end{array}\right]$

(C) $\begin{bmatrix} 1 & 1 & -1 \\ 1 & 2 & -1 \\ -1 & -1 & 2 \end{bmatrix}\begin{bmatrix} 3 & -1 & 1 \\ -1 & 1 & 0 \\ 1 & 0 & 1 \end{bmatrix} = \begin{bmatrix} 1 & 0 & 0 \\ 0 & 1 & 0 \\ 0 & 0 & 1 \end{bmatrix}$

3. $\begin{bmatrix} -1 & 3 \\ -\frac{1}{2} & 1 \end{bmatrix}$

4. Does not exist

5. WHO IS WILHELM JORDAN

4.6 Matrix Equations and Systems of Linear Equations

- Matrix Equations
- Matrix Equations and Systems of Linear Equations
- Application

The identity matrix and inverse matrix discussed in the preceding section can be put to immediate use in the solution of certain simple matrix equations. Being able to solve a matrix equation gives us another important method of solving systems of equations, provided that the system is independent and has the same number of variables as equations. If the system is dependent or if it has either fewer or more variables than equations, we must return to the Gauss–Jordan method of elimination.

Matrix Equations

Solving simple matrix equations is similar to solving real number equations but with two important differences:

1. there is *no* operation of division for matrices, and
2. matrix multiplication is *not* commutative.

Compare the real number equation $4x = 9$ and the matrix equation $AX = B$. The real number equation can be solved by dividing both sides of the equation by 4. However, that approach cannot be used for $AX = B$, because there is no operation of division for matrices. Instead, we note that $4x = 9$ can be solved by multiplying both sides of the equation by $\frac{1}{4}$, the multiplicative inverse of 4. So we solve $AX = B$ by multiplying both sides of the equation, *on the left*, by A^{-1}, the inverse of A. Because matrix multiplication is not commutative, multiplying both sides of an equation on the left by A^{-1} is different from multiplying both sides of an equation on the right by A^{-1}. In the case of $AX = B$, it is multiplication on the left that is required. The details are presented in Example 1.

In solving matrix equations, we will be guided by the properties of matrices summarized in Theorem 1.

THEOREM 1 Basic Properties of Matrices

Assuming that all products and sums are defined for the indicated matrices A, B, C, I, and 0, then

Addition Properties

Associative:	$(A + B) + C = A + (B + C)$
Commutative:	$A + B = B + A$
Additive identity:	$A + 0 = 0 + A = A$
Additive inverse:	$A + (-A) = (-A) + A = 0$

Multiplication Properties

Associative property:	$A(BC) = (AB)C$
Multiplicative identity:	$AI = IA = A$
Multiplicative inverse:	If A is a square matrix and A^{-1} exists, then $AA^{-1} = A^{-1}A = I$.

Combined Properties

Left distributive:	$A(B + C) = AB + AC$
Right distributive:	$(B + C)A = BA + CA$

Equality

Addition:	If $A = B$, then $A + C = B + C$.
Left multiplication:	If $A = B$, then $CA = CB$.
Right multiplication:	If $A = B$, then $AC = BC$.

EXAMPLE 1 **Solving a Matrix Equation** Given an $n \times n$ matrix A and $n \times 1$ column matrices B and X, solve $AX = B$ for X. Assume that all necessary inverses exist.

SOLUTION We are interested in finding a column matrix X that satisfies the matrix equation $AX = B$. To solve this equation, we multiply both sides on the left by A^{-1} to isolate X on the left side.

$$\begin{aligned} AX &= B && \text{Use the left multiplication property.} \\ A^{-1}(AX) &= A^{-1}B && \text{Use the associative property.} \\ (A^{-1}A)X &= A^{-1}B && A^{-1}A = I \\ IX &= A^{-1}B && IX = X \\ X &= A^{-1}B \end{aligned}$$

Matched Problem 1 Given an $n \times n$ matrix A and $n \times 1$ column matrices B, C, and X, solve $AX + C = B$ for X. Assume that all necessary inverses exist.

⚠ CAUTION Do not mix the left multiplication property and the right multiplication property. If $AX = B$, then

$$A^{-1}(AX) \neq BA^{-1}$$ ▲

Matrix Equations and Systems of Linear Equations

Now we show how independent systems of linear equations with the same number of variables as equations can be solved. First, convert the system into a matrix equation of the form $AX = B$, and then use $X = A^{-1}B$ as obtained in Example 1.

EXAMPLE 2 **Using Inverses to Solve Systems of Equations** Use matrix inverse methods to solve the system:

$$\begin{aligned} x_1 - x_2 + x_3 &= 1 \\ 2x_2 - x_3 &= 1 \\ 2x_1 + 3x_2 \quad &= 1 \end{aligned} \qquad (1)$$

SOLUTION The inverse of the coefficient matrix

$$A = \begin{bmatrix} 1 & -1 & 1 \\ 0 & 2 & -1 \\ 2 & 3 & 0 \end{bmatrix}$$

provides an efficient method for solving this system. To see how, we convert system (1) into a matrix equation:

$$\overset{A}{\begin{bmatrix} 1 & -1 & 1 \\ 0 & 2 & -1 \\ 2 & 3 & 0 \end{bmatrix}} \overset{X}{\begin{bmatrix} x_1 \\ x_2 \\ x_3 \end{bmatrix}} = \overset{B}{\begin{bmatrix} 1 \\ 1 \\ 1 \end{bmatrix}} \qquad (2)$$

Check that matrix equation (2) is equivalent to system (1) by finding the product of the left side and then equating corresponding elements on the left with those on the right.

We are interested in finding a column matrix X that satisfies the matrix equation $AX = B$. In Example 1 we found that if A^{-1} exists, then

$$X = A^{-1}B$$

The inverse of A was found in Example 2, Section 4.5, to be

$$A^{-1} = \begin{bmatrix} 3 & 3 & -1 \\ -2 & -2 & 1 \\ -4 & -5 & 2 \end{bmatrix}$$

Therefore,

$$\overset{X}{\begin{bmatrix} x_1 \\ x_2 \\ x_3 \end{bmatrix}} = \overset{A^{-1}}{\begin{bmatrix} 3 & 3 & -1 \\ -2 & -2 & 1 \\ -4 & -5 & 2 \end{bmatrix}} \overset{B}{\begin{bmatrix} 1 \\ 1 \\ 1 \end{bmatrix}} = \begin{bmatrix} 5 \\ -3 \\ -7 \end{bmatrix}$$

and we can conclude that $x_1 = 5, x_2 = -3$, and $x_3 = -7$. Check this result in system (1).

Matched Problem 2 Use matrix inverse methods to solve the system:

$$\begin{aligned} 3x_1 - x_2 + x_3 &= 1 \\ -x_1 + x_2 \quad\quad &= 3 \\ x_1 \quad\quad + x_3 &= 2 \end{aligned}$$

[*Note:* The inverse of the coefficient matrix was found in Matched Problem 2, Section 4.5.]

At first glance, using matrix inverse methods seems to require the same amount of effort as using Gauss–Jordan elimination. In either case, row operations must be applied to an augmented matrix involving the coefficients of the system. The advantage of the inverse matrix method becomes readily apparent when solving a number of systems with a common coefficient matrix and different constant terms.

EXAMPLE 3 **Using Inverses to Solve Systems of Equations** Use matrix inverse methods to solve each of the following systems:

(A)
$$\begin{aligned} x_1 - x_2 + x_3 &= 3 \\ 2x_2 - x_3 &= 1 \\ 2x_1 + 3x_2 \quad\quad &= 4 \end{aligned}$$

(B)
$$\begin{aligned} x_1 - x_2 + x_3 &= -5 \\ 2x_2 - x_3 &= 2 \\ 2x_1 + 3x_2 \quad\quad &= -3 \end{aligned}$$

SOLUTION Notice that both systems have the same coefficient matrix A as system (1) in Example 2. Only the constant terms have changed. We can use A^{-1} to solve these systems just as we did in Example 2.

(A)
$$\overset{X}{\begin{bmatrix} x_1 \\ x_2 \\ x_3 \end{bmatrix}} = \overset{A^{-1}}{\begin{bmatrix} 3 & 3 & -1 \\ -2 & -2 & 1 \\ -4 & -5 & 2 \end{bmatrix}} \overset{B}{\begin{bmatrix} 3 \\ 1 \\ 4 \end{bmatrix}} = \begin{bmatrix} 8 \\ -4 \\ -9 \end{bmatrix}$$

$x_1 = 8, x_2 = -4$, and $x_3 = -9$.

(B)
$$\overset{X}{\begin{bmatrix} x_1 \\ x_2 \\ x_3 \end{bmatrix}} = \overset{A^{-1}}{\begin{bmatrix} 3 & 3 & -1 \\ -2 & -2 & 1 \\ -4 & -5 & 2 \end{bmatrix}} \overset{B}{\begin{bmatrix} -5 \\ 2 \\ -3 \end{bmatrix}} = \begin{bmatrix} -6 \\ 3 \\ 4 \end{bmatrix}$$

$x_1 = -6, x_2 = 3$, and $x_3 = 4$.

Matched Problem 3 Use matrix inverse methods to solve each of the following systems (see Matched Problem 2):

(A) $$\begin{aligned} 3x_1 - x_2 + x_3 &= 3 \\ -x_1 + x_2 &= -3 \\ x_1 + x_3 &= 2 \end{aligned}$$

(B) $$\begin{aligned} 3x_1 - x_2 + x_3 &= -5 \\ -x_1 + x_2 &= 1 \\ x_1 + x_3 &= -4 \end{aligned}$$

As Examples 2 and 3 illustrate, inverse methods are very convenient for hand calculations because once the inverse is found, it can be used to solve any new system formed by changing only the constant terms. Since most graphing calculators and computers can compute the inverse of a matrix, this method also adapts readily to graphing calculator and spreadsheet solutions (Fig. 1). However, if your graphing calculator (or spreadsheet) also has a built-in procedure for finding the reduced form of an augmented matrix, it is just as convenient to use Gauss–Jordan elimination. Furthermore, Gauss–Jordan elimination can be used in all cases and, as noted below, matrix inverse methods cannot always be used.

	A	B	C	D	E	F	G	H	I	J	K
1			A			B	X		B	X	
2		1	-1	1		3	8		-5	-6	
3		0	2	-1		1	-4		2	3	
4		2	3	0		4	-9		-3	4	
5											

Figure 1 Using inverse methods on a spreadsheet: The values in G2:G4 are produced by the command MMULT (MINVERSE(B2:D4),F2:F4)

SUMMARY Using Inverse Methods to Solve Systems of Equations

If the number of equations in a system equals the number of variables and the coefficient matrix has an inverse, then the system will always have a unique solution that can be found by using the inverse of the coefficient matrix to solve the corresponding matrix equation.

Matrix equation	Solution
$AX = B$	$X = A^{-1}B$

CONCEPTUAL INSIGHT

There are two cases where inverse methods will not work:

Case 1. The coefficient matrix is singular.

Case 2. The number of variables is not the same as the number of equations.

In either case, use Gauss–Jordan elimination.

Application

The following application illustrates the usefulness of the inverse matrix method for solving systems of equations.

EXAMPLE 4 **Investment Analysis** An investment advisor currently has two types of investments available for clients: a conservative investment A that pays 5% per year and a higher risk investment B that pays 10% per year. Clients may divide their investments between the two to achieve any total return desired between 5% and 10%. However, the higher the desired return, the higher the risk. How should each client invest to achieve the indicated return?

	Client 1	Client 2	Client 3	Client k
Total investment	\$20,000	\$50,000	\$10,000	k_1
Annual return desired	\$ 1,200	\$ 3,750	\$ 900	k_2
	(6%)	(7.5%)	(9%)	

SOLUTION The answer to this problem involves six quantities, two for each client. Utilizing inverse matrices provides an efficient way to find these quantities. We will solve the problem for an arbitrary client k with unspecified amounts k_1 for the total investment and k_2 for the annual return. (Do not confuse k_1 and k_2 with variables. Their values are known—they just differ for each client.)

Let x_1 = amount invested in A by a given client
x_2 = amount invested in B by a given client

Then we have the following mathematical model:

$$\begin{aligned} x_1 + \quad x_2 &= k_1 & &\text{Total invested} \\ 0.05x_1 + 0.1x_2 &= k_2 & &\text{Total annual return desired} \end{aligned}$$

Write as a matrix equation:

$$\overset{A}{\begin{bmatrix} 1 & 1 \\ 0.05 & 0.1 \end{bmatrix}} \overset{X}{\begin{bmatrix} x_1 \\ x_2 \end{bmatrix}} = \overset{B}{\begin{bmatrix} k_1 \\ k_2 \end{bmatrix}}$$

If A^{-1} exists, then

$$X = A^{-1}B$$

We now find A^{-1} by starting with the augmented matrix $[A \mid I]$ and proceeding as discussed in Section 4.5:

$$\left[\begin{array}{cc|cc} 1 & 1 & 1 & 0 \\ 0.05 & 0.1 & 0 & 1 \end{array}\right] \quad 20R_2 \rightarrow R_2$$

$$\sim \left[\begin{array}{cc|cc} 1 & 1 & 1 & 0 \\ 1 & 2 & 0 & 20 \end{array}\right] \quad (-1)R_1 + R_2 \rightarrow R_2$$

$$\sim \left[\begin{array}{cc|cc} 1 & 1 & 1 & 0 \\ 0 & 1 & -1 & 20 \end{array}\right] \quad (-1)R_2 + R_1 \rightarrow R_1$$

$$\sim \left[\begin{array}{cc|cc} 1 & 0 & 2 & -20 \\ 0 & 1 & -1 & 20 \end{array}\right]$$

Therefore,

$$A^{-1} = \begin{bmatrix} 2 & -20 \\ -1 & 20 \end{bmatrix} \qquad \text{Check: } \overset{A^{-1}}{\begin{bmatrix} 2 & -20 \\ -1 & 20 \end{bmatrix}} \overset{A}{\begin{bmatrix} 1 & 1 \\ 0.1 & 0.2 \end{bmatrix}} = \overset{I}{\begin{bmatrix} 1 & 0 \\ 0 & 1 \end{bmatrix}}$$

and

$$\overset{X}{\begin{bmatrix} x_1 \\ x_2 \end{bmatrix}} = \overset{A^{-1}}{\begin{bmatrix} 2 & -20 \\ -1 & 20 \end{bmatrix}} \overset{B}{\begin{bmatrix} k_1 \\ k_2 \end{bmatrix}}$$

To solve each client's investment problem, we replace k_1 and k_2 with appropriate values from the table and multiply by A^{-1}:

Client 1

$$\begin{bmatrix} x_1 \\ x_2 \end{bmatrix} = \begin{bmatrix} 2 & -20 \\ -1 & 20 \end{bmatrix} \begin{bmatrix} 20{,}000 \\ 1{,}200 \end{bmatrix} = \begin{bmatrix} 16{,}000 \\ 4{,}000 \end{bmatrix}$$

Solution: $x_1 = \$16{,}000$ in investment A, $x_2 = \$4{,}000$ in investment B

Client 2

$$\begin{bmatrix} x_1 \\ x_2 \end{bmatrix} = \begin{bmatrix} 2 & -20 \\ -1 & 20 \end{bmatrix} \begin{bmatrix} 50{,}000 \\ 3{,}750 \end{bmatrix} = \begin{bmatrix} 25{,}000 \\ 25{,}000 \end{bmatrix}$$

Solution: $x_1 = \$25{,}000$ in investment A, $x_2 = \$25{,}000$ in investment B

Client 3

$$\begin{bmatrix} x_1 \\ x_2 \end{bmatrix} = \begin{bmatrix} 2 & -20 \\ -1 & 20 \end{bmatrix} \begin{bmatrix} 10{,}000 \\ 900 \end{bmatrix} = \begin{bmatrix} 2{,}000 \\ 8{,}000 \end{bmatrix}$$

Solution: $x_1 = \$2{,}000$ in investment A, $x_2 = \$8{,}000$ in investment B

Matched Problem 4 Repeat Example 4 with investment A paying 4% and investment B paying 12%.

Figure 2 illustrates a solution to Example 4 on a spreadsheet.

	A	B	C	D	E	F	G	H
1			Clients					
2		1	2	3		A		
3	Total Investment	\$ 20,000	\$ 50,000	\$ 10,000		1	1	
4	Annual Return	\$ 1,200	\$ 3,750	\$ 900		0.05	0.1	
5	Amount Invested in A	\$ 16,000	\$ 25,000	\$ 2,000				
6	Amount Invested in B	\$ 4,000	\$ 25,000	\$ 8,000				
7								

Figure 2

Explore and Discuss 1

Refer to the mathematical model in Example 4:

$$\overset{A}{\begin{bmatrix} 1 & 1 \\ 0.05 & 0.1 \end{bmatrix}} \overset{X}{\begin{bmatrix} x_1 \\ x_2 \end{bmatrix}} = \overset{B}{\begin{bmatrix} k_1 \\ k_2 \end{bmatrix}} \qquad (3)$$

(A) Does matrix equation (3) always have a solution for any constant matrix B?

(B) Do all these solutions make sense for the original problem? If not, give examples.

(C) If the total investment is $k_1 = \$10{,}000$, describe all possible annual returns k_2.

Exercises 4.6

Skills Warm-up Exercises

W *In Problems 1–8, solve each equation for x, where x represents a real number. (If necessary, review Section 1.1).*

1. $5x = -3$

2. $4x = 9$

3. $4x = 8x + 7$

4. $6x = -3x + 14$

5. $6x + 8 = -2x + 17$

6. $-4x + 3 = 5x + 12$

7. $10 - 3x = 7x + 9$

8. $2x + 7x + 1 = 8x + 3 - x$

A *Write Problems 9–12 as systems of linear equations without matrices.*

9. $\begin{bmatrix} 3 & 1 \\ 2 & -1 \end{bmatrix}\begin{bmatrix} x_1 \\ x_2 \end{bmatrix} = \begin{bmatrix} 5 \\ -4 \end{bmatrix}$

10. $\begin{bmatrix} -2 & 1 \\ -3 & 4 \end{bmatrix}\begin{bmatrix} x_1 \\ x_2 \end{bmatrix} = \begin{bmatrix} -5 \\ 7 \end{bmatrix}$

11. $\begin{bmatrix} -3 & 1 & 0 \\ 2 & 0 & 1 \\ -1 & 3 & -2 \end{bmatrix}\begin{bmatrix} x_1 \\ x_2 \\ x_3 \end{bmatrix} = \begin{bmatrix} 3 \\ -4 \\ 2 \end{bmatrix}$

12. $\begin{bmatrix} 2 & -1 & 0 \\ -2 & 3 & -1 \\ 4 & 0 & 3 \end{bmatrix}\begin{bmatrix} x_1 \\ x_2 \\ x_3 \end{bmatrix} = \begin{bmatrix} 6 \\ -4 \\ 7 \end{bmatrix}$

Write each system in Problems 13–16 as a matrix equation of the form $AX = B$.

13. $3x_1 - 4x_2 = 1$
$2x_1 + x_2 = 5$

14. $2x_1 + x_2 = 8$
$-5x_1 + 3x_2 = -4$

15. $x_1 - 3x_2 + 2x_3 = -3$
$-2x_1 + 3x_2 = 1$
$x_1 + x_2 + 4x_3 = -2$

16. $3x_1 + 2x_3 = 9$
$-x_1 + 4x_2 + x_3 = -7$
$-2x_1 + 3x_2 = 6$

Find x_1 and x_2 in Problems 17–20.

17. $\begin{bmatrix} x_1 \\ x_2 \end{bmatrix} = \begin{bmatrix} 3 & -2 \\ 1 & 4 \end{bmatrix}\begin{bmatrix} -2 \\ 1 \end{bmatrix}$

18. $\begin{bmatrix} x_1 \\ x_2 \end{bmatrix} = \begin{bmatrix} -2 & 1 \\ -1 & 2 \end{bmatrix}\begin{bmatrix} 3 \\ -2 \end{bmatrix}$

19. $\begin{bmatrix} x_1 \\ x_2 \end{bmatrix} = \begin{bmatrix} -2 & 3 \\ 2 & -1 \end{bmatrix}\begin{bmatrix} 3 \\ 2 \end{bmatrix}$

20. $\begin{bmatrix} x_1 \\ x_2 \end{bmatrix} = \begin{bmatrix} 3 & -1 \\ 0 & 2 \end{bmatrix}\begin{bmatrix} -2 \\ 1 \end{bmatrix}$

In Problems 21–24, find x_1 and x_2.

21. $\begin{bmatrix} 1 & -1 \\ 1 & -2 \end{bmatrix}\begin{bmatrix} x_1 \\ x_2 \end{bmatrix} = \begin{bmatrix} 5 \\ 7 \end{bmatrix}$

22. $\begin{bmatrix} 1 & 3 \\ 1 & 4 \end{bmatrix}\begin{bmatrix} x_1 \\ x_2 \end{bmatrix} = \begin{bmatrix} 9 \\ 6 \end{bmatrix}$

23. $\begin{bmatrix} 1 & 1 \\ 2 & -3 \end{bmatrix}\begin{bmatrix} x_1 \\ x_2 \end{bmatrix} = \begin{bmatrix} 15 \\ 10 \end{bmatrix}$

24. $\begin{bmatrix} 1 & 1 \\ 3 & -2 \end{bmatrix}\begin{bmatrix} x_1 \\ x_2 \end{bmatrix} = \begin{bmatrix} 10 \\ 20 \end{bmatrix}$

In Problems 25–30, solve for x_1 and x_2.

25. $\begin{bmatrix} 1 & 3 \\ 1 & 1 \end{bmatrix}\begin{bmatrix} x_1 \\ x_2 \end{bmatrix} + \begin{bmatrix} 5 \\ 2 \end{bmatrix} = \begin{bmatrix} 14 \\ 7 \end{bmatrix}$

26. $\begin{bmatrix} 2 & 1 \\ 1 & 3 \end{bmatrix}\begin{bmatrix} x_1 \\ x_2 \end{bmatrix} + \begin{bmatrix} 3 \\ 5 \end{bmatrix} = \begin{bmatrix} 10 \\ 16 \end{bmatrix}$

27. $\begin{bmatrix} 2 & 6 \\ -4 & -12 \end{bmatrix}\begin{bmatrix} x_1 \\ x_2 \end{bmatrix} + \begin{bmatrix} 5 \\ 2 \end{bmatrix} = \begin{bmatrix} 14 \\ 7 \end{bmatrix}$

28. $\begin{bmatrix} 2 & 1 \\ -6 & -3 \end{bmatrix}\begin{bmatrix} x_1 \\ x_2 \end{bmatrix} + \begin{bmatrix} 3 \\ 5 \end{bmatrix} = \begin{bmatrix} 10 \\ 16 \end{bmatrix}$

29. $\begin{bmatrix} 4 & 2 \\ 3 & 1 \end{bmatrix}\begin{bmatrix} x_1 \\ x_2 \end{bmatrix} - \begin{bmatrix} 4 \\ 3 \end{bmatrix} = \begin{bmatrix} 14 \\ 8 \end{bmatrix}$

30. $\begin{bmatrix} 3 & 2 \\ 2 & 1 \end{bmatrix}\begin{bmatrix} x_1 \\ x_2 \end{bmatrix} - \begin{bmatrix} 4 \\ 1 \end{bmatrix} = \begin{bmatrix} 4 \\ 3 \end{bmatrix}$

B *In Problems 31–38, write each system as a matrix equation and solve using inverses. [Note: The inverses were found in Problems 41–48, Exercises 4.5.]*

31. $x_1 + 2x_2 = k_1$
$x_1 + 3x_2 = k_2$
(A) $k_1 = 1, k_2 = 3$
(B) $k_1 = 3, k_2 = 5$
(C) $k_1 = -2, k_2 = 1$

32. $2x_1 + x_2 = k_1$
$5x_1 + 3x_2 = k_2$
(A) $k_1 = 2, k_2 = 13$
(B) $k_1 = 2, k_2 = 4$
(C) $k_1 = 1, k_2 = -3$

33. $x_1 + 3x_2 = k_1$
$2x_1 + 7x_2 = k_2$
(A) $k_1 = 2, k_2 = -1$
(B) $k_1 = 1, k_2 = 0$
(C) $k_1 = 3, k_2 = -1$

34. $2x_1 + x_2 = k_1$
$x_1 + x_2 = k_2$
(A) $k_1 = -1, k_2 = -2$
(B) $k_1 = 2, k_2 = 3$
(C) $k_1 = 2, k_2 = 0$

35. $x_1 - 3x_2 = k_1$
$x_2 + x_3 = k_2$
$2x_1 - x_2 + 4x_3 = k_3$
(A) $k_1 = 1, k_2 = 0, k_3 = 2$
(B) $k_1 = -1, k_2 = 1, k_3 = 0$
(C) $k_1 = 2, k_2 = -2, k_3 = 1$

36. $2x_1 + 3x_2 = k_1$
$x_1 + 2x_2 + 3x_3 = k_2$
$-x_2 - 5x_3 = k_3$
(A) $k_1 = 0, k_2 = 2, k_3 = 1$
(B) $k_1 = -2, k_2 = 0, k_3 = 1$
(C) $k_1 = 3, k_2 = 1, k_3 = 0$

37. $x_1 + x_2 = k_1$
$2x_1 + 3x_2 - x_3 = k_2$
$x_1 + 2x_3 = k_3$
(A) $k_1 = 2, k_2 = 0, k_3 = 4$
(B) $k_1 = 0, k_2 = 4, k_3 = -2$
(C) $k_1 = 4, k_2 = 2, k_3 = 0$

38. $x_1 - x_3 = k_1$
$2x_1 - x_2 = k_2$
$x_1 + x_2 - 2x_3 = k_3$
(A) $k_1 = 4, k_2 = 8, k_3 = 0$
(B) $k_1 = 4, k_2 = 0, k_3 = -4$
(C) $k_1 = 0, k_2 = 8, k_3 = -8$

In Problems 39–44, the matrix equation is not solved correctly. Explain the mistake and find the correct solution. Assume that the indicated inverses exist.

39. $AX = B, X = \dfrac{B}{A}$

40. $XA = B, X = \dfrac{B}{A}$

41. $XA = B, X = A^{-1}B$

42. $AX = B, X = BA^{-1}$

43. $AX = BA, X = A^{-1}BA, X = B$

44. $XA = AB, X = ABA^{-1}, X = B$

In Problems 45–50, explain why the system cannot be solved by matrix inverse methods. Discuss methods that could be used and then solve the system.

45. $-2x_1 + 4x_2 = -5$
$6x_1 - 12x_2 = 15$

46. $-2x_1 + 4x_2 = 5$
$6x_1 - 12x_2 = 15$

47. $x_1 - 3x_2 - 2x_3 = -1$
$-2x_1 + 6x_2 + 4x_3 = 3$

48. $x_1 - 3x_2 - 2x_3 = -1$
$-2x_1 + 7x_2 + 3x_3 = 3$

49. $x_1 - 2x_2 + 3x_3 = 1$
$2x_1 - 3x_2 - 2x_3 = 3$
$x_1 - x_2 - 5x_3 = 2$

50. $x_1 - 2x_2 + 3x_3 = 1$
$2x_1 - 3x_2 - 2x_3 = 3$
$x_1 - x_2 - 5x_3 = 4$

C *For $n \times n$ matrices A and B, and $n \times 1$ column matrices C, D, and X, solve each matrix equation in Problems 51–56 for X. Assume that all necessary inverses exist.*

51. $AX - BX = C$

52. $AX + BX = C$

53. $AX + X = C$

54. $AX - X = C$

55. $AX - C = D - BX$

56. $AX + C = BX + D$

In Problems 57 and 58, solve for x_1 and x_2.

57. $\begin{bmatrix} 5 & 10 \\ 4 & 7 \end{bmatrix}\begin{bmatrix} x_1 \\ x_2 \end{bmatrix} - \begin{bmatrix} 2 & -1 \\ 3 & 3 \end{bmatrix}\begin{bmatrix} x_1 \\ x_2 \end{bmatrix} = \begin{bmatrix} 97 \\ 35 \end{bmatrix}$

58. $\begin{bmatrix} 5 & 1 \\ 2 & 2 \end{bmatrix}\begin{bmatrix} x_1 \\ x_2 \end{bmatrix} + \begin{bmatrix} -3 & 2 \\ 1 & 3 \end{bmatrix}\begin{bmatrix} x_1 \\ x_2 \end{bmatrix} = \begin{bmatrix} 20 \\ 31 \end{bmatrix}$

In Problems 59–62, write each system as a matrix equation and solve using the inverse coefficient matrix. Use a graphing calculator or computer to perform the necessary calculations.

59. $x_1 + 5x_2 + 6x_3 = 76$
$2x_1 + 3x_2 + 8x_3 = 92$
$11x_1 + 9x_2 + 4x_3 = 181$

60. $7x_1 + 2x_2 + 7x_3 = 59$
$2x_1 + x_2 + x_3 = 15$
$3x_1 + 4x_2 + 9x_3 = 53$

61. $2x_1 + x_2 + 5x_3 + 5x_4 = 37$
$3x_1 - 4x_2 + 3x_3 + 2x_4 = 0$
$7x_1 + 3x_2 + 8x_3 + 4x_4 = 45$
$5x_1 + 9x_2 + 6x_3 + 7x_4 = 94$

62. $2x_1 + x_2 + 6x_3 + 5x_4 = 54$
$3x_1 - 4x_2 + 15x_3 + 2x_4 = 84$
$7x_1 + 3x_2 + 7x_3 + 4x_4 = 34$
$5x_1 + 9x_2 + 7x_3 + 7x_4 = 77$

Applications

Construct a mathematical model for each of the following problems. (The answers in the back of the book include both the mathematical model and the interpretation of its solution.) Use matrix inverse methods to solve the model and then interpret the solution.

63. Concert tickets. A concert hall has 10,000 seats and two categories of ticket prices, \$25 and \$35. Assume that all seats in each category can be sold.

	Concert		
	1	2	3
Tickets sold	10,000	10,000	10,000
Return required	\$275,000	\$300,000	\$325,000

(A) How many tickets of each category should be sold to bring in each of the returns indicated in the table?

(B) Is it possible to bring in a return of \$200,000? Of \$400,000? Explain.

(C) Describe all the possible returns.

64. Parking receipts. Parking fees at a zoo are \$5.00 for local residents and \$7.50 for all others. At the end of each day, the total number of vehicles parked that day and the gross receipts for the day are recorded, but the number of vehicles in each category is not. The following table contains the relevant information for a recent 4-day period:

	Day			
	1	2	3	4
Vehicles parked	1,200	1,550	1,740	1,400
Gross receipts	\$7,125	\$9,825	\$11,100	\$8,650

(A) How many vehicles in each category used the zoo's parking facilities each day?

(B) If 1,200 vehicles are parked in one day, is it possible to take in gross receipts of \$5,000? Of \$10,000? Explain.

(C) Describe all possible gross receipts on a day when 1,200 vehicles are parked.

65. Production scheduling. A supplier manufactures car and truck frames at two different plants. The production rates (in frames per hour) for each plant are given in the table:

Plant	Car Frames	Truck Frames
A	10	5
B	8	8

How many hours should each plant be scheduled to operate to exactly fill each of the orders in the following table?

	Orders		
	1	2	3
Car frames	3,000	2,800	2,600
Truck frames	1,600	2,000	2,200

66. Production scheduling. Labor and material costs for manufacturing two guitar models are given in the table:

Guitar Model	Labor Cost	Material Cost
A	\$30	\$20
B	\$40	\$30

(A) If a total of \$3,000 a week is allowed for labor and material, how many of each model should be produced each week to use exactly each of the allocations of the \$3,000 indicated in the following table?

	Weekly Allocation		
	1	2	3
Labor	\$1,800	\$1,750	\$1,720
Material	\$1,200	\$1,250	\$1,280

(B) Is it possible to use an allocation of \$1,600 for labor and \$1,400 for material? Of \$2,000 for labor and \$1,000 for material? Explain.

67. Incentive plan. A small company provides an incentive plan for its top executives. Each executive receives as a bonus a percentage of the portion of the annual profit that remains after the bonuses for the other executives have been deducted (see the table). If the company has an annual profit of $2 million, find the bonus for each executive. Round each bonus to the nearest hundred dollars.

Officer	Bonus
President	3%
Executive vice president	2.5%
Associate vice president	2%
Assistant vice president	1.5%

68. Incentive plan. Repeat Problem 67 if the company decides to include a 1% bonus for the sales manager in the incentive plan.

69. Diets. A biologist has available two commercial food mixes containing the percentage of protein and fat given in the table.

Mix	Protein(%)	Fat(%)
A	20	4
B	14	3

(A) How many ounces of each mix should be used to prepare each of the diets listed in the following table?

	Diet		
	1	2	3
Protein	80 oz	90 oz	100 oz
Fat	17 oz	18 oz	21 oz

(B) Is it possible to prepare a diet consisting of 100 ounces of protein and 22 ounces of fat? Of 80 ounces of protein and 15 ounces of fat? Explain.

70. Education. A state university system is planning to hire new faculty at the rank of lecturer or instructor for several of its two-year community colleges. The number of sections taught and the annual salary (in thousands of dollars) for each rank are given in the table.

	Rank	
	Lecturer	Instructor
Sections taught	3	4
Annual salary (thousand $)	20	25

The number of sections taught by new faculty and the amount budgeted for salaries (in thousands of dollars) at each of the colleges are given in the following table. How many faculty of each rank should be hired at each college to exactly meet the demand for sections and completely exhaust the salary budget?

	Community College		
	1	2	3
Demand for sections	30	33	35
Salary budget (thousand $)	200	210	220

Answers to Matched Problems

1.
$$\begin{aligned} AX + C &= B \\ (AX + C) - C &= B - C \\ AX + (C - C) &= B - C \\ AX + 0 &= B - C \\ AX &= B - C \\ A^{-1}(AX) &= A^{-1}(B - C) \\ (A^{-1}A)X &= A^{-1}(B - C) \\ IX &= A^{-1}(B - C) \\ X &= A^{-1}(B - C) \end{aligned}$$

2. $x_1 = 2, x_2 = 5, x_3 = 0$

3. (A) $x_1 = -2, x_2 = -5, x_3 = 4$

(B) $x_1 = 0, x_2 = 1, x_3 = -4$

4. $A^{-1} = \begin{bmatrix} 1.5 & -12.5 \\ -0.5 & 12.5 \end{bmatrix}$; client 1: $15,000 in *A* and $5,000 in *B*; client 2: $28,125 in *A* and $21,875 in *B*; client 3: $3,750 in *A* and $6,250 in *B*

4.7 Leontief Input–Output Analysis

- Two-Industry Model
- Three-Industry Model

An important application of matrices and their inverses is **input–output analysis**. Wassily Leontief (1905–1999), the primary force behind this subject, was awarded the Nobel Prize in economics in 1973 because of the significant impact his work had on economic planning for industrialized countries. Among other things, he conducted a comprehensive study of how 500 sectors of the U.S. economy interacted with each other. Of course, large-scale computers played a crucial role in this analysis.

Our investigation will be more modest. In fact, we start with an economy comprised of only two industries. From these humble beginnings, ideas and definitions will evolve that can be readily generalized for more realistic economies. Input–output analysis attempts to establish equilibrium conditions under which industries in an economy have just enough output to satisfy each other's demands in addition to final (outside) demands.

Two-Industry Model

We start with an economy comprised of only two industries, electric company E and water company W. Output for both companies is measured in dollars. The electric company uses both electricity and water (inputs) in the production of electricity (output), and the water company uses both electricity and water (inputs) in the production of water (output). Suppose that the production of each dollar's worth of electricity requires \$0.30 worth of electricity and \$0.10 worth of water, and the production of each dollar's worth of water requires \$0.20 worth of electricity and \$0.40 worth of water. If the final demand (the demand from all other users of electricity and water) is

$$d_1 = \$12 \text{ million for electricity}$$

$$d_2 = \$8 \text{ million for water}$$

how much electricity and water should be produced to meet this final demand?

To begin, suppose that the electric company produces \$12 million worth of electricity and the water company produces \$8 million worth of water. Then the production processes of the companies would require

$$\underset{\substack{\text{Electricity}\\\text{required to}\\\text{produce}\\\text{electricity}}}{0.3(12)} + \underset{\substack{\text{Electricity}\\\text{required to}\\\text{produce}\\\text{water}}}{0.2(8)} = \$5.2 \text{ million of electricity}$$

and

$$\underset{\substack{\text{Water}\\\text{required to}\\\text{produce}\\\text{electricity}}}{0.1(12)} + \underset{\substack{\text{Water}\\\text{required to}\\\text{produce}\\\text{water}}}{0.4(8)} = \$4.4 \text{ million of water}$$

leaving only \$6.8 million of electricity and \$3.6 million of water to satisfy the final demand. To meet the internal demands of both companies and to end up with enough electricity for the final outside demand, both companies must produce more than just the final demand. In fact, they must produce exactly enough to meet their own internal demands plus the final demand. To determine the total output that each company must produce, we set up a system of equations.

If

$$x_1 = \text{total output from electric company}$$

$$x_2 = \text{total output from water company}$$

then, reasoning as before, the internal demands are

$$0.3x_1 + 0.2x_2 \quad \text{Internal demand for electricity}$$

$$0.1x_1 + 0.4x_2 \quad \text{Internal demand for water}$$

Combining the internal demand with the final demand produces the following system of equations:

$$\begin{aligned} \underset{\substack{\text{Total}\\\text{output}}}{x_1} &= \underset{\substack{\text{Internal}\\\text{demand}}}{0.3x_1 + 0.2x_2} + \underset{\substack{\text{Final}\\\text{demand}}}{d_1} \\ x_2 &= 0.1x_1 + 0.4x_2 + d_2 \end{aligned} \qquad (1)$$

or, in matrix form,

$$\begin{bmatrix} x_1 \\ x_2 \end{bmatrix} = \begin{bmatrix} 0.3 & 0.2 \\ 0.1 & 0.4 \end{bmatrix} \begin{bmatrix} x_1 \\ x_2 \end{bmatrix} + \begin{bmatrix} d_1 \\ d_2 \end{bmatrix}$$

or

$$X = MX + D \qquad (2)$$

where

$$D = \begin{bmatrix} d_1 \\ d_2 \end{bmatrix} \quad \text{Final demand matrix}$$

$$X = \begin{bmatrix} x_1 \\ x_2 \end{bmatrix} \quad \text{Output matrix}$$

$$M = \begin{matrix} & \begin{matrix} E & W \end{matrix} \\ \begin{matrix} E \\ W \end{matrix} & \begin{bmatrix} 0.3 & 0.2 \\ 0.1 & 0.4 \end{bmatrix} \end{matrix} \quad \text{Technology matrix}$$

The **technology matrix** is the heart of input–output analysis. The elements in the technology matrix are determined as follows (read from left to right and then up):

$$\begin{matrix} & \text{Output} & \\ & E & W \\ & \uparrow & \uparrow \\ \text{Input} \quad \begin{matrix} E \rightarrow \\ \\ W \rightarrow \end{matrix} & \begin{bmatrix} \begin{pmatrix} \text{input from } E \\ \text{to produce \$1} \\ \text{of electricity} \end{pmatrix} & \begin{pmatrix} \text{input from } E \\ \text{to produce \$1} \\ \text{of water} \end{pmatrix} \\ \begin{pmatrix} \text{input from } W \\ \text{to produce \$1} \\ \text{of electricity} \end{pmatrix} & \begin{pmatrix} \text{input from } W \\ \text{to produce \$1} \\ \text{of water} \end{pmatrix} \end{bmatrix} & = M \end{matrix}$$

CONCEPTUAL INSIGHT

Labeling the rows and columns of the technology matrix with the first letter of each industry is an important part of the process. The same order must be used for columns as for rows, and that same order must be used for the entries of D (the final demand matrix) and the entries of X (the output matrix). In this book we normally label the rows and columns in alphabetical order.

Now we solve equation (2) for X. We proceed as in Section 4.6:

$$X = MX + D$$

$$X - MX = D$$

$$IX - MX = D \qquad I = \begin{bmatrix} 1 & 0 \\ 0 & 1 \end{bmatrix}$$

$$(I - M)X = D$$

$$X = (I - M)^{-1}D \qquad \text{Assuming } I - M \text{ has an inverse} \tag{3}$$

Omitting the details of the calculations, we find

$$I - M = \begin{bmatrix} 0.7 & -0.2 \\ -0.1 & 0.6 \end{bmatrix} \quad \text{and} \quad (I - M)^{-1} = \begin{bmatrix} 1.5 & 0.5 \\ 0.25 & 1.75 \end{bmatrix}$$

Then we have

$$\begin{bmatrix} x_1 \\ x_2 \end{bmatrix} = \begin{bmatrix} 1.5 & 0.5 \\ 0.25 & 1.75 \end{bmatrix} \begin{bmatrix} d_1 \\ d_2 \end{bmatrix} = \begin{bmatrix} 1.5 & 0.5 \\ 0.25 & 1.75 \end{bmatrix} \begin{bmatrix} 12 \\ 8 \end{bmatrix} = \begin{bmatrix} 22 \\ 17 \end{bmatrix} \tag{4}$$

Therefore, the electric company must produce an output of \$22 million and the water company must produce an output of \$17 million so that each company can meet both internal and final demands.

CHECK We use equation (2) to check our work:

$$X = MX + D$$

$$\begin{bmatrix} 22 \\ 17 \end{bmatrix} \stackrel{?}{=} \begin{bmatrix} 0.3 & 0.2 \\ 0.1 & 0.4 \end{bmatrix} \begin{bmatrix} 22 \\ 17 \end{bmatrix} + \begin{bmatrix} 12 \\ 8 \end{bmatrix}$$

$$\begin{bmatrix} 22 \\ 17 \end{bmatrix} \stackrel{?}{=} \begin{bmatrix} 10 \\ 9 \end{bmatrix} + \begin{bmatrix} 12 \\ 8 \end{bmatrix}$$

$$\begin{bmatrix} 22 \\ 17 \end{bmatrix} \stackrel{\checkmark}{=} \begin{bmatrix} 22 \\ 17 \end{bmatrix}$$

To solve this input–output problem on a graphing calculator, simply store matrices M, D, and I in memory; then use equation (3) to find X and equation (2) to check your results. Figure 1 illustrates this process on a graphing calculator.

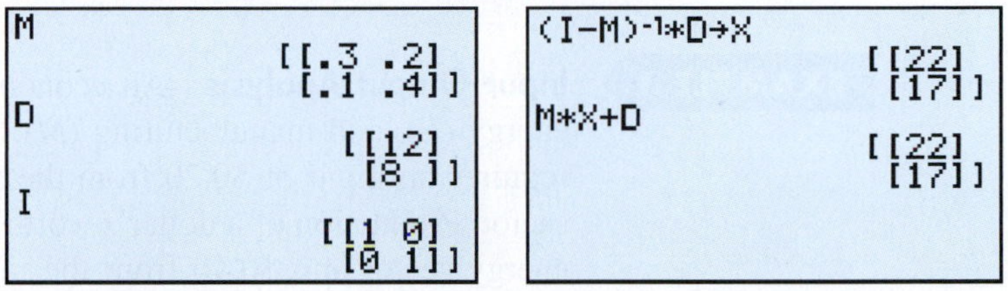

(A) Store M, D, and I in the graphing calculator's memory (B) Compute X and check in equation (2)

Figure 1

Actually, equation (4) solves the original problem for arbitrary final demands d_1 and d_2. This is very useful, since equation (4) gives a quick solution not only for the final demands stated in the original problem but also for various other projected final demands. If we had solved system (1) by Gauss–Jordan elimination, then we would have to start over for each new set of final demands.

Suppose that in the original problem the projected final demands 5 years from now are $d_1 = 24$ and $d_2 = 16$. To determine each company's output for this projection, we simply substitute these values into equation (4) and multiply:

$$\begin{bmatrix} x_1 \\ x_2 \end{bmatrix} = \begin{bmatrix} 1.5 & 0.5 \\ 0.25 & 1.75 \end{bmatrix} \begin{bmatrix} 24 \\ 16 \end{bmatrix} = \begin{bmatrix} 44 \\ 34 \end{bmatrix}$$

We summarize these results for convenient reference.

SUMMARY Solution to a Two-Industry Input–Output Problem

Given two industries, C_1 and C_2, with

$$\begin{array}{cccc} & \text{Technology matrix} & \text{Output matrix} & \text{Final demand matrix} \\ & \quad C_1 \quad C_2 & & \\ M = \begin{matrix} C_1 \\ C_2 \end{matrix} & \begin{bmatrix} a_{11} & a_{12} \\ a_{21} & a_{22} \end{bmatrix} & X = \begin{bmatrix} x_1 \\ x_2 \end{bmatrix} & D = \begin{bmatrix} d_1 \\ d_2 \end{bmatrix} \end{array}$$

where a_{ij} is the input required from C_i to produce a dollar's worth of output for C_j, the solution to the input–output matrix equation

$$\begin{array}{ccccc} \text{Total output} & & \text{Internal demand} & & \text{Final demand} \\ X & = & MX & + & D \end{array}$$

is

$$X = (I - M)^{-1}D \tag{3}$$

assuming that $I - M$ has an inverse.

Three-Industry Model

Equations (2) and (3) in the solution to a two-industry input–output problem are the same for a three-industry economy, a four-industry economy, or an economy with n industries (where n is any natural number). The steps we took going from equation (2) to equation (3) hold for arbitrary matrices as long as the matrices have the correct sizes and $(I - M)^{-1}$ exists.

Explore and Discuss 1

If equations (2) and (3) are valid for an economy with n industries, discuss the size of all the matrices in each equation.

The next example illustrates the application of equations (2) and (3) to a three-industry economy.

EXAMPLE 1 **Input–Output Analysis** An economy is based on three sectors, agriculture (A), energy (E), and manufacturing (M). Production of a dollar's worth of agriculture requires an input of \$0.20 from the agriculture sector and \$0.40 from the energy sector. Production of a dollar's worth of energy requires an input of \$0.20 from the energy sector and \$0.40 from the manufacturing sector. Production of a dollar's worth of manufacturing requires an input of \$0.10 from the agriculture sector, \$0.10 from the energy sector, and \$0.30 from the manufacturing sector. Find the output from each sector that is needed to satisfy a final demand of \$20 billion for agriculture, \$10 billion for energy, and \$30 billion for manufacturing.

SOLUTION Since this is a three-industry problem, the technology matrix will be a 3×3 matrix, and the output and final demand matrices will be 3×1 column matrices.

To begin, we form a blank 3×3 technology matrix and label the rows and columns in alphabetical order.

Technology matrix

$$\text{Input}\quad \begin{matrix} A \\ E \\ M \end{matrix}\overset{\displaystyle \text{Output} \atop \displaystyle A \quad E \quad M}{\begin{bmatrix} & & \\ & & \\ & & \end{bmatrix}} = M$$

Now we analyze the production information given in the problem, beginning with agriculture.

> **"Production of a dollar's worth of agriculture requires an input of \$0.20 from the agriculture sector and \$0.40 from the energy sector."**

We organize this information in a table and then insert it in the technology matrix. Since manufacturing is not mentioned in the agriculture production information, the input from manufacturing is \$0.

Agriculture

$$\begin{matrix} \text{Input} & & \text{Output} \\ A & \xrightarrow{0.2} & A \\ E & \xrightarrow{0.4} & A \\ M & \xrightarrow{0} & A \end{matrix} \qquad \begin{matrix} A \\ E \\ M \end{matrix}\overset{\displaystyle A \quad E \quad M}{\begin{bmatrix} 0.2 & & \\ 0.4 & & \\ 0 & & \end{bmatrix}}$$

"Production of a dollar's worth of energy requires an input of $0.20 from the energy sector and $0.40 from the manufacturing sector."

$$\begin{array}{ccc} \textbf{Energy} & & \\ \textbf{Input} & & \textbf{Output} \\ A & \xrightarrow{0} & E \\ E & \xrightarrow{0.2} & E \\ M & \xrightarrow{0.4} & E \end{array} \qquad \begin{array}{c} \\ A \\ E \\ M \end{array} \begin{array}{c} \begin{matrix} A & E & M \end{matrix} \\ \begin{bmatrix} 0.2 & 0 & \\ 0.4 & 0.2 & \\ 0 & 0.4 & \end{bmatrix} \end{array}$$

"Production of a dollar's worth of manufacturing requires an input of $0.10 from the agriculture sector, $0.10 from the energy sector, and $0.30 from the manufacturing sector."

$$\begin{array}{ccc} \textbf{Manufacturing} & & \\ \textbf{Input} & & \textbf{Output} \\ A & \xrightarrow{0.1} & M \\ E & \xrightarrow{0.1} & M \\ M & \xrightarrow{0.3} & M \end{array} \qquad \begin{array}{c} \\ A \\ E \\ M \end{array} \begin{array}{c} \begin{matrix} A & E & M \end{matrix} \\ \begin{bmatrix} 0.2 & 0 & 0.1 \\ 0.4 & 0.2 & 0.1 \\ 0 & 0.4 & 0.3 \end{bmatrix} \end{array}$$

Therefore,

$$\begin{array}{c}\text{Technology matrix}\\ M = \begin{array}{c} \\ A \\ E \\ M \end{array} \begin{array}{c} \begin{matrix} A & E & M \end{matrix} \\ \begin{bmatrix} 0.2 & 0 & 0.1 \\ 0.4 & 0.2 & 0.1 \\ 0 & 0.4 & 0.3 \end{bmatrix} \end{array}\end{array} \quad \begin{array}{c}\text{Final demand}\\ \text{matrix}\\ D = \begin{bmatrix} 20 \\ 10 \\ 30 \end{bmatrix}\end{array}, \quad \begin{array}{c}\text{Output}\\ \text{matrix}\\ X = \begin{bmatrix} x_1 \\ x_2 \\ x_3 \end{bmatrix}\end{array}$$

where M, X, and D satisfy the input–output equation $X = MX + D$. Since the solution to this equation is $X = (I - M)^{-1}D$, we must first find $I - M$ and then $(I - M)^{-1}$. Omitting the details of the calculations, we have

$$I - M = \begin{bmatrix} 0.8 & 0 & -0.1 \\ -0.4 & 0.8 & -0.1 \\ 0 & -0.4 & 0.7 \end{bmatrix}$$

and

$$(I - M)^{-1} = \begin{bmatrix} 1.3 & 0.1 & 0.2 \\ 0.7 & 1.4 & 0.3 \\ 0.4 & 0.8 & 1.6 \end{bmatrix}$$

So the output matrix X is given by

$$\overset{X}{\begin{bmatrix} x_1 \\ x_2 \\ x_3 \end{bmatrix}} = \overset{(I - M)^{-1}}{\begin{bmatrix} 1.3 & 0.1 & 0.2 \\ 0.7 & 1.4 & 0.3 \\ 0.4 & 0.8 & 1.6 \end{bmatrix}} \overset{D}{\begin{bmatrix} 20 \\ 10 \\ 30 \end{bmatrix}} = \begin{bmatrix} 33 \\ 37 \\ 64 \end{bmatrix}$$

An output of $33 billion for agriculture, $37 billion for energy, and $64 billion for manufacturing will meet the given final demands. You should check this result in equation (2).

Figure 2 illustrates a spreadsheet solution for Example 1.

	A	B	C	D	E	F	G	H	I	J	K	L	M
1		Technology Matrix M								Final		Output	
2		A	E	M			I-M			Demand			
3	A	0.2	0	0.1		0.8	0	-0.1		20		33	
4	E	0.4	0.2	0.1		-0.4	0.8	-0.1		10		37	
5	M	0	0.4	0.3		0	-0.4	0.7		30		64	
6													

Figure 2 The command MMULT(MINVERSE(F3:H5), J3:J5) produces the output in L3:L5

Matched Problem 1 An economy is based on three sectors, coal, oil, and transportation. Production of a dollar's worth of coal requires an input of \$0.20 from the coal sector and \$0.40 from the transportation sector. Production of a dollar's worth of oil requires an input of \$0.10 from the oil sector and \$0.20 from the transportation sector. Production of a dollar's worth of transportation requires an input of \$0.40 from the coal sector, \$0.20 from the oil sector, and \$0.20 from the transportation sector.

(A) Find the technology matrix M.

(B) Find $(I - M)^{-1}$.

(C) Find the output from each sector that is needed to satisfy a final demand of \$30 billion for coal, \$10 billion for oil, and \$20 billion for transportation.

Exercises 4.7

Skills Warm-up Exercises

W *In Problems 1–8, solve each equation for x, where x represents a real number. (If necessary, review Section 1.1.)*

1. $x = 3x + 6$ **2.** $x = 4x - 5$

3. $x = 0.9x + 10$ **4.** $x = 0.6x + 84$

5. $x = 0.2x + 3.2$ **6.** $x = 0.3x + 4.2$

7. $x = 0.68x + 2.56$ **8.** $x = 0.98x + 8.24$

A *Problems 9–14 pertain to the following input–output model: Assume that an economy is based on two industrial sectors, agriculture (A) and energy (E). The technology matrix M and final demand matrices (in billions of dollars) are*

$$\begin{array}{c} \\ A \\ E \end{array}\begin{array}{c} \begin{array}{cc} A & E \end{array} \\ \begin{bmatrix} 0.4 & 0.2 \\ 0.2 & 0.1 \end{bmatrix} \end{array} = M$$

$$D_1 = \begin{bmatrix} 6 \\ 4 \end{bmatrix} \quad D_2 = \begin{bmatrix} 8 \\ 5 \end{bmatrix} \quad D_3 = \begin{bmatrix} 12 \\ 9 \end{bmatrix}$$

9. How much input from A and E are required to produce a dollar's worth of output for A?

10. How much input from A and E are required to produce a dollar's worth of output for E?

11. Find $I - M$ and $(I - M)^{-1}$.

12. Find the output for each sector that is needed to satisfy the final demand D_1.

13. Repeat Problem 12 for D_2.

14. Repeat Problem 12 for D_3.

B *Problems 15–20 pertain to the following input–output model: Assume that an economy is based on three industrial sectors: agriculture (A), building (B), and energy (E). The technology matrix M and final demand matrices (in billions of dollars) are*

$$\begin{array}{c} \\ A \\ B \\ E \end{array}\begin{array}{c} \begin{array}{ccc} A & B & E \end{array} \\ \begin{bmatrix} 0.3 & 0.2 & 0.2 \\ 0.1 & 0.1 & 0.1 \\ 0.2 & 0.1 & 0.1 \end{bmatrix} \end{array} = M$$

$$D_1 = \begin{bmatrix} 5 \\ 10 \\ 15 \end{bmatrix} \quad D_2 = \begin{bmatrix} 20 \\ 15 \\ 10 \end{bmatrix}$$

15. How much input from A, B, and E are required to produce a dollar's worth of output for A?

16. How much of each of A's output dollars is required as input for each of the three sectors?

17. Find $I - M$.

18. Find $(I - M)^{-1}$.
Show that $(I - M)^{-1}(I - M) = I$.

19. Use $(I - M)^{-1}$ in Problem 18 to find the output for each sector that is needed to satisfy the final demand D_1.

20. Repeat Problem 19 for D_2.

In Problems 21–26, find $(I - M)^{-1}$ and X.

21. $M = \begin{bmatrix} 0.2 & 0.2 \\ 0.3 & 0.3 \end{bmatrix}$; $D = \begin{bmatrix} 10 \\ 25 \end{bmatrix}$

22. $M = \begin{bmatrix} 0.4 & 0.2 \\ 0.6 & 0.8 \end{bmatrix}$; $D = \begin{bmatrix} 30 \\ 50 \end{bmatrix}$

23. $M = \begin{bmatrix} 0.7 & 0.8 \\ 0.3 & 0.2 \end{bmatrix}$; $D = \begin{bmatrix} 25 \\ 75 \end{bmatrix}$

24. $M = \begin{bmatrix} 0.4 & 0.1 \\ 0.2 & 0.3 \end{bmatrix}$; $D = \begin{bmatrix} 15 \\ 20 \end{bmatrix}$

C **25.** $M = \begin{bmatrix} 0.3 & 0.1 & 0.3 \\ 0.2 & 0.1 & 0.2 \\ 0.1 & 0.1 & 0.1 \end{bmatrix}$; $D = \begin{bmatrix} 20 \\ 5 \\ 10 \end{bmatrix}$

26. $M = \begin{bmatrix} 0.3 & 0.2 & 0.3 \\ 0.1 & 0.1 & 0.1 \\ 0.1 & 0.2 & 0.1 \end{bmatrix}$; $D = \begin{bmatrix} 10 \\ 25 \\ 15 \end{bmatrix}$

27. The technology matrix for an economy based on agriculture (A) and manufacturing (M) is

$$M = \begin{array}{c} \\ A \\ M \end{array} \begin{array}{c} \begin{array}{cc} A & M \end{array} \\ \begin{bmatrix} 0.3 & 0.25 \\ 0.1 & 0.25 \end{bmatrix} \end{array}$$

(A) Find the output for each sector that is needed to satisfy a final demand of \$40 million for agriculture and \$40 million for manufacturing.

(B) Discuss the effect on the final demand if the agriculture output in part (A) is increased by \$20 million and manufacturing output remains unchanged.

28. The technology matrix for an economy based on energy (E) and transportation (T) is

$$M = \begin{array}{c} \\ E \\ T \end{array} \begin{array}{c} \begin{array}{cc} E & T \end{array} \\ \begin{bmatrix} 0.25 & 0.25 \\ 0.4 & 0.2 \end{bmatrix} \end{array}$$

(A) Find the output for each sector that is needed to satisfy a final demand of \$50 million for energy and \$50 million for transportation.

(B) Discuss the effect on the final demand if the transportation output in part (A) is increased by \$40 million and the energy output remains unchanged.

29. Refer to Problem 27. Fill in the elements in the following technology matrix.

$$T = \begin{array}{c} \\ M \\ A \end{array} \begin{array}{c} \begin{array}{cc} M & A \end{array} \\ \begin{bmatrix} & \\ & \end{bmatrix} \end{array}$$

Use this matrix to solve Problem 27. Discuss any differences in your calculations and in your answers.

30. Refer to Problem 28. Fill in the elements in the following technology matrix.

$$T = \begin{array}{c} \\ T \\ E \end{array} \begin{array}{c} \begin{array}{cc} T & E \end{array} \\ \begin{bmatrix} & \\ & \end{bmatrix} \end{array}$$

Use this matrix to solve Problem 28. Discuss any differences in your calculations and in your answers.

31. The technology matrix for an economy based on energy (E) and mining (M) is

$$M = \begin{array}{c} \\ E \\ M \end{array} \begin{array}{c} \begin{array}{cc} E & M \end{array} \\ \begin{bmatrix} 0.2 & 0.3 \\ 0.4 & 0.3 \end{bmatrix} \end{array}$$

The management of these two sectors would like to set the total output level so that the final demand is always 40% of the total output. Discuss methods that could be used to accomplish this objective.

32. The technology matrix for an economy based on automobiles (A) and construction (C) is

$$M = \begin{array}{c} \\ A \\ C \end{array} \begin{array}{c} \begin{array}{cc} A & C \end{array} \\ \begin{bmatrix} 0.1 & 0.4 \\ 0.1 & 0.1 \end{bmatrix} \end{array}$$

The management of these two sectors would like to set the total output level so that the final demand is always 70% of the total output. Discuss methods that could be used to accomplish this objective.

33. All the technology matrices in the text have elements between 0 and 1. Why is this the case? Would you ever expect to find an element in a technology matrix that is negative? That is equal to 0? That is equal to 1? That is greater than 1?

34. The sum of the elements in a column of any of the technology matrices in the text is less than 1. Why is this the case? Would you ever expect to find a column with a sum equal to 1? Greater than 1? How would you describe an economic system where the sum of the elements in every column of the technology matrix is 1?

Applications

35. Coal, steel. An economy is based on two industrial sectors, coal and steel. Production of a dollar's worth of coal requires an input of \$0.10 from the coal sector and \$0.20 from the steel sector. Production of a dollar's worth of steel requires an input of \$0.20 from the coal sector and \$0.40 from the steel sector. Find the output for each sector that is needed to satisfy a final demand of \$20 billion for coal and \$10 billion for steel.

36. Transportation, manufacturing. An economy is based on two sectors, transportation and manufacturing. Production of a dollar's worth of transportation requires an input of \$0.10 from each sector and production of a dollar's worth of manufacturing requires an input of \$0.40 from each sector. Find the output for each sector that is needed to satisfy a final demand of \$5 billion for transportation and \$20 billion for manufacturing.

37. Agriculture, tourism. The economy of a small island nation is based on two sectors, agriculture and tourism. Production of a dollar's worth of agriculture requires an input of \$0.20 from agriculture and \$0.15 from tourism. Production of a dollar's worth of tourism requires an input of \$0.40 from agriculture and \$0.30 from tourism. Find the output from each sector that is needed to satisfy a final demand of \$60 million for agriculture and \$80 million for tourism.

38. Agriculture, oil. The economy of a country is based on two sectors, agriculture and oil. Production of a dollar's worth of agriculture requires an input of \$0.40 from agriculture and \$0.35 from oil. Production of a dollar's worth of oil requires an input of $0.20 from agriculture and $0.05 from oil. Find the output from each sector that is needed to satisfy a final demand of $40 million for agriculture and $250 million for oil.

39. Agriculture, manufacturing, energy. An economy is based on three sectors, agriculture, manufacturing, and energy. Production of a dollar's worth of agriculture requires inputs of $0.20 from agriculture, $0.20 from manufacturing, and $0.20 from energy. Production of a dollar's worth of manufacturing requires inputs of $0.40 from agriculture, $0.10 from manufacturing, and $0.10 from energy. Production of a dollar's worth of energy requires inputs of $0.30 from agriculture, $0.10 from manufacturing, and $0.10 from energy. Find the output for each sector that is needed to satisfy a final demand of $10 billion for agriculture, $15 billion for manufacturing, and $20 billion for energy.

40. Electricity, natural gas, oil. A large energy company produces electricity, natural gas, and oil. The production of a dollar's worth of electricity requires inputs of $0.30 from electricity, $0.10 from natural gas, and $0.20 from oil. Production of a dollar's worth of natural gas requires inputs of $0.30 from electricity, $0.10 from natural gas, and $0.20 from oil. Production of a dollar's worth of oil requires inputs of $0.10 from each sector. Find the output for each sector that is needed to satisfy a final demand of $25 billion for electricity, $15 billion for natural gas, and $20 billion for oil.

41. Four sectors. An economy is based on four sectors, agriculture (A), energy (E), labor (L), and manufacturing (M). The table gives the input requirements for a dollar's worth of output for each sector, along with the projected final demand (in billions of dollars) for a 3-year period. Find the output for each sector that is needed to satisfy each of these final demands. Round answers to the nearest billion dollars.

		Output				Final Demand		
		A	E	L	M	1	2	3
Input	A	0.05	0.17	0.23	0.09	23	32	55
	E	0.07	0.12	0.15	0.19	41	48	62
	L	0.25	0.08	0.03	0.32	18	21	25
	M	0.11	0.19	0.28	0.16	31	33	35

42. Repeat Problem 41 with the following table:

		Output				Final Demand		
		A	E	L	M	1	2	3
Input	A	0.07	0.09	0.27	0.12	18	22	37
	E	0.14	0.07	0.21	0.24	26	31	42
	L	0.17	0.06	0.02	0.21	12	19	28
	M	0.15	0.13	0.31	0.19	41	45	49

Answers to Matched Problems

1. (A) $\begin{bmatrix} 0.2 & 0 & 0.4 \\ 0 & 0.1 & 0.2 \\ 0.4 & 0.2 & 0.2 \end{bmatrix}$ (B) $\begin{bmatrix} 1.7 & 0.2 & 0.9 \\ 0.2 & 1.2 & 0.4 \\ 0.9 & 0.4 & 1.8 \end{bmatrix}$

(C) $71 billion for coal, $26 billion for oil, and $67 billion for transportation

Chapter 4 Summary and Review

Important Terms, Symbols, and Concepts

4.1 Review: Systems of Linear Equations in Two Variables

EXAMPLES

- The **solution** of a system is an ordered pair of real numbers that satisfies each equation in the system. Solution by **graphing** is one method that can be used to find a solution. Ex. 1, p. 175 Ex. 2, p. 176
- A linear system is **consistent** and **independent** if it has a unique solution, **consistent** and **dependent** if it has more than one solution, and **inconsistent** if it has no solution. A linear system that is consistent and dependent actually has an infinite number of solutions.
- A **graphing calculator** provides accurate solutions to a linear system. Ex. 3, p. 178
- The **substitution** method can also be used to solve linear systems. Ex. 4, p. 179
- The **method of elimination by addition** is easily extended to larger systems. Ex. 5, p. 180

4.2 Systems of Linear Equations and Augmented Matrices

- A **matrix** is a rectangular array of real numbers. **Row operations** performed on an **augmented matrix** produce equivalent systems (Theorem 1, page 190). Ex. 1, p. 191 Ex. 2, p. 193
- There are only three possible final forms for the augmented matrix for a linear system of two equations in two variables (p. 195). Ex. 3, p. 193 Ex. 4, p. 195

4.3 Gauss–Jordan Elimination

- There are many possibilities for the final **reduced form** of the augmented matrix of a larger system of linear equations. Reduced form is defined on page 198.
- The **Gauss–Jordan elimination procedure** is described on pages 200 and 201.

Ex. 1, p. 198
Ex. 2, p. 199
Ex. 3, p. 201
Ex. 4, p. 202
Ex. 5, p. 203

4.4 Matrices: Basic Operations

- Two matrices are **equal** if they are the same size and their corresponding elements are equal. The **sum** of two matrices of the same size is the matrix with elements that are the sum of the corresponding elements of the two given matrices. Ex. 1, p. 211
- The **negative of a matrix** is the matrix with elements that are the negatives of the given matrix. If A and B are matrices of the same size, then B can be subtracted from A by adding the negative of B to A. Ex. 2, p. 212
- Matrix equations involving addition and subtraction are solved much like real number equations. Ex. 3, p. 212
- The product of a real number k and a matrix M is the matrix formed by multiplying each element of M by k. Ex. 4, p. 213
- The product of a row matrix and a column matrix is defined on page 214. Ex. 6, p. 214
- The matrix product of an $m \times p$ matrix with a $p \times n$ is defined on page 215. Ex. 8, p. 216

4.5 Inverse of a Square Matrix

- The **identity matrix** for multiplication is defined on page 224.
- The **inverse** of a square matrix is defined on page 226.

Ex. 1, p. 224
Ex. 2, p. 227
Ex. 3, p. 230
Ex. 4, p. 231

4.6 Matrix Equations and Systems of Linear Equations

- Basic properties of matrices are summarized in Theorem 1 on page 236.
- **Matrix inverse methods** for solving systems of equations are described in the Summary on page 239.

Ex. 1, p. 237
Ex. 2, p. 237
Ex. 3, p. 238

4.7 Leontief Input–Output Analysis

- Leontief **input–output** analysis is summarized on page 247. Ex. 1, p. 248

Review Exercises

Work through all the problems in this chapter review and check your answers in the back of the book. Answers to all problems are there along with section numbers in italics to indicate where each type of problem is discussed. Where weaknesses show up, review appropriate sections in the text.

1. Solve the following system by graphing:

$$\begin{aligned} 2x - y &= 4 \\ x - 2y &= -4 \end{aligned}$$

2. Solve the system in Problem 1 by substitution.

3. If a matrix is in reduced form, say so. If not, explain why and state the row operation(s) necessary to transform the matrix into reduced form.

(A) $\left[\begin{array}{cc|c} 0 & 1 & 2 \\ 1 & 0 & 3 \end{array}\right]$ (B) $\left[\begin{array}{cc|c} 1 & 0 & 2 \\ 0 & 3 & 3 \end{array}\right]$

(C) $\left[\begin{array}{ccc|c} 1 & 0 & 1 & 2 \\ 0 & 1 & 1 & 3 \end{array}\right]$ (D) $\left[\begin{array}{ccc|c} 1 & 1 & 0 & 2 \\ 0 & 1 & 1 & 3 \end{array}\right]$

4. Given matrices A and B,

$$A = \begin{bmatrix} 5 & 3 & -1 & 0 & 2 \\ -4 & 8 & 1 & 3 & 0 \end{bmatrix} \qquad B = \begin{bmatrix} -3 & 2 \\ 0 & 4 \\ -1 & 7 \end{bmatrix}$$

(A) What is the size of A? Of B?

(B) Find a_{24}, a_{15}, b_{31}, and b_{22}.

(C) Is AB defined? Is BA defined?

5. Find x_1 and x_2:

(A) $\begin{bmatrix} 1 & -2 \\ 1 & -3 \end{bmatrix} \begin{bmatrix} x_1 \\ x_2 \end{bmatrix} = \begin{bmatrix} 4 \\ 2 \end{bmatrix}$

(B) $\begin{bmatrix} 5 & 3 \\ 1 & 1 \end{bmatrix} \begin{bmatrix} x_1 \\ x_2 \end{bmatrix} + \begin{bmatrix} 25 \\ 14 \end{bmatrix} = \begin{bmatrix} 18 \\ 22 \end{bmatrix}$

In Problems 6–14, perform the operations that are defined, given the following matrices:

$$A = \begin{bmatrix} 1 & 2 \\ 3 & 1 \end{bmatrix} \quad B = \begin{bmatrix} 2 & 1 \\ 1 & 1 \end{bmatrix} \quad C = [2 \quad 3] \quad D = \begin{bmatrix} 1 \\ 2 \end{bmatrix}$$

6. $A + B$ **7.** $B + D$

8. $A - 2B$ **9.** AB

10. AC **11.** AD

12. DC **13.** CD

14. $C + D$

15. Find the inverse of the matrix A given below by appropriate row operations on $[A \mid I]$. Show that $A^{-1}A = I$.

$$A = \begin{bmatrix} 4 & 3 \\ 3 & 2 \end{bmatrix}$$

16. Solve the following system using elimination by addition:

$$\begin{aligned} 4x_1 + 3x_2 &= 3 \\ 3x_1 + 2x_2 &= 5 \end{aligned}$$

17. Solve the system in Problem 16 by performing appropriate row operations on the augmented matrix of the system.

18. Solve the system in Problem 16 by writing the system as a matrix equation and using the inverse of the coefficient matrix (see Problem 15). Also, solve the system if the constants 3 and 5 are replaced by 7 and 10, respectively. By 4 and 2, respectively.

In Problems 19–24, perform the operations that are defined, given the following matrices:

$$A = \begin{bmatrix} 2 & -2 \\ 1 & 0 \\ 3 & 2 \end{bmatrix} \quad B = \begin{bmatrix} -1 \\ 2 \\ 3 \end{bmatrix} \quad C = [2 \quad 1 \quad 3]$$

$$D = \begin{bmatrix} 3 & -2 & 1 \\ -1 & 1 & 2 \end{bmatrix} \quad E = \begin{bmatrix} 3 & -4 \\ -1 & 0 \end{bmatrix}$$

19. $A + D$ **20.** $E + DA$

21. $DA - 3E$ **22.** BC

23. CB **24.** $AD - BC$

25. Find the inverse of the matrix A given below by appropriate row operations on $[A \mid I]$. Show that $A^{-1}A = I$.

$$A = \begin{bmatrix} 1 & 2 & 3 \\ 2 & 3 & 4 \\ 1 & 2 & 1 \end{bmatrix}$$

26. Solve by Gauss–Jordan elimination:

(A)
$$\begin{aligned} x_1 + 2x_2 + 3x_3 &= 1 \\ 2x_1 + 3x_2 + 4x_3 &= 3 \\ x_1 + 2x_2 + x_3 &= 3 \end{aligned}$$

(B)
$$\begin{aligned} x_1 + 2x_2 - x_3 &= 2 \\ 2x_1 + 3x_2 + x_3 &= -3 \\ 3x_1 + 5x_2 \quad &= -1 \end{aligned}$$

(C)
$$\begin{aligned} x_1 + x_2 + x_3 &= 8 \\ 3x_1 + 2x_2 + 4x_3 &= 21 \end{aligned}$$

27. Solve the system in Problem 26A by writing the system as a matrix equation and using the inverse of the coefficient matrix (see Problem 25). Also, solve the system if the constants 1, 3, and 3 are replaced by 0, 0, and −2, respectively. By −3, −4, and 1, respectively.

28. Discuss the relationship between the number of solutions of the following system and the constant k.

$$\begin{aligned} 2x_1 - 6x_2 &= 4 \\ -x_1 + kx_2 &= -2 \end{aligned}$$

29. An economy is based on two sectors, agriculture and energy. Given the technology matrix M and the final demand matrix D (in billions of dollars), find $(I - M)^{-1}$ and the output matrix X:

$$M = \begin{matrix} & \begin{matrix} A & E \end{matrix} \\ \begin{matrix} A \\ E \end{matrix} & \begin{bmatrix} 0.2 & 0.15 \\ 0.4 & 0.3 \end{bmatrix} \end{matrix} \qquad D = \begin{matrix} A \\ E \end{matrix} \begin{bmatrix} 30 \\ 20 \end{bmatrix}$$

30. Use the matrix M in Problem 29 to fill in the elements in the following technology matrix.

$$T = \begin{matrix} & \begin{matrix} E & A \end{matrix} \\ \begin{matrix} E \\ A \end{matrix} & \begin{bmatrix} \quad & \quad \\ \quad & \quad \end{bmatrix} \end{matrix}$$

Use this matrix to solve Problem 29. Discuss any differences in your calculations and in your answers.

31. An economy is based on two sectors, coal and steel. Given the technology matrix M and the final demand matrix D (in billions of dollars), find $(I - M)^{-1}$ and the output matrix X:

$$M = \begin{matrix} & \begin{matrix} C & S \end{matrix} \\ \begin{matrix} C \\ S \end{matrix} & \begin{bmatrix} 0.45 & 0.65 \\ 0.55 & 0.35 \end{bmatrix} \end{matrix} \qquad D = \begin{matrix} C \\ S \end{matrix} \begin{bmatrix} 40 \\ 10 \end{bmatrix}$$

32. Use graphical approximation techniques on a graphing calculator to find the solution of the following system to two decimal places:

$$\begin{aligned} x - 5y &= -5 \\ 2x + 3y &= 12 \end{aligned}$$

33. Find the inverse of the matrix A given below. Show that $A^{-1}A = I$.

$$A = \begin{bmatrix} 4 & 5 & 6 \\ 4 & 5 & -4 \\ 1 & 1 & 1 \end{bmatrix}$$

34. Solve the system

$$\begin{aligned} 0.04x_1 + 0.05x_2 + 0.06x_3 &= 360 \\ 0.04x_1 + 0.05x_2 - 0.04x_3 &= 120 \\ x_1 + x_2 + x_3 &= 7{,}000 \end{aligned}$$

by writing it as a matrix equation and using the inverse of the coefficient matrix. (Before starting, multiply the first two equations by 100 to eliminate decimals. Also, see Problem 33.)

35. Solve Problem 34 by Gauss–Jordan elimination.

36. Given the technology matrix M and the final demand matrix D (in billions of dollars), find $(I - M)^{-1}$ and the output matrix X:

$$M = \begin{bmatrix} 0.2 & 0 & 0.4 \\ 0.1 & 0.3 & 0.1 \\ 0 & 0.4 & 0.2 \end{bmatrix} \qquad D = \begin{bmatrix} 40 \\ 20 \\ 30 \end{bmatrix}$$

37. Discuss the number of solutions for a system of n equations in n variables if the coefficient matrix

(A) Has an inverse.

(B) Does not have an inverse.

38. Discuss the number of solutions for the system corresponding to the reduced form shown below if

(A) $m \neq 0$

(B) $m = 0$ and $n \neq 0$

(C) $m = 0$ and $n = 0$

$$\left[\begin{array}{ccc|c} 1 & 0 & -2 & 5 \\ 0 & 1 & 3 & 3 \\ 0 & 0 & m & n \end{array}\right]$$

39. One solution to the input–output equation $X = MX + D$ is given by $X = (I - M)^{-1}D$. Discuss the validity of each step in the following solutions of this equation. (Assume that all necessary inverses exist.) Are both solutions correct?

(A)
$$\begin{aligned} X &= MX + D \\ X - MX &= D \\ X(I - M) &= D \\ X &= D(I - M)^{-1} \end{aligned}$$

(B)
$$\begin{aligned} X &= MX + D \\ -D &= MX - X \\ -D &= (M - I)X \\ X &= (M - I)^{-1}(-D) \end{aligned}$$

Applications

40. **Break-even analysis.** A cookware manufacturer is preparing to market a new pasta machine. The company's fixed costs for research, development, tooling, etc., are \$243,000 and the variable costs are \$22.45 per machine. The company sells the pasta machine for \$59.95.

(A) Find the cost and revenue equations.

(B) Find the break-even point.

(C) Graph both equations in the same coordinate system and show the break-even point. Use the graph to determine the production levels that will result in a profit and in a loss.

41. **Resource allocation.** An international mining company has two mines in Voisey's Bay and Hawk Ridge. The composition of the ore from each field is given in the table. How many tons of ore from each mine should be used to obtain exactly 6 tons of nickel and 8 tons of copper?

Mine	Nickel (%)	Copper (%)
Voisey's Bay	2	4
Hawk Ridge	3	2

42. **Resource allocation.**

(A) Set up Problem 41 as a matrix equation and solve using the inverse of the coefficient matrix.

(B) Solve Problem 41 as in part (A) if 7.5 tons of nickel and 7 tons of copper are needed.

43. **Business leases.** A grain company wants to lease a fleet of 20 covered hopper railcars with a combined capacity of 108,000 cubic feet. Hoppers with three different carrying capacities are available: 3,000 cubic feet, 4,500 cubic feet, and 6,000 cubic feet.

(A) How many of each type of hopper should they lease?

(B) The monthly rates for leasing these hoppers are \$180 for 3,000 cubic feet, \$225 for 4,500 cubic feet, and \$325 for 6,000 cubic feet. Which of the solutions in part (A) would minimize the monthly leasing costs?

44. **Material costs.** A manufacturer wishes to make two different bronze alloys in a metal foundry. The quantities of copper, tin, and zinc needed are indicated in matrix M. The costs for these materials (in dollars per pound) from two suppliers are summarized in matrix N. The company must choose one supplier or the other.

$$\begin{array}{cc} & \begin{array}{ccc} \text{Copper} & \text{Tin} & \text{Zinc} \end{array} \\ M = & \begin{bmatrix} 4{,}800\text{ lb} & 600\text{ lb} & 300\text{ lb} \\ 6{,}000\text{ lb} & 1{,}400\text{ lb} & 700\text{ lb} \end{bmatrix} \begin{array}{l} \text{Alloy 1} \\ \text{Alloy 2} \end{array} \end{array}$$

$$\begin{array}{cc} & \begin{array}{cc} \text{Supplier A} & \text{Supplier B} \end{array} \\ N = & \begin{bmatrix} \$0.75 & \$0.70 \\ \$6.50 & \$6.70 \\ \$0.40 & \$0.50 \end{bmatrix} \begin{array}{l} \text{Copper} \\ \text{Tin} \\ \text{Zinc} \end{array} \end{array}$$

(A) Discuss possible interpretations of the elements in the matrix products MN and NM.

(B) If either product MN or NM has a meaningful interpretation, find the product and label its rows and columns.

(C) Discuss methods of matrix multiplication that can be used to determine the supplier that will provide the necessary materials at the lowest cost.

45. **Labor costs.** A company with manufacturing plants in California and Texas has labor-hour and wage requirements for the manufacture of two inexpensive calculators as given in matrices M and N below:

Labor-hours per calculator

$$\begin{array}{cc} & \begin{array}{ccc} \text{Fabricating department} & \text{Assembly department} & \text{Packaging department} \end{array} \\ M = & \begin{bmatrix} 0.15\text{ hr} & 0.10\text{ hr} & 0.05\text{ hr} \\ 0.25\text{ hr} & 0.20\text{ hr} & 0.05\text{ hr} \end{bmatrix} \begin{array}{l} \text{Model A} \\ \text{Model B} \end{array} \end{array}$$

Hourly wages

$$\begin{array}{cc} & \begin{array}{cc} \text{California plant} & \text{Texas plant} \end{array} \\ N = & \begin{bmatrix} \$12 & \$10 \\ \$15 & \$12 \\ \$\ 7 & \$\ 6 \end{bmatrix} \begin{array}{l} \text{Fabricating department} \\ \text{Assembly department} \\ \text{Packaging department} \end{array} \end{array}$$

(A) Find the labor cost for producing one model B calculator at the California plant.

(B) Discuss possible interpretations of the elements in the matrix products MN and NM.

(C) If either product MN or NM has a meaningful interpretation, find the product and label its rows and columns.

46. **Investment analysis.** A person has \$5,000 to invest, part at 5% and the rest at 10%. How much should be invested at each rate to yield \$400 per year? Solve using augmented matrix methods.

47. **Investment analysis.** Solve Problem 46 by using a matrix equation and the inverse of the coefficient matrix.

48. **Investment analysis.** In Problem 46, is it possible to have an annual yield of \$200? Of \$600? Describe all possible annual yields.

49. **Ticket prices.** An outdoor amphitheater has 25,000 seats. Ticket prices are \$8, \$12, and \$20, and the number of tickets priced at \$8 must equal the number priced at \$20. How many tickets of each type should be sold (assuming that all seats can be sold) to bring in each of the returns indicated in the table? Solve using the inverse of the coefficient matrix.

	Concert		
	1	2	3
Tickets sold	25,000	25,000	25,000
Return required	\$320,000	\$330,000	\$340,000

50. **Ticket prices.** Discuss the effect on the solutions to Problem 49 if it is no longer required to have an equal number of \$8 tickets and \$20 tickets.

51. **Input–output analysis.** An economy is based on two industrial sectors, agriculture and fabrication. Production of a dollar's worth of agriculture requires an input of \$0.30 from the agriculture sector and \$0.20 from the fabrication sector. Production of a dollar's worth of fabrication requires \$0.10 from the agriculture sector and \$0.40 from the fabrication sector.

(A) Find the output for each sector that is needed to satisfy a final demand of \$50 billion for agriculture and \$20 billion for fabrication.

(B) Find the output for each sector that is needed to satisfy a final demand of \$80 billion for agriculture and \$60 billion for fabrication.

52. **Cryptography.** The following message was encoded with the matrix B shown below. Decode the message.

7	25	30	19	6	24	20	8	28	5	14	14
9	23	28	15	6	21	13	1	14	21	26	29

$$B = \begin{bmatrix} 1 & 1 & 0 \\ 1 & 0 & 1 \\ 1 & 1 & 1 \end{bmatrix}$$

53. **Traffic flow.** The rush-hour traffic flow (in vehicles per hour) for a network of four one-way streets is shown in the figure.

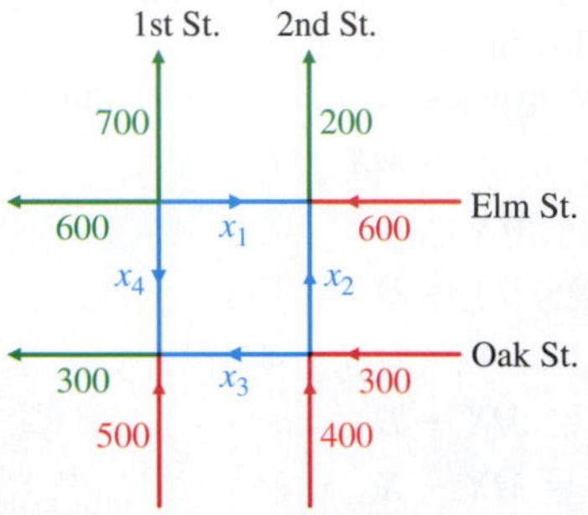

(A) Write the system of equations determined by the flow of traffic through the four intersections.

(B) Find the solution of the system in part (A).

(C) What is the maximum number of vehicles per hour that can travel from Oak Street to Elm Street on 1st Street? What is the minimum number?

(D) If traffic lights are adjusted so that 500 vehicles per hour travel from Oak Street to Elm Street on 1st Street, determine the flow around the rest of the network.

5 Linear Inequalities and Linear Programming

Introduction

Real-world problems often involve limitations on materials, time, and money. To express such constraints mathematically, we formulate systems of inequalities. In Chapter 5 we discuss systems of inequalities in two variables and introduce a relatively new mathematical tool called linear programming. Linear programming can be used in the textile industry, for example, to determine how blended yarns should be combined to produce a fabric of maximum strength (see Problems 59 and 60 in Section 5.1).

5.1 Linear Inequalities in Two Variables

- Graphing Linear Inequalities in Two Variables
- Application

Graphing Linear Inequalities in Two Variables

We know how to graph first-degree equations such as

$$y = 2x - 3 \quad \text{and} \quad 2x - 3y = 5$$

but how do we graph first-degree inequalities such as the following?

$$y \le 2x - 3 \quad \text{and} \quad 2x - 3y > 5$$

We will find that graphing these inequalities is similar to graphing the equations, but first we must discuss some important subsets of a plane in a rectangular coordinate system.

A line divides the plane into two regions called **half-planes**. A vertical line divides it into **left** and **right half-planes**; a nonvertical line divides it into **upper** and **lower half-planes**. In either case, the dividing line is called the **boundary line** of each half-plane, as indicated in Figure 1.

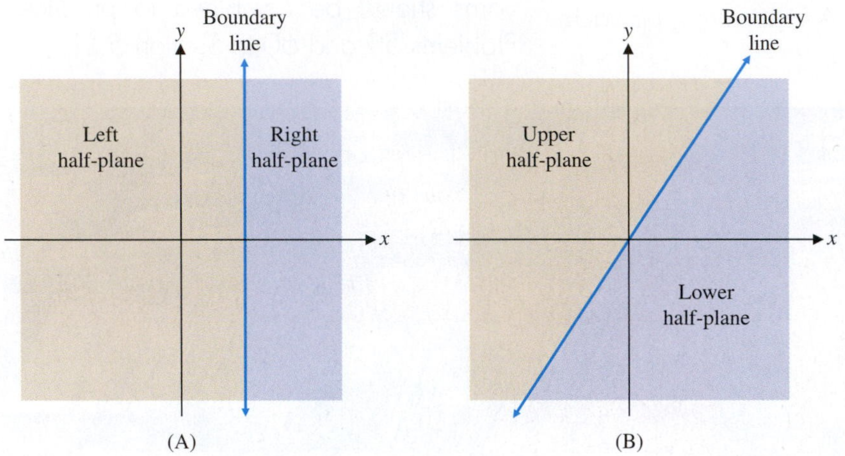

Figure 1

To find the half-planes determined by a linear equation such as $y - x = -2$, we rewrite the equation as $y = x - 2$. For any given value of x, there is exactly one value for y such that (x, y) lies on the line. For example, for $x = 4$, we have $y = 4 - 2 = 2$. For the same x and smaller values of y, the point (x, y) will lie below the line since $y < x - 2$. So the lower half-plane corresponds to the solution of the inequality $y < x - 2$. Similarly, the upper half-plane corresponds to $y > x - 2$, as shown in Figure 2.

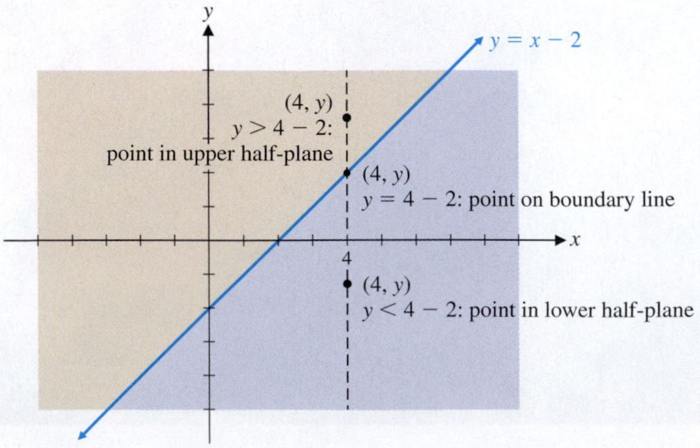

Figure 2

The four inequalities formed from $y = x - 2$, replacing the $=$ sign by $>$, $\geq$, $<$, and $\leq$, respectively, are

$$y > x - 2 \qquad y \geq x - 2 \qquad y < x - 2 \qquad y \leq x - 2$$

The graph of each is a half-plane, excluding the boundary line for $<$ and $>$ and including the boundary line for $\leq$ and $\geq$. In Figure 3, the half-planes are indicated with small arrows on the graph of $y = x - 2$ and then graphed as shaded regions. Excluded boundary lines are shown as dashed lines, and included boundary lines are shown as solid lines.

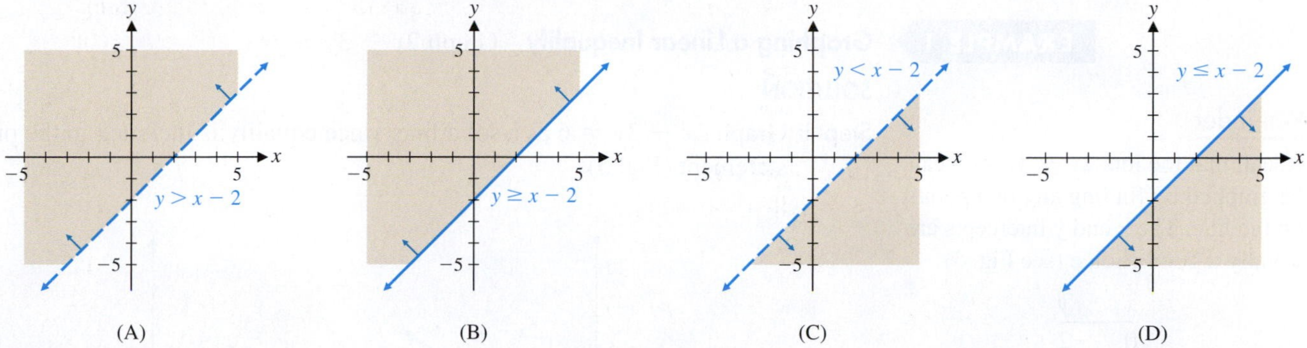

Figure 3

Figure 4 shows the graphs of Figures 3B and 3D on a graphing calculator. Note that it is impossible to show a dotted boundary line when using shading on a calculator.

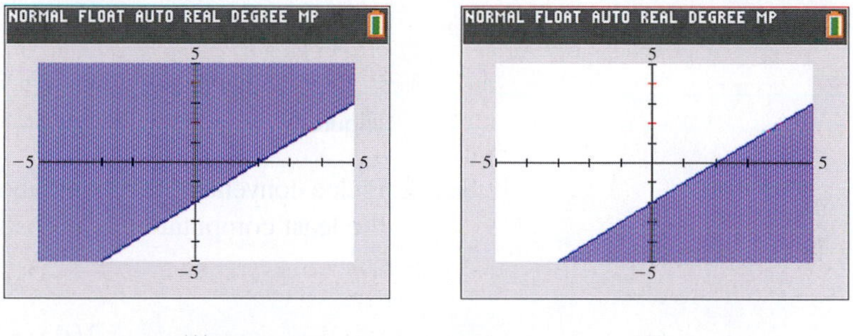

Figure 4

The preceding discussion suggests the following theorem, which is stated without proof:

THEOREM 1 Graphs of Linear Inequalities

The graph of the linear inequality

$$Ax + By < C \qquad \text{or} \qquad Ax + By > C$$

with $B \neq 0$, is either the upper half-plane or the lower half-plane (but not both) determined by the line $Ax + By = C$.

If $B = 0$ and $A \neq 0$, the graph of

$$Ax < C \qquad \text{or} \qquad Ax > C$$

is either the left half-plane or the right half-plane (but not both) determined by the line $Ax = C$.

As a consequence of this theorem, we state a simple and fast mechanical procedure for graphing linear inequalities.

PROCEDURE Graphing Linear Inequalities

Step 1 First graph $Ax + By = C$ as a dashed line if equality is not included in the original statement, or as a solid line if equality is included.

Step 2 Choose a test point anywhere in the plane not on the line [the origin $(0, 0)$ usually requires the least computation], and substitute the coordinates into the inequality.

Step 3 Does the test point satisfy the original inequality? If so, shade the half-plane that contains the test point. If not, shade the opposite half-plane.

EXAMPLE 1 **Graphing a Linear Inequality** Graph $2x - 3y \le 6$.

SOLUTION

Step 1 Graph $2x - 3y = 6$ as a solid line, since equality is included in the original statement (Fig. 5).

Reminder

Recall that the line $2x - 3y = 6$ can be graphed by finding any two points on the line. The x and y intercepts are usually a good choice (see Fig. 5).

x	y
0	-2
3	0

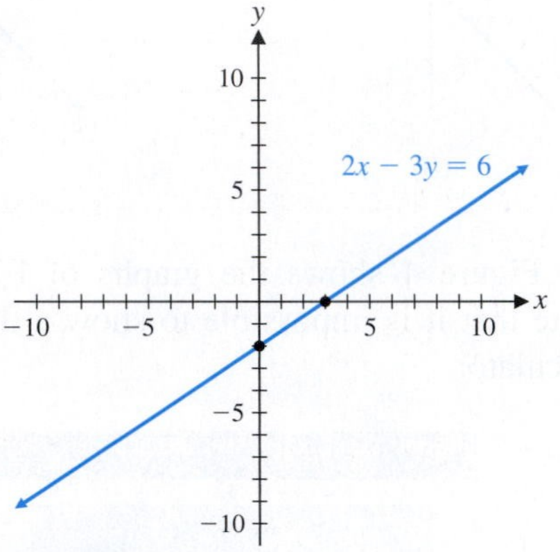

Figure 5

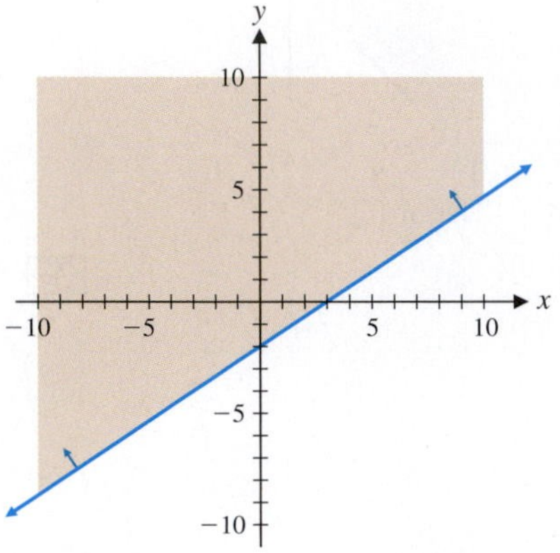

Figure 6

Step 2 Pick a convenient test point above or below the line. The origin $(0, 0)$ requires the least computation, so substituting $(0, 0)$ into the inequality, we get

$$2x - 3y \le 6$$

$$2(0) - 3(0) = 0 \le 6$$

This is a true statement; therefore, the point $(0, 0)$ is in the solution set.

Step 3 The line $2x - 3y = 6$ and the half-plane containing the origin form the graph of $2x - 3y \le 6$, as shown in Figure 6.

Matched Problem 1 Graph $6x - 3y > 18$.

Explore and Discuss 1

In Step 2 of Example 1, $(0, 0)$ was used as a test point in graphing a linear inequality. Describe those linear inequalities for which $(0, 0)$ is not a valid test point. In that case, how would you choose a test point to make calculation easy?

EXAMPLE 2 **Graphing Inequalities** Graph

(A) $y > -3$ (B) $2x \le 5$ (C) $x \le 3y$

SOLUTION

(A) Step 1 Graph the horizontal line $y = -3$ as a dashed line, since equality is not included in the original statement (Fig. 7).

Step 2 Substituting $x = 0$ and $y = 0$ in the inequality produces a true statement, so the point $(0, 0)$ is in the solution set.

Step 3 The graph of the solution set is the upper half-plane, excluding the boundary line (Fig. 8).

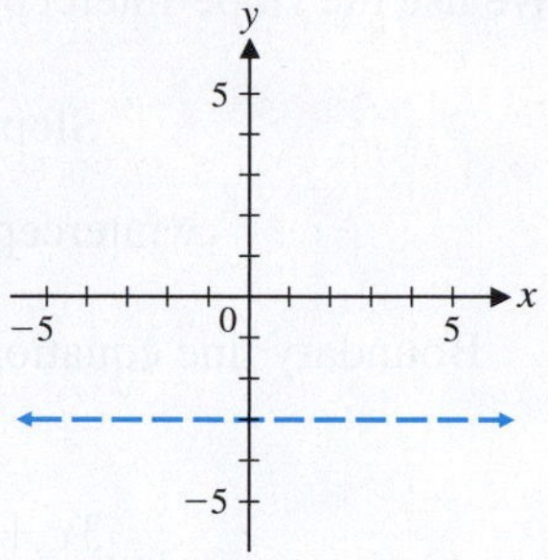

Figure 7

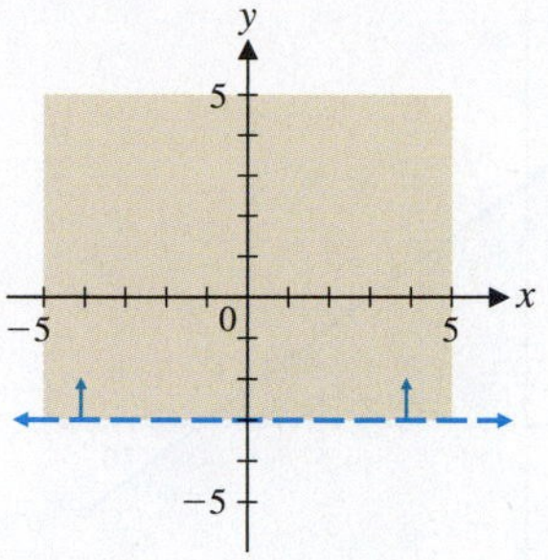

Figure 8

(B) Step 1 Graph the vertical line $2x = 5$ as a solid line, since equality is included in the original statement (Fig. 9).

Step 2 Substituting $x = 0$ and $y = 0$ in the inequality produces a true statement, so the point $(0, 0)$ is in the solution set.

Step 3 The graph of the solution set is the left half-plane, including the boundary line (Fig. 10).

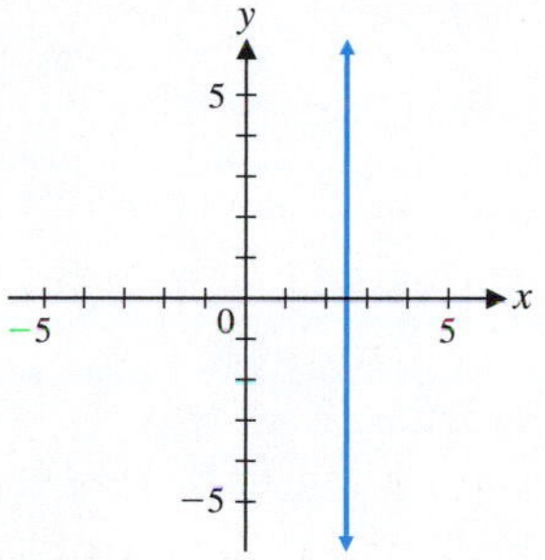

Figure 9

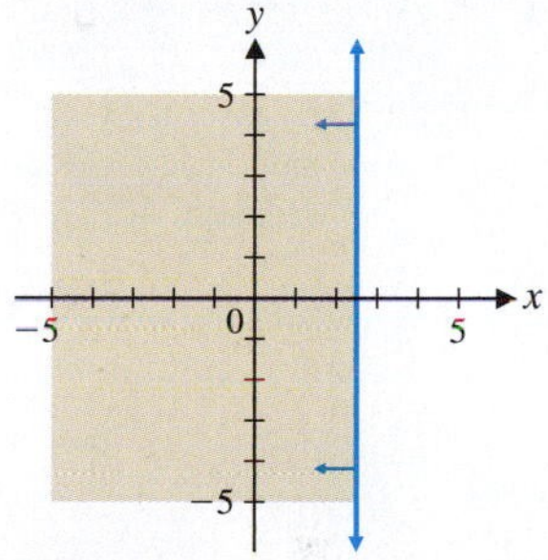

Figure 10

(C) Step 1 Graph the line $x = 3y$ as a solid line, since equality is included in the original statement (Fig. 11).

Step 2 Since the line passes through the origin, we must use a different test point. We choose $(0, 2)$ for a test point and conclude that this point is in the solution set.

Step 3 The graph of the solution set is the upper half-plane, including the boundary line (Fig. 12).

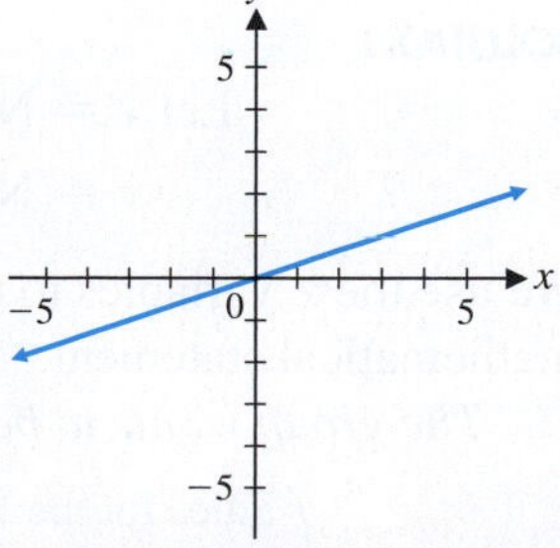

Figure 11

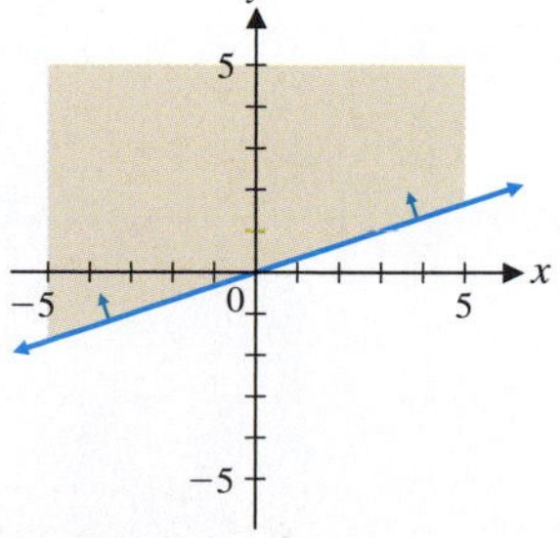

Figure 12

Matched Problem 2 Graph

(A) $y < 4$ (B) $4x \geq -9$ (C) $3x \geq 2y$

EXAMPLE 3 **Interpreting a Graph** Find the linear inequality whose graph is given in Figure 13. Write the boundary line equation in the form $Ax + By = C$, where A, B, and C are integers, before stating the inequality.

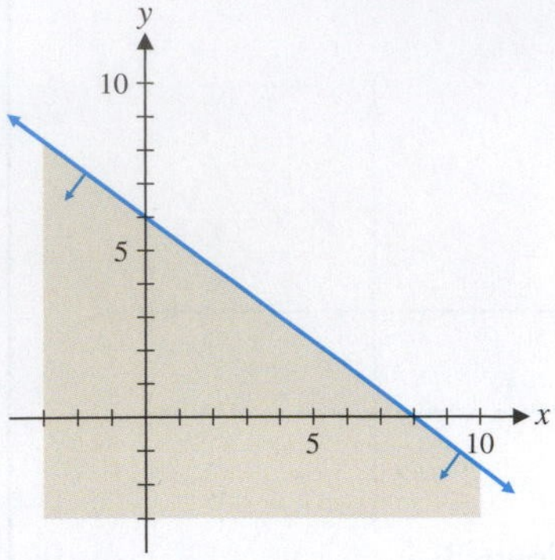

Figure 13

SOLUTION The boundary line (Fig. 13) passes through the points $(0, 6)$ and $(8, 0)$. We use the slope-intercept form to find the equation of this line:

$$\text{Slope: } m = \frac{0 - 6}{8 - 0} = -\frac{6}{8} = -\frac{3}{4}$$

$$y \text{ intercept: } b = 6$$

$$\text{Boundary line equation: } y = -\frac{3}{4}x + 6 \quad \text{Multiply both sides by 4.}$$

$$4y = -3x + 24 \quad \text{Add } 3x \text{ to both sides.}$$

$$3x + 4y = 24 \quad \text{Form: } Ax + By = C$$

Since $(0, 0)$ is in the shaded region in Figure 13 and the boundary line is solid, the graph in Figure 13 is the graph of $3x + 4y \le 24$.

Matched Problem 3 Find the linear inequality whose graph is given in Figure 14. Write the boundary line equation in the form $Ax + By = C$, where A, B, and C are integers, before stating the inequality.

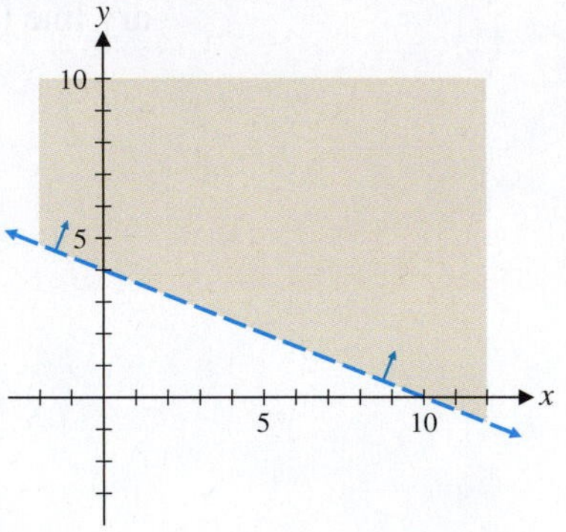

Figure 14

Application

EXAMPLE 4 **Sales** A concert promoter wants to book a rock group for a stadium concert. A ticket for admission to the stadium playing field will cost \$125, and a ticket for a seat in the stands will cost \$175. The group wants to be guaranteed total ticket sales of at least \$700,000. How many tickets of each type must be sold to satisfy the group's guarantee? Express the answer as a linear inequality and draw its graph.

SOLUTION

Let x = Number of tickets sold for the playing field

y = Number of tickets sold for seats in the stands

We use these variables to translate the following statement from the problem into a mathematical statement:

The group wants to be guaranteed total ticket sales of at least \$700,000.

$$\binom{\text{Sales for the}}{\text{playing field}} + \binom{\text{Sales for seats}}{\text{in the stands}} \quad \binom{\text{At}}{\text{least}} \quad \binom{\text{Total sales}}{\text{guaranteed}}$$

$$125x \quad + \quad 175y \quad \ge \quad 700{,}000$$

Dividing both sides of this inequality by 25, x, and y must satisfy

$$5x + 7y \ge 28{,}000$$

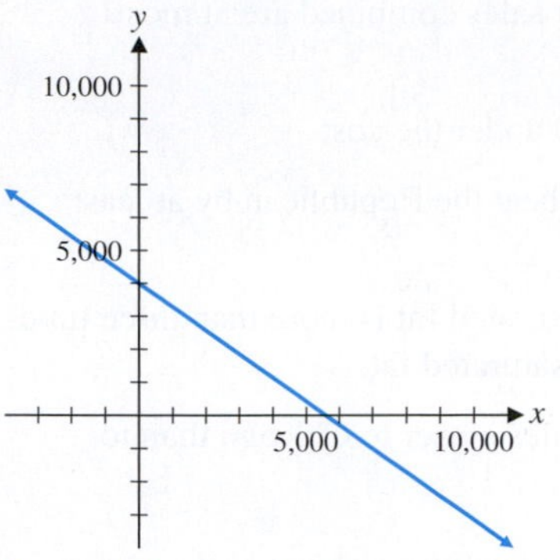

Figure 15

We use the three-step procedure to graph this inequality.

Step 1 Graph $5x + 7y = 28{,}000$ as a solid line (Fig. 15).

Step 2 Substituting $x = 0$ and $y = 0$ in the inequality produces a false statement, so the point $(0, 0)$ is not in the solution set.

Step 3 The graph of the inequality is the upper half-plane including the boundary line (Fig. 16), but does this graph really represent ticket sales?

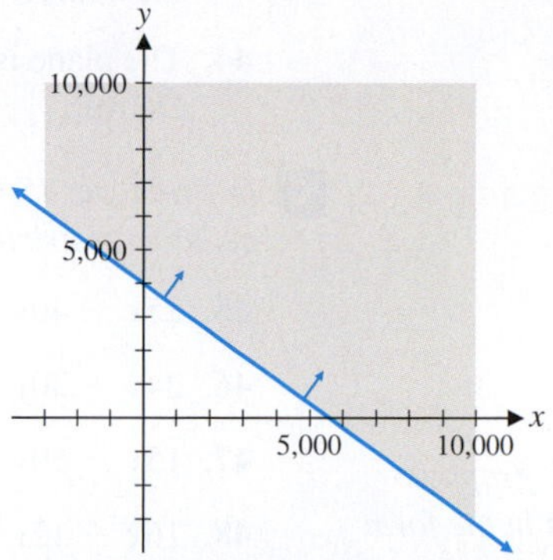

Figure 16

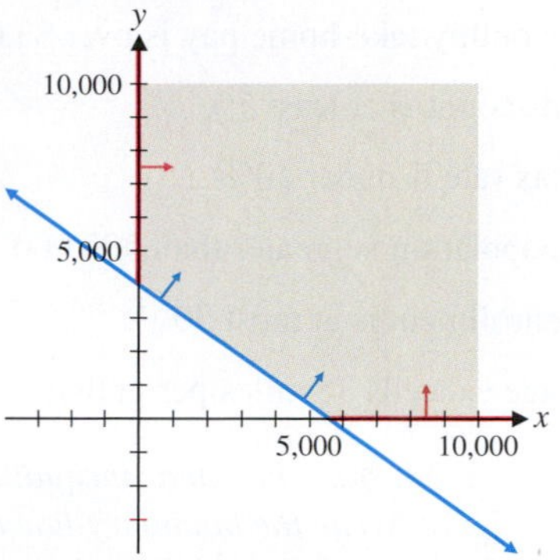

Figure 17

The shaded region in Figure 16 contains points in the second quadrant (where $x < 0$) and the fourth quadrant (where $y < 0$). It is not possible to sell a negative number of tickets, so we must restrict both x and y to the first quadrant. With this restriction, the solution becomes

$$5x + 7y \geq 28{,}000$$
$$x \geq 0, \; y \geq 0$$

and the graph is shown in Figure 17. There is yet another restriction on x and y. It is not possible to sell a fractional number of tickets, so both x and y must be integers. So the solutions of the original problem correspond to those points of the shaded region in Figure 17 that have integer coordinates. This restriction is not indicated in Figure 17, because the points with integer coordinates are too close together (about 9,000 such points per inch) to be visually distinguishable from other points.

Matched Problem 4 A food vendor at a rock concert sells hot dogs for \$4 and hamburgers for \$5. How many of these sandwiches must be sold to produce sales of at least \$1,000? Express the answer as a linear inequality and draw its graph.

Exercises 5.1

Skills Warm-up Exercises

For Problems 1–8, if necessary, review Section 1.2.

1. Is the point $(3, 5)$ on the line $y = 2x + 1$?

2. Is the point $(7, 9)$ on the line $y = 3x - 11$?

3. Is the point $(3, 5)$ in the solution set of $y \leq 2x + 1$?

4. Is the point $(7, 9)$ in the solution set of $y \leq 3x - 11$?

5. Is the point $(10, 12)$ on the line $13x - 11y = 2$?

6. Is the point $(21, 25)$ on the line $30x - 27y = 1$?

7. Is the point $(10, 12)$ in the solution set of $13x - 11y \geq 2$?

8. Is the point $(21, 25)$ in the solution set of $30x - 27y \leq 1$?

A *Graph each inequality in Problems 9–18.*

9. $y \leq x - 1$

10. $y > x + 1$

11. $3x - 2y > 6$

12. $2x - 5y \leq 10$

13. $x \geq -4$

14. $y < 5$

15. $6x + 4y \geq 24$

16. $4x + 8y \geq 32$

17. $5x \leq -2y$

18. $6x \geq 4y$

In Problems 19–22,

(A) *graph the set of points that satisfy the inequality.*

(B) *graph the set of points that do not satisfy the inequality.*

19. $2x + 3y < 18$

20. $3x + 4y > 24$

21. $5x - 2y \geq 20$

22. $3x - 5y \leq 30$

In Problems 23–32, define the variable and translate the sentence into an inequality.

23. There are fewer than 10 applicants.

24. She consumes no more than 900 calories per day.

25. He practices no less than 2.5 hours per day.

26. The average attendance is less than 15,000.

27. The monthly take-home pay is over \$3,000.

28. The discount is at least 5%.

29. The tax rate is under 40%.

30. The population is greater than 500,000.

31. The enrollment is at most 30.

32. Mileage exceeds 35 miles per gallon.

B *In Exercises 33–38, state the linear inequality whose graph is given in the figure. Write the boundary-line equation in the form $Ax + By = C$, where A, B, and C are integers, before stating the inequality.*

33.

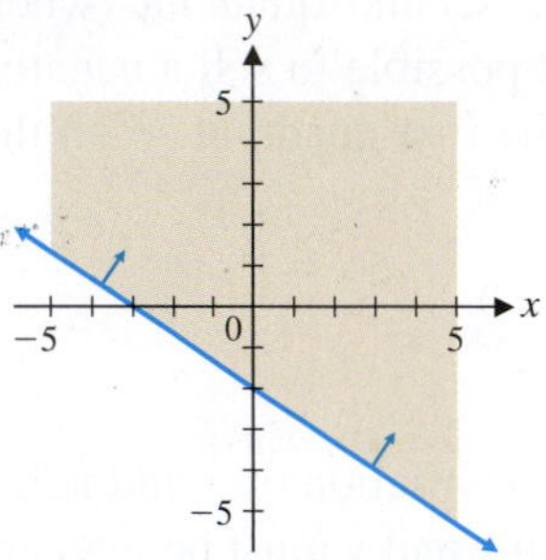

34.

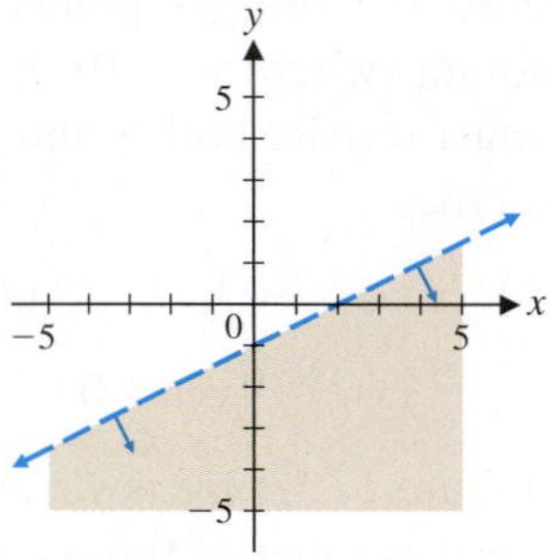

35.

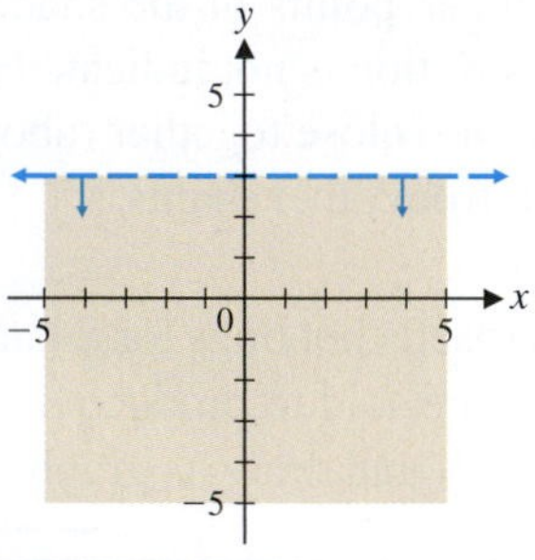

36.

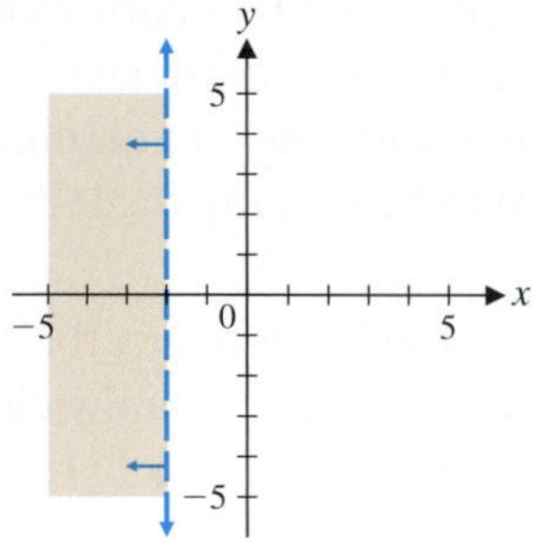

37.

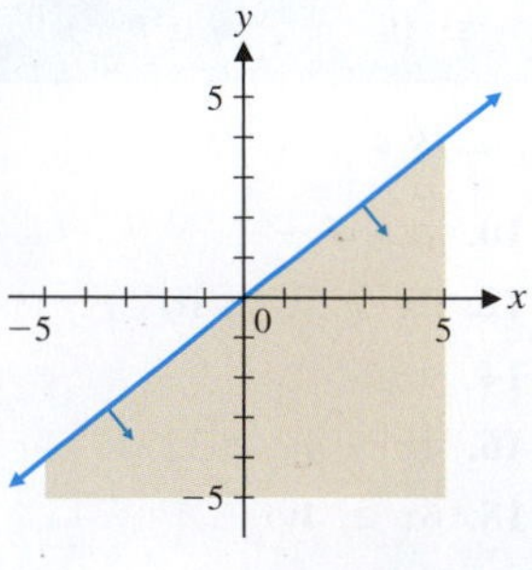

38.

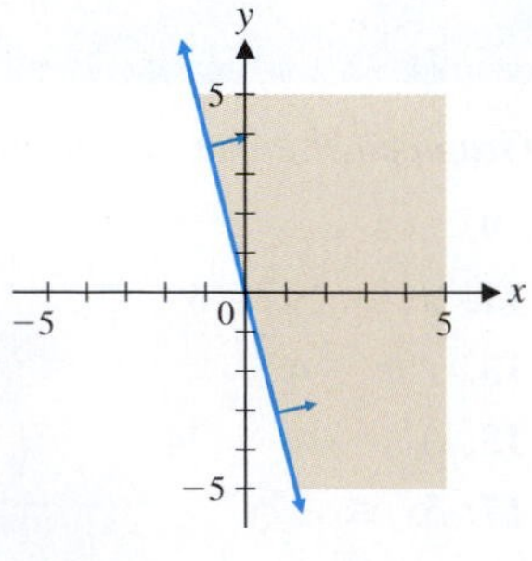

In Problems 39–44, define two variables and translate the sentence into an inequality.

39. Enrollment in finite mathematics plus enrollment in calculus is less than 300.

40. New-car sales and used-car sales combined are at most \$500,000.

41. Revenue is at least \$20,000 under the cost.

42. The Democratic candidate beat the Republican by at least seven percentage points.

43. The number of grams of saturated fat is more than three times the number of grams of unsaturated fat.

44. The plane is at least 500 miles closer to Chicago than to Denver.

C *In Problems 45–54, graph each inequality subject to the nonnegative restrictions.*

45. $25x + 40y \le 3{,}000,\ x \ge 0,\ y \ge 0$

46. $24x + 30y > 7{,}200,\ x \ge 0,\ y \ge 0$

47. $15x - 50y < 1{,}500,\ x \ge 0,\ y \ge 0$

48. $16x - 12y \ge 4{,}800,\ x \ge 0,\ y \ge 0$

49. $-18x + 30y \ge 2{,}700,\ x \ge 0,\ y \ge 0$

50. $-14x + 22y < 1{,}540,\ x \ge 0,\ y \ge 0$

51. $40x - 55y > 0,\ x \ge 0,\ y \ge 0$

52. $-35x + 75y \le 0,\ x \ge 0,\ y \ge 0$

53. $25x + 75y < -600,\ x \ge 0,\ y \ge 0$

54. $75x + 25y > -600,\ x \ge 0,\ y \ge 0$

Applications

In Problems 55–66, express your answer as a linear inequality with appropriate nonnegative restrictions and draw its graph.

55. Seed costs. Seed costs for a farmer are \$90 per acre for corn and \$70 per acre for soybeans. How many acres of each crop should the farmer plant if he wants to spend no more than \$11,000 on seed?

56. Labor costs. Labor costs for a farmer are \$120 per acre for corn and \$100 per acre for soybeans. How many acres of each crop should the farmer plant if he wants to spend no more than \$15,000 on labor?

57. Fertilizer. A farmer wants to use two brands of fertilizer for his corn crop. Brand *A* contains 26% nitrogen, 3% phosphate, and 3% potash. Brand *B* contains 16% nitrogen, 8% phosphate, and 8% potash.

(*Source:* Spectrum Analytic, Inc.)

(A) How many pounds of each brand of fertilizer should he add to each acre if he wants to add at least 120 pounds of nitrogen to each acre?

(B) How many pounds of each brand of fertilizer should he add to each acre if he wants to add at most 28 pounds of phosphate to each acre?

58. Fertilizer. A farmer wants to use two brands of fertilizer for his soybean crop. Brand A contains 18% nitrogen, 24% phosphate, and 12% potash. Brand B contains 5% nitrogen, 10% phosphate, and 15% potash.
(*Source:* Spectrum Analytic, Inc.)
(A) How many pounds of each brand of fertilizer should he add to each acre if he wants to add at least 50 pounds of phosphate to each acre?
(B) How many pounds of each brand of fertilizer should he add to each acre if he wants to add at most 60 pounds of potash to each acre?

59. Textiles. A textile mill uses two blended yarns—a standard blend that is 30% acrylic, 30% wool, and 40% nylon and a deluxe blend that is 9% acrylic, 39% wool, and 52% nylon—to produce various fabrics. How many pounds of each yarn should the mill use to produce a fabric that is at least 20% acrylic?

60. Textiles. Refer to Exercise 59. How many pounds of each yarn should the mill use to produce a fabric that is at least 45% nylon?

61. Customized vehicles. A company uses sedans and minivans to produce custom vehicles for transporting hotel guests to and from airports. Plant A can produce 10 sedans and 8 minivans per week, and Plant B can produce 8 sedans and 6 minivans per week. How many weeks should each plant operate in order to produce at least 400 sedans?

62. Customized vehicles. Refer to Exercise 61. How many weeks should each plant operate in order to produce at least 480 minivans?

63. Political advertising. A candidate has budgeted \$10,000 to spend on radio and television advertising. A radio ad costs \$200 per 30-second spot, and a television ad costs \$800 per 30-second spot. How many radio and television spots can the candidate purchase without exceeding the budget?

64. Political advertising. Refer to Problem 63. The candidate decides to replace the television ads with newspaper ads that cost \$500 per ad. How many radio spots and newspaper ads can the candidate purchase without exceeding the budget?

65. Mattresses. A company produces foam mattresses in two sizes: regular and king. It takes 5 minutes to cut the foam for a regular mattress and 6 minutes for a king mattress. If the cutting department has 50 labor-hours available each day, how many regular and king mattresses can be cut in one day?

66. Mattresses. Refer to Problem 65. It takes 15 minutes to cover a regular mattress and 20 minutes to cover a king mattress. If the covering department has 160 labor-hours available each day, how many regular and king mattresses can be covered in one day?

Answers to Matched Problems

1.

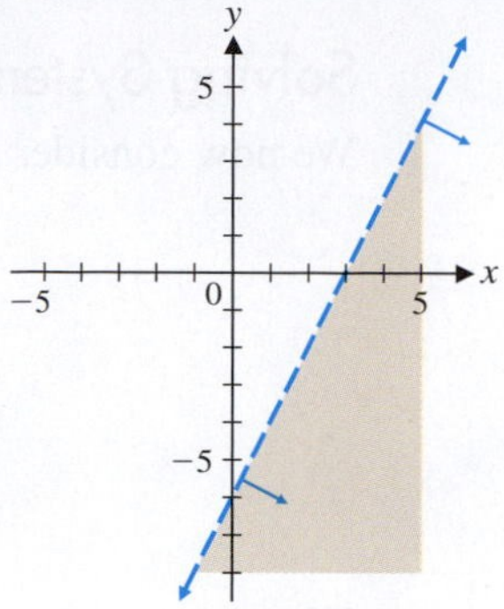

2. (A)

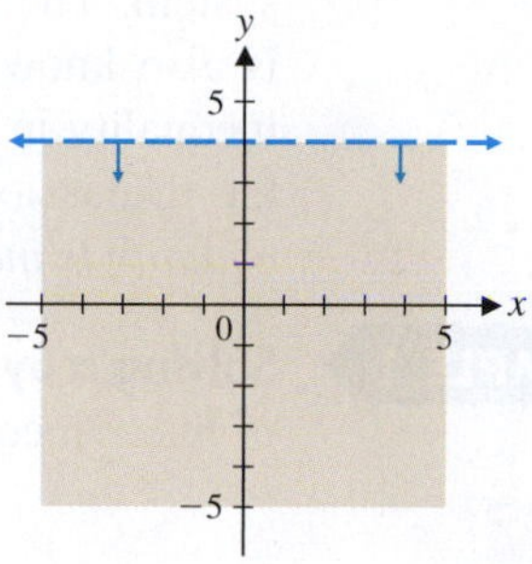

(B)

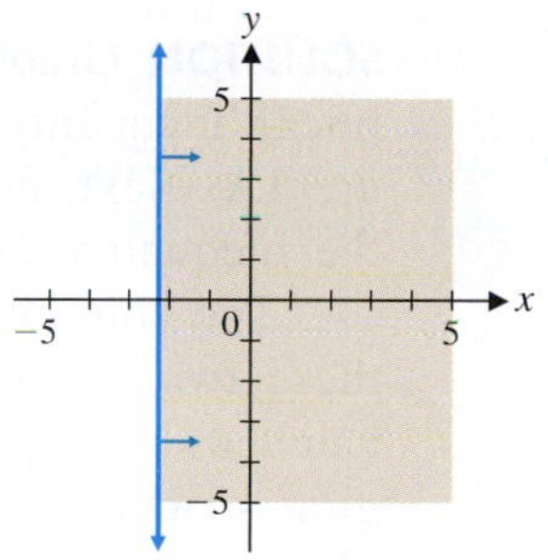

(C)

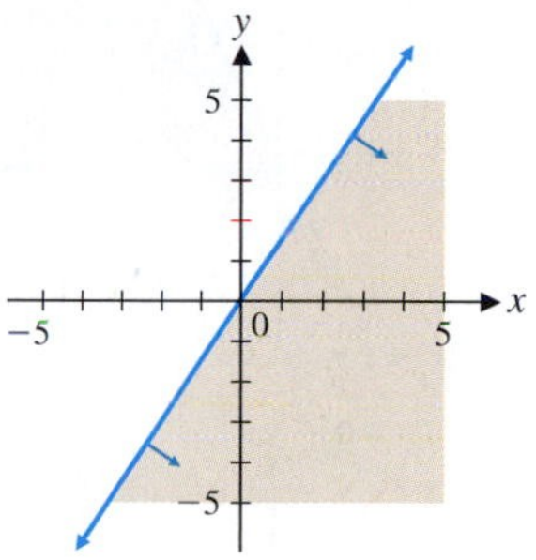

3. $2x + 5y > 20$

4. Let $x =$ Number of hot dogs sold
$y =$ Number of hamburgers sold
$4x + 5y \geq 1{,}000 \qquad x \geq 0, y \geq 0$

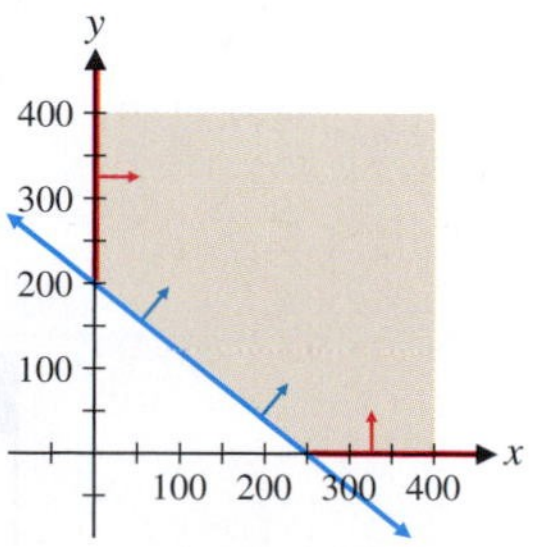

5.2 Systems of Linear Inequalities in Two Variables

- Solving Systems of Linear Inequalities Graphically
- Applications

Solving Systems of Linear Inequalities Graphically

We now consider systems of linear inequalities such as

$$\begin{aligned} x + y &\ge 6 \\ 2x - y &\ge 0 \end{aligned} \quad \text{and} \quad \begin{aligned} 2x + y &\le 22 \\ x + y &\le 13 \\ 2x + 5y &\le 50 \\ x &\ge 0 \\ y &\ge 0 \end{aligned}$$

We wish to **solve** such systems **graphically**—that is, to find the graph of all ordered pairs of real numbers (x, y) that simultaneously satisfy all the inequalities in the system. The graph is called the **solution region** for the system (the solution region is also known as the **feasible region**). To find the solution region, we graph each inequality in the system and then take the intersection of all the graphs. To simplify the discussion that follows, *we consider only systems of linear inequalities where equality is included in each statement in the system.*

EXAMPLE 1 **Solving a System of Linear Inequalities Graphically** Solve the following system of linear inequalities graphically:

$$\begin{aligned} x + y &\ge 6 \\ 2x - y &\ge 0 \end{aligned}$$

SOLUTION Graph the line $x + y = 6$ and shade the region that satisfies the linear inequality $x + y \ge 6$. This region is shaded with red lines in Figure 1A. Next, graph the line $2x - y = 0$ and shade the region that satisfies the inequality $2x - y \ge 0$. This region is shaded with blue lines in Figure 1A. The solution region for the system of inequalities is the intersection of these two regions. This is the region shaded in both red and blue (cross-hatched) in Figure 1A and redrawn in Figure 1B with only the solution region shaded. The coordinates of any point in the shaded region of Figure 1B specify a solution to the system. For example, the points $(2, 4)$, $(6, 3)$, and $(7.43, 8.56)$ are three of infinitely many solutions, as can be easily checked. The intersection point $(2, 4)$ is obtained by solving the equations $x + y = 6$ and $2x - y = 0$ simultaneously using any of the techniques discussed in Chapter 4.

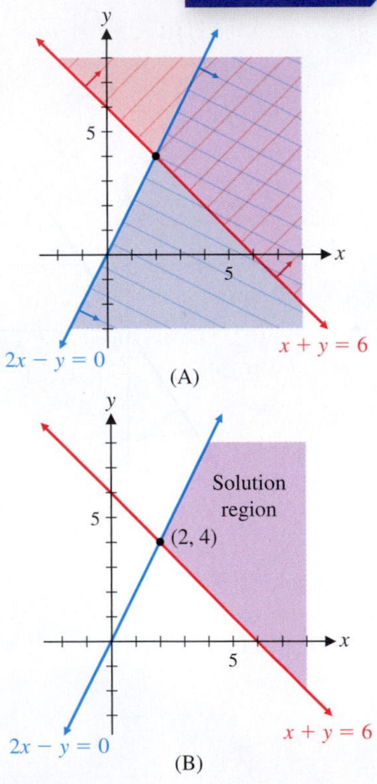

Figure 1

Matched Problem 1 Solve the following system of linear inequalities graphically:

$$\begin{aligned} 3x + y &\le 21 \\ x - 2y &\le 0 \end{aligned}$$

CONCEPTUAL INSIGHT

To check that you have shaded a solution region correctly, choose a test point in the region and check that it satisfies each inequality in the system. For example, choosing the point $(5, 4)$ in the shaded region in Figure 1B, we have

$$\begin{aligned} x + y &\ge 6 \\ 5 + 4 &\overset{?}{\ge} 6 \\ 9 &\overset{\checkmark}{\ge} 6 \end{aligned} \qquad \begin{aligned} 2x - y &\ge 0 \\ 10 - 4 &\overset{?}{\ge} 0 \\ 6 &\overset{\checkmark}{\ge} 0 \end{aligned}$$

The points of intersection of the lines that form the boundary of a solution region will play a fundamental role in the solution of linear programming problems, which are discussed in the next section.

DEFINITION Corner Point

A **corner point** of a solution region is a point in the solution region that is the intersection of two boundary lines.

For example, the point $(2, 4)$ is the only corner point of the solution region in Example 1 (Fig. 1B).

EXAMPLE 2 **Solving a System of Linear Inequalities Graphically** Solve the following system of linear inequalities graphically and find the corner points:

$$\begin{aligned} 2x + y &\le 22 \\ x + y &\le 13 \\ 2x + 5y &\le 50 \\ x &\ge 0 \\ y &\ge 0 \end{aligned}$$

SOLUTION The inequalities $x \ge 0$ and $y \ge 0$ indicate that the solution region will lie in the first quadrant. So we can restrict our attention to that portion of the plane. First, we graph the lines

$$\begin{aligned} 2x + y &= 22 \\ x + y &= 13 \\ 2x + 5y &= 50 \end{aligned}$$

Find the x and y intercepts of each line; then sketch the line through these points.

Next, choosing $(0, 0)$ as a test point, we see that the graph of each of the first three inequalities in the system consists of its corresponding line and the half-plane lying below the line, as indicated by the small arrows in Figure 2. The solution region of the system consists of the points in the first quadrant that simultaneously lie on or below all three of these lines (see the shaded region in Fig. 2).

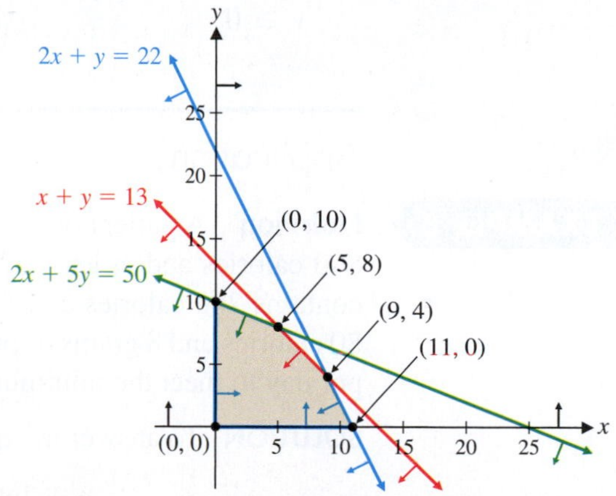

Figure 2

The corner points $(0, 0)$, $(0, 10)$, and $(11, 0)$ can be determined from the graph. The other two corner points are determined as follows:

Solve the system	Solve the system
$2x + 5y = 50$	$2x + y = 22$
$x + y = 13$	$x + y = 13$
to obtain $(5, 8)$.	to obtain $(9, 4)$.

Note that the lines $2x + 5y = 50$ and $2x + y = 22$ also intersect, but the intersection point is not part of the solution region and so is not a corner point.

Matched Problem 2 Solve the following system of linear inequalities graphically and find the corner points:

$$\begin{aligned} 5x + y &\geq 20 \\ x + y &\geq 12 \\ x + 3y &\geq 18 \\ x &\geq 0 \\ y &\geq 0 \end{aligned}$$

If we compare the solution regions of Examples 1 and 2, we see that there is a fundamental difference between these two regions. We can draw a circle around the solution region in Example 2; however, it is impossible to include all the points in the solution region in Example 1 in any circle, no matter how large we draw it. This leads to the following definition:

DEFINITION Bounded and Unbounded Solution Regions

A solution region of a system of linear inequalities is **bounded** if it can be enclosed within a circle. If it cannot be enclosed within a circle, it is **unbounded**.

The solution region for Example 2 is bounded, and the solution region for Example 1 is unbounded. This definition will be important in the next section.

Explore and Discuss 1

Determine whether the solution region of each system of linear inequalities is bounded or unbounded.

(A) $y \leq 1$, $x \geq 0$, $y \geq 0$

(B) $x \leq 100$, $y \leq 200$, $x \geq 0$, $y \geq 0$

(C) $x \leq y$, $y \leq x$, $x \geq 0$, $y \geq 0$

Applications

EXAMPLE 3 **Nutrition** A patient on a brown rice and skim milk diet is required to have at least 800 calories and at least 32 grams of protein per day. Each serving of brown rice contains 200 calories and 5 grams of protein. Each serving of skim milk contains 80 calories and 8 grams of protein. How many servings of each food should be eaten per day to meet the minimum daily requirements?

SOLUTION To answer the question, we need to solve for x and y, where

$$x = \text{number of daily servings of brown rice}$$

$$y = \text{number of daily servings of skim milk}$$

We arrange the information given in the problem in a table, with columns corresponding to x and y.

	Brown Rice	Skim Milk	Minimum Daily Requirement
Calories	200 cal/svg	80 cal/svg	800 cal
Protein	5 g/svg	8 g/svg	32 g

The number of calories in x servings of brown rice is $200x$, and the number of calories in y servings of skim milk is $80y$. So, to meet the minimum daily requirement for calories, $200x + 80y$ must be greater than or equal to 800. This gives the first of the inequalities below. The second inequality expresses the condition that the minimum daily requirement for protein is met. The last two inequalities express the fact that the number of servings of each food cannot be a negative number.

$$\begin{aligned} 200x + 80y &\geq 800 && \text{Requirement for calories} \\ 5x + 8y &\geq 32 && \text{Requirement for protein} \\ x &\geq 0 && \text{Nonnegative restriction on } x \\ y &\geq 0 && \text{Nonnegative restriction on } y \end{aligned}$$

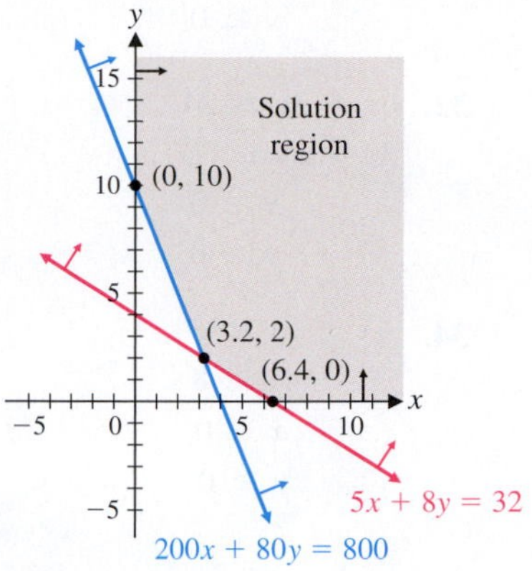

Figure 3

We graph this system of inequalities, and shade the solution region (Figure 3). Each point in the shaded area, including the straight-line boundaries, will meet the minimum daily requirements for calories and protein; any point outside the shaded area will not. For example, 4 servings of brown rice and 2 servings of skim milk will meet the minimum daily requirements, but 3 servings of brown rice and 2 servings of skim milk will not. Note that the solution region is unbounded.

Matched Problem 3 A manufacturing plant makes two types of inflatable boats—a two-person boat and a four-person boat. Each two-person boat requires 0.9 labor-hour in the cutting department and 0.8 labor-hour in the assembly department. Each four-person boat requires 1.8 labor-hours in the cutting department and 1.2 labor-hours in the assembly department. The maximum labor-hours available each month in the cutting and assembly departments are 864 and 672, respectively.

(A) Summarize this information in a table.

(B) If x two-person boats and y four-person boats are manufactured each month, write a system of linear inequalities that reflects the conditions indicated. Graph the feasible region.

Exercises 5.2

Skills Warm-up Exercises

W *For Problems 1–8, if necessary, review Section 1.2. Problems 1–4 refer to the following system of linear inequalities:*

$$\begin{aligned} 4x + y &\leq 20 \\ 3x + 5y &\leq 37 \\ x &\geq 0 \\ y &\geq 0 \end{aligned}$$

1. Is the point $(3, 5)$ in the solution region?

2. Is the point $(4, 5)$ in the solution region?

3. Is the point $(3, 6)$ in the solution region?

4. Is the point $(2, 6)$ in the solution region?

Problems 5–8 refer to the following system of linear inequalities:

$$\begin{aligned} 5x + y &\leq 32 \\ 7x + 4y &\geq 45 \\ x &\geq 0 \\ y &\geq 0 \end{aligned}$$

5. Is the point $(4, 3)$ in the solution region?

6. Is the point $(5, 3)$ in the solution region?

7. Is the point $(6, 2)$ in the solution region?

8. Is the point $(5, 2)$ in the solution region?

A *In Problems 9–12, match the solution region of each system of linear inequalities with one of the four regions shown in the figure.*

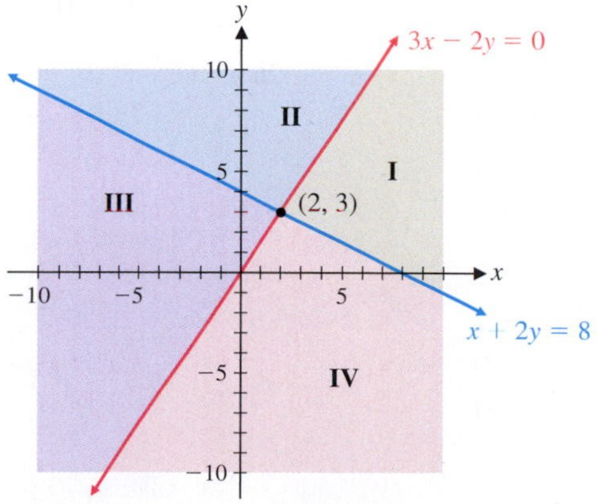

9. $x + 2y \leq 8$
$3x - 2y \geq 0$

10. $x + 2y \geq 8$
$3x - 2y \leq 0$

11. $x + 2y \geq 8$
$3x - 2y \geq 0$

12. $x + 2y \leq 8$
$3x - 2y \leq 0$

In Problems 13–16, solve each system of linear inequalities graphically.

13. $3x + y \ge 6$, $x \le 4$

14. $3x + 4y \le 12$, $y \ge -3$

15. $x - 2y \le 12$, $2x + y \ge 4$

16. $2x + 5y \le 20$, $x - 5y \ge -5$

B *In Problems 17–20, match the solution region of each system of linear inequalities with one of the four regions shown in the figure. Identify the corner points of each solution region.*

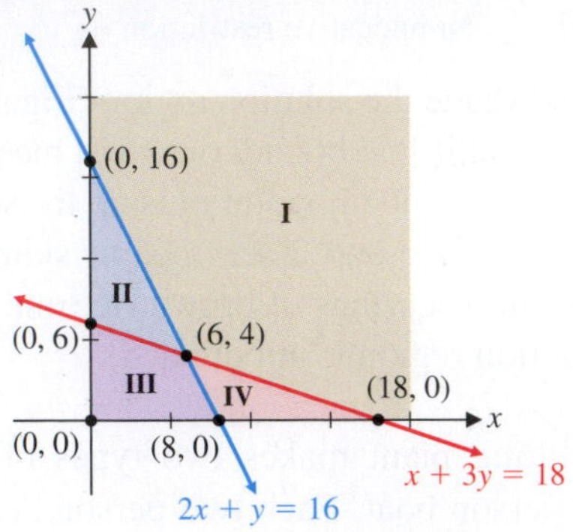

17. $x + 3y \le 18$, $2x + y \ge 16$, $x \ge 0$, $y \ge 0$

18. $x + 3y \le 18$, $2x + y \le 16$, $x \ge 0$, $y \ge 0$

19. $x + 3y \ge 18$, $2x + y \ge 16$, $x \ge 0$, $y \ge 0$

20. $x + 3y \ge 18$, $2x + y \le 16$, $x \ge 0$, $y \ge 0$

In Problems 21–28, is the solution region bounded or unbounded?

21. $3x + y \le 6$, $x \ge 0$, $y \ge 0$

22. $x + 2y \ge 4$, $x \ge 0$, $y \ge 0$

23. $5x - 2y \ge 10$, $x \ge 0$, $y \ge 0$

24. $4x - 3y \le 12$, $x \ge 0$, $y \ge 0$

25. $-x + y \le 4$, $x \le 10$, $x \ge 0$, $y \ge 0$

26. $x - y \le 3$, $x \le 9$, $x \ge 0$, $y \ge 0$

27. $-x + 2y \ge 2$, $2x - y \le 2$, $x \ge 0$, $y \ge 0$

28. $-x + 2y \le 2$, $2x - y \le 2$, $x \ge 0$, $y \ge 0$

Solve the systems in Problems 29–38 graphically and indicate whether each solution region is bounded or unbounded. Find the coordinates of each corner point.

29. $2x + 3y \le 12$, $x \ge 0$, $y \ge 0$

30. $3x + 4y \le 24$, $x \ge 0$, $y \ge 0$

31. $2x + y \le 10$, $x + 2y \le 8$, $x \ge 0$, $y \ge 0$

32. $6x + 3y \le 24$, $3x + 6y \le 30$, $x \ge 0$, $y \ge 0$

33. $2x + y \ge 10$, $x + 2y \ge 8$, $x \ge 0$, $y \ge 0$

34. $4x + 3y \ge 24$, $3x + 4y \ge 8$, $x \ge 0$, $y \ge 0$

35. $2x + y \le 10$, $x + y \le 7$, $x + 2y \le 12$, $x \ge 0$, $y \ge 0$

36. $3x + y \le 21$, $x + y \le 9$, $x + 3y \le 21$, $x \ge 0$, $y \ge 0$

37. $2x + y \ge 16$, $x + y \ge 12$, $x + 2y \ge 14$, $x \ge 0$, $y \ge 0$

38. $3x + y \ge 24$, $x + y \ge 16$, $x + 3y \ge 30$, $x \ge 0$, $y \ge 0$

C *Solve the systems in Problems 39–48 graphically and indicate whether each solution region is bounded or unbounded. Find the coordinates of each corner point.*

39. $x + 4y \le 32$, $3x + y \le 30$, $4x + 5y \ge 51$

40. $x + y \le 11$, $x + 5y \ge 15$, $2x + y \ge 12$

41. $4x + 3y \le 48$, $2x + y \ge 24$, $x \le 9$

42. $2x + 3y \ge 24$, $x + 3y \le 15$, $y \ge 4$

43. $x - y \le 0$, $2x - y \le 4$, $0 \le x \le 8$

44. $2x + 3y \ge 12$, $-x + 3y \le 3$, $0 \le y \le 5$

45. $-x + 3y \ge 1$, $5x - y \ge 9$, $x + y \le 9$, $x \le 5$

46. $x + y \le 10$, $5x + 3y \ge 15$, $-2x + 3y \le 15$, $2x - 5y \le 6$

47. $16x + 13y \le 120$, $3x + 4y \ge 25$, $-4x + 3y \le 11$

48. $2x + 2y \le 21$, $-10x + 5y \le 24$, $3x + 5y \ge 37$

Problems 49 and 50 introduce an algebraic process for finding the corner points of a solution region without drawing a graph. We will discuss this process later in the chapter.

49. Consider the following system of inequalities and corresponding boundary lines:

$$\begin{aligned} 3x + 4y &\le 36 & \quad 3x + 4y &= 36 \\ 3x + 2y &\le 30 & \quad 3x + 2y &= 30 \\ x &\ge 0 & \quad x &= 0 \\ y &\ge 0 & \quad y &= 0 \end{aligned}$$

(A) Use algebraic methods to find the intersection points (if any exist) for each possible pair of boundary lines. (There are six different possible pairs.)

(B) Test each intersection point in all four inequalities to determine which are corner points.

50. Repeat Problem 49 for

$$\begin{aligned} 2x + y &\le 16 & \quad 2x + y &= 16 \\ 2x + 3y &\le 36 & \quad 2x + 3y &= 36 \\ x &\ge 0 & \quad x &= 0 \\ y &\ge 0 & \quad y &= 0 \end{aligned}$$

Applications

51. Water skis. A manufacturing company makes two types of water skis, a trick ski and a slalom ski. The trick ski requires 6 labor-hours for fabricating and 1 labor-hour for finishing. The slalom ski requires 4 labor-hours for fabricating and 1 labor-hour for finishing. The maximum labor-hours available per day for fabricating and finishing are 108 and 24, respectively. If x is the number of trick skis and y is the number of slalom skis produced per day, write a system of linear inequalities that indicates appropriate restraints on x and y. Find the set of feasible solutions graphically for the number of each type of ski that can be produced.

52. Furniture. A furniture manufacturing company manufactures dining-room tables and chairs. A table requires 8 labor-hours for assembling and 2 labor-hours for finishing. A chair requires 2 labor-hours for assembling and 1 labor-hour for finishing. The maximum labor-hours available per day for assembly and finishing are 400 and 120, respectively. If x is the number of tables and y is the number of chairs produced per day, write a system of linear inequalities that indicates appropriate restraints on x and y. Find the set of feasible solutions graphically for the number of tables and chairs that can be produced.

53. Water skis. Refer to Problem 51. The company makes a profit of \$50 on each trick ski and a profit of \$60 on each slalom ski.

(A) If the company makes 10 trick skis and 10 slalom skis per day, the daily profit will be \$1,100. Are there other production schedules that will result in a daily profit of \$1,100? How are these schedules related to the graph of the line $50x + 60y = 1{,}100$?

(B) Find a production schedule that will produce a daily profit greater than \$1,100 and repeat part (A) for this schedule.

(C) Discuss methods for using lines like those in parts (A) and (B) to find the largest possible daily profit.

54. Furniture. Refer to Problem 52. The company makes a profit of \$50 on each table and a profit of \$15 on each chair.

(A) If the company makes 20 tables and 20 chairs per day, the daily profit will be \$1,300. Are there other production schedules that will result in a daily profit of \$1,300? How are these schedules related to the graph of the line $50x + 15y = 1{,}300$?

(B) Find a production schedule that will produce a daily profit greater than \$1,300 and repeat part (A) for this schedule.

(C) Discuss methods for using lines like those in parts (A) and (B) to find the largest possible daily profit.

55. Plant food. A farmer can buy two types of plant food, mix A and mix B. Each cubic yard of mix A contains 20 pounds of phosphoric acid, 30 pounds of nitrogen, and 5 pounds of potash. Each cubic yard of mix B contains 10 pounds of phosphoric acid, 30 pounds of nitrogen, and 10 pounds of potash. The minimum monthly requirements are 460 pounds of phosphoric acid, 960 pounds of nitrogen, and 220 pounds of potash. If x is the number of cubic yards of mix A used and y is the number of cubic yards of mix B used, write a system of linear inequalities that indicates appropriate restraints on x and y. Find the set of feasible solutions graphically for the amounts of mix A and mix B that can be used.

56. Nutrition. A dietitian in a hospital is to arrange a special diet using two foods. Each ounce of food M contains 30 units of calcium, 10 units of iron, and 10 units of vitamin A. Each ounce of food N contains 10 units of calcium, 10 units of iron, and 30 units of vitamin A. The minimum requirements in the diet are 360 units of calcium, 160 units of iron, and 240 units of vitamin A. If x is the number of ounces of food M used and y is the number of ounces of food N used, write a system of linear inequalities that reflects the conditions indicated. Find the set of feasible solutions graphically for the amount of each kind of food that can be used.

57. Psychology. A psychologist uses two types of boxes when studying mice and rats. Each mouse spends 10 minutes per day in box A and 20 minutes per day in box B. Each rat spends 20 minutes per day in box A and 10 minutes per day in box B. The total maximum time available per day is 800 minutes for box A and 640 minutes for box B. If x is the number of mice used and y the number of rats used, write a system of linear inequalities that indicates appropriate restrictions on x and y. Find the set of feasible solutions graphically.

Answers to Matched Problems

1.

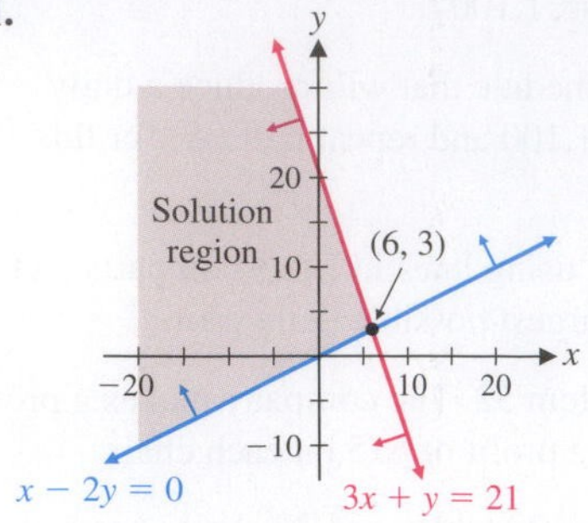

2.

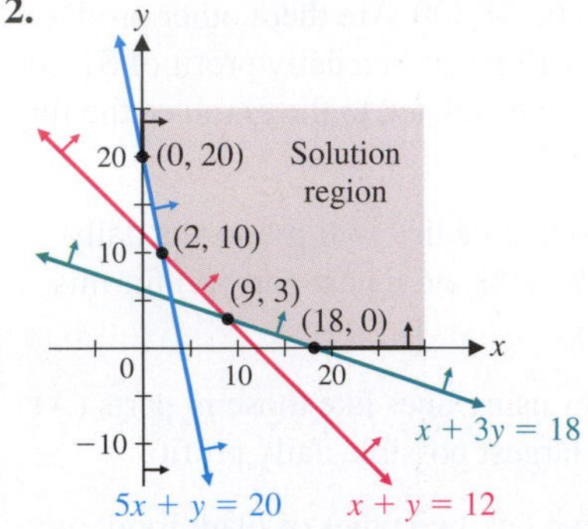

3. (A)

	Labor-Hours Required		Maximum Labor-Hours Available per Month
	Two-Person Boat	Four-Person Boat	
Cutting Department	0.9	1.8	864
Assembly Department	0.8	1.2	672

(B) $0.9x + 1.8y \le 864$
$0.8x + 1.2y \le 672$
$x \ge 0$
$y \ge 0$

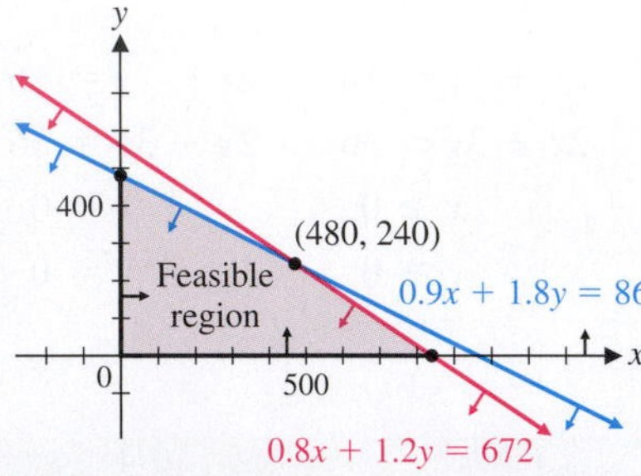

5.3 Linear Programming in Two Dimensions: A Geometric Approach

- A Linear Programming Problem
- General Description of Linear Programming
- Geometric Method for Solving Linear Programming Problems
- Applications

Several problems discussed in the preceding section are related to a more general type of problem called a *linear programming problem*. **Linear programming** is a mathematical process that has been developed to help management in decision making. We introduce this topic by considering an example in detail, using an intuitive geometric approach. Insight gained from this approach will prove invaluable when later we consider an algebraic approach that is less intuitive but necessary to solve most real-world problems.

A Linear Programming Problem

We begin our discussion with a concrete example. The solution method will suggest two important theorems and a simple general geometric procedure for solving linear programming problems in two variables.

EXAMPLE 1 **Production Scheduling** A manufacturer of lightweight mountain tents makes a standard model and an expedition model. Each standard tent requires 1 labor-hour from the cutting department and 3 labor-hours from the assembly department. Each expedition tent requires 2 labor-hours from the cutting department and 4 labor-hours from the assembly department. The maximum labor-hours available per day in the cutting and assembly departments are 32 and 84, respectively. If the company makes a profit of \$50 on each standard tent and \$80 on each expedition tent, how many tents of each type should be manufactured each day to maximize the total daily profit (assuming that all tents can be sold)?

SOLUTION This is an example of a linear programming problem. We begin by analyzing the question posed in this example.

According to the question, the *objective* of management is to maximize profit. Since the profits for standard and expedition tents differ, management must decide how many of each type of tent to manufacture. So it is reasonable to introduce the following **decision variables**:

$$\text{Let } x = \text{number of standard tents produced per day}$$
$$y = \text{number of expedition tents produced per day}$$

Now we summarize the manufacturing requirements, objectives, and restrictions in Table 1, with the decision variables related to the columns in the table.

Table 1

	Labor-Hours per Tent		Maximum Labor-Hours Available per Day
	Standard Model	Expedition Model	
Cutting department	1	2	32
Assembly department	3	4	84
Profit per tent	\$50	\$80	

Using the last line of Table 1, we form the **objective function**, in this case the profit P, in terms of the decision variables (we assume that all tents manufactured are sold):

$$P = 50x + 80y \quad \text{Objective function}$$

The **objective** is to find values of the decision variables that produce the **optimal value** (in this case, maximum value) of the objective function.

The form of the objective function indicates that the profit can be made as large as we like, simply by producing enough tents. But any manufacturing company has limits imposed by available resources, plant capacity, demand, and so on. These limits are referred to as **problem constraints**. Using the information in Table 1, we can determine two problem constraints.

$$\begin{pmatrix}\text{daily cutting}\\ \text{time for } x\\ \text{standard tents}\end{pmatrix} + \begin{pmatrix}\text{daily cutting}\\ \text{time for } y\\ \text{expedition tents}\end{pmatrix} \le \begin{pmatrix}\text{maximum labor-}\\ \text{hours available}\\ \text{per day}\end{pmatrix} \quad \text{Cutting department constraint}$$

$$1x \quad + \quad 2y \quad \le \quad 32$$

$$\begin{pmatrix}\text{daily assembly}\\ \text{time for } x\\ \text{standard tents}\end{pmatrix} + \begin{pmatrix}\text{daily assembly}\\ \text{time for } y\\ \text{expedition tents}\end{pmatrix} \le \begin{pmatrix}\text{maximum labor-}\\ \text{hours available}\\ \text{per day}\end{pmatrix} \quad \text{Assembly department constraint}$$

$$3x \quad + \quad 4y \quad \le \quad 84$$

It is not possible to manufacture a negative number of tents; thus, we have the **nonnegative constraints**

$$x \ge 0 \text{ and } y \ge 0$$

which we usually write in the form

$$x, y \ge 0 \quad \text{Nonnegative constraints}$$

We now have a **mathematical model** for the problem under consideration:

$$\begin{aligned} \text{Maximize} \quad & P = 50x + 80y && \text{Objective function} \\ \text{subject to} \quad & \left.\begin{aligned} x + 2y &\le 32 \\ 3x + 4y &\le 84 \end{aligned}\right\} && \text{Problem constraints} \\ & x, y \ge 0 && \text{Nonnegative constraints} \end{aligned}$$

Solving the set of linear inequality constraints **graphically**, we obtain the feasible region for production schedules (Fig. 1).

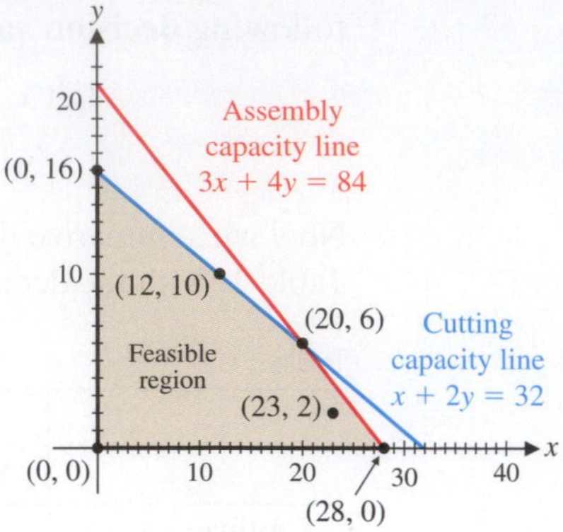

Figure 1

By choosing a production schedule (x, y) from the feasible region, a profit can be determined using the objective function

$$P = 50x + 80y$$

For example, if $x = 12$ and $y = 10$, the profit for the day would be

$$P = 50(12) + 80(10)$$
$$= \$1{,}400$$

Or if $x = 23$ and $y = 2$, the profit for the day would be

$$P = 50(23) + 80(2)$$
$$= \$1{,}310$$

Out of all possible production schedules (x, y) from the feasible region, which schedule(s) produces the *maximum* profit? This is a **maximization problem**. Since point-by-point checking is impossible (there are infinitely many points to check), we must find another way.

By assigning P in $P = 50x + 80y$ a particular value and plotting the resulting equation in the coordinate system shown in Figure 1, we obtain a **constant-profit line**. Every point in the feasible region on this line represents a production schedule that will produce the same profit. By doing this for a number of values for P, we obtain a family of constant-profit lines (Fig. 2) that are parallel to each other, since they all have the same slope. To see this, we write $P = 50x + 80y$ in the slope-intercept form

$$y = -\frac{5}{8}x + \frac{P}{80}$$

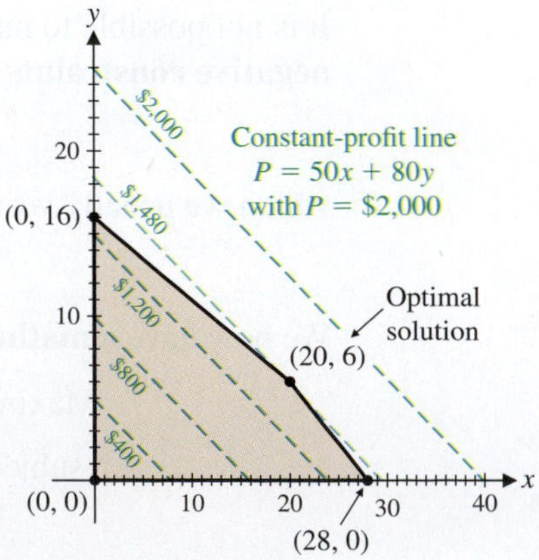

Figure 2 Constant-profit lines

and note that for any profit P, the constant-profit line has slope $-\frac{5}{8}$. We also observe that as the profit P increases, the y intercept ($P/80$) increases, and the line moves away from the origin.

Therefore, the maximum profit occurs at a point where a constant-profit line is the farthest from the origin but still in contact with the feasible region, in this example, at $(20, 6)$ (see Fig. 2). So profit is maximized if the manufacturer makes 20 standard tents and 6 expedition tents per day, and the maximum profit is

$$\begin{aligned} P &= 50(20) + 80(6) \\ &= \$1{,}480 \end{aligned}$$

The point $(20, 6)$ is called an **optimal solution** to the problem because it maximizes the objective (profit) function and is in the feasible region. In general, it appears that a maximum profit occurs at one of the corner points. We also note that the minimum profit $(P = 0)$ occurs at the corner point $(0, 0)$.

Matched Problem 1 A manufacturing plant makes two types of inflatable boats—a two-person boat and a four-person boat. Each two-person boat requires 0.9 labor-hour from the cutting department and 0.8 labor-hour from the assembly department. Each four-person boat requires 1.8 labor-hours from the cutting department and 1.2 labor-hours from the assembly department. The maximum labor-hours available per month in the cutting department and the assembly department are 864 and 672, respectively. The company makes a profit of \$25 on each two-person boat and \$40 on each four-person boat.

(A) Identify the decision variables.

(B) Summarize the relevant material in a table similar to Table 1 in Example 1.

(C) Write the objective function P.

(D) Write the problem constraints and nonnegative constraints.

(E) Graph the feasible region. Include graphs of the objective function for $P = \$5{,}000$, $P = \$10{,}000$, $P = \$15{,}000$, and $P = \$21{,}600$.

(F) From the graph and constant-profit lines, determine how many boats should be manufactured each month to maximize the profit. What is the maximum profit?

Before proceeding further, let's summarize the steps we used to form the model in Example 1.

PROCEDURE Constructing a Model for an Applied Linear Programming Problem

Step 1 Introduce decision variables.

Step 2 Summarize relevant material in table form, relating columns to the decision variables, if possible (see Table 1).

Step 3 Determine the objective and write a linear objective function.

Step 4 Write problem constraints using linear equations and/or inequalities.

Step 5 Write nonnegative constraints.

Explore and Discuss 1

Refer to the feasible region S shown in Figure 3.

(A) Let $P = x + y$. Graph the constant-profit lines through the points $(5, 5)$ and $(10, 10)$. Place a straightedge along the line with the smaller profit and slide it in the direction of increasing profit, without changing its slope. What is the maximum value of P? Where does this maximum value occur?

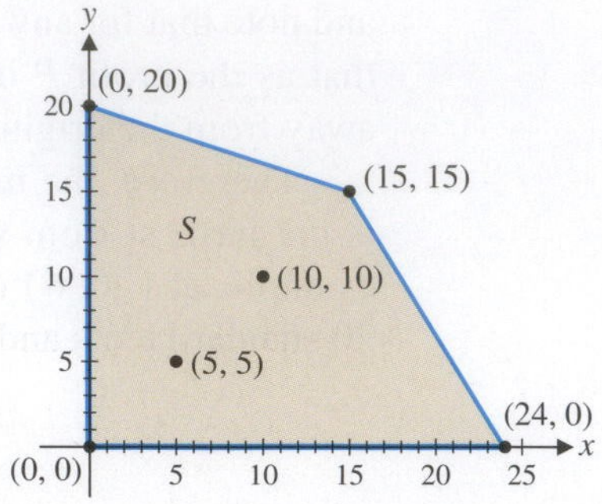

Figure 3

(B) Repeat part (A) for $P = x + 10y$.

(C) Repeat part (A) for $P = 10x + y$.

General Description of Linear Programming

In Example 1 and Matched Problem 1, the optimal solution occurs at a corner point of the feasible region. Is this always the case? The answer is a qualified yes, as we will see in Theorem 1. First, we give a few general definitions.

A **linear programming problem** is one that is concerned with finding the **optimal value** (maximum or minimum value) of a linear **objective function** z of the form

$$z = ax + by, \text{ where } a \text{ and } b \text{ do not both } = 0$$

and the **decision variables** x and y are subject to **problem constraints** in the form of $\le$ or $\ge$ linear inequalities and equations. In addition, the decision variables must satisfy the **nonnegative constraints** $x \ge 0, y \ge 0$. The set of points satisfying both the problem constraints and the nonnegative constraints is called the **feasible region** for the problem. Any point in the feasible region that produces the optimal value of the objective function over the feasible region is called an **optimal solution**.

THEOREM 1 Fundamental Theorem of Linear Programming

If the optimal value of the objective function in a linear programming problem exists, then that value must occur at one or more of the corner points of the feasible region.

Theorem 1 provides a simple procedure for solving a linear programming problem, *provided that the problem has an optimal solution—not all do*. In order to use Theorem 1, we must know that the problem under consideration has an optimal solution. Theorem 2 provides some conditions that will ensure that a linear programming problem has an optimal solution.

THEOREM 2 Existence of Optimal Solutions

(A) If the feasible region for a linear programming problem is bounded, then both the maximum value and the minimum value of the objective function always exist.

(B) If the feasible region is unbounded and the coefficients of the objective function are positive, then the minimum value of the objective function exists but the maximum value does not.

(C) If the feasible region is empty (that is, there are no points that satisfy all the constraints), then both the maximum value and the minimum value of the objective function do not exist.

Geometric Method for Solving Linear Programming Problems

The preceding discussion leads to the following procedure for the geometric solution of linear programming problems with two decision variables:

PROCEDURE Geometric Method for Solving a Linear Programming Problem with Two Decision Variables

Step 1 Graph the feasible region. Then, if an optimal solution exists according to Theorem 2, find the coordinates of each corner point.

Step 2 Construct a **corner point table** listing the value of the objective function at each corner point.

Step 3 Determine the optimal solution(s) from the table in Step 2.

Step 4 For an applied problem, interpret the optimal solution(s) in terms of the original problem.

Before we consider more applications, let's use this procedure to solve some linear programming problems where the model has already been determined.

EXAMPLE 2 **Solving a Linear Programming Problem**

(A) Minimize and maximize $z = 3x + y$

subject to
$$\begin{aligned} 2x + y &\le 20 \\ 10x + y &\ge 36 \\ 2x + 5y &\ge 36 \\ x, y &\ge 0 \end{aligned}$$

(B) Minimize and maximize $z = 10x + 20y$

subject to
$$\begin{aligned} 6x + 2y &\ge 36 \\ 2x + 4y &\ge 32 \\ y &\le 20 \\ x, y &\ge 0 \end{aligned}$$

SOLUTION

(A) Step 1 Graph the feasible region S (Fig. 4). Then, after checking Theorem 2 to determine whether an optimal solution exists, find the coordinates of each corner point. Since S is bounded, z will have both a maximum and a minimum value on S (Theorem 2A) and these will both occur at corner points (Theorem 1).

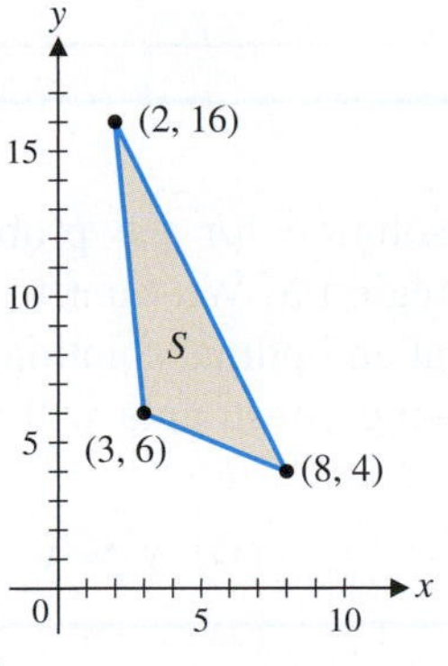

Figure 4

Step 2 Evaluate the objective function at each corner point, as shown in the table.

Corner Point	
(x, y)	$z = 3x + y$
(3, 6)	15
(2, 16)	22
(8, 4)	28

Step 3 Determine the optimal solutions from Step 2. Examining the values in the table, we see that the minimum value of z is 15 at (3, 6) and the maximum value of z is 28 at (8, 4).

(B) Step 1 Graph the feasible region S (Fig. 5). Then, after checking Theorem 2 to determine whether an optimal solution exists, find the coordinates of each corner point. Since S is unbounded and the coefficients of the

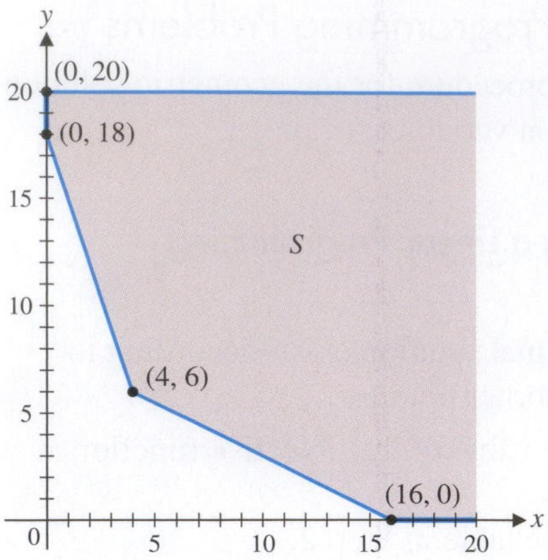

Figure 5

objective function are positive, z has a minimum value on S but no maximum value (Theorem 2B).

Step 2 Evaluate the objective function at each corner point, as shown in the table.

Corner Point (x, y)	$z = 10x + 20y$
$(0, 20)$	400
$(0, 18)$	360
$(4, 6)$	160
$(16, 0)$	160

Step 3 Determine the optimal solution from Step 2. The minimum value of z is 160 at $(4, 6)$ and at $(16, 0)$.

The solution to Example 2B is a **multiple optimal solution**. In general, if two corner points are both optimal solutions to a linear programming problem, then any point on the line segment joining them is also an optimal solution. This is the only way that optimal solutions can occur at noncorner points.

Matched Problem 2

(A) Maximize and minimize $z = 4x + 2y$ subject to the constraints given in Example 2A.

(B) Maximize and minimize $z = 20x + 5y$ subject to the constraints given in Example 2B.

CONCEPTUAL INSIGHT

Determining that an optimal solution exists is a critical step in the solution of a linear programming problem. If you skip this step, you may examine a corner point table like the one in the solution of Example 2B and erroneously conclude that the maximum value of the objective function is 400.

Explore and Discuss 2

In Example 2B we saw that there was no optimal solution for the problem of maximizing the objective function z over the feasible region S. We want to add an additional constraint to modify the feasible region so that an optimal solution for the maximization problem does exist. Which of the following constraints will accomplish this objective?

(A) $x \le 20$ (B) $y \ge 4$ (C) $x \le y$ (D) $y \le x$

For an illustration of Theorem 2C, consider the following:

$$\begin{aligned} \text{Maximize} \quad & P = 2x + 3y \\ \text{subject to} \quad & x + y \ge 8 \\ & x + 2y \le 8 \\ & 2x + y \le 10 \\ & x, y \ge 0 \end{aligned}$$

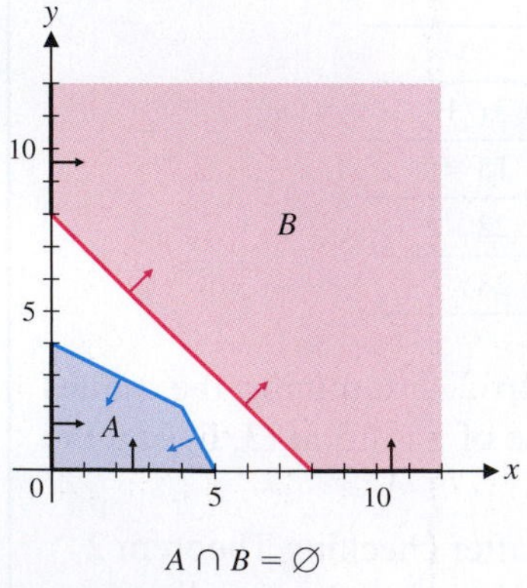

Figure 6

The intersection of the graphs of the constraint inequalities is the empty set (Fig. 6); so the *feasible region is empty*. If this happens, the problem should be reexamined to see if it has been formulated properly. If it has, the management may have to reconsider items such as labor-hours, overtime, budget, and supplies allocated to the project in order to obtain a nonempty feasible region and a solution to the original problem.

Applications

EXAMPLE 3 **Medication** A hospital patient is required to have at least 84 units of drug *A* and 120 units of drug *B* each day (assume that an overdose of either drug is harmless). Each gram of substance *M* contains 10 units of drug *A* and 8 units of drug *B*, and each gram of substance *N* contains 2 units of drug *A* and 4 units of drug *B*. Now, suppose that both *M* and *N* contain an undesirable drug *D*: 3 units per gram in *M* and 1 unit per gram in *N*. How many grams of each of substances *M* and *N* should be mixed to meet the minimum daily requirements and simultaneously minimize the intake of drug *D*? How many units of the undesirable drug *D* will be in this mixture?

SOLUTION First we construct the mathematical model.

Step 1 Introduce decision variables. According to the questions asked, we must decide how many grams of substances *M* and *N* should be mixed to form the daily dose of medication. These two quantities are the decision variables:

$$x = \text{number of grams of substance } M \text{ used}$$

$$y = \text{number of grams of substance } N \text{ used}$$

Step 2 Summarize relevant material in a table, relating the columns to substances *M* and *N*.

	Amount of Drug per Gram		Minimum Daily Requirement
	Substance *M*	Substance *N*	
Drug *A*	10 units/gram	2 units/gram	84 units
Drug *B*	8 units/gram	4 units/gram	120 units
Drug *D*	3 units/gram	1 unit/gram	

Step 3 Determine the objective and the objective function. The objective is to minimize the amount of drug *D* in the daily dose of medication. Using the decision variables and the information in the table, we form the linear objective function

$$C = 3x + y$$

Step 4 Write the problem constraints. The constraints in this problem involve minimum requirements, so the inequalities will take a different form:

$$10x + 2y \geq 84 \quad \text{Drug } A \text{ constraint}$$

$$8x + 4y \geq 120 \quad \text{Drug } B \text{ constraint}$$

Step 5 Add the nonnegative constraints and summarize the model.

$$\begin{aligned} \text{Minimize} \quad & C = 3x + y && \text{Objective function} \\ \text{subject to} \quad & 10x + 2y \geq 84 && \text{Drug } A \text{ constraint} \\ & 8x + 4y \geq 120 && \text{Drug } B \text{ constraint} \\ & x, y \geq 0 && \text{Nonnegative constraints} \end{aligned}$$

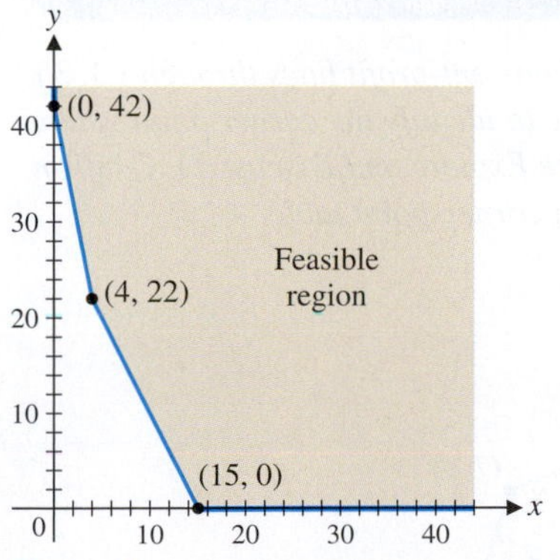

Figure 7

Corner Point (x, y)	$C = 3x + y$
(0, 42)	42
(4, 22)	34
(15, 0)	45

Now we use the geometric method to solve the problem.

Step 1 Graph the feasible region (Fig. 7). Then, after checking Theorem 2 to determine whether an optimal solution exists, find the coordinates of each corner point. Since the feasible region is unbounded and the coefficients of the objective function are positive, this minimization problem has a solution.

Step 2 Evaluate the objective function at each corner point, as shown in the table.

Step 3 Determine the optimal solution from Step 2. The optimal solution is $C = 34$ at the corner point $(4, 22)$.

Step 4 Interpret the optimal solution in terms of the original problem. If we use 4 grams of substance *M* and 22 grams of substance *N*, we will supply the minimum daily requirements for drugs A and B and minimize the intake of the undesirable drug D at 34 units. (Any other combination of M and N from the feasible region will result in a larger amount of the undesirable drug D.)

Matched Problem 3 A chicken farmer can buy a special food mix A at 20¢ per pound and a special food mix B at 40¢ per pound. Each pound of mix A contains 3,000 units of nutrient N_1 and 1,000 units of nutrient N_2; each pound of mix B contains 4,000 units of nutrient N_1 and 4,000 units of nutrient N_2. If the minimum daily requirements for the chickens collectively are 36,000 units of nutrient N_1 and 20,000 units of nutrient N_2, how many pounds of each food mix should be used each day to minimize daily food costs while meeting (or exceeding) the minimum daily nutrient requirements? What is the minimum daily cost? Construct a mathematical model and solve using the geometric method.

CONCEPTUAL INSIGHT

Refer to Example 3. If we change the minimum requirement for drug B from 120 to 125, the optimal solution changes to 3.6 grams of substance M and 24.1 grams of substance N, correct to one decimal place.

Now refer to Example 1. If we change the maximum labor-hours available per day in the assembly department from 84 to 79, the solution changes to 15 standard tents and 8.5 expedition tents.

We can measure 3.6 grams of substance M and 24.1 grams of substance N, but how can we make 8.5 tents? Should we make 8 tents? Or 9 tents? If the solutions to a problem must be integers and the optimal solution found graphically involves decimals, then rounding the decimal value to the nearest integer does not always produce the *optimal integer solution* (see Problem 44, Exercises 5.3). Finding optimal integer solutions to a linear programming problem is called *integer programming* and requires special techniques that are beyond the scope of this book. As mentioned earlier, if we encounter a solution like 8.5 tents per day, we will interpret this as an *average* value over many days of production.

Exercises 5.3

W **Skills Warm-up Exercises**

In Problems 1–8, if necessary, review Theorem 1. In Problems 1–4, the feasible region is the set of points on and inside the rectangle with vertices $(0, 0)$, $(12, 0)$, $(0, 5)$, and $(12, 5)$. Find the maximum and minimum values of the objective function Q over the feasible region.

1. $Q = 7x + 14y$

2. $Q = 3x + 15y$

3. $Q = 10x - 12y$

4. $Q = -9x + 20y$

In Problems 5–8, the feasible region is the set of points on and inside the triangle with vertices $(0, 0)$, $(8, 0)$, and $(0, 10)$. Find the maximum and minimum values of the objective function Q over the feasible region.

5. $Q = -4x - 3y$

6. $Q = 3x + 2y$

7. $Q = -6x + 4y$

8. $Q = 10x - 8y$

A *In Problems 9–12, graph the constant-profit lines through $(3, 3)$ and $(6, 6)$. Use a straightedge to identify the corner point where the maximum profit occurs (see Explore and Discuss 1). Confirm your answer by constructing a corner-point table.*

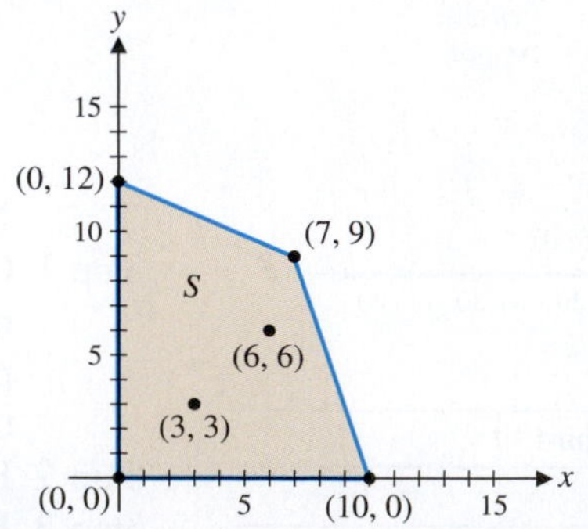

9. $P = x + y$

10. $P = 4x + y$

11. $P = 3x + 7y$

12. $P = 9x + 3y$

In Problems 13–16, graph the constant-cost lines through $(9, 9)$ *and* $(12, 12)$. *Use a straightedge to identify the corner point where the minimum cost occurs. Confirm your answer by constructing a corner-point table.*

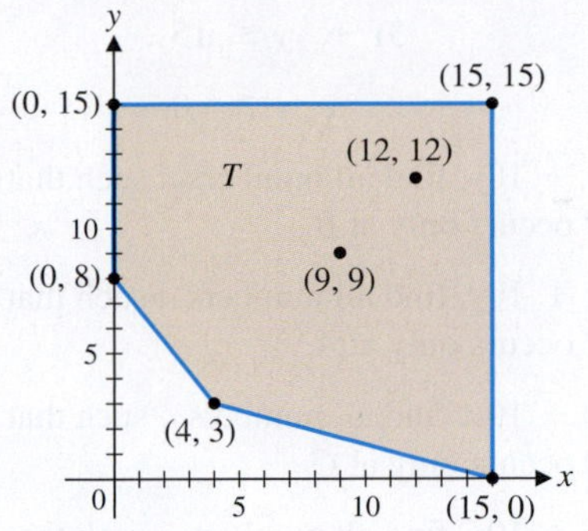

13. $C = 7x + 4y$ **14.** $C = 7x + 9y$

15. $C = 3x + 8y$ **16.** $C = 2x + 11y$

B *Solve the linear programming problems stated in Problems 17–38.*

17. Maximize $P = 10x + 75y$
subject to $x + 8y \le 24$
$x, y \ge 0$

18. Maximize $P = 30x + 12y$
subject to $3x + y \le 18$
$x, y \ge 0$

19. Minimize $C = 8x + 9y$
subject to $5x + 6y \ge 60$
$x, y \ge 0$

20. Minimize $C = 15x + 25y$
subject to $4x + 7y \ge 28$
$x, y \ge 0$

21. Maximize $P = 5x + 5y$
subject to $2x + y \le 10$
$x + 2y \le 8$
$x, y \ge 0$

22. Maximize $P = 3x + 2y$
subject to $6x + 3y \le 24$
$3x + 6y \le 30$
$x, y \ge 0$

23. Minimize and maximize
$z = 2x + 3y$
subject to $2x + y \ge 10$
$x + 2y \ge 8$
$x, y \ge 0$

24. Minimize and maximize
$z = 8x + 7y$
subject to $4x + 3y \ge 24$
$3x + 4y \ge 8$
$x, y \ge 0$

25. Maximize $P = 30x + 40y$
subject to $2x + y \le 10$
$x + y \le 7$
$x + 2y \le 12$
$x, y \ge 0$

26. Maximize $P = 20x + 10y$
subject to $3x + y \le 21$
$x + y \le 9$
$x + 3y \le 21$
$x, y \ge 0$

27. Minimize and maximize
$z = 10x + 30y$
subject to $2x + y \ge 16$
$x + y \ge 12$
$x + 2y \ge 14$
$x, y \ge 0$

28. Minimize and maximize
$z = 400x + 100y$
subject to $3x + y \ge 24$
$x + y \ge 16$
$x + 3y \ge 30$
$x, y \ge 0$

29. Minimize and maximize
$P = 30x + 10y$
subject to $2x + 2y \ge 4$
$6x + 4y \le 36$
$2x + y \le 10$
$x, y \ge 0$

30. Minimize and maximize
$P = 2x + y$
subject to $x + y \ge 2$
$6x + 4y \le 36$
$4x + 2y \le 20$
$x, y \ge 0$

31. Minimize and maximize
$P = 3x + 5y$
subject to $x + 2y \le 6$
$x + y \le 4$
$2x + 3y \ge 12$
$x, y \ge 0$

32. Minimize and maximize
$P = -x + 3y$
subject to $2x - y \ge 4$
$-x + 2y \le 4$
$y \le 6$
$x, y \ge 0$

33. Minimize and maximize

$$\begin{aligned} P &= 20x + 10y \\ \text{subject to}\quad 2x + 3y &\ge 30 \\ 2x + y &\le 26 \\ -2x + 5y &\le 34 \\ x, y &\ge 0 \end{aligned}$$

34. Minimize and maximize

$$\begin{aligned} P &= 12x + 14y \\ \text{subject to}\quad -2x + y &\ge 6 \\ x + y &\le 15 \\ 3x - y &\ge 0 \\ x, y &\ge 0 \end{aligned}$$

35. Maximize $P = 20x + 30y$

$$\begin{aligned} \text{subject to}\quad 0.6x + 1.2y &\le 960 \\ 0.03x + 0.04y &\le 36 \\ 0.3x + 0.2y &\le 270 \\ x, y &\ge 0 \end{aligned}$$

36. Minimize $C = 30x + 10y$

$$\begin{aligned} \text{subject to}\quad 1.8x + 0.9y &\ge 270 \\ 0.3x + 0.2y &\ge 54 \\ 0.01x + 0.03y &\ge 3.9 \\ x, y &\ge 0 \end{aligned}$$

37. Maximize $P = 525x + 478y$

$$\begin{aligned} \text{subject to}\quad 275x + 322y &\le 3{,}381 \\ 350x + 340y &\le 3{,}762 \\ 425x + 306y &\le 4{,}114 \\ x, y &\ge 0 \end{aligned}$$

38. Maximize $P = 300x + 460y$

$$\begin{aligned} \text{subject to}\quad 245x + 452y &\le 4{,}181 \\ 290x + 379y &\le 3{,}888 \\ 390x + 299y &\le 4{,}407 \\ x, y &\ge 0 \end{aligned}$$

C *In Problems 39 and 40, explain why Theorem 2 cannot be used to conclude that a maximum or minimum value exists. Graph the feasible regions and use graphs of the objective function $z = x - y$ for various values of z to discuss the existence of a maximum value and a minimum value.*

39. Minimize and maximize

$$\begin{aligned} z &= x - y \\ \text{subject to}\quad x - 2y &\le 0 \\ 2x - y &\le 6 \\ x, y &\ge 0 \end{aligned}$$

40. Minimize and maximize

$$\begin{aligned} z &= x - y \\ \text{subject to}\quad x - 2y &\ge -6 \\ 2x - y &\ge 0 \\ x, y &\ge 0 \end{aligned}$$

Problems 41–48 refer to the bounded feasible region with corner points $O = (0, 0)$, $A = (0, 5)$, $B = (4, 3)$, and $C = (5, 0)$ that is determined by the system of inequalities

$$\begin{aligned} x + 2y &\le 10 \\ 3x + y &\le 15 \\ x, y &\ge 0 \end{aligned}$$

41. If $P = ax + 10y$, find all numbers a such that the maximum value of P occurs only at B.

42. If $P = ax + 10y$, find all numbers a such that the maximum value of P occurs only at A.

43. If $P = ax + 10y$, find all numbers a such that the maximum value of P occurs only at C.

44. If $P = ax + 10y$, find all numbers a such that the maximum value of P occurs at both A and B.

45. If $P = ax + 10y$, find all numbers a such that the maximum value of P occurs at both B and C.

46. If $P = ax + 10y$, find all numbers a such that the minimum value of P occurs only at C.

47. If $P = ax + 10y$, find all numbers a such that the minimum value of P occurs at both O and C.

48. If $P = ax + 10y$, explain why the minimum value of P cannot occur at B.

Applications

In Problems 49–64, construct a mathematical model in the form of a linear programming problem. (The answers in the back of the book for these application problems include the model.) Then solve by the geometric method.

49. Water skis. A manufacturing company makes two types of water skis—a trick ski and a slalom ski. The relevant manufacturing data are given in the table below.

	Labor-Hours per Ski		Maximum Labor-Hours
Department	**Trick Ski**	**Slalom Ski**	**Available per Day**
Fabricating	6	4	108
Finishing	1	1	24

(A) If the profit on a trick ski is \$40 and the profit on a slalom ski is \$30, how many of each type of ski should be manufactured each day to realize a maximum profit? What is the maximum profit?

(B) Discuss the effect on the production schedule and the maximum profit if the profit on a slalom ski decreases to \$25.

(C) Discuss the effect on the production schedule and the maximum profit if the profit on a slalom ski increases to \$45.

50. Furniture. A furniture manufacturing company manufactures dining-room tables and chairs. The relevant manufacturing data are given in the table below.

Department	Labor-Hours per Unit		Maximum Labor-Hours Available per Day
	Table	Chair	
Assembly	8	2	400
Finishing	2	1	120
Profit per unit	\$90	\$25	

(A) How many tables and chairs should be manufactured each day to realize a maximum profit? What is the maximum profit?

(B) Discuss the effect on the production schedule and the maximum profit if the marketing department of the company decides that the number of chairs produced should be at least four times the number of tables produced.

51. **Production scheduling.** A furniture company has two plants that produce the lumber used in manufacturing tables and chairs. In 1 day of operation, plant *A* can produce the lumber required to manufacture 20 tables and 60 chairs, and plant *B* can produce the lumber required to manufacture 25 tables and 50 chairs. The company needs enough lumber to manufacture at least 200 tables and 500 chairs.

(A) If it costs \$1,000 to operate plant *A* for 1 day and \$900 to operate plant *B* for 1 day, how many days should each plant be operated to produce a sufficient amount of lumber at a minimum cost? What is the minimum cost?

(B) Discuss the effect on the operating schedule and the minimum cost if the daily cost of operating plant *A* is reduced to \$600 and all other data in part (A) remain the same.

(C) Discuss the effect on the operating schedule and the minimum cost if the daily cost of operating plant *B* is reduced to \$800 and all other data in part (A) remain the same.

52. **Computers.** An electronics firm manufactures two types of personal computers—a standard model and a portable model. The production of a standard computer requires a capital expenditure of \$400 and 40 hours of labor. The production of a portable computer requires a capital expenditure of \$250 and 30 hours of labor. The firm has \$20,000 capital and 2,160 labor-hours available for production of standard and portable computers.

(A) What is the maximum number of computers the company is capable of producing?

(B) If each standard computer contributes a profit of \$320 and each portable model contributes a profit of \$220, how much profit will the company make by producing the maximum number of computers determined in part (A)? Is this the maximum profit? If not, what is the maximum profit?

53. **Transportation.** The officers of a high school senior class are planning to rent buses and vans for a class trip. Each bus can transport 40 students, requires 3 chaperones, and costs \$1,200 to rent. Each van can transport 8 students, requires 1 chaperone, and costs \$100 to rent. Since there are 400 students in the senior class that may be eligible to go on the trip, the officers must plan to accommodate at least 400 students. Since only 36 parents have volunteered to serve as chaperones, the officers must plan to use at most 36 chaperones. How many vehicles of each type should the officers rent in order to minimize the transportation costs? What are the minimal transportation costs?

54. **Transportation.** Refer to Problem 53. If each van can transport 7 people and there are 35 available chaperones, show that the optimal solution found graphically involves decimals. Find all feasible solutions with integer coordinates and identify the one that minimizes the transportation costs. Can this optimal integer solution be obtained by rounding the optimal decimal solution? Explain.

55. **Investment.** An investor has \$60,000 to invest in a CD and a mutual fund. The CD yields 5% and the mutual fund yields an average of 9%. The mutual fund requires a minimum investment of \$10,000, and the investor requires that at least twice as much should be invested in CDs as in the mutual fund. How much should be invested in CDs and how much in the mutual fund to maximize the return? What is the maximum return?

56. **Investment.** An investor has \$24,000 to invest in bonds of AAA and B qualities. The AAA bonds yield an average of 6%, and the B bonds yield 10%. The investor requires that at least three times as much money should be invested in AAA bonds as in B bonds. How much should be invested in each type of bond to maximize the return? What is the maximum return?

57. **Pollution control.** Because of new federal regulations on pollution, a chemical plant introduced a new, more expensive process to supplement or replace an older process used in the production of a particular chemical. The older process emitted 20 grams of sulfur dioxide and 40 grams of particulate matter into the atmosphere for each gallon of chemical produced. The new process emits 5 grams of sulfur dioxide and 20 grams of particulate matter for each gallon produced. The company makes a profit of 60¢ per gallon and 20¢ per gallon on the old and new processes, respectively.

(A) If the government allows the plant to emit no more than 16,000 grams of sulfur dioxide and 30,000 grams of particulate matter daily, how many gallons of the chemical should be produced by each process to maximize daily profit? What is the maximum daily profit?

(B) Discuss the effect on the production schedule and the maximum profit if the government decides to restrict emissions of sulfur dioxide to 11,500 grams daily and all other data remain unchanged.

(C) Discuss the effect on the production schedule and the maximum profit if the government decides to restrict emissions of sulfur dioxide to 7,200 grams daily and all other data remain unchanged.

58. **Capital expansion.** A fast-food chain plans to expand by opening several new restaurants. The chain operates two types of restaurants, drive-through and full-service. A drive-through restaurant costs \$100,000 to construct, requires 5 employees, and has an expected annual revenue of \$200,000. A full-service restaurant costs \$150,000 to construct, requires 15 employees, and has an expected annual revenue of \$500,000. The chain has \$2,400,000 in capital available for expansion. Labor contracts require that they hire no more than 210 employees, and licensing restrictions require that they open no more than 20 new restaurants. How many

restaurants of each type should the chain open in order to maximize the expected revenue? What is the maximum expected revenue? How much of their capital will they use and how many employees will they hire?

59. Fertilizer. A fruit grower can use two types of fertilizer in his orange grove, brand A and brand B. The amounts (in pounds) of nitrogen, phosphoric acid, and chloride in a bag of each brand are given in the table. Tests indicate that the grove needs at least 1,000 pounds of phosphoric acid and at most 400 pounds of chloride.

	Pounds per Bag	
	Brand A	Brand B
Nitrogen	8	3
Phosphoric acid	4	4
Chloride	2	1

(A) If the grower wants to maximize the amount of nitrogen added to the grove, how many bags of each mix should be used? How much nitrogen will be added?

(B) If the grower wants to minimize the amount of nitrogen added to the grove, how many bags of each mix should be used? How much nitrogen will be added?

60. Nutrition. A dietitian is to arrange a special diet composed of two foods, M and N. Each ounce of food M contains 30 units of calcium, 10 units of iron, 10 units of vitamin A, and 8 units of cholesterol. Each ounce of food N contains 10 units of calcium, 10 units of iron, 30 units of vitamin A, and 4 units of cholesterol. If the minimum daily requirements are 360 units of calcium, 160 units of iron, and 240 units of vitamin A, how many ounces of each food should be used to meet the minimum requirements and at the same time minimize the cholesterol intake? What is the minimum cholesterol intake?

61. Plant food. A farmer can buy two types of plant food, mix A and mix B. Each cubic yard of mix A contains 20 pounds of phosphoric acid, 30 pounds of nitrogen, and 5 pounds of potash. Each cubic yard of mix B contains 10 pounds of phosphoric acid, 30 pounds of nitrogen, and 10 pounds of potash. The minimum monthly requirements are 460 pounds of phosphoric acid, 960 pounds of nitrogen, and 220 pounds of potash. If mix A costs \$30 per cubic yard and mix B costs \$35 per cubic yard, how many cubic yards of each mix should the farmer blend to meet the minimum monthly requirements at a minimal cost? What is this cost?

62. Animal food. A laboratory technician in a medical research center is asked to formulate a diet from two commercially packaged foods, food A and food B, for a group of animals. Each ounce of food A contains 8 units of fat, 16 units of carbohydrate, and 2 units of protein. Each ounce of food B contains 4 units of fat, 32 units of carbohydrate, and 8 units of protein. The minimum daily requirements are 176 units of fat, 1,024 units of carbohydrate, and 384 units of protein. If food A costs 5¢ per ounce and food B costs 5¢ per ounce, how many ounces of each food should be used to meet the minimum daily requirements at the least cost? What is the cost for this amount of food?

63. Psychology. A psychologist uses two types of boxes with mice and rats. The amount of time (in minutes) that each mouse and each rat spends in each box per day is given in the table. What is the maximum number of mice and rats that can be used in this experiment? How many mice and how many rats produce this maximum?

	Time		Maximum Time
	Mice	Rats	Available per Day
Box A	10 min	20 min	800 min
Box B	20 min	10 min	640 min

64. Sociology. A city council voted to conduct a study on inner-city community problems using sociologists and research assistants from a nearby university. Allocation of time and costs per week are given in the table. How many sociologists and how many research assistants should be hired to minimize the cost and meet the weekly labor-hour requirements? What is the minimum weekly cost?

	Labor-Hours		Minimum Labor-Hours Needed per
	Sociologist	Research Assistant	Week
Fieldwork	10	30	180
Research center	30	10	140
Costs per week	\$500	\$300	

Answers to Matched Problems

1. (A) x = number of two-person boats produced each month
y = number of four-person boats produced each month

(B)

	Labor-Hours Required		Maximum Labor-Hours Available
	Two-Person Boat	Four-Person Boat	per Month
Cutting department	0.9	1.8	864
Assembly department	0.8	1.2	672
Profit per boat	\$25	\$40	

(C) $P = 25x + 40y$

(D) $0.9x + 1.8y \le 864$
$0.8x + 1.2y \le 672$
$x, y \ge 0$

(E)

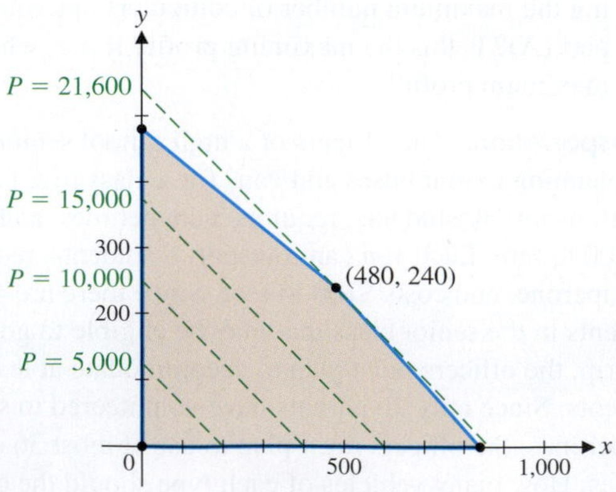

(F) 480 two-person boats, 240 four-person boats; Max $P = \$21{,}600$ per month

2. (A) Min $z = 24$ at $(3, 6)$; Max $z = 40$ at $(2, 16)$ and $(8, 4)$ (multiple optimal solution)

(B) Min $z = 90$ at $(0, 18)$; no maximum value

3. Min $C = 0.2x + 0.4y$

subject to $3{,}000x + 4{,}000y \geq 36{,}000$

$1{,}000x + 4{,}000y \geq 20{,}000$

$x, y \geq 0$

8 lb of mix *A*, 3 lb of mix *B*; Min $C = \$2.80$ per day

Chapter 5 Summary and Review

Important Terms, Symbols, and Concepts

5.1 Linear Inequalities in Two Variables

EXAMPLES

- A line divides the plane into two regions called **half-planes**. A vertical line divides the plane into **left** and **right half-planes**; a nonvertical line divides it into **upper** and **lower half-planes**. In either case, the dividing line is called the **boundary line** of each half-plane.
- The **graph of a linear inequality** is the half-plane obtained by following the procedure on page 260.
- The variables in an applied problem are often required to be nonnegative.

Ex. 1, p. 260
Ex. 2, p. 260
Ex. 3, p. 262
Ex. 4, p. 262

5.2 System of Linear Inequalities in Two Variables

- The **solution region** (also called the **feasible region**) of a system of linear inequalities is the graph of all ordered pairs that simultaneously satisfy all the inequalities in the system. Ex. 1, p. 266
- A **corner point** of a solution region is a point in the region that is the intersection of two boundary lines. Ex. 2, p. 267
- A solution region is **bounded** if it can be enclosed in a circle and **unbounded** if it cannot. Ex. 3, p. 268

5.3 Linear Programming in Two Dimensions: A Geometric Approach

- The problem of finding the optimal (maximum or minimum) value of a linear objective function on a feasible region is called a **linear programming problem**. Ex. 1, p. 272
- The optimal value (if it exists) of the objective function in a linear programming problem must occur at one (or more) of the corner points of the feasible region (Theorem 1, page 276). Existence criteria are described in Theorem 2, page 276, and a solution procedure is listed on page 277. Ex. 2, p. 277
Ex. 3, p. 279

Review Exercises

Work through all the problems in this chapter review and check answers in the back of the book. Answers to all review problems are there, and following each answer is a number in italics indicating the section in which that type of problem is discussed. Where weaknesses show up, review appropriate sections in the text.

A *Graph each inequality.*

1. $x > 2y - 3$

2. $3y - 5x \leq 30$

Graph the systems in Problems 3–6 and indicate whether each solution region is bounded or unbounded. Find the coordinates of each corner point.

3. $5x + 9y \leq 90$
$x, y \geq 0$

4. $15x + 16y \geq 1{,}200$
$x, y \geq 0$

5. $2x + y \leq 8$
$3x + 9y \leq 27$
$x, y \geq 0$

6. $3x + y \geq 9$
$2x + 4y \geq 16$
$x, y \geq 0$

B *In Exercises 7 and 8, state the linear inequality whose graph is given in the figure. Write the boundary line equation in the form $Ax + By = C$, with A, B, and C integers, before stating the inequality.*

7.

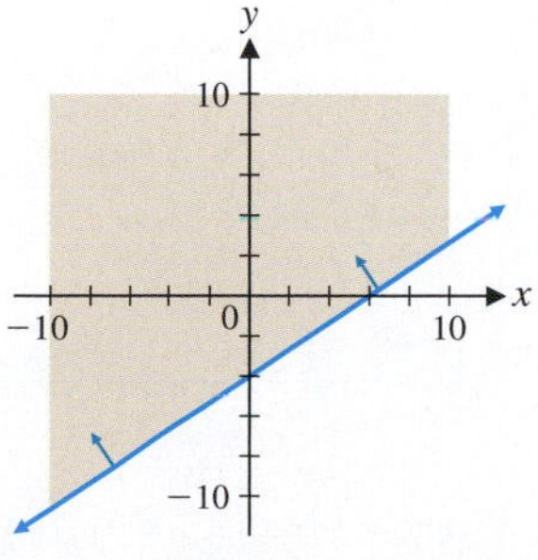

8.

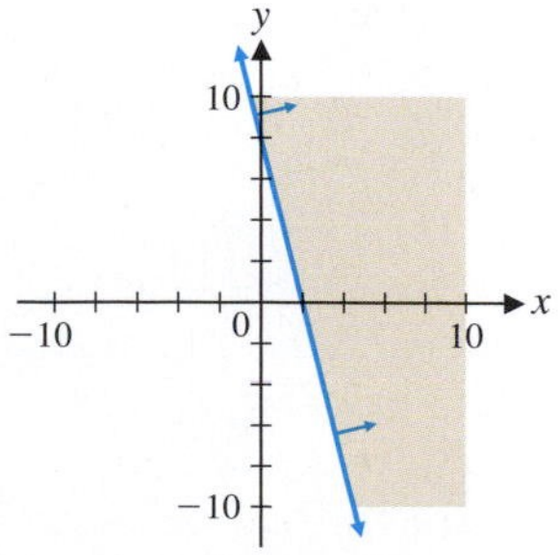

Solve the linear programming problems in Problems 9–13.

9. Maximize $P = 2x + 6y$

subject to $x + 2y \leq 8$

$2x + y \leq 10$

$x, y \geq 0$

10. Minimize $C = 5x + 2y$

subject to
$$\begin{aligned} x + 3y &\ge 15 \\ 2x + y &\ge 20 \\ x, y &\ge 0 \end{aligned}$$

C **11.** Maximize $P = 3x + 4y$

subject to
$$\begin{aligned} x + 2y &\le 12 \\ x + y &\le 7 \\ 2x + y &\le 10 \\ x, y &\ge 0 \end{aligned}$$

12. Minimize $C = 8x + 3y$

subject to
$$\begin{aligned} x + y &\ge 10 \\ 2x + y &\ge 15 \\ x &\ge 3 \\ x, y &\ge 0 \end{aligned}$$

13. Maximize $P = 3x + 2y$

subject to
$$\begin{aligned} 2x + y &\le 22 \\ x + 3y &\le 26 \\ x &\le 10 \\ y &\le 10 \\ x, y &\ge 0 \end{aligned}$$

Applications

14. Electronics. A company uses two machines to solder circuit boards, an oven and a wave soldering machine. A circuit board for a calculator needs 4 minutes in the oven and 2 minutes on the wave machine, while a circuit board for a toaster requires 3 minutes in the oven and 1 minute on the wave machine. (*Source*: Universal Electronics)

(A) How many circuit boards for calculators and toasters can be produced if the oven is available for 5 hours? Express your answer as a linear inequality with appropriate nonnegative restrictions and draw its graph.

(B) How many circuit boards for calculators and toasters can be produced if the wave machine is available for 2 hours? Express your answer as a linear inequality with appropriate nonnegative restrictions and draw its graph.

In Problems 15 and 16, construct a mathematical model in the form of a linear programming problem. (The answers in the back of the book for these application problems include the model.) Then solve the problem by the indicated method.

15. Sail manufacture. South Shore Sail Loft manufactures regular and competition sails. Each regular sail takes 2 hours to cut and 4 hours to sew. Each competition sail takes 3 hours to cut and 10 hours to sew. There are 150 hours available in the cutting department and 380 hours available in the sewing department.

(A) If the Loft makes a profit of \$100 on each regular sail and \$200 on each competition sail, how many sails of each type should the company manufacture to maximize its profit? What is the maximum profit?

(B) An increase in the demand for competition sails causes the profit on a competition sail to rise to \$260. Discuss the effect of this change on the number of sails manufactured and on the maximum profit.

(C) A decrease in the demand for competition sails causes the profit on a competition sail to drop to \$140. Discuss the effect of this change on the number of sails manufactured and on the maximum profit.

16. Animal food. A special diet for laboratory animals is to contain at least 850 units of vitamins, 800 units of minerals, and 1,150 calories. There are two feed mixes available, mix *A* and mix *B*. A gram of mix *A* contains 2 units of vitamins, 2 units of minerals, and 4 calories. A gram of mix *B* contains 5 units of vitamins, 4 units of minerals, and 5 calories.

(A) If mix *A* costs \$0.04 per gram and mix *B* costs \$0.09 per gram, how many grams of each mix should be used to satisfy the requirements of the diet at minimal cost? What is the minimum cost?

(B) If the price of mix *B* decreases to \$0.06 per gram, discuss the effect of this change on the solution in part (A).

(C) If the price of mix *B* increases to \$0.12 per gram, discuss the effect of this change on the solution in part (A).

6 Linear Programming: The Simplex Method

Introduction

The geometric method of solving linear programming problems (presented in Chapter 5) provides an overview of linear programming. But, practically speaking, the geometric method is useful only for problems involving two decision variables and relatively few problem constraints. What happens when we need more decision variables and more problem constraints? We use an algebraic method called the *simplex method*, developed by George B. Dantzig (1914–2005) in 1947. Ideally suited to computer use, the simplex method is used routinely on applied problems involving thousands of variables and problem constraints. In this chapter, we move from an introduction to the simplex method to the *big M method*, which can be used to solve linear programming problems with a large number of variables and constraints. We also explore many applications. A mining company, for example, can use the simplex method to schedule its operations so that production quotas are met at minimum cost (see Problem 50 in Section 6.3).

6.1 The Table Method: An Introduction to the Simplex Method

- Standard Maximization Problems in Standard Form
- Slack Variables
- The Table Method: Basic Solutions and Basic Feasible Solutions
- Basic and Nonbasic Variables
- Summary

In Chapter 5, we denoted variables by single letters such as x and y. In this chapter, we will use letters with subscripts, for example, x_1, x_2, x_3, to denote variables. With this new notation, we will lay the groundwork for solving linear programming problems algebraically, by means of the *simplex method*. In this section, we introduce the *table method* to provide an introduction to the simplex method. Both methods, the table method and the simplex method, solve linear programming problems without the necessity of drawing a graph of the feasible region.

Standard Maximization Problems in Standard Form

The tent production problem that we considered in Section 5.3 is an example of a *standard maximization problem in standard form*. We restate the tent production problem below using subscript notation for the variables.

$$\begin{aligned} \text{Maximize} \quad & P = 50x_1 + 80x_2 && \text{Objective function} \\ \text{subject to} \quad & x_1 + 2x_2 \le 32 && \text{Cutting department constraint} \\ & 3x_1 + 4x_2 \le 84 && \text{Assembly department constraint} \\ & x_1, x_2 \ge 0 && \text{Nonnegative constraints} \end{aligned} \tag{1}$$

The decision variables x_1 and x_2 are the number of standard and expedition tents, respectively, produced each day.

Notice that the problem constraints involve $\le$ inequalities with positive constants to the right of the inequality. Maximization problems that satisfy this condition are called *standard maximization problems*. In this and the next section, we restrict our attention to standard maximization problems.

DEFINITION Standard Maximization Problem in Standard Form

A linear programming problem is said to be a **standard maximization problem in standard form** if its mathematical model is of the following form:
Maximize the objective function

$$P = c_1x_1 + c_2x_2 + \cdots + c_kx_k$$

subject to problem constraints of the form

$$a_1x_1 + a_2x_2 + \cdots + a_kx_k \le b \quad b \ge 0$$

with nonnegative constraints

$$x_1, x_2, \ldots, x_k \ge 0$$

Note: Mathematical model (1) is a standard maximization problem in standard form. The coefficients of the objective function can be any real numbers.

Explore and Discuss 1

Find an example of a standard maximization problem in standard form involving two variables and one problem constraint such that

(A) The feasible region is bounded.

(B) The feasible region is unbounded.

Is it possible for a standard maximization problem to have no solution? Explain.

Slack Variables

To adapt a linear programming problem to the matrix methods used in the simplex process (as discussed in the next section), we convert the problem constraint inequalities into a system of linear equations using *slack variables*. In particular, to convert the system of inequalities from model (1),

$$\begin{aligned} x_1 + 2x_2 &\le 32 && \text{Cutting department constraint} \\ 3x_1 + 4x_2 &\le 84 && \text{Assembly department constraint} \\ x_1, x_2 &\ge 0 && \text{Nonnegative constraints} \end{aligned} \tag{2}$$

into a system of equations, we add variables s_1 and s_2 to the left sides of the problem constraint inequalities in (2) to obtain

$$\begin{aligned} x_1 + 2x_2 + s_1 &= 32 \\ 3x_1 + 4x_2 + s_2 &= 84 \end{aligned} \tag{3}$$

The variables s_1 and s_2 are called **slack variables** because each makes up the difference (takes up the slack) between the left and right sides of an inequality in system (2). For example, if we produced 20 standard tents $(x_1 = 20)$ and 5 expedition tents $(x_2 = 5)$, then the number of labor-hours used in the cutting department would be $20 + 2(5) = 30$, leaving a slack of 2 unused labor-hours out of the 32 available. So s_1 would be equal to 2.

Notice that if the decision variables x_1 and x_2 satisfy the system of constraint inequalities (2), then the slack variables s_1 and s_2 are nonnegative.

The Table Method: Basic Solutions and Basic Feasible Solutions

We call the system of inequalities of a linear programming problem an ***i*-system** ("*i*" for inequality), and we call the associated system of linear equations, obtained via slack variables, an ***e*-system** ("*e*" for equation).

The **solutions** of the *i*-system (2) are the points in the feasible region of Figure 1 (the feasible region was graphed earlier in Section 5.3). For example, $(20, 4)$ is a solution of the *i*-system, but $(20, 8)$ is not (Fig. 1).

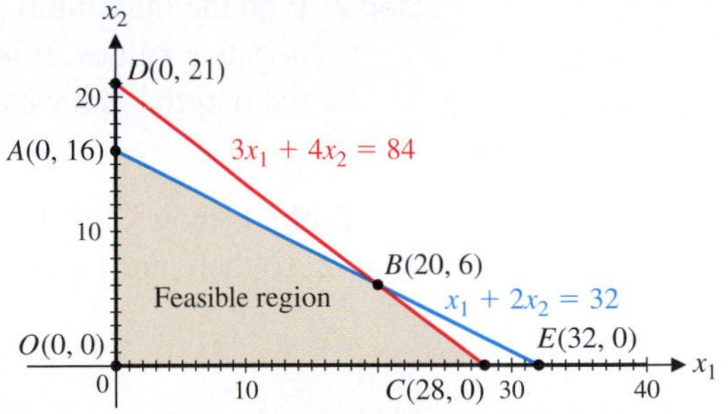

Figure 1

The solutions of the *e*-system (3) are quadruples (x_1, x_2, s_1, s_2). Any point $P = (x_1, x_2)$ in the plane corresponds to a unique solution of the *e*-system (3). For example, $(20, 4)$ corresponds to $(20, 4, 4, 8)$; we say that $(20, 4, 4, 8)$ are the **expanded coordinates** of $(20, 4)$. Similarly, the expanded coordinates of $(20, 8)$ are $(20, 8, -4, -8)$. Both $(20, 4, 4, 8)$ and $(20, 8, -4, -8)$ are solutions of (3).

Systems (2) and (3) do not have the same solutions. However, the solutions of the *i*-system (2) correspond to the solutions of the *e*-system (3) in which the variables x_1, x_2, s_1, and s_2 are all nonnegative. Such a solution of the *e*-system, in which the values of all decision variables and slack variables are nonnegative, is called a **feasible solution**. So $(20, 4, 4, 8)$ is a feasible solution of the *e*-system (3), but the solution $(20, 8, -4, -8)$ is not feasible. Note that $(20, 4)$ is in the feasible region (Fig. 1), but $(20, 8)$ is not. To summarize, the feasible solutions of (3) correspond to the points of the feasible region shown in Figure 1.

Any line segment that forms a boundary for the feasible region of (2) lies on one of the following four lines [replace "≤" by "=" in each inequality of *i*-system (2)]:

$$x_1 + 2x_2 = 32$$
$$3x_1 + 4x_2 = 84$$
$$x_1 \qquad = 0$$
$$x_2 = 0$$

So any corner point of the feasible region must lie on at least two of the four lines. If P is the intersection of the first two lines, then $s_1 = 0$ and $s_2 = 0$ in the expanded coordinates for P. If P is the intersection of the first and third lines, then $s_1 = 0$ and $x_1 = 0$ in the expanded coordinates for P. Continuing in this way, we conclude that for any corner point P of the feasible region, at least two of the variables x_1, x_2, s_1, and s_2 must equal 0 in the expanded coordinates of P.

The reasoning above gives a procedure, called the **table method**, for solving a standard maximization problem in standard form.

PROCEDURE The Table Method (Two Decision Variables)

Assume that a standard maximization problem in standard form has two decision variables x_1 and x_2 and m problem constraints.

Step 1 Use slack variables $s_1, s_2, \ldots, s_m$ to convert the *i*-system to an *e*-system.

Step 2 Form a table with $(m + 2)(m + 1)/2$ rows and $m + 2$ columns labeled $x_1, x_2, s_1, s_2, \ldots, s_m$. In the first row, assign 0 to x_1 and x_2. In the second row, assign 0 to x_1 and s_1. Continue until the rows contain all possible combinations of assigning two 0's to the variables.

Step 3 Complete each row to a solution of the *e*-system, if possible. Because two of the variables have the value 0, this involves solving a system of m linear equations in m variables. Use the Gauss–Jordan method, or another method if you find it easier. If the system has no solution or infinitely many solutions, do not complete the row.

Step 4 Find the maximum value of P over those completed rows that have no negative values. It is the optimal value of the objective function, provided the optimal value exists.

Table 1 shows the table of step 2 for the tent production problem. Note that $m = 2$, so there are $(m + 2)(m + 1)/2 = 6$ rows and $m + 2 = 4$ columns.

Table 1

x_1	x_2	s_1	s_2
0	0		
0		0	
0			0
	0	0	
	0		0
		0	0

Table 2 shows the table of step 3: Each row of Table 1 has been completed to a solution of the *e*-system (3). This involves solving the *e*-system six different times. Of course, the solution in the first row is easy: If x_1 and x_2 are both assigned the value 0 in (3), then clearly $s_1 = 32$ and $s_2 = 84$. The remaining rows are completed similarly: After two variables are assigned the value 0, the resulting system of two equations in two variables can be solved by the Gauss–Jordan method, or by substitution, or by elimination by addition. Such solutions of (3), in which two of the variables have been assigned the value 0, are called **basic solutions**. So there are six basic solutions in Table 2. Note that the six basic solutions correspond to the six intersection points O, A, B, C, D, and E of Figure 1.

Table 2 Basic Solutions

x_1	x_2	s_1	s_2
0	0	32	84
0	16	0	20
0	21	−10	0
32	0	0	−12
28	0	4	0
20	6	0	0

We use step 4 of the procedure to obtain the solution of the linear programming problem. The value of $P = 50x_1 + 80x_2$ is calculated for each row of Table 2 that has no negative values. See Table 3. We conclude that the solution of the linear programming problem is

$$\text{Max } P = \$1{,}480 \text{ at } x_1 = 20, x_2 = 6$$

Table 3 The Table Method

x_1	x_2	s_1	s_2	$P = 50x_1 + 80x_2$
0	0	32	84	0
0	16	0	20	1,280
0	21	−10	0	–
32	0	0	−12	–
28	0	4	0	1,400
20	6	0	0	1,480

We have ignored the third and fourth rows because $s_1 = -10$ in the third row and $s_2 = -12$ in the fourth. Those basic solutions are not feasible. The four remaining basic solutions are feasible. We call them **basic feasible solutions**. Note that the basic feasible solutions correspond to the four corner points O, A, B, C of the feasible region (Fig. 1).

Because basic feasible solutions correspond to the corner points of the feasible region, we can reformulate the fundamental theorem (Theorem 1 in Section 5.3):

THEOREM 1 Fundamental Theorem of Linear Programming: Version 2

If the optimal value of the objective function in a linear programming problem exists, then that value must occur at one or more of the basic feasible solutions.

EXAMPLE 1 **Slack Variables and Basic Solutions**

(A) Use slack variables s_1 and s_2 to convert the following i-system (system of inequalities) to an e-system (system of equations).

$$\begin{aligned} 3x_1 + 2x_2 &\le 21 \\ x_1 + 5x_2 &\le 20 \\ x_1, x_2 &\ge 0 \end{aligned}$$

(B) Find the basic solution for which $x_1 = 0$ and $s_1 = 0$.

(C) Find the basic solution for which $s_1 = 0$ and $s_2 = 0$.

SOLUTION

(A) Introduce slack variables s_1 and s_2 (one slack variable for each of the problem constraint inequalities):

$$\begin{aligned} 3x_1 + 2x_2 + s_1 \quad\quad &= 21 \\ x_1 + 5x_2 \quad\quad + s_2 &= 20 \end{aligned} \qquad (4)$$

(B) Substitute $x_1 = 0$ and $s_1 = 0$ into *e*-system (4):

$$\begin{aligned} 3(0) + 2x_2 + 0 \quad &= 21 \\ 0 + 5x_2 \quad + s_2 &= 20 \end{aligned}$$

Divide the first equation by 2 to get $x_2 = 10.5$. Substitute $x_2 = 10.5$ in the second equation and solve for s_2:

$$s_2 = 20 - 5(10.5) = -32.5$$

The basic solution is

$$(x_1, x_2, s_1, s_2) = (0, 10.5, 0, -32.5)$$

Note that this basic solution is not feasible because at least one of the variables (s_2 in this case) has a negative value.

(C) Substitute $s_1 = 0$ and $s_2 = 0$ into *e*-system (4):

$$\begin{aligned} 3x_1 + 2x_2 + 0 \quad &= 21 \\ x_1 + 5x_2 \quad + 0 &= 20 \end{aligned}$$

This system of two equations in two variables can be solved by Gauss–Jordan elimination or by another of our standard methods. Multiplying the second equation by -3 and adding the two equations gives $-13x_2 = -39$, so $x_2 = 3$. Substituting $x_2 = 3$ in the second equation gives $x_1 = 5$.

The basic solution is

$$(x_1, x_2, s_1, s_2) = (5, 3, 0, 0)$$

Note that this basic solution is feasible because none of the variables has a negative value.

Matched Problem 1 Refer to Example 1. Find the basic solution for which $x_2 = 0$ and $s_1 = 0$.

EXAMPLE 2 **The Table Method** Construct the table of basic solutions and use it to solve the following linear programming problem:

$$\begin{aligned} \text{Maximize } P &= 10x_1 + 25x_2 \\ \text{subject to } 3x_1 + 2x_2 &\le 21 \\ x_1 + 5x_2 &\le 20 \\ x_1, x_2 &\ge 0 \end{aligned}$$

SOLUTION The system of inequalities is identical to the *i*-system of Example 1. The number of problem constraints is $m = 2$, so there will be $(m + 2)(m + 1)/2 = 6$ rows in the table of basic solutions. Because there are two decision variables, x_1 and x_2, we assign two zeros to each row of the table in all possible combinations. We also include the basic solution that was found in Example 1B (row 2 of table) and the basic solution that was found in Example 1C (last row).

x_1	x_2	s_1	s_2
0	0		
0	10.5	0	−32.5
0			0
	0	0	
	0		0
5	3	0	0

We complete the table working one row at a time. We substitute 0's for the two variables indicated by the row, in the *e*-system

$$3x_1 + 2x_2 + s_1 = 21$$

$$x_1 + 5x_2 + s_2 = 20$$

The result is a system of two equations in two variables, which can be solved by Gauss–Jordan elimination or another of our standard methods. Table 4 shows all six basic solutions and the values of the objective function $P = 10x_1 + 25x_2$ at the four basic feasible solutions.

Table 4 The Table Method

x_1	x_2	s_1	s_2	$P = 10x_1 + 25x_2$
0	0	21	20	0
0	10.5	0	−32.5	–
0	4	13	0	100
7	0	0	13	70
20	0	−39	0	–
5	3	0	0	125

We conclude that

$$\text{Max } P = 125 \text{ at } x_1 = 5, x_2 = 3$$

Matched Problem 2 Construct the table of basic solutions and use it to solve the following linear programming problem:

$$\begin{aligned} \text{Maximize} \quad & P = 30x_1 + 40x_2 \\ \text{subject to} \quad & 2x_1 + 3x_2 \le 24 \\ & 4x_1 + 3x_2 \le 36 \\ & x_1, x_2 \ge 0 \end{aligned}$$

Explore and Discuss 1

The following linear programming problem has only one problem constraint:

$$\begin{aligned} \text{Maximize} \quad & P = 2x_1 + 3x_2 \\ \text{subject to} \quad & 4x_1 + 5x_2 \le 20 \\ & x_1, x_2 \ge 0 \end{aligned}$$

Solve it by the table method, then solve it by graphing, and compare the two solutions.

EXAMPLE 3 **The Table Method** Construct the table of basic solutions and use it to solve the following linear programming problem:

$$\begin{aligned} \text{Maximize} \quad & P = 40x_1 + 50x_2 \\ \text{subject to} \quad & x_1 + 6x_2 \le 72 \\ & x_1 + 3x_2 \le 45 \\ & 2x_1 + 3x_2 \le 72 \\ & x_1, x_2 \ge 0 \end{aligned}$$

SOLUTION There are $m = 3$ problem constraints, so we use slack variables s_1, s_2, and s_3 to convert the *i*-system to the *e*-system (5):

$$\begin{aligned} x_1 + 6x_2 + s_1 \qquad\qquad &= 72 \\ x_1 + 3x_2 \qquad + s_2 \qquad &= 45 \\ 2x_1 + 3x_2 \qquad\qquad + s_3 &= 72 \end{aligned} \tag{5}$$

There will be $(m + 2)(m + 1)/2 = 10$ rows in the table of basic solutions. Because there are two decision variables, x_1 and x_2, we assign two zeros to each row of the table in all possible combinations.

x_1	x_2	s_1	s_2	s_3
0	0			
0		0		
0			0	
0				0
	0	0		
	0		0	
	0			0
		0	0	
		0		0
			0	0

We complete the table working one row at a time. We substitute 0's for the two variables indicated by the row, in the *e*-system (5). The result is a system of three equations in three variables, which can be solved by Gauss–Jordan elimination, or by another of our standard methods. Table 5 shows all ten basic solutions, and the values of the objective function, $P = 40x_1 + 50x_2$, at the five basic feasible solutions.

Table 5 The Table Method

x_1	x_2	s_1	s_2	s_3	$P = 40x_1 + 50x_2$
0	0	72	45	72	0
0	12	0	9	36	600
0	15	−18	0	27	–
0	24	−72	−27	0	–
72	0	0	−27	−72	–
45	0	27	0	−18	–
36	0	36	9	0	1,440
18	9	0	0	9	1,170
24	8	0	−3	0	–
27	6	9	0	0	1,380

We conclude that

$$\text{Max } P = 1{,}440 \text{ at } x_1 = 36, x_2 = 0$$

Matched Problem 3 Construct the table of basic solutions and use it to solve the following linear programming problem:

$$\begin{aligned} \text{Maximize} \quad & P = 36x_1 + 24x_2 \\ \text{subject to} \quad & x_1 + 2x_2 \le 8 \\ & x_1 + x_2 \le 5 \\ & 2x_1 + x_2 \le 8 \\ & x_1, x_2 \ge 0 \end{aligned}$$

Basic and Nonbasic Variables

The basic solutions associated with a linear programming problem are found by assigning the value 0 to certain decision variables (the x_i's) and slack variables (the s_i's). Consider, for example, row 2 of Table 5. That row shows the basic solution $(x_1, x_2, s_1, s_2, s_3) = (0, 12, 0, 9, 36)$. It is customary to refer to the variables that are assigned the value 0 as **nonbasic variables** and to the others as **basic variables**. So for the basic solution of row 2, the basic variables are x_2, s_2, and s_3; the nonbasic variables are x_1 and s_1.

Note that the classification of variables as basic or nonbasic depends on the basic solution. Row 8 of Table 5 shows the basic solution $(x_1, x_2, s_1, s_2, s_3) = (18, 9, 0, 0, 9)$. For row 8, the basic variables are x_1, x_2, and s_3; the nonbasic variables are s_1 and s_2.

EXAMPLE 4 **Basic and Nonbasic Variables** Refer to Table 5. For the basic solution $(x_1, x_2, s_1, s_2, s_3) = (36, 0, 36, 9, 0)$ in row 7 of Table 5, classify the variables as basic or nonbasic.

SOLUTION The basic variables are x_1, s_1, and s_2. The other variables, x_2 and s_3, were assigned the value 0, and therefore are nonbasic.

Matched Problem 4 Refer to Table 5. For the basic solution $(x_1, x_2, s_1, s_2, s_3) = (27, 6, 9, 0, 0)$ of row 10, classify the variables as basic or nonbasic.

Explore and Discuss 2

Use the table method to solve the following linear programming problem, and explain why one of the rows in the table cannot be completed to a basic solution:

$$\begin{aligned} \text{Maximize} \quad & P = 10x_1 + 12x_2 \\ \text{subject to} \quad & x_1 + x_2 \le 2 \\ & x_1 + x_2 \le 3 \\ & x_1, x_2 \ge 0 \end{aligned}$$

Summary

The examples in this section illustrate the table method when there are two decision variables. But the method can be used when there are k decision variables, where k is any positive integer.

The number of ways in which r objects can be chosen from a set of n objects, without regard to order, is denoted by ${}_nC_r$ and given by the formula

$${}_nC_r = \frac{n!}{r!(n-r)!}$$

(The formula, giving the number of combinations of n distinct objects taken r at a time, is explained and derived in Chapter 7). If there are k decision variables and m problem constraints in a linear programming problem, then the number of rows in the table of basic solutions is ${}_{k+m}C_k$, because this is the number of ways of selecting k of the $k + m$ variables to be assigned the value 0.

Reminder

The expression 6!, read "6 factorial," stands for the product of the natural numbers from 1 through 6. So $6! = 6 \cdot 5 \cdot 4 \cdot 3 \cdot 2 \cdot 1 = 720$, and

$$\begin{aligned} {}_6C_2 &= \frac{6!}{2!(6-2)!} \\ &= \frac{6 \cdot 5 \cdot 4 \cdot 3 \cdot 2 \cdot 1}{2 \cdot 1 \cdot 4 \cdot 3 \cdot 2 \cdot 1} \\ &= \frac{6 \cdot 5}{2 \cdot 1} = 15. \end{aligned}$$

PROCEDURE The Table Method (*k* Decision Variables)

Assume that a standard maximization problem in standard form has k decision variables $x_1, x_2, \ldots, x_k$, and m problem constraints.

Step 1 Use slack variables $s_1, s_2, \ldots, s_m$ to convert the *i*-system to an *e*-system.

Step 2 Form a table with ${}_{k+m}C_k$ rows and $k + m$ columns labeled $x_1, x_2, \ldots, x_k, s_1, s_2, \ldots, s_m$. In the first row, assign 0 to $x_1, x_2, \ldots, x_k$. Continue until the rows contain all possible combinations of assigning k 0's to the variables.

Step 3 Complete each row to a solution of the *e*-system, if possible. Because k of the variables have the value 0, this involves solving a system of m linear equations in m variables. Use the Gauss–Jordan method, or another method if you find it easier. If the system has no solutions, or infinitely many solutions, do not complete the row.

Step 4 Find the maximum value of P over those completed rows that have no negative values (that is, over the basic feasible solutions). It is the optimal value of the objective function, provided the optimal value exists.

The benefit of the table method is that it gives a procedure for **finding all corner points of the feasible region without drawing a graph**.

Unfortunately, the number of rows in the table becomes too large to be practical, even for computers, when the number of decision variables and problem constraints is large. For example, with $k = 30$ decision variables and $m = 35$ problem constraints, the number of rows is

$${}_{65}C_{30} \approx 3 \times 10^{18}$$

We need a procedure for finding the optimal solution of a linear programming problem without having to find every corner point. The *simplex method*, discussed in the next section, is such a procedure. It gives a practical method for solving large linear programming problems.

Exercises 6.1

Skills Warm-up Exercises

W *In Problems 1–8, evaluate the expression. (If necessary, review Section B.3).*

1. $\dfrac{8!}{3!5!}$ **2.** $\dfrac{10!}{2!8!}$ **3.** $\dfrac{11!}{9!2!}$ **4.** $\dfrac{7!}{4!3!}$

5. In how many ways can two variables be chosen from $x_1, x_2, s_1, s_2, s_3, s_4, s_5$ and assigned the value 0?

6. In how many ways can three variables be chosen from $x_1, x_2, x_3, s_1, s_2, s_3$ and assigned the value 0?

7. In how many ways can four variables be chosen from $x_1, x_2, x_3, x_4, s_1, s_2, s_3, s_4$ and assigned the value 0?

8. In how many ways can three variables be chosen from $x_1, x_2, s_1, s_2, s_3, s_4, s_5, s_6$, and assigned the value 0?

A *Problems 9–12 refer to the system*

$$\begin{aligned} 2x_1 + 5x_2 + s_1 \quad &= 10 \\ x_1 + 3x_2 \quad + s_2 &= 8 \end{aligned}$$

9. Find the solution of the system for which $x_1 = 0, s_1 = 0$.

10. Find the solution of the system for which $x_1 = 0, s_2 = 0$.

11. Find the solution of the system for which $x_2 = 0, s_2 = 0$.

12. Find the solution of the system for which $x_2 = 0, s_1 = 0$.

In Problems 13–20, write the e-system obtained via slack variables for the given linear programming problem.

13. Maximize $P = 5x_1 + 7x_2$

subject to
$$\begin{aligned} 2x_1 + 3x_2 &\le 9 \\ 6x_1 + 7x_2 &\le 13 \\ x_1, x_2 &\ge 0 \end{aligned}$$

14. Maximize $P = 35x_1 + 25x_2$

subject to
$$\begin{aligned} 10x_1 + 15x_2 &\le 100 \\ 5x_1 + 20x_2 &\le 120 \\ x_1, x_2 &\ge 0 \end{aligned}$$

15. Maximize $P = 3x_1 + 5x_2$
subject to
$$12x_1 - 14x_2 \le 55$$
$$19x_1 + 5x_2 \le 40$$
$$-8x_1 + 11x_2 \le 64$$
$$x_1, x_2 \ge 0$$

16. Maximize $P = 13x_1 + 25x_2$
subject to
$$3x_1 + 5x_2 \le 27$$
$$8x_1 + 3x_2 \le 19$$
$$4x_1 + 9x_2 \le 34$$
$$x_1, x_2 \ge 0$$

17. Maximize $P = 4x_1 + 7x_2$
subject to
$$6x_1 + 5x_2 \le 18$$
$$x_1, x_2 \ge 0$$

18. Maximize $P = 13x_1 + 8x_2$
subject to
$$x_1 + 2x_2 \le 20$$
$$x_1, x_2 \ge 0$$

19. Maximize $P = x_1 + 2x_2$
subject to
$$4x_1 - 3x_2 \le 12$$
$$5x_1 + 2x_2 \le 25$$
$$-3x_1 + 7x_2 \le 32$$
$$2x_1 + x_2 \le 9$$
$$x_1, x_2 \ge 0$$

20. Maximize $P = 8x_1 + 9x_2$
subject to
$$30x_1 - 25x_2 \le 75$$
$$10x_1 + 13x_2 \le 30$$
$$5x_1 + 18x_2 \le 40$$
$$40x_1 + 36x_2 \le 85$$
$$x_1, x_2 \ge 0$$

B *Problems 21–30 refer to the table below of the six basic solutions to the e-system*

$$2x_1 + 3x_2 + s_1 = 24$$
$$4x_1 + 3x_2 + s_2 = 36$$

	x_1	x_2	s_1	s_2
(A)	0	0	24	36
(B)	0	8	0	12
(C)	0	12	−12	0
(D)	12	0	0	−12
(E)	9	0	6	0
(F)	6	4	0	0

21. In basic solution (A), which variables are basic?

22. In basic solution (B), which variables are nonbasic?

23. In basic solution (C), which variables are nonbasic?

24. In basic solution (D), which variables are basic?

25. Which of the six basic solutions are feasible? Explain.

26. Which of the basic solutions are not feasible? Explain.

27. Use the basic feasible solutions to maximize $P = 2x_1 + 5x_2$.

28. Use the basic feasible solutions to maximize $P = 8x_1 + 5x_2$.

29. Describe geometrically the set of all points in the plane such that $s_1 > 0$.

30. Describe geometrically the set of all points in the plane such that $s_2 < 0$.

Problems 31–40 refer to the partially completed table below of the 10 basic solutions to the e-system

$$x_1 + x_2 + s_1 = 24$$
$$2x_1 + x_2 + s_2 = 30$$
$$4x_1 + x_2 + s_3 = 48$$

	x_1	x_2	s_1	s_2	s_3
(A)	0	0	24	30	48
(B)	0	24	0	6	24
(C)	0	30	−6	0	18
(D)	0	48	−24	−18	0
(E)	24	0	0	−18	−48
(F)	15	0	9	0	−12
(G)		0			0
(H)			0	0	
(I)			0		0
(J)				0	0

31. In basic solution (C), which variables are basic?

32. In basic solution (E), which variables are nonbasic?

33. In basic solution (G), which variables are nonbasic?

34. In basic solution (I), which variables are basic?

35. Which of the basic solutions (A) through (F) are not feasible? Explain.

36. Which of the basic solutions (A) through (F) are feasible? Explain.

37. Find basic solution (G).

38. Find basic solution (H).

39. Find basic solution (I).

40. Find basic solution (J).

In Problems 41–48, convert the given i-system to an e-system using slack variables. Then construct a table of all basic solutions of the e-system. For each basic solution, indicate whether or not it is feasible.

41. $4x_1 + 5x_2 \le 20$
$x_1, x_2 \ge 0$

42. $3x_1 + 8x_2 \le 24$
$x_1, x_2 \ge 0$

43. $x_1 + x_2 \le 6$
$x_1 + 4x_2 \le 12$
$x_1, x_2 \ge 0$

44. $5x_1 + x_2 \le 15$
$x_1 + x_2 \le 7$
$x_1, x_2 \ge 0$

45. $2x_1 + 5x_2 \le 20$
$x_1 + 2x_2 \le 9$
$x_1, x_2 \ge 0$

46. $x_1 + 3x_2 \le 18$
$5x_1 + 4x_2 \le 35$
$x_1, x_2 \ge 0$

47. $x_1 + 2x_2 \le 24$
$x_1 + x_2 \le 15$
$2x_1 + x_2 \le 24$
$x_1, x_2 \ge 0$

48. $5x_1 + 4x_2 \le 240$
$5x_1 + 2x_2 \le 150$
$5x_1 + x_2 \le 120$
$x_1, x_2 \ge 0$

In Problems 49–54, graph the system of inequalities from the given problem, and list the corner points of the feasible region. Verify that the corner points of the feasible region correspond to the basic feasible solutions of the associated e-system.

49. Problem 41

50. Problem 42

51. Problem 43

52. Problem 44

53. Problem 45

54. Problem 46

C **55.** For a standard maximization problem in standard form, with two decision variables x_1 and x_2, explain why the feasible region is not empty.

56. For a standard maximization problem in standard form, with k decision variables, $x_1, x_2, \ldots, x_k$, explain why the feasible region is not empty.

57. If $5x_1 + 4x_2 \le 1{,}000$ is one of the problem constraints in a standard maximization problem in standard form with two decision variables, explain why the optimal value of the objective function exists. [*Hint:* See Theorem 2 in Section 5.3].

58. If $a_1x_1 + a_2x_2 \le b$ is one of the problem constraints in a standard maximization problem in standard form with two decision variables, and a_1 and a_2 are both positive, explain why the optimal value of the objective function exists. [*Hint:* See Theorem 2 in Section 5.3].

In Problems 59–66, solve the given linear programming problem using the table method (the table of basic solutions was constructed in Problems 41–48).

59. Maximize $P = 10x_1 + 9x_2$
subject to $4x_1 + 5x_2 \le 20$
$x_1, x_2 \ge 0$

60. Maximize $P = 4x_1 + 7x_2$
subject to $3x_1 + 8x_2 \le 24$
$x_1, x_2 \ge 0$

61. Maximize $P = 15x_1 + 20x_2$
subject to $x_1 + x_2 \le 6$
$x_1 + 4x_2 \le 12$
$x_1, x_2 \ge 0$

62. Maximize $P = 5x_1 + 20x_2$
subject to $5x_1 + x_2 \le 15$
$x_1 + x_2 \le 7$
$x_1, x_2 \ge 0$

63. Maximize $P = 25x_1 + 10x_2$
subject to $2x_1 + 5x_2 \le 20$
$x_1 + 2x_2 \le 9$
$x_1, x_2 \ge 0$

64. Maximize $P = 40x_1 + 50x_2$
subject to $x_1 + 3x_2 \le 18$
$5x_1 + 4x_2 \le 35$
$x_1, x_2 \ge 0$

65. Maximize $P = 30x_1 + 40x_2$
subject to $x_1 + 2x_2 \le 24$
$x_1 + x_2 \le 15$
$2x_1 + x_2 \le 24$
$x_1, x_2 \ge 0$

66. Maximize $P = x_1 + x_2$
subject to $5x_1 + 4x_2 \le 240$
$5x_1 + 2x_2 \le 150$
$5x_1 + x_2 \le 120$
$x_1, x_2 \ge 0$

In Problems 67–70, explain why the linear programming problem has no optimal solution.

67. Maximize $P = 8x_1 + 9x_2$
subject to $3x_1 - 7x_2 \le 42$
$x_1, x_2 \ge 0$

68. Maximize $P = 12x_1 + 8x_2$
subject to $-2x_1 + 10x_2 \le 30$
$x_1, x_2 \ge 0$

69. Maximize $P = 6x_1 + 13x_2$
subject to $-4x_1 + x_2 \le 4$
$4x_1 - 5x_2 \le 12$
$x_1, x_2 \ge 0$

70. Maximize $P = 18x_1 + 11x_2$
subject to $3x_1 - 2x_2 \le 6$
$-3x_1 + 2x_2 \le 6$
$x_1, x_2 \ge 0$

In Problems 71–72, explain why the linear programming problem has an optimal solution, and find it using the table method.

71. Maximize $P = 20x_1 + 25x_2$
subject to $-2x_1 + x_2 \le 50$
$x_1 \le 100$
$x_1, x_2 \ge 0$

72. Maximize $P = 15x_1 + 12x_2$
subject to $-2x_1 + 5x_2 \le 10$
$2x_1 - x_2 \le 6$
$x_1, x_2 \ge 0$

73. A linear programming problem has four decision variables x_1, x_2, x_3, x_4, and six problem constraints. How many rows are there in the table of basic solutions of the associated e-system?

74. A linear programming problem has five decision variables x_1, x_2, x_3, x_4, x_5 and six problem constraints. How many rows are there in the table of basic solutions of the associated e-system?

75. A linear programming problem has 30 decision variables $x_1, x_2, \ldots, x_{30}$ and 42 problem constraints. How many rows are there in the table of basic solutions of the associated e-system? (Write the answer using scientific notation.)

76. A linear programming problem has 40 decision variables $x_1, x_2, \ldots, x_{40}$ and 85 problem constraints. How many rows are there in the table of basic solutions of the associated e-system? (Write the answer using scientific notation.)

Answers to Matched Problems

1. $(x_1, x_2, s_1, s_2) = (7, 0, 0, 13)$

2.

x_1	x_2	s_1	s_2	$P = 30x + 40x$
0	0	24	36	0
0	8	0	12	320
0	12	−12	0	–
12	0	0	−12	–
9	0	6	0	270
6	4	0	0	340

Max $P = 340$ at $x_1 = 6, x_2 = 4$

3.

x_1	x_2	s_1	s_2	s_3	$P = 36x_1 + 24x_2$
0	0	8	5	8	0
0	4	0	1	4	96
0	5	−2	0	3	–
0	8	−8	−3	0	–
8	0	0	−3	−8	–
5	0	3	0	−2	–
4	0	4	1	0	144
2	3	0	0	1	144
8/3	8/3	0	−1/3	0	–
3	2	1	0	0	156

Max $P = 156$ at $x_1 = 3, x_2 = 2$

4. x_1, x_2, and s_1 are basic; s_2 and s_3 are nonbasic

6.2 The Simplex Method: Maximization with Problem Constraints of the Form ≤

- Initial System
- Simplex Tableau
- Pivot Operation
- Interpreting the Simplex Process Geometrically
- Simplex Method Summarized
- Application

Now we can develop the simplex method for a standard maximization problem. The simplex method is most useful when used with computers. Consequently, it is not intended that you become an expert in manually solving linear programming problems using the simplex method. But it is important that you become proficient in constructing the models for linear programming problems so that they can be solved using a computer, and it is also important that you develop skill in interpreting the results. One way to gain this proficiency and interpretive skill is to set up and manually solve a number of fairly simple linear programming problems using the simplex method. This is the main goal in this section and in Sections 6.3 and 6.4. To assist you in learning to develop the models, the answer sections for Exercises 6.2, 6.3, and 6.4 contain both the model and its solution.

Initial System

We will introduce the concepts and procedures involved in the simplex method through an example—the tent production example discussed earlier. We restate the problem here in standard form for convenient reference:

$$\begin{aligned} \text{Maximize} \quad & P = 50x_1 + 80x_2 && \text{Objective function} \\ \text{subject to} \quad & x_1 + 2x_2 \le 32 \\ & 3x_1 + 4x_2 \le 84 && \text{Problem constraints} \qquad (1) \\ & x_1, x_2 \ge 0 && \text{Nonnegative constraints} \end{aligned}$$

Introducing slack variables s_1 and s_2, we convert the problem constraint inequalities in problem (1) into the following system of problem constraint equations:

$$\begin{aligned} x_1 + 2x_2 + s_1 \quad\quad &= 32 \\ 3x_1 + 4x_2 \quad\quad + s_2 &= 84 \\ x_1, x_2, s_1, s_2 &\geq 0 \end{aligned} \tag{2}$$

Since a basic solution of system (2) is not feasible if it contains any negative values, we have also included the nonnegative constraints for both the decision variables x_1 and x_2 and the slack variables s_1 and s_2. From our discussion in Section 6.1, we know that out of the infinitely many solutions to system (2), an optimal solution is one of the basic feasible solutions, which correspond to the corner points of the feasible region.

As part of the simplex method we add the objective function equation $P = 50x_1 + 80x_2$ in the form $-50x_1 - 80x_2 + P = 0$ to system (2) to create what is called the **initial system**:

$$\begin{aligned} x_1 + 2x_2 + s_1 \quad\quad\quad\quad &= 32 \\ 3x_1 + 4x_2 \quad\quad + s_2 \quad\quad &= 84 \\ -50x_1 - 80x_2 \quad\quad\quad\quad + P &= 0 \\ x_1, x_2, s_1, s_2 &\geq 0 \end{aligned} \tag{3}$$

When we add the objective function equation to system (2), we must slightly modify the earlier definitions of basic solution and basic feasible solution so that they apply to the initial system (3).

DEFINITION Basic Solutions and Basic Feasible Solutions for Initial Systems

1. The objective function variable P is always selected as a basic variable.
2. Note that a basic solution of system (3) is also a basic solution of system (2) after P is deleted.
3. If a basic solution of system (3) is a basic feasible solution of system (2) after deleting P, then the basic solution of system (3) is called a **basic feasible solution** of system (3).
4. A basic feasible solution of system (3) can contain a negative number, but only if it is the value of P, the objective function variable.

These changes lead to a small change in the second version of the fundamental theorem (see Theorem 1, Section 6.1).

THEOREM 1 Fundamental Theorem of Linear Programming: Version 3

If the optimal value of the objective function in a linear programming problem exists, then that value must occur at one or more of the basic feasible solutions of the initial system.

With these adjustments understood, we start the simplex process with a basic feasible solution of the initial system (3), which we will refer to as an **initial basic feasible solution**. An initial basic feasible solution that is easy to find is the one associated with the origin.

Since system (3) has three equations and five variables, it has three basic variables and two nonbasic variables. Looking at the system, we see that x_1 and x_2 appear in all equations, and s_1, s_2, and P each appear only once and each in a different equation. A basic solution can be found by inspection by selecting s_1, s_2, and P as the basic variables (remember, P is always selected as a basic variable) and x_1 and x_2 as the nonbasic variables to be set equal to 0. Setting x_1 and x_2 equal to 0 and solving for the basic variables, we obtain the basic solution:

$$x_1 = 0, \quad x_2 = 0, \quad s_1 = 32, \quad s_2 = 84, \quad P = 0$$

This basic solution is feasible since none of the variables (excluding P) are negative. This is the initial basic feasible solution that we seek.

Now you can see why we wanted to add the objective function equation to system (2): A basic feasible solution of system (3) not only includes a basic feasible solution of system (2), but, in addition, it includes the value of P for that basic feasible solution of system (2).

The initial basic feasible solution we just found is associated with the origin. Of course, if we do not produce any tents, we do not expect a profit, so $P = \$0$. Starting with this easily obtained initial basic feasible solution, the simplex process moves through each iteration (repetition) to another basic feasible solution, each time improving the profit. The process continues until the maximum profit is reached, then the process stops.

Simplex Tableau

To facilitate the search for the optimal solution, we turn to matrix methods discussed in Chapter 4. Our first step is to write the augmented matrix for the initial system (3). This matrix is called the **initial simplex tableau**, and it is simply a tabulation of the coefficients in system (3).

$$\begin{array}{c} \\ s_1 \\ s_2 \\ P \end{array} \begin{array}{c} \begin{array}{ccccc|c} x_1 & x_2 & s_1 & s_2 & P & \end{array} \\ \left[\begin{array}{ccccc|c} 1 & 2 & 1 & 0 & 0 & 32 \\ 3 & 4 & 0 & 1 & 0 & 84 \\ \hdashline -50 & -80 & 0 & 0 & 1 & 0 \end{array}\right] \end{array} \quad \text{Initial simplex tableau} \qquad (4)$$

In tableau (4), the row below the dashed line always corresponds to the objective function. Each of the basic variables we selected above, s_1, s_2, and P, is also placed on the left of the tableau so that the intersection element in its row and column is not 0. For example, we place the basic variable s_1 on the left so that the intersection element of the s_1 row and the s_1 column is 1 and not 0. The basic variable s_2 is similarly placed. The objective function variable P is always placed at the bottom. The reason for writing the basic variables on the left in this way is that this placement makes it possible to read certain basic feasible solutions directly from the tableau. If $x_1 = 0$ and $x_2 = 0$, the basic variables on the left of tableau (4) are lined up with their corresponding values, 32, 84, and 0, to the right of the vertical line.

Looking at tableau (4) relative to the choice of s_1, s_2, and P as basic variables, we see that each basic variable is above a column that has all 0 elements except for a single 1 and that no two such columns contain 1's in the same row. These observations lead to a formalization of the process of selecting basic and nonbasic variables that is an important part of the simplex method:

PROCEDURE Selecting Basic and Nonbasic Variables for the Simplex Method

Given a simplex tableau,

Step 1 *Numbers of variables.* Determine the number of basic variables and the number of nonbasic variables. These numbers do not change during the simplex process.

Step 2 *Selecting basic variables.* A variable can be selected as a basic variable only if it corresponds to a column in the tableau that has exactly one nonzero element (usually 1) and the nonzero element in the column is not in the same row as the nonzero element in the column of another basic variable. This procedure always selects P as a basic variable, since the P column never changes during the simplex process.

Step 3 *Selecting nonbasic variables.* After the basic variables are selected in step 2, the remaining variables are selected as the nonbasic variables. The tableau columns under the nonbasic variables usually contain more than one nonzero element.

The earlier selection of s_1, s_2, and P as basic variables and x_1 and x_2 as nonbasic variables conforms to this prescribed convention of selecting basic and nonbasic variables for the simplex process.

Pivot Operation

The simplex method swaps one of the nonbasic variables, x_1 or x_2, for one of the basic variables, s_1 or s_2 (but not P), as a step toward improving the profit. For a nonbasic variable to be classified as a basic variable, we need to perform appropriate row operations on the tableau so that the newly selected basic variable will end up with exactly one nonzero element in its column. In this process, the old basic variable will usually gain additional nonzero elements in its column as it becomes nonbasic.

Which nonbasic variable should we select to become basic? It makes sense to select the nonbasic variable that will increase the profit the most per unit change in that variable. Looking at the objective function

$$P = 50x_1 + 80x_2$$

we see that if x_1 stays a nonbasic variable (set equal to 0) and if x_2 becomes a new basic variable, then

$$P = 50(0) + 80x_2 = 80x_2$$

and for each unit increase in x_2, P will increase \$80. If x_2 stays a nonbasic variable and x_1 becomes a new basic variable, then (reasoning in the same way) for each unit increase in x_1, P will increase only \$50. So we select the nonbasic variable x_2 to enter the set of basic variables, and call it the **entering variable**. (The basic variable leaving the set of basic variables to become a nonbasic variable is called the **exiting variable**, which will be discussed shortly.)

The column corresponding to the entering variable is called the **pivot column**. Looking at the bottom row in tableau (4)—the objective function row below the dashed line—we see that the pivot column is associated with the column to the left of the P column that has the most negative bottom element. In general, the most negative element in the bottom row to the left of the P column indicates the variable above it that will produce the greatest increase in P for a unit increase in that variable. For this reason, we call the elements in the bottom row of the tableau, to the left of the P column, **indicators**.

We illustrate the indicators, the pivot column, the entering variable, and the initial basic feasible solution below:

Entering variable ↓ (above the x_2 column)

$$\begin{array}{c} \\ s_1 \\ s_2 \\ P \end{array}\begin{array}{c} \begin{array}{cccccc} x_1 & x_2 & s_1 & s_2 & P & \end{array} \\ \left[\begin{array}{rrrrr|r} 1 & 2 & 1 & 0 & 0 & 32 \\ 3 & 4 & 0 & 1 & 0 & 84 \\ \hdashline -50 & -80 & 0 & 0 & 1 & 0 \end{array}\right] \end{array} \quad (5)$$

Initial simplex tableau

Indicators are shown in color.

↑ Pivot column (the x_2 column)

$$x_1 = 0, \quad x_2 = 0, \quad s_1 = 32, \quad s_2 = 84, \quad P = 0$$ Initial basic feasible solution

Now that we have chosen the nonbasic variable x_2 as the entering variable (the nonbasic variable to become basic), which of the two basic variables, s_1 or s_2, should we choose as the exiting variable (the basic variable to become nonbasic)? We saw above that for $x_1 = 0$, each unit increase in the entering variable x_2 results in an increase of \$80 for P. Can we increase x_2 without limit? No! A limit is imposed by the nonnegative requirements for s_1 and s_2. (Remember that if any of the basic variables except P become negative, we no longer have a feasible solution.) So we rephrase the question and ask: How much can x_2 be increased when $x_1 = 0$ without causing s_1 or s_2 to become negative? To see how much x_2 can be increased, we refer to tableau (5) or system (3) and write the two problem constraint equations with $x_1 = 0$:

$$2x_2 + s_1 = 32$$
$$4x_2 + s_2 = 84$$

Solving for s_1 and s_2, we have

$$s_1 = 32 - 2x_2$$
$$s_2 = 84 - 4x_2$$

For s_1 and s_2 to be nonnegative, x_2 must be chosen so that both $32 - 2x_2$ and $84 - 4x_2$ are nonnegative. That is, so that

$$32 - 2x_2 \geq 0 \qquad \text{and} \qquad 84 - 4x_2 \geq 0$$
$$-2x_2 \geq -32 \qquad\qquad -4x_2 \geq -84$$
$$x_2 \leq \tfrac{32}{2} = 16 \qquad\qquad x_2 \leq \tfrac{84}{4} = 21$$

For both inequalities to be satisfied, x_2 must be less than or equal to the smaller of the values, which is 16. So x_2 can increase to 16 without either s_1 or s_2 becoming negative. Now, observe how each value (16 and 21) can be obtained directly from the following tableau:

Entering variable ↓ (column x_2)

$$\begin{array}{c} \\ s_1 \\ s_2 \\ P \end{array}\begin{array}{c} \begin{array}{cccccc} x_1 & x_2 & s_1 & s_2 & P & \end{array} \\ \left[\begin{array}{rrrrr|r} 1 & 2 & 1 & 0 & 0 & 32 \\ 3 & 4 & 0 & 1 & 0 & 84 \\ \hline -50 & -80 & 0 & 0 & 1 & 0 \end{array}\right] \end{array} \begin{array}{l} \\ \tfrac{32}{2} = 16 \text{ (smallest)} \\ \\ \tfrac{84}{4} = 21 \end{array} \qquad (6)$$

↑ Pivot column

From tableau (6) we can determine the amount that the entering variable can increase by choosing the smallest of the quotients obtained by dividing each element in the last column above the dashed line by the corresponding *positive* element in the pivot column. The row with the smallest quotient is called the **pivot row**, and the variable to the left of the pivot row is the exiting variable. In this case, s_1 will be the exiting variable, and the roles of x_2 and s_1 will be interchanged. The element at the intersection of the pivot column and the pivot row is called the **pivot element**, and we circle this element for ease of recognition. Since a negative or 0 element in the pivot column places no restriction on the amount that an entering variable can increase, it is not necessary to compute the quotient for negative or 0 values in the pivot column.

A negative or 0 element is never selected for the pivot element.

The following tableau illustrates this process, which is summarized in the next box.

Entering variable ↓ (x_2); Pivot element (circled 2); Exiting variable → s_1; ← Pivot row; ↑ Pivot column

$$\begin{array}{c} \\ s_1 \\ s_2 \\ P \end{array}\begin{array}{c} \begin{array}{cccccc} x_1 & x_2 & s_1 & s_2 & P & \end{array} \\ \left[\begin{array}{rrrrr|r} 1 & \textcircled{2} & 1 & 0 & 0 & 32 \\ 3 & 4 & 0 & 1 & 0 & 84 \\ \hdashline -50 & -80 & 0 & 0 & 1 & 0 \end{array}\right] \end{array} \qquad (7)$$

PROCEDURE Selecting the Pivot Element

Step 1 Locate the most negative indicator in the bottom row of the tableau to the left of the P column (the negative number with the largest absolute value). The column containing this element is the *pivot column*. If there is a tie for the most negative indicator, choose either column.

Step 2 Divide each *positive* element in the pivot column above the dashed line into the corresponding element in the last column. The *pivot row* is the row corresponding to the smallest quotient obtained. If there is a tie for the smallest quotient, choose either row. If the pivot column above the dashed line has no positive elements, there is no solution, and we stop.

Step 3 The *pivot* (or *pivot element*) is the element at the intersection of the pivot column and pivot row.

Note: The pivot element is always positive and never appears in the bottom row.

Remember: The entering variable is at the top of the pivot column, and the exiting variable is at the left of the pivot row.

In order for x_2 to be classified as a basic variable, we perform row operations on tableau (7) so that the pivot element is transformed into 1 and all other elements in the column into 0's. This procedure for transforming a nonbasic variable into a basic variable is called a *pivot operation,* or *pivoting,* and is summarized in the following box.

PROCEDURE Performing a Pivot Operation

A **pivot operation**, or **pivoting**, consists of performing row operations as follows:

Step 1 Multiply the pivot row by the reciprocal of the pivot element to transform the pivot element into a 1. (If the pivot element is already a 1, omit this step.)

Step 2 Add multiples of the pivot row to other rows in the tableau to transform all other nonzero elements in the pivot column into 0's.

CONCEPTUAL INSIGHT

A pivot operation uses some of the same row operations as those used in Gauss–Jordan elimination, but there is one essential difference. **In a pivot operation, you can never interchange two rows.**

Performing a pivot operation has the following effects:

1. The (entering) nonbasic variable becomes a basic variable.
2. The (exiting) basic variable becomes a nonbasic variable.
3. The value of the objective function is increased, or, in some cases, remains the same.

We now carry out the pivot operation on tableau (7). (To facilitate the process, we do not repeat the variables after the first tableau, and we use "Enter" and "Exit" for "Entering variable" and "Exiting variable," respectively.)

$$
\begin{array}{r} \\ \\ \text{Exit} \rightarrow s_1 \\ s_2 \\ P \end{array}
\begin{array}{c}
\begin{array}{ccccc|c} & \text{Enter} & & & & \\ & \downarrow & & & & \\ x_1 & x_2 & s_1 & s_2 & P & \end{array} \\
\left[\begin{array}{ccccc|c} 1 & \textcircled{2} & 1 & 0 & 0 & 32 \\ 3 & 4 & 0 & 1 & 0 & 84 \\ \hdashline -50 & -80 & 0 & 0 & 1 & 0 \end{array}\right]
\end{array}
\quad \tfrac{1}{2}R_1 \rightarrow R_1
$$

$$
\sim \left[\begin{array}{ccccc|c} \frac{1}{2} & \textcircled{1} & \frac{1}{2} & 0 & 0 & 16 \\ 3 & 4 & 0 & 1 & 0 & 84 \\ \hdashline -50 & -80 & 0 & 0 & 1 & 0 \end{array}\right]
\quad \begin{array}{l} (-4)R_1 + R_2 \rightarrow R_2 \\ 80R_1 + R_3 \rightarrow R_3 \end{array}
$$

$$
\sim \left[\begin{array}{ccccc|c} \frac{1}{2} & 1 & \frac{1}{2} & 0 & 0 & 16 \\ 1 & 0 & -2 & 1 & 0 & 20 \\ \hdashline -10 & 0 & 40 & 0 & 1 & 1{,}280 \end{array}\right]
$$

We have completed the pivot operation, and now we must insert appropriate variables for this new tableau. Since x_2 replaced s_1, the basic variables are now x_2, s_2, and P, as indicated by the labels on the left side of the new tableau. Note that this selection of basic variables agrees with the procedure outlined on pages 301 and 302 for selecting basic variables. We write the new basic feasible solution by setting the nonbasic variables x_1 and s_1 equal to 0 and solving for the basic variables by inspection. (Remember, the values of the basic variables listed on the left are the corresponding numbers to the right of the vertical line. To see this, substitute $x_1 = 0$ and $s_1 = 0$ in the corresponding system shown next to the simplex tableau.)

$$
\begin{array}{c} \\ x_2 \\ s_2 \\ P \end{array}
\begin{array}{c}
\begin{array}{ccccc|c} x_1 & x_2 & s_1 & s_2 & P & \end{array} \\
\left[\begin{array}{ccccc|c} \frac{1}{2} & 1 & \frac{1}{2} & 0 & 0 & 16 \\ 1 & 0 & -2 & 1 & 0 & 20 \\ \hdashline -10 & 0 & 40 & 0 & 1 & 1{,}280 \end{array}\right]
\end{array}
\quad
\begin{array}{rl} \frac{1}{2}x_1 + x_2 + \frac{1}{2}s_1 & = 16 \\ x_1 - 2s_1 + s_2 & = 20 \\ -10x_1 + 40s_1 + P & = 1{,}280 \end{array}
$$

$$x_1 = 0, \quad x_2 = 16, \quad s_1 = 0, \quad s_2 = 20, \quad P = \$1{,}280$$

A profit of \$1,280 is a marked improvement over the \$0 profit produced by the initial basic feasible solution. But we can improve P still further, since a negative indicator still remains in the bottom row. To see why, we write out the objective function:

$$-10x_1 + 40s_1 + P = 1{,}280$$

or

$$P = 10x_1 - 40s_1 + 1{,}280$$

If s_1 stays a nonbasic variable (set equal to 0) and x_1 becomes a new basic variable, then

$$P = 10x_1 - 40(0) + 1{,}280 = 10x_1 + 1{,}280$$

and for each unit increase in x_1, P will increase \$10.

We now go through another iteration of the simplex process using another pivot element. The pivot element and the entering and exiting variables are shown in the following tableau:

$$
\begin{array}{r c}
 & \text{Enter} \\
 & \downarrow \\
\end{array}
\quad
\begin{array}{r}
 \\ x_2 \\ \text{Exit} \rightarrow s_2 \\ P
\end{array}
\begin{array}{c}
\begin{array}{ccccc|c} x_1 & x_2 & s_1 & s_2 & P & \end{array} \\
\left[\begin{array}{ccccc|c}
\frac{1}{2} & 1 & \frac{1}{2} & 0 & 0 & 16 \\
\textcircled{1} & 0 & -2 & 1 & 0 & 20 \\
\hline
-10 & 0 & 40 & 0 & 1 & 1{,}280
\end{array}\right]
\end{array}
\quad
\begin{array}{l}
\\ \frac{16}{1/2} = 32 \\ \frac{20}{1} = 20 \\ \\
\end{array}
$$

We now pivot on (the circled) 1. That is, we perform a pivot operation using this 1 as the pivot element. Since the pivot element is 1, we do not need to perform the first step in the pivot operation, so we proceed to the second step to get 0's above and below the pivot element 1. As before, to facilitate the process, we omit writing the variables, except for the first tableau.

$$
\begin{array}{r}
\\ x_2 \\ \text{Exit} \rightarrow s_2 \\ P
\end{array}
\begin{array}{c}
\begin{array}{ccccc|c} \overset{\text{Enter}}{\underset{\downarrow}{}} x_1 & x_2 & s_1 & s_2 & P & \end{array} \\
\left[\begin{array}{ccccc|c}
\frac{1}{2} & 1 & \frac{1}{2} & 0 & 0 & 16 \\
\textcircled{1} & 0 & -2 & 1 & 0 & 20 \\
\hline
-10 & 0 & 40 & 0 & 1 & 1{,}280
\end{array}\right]
\end{array}
\quad
\begin{array}{l}
(-\frac{1}{2})R_2 + R_1 \rightarrow R_1 \\ \\ 10R_2 + R_3 \rightarrow R_3
\end{array}
$$

$$
\sim \left[\begin{array}{ccccc|c}
0 & 1 & \frac{3}{2} & -\frac{1}{2} & 0 & 6 \\
1 & 0 & -2 & 1 & 0 & 20 \\
\hline
0 & 0 & 20 & 10 & 1 & 1{,}480
\end{array}\right]
$$

Since there are no more negative indicators in the bottom row, we are done. Let us insert the appropriate variables for this last tableau and write the corresponding basic feasible solution. The basic variables are now x_1, x_2, and P, so to get the corresponding basic feasible solution, we set the nonbasic variables s_1 and s_2 equal to 0 and solve for the basic variables by inspection.

$$
\begin{array}{r}
\\ x_2 \\ x_1 \\ P
\end{array}
\begin{array}{c}
\begin{array}{ccccc|c} x_1 & x_2 & s_1 & s_2 & P & \end{array} \\
\left[\begin{array}{ccccc|c}
0 & 1 & \frac{3}{2} & -\frac{1}{2} & 0 & 6 \\
1 & 0 & -2 & 1 & 0 & 20 \\
\hline
0 & 0 & 20 & 10 & 1 & 1{,}480
\end{array}\right]
\end{array}
$$

$$x_1 = 20, \quad x_2 = 6, \quad s_1 = 0, \quad s_2 = 0, \quad P = 1{,}480$$

To see why this is the maximum, we rewrite the objective function from the bottom row:

$$20s_1 + 10s_2 + P = 1{,}480$$

$$P = 1{,}480 - 20s_1 - 10s_2$$

Since s_1 and s_2 cannot be negative, any increase of either variable from 0 will make the profit smaller.

Finally, returning to our original problem, we conclude that a production schedule of 20 standard tents and 6 expedition tents will produce a maximum profit of \$1,480 per day, just as was found by the geometric method in Section 5.3. The fact that the slack variables are both 0 means that, for this production schedule, the plant will operate at full capacity—there is no slack in either the cutting department or the assembly department.

Interpreting the Simplex Process Geometrically

We can interpret the simplex process geometrically in terms of the feasible region graphed in the preceding section. Table 1 lists the three basic feasible solutions we just found using the simplex method (in the order they were found). Table 1 also includes the corresponding corner points of the feasible region illustrated in Figure 1.

Table 1 Basic Feasible Solution (obtained above)

x_1	x_2	s_1	s_2	$P(\$)$	Corner Point
0	0	32	84	0	$O(0, 0)$
0	16	0	20	1,280	$A(0, 16)$
20	6	0	0	1,480	$B(20, 6)$

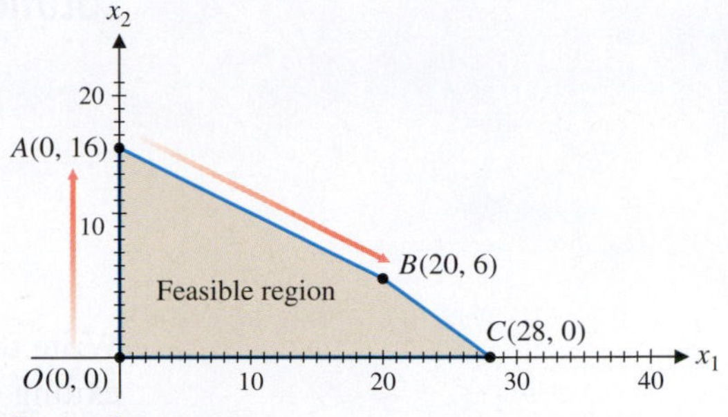

Figure 1

Looking at Table 1 and Figure 1, we see that the simplex process started at the origin, moved to the adjacent corner point $A(0, 16)$, and then to the optimal solution $B(20, 6)$ at the next adjacent corner point. This is typical of the simplex process.

Simplex Method Summarized

Before presenting additional examples, we summarize the important parts of the simplex method schematically in Figure 2.

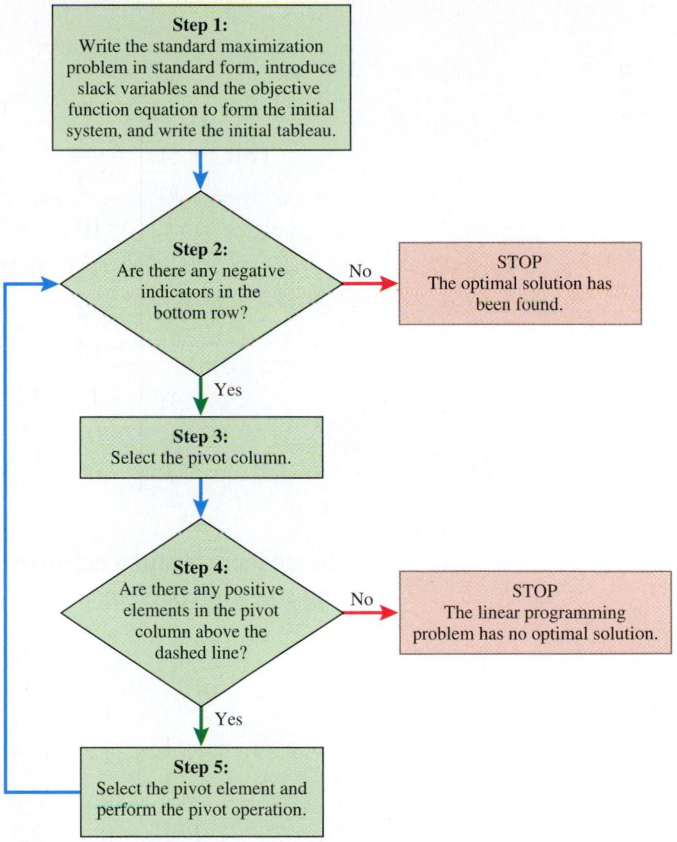

Figure 2 Simplex algorithm for standard maximization problems (Problem constraints are of the ≤ form with nonnegative constants on the right. The coefficients of the objective function can be any real numbers.)

EXAMPLE 1 **Using the Simplex Method** Solve the following linear programming problem using the simplex method:

$$\begin{aligned} \text{Maximize} \quad & P = 10x_1 + 5x_2 \\ \text{subject to} \quad & 4x_1 + x_2 \le 28 \\ & 2x_1 + 3x_2 \le 24 \\ & x_1, x_2 \ge 0 \end{aligned}$$

SOLUTION Introduce slack variables s_1 and s_2, and write the initial system:

$$4x_1 + x_2 + s_1 \qquad\qquad = 28$$
$$2x_1 + 3x_2 \qquad + s_2 \qquad = 24$$
$$-10x_1 - 5x_2 \qquad\qquad + P = 0$$
$$x_1, x_2, s_1, s_2 \geq 0$$

Write the simplex tableau, and identify the first pivot element and the entering and exiting variables:

$$
\begin{array}{c}
\text{Enter} \\ \downarrow
\end{array}
$$

$$
\begin{array}{cc}
 & \begin{array}{ccccc} x_1 & x_2 & s_1 & s_2 & P \end{array} \\
\begin{array}{c} \text{Exit} \to s_1 \\ s_2 \\ P \end{array} &
\left[\begin{array}{ccccc|c}
④ & 1 & 1 & 0 & 0 & 28 \\
2 & 3 & 0 & 1 & 0 & 24 \\
\hdashline
-10 & -5 & 0 & 0 & 1 & 0
\end{array} \right]
\end{array}
\quad
\begin{array}{l} \frac{28}{4} = 7 \\ \frac{24}{2} = 12 \end{array}
$$

Perform the pivot operation:

$$
\begin{array}{c}
\text{Enter} \\ \downarrow
\end{array}
$$

$$
\begin{array}{cc}
 & \begin{array}{ccccc} x_1 & x_2 & s_1 & s_2 & P \end{array} \\
\begin{array}{c} \text{Exit} \to s_1 \\ s_2 \\ P \end{array} &
\left[\begin{array}{ccccc|c}
④ & 1 & 1 & 0 & 0 & 28 \\
2 & 3 & 0 & 1 & 0 & 24 \\
\hdashline
-10 & -5 & 0 & 0 & 1 & 0
\end{array} \right]
\end{array}
\quad \frac{1}{4}R_1 \to R_1
$$

$$
\sim
\left[\begin{array}{ccccc|c}
① & 0.25 & 0.25 & 0 & 0 & 7 \\
2 & 3 & 0 & 1 & 0 & 24 \\
\hdashline
-10 & -5 & 0 & 0 & 1 & 0
\end{array} \right]
\quad
\begin{array}{l} (-2)R_1 + R_2 \to R_2 \\ 10R_1 + R_3 \to R_3 \end{array}
$$

$$
\begin{array}{c} x_1 \\ \sim\ s_2 \\ P \end{array}
\left[\begin{array}{ccccc|c}
1 & 0.25 & 0.25 & 0 & 0 & 7 \\
0 & 2.5 & -0.5 & 1 & 0 & 10 \\
\hdashline
0 & -2.5 & 2.5 & 0 & 1 & 70
\end{array} \right]
$$

Since there is still a negative indicator in the last row, we repeat the process by finding a new pivot element:

$$
\begin{array}{c}
\text{Enter} \\ \downarrow
\end{array}
$$

$$
\begin{array}{cc}
 & \begin{array}{ccccc} x_1 & x_2 & s_1 & s_2 & P \end{array} \\
\begin{array}{c} x_1 \\ \text{Exit} \to s_2 \\ P \end{array} &
\left[\begin{array}{ccccc|c}
1 & 0.25 & 0.25 & 0 & 0 & 7 \\
0 & ②.⑤ & -0.5 & 1 & 0 & 10 \\
\hdashline
0 & -2.5 & 2.5 & 0 & 1 & 70
\end{array} \right]
\end{array}
\quad
\begin{array}{l} \frac{7}{0.25} = 28 \\ \frac{10}{2.5} = 4 \end{array}
$$

Performing the pivot operation, we obtain

$$
\begin{array}{c}
\text{Enter} \\ \downarrow
\end{array}
$$

$$
\begin{array}{cc}
 & \begin{array}{ccccc} x_1 & x_2 & s_1 & s_2 & P \end{array} \\
\begin{array}{c} x_1 \\ \text{Exit} \to s_2 \\ P \end{array} &
\left[\begin{array}{ccccc|c}
1 & 0.25 & 0.25 & 0 & 0 & 7 \\
0 & ②.⑤ & -0.5 & 1 & 0 & 10 \\
\hdashline
0 & -2.5 & 2.5 & 0 & 1 & 70
\end{array} \right]
\end{array}
\quad \frac{1}{2.5}R_2 \to R_2
$$

$$\sim \left[\begin{array}{cccccc|c} 1 & 0.25 & 0.25 & 0 & 0 & & 7 \\ 0 & \textcircled{1} & -0.2 & 0.4 & 0 & & 4 \\ \hdashline 0 & -2.5 & 2.5 & 0 & 1 & & 70 \end{array}\right] \begin{array}{l} (-0.25)R_2 + R_1 \rightarrow R_1 \\ \\ 2.5R_2 + R_3 \rightarrow R_3 \end{array}$$

$$\sim \begin{array}{c} x_1 \\ x_2 \\ P \end{array} \left[\begin{array}{ccccc|c} 1 & 0 & 0.3 & -0.1 & 0 & 6 \\ 0 & 1 & -0.2 & 0.4 & 0 & 4 \\ \hdashline 0 & 0 & 2 & 1 & 1 & 80 \end{array}\right]$$

Since all the indicators in the last row are nonnegative, we stop and read the optimal solution:

$$\text{Max } P = 80 \quad \text{at} \quad x_1 = 6, \quad x_2 = 4, \quad s_1 = 0, \quad s_2 = 0$$

(To see why this makes sense, write the objective function corresponding to the last row to see what happens to P when you try to increase s_1 or s_2.)

Matched Problem 1 Solve the following linear programming problem using the simplex method:

$$\begin{aligned} \text{Maximize} \quad & P = 2x_1 + x_2 \\ \text{subject to} \quad & 5x_1 + x_2 \le 9 \\ & x_1 + x_2 \le 5 \\ & x_1, x_2 \le 0 \end{aligned}$$

Explore and Discuss 1

Graph the feasible region for the linear programming problem in Example 1 and trace the path to the optimal solution determined by the simplex method.

EXAMPLE 2 **Using the Simplex Method** Solve using the simplex method:

$$\begin{aligned} \text{Maximize} \quad & P = 6x_1 + 3x_2 \\ \text{subject to} \quad & -2x_1 + 3x_2 \le 9 \\ & -x_1 + 3x_2 \le 12 \\ & x_1, x_2 \ge 0 \end{aligned}$$

SOLUTION Write the initial system using the slack variables s_1 and s_2:

$$\begin{aligned} -2x_1 + 3x_2 + s_1 \qquad\qquad &= 9 \\ -x_1 + 3x_2 \qquad + s_2 \qquad &= 12 \\ -6x_1 - 3x_2 \qquad\qquad + P &= 0 \end{aligned}$$

Write the simplex tableau and identify the first pivot element:

$$\begin{array}{c} \\ s_1 \\ s_2 \\ P \end{array} \begin{array}{c} \begin{array}{ccccc} x_1 & x_2 & s_1 & s_2 & P \end{array} \\ \left[\begin{array}{ccccc|c} -2 & 3 & 1 & 0 & 0 & 9 \\ -1 & 3 & 0 & 1 & 0 & 12 \\ \hdashline -6 & -3 & 0 & 0 & 1 & 0 \end{array}\right] \end{array}$$

↑ Pivot column

Since both elements in the pivot column above the dashed line are negative, we are unable to select a pivot row. We stop and conclude that there is no optimal solution.

Matched Problem 2 Solve using the simplex method:

$$\begin{aligned} \text{Maximize} \quad & P = 2x_1 + 3x_2 \\ \text{subject to} \quad & -3x_1 + 4x_2 \le 12 \\ & x_2 \le 2 \\ & x_1, x_2 \ge 0 \end{aligned}$$

Refer to Examples 1 and 2. In Example 1 we concluded that we had found the optimal solution because we could not select a pivot column. In Example 2 we concluded that the problem had no optimal solution because we selected a pivot column and then could not select a pivot row. Notice that we do not try to continue with the simplex method by selecting a negative pivot element or using a different column for the pivot column. Remember:

If it is not possible to select a pivot column, then the simplex method stops and we conclude that the optimal solution has been found. If the pivot column has been selected and it is not possible to select a pivot row, then the simplex method stops and we conclude that there is no optimal solution.

Application

EXAMPLE 3 **Agriculture** A farmer owns a 100-acre farm and plans to plant at most three crops. The seed for crops *A*, *B*, and *C* costs \$40, \$20, and \$30 per acre, respectively. A maximum of \$3,200 can be spent on seed. Crops *A*, *B*, and *C* require one, two, and one work days per acre, respectively, and there are a maximum of 160 work days available. If the farmer can make a profit of \$100 per acre on crop *A*, \$300 per acre on crop *B*, and \$200 per acre on crop *C*, how many acres of each crop should be planted to maximize profit?

SOLUTION The farmer must decide on the number of acres of each crop that should be planted. So the decision variables are

$$x_1 = \text{number of acres of crop } A$$

$$x_2 = \text{number of acres of crop } B$$

$$x_3 = \text{number of acres of crop } C$$

The farmer's objective is to maximize profit:

$$P = 100x_1 + 300x_2 + 200x_3$$

The farmer is limited by the number of acres available for planting, the money available for seed, and the available work days. These limitations lead to the following constraints:

$$\begin{aligned} x_1 + x_2 + x_3 &\le 100 && \text{Acreage constraint} \\ 40x_1 + 20x_2 + 30x_3 &\le 3{,}200 && \text{Monetary constraint} \\ x_1 + 2x_2 + x_3 &\le 160 && \text{Labor constraint} \end{aligned}$$

Adding the nonnegative constraints, we have the following model for a linear programming problem:

$$\begin{aligned} \text{Maximize} \quad & P = 100x_1 + 300x_2 + 200x_3 && \text{Objective function} \\ \text{subject to} \quad & \left.\begin{aligned} x_1 + x_2 + x_3 &\le 100 \\ 40x_1 + 20x_2 + 30x_3 &\le 3{,}200 \\ x_1 + 2x_2 + x_3 &\le 160 \end{aligned}\right\} && \text{Problem constraints} \\ & x_1, x_2, x_3 \ge 0 && \text{Nonnegative constraints} \end{aligned}$$

Next, we introduce slack variables and form the initial system:

$$
\begin{array}{rcl}
x_1 + x_2 + x_3 + s_1 & = & 100 \\
40x_1 + 20x_2 + 30x_3 + s_2 & = & 3{,}200 \\
x_1 + 2x_2 + x_3 + s_3 & = & 160 \\
-100x_1 - 300x_2 - 200x_3 + P & = & 0 \\
\end{array}
$$

$$x_1, x_2, x_3, s_1, s_2, s_3 \ge 0$$

Notice that the initial system has $7 - 4 = 3$ nonbasic variables and 4 basic variables. Now we form the simplex tableau and solve by the simplex method:

$$
\begin{array}{r}
\\ s_1 \\ s_2 \\ \text{Exit} \rightarrow s_3 \\ P
\end{array}
\begin{array}{c}
\begin{array}{ccccccc|c}
x_1 & \overset{\text{Enter}}{\overset{\downarrow}{x_2}} & x_3 & s_1 & s_2 & s_3 & P & \\
\end{array}\\
\left[\begin{array}{ccccccc|c}
1 & 1 & 1 & 1 & 0 & 0 & 0 & 100 \\
40 & 20 & 30 & 0 & 1 & 0 & 0 & 3{,}200 \\
1 & \textcircled{2} & 1 & 0 & 0 & 1 & 0 & 160 \\
\hline
-100 & -300 & -200 & 0 & 0 & 0 & 1 & 0
\end{array}\right]
\end{array}
\quad
\begin{array}{l}
\\ \\ \\ 0.5R_3 \rightarrow R_3 \\ \\
\end{array}
$$

$$
\sim
\left[\begin{array}{ccccccc|c}
1 & 1 & 1 & 1 & 0 & 0 & 0 & 100 \\
40 & 20 & 30 & 0 & 1 & 0 & 0 & 3{,}200 \\
0.5 & \textcircled{1} & 0.5 & 0 & 0 & 0.5 & 0 & 80 \\
\hline
-100 & -300 & -200 & 0 & 0 & 0 & 1 & 0
\end{array}\right]
\quad
\begin{array}{l}
(-1)R_3 + R_1 \rightarrow R_1 \\
(-20)R_3 + R_2 \rightarrow R_2 \\
\\
300R_3 + R_4 \rightarrow R_4
\end{array}
$$

$$
\begin{array}{r}
\\ \text{Exit} \rightarrow s_1 \\ s_2 \\ x_2 \\ P
\end{array}
\begin{array}{c}
\begin{array}{ccccccc|c}
x_1 & x_2 & \overset{\text{Enter}}{\overset{\downarrow}{x_3}} & s_1 & s_2 & s_3 & P & \\
\end{array}\\
\left[\begin{array}{ccccccc|c}
0.5 & 0 & \textcircled{0.5} & 1 & 0 & -0.5 & 0 & 20 \\
30 & 0 & 20 & 0 & 1 & -10 & 0 & 1{,}600 \\
0.5 & 1 & 0.5 & 0 & 0 & 0.5 & 0 & 80 \\
\hline
50 & 0 & -50 & 0 & 0 & 150 & 1 & 24{,}000
\end{array}\right]
\end{array}
\quad
\begin{array}{l}
\\ 2R_1 \rightarrow R_1 \\ \\ \\ \\
\end{array}
$$

$$
\sim
\left[\begin{array}{ccccccc|c}
1 & 0 & \textcircled{1} & 2 & 0 & -1 & 0 & 40 \\
30 & 0 & 20 & 0 & 1 & -10 & 0 & 1{,}600 \\
0.5 & 1 & 0.5 & 0 & 0 & 0.5 & 0 & 80 \\
\hline
50 & 0 & -50 & 0 & 0 & 150 & 1 & 24{,}000
\end{array}\right]
\quad
\begin{array}{l}
\\
(-20)R_1 + R_2 \rightarrow R_2 \\
(-0.5)R_1 + R_3 \rightarrow R_3 \\
50R_1 + R_4 \rightarrow R_4
\end{array}
$$

$$
\sim
\begin{array}{r}
x_3 \\ s_2 \\ x_2 \\ P
\end{array}
\left[\begin{array}{ccccccc|c}
1 & 0 & 1 & 2 & 0 & -1 & 0 & 40 \\
10 & 0 & 0 & -40 & 1 & 10 & 0 & 800 \\
0 & 1 & 0 & -1 & 0 & 1 & 0 & 60 \\
\hline
100 & 0 & 0 & 100 & 0 & 100 & 1 & 26{,}000
\end{array}\right]
$$

All indicators in the bottom row are nonnegative, and now we can read the optimal solution:

$$x_1 = 0, \quad x_2 = 60, \quad x_3 = 40, \quad s_1 = 0, \quad s_2 = 800, \quad s_3 = 0, \quad P = \$26{,}000$$

So if the farmer plants 60 acres in crop B, 40 acres in crop C, and no crop A, the maximum profit of \$26,000 will be realized. The fact that $s_2 = 800$ tells us (look at the second row in the equations at the start) that this maximum profit is reached by using only \$2,400 of the \$3,200 available for seed; that is, we have a slack of \$800 that can be used for some other purpose.

There are many types of software that can be used to solve linear programming problems by the simplex method. Figure 3 illustrates a solution to Example 3 in Excel, a popular spreadsheet for personal computers.

	A	B	C	D	E	F
1	Resources	Crop A	Crop B	Crop C	Available	Used
2	Acres	1	1	1	100	100
3	Seed ($)	40	20	30	3,200	2,400
4	Workdays	1	2	1	160	160
5	Profit per acre	100	300	200	26,000	<-- Total profit
6	Acres to plant	0	60	40		

Figure 3

Matched Problem 3 Repeat Example 3 modified as follows:

	Investment per Acre			
	Crop *A*	**Crop *B***	**Crop *C***	**Maximum Available**
Seed cost	$24	$40	$30	$3,600
Work days	1	2	2	160
Profit	$140	$200	$160	

CONCEPTUAL INSIGHT

If you are solving a system of linear equations or inequalities, then you can multiply both sides of an equation by any nonzero number and both sides of an inequality by any positive number without changing the solution set. This is still the case for the simplex method, but you must be careful when you interpret the results. For example, consider the second problem constraint in the model for Example 3:

$$40x_1 + 20x_2 + 30x_3 \le 3{,}200$$

Multiplying both sides by $\frac{1}{10}$ before introducing slack variables simplifies subsequent calculations. However, performing this operation has a side effect—it changes the units of the slack variable from dollars to tens of dollars. Compare the following two equations:

$$40x_1 + 20x_2 + 30x_3 + s_2 = 3{,}200 \qquad s_2 \text{ represents dollars}$$

$$4x_1 + 2x_2 + 3x_3 + s_2' = 320 \qquad s_2' \text{ represents tens of dollars}$$

In general, if you multiply a problem constraint by a positive number, remember to take this into account when you interpret the value of the slack variable for that constraint.

The feasible region for the linear programming problem in Example 3 has eight corner points, but the simplex method found the solution in only two steps. In larger problems, the difference between the total number of corner points and the number of steps required by the simplex method is even more dramatic. A feasible region may have hundreds or even thousands of corner points, yet the simplex method will often find the optimal solution in 10 or 15 steps.

To simplify this introduction to the simplex method, we have purposely avoided certain degenerate cases that lead to difficulties. Discussion and resolution of these problems is left to a more advanced treatment of the subject.

Exercises 6.2

A *For the simplex tableaux in Problems 1–4,*

(A) Identify the basic and nonbasic variables.

(B) Find the corresponding basic feasible solution.

(C) Determine whether the optimal solution has been found, an additional pivot is required, or the problem has no optimal solution.

1.
$$\begin{array}{cccccc} x_1 & x_2 & s_1 & s_2 & P & \end{array}$$
$$\left[\begin{array}{ccccc|c} 2 & 1 & 0 & 3 & 0 & 12 \\ 3 & 0 & 1 & -2 & 0 & 15 \\ \hline -4 & 0 & 0 & 4 & 1 & 50 \end{array}\right]$$

2.
$$\begin{array}{cccccc} x_1 & x_2 & s_1 & s_2 & P & \end{array}$$
$$\left[\begin{array}{ccccc|c} 1 & 4 & -2 & 0 & 0 & 10 \\ 0 & 2 & 3 & 1 & 0 & 25 \\ \hline 0 & 5 & 6 & 0 & 1 & 35 \end{array}\right]$$

3.
$$\begin{array}{cccccccc} x_1 & x_2 & x_3 & s_1 & s_2 & s_3 & P & \end{array}$$
$$\left[\begin{array}{ccccccc|c} -2 & 0 & 1 & 3 & 1 & 0 & 0 & 5 \\ 0 & 1 & 0 & -2 & 0 & 0 & 0 & 15 \\ -1 & 0 & 0 & 4 & 1 & 1 & 0 & 12 \\ \hline -4 & 0 & 0 & 2 & 4 & 0 & 1 & 45 \end{array}\right]$$

4.
$$\begin{array}{cccccccc} x_1 & x_2 & x_3 & s_1 & s_2 & s_3 & P & \end{array}$$
$$\left[\begin{array}{ccccccc|c} 0 & 2 & -1 & 1 & 4 & 0 & 0 & 5 \\ 0 & 1 & 2 & 0 & -2 & 1 & 0 & 2 \\ 1 & 3 & 0 & 0 & 5 & 0 & 0 & 11 \\ \hline 0 & -5 & 4 & 0 & -3 & 0 & 1 & 27 \end{array}\right]$$

In Problems 5–8, find the pivot element, identify the entering and exiting variables, and perform one pivot operation.

5.
$$\begin{array}{cccccc} x_1 & x_2 & s_1 & s_2 & P & \end{array}$$
$$\left[\begin{array}{ccccc|c} 1 & 4 & 1 & 0 & 0 & 4 \\ 3 & 5 & 0 & 1 & 0 & 24 \\ \hline -8 & -5 & 0 & 0 & 1 & 0 \end{array}\right]$$

6.
$$\begin{array}{cccccc} x_1 & x_2 & s_1 & s_2 & P & \end{array}$$
$$\left[\begin{array}{ccccc|c} 1 & 6 & 1 & 0 & 0 & 36 \\ 3 & 1 & 0 & 1 & 0 & 5 \\ \hline -1 & -2 & 0 & 0 & 1 & 0 \end{array}\right]$$

7.
$$\begin{array}{ccccccc} x_1 & x_2 & s_1 & s_2 & s_3 & P & \end{array}$$
$$\left[\begin{array}{cccccc|c} 2 & 1 & 1 & 0 & 0 & 0 & 4 \\ 3 & 0 & 1 & 1 & 0 & 0 & 8 \\ 0 & 0 & 2 & 0 & 1 & 0 & 2 \\ \hline -4 & 0 & -3 & 0 & 0 & 1 & 5 \end{array}\right]$$

8.
$$\begin{array}{ccccccc} x_1 & x_2 & s_1 & s_2 & s_3 & P & \end{array}$$
$$\left[\begin{array}{cccccc|c} 0 & 0 & 2 & 1 & 1 & 0 & 2 \\ 1 & 0 & -4 & 0 & 1 & 0 & 3 \\ 0 & 1 & 5 & 0 & 2 & 0 & 11 \\ \hline 0 & 0 & -6 & 0 & -5 & 1 & 18 \end{array}\right]$$

In Problems 9–12,

(A) Using slack variables, write the initial system for each linear programming problem.

(B) Write the simplex tableau, circle the first pivot, and identify the entering and exiting variables.

(C) Use the simplex method to solve the problem.

9. Maximize $P = 15x_1 + 10x_2$
subject to
$$\begin{aligned} 2x_1 + x_2 &\le 10 \\ x_1 + 3x_2 &\le 10 \\ x_1, x_2 &\ge 0 \end{aligned}$$

10. Maximize $P = 3x_1 + 2x_2$
subject to
$$\begin{aligned} 5x_1 + 2x_2 &\le 20 \\ 3x_1 + 2x_2 &\le 16 \\ x_1, x_2 &\ge 0 \end{aligned}$$

11. Repeat Problem 9 with the objective function changed to $P = 30x_1 + x_2$.

12. Repeat Problem 10 with the objective function changed to $P = x_1 + 3x_2$.

B *Solve the linear programming problems in Problems 13–32 using the simplex method.*

13. Maximize $P = 30x_1 + 40x_2$
subject to
$$\begin{aligned} 2x_1 + x_2 &\le 10 \\ x_1 + x_2 &\le 7 \\ x_1 + 2x_2 &\le 12 \\ x_1, x_2 &\ge 0 \end{aligned}$$

14. Maximize $P = 15x_1 + 20x_2$
subject to
$$\begin{aligned} 2x_1 + x_2 &\le 9 \\ x_1 + x_2 &\le 6 \\ x_1 + 2x_2 &\le 10 \\ x_1, x_2 &\ge 0 \end{aligned}$$

15. Maximize $P = 2x_1 + 3x_2$
subject to
$$\begin{aligned} -2x_1 + x_2 &\le 2 \\ -x_1 + x_2 &\le 5 \\ x_2 &\le 6 \\ x_1, x_2 &\ge 0 \end{aligned}$$

16. Repeat Problem 15 with $P = -x_1 + 3x_2$.

17. Maximize $P = -x_1 + 2x_2$
subject to
$$\begin{aligned} -x_1 + x_2 &\le 2 \\ -x_1 + 3x_2 &\le 12 \\ x_1 - 4x_2 &\le 4 \\ x_1, x_2 &\ge 0 \end{aligned}$$

18. Repeat Problem 17 with $P = x_1 + 2x_2$.

19. Maximize $P = 15x_1 + 36x_2$
subject to $x_1 + 3x_2 \le 6$
$x_1, x_2 \ge 0$

20. Maximize $P = 8x_1 + 7x_2$
subject to $2x_1 - 5x_2 \le 10$
$x_1, x_2 \ge 0$

21. Maximize $P = 27x_1 + 64x_2$
subject to $8x_1 - 3x_2 \le 24$
$x_1, x_2 \ge 0$

22. Maximize $P = 45x_1 + 30x_2$
subject to $4x_1 + 3x_2 \le 12$
$x_1, x_2 \ge 0$

23. Maximize $P = 5x_1 + 2x_2 - x_3$
subject to $x_1 + x_2 - x_3 \le 10$
$2x_1 + 4x_2 + 3x_3 \le 30$
$x_1, x_2, x_3 \ge 0$

24. Maximize $P = 4x_1 - 3x_2 + 2x_3$
subject to $x_1 + 2x_2 - x_3 \le 5$
$3x_1 + 2x_2 + 2x_3 \le 22$
$x_1, x_2, x_3 \ge 0$

25. Maximize $P = 2x_1 + 3x_2 + 4x_3$
subject to $x_1 + x_3 \le 4$
$x_2 + x_3 \le 3$
$x_1, x_2, x_3 \ge 0$

26. Maximize $P = x_1 + x_2 + 2x_3$
subject to $x_1 - 2x_2 + x_3 \le 9$
$2x_1 + x_2 + 2x_3 \le 28$
$x_1, x_2, x_3 \ge 0$

27. Maximize $P = 4x_1 + 3x_2 + 2x_3$
subject to $3x_1 + 2x_2 + 5x_3 \le 23$
$2x_1 + x_2 + x_3 \le 8$
$x_1 + x_2 + 2x_3 \le 7$
$x_1, x_2, x_3 \ge 0$

28. Maximize $P = 4x_1 + 2x_2 + 3x_3$
subject to $x_1 + x_2 + x_3 \le 11$
$2x_1 + 3x_2 + x_3 \le 20$
$x_1 + 3x_2 + 2x_3 \le 20$
$x_1, x_2, x_3 \ge 0$

29. Maximize $P = 20x_1 + 30x_2$
subject to $0.6x_1 + 1.2x_2 \le 960$
$0.03x_1 + 0.04x_2 \le 36$
$0.3x_1 + 0.2x_2 \le 270$
$x_1, x_2 \ge 0$

30. Repeat Problem 29 with $P = 20x_1 + 20x_2$.

31. Maximize $P = x_1 + 2x_2 + 3x_3$
subject to $2x_1 + 2x_2 + 8x_3 \le 600$
$x_1 + 3x_2 + 2x_3 \le 600$
$3x_1 + 2x_2 + x_3 \le 400$
$x_1, x_2, x_3 \ge 0$

32. Maximize $P = 10x_1 + 50x_2 + 10x_3$
subject to $3x_1 + 3x_2 + 3x_3 \le 66$
$6x_1 - 2x_2 + 4x_3 \le 48$
$3x_1 + 6x_2 + 9x_3 \le 108$
$x_1, x_2, x_3 \ge 0$

C *In Problems 33 and 34, first solve the linear programming problem by the simplex method, keeping track of the basic feasible solutions at each step. Then graph the feasible region and illustrate the path to the optimal solution determined by the simplex method.*

33. Maximize $P = 2x_1 + 5x_2$
subject to $x_1 + 2x_2 \le 40$
$x_1 + 3x_2 \le 48$
$x_1 + 4x_2 \le 60$
$x_2 \le 14$
$x_1, x_2 \ge 0$

34. Maximize $P = 5x_1 + 3x_2$
subject to $5x_1 + 4x_2 \le 100$
$2x_1 + x_2 \le 28$
$4x_1 + x_2 \le 42$
$x_1 \le 10$
$x_1, x_2 \ge 0$

Solve Problems 35 and 36 by the simplex method and also by graphing (the geometric method). Compare and contrast the results.

35. Maximize $P = 2x_1 + 3x_2$
subject to $-2x_1 + x_2 \le 4$
$x_2 \le 10$
$x_1, x_2 \ge 0$

36. Maximize $P = 2x_1 + 3x_2$
subject to $-x_1 + x_2 \le 2$
$x_2 \le 4$
$x_1, x_2 \ge 0$

In Problems 37–40, there is a tie for the choice of the first pivot column. Use the simplex method to solve each problem two different ways: first by choosing column 1 as the first pivot column, and then by choosing column 2 as the first pivot column. Discuss the relationship between these two solutions.

37. Maximize $P = x_1 + x_2$
subject to $2x_1 + x_2 \le 16$
$x_1 \le 6$
$x_2 \le 10$
$x_1, x_2 \ge 0$

38. Maximize $P = x_1 + x_2$
subject to $x_1 + 2x_2 \le 10$
$x_1 \le 6$
$x_2 \le 4$
$x_1, x_2 \ge 0$

39. Maximize $P = 3x_1 + 3x_2 + 2x_3$
subject to $x_1 + x_2 + 2x_3 \le 20$
$2x_1 + x_2 + 4x_3 \le 32$
$x_1, x_2, x_3 \ge 0$

40. Maximize $P = 2x_1 + 2x_2 + x_3$
subject to $x_1 + x_2 + 3x_3 \le 10$
$2x_1 + 4x_2 + 5x_3 \le 24$
$x_1, x_2, x_3 \ge 0$

Applications

In Problems 41–56, construct a mathematical model in the form of a linear programming problem. (The answers in the back of the book for these application problems include the model.) Then solve the problem using the simplex method. Include an interpretation of any nonzero slack variables in the optimal solution.

41. Manufacturing: resource allocation. A small company manufactures three different electronic components for computers. Component *A* requires 2 hours of fabrication and 1 hour of assembly; component *B* requires 3 hours of fabrication and 1 hour of assembly; and component *C* requires 2 hours of fabrication and 2 hours of assembly. The company has up to 1,000 labor-hours of fabrication time and 800 labor-hours of assembly time available per week. The profit on each component, *A*, *B*, and *C*, is $7, $8, and $10, respectively. How many components of each type should the company manufacture each week in order to maximize its profit (assuming that all components manufactured can be sold)? What is the maximum profit?

42. Manufacturing: resource allocation. Solve Problem 41 with the additional restriction that the combined total number of components produced each week cannot exceed 420. Discuss the effect of this restriction on the solution to Problem 41.

43. Investment. An investor has at most $100,000 to invest in government bonds, mutual funds, and money market funds. The average yields for government bonds, mutual funds, and money market funds are 8%, 13%, and 15%, respectively. The investor's policy requires that the total amount invested in mutual and money market funds not exceed the amount invested in government bonds. How much should be invested in each type of investment in order to maximize the return? What is the maximum return?

44. Investment. Repeat Problem 43 under the additional assumption that no more than $30,000 can be invested in money market funds.

45. Advertising. A department store has up to $20,000 to spend on television advertising for a sale. All ads will be placed with one television station. A 30-second ad costs $1,000 on daytime TV and is viewed by 14,000 potential customers, $2,000 on prime-time TV and is viewed by 24,000 potential customers, and $1,500 on late-night TV and is viewed by 18,000 potential customers. The television station will not accept a total of more than 15 ads in all three time periods. How many ads should be placed in each time period in order to maximize the number of potential customers who will see the ads? How many potential customers will see the ads? (Ignore repeated viewings of the ad by the same potential customer.)

46. Advertising. Repeat Problem 45 if the department store increases its budget to $24,000 and requires that at least half of the ads be placed during prime-time.

47. Home construction. A contractor is planning a new housing development consisting of colonial, split-level, and ranch-style houses. A colonial house requires $\frac{1}{2}$ acre of land, $60,000 capital, and 4,000 labor-hours to construct, and returns a profit of $20,000. A split-level house requires $\frac{1}{2}$ acre of land, $60,000 capital, and 3,000 labor-hours to construct, and returns a profit of $18,000. A ranch house requires 1 acre of land, $80,000 capital, and 4,000 labor-hours to construct, and returns a profit of $24,000. The contractor has 30 acres of land, $3,200,000 capital, and 180,000 labor-hours available. How many houses of each type should be constructed to maximize the contractor's profit? What is the maximum profit?

48. Bicycle manufacturing. A company manufactures three-speed, five-speed, and ten-speed bicycles. Each bicycle passes through three departments: fabrication, painting & plating, and final assembly. The relevant manufacturing data are given in the table.

	Labor-Hours per Bicycle			Maximum Labor-Hours Available per Day
	Three-Speed	Five-Speed	Ten-Speed	
Fabrication	3	4	5	120
Painting & plating	5	3	5	130
Final assembly	4	3	5	120
Profit per bicycle ($)	80	70	100	

How many bicycles of each type should the company manufacture per day in order to maximize its profit? What is the maximum profit?

49. Home building. Repeat Problem 47 if the profit on a colonial house decreases from $20,000 to $17,000 and all other data remain the same. If the slack associated with any problem constraint is nonzero, find it.

50. Bicycle manufacturing. Repeat Problem 48 if the profit on a ten-speed bicycle increases from $100 to $110 and all other data remain the same. If the slack associated with any problem constraint is nonzero, find it.

51. Home building. Repeat Problem 47 if the profit on a colonial house increases from $20,000 to $25,000 and all other data remain the same. If the slack associated with any problem constraint is nonzero, find it.

52. Bicycle manufacturing. Repeat Problem 48 if the profit on a five-speed bicycle increases from \$70 to \$110 and all other data remain the same. If the slack associated with any problem constraint is nonzero, find it.

53. Animal nutrition. The natural diet of a certain animal consists of three foods: *A*, *B*, and *C*. The number of units of calcium, iron, and protein in 1 gram of each food and the average daily intake are given in the table. A scientist wants to investigate the effect of increasing the protein in the animal's diet while not allowing the units of calcium and iron to exceed its average daily intakes. How many grams of each food should be used to maximize the amount of protein in the diet? What is the maximum amount of protein?

	Units per Gram			
	Food *A*	Food *B*	Food *C*	Average Daily Intake (units)
Calcium	1	3	2	30
Iron	2	1	2	24
Protein	3	4	5	60

54. Animal nutrition. Repeat Problem 53 if the scientist wants to maximize the daily calcium intake while not allowing the intake of iron or protein to exceed the average daily intake.

55. Opinion survey. A political scientist received a grant to fund a research project on voting trends. The budget includes \$3,200 for conducting door-to-door interviews on the day before an election. Undergraduate students, graduate students, and faculty members will be hired to conduct the interviews. Each undergraduate student will conduct 18 interviews for \$100. Each graduate student will conduct 25 interviews for \$150. Each faculty member will conduct 30 interviews for \$200. Due to limited transportation facilities, no more than 20 interviewers can be hired. How many undergraduate students, graduate students, and faculty members should be hired in order to maximize the number of interviews? What is the maximum number of interviews?

56. Opinion survey. Repeat Problem 55 if one of the requirements of the grant is that at least 50% of the interviewers be undergraduate students.

Answers to Matched Problems

1. Max $P = 6$ when $x_1 = 1$ and $x_2 = 4$
2. No optimal solution
3. 40 acres of crop *A*, 60 acres of crop *B*, no crop *C*; max $P = \$17,600$ (since $s_2 = 240$, \$240 out of the \$3,600 will not be spent).

6.3 The Dual Problem: Minimization with Problem Constraints of the Form ≥

- Formation of the Dual Problem
- Solution of Minimization Problems
- Application: Transportation Problem
- Summary of Problem Types and Solution Methods

In the preceding section, we restricted attention to standard maximization problems (problem constraints of the form ≤, with nonnegative constants on the right and any real numbers as objective function coefficients). Now we will consider minimization problems with ≥ problem constraints. These two types of problems turn out to be very closely related.

Formation of the Dual Problem

Associated with each minimization problem with ≥ constraints is a maximization problem called the **dual problem**. To illustrate the procedure for forming the dual problem, consider the following minimization problem:

$$\begin{aligned} \text{Minimize} \quad & C = 16x_1 + 45x_2 \\ \text{subject to} \quad & 2x_1 + 5x_2 \geq 50 \\ & x_1 + 3x_2 \geq 27 \\ & x_1, x_2 \geq 0 \end{aligned} \tag{1}$$

The first step in forming the dual problem is to construct a matrix using the problem constraints and the objective function written in the following form:

$$\begin{aligned} 2x_1 + 5x_2 &\geq 50 \\ x_1 + 3x_2 &\geq 27 \\ 16x_1 + 45x_2 &= C \end{aligned} \qquad A = \left[\begin{array}{cc|c} 2 & 5 & 50 \\ 1 & 3 & 27 \\ \hline 16 & 45 & 1 \end{array}\right]$$

⚠ CAUTION Do not confuse matrix A with the simplex tableau. We use a solid horizontal line in matrix A to help distinguish the dual matrix from the simplex tableau. No slack variables are involved in matrix A, and the coefficient of C is in the same column as the constants from the problem constraints. ▲

Now we will form a second matrix called the *transpose of A*. In general, the **transpose** of a given matrix A is the matrix A^{T} formed by interchanging the rows and corresponding columns of A (first row with first column, second row with second column, and so on).

$$A = \left[\begin{array}{cc|c} 2 & 5 & 50 \\ 1 & 3 & 27 \\ \hline 16 & 45 & 1 \end{array}\right]$$

R_1 in $A = C_1$ in A^T
R_2 in $A = C_2$ in A^T
R_3 in $A = C_3$ in A^T

$$A^T = \left[\begin{array}{cc|c} 2 & 1 & 16 \\ 5 & 3 & 45 \\ \hline 50 & 27 & 1 \end{array}\right]$$

A^T is the transpose of A.

We can use the rows of A^{T} to define a new linear programming problem. This new problem will always be a maximization problem with ≤ problem constraints. To avoid confusion, we will use different variables in this new problem:

$$\begin{aligned} 2y_1 + y_2 &\le 16 \\ 5y_1 + 3y_2 &\le 45 \\ 50y_1 + 27y_2 &= P \end{aligned} \qquad A^{\mathrm{T}} = \begin{array}{c} \begin{array}{cc} y_1 & y_2 \end{array} \\ \left[\begin{array}{cc|c} 2 & 1 & 16 \\ 5 & 3 & 45 \\ \hline 50 & 27 & 1 \end{array}\right] \end{array}$$

The dual of the minimization problem (1) is the following maximization problem:

$$\begin{aligned} \text{Maximize} \quad & P = 50y_1 + 27y_2 \\ \text{subject to} \quad & 2y_1 + y_2 \le 16 \\ & 5y_1 + 3y_2 \le 45 \\ & y_1, y_2 \ge 0 \end{aligned} \tag{2}$$

Explore and Discuss 1

Excluding the nonnegative constraints, the components of a linear programming problem can be divided into three categories: the coefficients of the objective function, the coefficients of the problem constraints, and the constants on the right side of the problem constraints. Write a verbal description of the relationship between the components of the original minimization problem (1) and the dual maximization problem (2).

The procedure for forming the dual problem is summarized in the following box:

PROCEDURE Formation of the Dual Problem

Given a minimization problem with ≥ problem constraints,

Step 1 Use the coefficients and constants in the problem constraints and the objective function to form a matrix A with the coefficients of the objective function in the last row.

Step 2 Interchange the rows and columns of matrix A to form the matrix A^{T}, the transpose of A.

Step 3 Use the rows of A^{T} to form a maximization problem with ≤ problem constraints.

EXAMPLE 1 **Forming the Dual Problem** Form the dual problem:

$$\begin{aligned} \text{Minimize} \quad & C = 40x_1 + 12x_2 + 40x_3 \\ \text{subject to} \quad & 2x_1 + x_2 + 5x_3 \ge 20 \\ & 4x_1 + x_2 + x_3 \ge 30 \\ & x_1, x_2, x_3 \ge 0 \end{aligned}$$

SOLUTION

Step 1 Form the matrix A:

$$A = \left[\begin{array}{ccc|c} 2 & 1 & 5 & 20 \\ 4 & 1 & 1 & 30 \\ \hline 40 & 12 & 40 & 1 \end{array}\right]$$

Step 2 Form the matrix A^{T}, the transpose of A:

$$A^{\mathrm{T}} = \left[\begin{array}{cc|c} 2 & 4 & 40 \\ 1 & 1 & 12 \\ 5 & 1 & 40 \\ \hline 20 & 30 & 1 \end{array}\right]$$

Step 3 State the dual problem:

$$\begin{aligned} \text{Maximize} \quad & P = 20y_1 + 30y_2 \\ \text{subject to} \quad & 2y_1 + 4y_2 \le 40 \\ & y_1 + y_2 \le 12 \\ & 5y_1 + y_2 \le 40 \\ & y_1, y_2 \ge 0 \end{aligned}$$

Matched Problem 1 Form the dual problem:

$$\begin{aligned} \text{Minimize} \quad & C = 16x_1 + 9x_2 + 21x_3 \\ \text{subject to} \quad & x_1 + x_2 + 3x_3 \ge 12 \\ & 2x_1 + x_2 + x_3 \ge 16 \\ & x_1, x_2, x_3 \ge 0 \end{aligned}$$

Solution of Minimization Problems

The following theorem establishes the relationship between the solution of a minimization problem and the solution of its dual problem:

THEOREM 1 Fundamental Principle of Duality

A minimization problem has a solution if and only if its dual problem has a solution. If a solution exists, then the optimal value of the minimization problem is the same as the optimal value of the dual problem.

The proof of Theorem 1 is beyond the scope of this text. However, we can illustrate Theorem 1 by solving minimization problem (1) and its dual maximization problem (2) geometrically.

ORIGINAL PROBLEM (1)

$$\begin{aligned} \text{Minimize} \quad & C = 16x_1 + 45x_2 \\ \text{subject to} \quad & 2x_1 + 5x_2 \ge 50 \\ & x_1 + 3x_2 \ge 27 \\ & x_1, x_2 \ge 0 \end{aligned}$$

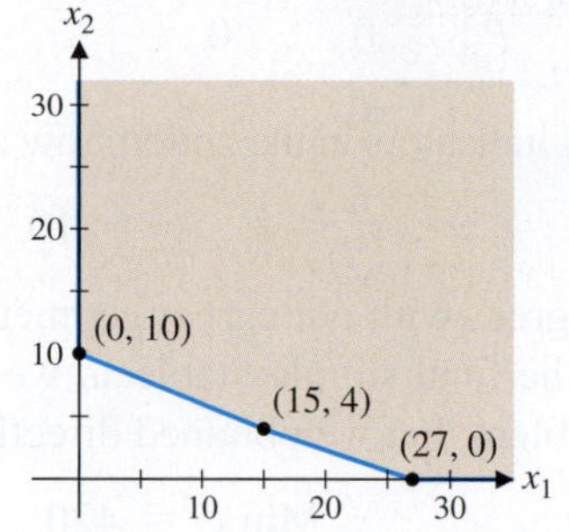

Corner Point

(x_1, x_2)	$C = 16x_1 + 45x_2$
$(0, 10)$	450
$(15, 4)$	420
$(27, 0)$	432

Min $C = 420$ at $(15, 4)$

DUAL PROBLEM (2)

$$\begin{aligned} \text{Maximize} \quad & P = 50y_1 + 27y_2 \\ \text{subject to} \quad & 2y_1 + y_2 \le 16 \\ & 5y_1 + 3y_2 \le 45 \\ & y_1, y_2 \ge 0 \end{aligned}$$

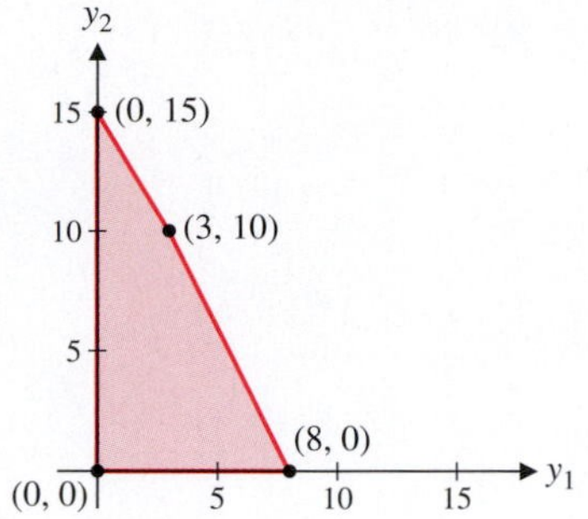

Corner Point

(y_1, y_2)	$P = 50y_1 + 27y_2$
$(0, 0)$	0
$(0, 15)$	405
$(3, 10)$	420
$(8, 0)$	400

Max $P = 420$ at $(3, 10)$

Note that the minimum value of C in problem (1) is the same as the maximum value of P in problem (2). The optimal solutions producing this optimal value are different: $(15, 4)$ is the optimal solution for problem (1), and $(3, 10)$ is the optimal solution for problem (2). Theorem 1 only guarantees that the optimal values of a minimization problem and its dual are equal, not that the optimal solutions are the same. In general, it is not possible to determine an optimal solution for a minimization problem by examining the feasible set for the dual problem. However, it is possible to apply the simplex method to the dual problem and find both the optimal value and an optimal solution to the original minimization problem. To see how this is done, we will solve problem (2) using the simplex method.

For reasons that will become clear later, we will use the variables x_1 and x_2 from the original problem as the slack variables in the dual problem:

$$\begin{aligned} 2y_1 + \quad y_2 + x_1 \qquad\qquad &= 16 \\ 5y_1 + \quad 3y_2 \qquad + x_2 \qquad &= 45 \\ -50y_1 - 27y_2 \qquad\qquad + P &= 0 \end{aligned}$$

Initial system for the dual problem

$$\begin{matrix} & y_1 & y_2 & x_1 & x_2 & P & \\ x_1 & \textcircled{2} & 1 & 1 & 0 & 0 & 16 \\ x_2 & 5 & 3 & 0 & 1 & 0 & 45 \\ P & -50 & -27 & 0 & 0 & 1 & 0 \end{matrix} \qquad 0.5R_1 \to R_1$$

$$\sim \left[\begin{array}{rrrrr|r} \textcircled{1} & 0.5 & 0.5 & 0 & 0 & 8 \\ 5 & 3 & 0 & 1 & 0 & 45 \\ -50 & -27 & 0 & 0 & 1 & 0 \end{array}\right] \qquad \begin{matrix} (-5)R_1 + R_2 \to R_2 \\ 50R_1 + R_3 \to R_3 \end{matrix}$$

$$\sim \begin{matrix} y_1 \\ x_2 \\ P \end{matrix} \left[\begin{array}{rrrrr|r} 1 & 0.5 & 0.5 & 0 & 0 & 8 \\ 0 & \textcircled{0.5} & -2.5 & 1 & 0 & 5 \\ 0 & -2 & 25 & 0 & 1 & 400 \end{array}\right] \qquad 2R_2 \to R_2$$

$$\sim \left[\begin{array}{ccccc|c} 1 & 0.5 & 0.5 & 0 & 0 & 8 \\ 0 & \textcircled{1} & -5 & 2 & 0 & 10 \\ \hdashline 0 & -2 & 25 & 0 & 1 & 400 \end{array}\right] \begin{array}{l} (-0.5)R_2 + R_1 \rightarrow R_1 \\ \\ 2R_2 + R_3 \rightarrow R_3 \end{array}$$

$$\sim \begin{array}{c} y_1 \\ y_2 \\ P \end{array} \left[\begin{array}{ccccc|c} 1 & 0 & 3 & -1 & 0 & 3 \\ 0 & 1 & -5 & 2 & 0 & 10 \\ \hdashline 0 & 0 & 15 & 4 & 1 & 420 \end{array}\right]$$

Since all indicators in the bottom row are nonnegative, the solution to the dual problem is

$$y_1 = 3, \quad y_2 = 10, \quad x_1 = 0, \quad x_2 = 0, \quad P = 420$$

which agrees with our earlier geometric solution. Furthermore, examining the bottom row of the final simplex tableau, we see the same optimal solution to the minimization problem that we obtained directly by the geometric method:

$$\text{Min } C = 420 \qquad \text{at} \qquad x_1 = 15, \ x_2 = 4$$

This is no accident.

An optimal solution to a minimization problem always can be obtained from the bottom row of the final simplex tableau for the dual problem.

We can see that using x_1 and x_2 as slack variables in the dual problem makes it easy to identify the solution of the original problem.

Explore and Discuss 2

The simplex method can be used to solve any standard maximization problem. Which of the following minimization problems have dual problems that are standard maximization problems? (Do not solve the problems.)

(A) Minimize $C = 2x_1 + 3x_2$

subject to
$$\begin{aligned} 2x_1 - 5x_2 &\ge 4 \\ x_1 - 3x_2 &\ge -6 \\ x_1, x_2 &\ge 0 \end{aligned}$$

(B) Minimize $C = 2x_1 - 3x_2$

subject to
$$\begin{aligned} -2x_1 + 5x_2 &\ge 4 \\ -x_1 + 3x_2 &\ge 6 \\ x_1, x_2 &\ge 0 \end{aligned}$$

What conditions must a minimization problem satisfy so that its dual problem is a standard maximization problem?

The procedure for solving a minimization problem by applying the simplex method to its dual problem is summarized in the following box:

PROCEDURE Solution of a Minimization Problem

Given a minimization problem with nonnegative coefficients in the objective function,

Step 1 Write all problem constraints as $\ge$ inequalities. (This may introduce negative numbers on the right side of some problem constraints.)

Step 2 Form the dual problem.

Step 3 Write the initial system of the dual problem, using the variables from the minimization problem as slack variables.

Step 4 Use the simplex method to solve the dual problem.

Step 5 Read the solution of the minimization problem from the bottom row of the final simplex tableau in step 4.

Note: If the dual problem has no optimal solution, the minimization problem has no optimal solution.

EXAMPLE 2 **Solving a Minimization Problem** Solve the following minimization problem by maximizing the dual problem:

$$\begin{aligned} \text{Minimize} \quad & C = 40x_1 + 12x_2 + 40x_3 \\ \text{subject to} \quad & 2x_1 + x_2 + 5x_3 \geq 20 \\ & 4x_1 + x_2 + x_3 \geq 30 \\ & x_1, x_2, x_3 \geq 0 \end{aligned}$$

SOLUTION From Example 1, the dual problem is

$$\begin{aligned} \text{Maximize} \quad & P = 20y_1 + 30y_2 \\ \text{subject to} \quad & 2y_1 + 4y_2 \leq 40 \\ & y_1 + y_2 \leq 12 \\ & 5y_1 + y_2 \leq 40 \\ & y_1, y_2 \geq 0 \end{aligned}$$

Using x_1, x_2, and x_3 for slack variables, we obtain the initial system for the dual problem:

$$\begin{aligned} 2y_1 + 4y_2 + x_1 \qquad\qquad\qquad &= 40 \\ y_1 + y_2 \qquad + x_2 \qquad\qquad &= 12 \\ 5y_1 + y_2 \qquad\qquad + x_3 \qquad &= 40 \\ -20y_1 - 30y_2 \qquad\qquad\qquad + P &= 0 \end{aligned}$$

Now we form the simplex tableau and solve the dual problem:

$$\begin{array}{c} \\ x_1 \\ x_2 \\ x_3 \\ P \end{array} \begin{array}{c} \begin{matrix} y_1 & y_2 & x_1 & x_2 & x_3 & P \end{matrix} \\ \left[\begin{array}{cccccc|c} 2 & \textcircled{4} & 1 & 0 & 0 & 0 & 40 \\ 1 & 1 & 0 & 1 & 0 & 0 & 12 \\ 5 & 1 & 0 & 0 & 1 & 0 & 40 \\ \hdashline -20 & -30 & 0 & 0 & 0 & 1 & 0 \end{array}\right] \end{array} \quad \tfrac{1}{4}R_1 \to R_1$$

$$\sim \left[\begin{array}{cccccc|c} \frac{1}{2} & \textcircled{1} & \frac{1}{4} & 0 & 0 & 0 & 10 \\ 1 & 1 & 0 & 1 & 0 & 0 & 12 \\ 5 & 1 & 0 & 0 & 1 & 0 & 40 \\ \hdashline -20 & -30 & 0 & 0 & 0 & 1 & 0 \end{array}\right] \quad \begin{array}{l} \\ (-1)R_1 + R_2 \to R_2 \\ (-1)R_1 + R_3 \to R_3 \\ 30R_1 + R_4 \to R_4 \end{array}$$

$$\sim \begin{array}{c} y_2 \\ x_2 \\ x_3 \\ P \end{array} \left[\begin{array}{cccccc|c} \frac{1}{2} & 1 & \frac{1}{4} & 0 & 0 & 0 & 10 \\ \textcircled{\frac{1}{2}} & 0 & -\frac{1}{4} & 1 & 0 & 0 & 2 \\ \frac{9}{2} & 0 & -\frac{1}{4} & 0 & 1 & 0 & 30 \\ \hdashline -5 & 0 & \frac{15}{2} & 0 & 0 & 1 & 300 \end{array}\right] \quad \begin{array}{l} \\ 2R_2 \to R_2 \\ \\ \\ \end{array}$$

$$\sim\left[\begin{array}{cccccc|c} \frac{1}{2} & 1 & \frac{1}{4} & 0 & 0 & 0 & 10 \\ \textcircled{1} & 0 & -\frac{1}{2} & 2 & 0 & 0 & 4 \\ \frac{9}{2} & 0 & -\frac{1}{4} & 0 & 1 & 0 & 30 \\ \hdashline -5 & 0 & \frac{15}{2} & 0 & 0 & 1 & 300 \end{array}\right] \begin{array}{l} \left(-\frac{1}{2}\right)R_2 + R_1 \to R_1 \\ \\ \left(-\frac{9}{2}\right)R_2 + R_3 \to R_3 \\ 5R_2 + R_4 \to R_4 \end{array}$$

$$\sim \begin{array}{c} y_2 \\ y_1 \\ x_3 \\ P \end{array}\left[\begin{array}{cccccc|c} 0 & 1 & \frac{1}{2} & -1 & 0 & 0 & 8 \\ 1 & 0 & -\frac{1}{2} & 2 & 0 & 0 & 4 \\ 0 & 0 & 2 & -9 & 1 & 0 & 12 \\ \hdashline 0 & 0 & 5 & 10 & 0 & 1 & 320 \end{array}\right]$$

From the bottom row of this tableau, we see that

$$\text{Min } C = 320 \quad \text{at} \quad x_1 = 5, \quad x_2 = 10, \quad x_3 = 0$$

Matched Problem 2 Solve the following minimization problem by maximizing the dual problem (see Matched Problem 1):

$$\begin{aligned} \text{Minimize } C &= 16x_1 + 9x_2 + 21x_3 \\ \text{subject to } \quad x_1 + x_2 + 3x_3 &\ge 12 \\ 2x_1 + x_2 + x_3 &\ge 16 \\ x_1, x_2, x_3 &\ge 0 \end{aligned}$$

CONCEPTUAL INSIGHT

In Section 6.2, we noted that multiplying a problem constraint by a number changes the units of the slack variable. This requires special interpretation of the value of the slack variable in the optimal solution, but causes no serious problems. However, when using the dual method, multiplying a problem constraint in the dual problem by a number can have serious consequences—the bottom row of the final simplex tableau may no longer give the correct solution to the minimization problem. To see this, refer to the first problem constraint of the dual problem in Example 2:

$$2y_1 + 4y_2 \le 40$$

If we multiply this constraint by $\frac{1}{2}$ and then solve, the final tableau is:

$$\begin{array}{cccccc} y_1 & y_2 & x_1 & x_2 & x_3 & P \end{array}$$
$$\left[\begin{array}{cccccc|c} 0 & 1 & 1 & -1 & 0 & 0 & 8 \\ 1 & 0 & -1 & 2 & 0 & 0 & 4 \\ 0 & 0 & 4 & -9 & 1 & 0 & 12 \\ \hdashline 0 & 0 & 10 & 10 & 0 & 1 & 320 \end{array}\right]$$

The bottom row of this tableau indicates that the optimal solution to the minimization problem is $C = 320$ at $x_1 = 10$ and $x_2 = 10$. This is not the correct answer ($x_1 = 5$ is the correct answer). Thus, **you should never multiply a problem constraint in a maximization problem by a number if that maximization problem is being used to solve a minimization problem**. You may still simplify problem constraints in a minimization problem before forming the dual problem.

EXAMPLE 3 **Solving a Minimization Problem** Solve the following minimization problem by maximizing the dual problem:

$$\begin{aligned} \text{Minimize} \quad & C = 5x_1 + 10x_2 \\ \text{subject to} \quad & x_1 - x_2 \ge 1 \\ & -x_1 + x_2 \ge 2 \\ & x_1, x_2 \ge 0 \end{aligned}$$

SOLUTION $A = \left[\begin{array}{rr|r} 1 & -1 & 1 \\ -1 & 1 & 2 \\ \hline 5 & 10 & 1 \end{array}\right] \qquad A^{\mathrm{T}} = \left[\begin{array}{rr|r} 1 & -1 & 5 \\ -1 & 1 & 10 \\ \hline 1 & 2 & 1 \end{array}\right]$

The dual problem is

$$\begin{aligned} \text{Maximize} \quad & P = y_1 + 2y_2 \\ \text{subject to} \quad & y_1 - y_2 \le 5 \\ & -y_1 + y_2 \le 10 \\ & y_1, y_2 \ge 0 \end{aligned}$$

Introduce slack variables x_1 and x_2, and form the initial system for the dual problem:

$$\begin{aligned} y_1 - y_2 + x_1 \qquad\quad &= 5 \\ -y_1 + y_2 \qquad + x_2 \quad &= 10 \\ -y_1 - 2y_2 \qquad\qquad + P &= 0 \end{aligned}$$

Form the simplex tableau and solve:

$$\begin{array}{c} \\ x_1 \\ x_2 \\ P \end{array} \begin{array}{c} \begin{array}{ccccc|c} y_1 & y_2 & x_1 & x_2 & P & \end{array} \\ \left[\begin{array}{rrrrr|r} 1 & -1 & 1 & 0 & 0 & 5 \\ -1 & \textcircled{1} & 0 & 1 & 0 & 10 \\ \hline -1 & -2 & 0 & 0 & 1 & 0 \end{array}\right] \end{array} \quad \begin{array}{l} R_2 + R_1 \to R_1 \\ \\ 2R_2 + R_3 \to R_3 \end{array}$$

$$\sim \left[\begin{array}{rrrrr|r} 0 & 0 & 1 & 1 & 0 & 15 \\ -1 & 1 & 0 & 1 & 0 & 10 \\ \hline -3 & 0 & 0 & 2 & 1 & 20 \end{array}\right]$$

No positive elements above dashed line in pivot column

↑ Pivot column

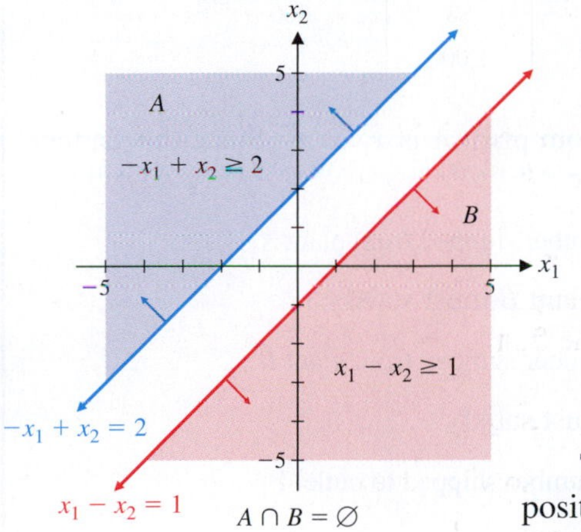

Figure 1

The −3 in the bottom row indicates that column 1 is the pivot column. Since no positive elements appear in the pivot column above the dashed line, we are unable to select a pivot row. We stop the pivot operation and conclude that this maximization problem has no optimal solution (see Fig. 2, Section 6.2). Theorem 1 now implies that the original minimization problem has no solution. The graph of the inequalities in the minimization problem (Fig. 1) shows that the feasible region is empty; so it is not surprising that an optimal solution does not exist.

Matched Problem 3 Solve the following minimization problem by maximizing the dual problem:

$$\begin{aligned} \text{Minimize} \quad & C = 2x_1 + 3x_2 \\ \text{subject to} \quad & x_1 - 2x_2 \ge 2 \\ & -x_1 + x_2 \ge 1 \\ & x_1, x_2 \ge 0 \end{aligned}$$

Application: Transportation Problem

One of the first applications of linear programming was to minimize the cost of transporting materials. Problems of this type are referred to as **transportation problems**.

EXAMPLE 4 **Transportation Problem** A computer manufacturing company has two assembly plants, plant *A* and plant *B*, and two distribution outlets, outlet I and outlet II. Plant *A* can assemble at most 700 computers a month, and plant *B* can assemble at most 900 computers a month. Outlet I must have at least 500 computers a month, and outlet II must have at least 1,000 computers a month. Transportation costs for shipping one computer from each plant to each outlet are as follows: \$6 from plant *A* to outlet I, \$5 from plant *A* to outlet II, \$4 from plant *B* to outlet I, \$8 from plant *B* to outlet II. Find a shipping schedule that minimizes the total cost of shipping the computers from the assembly plants to the distribution outlets. What is this minimum cost?

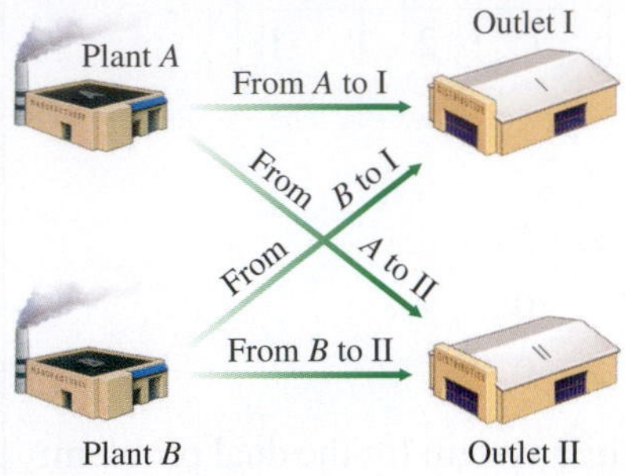

Figure 2

SOLUTION To form a shipping schedule, we must decide how many computers to ship from either plant to either outlet (Fig. 2). This will involve four decision variables:

$x_1 =$ number of computers shipped from plant *A* to outlet I

$x_2 =$ number of computers shipped from plant *A* to outlet II

$x_3 =$ number of computers shipped from plant *B* to outlet I

$x_4 =$ number of computers shipped from plant *B* to outlet II

Next, we summarize the relevant data in a table. Note that we do not follow the usual technique of associating each variable with a column of the table. Instead, sources are associated with the rows, and destinations are associated with the columns.

	Distribution Outlet		Assembly
	I	II	Capacity
Plant *A*	\$6	\$5	700
Plant *B*	\$4	\$8	900
Minimum required	500	1,000	

The total number of computers shipped from plant *A* is $x_1 + x_2$. Since this cannot exceed the assembly capacity at *A*, we have

$$x_1 + x_2 \le 700 \qquad \text{Number shipped from plant } A$$

Similarly, the total number shipped from plant *B* must satisfy

$$x_3 + x_4 \le 900 \qquad \text{Number shipped from plant } B$$

The total number shipped to each outlet must satisfy

$$x_1 + x_3 \ge 500 \qquad \text{Number shipped to outlet I}$$

and

$$x_2 + x_4 \ge 1{,}000 \qquad \text{Number shipped to outlet II}$$

Using the shipping charges in the table, the total shipping charges are

$$C = 6x_1 + 5x_2 + 4x_3 + 8x_4$$

We must solve the following linear programming problem:

$$\begin{array}{lrcll}
\text{Minimize} & C = 6x_1 + 5x_2 + 4x_3 + 8x_4 & & & \\
\text{subject to} & x_1 + x_2 \qquad\qquad & \le & 700 & \text{Available from } A \\
& x_3 + x_4 & \le & 900 & \text{Available from } B \\
& x_1 \qquad + x_3 \qquad & \ge & 500 & \text{Required at I} \\
& x_2 \qquad + x_4 & \ge & 1{,}000 & \text{Required at II} \\
& x_1, x_2, x_3, x_4 & \ge & 0 &
\end{array}$$

Before we can solve this problem, we must multiply the first two constraints by -1 so that all the problem constraints are of the $\geq$ type. This will introduce negative constants into the minimization problem but not into the dual problem. Since the coefficients of C are nonnegative, the constants in the dual problem will be nonnegative and the dual will be a standard maximization problem. The problem can now be stated as

$$\begin{aligned}
\text{Minimize} \quad & C = 6x_1 + 5x_2 + 4x_3 + 8x_4 \\
\text{subject to} \quad & -x_1 - x_2 \geq -700 \\
& -x_3 - x_4 \geq -900 \\
& x_1 + x_3 \geq 500 \\
& x_2 + x_4 \geq 1{,}000 \\
& x_1, x_2, x_3, x_4 \geq 0
\end{aligned}$$

$$A = \left[\begin{array}{cccc|c} -1 & -1 & 0 & 0 & -700 \\ 0 & 0 & -1 & -1 & -900 \\ 1 & 0 & 1 & 0 & 500 \\ 0 & 1 & 0 & 1 & 1{,}000 \\ \hline 6 & 5 & 4 & 8 & 1 \end{array}\right]$$

$$A^{T} = \left[\begin{array}{cccc|c} -1 & 0 & 1 & 0 & 6 \\ -1 & 0 & 0 & 1 & 5 \\ 0 & -1 & 1 & 0 & 4 \\ 0 & -1 & 0 & 1 & 8 \\ \hline -700 & -900 & 500 & 1{,}000 & 1 \end{array}\right]$$

The dual problem is

$$\begin{aligned}
\text{Maximize} \quad & P = -700y_1 - 900y_2 + 500y_3 + 1{,}000y_4 \\
\text{subject to} \quad & -y_1 + y_3 \leq 6 \\
& -y_1 + y_4 \leq 5 \\
& -y_2 + y_3 \leq 4 \\
& -y_2 + y_4 \leq 8 \\
& y_1, y_2, y_3, y_4 \geq 0
\end{aligned}$$

Introduce slack variables x_1, x_2, x_3, and x_4, and form the initial system for the dual problem:

$$\begin{aligned}
-y_1 + y_3 + x_1 &= 6 \\
-y_1 + y_4 + x_2 &= 5 \\
-y_2 + y_3 + x_3 &= 4 \\
-y_2 + y_4 + x_4 &= 8 \\
700y_1 + 900y_2 - 500y_3 - 1{,}000y_4 + P &= 0
\end{aligned}$$

Form the simplex tableau and solve:

$$\begin{array}{c} \\ x_1 \\ x_2 \\ x_3 \\ x_4 \\ P \end{array}
\begin{array}{c} \begin{array}{ccccccccc} y_1 & y_2 & y_3 & y_4 & x_1 & x_2 & x_3 & x_4 & P \end{array} \\
\left[\begin{array}{ccccccccc|c} -1 & 0 & 1 & 0 & 1 & 0 & 0 & 0 & 0 & 6 \\ -1 & 0 & 0 & \textcircled{1} & 0 & 1 & 0 & 0 & 0 & 5 \\ 0 & -1 & 1 & 0 & 0 & 0 & 1 & 0 & 0 & 4 \\ 0 & -1 & 0 & 1 & 0 & 0 & 0 & 1 & 0 & 8 \\ \hline 700 & 900 & -500 & -1{,}000 & 0 & 0 & 0 & 0 & 1 & 0 \end{array}\right] \end{array}
\begin{array}{l} \\ \\ \\ \\ (-1)R_2 + R_4 \rightarrow R_4 \\ 1{,}000R_2 + R_5 \rightarrow R_5 \end{array}$$

$$
\begin{array}{c} \\ x_1 \\ y_4 \\ \sim x_3 \\ x_4 \\ P \end{array}
\begin{array}{c}
\begin{array}{ccccccccc} y_1 & y_2 & y_3 & y_4 & x_1 & x_2 & x_3 & x_4 & P \end{array} \\
\left[\begin{array}{rrrrrrrrr|r}
-1 & 0 & 1 & 0 & 1 & 0 & 0 & 0 & 0 & 6 \\
-1 & 0 & 0 & 1 & 0 & 1 & 0 & 0 & 0 & 5 \\
0 & -1 & \textcircled{1} & 0 & 0 & 0 & 1 & 0 & 0 & 4 \\
1 & -1 & 0 & 0 & 0 & -1 & 0 & 1 & 0 & 3 \\
\hline
-300 & 900 & -500 & 0 & 0 & 1{,}000 & 0 & 0 & 1 & 5{,}000
\end{array}\right]
\end{array}
\begin{array}{l} \\ (-1)R_3 + R_1 \rightarrow R_1 \\ \\ \\ \\ 500R_3 + R_5 \rightarrow R_5 \end{array}
$$

$$
\begin{array}{c} x_1 \\ y_4 \\ \sim y_3 \\ x_4 \\ P \end{array}
\left[\begin{array}{rrrrrrrrr|r}
-1 & 1 & 0 & 0 & 1 & 0 & -1 & 0 & 0 & 2 \\
-1 & 0 & 0 & 1 & 0 & 1 & 0 & 0 & 0 & 5 \\
0 & -1 & 1 & 0 & 0 & 0 & 1 & 0 & 0 & 4 \\
\textcircled{1} & -1 & 0 & 0 & 0 & -1 & 0 & 1 & 0 & 3 \\
\hline
-300 & 400 & 0 & 0 & 0 & 1{,}000 & 500 & 0 & 1 & 7{,}000
\end{array}\right]
\begin{array}{l} R_4 + R_1 \rightarrow R_1 \\ R_4 + R_2 \rightarrow R_2 \\ \\ \\ 300R_4 + R_5 \rightarrow R_5 \end{array}
$$

$$
\begin{array}{c} x_1 \\ y_4 \\ \sim y_3 \\ y_1 \\ P \end{array}
\left[\begin{array}{rrrrrrrrr|r}
0 & 0 & 0 & 0 & 1 & -1 & -1 & 1 & 0 & 5 \\
0 & -1 & 0 & 1 & 0 & 0 & 0 & 1 & 0 & 8 \\
0 & -1 & 1 & 0 & 0 & 0 & 1 & 0 & 0 & 4 \\
1 & -1 & 0 & 0 & 0 & -1 & 0 & 1 & 0 & 3 \\
\hline
0 & 100 & 0 & 0 & 0 & 700 & 500 & 300 & 1 & 7{,}900
\end{array}\right]
$$

From the bottom row of this tableau, we have

$$\text{Min } C = 7{,}900 \quad \text{at} \quad x_1 = 0, \quad x_2 = 700, \quad x_3 = 500, \quad x_4 = 300$$

The shipping schedule that minimizes the shipping charges is 700 from plant *A* to outlet II, 500 from plant *B* to outlet I, and 300 from plant *B* to outlet II. The total shipping cost is \$7,900.

Figure 3 shows a solution to Example 4 in Excel. Notice that Excel permits the user to organize the original data and the solution in a format that is clear and easy to read. This is one of the main advantages of using spreadsheets to solve linear programming problems.

	A	B	C	D	E	F	G	H	I
1		DATA					SHIPPING SCHEDULE		
2		Distribution					Distribution		
3		outlet		Assembly			outlet		
4		I	II	capacity			I	II	Total
5	Plant A	\$6	\$5	700		Plant A	0	700	700
6	Plant B	\$4	\$8	900		Plant B	500	300	800
7	Minimum					Total	500	1,000	
8	required	500	1,000					Total cost	\$7,900

Figure 3

Matched Problem 4 Repeat Example 4 if the shipping charge from plant *A* to outlet I is increased to \$7 and the shipping charge from plant *B* to outlet II is decreased to \$3.

Summary of Problem Types and Solution Methods

In this and the preceding sections, we have solved both maximization and minimization problems, but with certain restrictions on the problem constraints, constants on the right, and/or objective function coefficients. Table 1 summarizes the types of problems and methods of solution we have considered so far.

Table 1 Summary of Problem Types and Simplex Solution Methods

Problem Type	Problem Constraints	Right-Side Constants	Coefficients of Objective Function	Method of Solution
1. Maximization	$\le$	Nonnegative	Any real numbers	Simplex method with slack variables
2. Minimization	$\ge$	Any real numbers	Nonnegative	Form dual problem and solve by simplex method with slack variables

The next section develops a generalized version of the simplex method that can handle both maximization and minimization problems with any combination of $\le$, $\ge$, and $=$ problem constraints.

Exercises 6.3

In Problems 1–8, find the transpose of each matrix.

A

1. $\begin{bmatrix} -5 & 3 \\ 2 & 7 \end{bmatrix}$

2. $\begin{bmatrix} 4 & -1 \\ 1 & 6 \end{bmatrix}$

3. $\begin{bmatrix} 1 & 7 & 8 \\ 6 & 3 & 5 \\ 9 & 4 & 2 \end{bmatrix}$

4. $\begin{bmatrix} 9 & 1 & 3 \\ 1 & 8 & 2 \\ 3 & 2 & 7 \end{bmatrix}$

5. $\begin{bmatrix} 2 & 1 & -6 & 0 & -1 \\ 5 & 2 & 0 & 1 & 3 \end{bmatrix}$

6. $\begin{bmatrix} 7 & 3 & -1 & 3 \\ -6 & 1 & 0 & -9 \end{bmatrix}$

7. $\begin{bmatrix} 1 & 2 & -1 \\ 0 & 2 & -7 \\ 8 & 0 & 1 \\ 4 & -1 & 3 \end{bmatrix}$

8. $\begin{bmatrix} 1 & -1 & 3 & 2 \\ 1 & -4 & 0 & 2 \\ 4 & -5 & 6 & 1 \\ -3 & 8 & 0 & -1 \\ 2 & 7 & -3 & 1 \end{bmatrix}$

In Problems 9 and 10,

(A) Form the dual problem.

(B) Write the initial system for the dual problem.

(C) Write the initial simplex tableau for the dual problem and label the columns of the tableau.

9. Minimize $C = 8x_1 + 9x_2$
subject to
$$x_1 + 3x_2 \ge 4$$
$$2x_1 + x_2 \ge 5$$
$$x_1, x_2 \ge 0$$

10. Minimize $C = 12x_1 + 5x_2$
subject to
$$2x_1 + x_2 \ge 7$$
$$3x_1 + x_2 \ge 9$$
$$x_1, x_2 \ge 0$$

In Problems 11 and 12, a minimization problem, the corresponding dual problem, and the final simplex tableau in the solution of the dual problem are given.

(A) Find the optimal solution of the dual problem.

(B) Find the optimal solution of the minimization problem.

11. Minimize $C = 21x_1 + 50x_2$
subject to
$$2x_1 + 5x_2 \ge 12$$
$$3x_1 + 7x_2 \ge 17$$
$$x_1, x_2 \ge 0$$

Maximize $P = 12y_1 + 17y_2$
subject to
$$2y_1 + 3y_2 \le 21$$
$$5y_1 + 7y_2 \le 50$$
$$y_1, y_2 \ge 0$$

$$\begin{array}{c} \\ \\ \\ \end{array}\begin{matrix} y_1 & y_2 & x_1 & x_2 & P & \\ \end{matrix}$$
$$\left[\begin{array}{ccccc|c} 0 & 1 & 5 & -2 & 0 & 5 \\ 1 & 0 & -7 & 3 & 0 & 3 \\ \hline 0 & 0 & 1 & 2 & 1 & 121 \end{array}\right]$$

12. Minimize $C = 16x_1 + 25x_2$
subject to
$$3x_1 + 5x_2 \ge 30$$
$$2x_1 + 3x_2 \ge 19$$
$$x_1, x_2 \ge 0$$

Maximize $P = 30y_1 + 19y_2$
subject to
$$3y_1 + 2y_2 \le 16$$
$$5y_1 + 3y_2 \le 25$$
$$y_1, y_2 \ge 0$$

$$\begin{matrix} y_1 & y_2 & x_1 & x_2 & P & \\ \end{matrix}$$
$$\left[\begin{array}{ccccc|c} 0 & 1 & 5 & -3 & 0 & 5 \\ 1 & 0 & -3 & 2 & 0 & 2 \\ \hline 0 & 0 & 5 & 3 & 1 & 155 \end{array}\right]$$

In Problems 13–20,

(A) Form the dual problem.

(B) Find the solution to the original problem by applying the simplex method to the dual problem.

13. Minimize $C = 9x_1 + 2x_2$
subject to
$$4x_1 + x_2 \ge 13$$
$$3x_1 + x_2 \ge 12$$
$$x_1, x_2 \ge 0$$

14. Minimize $C = x_1 + 4x_2$
subject to $x_1 + 2x_2 \geq 5$
$x_1 + 3x_2 \geq 6$
$x_1, x_2 \geq 0$

15. Minimize $C = 7x_1 + 12x_2$
subject to $2x_1 + 3x_2 \geq 15$
$x_1 + 2x_2 \geq 8$
$x_1, x_2 \geq 0$

16. Minimize $C = 3x_1 + 5x_2$
subject to $2x_1 + 3x_2 \geq 7$
$x_1 + 2x_2 \geq 4$
$x_1, x_2 \geq 0$

17. Minimize $C = 11x_1 + 4x_2$
subject to $2x_1 + x_2 \geq 8$
$-2x_1 + 3x_2 \geq 4$
$x_1, x_2 \geq 0$

18. Minimize $C = 40x_1 + 10x_2$
subject to $2x_1 + x_2 \geq 12$
$3x_1 - x_2 \geq 3$
$x_1, x_2 \geq 0$

19. Minimize $C = 7x_1 + 9x_2$
subject to $-3x_1 + x_2 \geq 6$
$x_1 - 2x_2 \geq 4$
$x_1, x_2 \geq 0$

20. Minimize $C = 10x_1 + 15x_2$
subject to $-4x_1 + x_2 \geq 12$
$12x_1 - 3x_2 \geq 10$
$x_1, x_2 \geq 0$

B *Solve the linear programming problems in Problems 21–36 by applying the simplex method to the dual problem.*

21. Minimize $C = 3x_1 + 9x_2$
subject to $2x_1 + x_2 \geq 8$
$x_1 + 2x_2 \geq 8$
$x_1, x_2 \geq 0$

22. Minimize $C = 2x_1 + x_2$
subject to $x_1 + x_2 \geq 8$
$x_1 + 2x_2 \geq 4$
$x_1, x_2 \geq 0$

23. Minimize $C = 7x_1 + 5x_2$
subject to $x_1 + x_2 \geq 4$
$x_1 - 2x_2 \geq -8$
$-2x_1 + x_2 \geq -8$
$x_1, x_2 \geq 0$

24. Minimize $C = 10x_1 + 4x_2$
subject to $2x_1 + x_2 \geq 6$
$x_1 - 4x_2 \geq -24$
$-8x_1 + 5x_2 \geq -24$
$x_1, x_2 \geq 0$

25. Minimize $C = 10x_1 + 30x_2$
subject to $2x_1 + x_2 \geq 16$
$x_1 + x_2 \geq 12$
$x_1 + 2x_2 \geq 14$
$x_1, x_2 \geq 0$

26. Minimize $C = 40x_1 + 10x_2$
subject to $3x_1 + x_2 \geq 24$
$x_1 + x_2 \geq 16$
$x_1 + 4x_2 \geq 30$
$x_1, x_2 \geq 0$

27. Minimize $C = 5x_1 + 7x_2$
subject to $x_1 \geq 4$
$x_1 + x_2 \geq 8$
$x_1 + 2x_2 \geq 10$
$x_1, x_2 \geq 0$

28. Minimize $C = 4x_1 + 5x_2$
subject to $2x_1 + x_2 \geq 12$
$x_1 + x_2 \geq 9$
$x_2 \geq 4$
$x_1, x_2 \geq 0$

29. Minimize $C = 60x_1 + 25x_2$
subject to $2x_1 + x_2 \geq 4$
$x_1, x_2 \geq 0$

30. Minimize $C = 35x_1 + 90x_2$
subject to $2x_1 + 5x_2 \geq 10$
$x_1, x_2 \geq 0$

31. Minimize $C = 10x_1 + 25x_2 + 12x_3$
subject to $2x_1 + 6x_2 + 3x_3 \geq 6$
$x_1, x_2, x_3 \geq 0$

32. Minimize $C = 10x_1 + 5x_2 + 15x_3$
subject to $7x_1 + 3x_2 + 6x_3 \geq 42$
$x_1, x_2, x_3 \geq 0$

33. Minimize $C = 10x_1 + 7x_2 + 12x_3$
subject to $x_1 + x_2 + 2x_3 \geq 7$
$2x_1 + x_2 + x_3 \geq 4$
$x_1, x_2, x_3 \geq 0$

34. Minimize $C = 14x_1 + 8x_2 + 20x_3$
subject to $x_1 + x_2 + 3x_3 \geq 6$
$2x_1 + x_2 + x_3 \geq 9$
$x_1, x_2, x_3 \geq 0$

35. Minimize $C = 5x_1 + 2x_2 + 2x_3$
subject to $x_1 - 4x_2 + x_3 \geq 6$
$-x_1 + x_2 - 2x_3 \geq 4$
$x_1, x_2, x_3 \geq 0$

36. Minimize $C = 6x_1 + 8x_2 + 3x_3$
subject to $-3x_1 - 2x_2 + x_3 \geq 4$
$x_1 + x_2 - x_3 \geq 2$
$x_1, x_2, x_3 \geq 0$

37. A minimization problem has 4 variables and 2 problem constraints. How many variables and problem constraints are in the dual problem?

38. A minimization problem has 3 variables and 5 problem constraints. How many variables and problem constraints are in the dual problem?

39. If you want to solve a minimization problem by applying the geometric method to the dual problem, how many variables and problem constraints must be in the original problem?

40. If you want to solve a minimization problem by applying the geometric method to the original problem, how many variables and problem constraints must be in the original problem?

In Problems 41 and 42,

(A) Form the dual problem.

(B) Is the dual problem a standard maximization problem in standard form? Explain.

41. Minimize $C = 4x_1 - x_2$
subject to
$$5x_1 + 2x_2 \ge 7$$
$$4x_1 + 6x_2 \ge 10$$
$$x_1, x_2 \ge 0$$

42. Minimize $C = -2x_1 + 9x_2$
subject to
$$3x_1 - x_2 \ge -8$$
$$x_1 + 4x_2 \ge 4$$
$$x_1, x_2 \ge 0$$

In Problems 43 and 44,

(A) Form an equivalent minimization problem with ≥ problem constraints (multiply inequalities by −1 if necessary),

(B) Form the dual of the equivalent problem.

(C) Is the dual problem a standard maximization problem in standard form? Explain.

43. Minimize $C = 3x_1 + x_2 + 5x_3$
subject to
$$2x_1 - 6x_2 - x_3 \le 10$$
$$-5x_1 + x_2 + 4x_3 \ge 15$$
$$x_1, x_2, x_3 \ge 0$$

44. Minimize $C = 25x_1 + 30x_2 + 50x_3$
subject to
$$-x_1 + 3x_2 - 8x_3 \le 12$$
$$6x_1 - 4x_2 + 5x_3 \le 20$$
$$x_1, x_2, x_3 \ge 0$$

C *Solve the linear programming problems in Problems 45–48 by applying the simplex method to the dual problem.*

45. Minimize $C = 16x_1 + 8x_2 + 4x_3$
subject to
$$3x_1 + 2x_2 + 2x_3 \ge 16$$
$$4x_1 + 3x_2 + x_3 \ge 14$$
$$5x_1 + 3x_2 + x_3 \ge 12$$
$$x_1, x_2, x_3 \ge 0$$

46. Minimize $C = 6x_1 + 8x_2 + 12x_3$
subject to
$$x_1 + 3x_2 + 3x_3 \ge 6$$
$$x_1 + 5x_2 + 5x_3 \ge 4$$
$$2x_1 + 2x_2 + 3x_3 \ge 8$$
$$x_1, x_2, x_3 \ge 0$$

47. Minimize $C = 5x_1 + 4x_2 + 5x_3 + 6x_4$
subject to
$$x_1 + x_2 \le 12$$
$$x_3 + x_4 \le 25$$
$$x_1 + x_3 \ge 20$$
$$x_2 + x_4 \ge 15$$
$$x_1, x_2, x_3, x_4 \ge 0$$

48. Repeat Problem 47 with $C = 4x_1 + 7x_2 + 5x_3 + 6x_4$.

Applications

In Problems 49–58, construct a mathematical model in the form of a linear programming problem. (The answers in the back of the book for these application problems include the model.) Then solve the problem by applying the simplex method to the dual problem.

49. Ice cream. A food processing company produces regular and deluxe ice cream at three plants. Per hour of operation, the Cedarburg plant produces 20 gallons of regular ice cream and 10 gallons of deluxe ice cream. The Grafton plant produces 10 gallons of regular and 20 gallons of deluxe, and the West Bend plant produces 20 gallons of regular and 20 gallons of deluxe. It costs \$70 per hour to operate the Cedarburg plant, \$75 per hour to operate the Grafton plant, and \$90 per hour to operate the West Bend plant. The company needs to produce at least 300 gallons of regular ice cream and at least 200 gallons of deluxe ice cream each day. How many hours per day should each plant operate in order to produce the required amounts of ice cream and minimize the cost of production? What is the minimum production cost?

50. Mining. A mining company operates two mines, each producing three grades of ore. The West Summit mine can produce 2 tons of low-grade ore, 3 tons of medium-grade ore, and 1 ton of high-grade ore in one hour of operation. The North Ridge mine can produce 2 tons of low-grade ore, 1 ton of medium-grade ore, and 2 tons of high-grade ore in one hour of operation. To satisfy existing orders, the company needs to produce at least 100 tons of low-grade ore, 60 tons of medium-grade ore, and 80 tons of high-grade ore. The cost of operating each mine varies, depending on conditions while extracting the ore. If it costs \$400 per hour to operate the West Summit mine and \$600 per hour to operate the North Ridge mine, how many hours should each mine operate to supply the required amounts of ore and minimize the cost of production? What is the minimum production cost?

51. Ice cream. Repeat Problem 49 if the demand for deluxe ice cream increases from 200 gallons to 300 gallons per day and all other data remain the same.

52. **Mining.** Repeat Problem 50 if it costs \$300 per hour to operate the West Summit mine and \$700 per hour to operate the North Ridge mine and all other data remain the same.

53. **Ice cream.** Repeat Problem 49 if the demand for deluxe ice cream increases from 200 gallons to 400 gallons per day and all other data remain the same.

54. **Mining.** Repeat Problem 50 if it costs \$800 per hour to operate the West Summit mine and \$200 per hour to operate the North Ridge mine and all other data remain the same.

55. **Human nutrition.** A dietitian arranges a special diet using three foods: *L*, *M*, and *N*. Each ounce of food *L* contains 20 units of calcium, 10 units of iron, 10 units of vitamin A, and 20 units of cholesterol. Each ounce of food *M* contains 10 units of calcium, 10 units of iron, 15 units of vitamin A, and 24 units of cholesterol. Each ounce of food *N* contains 10 units of calcium, 10 units of iron, 10 units of vitamin A, and 18 units of cholesterol. If the minimum daily requirements are 300 units of calcium, 200 units of iron, and 240 units of vitamin A, how many ounces of each food should be used to meet the minimum requirements and simultaneously minimize the cholesterol intake? What is the minimum cholesterol intake?

56. **Plant food.** A farmer can buy three types of plant food: mix *A*, mix *B*, and mix *C*. Each cubic yard of mix *A* contains 20 pounds of phosphoric acid, 10 pounds of nitrogen, and 10 pounds of potash. Each cubic yard of mix *B* contains 10 pounds of phosphoric acid, 10 pounds of nitrogen, and 15 pounds of potash. Each cubic yard of mix *C* contains 20 pounds of phosphoric acid, 20 pounds of nitrogen, and 5 pounds of potash. The minimum monthly requirements are 480 pounds of phosphoric acid, 320 pounds of nitrogen, and 225 pounds of potash. If mix *A* costs \$30 per cubic yard, mix *B* costs \$36 per cubic yard, and mix *C* costs \$39 per cubic yard, how many cubic yards of each mix should the farmer blend to meet the minimum monthly requirements at a minimal cost? What is the minimum cost?

57. **Education: resource allocation.** A metropolitan school district has two overcrowded high schools and two underenrolled high schools. To balance the enrollment, the school board decided to bus students from the overcrowded schools to the underenrolled schools. North Division High School has 300 more students than normal, and South Division High School has 500 more students than normal. Central High School can accommodate 400 additional students, and Washington High School can accommodate 500 additional students. The weekly cost of busing a student from North Division to Central is \$5, from North Division to Washington is \$2, from South Division to Central is \$3, and from South Division to Washington is \$4. Determine the number of students that should be bused from each overcrowded school to each underenrolled school in order to balance the enrollment and minimize the cost of busing the students. What is the minimum cost?

58. **Education: resource allocation.** Repeat Problem 57 if the weekly cost of busing a student from North Division to Washington is \$7 and all other data remain the same.

Answers to Matched Problems

1. Maximize $P = 12y_1 + 16y_2$
   subject to
   $$\begin{aligned} y_1 + 2y_2 &\le 16 \\ y_1 + y_2 &\le 9 \\ 3y_1 + y_2 &\le 21 \\ y_1, y_2 &\ge 0 \end{aligned}$$

2. Min $C = 136$ at $x_1 = 4, x_2 = 8, x_3 = 0$

3. Dual problem:
   Maximize $P = 2y_1 + y_2$
   subject to
   $$\begin{aligned} y_1 - y_2 &\le 2 \\ -2y_1 + y_2 &\le 3 \\ y_1, y_2 &\ge 0 \end{aligned}$$
   No optimal solution

4. 600 from plant *A* to outlet II, 500 from plant *B* to outlet I, 400 from plant *B* to outlet II; total shipping cost is \$6,200.

6.4 Maximization and Minimization with Mixed Problem Constraints

- Introduction to the Big *M* Method
- Big *M* Method
- Minimization by the Big *M* Method
- Summary of Solution Methods
- Larger Problems: Refinery Application

In the preceding two sections, we have seen how to solve both maximization and minimization problems, but with rather severe restrictions on problem constraints, right-side constants, and/or objective function coefficients (see the summary in Table 1 of Section 6.3). In this section we present a generalized version of the simplex method that will solve both maximization and minimization problems with any combination of $\le$, $\ge$, and $=$ problem constraints. The only requirement is that each problem constraint must have a nonnegative constant on the right side. (This restriction is easily accommodated, as you will see.)

Introduction to the Big *M* Method

We introduce the *big M method* through a simple maximization problem with mixed problem constraints. The key parts of the method will then be summarized and applied to more complex problems.

Consider the following problem:

$$\begin{aligned} \text{Maximize} \quad & P = 2x_1 + x_2 \\ \text{subject to} \quad & x_1 + x_2 \le 10 \\ & -x_1 + x_2 \ge 2 \\ & x_1, x_2 \ge 0 \end{aligned} \tag{1}$$

To form an equation out of the first inequality, we introduce a slack variable s_1, as before, and write

$$x_1 + x_2 + s_1 = 10$$

How can we form an equation out of the second inequality? We introduce a second variable s_2 and subtract it from the left side:

$$-x_1 + x_2 - s_2 = 2$$

The variable s_2 is called a **surplus variable** because it is the amount (surplus) by which the left side of the inequality exceeds the right side.

Next, we express the linear programming problem (1) as a system of equations:

$$\begin{aligned} x_1 + x_2 + s_1 \qquad\qquad &= 10 \\ -x_1 + x_2 \qquad - s_2 \qquad &= 2 \\ -2x_1 - x_2 \qquad\qquad + P &= 0 \\ x_1, x_2, s_1, s_2 &\ge 0 \end{aligned} \tag{2}$$

It can be shown that a basic solution of system (2) is not feasible if any of the variables (excluding P) are negative. So **a surplus variable is required to satisfy the nonnegative constraint.**

The basic solution found by setting the nonbasic variables x_1 and x_2 equal to 0 is

$$x_1 = 0, \quad x_2 = 0, \quad s_1 = 10, \quad s_2 = -2, \quad P = 0$$

But this basic solution is not feasible, since the surplus variable s_2 is negative (which is a violation of the nonnegative requirements of all variables except P). The simplex method works only when the basic solution for a tableau is feasible, so we cannot solve this problem simply by writing the tableau for (2) and starting pivot operations.

To use the simplex method on problems with mixed constraints, we turn to an ingenious device called an *artificial variable*. This variable has no physical meaning in the original problem. It is introduced solely for the purpose of obtaining a basic feasible solution so that we can apply the simplex method. An **artificial variable** is a variable introduced into each equation that has a surplus variable. As before, to ensure that we consider only basic feasible solutions, **an artificial variable is required to satisfy the nonnegative constraint**. (As we will see later, artificial variables are also used to augment equality problem constraints when they are present.)

Returning to the problem at hand, we introduce an artificial variable a_1 into the equation involving the surplus variable s_2:

$$-x_1 + x_2 - s_2 + a_1 = 2$$

To prevent an artificial variable from becoming part of an optimal solution to the original problem, a very large "penalty" is introduced into the objective function. This penalty is created by choosing a positive constant M so large that the artificial variable is forced to be 0 in any final optimal solution of the original problem. (Since the constant M can be as large as we wish in computer solutions, M is often selected as the largest number the computer can hold!) Then we add the term $-Ma_1$ to the objective function:

$$P = 2x_1 + x_2 - Ma_1$$

We now have a new problem, which we call the **modified problem**:

$$\begin{aligned} \text{Maximize} \quad & P = 2x_1 + x_2 - Ma_1 \\ \text{subject to} \quad & x_1 + x_2 + s_1 = 10 \\ & -x_1 + x_2 - s_2 + a_1 = 2 \\ & x_1, x_2, s_1, s_2, a_1 \ge 0 \end{aligned} \tag{3}$$

The initial system for the modified problem (3) is

$$\begin{aligned} x_1 + x_2 + s_1 \quad & = 10 \\ -x_1 + x_2 - s_2 + a_1 \quad & = 2 \\ -2x_1 - x_2 + Ma_1 + P & = 0 \\ x_1, x_2, s_1, s_2, a_1 & \ge 0 \end{aligned} \tag{4}$$

We next write the augmented matrix for system (4), which we call the **preliminary simplex tableau** for the modified problem. (The reason we call it the "preliminary" simplex tableau instead of the "initial" simplex tableau will be made clear shortly.)

$$\begin{array}{c} \\ \end{array} \begin{matrix} x_1 & x_2 & s_1 & s_2 & a_1 & P & \end{matrix}$$
$$\left[\begin{array}{rrrrrr|r} 1 & 1 & 1 & 0 & 0 & 0 & 10 \\ -1 & 1 & 0 & -1 & 1 & 0 & 2 \\ \hline -2 & -1 & 0 & 0 & M & 1 & 0 \end{array}\right] \tag{5}$$

To start the simplex process, including any necessary pivot operations, the preliminary simplex tableau should either meet the two requirements given in the following box or be transformed by row operations into a tableau that meets these two requirements.

DEFINITION Initial Simplex Tableau

For a system tableau to be considered an **initial simplex tableau**, it must satisfy the following two requirements:

1. The requisite number of basic variables must be selectable by the process described in Section 6.2. That is, a variable can be selected as a basic variable only if it corresponds to a column in the tableau that has exactly one nonzero element and the nonzero element in the column is not in the same row as the nonzero element in the column of another basic variable. The remaining variables are then selected as nonbasic variables to be set equal to 0 in determining a basic solution.
2. The basic solution found by setting the nonbasic variables equal to 0 is feasible.

Tableau (5) satisfies the first initial simplex tableau requirement, since s_1, s_2, and P can be selected as basic variables according to the first requirement. (Not all preliminary simplex tableaux satisfy the first requirement; see Example 2.) However, tableau (5) does not satisfy the second initial simplex tableau requirement since the basic solution is not feasible $(s_2 = -2)$. To use the simplex method, we must use row operations to transform tableau (5) into an equivalent matrix that satisfies both initial simplex tableau requirements. **Note that this transformation is not a pivot operation.**

To get an idea of how to proceed, notice in tableau (5) that -1 in the s_2 column is in the same row as 1 in the a_1 column. This is not an accident! The artificial variable a_1 was introduced so that this would happen. If we eliminate M from the bottom of the a_1 column, the nonbasic variable a_1 will become a basic variable and the troublesome basic variable s_2 will become a nonbasic variable. We proceed to eliminate M from the a_1 column using row operations:

$$\begin{array}{c} \\ \\ \\ \\ \sim \\ \\ \end{array}\begin{array}{c} \begin{matrix} x_1 & x_2 & s_1 & s_2 & a_1 & P & \end{matrix} \\ \left[\begin{array}{cccccc|c} 1 & 1 & 1 & 0 & 0 & 0 & 10 \\ -1 & 1 & 0 & -1 & 1 & 0 & 2 \\ \hdashline -2 & -1 & 0 & 0 & M & 1 & 0 \end{array}\right] \quad (-M)R_2 + R_3 \rightarrow R_3 \\ \left[\begin{array}{cccccc|c} 1 & 1 & 1 & 0 & 0 & 0 & 10 \\ -1 & 1 & 0 & -1 & 1 & 0 & 2 \\ \hdashline M-2 & -M-1 & 0 & M & 0 & 1 & -2M \end{array}\right] \end{array}$$

From this last matrix we see that the basic variables are s_1, a_1, and P. The basic solution found by setting the nonbasic variables x_1, x_2, and s_2 equal to 0 is

$$x_1 = 0, \quad x_2 = 0, \quad s_1 = 10, \quad s_2 = 0, \quad a_1 = 2, \quad P = -2M$$

The basic solution is feasible (P can be negative), and both requirements for an initial simplex tableau are met. We perform pivot operations to find an optimal solution.

The pivot column is determined by the most negative indicator in the bottom row of the tableau. Since M is a positive number, $-M - 1$ is certainly a negative indicator. What about the indicator $M - 2$? Remember that M is a very large positive number. We will assume that M is so large that any expression of the form $M - k$ is positive. So the only negative indicator in the bottom row is $-M - 1$.

$$\begin{array}{c} \\ \\ \text{Pivot row} \rightarrow \\ \\ \end{array}\begin{array}{c} \begin{matrix} x_1 & x_2 & s_1 & s_2 & a_1 & P & \end{matrix} \\ \left[\begin{array}{cccccc|c} 1 & 1 & 1 & 0 & 0 & 0 & 10 \\ -1 & \textcircled{1} & 0 & -1 & 1 & 0 & 2 \\ \hdashline M-2 & -M-1 & 0 & M & 0 & 1 & -2M \end{array}\right] \begin{array}{l} \frac{10}{1} = 10 \\ \frac{2}{1} = 2 \\ \\ \end{array} \\ \uparrow \\ \text{Pivot column} \end{array}$$

Having identified the pivot element, we now begin pivoting:

$$\begin{array}{c} \begin{matrix} & x_1 & x_2 & s_1 & s_2 & a_1 & P & \end{matrix} \\ \begin{array}{c} s_1 \\ a_1 \\ P \end{array}\left[\begin{array}{cccccc|c} 1 & 1 & 1 & 0 & 0 & 0 & 10 \\ -1 & \textcircled{1} & 0 & -1 & 1 & 0 & 2 \\ \hdashline M-2 & -M-1 & 0 & M & 0 & 1 & -2M \end{array}\right] \begin{array}{l} (-1)R_2 + R_1 \rightarrow R_1 \\ \\ (M+1)R_2 + R_3 \rightarrow R_3 \end{array} \\ \sim \begin{array}{c} s_1 \\ x_2 \\ P \end{array}\left[\begin{array}{cccccc|c} \textcircled{2} & 0 & 1 & 1 & -1 & 0 & 8 \\ -1 & 1 & 0 & -1 & 1 & 0 & 2 \\ \hdashline -3 & 0 & 0 & -1 & M+1 & 1 & 2 \end{array}\right] \begin{array}{l} \frac{1}{2}R_1 \rightarrow R_1 \\ \\ \\ \end{array} \\ \sim \left[\begin{array}{cccccc|c} \textcircled{1} & 0 & \frac{1}{2} & \frac{1}{2} & -\frac{1}{2} & 0 & 4 \\ -1 & 1 & 0 & -1 & 1 & 0 & 2 \\ \hdashline -3 & 0 & 0 & -1 & M+1 & 1 & 2 \end{array}\right] \begin{array}{l} \\ R_1 + R_2 \rightarrow R_2 \\ 3R_1 + R_3 \rightarrow R_3 \end{array} \\ \sim \begin{array}{c} x_1 \\ x_2 \\ P \end{array}\left[\begin{array}{cccccc|c} 1 & 0 & \frac{1}{2} & \frac{1}{2} & -\frac{1}{2} & 0 & 4 \\ 0 & 1 & \frac{1}{2} & -\frac{1}{2} & \frac{1}{2} & 0 & 6 \\ \hdashline 0 & 0 & \frac{3}{2} & \frac{1}{2} & M-\frac{1}{2} & 1 & 14 \end{array}\right] \end{array}$$

Since all the indicators in the last row are nonnegative ($M - \frac{1}{2}$ is nonnegative because M is a very large positive number), we can stop and write the optimal solution:

$$\text{Max } P = 14 \quad \text{at} \quad x_1 = 4, \quad x_2 = 6, \quad s_1 = 0, \quad s_2 = 0, \quad a_1 = 0$$

This is an optimal solution to the modified problem (3). How is it related to the original problem (2)? Since $a_1 = 0$ in this solution,

$$x_1 = 4, \quad x_2 = 6, \quad s_1 = 0, \quad s_2 = 0, \quad P = 14 \qquad (6)$$

is certainly a feasible solution for system (2). [You can verify this by direct substitution into system (2).] Surprisingly, it turns out that solution (6) is an optimal solution to the original problem. To see that this is true, suppose we were able to find feasible values of x_1, x_2, s_1, and s_2 that satisfy the original system (2) and produce a value of $P > 14$. Then by using these same values in problem (3) along with $a_1 = 0$, we would have a feasible solution of problem (3) with $P > 14$. This contradicts the fact that $P = 14$ is the maximum value of P for the modified problem. Solution (6) is an optimal solution for the original problem.

As this example illustrates, if $a_1 = 0$ is an optimal solution for the modified problem, then deleting a_1 produces an optimal solution for the original problem. What happens if $a_1 \neq 0$ in the optimal solution for the modified problem? In this case, it can be shown that the original problem has no optimal solution because its feasible set is empty.

In larger problems, each $\geq$ problem constraint will require the introduction of a surplus variable and an artificial variable. If one of the problem constraints is an equation rather than an inequality, then there is no need to introduce a slack or surplus variable. However, each = problem constraint will require the introduction of another artificial variable to prevent the initial basic solution from violating the equality constraint—the decision variables are often 0 in the initial basic solution (see Example 2). Finally, each artificial variable also must be included in the objective function for the modified problem. The same constant M can be used for each artificial variable. Because of the role that the constant M plays in this approach, this method is often called the ***big M method***.

Big M Method

We summarize the key steps of the big M method and use them to solve several problems.

PROCEDURE Big M Method: Introducing Slack, Surplus, and Artificial Variables to Form the Modified Problem

Step 1 If any problem constraints have negative constants on the right side, multiply both sides by -1 to obtain a constraint with a nonnegative constant. (If the constraint is an inequality, this will reverse the direction of the inequality.)

Step 2 Introduce a slack variable in each $\leq$ constraint.

Step 3 Introduce a surplus variable and an artificial variable in each $\geq$ constraint.

Step 4 Introduce an artificial variable in each = constraint.

Step 5 For each artificial variable a_i, add $-Ma_i$ to the objective function. Use the same constant M for all artificial variables.

EXAMPLE 1 **Finding the Modified Problem** Find the modified problem for the following linear programming problem. (Do not attempt to solve the problem.)

$$\begin{aligned}
\text{Maximize} \quad & P = 2x_1 + 5x_2 + 3x_3 \\
\text{subject to} \quad & x_1 + 2x_2 - x_3 \leq 7 \\
& -x_1 + x_2 - 2x_3 \leq -5 \\
& x_1 + 4x_2 + 3x_3 \geq 1 \\
& 2x_1 - x_2 + 4x_3 = 6 \\
& x_1, x_2, x_3 \geq 0
\end{aligned}$$

SOLUTION First, we multiply the second constraint by -1 to change -5 to 5:

$$(-1)(-x_1 + x_2 - 2x_3) \geq (-1)(-5)$$

$$x_1 - x_2 + 2x_3 \geq 5$$

Next, we introduce the slack, surplus, and artificial variables according to the procedure stated in the box:

$$\begin{aligned}
x_1 + 2x_2 - x_3 + s_1 &= 7\\
x_1 - x_2 + 2x_3 - s_2 + a_1 &= 5\\
x_1 + 4x_2 + 3x_3 - s_3 + a_2 &= 1\\
2x_1 - x_2 + 4x_3 + a_3 &= 6
\end{aligned}$$

Finally, we add $-Ma_1$, $-Ma_2$, and $-Ma_3$ to the objective function:

$$P = 2x_1 + 5x_2 + 3x_3 - Ma_1 - Ma_2 - Ma_3$$

The modified problem is

$$\begin{aligned}
\text{Maximize} \quad & P = 2x_1 + 5x_2 + 3x_3 - Ma_1 - Ma_2 - Ma_3\\
\text{subject to} \quad & x_1 + 2x_2 - x_3 + s_1 = 7\\
& x_1 - x_2 + 2x_3 - s_2 + a_1 = 5\\
& x_1 + 4x_2 + 3x_3 - s_3 + a_2 = 1\\
& 2x_1 - x_2 + 4x_3 + a_3 = 6\\
& x_1, x_2, x_3, s_1, s_2, s_3, a_1, a_2, a_3 \geq 0
\end{aligned}$$

Matched Problem 1 Repeat Example 1 for

$$\begin{aligned}
\text{Maximize} \quad & P = 3x_1 - 2x_2 + x_3\\
\text{subject to} \quad & x_1 - 2x_2 + x_3 \geq 5\\
& -x_1 - 3x_2 + 4x_3 \leq -10\\
& 2x_1 + 4x_2 + 5x_3 \leq 20\\
& 3x_1 - x_2 - x_3 = -15\\
& x_1, x_2, x_3 \geq 0
\end{aligned}$$

Now we can list the key steps for solving a problem using the big M method. The various steps and remarks are based on a number of important theorems, which we assume without proof. In particular, step 2 is based on the fact that (except for some degenerate cases not considered here) if the modified linear programming problem has an optimal solution, then the preliminary simplex tableau will be transformed into an initial simplex tableau by eliminating the M's from the columns corresponding to the artificial variables in the preliminary simplex tableau. Having obtained an initial simplex tableau, we perform pivot operations.

PROCEDURE Big *M* Method: Solving the Problem

Step 1 Form the preliminary simplex tableau for the modified problem.

Step 2 Use row operations to eliminate the M's in the bottom row of the preliminary simplex tableau in the columns corresponding to the artificial variables. The resulting tableau is the initial simplex tableau.

Step 3 Solve the modified problem by applying the simplex method to the initial simplex tableau found in step 2.

Step 4 Relate the optimal solution of the modified problem to the original problem.

(A) If the modified problem has no optimal solution, then the original problem has no optimal solution.

(B) If all artificial variables are 0 in the optimal solution to the modified problem, then delete the artificial variables to find an optimal solution to the original problem.

(C) If any artificial variables are nonzero in the optimal solution to the modified problem, then the original problem has no optimal solution.

EXAMPLE 2 **Using the Big *M* Method** Solve the following linear programming problem using the big M method:

$$\begin{aligned} \text{Maximize} \quad & P = x_1 - x_2 + 3x_3 \\ \text{subject to} \quad & x_1 + x_2 \le 20 \\ & x_1 + x_3 = 5 \\ & x_2 + x_3 \ge 10 \\ & x_1, x_2, x_3 \ge 0 \end{aligned}$$

SOLUTION State the modified problem:

$$\begin{aligned} \text{Maximize} \quad & P = x_1 - x_2 + 3x_3 - Ma_1 - Ma_2 \\ \text{subject to} \quad & x_1 + x_2 + s_1 = 20 \\ & x_1 + x_3 + a_1 = 5 \\ & x_2 + x_3 - s_2 + a_2 = 10 \\ & x_1, x_2, x_3, s_1, s_2, a_1, a_2 \ge 0 \end{aligned}$$

Write the preliminary simplex tableau for the modified problem, and find the initial simplex tableau by eliminating the M's from the artificial variable columns:

$$\begin{array}{c} \\ \\ \end{array}\begin{matrix} x_1 & x_2 & x_3 & s_1 & a_1 & s_2 & a_2 & P & & \end{matrix}$$

$$\left[\begin{array}{cccccccc|c} 1 & 1 & 0 & 1 & 0 & 0 & 0 & 0 & 20 \\ 1 & 0 & 1 & 0 & 1 & 0 & 0 & 0 & 5 \\ 0 & 1 & 1 & 0 & 0 & -1 & 1 & 0 & 10 \\ \hline -1 & 1 & -3 & 0 & M & 0 & M & 1 & 0 \end{array}\right] \quad \begin{array}{l} \text{Eliminate } M \text{ from the } a_1 \text{ column} \\ \\ \\ (-M)R_2 + R_4 \rightarrow R_4 \end{array}$$

$$\sim \left[\begin{array}{cccccccc|c} 1 & 1 & 0 & 1 & 0 & 0 & 0 & 0 & 20 \\ 1 & 0 & 1 & 0 & 1 & 0 & 0 & 0 & 5 \\ 0 & 1 & 1 & 0 & 0 & -1 & 1 & 0 & 10 \\ \hline -M-1 & 1 & -M-3 & 0 & 0 & 0 & M & 1 & -5M \end{array}\right] \quad \begin{array}{l} \text{Eliminate } M \text{ from the } a_2 \text{ column} \\ \\ \\ (-M)R_3 + R_4 \rightarrow R_4 \end{array}$$

$$\sim \left[\begin{array}{cccccccc|c} 1 & 1 & 0 & 1 & 0 & 0 & 0 & 0 & 20 \\ 1 & 0 & 1 & 0 & 1 & 0 & 0 & 0 & 5 \\ 0 & 1 & 1 & 0 & 0 & -1 & 1 & 0 & 10 \\ \hline -M-1 & -M+1 & -2M-3 & 0 & 0 & M & 0 & 1 & -15M \end{array}\right]$$

From this last matrix we see that the basic variables are s_1, a_1, a_2, and P. The basic solution found by setting the nonbasic variables x_1, x_2, x_3, and s_2 equal to 0 is

$$x_1 = 0, \quad x_2 = 0, \quad x_3 = 0, \quad s_1 = 20, \quad a_1 = 5, \quad s_2 = 0, \quad a_2 = 10, \quad P = -15M$$

The basic solution is feasible, and both requirements for an initial simplex tableau are met. We perform pivot operations to find the optimal solution.

$$
\begin{array}{c} \\ s_1 \\ a_1 \\ a_2 \\ P \end{array}
\begin{array}{cccccccc|c}
x_1 & x_2 & x_3 & s_1 & a_1 & s_2 & a_2 & P & \\
1 & 1 & 0 & 1 & 0 & 0 & 0 & 0 & 20 \\
1 & 0 & \textcircled{1} & 0 & 1 & 0 & 0 & 0 & 5 \\
0 & 1 & 1 & 0 & 0 & -1 & 1 & 0 & 10 \\
\hdashline
-M-1 & -M+1 & -2M-3 & 0 & 0 & M & 0 & 1 & -15M
\end{array}
\quad
\begin{array}{l} \\ \\ \\ (-1)R_2 + R_3 \rightarrow R_3 \\ (2M+3)R_2 + R_4 \rightarrow R_4 \end{array}
$$

$$
\sim
\begin{array}{c} s_1 \\ x_3 \\ a_2 \\ P \end{array}
\left[\begin{array}{cccccccc|c}
1 & 1 & 0 & 1 & 0 & 0 & 0 & 0 & 20 \\
1 & 0 & 1 & 0 & 1 & 0 & 0 & 0 & 5 \\
-1 & \textcircled{1} & 0 & 0 & -1 & -1 & 1 & 0 & 5 \\
\hdashline
M+2 & -M+1 & 0 & 0 & 2M+3 & M & 0 & 1 & -5M+15
\end{array}\right]
\quad
\begin{array}{l} (-1)R_3 + R_1 \rightarrow R_1 \\ \\ \\ (M-1)R_3 + R_4 \rightarrow R_4 \end{array}
$$

$$
\sim
\begin{array}{c} s_1 \\ x_3 \\ x_2 \\ P \end{array}
\left[\begin{array}{cccccccc|c}
2 & 0 & 0 & 1 & 1 & 1 & -1 & 0 & 15 \\
1 & 0 & 1 & 0 & 1 & 0 & 0 & 0 & 5 \\
-1 & 1 & 0 & 0 & -1 & -1 & 1 & 0 & 5 \\
\hdashline
3 & 0 & 0 & 0 & M+4 & 1 & M-1 & 1 & 10
\end{array}\right]
$$

Since the bottom row has no negative indicators, we can stop and write the optimal solution to the modified problem:

$$x_1 = 0, \quad x_2 = 5, \quad x_3 = 5, \quad s_1 = 15, \quad a_1 = 0, \quad s_2 = 0, \quad a_2 = 0, \quad P = 10$$

Since $a_1 = 0$ and $a_2 = 0$, the solution to the original problem is

$$\text{Max } P = 10 \quad \text{at} \quad x_1 = 0, \quad x_2 = 5, \quad x_3 = 5$$

Matched Problem 2 Solve the following linear programming problem using the big M method:

$$
\begin{array}{ll}
\text{Maximize} & P = x_1 + 4x_2 + 2x_3 \\
\text{subject to} & \quad\; x_2 + x_3 \le 4 \\
& x_1 \qquad - x_3 = 6 \\
& x_1 - x_2 - x_3 \ge 1 \\
& x_1, x_2, x_3 \ge 0
\end{array}
$$

EXAMPLE 3 **Using the Big M Method** Solve the following linear programming problem using the big M method:

$$
\begin{array}{ll}
\text{Maximize} & P = 3x_1 + 5x_2 \\
\text{subject to} & 2x_1 + x_2 \le 4 \\
& x_1 + 2x_2 \ge 10 \\
& x_1, x_2 \ge 0
\end{array}
$$

SOLUTION Introducing slack, surplus, and artificial variables, we obtain the modified problem:

$$
\begin{array}{rcl}
2x_1 + x_2 + s_1 & = & 4 \\
x_1 + 2x_2 \qquad - s_2 + a_1 & = & 10 \\
-3x_1 - 5x_2 \qquad\qquad + Ma_1 + P & = & 0
\end{array}
\qquad \text{Modified problem}
$$

Preliminary simplex tableau

$$
\begin{array}{c} \\ \\ \\ \\ \end{array}
\begin{array}{cccccc|c}
x_1 & x_2 & s_1 & s_2 & a_1 & P & \\
\hline
2 & 1 & 1 & 0 & 0 & 0 & 4 \\
1 & 2 & 0 & -1 & 1 & 0 & 10 \\
\hdashline
-3 & -5 & 0 & 0 & M & 1 & 0
\end{array}
$$

Eliminate M in the a_1 column.

$(-M)R_2 + R_3 \rightarrow R_3$

Initial simplex tableau

$$
\sim
\begin{array}{c}
s_1 \\ a_1 \\ P
\end{array}
\left[
\begin{array}{cccccc|c}
2 & \textcircled{1} & 1 & 0 & 0 & 0 & 4 \\
1 & 2 & 0 & -1 & 1 & 0 & 10 \\
\hdashline
-M-3 & -2M-5 & 0 & M & 0 & 1 & -10M
\end{array}
\right]
$$

Begin pivot operations.

$(-2)R_1 + R_2 \rightarrow R_2$

$(2M+5)R_1 + R_3 \rightarrow R_3$

$$
\sim
\begin{array}{c}
x_2 \\ a_1 \\ P
\end{array}
\left[
\begin{array}{cccccc|c}
2 & 1 & 1 & 0 & 0 & 0 & 4 \\
-3 & 0 & -2 & -1 & 1 & 0 & 2 \\
\hdashline
3M+7 & 0 & 2M+5 & M & 0 & 1 & -2M+20
\end{array}
\right]
$$

The optimal solution of the modified problem is

$$
\begin{aligned}
x_1 = 0, \quad x_2 = 4, \quad s_1 = 0, \quad s_2 = 0, \\
a_1 = 2, \quad P = -2M + 20
\end{aligned}
$$

Since $a_1 \neq 0$, the original problem has no optimal solution. Figure 1 shows that the feasible region for the original problem is empty.

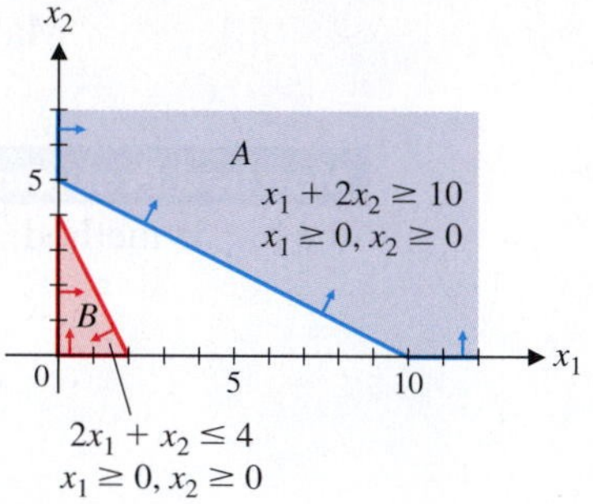

Figure 1

$A \cap B = \varnothing$

Matched Problem 3 Solve the following linear programming problem using the big M method:

$$
\begin{aligned}
\text{Maximize} \quad & P = 3x_1 + 2x_2 \\
\text{subject to} \quad & x_1 + 5x_2 \leq 5 \\
& 2x_1 + x_2 \geq 12 \\
& x_1, x_2 \geq 0
\end{aligned}
$$

Minimization by the Big M Method

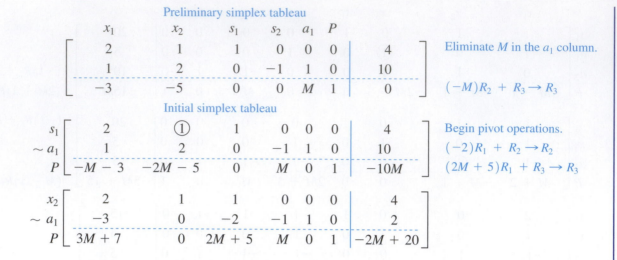

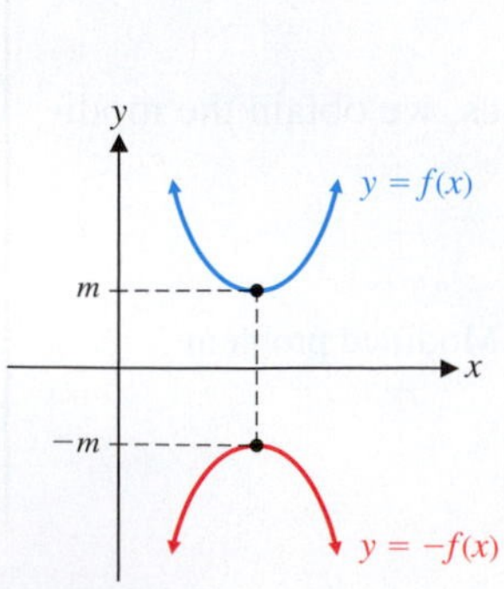

Figure 2

In addition to solving any maximization problem, the big M method can be used to solve minimization problems. To minimize an objective function, we need only to maximize its negative. Figure 2 illustrates the fact that the minimum value of a function f occurs at the same point as the maximum value of the function $-f$. Furthermore, if m is the minimum value of f, then $-m$ is the maximum value of $-f$, and conversely. So we can find the minimum value of a function f by finding the maximum value of $-f$ and then changing the sign of the maximum value.

EXAMPLE 4 **Production Scheduling: Minimization Problem** A small jewelry manufacturer hires a highly skilled gem cutter to work at least 6 hours per day. On the other hand, the polishing facilities can be used for at most 10 hours per day. The company specializes in three kinds of semiprecious gemstones, *J*, *K*, and *L*. Relevant cutting, polishing, and cost requirements are listed in the table. How many gemstones of each type should be processed each day to minimize the cost of the finished stones? What is the minimum cost?

	J	*K*	*L*
Cutting	1 hr	1 hr	1 hr
Polishing	2 hr	1 hr	2 hr
Cost per stone	\$30	\$30	\$10

SOLUTION Since we must decide how many gemstones of each type should be processed each day, the decision variables are

$$x_1 = \text{number of type } J \text{ gemstones processed each day}$$
$$x_2 = \text{number of type } K \text{ gemstones processed each day}$$
$$x_3 = \text{number of type } L \text{ gemstones processed each day}$$

Since the data is already summarized in a table, we can proceed directly to the model:

$$\text{Minimize} \quad C = 30x_1 + 30x_2 + 10x_3 \qquad \text{Objective function}$$
$$\text{subject to} \quad \left.\begin{aligned} x_1 + x_2 + x_3 &\ge 6 \\ 2x_1 + x_2 + 2x_3 &\le 10 \end{aligned}\right\} \qquad \text{Problem constraints}$$
$$x_1, x_2, x_3 \ge 0 \qquad \text{Nonnegative constraints}$$

We convert this to a maximization problem by letting

$$P = -C = -30x_1 - 30x_2 - 10x_3$$

We get

$$\begin{aligned} \text{Maximize} \quad P &= -30x_1 - 30x_2 - 10x_3 \\ \text{subject to} \quad x_1 + x_2 + x_3 &\ge 6 \\ 2x_1 + x_2 + 2x_3 &\le 10 \\ x_1, x_2, x_3 &\ge 0 \end{aligned}$$

and Min $C = -$Max P. To solve, we first state the modified problem:

$$\begin{aligned} x_1 + x_2 + x_3 - s_1 + a_1 \qquad &= 6 \\ 2x_1 + x_2 + 2x_3 \qquad + s_2 &= 10 \\ 30x_1 + 30x_2 + 10x_3 \qquad + Ma_1 \qquad + P &= 0 \\ x_1, x_2, x_3, s_1, s_2, a_1 &\ge 0 \end{aligned}$$

$$\begin{array}{c} \\ \\ \\ \end{array}\begin{matrix} x_1 & x_2 & x_3 & s_1 & a_1 & s_2 & P & \\ \end{matrix}$$
$$\left[\begin{array}{ccccccc|c} 1 & 1 & 1 & -1 & 1 & 0 & 0 & 6 \\ 2 & 1 & 2 & 0 & 0 & 1 & 0 & 10 \\ \hdashline 30 & 30 & 10 & 0 & M & 0 & 1 & 0 \end{array}\right] \quad \begin{array}{l} \text{Eliminate } M \text{ in the } a_1 \text{ column} \\ (-M)R_1 + R_3 \to R_3 \end{array}$$

Begin pivot operations. Assume M is so large that $-M + 30$ and $-M + 10$ are negative

$$\begin{array}{r} a_1 \\ \sim s_2 \\ P \end{array}\left[\begin{array}{ccccccc|c} 1 & 1 & 1 & -1 & 1 & 0 & 0 & 6 \\ 2 & 1 & \textcircled{2} & 0 & 0 & 1 & 0 & 10 \\ \hdashline -M+30 & -M+30 & -M+10 & M & 0 & 0 & 1 & -6M \end{array}\right] \quad 0.5R_2 \rightarrow R_2$$

$$\sim\left[\begin{array}{ccccccc|c} 1 & 1 & 1 & -1 & 1 & 0 & 0 & 6 \\ 1 & 0.5 & \textcircled{1} & 0 & 0 & 0.5 & 0 & 5 \\ \hdashline -M+30 & -M+30 & -M+10 & M & 0 & 0 & 1 & -6M \end{array}\right] \quad \begin{array}{l} (-1)R_2 + R_1 \rightarrow R_1 \\ \\ (M-10)R_2 + R_3 \rightarrow R_3 \end{array}$$

$$\begin{array}{r} a_1 \\ \sim x_3 \\ P \end{array}\left[\begin{array}{ccccccc|c} 0 & \textcircled{0.5} & 0 & -1 & 1 & -0.5 & 0 & 1 \\ 1 & 0.5 & 1 & 0 & 0 & 0.5 & 0 & 5 \\ \hdashline 20 & -0.5M+25 & 0 & M & 0 & 0.5M-5 & 1 & -M-50 \end{array}\right] \quad 2R_1 \rightarrow R_1$$

$$\sim\left[\begin{array}{ccccccc|c} 0 & \textcircled{1} & 0 & -2 & 2 & -1 & 0 & 2 \\ 1 & 0.5 & 1 & 0 & 0 & 0.5 & 0 & 5 \\ \hdashline 20 & -0.5M+25 & 0 & M & 0 & 0.5M-5 & 1 & -M-50 \end{array}\right] \quad \begin{array}{l} (-0.5)R_1 + R_2 \rightarrow R_2 \\ (0.5M-25)R_1 + R_3 \rightarrow R_3 \end{array}$$

$$\begin{array}{r} x_2 \\ \sim x_3 \\ P \end{array}\left[\begin{array}{ccccccc|c} 0 & 1 & 0 & -2 & 2 & -1 & 0 & 2 \\ 1 & 0 & 1 & 1 & -1 & 1 & 0 & 4 \\ \hdashline 20 & 0 & 0 & 50 & M-50 & 20 & 1 & -100 \end{array}\right]$$

The bottom row has no negative indicators, so the optimal solution for the modified problem is

$$x_1 = 0, \quad x_2 = 2, \quad x_3 = 4, \quad s_1 = 0, \quad a_1 = 0, \quad s_2 = 0, \quad P = -100$$

Since $a_1 = 0$, deleting a_1 produces the optimal solution to the original maximization problem and also to the minimization problem. Thus,

$$\text{Min } C = -\text{Max } P = -(-100) = 100 \quad \text{at} \quad x_1 = 0, \quad x_2 = 2, \quad x_3 = 4$$

That is, a minimum cost of \$100 for gemstones will be realized if no type J, 2 type K, and 4 type L stones are processed each day.

Matched Problem 4 Repeat Example 4 if the gem cutter works at least 8 hours a day and all other data remain the same.

Summary of Solution Methods

The big M method can be used to solve any minimization problems, including those that can be solved by the dual method. (Note that Example 4 could have been solved by the dual method.) Both methods of solving minimization problems are important. You will be instructed to solve most minimization problems in Exercise 6.4 by the big M method in order to gain more experience with this method. However, if the method of solution is not specified, the dual method is usually easier.

Table 1 should help you select the proper method of solution for any linear programming problem.

Table 1 Summary of Problem Types and Simplex Solution Methods

Problem Type	Problem Constraints	Right-Side Constants	Coefficients of Objective Function	Method of Solution
1. Maximization	$\le$	Nonnegative	Any real numbers	Simplex method with slack variables
2. Minimization	$\ge$	Any real numbers	Nonnegative	Form dual problem and solve by the preceding method
3. Maximization	Mixed $(\le, \ge, =)$	Nonnegative	Any real numbers	Form modified problem with slack, surplus, and artificial variables, and solve by the big M method
4. Minimization	Mixed $(\le, \ge, =)$	Nonnegative	Any real numbers	Maximize negative of objective function by the preceding method

Larger Problems: Refinery Application

Up to this point, all the problems we have considered could be solved by hand. However, the real value of the simplex method lies in its ability to solve problems with a large number of variables and constraints, where a computer is generally used to perform the actual pivot operations. As a final application, we consider a problem that would require the use of a computer to complete the solution.

EXAMPLE 5 **Petroleum Blending** A refinery produces two grades of gasoline, regular and premium, by blending together two components, A and B. Component A has an octane rating of 90 and costs $28 a barrel. Component B has an octane rating of 110 and costs $32 a barrel. The octane rating for regular gasoline must be at least 95, and the octane rating for premium must be at least 105. Regular gasoline sells for $34 a barrel and premium sells for $40 a barrel. Currently, the company has 30,000 barrels of component A and 20,000 barrels of component B. It also has orders for 20,000 barrels of regular and 10,000 barrels of premium that must be filled. Assuming that all the gasoline produced can be sold, determine the maximum possible profit.

SOLUTION This problem is similar to the transportation problem in Section 6.3. That is, to maximize the profit, we must decide how much of each component must be used to produce each grade of gasoline. Thus, the decision variables are

$$x_1 = \text{number of barrels of component } A \text{ used in regular gasoline}$$

$$x_2 = \text{number of barrels of component } A \text{ used in premium gasoline}$$

$$x_3 = \text{number of barrels of component } B \text{ used in regular gasoline}$$

$$x_4 = \text{number of barrels of component } B \text{ used in premium gasoline}$$

Table 2

Component	Octane Rating	Cost ($)	Available Supply
A	90	28	30,000 barrels
B	110	32	20,000 barrels
Grade	**Minimum Octane Rating**	**Selling Price($)**	**Existing Orders**
Regular	95	34	20,000 barrels
Premium	105	40	10,000 barrels

Next, we summarize the data in table form (Table 2). Once again, we have to adjust the form of the table to fit the data.

The total amount of component A used is $x_1 + x_2$. This cannot exceed the available supply. Thus, one constraint is

$$x_1 + x_2 \le 30{,}000$$

The corresponding inequality for component B is

$$x_3 + x_4 \le 20{,}000$$

The amounts of regular and premium gasoline produced must be sufficient to meet the existing orders:

$$x_1 + x_3 \ge 20{,}000 \quad \text{Regular}$$

$$x_2 + x_4 \ge 10{,}000 \quad \text{Premium}$$

Now consider the octane ratings. The octane rating of a blend is simply the proportional average of the octane ratings of the components. So the octane rating for regular gasoline is

$$90\frac{x_1}{x_1 + x_3} + 110\frac{x_3}{x_1 + x_3}$$

where $x_1/(x_1 + x_3)$ is the percentage of component A used in regular gasoline and $x_3/(x_1 + x_3)$ is the percentage of component B. The final octane rating of regular gasoline must be at least 95; so

$$90\frac{x_1}{x_1 + x_3} + 110\frac{x_3}{x_1 + x_3} \geq 95 \qquad \text{Multiply by } x_1 + x_3.$$

$$90x_1 + 110x_3 \geq 95(x_1 + x_3) \qquad \text{Collect like terms on the right side.}$$

$$0 \geq 5x_1 - 15x_3 \qquad \text{Octane rating for regular}$$

The corresponding inequality for premium gasoline is

$$90\frac{x_2}{x_2 + x_4} + 110\frac{x_4}{x_2 + x_4} \geq 105$$

$$90x_2 + 110x_4 \geq 105(x_2 + x_4)$$

$$0 \geq 15x_2 - 5x_4 \qquad \text{Octane rating for premium}$$

The cost of the components used is

$$C = 28(x_1 + x_2) + 32(x_3 + x_4)$$

The revenue from selling all the gasoline is

$$R = 34(x_1 + x_3) + 40(x_2 + x_4)$$

and the profit is

$$\begin{aligned} P &= R - C \\ &= 34(x_1 + x_3) + 40(x_2 + x_4) - 28(x_1 + x_2) - 32(x_3 + x_4) \\ &= (34 - 28)x_1 + (40 - 28)x_2 + (34 - 32)x_3 + (40 - 32)x_4 \\ &= 6x_1 + 12x_2 + 2x_3 + 8x_4 \end{aligned}$$

To find the maximum profit, we must solve the following linear programming problem:

$$\begin{array}{lrll} \text{Maximize} & P = 6x_1 + 12x_2 + 2x_3 + 8x_4 & & \text{Profit} \\ \text{subject to} & x_1 + x_2 & \leq 30{,}000 & \text{Available } A \\ & x_3 + x_4 & \leq 20{,}000 & \text{Available } B \\ & x_1 + x_3 & \geq 20{,}000 & \text{Required regular} \\ & x_2 + x_4 & \geq 10{,}000 & \text{Required premium} \\ & 5x_1 - 15x_3 & \leq 0 & \text{Octane for regular} \\ & 15x_2 - 5x_4 & \leq 0 & \text{Octane for premium} \\ & x_1, x_2, x_3, x_4 & \geq 0 & \end{array}$$

We will use technology to solve this large problem. There are many types of software that use the big M method to solve linear programming problems, including Java applets, graphing calculator programs, and spreadsheets. Because you are likely to use different software than we did, we will simply display the initial and final tableaux. Notice that in the last row of the initial tableau, we entered a large number, 10^6, instead of the symbol M. This is typical of software implementations of the big M method.

$$\begin{array}{c} \begin{matrix} x_1 & x_2 & x_3 & x_4 & s_1 & s_2 & s_3 & a_1 & s_4 & a_2 & s_5 & s_6 & P & \end{matrix} \\ \left[\begin{array}{rrrrrrrrrrrrr|r} 1 & 1 & 0 & 0 & 1 & 0 & 0 & 0 & 0 & 0 & 0 & 0 & 0 & 30{,}000 \\ 0 & 0 & 1 & 1 & 0 & 1 & 0 & 0 & 0 & 0 & 0 & 0 & 0 & 20{,}000 \\ 1 & 0 & 1 & 0 & 0 & 0 & -1 & 1 & 0 & 0 & 0 & 0 & 0 & 20{,}000 \\ 0 & 1 & 0 & 1 & 0 & 0 & 0 & 0 & -1 & 1 & 0 & 0 & 0 & 10{,}000 \\ 5 & 0 & -15 & 0 & 0 & 0 & 0 & 0 & 0 & 0 & 1 & 0 & 0 & 0 \\ 0 & 15 & 0 & -5 & 0 & 0 & 0 & 0 & 0 & 0 & 0 & 1 & 0 & 0 \\ \hdashline -6 & -12 & -2 & -8 & 0 & 0 & 0 & 10^6 & 0 & 10^6 & 0 & 0 & 1 & 0 \end{array}\right] \end{array}$$

The final table produced by the software is

x_1	x_2	x_3	x_4	s_1	s_2	s_3	a_1	s_4	a_2	s_5	s_6	P	
0	0	0	0	1.5	−0.5	1	−1	0	0	−0.1	−0.1	0	15,000
0	0	0	0	−0.5	1.5	0	0	1	−1	0.1	0.1	0	5,000
0	0	1	0	0.375	−0.125	0	0	0	0	−0.075	−0.025	0	8,750
0	0	0	1	−0.375	1.125	0	0	0	0	0.075	0.025	0	11,250
1	0	0	0	1.125	−0.375	0	0	0	0	−0.025	−0.075	0	26,250
0	1	0	0	−0.125	0.375	0	0	0	0	0.025	0.075	0	3,750
0	0	0	0	3	11	0	10^6	0	10^6	0.6	0.6	1	310,000

From the final tableau, we see that the refinery should blend 26,250 barrels of component *A* and 8,750 barrels of component *B* to produce 35,000 barrels of regular gasoline. They should blend 3,750 barrels of component *A* and 11,250 barrels of component *B* to produce 15,000 barrels of premium gasoline. This will result in a maximum profit of $310,000.

Explore and Discuss 1

Interpret the values of the slack and surplus variables in the computer solution to Example 5.

Matched Problem 5 Suppose that the refinery in Example 5 has 35,000 barrels of component *A*, which costs $25 a barrel, and 15,000 barrels of component *B*, which costs $35 a barrel. If all other data remain the same, formulate a linear programming problem to find the maximum profit. Do not attempt to solve the problem (unless you have access to software that can solve linear programming problems).

Exercises 6.4

A *In Problems 1–8,*

(A) Introduce slack, surplus, and artificial variables and form the modified problem.

(B) Write the preliminary simplex tableau for the modified problem and find the initial simplex tableau.

(C) Find the optimal solution of the modified problem by applying the simplex method to the initial simplex tableau.

(D) Find the optimal solution of the original problem, if it exists.

1. Maximize $P = 5x_1 + 2x_2$
subject to
$$x_1 + 2x_2 \le 12$$
$$x_1 + x_2 \ge 4$$
$$x_1, x_2 \ge 0$$

2. Maximize $P = 3x_1 + 7x_2$
subject to
$$2x_1 + x_2 \le 16$$
$$x_1 + x_2 \ge 6$$
$$x_1, x_2 \ge 0$$

3. Maximize $P = 3x_1 + 5x_2$
subject to
$$2x_1 + x_2 \le 8$$
$$x_1 + x_2 = 6$$
$$x_1, x_2 \ge 0$$

4. Maximize $P = 4x_1 + 3x_2$
subject to
$$x_1 + 3x_2 \le 24$$
$$x_1 + x_2 = 12$$
$$x_1, x_2 \ge 0$$

5. Maximize $P = 4x_1 + 3x_2$
subject to
$$-x_1 + 2x_2 \le 2$$
$$x_1 + x_2 \ge 4$$
$$x_1, x_2 \ge 0$$

6. Maximize $P = 3x_1 + 4x_2$
subject to
$$x_1 - 2x_2 \le 2$$
$$x_1 + x_2 \ge 5$$
$$x_1, x_2 \ge 0$$

7. Maximize $P = 5x_1 + 10x_2$
subject to
$x_1 + x_2 \le 3$
$2x_1 + 3x_2 \ge 12$
$x_1, x_2 \ge 0$

8. Maximize $P = 4x_1 + 6x_2$
subject to
$x_1 + x_2 \le 2$
$3x_1 + 5x_2 \ge 15$
$x_1, x_2 \ge 0$

B *Use the big M method to solve Problems 9–22.*

9. Minimize and maximize $P = 2x_1 - x_2$
subject to
$x_1 + x_2 \le 8$
$5x_1 + 3x_2 \ge 30$
$x_1, x_2 \ge 0$

10. Minimize and maximize $P = -4x_1 + 16x_2$
subject to
$3x_1 + x_2 \le 28$
$x_1 + 2x_2 \ge 16$
$x_1, x_2 \ge 0$

11. Maximize $P = 2x_1 + 5x_2$
subject to
$x_1 + 2x_2 \le 18$
$2x_1 + x_2 \le 21$
$x_1 + x_2 \ge 10$
$x_1, x_2 \ge 0$

12. Maximize $P = 6x_1 + 2x_2$
subject to
$x_1 + 2x_2 \le 20$
$2x_1 + x_2 \le 16$
$x_1 + x_2 \ge 9$
$x_1, x_2 \ge 0$

13. Maximize $P = 10x_1 + 12x_2 + 20x_3$
subject to
$3x_1 + x_2 + 2x_3 \ge 12$
$x_1 - x_2 + 2x_3 = 6$
$x_1, x_2, x_3 \ge 0$

14. Maximize $P = 5x_1 + 7x_2 + 9x_3$
subject to
$x_1 - x_2 + x_3 \ge 20$
$2x_1 + x_2 + 5x_3 = 35$
$x_1, x_2, x_3 \ge 0$

15. Minimize $C = -5x_1 - 12x_2 + 16x_3$
subject to
$x_1 + 2x_2 + x_3 \le 10$
$2x_1 + 3x_2 + x_3 \ge 6$
$2x_1 + x_2 - x_3 = 1$
$x_1, x_2, x_3 \ge 0$

16. Minimize $C = -3x_1 + 15x_2 - 4x_3$
subject to
$2x_1 + x_2 + 3x_3 \le 24$
$x_1 + 2x_2 + x_3 \ge 6$
$x_1 - 3x_2 + x_3 = 2$
$x_1, x_2, x_3 \ge 0$

17. Maximize $P = 3x_1 + 5x_2 + 6x_3$
subject to
$2x_1 + x_2 + 2x_3 \le 8$
$2x_1 + x_2 - 2x_3 = 0$
$x_1, x_2, x_3 \ge 0$

18. Maximize $P = 3x_1 + 6x_2 + 2x_3$
subject to
$2x_1 + 2x_2 + 3x_3 \le 12$
$2x_1 - 2x_2 + x_3 = 0$
$x_1, x_2, x_3 \ge 0$

19. Maximize $P = 2x_1 + 3x_2 + 4x_3$
subject to
$x_1 + 2x_2 + x_3 \le 25$
$2x_1 + x_2 + 2x_3 \le 60$
$x_1 + 2x_2 - x_3 \ge 10$
$x_1, x_2, x_3 \ge 0$

20. Maximize $P = 5x_1 + 2x_2 + 9x_3$
subject to
$2x_1 + 4x_2 + x_3 \le 150$
$3x_1 + 3x_2 + x_3 \le 90$
$-x_1 + 5x_2 + x_3 \ge 120$
$x_1, x_2, x_3 \ge 0$

21. Maximize $P = x_1 + 2x_2 + 5x_3$
subject to
$x_1 + 3x_2 + 2x_3 \le 60$
$2x_1 + 5x_2 + 2x_3 \ge 50$
$x_1 - 2x_2 + x_3 \ge 40$
$x_1, x_2, x_3 \ge 0$

22. Maximize $P = 2x_1 + 4x_2 + x_3$
subject to
$2x_1 + 3x_2 + 5x_3 \le 280$
$2x_1 + 2x_2 + x_3 \ge 140$
$2x_1 + x_2 \ge 150$
$x_1, x_2, x_3 \ge 0$

23. Solve Problems 5 and 7 by graphing (the geometric method).

24. Solve Problems 6 and 8 by graphing (the geometric method).

C *Problems 25–32 are mixed. Some can be solved by the methods presented in Sections 6.2 and 6.3 while others must be solved by the big M method.*

25. Minimize $C = 10x_1 - 40x_2 - 5x_3$
subject to
$x_1 + 3x_2 \le 6$
$4x_2 + x_3 \le 3$
$x_1, x_2, x_3 \ge 0$

26. Maximize $P = 7x_1 - 5x_2 + 2x_3$
subject to
$x_1 - 2x_2 + x_3 \ge -8$
$x_1 - x_2 + x_3 \le 10$
$x_1, x_2, x_3 \ge 0$

27. Maximize $P = -5x_1 + 10x_2 + 15x_3$
subject to
$2x_1 + 3x_2 + x_3 \le 24$
$x_1 - 2x_2 - 2x_3 \ge 1$
$x_1, x_2, x_3 \ge 0$

28. Minimize $C = -5x_1 + 10x_2 + 15x_3$

subject to $2x_1 + 3x_2 + x_3 \le 24$

$x_1 - 2x_2 - 2x_3 \ge 1$

$x_1, x_2, x_3 \ge 0$

29. Minimize $C = 10x_1 + 40x_2 + 5x_3$

subject to $x_1 + 3x_2 \ge 6$

$4x_2 + x_3 \ge 3$

$x_1, x_2, x_3 \ge 0$

30. Maximize $P = 8x_1 + 2x_2 - 10x_3$

subject to $x_1 + x_2 - 3x_3 \le 6$

$4x_1 - x_2 + 2x_3 \le -7$

$x_1, x_2, x_3 \ge 0$

31. Maximize $P = 12x_1 + 9x_2 + 5x_3$

subject to $x_1 + 3x_2 + x_3 \le 40$

$2x_1 + x_2 + 3x_3 \le 60$

$x_1, x_2, x_3 \ge 0$

32. Minimize $C = 10x_1 + 12x_2 + 28x_3$

subject to $4x_1 + 2x_2 + 3x_3 \ge 20$

$3x_1 - x_2 - 4x_3 \le 10$

$x_1, x_2, x_3 \ge 0$

Applications

In Problems 33–38, construct a mathematical model in the form of a linear programming problem. (The answers in the back of the book for these application problems include the model.) Then solve the problem using the big M method.

33. Advertising. An advertising company wants to attract new customers by placing a total of at most 10 ads in 3 newspapers. Each ad in the *Sentinel* costs \$200 and will be read by 2,000 people. Each ad in the *Journal* costs \$200 and will be read by 500 people. Each ad in the *Tribune* costs \$100 and will be read by 1,500 people. The company wants at least 16,000 people to read its ads. How many ads should it place in each paper in order to minimize the advertising costs? What is the minimum cost?

34. Advertising. Discuss the effect on the solution to Problem 33 if the *Tribune* will not accept more than 4 ads from the company.

35. Human nutrition. A person on a high-protein, low-carbohydrate diet requires at least 100 units of protein and at most 24 units of carbohydrates daily. The diet will consist entirely of three special liquid diet foods: *A*, *B*, and *C*. The contents and costs of the diet foods are given in the table. How many bottles of each brand of diet food should be consumed daily in order to meet the protein and carbohydrate requirements at minimal cost? What is the minimum cost?

	Units per Bottle		
	A	*B*	*C*
Protein	10	10	20
Carbohydrates	2	3	4
Cost per bottle (\$)	0.60	0.40	0.90

36. Human nutrition. Discuss the effect on the solution to Problem 35 if the cost of brand *C* liquid diet food increases to \$1.50 per bottle.

37. Plant food. A farmer can use three types of plant food: mix *A*, mix *B*, and mix *C*. The amounts (in pounds) of nitrogen, phosphoric acid, and potash in a cubic yard of each mix are given in the table. Tests performed on the soil indicate that the field needs at least 800 pounds of potash. The tests also indicate that no more than 700 pounds of phosphoric acid should be added to the field. The farmer plans to plant a crop that requires a great deal of nitrogen. How many cubic yards of each mix should be added to the field in order to satisfy the potash and phosphoric acid requirements and maximize the amount of nitrogen? What is the maximum amount of nitrogen?

	Pounds per Cubic Yard		
	A	*B*	*C*
Nitrogen	12	16	8
Phosphoric acid	12	8	16
Potash	16	8	16

38. Plant food. Discuss the effect on the solution to Problem 37 if the limit on phosphoric acid is increased to 1,000 pounds.

In Problems 39–47, construct a mathematical model in the form of a linear programming problem. Do not solve.

39. Manufacturing. A company manufactures car and truck frames at plants in Milwaukee and Racine. The Milwaukee plant has a daily operating budget of \$50,000 and can produce at most 300 frames daily in any combination. It costs \$150 to manufacture a car frame and \$200 to manufacture a truck frame at the Milwaukee plant. The Racine plant has a daily operating budget of \$35,000, and can produce a maximum combined total of 200 frames daily. It costs \$135 to manufacture a car frame and \$180 to manufacture a truck frame at the Racine plant. Based on past demand, the company wants to limit production to a maximum of 250 car frames and 350 truck frames per day. If the company realizes a profit of \$50 on each car frame and \$70 on each truck frame, how many frames of each type should be produced at each plant to maximize the daily profit?

40. Loan distributions. A savings and loan company has \$3 million to lend. The types of loans and annual returns offered are given in the table. State laws require that at least 50% of the money loaned for mortgages must be for first mortgages and that at least 30% of the total amount loaned must be for either first or second mortgages. Company policy requires that the amount of signature and automobile loans cannot exceed 25% of the total amount loaned and that signature loans cannot exceed 15% of the total amount loaned. How much money should be allocated to each type of loan in order to maximize the company's return?

Type of Loan	Annual Return (%)
Signature	18
First mortgage	12
Second mortgage	14
Automobile	16

41. Oil refining. A refinery produces two grades of gasoline, regular and premium, by blending together three components: *A*, *B*, and *C*. Component *A* has an octane rating of 90 and costs $28 a barrel, component *B* has an octane rating of 100 and costs $30 a barrel, and component *C* has an octane rating of 110 and costs $34 a barrel. The octane rating for regular must be at least 95 and the octane rating for premium must be at least 105. Regular gasoline sells for $38 a barrel and premium sells for $46 a barrel. The company has 40,000 barrels of component *A*, 25,000 barrels of component *B*, and 15,000 barrels of component *C*. It must produce at least 30,000 barrels of regular and 25,000 barrels of premium. How should the components be blended in order to maximize profit?

42. Trail mix. A company makes two brands of trail mix, regular and deluxe, by mixing dried fruits, nuts, and cereal. The recipes for the mixes are given in the table. The company has 1,200 pounds of dried fruits, 750 pounds of nuts, and 1,500 pounds of cereal for the mixes. The company makes a profit of $0.40 on each pound of regular mix and $0.60 on each pound of deluxe mix. How many pounds of each ingredient should be used in each mix in order to maximize the company's profit?

Type of Mix	Ingredients
Regular	At least 20% nuts
	At most 40% cereal
Deluxe	At least 30% nuts
	At most 25% cereal

43. Investment strategy. An investor is planning to divide her investments among high-tech mutual funds, global mutual funds, corporate bonds, municipal bonds, and CDs. Each of these investments has an estimated annual return and a risk factor (see the table). The risk level for each choice is the product of its risk factor and the percentage of the total funds invested in that choice. The total risk level is the sum of the risk levels for all the investments. The investor wants at least 20% of her investments to be in CDs and does not want the risk level to exceed 1.8. What percentage of her total investments should be invested in each choice to maximize the return?

Investment	Annual Return (%)	Risk Factor
High-tech funds	11	2.7
Global funds	10	1.8
Corporate bonds	9	1.2
Muncipal bonds	8	0.5
CDs	5	0

44. Investment strategy. Refer to Problem 43. Suppose the investor decides that she would like to minimize the total risk factor, as long as her return does not fall below 9%. What percentage of her total investments should be invested in each choice to minimize the total risk level?

45. Human nutrition. A dietitian arranges a special diet using foods *L*, *M*, and *N*. The table gives the nutritional contents and cost of 1 ounce of each food. The diet's daily requirements are at least 400 units of calcium, at least 200 units of iron, at least 300 units of vitamin A, at most 150 units of cholesterol, and at most 900 calories. How many ounces of each food should be used in order to meet the diet's requirements at a minimal cost?

	Units per Bottle		
	L	*M*	*N*
Calcium	30	10	30
Iron	10	10	10
Vitamin A	10	30	20
Cholesterol	8	4	6
Calories	60	40	50
Cost per ounce ($)	0.40	0.60	0.80

46. Mixing feed. A farmer grows three crops: corn, oats, and soybeans. He mixes them to feed his cows and pigs. At least 40% of the feed mix for the cows must be corn. The feed mix for the pigs must contain at least twice as much soybeans as corn. He has harvested 1,000 bushels of corn, 500 bushels of oats, and 1,000 bushels of soybeans. He needs 1,000 bushels of each feed mix for his livestock. The unused corn, oats, and soybeans can be sold for $4, $3.50, and $3.25 a bushel, respectively (thus, these amounts also represent the cost of the crops used to feed the livestock). How many bushels of each crop should be used in each feed mix in order to produce sufficient food for the livestock at a minimal cost?

47. Transportation. Three towns are forming a consolidated school district with two high schools. Each high school has a maximum capacity of 2,000 students. Town *A* has 500 high school students, town *B* has 1,200, and town *C* has 1,800. The weekly costs of transporting a student from each town to each school are given in the table. In order to balance the enrollment, the school board decided that each high school must enroll at least 40% of the total student population. Furthermore, no more than 60% of the students in any town should be sent to the same high school. How many students from each town should be enrolled in each school in order to meet these requirements and minimize the cost of transporting the students?

	Weekly Transportation Cost per Student ($)	
	School I	School II
Town *A*	4	8
Town *B*	6	4
Town *C*	3	9

Answers to Matched Problems

1. Maximize $P = 3x_1 - 2x_2 + x_3 - Ma_1 - Ma_2 - Ma_3$
subject to
$$\begin{aligned} x_1 - 2x_2 + x_3 - s_1 + a_1 &= 5 \\ x_1 + 3x_2 - 4x_3 - s_2 + a_2 &= 10 \\ 2x_1 + 4x_2 + 5x_3 + s_3 &= 20 \\ -3x_1 + x_2 + x_3 + a_3 &= 15 \\ x_1, x_2, x_3, s_1, a_1, s_2, a_2, s_3, a_3 &\geq 0 \end{aligned}$$
2. Max $P = 22$ at $x_1 = 6, x_2 = 4, x_3 = 0$
3. No optimal solution
4. A minimum cost of \$200 is realized when no type J, 6 type K, and 2 type L stones are processed each day.
5. Maximize $P = 9x_1 + 15x_2 - x_3 + 5x_4$
subject to
$$\begin{aligned} x_1 + x_2 &\leq 35{,}000 \\ x_3 + x_4 &\leq 15{,}000 \\ x_1 + x_3 &\geq 20{,}000 \\ x_2 + x_4 &\geq 10{,}000 \\ 5x_1 - 15x_3 &\leq 0 \\ 15x_2 - 5x_4 &\leq 0 \\ x_1, x_2, x_3, x_4 &\geq 0 \end{aligned}$$

Chapter 6 Summary and Review

Important Terms, Symbols, and Concepts

6.1 The Table Method: An Introduction to the Simplex Method

EXAMPLES

- A linear programming problem is said to be a **standard maximization problem in standard form** if its mathematical model is of the following form: Maximize the objective function
$$P = c_1x_1 + c_2x_2 + \cdots + c_kx_k$$
subject to problem constraints of the form
$$a_1x_1 + a_2x_2 + \cdots + a_kx_k \leq b \qquad b \geq 0$$
with nonnegative constraints
$$x_1, x_2, \ldots, x_k \geq 0$$
- The system of inequalities (***i*-system**) of a linear programming problem is converted to a system of equations (***e*-system**) by means of **slack variables**. A solution of the e-system is a **feasible solution** if the values of all decision variables and slack variables are nonnegative. The feasible solutions of the e-system correspond to the points in the feasible region of the i-system. A **basic solution** of the e-system is found by setting k of the variables equal to 0, where k is the number of decision variables $x_1, x_2, \ldots, x_k$. A solution of the e-system that is both basic and feasible is called a **basic feasible solution**. The **table method** for solving a linear programming problem consists of constructing a table of all basic solutions, determining which of the basic solutions are feasible, and then maximizing the objective function over the basic feasible solutions. A procedure for carrying out the table method in the case of $k = 2$ decision variables is given on page 290. For an arbitrary number of decision variables, see the procedure on page 296. Ex. 1, p. 291 Ex. 2, p. 292 Ex. 3, p. 293
- The **fundamental theorem of linear programming** can be formulated in terms of basic feasible solutions. It states that an optimal solution to the linear programming problem, if one exists, must occur at one or more of the basic feasible solutions.
- The k variables that are assigned the value 0, in order to generate a basic solution, are called **nonbasic variables**. The remaining variables are called **basic variables**. So the classification of variables as basic or nonbasic depends on the basic solution under consideration. Ex. 4, p. 295
- The benefit of the table method is that it gives a procedure for **finding all corner points of the feasible region without drawing a graph**. But the table method has a drawback: If the number of decision variables and problem constraints is large, then the number of rows in the table (that is, the number of basic solutions) becomes too large for practical implementation. The *simplex method,* discussed in Section 6.2, gives a practical method for solving large linear programming problems.

6.2 The Simplex Method: Maximization with Problem Constraints of the Form ≤

- Adding the objective function to the system of constraint equations produces the **initial system**. Negative values of the objective function variable are permitted in a basic feasible solution as long as all other variables are nonnegative. The fundamental theorem of linear programming also applies to initial systems.
- The augmented matrix of the initial system is called the **initial simplex tableau**. The **simplex method** consists of performing **pivot operations**, starting with the initial simplex tableau, until an optimal solution is found (if one exists). The procedure is illustrated in Figure 2 (p. 307). Ex. 1, p. 307 Ex. 2, p. 309 Ex. 3, p. 310

6.3 The Dual Problem: Minimization with Problem Constraints of the Form ≥

- By the **Fundamental Principle of Duality**, a linear programming problem that asks for the minimum of the objective function over a region described by ≥ problem constraints can be solved by first forming the **dual problem** and then using the simplex method. Ex. 1, p. 318 Ex. 2, p. 321 Ex. 3, p. 323

6.4 Maximization and Minimization with Mixed Problem Constraints

- The **big *M* method** can be used to find the maximum of any objective function on any feasible region. The solution process involves the introduction of two new types of variables, **surplus variables** and **artificial variables**, and a modification of the objective function. The result is an initial tableau that can be transformed into the tableau of a **modified problem**. Ex. 1, p. 334
- Applying the simplex method to the modified problem produces a solution to the original problem, if one exists. Ex. 2, p. 336 Ex. 3, p. 337
- The dual method can be used to solve *only* certain minimization problems. But *all* minimization problems can be solved by using the big *M* method to find the maximum of the negative of the objective function. The big *M* method also lends itself to computer implementation. Ex. 4, p. 339 Ex. 5, p. 341

Review Exercises

Work through all the problems in this chapter review and check your answers in the back of the book. Answers to all review problems are there along with section numbers in italics to indicate where each type of problem is discussed. Where weaknesses show up, review appropriate sections in the text.

A *Problems 1–7 refer to the partially completed table of the six basic solutions to the e-system*

$$2x_1 + 5x_2 + s_1 = 32$$

$$x_1 + 2x_2 + s_2 = 14$$

	x_1	x_2	s_1	s_2
(A)	0	0	32	14
(B)	0	6.4	0	1.2
(C)	0	7	−3	0
(D)	16	0	0	−2
(E)		0		0
(F)			0	0

1. In basic solution (B), which variables are basic?

2. In basic solution (D), which variables are nonbasic?

3. Find basic solution (E).

4. Find basic solution (F).

5. Which of the six basic solutions are feasible?

6. Describe geometrically the set of points in the plane such that $s_1 < 0$.

7. Use the basic feasible solutions to maximize $P = 50x_1 + 60x_2$.

8. A linear programming problem has 6 decision variables and 3 problem constraints. How many rows are there in the table of basic solutions of the corresponding *e*-system?

9. Given the linear programming problem

$$\begin{aligned} \text{Maximize} \quad & P = 6x_1 + 2x_2 \\ \text{subject to} \quad & 2x_1 + x_2 \le 8 \\ & x_1 + 2x_2 \le 10 \\ & x_1, x_2 \ge 0 \end{aligned}$$

convert the problem constraints into a system of equations using slack variables.

10. How many basic variables and how many nonbasic variables are associated with the system in Problem 9?

11. Find all basic solutions for the system in Problem 9, and determine which basic solutions are feasible.

12. Write the simplex tableau for Problem 9, and circle the pivot element. Indicate the entering and exiting variables.

13. Solve Problem 9 using the simplex method.

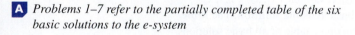

14. For the simplex tableau below, identify the basic and nonbasic variables. Find the pivot element, the entering and exiting variables, and perform one pivot operation.

$$\begin{array}{c} \\ \\ \\ \\ \\ \end{array}\begin{array}{ccccccc|c} x_1 & x_2 & x_3 & s_1 & s_2 & s_3 & P & \\ 2 & 1 & 3 & -1 & 0 & 0 & 0 & 20 \\ 3 & 0 & 4 & 1 & 1 & 0 & 0 & 30 \\ 2 & 0 & 5 & 2 & 0 & 1 & 0 & 10 \\ \hdashline -8 & 0 & -5 & 3 & 0 & 0 & 1 & 50 \end{array}$$

15. Find the basic solution for each tableau. Determine whether the optimal solution has been reached, additional pivoting is required, or the problem has no optimal solution.

(A)
$$\begin{array}{ccccc|c} x_1 & x_2 & s_1 & s_2 & P & \\ 4 & 1 & 0 & 0 & 0 & 2 \\ 2 & 0 & 1 & 1 & 0 & 5 \\ \hdashline -2 & 0 & 3 & 0 & 1 & 12 \end{array}$$

(B)
$$\begin{array}{ccccc|c} x_1 & x_2 & s_1 & s_2 & P & \\ -1 & 3 & 0 & 1 & 0 & 7 \\ 0 & 2 & 1 & 0 & 0 & 0 \\ \hdashline -2 & 1 & 0 & 0 & 1 & 22 \end{array}$$

(C)
$$\begin{array}{ccccc|c} x_1 & x_2 & s_1 & s_2 & P & \\ 1 & -2 & 0 & 4 & 0 & 6 \\ 0 & 2 & 1 & 6 & 0 & 15 \\ \hdashline 0 & 3 & 0 & 2 & 1 & 10 \end{array}$$

16. Form the dual problem of

$$\begin{aligned} \text{Minimize} \quad & C = 5x_1 + 2x_2 \\ \text{subject to} \quad & x_1 + 3x_2 \ge 15 \\ & 2x_1 + x_2 \ge 20 \\ & x_1, x_2 \ge 0 \end{aligned}$$

17. Write the initial system for the dual problem in Problem 16.

18. Write the first simplex tableau for the dual problem in Problem 16 and label the columns.

19. Use the simplex method to find the optimal solution of the dual problem in Problem 16.

20. Use the final simplex tableau from Problem 19 to find the optimal solution of the linear programming problem in Problem 16.

B **21.** Solve the linear programming problem using the simplex method.

$$\begin{aligned} \text{Maximize} \quad & P = 3x_1 + 4x_2 \\ \text{subject to} \quad & 2x_1 + 4x_2 \le 24 \\ & 3x_1 + 3x_2 \le 21 \\ & 4x_1 + 2x_2 \le 20 \\ & x_1, x_2 \ge 0 \end{aligned}$$

22. Form the dual problem of the linear programming problem

$$\begin{aligned} \text{Minimize} \quad & C = 3x_1 + 8x_2 \\ \text{subject to} \quad & x_1 + x_2 \ge 10 \\ & x_1 + 2x_2 \ge 15 \\ & x_2 \ge 3 \\ & x_1, x_2 \ge 0 \end{aligned}$$

23. Solve Problem 22 by applying the simplex method to the dual problem.

Solve the linear programming Problems 24 and 25.

24.
$$\begin{aligned} \text{Maximize} \quad & P = 5x_1 + 3x_2 - 3x_3 \\ \text{subject to} \quad & x_1 - x_2 - 2x_3 \le 3 \\ & 2x_1 + 2x_2 - 5x_3 \le 10 \\ & x_1, x_2, x_3 \ge 0 \end{aligned}$$

25.
$$\begin{aligned} \text{Maximize} \quad & P = 5x_1 + 3x_2 - 3x_3 \\ \text{subject to} \quad & x_1 - x_2 - 2x_3 \le 3 \\ & x_1 + x_2 \le 5 \\ & x_1, x_2, x_3 \ge 0 \end{aligned}$$

26. Solve the linear programming problem using the table method:

$$\begin{aligned} \text{Maximize} \quad & P = 10x_1 + 7x_2 + 8x_3 \\ \text{subject to} \quad & 2x_1 + x_2 + 3x_3 \le 12 \\ & x_1, x_2 \ge 0 \end{aligned}$$

27. Refer to Problem 26. How many pivot operations are required to solve the linear programming problem using the simplex method?

In Problems 28 and 29,

(A) Introduce slack, surplus, and artificial variables and form the modified problem.

(B) Write the preliminary simplex tableau for the modified problem and find the initial simplex tableau.

(C) Find the optimal solution of the modified problem by applying the simplex method to the initial simplex tableau.

(D) Find the optimal solution of the original problem, if it exists.

28.
$$\begin{aligned} \text{Maximize} \quad & P = x_1 + 3x_2 \\ \text{subject to} \quad & x_1 + x_2 \ge 6 \\ & x_1 + 2x_2 \le 8 \\ & x_1, x_2 \ge 0 \end{aligned}$$

29.
$$\begin{aligned} \text{Maximize} \quad & P = x_1 + x_2 \\ \text{subject to} \quad & x_1 + x_2 \ge 5 \\ & x_1 + 2x_2 \le 4 \\ & x_1, x_2 \ge 0 \end{aligned}$$

30. Find the modified problem for the following linear programming problem. (Do not solve.)

$$\begin{aligned} \text{Maximize} \quad & P = 2x_1 + 3x_2 + x_3 \\ \text{subject to} \quad & x_1 - 3x_2 + x_3 \le 7 \\ & -x_1 - x_2 + 2x_3 \le -2 \\ & 3x_1 + 2x_2 - x_3 = 4 \\ & x_1, x_2, x_3 \ge 0 \end{aligned}$$

Write a brief verbal description of the type of linear programming problem that can be solved by the method indicated in Problems 31–33. Include the type of optimization, the number of variables, the type of constraints, and any restrictions on the coefficients and constants.

31. Basic simplex method with slack variables

32. Dual problem method

33. Big M method

C **34.** Solve the following linear programming problem by the simplex method, keeping track of the obvious basic solution at each step. Then graph the feasible region and illustrate the path to the optimal solution determined by the simplex method.

$$\begin{aligned} \text{Maximize} \quad & P = 2x_1 + 3x_2 \\ \text{subject to} \quad & x_1 + 2x_2 \le 22 \\ & 3x_1 + x_2 \le 26 \\ & x_1 \le 8 \\ & x_2 \le 10 \\ & x_1, x_2 \ge 0 \end{aligned}$$

35. Solve by the dual problem method:

$$\begin{aligned} \text{Minimize} \quad & C = 3x_1 + 2x_2 \\ \text{subject to} \quad & 2x_1 + x_2 \le 20 \\ & 2x_1 + x_2 \ge 9 \\ & x_1 + x_2 \ge 6 \\ & x_1, x_2 \ge 0 \end{aligned}$$

36. Solve Problem 35 by the big M method.

37. Solve by the dual problem method:

$$\begin{aligned} \text{Minimize} \quad & C = 15x_1 + 12x_2 + 15x_3 + 18x_4 \\ \text{subject to} \quad & x_1 + x_2 \le 240 \\ & x_3 + x_4 \le 500 \\ & x_1 + x_3 \ge 400 \\ & x_2 + x_4 \ge 300 \\ & x_1, x_2, x_3, x_4 \ge 0 \end{aligned}$$

Applications

In Problems 38–41, construct a mathematical model in the form of a linear programming problem. (The answers in the back of the book for these application problems include the model.) Then solve the problem by the simplex, dual problem, or big M methods.

38. Investment. An investor has \$150,000 to invest in oil stock, steel stock, and government bonds. The bonds are guaranteed to yield 5%, but the yield for each stock can vary. To protect against major losses, the investor decides that the amount invested in oil stock should not exceed \$50,000 and that the total amount invested in stock cannot exceed the amount invested in bonds by more than \$25,000.

(A) If the oil stock yields 12% and the steel stock yields 9%, how much money should be invested in each alternative in order to maximize the return? What is the maximum return?

(B) Repeat part (A) if the oil stock yields 9% and the steel stock yields 12%.

39. Manufacturing. A company manufactures outdoor furniture consisting of regular chairs, rocking chairs, and chaise lounges. Each piece of furniture passes through three different production departments: fabrication, assembly, and finishing. Each regular chair takes 1 hour to fabricate, 2 hours to assemble, and 3 hours to finish. Each rocking chair takes 2 hours to fabricate, 2 hours to assemble, and 3 hours to finish. Each chaise lounge takes 3 hours to fabricate, 4 hours to assemble, and 2 hours to finish. There are 2,500 labor-hours available in the fabrication department, 3,000 in the assembly department, and 3,500 in the finishing department. The company makes a profit of \$17 on each regular chair, \$24 on each rocking chair, and \$31 on each chaise lounge.

(A) How many chairs of each type should the company produce in order to maximize profit? What is the maximum profit?

(B) Discuss the effect on the optimal solution in part (A) if the profit on a regular chair is increased to \$25 and all other data remain the same.

(C) Discuss the effect on the optimal solution in part (A) if the available hours on the finishing department are reduced to 3,000 and all other data remain the same.

40. Shipping schedules. A company produces motors for washing machines at factory A and factory B. The motors are then shipped to either plant X or plant Y, where the washing machines are assembled. The maximum number of motors that can be produced at each factory monthly, the minimum number required monthly for each plant to meet anticipated demand, and the shipping charges for one motor are given in the table. Determine a shipping schedule that will minimize the cost of transporting the motors from the factories to the assembly plants.

	Plant X	Plant Y	Maximum Production
Factory A	\$5	\$8	1,500
Factory B	\$9	\$7	1,000
Minimum Requirement	900	1,200	

41. Blending–food processing. A company blends long-grain rice and wild rice to produce two brands of rice mixes: brand A, which is marketed under the company's name, and brand B, which is marketed as a generic brand. Brand A must contain at least 10% wild rice, and brand B must contain at least 5% wild rice. Long-grain rice costs \$0.70 per pound, and wild rice costs \$3.40 per pound. The company sells brand A for \$1.50 a pound and brand B for \$1.20 a pound. The company has 8,000 pounds of long-grain rice and 500 pounds of wild rice on hand. How should the company use the available rice to maximize its profit? What is the maximum profit?

7 Logic, Sets, and Counting

Introduction

Quality control is crucial in manufacturing to insure product safety and reliability. One method of quality control is to test a sample of manufactured parts (see Problems 71 and 72 in Section 7.4). If the sample does not pass the test, there is a problem in production that must be rectified. The mathematics behind this method of quality control involves logic, sets, and counting, the key concepts explored in Chapter 7.

Logic and sets form the foundation of mathematics. That foundation remained largely out of view in Chapters 1 through 6. In Chapter 7, however, we study logic and sets explicitly with an eye toward the topic of probability in Chapter 8. We introduce the symbolic logic of propositions in Section 7.1, study sets in Section 7.2, and consider various counting techniques in Section 7.4.

7.1 Logic

- Propositions and Connectives
- Truth Tables
- Logical Implications and Equivalences

Consider the kind of logical reasoning that is used in everyday language. For example, suppose that the following two statements are true:

"If today is Saturday, then Derek plays soccer" and

"Today is Saturday."

From the truth of these two statements, we can conclude that

"Derek plays soccer."

Similarly, suppose that the following two mathematical statements are true:

"If the sum of the digits of 71,325 is divisible by 9, then 71,325 is divisible by 9" and

"The sum of the digits of 71,325 is divisible by 9."

From the truth of these two statements, we can conclude that

"71,325 is divisible by 9."

Logic is the study of the form of arguments. Each of the preceding arguments, the first about Derek and the second about divisibility by 9, has the same form, namely,

$$[(p \rightarrow q) \wedge p] \Rightarrow q$$

In this section, we introduce the notation that is required to represent an argument in compact form, and we establish precision in the use of logical deduction that forms the foundation for the proof of mathematical theorems.

Propositions and Connectives

A **proposition** is a statement (not a question or command) that is either true or false. So the statement

"There is a prime number between 2,312 and 2,325"

is a proposition. It is a proposition even if we do not know or cannot determine whether it is true. We use lowercase letters such as p, q, and r to denote propositions.

If p and q are propositions, then the compound propositions

$$\neg p, \quad p \vee q, \quad p \wedge q, \quad \text{and} \quad p \rightarrow q$$

can be formed using the negation symbol $\neg$ and the connectives $\vee$, $\wedge$, and $\rightarrow$. These propositions are called "not p," "p or q," "p and q," and "if p then q," respectively. We use a *truth table* to specify each of these compound propositions. A truth table gives the proposition's truth value, T (true) or F (false), for all possible values of its variables.

DEFINITION Negation

If p is a proposition, then the proposition $\neg p$, read **not** $\boldsymbol{p}$, or the **negation** of p, is false if p is true and true if p is false.

p	$\neg p$
T	F
F	T

DEFINITION Disjunction

If p and q are propositions, then the proposition $p \vee q$, read **p or q**, or the **disjunction** of p and q, is true if p is true, or if q is true, or if both are true, and is false otherwise.

p	q	$p \vee q$
T	T	T
T	F	T
F	T	T
F	F	F

Note that *or* is used in the inclusive sense; that is, it includes the possibility that both p and q are true. This mathematical usage differs from the way that *or* is sometimes used in everyday language ("I will order chicken or I will order fish") when we intend to exclude the possibility that both are true.

DEFINITION Conjunction

If p and q are propositions, then the proposition $p \wedge q$, read **p and q**, or the **conjunction** of p and q, is true if both p and q are true and is false otherwise.

p	q	$p \wedge q$
T	T	T
T	F	F
F	T	F
F	F	F

The truth table for $p \wedge q$ is just what you would expect, based on the use of "and" in everyday language. Note that there is just one T in the third column of the truth table for conjunction; both p and q must be true in order for $p \wedge q$ to be true.

DEFINITION Conditional

If p and q are propositions, then the proposition $p \rightarrow q$, read **if p then q**, or the **conditional with hypothesis p and conclusion q**, is false if p is true and q is false, but is true otherwise.

p	q	$p \rightarrow q$
T	T	T
T	F	F
F	T	T
F	F	T

Note that the definition of $p \rightarrow q$ differs somewhat from the use of "if p then q" in everyday language. For example, we might question whether the proposition

"If Paris is in Switzerland, then Queens is in New York"

is true on the grounds that there is no apparent connection between p ("Paris is in Switzerland") and q ("Queens is in New York"). We consider it to be true, however, in accordance with the definition of $p \rightarrow q$, because p is false. Whenever the hypothesis p is false, we say that the conditional $p \rightarrow q$ is **vacuously true**.

CONCEPTUAL INSIGHT

It is helpful to think of the conditional as a guarantee. For example, an instructor of a mathematics course might give a student the guarantee: "If you score at least 90%, then you will get an A." Suppose the student scores less than 90%, so the hypothesis is false. The guarantee remains in effect even though it is not applicable. We say that the conditional statement is *vacuously true*. In fact, there is only one circumstance in which the conditional statement could be false: The student scores at least 90% (that is, the hypothesis is true), but the grade is not an A (the conclusion is false).

EXAMPLE 1 **Compound Propositions** Consider the propositions p and q:

p: "$4 + 3$ is even."

q: "$4^2 + 3^2$ is odd."

Express each of the following propositions as an English sentence and determine whether it is true or false.

(A) $\neg p$ (B) $\neg q$ (C) $p \vee q$ (D) $p \wedge q$ (E) $p \rightarrow q$

SOLUTION

(A) $\neg p$: "$4 + 3$ is not even"
(Note that we modified the wording of "Not $4 + 3$ is even" to standard English usage.) Because $4 + 3 = 7$ is odd, p is false, and therefore $\neg p$ is true.

(B) $\neg q$: "$4^2 + 3^2$ is not odd"
Because $4^2 + 3^2 = 25$ is odd, q is true, and therefore $\neg q$ is false.

(C) $p \vee q$: "$4 + 3$ is even or $4^2 + 3^2$ is odd"
Because q is true, $p \vee q$ is true.

(D) $p \wedge q$: "$4 + 3$ is even and $4^2 + 3^2$ is odd"
Because p is false, $p \wedge q$ is false.

(E) $p \rightarrow q$: "if $4 + 3$ is even, then $4^2 + 3^2$ is odd"
Because p is false, $p \rightarrow q$ is (vacuously) true.

Matched Problem 1 Consider the propositions p and q:

p: "$14^2 < 200$"

q: "$23^2 < 500$"

Express each of the following propositions as an English sentence and determine whether it is true or false.

(A) $\neg p$ (B) $\neg q$ (C) $p \vee q$ (D) $p \wedge q$ (E) $p \rightarrow q$

With any conditional $p \rightarrow q$, we associate two other propositions: the *converse* of $p \rightarrow q$ and the *contrapositive* of $p \rightarrow q$.

DEFINITION Converse and Contrapositive

Let $p \rightarrow q$ be a conditional proposition. The proposition $q \rightarrow p$ is called the **converse** of $p \rightarrow q$. The proposition $\neg q \rightarrow \neg p$ is called the **contrapositive** of $p \rightarrow q$.

EXAMPLE 2 **Converse and Contrapositive** Consider the propositions p and q:

p: "$2 + 2 = 4$"

q: "9 is a prime"

Express each of the following propositions as an English sentence and determine whether it is true or false.

(A) $p \rightarrow q$ (B) The converse of $p \rightarrow q$ (C) The contrapositive of $p \rightarrow q$

SOLUTION

(A) $p \rightarrow q$: "if 2 + 2 = 4, then 9 is a prime"
Because p is true and q is false, $p \rightarrow q$ is false.

(B) $q \rightarrow p$: "if 9 is a prime, then 2 + 2 = 4"
Because q is false, $q \rightarrow p$ is vacuously true.

(C) $\neg q \rightarrow \neg p$: "if 9 is not prime, then 2 + 2 is not equal to 4"
Because q is false and p is true, $\neg q$ is true and $\neg p$ is false, so $\neg q \rightarrow \neg p$ is false.

Matched Problem 2 Consider the propositions p and q:

p: "$5^2 + 12^2 = 13^2$"
q: "$7^2 + 24^2 = 25^2$"

Express each of the following propositions as an English sentence and determine whether it is true or false.

(A) $p \rightarrow q$ (B) The converse of $p \rightarrow q$ (C) The contrapositive of $p \rightarrow q$

Truth Tables

A **truth table** for a compound proposition specifies whether it is true or false for any assignment of truth values to its variables. Such a truth table can be constructed for any compound proposition by referring to the truth tables in the definitions of $\neg$, $\vee$, $\wedge$, and $\rightarrow$.

EXAMPLE 3 **Constructing Truth Tables** Construct the truth table for $\neg p \vee q$.

p	q	$\neg p$	$\neg p \vee q$
T	T	F	T
T	F	F	F
F	T	T	T
F	F	T	T

SOLUTION The proposition contains two variables, p and q, so the truth table will consist of four rows, one for each possible assignment of truth values to two variables (TT, TF, FT, FF). Although the truth table itself consists of three columns, one labeled p, another labeled q, and the third labeled $\neg p \vee q$, it is helpful to insert an additional column labeled $\neg p$. The entries in that additional column are obtained from the first column, changing any T to F, and vice versa, in accordance with the definition of $\neg$. The entries in the last column are obtained from the third and second columns, in accordance with the definition of $\vee$ (in a given row, if either entry in those columns is a T, then the entry in the last column is T; if both are F, the entry in the last column is F).

Note that the truth table for $\neg p \vee q$ is identical to the truth table for $p \rightarrow q$ (see the definition of a conditional).

Matched Problem 3 Construct the truth table for $p \wedge \neg q$.

EXAMPLE 4 **Constructing a Truth Table** Construct the truth table for $[(p \rightarrow q) \wedge p] \rightarrow q$.

SOLUTION It is helpful to insert a third column labeled $p \rightarrow q$ and a fourth labeled $(p \rightarrow q) \wedge p$. We complete the first three columns. Then we use the third and first columns to complete the fourth, and we use the fourth and second columns to complete the last column.

p	q	$p \rightarrow q$	$(p \rightarrow q) \wedge p$	$[(p \rightarrow q) \wedge p] \rightarrow q$
T	T	T	T	T
T	F	F	F	T
F	T	T	F	T
F	F	T	F	T

Note that $[(p \rightarrow q) \wedge p] \rightarrow q$ is always true, regardless of the truth values of p and q.

Matched Problem 4 Construct the truth table for $[(p \rightarrow q) \wedge \neg q] \rightarrow \neg p$.

Any proposition is either a *tautology*, a *contradiction*, or a *contingency*. The proposition $[(p \rightarrow q) \wedge p] \rightarrow q$ of Example 4, which is always true, is a tautology. The proposition $\neg p \vee q$ of Example 3, which may be true or false, is a contingency.

DEFINITION Tautology, Contradiction, and Contingency

A proposition is a **tautology** if each entry in its column of the truth table is T, a **contradiction** if each entry is F, and a **contingency** if at least one entry is T and at least one entry is F.

EXAMPLE 5 **Constructing a Truth Table** Construct the truth table for $p \wedge \neg(p \vee q)$.

SOLUTION It is helpful to insert a third column labeled $p \vee q$ and a fourth labeled $\neg(p \vee q)$. We complete the first three columns. Then we use the third column to complete the fourth, and we use the first and fourth columns to complete the last column.

p	q	$p \vee q$	$\neg(p \vee q)$	$p \wedge \neg(p \vee q)$
T	T	T	F	F
T	F	T	F	F
F	T	T	F	F
F	F	F	T	F

Note that $p \wedge \neg(p \vee q)$ is a contradiction; it is always false, regardless of the truth values of p and q.

Matched Problem 5 Construct the truth table for $(p \rightarrow q) \wedge (p \wedge \neg q)$.

Explore and Discuss 1

The LOGIC menu on the TI-84 Plus CE contains the operators "and," "or," "xor," and "not." The truth table for $p \wedge q$, for example, can be displayed by entering all combinations of truth values for p and q in lists L_1 and L_2 (using 1 to represent T and 0 to represent F), and then, for L_3, entering "L_1 and L_2" (see Figure 1).

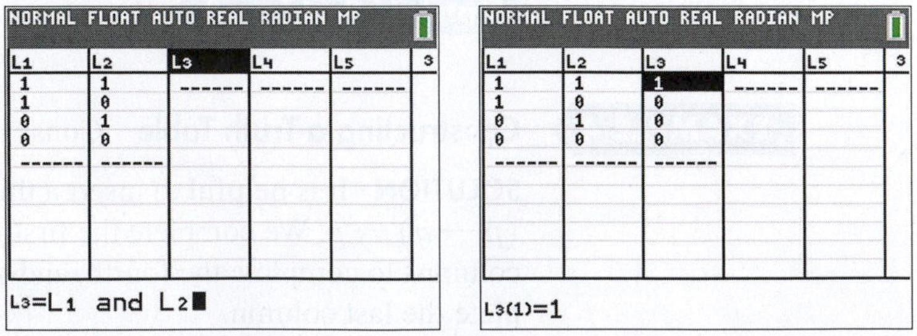

Figure 1

(A) Use a calculator to display the truth table for the "exclusive or" operator "xor."

(B) Show that the truth table for "xor" is identical to the truth table for the proposition $(p \vee q) \wedge \neg(p \wedge q)$.

Logical Implications and Equivalences

At the beginning of the section, we noted that the notation $[(p \rightarrow q) \wedge p] \Rightarrow q$ could be used to represent the form of a familiar logical deduction. The notion of such a deduction can be defined in terms of truth tables. Consider the truth tables for the propositions $(p \rightarrow q) \wedge p$ and q:

Table 1

p	q	$p \rightarrow q$	$(p \rightarrow q) \wedge p$
T	T	T	T
T	F	F	F
F	T	T	F
F	F	T	F

Whenever the proposition $(p \rightarrow q) \wedge p$ is true (in this case, in the first row only), the proposition q is also true. We say that $(p \rightarrow q) \wedge p$ *implies* q, or that $[(p \rightarrow q) \wedge p] \Rightarrow q$ is a *logical implication*.

DEFINITION Logical Implication

Consider the rows of the truth tables for the compound propositions P and Q. If whenever P is true, Q is also true, we say that P **logically implies** Q and write $P \Rightarrow Q$. We call $P \Rightarrow Q$ a **logical implication**.

CONCEPTUAL INSIGHT

If $P \Rightarrow Q$, then whenever P is true, so is Q. So to say that P implies Q is the same as saying that the proposition $P \rightarrow Q$ is a tautology. This gives us two methods of verifying that P implies Q: We can check the rows of the truth tables for P and Q as we did previously, or we can construct the truth table for $P \rightarrow Q$ to check that it is a tautology. Compare the truth tables in Table 1 with the truth table of Example 4 to decide which of the two methods you prefer.

EXAMPLE 6 **Verifying a Logical Implication** Show that $[(p \rightarrow q) \wedge \neg q] \Rightarrow \neg p$.

SOLUTION To construct the truth table for $(p \rightarrow q) \wedge \neg q$, it is helpful to insert a third column labeled $p \rightarrow q$ and a fourth labeled $\neg q$. We complete the first three columns. Then we use the second column to complete the fourth, and we use the third and fourth columns to complete the fifth column.

p	q	$p \rightarrow q$	$\neg q$	$(p \rightarrow q) \wedge \neg q$	$\neg p$
T	T	T	F	F	F
T	F	F	T	F	F
F	T	T	F	F	T
F	F	T	T	T	T

Now we compare the fifth column to the sixth: Whenever $(p \rightarrow q) \wedge \neg q$ is true (in the fourth row only), $\neg p$ is also true. We conclude that $[(p \rightarrow q) \wedge \neg q] \Rightarrow \neg p$.

Matched Problem 6 Show that $[(p \rightarrow q) \rightarrow p)] \Rightarrow (q \rightarrow p)$.

If the compound propositions P and Q have identical truth tables, then $P \Rightarrow Q$ and $Q \Rightarrow P$. In this case, we say that P and Q are *logically equivalent*.

DEFINITION Logical Equivalence

If the compound propositions P and Q have identical truth tables, we say that P and Q are **logically equivalent** and write $P \equiv Q$. We call $P \equiv Q$ a **logical equivalence**.

In Example 3, we noted that $p \rightarrow q$ and $\neg p \vee q$ have identical truth tables. Therefore, $p \rightarrow q \equiv \neg p \vee q$. This is formula (4) in Table 2, which lists several logical equivalences. The first three equivalences of Table 2 are obvious, and the last three equivalences are established in Example 7, Matched Problem 7, and Example 8, respectively.

Table 2 Some Logical Equivalences

$\neg(\neg p) \equiv p$	(1)
$p \vee q \equiv q \vee p$	(2)
$p \wedge q \equiv q \wedge p$	(3)
$p \rightarrow q \equiv \neg p \vee q$	(4)
$\neg(p \vee q) \equiv \neg p \wedge \neg q$	(5)
$\neg(p \wedge q) \equiv \neg p \vee \neg q$	(6)
$p \rightarrow q \equiv \neg q \rightarrow \neg p$	(7)

CONCEPTUAL INSIGHT

Formulas (5) and (6) in Table 2 are known as **De Morgan's laws**. They may remind you of the way a negative sign is distributed over a binomial in algebra:

$$-(a + b) = (-a) + (-b)$$

But there is an important difference: Formula (5) has a disjunction $(\vee)$ on the left side but a conjunction $(\wedge)$ on the right side. Similarly, formula (6) has a conjunction on the left side but a disjunction on the right side.

EXAMPLE 7 Verifying a Logical Equivalence Show that $\neg(p \vee q) \equiv \neg p \wedge \neg q$.

SOLUTION We construct truth tables for $\neg(p \vee q)$ and $\neg p \wedge \neg q$.

p	q	$p \vee q$	$\neg(p \vee q)$	$\neg p$	$\neg q$	$\neg p \wedge \neg q$
T	T	T	F	F	F	F
T	F	T	F	F	T	F
F	T	T	F	T	F	F
F	F	F	T	T	T	T

The fourth and seventh columns are identical, so $\neg(p \vee q) \equiv \neg p \wedge \neg q$.

Matched Problem 7 Show that $\neg(p \wedge q) \equiv \neg p \vee \neg q$.

One way to show that two propositions are logically equivalent is to check that their truth tables are identical (as in Example 7). Another way is to convert one to the other by a sequence of steps, where a known logical equivalence is used at each step to replace part or all of the proposition by an equivalent proposition. This procedure, analogous to simplifying an algebraic expression, is illustrated in Example 8 to show that **any conditional proposition is logically equivalent to its contrapositive**.

EXAMPLE 8 **Any Conditional and Its Contrapositive Are Logically Equivalent** Show that $p \rightarrow q \equiv \neg q \rightarrow \neg p$.

SOLUTION Each step is justified by a reference to one of the formulas in Table 2:

$$\begin{aligned} p \rightarrow q &\equiv \neg p \vee q && \text{By (4)} \\ &\equiv q \vee \neg p && \text{By (2)} \\ &\equiv \neg(\neg q) \vee \neg p && \text{By (1)} \\ &\equiv \neg q \rightarrow \neg p && \text{By (4)} \end{aligned}$$

Therefore, $p \rightarrow q \equiv \neg q \rightarrow \neg p$.

Matched Problem 8 Use equivalences from Table 2 to show that $p \rightarrow q \equiv \neg(\neg q \wedge p)$.

Explore and Discuss 2

If a compound proposition contains three variables p, q, and r, then its truth table will have eight rows, one for each of the eight ways of assigning truth values to p, q, and r (TTT, TTF, TFT, TFF, FTT, FTF, FFT, FFF). Construct truth tables to verify the following logical implication and equivalences:

(A) $(p \rightarrow q) \wedge (q \rightarrow r) \Rightarrow (p \rightarrow r)$

(B) $p \wedge (q \vee r) \equiv (p \wedge q) \vee (p \wedge r)$

(C) $p \vee (q \wedge r) \equiv (p \vee q) \wedge (p \vee r)$

Exercises 7.1

W **Skills Warm-up Exercises**

*In Problems 1–6, refer to the footnote for the definitions of divisor, multiple, prime, even, and odd.**

1. List the positive integers that are divisors of 20.
2. List the positive integers that are divisors of 24.
3. List the positive multiples of 11 that are less than 60.
4. List the positive multiples of 9 that are less than 50.
5. List the primes between 20 and 30.
6. List the primes between 10 and 20.
7. Explain why the sum of any two odd integers is even.
8. Explain why the product of any two odd integers is odd.

A *In Problems 9–14, express each proposition as an English sentence and determine whether it is true or false, where p and q are the propositions*

p: "91 is prime" q: "91 is odd"

9. $\neg q$
10. $p \vee q$
11. $p \wedge q$
12. $p \rightarrow q$
13. The converse of $p \rightarrow q$
14. The contrapositive of $p \rightarrow q$

In Problems 15–20, express each proposition as an English sentence and determine whether it is true or false, where r and s are the propositions

r: "the moon is a cube"
s: "rain is wet"

15. $r \rightarrow s$
16. $r \wedge s$
17. $r \vee s$
18. $\neg r$
19. The contrapositive of $r \rightarrow s$
20. The converse of $r \rightarrow s$

In Problems 21–28, describe each proposition as a negation, disjunction, conjunction, or conditional, and determine whether the proposition is true or false.

21. $-3 < 0$ or $-3 > 0$
22. $-3 < 0$ and $-3 > 0$
23. If $-3 < 0$, then $(-3)^2 < 0$
24. -3 is not greater than 0
25. 11 is not prime
26. 9 is even or 9 is prime

*An integer d is a ***divisor*** of an integer n (and n is a ***multiple*** of d) if $n = kd$ for some integer k. An integer n is ***even*** if 2 is a divisor of n; otherwise, n is odd. An integer $p > 1$ is ***prime*** if its only positive divisors are 1 and p.

27. 7 is odd and 7 is prime

28. If 4 is even, then 4 is prime

In Problems 29–34, state the converse and the contrapositive of the given proposition.

29. If triangle ABC is equilateral, then triangle ABC is equiangular.

30. If triangle ABC is isosceles, then the base angles of triangle ABC are congruent.

31. If $f(x)$ is a linear function with positive slope, then $f(x)$ is an increasing function.

32. If $g(x)$ is a quadratic function, then $g(x)$ is a function that is neither increasing nor decreasing.

33. If n is an integer that is a multiple of 8, then n is an integer that is a multiple of 2 and a multiple of 4.

34. If n is an integer that is a multiple of 6, then n is an integer that is a multiple of 2 and a multiple of 3.

B *In Problems 35–52, construct a truth table for the proposition and determine whether the proposition is a contingency, tautology, or contradiction.*

35. $\neg p \wedge q$

36. $p \vee \neg q$

37. $\neg p \rightarrow q$

38. $p \rightarrow \neg q$

39. $q \wedge (p \vee q)$

40. $q \vee (p \wedge q)$

41. $p \vee (p \rightarrow q)$

42. $p \wedge (p \rightarrow q)$

43. $p \rightarrow (p \wedge q)$

44. $p \rightarrow (p \vee q)$

45. $(p \rightarrow q) \rightarrow \neg p$

46. $(p \rightarrow q) \rightarrow \neg q$

47. $\neg p \rightarrow (p \vee q)$

48. $\neg p \rightarrow (p \wedge q)$

49. $q \rightarrow (\neg p \wedge q)$

50. $q \rightarrow (p \vee \neg q)$

51. $(\neg p \wedge q) \wedge (q \rightarrow p)$

52. $(p \rightarrow \neg q) \wedge (p \wedge q)$

C *In Problems 53–58, construct a truth table to verify each implication.*

53. $p \Rightarrow p \vee q$

54. $\neg p \Rightarrow p \rightarrow q$

55. $\neg p \wedge q \Rightarrow p \vee q$

56. $p \wedge q \Rightarrow p \rightarrow q$

57. $\neg p \rightarrow (q \wedge \neg q) \Rightarrow p$

58. $(p \wedge \neg p) \Rightarrow q$

In Problems 59–64, construct a truth table to verify each equivalence.

59. $\neg p \rightarrow (p \vee q) \equiv p \vee q$

60. $q \rightarrow (\neg p \wedge q) \equiv \neg(p \wedge q)$

61. $q \wedge (p \vee q) \equiv q \vee (p \wedge q)$

62. $p \wedge (p \rightarrow q) \equiv p \wedge q$

63. $p \vee (p \rightarrow q) \equiv p \rightarrow (p \vee q)$

64. $p \rightarrow (p \wedge q) \equiv p \rightarrow q$

In Problems 65–68, verify each equivalence using formulas from Table 2.

65. $p \rightarrow \neg q \equiv \neg(p \wedge q)$

66. $\neg p \rightarrow q \equiv p \vee q$

67. $\neg(p \rightarrow q) \equiv p \wedge \neg q$

68. $\neg(\neg p \rightarrow \neg q) \equiv q \wedge \neg p$

69. If p is the proposition "I want pickles" and q is the proposition "I want tomatoes," rewrite the sentence "I want tomatoes, but I do not want pickles" using symbols.

70. Let p be the proposition "every politician is honest." Explain why the statement "every politician is dishonest" is not equivalent to $\neg p$. Express $\neg p$ as an English sentence without using the word *not*.

71. If the conditional proposition p is a contingency, is $\neg p$ a contingency, a tautology, or a contradiction? Explain.

72. If the conditional proposition p is a contradiction, is $\neg p$ a contingency, a tautology, or a contradiction? Explain.

73. Can a conditional proposition be false if its contrapositive is true? Explain.

74. Can a conditional proposition be false if its converse is true? Explain.

Answers to Matched Problems

1. (A) "14^2 is not less than 200"; false
(B) "23^2 is not less than 500"; true
(C) "14^2 is less than 200 or 23^2 is less than 500"; true
(D) "14^2 is less than 200 and 23^2 is less than 500"; false
(E) "If 14^2 is less than 200, then 23^2 is less than 500"; false

2. (A) "If $5^2 + 12^2 = 13^2$, then $7^2 + 24^2 = 25^2$"; true
(B) "If $7^2 + 24^2 = 25^2$, then $5^2 + 12^2 = 13^2$; true
(C) "If $7^2 + 24^2 \neq 25^2$, then $5^2 + 12^2 \neq 13^2$; true

3.

p	q	$p \wedge \neg q$
T	T	F
T	F	T
F	T	F
F	F	F

4.

p	q	$[(p \rightarrow q) \wedge \neg q] \rightarrow \neg p$
T	T	T
T	F	T
F	T	T
F	F	T

5.

p	q	$(p \rightarrow q) \wedge (p \wedge \neg q)$
T	T	F
T	F	F
F	T	F
F	F	F

6.

p	q	$(p \rightarrow q) \rightarrow p$	$q \rightarrow p$
T	T	T	T
T	F	T	T
F	T	F	F
F	F	F	T

Whenever $(p \rightarrow q) \rightarrow p$ is true, so is $q \rightarrow p$, so $[(p \rightarrow q) \rightarrow p)] \Rightarrow (q \rightarrow p)$.

7.

p	q	$\neg(p \wedge q)$	$\neg p \vee \neg q$
T	T	F	F
T	F	T	T
F	T	T	T
F	F	T	T

The third and fourth columns are identical, so $\neg(p \wedge q) \equiv \neg p \vee \neg q$.

8. $p \rightarrow q \equiv \neg p \vee q$ By (4)

$\equiv q \vee \neg p$ By (2)

$\equiv \neg(\neg q) \vee \neg p$ By (1)

$\equiv \neg(\neg q \wedge p)$ By (6)

7.2 Sets

- Set Properties and Set Notation
- Venn Diagrams and Set Operations
- Application

In this section, we review a few key ideas about sets. Set concepts and notation help us talk about certain mathematical ideas with greater clarity and precision, and they are indispensable to a clear understanding of probability.

Set Properties and Set Notation

We can think of a **set** as any collection of objects specified in such a way that we can tell whether any given object is or is not in the collection. Capital letters, such as A, B, and C, are often used to designate particular sets. Each object in a set is called a **member**, or **element**, of the set. Symbolically,

$a \in A$ means "a is an element of set A"

$a \notin A$ means "a is not an element of set A"

A set without any elements is called the **empty**, or **null**, **set**. For example, the set of all people over 20 feet tall is an empty set. Symbolically,

$\varnothing$ denotes the empty set

A set is described either by listing all its elements between braces { } (the listing method) or by enclosing a rule within braces that determines the elements of the set (the rule method). So if $P(x)$ is a statement about x, then

$S = \{x|P(x)\}$ means "S is the set of all x such that $P(x)$ is true"

Recall that the vertical bar within the braces is read "such that." The following example illustrates the rule and listing methods of representing sets.

EXAMPLE 1 **Representing Sets**

Rule method		Listing method
$\{x \mid x \text{ is a weekend day}\}$	=	{Saturday, Sunday}
$\{x \mid x^2 = 4\}$	=	$\{-2, 2\}$
$\{x \mid x \text{ is an odd positive counting number}\}$	=	$\{1, 3, 5, \ldots\}$

The three dots $(\ldots)$ in the last set of Example 1 indicate that the pattern established by the first three entries continues indefinitely. The first two sets in Example 1 are **finite sets** (the elements can be counted, and there is an end); the last set is an **infinite set** (there is no end when counting the elements). When listing the elements in a set, we do not list an element more than once, and the order in which the elements are listed does not matter.

Matched Problem 1 Let G be the set of all numbers such that $x^2 = 9$.

(A) Denote G by the rule method.

(B) Denote G by the listing method.

(C) Indicate whether the following are true or false: $3 \in G$, $9 \notin G$.

If each element of a set A is also an element of set B, we say that A is a **subset** of B. For example, the set of all women students in a class is a subset of the whole class. Note that the definition implies that every set is a subset of itself. If set A and set B have exactly the same elements, then the two sets are said to be **equal**. Symbolically,

$A \subset B$	means	"A is a subset of B"
$A = B$	means	"A and B have exactly the same elements"
$A \not\subset B$	means	"A is not a subset of B"
$A \neq B$	means	"A and B do not have exactly the same elements"

From the definition of subset, we conclude that

$\varnothing$ is a subset of every set, and
if $A \subset B$ and $B \subset A$, then $A = B$.

CONCEPTUAL INSIGHT

To conclude that $\varnothing$ is a subset of any set A, we must show that the conditional proposition "if $x \in \varnothing$, then $x \in A$" is true. But the set $\varnothing$ has no element, so the hypothesis of the conditional is false, and therefore the conditional itself is vacuously true (see Section 7.1). We correctly conclude that $\varnothing$ is a subset of every set. Note, however, that $\varnothing$ is *not* an *element* of every set.

EXAMPLE 2 **Set Notation** If $A = \{-3, -1, 1, 3\}$, $B = \{3, -3, 1, -1\}$, and $C = \{-3, -2, -1, 0, 1, 2, 3\}$, then each of the following statements is true:

$$A = B \qquad A \subset C \qquad A \subset B$$
$$C \neq A \qquad C \not\subset A \qquad B \subset A$$
$$\varnothing \subset A \qquad \varnothing \subset C \qquad \varnothing \notin A$$

Matched Problem 2 Given $A = \{0, 2, 4, 6\}$, $B = \{0, 1, 2, 3, 4, 5, 6\}$, and $C = \{2, 6, 0, 4\}$, indicate whether the following relationships are true (T) or false (F):

(A) $A \subset B$ (B) $A \subset C$ (C) $A = C$
(D) $C \subset B$ (E) $B \not\subset A$ (F) $\varnothing \subset B$

EXAMPLE 3 **Subsets** List all subsets of the set $\{a, b, c\}$.

SOLUTION

$$\{a, b, c\}, \quad \{a, b\}, \quad \{a, c\}, \quad \{b, c\}, \quad \{a\}, \quad \{b\}, \quad \{c\}, \quad \varnothing$$

Matched Problem 3 List all subsets of the set $\{1, 2\}$.

Explore and Discuss 1

Which of the following statements are true?
(A) $\varnothing \subset \varnothing$
(B) $\varnothing \in \varnothing$
(C) $\varnothing = \{0\}$
(D) $\varnothing \subset \{0\}$

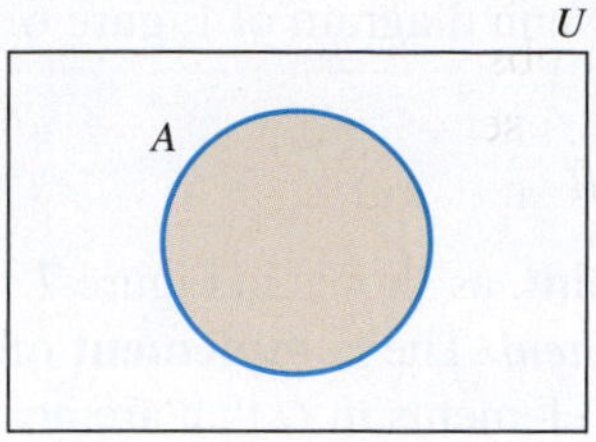

Figure 1 A is the shaded region

Venn Diagrams and Set Operations

If the set U is the set of all rental units in a city, and A is the set of all rental units within one mile of the college campus, then it is natural to picture A by Figure 1, called a *Venn diagram.*

The circle is an imaginary boundary that separates the elements of A (inside the circle) from the elements of U that are not in A (outside the circle).

The Venn diagram of Figure 1 can be used to picture any set A and *universal set* U. The **universal set** is the set of all elements under consideration. It is customary to place the label A near the circle itself, but the elements of U are imagined to be arranged so that the elements of U that are in A lie inside the circle and the elements of U that are not in A lie outside the circle.

Let U be the set of all 26 lowercase letters of the English alphabet, let $A = \{a, b, c, d, e\}$, and let $B = \{d, e, f, g, h, i\}$. Sets A and B are pictured in the Venn diagram of Figure 2. Note that the elements d and e belong to both A and B and that the elements a, b, c belong to A but not to B. Of course sets may have hundreds or thousands of elements, so we seldom write the names of the elements as in Figure 2; instead, we represent the situation by the Venn diagram of Figure 3 or by the Venn diagram of Figure 4.

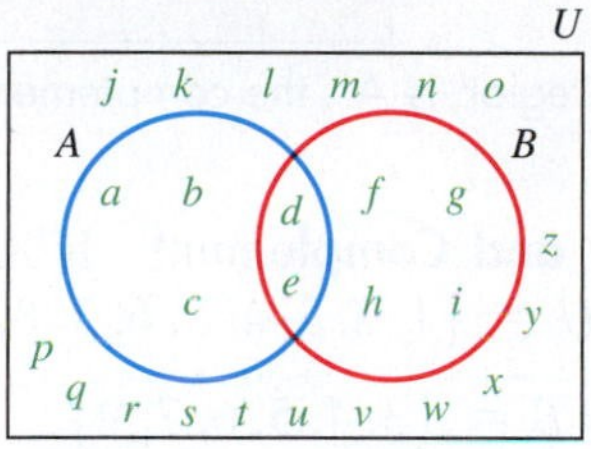

Figure 2

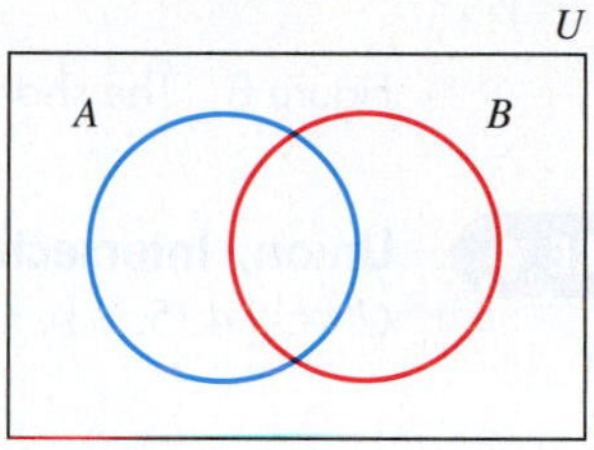

Figure 3

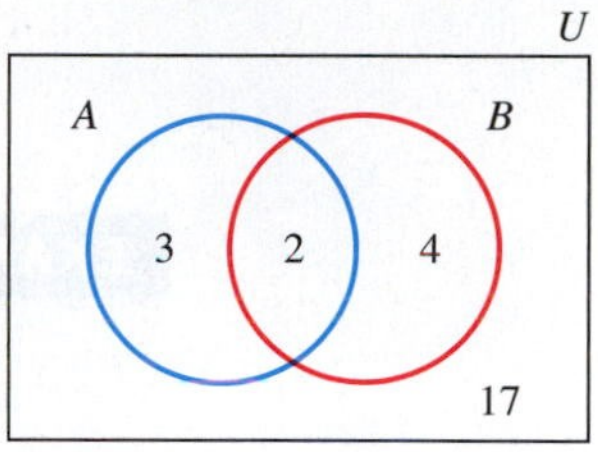

Figure 4

The numbers in Figure 4 are interpreted as follows: There are 2 elements in the overlap of A and B, 3 elements that belong to A but not B, 4 elements that belong to B but not A, and 17 elements that belong to neither A nor B. Using such a diagram, it is easy to calculate the number of elements in A (just add 2 and 3) or the number of elements in B (just add 2 and 4).

The **union** of sets A and B, denoted by $A \cup B$, is the set of elements formed by combining all the elements of A and all the elements of B into one set.

DEFINITION Union

$$A \cup B = \{x | x \in A \textbf{ or } x \in B\}$$

Here we use the word **or** in the way it is always used in mathematics; that is, x may be an element of set A or set B or both.

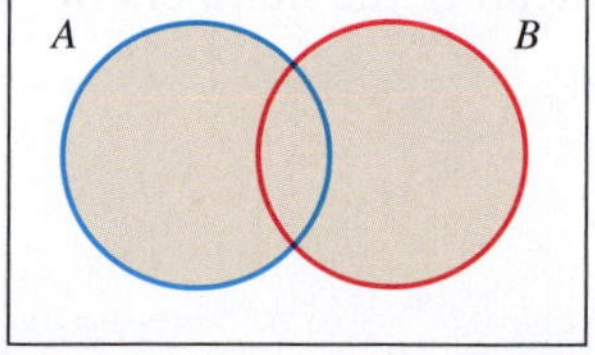

Figure 5 $A \cup B$ is the shaded region.

The union of two sets A and B is pictured in the Venn diagram of Figure 5. Note that

$$A \subset A \cup B \quad \text{and} \quad B \subset A \cup B$$

The **intersection** of sets A and B, denoted by $A \cap B$, is the set of elements in set A that are also in set B.

DEFINITION Intersection

$$A \cap B = \{x | x \in A \textbf{ and } x \in B\}$$

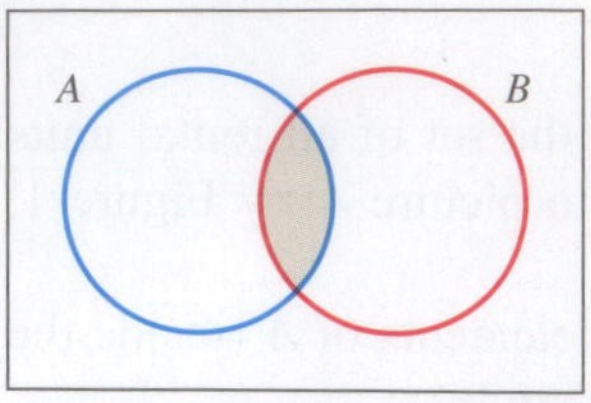

Figure 6 $A \cap B$ is the shaded region.

The intersection of two sets A and B is pictured in the Venn diagram of Figure 6. Note that

$$A \cap B \subset A \quad \text{and} \quad A \cap B \subset B$$

If $A \cap B = \varnothing$, then the sets A and B are said to be **disjoint**, as shown in Figure 7. We now define an operation on sets called the *complement*. The **complement** of A (relative to a universal set U), denoted by A', is the set of elements in U that are not in A (Fig. 8).

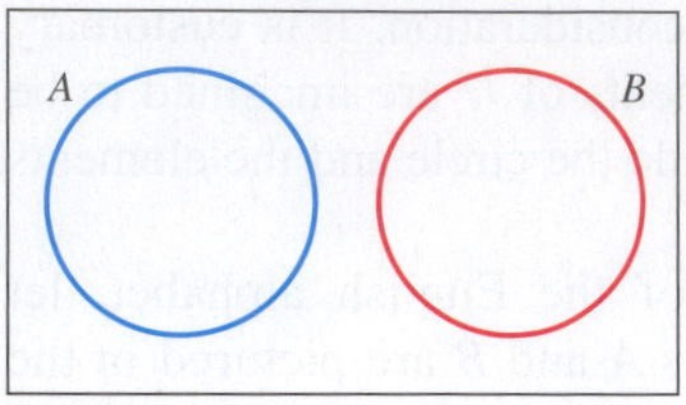

Figure 7 $A \cap B = \varnothing$; A and B are disjoint.

DEFINITION Complement

$$A' = \{x \in U \mid x \notin A\}$$

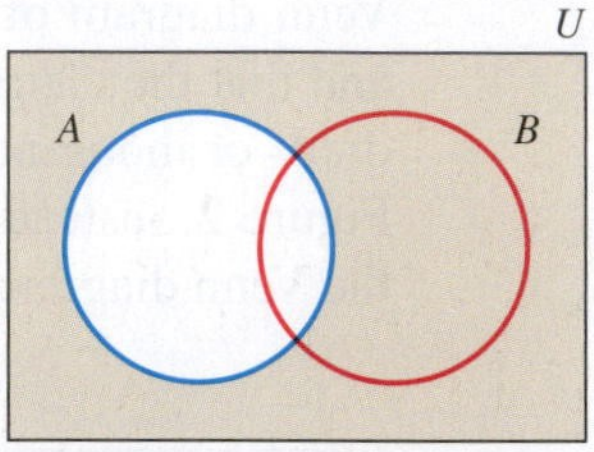

Figure 8 The shaded region is A', the complement of A.

EXAMPLE 4 **Union, Intersection, and Complement** If $A = \{3, 6, 9\}$, $B = \{3, 4, 5, 6, 7\}$, $C = \{4, 5, 7\}$, and $U = \{1, 2, 3, 4, 5, 6, 7, 8, 9\}$, then

$$A \cup B = \{3, 4, 5, 6, 7, 9\}$$
$$A \cap B = \{3, 6\}$$
$$A \cap C = \varnothing \qquad \text{A and C are disjoint.}$$
$$B' = \{1, 2, 8, 9\}$$

Matched Problem 4 If $R = \{1, 2, 3, 4\}$, $S = \{1, 3, 5, 7\}$, $T = \{2, 4\}$, and $U = \{1, 2, 3, 4, 5, 6, 7, 8, 9\}$, find

(A) $R \cup S$ (B) $R \cap S$ (C) $S \cap T$ (D) S'

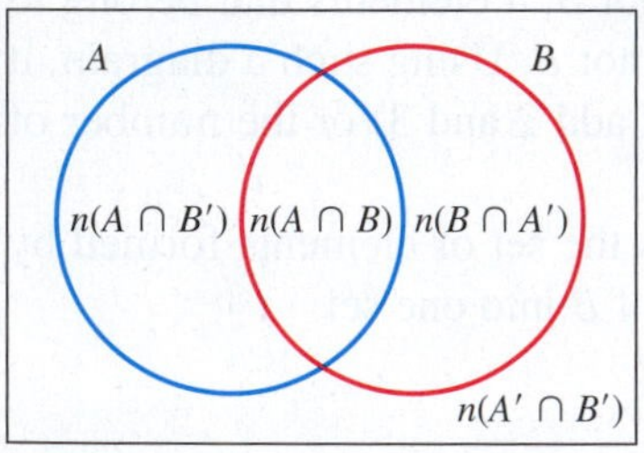

Figure 9

The **number of elements** in a set A is denoted by $n(A)$. So if A and B are sets, then the numbers that are often shown in a Venn diagram, as in Figure 4, are $n(A \cap B')$, $n(A \cap B)$, $n(B \cap A')$, and $n(A' \cap B')$ (see Fig. 9).

EXAMPLE 5 **Numbers of Elements** Let the universal set U be the set of positive integers less than or equal to 100. Let A be the set of multiples of 3 in U, and let B be the set of multiples of 5 in U.

(A) Find $n(A \cap B)$, $n(A \cap B')$, $n(B \cap A')$, and $n(A' \cap B')$.

(B) Draw a Venn diagram with circles labeled A and B, indicating the numbers of elements in the subsets of part (A).

SOLUTION

(A) $A = \{3, 6, 9, \ldots, 99\}$, so $n(A) = 33$.

$B = \{5, 10, 15, \ldots, 100\}$, so $n(B) = 20$.

$A \cap B = \{15, 30, 45, \ldots, 90\}$, so $n(A \cap B) = 6$.

$n(A \cap B') = 33 - 6 = 27$

$n(B \cap A') = 20 - 6 = 14$

$n(A' \cap B') = 100 - (6 + 27 + 14) = 53$

(B)

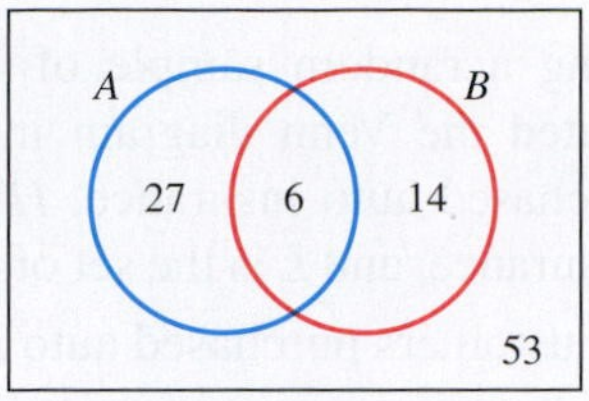

Matched Problem 5 Let the universal set U be the set of positive integers less than or equal to 100. Let A be the set of multiples of 3 in U, and let B be the set of multiples of 7 in U.

(A) Find $n(A \cap B)$, $n(A \cap B')$, $n(B \cap A')$, and $n(A' \cap B')$.

(B) Draw a Venn diagram with circles labeled A and B, indicating the numbers of elements in the subsets of part (A).

Application

EXAMPLE 6 **Exit Polling** In the 2016 presidential election, an exit poll of 100 voters produced the results in the table (23 men voted for Clinton, 29 women for Clinton, 26 men for Trump, and 22 women for Trump).

	Men	Women
Clinton	23	29
Trump	26	22

Let the universal set U be the set of 100 voters, C the set of voters for Clinton, T the set of voters for Trump, M the set of male voters, and W the set of female voters.

(A) Find $n(C \cap M)$, $n(C \cap M')$, $n(M \cap C')$, and $n(C' \cap M')$.

(B) Draw a Venn diagram with circles labeled C and M, indicating the numbers of elements in the subsets of part (A).

SOLUTION

(A) The set C' is equal to the set T, and the set M' is equal to the set W.

$$n(C \cap M) = 23, n(C \cap M') = 29$$
$$n(M \cap C') = 26, n(C' \cap M') = 22$$

(B)

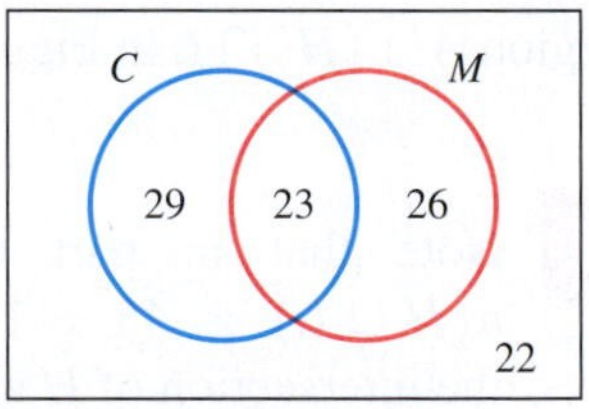

Matched Problem 6 Refer to Example 6.

(A) Find $n(T \cap W)$, $n(T \cap W')$, $n(W \cap T')$, and $n(T' \cap W')$.

(B) Draw a Venn diagram with circles labeled T and W, indicating the numbers of elements in the subsets of part (A).

Explore and Discuss 2

In Example 6, find the number of voters in the set $(C \cup M) \cap M'$. Describe this set verbally and with a Venn diagram.

EXAMPLE 7 **Insurance** Using a random sample of 100 insurance customers, an insurance company generated the Venn diagram in Figure 10 where A is the set of customers who purchased auto insurance, H is the set of customers who purchased homeowner's insurance, and L is the set of customers who purchased life insurance.

(A) How many customers purchased auto insurance?

(B) Shade the region $H \cup L$ in Figure 10. Find $n(H \cup L)$.

(C) Shade the region $A \cap H \cap L'$ in Figure 10. Find $n(A \cap H \cap L')$.

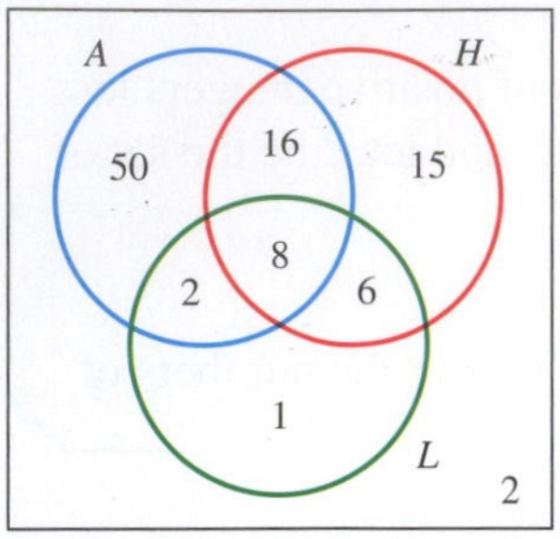

Figure 10

SOLUTION

(A) The number of customers who purchased auto insurance is the number of customers in the set A. We compute $n(A)$ by adding all of the numbers in the Venn diagram that lie inside the circle labeled A. This gives

$$50 + 16 + 8 + 2 = 76.$$

(B) We compute $n(H \cup L)$ by adding all of the numbers in Figure 10 that lie inside the circle labeled H or the circle labeled L; that is, $n(H \cup L) = 15 + 16 + 8 + 6 + 2 + 1 = 48$. Out of the 100 surveyed customers, 48 purchased homeowner's insurance or life insurance.

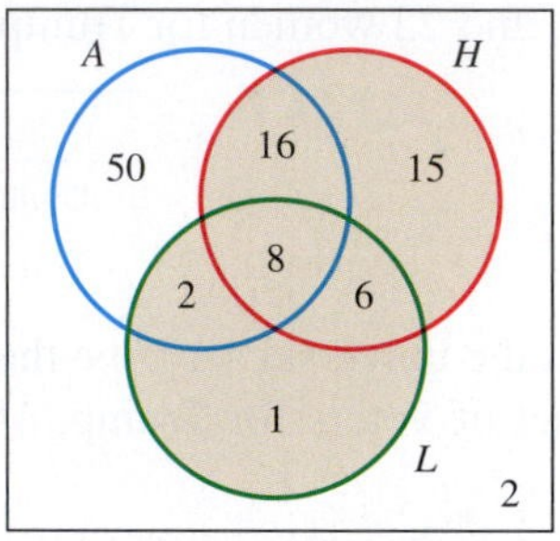

$H \cup L$

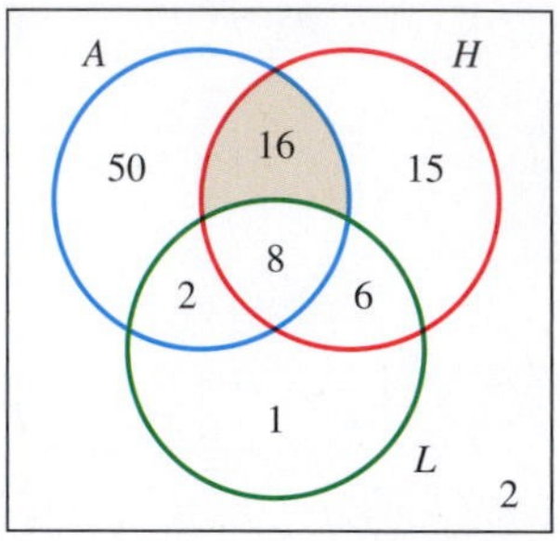

$A \cap H \cap L'$

(C) Using Figure 10, we get that $n(A \cap H \cap L') = 16$. Out of the 100 surveyed customers, there were 16 who purchased auto insurance and homeowner's insurance and did not purchase life insurance.

Matched Problem 7 Refer to Example 7.

(A) How many customers purchased homeowner's insurance?

(B) Shade the region $A \cup L$ in Figure 10. Find $n(A \cup L)$.

(C) Shade the region $A' \cap H' \cap L$ in Figure 10. Find $n(A' \cap H' \cap L)$.

CAUTION Note that in part (B), $n(H) = 45$ and $n(L) = 17$, but $n(H \cup L) \neq 45 + 17 = 62$. We do not count the customers in the intersection of H and L twice. ▲

Exercises 7.2

Skills Warm-up Exercises

W *In Problems 1–6, answer yes or no. (If necessary, review Section A.1).*

1. Is the set of even integers a subset of the set of odd integers?

2. Is the set of rational numbers a subset of the set of integers?

3. Is the set of integers the intersection of the set of even integers and the set of odd integers?

4. Is the set of integers the union of the set of even integers and the set of odd integers?

5. If the universal set is the set of integers, is the set of positive integers the complement of the set of negative integers?

6. If the universal set is the set of integers, is the set of even integers the complement of the set of odd integers?

A *In Problems 7–14, indicate true (T) or false (F).*

7. $\{1, 2\} \subset \{2, 1\}$

8. $\{3, 2, 1\} \subset \{1, 2, 3, 4\}$

9. $\{5, 10\} = \{10, 5\}$

10. $1 \in \{10, 11\}$

11. $\{0\} \in \{0, \{0\}\}$

12. $\{0, 6\} = \{6\}$

13. $8 \in \{1, 2, 4\}$

14. $\emptyset \subset \{1, 2, 3\}$

In Problems 15–28, write the resulting set using the listing method.

15. $\{1, 2, 3\} \cap \{2, 3, 4\}$

16. $\{1, 2, 4\} \cup \{4, 8, 16\}$

17. $\{1, 2, 3\} \cup \{2, 3, 4\}$

18. $\{1, 2, 4\} \cap \{4, 8, 16\}$

19. $\{1, 4, 7\} \cup \{10, 13\}$

20. $\{-3, -1\} \cap \{1, 3\}$

21. $\{1, 4, 7\} \cap \{10, 13\}$

22. $\{-3, -1,\} \cup \{1, 3\}$

B **23.** $\{x | x^2 = 25\}$

24. $\{x | x^2 = 36\}$

25. $\{x | x^3 = -27\}$

26. $\{x | x^4 = 16\}$

27. $\{x | x$ is an odd number between 1 and 9, inclusive$\}$

28. $\{x | x$ is a month starting with $M\}$

29. For $U = \{1, 2, 3, 4, 5\}$ and $A = \{2, 3, 4\}$, find A'.

30. For $U = \{7, 8, 9, 10, 11\}$ and $A = \{7, 11\}$, find A'.

In Problems 31–44, refer to the Venn diagram below and find the indicated number of elements.

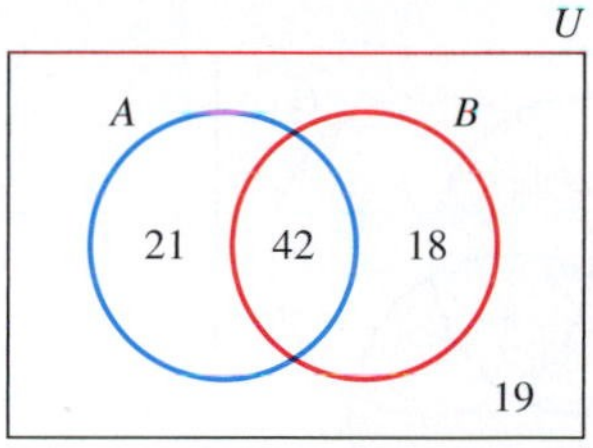

31. $n(U)$

32. $n(A)$

33. $n(B)$

34. $n(A \cap B)$

35. $n(A \cup B)$

36. $n(B')$

37. $n(A')$

38. $n(A \cap B')$

39. $n(B \cap A')$

40. $n((A \cap B)')$

41. $n((A \cup B)')$

42. $n(A' \cap B')$

43. $n(A \cup A')$

44. $n(A \cap A')$

45. If $R = \{1, 2, 3, 4\}$ and $T = \{2, 4, 6\}$, find

(A) $\{x | x \in R$ **or** $x \in T\}$

(B) $R \cup T$

46. If $R = \{1, 3, 4\}$ and $T = \{2, 4, 6\}$, find

(A) $\{x | x \in R$ **and** $x \in T\}$

(B) $R \cap T$

47. For $P = \{1, 2, 3, 4\}$, $Q = \{2, 4, 6\}$, and $R = \{3, 4, 5, 6\}$, find $P \cup (Q \cap R)$.

48. For P, Q, and R in Problem 47, find $P \cap (Q \cup R)$.

C *In Problems 49–52, determine whether the given set is finite or infinite. Consider the set N of positive integers to be the universal set.*

49. $\{n \in N | n > 100\}$

50. $\{n \in N | n < 1000\}$

51. $\{2, 3, 5, 7, 11, 13\}$

52. $\{2, 4, 6, 8, 10, \dots\}$

In Problems 53–58, draw a Venn diagram for sets A, B, and C and shade the given region.

53. $A \cap B' \cap C$

54. $A' \cap B' \cap C$

55. $(A \cap B)'$

56. $(A \cup B)'$

57. $A' \cup (B' \cap C)$

58. $(A \cap B)' \cup C$

In Problems 59–62, are the given sets disjoint? Let H, T, P, and E denote the sets in Problems 49, 50, 51, and 52, respectively.

59. H' and T'

60. E and P

61. P' and H

62. E and E'

In Problems 63–72, discuss the validity of each statement. Venn diagrams may be helpful. If the statement is true, explain why. If not, give a counterexample.

63. If $A \subset B$, then $A \cap B = A$.

64. If $A \subset B$, then $A \cup B = A$.

65. If $A \cup B = A$, then $A \subset B$.

66. If $A \cap B = A$, then $A \subset B$.

67. If $A \cap B = \emptyset$, then $A = \emptyset$.

68. If $A = \emptyset$, then $A \cap B = \emptyset$.

69. If $A \subset B$, then $A' \subset B'$.

70. If $A \subset B$, then $B' \subset A'$.

71. The empty set is an element of every set.

72. The empty set is a subset of the empty set.

73. How many subsets does each of the following sets contain?

(A) $\{a\}$

(B) $\{a, b\}$

(C) $\{a, b, c\}$

(D) $\{a, b, c, d\}$

74. Let A be a set that contains exactly n elements. Find a formula in terms of n for the number of subsets of A.

Applications

Enrollments. *In Problems 75–88, find the indicated number of elements by referring to the following table of enrollments in a finite mathematics class:*

	Freshmen	Sophomores
Arts & Sciences	19	14
Business	66	21

Let the universal set U be the set of all 120 students in the class, A the set of students from the College of Arts & Sciences, B the set of students from the College of Business, F the set of freshmen, and S the set of sophomores.

75. $n(F)$

76. $n(S)$

77. $n(A)$

78. $n(B)$

79. $n(A \cap S)$

80. $n(A \cap F)$

81. $n(B \cap F)$

82. $n(B \cap S)$

83. $n(A \cup S)$

84. $n(A \cup F)$

85. $n(B \cup F)$

86. $n(B \cup S)$

87. $n(A \cap B)$

88. $n(F \cup S)$

89. Committee selection. A company president and three vice-presidents, denoted by the set $\{P, V_1, V_2, V_3\}$, wish to select a committee of 2 people from among themselves. How many ways can this committee be formed? That is, how many 2-person subsets can be formed from a set of 4 people?

90. Voting coalition. The company's leaders in Problem 89 decide for or against certain measures as follows: The president has 2 votes and each vice-president has 1 vote. Three favorable votes are needed to pass a measure. List all minimal winning coalitions; that is, list all subsets of $\{P, V_1, V_2, V_3\}$ that represent exactly 3 votes.

Blood types. *When receiving a blood transfusion, a recipient must have all the antigens of the donor. A person may have one or more of the three antigens A, B, and Rh or none at all. Eight blood types are possible, as indicated in the following Venn diagram, where U is the set of all people under consideration:*

An A− person has A antigens but no B or Rh antigens, an O+ person has Rh but neither A nor B, an AB− person has A and B but no Rh, and so on.

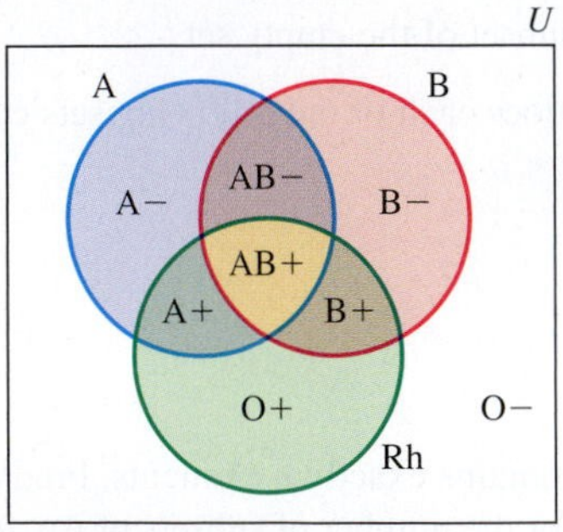

Blood Types
A−
A+
B−
B+
AB−
AB+
O−
O+

In Problems 91–98, use the Venn diagram to indicate which of the eight blood types are included in each set.

91. A ∩ Rh

92. A ∩ B

93. A ∪ Rh

94. A ∪ B

95. (A ∪ B)′

96. (A ∪ B ∪ Rh)′

97. A′ ∩ B

98. Rh′ ∩ A

Answers to Matched Problems

1. (A) $\{x \mid x^2 = 9\}$ (B) $\{-3, 3\}$ (C) True; true

2. All are true.

3. $\{1, 2\}, \{1\}, \{2\}, \varnothing$

4. (A) $\{1, 2, 3, 4, 5, 7\}$ (B) $\{1, 3\}$
(C) $\varnothing$ (D) $\{2, 4, 6, 8, 9\}$

5. (A) $n(A \cap B) = 4; n(A \cap B') = 29; n(B \cap A') = 10;$
$n(A' \cap B') = 57$

(B)

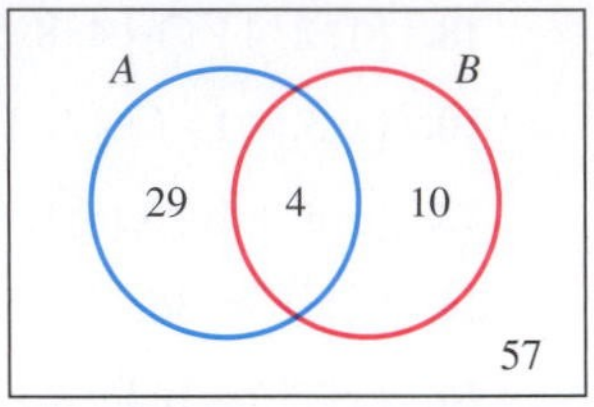

6. (A) $n(T \cap W) = 22; n(T \cap W') = 26; n(W \cap T') = 29;$
$n(T' \cap W') = 23$

(B)

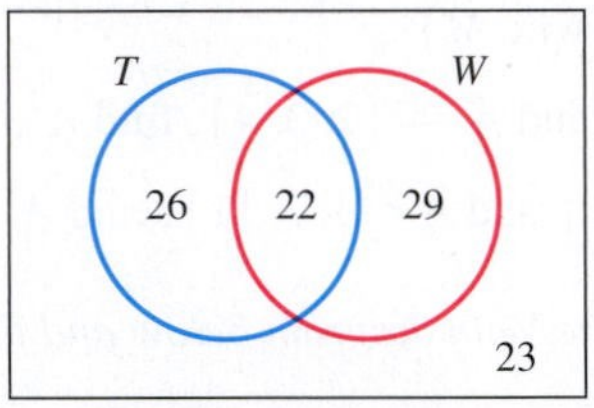

7. (A) $n(H) = 45$

(B)

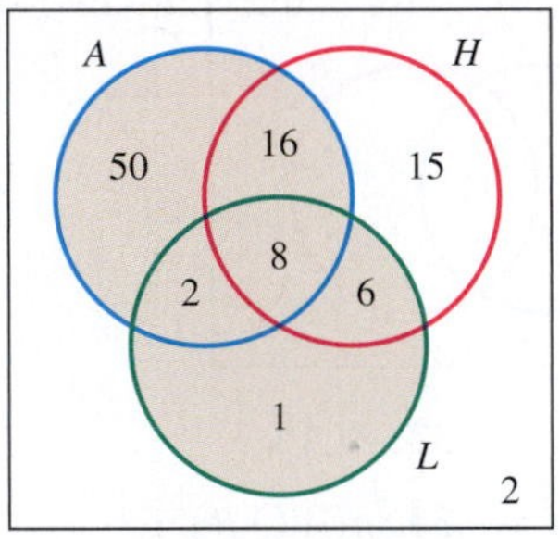

$n(A \cup L) = 83$

(C)

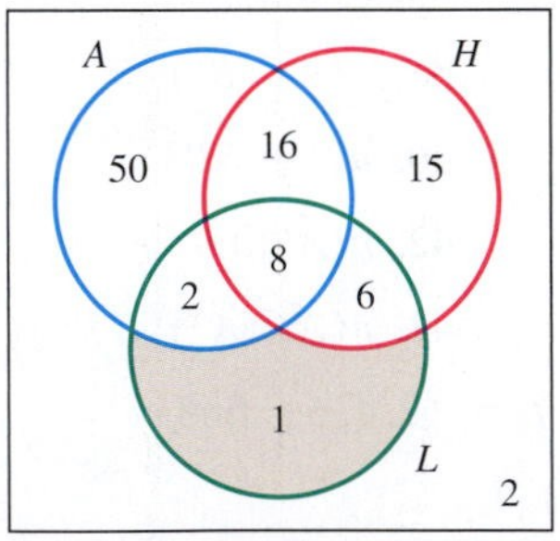

$n(A' \cap H' \cap L) = 1$

7.3 Basic Counting Principles

- Addition Principle
- Multiplication Principle

Addition Principle

If the enrollment in a college chemistry class consists of 13 males and 15 females, then there are a total of 28 students enrolled in the class. This is a simple example of a **counting technique**, a method for determining the number of elements in a set without actually enumerating the elements one by one. Set operations play a fundamental role in many counting techniques. For example, if M is the set of male students in the chemistry class and F is the set of female students, then the *union* of sets M and F, denoted $M \cup F$, is the set of all students in the class. Since these sets have no elements in common, the *intersection* of sets M and F, denoted $M \cap F$, is the *empty set* $\varnothing$; we then say that M and F are *disjoint sets*. The total number of students enrolled in the class is the number of elements in $M \cup F$, denoted by $n(M \cup F)$ and given by

$$\begin{aligned} n(M \cup F) &= n(M) + n(F) \\ &= 13 + 15 = 28 \end{aligned}$$

In this example, the number of elements in the union of sets M and F is the sum of the number of elements in M and in F. However, this does not work for all pairs of sets. To see why, consider another example. Suppose that the enrollment in a mathematics class consists of 22 math majors and 16 physics majors, and that 7 of these students have majors in both subjects. If M represents the set of math majors and P represents the set of physics majors, then $M \cap P$ represents the set of double majors. It is tempting to proceed as before and conclude that there are $22 + 16 = 38$ students in the class, but this is incorrect. We have counted the double majors twice, once as math majors and again as physics majors. To correct for this double counting, we subtract the number of double majors from this sum. Thus, the total number of students enrolled in this class is given by

$$\begin{aligned} n(M \cup P) &= n(M) + n(P) - n(M \cap P) \\ &= 22 + 16 - 7 = 31 \end{aligned} \qquad (1)$$

Equation (1) illustrates the *addition principle* for counting the elements in the union of two sets.

THEOREM 1 Addition Principle (for Counting)

For any two sets A and B,

$$n(A \cup B) = n(A) + n(B) - n(A \cap B) \qquad (2)$$

Note that if A and B are disjoint, then $n(A \cap B) = 0$, and equation (2) becomes $n(A \cup B) = n(A) + n(B)$.

EXAMPLE 1 **Employee Benefits** According to a survey of business firms in a certain city, 750 firms offer their employees health insurance, 640 offer dental insurance, and 280 offer health insurance and dental insurance. How many firms offer their employees health insurance or dental insurance?

SOLUTION If H is the set of firms that offer their employees health insurance and D is the set that offer dental insurance, then

$H \cap D =$ set of firms that offer health insurance **and** dental insurance

$H \cup D =$ set of firms that offer health insurance **or** dental insurance

So

$$n(H) = 750 \qquad n(D) = 640 \qquad n(H \cap D) = 280$$

and

$$\begin{aligned} n(H \cup D) &= n(H) + n(D) - n(H \cap D) \\ &= 750 + 640 - 280 = 1{,}110 \end{aligned}$$

Therefore, 1,110 firms offer their employees health insurance or dental insurance.

Matched Problem 1 The survey in Example 1 also indicated that 345 firms offer their employees group life insurance, 285 offer long-term disability insurance, and 115 offer group life insurance and long-term disability insurance. How many firms offer their employees group life insurance or long-term disability insurance?

EXAMPLE 2 **Market Research** A survey of 100 families found that 35 families subscribe to the video streaming service Webfilms, 60 families subscribe to the video streaming service Nile Prime, and 20 families subscribe to both video streaming services.

(A) How many families subscribe to Webfilms but not Nile Prime?

(B) How many subscribe to Nile Prime but not Webfilms?

(C) How many do not subscribe to either video streaming service?

(D) Organize this information in a table.

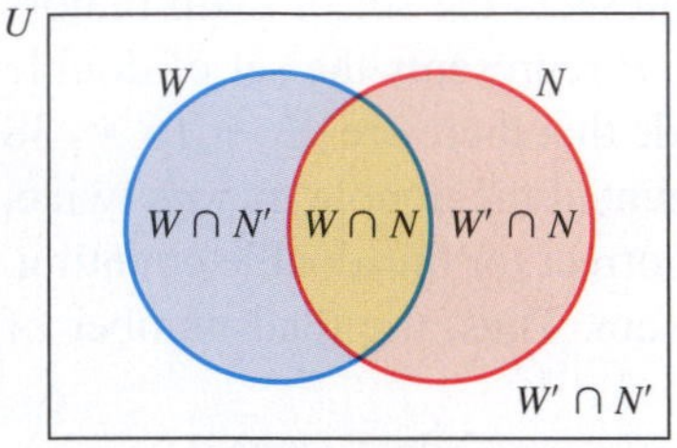

Figure 1 Venn diagram for the video streaming survey

SOLUTION Let U be the group of families surveyed. Let W be the set of families that subscribe to Webfilms, and let N be the set of families that subscribe to Nile Prime. Since U contains all the elements under consideration, it is the *universal set* for this problem. The *complement* of W, denoted by W', is the set of families in the survey group U that do not subscribe to Webfilms. Similarly, N' is the set of families in the group that do not subscribe to Nile Prime. Using the sets W and N, their complements, and set intersection, we can divide U into the four disjoint subsets defined below and illustrated as a Venn diagram in Figure 1.

$W \cap N$ = set of families that subscribe to both services

$W \cap N'$ = set of families that subscribe to Webfilms but not Nile Prime

$W' \cap N$ = set of families that subscribe to Nile Prime but not Webfilms

$W' \cap N'$ = set of families that do not subscribe to either service

The given survey information can be expressed in terms of set notation as

$$n(U) = 100 \qquad n(W) = 35 \qquad n(N) = 60 \qquad n(W \cap N) = 20$$

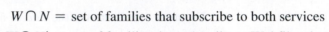

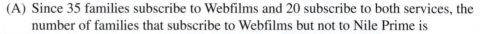

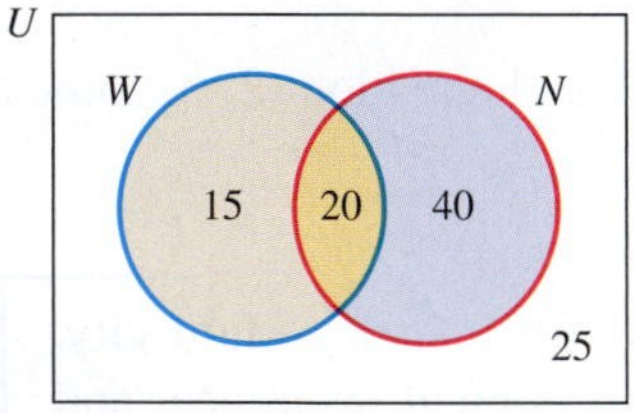

Figure 2 Survey results

We can use this information and a Venn diagram to answer parts (A)–(C). First, we place 20 in $W \cap N$ in the diagram (see Fig. 2). As we proceed with parts (A) through (C), we add each answer to the diagram.

(A) Since 35 families subscribe to Webfilms and 20 subscribe to both services, the number of families that subscribe to Webfilms but not to Nile Prime is

$$N(W \cap N') = 35 - 20 = 15.$$

(B) In a similar manner, the number of families that subscribe to Nile Prime but not Webfilms is

$$N(W' \cap N) = 60 - 20 = 40.$$

(C) The total number of subscribers is $20 + 15 + 40 = 75$. The total number of families that do not subscribe to either video streaming service is

$$n(W' \cap N') = 100 - 75 = 25.$$

(D) Venn diagrams are useful tools for determining the number of elements in the various sets of a survey, but often the results must be presented in the form of a table, rather than a diagram. Table 1 contains the information in Figure 2 and also includes totals that give the numbers of elements in the sets W, W', N, N', and U.

Table 1

		Nile Prime		
		Subscriber, N	**Nonsubscriber, N'**	**Totals**
Webfilms	**Subscriber, W**	20	15	35
	Nonsubscriber, W'	40	25	65
	Totals	60	40	100

CONCEPTUAL INSIGHT

Carefully compare Table 1 and Figure 2 of Example 2, and note that we do *not* include any of the totals in Table 1 in the Venn diagram. Instead, the numbers in the Venn diagram give the numbers of elements in the four disjoint sets $W \cap N$, $W \cap N'$, $W' \cap N$, and $W' \cap N'$. From Figure 2, it is easy to construct Table 1 (the totals are easily calculated from the Venn diagram). And from Table 1, it is easy to construct the Venn diagram of Figure 2 (simply disregard the totals).

Matched Problem 2 Students at a university have the option to stream TV shows over the internet or to watch cable TV. A survey of 100 college students produced the following results: In the past 30 days, 65 people have streamed TV shows, 45 have watched cable TV, and 30 have done both.

(A) During this 30-day period, how many people in the survey have streamed TV shows but not watched cable TV?

(B) How many have watched cable TV but not streamed TV shows?

(C) How many have done neither?

(D) Organize this information in a table.

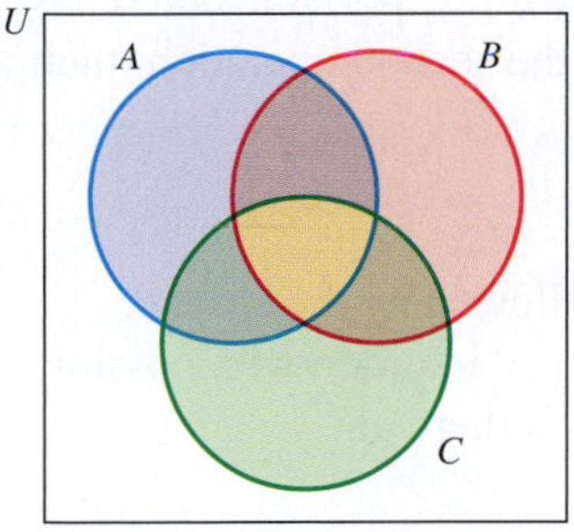

Figure 3

Explore and Discuss 1

Let A, B, and C be three sets. Use a Venn diagram (Fig. 3) to explain the following equation:

$$n(A \cup B \cup C) = n(A) + n(B) + n(C) - n(A \cap B) - n(A \cap C) - n(B \cap C) + n(A \cap B \cap C)$$

Multiplication Principle

As we have just seen, if the elements of a set are determined by the union operation, addition and subtraction are used to count the number of elements in the set. Now we want to consider sets whose elements are determined by a sequence of operations. We will see that multiplication is used to count the number of elements in sets formed this way.

EXAMPLE 3 **Product Mix** A retail store stocks windbreaker jackets in small, medium, large, and extra large. All are available in blue or red. What are the combined choices, and how many combined choices are there?

SOLUTION To solve the problem, we use a tree diagram:

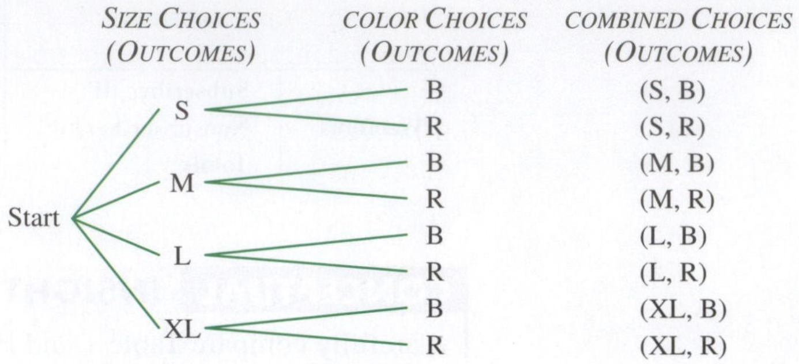

There are 8 possible combined choices (outcomes). There are 4 ways that a size can be chosen and 2 ways that a color can be chosen. The first element in the ordered pair represents a size choice, and the second element represents a color choice.

Matched Problem 3 A company offers its employees health plans from three different companies: *R*, *S*, and *T*. Each company offers two levels of coverage, *A* and *B*, with one level requiring additional employee contributions. What are the combined choices, and how many choices are there? Solve using a tree diagram.

Suppose that you asked, "From the 26 letters in the alphabet, how many ways can 3 letters appear on a license plate if no letter is repeated?" To try to count the possibilities using a tree diagram would be extremely tedious. The following **multiplication principle** will enable us to solve this problem easily. In addition, it forms the basis for developing other counting devices in the next section.

THEOREM 2 Multiplication Principle (for Counting)

1. If two operations O_1 and O_2 are performed in order, with N_1 possible outcomes for the first operation and N_2 possible outcomes for the second operation, then there are

$$N_1 \cdot N_2$$

possible combined outcomes of the first operation followed by the second.

2. In general, if n operations $O_1, O_2, \ldots, O_n$ are performed in order, with possible number of outcomes $N_1, N_2, \ldots, N_n$, respectively, then there are

$$N_1 \cdot N_2 \cdot \cdots \cdot N_n$$

possible combined outcomes of the operations performed in the given order.

In Example 3, we see that there are 4 possible outcomes in choosing a size (the first operation) and 2 possible outcomes in choosing a color (the second operation). So by the multiplication principle, there are $4 \cdot 2 = 8$ possible combined outcomes. Use the multiplication principle to solve Matched Problem 3. [*Answer:* $3 \cdot 2 = 6$]

To answer the license plate question: There are 26 ways the first letter can be chosen; after a first letter is chosen, there are 25 ways a second letter can be chosen; and after 2 letters are chosen, there are 24 ways a third letter can be chosen. So, using the multiplication principle, there are $26 \cdot 25 \cdot 24 = 15{,}600$ possible 3-letter license plates if no letter is repeated.

EXAMPLE 4 **Computer-Assisted Testing** Many colleges and universities use computer-assisted testing. Suppose that a screening test is to consist of 5 questions, and a computer stores 5 comparable questions for the first test question, 8 for the second, 6 for the third, 5 for the fourth, and 10 for the fifth. How many different 5-question tests can the computer select? (Two tests are considered different if they differ in one or more questions.)

SOLUTION

O_1:	Selecting the first question	N_1:	5 ways
O_2:	Selecting the second question	N_2:	8 ways
O_3:	Selecting the third question	N_3:	6 ways
O_4:	Selecting the fourth question	N_4:	5 ways
O_5:	Selecting the fifth question	N_5:	10 ways

The computer can generate

$$5 \cdot 8 \cdot 6 \cdot 5 \cdot 10 = 12{,}000 \text{ different tests}$$

Matched Problem 4 Each question on a multiple-choice test has 5 choices. If there are 5 such questions on a test, how many different responses are possible if only 1 choice is marked for each question?

EXAMPLE 5 **Code Words** How many 3-letter code words are possible using the first 8 letters of the alphabet if

(A) No letter can be repeated?

(B) Letters can be repeated?

(C) Adjacent letters cannot be alike?

SOLUTION To form 3-letter code words from the 8 letters available, we select a letter for the first position, one for the second position, and one for the third position. Altogether, there are three operations.

(A) No letter can be repeated:

O_1:	Selecting the first letter	N_1:	8 ways	
O_2:	Selecting the second letter	N_2:	7 ways	Since 1 letter has been used
O_3:	Selecting the third letter	N_3:	6 ways	Since 2 letters have been used

There are

$$8 \cdot 7 \cdot 6 = 336 \text{ possible code words} \qquad \text{Possible combined operations}$$

(B) Letters can be repeated:

O_1:	Selecting the first letter	N_1:	8 ways	
O_2:	Selecting the second letter	N_2:	8 ways	Repeats allowed
O_3:	Selecting the third letter	N_3:	8 ways	Repeats allowed

There are

$$8 \cdot 8 \cdot 8 = 8^3 = 512 \text{ possible code words}$$

(C) Adjacent letters cannot be alike:

O_1:	Selecting the first letter	N_1:	8 ways	
O_2:	Selecting the second letter	N_2:	7 ways	Cannot be the same as the first
O_3:	Selecting the third letter	N_3:	7 ways	Cannot be the same as the second, but can be the same as the first

There are

$$8 \cdot 7 \cdot 7 = 392 \text{ possible code words}$$

Matched Problem 5 How many 4-letter code words are possible using the first 10 letters of the alphabet under the three different conditions stated in Example 5?

Exercises 7.3

Skills Warm-up Exercises

W *In Problems 1–6, Solve for x. (If necessary, review Section 1.1).*

1. $50 = 34 + 29 - x$

2. $124 = 73 + 87 - x$

3. $4x = 23 + 42 - x$

4. $7x = 51 + 45 - x$

5. $3(x + 11) = 65 + x - 14$

6. $12(x + 5) = x + 122 - 29$

A *Solve Problems 7–10 two ways: (A) using a tree diagram, and (B) using the multiplication principle.*

7. How many ways can 2 coins turn up—heads, H, or tails, T—if the combined outcome (H, T) is to be distinguished from the outcome (T, H)?

8. How many 2-letter code words can be formed from the first 3 letters of the alphabet if no letter can be used more than once?

9. A coin is tossed with possible outcomes of heads H, or tails T. Then a single die is tossed with possible outcomes 1, 2, 3, 4, 5, or 6. How many combined outcomes are there?

10. In how many ways can 3 coins turn up—heads H, or tails T—if combined outcomes such as (H, T, H), (H, H, T), and (T, H, H) are considered as different?

11. An entertainment guide recommends 6 restaurants and 3 plays that appeal to a couple.

(A) If the couple goes to dinner or a play, but not both, how many selections are possible?

(B) If the couple goes to dinner and then to a play, how many combined selections are possible?

12. A college offers 2 introductory courses in history, 3 in science, 2 in mathematics, 2 in philosophy, and 1 in English.

(A) If a freshman takes one course in each area during her first semester, how many course selections are possible?

(B) If a part-time student can afford to take only one introductory course, how many selections are possible?

13. How many 3-letter code words can be formed from the letters *A, B, C, D, E* if no letter is repeated? If letters can be repeated? If adjacent letters must be different?

14. How many 4-letter code words can be formed from the letters *A, B, C, D, E, F, G* if no letter is repeated? If letters can be repeated? If adjacent letters must be different?

B **15.** A county park system rates its 20 golf courses in increasing order of difficulty as bronze, silver, or gold. There are only two gold courses and twice as many bronze as silver courses.

(A) If a golfer decides to play a round at a silver or gold course, how many selections are possible?

(B) If a golfer decides to play one round per week for 3 weeks, first on a bronze course, then silver, then gold, how many combined selections are possible?

16. The 14 colleges of interest to a high school senior include 6 that are expensive (tuition more than \$30,000 per year), 7 that are far from home (more than 200 miles away), and 2 that are both expensive and far from home.

(A) If the student decides to select a college that is not expensive and within 200 miles of home, how many selections are possible?

(B) If the student decides to attend a college that is not expensive and within 200 miles from home during his first two years of college, and then will transfer to a college that is not expensive but is far from home, how many selections of two colleges are possible?

In Problems 17–24, use the given information to determine the number of elements in each of the four disjoint subsets in the following Venn diagram.

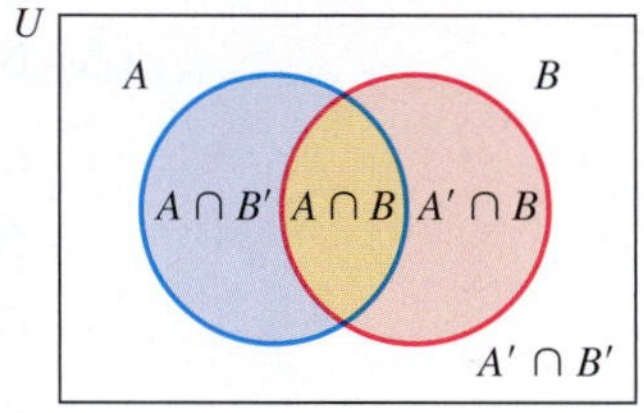

17. $n(A) = 100, n(B) = 90, n(A \cap B) = 50, n(U) = 200$

18. $n(A) = 40, n(B) = 60, n(A \cap B) = 20, n(U) = 100$

19. $n(A) = 35, n(B) = 85, n(A \cup B) = 90, n(U) = 100$

20. $n(A) = 65, n(B) = 150, n(A \cup B) = 175, n(U) = 200$

21. $n(A') = 110, n(B') = 220, n(A' \cap B') = 60, n(U) = 300$

22. $n(A') = 70, n(B') = 170, n(A' \cap B') = 40, n(U) = 300$

23. $n(A') = 20, n(B') = 40, n(A' \cup B') = 50, n(U) = 80$

24. $n(A') = 30, n(B') = 10, n(A' \cup B') = 35, n(U) = 60$

In Problems 25–32, use the given information to complete the following table.

	A	A'	Totals
B	?	?	?
B'	?	?	?
Totals	?	?	?

25. $n(A) = 70, \quad n(B) = 90,$
$n(A \cap B) = 30, \quad n(U) = 200$

26. $n(A) = 55, \quad n(B) = 65,$
$n(A \cap B) = 35, \quad n(U) = 100$

27. $n(A) = 45, \quad n(B) = 55,$
$n(A \cup B) = 80, \quad n(U) = 100$

28. $n(A) = 80, \quad n(B) = 70,$
$n(A \cup B) = 110, \quad n(U) = 200$

29. $n(A') = 15, \quad n(B') = 24,$
$n(A' \cup B') = 32, \quad n(U) = 90$

30. $n(A') = 81, \quad n(B') = 90,$
$n(A' \cap B') = 63, \quad n(U) = 180$

31. $n(A) = 110, \quad n(B) = 145,$
$n(A \cup B) = 255, \quad n(U) = 300$

32. $n(A) = 175, \quad n(B) = 125,$
$n(A \cup B) = 300, \quad n(U) = 300$

In Problems 33 and 34, discuss the validity of each statement. If the statement is always true, explain why. If not, give a counterexample.

33. (A) If A or B is the empty set, then A and B are disjoint.

(B) If A and B are disjoint, then A or B is the empty set.

34. (A) If A and B are disjoint, then $n(A \cap B) = n(A) + n(B)$.

(B) If $n(A \cup B) = n(A) + n(B)$, then A and B are disjoint.

35. A particular new car model is available with 5 choices of color, 3 choices of transmission, 4 types of interior, and 2 types of engine. How many different variations of this model are possible?

36. A delicatessen serves meat sandwiches with the following options: 3 kinds of bread, 5 kinds of meat, and lettuce or sprouts. How many different sandwiches are possible, assuming that one item is used out of each category?

37. Using the English alphabet, how many 5-character case-sensitive passwords are possible?

38. Using the English alphabet, how many 5-character case-sensitive passwords are possible if each character is a letter or a digit?

39. A combination lock has 5 wheels, each labeled with the 10 digits from 0 to 9. How many 5-digit opening combinations are possible if no digit is repeated? If digits can be repeated? If successive digits must be different?

40. A small combination lock has 3 wheels, each labeled with the 10 digits from 0 to 9. How many 3-digit combinations are possible if no digit is repeated? If digits can be repeated? If successive digits must be different?

41. How many different license plates are possible if each contains 3 letters (out of the alphabet's 26 letters) followed by 3 digits (from 0 to 9)? How many of these license plates contain no repeated letters and no repeated digits?

42. How many 5-digit ZIP code numbers are possible? How many of these numbers contain no repeated digits?

43. In Example 3, does it make any difference in which order the selection operations are performed? That is, if we select a jacket color first and then select a size, are there as many combined choices available as selecting a size first and then a color? Justify your answer using tree diagrams and the multiplication principle.

44. Explain how three sets, A, B, and C, can be related to each other in order for the following equation to hold true (Venn diagrams may be helpful):

$$n(A \cup B \cup C) = n(A) + n(B) + n(C) - n(A \cap C) - n(B \cap C)$$

C *Problems 45–48 refer to the following Venn diagram.*

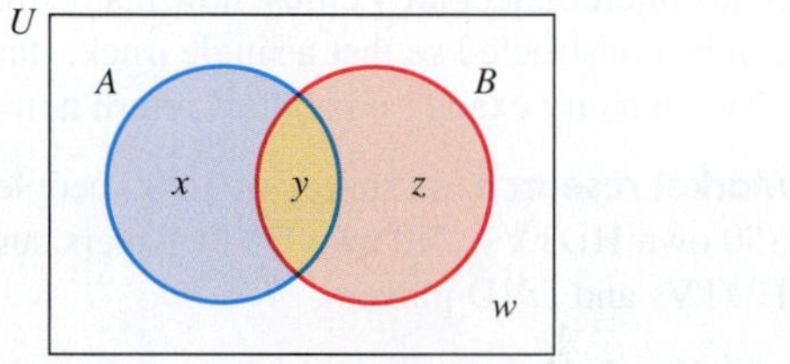

45. Which of the numbers x, y, z, or w must equal 0 if $B \subset A$?

46. Which of the numbers x, y, z, or w must equal 0 if A and B are disjoint?

47. Which of the numbers x, y, z, or w must equal 0 if $A \cup B = A \cap B$?

48. Which of the numbers x, y, z, or w must equal 0 if $A \cup B = U$?

49. A group of 75 people includes 32 who play tennis, 37 who play golf, and 8 who play both tennis and golf. How many people in the group play neither sport?

50. A class of 30 music students includes 13 who play the piano, 16 who play the guitar, and 5 who play both the piano and the guitar. How many students in the class play neither instrument?

51. A group of 100 people touring Europe includes 42 people who speak French, 55 who speak German, and 17 who speak neither language. How many people in the group speak both French and German?

52. A high school football team with 40 players includes 16 players who played offense last year, 17 who played defense, and 12 who were not on last year's team. How many players from last year played both offense and defense?

Applications

53. Management. A management selection service classifies its applicants (using tests and interviews) as high-IQ, middle-IQ, or low-IQ and as aggressive or passive. How many combined classifications are possible?

(A) Solve using a tree diagram.

(B) Solve using the multiplication principle.

54. **Management.** A corporation plans to fill 2 different positions for vice-president, V_1 and V_2, from administrative officers in 2 of its manufacturing plants. Plant A has 6 officers and plant B has 8. How many ways can these 2 positions be filled if the V_1 position is to be filled from plant A and the V_2 position from plant B? How many ways can the 2 positions be filled if the selection is made without regard to plant?

55. **Transportation.** A sales representative who lives in city A wishes to start from home and visit 3 different cities: B, C, and D. She must choose whether to drive her own car or to fly. If all cities are interconnected by both roads and airlines, how many travel plans can be constructed to visit each city exactly once and return home?

56. **Transportation.** A manufacturing company in city A wishes to truck its product to 4 different cities: B, C, D, and E. If roads interconnect all 4 cities, how many different route plans can be constructed so that a single truck, starting from A, will visit each city exactly once, then return home?

57. **Market research.** A survey of 1,200 people indicates that 850 own HDTVs, 740 own DVD players, and 580 own HDTVs and DVD players.

(A) How many people in the survey own either an HDTV or a DVD player?

(B) How many own neither an HDTV nor a DVD player?

(C) How many own an HDTV and do not own a DVD player?

58. **Market research.** A survey of 800 small businesses indicates that 250 own a video conferencing system, 420 own projection equipment, and 180 own a video conferencing system and projection equipment.

(A) How many businesses in the survey own either a video conferencing system or projection equipment?

(B) How many own neither a video conferencing system nor projection equipment?

(C) How many own projection equipment and do not own a video conferencing system?

59. **Communications.** A cable television company has 8,000 subscribers in a suburban community. The company offers two premium channels: HBO and Showtime. If 2,450 subscribers receive HBO, 1,940 receive Showtime, and 5,180 do not receive any premium channel, how many subscribers receive both HBO and Showtime?

60. **Communications.** A cable company offers its 10,000 customers two special services: high-speed internet and digital phone. If 3,770 customers use high-speed internet, 3,250 use digital phone, and 4,530 do not use either of these services, how many customers use both high-speed internet and digital phone?

61. **Minimum wage.** The table gives the number of male and female workers earning at or below the minimum wage for several age categories.

	Workers per Age Group (thousands)			
	16–19	20–24	25 +	Totals
Males at Minimum Wage	343	154	237	734
Males below Minimum Wage	118	102	159	379
Females at Minimum Wage	367	186	503	1,056
Females below Minimum Wage	251	202	540	993
Totals	1,079	644	1,439	3,162

(A) How many males are of age 20–24 and earn below minimum wage?

(B) How many females are of age 20 or older and earn minimum wage?

(C) How many workers are of age 16–19 or males earning minimum wage?

(D) How many workers earn below minimum wage?

62. **Minimum wage.** Refer to the table in Problem 61.

(A) How many females are of age 16–19 and earn minimum wage?

(B) How many males are of age 16–24 and earn below minimum wage?

(C) How many workers are of age 20–24 or females earning below minimum wage?

(D) How many workers earn minimum wage?

63. **Medicine.** A medical researcher classifies subjects according to male or female; smoker or nonsmoker; and underweight, average weight, or overweight. How many combined classifications are possible?

(A) Solve using a tree diagram.

(B) Solve using the multiplication principle.

64. **Family planning.** A couple is planning to have 3 children. How many boy–girl combinations are possible? Distinguish between combined outcomes such as (B, B, G), (B, G, B), and (G, B, B).

(A) Solve using a tree diagram.

(B) Solve using the multiplication principle.

65. **Politics.** A politician running for a third term is planning to contact all contributors to her first two campaigns. If 1,475 individuals contributed to the first campaign, 2,350 contributed to the second campaign, and 920 contributed to the first and second campaigns, how many individuals have contributed to the first or second campaign?

66. **Politics.** If 12,457 people voted for a politician in his first election, 15,322 voted for him in his second election, and 9,345 voted for him in the first and second elections, how many people voted for this politician in the first or second election?

Answers to Matched Problems

1. 515
2. (A) 35
   (B) 15
   (C) 20
   (D)

		Cable TV		
		Viewer	Nonviewer	Totals
Internet	Streamer	30	35	65
	Nonstreamer	15	20	35
	Totals	45	55	100

3. There are 6 combined choices:

COMPANY CHOICES (OUTCOMES)	COVERAGE CHOICES (OUTCOMES)	COMBINED CHOICES (OUTCOMES)

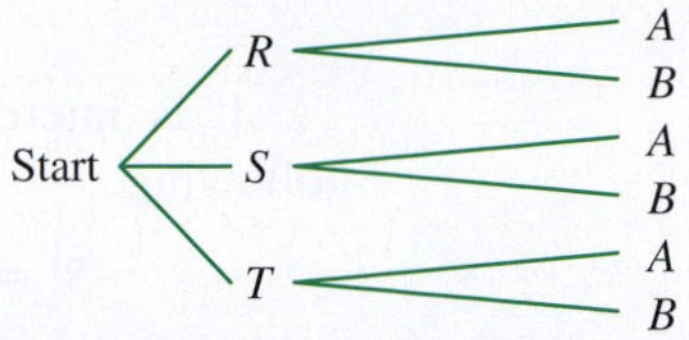

(R, A)
(R, B)
(S, A)
(S, B)
(T, A)
(T, B)

4. 5^5, or 3,125
5. (A) $10 \cdot 9 \cdot 8 \cdot 7 = 5{,}040$
   (B) $10 \cdot 10 \cdot 10 \cdot 10 = 10{,}000$
   (C) $10 \cdot 9 \cdot 9 \cdot 9 = 7{,}290$

7.4 Permutations and Combinations

- Factorials
- Permutations
- Combinations
- Applications

The multiplication principle discussed in the preceding section can be used to develop two additional counting devices that are extremely useful in more complicated counting problems. Both of these devices use *factorials*.

Factorials

When using the multiplication principle, we encountered expressions such as

$$26 \cdot 25 \cdot 24 \quad \text{or} \quad 8 \cdot 7 \cdot 6$$

where each natural number factor is decreased by 1 as we move from left to right. The factors in the following product continue to decrease by 1 until a factor of 1 is reached:

$$5 \cdot 4 \cdot 3 \cdot 2 \cdot 1$$

Products like this are encountered so frequently in counting problems that it is useful to express them in a concise notation. The product of the first n natural numbers is called ***n* factorial** and is denoted by $n!$ Also, we define **zero factorial, 0!**, to be 1.

DEFINITION Factorial*

For a natural number n,

$$n! = n(n-1)(n-2) \cdot \cdots \cdot 2 \cdot 1 \qquad 4! = 4 \cdot 3 \cdot 2 \cdot 1$$
$$0! = 1$$
$$n! = n \cdot (n-1)!$$

EXAMPLE 1 Computing Factorials

(A) $5! = 5 \cdot 4 \cdot 3 \cdot 2 \cdot 1 = 120$

(B) $\dfrac{7!}{6!} = \dfrac{7 \cdot \cancel{6!}}{\cancel{6!}} = 7$

(C) $\dfrac{8!}{5!} = \dfrac{8 \cdot 7 \cdot 6 \cdot \cancel{5!}}{\cancel{5!}} = 8 \cdot 7 \cdot 6 = 336$

(D) $\dfrac{52!}{5!47!} = \dfrac{52 \cdot 51 \cdot 50 \cdot 49 \cdot 48 \cdot \cancel{47!}}{5 \cdot 4 \cdot 3 \cdot 2 \cdot 1 \cdot \cancel{47!}} = 2{,}598{,}960$

*Many calculators have an n! key or its equivalent.

Matched Problem 1 Find

(A) $6!$ (B) $\frac{10!}{9!}$ (C) $\frac{10!}{7!}$ (D) $\frac{5!}{0!3!}$ (E) $\frac{20!}{3!17!}$

It is interesting and useful to note that $n!$ grows very rapidly. Compare the following:

$$5! = 120 \qquad 10! = 3{,}628{,}800 \qquad 15! = 1{,}307{,}674{,}368{,}000$$

Try 69!, 70!, and 71! on your calculator.

Permutations

A particular (horizontal or vertical) arrangement of a set of paintings on a wall is called a *permutation* of the set of paintings.

DEFINITION Permutation of a Set of Objects

A **permutation** of a set of distinct objects is an arrangement of the objects in a specific order without repetition.

Suppose that 4 pictures are to be arranged from left to right on one wall of an art gallery. How many permutations (ordered arrangements) are possible? Using the multiplication principle, there are 4 ways of selecting the first picture; after the first picture is selected, there are 3 ways of selecting the second picture. After the first 2 pictures are selected, there are 2 ways of selecting the third picture, and after the first 3 pictures are selected, there is only 1 way to select the fourth. So the number of permutations (ordered arrangements) of the set of 4 pictures is

$$4 \cdot 3 \cdot 2 \cdot 1 = 4! = 24$$

In general, how many permutations of a set of n distinct objects are possible? Reasoning as above, there are n ways in which the first object can be chosen, there are $n - 1$ ways in which the second object can be chosen, and so on. Using the multiplication principle, we have the following:

THEOREM 1 Number of Permutations of n Objects

The number of permutations of n distinct objects without repetition, denoted by ${}_nP_n$, is

$${}_nP_n = n(n-1) \cdot \cdots \cdot 2 \cdot 1 = n! \qquad n \text{ factors}$$

Example: The number of permutations of 7 objects is

$${}_7P_7 = 7 \cdot 6 \cdot 5 \cdot 4 \cdot 3 \cdot 2 \cdot 1 = 7! \qquad 7 \text{ factors}$$

Now suppose that the director of the art gallery decides to use only 2 of the 4 available paintings, and they will be arranged on the wall from left to right. We are now talking about a particular arrangement of 2 paintings out of the 4, which is called a *permutation of 4 objects taken 2 at a time*. In general,

DEFINITION Permutation of n Objects Taken r at a Time

A permutation of a set of n distinct objects taken r at a time without repetition is an arrangement of r of the n objects in a specific order.

How many ordered arrangements of 2 pictures can be formed from the 4? That is, how many permutations of 4 objects taken 2 at a time are there? There are 4 ways that the first picture can be selected; after selecting the first picture, there are 3 ways that the second picture can be selected. So the number of permutations of a set of 4 objects taken 2 at a time, which is denoted by ${}_4P_2$, is given by

$${}_4P_2 = 4 \cdot 3$$

In terms of factorials, we have

$${}_4P_2 = 4 \cdot 3 = \frac{4 \cdot 3 \cdot \mathbf{2!}}{\mathbf{2!}} = \frac{4!}{2!} \quad \text{Multiplying } 4 \cdot 3 \text{ by 1 in the form } 2!/2!$$

Reasoning in the same way as in the example, we find that the number of permutations of n distinct objects taken r at a time without repetition $(0 \le r \le n)$ is given by

$$\begin{aligned} {}_nP_r &= n(n-1)(n-2) \cdot \cdots \cdot (n-r+1) && r \text{ factors} \\ {}_9P_6 &= 9(9-1)(9-2) \cdot \cdots \cdot (9-6+1) && 6 \text{ factors} \\ &= 9 \cdot 8 \cdot 7 \cdot 6 \cdot 5 \cdot 4 \end{aligned}$$

Multiplying the right side of the equation for ${}_nP_r$ by 1 in the form $(n-r)!/(n-r)!$, we obtain a factorial form for ${}_nP_r$:

$${}_nP_r = n(n-1)(n-2) \cdot \cdots \cdot (n-r+1)\frac{(n-r)!}{(n-r)!}$$

But, since

$$n(n-1)(n-2) \cdot \cdots \cdot (n-r+1)(n-r)! = n!$$

the expression above simplifies to

$${}_nP_r = \frac{n!}{(n-r)!}$$

We summarize these results in Theorem 2.

THEOREM 2 Number of Permutations of *n* Objects Taken *r* at a Time

The number of permutations of n distinct objects taken r at a time without repetition is given by*

$${}_nP_r = n(n-1)(n-2) \cdot \cdots \cdot (n-r+1) \qquad r \text{ factors}$$

$${}_5P_2 = 5 \cdot 4 \text{ factors}$$

or

$${}_nP_r = \frac{n!}{(n-r)!} \qquad 0 \le r \le n \qquad {}_5P_2 = \frac{5!}{(5-2)!} = \frac{5!}{3!}$$

Note: ${}_nP_n = \dfrac{n!}{(n-n)!} = \dfrac{n!}{0!} = n!$ permutations of n objects taken n at a time.

Remember, by definition, $0! = 1$.

---

* In place of the symbol ${}_nP_r$, the symbols P_r^n, $P_{n,r}$, and $P(n, r)$ are often used.

EXAMPLE 2 **Permutations** Given the set $\{A, B, C\}$, how many permutations are possible for this set of 3 objects taken 2 at a time? Answer the question

(A) Using a tree diagram

(B) Using the multiplication principle

(C) Using the two formulas for ${}_nP_r$

SOLUTION

(A) Using a tree diagram:

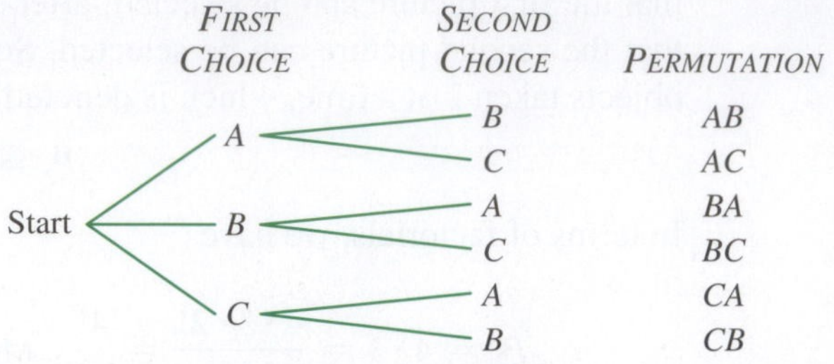

There are 6 permutations of 3 objects taken 2 at a time.

(B) Using the multiplication principle:

O_1: Fill the first position N_1: 3 ways

O_2: Fill the second position N_2: 2 ways

There are

$$3 \cdot 2 = 6 \text{ permutations of 3 objects taken 2 at a time}$$

(C) Using the two formulas for ${}_nP_r$:

$$\overbrace{{}_3P_2 = 3 \cdot 2}^{\text{2 factors}} = 6 \quad \text{or} \quad {}_3P_2 = \frac{3!}{(3-2)!} = \frac{3 \cdot 2 \cdot 1}{1} = 6$$

There are 6 permutations of 3 objects taken 2 at a time. Of course, all three methods produce the same answer.

Matched Problem 2 Given the set $\{A, B, C, D\}$, how many permutations are possible for this set of 4 objects taken 2 at a time? Answer the question

(A) Using a tree diagram

(B) Using the multiplication principle

(C) Using the two formulas for ${}_nP_r$

In Example 2 you probably found the multiplication principle to be the easiest method to use. But for large values of n and r, you will find that the factorial formula is more convenient. In fact, many calculators have functions that compute $n!$ and ${}_nP_r$ directly.

EXAMPLE 3 **Permutations** Find the number of permutations of 13 objects taken 8 at a time. Compute the answer using a calculator.

SOLUTION We use the factorial formula for ${}_nP_r$:

$${}_{13}P_8 = \frac{13!}{(13-8)!} = \frac{13!}{5!} = 51{,}891{,}840$$

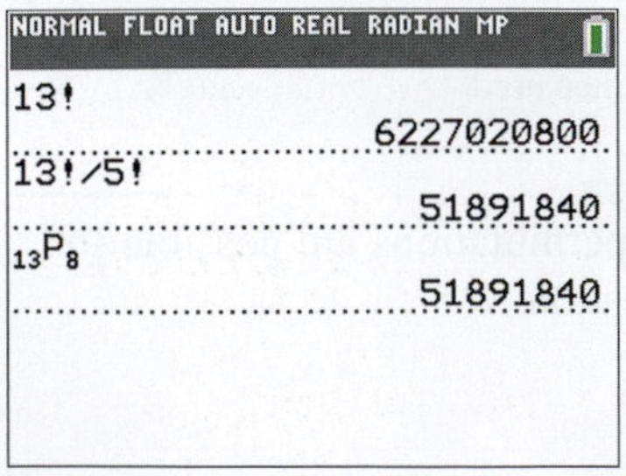

Figure 1

Using a tree diagram to solve this problem would involve a monumental effort. Using the multiplication principle would mean multiplying $13 \cdot 12 \cdot 11 \cdot 10 \cdot 9 \cdot 8 \cdot 7 \cdot 6$ (8 factors), which is not too bad. However, a calculator can provide instant results (see Fig. 1).

Matched Problem 3 Find the number of permutations of 30 objects taken 4 at a time. Compute the answer using a calculator.

Combinations

Suppose that an art museum owns 8 paintings by a given artist and another art museum wishes to borrow 3 of these paintings for a special show. In selecting 3 of the 8 paintings for shipment, the order would not matter, and we would simply be selecting a 3-element subset from the set of 8 paintings. That is, we would be selecting what is called *a combination of 8 objects taken 3 at a time.*

DEFINITION Combination of *n* Objects Taken *r* at a Time

A **combination** of a set of n distinct objects taken r at a time without repetition is an r-element subset of the set of n objects. The arrangement of the elements in the subset does not matter.

How many ways can the 3 paintings be selected out of the 8 available? That is, what is the number of combinations of 8 objects taken 3 at a time? To answer this question, and to get a better insight into the general problem, we return to Example 2.

In Example 2, we were given the set $\{A, B, C\}$ and found the number of permutations of 3 objects taken 2 at a time using a tree diagram. From this tree diagram, we can determine the number of combinations of 3 objects taken 2 at a time (the number of 2-element subsets from a 3-element set), and compare it with the number of permutations (see Fig. 2).

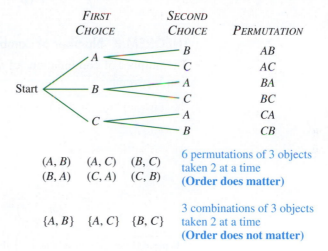

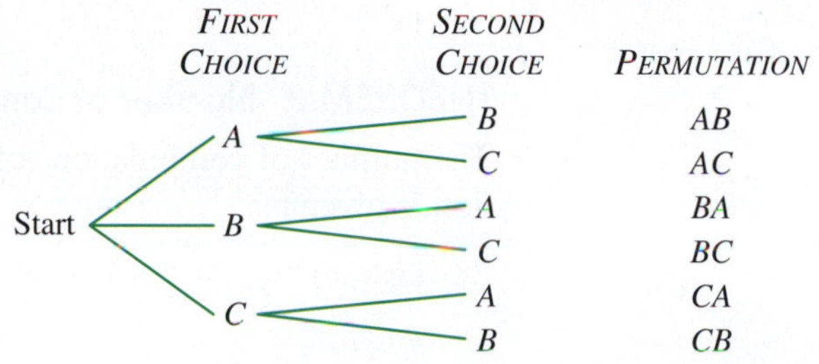

(A, B) (A, C) (B, C)
(B, A) (C, A) (C, B)

6 permutations of 3 objects taken 2 at a time
(Order does matter)

$\{A, B\}$ $\{A, C\}$ $\{B, C\}$

3 combinations of 3 objects taken 2 at a time
(Order does not matter)

Figure 2

There are fewer combinations than permutations, as we would expect. To each subset (combination), there corresponds two ordered pairs (permutations). We denote the number of combinations in Figure 2 by

$$_3C_2 \quad \text{or} \quad \binom{3}{2}$$

Our final goal is to find a factorial formula for $_nC_r$, the number of combinations of n objects taken r at a time. But first, we will develop a formula for $_3C_2$, and then we will generalize from this experience.

We know the number of permutations of 3 objects taken 2 at a time is given by $_3P_2$, and we have a formula for computing this number. Now, suppose we think of $_3P_2$ in terms of two operations:

O_1: Selecting a subset of 2 elements $\qquad N_1$: $_3C_2$ ways

O_2: Arranging the subset in a given order $\qquad N_2$: $2!$ ways

The combined operation, O_1 followed by O_2, produces a permutation of 3 objects taken 2 at a time. Thus,

$$_3P_2 = {_3C_2} \cdot 2! \quad \text{or} \quad {_3C_2} = \frac{_3P_2}{2!}$$

To find $_3C_2$, the number of combinations of 3 objects taken 2 at a time, we substitute

$$_3P_2 = \frac{3!}{(3-2)!}$$

and solve for $_3C_2$:

$$_3C_2 = \frac{3!}{2!(3-2)!} = \frac{3 \cdot 2 \cdot 1}{(2 \cdot 1)(1)} = 3$$

This result agrees with the result obtained by using a tree diagram. Note that the number of combinations of 3 objects taken 2 at a time is the same as the number of permutations of 3 objects taken 2 at a time divided by the number of permutations of the elements in a 2-element subset. Figure 2 also reveals this observation.

Reasoning the same way as in the example, the number of combinations of n objects taken r at a time $(0 \le r \le n)$ is given by

$$_nC_r = \frac{_nP_r}{r!} \qquad \text{Substitute } {_nP_r} = \frac{n!}{(n-r)!}.$$

$$= \frac{n!}{r!(n-r)!}$$

THEOREM 3 Number of Combinations of *n* Objects Taken *r* at a Time

The number of combinations of n distinct objects taken r at a time without repetition is given by*

$$_nC_r = \binom{n}{r} \qquad\qquad {_{52}C_5} = \binom{52}{5}$$

$$= \frac{_nP_r}{r!} \qquad\qquad = \frac{_{52}P_5}{5!}$$

$$= \frac{n!}{r!(n-r)!} \quad 0 \le r \le n \qquad = \frac{52!}{5!(52-5)!}$$

* In place of the symbols $_nC_r$ and $\binom{n}{r}$, the symbols C_r^n, $C_{n,r}$ and $C(n, r)$ are often used.

Now we can answer the question posed earlier in the museum example. There are

$$_8C_3 = \frac{8!}{3!(8-3)!} = \frac{8!}{3!5!} = \frac{8 \cdot 7 \cdot 6 \cdot 5!}{3 \cdot 2 \cdot 1 \cdot 5!} = 56$$

ways that 3 paintings can be selected for shipment. That is, there are 56 combinations of 8 objects taken 3 at a time.

EXAMPLE 4 **Permutations and Combinations** From a committee of 10 people,

(A) In how many ways can we choose a chairperson, a vice-chairperson, and a secretary, assuming that one person cannot hold more than one position?

(B) In how many ways can we choose a subcommittee of 3 people?

SOLUTION Note how parts (A) and (B) differ. In part (A), order of choice makes a difference in the selection of the officers. In part (B), the ordering does not matter in choosing a 3-person subcommittee. In part (A), we are interested in the number of *permutations* of 10 objects taken 3 at a time; and in part (B), we are interested in the number of *combinations* of 10 objects taken 3 at a time. These quantities are computed as follows (and since the numbers are not large, we do not need to use a calculator):

$$\text{(A)}\ {}_{10}P_3 = \frac{10!}{(10-3)!} = \frac{10!}{7!} = \frac{10 \cdot 9 \cdot 8 \cdot \cancel{7!}}{\cancel{7!}} = 720 \text{ ways}$$

$$\text{(B)}\ {}_{10}C_3 = \frac{10!}{3!(10-3)!} = \frac{10!}{3!7!} = \frac{10 \cdot 9 \cdot 8 \cdot \cancel{7!}}{3 \cdot 2 \cdot 1 \cdot \cancel{7!}} = 120 \text{ ways}$$

Matched Problem 4 From a committee of 12 people,

(A) In how many ways can we choose a chairperson, a vice-chairperson, a secretary, and a treasurer, assuming that one person cannot hold more than one position?

(B) In how many ways can we choose a subcommittee of 4 people?

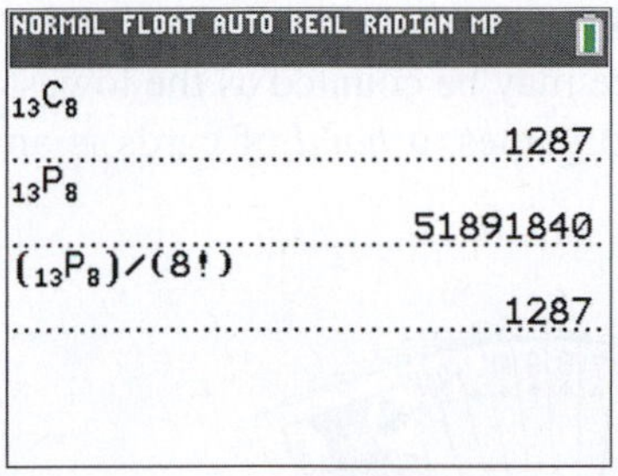

Figure 3

If n and r are large numbers, a calculator is useful in evaluating expressions involving factorials. Many calculators have a function that computes ${}_nC_r$ directly (see Fig. 3).

EXAMPLE 5 **Combinations** Find the number of combinations of 13 objects taken 8 at a time. Compute the answer using a calculator.

SOLUTION

$${}_{13}C_8 = \binom{13}{8} = \frac{13!}{8!(13-8)!} = \frac{13!}{8!5!} = 1{,}287$$

Compare the result in Example 5 with that obtained in Example 3, and note that ${}_{13}C_8$ is substantially smaller than ${}_{13}P_8$ (see Fig. 3).

Matched Problem 5 Find the number of combinations of 30 objects taken 4 at a time. Compute the answer using a calculator.

CONCEPTUAL INSIGHT

Permutations and combinations are similar in that both are selections in which repetition is *not* allowed. But there is a crucial distinction between the two:

In a permutation, order is vital.

In a combination, order is irrelevant.

To determine whether a given selection is a permutation or combination, see if rearranging the order in which the elements are selected would produce a different object. If so, the selection is a permutation; if not, the selection is a combination.

Explore and Discuss 1

(A) List alphabetically by the first letter, all 3-letter license plate codes consisting of 3 different letters chosen from M, A, T, H. Discuss how this list relates to ${}_nP_r$.

(B) Reorganize the list from part (A) so that all codes without M come first, then all codes without A, then all codes without T, and finally all codes without H. Discuss how this list illustrates the formula ${}_nP_r = r!\,{}_nC_r$.

Applications

We now consider some applications of permutations and combinations. Several applications in this section involve a standard 52-card deck of playing cards.

Standard 52-Card Deck of Playing Cards

A standard deck of 52 cards (see Fig. 4) has four 13-card suits: diamonds, hearts, clubs, and spades. The diamonds and hearts are red, and the clubs and spades are black. Each 13-card suit contains cards numbered from 2 to 10, a jack, a queen, a king, and an ace. The number or letter on a card indicates its *rank*. So there are 13 ranks and 4 cards of each rank. The jack, queen, and king are called *face cards*. (The ace is *not* a face card.) Depending on the game, the ace may be counted as the lowest and/or the highest card in the suit. In traditional card games, a *hand* of cards is an unordered subset of the deck.

Figure 4

EXAMPLE 6 **Counting Techniques** How many 5-card hands have 3 aces and 2 kings?

SOLUTION The solution involves both the multiplication principle and combinations. Think of selecting the 5-card hand in terms of the following two operations:

O_1: Choosing 3 aces out of 4 possible (order is not important) $\quad N_1$: ${}_4C_3$

O_2: Choosing 2 kings out of 4 possible (order is not important) $\quad N_2$: ${}_4C_2$

Using the multiplication principle, we have

$$\begin{aligned} \text{number of hands} &= {}_4C_3 \cdot {}_4C_2 \\ &= \frac{4!}{3!(4-3)!} \cdot \frac{4!}{2!(4-2)!} \\ &= 4 \cdot 6 = 24 \end{aligned}$$

Matched Problem 6 How many 5-card hands have 3 hearts and 2 spades?

EXAMPLE 7 **Counting Techniques** Serial numbers for a product are made using 2 letters followed by 3 numbers. If the letters are taken from the first 8 letters of the alphabet with no repeats and the numbers are taken from the 10 digits (0–9) with no repeats, how many serial numbers are possible?

SOLUTION The solution involves both the multiplication principle and permutations. Think of selecting a serial number in terms of the following two operations:

O_1: Choosing 2 letters out of 8 available (order is important) $\quad N_1$: ${}_8P_2$

O_2: Choosing 3 numbers out of 10 available (order is important) $\quad N_2$: ${}_{10}P_3$

Using the multiplication principle, we have

$$\begin{aligned}\text{number of serial numbers} &= {}_8P_2 \cdot {}_{10}P_3\\ &= \frac{8!}{(8-2)!}\cdot\frac{10!}{(10-3)!}\\ &= 56 \cdot 720 = 40{,}320\end{aligned}$$

Matched Problem 7 Repeat Example 7 under the same conditions, except that the serial numbers will now have 3 letters followed by 2 digits (no repeats).

EXAMPLE 8 **Counting Techniques** A company has 7 senior and 5 junior officers. It wants to form an ad hoc legislative committee. In how many ways can a 4-officer committee be formed so that it is composed of

(A) Any 4 officers?

(B) 4 senior officers?

(C) 3 senior officers and 1 junior officer?

(D) 2 senior and 2 junior officers?

(E) At least 2 senior officers?

SOLUTION

(A) Since there are a total of 12 officers in the company, the number of different 4-member committees is

$$ {}_{12}C_4 = \frac{12!}{4!(12-4)!} = \frac{12!}{4!8!} = 495$$

(B) If only senior officers can be on the committee, the number of different committees is

$$ {}_7C_4 = \frac{7!}{4!(7-4)!} = \frac{7!}{4!3!} = 35$$

(C) The 3 senior officers can be selected in ${}_7C_3$ ways, and the 1 junior officer can be selected in ${}_5C_1$ ways. Applying the multiplication principle, the number of ways that 3 senior officers and 1 junior officer can be selected is

$$ {}_7C_3 \cdot {}_5C_1 = \frac{7!}{3!(7-3)!}\cdot\frac{5!}{1!(5-1)!} = \frac{7!5!}{3!4!1!4!} = 175$$

(D) $${}_7C_2 \cdot {}_5C_2 = \frac{7!}{2!(7-2)!}\cdot\frac{5!}{2!(5-2)!} = \frac{7!5!}{2!5!2!3!} = 210$$

(E) The committees with *at least* 2 senior officers can be divided into three disjoint collections:

1. Committees with 4 senior officers and 0 junior officers

2. Committees with 3 senior officers and 1 junior officer

3. Committees with 2 senior officers and 2 junior officers

The number of committees of types 1, 2, and 3 is computed in parts (B), (C), and (D), respectively. The total number of committees of all three types is the sum of these quantities:

$$\underbrace{{}_7C_4}_{\text{Type 1}} + \underbrace{{}_7C_3 \cdot {}_5C_1}_{\text{Type 2}} + \underbrace{{}_7C_2 \cdot {}_5C_2}_{\text{Type 3}} = 35 + 175 + 210 = 420$$

Matched Problem 8 Given the information in Example 8, answer the following questions:

(A) How many 4-officer committees with 1 senior officer and 3 junior officers can be formed?

(B) How many 4-officer committees with 4 junior officers can be formed?

(C) How many 4-officer committees with at least 2 junior officers can be formed?

EXAMPLE 9 **Counting Techniques** From a standard 52-card deck, how many 3-card hands have all cards from the same suit?

SOLUTION There are 13 cards in each suit, so the number of 3-card hands having all hearts, for example, is

$${}_{13}C_3 = \frac{13!}{3!(13-3)!} = \frac{13!}{3!10!} = 286$$

Similarly, there are 286 3-card hands having all diamonds, 286 having all clubs, and 286 having all spades. So the total number of 3-card hands having all cards from the same suit is

$$4 \cdot {}_{13}C_3 = 1{,}144$$

Matched Problem 9 From a standard 52-card deck, how many 5-card hands have all cards from the same suit?

Exercises 7.4

Skills Warm-up Exercises

W *In Problems 1–6, evaluate the given expression without using a calculator. (If necessary, review Section A.4).*

1. $\dfrac{12 \cdot 11 \cdot 10}{3 \cdot 2 \cdot 1}$

2. $\dfrac{12 \cdot 10 \cdot 8}{6 \cdot 4 \cdot 2}$

3. $\dfrac{10 \cdot 9 \cdot 8 \cdot 7 \cdot 6}{5 \cdot 4 \cdot 3 \cdot 2 \cdot 1}$

4. $\dfrac{8 \cdot 7 \cdot 6 \cdot 5}{4 \cdot 3 \cdot 2 \cdot 1}$

5. $\dfrac{100 \cdot 99 \cdot 98 \cdot \ldots \cdot 3 \cdot 2 \cdot 1}{98 \cdot 97 \cdot 96 \cdot \ldots \cdot 3 \cdot 2 \cdot 1}$

6. $\dfrac{11 \cdot 10 \cdot 9 \cdot \ldots \cdot 3 \cdot 2 \cdot 1}{8 \cdot 7 \cdot 6 \cdot 5 \cdot 4 \cdot 3 \cdot 2 \cdot 1}$

A *In Problems 7–26, evaluate the expression. If the answer is not an integer, round to four decimal places.*

7. $8!$

8. $9!$

9. $(4 + 3)!$

10. $(7 + 3)!$

11. $(23 - 17)!$

12. $(52 - 47)!$

13. $\dfrac{11!}{8!}$

14. $\dfrac{20!}{18!}$

15. $\dfrac{8!}{4!(8-4)!}$

16. $\dfrac{10!}{5!(10-5)!}$

17. $\dfrac{500!}{498!}$

18. $\dfrac{601!}{599!}$

19. ${}_{13}C_8$

20. ${}_{15}C_{10}$

21. ${}_{18}P_6$

22. ${}_{10}P_7$

23. $\dfrac{{}_{12}P_7}{12^7}$

24. $\dfrac{{}_{365}P_{25}}{365^{25}}$

25. $\dfrac{{}_{39}C_5}{{}_{52}C_5}$

26. $\dfrac{{}_{26}C_4}{{}_{52}C_4}$

In Problems 27–30, simplify each expression assuming that n is an integer and $n \geq 2$.

27. $\dfrac{n!}{(n-2)!}$

28. $\dfrac{(n+1)!}{3!(n-2)!}$

29. $\dfrac{(n+1)!}{2!(n-1)!}$

30. $\dfrac{(n+3)!}{(n+1)!}$

In Problems 31–36, would you consider the selection to be a permutation, a combination, or neither? Explain your reasoning.

31. The university president named 3 new officers: a vice-president of finance, a vice-president of academic affairs, and a vice-president of student affairs.

32. The university president selected 2 of her vice-presidents to attend the dedication ceremony of a new branch campus.

33. A student checked out 4 novels from the library.

34. A student bought 4 books: 1 for his father, 1 for his mother, 1 for his younger sister, and 1 for his older brother.

35. A father ordered an ice cream cone (chocolate, vanilla, or strawberry) for each of his 4 children.

36. A book club meets monthly at the home of one of its 10 members. In December, the club selects a host for each meeting of the next year.

37. In a horse race, how many different finishes among the first 3 places are possible if 10 horses are running? (Exclude ties.)

38. In a long-distance foot race, how many different finishes among the first 5 places are possible if 50 people are running? (Exclude ties.)

39. How many ways can a 3-person subcommittee be selected from a committee of 7 people? How many ways can a president, vice-president, and secretary be chosen from a committee of 7 people?

40. Nine cards are numbered with the digits from 1 to 9. A 3-card hand is dealt, 1 card at a time. How many hands are possible in which

(A) Order is taken into consideration?

(B) Order is not taken into consideration?

B **41.** From a standard 52-card deck, how many 6-card hands consist entirely of red cards?

42. From a standard 52-card deck, how many 6-card hands consist entirely of clubs?

43. From a standard 52-card deck, how many 5-card hands consist entirely of face cards?

44. From a standard 52-card deck, how many 5-card hands consist entirely of queens?

45. From a standard 52-card deck, how many 7-card hands contain four kings?

46. From a standard 52-card deck, how many 7-card hands consist of 3 hearts and 4 diamonds?

47. From a standard 52-card deck, how many 4-card hands contain a card from each suit?

48. From a standard 52-card deck, how many 4-card hands consist of cards from the same suit?

49. From a standard 52-card deck, how many 5-card hands contain 3 cards of one rank and 2 cards of a different rank?

50. From a standard 52-card deck, how many 5-card hands contain 5 different ranks of cards?

51. A catering service offers 8 appetizers, 10 main courses, and 7 desserts. A banquet committee selects 3 appetizers, 4 main courses, and 2 desserts. How many ways can this be done?

52. Three departments have 12, 15, and 18 members, respectively. If each department selects a delegate and an alternate to represent the department at a conference, how many ways can this be done?

In Problems 53 and 54, refer to the table in the graphing calculator display below, which shows $y_1 = {}_nP_r$ and $y_2 = {}_nC_r$ for $n = 6$.

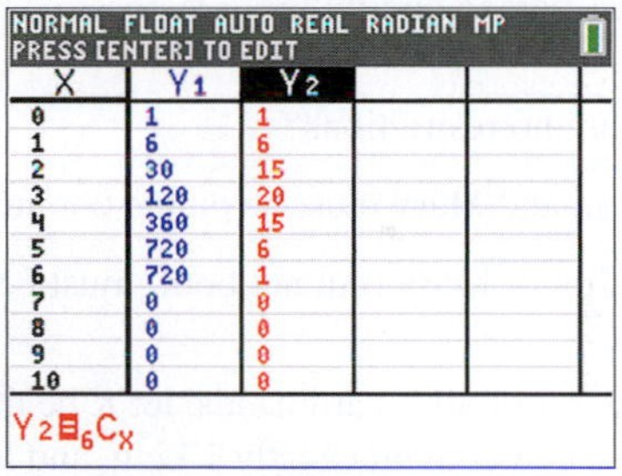

X	Y1	Y2
0	1	1
1	6	6
2	30	15
3	120	20
4	360	15
5	720	6
6	720	1
7	0	0
8	0	0
9	0	0
10	0	0

53. Discuss and explain the symmetry of the numbers in the y_2 column of the table.

54. Explain how the table illustrates the formula

$$ {}_nP_r = r!\,{}_nC_r $$

In Problems 55–60, discuss the validity of each statement. If the statement is true, explain why. If not, give a counterexample.

55. If n is a positive integer, then $n! < (n+1)!$

56. If n is a positive integer greater than 3, then $n! > 2^n$.

57. If n and r are positive integers and $1 < r < n$, then ${}_nP_r < {}_nP_{r+1}$.

58. If n and r are positive integers and $1 < r < n$, then ${}_nC_r < {}_nC_{r+1}$.

59. If n and r are positive integers and $1 < r < n$, then ${}_nC_r = {}_nC_{n-r}$.

60. If n and r are positive integers and $1 < r < n$, then ${}_nP_r = {}_nP_{n-r}$.

61. Eight distinct points are selected on the circumference of a circle.

(A) How many line segments can be drawn by joining the points in all possible ways?

(B) How many triangles can be drawn using these 8 points as vertices?

(C) How many quadrilaterals can be drawn using these 8 points as vertices?

62. Five distinct points are selected on the circumference of a circle.

(A) How many line segments can be drawn by joining the points in all possible ways?

(B) How many triangles can be drawn using these 5 points as vertices?

(C) How many quadrilaterals can be drawn using these 5 points as vertices?

63. In how many ways can 4 people sit in a row of 6 chairs?

64. In how many ways can 3 people sit in a row of 7 chairs?

65. A basketball team has 5 distinct positions. Out of 8 players, how many starting teams are possible if

(A) The distinct positions are taken into consideration?

(B) The distinct positions are not taken into consideration?

(C) The distinct positions are not taken into consideration, but either Mike or Ken (but not both) must start?

66. How many 4-person committees are possible from a group of 9 people if

(A) There are no restrictions?

(B) Both Jim and Mary must be on the committee?

(C) Either Jim or Mary (but not both) must be on the committee?

67. Let U be the set of all 2-card hands, let K be the set of all 2-card hands that contain exactly 1 king, and let H be the set of all 2-card hands that contain exactly 1 heart. Find $n(K \cap H')$, $n(K \cap H)$, $n(K' \cap H)$, and $n(K' \cap H')$.

68. Let U be the set of all 2-card hands, let K be the set of all 2-card hands that contain exactly 1 king, and let Q be the set of all 2-card hands that contain exactly 1 queen. Find $n(K \cap Q')$, $n(K \cap Q)$, $n(K' \cap Q)$, and $n(K' \cap Q')$.

69. Note from the table in the graphing calculator display below that the largest value of ${}_nC_r$ when $n = 20$ is ${}_{20}C_{10} = 184{,}756$. Use a similar table to find the largest value of ${}_nC_r$ when $n = 24$.

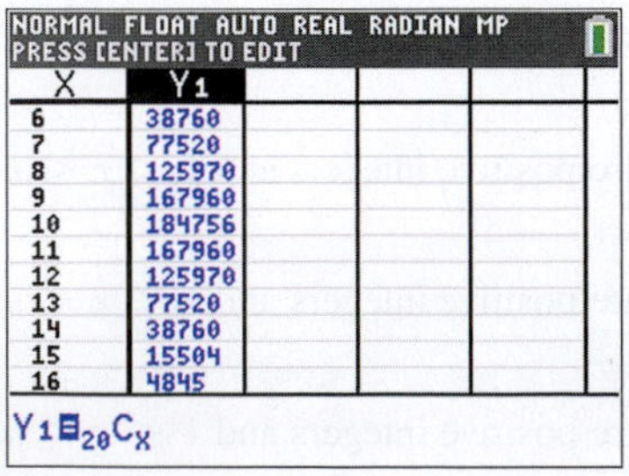
NORMAL FLOAT AUTO REAL RADIAN MP
PRESS [ENTER] TO EDIT

X	Y1
6	38760
7	77520
8	125970
9	167960
10	184756
11	167960
12	125970
13	77520
14	38760
15	15504
16	4845

$Y_1 = {}_{20}C_X$

70. Note from the table in the graphing calculator display that the largest value of ${}_nC_r$ when $n = 21$ is ${}_{21}C_{10} = {}_{21}C_{11} = 352{,}716$. Use a similar table to find the largest value of ${}_nC_r$ when $n = 17$.

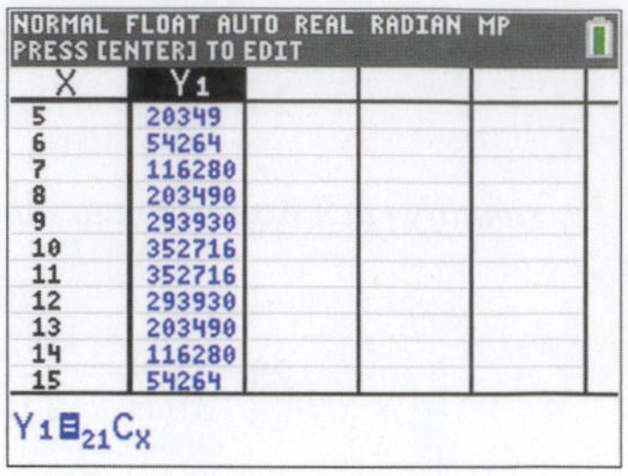
NORMAL FLOAT AUTO REAL RADIAN MP
PRESS [ENTER] TO EDIT

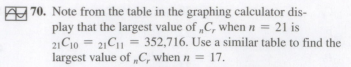

X	Y1
5	20349
6	54264
7	116280
8	203490
9	293930
10	352716
11	352716
12	293930
13	203490
14	116280
15	54264

$Y_1 = {}_{21}C_X$

Applications

71. Quality control. An office supply store receives a shipment of 24 high-speed printers, including 5 that are defective. Three of these printers are selected for a store display.

(A) How many selections can be made?

(B) How many of these selections will contain no defective printers?

72. Quality control. An electronics store receives a shipment of 30 graphing calculators, including 6 that are defective. Four of these calculators are selected for a local high school.

(A) How many selections can be made?

(B) How many of these selections will contain no defective calculators?

73. Business closings. A jewelry store chain with 8 stores in Georgia, 12 in Florida, and 10 in Alabama is planning to close 10 of these stores.

(A) How many ways can this be done?

(B) The company decides to close 2 stores in Georgia, 5 in Florida, and 3 in Alabama. In how many ways can this be done?

74. Employee layoffs. A real estate company with 14 employees in its central office, 8 in its north office, and 6 in its south office is planning to lay off 12 employees.

(A) How many ways can this be done?

(B) The company decides to lay off 5 employees from the central office, 4 from the north office, and 3 from the south office. In how many ways can this be done?

75. Personnel selection. Suppose that 6 female and 5 male applicants have been successfully screened for 5 positions. In how many ways can the following compositions be selected?

(A) 3 females and 2 males

(B) 4 females and 1 male

(C) 5 females

(D) 5 people regardless of sex

(E) At least 4 females

76. Committee selection. A 4-person grievance committee is selected out of 2 departments A and B, with 15 and 20 people, respectively. In how many ways can the following committees be selected?

(A) 3 from A and 1 from B

(B) 2 from A and 2 from B

(C) All from A

(D) 4 people regardless of department

(E) At least 3 from department A

77. Medicine. There are 8 standard classifications of blood type. An examination for prospective laboratory technicians consists of having each candidate determine the type for 3 blood samples. How many different examinations can be given if no 2 of the samples provided for the candidate have the same type? If 2 or more samples have the same type?

78. Medical research. Because of limited funds, 5 research centers are chosen out of 8 suitable ones for a study on heart disease. How many choices are possible?

79. Politics. A nominating convention will select a president and vice-president from among 4 candidates. Campaign buttons, listing a president and a vice-president, will be designed for each possible outcome before the convention. How many different kinds of buttons should be designed?

80. Politics. In how many different ways can 6 candidates for an office be listed on a ballot?

Answers to Matched Problems

1. (A) 720 (B) 10 (C) 720 (D) 20 (E) 1,140

2. (A)

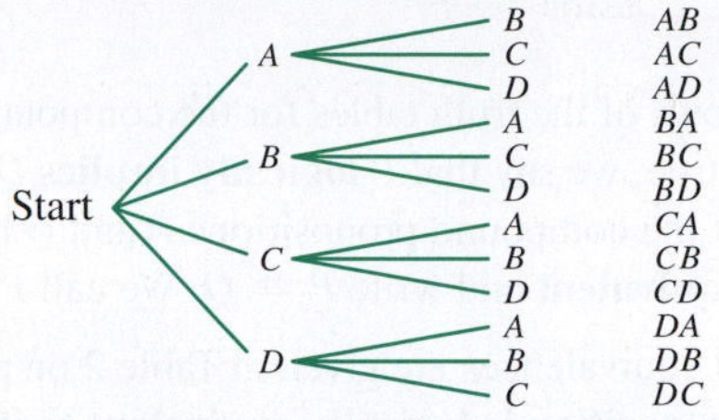

12 permutations of 4 objects taken 2 at a time

(B) O_1: Fill first position N_1: 4 ways
O_2: Fill second position N_2: 3 ways
$4 \cdot 3 = 12$

(C) ${}_4P_2 = 4 \cdot 3 = 12$; ${}_4P_2 = \dfrac{4!}{(4-2)!} = 12$

3. ${}_{30}P_4 = \dfrac{30!}{(30-4)!} = 657{,}720$

4. (A) ${}_{12}P_4 = \dfrac{12!}{(12-4)!} = 11{,}880$ ways

(B) ${}_{12}C_4 = \dfrac{12!}{4!(12-4)!} = 495$ ways

5. ${}_{30}C_4 = \dfrac{30!}{4!(30-4)!} = 27{,}405$

6. ${}_{13}C_3 \cdot {}_{13}C_2 = 22{,}308$

7. ${}_8P_3 \cdot {}_{10}P_2 = 30{,}240$

8. (A) ${}_7C_1 \cdot {}_5C_3 = 70$ (B) ${}_5C_4 = 5$

(C) ${}_7C_2 \cdot {}_5C_2 + {}_7C_1 \cdot {}_5C_3 + {}_5C_4 = 285$

9. $4 \cdot {}_{13}C_5 = 5{,}148$

Chapter 7 Summary and Review

Important Terms, Symbols, and Concepts

7.1 Logic

EXAMPLES

- A **proposition** is a statement (not a question or command) that is either true or false.
- If p and q are propositions, then the compound propositions, Ex. 1, p. 354

$$\neg p, p \vee q, p \wedge q, \text{ and } p \rightarrow q$$

can be formed using the negation symbol $\neg$ and the connectives $\vee$, $\wedge$, and $\rightarrow$. These propositions are called **not *p*, *p* or *q*, *p* and *q***, and **if *p* then *q***, respectively (or **negation, disjunction, conjunction**, and **conditional**, respectively). Each of these compound propositions is specified by a truth table (see pages 352 and 353).

- Given any conditional proposition $p \rightarrow q$, the proposition $q \rightarrow p$ is called the **converse** of $p \rightarrow q$, and the proposition $\neg q \rightarrow \neg p$, is called the **contrapositive** of $p \rightarrow q$. Ex. 2, p. 354

- A **truth table** for a compound proposition specifies whether it is true or false for any assignment of truth values to its variables. A proposition is a **tautology** if each entry in its column of the truth table is T, a **contradiction** if each entry is F, and a **contingency** if at least one entry is T and at least one entry is F. Ex. 3, p. 355 Ex. 4, p. 355 Ex. 5, p. 356
- Consider the rows of the truth tables for the compound propositions P and Q. If whenever P is true, Q is also true, we say that P **logically implies** Q and write $P \Rightarrow Q$. We call $P \Rightarrow Q$ a **logical implication**. If the compound propositions P and Q have identical truth tables, we say that P and Q are **logically equivalent** and write $P \equiv Q$. We call $P \equiv Q$ a **logical equivalence**. Ex. 6, p. 357 Ex. 7, p. 358
- Several logical equivalences are given in Table 2 on page 358. The last of these implies that **any conditional proposition is logically equivalent to its contrapositive**.

7.2 Sets

- A **set** is a collection of objects specified in such a way that we can tell whether any given object is or is not in the collection.
- Each object in a set is called a **member**, or **element**, of the set. If a is an element of the set A, we write $a \in A$.
- A set without any elements is called the **empty**, or **null**, set, denoted by $\varnothing$.
- A set can be described by listing its elements, or by giving a rule that determines the elements of the set. If $P(x)$ is a statement about x, then $\{x|P(x)\}$ denotes the set of all x such that $P(x)$ is true. Ex. 1, p. 361
- A set is **finite** if its elements can be counted and there is an end; a set such as the positive integers, in which there is no end in counting its elements, is **infinite**.
- We write $A \subset B$, and say that A is a **subset** of B, if each element of A is an element of B. We write $A = B$, and say that sets A and B are equal, if they have exactly the same elements. The empty set $\varnothing$ is a subset of every set. Ex. 2, p. 362 Ex. 3, p. 362
- If A and B are sets, then Ex. 4, p. 364

$$A \cup B = \{x|x \in A \text{ or } x \in B\}$$

is called the **union** of A and B, and

$$A \cap B = \{x|x \in A \text{ and } x \in B\}$$

is called the **intersection** of A and B.

- **Venn diagrams** are useful in visualizing set relationships. Ex. 5, p. 364
- If $A \cap B = \varnothing$, the sets A and B are said to be **disjoint**.
- The set of all elements under consideration in a given discussion is called the **universal set** U. The set $A' = \{x \in U | x \notin A\}$ is called the **complement** of A (relative to U).
- The **number of elements** in set A is denoted by $n(A)$. So if A and B are sets, then the numbers that are often shown in a Venn diagram, as in Figure 4 on page 363, are $n(A \cap B')$, $n(A \cap B)$, $n(B \cap A')$, and $n(A' \cap B')$ (see Fig. 9 on page 364).

7.3 Basic Counting Principles

- If A and B are sets, then the number of elements in the union of A and B is given by the **addition principle** for counting (Theorem 1, page 369). Ex. 1, p. 369 Ex. 2, p. 370
- If the elements of a set are determined by a sequence of operations, tree diagrams can be used to list all combined outcomes. To count the number of combined outcomes without using a tree diagram, use the **multiplication principle** for counting (Theorem 2, page 372). Ex. 3, p. 371 Ex. 4, p. 373 Ex. 5, p. 373

7.4 Permutations and Combinations

- The product of the first n natural numbers, denoted $n!$, is called ***n* factorial**: Ex. 1, p. 377

$$n! = n(n-1)(n-2)\cdot \cdots \cdot 2 \cdot 1$$

$$0! = 1$$

$$n! = n \cdot (n-1)!$$

- A **permutation** of a set of distinct objects is an arrangement of the objects in a specific order without repetition. The number of permutations of a set of n distinct objects is given by ${}_nP_n = n!$. A permutation of a set of n distinct objects taken r at a time without repetition is an arrangement of r of the n objects in a specific order. The number of permutations of n distinct objects taken r at a time without repetition is given by

$$ {}_nP_r = \frac{n!}{(n-r)!} \qquad 0 \le r \le n $$

Ex. 2, p. 379
Ex. 3, p. 380

- A **combination** of a set of n distinct objects taken r at a time without repetition is an r-element subset of the set of n objects. The arrangement of the elements in the subset is irrelevant. The number of combinations of n distinct objects taken r at a time without repetition is given by

$$ {}_nC_r = \binom{n}{r} = \frac{{}_nP_r}{r!} = \frac{n!}{r!(n-r)!} \qquad 0 \le r \le n $$

Ex. 4, p. 382
Ex. 5, p. 383
Ex. 6, p. 384
Ex. 7, p. 385
Ex. 8, p. 385
Ex. 9, p. 386

Review Exercises

Work through all the problems in this chapter review and check your answers in the back of the book. Answers to all problems are there along with section numbers in italics to indicate where each type of problem is discussed. Where weaknesses show up, review appropriate sections in the text.

In Problems 1–6, express each proposition in an English sentence and determine whether it is true or false, where p and q are the propositions

$$ p: \text{“}2^3 < 3^2\text{”} \qquad q: \text{“}3^4 < 4^3\text{”} $$

1. $\neg q$

2. $p \lor q$

3. $p \land q$

4. $p \to q$

5. The converse of $p \to q$

6. The contrapositive of $p \to q$

In Problems 7–10, indicate true (T) or false (F).

7. $\{5, 6, 7\} = \{6, 7, 5\}$

8. $5 \in \{55, 555\}$

9. $\{9, 27\} \subset \{3, 9, 27, 81\}$

10. $\{1, 2\} \subset \{1, \{1, 2\}\}$

In Problems 11–14, describe each proposition as a negation, disjunction, conjunction, or conditional, and determine whether the proposition is true or false.

11. If 9 is prime, then 10 is odd.

12. 7 is even or 8 is odd.

13. 53 is prime and 57 is prime.

14. 51 is not prime.

In Problems 15–16, state the converse and the contrapositive of the given proposition.

15. If the square matrix A has a row of zeros, then the square matrix A has no inverse.

16. If the square matrix A is an identity matrix, then the square matrix A has an inverse.

In Problems 17–19, write the resulting set using the listing method.

17. $\{1, 2, 3, 4\} \cup \{2, 3, 4, 5\}$

18. $\{1, 2, 3, 4\} \cap \{2, 3, 4, 5\}$

19. $\{1, 2, 3, 4\} \cap \{5, 6\}$

20. A single die is rolled, and a coin is flipped. How many combined outcomes are possible? Solve:

(A) Using a tree diagram

(B) Using the multiplication principle

21. Use the Venn diagram to find the number of elements in each of the following sets:

(A) A (B) B (C) U (D) A'

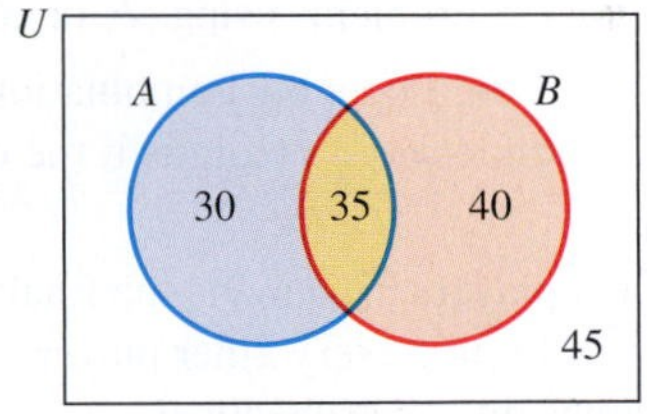

Figure for Problems 21 and 22

22. Use the Venn diagram to find the number of elements in each of the following sets:

(A) $A \cap B$ (B) $A \cup B$

(C) $(A \cap B)'$ (D) $(A \cup B)'$

Evaluate the expressions in Problems 23–28.

23. $(10 - 6)!$

24. $\dfrac{15!}{10!}$

25. $\dfrac{15!}{10!5!}$

26. ${}_8C_5$

27. ${}_8P_5$

28. ${}_{13}C_4 \cdot {}_{13}C_1$

29. How many seating arrangements are possible with 6 people and 6 chairs in a row? Solve using the multiplication principle.

30. Solve Problem 29 using permutations or combinations, whichever is applicable.

In Problems 31–36, construct a truth table for the proposition and determine whether the proposition is a contingency, tautology, or contradiction.

31. $(p \rightarrow q) \wedge (q \rightarrow p)$

32. $p \vee (q \rightarrow p)$

33. $(p \vee \neg p) \rightarrow (q \wedge \neg q)$

34. $\neg q \wedge (p \rightarrow q)$

35. $\neg p \rightarrow (p \rightarrow q)$

36. $\neg(p \vee \neg q)$

In Problems 37–40, determine whether the given set is finite or infinite. Consider the set Z of integers to be the universal set, and let

$$M = \{n \in Z \mid n < 10^6\}$$

$$K = \{n \in Z \mid n > 10^3\}$$

$$E = \{n \in Z \mid n \text{ is even}\}$$

37. $E \cup K$

38. $M \cap K$

39. K'

40. $E \cap M$

In Problems 41–42, determine whether or not the given sets are disjoint. For the definitions of M, K, and E, refer to the instructions for Problems 37–40.

41. M' and K'

42. M and E'

43. Draw a Venn diagram for sets A, B, and C and shade the region described by $A' \cap (B \cup C)$.

44. A man has 5 children. Each of those children has 3 children, who in turn each have 2 children. Discuss the number of descendants that the man has.

45. How many 3-letter code words are possible using the first 8 letters of the alphabet if no letter can be repeated? If letters can be repeated? If adjacent letters cannot be alike?

46. Solve the following problems using ${}_nP_r$ or ${}_nC_r$:

(A) How many 3-digit opening combinations are possible on a combination lock with 6 digits if the digits cannot be repeated?

(B) Five tennis players have made the finals. If each of the 5 players is to play every other player exactly once, how many games must be scheduled?

47. Use graphical techniques on a graphing calculator to find the largest value of ${}_nC_r$ when $n = 25$.

48. If 3 operations O_1, O_2, O_3 are performed in order, with possible number of outcomes N_1, N_2, N_3, respectively, determine the number of branches in the corresponding tree diagram.

In Problems 49–51, write the resulting set using the listing method.

49. $\{x \mid x^3 - x = 0\}$

50. $\{x \mid x \text{ is a positive integer and } x! < 100\}$

51. $\{x \mid x \text{ is a positive integer that is a perfect square and } x < 50\}$

52. A software development department consists of 6 women and 4 men.

(A) How many ways can the department select a chief programmer, a backup programmer, and a programming librarian?

(B) How many of the selections in part (A) consist entirely of women?

(C) How many ways can the department select a team of 3 programmers to work on a particular project?

53. A group of 150 people includes 52 who play chess, 93 who play checkers, and 28 who play both chess and checkers. How many people in the group play neither game?

Problems 54 and 55 refer to the following Venn diagram.

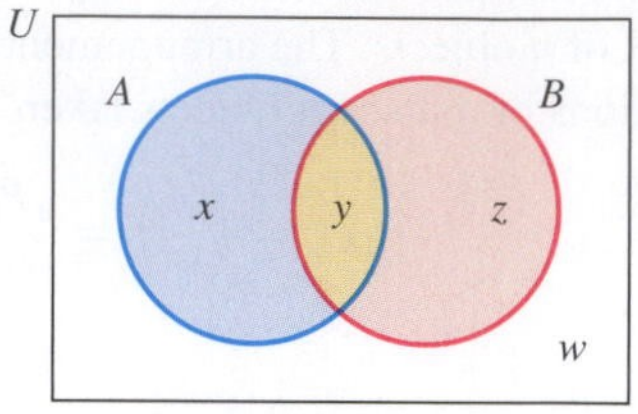

54. Which of the numbers x, y, z, or w must equal 0 if $A \subset B$?

55. Which of the numbers x, y, z, or w must equal 0 if $A \cap B = U$?

In Problems 56–58, discuss the validity of each statement. If the statement is true, explain why. If not, give a counterexample.

56. If n and r are positive integers and $1 < r < n$, then ${}_nC_r < {}_nP_r$.

57. If n and r are positive integers and $1 < r < n$, then ${}_nP_r < n!$.

58. If n and r are positive integers and $1 < r < n$, then ${}_nC_r < n!$.

In Problems 59–64, construct a truth table to verify the implication or equivalence.

59. $p \wedge q \Rightarrow p$

60. $q \Rightarrow p \rightarrow q$

61. $\neg p \rightarrow (q \wedge \neg q) \equiv p$

62. $p \vee q \equiv \neg p \rightarrow q$

63. $p \wedge (p \rightarrow q) \Rightarrow q$

64. $\neg(p \wedge \neg q) \equiv p \rightarrow q$

65. How many different 5-child families are possible where the gender of the children in the order of their births is taken into consideration [that is, birth sequences such as (B, G, G, B, B) and (G, B, G, B, B) produce different families]? How many families are possible if the order pattern is not taken into account?

66. Can a selection of r objects from a set of n distinct objects, where n is a positive integer, be a combination and a permutation simultaneously? Explain.

Applications

67. Transportation. A distribution center A wishes to send its products to five different retail stores: B, C, D, E, and F. How many different route plans can be constructed so that a single truck, starting from A, will deliver to each store exactly once and then return to the center?

68. Market research. A survey of 1,000 people indicates that 340 have invested in stocks, 480 have invested in bonds, and 210 have invested in stocks and bonds.

(A) How many people in the survey have invested in stocks or bonds?

(B) How many have invested in neither stocks nor bonds?

(C) How many have invested in bonds and not stocks?

69. Medical research. In a study of twins, a sample of 6 pairs of identical twins will be selected for medical tests from a group of 40 pairs of identical twins. In how many ways can this be done?

70. Elections. In an unusual recall election, there are 67 candidates to replace the governor of a state. To negate the advantage that might accrue to candidates whose names appear near the top of the ballot, it is proposed that equal numbers of ballots be printed for each possible order in which the candidates' names can be listed.

(A) In how many ways can the candidates' names be listed?

(B) Explain why the proposal is not feasible, and discuss possible alternatives.

8 Probability

Introduction

Like other branches of mathematics, probability evolved out of practical considerations. Girolamo Cardano (1501–1576), a gambler and physician, produced some of the best mathematics of his time, including a systematic analysis of gambling problems. In 1654, another gambler, Chevalier de Méré, approached the well-known French philosopher and mathematician Blaise Pascal (1623–1662) regarding certain dice problems. Pascal became interested in these problems, studied them, and discussed them with Pierre de Fermat (1601–1665), another French mathematician. So out of the gaming rooms of western Europe, the study of probability was born.

Despite this lowly birth, probability has matured into a highly respected and immensely useful branch of mathematics. It is used in practically every field. In particular, probability plays a critical role in the management of risk, from insuring a painting against theft (see Problem 49 in Section 8.5) to diversifying an investment portfolio to designing a health care system.

8.1 Sample Spaces, Events, and Probability

- Experiments
- Sample Spaces and Events
- Probability of an Event
- Equally Likely Assumption

This section provides a brief and relatively informal introduction to probability. It assumes a familiarity with the basics of set theory as presented in Chapter 7, including the union, intersection, and complement of sets, as well as various techniques for counting the number of elements in a set. Probability studies involve many subtle ideas, and care must be taken at the beginning to understand the fundamental concepts.

Experiments

Some experiments do not yield the same results each time that they are performed, no matter how carefully they are repeated under the same conditions. These experiments are called **random experiments**. Familiar examples of random experiments are flipping coins, rolling dice, observing the frequency of defective items from an assembly line, or observing the frequency of deaths in a certain age group.

Probability theory is a branch of mathematics that has been developed to deal with outcomes of random experiments, both real and conceptual. In the work that follows, we simply use the word **experiment** to mean a random experiment.

Figure 1

Sample Spaces and Events

Associated with outcomes of experiments are *sample spaces* and *events*. Consider the experiment, "A wheel with 18 numbers on the perimeter (Fig. 1) spins and comes to rest so that a pointer points within a numbered sector."

What outcomes might we observe? When the wheel stops, we might be interested in which number is next to the pointer, or whether that number is an odd number, or whether that number is divisible by 5, or whether that number is prime, or whether the pointer is in a red or green sector, and so on. The list of possible outcomes appears endless. In general, there is no unique method of analyzing all possible outcomes of an experiment. Therefore, before conducting an experiment, it is important to decide just what outcomes are of interest.

Suppose we limit our interest to the set of numbers on the wheel and to various subsets of these numbers, such as the set of prime numbers or the set of odd numbers on the wheel. Having decided what to observe, we make a list of outcomes of the experiment, called *simple outcomes* or *simple events*, such that in each trial of the experiment (each spin of the wheel), one and only one of the outcomes on the list will occur. For our stated interests, we choose each number on the wheel as a simple event and form the set

$$S = \{1, 2, 3, \ldots, 17, 18\}$$

The set of simple events S for the experiment is called a *sample space* for the experiment.

Now consider the outcome, "When the wheel comes to rest, the number next to the pointer is divisible by 4." This outcome is not a simple outcome (or simple event) since it is not associated with one and only one element in the sample space S. The outcome will occur whenever any one of the simple events 4, 8, 12, or 16 occurs, that is, whenever an element in the subset

$$E = \{4, 8, 12, 16\}$$

occurs. Subset E is called a *compound event* (and the outcome, a *compound outcome*).

DEFINITION Sample Spaces and Events

If we formulate a set S of outcomes (events) of an experiment in such a way that in each trial of the experiment one and only one of the outcomes (events) in the set will occur, we call the set S a **sample space** for the experiment. Each element in S is called a **simple outcome**, or **simple event**.

An **event** E is defined to be any subset of S (including the empty set $\varnothing$ and the sample space S). Event E is a **simple event** if it contains only one element and a **compound event** if it contains more than one element. We say that **an event** E **occurs** if any of the simple events in E occurs.

We use the terms *event* and *outcome of an experiment* interchangeably. Technically, an event is the mathematical counterpart of an outcome of an experiment, but we will not insist on strict adherence to this distinction in our development of probability.

Real World	Mathematical Model
Experiment (real or conceptual)	Sample space (set S)
Outcome (simple or compound)	Event (subset of S; simple or compound)

EXAMPLE 1 **Simple and Compound Events** Relative to the number wheel experiment (Fig. 1) and the sample space

$$S = \{1, 2, 3, \ldots, 17, 18\}$$

what is the event E (subset of the sample space S) that corresponds to each of the following outcomes? Indicate whether the event is a simple event or a compound event.

(A) The outcome is a prime number. (B) The outcome is the square of 4.

SOLUTION

(A) The outcome is a prime number if any of the simple events 2, 3, 5, 7, 11, 13, or 17 occurs.* To say "A prime number occurs" is the same as saying that the experiment has an outcome in the set

$$E = \{2, 3, 5, 7, 11, 13, 17\}$$

Since event E has more than one element, it is a compound event.

(B) The outcome is the square of 4 if 16 occurs. To say "The square of 4 occurs" is the same as saying that the experiment has an outcome in the set

$$E = \{16\}$$

Since E has only one element, it is a simple event.

Matched Problem 1 Repeat Example 1 for

(A) The outcome is a number divisible by 12.

(B) The outcome is an even number greater than 15.

*Technically, we should write $\{2\}$, $\{3\}$, $\{5\}$, $\{7\}$, $\{11\}$, $\{13\}$, and $\{17\}$ for the simple events since there is a logical distinction between an element of a set and a subset consisting of only that element. But we will keep this in mind and drop the braces for simple events to simplify the notation.

EXAMPLE 2 **Sample Spaces** A nickel and a dime are tossed. How do we identify a sample space for this experiment? There are a number of possibilities, depending on our interest. We will consider three.

(A) If we are interested in whether each coin falls heads (H) or tails (T), then, using a tree diagram, we can easily determine an appropriate sample space for the experiment:

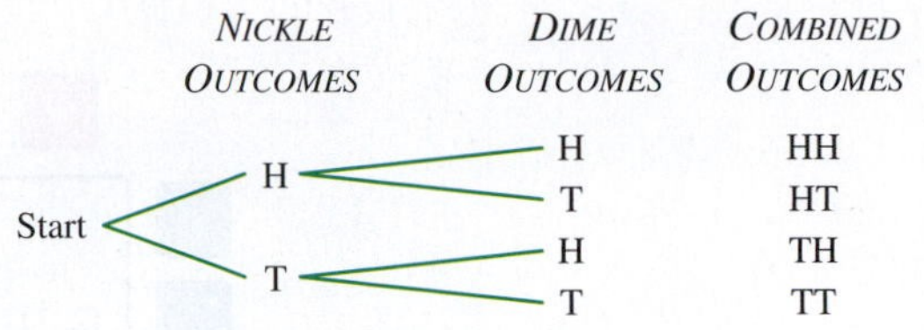

$$S_1 = \{HH, HT, TH, TT\}$$

and there are 4 simple events in the sample space.

(B) If we are interested only in the number of heads that appear on a single toss of the two coins, we can let

$$S_2 = \{0, 1, 2\}$$

and there are 3 simple events in the sample space.

(C) If we are interested in whether the coins match (*M*) or do not match (*D*), we can let

$$S_3 = \{M, D\}$$

and there are only 2 simple events in the sample space.

CONCEPTUAL INSIGHT

There is no single correct sample space for a given experiment. When specifying a sample space for an experiment, we include as much detail as necessary to answer all questions of interest regarding the outcomes of the experiment. If in doubt, we choose a sample space that contains more elements rather than fewer.

In Example 2, which sample space would be appropriate for all three interests? Sample space S_1 contains more information than either S_2 or S_3. If we know which outcome has occurred in S_1, then we know which outcome has occurred in S_2 and S_3. However, the reverse is not true. (Note that the simple events in S_2 and S_3 are simple or compound events in S_1.) In this sense, we say that S_1 is a more **fundamental sample space** than either S_2 or S_3. Thus, we would choose S_1 as an appropriate sample space for all three expressed interests.

Matched Problem 2 An experiment consists of recording the boy–girl composition of a two-child family. What would be an appropriate sample space

(A) If we are interested in the genders of the children in the order of their births? Draw a tree diagram.

(B) If we are interested only in the number of girls in a family?

(C) If we are interested only in whether the genders are alike (*A*) or different (*D*)?

(D) For all three interests expressed in parts (*A*) to (*C*)?

EXAMPLE 3 **Sample Spaces and Events** Consider an experiment of rolling two dice. Figure 2 shows a convenient sample space that will enable us to answer many questions about interesting events. Let S be the set of all ordered pairs in the figure. The simple event $(3, 2)$ is distinguished from the simple event $(2, 3)$. The former indicates that a 3 turned up on the first die and a 2 on the second, while the latter indicates that a 2 turned up on the first die and a 3 on the second.

Second Die

	(1, 1)	(1, 2)	(1, 3)	(1, 4)	(1, 5)	(1, 6)
	(2, 1)	(2, 2)	(2, 3)	(2, 4)	(2, 5)	(2, 6)
First Die	(3, 1)	(3, 2)	(3, 3)	(3, 4)	(3, 5)	(3, 6)
	(4, 1)	(4, 2)	(4, 3)	(4, 4)	(4, 5)	(4, 6)
	(5, 1)	(5, 2)	(5, 3)	(5, 4)	(5, 5)	(5, 6)
	(6, 1)	(6, 2)	(6, 3)	(6, 4)	(6, 5)	(6, 6)

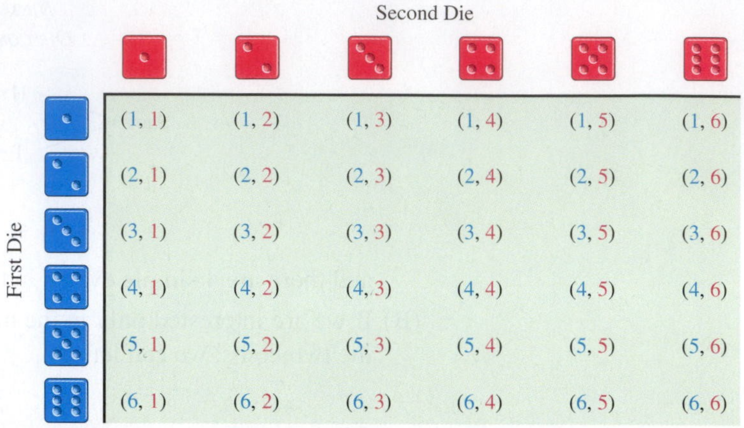

Figure 2

What is the event (subset of the sample space S) that corresponds to each of the following outcomes?

(A) A sum of 7 turns up. (B) A sum of 11 turns up.

(C) A sum less than 4 turns up. (D) A sum of 12 turns up.

SOLUTION

(A) By "A sum of 7 turns up," we mean that the sum of all dots on both turned-up faces is 7. This outcome corresponds to the event

$$\{(6, 1), (5, 2), (4, 3), (3, 4), (2, 5), (1, 6)\}$$

(B) "A sum of 11 turns up" corresponds to the event

$$\{(6, 5), (5, 6)\}$$

(C) "A sum less than 4 turns up" corresponds to the event

$$\{(1, 1), (2, 1), (1, 2)\}$$

(D) "A sum of 12 turns up" corresponds to the event

$$\{(6, 6)\}$$

Matched Problem 3 Refer to the sample space shown in Figure 2. What is the event that corresponds to each of the following outcomes?

(A) A sum of 5 turns up.

(B) A sum that is a prime number greater than 7 turns up.

As indicated earlier, we often use the terms *event* and *outcome of an experiment* interchangeably. In Example 3, we might say "the event, 'A sum of 11 turns up'" in place of "the outcome, 'A sum of 11 turns up,'" or even write

$$E = \text{a sum of 11 turns up} = \{(6, 5), (5, 6)\}$$

Probability of an Event

The next step in developing our mathematical model for probability studies is the introduction of a *probability function*. This is a function that assigns to an arbitrary event associated with a sample space a real number between 0 and 1, inclusive. We start by discussing ways in which probabilities are assigned to simple events in the sample space S.

DEFINITION Probabilities for Simple Events

Given a sample space

$$S = \{e_1, e_2, \ldots, e_n\}$$

with n simple events, to each simple event e_i we assign a real number, denoted by $P(e_i)$, called the **probability of the event e_i**. These numbers can be assigned in an arbitrary manner as long as the following two conditions are satisfied:

Condition 1. The probability of a simple event is a number between 0 and 1, inclusive. That is,

$$0 \le P(e_i) \le 1$$

Condition 2. The sum of the probabilities of all simple events in the sample space is 1. That is,

$$P(e_1) + P(e_2) + \cdots + P(e_n) = 1$$

Any probability assignment that satisfies Conditions 1 and 2 is said to be an **acceptable probability assignment**.

Our mathematical theory does not explain how acceptable probabilities are assigned to simple events. These assignments are generally based on the expected or actual percentage of times that a simple event occurs when an experiment is repeated a large number of times. Assignments based on this principle are called **reasonable**.

Let an experiment be the flipping of a single coin, and let us choose a sample space S to be

$$S = \{\text{H}, \text{T}\}$$

If a coin appears to be fair, we are inclined to assign probabilities to the simple events in S as follows:

$$P(\text{H}) = \frac{1}{2} \quad \text{and} \quad P(\text{T}) = \frac{1}{2}$$

These assignments are based on reasoning that, since there are 2 ways a coin can land, in the long run, a head will turn up half the time and a tail will turn up half the time. These probability assignments are acceptable since both conditions for acceptable probability assignments stated in the preceding box are satisfied:

1. $0 \le P(\text{H}) \le 1, \quad 0 \le P(\text{T}) \le 1$
2. $P(\text{H}) + P(\text{T}) = \frac{1}{2} + \frac{1}{2} = 1$

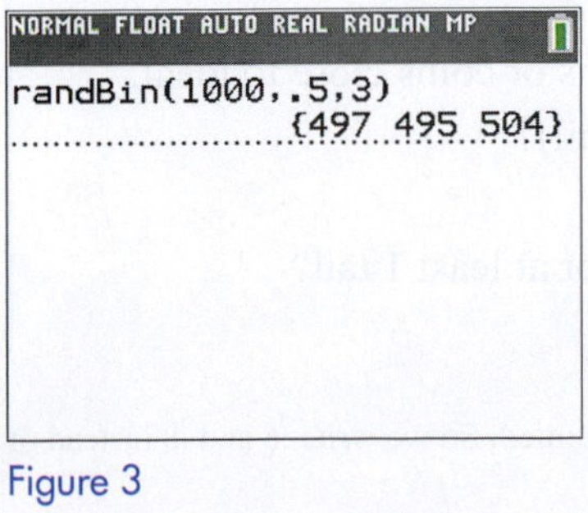

Figure 3

If we were to flip a coin 1,000 times, we would expect a head to turn up approximately, but not exactly, 500 times. The random number feature on a graphing calculator can be used to simulate 1,000 flips of a coin. Figure 3 shows the results of 3 such simulations: 497 heads the first time, 495 heads the second, and 504 heads the third.

If, however, we get only 376 heads in 1,000 flips of a coin, we might suspect that the coin is not fair. Then we might assign the simple events in the sample space S the following probabilities, based on our experimental results:

$$P(\text{H}) = .376 \quad \text{and} \quad P(\text{T}) = .624$$

This is also an acceptable assignment. However, the probability assignment

$$P(\text{H}) = 1 \quad \text{and} \quad P(\text{T}) = 0$$

although acceptable, is not reasonable (unless the coin has 2 heads). And the assignment

$$P(\text{H}) = .6 \quad \text{and} \quad P(\text{T}) = .8$$

is not acceptable, since $.6 + .8 = 1.4$, which violates Condition 2 in the box on page 399.*

It is important to keep in mind that out of the infinitely many possible acceptable probability assignments to simple events in a sample space, we are generally inclined to choose one assignment over another based on reasoning or experimental results.

Given an acceptable probability assignment for simple events in a sample space S, how do we define the probability of an arbitrary event E associated with S?

DEFINITION Probability of an Event *E*

Given an acceptable probability assignment for the simple events in a sample space S, we define the **probability of an arbitrary event *E*, denoted by $P(E)$**, as follows:

(A) If E is the empty set, then $P(E) = 0$.
(B) If E is a simple event, then $P(E)$ has already been assigned.
(C) If E is a compound event, then $P(E)$ is the sum of the probabilities of all the simple events in E.
(D) If E is the sample space S, then $P(E) = P(S) = 1$ [this is a special case of part (C)].

EXAMPLE 4 **Probabilities of Events** Let us return to Example 2, the tossing of a nickel and a dime, and the sample space

$$S = \{\text{HH, HT, TH, TT}\}$$

Since there are 4 simple outcomes and the coins are assumed to be fair, it would appear that each outcome would occur 25% of the time, in the long run. Let us assign the same probability of $\frac{1}{4}$ to each simple event in S:

Simple Event e_i	HH	HT	TH	TT
$P(e_i)$	$\frac{1}{4}$	$\frac{1}{4}$	$\frac{1}{4}$	$\frac{1}{4}$

This is an acceptable assignment according to Conditions 1 and 2, and it is a reasonable assignment for ideal (perfectly balanced) coins or coins close to ideal.

(A) What is the probability of getting 1 head (and 1 tail)?
(B) What is the probability of getting at least 1 head?
(C) What is the probability of getting at least 1 head or at least 1 tail?
(D) What is the probability of getting 3 heads?

*In probability studies, the 0 to the left of the decimal is usually omitted. So we write .6 and .8 instead of 0.6 and 0.8.

SOLUTION

(A) E_1 = getting 1 head = {HT, TH}

Since E_1 is a compound event, we use part (C) in the box and find $P(E_1)$ by adding the probabilities of the simple events in E_1:

$$P(E_1) = P(\text{HT}) + P(\text{TH}) = \frac{1}{4} + \frac{1}{4} = \frac{1}{2}$$

(B) E_2 = getting at least 1 head = {HH, HT, TH}

$$P(E_2) = P(\text{HH}) + P(\text{HT}) + P(\text{TH}) = \frac{1}{4} + \frac{1}{4} + \frac{1}{4} = \frac{3}{4}$$

(C) E_3 = {HH, HT, TH, TT} = S

$$P(E_3) = P(S) = 1 \quad \tfrac{1}{4} + \tfrac{1}{4} + \tfrac{1}{4} + \tfrac{1}{4} = 1$$

(D) E_4 = getting 3 heads = $\varnothing$ Empty set

$$P(\varnothing) = 0$$

PROCEDURE Steps for Finding the Probability of an Event *E*

Step 1 Set up an appropriate sample space S for the experiment.

Step 2 Assign acceptable probabilities to the simple events in S.

Step 3 To obtain the probability of an arbitrary event E, add the probabilities of the simple events in E.

The function P defined in Steps 2 and 3 is a **probability function** whose domain is all possible events (subsets) in the sample space S and whose range is a set of real numbers between 0 and 1, inclusive.

Matched Problem 4 Suppose in Example 4 that after flipping the nickel and dime 1,000 times, we find that HH turns up 273 times, HT turns up 206 times, TH turns up 312 times, and TT turns up 209 times. On the basis of this evidence, we assign probabilities to the simple events in S as follows:

Simple Event				
e_i	**HH**	**HT**	**TH**	**TT**
$P(e_i)$	.273	.206	.312	.209

This is an acceptable and reasonable probability assignment for the simple events in S. What are the probabilities of the following events?

(A) E_1 = getting at least 1 tail

(B) E_2 = getting 2 tails

(C) E_3 = getting at least 1 head or at least 1 tail

Example 4 and Matched Problem 4 illustrate two important ways in which acceptable and reasonable probability assignments are made for simple events in a sample space S. Each approach has its advantage in certain situations:

1. *Theoretical Approach.* We use assumptions and a deductive reasoning process to assign probabilities to simple events. No experiments are actually conducted. This is what we did in Example 4.
2. *Empirical Approach.* We assign probabilities to simple events based on the results of actual experiments. This is what we did in Matched Problem 4.

Empirical probability concepts are stated more precisely as follows: If we conduct an experiment n times and event E occurs with **frequency $f(E)$**, then the ratio $f(E)/n$ is called the **relative frequency** of the occurrence of event E in n trials. We define the **empirical probability** of E, denoted by **$P(E)$**, by the number (if it exists) that the relative frequency $f(E)/n$ approaches as n gets larger and larger. Therefore,

$$P(E) \approx \frac{\text{frequency of occurrence of } E}{\text{total number of trials}} = \frac{f(E)}{n}$$

For any particular n, the relative frequency $f(E)/n$ is also called the **approximate empirical probability** of event E.

For most of this section, we emphasize the theoretical approach. In the next section, we return to the empirical approach.

Equally Likely Assumption

In tossing a nickel and a dime (Example 4), we assigned the same probability, $\frac{1}{4}$, to each simple event in the sample space $S = \{\text{HH, HT, TH, TT}\}$. By assigning the same probability to each simple event in S, we are actually making the assumption that each simple event is as likely to occur as any other. We refer to this as an **equally likely assumption**. If, in a sample space

$$S = \{e_1, e_2, \ldots, e_n\}$$

with n elements, we assume that each simple event e_i is as likely to occur as any other, then we assign the probability $1/n$ to each. That is,

$$P(e_i) = \frac{1}{n}$$

Under an equally likely assumption, we can develop a very useful formula for finding probabilities of arbitrary events associated with a sample space S. Consider the following example:

If a single die is rolled and we assume that each face is as likely to come up as any other, then for the sample space

$$S = \{1, 2, 3, 4, 5, 6\}$$

we assign a probability of $\frac{1}{6}$ to each simple event since there are 6 simple events. The probability of

$$E = \text{rolling a prime number} = \{2, 3, 5\}$$

is

Number of elements in E ↓

$$P(E) = P(2) + P(3) + P(5) = \frac{1}{6} + \frac{1}{6} + \frac{1}{6} = \frac{3}{6} = \frac{1}{2}$$

↑ Number of elements in S

Under the assumption that each simple event is as likely to occur as any other, the computation of the probability of the occurrence of any event E in a sample space S is the number of elements in E divided by the number of elements in S.

THEOREM 1 Probability of an Arbitrary Event under an Equally Likely Assumption

If we assume that each simple event in sample space S is as likely to occur as any other, then the probability of an arbitrary event E in S is given by

$$P(E) = \frac{\text{number of elements in } E}{\text{number of elements in } S} = \frac{n(E)}{n(S)}$$

EXAMPLE 5 **Probabilities and Equally Likely Assumptions** Let us again consider rolling two dice, and assume that each simple event in the sample space shown in Figure 2 (page 398) is as likely as any other. Find the probabilities of the following events:

(A) E_1 = a sum of 7 turns up

(B) E_2 = a sum of 11 turns up

(C) E_3 = a sum less than 4 turns up

(D) E_4 = a sum of 12 turns up

SOLUTION Referring to Figure 2 (page 398) and the results found in Example 3, we find

(A) $P(E_1) = \frac{n(E_1)}{n(S)} = \frac{6}{36} = \frac{1}{6}$ $\quad E_1 = \{(6,1),(5,2),(4,3),(3,4),(2,5),(1,6)\}$

(B) $P(E_2) = \frac{n(E_2)}{n(S)} = \frac{2}{36} = \frac{1}{18}$ $\quad E_2 = \{(6,5),(5,6)\}$

(C) $P(E_3) = \frac{n(E_3)}{n(S)} = \frac{3}{36} = \frac{1}{12}$ $\quad E_3 = \{(1,1),(2,1),(1,2)\}$

(D) $P(E_4) = \frac{n(E_4)}{n(S)} = \frac{1}{36}$ $\quad E_4 = \{(6,6)\}$

Matched Problem 5 Under the conditions in Example 5, find the probabilities of the following events (each event refers to the sum of the dots facing up on both dice):

(A) E_5 = a sum of 5 turns up

(B) E_6 = a sum that is a prime number greater than 7 turns up

EXAMPLE 6 **Simulation and Empirical Probabilities** Use output from the random number feature of a graphing calculator to simulate 100 rolls of two dice. Determine the empirical probabilities of the following events, and compare with the theoretical probabilities:

(A) E_1 = a sum of 7 turns up

(B) E_2 = a sum of 11 turns up

SOLUTION A graphing calculator can be used to select a random integer from 1 to 6. Each of the six integers in the given range is equally likely to be selected. Therefore, by selecting a random integer from 1 to 6 and adding it to a second random integer from 1 to 6, we simulate rolling two dice and recording the sum (see the first command in Figure 4A). The second command in Figure 4A simulates 100 rolls of two dice; the sums are stored in list L_1. From the statistical plot of L_1 in Figure 4B we obtain the empirical probabilities.*

(A) The empirical probability of E_1 is $\frac{15}{100} = .15$; the theoretical probability of E_1 (see Example 5A) is $\frac{6}{36} = .167$.

(B) The empirical probability of E_2 is $\frac{6}{100} = .06$; the theoretical probability of E_2 (see Example 5B) is $\frac{2}{36} = .056$.

*If you simulate this experiment on your graphing calculator, you should not expect to get the same empirical probabilities.

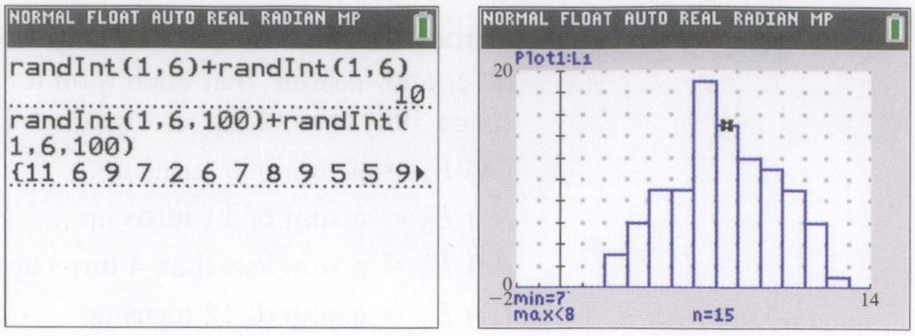

Figure 4 (A) (B)

Matched Problem 6 Use the graphing calculator output in Figure 4B to determine the empirical probabilities of the following events, and compare with the theoretical probabilities:

(A) E_3 = a sum less than 4 turns up

(B) E_4 = a sum of 12 turns up

Explore and Discuss 1

A shipment box contains 12 graphing calculators, out of which 2 are defective. A calculator is drawn at random from the box and then, without replacement, a second calculator is drawn. Discuss whether the equally likely assumption would be appropriate for the sample space $S = \{GG, GD, DG, DD\}$, where G is a good calculator and D is a defective one.

We now turn to some examples that make use of the counting techniques developed in Chapter 7.

EXAMPLE 7 **Probability and Equally Likely Assumption** In drawing 5 cards from a 52-card deck without replacement, what is the probability of getting 5 spades?

SOLUTION Let the sample space S be the set of all 5-card hands from a 52-card deck. Since the order in a hand does not matter, $n(S) = {}_{52}C_5$. Let event E be the set of all 5-card hands from 13 spades. Again, the order does not matter and $n(E) = {}_{13}C_5$. Assuming that each 5-card hand is as likely as any other,

$$P(E) = \frac{n(E)}{n(S)} = \frac{{}_{13}C_5}{{}_{52}C_5} = \frac{1{,}287}{2{,}598{,}960} \approx .0005$$

(Some calculators display the answer as 4.951980792E–4. This means the same thing as the scientific notation $4.951980792 \times 10^{-4}$, so the answer, rounded to 4 decimal places, is .0005).

Matched Problem 7 In drawing 7 cards from a 52-card deck without replacement, what is the probability of getting 7 hearts?

EXAMPLE 8 **Probability and Equally Likely Assumption** The board of regents of a university is made up of 12 men and 16 women. If a committee of 6 is chosen at random, what is the probability that it will contain 3 men and 3 women?

SOLUTION Let S be the set of all 6-person committees out of 28 people. Then

$$n(S) = {}_{28}C_6$$

Let E be the set of all 6-person committees with 3 men and 3 women. To find $n(E)$, we use the multiplication principle and the following two operations:

O_1: Select 3 men out of the 12 available $\quad N_1$: ${}_{12}C_3$

O_2: Select 3 women out of the 16 available $\quad N_2$: ${}_{16}C_3$

Therefore,

$$n(E) = N_1 \cdot N_2 = {}_{12}C_3 \cdot {}_{16}C_3$$

and

$$P(E) = \frac{n(E)}{n(S)} = \frac{{}_{12}C_3 \cdot {}_{16}C_3}{{}_{28}C_6} \approx .327$$

Matched Problem 8 What is the probability that the committee in Example 8 will have 4 men and 2 women?

There are many counting problems for which it is not possible to produce a simple formula that will yield the number of possible cases. In situations of this type, we often revert back to tree diagrams and counting branches.

Exercises 8.1

Skills Warm-up Exercises

In Problems 1–6, without using a calculator, determine which event, E or F, is more likely to occur. (If necessary, review Section A.1.)

1. $P(E) = \frac{5}{6}; P(F) = \frac{4}{5}$

2. $P(E) = \frac{2}{7}; P(F) = \frac{1}{3}$

3. $P(E) = \frac{3}{8}; P(F) = .4$

4. $P(E) = .9; P(F) = \frac{7}{8}$

5. $P(E) = .15; P(F) = \frac{1}{6}$

6. $P(E) = \frac{5}{7}; P(F) = \frac{6}{11}$

A *A circular spinner is divided into 15 sectors of equal area: 6 red sectors, 5 blue, 3 yellow, and 1 green. In Problems 7–14, consider the experiment of spinning the spinner once. Find the probability that the spinner lands on:*

7. Blue

8. Yellow

9. Yellow or green

10. Red or blue

11. Orange

12. Yellow, red, or green

13. Blue, red, yellow, or green

14. Purple

Refer to the description of a standard deck of 52 cards and Figure 4 on page 384. An experiment consists of drawing 1 card from a standard 52-card deck. In Problems 15–24, what is the probability of drawing

15. A club

16. A black card

17. A heart or diamond

18. A numbered card

19. The jack of clubs

20. An ace

21. A king or spade

22. A red queen

23. A black diamond

24. A six or club

25. In a family with 2 children, excluding multiple births, what is the probability of having 2 children of the opposite gender? Assume that a girl is as likely as a boy at each birth.

26. In a family with 2 children, excluding multiple births, what is the probability of having 2 girls? Assume that a girl is as likely as a boy at each birth.

27. A store carries four brands of DVD players: J, G, P, and S. From past records, the manager found that the relative frequency of brand choice among customers varied. Which of the following probability assignments for a particular customer choosing a particular brand of DVD player would have to be rejected? Why?

(A) $P(J) = .15, P(G) = -.35, P(P) = .50, P(S) = .70$

(B) $P(J) = .32, P(G) = .28, P(P) = .24, P(S) = .30$

(C) $P(J) = .26, P(G) = .14, P(P) = .30, P(S) = .30$

28. Using the probability assignments in Problem 27C, what is the probability that a random customer will not choose brand S?

29. Using the probability assignments in Problem 27C, what is the probability that a random customer will choose brand J or brand P?

30. Using the probability assignments in Problem 27C, what is the probability that a random customer will not choose brand J or brand P?

B **31.** In a family with 3 children, excluding multiple births, what is the probability of having 2 boys and 1 girl, in that order? Assume that a boy is as likely as a girl at each birth.

32. In a family with 3 children, excluding multiple births, what is the probability of having 2 boys and 1 girl, in any order? Assume that a boy is as likely as a girl at each birth.

33. A keypad at the entrance of a building has 10 buttons labeled 0 through 9. What is the probability of a person correctly guessing a 4-digit entry code?

34. A keypad at the entrance of a building has 10 buttons labeled 0 through 9. What is the probability of a person correctly guessing a 4-digit entry code if they know that no digits repeat?

Refer to the description of a standard deck of 52 cards and Figure 4 on page 384. An experiment consists of dealing 5 cards from a standard 52-card deck. In Problems 35–38, what is the probability of being dealt

35. 5 black cards?

36. 5 hearts?

37. 5 face cards?

38. 5 nonface cards?

39. Twenty thousand students are enrolled at a state university. A student is selected at random, and his or her birthday (month and day, not year) is recorded. Describe an appropriate sample space for this experiment and assign acceptable probabilities to the simple events. What are your assumptions in making this assignment?

40. In a three-way race for the U.S. Senate, polls indicate that the two leading candidates are running neck-and-neck, while the third candidate is receiving half the support of either of the others. Registered voters are chosen at random and asked which of the three will get their vote. Describe an appropriate sample space for this random survey experiment and assign acceptable probabilities to the simple events.

41. Suppose that 5 thank-you notes are written and 5 envelopes are addressed. Accidentally, the notes are randomly inserted into the envelopes and mailed without checking the addresses. What is the probability that all the notes will be inserted into the correct envelopes?

42. Suppose that 6 people check their coats in a checkroom. If all claim checks are lost and the 6 coats are randomly returned, what is the probability that all the people will get their own coats back?

An experiment consists of rolling two fair dice and adding the dots on the two sides facing up. Using the sample space shown in Figure 2 (page 398) and, assuming each simple event is as likely as any other, find the probability of the sum of the dots indicated in Problems 43–56.

43. Sum is 6.

44. Sum is 8.

45. Sum is less than 5.

46. Sum is greater than 8.

47. Sum is not 7 or 11.

48. Sum is not 2, 4, or 6.

49. Sum is 1.

50. Sum is 13.

51. Sum is divisible by 3.

52. Sum is divisible by 4.

53. Sum is 7 or 11 (a "natural").

54. Sum is 2, 3, or 12 ("craps").

55. Sum is divisible by 2 or 3.

56. Sum is divisible by 2 and 3.

An experiment consists of tossing three fair (not weighted) coins, except that one of the three coins has a head on both sides. Compute the probability of obtaining the indicated results in Problems 57–62.

57. 1 head

58. 2 heads

59. 3 heads

60. 0 heads

61. More than 1 head

62. More than 1 tail

In Problems 63–68, a sample space S is described. Would it be reasonable to make the equally likely assumption? Explain.

63. A single card is drawn from a standard deck. We are interested in whether or not the card drawn is a heart, so an appropriate sample space is $S = \{H, N\}$.

64. A single fair coin is tossed. We are interested in whether the coin falls heads or tails, so an appropriate sample space is $S = \{H, T\}$.

65. A single fair die is rolled. We are interested in whether or not the number rolled is even or odd, so an appropriate sample space is $S = \{E, O\}$.

66. A nickel and dime are tossed. We are interested in the number of heads that appear, so an appropriate sample space is $S = \{0, 1, 2\}$.

67. A wheel of fortune has seven sectors of equal area colored red, orange, yellow, green, blue, indigo, and violet. We are interested in the color that the pointer indicates when the wheel stops, so an appropriate sample space is $S = \{R, O, Y, G, B, I, V\}$.

68. A wheel of fortune has seven sectors of equal area colored red, orange, yellow, red, orange, yellow, and red. We are interested in the color that the pointer indicates when the wheel stops, so an appropriate sample space is $S = \{R, O, Y\}$.

69. (A) Is it possible to get 19 heads in 20 flips of a fair coin? Explain.

(B) If you flipped a coin 40 times and got 37 heads, would you suspect that the coin was unfair? Why or why not? If you suspect an unfair coin, what empirical probabilities would you assign to the simple events of the sample space?

70. (A) Is it possible to get 7 double 6's in 10 rolls of a pair of fair dice? Explain.

(B) If you rolled a pair of dice 36 times and got 11 double 6's, would you suspect that the dice were unfair? Why or why not? If you suspect loaded dice, what empirical probability would you assign to the event of rolling a double 6?

C *An experiment consists of rolling two fair (not weighted) 4-sided dice and adding the dots on the two sides facing up. Each die is numbered 1–4. Compute the probability of obtaining the indicated sums in Problems 71–78.*

71. 2 **72.** 3 **73.** 4

74. 5 **75.** 6 **76.** 7

77. An odd sum **78.** An even sum

In Problems 79–86, find the probability of being dealt the given hand from a standard 52-card deck. Refer to the description of a standard 52-card deck on page 384.

79. A 5-card hand that consists entirely of red cards

80. A 5-card hand that consists entirely of face cards

81. A 6-card hand that contains exactly two face cards

82. A 6-card hand that contains exactly two clubs

83. A 4-card hand that contains no aces

84. A 4-card hand that contains no face cards

85. A 7-card hand that contains exactly 2 diamonds and exactly 2 spades

86. A 7-card hand that contains exactly 1 king and exactly 2 jacks

In Problems 87–90, several experiments are simulated using the random number feature on a graphing calculator. For example, the roll of a fair die can be simulated by selecting a random integer from 1 to 6, and 50 rolls of a fair die by selecting 50 random integers from 1 to 6 (see Fig. A for Problem 87 and your user's manual).

87. From the statistical plot of the outcomes of rolling a fair die 50 times (see Fig. B), we see, for example, that the number 4 was rolled exactly 5 times.

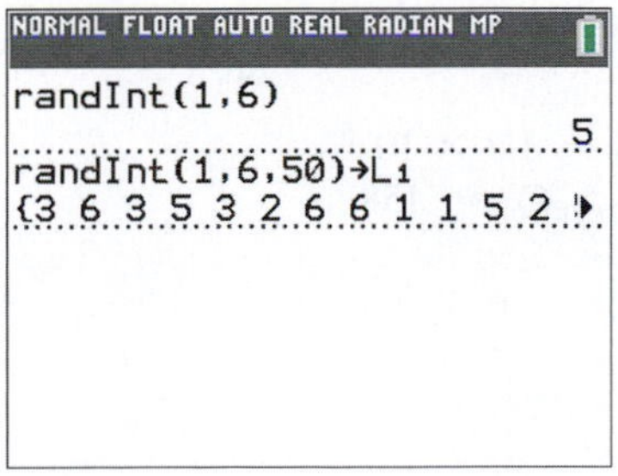

(A)

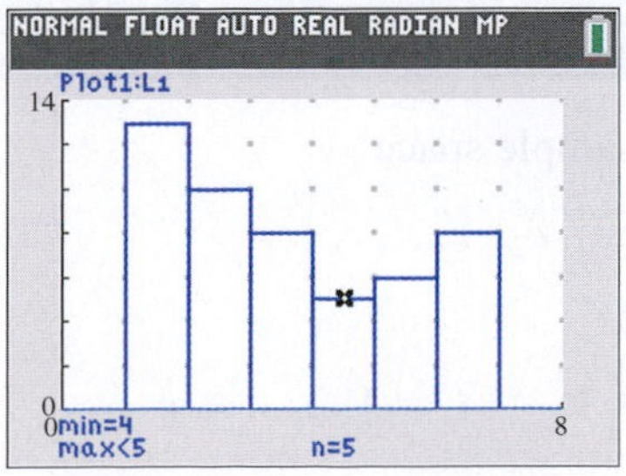

(B)

(A) What is the empirical probability that the number 6 was rolled?

(B) What is the probability that a 6 is rolled under the equally likely assumption?

(C) Use a graphing calculator to simulate 100 rolls of a fair die and determine the empirical probabilities of the six outcomes.

88. Use a graphing calculator to simulate 200 tosses of a nickel and dime, representing the outcomes HH, HT, TH, and TT by 1, 2, 3, and 4, respectively.

(A) Find the empirical probabilities of the four outcomes.

(B) What is the probability of each outcome under the equally likely assumption?

89. (A) Explain how a graphing calculator can be used to simulate 500 tosses of a coin.

(B) Carry out the simulation and find the empirical probabilities of the two outcomes.

(C) What is the probability of each outcome under the equally likely assumption?

90. From a box containing 12 balls numbered 1 through 12, one ball is drawn at random.

(A) Explain how a graphing calculator can be used to simulate 400 repetitions of this experiment.

(B) Carry out the simulation and find the empirical probability of drawing the 8 ball.

(C) What is the probability of drawing the 8 ball under the equally likely assumption?

Applications

91. Consumer testing. Twelve popular brands of beer are used in a blind taste study for consumer recognition.

(A) If 4 distinct brands are chosen at random from the 12 and if a consumer is not allowed to repeat any answers, what is the probability that all 4 brands could be identified by just guessing?

(B) If repeats are allowed in the 4 brands chosen at random from the 12 and if a consumer is allowed to repeat answers, what is the probability that all 4 brands are identified correctly by just guessing?

92. Consumer testing. Six popular brands of cola are to be used in a blind taste study for consumer recognition.

(A) If 3 distinct brands are chosen at random from the 6 and if a consumer is not allowed to repeat any answers, what is the probability that all 3 brands could be identified by just guessing?

(B) If repeats are allowed in the 3 brands chosen at random from the 6 and if a consumer is allowed to repeat answers, what is the probability that all 3 brands are identified correctly by just guessing?

93. Personnel selection. Suppose that 6 female and 5 male applicants have been successfully screened for 5 positions.

If the 5 positions are filled at random from the 11 finalists, what is the probability of selecting

(A) 3 females and 2 males?

(B) 4 females and 1 male?

(C) 5 females?

(D) At least 4 females?

94. Committee selection. A 4-person grievance committee is to include employees in 2 departments, A and B, with 15 and 20 employees, respectively. If the 4 people are selected at random from the 35 employees, what is the probability of selecting

(A) 3 from A and 1 from B?

(B) 2 from A and 2 from B?

(C) All 4 from A?

(D) At least 3 from A?

95. Medicine. A laboratory technician is to be tested on identifying blood types from 8 standard classifications.

(A) If 3 distinct samples are chosen at random from the 8 types and if the technician is not allowed to repeat any answers, what is the probability that all 3 could be correctly identified by just guessing?

(B) If repeats are allowed in the 3 blood types chosen at random from the 8 and if the technician is allowed to repeat answers, what is the probability that all 3 are identified correctly by just guessing?

96. Medical research. Because of limited funds, 5 research centers are to be chosen out of 8 suitable ones for a study on heart disease. If the selection is made at random, what is the probability that 5 particular research centers will be chosen?

97. Politics. A town council has 9 members: 5 Democrats and 4 Republicans. A 3-person zoning committee is selected at random.

(A) What is the probability that all zoning committee members are Democrats?

(B) What is the probability that a majority of zoning committee members are Democrats?

98. Politics. There are 10 senators (half Democrats, half Republicans) and 16 representatives (half Democrats, half Republicans) who wish to serve on a joint congressional committee on tax reform. An 8-person committee is chosen at random from those who wish to serve.

(A) What is the probability that the joint committee contains equal numbers of senators and representatives?

(B) What is the probability that the joint committee contains equal numbers of Democrats and Republicans?

Answers to Matched Problems

1. (A) $E = \{12\}$; simple event
(B) $E = \{16, 18\}$; compound event

2. (A) $S_1 = \{BB, BG, GB, GG\}$;

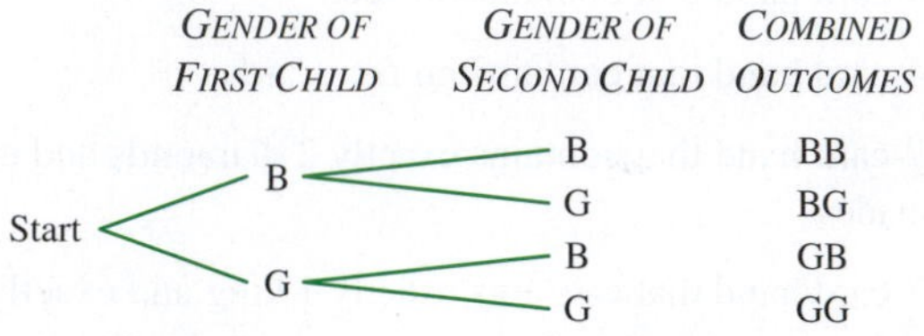

(B) $S_2 = \{0, 1, 2\}$ (C) $S_3 = \{A, D\}$ (D) S_1

3. (A) $\{(4, 1), (3, 2), (2, 3), (1, 4)\}$
(B) $\{(6, 5), (5, 6)\}$

4. (A) .727 (B) .209 (C) 1

5. (A) $P(E_5) = \frac{1}{9}$ (B) $P(E_6) = \frac{1}{18}$

6. (A) $\frac{9}{100} = .09$ (empirical); $\frac{3}{36} = .083$ (theoretical)
(B) $\frac{1}{100} = .01$ (empirical); $\frac{1}{36} = .028$ (theoretical)

7. ${}_{13}C_7/{}_{52}C_7 \approx 1.3 \times 10^{-5}$

8. ${}_{12}C_4 \cdot {}_{16}C_2/{}_{28}C_6 \approx .158$

8.2 Union, Intersection, and Complement of Events; Odds

- Union and Intersection
- Complement of an Event
- Odds
- Applications to Empirical Probability

Recall from Section 8.1 that given a sample space

$$S = \{e_1, e_2, \ldots, e_n\}$$

any function P defined on S such that

$$0 \le P(e_i) \le 1 \qquad i = 1, 2, \ldots, n$$

and

$$P(e_1) + P(e_2) + \cdots + P(e_n) = 1$$

is called a *probability function*. In addition, any subset of S is called an *event E*, and the probability of E is the sum of the probabilities of the simple events in E.

Union and Intersection

Since events are subsets of a sample space, the union and intersection of events are simply the union and intersection of sets as defined in the following box. In this section, we concentrate on the union of events and consider only simple cases of intersection. More complicated cases of intersection will be investigated in the next section.

DEFINITION Union and Intersection of Events*

If A and B are two events in a sample space S, then the **union** of A and B, denoted by $\mathbf{A \cup B}$, and the **intersection** of A and B, denoted by $\mathbf{A \cap B}$, are defined as follows:

$$A \cup B = \{e \in S \mid e \in A \textbf{ or } e \in B\} \qquad A \cap B = \{e \in S \mid e \in A \textbf{ and } e \in B\}$$

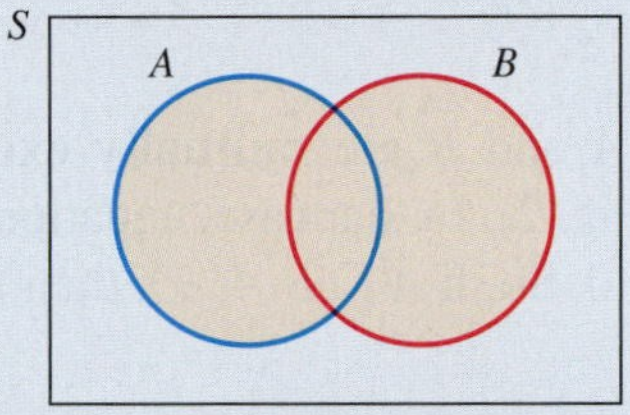

$A \cup B$ is shaded

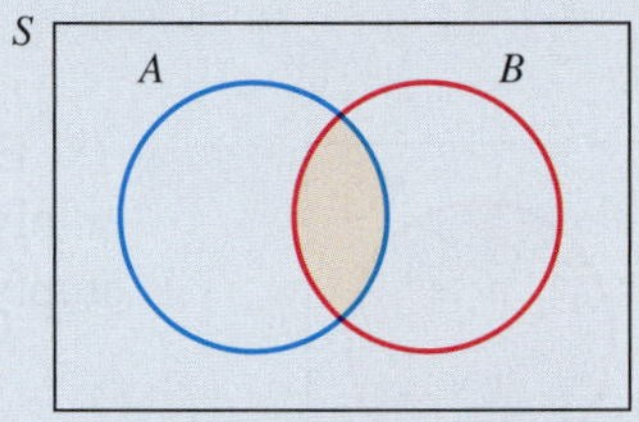

$A \cap B$ is shaded

Furthermore, we define

The **event A or B** to be $A \cup B$

The **event A and B** to be $A \cap B$

*See Section 7.2 for a discussion of set notation.

EXAMPLE 1 **Probability Involving Union and Intersection** Consider the sample space of equally likely events for the rolling of a single fair die:

$$S = \{1, 2, 3, 4, 5, 6\}$$

(A) What is the probability of rolling a number that is odd **and** exactly divisible by 3?

(B) What is the probability of rolling a number that is odd **or** exactly divisible by 3?

SOLUTION

(A) Let A be the event of rolling an odd number, B the event of rolling a number divisible by 3, and F the event of rolling a number that is odd **and** divisible by 3. Then (Fig. 1A)

$$A = \{1, 3, 5\} \qquad B = \{3, 6\} \qquad F = A \cap B = \{3\}$$

The probability of rolling a number that is odd **and** exactly divisible by 3 is

$$P(F) = P(A \cap B) = \frac{n(A \cap B)}{n(S)} = \frac{1}{6}$$

(B) Let A and B be the same events as in part (A), and let E be the event of rolling a number that is odd **or** divisible by 3. Then (Fig. 1B)

$$A = \{1, 3, 5\} \qquad B = \{3, 6\} \qquad E = A \cup B = \{1, 3, 5, 6\}$$

The probability of rolling a number that is odd **or** exactly divisible by 3 is

$$P(E) = P(A \cup B) = \frac{n(A \cup B)}{n(S)} = \frac{4}{6} = \frac{2}{3}$$

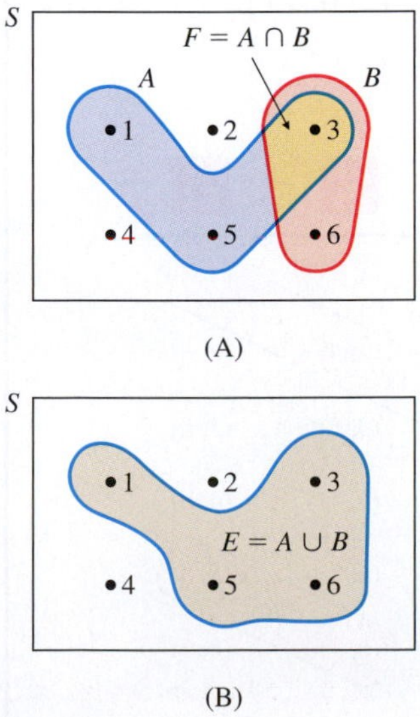

Figure 1

Matched Problem 1 Use the sample space in Example 1 to answer the following:

(A) What is the probability of rolling an odd number **and** a prime number?

(B) What is the probability of rolling an odd number **or** a prime number?

Suppose that A and B are events in a sample space S. How is the probability of $A \cup B$ related to the individual probabilities of A and of B? Think of the probability of an element of S as being its weight. To find the total weight of the elements of $A \cup B$, we weigh all of the elements of A, then weigh all of the elements of B, and subtract from the sum the weight of all of the elements that were weighed twice. This gives the formula for $P(A \cup B)$ in Theorem 1.

THEOREM 1 Probability of the Union of Two Events

For any events A and B,

$$P(A \cup B) = P(A) + P(B) - P(A \cap B) \tag{1}$$

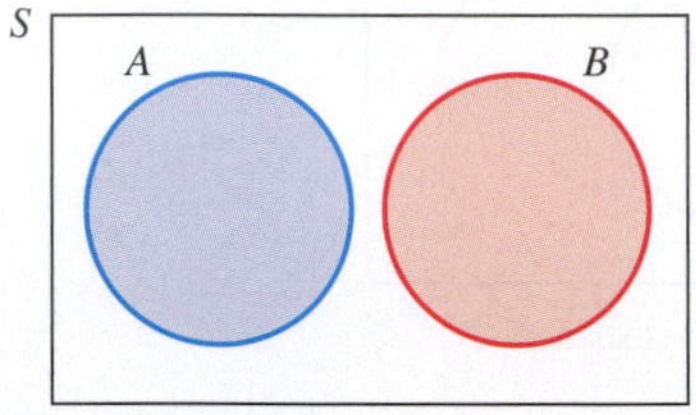

Figure 2 Mutually exclusive: $A \cap B = \varnothing$

Events A and B are **mutually exclusive (disjoint)** if their intersection is the empty set (Fig. 2). In that case, equation (1) simplifies, because the probability of the empty set is 0. So, if $A \cap B = \varnothing$, then $P(A \cup B) = P(A) + P(B)$.

CONCEPTUAL INSIGHT

Note the similarity between equation (1) in Theorem 1 and the formula obtained in Section 7.3 for counting the number of elements in $A \cup B$:

$$n(A \cup B) = n(A) + n(B) - n(A \cap B)$$

To find the number of elements in $A \cup B$, we count all of the elements of A, then count all of the elements of B, and subtract from the sum the number of elements that were counted twice.

EXAMPLE 2 **Probability Involving Union and Intersection** Suppose that two fair dice are rolled.

(A) What is the probability that a sum of 7 or 11 turns up?

(B) What is the probability that both dice turn up the same or that a sum less than 5 turns up?

SOLUTION

(A) If A is the event that a sum of 7 turns up and B is the event that a sum of 11 turns up, then (Fig. 3) the event that a sum of 7 or 11 turns up is $A \cup B$, where

$$A = \{(1, 6), (2, 5), (3, 4), (4, 3), (5, 2), (6, 1)\}$$
$$B = \{(5, 6), (6, 5)\}$$

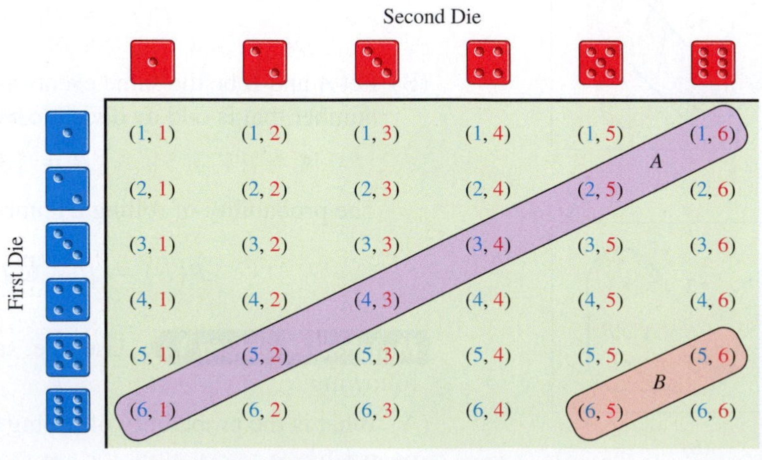

Figure 3

Since events A and B are mutually exclusive, we can use the simplified version of equation (1) to calculate $P(A \cup B)$:

$$P(A \cup B) = P(A) + P(B) = \frac{6}{36} + \frac{2}{36} = \frac{8}{36} = \frac{2}{9}$$

In this equally likely sample space, $n(A) = 6$, $n(B) = 2$, and $n(S) = 36$.

(B) If A is the event that both dice turn up the same and B is the event that the sum is less than 5, then (Fig. 4) the event that both dice turn up the same or the sum is less than 5 is $A \cup B$, where

$$A = \{(1,1), (2,2), (3,3), (4,4), (5,5), (6,6)\}$$
$$B = \{(1,1), (1,2), (1,3), (2,1), (2,2), (3,1)\}$$

Since $A \cap B = \{(1,1), (2,2)\}$, A and B are not mutually exclusive, and we use equation (1) to calculate $P(A \cup B)$:

$$P(A \cup B) = P(A) + P(B) - P(A \cap B) = \frac{6}{36} + \frac{6}{36} - \frac{2}{36} = \frac{10}{36} = \frac{5}{18}$$

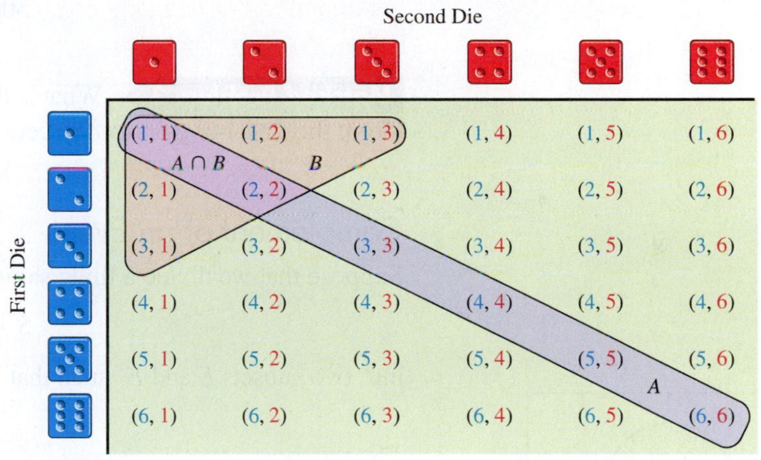

Figure 4

Matched Problem 2 Use the sample space in Example 2 to answer the following:

(A) What is the probability that a sum of 2 or 3 turns up?

(B) What is the probability that both dice turn up the same or that a sum greater than 8 turns up?

You no doubt noticed in Example 2 that we actually did not have to use equation (1). We could have proceeded as in Example 1 and simply counted sample points in $A \cup B$. The following example illustrates the use of equation (1) in a situation where visual representation of sample points is not practical.

EXAMPLE 3 **Probability Involving Union and Intersection** What is the probability that a number selected at random from the first 500 positive integers is (exactly) divisible by 3 or 4?

SOLUTION Let A be the event that a drawn integer is divisible by 3 and B the event that a drawn integer is divisible by 4. Note that events A and B are not mutually exclusive because multiples of 12 are divisible by both 3 and 4. Since each of the positive integers from 1 to 500 is as likely to be drawn as any other, we can use $n(A)$, $n(B)$, and $n(A \cap B)$ to determine $P(A \cup B)$, where

$$n(A) = \text{the largest integer less than or equal to } \frac{500}{3} = 166$$

$$n(B) = \text{the largest integer less than or equal to } \frac{500}{4} = 125$$

$$n(A \cap B) = \text{the largest integer less than or equal to } \frac{500}{12} = 41$$

Now we can compute $P(A \cup B)$:

$$\begin{aligned} P(A \cup B) &= P(A) + P(B) - P(A \cap B) \\ &= \frac{n(A)}{n(S)} + \frac{n(B)}{n(S)} - \frac{n(A \cap B)}{n(S)} \\ &= \frac{166}{500} + \frac{125}{500} - \frac{41}{500} = \frac{250}{500} = .5 \end{aligned}$$

Matched Problem 3 What is the probability that a number selected at random from the first 140 positive integers is (exactly) divisible by 4 or 6?

Complement of an Event

Suppose that we divide a finite sample space

$$S = \{e_1, \ldots, e_n\}$$

into two subsets E and E' such that

$$E \cap E' = \varnothing$$

that is, E and E' are mutually exclusive and

$$E \cup E' = S$$

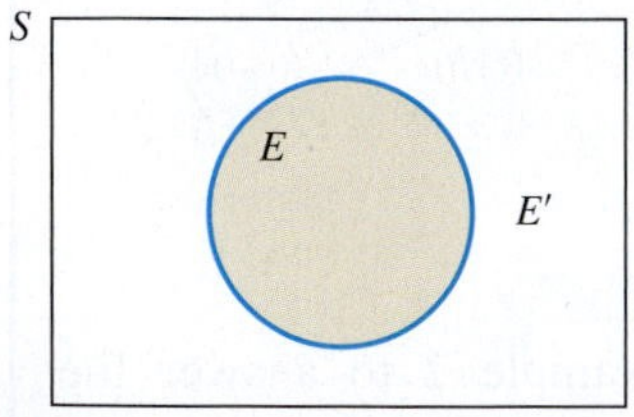

Figure 5

Then E' is called the **complement of E** relative to S. Note that E' contains all the elements of S that are not in E (Fig. 5). Furthermore,

$$\begin{aligned} P(S) &= P(E \cup E') \\ &= P(E) + P(E') = 1 \end{aligned}$$

Therefore,

$$\boldsymbol{P(E) = 1 - P(E') \qquad P(E') = 1 - P(E)} \tag{2}$$

If the probability of rain is .67, then the probability of no rain is $1 - .67 = .33$; if the probability of striking oil is .01, then the probability of not striking oil is .99. If the probability of having at least 1 boy in a 2-child family is .75, what is the probability of having 2 girls? [*Answer*:.25.]

Explore and Discuss 1

(A) Suppose that E and F are complementary events. Are E and F necessarily mutually exclusive? Explain why or why not.

(B) Suppose that E and F are mutually exclusive events. Are E and F necessarily complementary? Explain why or why not.

In looking for $P(E)$, there are situations in which it is easier to find $P(E')$ first, and then use equations (2) to find $P(E)$. The next two examples illustrate two such situations.

EXAMPLE 4 **Quality Control** A shipment of 45 precision parts, including 9 that are defective, is sent to an assembly plant. The quality control division selects 10 at random for testing and rejects the entire shipment if 1 or more in the sample are found to be defective. What is the probability that the shipment will be rejected?

SOLUTION If E is the event that 1 or more parts in a random sample of 10 are defective, then E', the complement of E, is the event that no parts in a random sample of 10 are defective. It is easier to compute $P(E')$ than to compute $P(E)$ directly. Once $P(E')$ is found, we will use $P(E) = 1 - P(E')$ to find $P(E)$.

The sample space S for this experiment is the set of all subsets of 10 elements from the set of 45 parts shipped. Thus, since there are $45 - 9 = 36$ nondefective parts,

$$P(E') = \frac{n(E')}{n(S)} = \frac{{}_{36}C_{10}}{{}_{45}C_{10}} \approx .08$$

and

$$P(E) = 1 - P(E') \approx 1 - .08 = .92$$

Matched Problem 4 A shipment of 40 precision parts, including 8 that are defective, is sent to an assembly plant. The quality control division selects 10 at random for testing and rejects the entire shipment if 1 or more in the sample are found to be defective. What is the probability that the shipment will be rejected?

EXAMPLE 5 **Birthday Problem** In a group of n people, what is the probability that at least 2 people have the same birthday (same month and day, excluding February 29)? Make a guess for a class of 40 people, and check your guess with the conclusion of this example.

SOLUTION If we form a list of the birthdays of all the people in the group, we have a simple event in the sample space

$$S = \text{set of all lists of } n \text{ birthdays}$$

For any person in the group, we will assume that any birthday is as likely as any other, so that the simple events in S are equally likely. How many simple events are in the set S? Since any person could have any one of 365 birthdays (excluding February 29), the multiplication principle implies that the number of simple events in S is

$$n(S) = \underset{\text{1st person}}{365} \cdot \underset{\text{2nd person}}{365} \cdot \underset{\text{3rd person}}{365} \cdot \cdots \cdot \underset{n\text{th person}}{365}$$
$$= 365^n$$

Now, let E be the event that at least 2 people in the group have the same birthday. Then E' is the event that no 2 people have the same birthday. The multiplication principle can be used to determine the number of simple events in E':

$$\begin{aligned} n(E') &= \underset{\text{1st person}}{365} \cdot \underset{\text{2nd person}}{364} \cdot \underset{\text{3rd person}}{363} \cdot \cdots \cdot \underset{n\text{th person}}{(366 - n)} \\ &= \frac{[365 \cdot 364 \cdot 363 \cdot \cdots \cdot (366 - n)](365 - n)!}{(365 - n)!} \quad \text{Multiply numerator and denominator by } (365 - n)!. \\ &= \frac{365!}{(365 - n)!} \end{aligned}$$

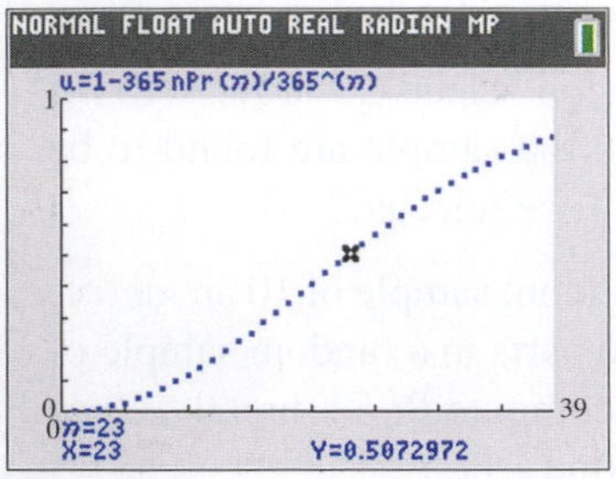

Figure 6

Since we have assumed that S is an equally likely sample space,

$$P(E') = \frac{n(E')}{n(S)} = \frac{\dfrac{365!}{(365 - n)!}}{365^n} = \frac{365!}{365^n(365 - n)!}$$

Therefore,

$$\begin{aligned} P(E) &= 1 - P(E') \\ &= 1 - \frac{365!}{365^n(365 - n)!} \end{aligned} \qquad (3)$$

Table 1 Birthday Problem

Number of People in Group	Probability That 2 or More Have Same Birthday
n	$P(E)$
5	.027
10	.117
15	.253
20	.411
23	.507
30	.706
40	.891
50	.970
60	.994
70	.999

Equation (3) is valid for any n satisfying $1 \le n \le 365$. [What is $P(E)$ if $n > 365$?] For example, in a group of 5 people,

$$\begin{aligned} P(E) &= 1 - \frac{365!}{(365)^5\, 360!} \\ &= 1 - \frac{\cancel{365} \cdot 364 \cdot 363 \cdot 362 \cdot 361 \cdot \cancel{360!}}{\cancel{365} \cdot 365 \cdot 365 \cdot 365 \cdot 365 \cdot \cancel{360!}} \\ &= .027 \end{aligned}$$

It is interesting to note that as the size of the group increases, $P(E)$ increases more rapidly than you might expect. Figure 6* shows the graph of $P(E)$ for $1 \le n \le 39$. Table 1 gives the value of $P(E)$ for selected values of n. If $n = 5$, Table 1 gives $P(E) = .027$, as calculated above. Notice that for a group of only 23 people, the probability that 2 or more have the same birthday is greater than $\frac{1}{2}$.

Matched Problem 5 Use equation (3) to evaluate $P(E)$ for $n = 4$.

Explore and Discuss 2

Determine the smallest number n such that in a group of n people, the probability that 2 or more have a birthday in the same month is greater than .5. Discuss the assumptions underlying your computation.

Odds

When the probability of an event E is known, it is often customary (especially in gaming situations) to speak of odds for or against E rather than the probability of E. For example, if you roll a fair die once, then the *odds for* rolling a 2 are 1 to 5 (also written 1 : 5), and the *odds against* rolling a 2 are 5 to 1 (or 5 : 1). This is consistent with the following instructions for converting probabilities to odds.

*See Problem 71 in Exercises 8.2 for a discussion of the form of equation (3) used to produce the graph in Figure 6.

DEFINITION From Probabilities to Odds

If $P(E)$ is the probability of the event E, then

(A) **Odds for E** $= \dfrac{P(E)}{1 - P(E)} = \dfrac{P(E)}{P(E')}$ $\quad P(E) \neq 1$

(B) **Odds against E** $= \dfrac{P(E')}{P(E)}$ $\quad P(E) \neq 0$

The ratio $\dfrac{P(E)}{P(E')}$, giving odds for E, is usually expressed as an equivalent ratio $\dfrac{a}{b}$ of whole numbers (by multiplying numerator and denominator by the same number), and written "a to b" or "$a : b$." In this case, the odds against E are written "b to a" or "$b : a$."

Odds have a natural interpretation in terms of **fair games**. Let's return to the experiment of rolling a fair die once. Recall that the odds for rolling a 2 are 1 to 5. Turn the experiment into a fair game as follows: If you bet \$1 on rolling a 2, then the house pays you \$5 (and returns your \$1 bet) if you roll a 2; otherwise, you lose the \$1 bet.

More generally, consider any experiment and an associated event E. If the odds for E are a to b, then the experiment can be turned into a fair game as follows: If you bet \$$a$ on event E, then the house pays you \$$b$ (and returns your \$$a$ bet) if E occurs; otherwise, you lose the \$$a$ bet.

EXAMPLE 6 **Probability and Odds**

(A) What are the odds for rolling a sum of 7 in a single roll of two fair dice?

(B) If you bet \$1 on rolling a sum of 7, what should the house pay (plus returning your \$1 bet) if you roll a sum of 7 in order for the game to be fair?

SOLUTION

(A) Let E denote the event of rolling a sum of 7. Then $P(E) = \frac{6}{36} = \frac{1}{6}$. So the odds for E are

$$\frac{P(E)}{P(E')} = \frac{\frac{1}{6}}{\frac{5}{6}} = \frac{1}{5} \qquad \text{Also written as "1 to 5" or "1 : 5."}$$

(B) The odds for rolling a sum of 7 are 1 to 5. The house should pay \$5 (and return your \$1 bet) if you roll a sum of 7 for the game to be fair.

Matched Problem 6

(A) What are the odds for rolling a sum of 8 in a single roll of two fair dice?

(B) If you bet \$5 that a sum of 8 will turn up, what should the house pay (plus returning your \$5 bet) if a sum of 8 does turn up in order for the game to be fair?

Now we will go in the other direction: If we are given the odds for an event, what is the probability of the event? (The verification of the following formula is left to Problem 75 in Exercises 8.2.)

If the odds for event E are a/b, then the probability of E is

$$P(E) = \frac{a}{a + b}$$

EXAMPLE 7 **Odds and Probability** If in repeated rolls of two fair dice, the odds for rolling a 5 before rolling a 7 are 2 to 3, then the probability of rolling a 5 before rolling a 7 is

$$P(E) = \frac{a}{a+b} = \frac{2}{2+3} = \frac{2}{5}$$

Matched Problem 7 If in repeated rolls of two fair dice, the odds against rolling a 6 before rolling a 7 are 6 to 5, then what is the probability of rolling a 6 before rolling a 7?

Applications to Empirical Probability

In the following discussions, the term *empirical probability* will mean the probability of an event determined by a sample that is used to approximate the probability of the corresponding event in the total population. How does the approximate empirical probability of an event determined from a sample relate to the actual probability of an event relative to the total population? In mathematical statistics an important theorem called the **law of large numbers** (or the **law of averages**) is proved. Informally, it states that the approximate empirical probability can be made as close to the actual probability as we please by making the sample sufficiently large.

EXAMPLE 8 **Market Research** From a survey of 1,000 people in Springfield, it was found that 500 people had tried a certain brand of diet cola, 600 had tried a certain brand of regular cola, and 200 had tried both types of cola. If a person from Springfield is selected at random, what is the (empirical) probability that

(A) He or she has tried the diet cola or the regular cola? What are the (empirical) odds for this event?

(B) He or she has tried one of the colas but not both? What are the (empirical) odds against this event?

SOLUTION Let D be the event that a person has tried the diet cola and R the event that a person has tried the regular cola. The events D and R can be used to partition the population of Springfield into four mutually exclusive subsets (a collection of subsets is mutually exclusive if the intersection of any two of them is the empty set):

$D \cap R$ = set of people who have tried both colas

$D \cap R'$ = set of people who have tried the diet cola but not the regular cola

$D' \cap R$ = set of people who have tried the regular cola but not the diet cola

$D' \cap R'$ = set of people who have not tried either cola

These sets are displayed in the Venn diagram in Figure 7.

The sample population of 1,000 persons is also partitioned into four mutually exclusive sets, with $n(D) = 500$, $n(R) = 600$, and $n(D \cap R) = 200$. By using a Venn diagram (Fig. 8), we can determine the number of sample points in the sets $D \cap R'$, $D' \cap R$, and $D' \cap R'$ (see Example 2, Section 7.3).

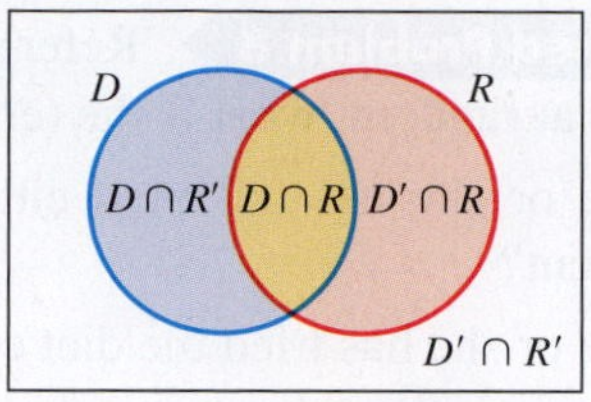

Figure 7 Total population

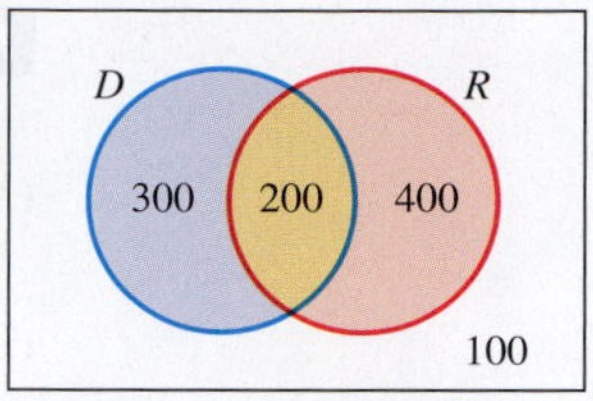

Figure 8 Sample population

These numbers are displayed in a table:

		Regular R	No Regular R'	Totals
Diet	D	200	300	500
No Diet	D'	400	100	500
Totals		600	400	1,000

Assuming that each sample point is equally likely, we form a probability table by dividing each entry in this table by 1,000, the total number surveyed. These are empirical probabilities for the sample population, which we can use to approximate the corresponding probabilities for the total population.

		Regular R	No Regular R'	Totals
Diet	D	.2	.3	.5
No Diet	D'	.4	.1	.5
Totals		.6	.4	1.0

Now we are ready to compute the required probabilities.

(A) The event that a person has tried the diet cola or the regular cola is $E = D \cup R$. We compute $P(E)$ two ways:

Method 1. Directly:

$$\begin{aligned} P(E) &= P(D \cup R) \\ &= P(D) + P(R) - P(D \cap R) \\ &= .5 + .6 - .2 = .9 \end{aligned}$$

Method 2. Using the complement of E:

$$\begin{aligned} P(E) &= 1 - P(E') \\ &= 1 - P(D' \cap R') \qquad E' = (D \cup R)' = D' \cap R' \text{ (see Fig. 8)} \\ &= 1 - .1 = .9 \end{aligned}$$

In either case,

$$\text{odds for } E = \frac{P(E)}{P(E')} = \frac{.9}{.1} = \frac{9}{1} \quad \text{or} \quad 9:1$$

(B) The event that a person has tried one cola but not both is the event that the person has tried diet cola and not regular cola or has tried regular cola and not diet cola. In terms of sets, this is event $E = (D \cap R') \cup (D' \cap R)$. Since $D \cap R'$ and $D' \cap R$ are mutually exclusive (Fig. 6),

$$\begin{aligned} P(E) &= P[(D \cap R') \cup (D' \cap R)] \\ &= P(D \cap R') + P(D' \cap R) \\ &= .3 + .4 = .7 \end{aligned}$$

$$\text{odds against } E = \frac{P(E')}{P(E)} = \frac{.3}{.7} = \frac{3}{7} \quad \text{or} \quad 3:7$$

Matched Problem 8 Refer to Example 8. If a person from Springfield is selected at random, what is the (empirical) probability that

(A) He or she has not tried either cola? What are the (empirical) odds for this event?

(B) He or she has tried the diet cola or has not tried the regular cola? What are the (empirical) odds against this event?

Exercises 8.2

Skills Warm-up Exercises

W *In Problems 1–6, write the expression as a quotient of integers, reduced to lowest terms. (If necessary, review Section A.1).*

1. $\dfrac{\frac{3}{10}}{\frac{9}{10}}$ **2.** $\dfrac{\frac{5}{12}}{\frac{7}{12}}$ **3.** $\dfrac{\frac{1}{8}}{\frac{3}{7}}$

4. $\dfrac{\frac{4}{5}}{\frac{5}{6}}$ **5.** $\dfrac{\frac{2}{9}}{1-\frac{2}{9}}$ **6.** $\dfrac{\frac{3}{16}}{1-\frac{3}{16}}$

A *Problems 7–12 refer to the Venn diagram below for events A and B in an equally likely sample space S. Find each of the indicated probabilities.*

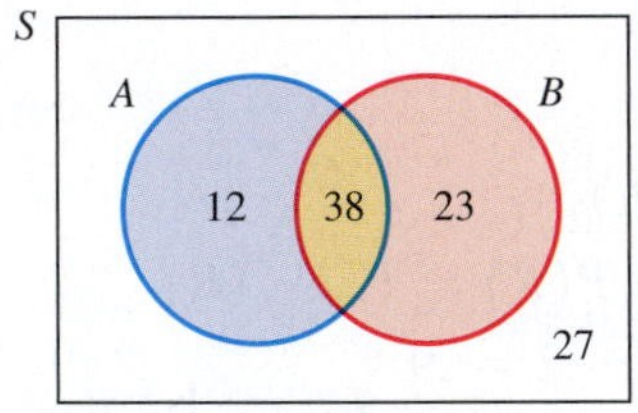

7. $P(A \cap B)$ **8.** $P(A \cup B)$

9. $P(A' \cup B)$ **10.** $P(A \cap B')$

11. $P((A \cup B)')$ **12.** $P((A \cap B)')$

A single card is drawn from a standard 52-card deck. Let D be the event that the card drawn is a diamond, and let F be the event that the card drawn is a face card. In Problems 13–24, find the indicated probabilities.

13. $P(D)$ **14.** $P(F)$

15. $P(F')$ **16.** $P(D')$

17. $P(D \cap F)$ **18.** $P(D' \cap F)$

19. $P(D \cup F)$ **20.** $P(D' \cup F)$

21. $P(D \cap F')$ **22.** $P(D' \cap F')$

23. $P(D \cup F')$ **24.** $P(D' \cup F')$

In a lottery game, a single ball is drawn at random from a container that contains 25 identical balls numbered from 1 through 25. In Problems 25–32, use equation (1) to compute the probability that the number drawn is

25. Odd or a multiple of 4

26. Even or a multiple of 7

27. Prime or greater than 20

28. Less than 10 or greater than 10

29. A multiple of 2 or a multiple of 5

30. A multiple of 3 or a multiple of 4

31. Less than 5 or greater than 20

32. Prime or less than 14

33. If the probability is .51 that a candidate wins the election, what is the probability that he loses?

34. If the probability is .03 that an automobile tire fails in less than 50,000 miles, what is the probability that the tire does not fail in 50,000 miles?

In Problems 35–38, use the equally likely sample space in Example 2 to compute the probability of the following events:

35. A sum that is less than or equal to 5

36. A sum that is greater than 9

37. The number on the first die is a 6 or the number on the second die is a 3.

38. The number on the first die is even or the number on the second die is even.

39. Given the following probabilities for an event E, find the odds for and against E:

(A) $\frac{3}{8}$ (B) $\frac{1}{4}$ (C) .4 (D) .55

40. Given the following probabilities for an event E, find the odds for and against E:

(A) $\frac{3}{5}$ (B) $\frac{1}{7}$ (C) .6 (D) .35

41. Compute the probability of event E if the odds in favor of E are

(A) $\frac{3}{8}$ (B) $\frac{11}{7}$ (C) $\frac{4}{1}$ (D) $\frac{49}{51}$

42. Compute the probability of event E if the odds in favor of E are

(A) $\frac{5}{9}$ (B) $\frac{4}{3}$ (C) $\frac{3}{7}$ (D) $\frac{23}{77}$

In Problems 43–48, discuss the validity of each statement. If the statement is always true, explain why. If not, give a counterexample.

43. If the odds for E equal the odds against E', then $P(E) = \frac{1}{2}$.

44. If the odds for E are a: b, then the odds against E are b: a.

45. If $P(E) + P(F) = P(E \cup F) + P(E \cap F)$, then E and F are mutually exclusive events.

46. The theoretical probability of an event is less than or equal to its empirical probability.

47. If E and F are complementary events, then E and F are mutually exclusive.

48. If E and F are mutually exclusive events, then E and F are complementary.

B *In Problems 49–52, compute the odds in favor of obtaining*

49. A head in a single toss of a coin

50. A number divisible by 3 in a single roll of a die

51. At least 1 head when a single coin is tossed 3 times

52. 1 head when a single coin is tossed twice

In Problems 53–56, compute the odds against obtaining

53. A number greater than 4 in a single roll of a die

54. 2 heads when a single coin is tossed twice

55. A 3 or an even number in a single roll of a die

56. An odd number or a number divisible by 3 in a single roll of a die

57. (A) What are the odds for rolling a sum of 5 in a single roll of two fair dice?

(B) If you bet \$1 that a sum of 5 will turn up, what should the house pay (plus returning your \$1 bet) if a sum of 5 turns up in order for the game to be fair?

58. (A) What are the odds for rolling a sum of 10 in a single roll of two fair dice?

(B) If you bet \$1 that a sum of 10 will turn up, what should the house pay (plus returning your \$1 bet) if a sum of 10 turns up in order for the game to be fair?

A pair of dice are rolled 1,000 times with the following frequencies of outcomes:

Sum	2	3	4	5	6	7	8	9	10	11	12
Frequency	10	30	50	70	110	150	170	140	120	80	70

Use these frequencies to calculate the approximate empirical probabilities and odds for the events in Problems 59 and 60.

59. (A) The sum is less than 4 or greater than 9.

(B) The sum is even or exactly divisible by 5.

60. (A) The sum is a prime number or is exactly divisible by 4.

(B) The sum is an odd number or exactly divisible by 3.

In Problems 61–64, a single card is drawn from a standard 52-card deck. Calculate the probability of each event.

61. A face card or a club is drawn.

62. A king or a heart is drawn.

63. A black card or an ace is drawn.

64. A heart or a number less than 7 (count an ace as 1) is drawn.

65. What is the probability of getting at least 1 diamond in a 5-card hand dealt from a standard 52-card deck?

66. What is the probability of getting at least 1 black card in a 7-card hand dealt from a standard 52-card deck?

67. What is the probability that a number selected at random from the first 100 positive integers is (exactly) divisible by 6 or 8?

68. What is the probability that a number selected at random from the first 60 positive integers is (exactly) divisible by 6 or 9?

C **69.** Explain how the three events A, B, and C from a sample space S are related to each other in order for the following equation to hold true:

$$P(A \cup B \cup C) = P(A) + P(B) + P(C) - P(A \cap B)$$

70. Explain how the three events A, B, and C from a sample space S are related to each other in order for the following equation to hold true:

$$P(A \cup B \cup C) = P(A) + P(B) + P(C)$$

71. Show that the solution to the birthday problem in Example 5 can be written in the form

$$P(E) = 1 - \frac{{}_{365}P_n}{365^n}$$

For a calculator that has a ${}_nP_r$ function, explain why this form may be better for direct evaluation than the other form used in the solution to Example 5. Try direct evaluation of both forms on a calculator for $n = 25$.

72. Many (but not all) calculators experience an overflow error when computing ${}_{365}P_n$ for $n > 39$ and when computing 365^n. Explain how you would evaluate $P(E)$ for any $n > 39$ on such a calculator.

73. In a group of n people $(n \le 12)$, what is the probability that at least 2 of them have the same birth month? (Assume that any birth month is as likely as any other.)

74. In a group of n people $(n \le 100)$, each person is asked to select a number between 1 and 100, write the number on a slip of paper and place the slip in a hat. What is the probability that at least 2 of the slips in the hat have the same number written on them?

75. If the odds in favor of an event E occurring are a to b, show that

$$P(E) = \frac{a}{a+b}$$

[*Hint:* Solve the equation $P(E)/P(E') = a/b$ for $P(E)$.]

76. If $P(E) = c/d$, show that odds in favor of E occurring are c to $d - c$.

77. The command in Figure A was used on a graphing calculator to simulate 50 repetitions of rolling a pair of dice and recording their sum. A statistical plot of the results is shown in Figure B.

(A) Use Figure B to find the empirical probability of rolling a 7 or 8.

(B) What is the theoretical probability of rolling a 7 or 8?

(C) Use a graphing calculator to simulate 200 repetitions of rolling a pair of dice and recording their sum, and find the empirical probability of rolling a 7 or 8.

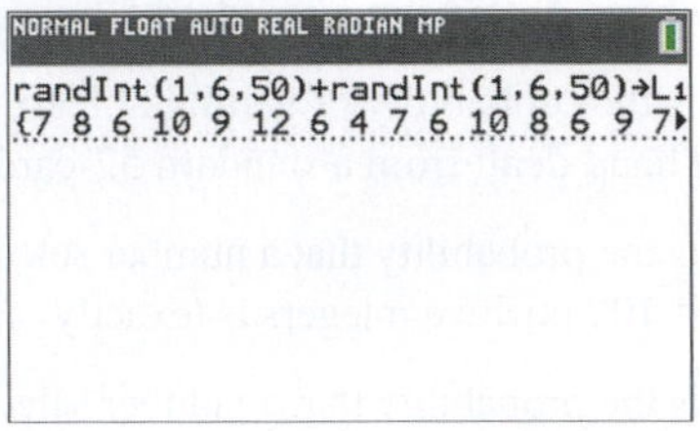

(A)

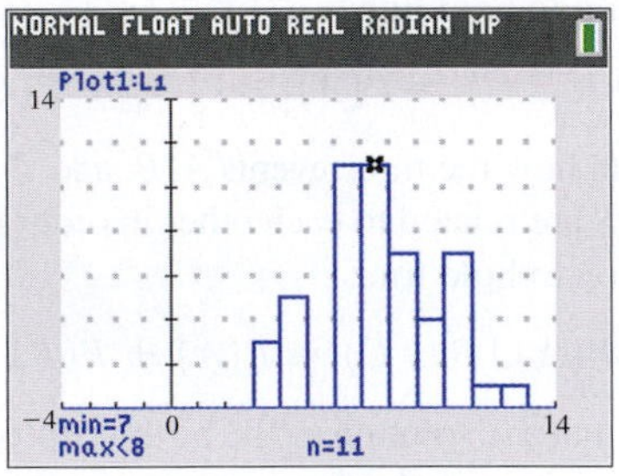

(B)

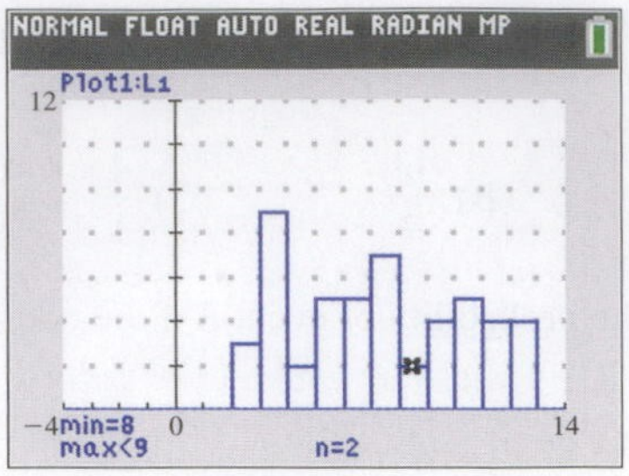

(B)

78. Consider the command in Figure A and the associated statistical plot in Figure B.

(A) Explain why the command does not simulate 50 repetitions of rolling a pair of dice and recording their sum.

(B) Describe an experiment that is simulated by this command.

(C) Simulate 200 repetitions of the experiment you described in part (B). Find the empirical probability of recording a 7 or 8, and the theoretical probability of recording a 7 or 8.

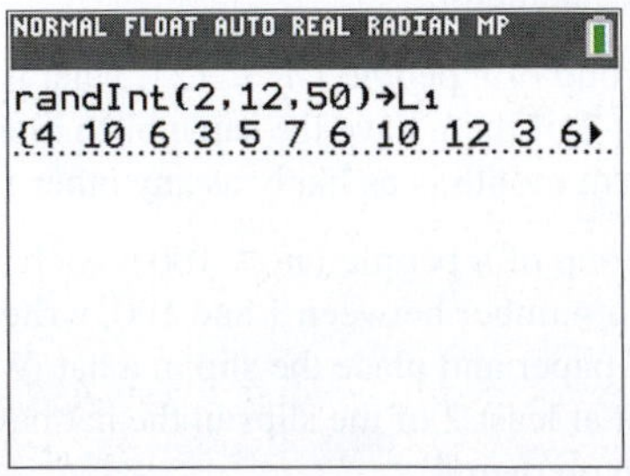

(A)

Applications

79. Market research. From a survey involving 1,000 university students, a market research company found that 750 students owned laptops, 450 owned cars, and 350 owned cars and laptops. If a university student is selected at random, what is the (empirical) probability that

(A) The student owns either a car or a laptop?

(B) The student owns neither a car nor a laptop?

80. Market research. Refer to Problem 79. If a university student is selected at random, what is the (empirical) probability that

(A) The student does not own a car?

(B) The student owns a car but not a laptop?

81. Insurance. By examining the past driving records of city drivers, an insurance company has determined the following (empirical) probabilities:

	Miles Driven per Year			
	Less Than 10,000, M_1	10,000–15,000, Inclusive, M_2	More Than 15,000, M_3	Totals
Accident A	.05	.1	.15	.3
No Accident A'	.15	.2	.35	.7
Totals	.2	.3	.5	1.0

If a city driver is selected at random, what is the probability that

(A) He or she drives less than 10,000 miles per year or has an accident?

(B) He or she drives 10,000 or more miles per year and has no accidents?

82. Insurance. Use the (empirical) probabilities in Problem 81 to find the probability that a city driver selected at random

(A) Drives more than 15,000 miles per year or has an accident

(B) Drives 15,000 or fewer miles per year and has an accident

83. Quality control. A shipment of 60 game systems, including 9 that are defective, is sent to a retail store. The receiving department selects 10 at random for testing and rejects the whole shipment if 1 or more in the sample are found to be defective. What is the probability that the shipment will be rejected?

84. **Quality control.** An assembly plant produces 40 outboard motors, including 7 that are defective. The quality control department selects 10 at random (from the 40 produced) for testing and will shut down the plant for troubleshooting if 1 or more in the sample are found to be defective. What is the probability that the plant will be shut down?

85. **Medicine.** In order to test a new drug for adverse reactions, the drug was administered to 1,000 test subjects with the following results: 60 subjects reported that their only adverse reaction was a loss of appetite, 90 subjects reported that their only adverse reaction was a loss of sleep, and 800 subjects reported no adverse reactions at all. If this drug is released for general use, what is the (empirical) probability that a person using the drug will suffer both a loss of appetite and a loss of sleep?

86. **Product testing.** To test a new car, an automobile manufacturer wants to select 4 employees to test-drive the car for 1 year. If 12 management and 8 union employees volunteer to be test drivers and the selection is made at random, what is the probability that at least 1 union employee is selected?

Problems 87 and 88 refer to the data in the following table, obtained from a random survey of 1,000 residents of a state. The participants were asked their political affiliations and their preferences in an upcoming election. (In the table, D = Democrat, R = Republican, and U = Unaffiliated.)

		D	R	U	Totals
Candidate A	A	200	100	85	385
Candidate B	B	250	230	50	530
No Preference	N	50	20	15	85
Totals		500	350	150	1,000

87. **Politics.** If a state resident is selected at random, what is the (empirical) probability that the resident is

(A) Not affiliated with a political party or has no preference? What are the odds for this event?

(B) Affiliated with a political party and prefers candidate A? What are the odds against this event?

88. **Politics.** If a state resident is selected at random, what is the (empirical) probability that the resident is

(A) A Democrat or prefers candidate B? What are the odds for this event?

(B) Not a Democrat and has no preference? What are the odds against this event?

Answers to Matched Problems

1. (A) $\frac{1}{3}$ (B) $\frac{2}{3}$ 2. (A) $\frac{1}{12}$ (B) $\frac{7}{18}$

3. $\frac{47}{140} \approx .336$ 4. .92 5. .016

6. (A) 5 : 31 (B) \$31 7. $\frac{5}{11} \approx .455$

8. (A) $P(D' \cap R') = .1$; odds for $D' \cap R' = \frac{1}{9}$ or 1 : 9

(B) $P(D \cup R') = .6$; odds against $D \cup R' = \frac{2}{3}$ or 2 : 3

8.3 Conditional Probability, Intersection, and Independence

- Conditional Probability
- Intersection of Events: Product Rule
- Probability Trees
- Independent Events
- Summary

In Section 8.2, we learned that the probability of the union of two events is related to the sum of the probabilities of the individual events (Theorem 1, p. 410):

$$P(A \cup B) = P(A) + P(B) - P(A \cap B)$$

In this section, we will learn how the probability of the intersection of two events is related to the product of the probabilities of the individual events. But first we must investigate the concept of *conditional probability*.

Conditional Probability

The probability of an event may change if we are told of the occurrence of another event. For example, if an adult (21 years or older) is selected at random from all adults in the United States, the probability of that person having lung cancer would not be high. However, if we are told that the person is also a heavy smoker, we would want to revise the probability upward.

In general, the probability of the occurrence of an event A, given the occurrence of another event B, is called a **conditional probability** and is denoted by $P(A|B)$.

In the preceding situation, events A and B would be

$$A = \text{adult has lung cancer}$$
$$B = \text{adult is a heavy smoker}$$

and $P(A|B)$ would represent the probability of an adult having lung cancer, given that he or she is a heavy smoker.

Our objective now is to try to formulate a precise definition of $P(A|B)$. It is helpful to start with a relatively simple problem, solve it intuitively, and then generalize from this experience.

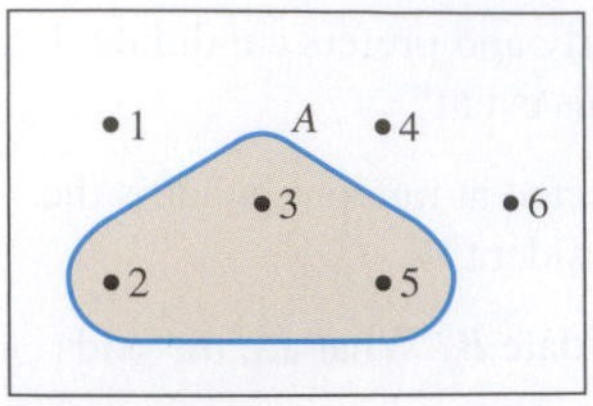

Figure 1

What is the probability of rolling a prime number (2, 3, or 5) in a single roll of a fair die? Let

$$S = \{1, 2, 3, 4, 5, 6\}$$

Then the event of rolling a prime number is (Fig. 1)

$$A = \{2, 3, 5\}$$

Thus, since we assume that each simple event in the sample space is equally likely,

$$P(A) = \frac{n(A)}{n(S)} = \frac{3}{6} = \frac{1}{2}$$

Now suppose you are asked, "In a single roll of a fair die, what is the probability that a prime number has turned up if we are given the additional information that an odd number has turned up?" The additional knowledge that another event has occurred, namely,

$$B = \text{odd number turns up}$$

puts the problem in a new light. We are now interested only in the part of event A (rolling a prime number) that is in event B (rolling an odd number). Event B, since we know it has occurred, becomes the new sample space. The Venn diagrams in Figure 2 illustrate the various relationships. Thus, the probability of A given B is the number of A elements in B divided by the total number of elements in B. Symbolically,

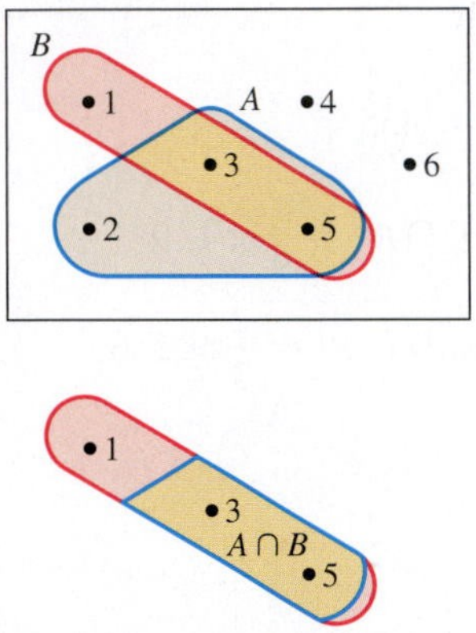

Figure 2 *B* is the new sample space.

$$P(A|B) = \frac{n(A \cap B)}{n(B)} = \frac{2}{3}$$

Dividing the numerator and denominator of $n(A \cap B)/n(B)$ by $n(S)$, the number of elements in the original sample space, we can express $P(A|B)$ in terms of $P(A \cap B)$ and $P(B)$:*

$$P(A|B) = \frac{n(A \cap B)}{n(B)} = \frac{\dfrac{n(A \cap B)}{n(S)}}{\dfrac{n(B)}{n(S)}} = \frac{P(A \cap B)}{P(B)}$$

Using the right side to compute $P(A|B)$ for the preceding example, we obtain the same result:

$$P(A|B) = \frac{P(A \cap B)}{P(B)} = \frac{\frac{2}{6}}{\frac{3}{6}} = \frac{2}{3}$$

We use the formula above to motivate the following definition of *conditional probability,* which applies to any sample space, including those having simple events that are not equally likely (see Example 1).

*Note that $P(A|B)$ is a probability based on the new sample space B, while $P(A \cap B)$ and $P(B)$ are both probabilities based on the original sample space S.

DEFINITION Conditional Probability

For events A and B in an arbitrary sample space S, we define the **conditional probability of A given B** by

$$P(A|B) = \frac{P(A \cap B)}{P(B)} \qquad P(B) \neq 0 \tag{1}$$

EXAMPLE 1 **Conditional Probability** A pointer is spun once on a circular spinner (Fig. 3). The probability assigned to the pointer landing on a given integer (from 1 to 6) is the ratio of the area of the corresponding circular sector to the area of the whole circle, as given in the table:

e_i	1	2	3	4	5	6
$P(e_i)$	.1	.2	.1	.1	.3	.2

$S = \{1, 2, 3, 4, 5, 6\}$

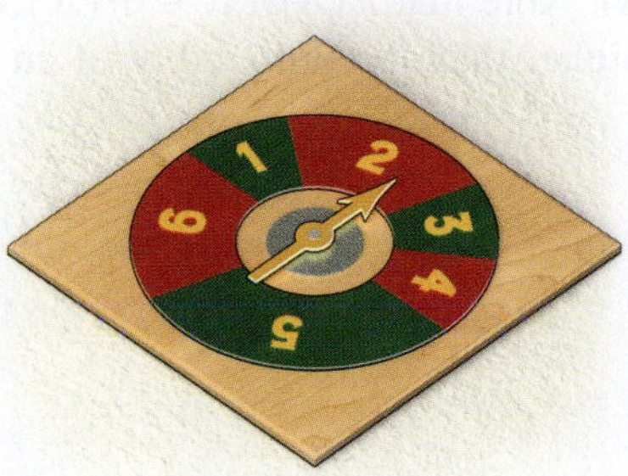

Figure 3

(A) What is the probability of the pointer landing on a prime number?

(B) What is the probability of the pointer landing on a prime number, given that it landed on an odd number?

SOLUTION Let the events E and F be defined as follows:

$$E = \text{pointer lands on a prime number} = \{2, 3, 5\}$$
$$F = \text{pointer lands on an odd number} = \{1, 3, 5\}$$

(A) $P(E) = P(2) + P(3) + P(5)$
$= .2 + .1 + .3 = .6$

(B) First note that $E \cap F = \{3, 5\}$.

$$P(E|F) = \frac{P(E \cap F)}{P(F)} = \frac{P(3) + P(5)}{P(1) + P(3) + P(5)}$$
$$= \frac{.1 + .3}{.1 + .1 + .3} = \frac{.4}{.5} = .8$$

Matched Problem 1 Refer to Example 1.

(A) What is the probability of the pointer landing on a number greater than 4?

(B) What is the probability of the pointer landing on a number greater than 4, given that it landed on an even number?

EXAMPLE 2 **Safety Research** Suppose that city records produced the following probability data on a driver being in an accident on the last day of a Memorial Day weekend:

	Accident A	No Accident A'	Totals
Rain R	.025	.335	.360
No Rain R'	.015	.625	.640
Totals	.040	.960	1.000

$S = \{RA, RA', R'A, R'A'\}$

(A) Find the probability of an accident, rain or no rain.

(B) Find the probability of rain, accident or no accident.

(C) Find the probability of an accident and rain.

(D) Find the probability of an accident, given rain.

SOLUTION

(A) Let $A = \{RA, R'A\}$ Event: "accident"

$$P(A) = P(RA) + P(R'A) = .025 + .015 = .040$$

(B) Let $R = \{RA, RA'\}$ Event: "rain"

$$P(R) = P(RA) + P(RA') = .025 + .335 = .360$$

(C) $A \cap R = \{RA\}$ Event: "accident and rain"

$$P(A \cap R) = P(RA) = .025$$

(D) $P(A|R) = \dfrac{P(A \cap R)}{P(R)} = \dfrac{.025}{.360} = .069$ Event: "accident, given rain"

Compare the result in part (D) with that in part (A). Note that $P(A|R) \neq P(A)$, and the probability of an accident, given rain, is higher than the probability of an accident without the knowledge of rain.

Matched Problem 2 Referring to the table in Example 2, determine the following:

(A) Probability of no rain

(B) Probability of an accident and no rain

(C) Probability of an accident, given no rain [Use formula (1) and the results of parts (A) and (B).]

Intersection of Events: Product Rule

Let's return to the original problem of this section, that is, representing the probability of an intersection of two events in terms of the probabilities of the individual events. If $P(A) \neq 0$ and $P(B) \neq 0$, then using formula (1), we can write

$$P(A|B) = \frac{P(A \cap B)}{P(B)} \quad \text{and} \quad P(B|A) = \frac{P(B \cap A)}{P(A)}$$

Solving the first equation for $P(A \cap B)$ and the second equation for $P(B \cap A)$, we have

$$P(A \cap B) = P(B)P(A|B) \quad \text{and} \quad P(B \cap A) = P(A)P(B|A)$$

Since $A \cap B = B \cap A$ for any sets A and B, it follows that

$$P(A \cap B) = P(B)P(A|B) = P(A)P(B|A)$$

and we have the **product rule:**

THEOREM 1 Product Rule

For events A and B with nonzero probabilities in a sample space S,

$$P(A \cap B) = P(A)P(B|A) = P(B)P(A|B) \tag{2}$$

and we can use either $P(A)P(B|A)$ or $P(B)P(A|B)$ to compute $P(A \cap B)$.

EXAMPLE 3 **Consumer Survey** If 60% of a department store's customers are female and 75% of the female customers have credit cards at the store, what is the probability that a customer selected at random is a female and has a store credit card?

SOLUTION Let

$$S = \text{all store customers}$$
$$F = \text{female customers}$$
$$C = \text{customers with a store credit card}$$

If 60% of the customers are female, then the probability that a customer selected at random is a female is

$$P(F) = .60$$

Since 75% of the female customers have store credit cards, the probability that a customer has a store credit card, given that the customer is a female, is

$$P(C|F) = .75$$

Using equation (2), the probability that a customer is a female and has a store credit card is

$$P(F \cap C) = P(F)P(C|F) = (.60)(.75) = .45$$

Matched Problem 3 If 80% of the male customers of the department store in Example 3 have store credit cards, what is the probability that a customer selected at random is a male and has a store credit card?

Figure 4

Probability Trees

We used tree diagrams in Section 7.3 to help us count the number of combined outcomes in a sequence of experiments. In a similar way, we will use probability trees to help us compute the probabilities of combined outcomes in a sequence of experiments.

EXAMPLE 4 **Probability Tree** Two balls are drawn in succession, without replacement, from a box containing 3 blue and 2 white balls (Fig. 4). What is the probability of drawing a white ball on the second draw?

SOLUTION We start with a tree diagram (Fig. 5) showing the combined outcomes of the two experiments (first draw and second draw). Then we assign a probability to each branch of the tree (Fig. 6). For example, we assign the probability $\frac{2}{5}$ to the branch Sw_1, since this is the probability of drawing a white ball on the first draw (there are 2 white balls and 3 blue balls in the box). What probability should be assigned to the branch w_1w_2? This is the conditional probability $P(w_2|w_1)$, that is, the probability of drawing a white ball on the second draw given that a white ball was drawn on the first draw and not replaced. Since the box now contains 1 white ball and 3 blue balls, the probability is $\frac{1}{4}$. Continuing in the same way, we assign probabilities to the other branches of the tree and obtain Figure 6.

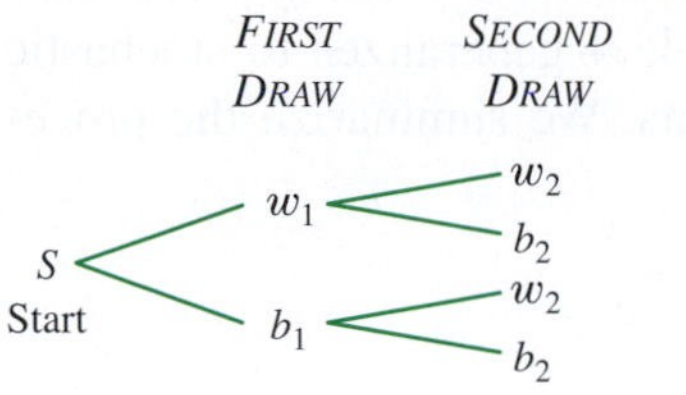

Figure 5

What is the probability of the combined outcome $w_1 \cap w_2$, that is, the probability of drawing a white ball on the first draw and a white ball on the second draw?* Using the product rule (2), we have

$$P(w_1 \cap w_2) = P(w_1)P(w_2|w_1)$$
$$= \left(\frac{2}{5}\right)\left(\frac{1}{4}\right) = \frac{1}{10}$$

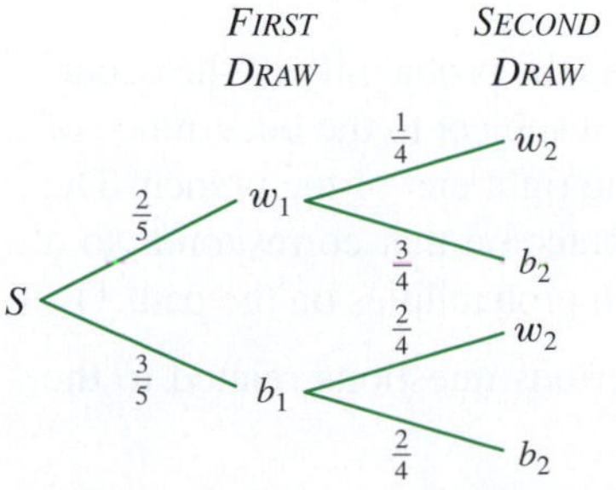

Figure 6

The combined outcome $w_1 \cap w_2$ corresponds to the unique path Sw_1w_2 in the tree diagram, and we see that the probability of reaching w_2 along this path is the product of the probabilities assigned to the branches on the path. Reasoning in this way, we

*The sample space for the combined outcomes is $S = \{w_1w_2, w_1b_2, b_1w_2, b_1b_2\}$. If we let $w_1 = \{w_1w_2, w_1b_2\}$ and $w_2 = \{w_1w_2, b_1w_2\}$, then $w_1 \cap w_2 = \{w_1w_2\}$.

obtain the probability of each remaining combined outcome by multiplying the probabilities assigned to the branches on the path corresponding to the given combined outcomes. These probabilities are often written at the ends of the paths to which they correspond (Fig. 7).

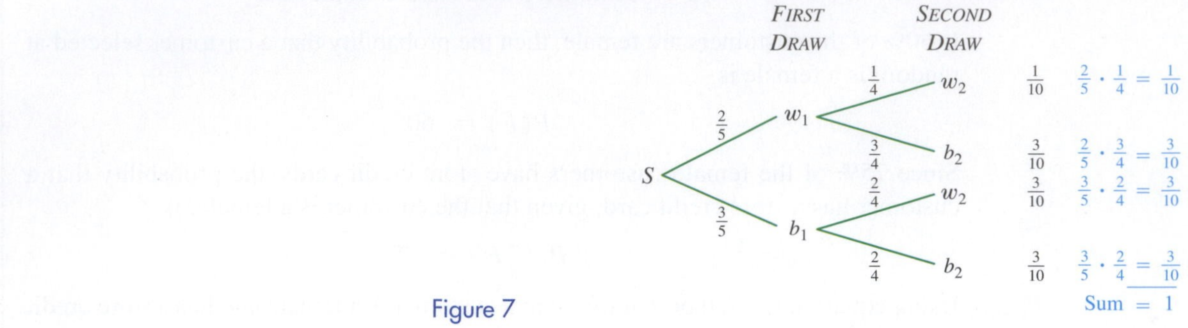

Figure 7

Now we can complete the problem. A white ball drawn on the second draw corresponds to either the combined outcome $w_1 \cap w_2$ or $b_1 \cap w_2$ occurring. Thus, since these combined outcomes are mutually exclusive,

$$P(w_2) = P(w_1 \cap w_2) + P(b_1 \cap w_2)$$
$$= \frac{1}{10} + \frac{3}{10} = \frac{4}{10} = \frac{2}{5}$$

which is the sum of the probabilities listed at the ends of the two paths terminating in w_2.

Matched Problem 4 Two balls are drawn in succession without replacement from a box containing 4 red and 2 white balls. What is the probability of drawing a red ball on the second draw?

The sequence of two experiments in Example 4 is an example of a *stochastic process*. In general, a **stochastic process** involves a sequence of experiments where the outcome of each experiment is not certain. Our interest is in making predictions about the process as a whole. The analysis in Example 4 generalizes to stochastic processes involving any finite sequence of experiments. We summarize the procedures used in Example 4 for general application:

PROCEDURE Constructing Probability Trees

Step 1 Draw a tree diagram corresponding to all combined outcomes of the sequence of experiments.

Step 2 Assign a probability to each tree branch. (This is the probability of the occurrence of the event on the right end of the branch subject to the occurrence of all events on the path leading to the event on the right end of the branch. The probability of the occurrence of a combined outcome that corresponds to a path through the tree is the product of all branch probabilities on the path.*)

Step 3 Use the results in Steps 1 and 2 to answer various questions related to the sequence of experiments as a whole.

*If we form a sample space S such that each simple event in S corresponds to one path through the tree, and if the probability assigned to each simple event in S is the product of the branch probabilities on the corresponding path, then it can be shown that this is not only an acceptable assignment (all probabilities for the simple events in S are nonnegative and their sum is 1), but it is the only assignment consistent with the method used to assign branch probabilities within the tree.

Explore and Discuss 1

Refer to the table on rain and accidents in Example 2 and use formula (1), where appropriate, to complete the following probability tree:

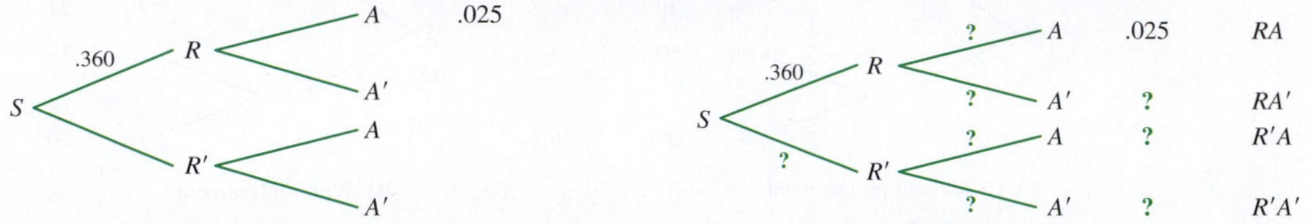

Discuss the difference between $P(R \cap A)$ and $P(A|R)$.

EXAMPLE 5 **Product Defects** An auto company A subcontracts the manufacturing of its onboard computers to two companies: 40% to company B and 60% to company C. Company B in turn subcontracts 70% of the orders it receives from company A to company D and the remaining 30% to company E, both subsidiaries of company B. When the onboard computers are completed by companies D, E, and C, they are shipped to company A to be used in various car models. It has been found that 1.5%, 1%, and .5% of the boards from D, E, and C, respectively, prove defective during the 3-year warranty period after a car is first sold. What is the probability that a given onboard computer will be defective during the 3-year warranty period?

SOLUTION Draw a tree diagram and assign probabilities to each branch (Fig. 8):

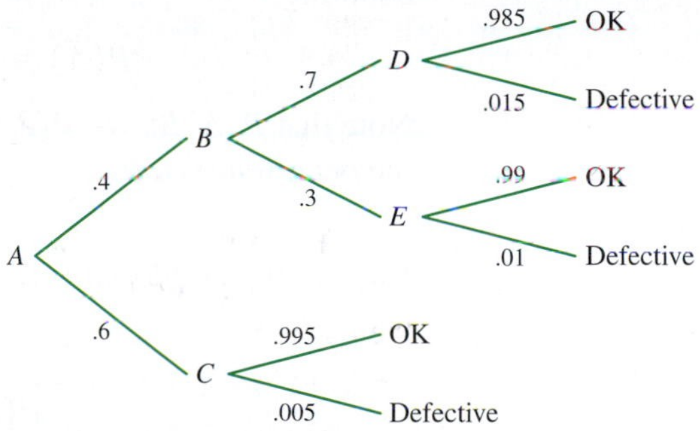

Figure 8

There are three paths leading to defective (the onboard computer will be defective within the 3-year warranty period). We multiply the branch probabilities on each path and add the three products:

$$\begin{aligned} P(\text{defective}) &= (.4)(.7)(.015) + (.4)(.3)(.01) + (.6)(.005) \\ &= .0084 \end{aligned}$$

Matched Problem 5 In Example 5, what is the probability that a given onboard computer came from company E or C?

Independent Events

We return to Example 4, which involved drawing two balls in succession without replacement from a box of 3 blue and 2 white balls. What difference does "without replacement" and "with replacement" make? Figure 9 shows probability trees corresponding to each case. Go over the probability assignments for the branches in Figure 9B to convince yourself of their correctness.

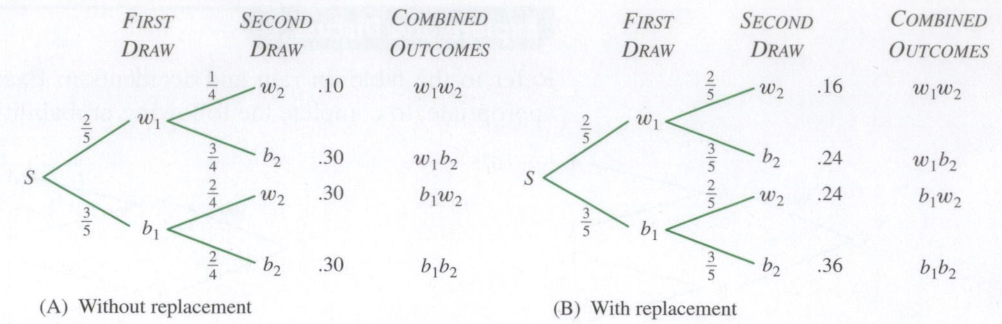

Figure 9 $S = \{w_1w_2, w_1b_2, b_1w_2, b_1b_2\}$

Let

$$A = \text{white ball on second draw} = \{w_1w_2, b_1w_2\}$$
$$B = \text{white ball on first draw} = \{w_1w_2, w_1b_2\}$$

We now compute $P(A|B)$ and $P(A)$ for each case in Figure 9.

Case 1. *Without replacement:*

$$P(A|B) = \frac{P(A \cap B)}{P(B)} = \frac{P\{w_1w_2\}}{P\{w_1w_2, w_1b_2\}} = \frac{.10}{.10 + .30} = .25$$

(This is the assignment to branch w_1w_2 that we made by looking in the box and counting.)

$$P(A) = P\{w_1w_2, b_1w_2\} = .10 + .30 = .40$$

Note that $P(A|B) \neq P(A)$, and we conclude that the probability of A is affected by the occurrence of B.

Case 2. *With replacement:*

$$P(A|B) = \frac{P(A \cap B)}{P(B)} = \frac{P\{w_1w_2\}}{P\{w_1w_2, w_1b_2\}} = \frac{.16}{.16 + .24} = .40$$

(This is the assignment to branch w_1w_2 that we made by looking in the box and counting.)

$$P(A) = P\{w_1w_2, b_1w_2\} = .16 + .24 = .40$$

Note that $P(A|B) = P(A)$, and we conclude that the probability of A is not affected by the occurrence of B.

Intuitively, if $P(A|B) = P(A)$, then it appears that event A is "independent" of B. Let us pursue this further. If events A and B are such that

$$P(A|B) = P(A)$$

then replacing the left side by its equivalent from formula (1), we obtain

$$\frac{P(A \cap B)}{P(B)} = P(A)$$

After multiplying both sides by $P(B)$, the last equation becomes

$$P(A \cap B) = P(A)P(B)$$

This result motivates the following definition of *independence:*

DEFINITION Independence

If A and B are any events in a sample space S, we say that **A and B are independent** if

$$P(A \cap B) = P(A)P(B) \tag{3}$$

Otherwise, A and B are said to be **dependent**.

From the definition of independence one can prove (see Problems 75 and 76, Exercises 8.3) the following theorem:

THEOREM 2 On Independence

If A and B are independent events with nonzero probabilities in a sample space S, then

$$P(A|B) = P(A) \quad \text{and} \quad P(B|A) = P(B) \tag{4}$$

If either equation in (4) holds, then A and B are independent.

CONCEPTUAL INSIGHT

Sometimes intuitive reasoning can be helpful in deciding whether or not two events are independent. Suppose that a fair coin is tossed five times. What is the probability of a head on the fifth toss, given that the first four tosses are all heads? Our intuition tells us that a coin has no memory, so the probability of a head on the fifth toss given four previous heads should be equal to the probability of a head on the fifth toss, namely, 1/2. In other words, the first equation of Theorem 2 holds intuitively, so "heads on the fifth toss" and "heads on the first four tosses" are independent events.

Often, unfortunately, intuition is *not* a reliable guide to the notion of independence. Independence is a technical concept. So in all cases, an appropriate sample space should be chosen, and either equation (3) or equation (4) should be tested, to confirm that two events are (or are not) independent.

EXAMPLE 6 **Testing for Independence** In two tosses of a single fair coin, show that the events "A head on the first toss" and "A head on the second toss" are independent.

SOLUTION Consider the sample space of equally likely outcomes for the tossing of a fair coin twice,

$$S = \{\text{HH, HT, TH, TT}\}$$

and the two events,

$$A = \text{a head on the first toss} = \{\text{HH, HT}\}$$
$$B = \text{a head on the second toss} = \{\text{HH, TH}\}$$

Then

$$P(A) = \frac{2}{4} = \frac{1}{2} \qquad P(B) = \frac{2}{4} = \frac{1}{2} \qquad P(A \cap B) = \frac{1}{4}$$

Thus,

$$P(A \cap B) = \frac{1}{4} = \frac{1}{2} \cdot \frac{1}{2} = P(A)P(B)$$

and the two events are independent. (The theory agrees with our intuition—a coin has no memory.)

Matched Problem 6 In Example 6, compute $P(B|A)$ and compare with $P(B)$.

EXAMPLE 7 **Testing for Independence** A single card is drawn from a standard 52-card deck. Test the following events for independence (try guessing the answer to each part before looking at the solution):

(A) $E =$ the drawn card is a spade.
$F =$ the drawn card is a face card.

(B) $G =$ the drawn card is a club.
$H =$ the drawn card is a heart.

SOLUTION

(A) To test E and F for independence, we compute $P(E \cap F)$ and $P(E)P(F)$. If they are equal, then events E and F are independent; if they are not equal, then events E and F are dependent.

$$P(E \cap F) = \frac{3}{52} \qquad P(E)P(F) = \left(\frac{13}{52}\right)\left(\frac{12}{52}\right) = \frac{3}{52}$$

Events E and F are independent. (Did you guess this?)

(B) Proceeding as in part (A), we see that

$$P(G \cap H) = P(\varnothing) = 0 \qquad P(G)P(H) = \left(\frac{13}{52}\right)\left(\frac{13}{52}\right) = \frac{1}{16}$$

Events G and H are dependent. (Did you guess this?)

⚠ CAUTION Students often confuse *mutually exclusive (disjoint) events* with *independent events*. One does not necessarily imply the other. In fact, it is not difficult to show (see Problem 79, Exercises 8.3) that any two mutually exclusive events A and B, with nonzero probabilities, are always dependent. ▲

Matched Problem 7 A single card is drawn from a standard 52-card deck. Test the following events for independence:

(A) $E =$ the drawn card is a red card
$F =$ the drawn card's number is divisible by 5 (face cards are not assigned values)

(B) $G =$ the drawn card is a king
$H =$ the drawn card is a queen

Explore and Discuss 2

In college basketball, would it be reasonable to assume that the following events are independent? Explain why or why not.

$A =$ the Golden Eagles win in the first round of the NCAA tournament.

$B =$ the Golden Eagles win in the second round of the NCAA tournament.

The notion of independence can be extended to more than two events:

DEFINITION Independent Set of Events

A set of events is said to be **independent** if for each finite subset $\{E_1, E_2, \ldots, E_k\}$

$$P(E_1 \cap E_2 \cap \cdots \cap E_k) = P(E_1)P(E_2) \cdot \cdots \cdot P(E_k) \qquad (5)$$

The next example makes direct use of this definition.

EXAMPLE 8 **Computer Control Systems** A space shuttle has four independent computer control systems. If the probability of failure (during flight) of any one system is .001, what is the probability of failure of all four systems?

SOLUTION Let

E_1 = failure of system 1 $\qquad$ E_3 = failure of system 3

E_2 = failure of system 2 $\qquad$ E_4 = failure of system 4

Then, since events E_1, E_2, E_3, and E_4 are given to be independent,

$$\begin{aligned} P(E_1 \cap E_2 \cap E_3 \cap E_4) &= P(E_1)P(E_2)P(E_3)P(E_4) \\ &= (.001)^4 \\ &= .000\ 000\ 000\ 001 \end{aligned}$$

Matched Problem 8 A single die is rolled 6 times. What is the probability of getting the sequence 1, 2, 3, 4, 5, 6?

Summary

The key results in this section are summarized in the following box:

SUMMARY Key Concepts

Conditional Probability

$$P(A|B) = \frac{P(A \cap B)}{P(B)} \qquad P(B|A) = \frac{P(B \cap A)}{P(A)}$$

Note: $P(A|B)$ is a probability based on the new sample space B, while $P(A \cap B)$ and $P(B)$ are probabilities based on the original sample space S.

Product Rule

$$P(A \cap B) = P(A)P(B|A) = P(B)P(A|B)$$

Independent Events

- A and B are **independent** if

$$P(A \cap B) = P(A)P(B)$$

- If A and B are independent events with nonzero probabilities, then

$$P(A|B) = P(A) \quad \text{and} \quad P(B|A) = P(B)$$

- If A and B are events with nonzero probabilities and either $P(A|B) = P(A)$ or $P(B|A) = P(B)$, then A and B are independent.
- If $E_1, E_2, \ldots, E_n$ are independent, then

$$P(E_1 \cap E_2 \cap \cdots \cap E_n) = P(E_1)P(E_2) \cdot \cdots \cdot P(E_n)$$

Exercises 8.3

Skills Warm-up Exercises

W *In Problems 1–6, use a tree diagram to represent a factorization of the given integer into primes, so that there are two branches at each number that is not prime. For example, the factorization $24 = 4 \cdot 6 = (2 \cdot 2) \cdot (2 \cdot 3)$ is represented by:*

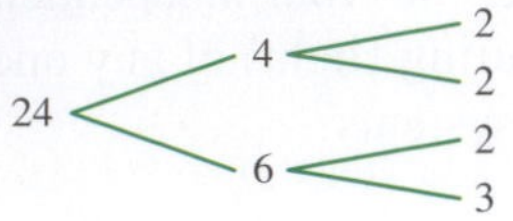

(If necessary, review Section A.3.)

1. 100 **2.** 120

3. 180 **4.** 225

5. 315 **6.** 360

A *A single card is drawn from a standard 52-card deck. In Problems 7–14, find the conditional probability that*

7. The card is an ace, given that it is a heart.

8. The card is red, given that it is a face card.

9. The card is a heart, given that it is an ace.

10. The card is a face card, given that it is red.

11. The card is black, given that it is a club.

12. The card is a jack, given that it is red.

13. The card is a club, given that it is black.

14. The card is red, given that it is a jack.

In Problems 15–22, find the conditional probability, in a single roll of two fair dice, that

15. The sum is less than 6, given that the sum is even.

16. The sum is 10, given that the roll is doubles.

17. The sum is even, given that the sum is less than 6.

18. The roll is doubles, given that the sum is 10.

19. The sum is greater than 7, given that neither die is a six.

20. The sum is odd, given that at least one die is a six.

21. Neither die is a six, given that the sum is greater than 7.

22. At least one die is a six, given that the sum is odd.

In Problems 23–42, use the table below. Events A, B, and C are mutually exclusive; so are D, E, and F.

	A	B	C	Totals
D	.20	.03	.07	.30
E	.28	.05	.07	.40
F	.22	.02	.06	.30
Totals	.70	.10	.20	1.00

In Problems 23–26, find each probability directly from the table.

23. $P(B)$ **24.** $P(E)$

25. $P(B \cap D)$ **26.** $P(C \cap E)$

In Problems 27–34, compute each probability using formula (1) on page 423 and appropriate table values.

27. $P(D|B)$ **28.** $P(C|E)$

29. $P(B|D)$ **30.** $P(E|C)$

31. $P(D|C)$ **32.** $P(E|A)$

33. $P(A|C)$ **34.** $P(B|B)$

In Problems 35–42, test each pair of events for independence.

35. *A* and *D* **36.** *A* and *E*

37. *B* and *D* **38.** *B* and *E*

39. *B* and *F* **40.** *C* and *F*

41. *A* and *B* **42.** *D* and *F*

B **43.** A fair coin is tossed 8 times.

(A) What is the probability of tossing a head on the 8th toss, given that the preceding 7 tosses were heads?

(B) What is the probability of getting 8 heads or 8 tails?

44. A fair die is rolled 5 times.

(A) What is the probability of getting a 6 on the 5th roll, given that a 6 turned up on the preceding 4 rolls?

(B) What is the probability that the same number turns up every time?

45. A pointer is spun once on the circular spinner shown below. The probability assigned to the pointer landing on a given integer (from 1 to 5) is the ratio of the area of the corresponding circular sector to the area of the whole circle, as given in the table:

e_i	1	2	3	4	5
$P(e_i)$	.3	.1	.2	.3	.1

Given the events

E = pointer lands on an even number

F = pointer lands on a number less than 4

(A) Find $P(F|E)$.

(B) Test events E and F for independence.

46. Repeat Problem 45 with the following events:

E = pointer lands on an odd number

F = pointer lands on a prime number

Compute the indicated probabilities in Problems 47 and 48 by referring to the following probability tree:

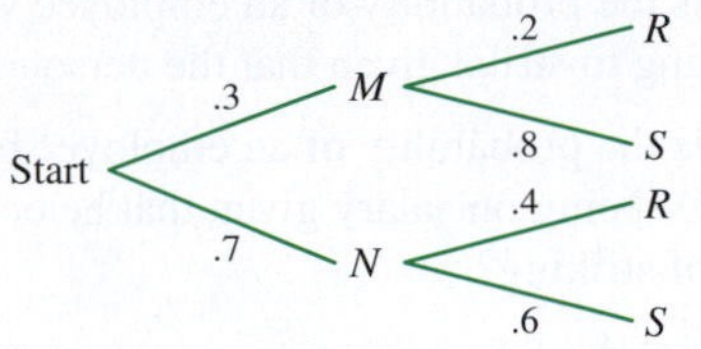

47. (A) $P(M \cap S)$ (B) $P(R)$

48. (A) $P(N \cap R)$ (B) $P(S)$

49. A fair coin is tossed twice. Consider the sample space $S = \{HH, HT, TH, TT\}$ of equally likely simple events. We are interested in the following events:

E_1 = a head on the first toss

E_2 = a tail on the first toss

E_3 = a tail on the second toss

E_4 = a head on the second toss

For each pair of events, discuss whether they are independent and whether they are mutually exclusive.

(A) E_1 and E_4 (B) E_1 and E_2

50. For each pair of events (see Problem 49), discuss whether they are independent and whether they are mutually exclusive.

(A) E_1 and E_3 (B) E_3 and E_4

51. In 2 throws of a fair die, what is the probability that you will get an even number on each throw? An even number on the first or second throw?

52. In 2 throws of a fair die, what is the probability that you will get at least 5 on each throw? At least 5 on the first or second throw?

53. Two cards are drawn in succession from a standard 52-card deck. What is the probability that the first card is a club and the second card is a heart

(A) If the cards are drawn without replacement?

(B) If the cards are drawn with replacement?

54. Two cards are drawn in succession from a standard 52-card deck. What is the probability that both cards are red

(A) If the cards are drawn without replacement?

(B) If the cards are drawn with replacement?

55. A card is drawn at random from a standard 52-card deck. Events G and H are

G = the drawn card is black.

H = the drawn card is divisible by 3 (face cards are not valued).

(A) Find $P(H|G)$.

(B) Test H and G for independence.

56. A card is drawn at random from a standard 52-card deck. Events M and N are

M = the drawn card is a diamond.

N = the drawn card is even (face cards are not valued).

(A) Find $P(N|M)$.

(B) Test M and N for independence.

57. Let A be the event that all of a family's children are the same gender, and let B be the event that the family has at most 1 boy. Assuming the probability of having a girl is the same as the probability of having a boy (both .5), test events A and B for independence if

(A) The family has 2 children.

(B) The family has 3 children.

58. An experiment consists of tossing n coins. Let A be the event that at least 2 heads turn up, and let B be the event that all the coins turn up the same. Test A and B for independence if

(A) 2 coins are tossed.

(B) 3 coins are tossed.

Problems 59–62 refer to the following experiment: 2 balls are drawn in succession out of a box containing 2 red and 5 white balls. Let R_i be the event that the ith ball is red, and let W_i be the event that the ith ball is white.

59. Construct a probability tree for this experiment and find the probability of each of the events $R_1 \cap R_2$, $R_1 \cap W_2$, $W_1 \cap R_2$, $W_1 \cap W_2$, given that the first ball drawn was

(A) Replaced before the second draw

(B) Not replaced before the second draw

60. Find the probability that the second ball was red, given that the first ball was

(A) Replaced before the second draw

(B) Not replaced before the second draw

61. Find the probability that at least 1 ball was red, given that the first ball was

(A) Replaced before the second draw

(B) Not replaced before the second draw

62. Find the probability that both balls were the same color, given that the first ball was

(A) Replaced before the second draw

(B) Not replaced before the second draw

C *In Problems 63–70, discuss the validity of each statement. If the statement is always true, explain why. If not, give a counterexample.*

63. If $P(A|B) = P(B)$, then A and B are independent.

64. If A and B are independent, then $P(A|B) = P(B|A)$.

65. If A is nonempty and $A \subset B$, then $P(A|B) \geq P(A)$.

66. If A and B are events, then $P(A|B) \leq P(B)$.

67. If A and B are mutually exclusive, then A and B are independent.

68. If A and B are independent, then A and B are mutually exclusive.

69. If two balls are drawn in succession, with replacement, from a box containing m red and n white balls $(m \geq 1$ and $n \geq 1)$, then

$$P(W_1 \cap R_2) = P(R_1 \cap W_2)$$

70. If two balls are drawn in succession, without replacement, from a box containing m red and n white balls $(m \geq 1$ and $n \geq 1)$, then

$$P(W_1 \cap R_2) = P(R_1 \cap W_2)$$

71. A box contains 2 red, 3 white, and 4 green balls. Two balls are drawn out of the box in succession without replacement. What is the probability that both balls are the same color?

72. For the experiment in Problem 71, what is the probability that no white balls are drawn?

73. An urn contains 2 one-dollar bills, 1 five-dollar bill, and 1 ten-dollar bill. A player draws bills one at a time without replacement from the urn until a ten-dollar bill is drawn. Then the game stops. All bills are kept by the player.

(A) What is the probability of winning \$16?

(B) What is the probability of winning all bills in the urn?

(C) What is the probability of the game stopping at the second draw?

74. Ann and Barbara are playing a tennis match. The first player to win 2 sets wins the match. For any given set, the probability that Ann wins that set is $\frac{2}{3}$. Find the probability that

(A) Ann wins the match.

(B) 3 sets are played.

(C) The player who wins the first set goes on to win the match.

75. Show that if A and B are independent events with nonzero probabilities in a sample space S, then

$$P(A|B) = P(A) \quad \text{and} \quad P(B|A) = P(B)$$

76. Show that if A and B are events with nonzero probabilities in a sample space S, and either $P(A|B) = P(A)$ or $P(B|A) = P(B)$, then events A and B are independent.

77. Show that $P(A|A) = 1$ when $P(A) \neq 0$.

78. Show that $P(A|B) + P(A'|B) = 1$.

79. Show that A and B are dependent if A and B are mutually exclusive and $P(A) \neq 0, P(B) \neq 0$.

80. Show that $P(A|B) = 1$ if B is a subset of A and $P(B) \neq 0$.

Applications

81. Labor relations. In a study to determine employee voting patterns in a recent strike election, 1,000 employees were selected at random and the following tabulation was made:

		Salary Classification		
		Hourly (H)	Salary (S)	Totals
To	Yes (Y)	400	200	600
Strike	No (N)	150	250	400
Totals		550	450	1,000

(A) Convert this table to a probability table by dividing each entry by 1,000.

(B) What is the probability of an employee voting to strike? Of voting to strike given that the person is paid hourly?

(C) What is the probability of an employee being on salary (S)? Of being on salary given that he or she voted in favor of striking?

82. Quality control. An automobile manufacturer produces 37% of its cars at plant A. If 5% of the cars manufactured at plant A have defective emission control devices, what is the probability that one of this manufacturer's cars was manufactured at plant A and has a defective emission control device?

83. Bonus incentives. If a salesperson has gross sales of over \$600,000 in a year, then he or she is eligible to play the company's bonus game: A black box contains 1 twenty-dollar bill, 2 five-dollar bills, and 1 one-dollar bill. Bills are drawn out of the box one at a time without replacement until a twenty-dollar bill is drawn. Then the game stops. The salesperson's bonus is 1,000 times the value of the bills drawn.

(A) What is the probability of winning a \$26,000 bonus?

(B) What is the probability of winning the maximum bonus, \$31,000, by drawing out all bills in the box?

(C) What is the probability of the game stopping at the third draw?

84. Personnel selection. To transfer into a particular technical department, a company requires an employee to pass a screening test. A maximum of 3 attempts are allowed at 6-month intervals between trials. From past records it is found that 40% pass on the first trial; of those that fail the first trial and take the test a second time, 60% pass; and of those that fail on the second trial and take the test a third time, 20% pass. For an employee wishing to transfer:

(A) What is the probability of passing the test on the first or second try?

(B) What is the probability of failing on the first 2 trials and passing on the third?

(C) What is the probability of failing on all 3 attempts?

85. U.S. Food and Drug Administration. An ice cream company wishes to use a new red dye to enhance the color in its strawberry ice cream. The U.S. Food and Drug Administration (FDA) requires the dye to be tested for cancer-producing potential using laboratory rats. The results of one test on 1,000 rats are summarized in the following table:

		Cancer C	No Cancer C'	Totals
Ate Red Dye	R	60	440	500
No Red Dye	R'	20	480	500
Totals		80	920	1,000

(A) Convert the table into a probability table by dividing each entry by 1,000.

(B) Are "developing cancer" and "eating red dye" independent events?

(C) Should the FDA approve or ban the use of the dye? Explain why or why not using $P(C|R)$ and $P(C)$.

86. Genetics. In a study to determine frequency and dependency of color-blindness relative to females and males, 1,000 people were chosen at random, and the following results were recorded:

	Female F	Male F'	Totals
Color-Blind C	2	24	26
Normal C'	518	456	974
Totals	520	480	1,000

(A) Convert this table to a probability table by dividing each entry by 1,000.

(B) What is the probability that a person is a woman, given that the person is color-blind? Are the events color-blindness and female independent?

(C) What is the probability that a person is color-blind, given that the person is a male? Are the events color-blindness and male independent?

Problems 87 and 88 refer to the data in the following table, obtained in a study to determine the frequency and dependency of IQ ranges relative to males and females. 1,000 people were chosen at random and the following results were recorded:

	IQ			
	Below 90 (A)	90–120 (B)	Above 120 (C)	Totals
Female F	130	286	104	520
Male F'	120	264	96	480
Totals	250	550	200	1,000

87. Psychology.

(A) What is the probability of a person having an IQ below 90, given that the person is a female? A male?

(B) What is the probability of a person having an IQ below 90?

(C) Are events A and F dependent? A and F'?

88. Psychology.

(A) What is the probability of a person having an IQ above 120, given that the person is a female? A male?

(B) What is the probability of a person being female and having an IQ above 120?

(C) Are events C and F dependent? C and F'?

89. Voting patterns. A survey of a precinct's residents revealed that 55% of the residents were members of the Democratic party and 60% of the Democratic party members voted in the last election. What is the probability that a person selected at random from this precinct is a member of the Democratic party and voted in the last election?

Answers to Matched Problems

1. (A) .5 (B) .4

2. (A) $P(R') = .640$ (B) $P(A \cap R') = .015$
(C) $P(A|R') = \dfrac{P(A \cap R')}{P(R')} = .023$

3. $P(M \cap C) = P(M)P(C|M) = .32$

4. $\frac{2}{3}$ 5. .72

6. $P(B|A) = \dfrac{P(A \cap B)}{P(A)} = \dfrac{\frac{1}{4}}{\frac{1}{2}} = \dfrac{1}{2} = P(B)$

7. (A) E and F are independent.
(B) G and H are dependent.

8. $\left(\frac{1}{6}\right)^6 \approx .000\,021\,4$

8.4 Bayes' Formula

In the preceding section, we discussed the conditional probability of the occurrence of an event, given the occurrence of an earlier event. Now we will reverse the problem and try to find the probability of an earlier event conditioned on the occurrence of a later event. As you will see, a number of practical problems have this form. First, let us consider a relatively simple problem that will provide the basis for a generalization.

EXAMPLE 1 **Probability of an Earlier Event Given a Later Event** One urn has 3 blue and 2 white balls; a second urn has 1 blue and 3 white balls (Fig. 1). A single fair die is rolled and if 1 or 2 comes up, a ball is drawn out of the first urn; otherwise, a ball is drawn out of the second urn. If the drawn ball is blue, what is the probability that it came out of the first urn? Out of the second urn?

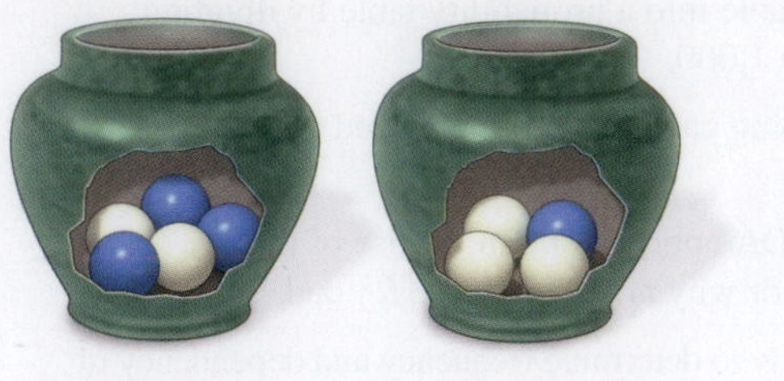

Figure 1

SOLUTION We form a probability tree, letting U_1 represent urn 1, U_2 urn 2, B a blue ball, and W a white ball. Then, on the various outcome branches, we assign appropriate probabilities. For example, $P(U_1) = \frac{1}{3}$, $P(B|U_1) = \frac{3}{5}$, and so on:

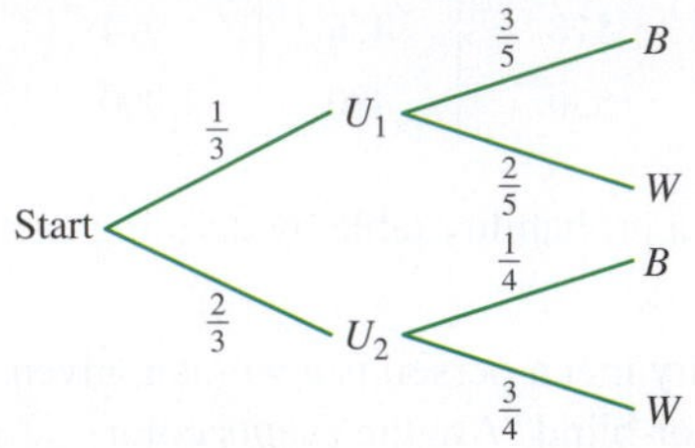

Now we are interested in finding $P(U_1|B)$, that is, the probability that the ball came out of urn 1, given that the drawn ball is blue. Using equation (1) from Section 8.3, we can write

$$P(U_1|B) = \frac{P(U_1 \cap B)}{P(B)} \tag{1}$$

If we look at the tree diagram, we can see that B is at the end of two different branches; thus,

$$P(B) = P(U_1 \cap B) + P(U_2 \cap B) \tag{2}$$

After substituting equation (2) into equation (1), we get

$$P(U_1|B) = \frac{P(U_1 \cap B)}{P(U_1 \cap B) + P(U_2 \cap B)} \qquad P(A \cap B) = P(A)P(B|A)$$

$$= \frac{P(U_1)P(B|U_1)}{P(U_1)P(B|U_1) + P(U_2)P(B|U_2)}$$

$$= \frac{P(B|U_1)P(U_1)}{P(B|U_1)P(U_1) + P(B|U_2)P(U_2)} \tag{3}$$

Equation (3) is really a lot simpler to use than it looks. You do not need to memorize it; you simply need to understand its form relative to the probability tree above. Referring to the probability tree, we see that

$$P(B|U_1)P(U_1) = \text{product of branch probabilities leading to } B \text{ through } U_1$$

$$= \left(\frac{3}{5}\right)\left(\frac{1}{3}\right) \qquad \text{We usually start at } B \text{ and work back through } U_1.$$

$$P(B|U_2)P(U_2) = \text{product of branch probabilities leading to } B \text{ through } U_2$$

$$= \left(\frac{1}{4}\right)\left(\frac{2}{3}\right) \qquad \text{We usually start at } B \text{ and work back through } U_2.$$

Equation (3) now can be interpreted in terms of the probability tree as follows:

$$P(U_1|B) = \frac{\text{product of branch probabilities leading to } B \text{ through } U_1}{\text{sum of all branch products leading to } B}$$

$$= \frac{\left(\frac{3}{5}\right)\left(\frac{1}{3}\right)}{\left(\frac{3}{5}\right)\left(\frac{1}{3}\right) + \left(\frac{1}{4}\right)\left(\frac{2}{3}\right)} = \frac{6}{11} \approx .55$$

Similarly,

$$P(U_2|B) = \frac{\text{product of branch probabilities leading to } B \text{ through } U_2}{\text{sum of all branch products leading to } B}$$

$$= \frac{\left(\frac{1}{4}\right)\left(\frac{2}{3}\right)}{\left(\frac{3}{5}\right)\left(\frac{1}{3}\right) + \left(\frac{1}{4}\right)\left(\frac{2}{3}\right)} = \frac{5}{11} \approx .45$$

Note: We could have obtained $P(U_2|B)$ by subtracting $P(U_1|B)$ from 1. Why?

Matched Problem 1 Repeat Example 1, but find $P(U_1|W)$ and $P(U_2|W)$.

Explore and Discuss 1

Study the probability tree below:

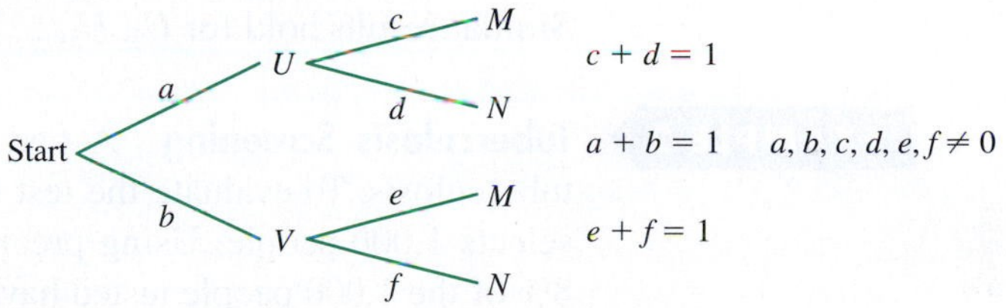

(A) Discuss the difference between $P(M|U)$ and $P(U|M)$, and between $P(N|V)$ and $P(V|N)$, in terms of a, b, c, d, e, and f.

(B) Show that $ac + ad + be + bf = 1$. What is the significance of this result?

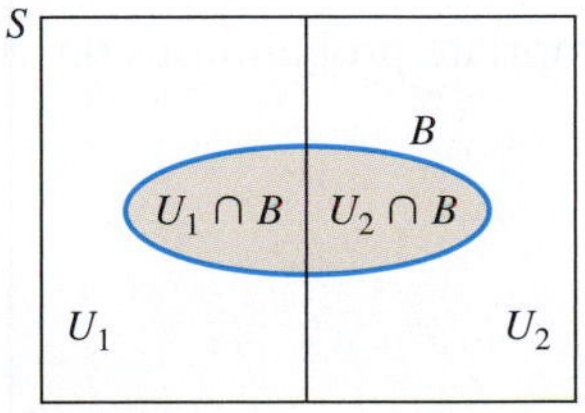

Figure 2

In generalizing the results in Example 1, it is helpful to look at its structure in terms of the Venn diagram shown in Figure 2. We note that U_1 and U_2 are mutually exclusive (disjoint), and their union forms S. The following two equations can now be interpreted in terms of this diagram:

$$P(U_1|B) = \frac{P(U_1 \cap B)}{P(B)} = \frac{P(U_1 \cap B)}{P(U_1 \cap B) + P(U_2 \cap B)}$$

$$P(U_2|B) = \frac{P(U_2 \cap B)}{P(B)} = \frac{P(U_2 \cap B)}{P(U_1 \cap B) + P(U_2 \cap B)}$$

Look over the equations and the diagram carefully.

Of course, there is no reason to stop here. Suppose that U_1, U_2, and U_3 are three mutually exclusive events whose union is the whole sample space S. Then, for an

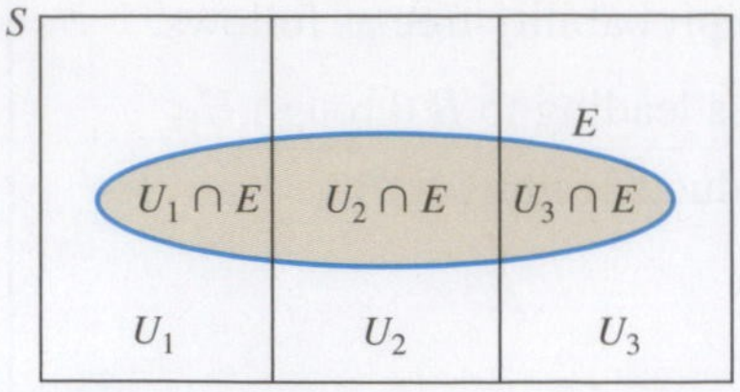

Figure 3

arbitrary event E in S, with $P(E) \neq 0$, the corresponding Venn diagram looks like Figure 3, and

$$P(U_1|E) = \frac{P(U_1 \cap E)}{P(E)} = \frac{P(U_1 \cap E)}{P(U_1 \cap E) + P(U_2 \cap E) + P(U_3 \cap E)}$$

Similar results hold for U_2 and U_3.

Using the same reasoning, we arrive at the following famous theorem, which was first stated by Thomas Bayes (1702–1763):

THEOREM 1 Bayes' Formula

Let $U_1, U_2, \ldots, U_n$ be n mutually exclusive events whose union is the sample space S. Let E be an arbitrary event in S such that $P(E) \neq 0$. Then,

$$P(U_1|E) = \frac{P(U_1 \cap E)}{P(E)} = \frac{P(U_1 \cap E)}{P(U_1 \cap E) + P(U_2 \cap E) + \cdots + P(U_n \cap E)}$$

$$= \frac{P(E|U_1)P(U_1)}{P(E|U_1)P(U_1) + P(E|U_2)P(U_2) + \cdots + P(E|U_n)P(U_n)}$$

Similar results hold for $U_2, U_3, \ldots, U_n$.

You do not need to memorize Bayes' formula. In practice, it is often easier to draw a probability tree and use the following:

$$P(U_1|E) = \frac{\text{product of branch probabilities leading to } E \text{ through } U_1}{\text{sum of all branch products leading to } E}$$

Similar results hold for $U_2, U_3, \ldots, U_n$.

EXAMPLE 2 **Tuberculosis Screening** A new, inexpensive skin test is devised for detecting tuberculosis. To evaluate the test before it is used, a medical researcher randomly selects 1,000 people. Using precise but more expensive methods, it is found that 8% of the 1,000 people tested have tuberculosis. Now each of the 1,000 subjects is given the new skin test, and the following results are recorded: The test indicates tuberculosis in 96% of those who have it and in 2% of those who do not. Based on these results, what is the probability of a randomly chosen person having tuberculosis given that the skin test indicates the disease? What is the probability of a person not having tuberculosis given that the skin test indicates the disease? (That is, what is the probability of the skin test giving a *false positive result*?)

SOLUTION To start, we form a tree diagram and place appropriate probabilities on each branch:

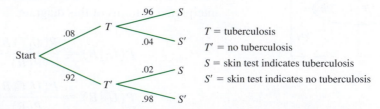

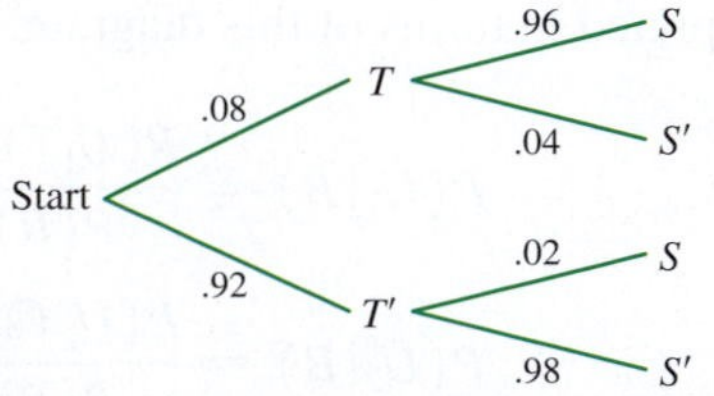

T = tuberculosis
T' = no tuberculosis
S = skin test indicates tuberculosis
S' = skin test indicates no tuberculosis

We are interested in finding $P(T|S)$, that is, the probability of a person having tuberculosis given that the skin test indicates the disease. Bayes' formula for this case is

$$P(T|S) = \frac{\text{product of branch probabilities leading to } S \text{ through } T}{\text{sum of all branch products leading to } S}$$

Substituting appropriate values from the probability tree, we obtain

$$P(T \mid S) = \frac{(.08)(.96)}{(.08)(.96) + (.92)(.02)} = .81$$

The probability of a person not having tuberculosis given that the skin test indicates the disease, denoted by $P(T' \mid S)$, is

$$P(T' \mid S) = 1 - P(T \mid S) = 1 - .81 = .19 \quad P(T \mid S) + P(T' \mid S) = 1$$

Matched Problem 2 What is the probability that a person has tuberculosis given that the test indicates no tuberculosis is present? (That is, what is the probability of the skin test giving a *false negative result*?) What is the probability that a person does not have tuberculosis given that the test indicates no tuberculosis is present?

CONCEPTUAL INSIGHT

From a public health standpoint, which is the more serious error in Example 2 and Matched Problem 2: a false positive result or a false negative result? A false positive result will certainly be unsettling to the subject, who might lose sleep thinking that she has tuberculosis. But she will be sent for further testing and will be relieved to learn it was a false alarm; she does not have the disease. A false negative result, on the other hand, is more serious. A person who has tuberculosis will be unaware that he has a communicable disease and so may pose a considerable risk to public health.

In designing an inexpensive skin test, we would expect a tradeoff between cost and false results. We might be willing to tolerate a moderate number of false positives if we could keep the number of false negatives at a minimum.

EXAMPLE 3 **Product Defects** A company produces 1,000 refrigerators a week at three plants. Plant *A* produces 350 refrigerators a week, plant *B* produces 250 refrigerators a week, and plant *C* produces 400 refrigerators a week. Production records indicate that 5% of the refrigerators produced at plant *A* will be defective, 3% of those produced at plant *B* will be defective, and 7% of those produced at plant *C* will be defective. All the refrigerators are shipped to a central warehouse. If a refrigerator at the warehouse is found to be defective, what is the probability that it was produced at plant *A*?

SOLUTION We begin by constructing a tree diagram:

A = produced at plant A
B = produced at plant B
C = produced at plant C
D = defective
D' = not defective

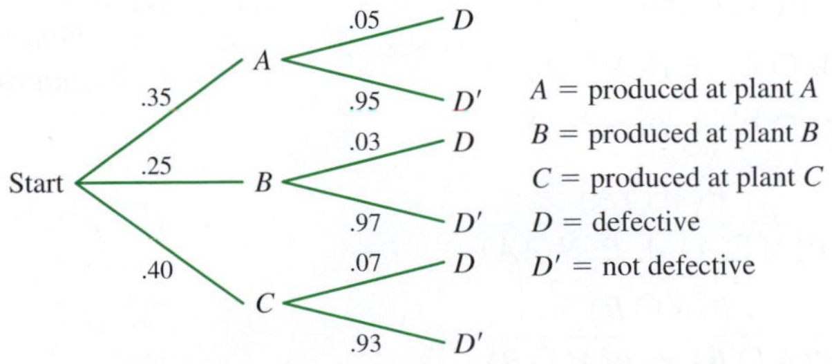

The probability that a defective refrigerator was produced at plant *A* is $P(A \mid D)$. Bayes' formula for this case is

$$P(A \mid D) = \frac{\text{product of branch probabilities leading to } D \text{ through } A}{\text{sum of all branch products leading to } D}$$

Using the values from the probability tree, we have

$$P(A|D) = \frac{(.35)(.05)}{(.35)(.05) + (.25)(.03) + (.40)(.07)}$$
$$\approx .33$$

Matched Problem 3 In Example 3, what is the probability that a defective refrigerator in the warehouse was produced at plant B? At plant C?

Exercises 8.4

Skills Warm-up Exercises

W *In Problems 1–6, write each expression as a quotient of integers, reduced to lowest terms. (If necessary, review Section A.1.)*

1. $\dfrac{\frac{1}{3}}{\frac{1}{3} + \frac{1}{2}}$

2. $\dfrac{\frac{2}{7}}{\frac{1}{4} + \frac{2}{7}}$

3. $\frac{1}{3} \div \frac{1}{3} + \frac{1}{2}$

4. $\frac{2}{7} \div \frac{1}{4} + \frac{2}{7}$

5. $\dfrac{\frac{4}{5} \cdot \frac{3}{4}}{\frac{1}{5} \cdot \frac{1}{3} + \frac{4}{5} \cdot \frac{3}{4}}$

6. $\dfrac{\frac{1}{5} \cdot \frac{2}{3}}{\frac{1}{5} \cdot \frac{2}{3} + \frac{4}{5} \cdot \frac{1}{4}}$

A *Find the probabilities in Problems 7–12 by referring to the tree diagram below.*

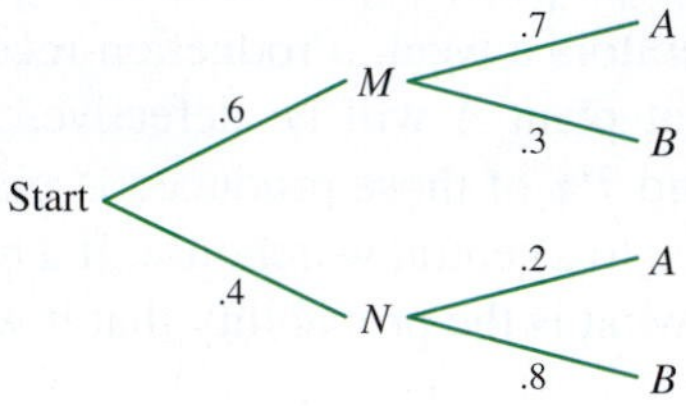

7. $P(M \cap A) = P(M)P(A|M)$

8. $P(N \cap B) = P(N)P(B|N)$

9. $P(A) = P(M \cap A) + P(N \cap A)$

10. $P(B) = P(M \cap B) + P(N \cap B)$

11. $P(M|A) = \dfrac{P(M \cap A)}{P(M \cap A) + P(N \cap A)}$

12. $P(N|B) = \dfrac{P(N \cap B)}{P(N \cap B) + P(M \cap B)}$

Find the probabilities in Problems 13–16 by referring to the following Venn diagram and using Bayes' formula (assume that the simple events in S are equally likely):

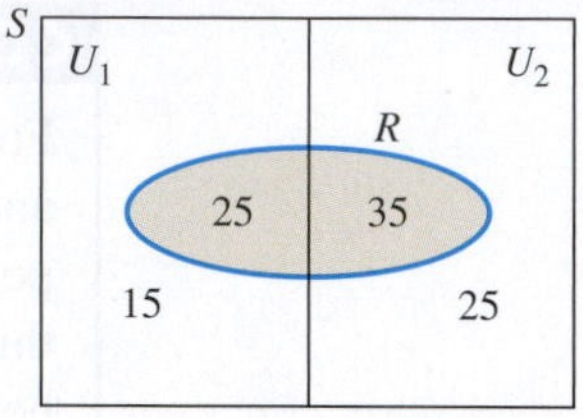

13. $P(U_1|R)$

14. $P(U_2|R)$

15. $P(U_1|R')$

16. $P(U_2|R')$

B *Find the probabilities in Problems 17–22 by referring to the following tree diagram and using Bayes' formula. Round answers to three decimal places.*

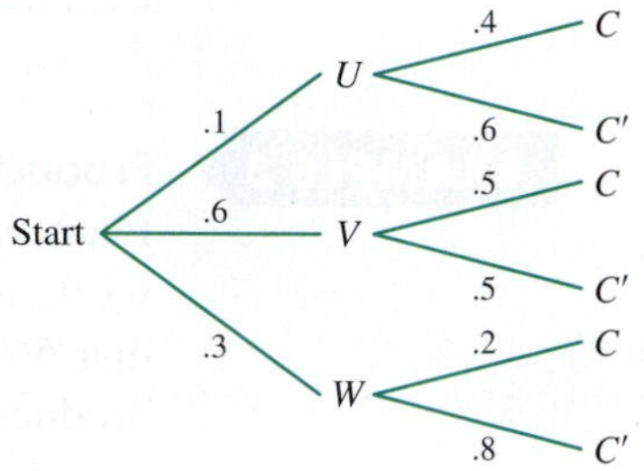

17. $P(U|C)$

18. $P(V|C')$

19. $P(W|C)$

20. $P(U|C')$

21. $P(V|C)$

22. $P(W|C')$

Find the probabilities in Problems 23–28 by referring to the tree diagram below.

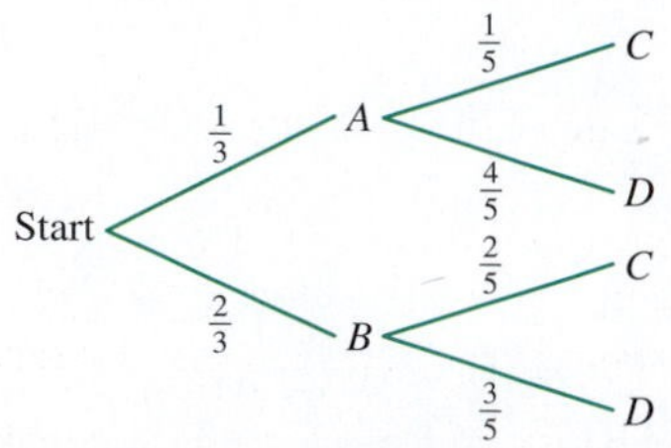

23. $P(A)$

24. $P(B)$

25. $P(C)$

26. $P(D)$

27. $P(A|C)$

28. $P(B|D)$

In Problems 29 and 30, use the probabilities in the first tree diagram to find the probability of each branch of the second tree diagram.

29.

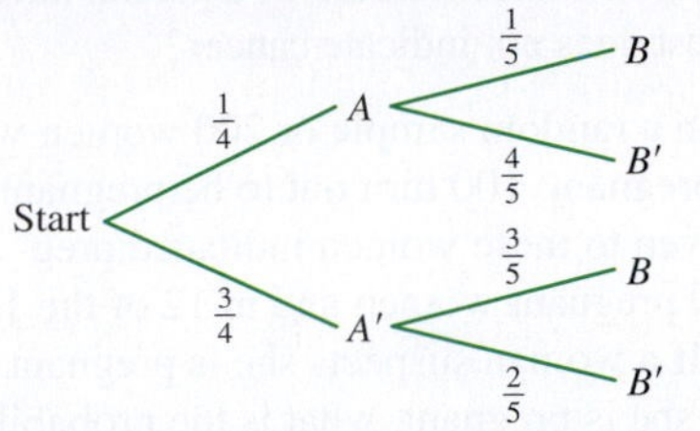

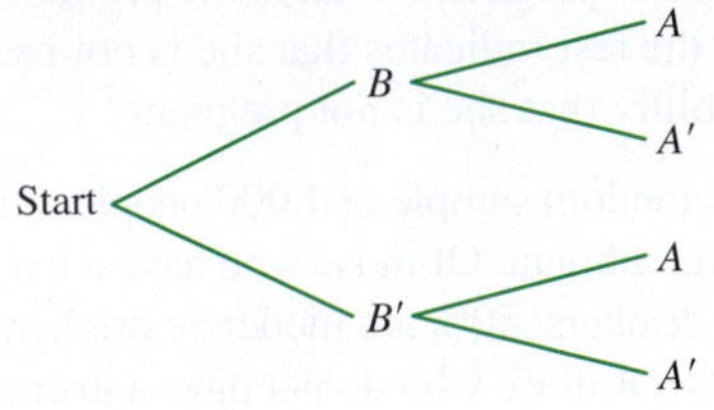

30.

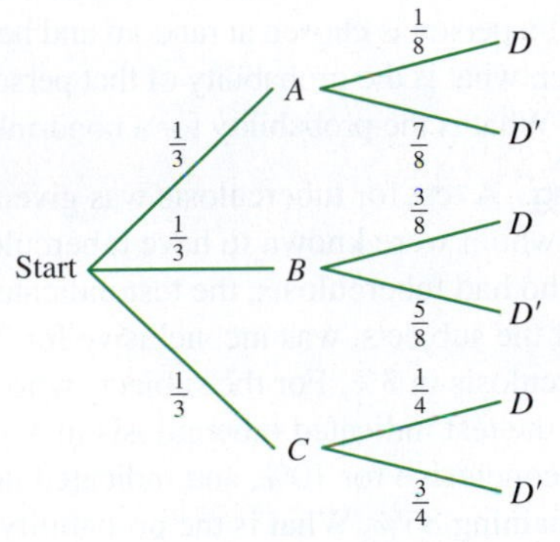

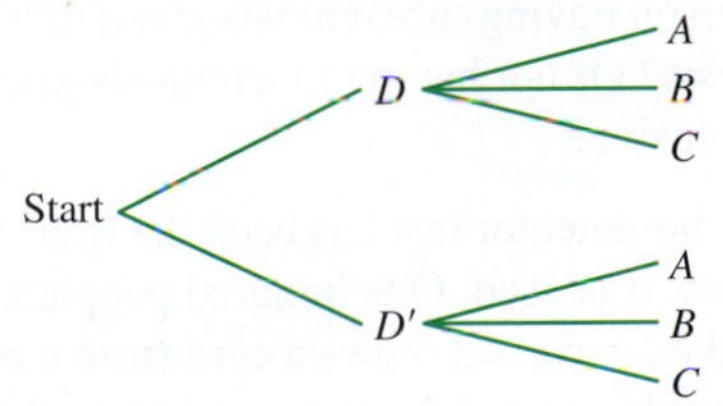

In Problems 31–34, one of two urns is chosen at random, with one as likely to be chosen as the other. Then a ball is withdrawn from the chosen urn. Urn 1 contains 1 white and 4 red balls, and urn 2 has 3 white and 2 red balls.

31. If a white ball is drawn, what is the probability that it came from urn 1?

32. If a white ball is drawn, what is the probability that it came from urn 2?

33. If a red ball is drawn, what is the probability that it came from urn 2?

34. If a red ball is drawn, what is the probability that it came from urn 1?

In Problems 35 and 36, an urn contains 4 red and 5 white balls. Two balls are drawn in succession without replacement.

35. If the second ball is white, what is the probability that the first ball was white?

36. If the second ball is red, what is the probability that the first ball was red?

In Problems 37 and 38, urn 1 contains 7 red and 3 white balls. Urn 2 contains 4 red and 5 white balls. A ball is drawn from urn 1 and placed in urn 2. Then a ball is drawn from urn 2.

37. If the ball drawn from urn 2 is red, what is the probability that the ball drawn from urn 1 was red?

38. If the ball drawn from urn 2 is white, what is the probability that the ball drawn from urn 1 was white?

In Problems 39 and 40, refer to the following probability tree:

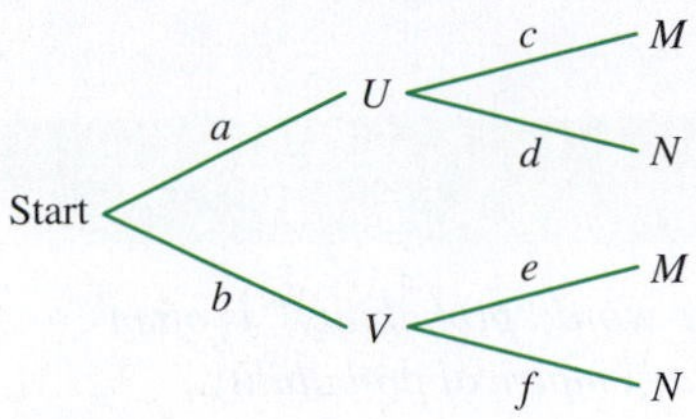

$c + d = 1$

$a + b = 1 \quad a, b, c, d, e, f \neq 0$

$e + f = 1$

39. Suppose that $c = e$. Discuss the dependence or independence of events U and M.

40. Suppose that $c = d = e = f$. Discuss the dependence or independence of events M and N.

In Problems 41 and 42, two balls are drawn in succession from an urn containing m blue balls and n white balls ($m \geq 2$ and $n \geq 2$). Discuss the validity of each statement. If the statement is always true, explain why. If not, give a counterexample.

41. (A) If the two balls are drawn with replacement, then $P(B_1|B_2) = P(B_2|B_1)$.

(B) If the two balls are drawn without replacement, then $P(B_1|B_2) = P(B_2|B_1)$.

42. (A) If the two balls are drawn with replacement, then $P(B_1|W_2) = P(W_2|B_1)$.

(B) If the two balls are drawn without replacement, then $P(B_1|W_2) = P(W_2|B_1)$.

C **43.** If 2 cards are drawn in succession from a standard 52-card deck without replacement and the second card is a heart, what is the probability that the first card is a heart?

44. A box contains 10 balls numbered 1 through 10. Two balls are drawn in succession without replacement. If the second ball drawn has the number 4 on it, what is the probability that the first ball had a smaller number on it? An even number on it?

In Problems 45–50, a player is dealt two cards from a 52-card deck. If the first card is black, the player returns it to the deck before drawing the second card. If the first card is red, the player sets it aside and then draws the second card.

45. What is the probability of drawing a red card on the second draw?

46. What is the probability of drawing a black card on the second draw?

47. If the second card drawn is red, what is the probability that the first card drawn was red?

48. If the second card drawn is black, what is the probability that the first card drawn was red?

49. If the second card drawn is red, what is the probability that the first card drawn was black?

50. If the second card drawn is black, what is the probability that the first card drawn was black?

51. Show that $P(U_1|R) + P(U_1'|R) = 1$.

52. If U_1 and U_2 are two mutually exclusive events whose union is the equally likely sample space S and if E is an arbitrary event in S such that $P(E) \neq 0$, show that

$$P(U_1|E) = \frac{n(U_1 \cap E)}{n(U_1 \cap E) + n(U_2 \cap E)}$$

Applications

In the following applications, the word "probability" is often understood to mean "approximate empirical probability."

53. Employee screening. The management of a company finds that 30% of the administrative assistants hired are unsatisfactory. The personnel director is instructed to devise a test that will improve the situation. One hundred employed administrative assistants are chosen at random and are given the newly constructed test. Out of these, 90% of the satisfactory administrative assistants pass the test and 20% of the unsatisfactory administrative assistants pass. Based on these results, if a person applies for a job, takes the test, and passes it, what is the probability that he or she is a satisfactory administrative assistant? If the applicant fails the test, what is the probability that he or she is a satisfactory administrative assistant?

54. Employee rating. A company has rated 75% of its employees as satisfactory and 25% as unsatisfactory. Personnel records indicate that 80% of the satisfactory workers had previous work experience, while only 40% of the unsatisfactory workers had any previous work experience. If a person with previous work experience is hired, what is the probability that this person will be a satisfactory employee? If a person with no previous work experience is hired, what is the probability that this person will be a satisfactory employee?

55. Product defects. A manufacturer obtains GPS systems from three different subcontractors: 20% from A, 40% from B, and 40% from C. The defective rates for these subcontractors are 1%, 3%, and 2%, respectively. If a defective GPS system is returned by a customer, what is the probability that it came from subcontractor A? From B? From C?

56. Product defects. A store sells three types of flash drives: brand A, brand B, and brand C. Of the flash drives it sells, 60% are brand A, 25% are brand B, and 15% are brand C. The store has found that 20% of the brand A flash drives, 15% of the brand B flash drives, and 5% of the brand C flash drives are returned as defective. If a flash drive is returned as defective, what is the probability that it is a brand A flash drive? A brand B flash drive? A brand C flash drive?

57. Cancer screening. A new, simple test has been developed to detect a particular type of cancer. The test must be evaluated before it is used. A medical researcher selects a random sample of 1,000 adults and finds (by other means) that 2% have this type of cancer. Each of the 1,000 adults is given the test, and it is found that the test indicates cancer in 98% of those who have it and in 1% of those who do not. Based on these results, what is the probability of a randomly chosen person having cancer given that the test indicates cancer? Of a person having cancer given that the test does not indicate cancer?

58. Pregnancy testing. In a random sample of 200 women who suspect that they are pregnant, 100 turn out to be pregnant. A new pregnancy test given to these women indicated pregnancy in 92 of the 100 pregnant women and in 12 of the 100 nonpregnant women. If a woman suspects she is pregnant and this test indicates that she is pregnant, what is the probability that she is pregnant? If the test indicates that she is not pregnant, what is the probability that she is not pregnant?

59. Medical research. In a random sample of 1,000 people, it is found that 7% have a liver ailment. Of those who have a liver ailment, 40% are heavy drinkers, 50% are moderate drinkers, and 10% are nondrinkers. Of those who do not have a liver ailment, 10% are heavy drinkers, 70% are moderate drinkers, and 20% are nondrinkers. If a person is chosen at random and he or she is a heavy drinker, what is the probability of that person having a liver ailment? What is the probability for a nondrinker?

60. Tuberculosis screening. A test for tuberculosis was given to 1,000 subjects, 8% of whom were known to have tuberculosis. For the subjects who had tuberculosis, the test indicated tuberculosis in 90% of the subjects, was inconclusive for 7%, and indicated no tuberculosis in 3%. For the subjects who did not have tuberculosis, the test indicated tuberculosis in 5% of the subjects, was inconclusive for 10%, and indicated no tuberculosis in the remaining 85%. What is the probability of a randomly selected person having tuberculosis given that the test indicates tuberculosis? Of not having tuberculosis given that the test was inconclusive?

61. Police science. A new lie-detector test has been devised and must be tested before it is used. One hundred people are selected at random, and each person draws a card from a box of 100 cards. Half the cards instruct the person to lie, and the others instruct the person to tell the truth. Of those who lied, 80% fail the new lie-detector test (that is, the test indicates lying). Of those who told the truth, 5% failed the test. What is the probability that a randomly chosen subject will have lied given that the subject failed the test? That the subject will not have lied given that the subject failed the test?

62. Politics. In a given county, records show that of the registered voters, 45% are Democrats, 35% are Republicans, and 20% are independents. In an election, 70% of the Democrats, 40% of the Republicans, and 80% of the independents voted in favor of a parks and recreation bond proposal. If a registered voter chosen at random is found to have voted in favor of the bond, what is the probability that the voter is a Republican? An independent? A Democrat?

Answers to Matched Problems

1. $P(U_1|W) = \frac{4}{19} \approx .21$; $P(U_2|W) = \frac{15}{19} \approx .79$

2. $P(T|S') = .004$; $P(T'|S') = .996$

3. $P(B|D) \approx .14$; $P(C|D) \approx .53$

8.5 Random Variable, Probability Distribution, and Expected Value

- Random Variable and Probability Distribution
- Expected Value of a Random Variable
- Decision Making and Expected Value

Random Variable and Probability Distribution

When performing a random experiment, a sample space S is selected in such a way that all probability problems of interest relative to the experiment can be solved. In many situations we may not be interested in each simple event in the sample space S but in some numerical value associated with the event. For example, if 3 coins are tossed, we may be interested in the number of heads that turn up rather than in the particular pattern that turns up. Or, in selecting a random sample of students, we may be interested in the proportion that are women rather than which particular students are women. In the same way, a "craps" player is usually interested in the sum of the dots on the showing faces of the dice rather than the pattern of dots on each face.

In each of these examples, there is a rule that assigns to each simple event in S a single real number. Mathematically speaking, we are dealing with a function (see Section 2.1). Historically, this particular type of function has been called a "random variable."

DEFINITION Random Variable

A **random variable** is a function that assigns a numerical value to each simple event in a sample space S.

The term *random variable* is an unfortunate choice, since it is neither random nor a variable—it is a function with a numerical value, and it is defined on a sample space. But the terminology has stuck and is now standard. Capital letters, such as X, are used to represent random variables.

Let us return to the experiment of tossing 3 coins. A sample space S of equally likely simple events is indicated in Table 1. Suppose that we are interested in the number of heads $(0, 1, 2, \text{or } 3)$ appearing on each toss of the 3 coins and the probability of each of these events. We introduce a random variable X (a function) that indicates the number of heads for each simple event in S (see the second column in Table 1). For example, $X(e_1) = 0$, $X(e_2) = 1$, and so on. The random variable X assigns a numerical value to each simple event in the sample space S.

Table 1 Number of Heads in the Toss of 3 Coins

Sample Space S	Number of Heads $X(e_i)$
e_1: TTT	0
e_2: TTH	1
e_3: THT	1
e_4: HTT	1
e_5: THH	2
e_6: HTH	2
e_7: HHT	2
e_8: HHH	3

We are interested in the probability of the occurrence of each image or range value of X, that is, in the probability of the occurrence of 0 heads, 1 head, 2 heads, or 3 heads in the single toss of 3 coins. We indicate this probability by

$$p(x) \quad \text{where} \quad x \in \{0, 1, 2, 3\}$$

The function p is called the **probability distribution* of the random variable X.**

What is $p(2)$, the probability of getting exactly 2 heads on the single toss of 3 coins? "Exactly 2 heads occur" is the event

$$E = \{\text{THH, HTH, HHT}\}$$

Thus,

$$p(2) = \frac{n(E)}{n(S)} = \frac{3}{8}$$

Proceeding similarly for $p(0)$, $p(1)$, and $p(3)$, we obtain the probability distribution of the random variable X presented in Table 2. Probability distributions are also represented graphically, as shown in Figure 1. The graph of a probability distribution is often called a **histogram**.

Table 2 Probability Distribution

Number of Heads x	0	1	2	3
Probability $p(x)$	$\frac{1}{8}$	$\frac{3}{8}$	$\frac{3}{8}$	$\frac{1}{8}$

*The probability distribution p of the random variable X is defined by $p(x) = P(\{e_i \in S \mid X(e_i) = x\})$, which, because of its cumbersome nature, is usually simplified to $p(x) = P(X = x)$ or simply $p(x)$. We will use the simplified notation.

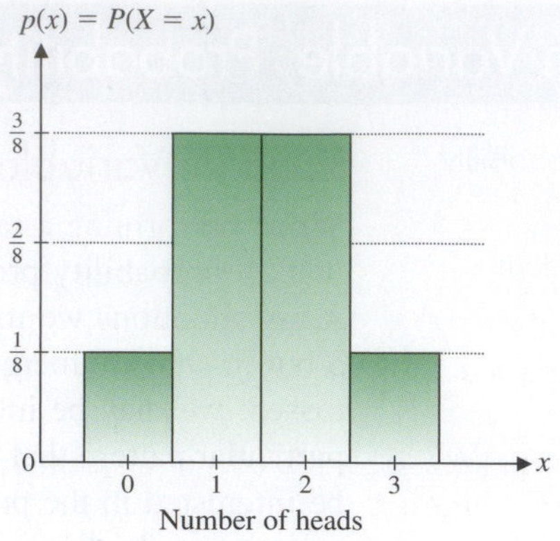

Figure 1 Histogram for a probability distribution

Note from Table 2 or Figure 1 that

1. $0 \le p(x) \le 1, \quad x \in \{0, 1, 2, 3\}$
2. $p(0) + p(1) + p(2) + p(3) = \frac{1}{8} + \frac{3}{8} + \frac{3}{8} + \frac{1}{8} = 1$

These are general properties that any probability distribution of a random variable X associated with a finite sample space must have.

THEOREM 1 Probability Distribution of a Random Variable X

The **probability distribution of a random variable** X, denoted by $P(X = x) = p(x)$, satisfies

1. $0 \le p(x) \le 1, \quad x \in \{x_1, x_2, \ldots, x_n\}$
2. $p(x_1) + p(x_2) + \cdots + p(x_n) = 1$

where $\{x_1, x_2, \ldots, x_n\}$ are the (range) values of X (see Fig. 2).

Figure 2 illustrates the process of forming a probability distribution of a random variable.

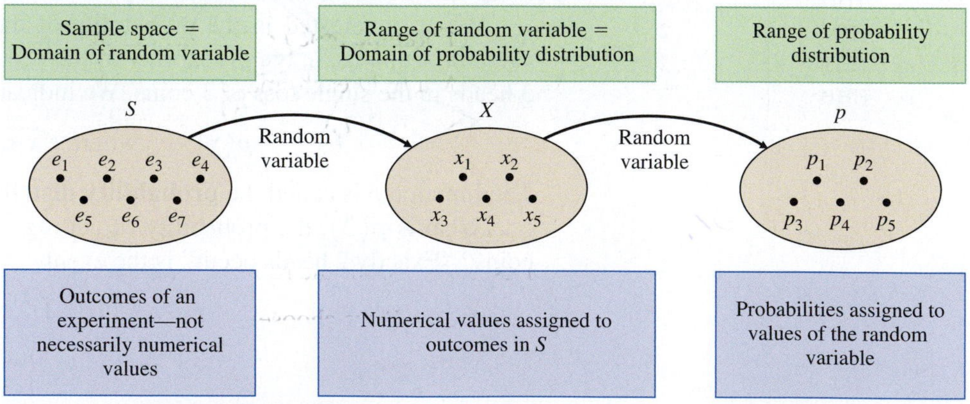

Figure 2 Probability distribution of a random variable for a finite sample space

Expected Value of a Random Variable

Suppose that the experiment of tossing 3 coins was repeated many times. What would be the average number of heads per toss (the total number of heads in all tosses divided by the total number of tosses)? Consulting the probability distribution in Table 2 or Figure 1, we would expect to toss 0 heads $\frac{1}{8}$ of the time, 1 head $\frac{3}{8}$ of the time, 2 heads $\frac{3}{8}$ of the time, and 3 heads $\frac{1}{8}$ of the time. In the long run, we would expect the average number of heads per toss of the 3 coins, or the *expected value* $E(X)$, to be given by

$$E(X) = 0\left(\frac{1}{8}\right) + 1\left(\frac{3}{8}\right) + 2\left(\frac{3}{8}\right) + 3\left(\frac{1}{8}\right) = \frac{12}{8} = 1.5$$

It is important to note that the expected value is not a value that will necessarily occur in a single experiment (1.5 heads cannot occur in the toss of 3 coins), but it is an average of what occurs over a large number of experiments. Sometimes we will toss more than 1.5 heads and sometimes less, but if the experiment is repeated many times, the average number of heads per experiment should be close to 1.5.

We now make the preceding discussion more precise through the following definition of expected value:

DEFINITION Expected Value of a Random Variable *X*

Given the probability distribution for the random variable X,

x_i	x_1	x_2	...	x_n
p_i	p_1	p_2	...	p_n

where $p_i = p(x_i)$, we define the **expected value of** X, denoted $E(X)$, by the formula

$$E(X) = x_1p_1 + x_2p_2 + \cdots + x_np_n$$

We again emphasize that the expected value is not the outcome of a single experiment, but a long-run average of outcomes of repeated experiments. The expected value is the weighted average of the possible outcomes, each weighted by its probability.

PROCEDURE Steps for Computing the Expected Value of a Random Variable *X*

Step 1 Form the probability distribution of the random variable X.

Step 2 Multiply each image value of X, x_i, by its corresponding probability of occurrence p_i; then add the results.

EXAMPLE 1

Expected Value What is the expected value (long-run average) of the number of dots facing up for the roll of a single die?

SOLUTION If we choose

$$S = \{1, 2, 3, 4, 5, 6\}$$

as our sample space, then each simple event is a numerical outcome reflecting our interest, and each is equally likely. The random variable X in this case is just the identity function (each number is associated with itself). The probability distribution for X is

x_i	1	2	3	4	5	6
p_i	$\frac{1}{6}$	$\frac{1}{6}$	$\frac{1}{6}$	$\frac{1}{6}$	$\frac{1}{6}$	$\frac{1}{6}$

Therefore,

$$E(X) = 1\left(\frac{1}{6}\right) + 2\left(\frac{1}{6}\right) + 3\left(\frac{1}{6}\right) + 4\left(\frac{1}{6}\right) + 5\left(\frac{1}{6}\right) + 6\left(\frac{1}{6}\right)$$

$$= \frac{21}{6} = 3.5$$

Matched Problem 1 Suppose that the die in Example 1 is not fair and we obtain (empirically) the following probability distribution for X:

x_i	1	2	3	4	5	6
p_i	.14	.13	.18	.20	.11	.24

[Note: Sum = 1.]

What is the expected value of X?

Explore and Discuss 1

From Example 1 we can conclude that the probability is 0 that a single roll of a fair die will equal the expected value for a roll of a die (the number of dots facing up is never 3.5). What is the probability that the sum for a single roll of a pair of dice will equal the expected value of the sum for a roll of a pair of dice?

EXAMPLE 2 **Expected Value** A carton of 20 laptop batteries contains 2 defective ones. A random sample of 3 is selected from the 20 and tested. Let X be the random variable associated with the number of defective batteries found in a sample.

(A) Find the probability distribution of X.

(B) Find the expected number of defective batteries in a sample.

SOLUTION

(A) The number of ways of selecting a sample of 3 from 20 (order is not important) is ${}_{20}C_3$. This is the number of simple events in the experiment, each as likely as the other. A sample will have either 0, 1, or 2 defective batteries. These are the values of the random variable in which we are interested. The probability distribution is computed as follows:

$$p(0) = \frac{{}_{18}C_3}{{}_{20}C_3} \approx .716 \quad p(1) = \frac{{}_2C_1 \cdot {}_{18}C_2}{{}_{20}C_3} \approx .268 \quad p(2) = \frac{{}_2C_2 \cdot {}_{18}C_1}{{}_{20}C_3} \approx .016$$

We summarize these results in a table:

x_i	0	1	2
p_i	.716	.268	.016

[Note: .716 + .268 + .016 = 1.]

(B) The expected number of defective batteries in a sample is readily computed as follows:

$$E(X) = (0)(.716) + (1)(.268) + (2)(.016) = .3$$

The expected value is not one of the random variable values; rather, it is a number that the average number of defective batteries in a sample would approach as the experiment is repeated without end.

Matched Problem 2 Repeat Example 2 using a random sample of 4.

EXAMPLE 3 **Expected Value of a Game** A spinner device is numbered from 0 to 5, and each of the 6 numbers is as likely to come up as any other. A player who bets $1 on any given number wins $4 (and gets the $1 bet back) if the pointer comes to rest on the chosen number; otherwise, the $1 bet is lost. What is the expected value of the game (long-run average gain or loss per game)?

SOLUTION The sample space of equally likely events is

$$S = \{0, 1, 2, 3, 4, 5\}$$

Each sample point occurs with a probability of $\frac{1}{6}$. The random variable X assigns $4 to the winning number and −$1 to each of the remaining numbers. So the probability of winning $4 is $\frac{1}{6}$ and of losing $1 is $\frac{5}{6}$. We form the probability distribution for X, called a **payoff table**, and compute the expected value of the game:

Payoff Table (Probability Distribution for X)

x_i	$4	−$1
p_i	$\frac{1}{6}$	$\frac{5}{6}$

$$E(X) = \$4\left(\frac{1}{6}\right) + (-\$1)\left(\frac{5}{6}\right) = -\$\frac{1}{6} \approx -\$0.1667 \approx -17¢ \text{ per game}$$

In the long run, the player will lose an average of about 17¢ per game.

Matched Problem 3 Repeat Example 3 with the player winning $5 instead of $4 if the chosen number turns up. The loss is still $1 if any other number turns up. Is this a fair game?

The game in Example 3 is *not* fair: The player tends to lose money in the long run. A game is **fair** if the expected value $E(X)$ is equal to 0; that is, the player neither wins nor loses money in the long run. The fair games discussed in Section 8.2 are fair according to this definition, because their payoff tables have the following form:

Payoff Table

x_i	$\$b$	$-\$a$
p_i	$\frac{a}{a+b}$	$\frac{b}{a+b}$

So $E(X) = b\left(\frac{a}{a+b}\right) + (-a)\frac{b}{a+b} = 0$

EXAMPLE 4 **Expected Value and Insurance** Suppose you are interested in insuring a car video system for $2,000 against theft. An insurance company charges a premium of $225 for coverage for 1 year, claiming an empirically determined probability of .1 that the system will be stolen sometime during the year. What is your expected return from the insurance company if you take out this insurance?

SOLUTION This is actually a game of chance in which your stake is $225. You have a .1 chance of receiving $1,775 from the insurance company ($2,000 minus your stake of $225) and a .9 chance of losing your stake of $225. What is the expected value of this "game"? We form a payoff table (the probability distribution for X) and compute the expected value:

Payoff Table

x_i	\$1,775	−\$225
p_i	.1	.9

$$E(X) = (\$1{,}775)(.1) + (-\$225)(.9) = -\$25$$

This means that if you insure with this company over many years and circumstances remain the same, you would have an average net loss to the insurance company of \$25 per year.

Matched Problem 4 Find the expected value in Example 4 from the insurance company's point of view.

CONCEPTUAL INSIGHT

Suppose that in a class of 10 students, the scores on the first exam are 85, 73, 82, 65, 95, 85, 73, 75, 85, and 75. To compute the class average (mean), we add the scores and divide by the number of scores:

$$\frac{85 + 73 + 82 + 65 + 95 + 85 + 73 + 75 + 85 + 75}{10} = \frac{793}{10} = 79.3$$

Because 1 student scored 95, 3 scored 85, 1 scored 82, 2 scored 75, 2 scored 73, and 1 scored 65, the probability distribution of an exam score, for a student chosen at random from the class, is as follows:

x_i	65	73	75	82	85	95
p_i	.1	.2	.2	.1	.3	.1

The expected value of the probability distribution is

$$65\left(\frac{1}{10}\right) + 73\left(\frac{2}{10}\right) + 75\left(\frac{2}{10}\right) + 82\left(\frac{1}{10}\right) + 85\left(\frac{3}{10}\right) + 95\left(\frac{1}{10}\right) = \frac{793}{10} = 79.3$$

By comparing the two computations, we see that the mean of a data set is just the expected value of the corresponding probability distribution.

Decision Making and Expected Value

We conclude this section with an example in decision making.

EXAMPLE 5 **Decision Analysis** An outdoor concert featuring a popular musical group is scheduled for a Sunday afternoon in a large open stadium. The promoter, worrying about being rained out, contacts a long-range weather forecaster who predicts the chance of rain on that Sunday to be .24. If it does not rain, the promoter is certain to net \$100,000; if it does rain, the promoter estimates that the net will be only \$10,000. An insurance company agrees to insure the concert for \$100,000 against rain at a premium of \$20,000. Should the promoter buy the insurance?

SOLUTION The promoter has a choice between two courses of action; A_1: Insure and A_2: Do not insure. As an aid in making a decision, the expected value is computed for each course of action. Probability distributions are indicated in the payoff table (read vertically):

Payoff Table

	A_1: Insure	A_2: Do Not Insure
p_i	x_i	x_i
.24 (rain)	\$90,000	\$10,000
.76 (no rain)	\$80,000	\$100,000

Note that the \$90,000 entry comes from the insurance company's payoff (\$100,000) minus the premium (\$20,000) plus gate receipts (\$10,000). The reasons for the other entries should be obvious. The expected value for each course of action is computed as follows:

A_1: Insure	A_2: Do Not Insure
$E(X) = x_1p_1 + x_2p_2$ $= (\$90{,}000)(.24) + (\$80{,}000)(.76)$ $= \$82{,}400$	$E(X) = (\$10{,}000)(.24) + (\$100{,}000)(.76)$ $= \$78{,}400$

It appears that the promoter's best course of action is to buy the insurance at \$20,000. The promoter is using a long-run average to make a decision about a single event—a common practice in making decisions in areas of uncertainty.

Matched Problem 5 In Example 5, what is the insurance company's expected value if it writes the policy?

Exercises 8.5

Skills Warm-up Exercises

W *In Problems 1–8, if necessary, review Section B.1.*

1. Find the average (mean) of the exam scores 73, 89, 45, 82, and 66.

2. Find the average (mean) of the exam scores 78, 64, 97, 60, 86, and 83.

3. Find the average (mean) of the exam scores in Problem 1, if 4 points are added to each score.

4. Find the average (mean) of the exam scores in Problem 2, if 3 points are subtracted from each score.

5. Find the average (mean) of the exam scores in Problem 1, if each score is multiplied by 2.

6. Find the average (mean) of the exam scores in Problem 2, if each score is divided by 2.

A 7. If the probability distribution for the random variable X is given in the table, what is the expected value of X?

x_i	−3	0	4
p_i	.3	.5	.2

8. If the probability distribution for the random variable X is given in the table, what is the expected value of X?

x_i	−2	−1	0	1	2
p_i	.1	.2	.4	.2	.1

9. You draw and keep a single bill from a hat that contains a \$5, \$20, \$50, and \$100 bill. What is the expected value of the game to you?

10. You draw and keep a single bill from a hat that contains a \$1, \$10, \$20, \$50, and \$100 bill. What is the expected value of the game to you?

11. You draw and keep a single coin from a bowl that contains 15 pennies, 10 dimes, and 25 quarters. What is the expected value of the game to you?

12. You draw and keep a single coin from a bowl that contains 120 nickels and 80 quarters. What is the expected value of the game to you?

13. You draw a single card from a standard 52-card deck. If it is red, you win \$50. Otherwise you get nothing. What is the expected value of the game to you?

14. You draw a single card from a standard 52-card deck. If it is an ace, you win \$104. Otherwise you get nothing. What is the expected value of the game to you?

15. In tossing 2 fair coins, what is the expected number of heads?

16. In a family with 2 children, excluding multiple births and assuming that a boy is as likely as a girl at each birth, what is the expected number of boys?

17. A fair coin is flipped. If a head turns up, you win \$1. If a tail turns up, you lose \$1. What is the expected value of the game? Is the game fair?

18. Repeat Problem 17, assuming an unfair coin with the probability of a head being .55 and a tail being .45.

B 19. After paying \$4 to play, a single fair die is rolled, and you are paid back the number of dollars corresponding to the number of dots facing up. For example, if a 5 turns up, \$5 is returned to you for a net gain, or payoff, of \$1; if a 1 turns up, \$1 is returned for a net gain of −\$3; and so on. What is the expected value of the game? Is the game fair?

20. Repeat Problem 19 with the same game costing \$3.50 for each play.

21. Two coins are flipped. You win \$2 if either 2 heads or 2 tails turn up; you lose \$3 if a head and a tail turn up. What is the expected value of the game?

22. In Problem 21, for the game to be fair, how much *should* you lose if a head and a tail turn up?

23. A friend offers the following game: She wins \$1 from you if, on four rolls of a single die, a 6 turns up at least once; otherwise, you win \$1 from her. What is the expected value of the game to you? To her?

24. On three rolls of a single die, you will lose \$10 if a 5 turns up at least once, and you will win \$7 otherwise. What is the expected value of the game?

25. A single die is rolled once. You win \$5 if a 1 or 2 turns up and \$10 if a 3, 4, or 5 turns up. How much should you lose if a 6 turns up in order for the game to be fair? Describe the steps you took to arrive at your answer.

26. A single die is rolled once. You lose \$12 if a number divisible by 3 turns up. How much should you win if a number not divisible by 3 turns up in order for the game to be fair? Describe the process and reasoning used to arrive at your answer.

27. A pair of dice is rolled once. Suppose you lose \$10 if a 7 turns up and win \$11 if an 11 or 12 turns up. How much should you win or lose if any other number turns up in order for the game to be fair?

28. A coin is tossed three times. Suppose you lose \$3 if 3 heads appear, lose \$2 if 2 heads appear, and win \$3 if 0 heads appear. How much should you win or lose if 1 head appears in order for the game to be fair?

29. A card is drawn from a standard 52-card deck. If the card is a king, you win \$10; otherwise, you lose \$1. What is the expected value of the game?

30. A card is drawn from a standard 52-card deck. If the card is a diamond, you win \$10; otherwise, you lose \$4. What is the expected value of the game?

31. A 5-card hand is dealt from a standard 52-card deck. If the hand contains at least one king, you win \$10; otherwise, you lose \$1. What is the expected value of the game?

32. A 5-card hand is dealt from a standard 52-card deck. If the hand contains at least one diamond, you win \$10; otherwise, you lose \$4. What is the expected value of the game?

33. The payoff table for two courses of action, A_1 or A_2, is given below. Which of the two actions will produce the largest expected value? What is it?

	A_1	A_2
p_i	x_i	x_i
.1	−\$200	−\$100
.2	\$100	\$200
.4	\$400	\$300
.3	\$100	\$200

34. The payoff table for three possible courses of action is given below. Which of the three actions will produce the largest expected value? What is it?

	A_1	A_2	A_3
P_i	x_i	x_i	x_i
.2	\$ 500	\$ 400	\$ 300
.4	\$1,200	\$1,100	\$1,000
.3	\$1,200	\$1,800	\$1,700
.1	\$1,200	\$1,800	\$2,400

35. Roulette wheels in Nevada generally have 38 equally spaced slots numbered 00, 0, 1, 2, . . . , 36. A player who bets \$1 on any given number wins \$35 (and gets the bet back) if the ball comes to rest on the chosen number; otherwise, the \$1 bet is lost. What is the expected value of this game?

36. In roulette (see Problem 35), the numbers from 1 to 36 are evenly divided between red and black. A player who bets \$1 on black wins \$1 (and gets the \$1 bet back) if the ball comes to rest on black; otherwise (if the ball lands on red, 0, or 00), the \$1 bet is lost. What is the expected value of the game?

37. A game has an expected value to you of \$100. It costs \$100 to play, but if you win, you receive \$100,000 (including your \$100 bet) for a net gain of \$99,900. What is the probability of winning? Would you play this game? Discuss the factors that would influence your decision.

38. A game has an expected value to you of −\$0.50. It costs \$2 to play, but if you win, you receive \$20 (including your \$2 bet) for a net gain of \$18. What is the probability of winning? Would you play this game? Discuss the factors that would influence your decision.

C **39.** Five thousand tickets are sold at \$1 each for a charity raffle. Tickets will be drawn at random and monetary prizes awarded as follows: 1 prize of \$500; 3 prizes of \$100, 5 prizes of \$20, and 20 prizes of \$5. What is the expected value of this raffle if you buy 1 ticket?

40. Ten thousand raffle tickets are sold at \$2 each for a local library benefit. Prizes are awarded as follows: 2 prizes of \$1,000, 4 prizes of \$500, and 10 prizes of \$100. What is the expected value of this raffle if you purchase 1 ticket?

41. A box of 10 flashbulbs contains 3 defective bulbs. A random sample of 2 is selected and tested. Let X be the random variable associated with the number of defective bulbs in the sample.

(A) Find the probability distribution of X.

(B) Find the expected number of defective bulbs in a sample.

42. A box of 8 flashbulbs contains 3 defective bulbs. A random sample of 2 is selected and tested. Let X be the random variable associated with the number of defective bulbs in a sample.

(A) Find the probability distribution of X.

(B) Find the expected number of defective bulbs in a sample.

43. One thousand raffle tickets are sold at $1 each. Three tickets will be drawn at random (without replacement), and each will pay $200. Suppose you buy 5 tickets.

(A) Create a payoff table for 0, 1, 2, and 3 winning tickets among the 5 tickets you purchased. (If you do not have any winning tickets, you lose $5; if you have 1 winning ticket, you net $195 since your initial $5 will not be returned to you; and so on.)

(B) What is the expected value of the raffle to you?

44. Repeat Problem 43 with the purchase of 10 tickets.

45. To simulate roulette on a graphing calculator, a random integer between −1 and 36 is selected (−1 represents 00; see Problem 35). The command in Figure A simulates 200 games.

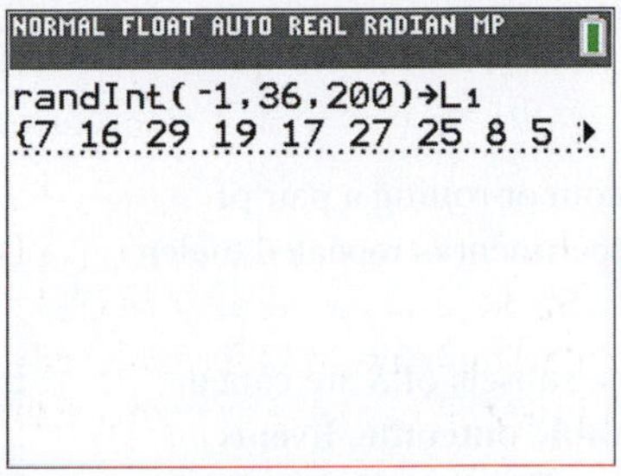

(A)

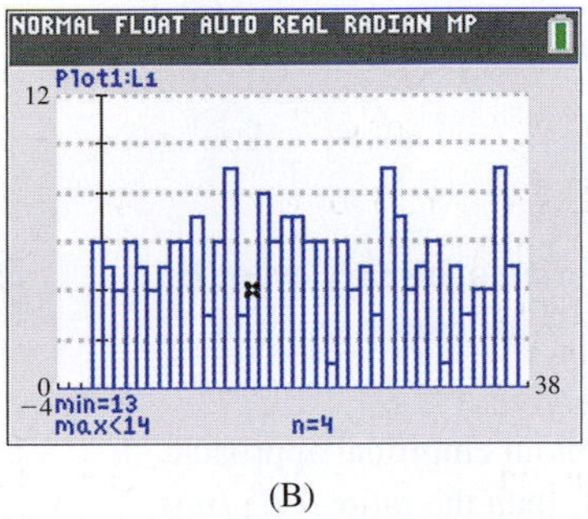

(B)

(A) Use the statistical plot in Figure B to determine the net gain or loss of placing a $1 bet on the number 13 in each of the 200 games.

(B) Compare the results of part (A) with the expected value of the game.

(C) Use a graphing calculator to simulate betting $1 on the number 7 in each of 500 games of roulette and compare the simulated and expected gains or losses.

46. Use a graphing calculator to simulate the results of placing a $1 bet on black in each of 400 games of roulette (see Problems 36 and 45) and compare the simulated and expected gains or losses.

47. A 3-card hand is dealt from a standard deck. You win $20 for each diamond in the hand. If the game is fair, how much should you lose if the hand contains no diamonds?

48. A 3-card hand is dealt from a standard deck. You win $100 for each king in the hand. If the game is fair, how much should you lose if the hand contains no kings?

Applications

49. Insurance. The annual premium for a $5,000 insurance policy against the theft of a painting is $150. If the (empirical) probability that the painting will be stolen during the year is .01, what is your expected return from the insurance company if you take out this insurance?

50. Insurance. An insurance company charges an annual premium of $75 for a $200,000 insurance policy against a house burning down. If the (empirical) probability that a house burns down in a given year is .0003, what is the expected value of the policy to the insurance company?

51. Decision analysis. After careful testing and analysis, an oil company is considering drilling in two different sites. It is estimated that site A will net $30 million if successful (probability .2) and lose $3 million if not (probability .8); site B will net $70 million if successful (probability .1) and lose $4 million if not (probability .9). Which site should the company choose according to the expected return for each site?

52. Decision analysis. Repeat Problem 51, assuming that additional analysis caused the estimated probability of success in field B to be changed from .1 to .11.

53. Genetics. Suppose that at each birth, having a girl is not as likely as having a boy. The probability assignments for the number of boys in a 3-child family are approximated empirically from past records and are given in the table. What is the expected number of boys in a 3-child family?

Number of Boys

x_i	p_i
0	.12
1	.36
2	.38
3	.14

54. Genetics. A pink-flowering plant is of genotype RW. If two such plants are crossed, we obtain a red plant (RR) with probability .25, a pink plant (RW or WR) with probability .50, and a white plant (WW) with probability .25, as shown in the table. What is the expected number of W genes present in a crossing of this type?

Number of W Genes Present

x_i	p_i
0	.25
1	.50
2	.25

55. Lottery. A \$2 Powerball lottery ticket has a 1/27.05 probability of winning \$4, a 1/317.39 probability of winning \$7, a 1/10,376.47 probability of winning \$100, a 1/913,129.18 probability of winning \$50,000, a 1/11,688,053.52 probability of winning \$1,000,000, and a 1/292,201,338 probability of winning the Grand Prize. If the Grand Prize is currently \$100,000,000, what is the expected value of a single Powerball lottery ticket?

56. Lottery. Repeat Problem 55, assuming that the Grand Prize is currently \$400,000,000.

Answers to Matched Problems

1. $E(X) = 3.73$

2. (A)

x_i	0	1	2
p_i	.632	.337	.032*

**Note:* Due to roundoff error, sum $= 1.001 \approx 1$.

(B) .4

3. $E(X) = \$0$; the game is fair

4. $E(X) = (-\$1,775)(.1) + (\$225)(.9) = \$25$ (This amount, of course, is necessary to cover expenses and profit.)

5. $E(X) = (-\$80,000)(.24) + (\$20,000)(.76) = -\$4,000$ (This means that the insurance company had other information regarding the weather than the promoter had; otherwise, the company would not have written this policy.)

Chapter 8 Summary and Review

Important Terms, Symbols, and Concepts

8.1 Sample Spaces, Events, and Probability

EXAMPLES

- Probability theory is concerned with **random experiments** (such as tossing a coin or rolling a pair of dice) for which different outcomes are obtained no matter how carefully the experiment is repeated under the same conditions.
- The set S of all outcomes of a random experiment is called a **sample space**. The subsets of S are called **events**. An event that contains only one outcome is called a **simple event** or **simple outcome**. Events that contain more than one outcome are **compound events**. We say that **an event *E* occurs** if any of the simple events in E occurs. Ex. 1, p. 396 Ex. 2, p. 397 Ex. 3, p. 398
- If $S = \{e_1, e_2, \dots, e_n\}$ is a sample space for an experiment, an **acceptable probability assignment** is an assignment of real numbers $P(e_i)$ to simple events such that

$$0 \le P(e_i) \le 1 \quad \text{and} \quad P(e_1) + P(e_2) + \cdots + P(e_n) = 1$$

- Each number $P(e_i)$ is called the **probability of the event e_i**. The **probability of an arbitrary event *E***, denoted by $P(E)$, is the sum of the probabilities of the simple events in E. If E is the empty set, then $P(E) = 0$. Ex. 4, p. 400
- Acceptable probability assignments can be made using a theoretical approach or an empirical approach. If an experiment is conducted n times and event E occurs with frequency $f(E)$, then the ratio $f(E)/n$ is called the **relative frequency** of the occurrence of E in n trials, or the **approximate empirical probability of *E***. The **empirical probability** of E is the number (if it exists) that $f(E)/n$ approaches as n gets larger and larger. Ex. 6, p. 403
- If the **equally likely assumption** is made, each simple event of the sample space $S = \{e_1, e_2, \dots, e_n\}$ is assigned the same probability, namely, $1/n$. Theorem 1 (p. 402) gives the probability of arbitrary events under the equally likely assumption. Ex. 5, p. 403 Ex. 7, p. 404 Ex. 8, p. 404

8.2 Union, Intersection, and Complement of Events; Odds

- Let A and B be two events in a sample space. Then $A \cup B = \{x \mid x \in A \text{ or } x \in B\}$ is the **union** of A and B; $A \cap B = \{x \mid x \in A \text{ and } x \in B\}$ is the **intersection** of A and B. Ex. 1, p. 409
- Events whose intersection is the empty set are said to be **mutually exclusive** or **disjoint**.

- The probability of the union of two events is given by

$$P(A \cup B) = P(A) + P(B) - P(A \cap B)$$

Ex. 2, p. 410
Ex. 3, p. 412

- The **complement** of event E, denoted by E', consists of those elements of S that do not belong to E:

$$P(E') = 1 - P(E)$$

Ex. 4, p. 413
Ex. 5, p. 413

- The language of **odds** is sometimes used, as an alternative to the language of probability, to describe the likelihood of an event. If $P(E)$ is the probability of E, then the **odds for E** are $P(E)/P(E')$ [usually expressed as a ratio of whole numbers and read as "$P(E)$ to $P(E')$"], and the **odds against E** are $P(E')/P(E)$.

Ex. 6, p. 415
Ex. 7, p. 416
Ex. 8, p. 416

- If the odds for an event E are a/b, then

$$P(E) = \frac{a}{a + b}$$

8.3 Conditional Probability, Intersection, and Independence

- If A and B are events in a sample space S, and $P(B) \neq 0$, then the **conditional probability of A given B** is defined by

$$P(A|B) = \frac{P(A \cap B)}{P(B)}$$

Ex. 1, p. 423
Ex. 2, p. 423

- By solving this equation for $P(A \cap B)$ we obtain the **product rule** (Theorem 1, p. 424):

$$P(A \cap B) = P(B)P(A|B) = P(A)P(B|A)$$

- Events A and B are **independent** if $P(A \cap B) = P(A)P(B)$.
- Theorem 2 (p. 429) gives a test for independence.

Ex. 3, p. 424
Ex. 4, p. 425
Ex. 5, p. 427
Ex. 6, p. 429
Ex. 7, p. 430
Ex. 8, p. 431

8.4 Bayes' Formula

- Let $U_1, U_2, \ldots, U_n$ be n mutually exclusive events whose union is the sample space S. Let E be an arbitrary event in S such that $P(E) \neq 0$. Then

$$P(U_1|E) = \frac{P(E|U_1)P(U_1)}{P(E|U_1)P(U_1) + P(E|U_2)P(U_2) + \cdots + P(E|U_n)P(U_n)}$$

$$= \frac{\text{product of branch probabilities leading to } E \text{ through } U_1}{\text{sum of all branch products leading to } E}$$

Similar results hold for $U_2, U_3, \ldots, U_n$. This formula is called **Bayes' formula**.

Ex. 1, p. 435
Ex. 2, p. 438
Ex. 3, p. 439

8.5 Random Variable, Probability Distribution, and Expected Value

- A **random variable** X is a function that assigns a numerical value to each simple event in a sample space S.
- The **probability distribution of** X assigns a probability $p(x)$ to each range element x of X: $p(x)$ is the sum of the probabilities of the simple events in S that are assigned the numerical value x.
- If a random variable X has range values $x_1, x_2, \ldots, x_n$ that have probabilities $p_1, p_2, \ldots, p_n$, respectively, then the **expected value** of X, denoted $E(X)$, is defined by

$$E(X) = x_1p_1 + x_2p_2 + \cdots + x_np_n$$

- Suppose the x_i's are payoffs in a game of chance. If the game is played a large number of times, the expected value approximates the average win per game.

Ex. 1, p. 445
Ex. 2, p. 446
Ex. 3, p. 447
Ex. 4, p. 447
Ex. 5, p. 448

Review Exercises

Work through all the problems in this chapter review and check your answers in the back of the book. Answers to all problems are there along with section numbers in italics to indicate where each type of problem is discussed. Where weaknesses show up, review appropriate sections in the text.

1. In a single deal of 5 cards from a standard 52-card deck, what is the probability of being dealt 5 clubs?

2. Brittani and Ramon are members of a 15-person ski club. If the president and treasurer are selected by lottery, what is the probability that Brittani will be president and Ramon will be treasurer? (A person cannot hold more than one office.)

3. Each of the first 10 letters of the alphabet is printed on a separate card. What is the probability of drawing 3 cards and getting the code word *dig* by drawing *d* on the first draw, *i* on the second draw, and *g* on the third draw? What is the probability of being dealt a 3-card hand containing the letters *d*, *i*, and *g* in any order?

4. A drug has side effects for 50 out of 1,000 people in a test. What is the approximate empirical probability that a person using the drug will have side effects?

5. A spinning device has 5 numbers, 1, 2, 3, 4, and 5, each as likely to turn up as the other. A person pays \$3 and then receives back the dollar amount corresponding to the number turning up on a single spin. What is the expected value of the game? Is the game fair?

6. If A and B are events in a sample space S and $P(A) = .3$, $P(B) = .4$, and $P(A \cap B) = .1$, find

(A) $P(A')$

(B) $P(A \cup B)$

7. A spinner lands on R with probability .3, on G with probability .5, and on B with probability .2. Find the probability and odds for the spinner landing on either R or G.

8. If in repeated rolls of two fair dice the odds for rolling a sum of 8 before rolling a sum of 7 are 5 to 6, then what is the probability of rolling a sum of 8 before rolling a sum of 7?

Answer Problems 9–17 using the table of probabilities shown below.

	X	Y	Z	Totals
S	.10	.25	.15	.50
T	.05	.20	.02	.27
R	.05	.15	.03	.23
Totals	.20	.60	.20	1.00

9. Find $P(T)$.

10. Find $P(Z)$.

11. Find $P(T \cap Z)$.

12. Find $P(R \cap Z)$.

13. Find $P(R|Z)$.

14. Find $P(Z|R)$.

15. Find $P(T|Z)$.

16. Are T and Z independent?

17. Are S and X independent?

Answer Problems 18–25 using the following probability tree:

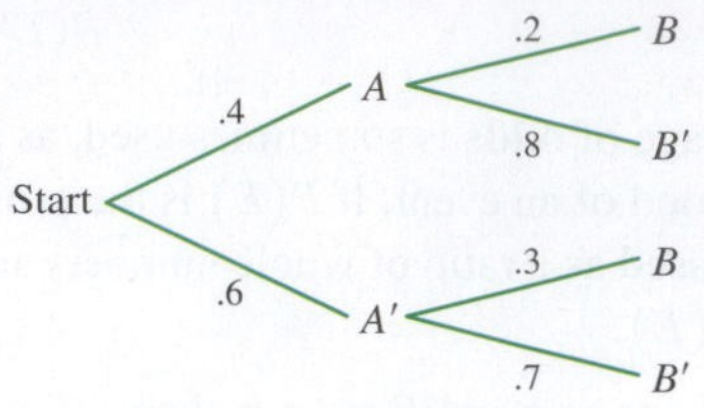

18. $P(A)$

19. $P(B|A)$

20. $P(B|A')$

21. $P(A \cap B)$

22. $P(A' \cap B)$

23. $P(B)$

24. $P(A|B)$

25. $P(A|B')$

26. (A) If 10 out of 32 students in a class were born in June, July, or August, what is the approximate empirical probability of any student being born in June, July, or August?

(B) If one is as likely to be born in any of the 12 months of a year as any other, what is the theoretical probability of being born in either June, July, or August?

(C) Discuss the discrepancy between the answers to parts (A) and (B).

In Problems 27 and 28, a sample space S is described. Would it be reasonable to make the equally likely assumption? Explain.

27. A 3-card hand is dealt from a standard deck. We are interested in the number of red cards in the hand, so an appropriate sample space is $S = \{0, 1, 2, 3\}$.

28. A 3-card hand is dealt from a standard deck. We are interested in whether there are more red cards or more black cards in the hand, so an appropriate sample space is $S = \{R, B\}$.

29. A player tosses two coins and receives \$5 if 2 heads turn up, loses \$4 if 1 head turns up, and wins \$2 if 0 heads turn up. Compute the expected value of the game. Is the game fair?

30. A spinning device has 3 numbers, 1, 2, and 3, each as likely to turn up as the other. If the device is spun twice, what is the probability that

(A) The same number turns up both times?

(B) The sum of the numbers turning up is 5?

31. In a single draw from a standard 52-card deck, what are the probability and odds for drawing

(A) A jack or a queen?

(B) A jack or a spade?

(C) A card other than an ace?

32. (A) What are the odds for rolling a sum of 5 on the single roll of two fair dice?

(B) If you bet \$1 that a sum of 5 will turn up, what should the house pay (plus return your \$1 bet) in order for the game to be fair?

33. Two coins are flipped 1,000 times with the following frequencies:

2 heads	210
1 head	480
0 heads	310

(A) Compute the empirical probability for each outcome.

(B) Compute the theoretical probability for each outcome.

(C) Using the theoretical probabilities computed in part (B), compute the expected frequency of each outcome, assuming fair coins.

34. A fair coin is tossed 10 times. On each of the first 9 tosses the outcome is heads. Discuss the probability of a head on the 10th toss.

35. An experiment consists of rolling a pair of fair dice. Let X be the random variable associated with the sum of the values that turn up.

(A) Find the probability distribution for X.

(B) Find the expected value of X.

36. Two dice are rolled. The sample space is chosen as the set of all ordered pairs of integers taken from $\{1, 2, 3, 4, 5, 6\}$. What is the event A that corresponds to the sum being divisible by 4? What is the event B that corresponds to the sum being divisible by 6? What are $P(A)$, $P(B)$, $P(A \cap B)$, and $P(A \cup B)$?

37. A person tells you that the following approximate empirical probabilities apply to the sample space $\{e_1, e_2, e_3, e_4\}$: $P(e_1) \approx .1$, $P(e_2) \approx -.2$, $P(e_3) \approx .6$, $P(e_4) \approx 2$. There are three reasons why P cannot be a probability function. Name them.

38. Use the following information to complete the frequency table below:

$$n(A) = 50, \quad n(B) = 45,$$
$$n(A \cup B) = 80, \quad n(U) = 100$$

	A	A'	Totals
B			
B'			
Totals			

39. A pointer is spun on a circular spinner. The probabilities of the pointer landing on the integers from 1 to 5 are given in the table below.

e_i	1	2	3	4	5
p_i	.2	.1	.3	.3	.1

(A) What is the probability of the pointer landing on an odd number?

(B) What is the probability of the pointer landing on a number less than 4 given that it landed on an odd number?

40. A card is drawn at random from a standard 52-card deck. If E is the event "The drawn card is red" and F is the event "The drawn card is an ace," then

(A) Find $P(F \mid E)$. (B) Test E and F for independence.

In Problems 41–45, urn U_1 contains 2 white balls and 3 red balls; urn U_2 contains 2 white balls and 1 red ball.

41. Two balls are drawn out of urn U_1 in succession. What is the probability of drawing a white ball followed by a red ball if the first ball is

(A) Replaced? (B) Not replaced?

42. Which of the two parts in Problem 41 involve dependent events?

43. In Problem 41, what is the expected number of red balls if the first ball is

(A) Replaced? (B) Not replaced?

44. An urn is selected at random by flipping a fair coin; then a ball is drawn from the urn. Compute:

(A) $P(R \mid U_1)$ (B) $P(R \mid U_2)$

(C) $P(U_2 \mid W)$ (D) $P(U_1 \mid R)$

45. In Problem 44, are the events "Selecting urn U_1" and "Drawing a red ball" independent?

46. From a standard deck of 52 cards, what is the probability of obtaining a 5-card hand

(A) Of all diamonds?

(B) Of 3 diamonds and 2 spades?

Write answers in terms of ${}_nC_r$ or ${}_nP_r$; do not evaluate.

47. A group of 10 people includes one married couple. If 4 people are selected at random, what is the probability that the married couple is selected?

48. A 5-card hand is drawn from a standard deck. Discuss how you can tell that the following two events are dependent without any computation.

S = hand consists entirely of spades

H = hand consists entirely of hearts

49. The command in Figure A was used on a graphing calculator to simulate 50 repetitions of rolling a pair of dice and recording the minimum of the two numbers. A statistical plot of the results is shown in Figure B.

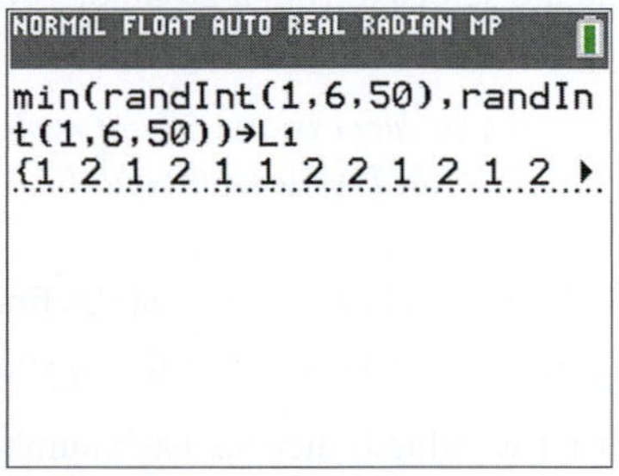

(A)

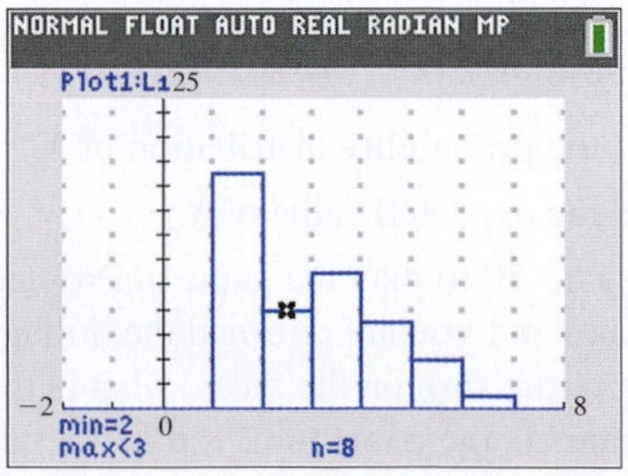

(B)

(A) Use Figure B to find the empirical probability that the minimum is 2.

(B) What is the theoretical probability that the minimum is 2?

(C) Using a graphing calculator to simulate 200 rolls of a pair of dice, determine the empirical probability that the minimum is 4 and compare with the theoretical probability.

50. A card is drawn at random from a standard 52-card deck. Using a graphing calculator to simulate 800 such draws, determine the empirical probability that the card is a black jack and compare with the theoretical probability.

In Problems 51–56, discuss the validity of each statement. If the statement is always true, explain why. If not, give a counterexample.

51. If $P(E) = 1$, then the odds for E are $1 : 1$.

52. If $E = F'$, then $P(E \cup F) = P(E) + P(F)$.

53. If E and F are complementary events, then E and F are independent.

54. If $P(E \cup F) = 1$, then E and F are complementary events.

55. If E and F are independent events, then $P(E)P(F) = P(E \cap F)$.

56. If E and F are mutually exclusive events, then $P(E)P(F) = P(E \cap F)$.

57. Three fair coins are tossed 1,000 times with the following frequencies of outcomes:

Number of Heads	0	1	2	3
Frequency	120	360	350	170

(A) What is the empirical probability of obtaining 2 heads?

(B) What is the theoretical probability of obtaining 2 heads?

(C) What is the expected frequency of obtaining 2 heads?

58. You bet a friend \$1 that you will get 1 or more double 6's on 24 rolls of a pair of fair dice. What is your expected value for this game? What is your friend's expected value? Is the game fair?

59. If 3 people are selected from a group of 7 men and 3 women, what is the probability that at least 1 woman is selected?

Two cards are drawn in succession without replacement from a standard 52-card deck. In Problems 60 and 61, compute the indicated probabilities.

60. The second card is a heart given that the first card is a heart.

61. The first card is a heart given that the second card is a heart.

62. Two fair (not weighted) dice are each numbered with a 3 on one side, a 2 on two sides, and a 1 on three sides. The dice are rolled, and the numbers on the two up faces are added. If X is the random variable associated with the sample space $S = \{2, 3, 4, 5, 6\}$:

(A) Find the probability distribution of X.

(B) Find the expected value of X.

63. If you pay \$3.50 to play the game in Problem 62 (the dice are rolled once) and you are returned the dollar amount corresponding to the sum on the faces, what is the expected value of the game? Is the game fair? If it is not fair, how much should you pay in order to make the game fair?

64. Suppose that 3 white balls and 1 black ball are placed in a box. Balls are drawn in succession without replacement until a black ball is drawn, and then the game is over. You win if the black ball is drawn on the fourth draw.

(A) What are the probability and odds for winning?

(B) If you bet \$1, what should the house pay you for winning (plus return your \$1 bet) if the game is to be fair?

65. If each of 5 people is asked to identify his or her favorite book from a list of 10 best-sellers, what is the probability that at least 2 of them identify the same book?

66. Let A and B be events with nonzero probabilities in a sample space S. Under what conditions is $P(A|B)$ equal to $P(B|A)$?

Applications

67. Market research. From a survey of 100 city residents, it was found that 40 read the daily newspaper, 70 watch the evening news, and 30 do both. What is the (empirical) probability that a resident selected at random

(A) Reads the daily paper or watches the evening news?

(B) Does neither?

(C) Does one but not the other?

68. Market research. A market research firm has determined that 40% of the people in a certain area have seen the advertising for a new product and that 85% of those who have seen the advertising have purchased the product. What is the probability that a person in this area has seen the advertising and purchased the product?

69. Market analysis. A clothing company selected 1,000 persons at random and surveyed them to determine a relationship between the age of the purchaser and the annual purchases of jeans. The results are given in the table.

	Jeans Purchased Annually				
Age	**0**	**1**	**2**	**Above 2**	**Totals**
Under 12	60	70	30	10	170
12–18	30	100	100	60	290
19–25	70	110	120	30	330
Over 25	100	50	40	20	210
Totals	260	330	290	120	1,000

Given the events

A = person buys 2 pairs of jeans

B = person is between 12 and 18 years old

C = person does not buy more than 2 pairs of jeans

D = person buys more than 2 pairs of jeans

(A) Find $P(A)$, $P(B)$, $P(A \cap B)$, $P(A|B)$, and $P(B|A)$.

(B) Are events A and B independent? Explain.

(C) Find $P(C)$, $P(D)$, $P(C \cap D)$, $P(C|D)$, and $P(D|C)$.

(D) Are events C and D mutually exclusive? Independent? Explain.

70. Decision analysis. A company sales manager, after careful analysis, presents two sales plans. It is estimated that plan A will net \$10 million if successful (probability .8) and lose \$2 million if not (probability .2); plan B will net \$12 million if successful (probability .7) and lose \$2 million if not (probability .3). What is the expected return for each plan? Which plan should be chosen based on the expected return?

71. Insurance. A \$2,000 bicycle is insured against theft for an annual premium of \$170. If the probability that the bicycle will be stolen during the year is .08 (empirically determined), what is the expected value of the policy?

72. Quality control. Twelve precision parts, including 2 that are substandard, are sent to an assembly plant. The plant will select 4 at random and will return the entire shipment if 1 or more of the sample are found to be substandard. What is the probability that the shipment will be returned?

73. Quality control. A dozen tablet computers, including 2 that are defective, are sent to a computer service center. A random sample of 3 is selected and tested. Let X be the random variable associated with the number of tablet computers in a sample that are defective.

(A) Find the probability distribution of X.

(B) Find the expected number of defective tablet computers in a sample.

74. Medicine: cardiogram test. By testing a large number of individuals, it has been determined that 82% of the population have normal hearts, 11% have some minor heart problems, and 7% have severe heart problems. Ninety-five percent of the persons with normal hearts, 30% of those with minor problems, and 5% of those with severe problems will pass a cardiogram test. What is the probability that a person who passes the cardiogram test has a normal heart?

75. Genetics. Six men in 100 and 1 woman in 100 are color-blind. A person is selected at random and is found to be color-blind. What is the probability that this person is a man? (Assume that the total population contains the same number of women as men.)

76. Voter preference. In a straw poll, 30 students in a mathematics class are asked to indicate their preference for president of student government. Approximate empirical probabilities are assigned on the basis of the poll: candidate A should receive 53% of the vote, candidate B should receive 37%, and candidate C should receive 10%. One week later, candidate B wins the election. Discuss the factors that may account for the discrepancy between the poll and the election results.

9 Markov Chains

Introduction

In this chapter, we consider a mathematical model that combines probability and matrices to analyze certain sequences. The model is called a *Markov chain*, after the Russian mathematician Andrei Markov (1856–1922). Recent applications of Markov chains involve a wide variety of topics, including finance, market research, genetics, medicine, demographics, psychology, and political science. Problem 92 in Section 9.1, for example, uses a Markov chain to model a training program for apprentice welders.

In Section 9.1 we introduce the basic properties of Markov chains. In the remaining sections, we discuss the long-term behavior of two different types of Markov chains.

9.1 Properties of Markov Chains

- Introduction
- Transition and State Matrices
- Powers of Transition Matrices
- Application

Introduction

In this section, we explore physical *systems* and their possible *states*. To understand what this means, consider the following examples:

1. A stock listed on the New York Stock Exchange either increases, decreases, or does not change in price each day that the exchange is open. The stock can be thought of as a physical system with three possible states: increase, decrease, or no change.
2. A commuter, relative to a rapid transit system, can be thought of as a physical system with two states, a user or a nonuser.
3. During each congressional election, a voting precinct casts a simple majority vote for a Republican candidate, a Democratic candidate, or a third-party candidate. The precinct, relative to all congressional elections past, present, and future, constitutes a physical system that is in one (and only one) of three states after each election: Republican, Democratic, or other.

If a system evolves from one state to another in such a way that chance elements are involved, then the system's progression through a sequence of states is called a **stochastic process** (*stochos* is the Greek word for "guess"). We will consider a simple example of a stochastic process, and out of it will arise further definitions and methodology.

A toothpaste company markets a product (brand A) that currently has 10% of the toothpaste market. The company hires a market research firm to estimate the percentage of the market that it might acquire in the future if it launches an aggressive sales campaign. The research firm uses test marketing and extensive surveys to predict the effect of the campaign. They find that if a person is using brand A, the probability is .8 that this person will buy it again when he or she runs out of toothpaste. On the other hand, a person using another brand will switch to brand A with a probability of .6 when he or she runs out of toothpaste. So each toothpaste consumer can be considered to be in one of two possible states:

$$A = \text{uses brand } A \qquad \text{or} \qquad A' = \text{uses another brand}$$

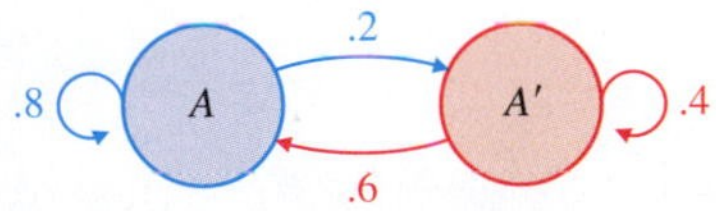

Figure 1 Transition diagram

The probabilities determined by the market research firm can be represented graphically in a **transition diagram** (Fig. 1).

We can also represent this information numerically in a **transition probability matrix**:

$$\text{Current state} \quad \begin{matrix} & \text{Next state} \\ & \begin{matrix} A & A' \end{matrix} \\ \begin{matrix} A \\ A' \end{matrix} & \begin{bmatrix} .8 & .2 \\ .6 & .4 \end{bmatrix} \end{matrix} = P$$

Explore and Discuss 1

(A) Refer to the transition diagram in Figure 1. What is the probability that a person using brand A will switch to another brand when he or she runs out of toothpaste?

(B) Refer to transition probability matrix P. What is the probability that a person who is not using brand A will not switch to brand A when he or she runs out of toothpaste?

(C) In Figure 1, the sum of the probabilities on the arrows leaving each state is 1. Will this be true for any transition diagram? Explain your answer.

(D) In transition probability matrix P, the sum of the probabilities in each row is 1. Will this be true for any transition probability matrix? Explain your answer.

The toothpaste company's 10% share of the market at the beginning of the sales campaign can be represented as an **initial-state distribution matrix**:

$$S_0 = \begin{matrix} A & A' \\ [.1 & .9] \end{matrix}$$

If a person is chosen at random, the probability that this person uses brand A (state A) is .1, and the probability that this person does not use brand A (state A') is .9. Thus, S_0 also can be interpreted as an **initial-state probability matrix**.

What are the probabilities of a person being in state A or A' on the first purchase after the start of the sales campaign? Let us look at the probability tree given below.

Note: A_0 represents state A at the beginning of the campaign, A_1' represents state A' on the first purchase after the campaign, and so on.

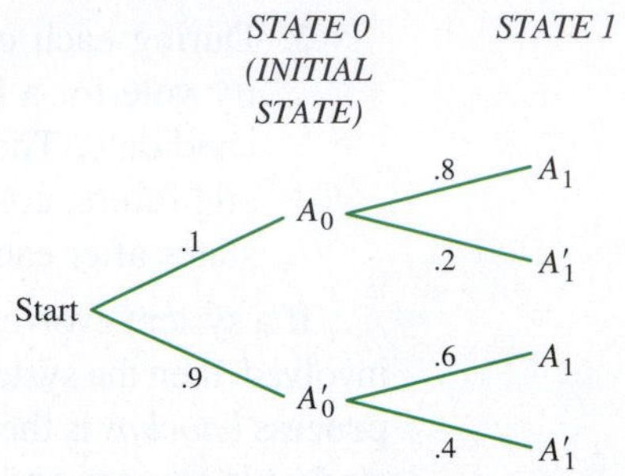

Proceeding as in Chapter 8, we can read the required probabilities directly from the tree:

$$\begin{aligned} P(A_1) &= P(A_0 \cap A_1) + P(A_0' \cap A_1) \\ &= (.1)(.8) + (.9)(.6) = .62 \\ P(A_1') &= P(A_0 \cap A_1') + P(A_0' \cap A_1') \\ &= (.1)(.2) + (.9)(.4) = .38 \end{aligned}$$

Note: $P(A_1) + P(A_1') = 1$, as expected.

The **first-state matrix** is

$$S_1 = \begin{matrix} A & A' \\ [.62 & .38] \end{matrix}$$

This matrix gives us the probabilities of a randomly chosen person being in state A or A' on the first purchase after the start of the campaign. We see that brand A's market share has increased from 10% to 62%.

Now, if you were asked to find the probabilities of a person being in state A or state A' on the tenth purchase after the start of the campaign, you might start to draw additional branches on the probability tree. However, you would soon become discouraged because the number of branches doubles for each successive purchase. By the tenth purchase, there would be $2^{11} = 2{,}048$ branches! Fortunately, we can convert the summing of branch products to matrix multiplication. In particular, if we multiply the initial-state matrix S_0 by the transition matrix P, we obtain the first-state matrix S_1:

$$S_0P = \underbrace{\begin{matrix} A & A' \\ [.1 & .9] \end{matrix}}_{\text{Initial state}} \underbrace{\begin{bmatrix} .8 & .2 \\ .6 & .4 \end{bmatrix}}_{\text{Transition matrix}} = [(.1)(.8) + (.9)(.6) \quad (.1)(.2) + (.9)(.4)] = \underbrace{\begin{matrix} A & A' \\ [.62 & .38] \end{matrix}}_{\text{First state}} = S_1$$

Compare with the tree computations above

As you might guess, we can get the second-state matrix S_2 (for the second purchase) by multiplying the first-state matrix by the transition matrix:

$$S_1 P = \begin{array}{cc} A & A' \\ [.62 & .38] \end{array} \begin{bmatrix} .8 & .2 \\ .6 & .4 \end{bmatrix} = \begin{array}{cc} A & A' \\ [.724 & .276] \end{array} = S_2$$

First state Second state

The third-state matrix S_3 is computed in a similar manner:

$$S_2 P = \begin{array}{cc} A & A' \\ [.724 & .276] \end{array} \begin{bmatrix} .8 & .2 \\ .6 & .4 \end{bmatrix} = \begin{array}{cc} A & A' \\ [.7448 & .2552] \end{array} = S_3$$

Second state Third state

Examining the values in the first three state matrices, we see that brand A's market share increases after each toothpaste purchase. Will the market share for brand A continue to increase until it approaches 100%, or will it level off at some value less than 100%? These questions are answered in Section 9.2 when we develop techniques for determining the long-run behavior of state matrices.

Transition and State Matrices

The sequence of trials (toothpaste purchases) with the constant transition matrix P is a special kind of stochastic process called a *Markov chain*. In general, a **Markov chain** is a sequence of experiments, trials, or observations such that the transition probability matrix from one state to the next is constant. A Markov chain has no memory. The various matrices associated with a Markov chain are defined in the next box.

DEFINITION Markov Chains

Given a Markov chain with n states, a **kth-state matrix** is a matrix of the form

$$S_k = \begin{bmatrix} s_{k1} & s_{k2} & \cdots & s_{kn} \end{bmatrix}$$

such that no entry is negative and the sum of the entries is 1.

Each entry s_{ki} is the proportion of the population that is in state i after the kth trial, or, equivalently, the probability of a randomly selected element of the population being in state i after the kth trial.

A **transition matrix** is a constant square matrix P of order n such that the entry in the ith row and jth column indicates the probability of the system moving from the ith state to the jth state on the next observation or trial. The sum of the entries in each row must be 1.

CONCEPTUAL INSIGHT

1. Since the entries in a kth-state matrix or transition matrix are probabilities, they must be real numbers between 0 and 1, inclusive.

2. Rearranging the various states and corresponding transition probabilities in a transition matrix will produce a different, but equivalent, transition matrix. For example, both of the following matrices are transition matrices for the toothpaste company discussed earlier:

$$P = \begin{array}{c} \\ A \\ A' \end{array} \begin{array}{cc} A & A' \\ \begin{bmatrix} .8 & .2 \\ .6 & .4 \end{bmatrix} \end{array} \qquad P' = \begin{array}{c} \\ A \\ A' \end{array} \begin{array}{cc} A & A' \\ \begin{bmatrix} .4 & .6 \\ .2 & .8 \end{bmatrix} \end{array}$$

Such rearrangements will affect the form of the matrices used in the solution of a problem but will not affect any of the information obtained from these matrices. In Section 9.3, we encounter situations where it will be helpful to select a transition matrix that has a special form. For now, you can choose any order for the states in a transition matrix.

As we indicated in the preceding discussion, matrix multiplication can be used to compute the various state matrices of a Markov chain:

If S_0 is the initial-state matrix and P is the transition matrix for a Markov chain, then the subsequent state matrices are given by

$$S_1 = S_0P \quad \text{First-state matrix}$$
$$S_2 = S_1P \quad \text{Second-state matrix}$$
$$S_3 = S_2P \quad \text{Third-state matrix}$$
$$\vdots$$
$$S_k = S_{k-1}P \quad k\text{th-state matrix}$$

EXAMPLE 1 **Insurance** An insurance company found that on average, over a period of 10 years, 23% of the drivers in a particular community who were involved in an accident one year were also involved in an accident the following year. They also found that only 11% of the drivers who were not involved in an accident one year were involved in an accident the following year. Use these percentages as approximate empirical probabilities for the following:

(A) Draw a transition diagram.

(B) Find the transition matrix P.

(C) If 5% of the drivers in the community are involved in an accident this year, what is the probability that a driver chosen at random from the community will be involved in an accident next year? Year after next?

SOLUTION

(A)

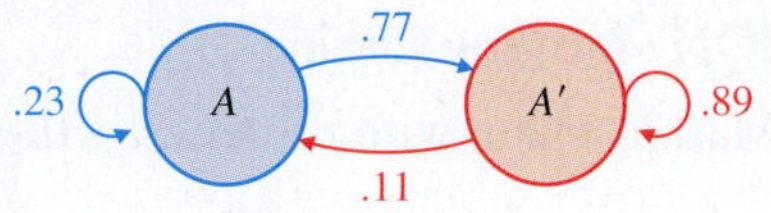

Next year

A = accident
A' = no accident

(B)

$$\begin{matrix} & & A & A' \\ \text{This} & A & .23 & .77 \\ \text{year} & A' & .11 & .89 \end{matrix} = P \quad \text{Transition matrix}$$

(C) The initial-state matrix S_0 is

$$S_0 = \begin{matrix} A & A' \\ [.05 & .95] \end{matrix} \quad \text{Initial-state matrix}$$

Thus,

$$S_0P = \underset{\substack{\text{This year} \\ \text{(initial state)}}}{\begin{matrix} A & A' \\ [.05 & .95] \end{matrix}} \begin{bmatrix} .23 & .77 \\ .11 & .89 \end{bmatrix} = \underset{\substack{\text{Next year} \\ \text{(first state)}}}{\begin{matrix} A & A' \\ [.116 & .884] \end{matrix}} = S_1$$

$$S_1P = \underset{\substack{\text{Next year} \\ \text{(first state)}}}{\begin{matrix} A & A' \\ [.116 & .884] \end{matrix}} \begin{bmatrix} .23 & .77 \\ .11 & .89 \end{bmatrix} = \underset{\substack{\text{Year after next} \\ \text{(second state)}}}{\begin{matrix} A & A' \\ [.12392 & .87608] \end{matrix}} = S_2$$

The probability that a driver chosen at random from the community will have an accident next year is .116, and the year after next is .12392. That is, it is expected that 11.6% of the drivers in the community will have an accident next year and 12.392% the year after.

Matched Problem 1 An insurance company classifies drivers as low-risk if they are accident-free for one year. Past records indicate that 98% of the drivers in the low-risk category (L) one year will remain in that category the next year, and 78% of the drivers who are not in the low-risk category (L') one year will be in the low-risk category the next year.

(A) Draw a transition diagram.

(B) Find the transition matrix P.

(C) If 90% of the drivers in the community are in the low-risk category this year, what is the probability that a driver chosen at random from the community will be in the low-risk category next year? Year after next?

Powers of Transition Matrices

Next we investigate the powers of a transition matrix.

The state matrices for a Markov chain are defined **recursively**; that is, each state matrix is defined in terms of the preceding state matrix. For example, to find the fourth-state matrix S_4, it is necessary to compute the preceding three state matrices:

$$S_1 = S_0P \qquad S_2 = S_1P \qquad S_3 = S_2P \qquad S_4 = S_3P$$

Is there any way to compute a given state matrix directly without first computing all the preceding state matrices? If we substitute the equation for S_1 into the equation for S_2, substitute this new equation for S_2 into the equation for S_3, and so on, a definite pattern emerges:

$$\begin{aligned} S_1 &= S_0P \\ S_2 &= S_1P = (S_0P)P = S_0P^2 \\ S_3 &= S_2P = (S_0P^2)P = S_0P^3 \\ S_4 &= S_3P = (S_0P^3)P = S_0P^4 \\ &\vdots \end{aligned}$$

In general, it can be shown that the kth-state matrix is given by $S_k = S_0P^k$. We summarize this important result in Theorem 1.

THEOREM 1 Powers of a Transition Matrix

If P is the transition matrix and S_0 is an initial-state matrix for a Markov chain, then the kth-state matrix is given by

$$S_k = S_0P^k$$

The entry in the ith row and jth column of P^k indicates the probability of the system moving from the ith state to the jth state in k observations or trials. The sum of the entries in each row of P^k is 1.

EXAMPLE 2 **Using P^k to Compute S_k** Find P^4 and use it to find S_4 for

$$P = \begin{matrix} & \begin{matrix} A & A' \end{matrix} \\ \begin{matrix} A \\ A' \end{matrix} & \begin{bmatrix} .1 & .9 \\ .6 & .4 \end{bmatrix} \end{matrix} \quad \text{and} \quad S_0 = \begin{matrix} \begin{matrix} A & A' \end{matrix} \\ \begin{bmatrix} .2 & .8 \end{bmatrix} \end{matrix}$$

SOLUTION $P^2 = PP = \begin{bmatrix} .1 & .9 \\ .6 & .4 \end{bmatrix}\begin{bmatrix} .1 & .9 \\ .6 & .4 \end{bmatrix} = \begin{bmatrix} .55 & .45 \\ .3 & .7 \end{bmatrix}$

$$P^4 = P^2P^2 = \begin{bmatrix} .55 & .45 \\ .3 & .7 \end{bmatrix}\begin{bmatrix} .55 & .45 \\ .3 & .7 \end{bmatrix} = \begin{bmatrix} .4375 & .5625 \\ .375 & .625 \end{bmatrix}$$

$$S_4 = S_0P^4 = \begin{bmatrix} .2 & .8 \end{bmatrix}\begin{bmatrix} .4375 & .5625 \\ .375 & .625 \end{bmatrix} = \begin{bmatrix} .3875 & .6125 \end{bmatrix}$$

Matched Problem 2 Find P^4 and use it to find S_4 for

$$P = \begin{matrix} & \begin{matrix} A & A' \end{matrix} \\ \begin{matrix} A \\ A' \end{matrix} & \begin{bmatrix} .8 & .2 \\ .3 & .7 \end{bmatrix} \end{matrix} \quad \text{and} \quad S_0 = \begin{matrix} \begin{matrix} A & A' \end{matrix} \\ \begin{bmatrix} .8 & .2 \end{bmatrix} \end{matrix}$$

If a graphing calculator or a computer is available for computing matrix products and powers of a matrix, finding state matrices for any number of trials becomes a routine calculation.

EXAMPLE 3 **Using a Graphing Calculator and P^k to Compute S^k** Use P^8 and a graphing calculator to find S_8 for P and S_0 as given in Example 2. Round values in S_8 to six decimal places.

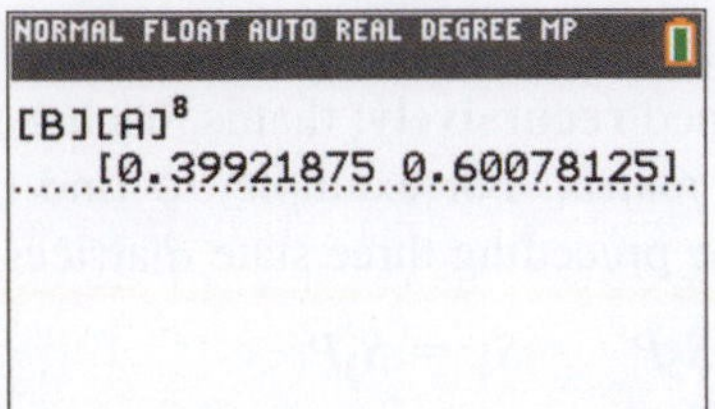

Figure 2

SOLUTION After storing the matrices P and S_0 in the graphing calculator's memory, we use the equation

$$S_8 = S_0 P^8$$

to compute S_8. Figure 2 shows the result on a typical graphing calculator, where the matrix names [A] and [B] denote P and S_0, respectively. We see that (to six decimal places)

$$S_8 = [\,.399219 \quad .600781\,]$$

Matched Problem 3 Use P^8 and a graphing calculator to find S_8 for P and S_0 as given in Matched Problem 2. Round values in S_8 to six decimal places.

Application

The next example illustrates the use of Theorem 1 in an application.

EXAMPLE 4 **Student Retention** Part-time students in a university MBA program are considered to be entry-level students until they complete 15 credits successfully. Then they are classified as advanced-level students and can take more advanced courses and work on the thesis required for graduation. Past records indicate that at the end of each year, 10% of the entry-level students (E) drop out of the program (D) and 30% become advanced-level students (A). Also, 10% of the advanced-level students drop out of the program and 40% graduate (G) each year. Students that graduate or drop out never return to the program.

(A) Draw a transition diagram. (B) Find the transition matrix P.

(C) What is the probability that an entry-level student graduates within 4 years? Drops out within 4 years?

SOLUTION

(A) If 10% of entry-level students drop out and 30% become advanced-level students, then the remaining 60% must continue as entry-level students for another year (see the diagram). Similarly, 50% of advanced-level students must continue as advanced-level students for another year. Since students who drop out never return, all students in state D in one year will continue in that state the next year. We indicate this by placing a 1 on the arrow from D back to D. State G is labeled in the same manner.

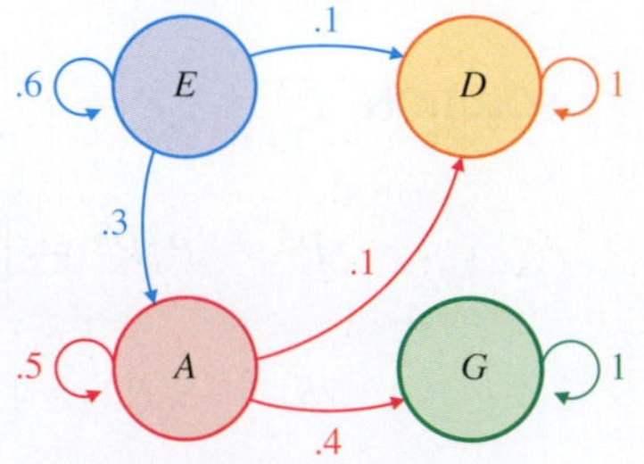

Transition diagram

(B)
$$P = \begin{array}{c} \\ E \\ D \\ A \\ G \end{array} \begin{array}{c} \begin{array}{cccc} E & D & A & G \end{array} \\ \begin{bmatrix} .6 & .1 & .3 & 0 \\ 0 & 1 & 0 & 0 \\ 0 & .1 & .5 & .4 \\ 0 & 0 & 0 & 1 \end{bmatrix} \end{array}$$
Transition matrix

(C) The probability that an entry-level student moves from state E to state G within 4 years is the entry in row 1 and column 4 of P^4 (Theorem 1). Hand computation of P^4 requires two multiplications:

$$P^2 = \begin{bmatrix} .6 & .1 & .3 & 0 \\ 0 & 1 & 0 & 0 \\ 0 & .1 & .5 & .4 \\ 0 & 0 & 0 & 1 \end{bmatrix} \begin{bmatrix} .6 & .1 & .3 & 0 \\ 0 & 1 & 0 & 0 \\ 0 & .1 & .5 & .4 \\ 0 & 0 & 0 & 1 \end{bmatrix} = \begin{bmatrix} .36 & .19 & .33 & .12 \\ 0 & 1 & 0 & 0 \\ 0 & .15 & .25 & .6 \\ 0 & 0 & 0 & 1 \end{bmatrix}$$

$$P^4 = P^2P^2 = \begin{bmatrix} .36 & .19 & .33 & .12 \\ 0 & 1 & 0 & 0 \\ 0 & .15 & .25 & .6 \\ 0 & 0 & 0 & 1 \end{bmatrix} \begin{bmatrix} .36 & .19 & .33 & .12 \\ 0 & 1 & 0 & 0 \\ 0 & .15 & .25 & .6 \\ 0 & 0 & 0 & 1 \end{bmatrix}$$

$$= \begin{bmatrix} .1296 & .3079 & .2013 & .3612 \\ 0 & 1 & 0 & 0 \\ 0 & .1875 & .0625 & .75 \\ 0 & 0 & 0 & 1 \end{bmatrix}$$

The probability that an entry-level student has graduated within 4 years is .3612. Similarly, the probability that an entry-level student has dropped out within 4 years is .3079 (the entry in row 1 and column 2 of P^4).

Matched Problem 4 Refer to Example 4. At the end of each year the faculty examines the progress that each advanced-level student has made on the required thesis. Past records indicate that 30% of advanced-level students (A) complete the thesis requirement (C) and 10% are dropped from the program for insufficient progress (D), never to return. The remaining students continue to work on their theses.

(A) Draw a transition diagram.

(B) Find the transition matrix P.

(C) What is the probability that an advanced-level student completes the thesis requirement within 4 years? Is dropped from the program for insufficient progress within 4 years?

Explore and Discuss 2

Refer to Example 4. States D and G are referred to as *absorbing states* because a student who enters either one of these states never leaves it. Absorbing states are discussed in detail in Section 9.3.

(A) How can absorbing states be recognized from a transition diagram? Draw a transition diagram with two states, one that is absorbing and one that is not, to illustrate.

(B) How can absorbing states be recognized from a transition matrix? Write the transition matrix for the diagram you drew in part (A) to illustrate.

Exercises 9.1

Skills Warm-up Exercises

W *In Problems 1–8, find the matrix product, if it is defined. (If necessary, review Section 4.4.)*

1. $\begin{bmatrix} 2 & 5 \\ 4 & 1 \end{bmatrix}\begin{bmatrix} 3 \\ 2 \end{bmatrix}$

2. $\begin{bmatrix} 4 & 5 \end{bmatrix}\begin{bmatrix} 6 & 9 \\ 3 & 7 \end{bmatrix}$

3. $\begin{bmatrix} 3 \\ 2 \end{bmatrix}\begin{bmatrix} 2 & 5 \\ 4 & 1 \end{bmatrix}$

4. $\begin{bmatrix} 6 & 9 \\ 3 & 7 \end{bmatrix}\begin{bmatrix} 4 & 5 \end{bmatrix}$

5. $\begin{bmatrix} 3 & 2 \end{bmatrix}\begin{bmatrix} 2 & 5 \\ 4 & 1 \end{bmatrix}$

6. $\begin{bmatrix} 4 \\ 5 \end{bmatrix}\begin{bmatrix} 6 & 9 \\ 3 & 7 \end{bmatrix}$

7. $\begin{bmatrix} 2 & 5 \\ 4 & 1 \end{bmatrix}\begin{bmatrix} 3 & 2 \end{bmatrix}$

8. $\begin{bmatrix} 6 & 9 \\ 3 & 7 \end{bmatrix}\begin{bmatrix} 4 \\ 5 \end{bmatrix}$

A *In Problems 9–14, use the transition matrix*

$$P = \begin{array}{c} \\ A \\ B \end{array}\begin{array}{c} \begin{array}{cc} A & B \end{array} \\ \begin{bmatrix} .7 & .3 \\ .1 & .9 \end{bmatrix} \end{array}$$

to find S_1 and S_2 for the indicated initial state matrix S_0.

9. $S_0 = [0 \quad 1]$

10. $S_0 = [1 \quad 0]$

11. $S_0 = [.6 \quad .4]$

12. $S_0 = [.2 \quad .8]$

13. $S_0 = [.25 \quad .75]$

14. $S_0 = [.75 \quad .25]$

In Problems 15–20, use the transition diagram

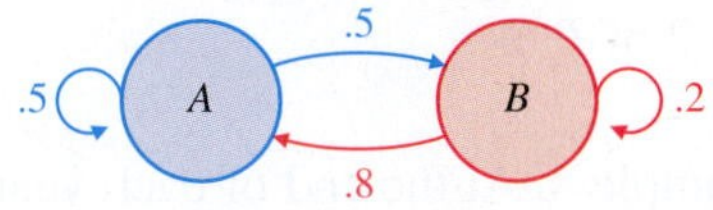

to find S_1 and S_2 for the indicated initial state matrix S_0.

15. $S_0 = [1 \quad 0]$

16. $S_0 = [0 \quad 1]$

17. $S_0 = [.3 \quad .7]$

18. $S_0 = [.9 \quad .1]$

19. $S_0 = [.5 \quad .5]$

20. $S_0 = [.2 \quad .8]$

In Problems 21–26, use the transition matrix

$$P = \begin{array}{c} \\ A \\ B \\ C \end{array}\begin{array}{c} \begin{array}{ccc} A & B & C \end{array} \\ \begin{bmatrix} .2 & .4 & .4 \\ .7 & .2 & .1 \\ .5 & .3 & .2 \end{bmatrix} \end{array}$$

to find S_1 and S_2 for the indicated initial state matrix S_0.

21. $S_0 = [0 \quad 1 \quad 0]$

22. $S_0 = [0 \quad 0 \quad 1]$

23. $S_0 = [.5 \quad 0 \quad .5]$

24. $S_0 = [.5 \quad .5 \quad 0]$

25. $S_0 = [.1 \quad .3 \quad .6]$

26. $S_0 = [.4 \quad .3 \quad .3]$

In Problems 27–32, use the transition diagram

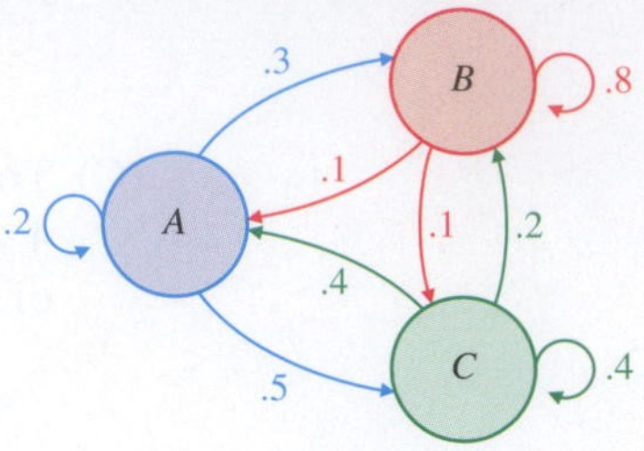

to find S_1 and S_2 for the indicated initial state matrix S_0.

27. $S_0 = [1 \quad 0 \quad 0]$

28. $S_0 = [0 \quad 1 \quad 0]$

29. $S_0 = [0 \quad .4 \quad .6]$

30. $S_0 = [.8 \quad 0 \quad .2]$

31. $S_0 = [.5 \quad .2 \quad .3]$

32. $S_0 = [.2 \quad .7 \quad .1]$

33. Draw the transition diagram that corresponds to the transition matrix of Problem 9.

34. Find the transition matrix that corresponds to the transition diagram of Problem 15.

35. Draw the transition matrix that corresponds to the transition diagram of Problem 27.

36. Find the transition diagram that corresponds to the transition matrix of Problem 21.

In Problems 37–44, could the given matrix be the transition matrix of a Markov chain?

37. $\begin{bmatrix} .3 & .7 \\ 1 & 0 \end{bmatrix}$

38. $\begin{bmatrix} .9 & .1 \\ .4 & .8 \end{bmatrix}$

39. $\begin{bmatrix} .5 & .5 \\ .7 & -.3 \end{bmatrix}$

40. $\begin{bmatrix} 0 & 1 \\ 1 & 0 \end{bmatrix}$

41. $\begin{bmatrix} .1 & .3 & .6 \\ .2 & .4 & .4 \end{bmatrix}$

42. $\begin{bmatrix} .2 & .8 \\ .5 & .5 \\ .9 & .1 \end{bmatrix}$

43. $\begin{bmatrix} .5 & .1 & .4 \\ 0 & .5 & .5 \\ .2 & .1 & .7 \end{bmatrix}$

44. $\begin{bmatrix} .3 & .3 & .4 \\ .7 & .2 & .2 \\ .1 & .8 & .1 \end{bmatrix}$

In Problems 45–50, is there a unique way of filling in the missing probabilities in the transition diagram? If so, complete the transition diagram and write the corresponding transition matrix. If not, explain why.

45.

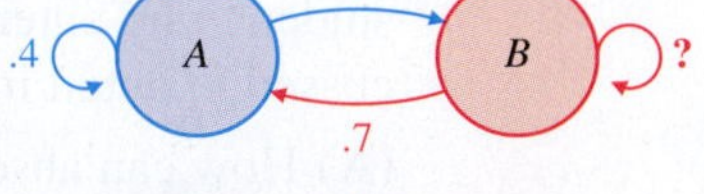

46.

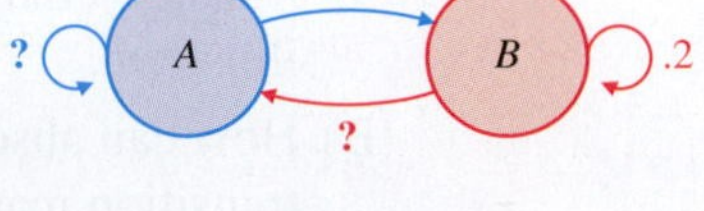

47.

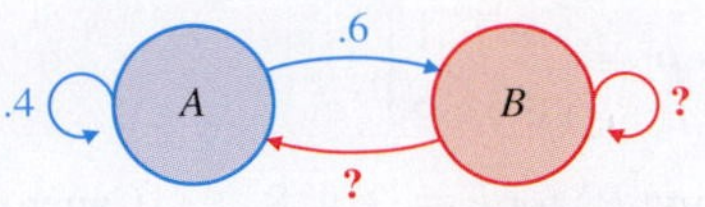

48.

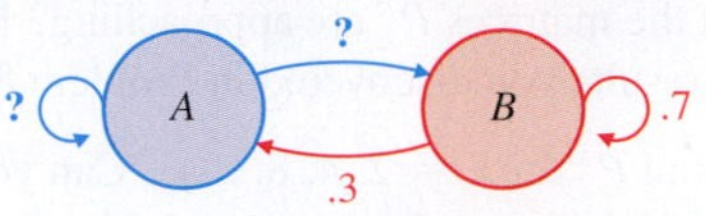

49.

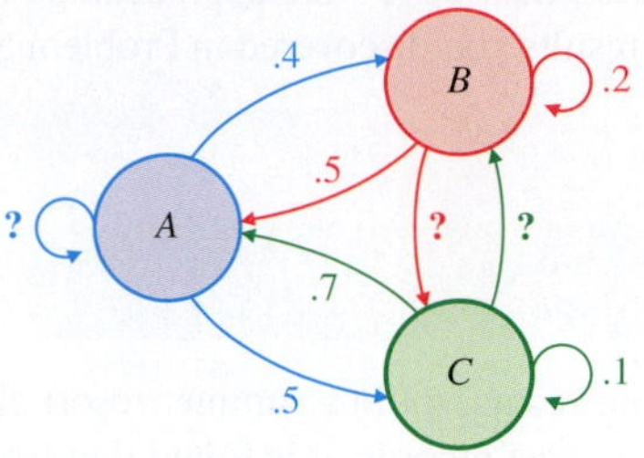

50.

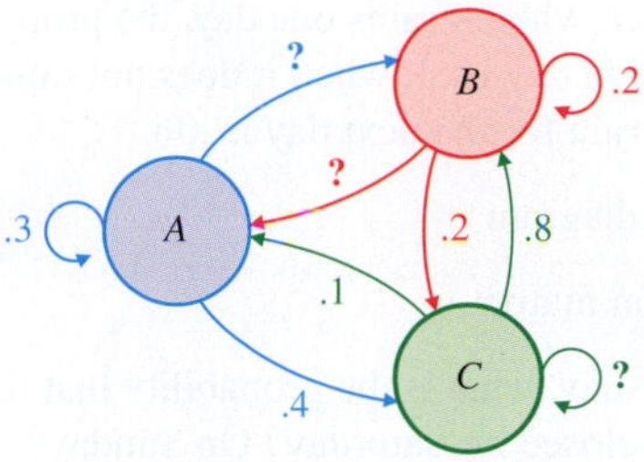

In Problems 51–56, are there unique values of a, b, and c that make P a transition matrix? If so, complete the transition matrix and draw the corresponding transition diagram. If not, explain why.

51. $P = \begin{matrix} & A & B & C \\ A & 0 & .5 & a \\ B & b & 0 & .4 \\ C & .2 & c & .1 \end{matrix}$

52. $P = \begin{matrix} & A & B & C \\ A & a & 0 & .9 \\ B & .2 & .3 & b \\ C & .6 & c & 0 \end{matrix}$

53. $P = \begin{matrix} & A & B & C \\ A & 0 & a & .3 \\ B & 0 & b & 0 \\ C & c & .8 & 0 \end{matrix}$

54. $P = \begin{matrix} & A & B & C \\ A & 0 & 1 & a \\ B & 0 & 0 & b \\ C & c & .5 & 0 \end{matrix}$

55. $P = \begin{matrix} & A & B & C \\ A & .2 & .1 & .7 \\ B & a & .4 & c \\ C & .5 & b & .4 \end{matrix}$

56. $P = \begin{matrix} & A & B & C \\ A & a & .8 & .1 \\ B & .3 & b & .4 \\ C & .6 & .5 & c \end{matrix}$

B *In Problems 57–60, use the given information to draw the transition diagram and find the transition matrix.*

57. A Markov chain has two states, A and B. The probability of going from state A to state B in one trial is .7, and the probability of going from state B to state A in one trial is .9.

58. A Markov chain has two states, A and B. The probability of going from state A to state A in one trial is .6, and the probability of going from state B to state B in one trial is .2.

59. A Markov chain has three states, A, B, and C. The probability of going from state A to state B in one trial is .1, and the probability of going from state A to state C in one trial is .3. The probability of going from state B to state A in one trial is .2, and the probability of going from state B to state C in one trial is .5. The probability of going from state C to state C in one trial is 1.

60. A Markov chain has three states, A, B, and C. The probability of going from state A to state B in one trial is 1. The probability of going from state B to state A in one trial is .5, and the probability of going from state B to state C in one trial is .5. The probability of going from state C to state A in one trial is 1.

Problems 61–70 refer to the following transition matrix P and its powers:

$$P = \begin{matrix} & A & B & C \\ A & .6 & .3 & .1 \\ B & .2 & .5 & .3 \\ C & .1 & .2 & .7 \end{matrix} \qquad P^2 = \begin{matrix} & A & B & C \\ A & .43 & .35 & .22 \\ B & .25 & .37 & .38 \\ C & .17 & .27 & .56 \end{matrix}$$

$$P^3 = \begin{matrix} & A & B & C \\ A & .35 & .348 & .302 \\ B & .262 & .336 & .402 \\ C & .212 & .298 & .49 \end{matrix}$$

61. Find the probability of going from state A to state B in two trials.

62. Find the probability of going from state B to state C in two trials.

63. Find the probability of going from state C to state A in three trials.

64. Find the probability of going from state B to state B in three trials.

65. Find S_2 for $S_0 = [1 \quad 0 \quad 0]$ and explain what it represents.

66. Find S_2 for $S_0 = [0 \quad 1 \quad 0]$ and explain what it represents.

67. Find S_3 for $S_0 = [0 \quad 0 \quad 1]$ and explain what it represents.

68. Find S_3 for $S_0 = [1 \quad 0 \quad 0]$ and explain what it represents.

69. Using a graphing calculator to compute powers of P, find the smallest positive integer n such that the corresponding entries in P^n and P^{n+1} are equal when rounded to two decimal places.

70. Using a graphing calculator to compute powers of P, find the smallest positive integer n such that the corresponding entries in P^n and P^{n+1} are equal when rounded to three decimal places.

In Problems 71–74, given the transition matrix P and initial-state matrix S_0, find P^4 and use P^4 to find S_4.

71. $P = \begin{matrix} & A & B \\ A & .1 & .9 \\ B & .6 & .4 \end{matrix}; \quad S_0 = [.8 \quad .2]$

72. $P = \begin{matrix} & A & B \\ A & .8 & .2 \\ B & .3 & .7 \end{matrix}; \quad S_0 = [.4 \quad .6]$

73. $P = \begin{matrix} & A & B & C \\ A & 0 & .4 & .6 \\ B & 0 & 0 & 1 \\ C & 1 & 0 & 0 \end{matrix}; \quad S_0 = [.2 \quad .3 \quad .5]$

74. $P = \begin{matrix} & A & B & C \\ A & 0 & 1 & 0 \\ B & .8 & 0 & .2 \\ C & 1 & 0 & 0 \end{matrix}; \quad S_0 = [.4 \quad .2 \quad .4]$

75. A Markov chain with two states has transition matrix P. If the initial-state matrix is $S_0 = [1 \quad 0]$, discuss the relationship between the entries in the kth-state matrix and the entries in the kth power of P.

76. Repeat Problem 75 if the initial-state matrix is $S_0 = [0 \quad 1]$.

77. Given the transition matrix

$$P = \begin{matrix} & \begin{matrix} A & B & C & D \end{matrix} \\ \begin{matrix} A \\ B \\ C \\ D \end{matrix} & \begin{bmatrix} .2 & .2 & .3 & .3 \\ 0 & 1 & 0 & 0 \\ .2 & .2 & .1 & .5 \\ 0 & 0 & 0 & 1 \end{bmatrix} \end{matrix}$$

(A) Find P^4.

(B) Find the probability of going from state A to state D in four trials.

(C) Find the probability of going from state C to state B in four trials.

(D) Find the probability of going from state B to state A in four trials.

78. Repeat Problem 77 for the transition matrix

$$P = \begin{matrix} & \begin{matrix} A & B & C & D \end{matrix} \\ \begin{matrix} A \\ B \\ C \\ D \end{matrix} & \begin{bmatrix} .5 & .3 & .1 & .1 \\ 0 & 1 & 0 & 0 \\ 0 & 0 & 1 & 0 \\ .1 & .2 & .3 & .4 \end{bmatrix} \end{matrix}$$

A matrix is called a **probability matrix** *if all its entries are real numbers between 0 and 1, inclusive, and the sum of the entries in each row is 1. So transition matrices are square probability matrices and state matrices are probability matrices with one row.*

79. Show that if

$$P = \begin{bmatrix} a & 1-a \\ 1-b & b \end{bmatrix}$$

is a probability matrix, then P^2 is a probability matrix.

80. Show that if

$$P = \begin{bmatrix} a & 1-a \\ 1-b & b \end{bmatrix} \quad \text{and} \quad S = [c \quad 1-c]$$

are probability matrices, then SP is a probability matrix.

Use a graphing calculator and the formula $S_k = S_0 P^k$ (Theorem 1) to compute the required state matrices in Problems 81–84.

81. The transition matrix for a Markov chain is

$$P = \begin{bmatrix} .4 & .6 \\ .2 & .8 \end{bmatrix}$$

(A) If $S_0 = [0 \quad 1]$, find $S_2, S_4, S_8, \ldots$. Can you identify a state matrix S that the matrices S_k seem to be approaching?

(B) Repeat part (A) for $S_0 = [1 \quad 0]$.

(C) Repeat part (A) for $S_0 = [.5 \quad .5]$.

(D) Find SP for any matrix S you identified in parts (A)–(C).

(E) Write a brief verbal description of the long-term behavior of the state matrices of this Markov chain based on your observations in parts (A)–(D).

82. Repeat Problem 81 for $P = \begin{bmatrix} .9 & .1 \\ .4 & .6 \end{bmatrix}$.

83. Refer to Problem 81. Find P^k for $k = 2, 4, 8, \ldots$. Can you identify a matrix Q that the matrices P^k are approaching? If so, how is Q related to the results you discovered in Problem 81?

84. Refer to Problem 82. Find P^k for $k = 2, 4, 8, \ldots$. Can you identify a matrix Q that the matrices P^k are approaching? If so, how is Q related to the results you discovered in Problem 82?

Applications

85. Scheduling. An outdoor restaurant in a summer resort closes only on rainy days. From past records, it is found that from May through September, when it rains one day, the probability of rain for the next day is .4; when it does not rain one day, the probability of rain for the next day is .06.

(A) Draw a transition diagram.

(B) Write the transition matrix.

(C) If it rains on Thursday, what is the probability that the restaurant will be closed on Saturday? On Sunday?

86. Scheduling. Repeat Problem 85 if the probability of rain following a rainy day is .6 and the probability of rain following a nonrainy day is .1.

87. Advertising. A television advertising campaign is conducted during the football season to promote a well-known brand X shaving cream. For each of several weeks, a survey is made, and it is found that each week, 80% of those using brand X continue to use it and 20% switch to another brand. It is also found that of those not using brand X, 20% switch to brand X while the other 80% continue using another brand.

(A) Draw a transition diagram.

(B) Write the transition matrix.

(C) If 20% of the people are using brand X at the start of the advertising campaign, what percentage will be using it 1 week later? 2 weeks later?

88. Car rental. A car rental agency has facilities at both JFK and LaGuardia airports. Assume that a car rented at either airport must be returned to one or the other airport. If a car is rented at LaGuardia, the probability that it will be returned there is .8; if a car is rented at JFK, the probability that it will be returned there is .7. Assume that the company rents all its 100 cars each day and that each car is rented (and returned) only once a day. If we start with 50 cars at each airport, then

(A) What is the expected distribution on the next day?

(B) What is the expected distribution 2 days later?

89. Homeowner's insurance. In a given city, the market for homeowner's insurance is dominated by two companies: National Property and United Family. Currently, National Property insures 50% of homes in the city, United Family insures 30%, and the remainder are insured by a collection of smaller companies. United Family decides to offer rebates to increase its market share. This has the following effects on insurance purchases for the next several years: each year 25% of National Property's customers switch to United Family and 10% switch to other companies; 10% of United Family's customers switch to National Property and 5% switch to other companies; 15% of the customers of other companies switch to National Property and 35% switch to United Family.

(A) Draw a transition diagram.

(B) Write the transition matrix.

(C) What percentage of homes will be insured by National Property next year? The year after next?

(D) What percentage of homes will be insured by United Family next year? The year after next?

90. Service contracts. A small community has two heating services that offer annual service contracts for home heating: Alpine Heating and Badger Furnaces. Currently, 25% of homeowners have service contracts with Alpine, 30% have service contracts with Badger, and the remainder do not have service contracts. Both companies launch aggressive advertising campaigns to attract new customers, with the following effects on service contract purchases for the next several years: each year 35% of homeowners with no current service contract decide to purchase a contract from Alpine and 40% decide to purchase one from Badger. In addition, 10% of the previous customers at each company decide to switch to the other company, and 5% decide they do not want a service contract.

(A) Draw a transition diagram.

(B) Write the transition matrix.

(C) What percentage of homes will have service contracts with Alpine next year? The year after next?

(D) What percentage of homes will have service contracts with Badger next year? The year after next?

91. Travel agent training. A chain of travel agencies maintains a training program for new travel agents. Initially, all new employees are classified as beginning agents requiring extensive supervision. Every 6 months, the performance of each agent is reviewed. Past records indicate that after each semiannual review, 40% of the beginning agents are promoted to intermediate agents requiring only minimal supervision, 10% are terminated for unsatisfactory performance, and the remainder continue as beginning agents. Furthermore, 30% of the intermediate agents are promoted to qualified travel agents requiring no supervision, 10% are terminated for unsatisfactory performance, and the remainder continue as intermediate agents.

(A) Draw a transition diagram.

(B) Write the transition matrix.

(C) What is the probability that a beginning agent is promoted to qualified agent within 1 year? Within 2 years?

92. Welder training. All welders in a factory begin as apprentices. Every year the performance of each apprentice is reviewed. Past records indicate that after each review, 10% of the apprentices are promoted to professional welder, 20% are terminated for unsatisfactory performance, and the remainder continue as apprentices.

(A) Draw a transition diagram.

(B) Write the transition matrix.

(C) What is the probability that an apprentice is promoted to professional welder within 2 years? Within 4 years?

93. Health plans. A midwestern university offers its employees three choices for health care: a clinic-based health maintenance organization (HMO), a preferred provider organization (PPO), and a traditional fee-for-service program (FFS). Each year, the university designates an open enrollment period during which employees may change from one health plan to another. Prior to the last open enrollment period, 20% of employees were enrolled in the HMO, 25% in the PPO, and the remainder in the FFS. During the open enrollment period, 15% of employees in the HMO switched to the PPO and 5% switched to the FFS, 20% of the employees in the PPO switched to the HMO and 10% to the FFS, and 25% of the employees in the FFS switched to the HMO and 30% switched to the PPO.

(A) Write the transition matrix.

(B) What percentage of employees were enrolled in each health plan after the last open enrollment period?

(C) If this trend continues, what percentage of employees will be enrolled in each plan after the next open enrollment period?

94. Dental insurance. Refer to Problem 93. During the open enrollment period, university employees can switch between two available dental care programs: the low-option plan (LOP) and the high-option plan (HOP). Prior to the last open enrollment period, 40% of employees were enrolled in the LOP and 60% in the HOP. During the open enrollment program, 30% of employees in the LOP switched to the HOP and 10% of employees in the HOP switched to the LOP.

(A) Write the transition matrix.

(B) What percentage of employees were enrolled in each dental plan after the last open enrollment period?

(C) If this trend continues, what percentage of employees will be enrolled in each dental plan after the next open enrollment period?

95. Housing trends. The 2000 census reported that 41.9% of the households in the District of Columbia were homeowners and the remainder were renters. During the next decade, 15.3% of homeowners became renters, and the rest continued to be homeowners. Similarly, 17.4% of renters became homeowners, and the rest continued to rent.

(A) Write the appropriate transition matrix.

(B) According to this transition matrix, what percentage of households were homeowners in 2010?

(C) If the transition matrix remains the same, what percentage of households will be homeowners in 2030?

96. Housing trends. The 2000 census reported that 66.4% of the households in Alaska were homeowners, and the remainder were renters. During the next decade, 37.2% of the homeowners became renters, and the rest continued to be homeowners. Similarly, 71.5% of the renters became homeowners, and the rest continued to rent.

(A) Write the appropriate transition matrix.

(B) According to this transition matrix, what percentage of households were homeowners in 2010?

(C) If the transition matrix remains the same, what percentage of households will be homeowners in 2030?

Answers to Matched Problems

1. (A)

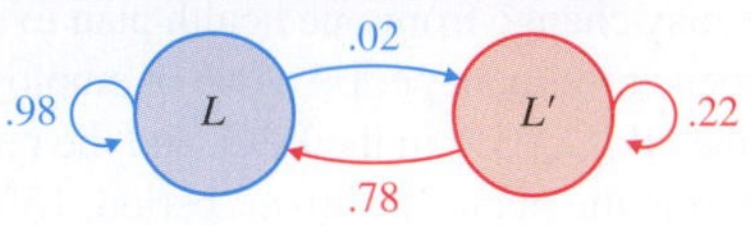

L = Low-risk
L' = Not low-risk

(B)

$$\begin{array}{c} \text{Next year} \\ \begin{array}{cc} L & L' \end{array} \end{array}$$

$$\begin{array}{c}\text{This}\\\text{year}\end{array}\begin{array}{c} L \\ L' \end{array}\begin{bmatrix} .98 & .02 \\ .78 & .22 \end{bmatrix} = P$$

(C) Next year: .96; year after next: .972

2. $P^4 = \begin{bmatrix} .625 & .375 \\ .5625 & .4375 \end{bmatrix}$; $S_4 = [.6125 \quad .3875]$

3. $S_8 = [.600781 \quad .399219]$

4. (A)

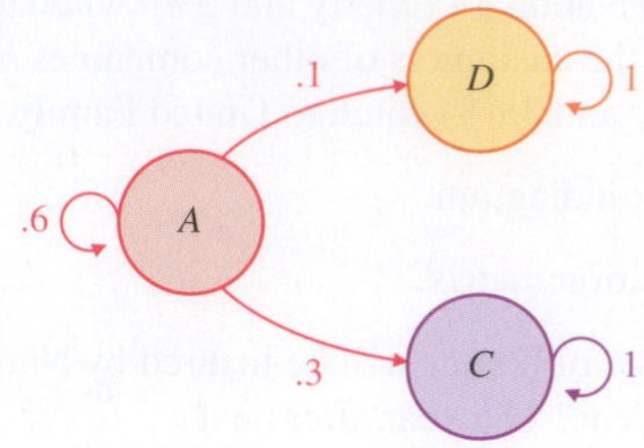

(B) $P = \begin{array}{c} \\ A \\ C \\ D \end{array}\begin{array}{c} \begin{array}{ccc} A & C & D \end{array} \\ \begin{bmatrix} .6 & .3 & .1 \\ 0 & 1 & 0 \\ 0 & 0 & 1 \end{bmatrix} \end{array}$

(C) .6528; .2176

9.2 Regular Markov Chains

- Stationary Matrices
- Regular Markov Chains
- Applications
- Graphing Calculator Approximations

Given a Markov chain with transition matrix P and initial-state matrix S_0, the entries in the state matrix S_k are the probabilities of being in the corresponding states after k trials. What happens to these probabilities as the number of trials k increases? In this section, we establish conditions on the transition matrix P that enable us to determine the long-run behavior of both the state matrices S_k and the powers of the transition matrix P^k.

Stationary Matrices

We begin by considering a concrete example—the toothpaste company discussed earlier. Recall that the transition matrix was given by

$$P = \begin{array}{c} \\ A \\ A' \end{array}\begin{array}{c} \begin{array}{cc} A & A' \end{array} \\ \begin{bmatrix} .8 & .2 \\ .6 & .4 \end{bmatrix} \end{array} \quad \begin{array}{l} A = \text{uses brand } A \text{ toothpaste} \\ A' = \text{uses another brand} \end{array}$$

Initially, this company had a 10% share of the toothpaste market. If the probabilities in the transition matrix P remain valid over a long period of time, what will happen to the company's market share? Examining the first several state matrices will give us some insight into this situation (matrix multiplication details are omitted):

$$\begin{aligned} S_0 &= [.1 \quad .9] \\ S_1 = S_0P &= [.62 \quad .38] \\ S_2 = S_1P &= [.724 \quad .276] \\ S_3 = S_2P &= [.7448 \quad .2552] \\ S_4 = S_3P &= [.74896 \quad .25104] \\ S_5 = S_4P &= [.749792 \quad .250208] \\ S_6 = S_5P &= [.7499584 \quad .2500416] \end{aligned}$$

It appears that the state matrices are getting closer and closer to $S = [.75 \quad .25]$ as we proceed to higher states. Let us multiply the matrix S (the matrix that the other state matrices appear to be approaching) by the transition matrix:

$$SP = [.75 \quad .25]\begin{bmatrix} .8 & .2 \\ .6 & .4 \end{bmatrix} = [.75 \quad .25] = S$$

No change occurs! The matrix $[.75 \quad .25]$ is called a **stationary matrix**. If we reach this state or are very close to it, the system is said to be at a steady state; that is, later states either will not change or will not change very much. In terms of this example, this means that in the long run a person will purchase brand A with a probability of .75. In other words, the company can expect to capture 75% of the market, assuming that the transition matrix does not change.

DEFINITION Stationary Matrix for a Markov Chain

The state matrix $S = [s_1 \quad s_2 \quad \ldots \quad s_n]$ is a **stationary matrix** for a Markov chain with transition matrix P if

$$SP = S$$

where $s_i \geq 0, i = 1, \ldots, n$, and $s_1 + s_2 + \cdots + s_n = 1$.

Explore and Discuss 1

(A) Suppose that the toothpaste company started with only 5% of the market instead of 10%. Write the initial-state matrix and find the next six state matrices. Discuss the behavior of these state matrices as you proceed to higher states.

(B) Repeat part (A) if the company started with 90% of the toothpaste market.

Regular Markov Chains

Does every Markov chain have a unique stationary matrix? And if a Markov chain has a unique stationary matrix, will the successive state matrices always approach this stationary matrix? Unfortunately, the answer to both these questions is no (see Problems 47–50, Exercises 9.2). However, there is one important type of Markov chain for which both questions always can be answered in the affirmative. These are called *regular Markov chains*.

DEFINITION Regular Markov Chains

A transition matrix P is **regular** if some power of P has only positive entries. A Markov chain is a **regular Markov chain** if its transition matrix is regular.

EXAMPLE 1 **Recognizing Regular Matrices** Which of the following matrices are regular?

(A) $P = \begin{bmatrix} .8 & .2 \\ .6 & .4 \end{bmatrix}$ (B) $P = \begin{bmatrix} 0 & 1 \\ 1 & 0 \end{bmatrix}$ (C) $P = \begin{bmatrix} .5 & .5 & 0 \\ 0 & .5 & .5 \\ 1 & 0 & 0 \end{bmatrix}$

SOLUTION

(A) This is the transition matrix for the toothpaste company. Since all the entries in P are positive, we can immediately conclude that P is regular.

(B) P has two 0 entries, so we must examine higher powers of P:

$$P^2 = \begin{bmatrix} 1 & 0 \\ 0 & 1 \end{bmatrix} \quad P^3 = \begin{bmatrix} 0 & 1 \\ 1 & 0 \end{bmatrix} \quad P^4 = \begin{bmatrix} 1 & 0 \\ 0 & 1 \end{bmatrix} \quad P^5 = \begin{bmatrix} 0 & 1 \\ 1 & 0 \end{bmatrix}$$

Since the powers of P oscillate between P and I, the 2×2 identity, all powers of P will contain 0 entries. Hence, P is not regular.

(C) Again, we examine higher powers of P:

$$P^2 = \begin{bmatrix} .25 & .5 & .25 \\ .5 & .25 & .25 \\ .5 & .5 & 0 \end{bmatrix} \quad P^3 = \begin{bmatrix} .375 & .375 & .25 \\ .5 & .375 & .125 \\ .25 & .5 & .25 \end{bmatrix}$$

Since all the entries in P^3 are positive, P is regular.

Matched Problem 1 Which of the following matrices are regular?

$$\text{(A) } P = \begin{bmatrix} .3 & .7 \\ 1 & 0 \end{bmatrix} \quad \text{(B) } P = \begin{bmatrix} 1 & 0 \\ 1 & 0 \end{bmatrix} \quad \text{(C) } P = \begin{bmatrix} 0 & 1 & 0 \\ .5 & 0 & .5 \\ .5 & 0 & .5 \end{bmatrix}$$

The relationships among successive state matrices, powers of the transition matrix, and the stationary matrix for a regular Markov chain are given in Theorem 1. The proof of this theorem is left to more advanced courses.

THEOREM 1 Properties of Regular Markov Chains

Let P be the transition matrix for a regular Markov chain.

(A) There is a unique stationary matrix S that can be found by solving the equation

$$SP = S$$

(B) Given any initial-state matrix S_0, the state matrices S_k approach the stationary matrix S.

(C) The matrices P^k approach a **limiting matrix** $\overline{P}$, where each row of $\overline{P}$ is equal to the stationary matrix S.

EXAMPLE 2 **Finding the Stationary Matrix** The transition matrix for a Markov chain is

$$P = \begin{bmatrix} .7 & .3 \\ .2 & .8 \end{bmatrix}$$

(A) Find the stationary matrix S.

(B) Discuss the long-run behavior of S_k and P^k.

SOLUTION

(A) Since P is regular, the stationary matrix S must exist. To find it, we must solve the equation $SP = S$. Let

$$S = [s_1 \quad s_2]$$

and write

$$[s_1 \quad s_2]\begin{bmatrix} .7 & .3 \\ .2 & .8 \end{bmatrix} = [s_1 \quad s_2]$$

After multiplying the left side, we obtain

$$[(.7s_1 + .2s_2) \quad (.3s_1 + .8s_2)] = [s_1 \quad s_2]$$

which is equivalent to the system

$$\begin{aligned} .7s_1 + .2s_2 &= s_1 \quad \text{or} \quad -.3s_1 + .2s_2 = 0 \\ .3s_1 + .8s_2 &= s_2 \quad \text{or} \quad .3s_1 - .2s_2 = 0 \end{aligned} \tag{1}$$

System (1) is dependent and has an infinite number of solutions. However, we are looking for a solution that is also a state matrix. This gives us another equation that we can add to system (1) to obtain a system with a unique solution.

$$\begin{aligned} -.3s_1 + .2s_2 &= 0 \\ .3s_1 - .2s_2 &= 0 \\ s_1 + s_2 &= 1 \end{aligned} \tag{2}$$

System (2) can be solved using matrix methods or elimination to obtain

$$s_1 = .4 \quad \text{and} \quad s_2 = .6$$

Therefore,

$$S = [.4 \quad .6]$$

is the stationary matrix.

CHECK

$$SP = [.4 \quad .6]\begin{bmatrix} .7 & .3 \\ .2 & .8 \end{bmatrix} = [.4 \quad .6] = S$$

(B) Given any initial-state matrix S_0, Theorem 1 guarantees that the state matrices S_k will approach the stationary matrix S. Furthermore,

$$P^k = \begin{bmatrix} .7 & .3 \\ .2 & .8 \end{bmatrix}^k \quad \text{approaches the limiting matrix} \quad \overline{P} = \begin{bmatrix} .4 & .6 \\ .4 & .6 \end{bmatrix}$$

Matched Problem 2 The transition matrix for a Markov chain is

$$P = \begin{bmatrix} .6 & .4 \\ .1 & .9 \end{bmatrix}$$

Find the stationary matrix S and the limiting matrix $\overline{P}$.

Applications

EXAMPLE 3 **Insurance** Refer to Example 1 in Section 9.1, where we found the following transition matrix for an insurance company:

$$P = \begin{array}{c} \\ A \\ A' \end{array}\!\!\begin{array}{c} \begin{array}{cc} A & A' \end{array} \\ \begin{bmatrix} .23 & .77 \\ .11 & .89 \end{bmatrix} \end{array} \quad \begin{aligned} A &= \text{accident} \\ A' &= \text{no accident} \end{aligned}$$

If these probabilities remain valid over a long period of time, what percentage of drivers are expected to have an accident during any given year?

SOLUTION To determine what happens in the long run, we find the stationary matrix by solving the following system:

$$[s_1 \quad s_2]\begin{bmatrix} .23 & .77 \\ .11 & .89 \end{bmatrix} = [s_1 \quad s_2] \quad \text{and} \quad s_1 + s_2 = 1$$

which is equivalent to

$$\begin{aligned} .23s_1 + .11s_2 &= s_1 \\ .77s_1 + .89s_2 &= s_2 \\ s_1 + s_2 &= 1 \end{aligned} \quad \text{or} \quad \begin{aligned} -.77s_1 + .11s_2 &= 0 \\ .77s_1 - .11s_2 &= 0 \\ s_1 + s_2 &= 1 \end{aligned}$$

Solving this system, we obtain

$$s_1 = .125 \quad \text{and} \quad s_2 = .875$$

The stationary matrix is $[.125 \quad .875]$, which means that in the long run, assuming that the transition matrix does not change, about 12.5% of drivers in the community will have an accident during any given year.

Matched Problem 3 Refer to Matched Problem 1 in Section 9.1, where we found the following transition matrix for an insurance company:

$$P = \begin{matrix} & \begin{matrix} L & L' \end{matrix} \\ \begin{matrix} L \\ L' \end{matrix} & \begin{bmatrix} .98 & .02 \\ .78 & .22 \end{bmatrix} \end{matrix} \qquad \begin{aligned} L &= \text{low-risk} \\ L' &= \text{not low-risk} \end{aligned}$$

If these probabilities remain valid for a long period of time, what percentage of drivers are expected to be in the low-risk category during any given year?

EXAMPLE 4 **Employee Evaluation** A company rates every employee as below average, average, or above average. Past performance indicates that each year, 10% of the below-average employees will raise their rating to average, and 25% of the average employees will raise their rating to above average. On the other hand, 15% of the average employees will lower their rating to below average, and 15% of the above-average employees will lower their rating to average. Company policy prohibits rating changes from below average to above average, or conversely, in a single year. Over the long run, what percentage of employees will receive below-average ratings? Average ratings? Above-average ratings?

SOLUTION First we find the transition matrix:

$$\text{This year} \quad \begin{matrix} & \text{Next year} \\ & \begin{matrix} A^- & A & A^+ \end{matrix} \\ \begin{matrix} A^- \\ A \\ A^+ \end{matrix} & \begin{bmatrix} .9 & .1 & 0 \\ .15 & .6 & .25 \\ 0 & .15 & .85 \end{bmatrix} \end{matrix} \qquad \begin{aligned} A^- &= \text{below average} \\ A &= \text{average} \\ A^+ &= \text{above average} \end{aligned}$$

To determine what happens over the long run, we find the stationary matrix by solving the following system:

$$[s_1 \quad s_2 \quad s_3] \begin{bmatrix} .9 & .1 & 0 \\ .15 & .6 & .25 \\ 0 & .15 & .85 \end{bmatrix} = [s_1 \quad s_2 \quad s_3] \quad \text{and} \quad s_1 + s_2 + s_3 = 1$$

which is equivalent to

$$\begin{aligned} .9s_1 + .15s_2 &= s_1 \\ .1s_1 + .6s_2 + .15s_3 &= s_2 \\ .25s_2 + .85s_3 &= s_3 \\ s_1 + s_2 + s_3 &= 1 \end{aligned} \quad \text{or} \quad \begin{aligned} -.1s_1 + .15s_2 &= 0 \\ .1s_1 - .4s_2 + .15s_3 &= 0 \\ .25s_2 - .15s_3 &= 0 \\ s_1 + s_2 + s_3 &= 1 \end{aligned}$$

Using Gauss–Jordan elimination to solve this system of four equations with three variables, we obtain

$$s_1 = .36 \qquad s_2 = .24 \qquad s_3 = .4$$

In the long run, 36% of employees will be rated as below average, 24% as average, and 40% as above average.

Matched Problem 4 A mail-order company classifies its customers as preferred, standard, or infrequent depending on the number of orders placed in a year. Past records indicate that each year, 5% of preferred customers are reclassified as standard and 12% as infrequent, 5% of standard customers are reclassified as preferred and 5% as infrequent, and 9% of infrequent customers are reclassified as preferred and 10% as standard. Assuming that these percentages remain valid, what percentage of customers are expected to be in each category in the long run?

Graphing Calculator Approximations

If P is the transition matrix for a regular Markov chain, then the powers of P approach the limiting matrix $\overline{P}$, where each row of $\overline{P}$ is equal to the stationary matrix S (Theorem 1C). We can use this result to approximate S by computing P^k for sufficiently large values of k. The next example illustrates this approach on a graphing calculator.

EXAMPLE 5 **Approximating the Stationary Matrix** Compute powers of the transition matrix P to approximate $\overline{P}$ and S to four decimal places. Check the approximation in the equation $SP = S$.

$$P = \begin{bmatrix} .5 & .2 & .3 \\ .7 & .1 & .2 \\ .4 & .1 & .5 \end{bmatrix}$$

SOLUTION To approximate $\overline{P}$ to four decimal places, we store P in a graphing calculator using the matrix name [A] (Fig. 1A), set the decimal display to four places, and compute powers of P until all three rows of P^k are identical. Examining the output in Figure 1B, we conclude that

$$\overline{P} = \begin{bmatrix} .4943 & .1494 & .3563 \\ .4943 & .1494 & .3563 \\ .4943 & .1494 & .3563 \end{bmatrix} \quad \text{and} \quad S = [.4943 \quad .1494 \quad .3563]$$

Entering S in the graphing calculator using the matrix name [B], and computing SP shows that these matrices are correct to four decimal places (Fig. 1C).

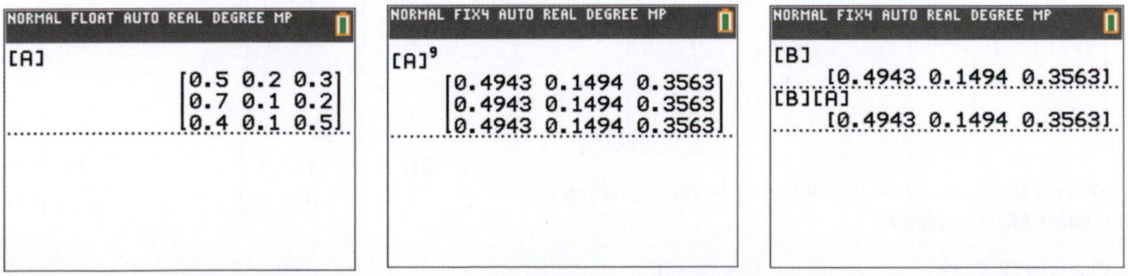

(A) P (B) P^9 (C) Check: $SP = S$

Figure 1

Matched Problem 5 Repeat Example 5 for

$$P = \begin{bmatrix} .3 & .6 & .1 \\ .2 & .3 & .5 \\ .1 & .2 & .7 \end{bmatrix}$$

CONCEPTUAL INSIGHT

1. We used a relatively small value of k to approximate $\overline{P}$ in Example 5. Many graphing calculators will compute P^k for large values of k almost as rapidly as for small values. However, round-off errors can occur in these calculations. A safe procedure is to start with a relatively small value of k, such as $k = 8$, and then keep doubling k until the rows of P^k are identical to the specified number of decimal places.
2. If any of the entries of P^k are approaching 0, then the graphing calculator may use scientific notation to display these entries as very small numbers. Figure 2 shows the 100th power of a transition matrix P using the matrix name [A]. The entry in row 2 and column 2 of P^{100} is approaching 0, but the graphing calculator displays it as 5.1538×10^{-53}. If this occurs, simply change this value to 0 in the corresponding entry in $\overline{P}$. Thus, from the output in Figure 2 we conclude that

$$P^k = \begin{bmatrix} 1 & 0 \\ .7 & .3 \end{bmatrix}^k \quad \text{approaches} \quad \overline{P} = \begin{bmatrix} 1 & 0 \\ 1 & 0 \end{bmatrix}$$

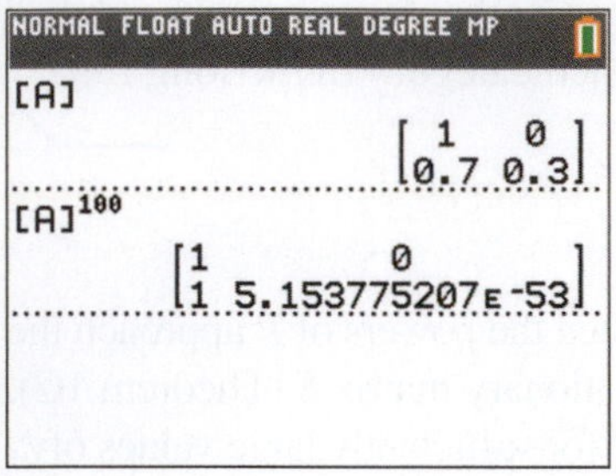

Figure 2

Exercises 9.2

Skills Warm-up Exercises

In Problems 1–8, without using a calculator, find P^{100}. (If necessary, review Section 4.4.) [Hint: First find P^2.]

1. $\begin{bmatrix} 1 & 0 \\ 0 & 0 \end{bmatrix}$
2. $\begin{bmatrix} 0 & 0 \\ 1 & 0 \end{bmatrix}$
3. $\begin{bmatrix} 1 & 0 \\ 0 & 1 \end{bmatrix}$
4. $\begin{bmatrix} 0 & 1 \\ 1 & 0 \end{bmatrix}$
5. $\begin{bmatrix} 1 & 0 & 0 \\ 0 & 0 & 0 \\ 0 & 0 & 1 \end{bmatrix}$
6. $\begin{bmatrix} 1 & 0 & 0 \\ 0 & 1 & 0 \\ 0 & 0 & 1 \end{bmatrix}$
7. $\begin{bmatrix} 0 & 0 & 1 \\ 0 & 0 & 0 \\ 0 & 0 & 0 \end{bmatrix}$
8. $\begin{bmatrix} 0 & 1 & 1 \\ 0 & 0 & 1 \\ 0 & 0 & 0 \end{bmatrix}$

A *In Problems 9–22, could the given matrix be the transition matrix of a regular Markov chain?*

9. $\begin{bmatrix} .6 & .4 \\ .4 & .6 \end{bmatrix}$
10. $\begin{bmatrix} .3 & .7 \\ .2 & .6 \end{bmatrix}$
11. $\begin{bmatrix} .1 & .9 \\ .5 & .4 \end{bmatrix}$
12. $\begin{bmatrix} .5 & .5 \\ .8 & .2 \end{bmatrix}$
13. $\begin{bmatrix} .4 & .6 \\ 0 & 1 \end{bmatrix}$
14. $\begin{bmatrix} .4 & .6 \\ 1 & 0 \end{bmatrix}$
15. $\begin{bmatrix} 0 & 1 \\ .8 & .2 \end{bmatrix}$
16. $\begin{bmatrix} .3 & .7 \\ .2 & .6 \end{bmatrix}$
17. $\begin{bmatrix} .6 & .4 \\ .1 & .9 \\ .3 & .7 \end{bmatrix}$
18. $\begin{bmatrix} .2 & .5 & .3 \\ .6 & .3 & .1 \end{bmatrix}$
19. $\begin{bmatrix} 0 & 1 & 0 \\ 0 & 0 & 1 \\ .5 & .5 & 0 \end{bmatrix}$
20. $\begin{bmatrix} .2 & 0 & .8 \\ 0 & 0 & 1 \\ .7 & 0 & .3 \end{bmatrix}$
21. $\begin{bmatrix} .1 & .3 & .6 \\ .8 & .1 & .1 \\ 0 & 0 & 1 \end{bmatrix}$
22. $\begin{bmatrix} 0 & 0 & 1 \\ .9 & 0 & .1 \\ 0 & 1 & 0 \end{bmatrix}$

B *For each transition matrix P in Problems 23–30, solve the equation SP = S to find the stationary matrix S and the limiting matrix $\overline{P}$.*

23. $P = \begin{bmatrix} .1 & .9 \\ .6 & .4 \end{bmatrix}$ **24.** $P = \begin{bmatrix} .8 & .2 \\ .3 & .7 \end{bmatrix}$

25. $P = \begin{bmatrix} .5 & .5 \\ .3 & .7 \end{bmatrix}$ **26.** $P = \begin{bmatrix} .9 & .1 \\ .7 & .3 \end{bmatrix}$

27. $P = \begin{bmatrix} .5 & .1 & .4 \\ .3 & .7 & 0 \\ 0 & .6 & .4 \end{bmatrix}$ **28.** $P = \begin{bmatrix} .4 & .1 & .5 \\ .2 & .8 & 0 \\ 0 & .5 & .5 \end{bmatrix}$

29. $P = \begin{bmatrix} .8 & .2 & 0 \\ .5 & .1 & .4 \\ 0 & .6 & .4 \end{bmatrix}$ **30.** $P = \begin{bmatrix} .2 & .8 & 0 \\ .6 & .1 & .3 \\ 0 & .9 & .1 \end{bmatrix}$

Problems 31–34 refer to the regular Markov chain with transition matrix

$$P = \begin{bmatrix} .5 & .5 \\ .2 & .8 \end{bmatrix}$$

31. For $S = [.2 \quad .5]$, calculate SP. Is S a stationary matrix? Explain.

32. For $S = [.6 \quad 1.5]$, calculate SP. Is S a stationary matrix? Explain.

33. For $S = [0 \quad 0]$, calculate SP. Is S a stationary matrix? Explain.

34. For $S = \begin{bmatrix} \frac{2}{7} & \frac{5}{7} \end{bmatrix}$, calculate SP. Is S a stationary matrix? Explain.

In Problems 35–40, discuss the validity of each statement. If the statement is always true, explain why. If not, give a counterexample.

35. The $n \times n$ identity matrix is the transition matrix for a regular Markov chain.

36. The $n \times n$ matrix in which each entry equals $\frac{1}{n}$ is the transition matrix for a regular Markov chain.

37. If the 2×2 matrix P is the transition matrix for a regular Markov chain, then, at most, one of the entries of P is equal to 0.

38. If the 3×3 matrix P is the transition matrix for a regular Markov chain, then, at most, two of the entries of P are equal to 0.

39. If a transition matrix P for a Markov chain has a stationary matrix S, then P is regular.

40. If P is the transition matrix for a Markov chain, then P has a unique stationary matrix.

In Problems 41–44, approximate the stationary matrix S for each transition matrix P by computing powers of the transition matrix P. Round matrix entries to four decimal places.

41. $P = \begin{bmatrix} .51 & .49 \\ .27 & .73 \end{bmatrix}$ **42.** $P = \begin{bmatrix} .68 & .32 \\ .19 & .81 \end{bmatrix}$

43. $P = \begin{bmatrix} .5 & .5 & 0 \\ 0 & .5 & .5 \\ .8 & .1 & .1 \end{bmatrix}$ **44.** $P = \begin{bmatrix} .2 & .2 & .6 \\ .5 & 0 & .5 \\ .5 & 0 & .5 \end{bmatrix}$

C **45.** A red urn contains 2 red marbles and 3 blue marbles, and a blue urn contains 1 red marble and 4 blue marbles. A marble is selected from an urn, the color is noted, and the marble is returned to the urn from which it was drawn. The next marble is drawn from the urn whose color is the same as the marble just drawn. This is a Markov process with two states: draw from the red urn or draw from the blue urn.

(A) Draw a transition diagram for this process.

(B) Write the transition matrix.

(C) Find the stationary matrix and describe the long-run behavior of this process.

46. Repeat Problem 45 if the red urn contains 5 red and 3 blue marbles, and the blue urn contains 1 red and 3 blue marbles.

47. Given the transition matrix

$$P = \begin{bmatrix} 0 & 1 \\ 1 & 0 \end{bmatrix}$$

(A) Discuss the behavior of the state matrices $S_1, S_2, S_3, \ldots$ for the initial-state matrix $S_0 = [.2 \quad .8]$.

(B) Repeat part (A) for $S_0 = [.5 \quad .5]$.

(C) Discuss the behavior of P^k, $k = 2, 3, 4, \ldots$.

(D) Which of the conclusions of Theorem 1 are not valid for this matrix? Why is this not a contradiction?

48. Given the transition matrix

$$P = \begin{bmatrix} 0 & 1 & 0 \\ 0 & 0 & 1 \\ 1 & 0 & 0 \end{bmatrix}$$

(A) Discuss the behavior of the state matrices $S_1, S_2, S_3, \ldots$ for the initial-state matrix $S_0 = [.2 \quad .3 \quad .5]$.

(B) Repeat part (A) for $S_0 = \begin{bmatrix} \frac{1}{3} & \frac{1}{3} & \frac{1}{3} \end{bmatrix}$.

(C) Discuss the behavior of P^k, $k = 2, 3, 4, \ldots$.

(D) Which of the conclusions of Theorem 1 are not valid for this matrix? Why is this not a contradiction?

49. The transition matrix for a Markov chain is

$$P = \begin{bmatrix} 1 & 0 & 0 \\ .2 & .2 & .6 \\ 0 & 0 & 1 \end{bmatrix}$$

(A) Show that $R = [1 \quad 0 \quad 0]$ and $S = [0 \quad 0 \quad 1]$ are both stationary matrices for P. Explain why this does not contradict Theorem 1A.

(B) Find another stationary matrix for P. [*Hint*: Consider $T = aR + (1-a)S$, where $0 < a < 1$.]

(C) How many different stationary matrices does P have?

50. The transition matrix for a Markov chain is

$$P = \begin{bmatrix} .7 & 0 & .3 \\ 0 & 1 & 0 \\ .2 & 0 & .8 \end{bmatrix}$$

(A) Show that $R = [.4 \quad 0 \quad .6]$ and $S = [0 \quad 1 \quad 0]$ are both stationary matrices for P. Explain why this does not contradict Theorem 1A.

(B) Find another stationary matrix for P. [*Hint*: Consider $T = aR + (1-a)S$, where $0 < a < 1$.]

(C) How many different stationary matrices does P have?

Problems 51 and 52 require the use of a graphing calculator.

51. Refer to the transition matrix P in Problem 49. What matrix $\overline{P}$ do the powers of P appear to be approaching? Are the rows of $\overline{P}$ stationary matrices for P?

52. Refer to the transition matrix P in Problem 50. What matrix $\overline{P}$ do the powers of P appear to be approaching? Are the rows of $\overline{P}$ stationary matrices for P?

53. The transition matrix for a Markov chain is

$$P = \begin{bmatrix} .1 & .5 & .4 \\ .3 & .2 & .5 \\ .7 & .1 & .2 \end{bmatrix}$$

Let M_k denote the maximum entry in the second column of P^k. Note that $M_1 = .5$.

(A) Find M_2, M_3, M_4, and M_5 to three decimal places.

(B) Explain why $M_k \geq M_{k+1}$ for all positive integers k.

54. The transition matrix for a Markov chain is

$$P = \begin{bmatrix} 0 & .2 & .8 \\ .3 & .3 & .4 \\ .6 & .1 & .3 \end{bmatrix}$$

Let m_k denote the minimum entry in the third column of P^k. Note that $m_1 = .3$.

(A) Find m_2, m_3, m_4, and m_5 to three decimal places.

(B) Explain why $m_k \leq m_{k+1}$ for all positive integers k.

Applications

55. Transportation. Most railroad cars are owned by individual railroad companies. When a car leaves its home railroad's tracks, it becomes part of a national pool of cars and can be used by other railroads. The rules governing the use of these pooled cars are designed to eventually return the car to the home tracks. A particular railroad found that each month, 11% of its boxcars on the home tracks left to join the national pool, and 29% of its boxcars in the national pool were returned to the home tracks. If these percentages remain valid for a long period of time, what percentage of its boxcars can this railroad expect to have on its home tracks in the long run?

56. Transportation. The railroad in Problem 55 also has a fleet of tank cars. If 14% of the tank cars on the home tracks enter the national pool each month, and 26% of the tank cars in the national pool are returned to the home tracks each month, what percentage of its tank cars can the railroad expect to have on its home tracks in the long run?

57. Labor force. Table 1 gives the percentage of the U.S. female population who were members of the civilian labor force in the indicated years. The following transition matrix P is proposed as a model for the data, where L represents females who are in the labor force and L' represents females who are not in the labor force:

$$\begin{array}{c} \text{Next decade} \\ \begin{array}{cc} & \begin{array}{cc} L & L' \end{array} \\ \text{Current decade} \begin{array}{c} L \\ L' \end{array} & \begin{bmatrix} .92 & .08 \\ .2 & .8 \end{bmatrix} = P \end{array} \end{array}$$

(A) Let $S_0 = [.433 \quad .567]$, and find S_1, S_2, S_3, and S_4. Compute the matrices exactly and then round entries to three decimal places.

(B) Construct a new table comparing the results from part (A) with the data in Table 1.

(C) According to this transition matrix, what percentage of the U.S. female population will be in the labor force in the long run?

Table 1

Year	Percent
1970	43.3
1980	51.5
1990	57.5
2000	59.8
2010	58.5

58. Home ownership. The U.S. Census Bureau published the home ownership rates given in Table 2.

Table 2

Year	Percent
1996	65.4
2000	67.4
2004	69.0
2008	67.8

The following transition matrix P is proposed as a model for the data, where H represents the households that own their home.

$$\begin{array}{c} \text{Four years later} \\ \begin{array}{cc} & \begin{array}{cc} H & H' \end{array} \\ \text{Current year} \begin{array}{c} H \\ H' \end{array} & \begin{bmatrix} .95 & .05 \\ .15 & .85 \end{bmatrix} = P \end{array} \end{array}$$

(A) Let $S_0 = [.654 \quad .346]$ and find S_1, S_2, and S_3. Compute both matrices exactly and then round entries to three decimal places.

(B) Construct a new table comparing the results from part (A) with the data in Table 2.

(C) According to this transition matrix, what percentage of households will own their home in the long run?

59. Market share. Consumers can choose between three long-distance telephone services: GTT, NCJ, and Dash. Aggressive marketing by all three companies results in a continual shift of customers among the three services. Each year, GTT loses 5% of its customers to NCJ and 20% to Dash, NCJ loses 15% of its customers to GTT and 10% to Dash, and Dash loses 5% of its customers to GTT and 10% to NCJ. Assuming that these percentages remain valid over a long period of time, what is each company's expected market share in the long run?

60. Market share. Consumers in a certain area can choose between three package delivery services: APS, GX, and WWP. Each week, APS loses 10% of its customers to GX and 20% to WWP, GX loses 15% of its customers to APS and 10% to WWP, and WWP loses 5% of its customers to APS and 5% to GX. Assuming that these percentages remain valid over a long period of time, what is each company's expected market share in the long run?

61. Insurance. An auto insurance company classifies its customers in three categories: poor, satisfactory, and preferred. Each year, 40% of those in the poor category are moved to satisfactory, and 20% of those in the satisfactory category are moved to preferred. Also, 20% in the preferred category are moved to the satisfactory category, and 20% in the satisfactory category are moved to the poor category. Customers are never moved from poor to preferred, or conversely, in a single year. Assuming that these percentages remain valid over a long period of time, how many customers are expected in each category in the long run?

62. Insurance. Repeat Problems 61 if 40% of preferred customers are moved to the satisfactory category each year, and all other information remains the same.

Problems 63 and 64 require the use of a graphing calculator.

63. Market share. Acme Soap Company markets one brand of soap, called Standard Acme (*SA*), and Best Soap Company markets two brands, Standard Best (*SB*) and Deluxe Best (*DB*). Currently, Acme has 40% of the market, and the remainder is divided equally between the two Best brands. Acme is considering the introduction of a second brand to get a larger share of the market. A proposed new brand, called brand *X*, was test-marketed in several large cities, producing the following transition matrix for the consumers' weekly buying habits:

$$P = \begin{matrix} & \begin{matrix} SB & DB & SA & X \end{matrix} \\ \begin{matrix} SB \\ DB \\ SA \\ X \end{matrix} & \begin{bmatrix} .4 & .1 & .3 & .2 \\ .3 & .2 & .2 & .3 \\ .1 & .2 & .2 & .5 \\ .3 & .3 & .1 & .3 \end{bmatrix} \end{matrix}$$

Assuming that *P* represents the consumers' buying habits over a long period of time, use this transition matrix and the initial-state matrix $S_0 = [.3 \quad .3 \quad .4 \quad 0]$ to compute successive state matrices in order to approximate the elements in the stationary matrix correct to two decimal places. If Acme decides to market this new soap, what is the long-run expected total market share for their two soaps?

64. Market share. Refer to Problem 63. The chemists at Acme Soap Company have developed a second new soap, called brand *Y*. Test-marketing this soap against the three established brands produces the following transition matrix:

$$P = \begin{matrix} & \begin{matrix} SB & DB & SA & Y \end{matrix} \\ \begin{matrix} SB \\ DB \\ SA \\ Y \end{matrix} & \begin{bmatrix} .3 & .2 & .2 & .3 \\ .2 & .2 & .2 & .4 \\ .2 & .2 & .4 & .2 \\ .1 & .2 & .3 & .4 \end{bmatrix} \end{matrix}$$

Proceed as in Problem 63 to approximate the elements in the stationary matrix correct to two decimal places. If Acme decides to market brand *Y*, what is the long-run expected total market share for Standard Acme and brand *Y*? Should Acme market brand *X* or brand *Y*?

65. Genetics. A given plant species has red, pink, or white flowers according to the genotypes RR, RW, and WW, respectively. If each of these genotypes is crossed with a pink-flowering plant (genotype RW), then the transition matrix is

$$\begin{matrix} & & \text{Next generation} \\ & & \begin{matrix} Red & Pink & White \end{matrix} \\ \text{This generation} & \begin{matrix} Red \\ Pink \\ White \end{matrix} & \begin{bmatrix} .5 & .5 & 0 \\ .25 & .5 & .25 \\ 0 & .5 & .5 \end{bmatrix} \end{matrix}$$

Assuming that the plants of each generation are crossed only with pink plants to produce the next generation, show that regardless of the makeup of the first generation, the genotype composition will eventually stabilize at 25% red, 50% pink, and 25% white. (Find the stationary matrix.)

66. Gene mutation. Suppose that a gene in a chromosome is of type *A* or type *B*. Assume that the probability that a gene of type *A* will mutate to type *B* in one generation is 10^{-4} and that a gene of type *B* will mutate to type *A* is 10^{-6}.

(A) What is the transition matrix?

(B) After many generations, what is the probability that the gene will be of type *A*? Of type *B*? (Find the stationary matrix.)

67. Rapid transit. A new rapid transit system has just started operating. In the first month of operation, it is found that 25% of commuters are using the system, while 75% still travel by car. The following transition matrix was determined from records of other rapid transit systems:

$$\begin{matrix} & & \text{Next month} \\ & & \begin{matrix} \text{Rapid transit} & Car \end{matrix} \\ \text{Current month} & \begin{matrix} \text{Rapid transit} \\ Car \end{matrix} & \begin{bmatrix} .8 & .2 \\ .3 & .7 \end{bmatrix} \end{matrix}$$

(A) What is the initial-state matrix?

(B) What percentage of commuters will be using the new system after 1 month? After 2 months?

(C) Find the percentage of commuters using each type of transportation after the new system has been in service for a long time.

68. Politics: filibuster. The Senate is in the middle of a floor debate, and a filibuster is threatened. Senator Hanks, who is still vacillating, has a probability of .1 of changing his mind during the next 5 minutes. If this pattern continues for each 5 minutes that the debate continues and if a 24-hour filibuster takes place before a vote is taken, what is the probability that Senator Hanks will cast a yes vote? A no vote?

(A) Complete the following transition matrix:

$$\begin{array}{c} \text{Current} \\ \text{5 minutes} \end{array} \begin{array}{c} \\ Yes \\ No \end{array} \overset{\text{Next 5 minutes}}{\overset{\begin{array}{cc} Yes & No \end{array}}{\begin{bmatrix} .9 & .1 \\ & \end{bmatrix}}}$$

(B) Find the stationary matrix and answer the two questions.

(C) What is the stationary matrix if the probability of Senator Hanks changing his mind (.1) is replaced with an arbitrary probability p?

The population center of the 48 contiguous states of the United States is the point where a flat, rigid map of the contiguous states would balance if the location of each person was represented on the map by a weight of equal measure. In 1790, the population center was 23 miles east of Baltimore, Maryland. By 1990, the center had shifted about 800 miles west and 100 miles south to a point in southeast Missouri. To study this shifting population, the U.S. Census Bureau divides the states into four regions as shown in the figure. Problems 69 and 70 deal with population shifts among these regions.

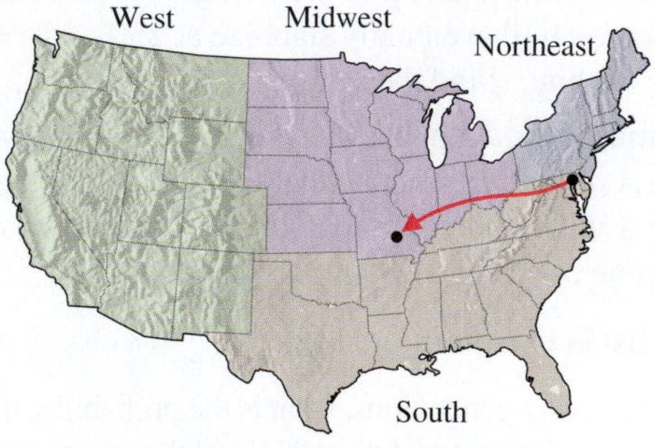

Figure for 69 and 70.: Regions of the United States and the center of population

69. Population shifts. Table 3 gives the percentage of the U.S. population living in the south region during the indicated years.

Table 3

Year	Percent
1970	30.9
1980	33.3
1990	34.4
2000	35.6
2010	37.1

The following transition matrix P is proposed as a model for the data, where S represents the population that lives in the south region:

$$\begin{array}{c} \text{Current} \\ \text{decade} \end{array} \begin{array}{c} \\ S \\ S' \end{array} \overset{\text{Next decade}}{\overset{\begin{array}{cc} S & S' \end{array}}{\begin{bmatrix} .61 & .39 \\ .21 & .79 \end{bmatrix}}} = P$$

(A) Let $S_0 = [.309 \quad .691]$ and find S_1, S_2, S_3, and S_4. Compute the matrices exactly and then round entries to three decimal places.

(B) Construct a new table comparing the results from part (A) with the data in Table 3.

(C) According to this transition matrix, what percentage of the population will live in the south region in the long run?

70. Population shifts. Table 4 gives the percentage of the U.S. population living in the northeast region during the indicated years.

Table 4

Year	Percent
1970	24.1
1980	21.7
1990	20.4
2000	19.0
2010	17.9

The following transition matrix P is proposed as a model for the data, where N represents the population that lives in the northeast region:

$$\begin{array}{c} \text{Current} \\ \text{decade} \end{array} \begin{array}{c} \\ N \\ N' \end{array} \overset{\text{Next decade}}{\overset{\begin{array}{cc} N & N' \end{array}}{\begin{bmatrix} .61 & .39 \\ .09 & .91 \end{bmatrix}}} = P$$

(A) Let $S_0 = [.241 \quad .759]$ and find S_1, S_2, S_3, and S_4. Compute the matrices exactly and then round entries to three decimal places.

(B) Construct a new table comparing the results from part (A) with the data in Table 4.

(C) According to this transition matrix, what percentage of the population will live in the northeast region in the long run?

Answers to Matched Problems

1. (A) Regular (B) Not regular (C) Regular

2. $S = [.2 \quad .8]$; $\overline{P} = \begin{bmatrix} .2 & .8 \\ .2 & .8 \end{bmatrix}$

3. 97.5%

4. 28% preferred, 43% standard, 29% infrequent

5. $\overline{P} = \begin{bmatrix} .1618 & .2941 & .5441 \\ .1618 & .2941 & .5441 \\ .1618 & .2941 & .5441 \end{bmatrix}$;

$S = [.1618 \quad .2941 \quad .5441]$

9.3 Absorbing Markov Chains

- Absorbing States and Absorbing Chains
- Standard Form
- Limiting Matrix
- Graphing Calculator Approximations

In Section 9.2, we saw that the powers of a regular transition matrix always approach a limiting matrix. Not all transition matrices have this property. In this section, we discuss another type of Markov chain, called an *absorbing Markov chain*. Although regular and absorbing Markov chains have some differences, they have one important similarity: the powers of the transition matrix for an absorbing Markov chain also approach a limiting matrix. After introducing basic concepts, we develop methods for finding the limiting matrix and discuss the relationship between the states in the Markov chain and the entries in the limiting matrix.

Absorbing States and Absorbing Chains

A state in a Markov chain is called an **absorbing state** if, once the state is entered, it is impossible to leave.

EXAMPLE 1 **Recognizing Absorbing States** Identify any absorbing states for the following transition matrices:

(A)
$$P = \begin{matrix} & \begin{matrix} A & B & C \end{matrix} \\ \begin{matrix} A \\ B \\ C \end{matrix} & \begin{bmatrix} 1 & 0 & 0 \\ .5 & .5 & 0 \\ 0 & .5 & .5 \end{bmatrix} \end{matrix}$$

(B)
$$P = \begin{matrix} & \begin{matrix} A & B & C \end{matrix} \\ \begin{matrix} A \\ B \\ C \end{matrix} & \begin{bmatrix} 0 & 0 & 1 \\ 0 & 1 & 0 \\ 1 & 0 & 0 \end{bmatrix} \end{matrix}$$

SOLUTION

(A) The probability of going from state A to state A is 1, and the probability of going from state A to either state B or state C is 0. Once state A is entered, it is impossible to leave, so A is an absorbing state. Since the probability of going from state B to state A is nonzero, it is possible to leave B, and B is not an absorbing state. Similarly, the probability of going from state C to state B is nonzero, so C is not an absorbing state.

(B) Reasoning as before, the 1 in row 2 and column 2 indicates that state B is an absorbing state. The probability of going from state A to state C and the probability of going from state C to state A are both nonzero. So A and C are not absorbing states.

Matched Problem 1 Identify any absorbing states for the following transition matrices:

(A)
$$P = \begin{matrix} & \begin{matrix} A & B & C \end{matrix} \\ \begin{matrix} A \\ B \\ C \end{matrix} & \begin{bmatrix} .5 & 0 & .5 \\ 0 & 1 & 0 \\ 0 & .5 & .5 \end{bmatrix} \end{matrix}$$

(B)
$$P = \begin{matrix} & \begin{matrix} A & B & C \end{matrix} \\ \begin{matrix} A \\ B \\ C \end{matrix} & \begin{bmatrix} 0 & 1 & 0 \\ 1 & 0 & 0 \\ 0 & 0 & 1 \end{bmatrix} \end{matrix}$$

The reasoning used to identify absorbing states in Example 1 is generalized in Theorem 1.

THEOREM 1 Absorbing States and Transition Matrices

A state in a Markov chain is **absorbing** if and only if the row of the transition matrix corresponding to the state has a 1 on the main diagonal and 0's elsewhere.

The presence of an absorbing state in a transition matrix does not guarantee that the powers of the matrix approach a limiting matrix nor that the state matrices in the corresponding Markov chain approach a stationary matrix. For example, if we square the matrix P from Example 1B, we obtain

$$P^2 = \begin{bmatrix} 0 & 0 & 1 \\ 0 & 1 & 0 \\ 1 & 0 & 0 \end{bmatrix}\begin{bmatrix} 0 & 0 & 1 \\ 0 & 1 & 0 \\ 1 & 0 & 0 \end{bmatrix} = \begin{bmatrix} 1 & 0 & 0 \\ 0 & 1 & 0 \\ 0 & 0 & 1 \end{bmatrix} = I$$

Since $P^2 = I$, the 3×3 identity matrix, it follows that

$$P^3 = PP^2 = PI = P \quad \text{Since } P^2 = I$$

$$P^4 = PP^3 = PP = I \quad \text{Since } P^3 = P \text{ and } PP = P^2 = I$$

In general, the powers of this transition matrix P oscillate between P and I and do not approach a limiting matrix.

Explore and Discuss 1

(A) For the initial-state matrix $S_0 = [a \quad b \quad c]$, find the first four state matrices, S_1, S_2, S_3, and S_4, in the Markov chain with transition matrix

$$P = \begin{bmatrix} 0 & 0 & 1 \\ 0 & 1 & 0 \\ 1 & 0 & 0 \end{bmatrix}$$

(B) Do the state matrices appear to be approaching a stationary matrix? Discuss.

To ensure that transition matrices for Markov chains with one or more absorbing states have limiting matrices, it is necessary to require the chain to satisfy one additional condition, as stated in the following definition.

DEFINITION Absorbing Markov Chains

A Markov chain is an **absorbing chain** if

1. There is at least one absorbing state; and
2. It is possible to go from each nonabsorbing state to at least one absorbing state in a finite number of steps.

As we saw earlier, absorbing states are identified easily by examining the rows of a transition matrix. It is also possible to use a transition matrix to determine whether a Markov chain is an absorbing chain, but this can be a difficult task, especially if the matrix is large. A transition diagram is often a more appropriate tool for determining whether a Markov chain is absorbing. The next example illustrates this approach for the two matrices discussed in Example 1.

EXAMPLE 2 **Recognizing Absorbing Markov Chains** Use a transition diagram to determine whether P is the transition matrix for an absorbing Markov chain.

(A)
$$P = \begin{array}{c} \\ A \\ B \\ C \end{array}\begin{array}{c} \begin{array}{ccc} A & B & C \end{array} \\ \begin{bmatrix} 1 & 0 & 0 \\ .5 & .5 & 0 \\ 0 & .5 & .5 \end{bmatrix} \end{array}$$

(B)
$$P = \begin{array}{c} \\ A \\ B \\ C \end{array}\begin{array}{c} \begin{array}{ccc} A & B & C \end{array} \\ \begin{bmatrix} 0 & 0 & 1 \\ 0 & 1 & 0 \\ 1 & 0 & 0 \end{bmatrix} \end{array}$$

SOLUTION

(A) From Example 1A, we know that A is the only absorbing state. The second condition in the definition of an absorbing Markov chain is satisfied if we can show that it is possible to go from the nonabsorbing states B and C to the absorbing state A in a finite number of steps. This is easily determined by drawing a transition diagram (Fig. 1). Examining the diagram, we see that it is possible to go from state B to the absorbing state A in one step and from state C to the absorbing state A in two steps. So P is the transition matrix for an absorbing Markov chain.

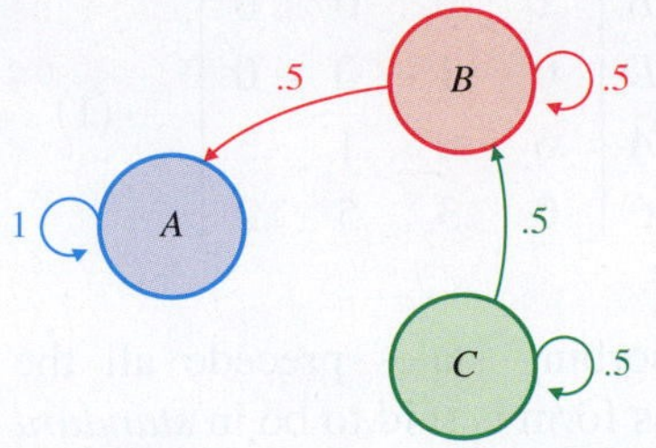

Figure 1

(B) Again, we draw the transition diagram (Fig. 2). From this diagram it is clear that it is impossible to go from either state A or state C to the absorbing state B. So P is not the transition matrix for an absorbing Markov chain.

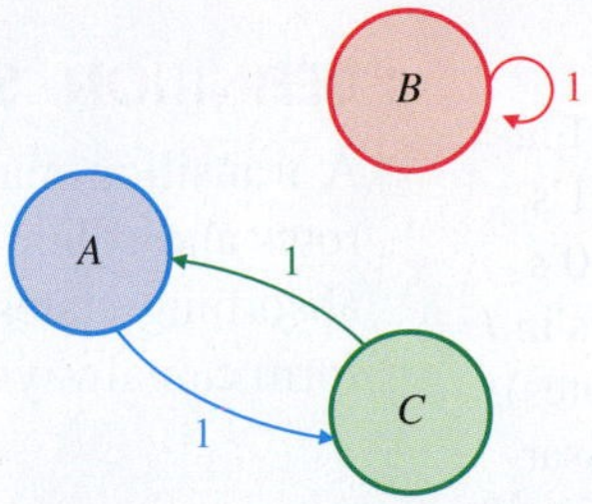

Figure 2

Matched Problem 2 Use a transition diagram to determine whether P is the transition matrix for an absorbing Markov chain.

(A)
$$P = \begin{matrix} & \begin{matrix} A & B & C \end{matrix} \\ \begin{matrix} A \\ B \\ C \end{matrix} & \begin{bmatrix} .5 & 0 & .5 \\ 0 & 1 & 0 \\ 0 & .5 & .5 \end{bmatrix} \end{matrix}$$

(B)
$$P = \begin{matrix} & \begin{matrix} A & B & C \end{matrix} \\ \begin{matrix} A \\ B \\ C \end{matrix} & \begin{bmatrix} 0 & 1 & 0 \\ 1 & 0 & 0 \\ 0 & 0 & 1 \end{bmatrix} \end{matrix}$$

Explore and Discuss 2

Determine whether each statement is true or false. Use examples and verbal arguments to support your conclusions.

(A) A Markov chain with two states, one nonabsorbing and one absorbing, is always an absorbing chain.

(B) A Markov chain with two states, both of which are absorbing, is always an absorbing chain.

(C) A Markov chain with three states, one nonabsorbing and two absorbing, is always an absorbing chain.

Standard Form

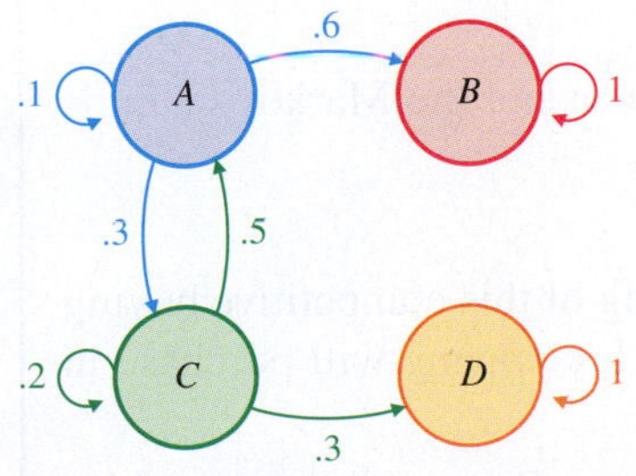

Figure 3

The transition matrix for a Markov chain is not unique. Consider the transition diagram in Figure 3. Since there are $4! = 24$ different ways to arrange the four states in this diagram, there are 24 different ways to write a transition matrix. (Some of these matrices may have identical entries, but all are different when the row and column labels are taken into account.) For example, the following matrices M, N, and P are three different transition matrices for this diagram.

$$M = \begin{array}{c} \\ A \\ B \\ C \\ D \end{array} \begin{array}{c} \begin{array}{cccc} A & B & C & D \end{array} \\ \begin{bmatrix} .1 & .6 & .3 & 0 \\ 0 & 1 & 0 & 0 \\ .5 & 0 & .2 & .3 \\ 0 & 0 & 0 & 1 \end{bmatrix} \end{array} \quad N = \begin{array}{c} \\ D \\ B \\ C \\ A \end{array} \begin{array}{c} \begin{array}{cccc} D & B & C & A \end{array} \\ \begin{bmatrix} 1 & 0 & 0 & 0 \\ 0 & 1 & 0 & 0 \\ .3 & 0 & .2 & .5 \\ 0 & .6 & .3 & .1 \end{bmatrix} \end{array} \quad P = \begin{array}{c} \\ B \\ D \\ A \\ C \end{array} \begin{array}{c} \begin{array}{cccc} B & D & A & C \end{array} \\ \begin{bmatrix} 1 & 0 & 0 & 0 \\ 0 & 1 & 0 & 0 \\ .6 & 0 & .1 & .3 \\ 0 & .3 & .5 & .2 \end{bmatrix} \end{array} \qquad (1)$$

In matrices N and P, notice that all the absorbing states precede all the nonabsorbing states. A transition matrix written in this form is said to be in *standard form*. We will find standard forms very useful in determining limiting matrices for absorbing Markov chains.

Reminder

The identity matrix I in the definition is a square matrix: It has 1's on the principal diagonal and 0's elsewhere (the number of rows in I is the number of absorbing states). The zero matrix 0 is not necessarily square (the number of rows in 0 is the number of absorbing states, and the number of columns is the number of nonabsorbing states).

DEFINITION Standard Forms for Absorbing Markov Chains

A transition matrix for an absorbing Markov chain is in **standard form** if the rows and columns are labeled so that all the absorbing states precede all the nonabsorbing states. (There may be more than one standard form.) Any standard form can always be partitioned into four submatrices:

$$\begin{array}{c} \\ A \\ N \end{array} \begin{array}{c} \begin{array}{cc} A & N \end{array} \\ \left[\begin{array}{c:c} I & 0 \\ \hdashline R & Q \end{array}\right] \end{array} \quad \begin{array}{l} A = \text{all absorbing states} \\ N = \text{all nonabsorbing states} \end{array}$$

where I is an identity matrix and 0 is a zero matrix.

Referring to the matrix P in (1), we see that the submatrices in this standard form are

$$I = \begin{bmatrix} 1 & 0 \\ 0 & 1 \end{bmatrix} \quad 0 = \begin{bmatrix} 0 & 0 \\ 0 & 0 \end{bmatrix} \quad P = \begin{array}{c} \\ B \\ D \\ A \\ C \end{array} \begin{array}{c} \begin{array}{cccc} B & D & A & C \end{array} \\ \left[\begin{array}{cc:cc} 1 & 0 & 0 & 0 \\ 0 & 1 & 0 & 0 \\ \hdashline .6 & 0 & .1 & .3 \\ 0 & .3 & .5 & .2 \end{array}\right] \end{array}$$

$$R = \begin{bmatrix} .6 & 0 \\ 0 & .3 \end{bmatrix} \quad Q = \begin{bmatrix} .1 & .3 \\ .5 & .2 \end{bmatrix}$$

Limiting Matrix

We will now discuss the long-run behavior of absorbing Markov chains.

EXAMPLE 3 **Real Estate Development** Two competing real estate companies are trying to buy all the farms in a particular area for future housing development. Each year, 20% of the farmers decide to sell to company A, 30% decide to sell to company B, and the rest continue to farm their land. Neither company ever sells any of the farms they purchase.

(A) Draw a transition diagram and determine whether or not the Markov chain is absorbing.

(B) Write a transition matrix that is in standard form.

(C) If neither company owns any farms at the beginning of this competitive buying process, estimate the percentage of farms that each company will purchase in the long run.

(D) If company A buys 50% of the farms before company B enters the competitive buying process, estimate the percentage of farms that each company will purchase in the long run.

SOLUTION

(A)

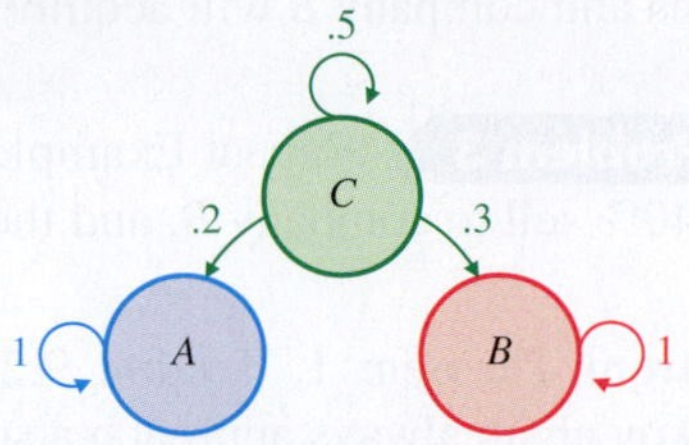

A = sells to company A
B = sells to company B
C = continues farming

The associated Markov chain is absorbing since there are two absorbing states, A and B. It is possible to go from the nonabsorbing state C to either A or B in one step.

(B) We use the transition diagram to write a transition matrix that is in standard form:

$$P = \begin{matrix} & \begin{matrix} A & B & C \end{matrix} \\ \begin{matrix} A \\ B \\ C \end{matrix} & \begin{bmatrix} 1 & 0 & 0 \\ 0 & 1 & 0 \\ .2 & .3 & .5 \end{bmatrix} \end{matrix}$$ Standard form

(C) At the beginning of the competitive buying process all the farmers are in state C (own a farm). Thus, $S_0 = [0 \quad 0 \quad 1]$. The successive state matrices are (multiplication details omitted):

$$\begin{aligned}
S_1 &= S_0P = [.2 \quad .3 \quad .5] \\
S_2 &= S_1P = [.3 \quad .45 \quad .25] \\
S_3 &= S_2P = [.35 \quad .525 \quad .125] \\
S_4 &= S_3P = [.375 \quad .5625 \quad .0625] \\
S_5 &= S_4P = [.3875 \quad .58125 \quad .03125] \\
S_6 &= S_5P = [.39375 \quad .590625 \quad .015625] \\
S_7 &= S_6P = [.396875 \quad .5953125 \quad .0078125] \\
S_8 &= S_7P = [.3984375 \quad .59765625 \quad .00390625] \\
S_9 &= S_8P = [.39921875 \quad .598828125 \quad .001953125]
\end{aligned}$$

It appears that these state matrices are approaching the matrix

$$S = \begin{matrix} A & B & C \\ [.4 & .6 & 0] \end{matrix}$$

This indicates that in the long run, company A will acquire approximately 40% of the farms and company B will acquire the remaining 60%.

(D) This time, at the beginning of the competitive buying process 50% of farmers are already in state A and the rest are in state C. So $S_0 = [.5 \quad 0 \quad .5]$. The successive state matrices are (multiplication details omitted):

$$\begin{aligned}
S_1 &= S_0P = [.6 \quad .15 \quad .25] \\
S_2 &= S_1P = [.65 \quad .225 \quad .125] \\
S_3 &= S_2P = [.675 \quad .2625 \quad .0625] \\
S_4 &= S_3P = [.6875 \quad .28125 \quad .03125] \\
S_5 &= S_4P = [.69375 \quad .290625 \quad .015625] \\
S_6 &= S_5P = [.696875 \quad .2953125 \quad .0078125] \\
S_7 &= S_6P = [.6984375 \quad .29765625 \quad .00390625] \\
S_8 &= S_7P = [.69921875 \quad .298828125 \quad .001953125]
\end{aligned}$$

These state matrices approach a matrix different from the one in part (C):

$$S' = \begin{matrix} A & B & C \\ [.7 & .3 & 0] \end{matrix}$$

Because of its head start, company A will now acquire approximately 70% of the farms and company B will acquire the remaining 30%.

Matched Problem 3 Repeat Example 3 if 10% of farmers sell to company A each year, 40% sell to company B, and the remainder continue farming.

Recall from Theorem 1, Section 9.2, that the successive state matrices of a regular Markov chain always approach a stationary matrix. Furthermore, this stationary matrix is unique. That is, changing the initial-state matrix does not change the stationary matrix. The successive state matrices for an absorbing Markov chain also approach a stationary matrix, but this matrix is not unique. To confirm this, consider the transition matrix P and the state matrices S and S' from Example 3:

$$P = \begin{matrix} & \begin{matrix} A & B & C \end{matrix} \\ \begin{matrix} A \\ B \\ C \end{matrix} & \begin{bmatrix} 1 & 0 & 0 \\ 0 & 1 & 0 \\ .2 & .3 & .5 \end{bmatrix} \end{matrix} \qquad S = \overset{\begin{matrix} A & B & C \end{matrix}}{\begin{bmatrix} .4 & .6 & 0 \end{bmatrix}} \qquad S' = \overset{\begin{matrix} A & B & C \end{matrix}}{\begin{bmatrix} .7 & .3 & 0 \end{bmatrix}}$$

It turns out that S and S' are both stationary matrices, as the following multiplications verify:

$$SP = \begin{bmatrix} .4 & .6 & 0 \end{bmatrix} \begin{bmatrix} 1 & 0 & 0 \\ 0 & 1 & 0 \\ .2 & .3 & .5 \end{bmatrix} = \begin{bmatrix} .4 & .6 & 0 \end{bmatrix} = S$$

$$S'P = \begin{bmatrix} .7 & .3 & 0 \end{bmatrix} \begin{bmatrix} 1 & 0 & 0 \\ 0 & 1 & 0 \\ .2 & .3 & .5 \end{bmatrix} = \begin{bmatrix} .7 & .3 & 0 \end{bmatrix} = S'$$

In fact, this absorbing Markov chain has an infinite number of stationary matrices (see Problems 57 and 58, Exercises 9.3).

Changing the initial-state matrix for an absorbing Markov chain can cause the successive state matrices to approach a different stationary matrix.

In Section 9.2, we used the unique stationary matrix for a regular Markov chain to find the limiting matrix $\overline{P}$. Since an absorbing Markov chain can have many different stationary matrices, we cannot expect this approach to work for absorbing chains. However, it turns out that transition matrices for absorbing chains do have limiting matrices, and they are not very difficult to find. Theorem 2 gives us the necessary tools. The proof of this theorem is left for more advanced courses.

THEOREM 2 Limiting Matrices for Absorbing Markov Chains

If a standard form P for an absorbing Markov chain is partitioned as

$$P = \left[\begin{array}{c|c} I & 0 \\ \hline R & Q \end{array}\right]$$

then P^k approaches a limiting matrix $\overline{P}$ as k increases, where

$$\overline{P} = \left[\begin{array}{c|c} I & 0 \\ \hline FR & 0 \end{array}\right]$$

The matrix F is given by $F = (I - Q)^{-1}$ and is called the **fundamental matrix** for P.

The identity matrix used to form the fundamental matrix F must be the same size as the matrix Q.

EXAMPLE 4 **Finding the Limiting Matrix**

(A) Find the limiting matrix $\overline{P}$ for the standard form P found in Example 3.

(B) Use $\overline{P}$ to find the limit of the successive state matrices for $S_0 = [0 \quad 0 \quad 1]$.

(C) Use $\overline{P}$ to find the limit of the successive state matrices for $S_0 = [.5 \quad 0 \quad .5]$.

SOLUTION

(A) From Example 3, we have

$$P = \left[\begin{array}{cc|c} 1 & 0 & 0 \\ 0 & 1 & 0 \\ \hline .2 & .3 & .5 \end{array}\right] \qquad \left[\begin{array}{c|c} I & 0 \\ \hline R & Q \end{array}\right]$$

where

$$I = \begin{bmatrix} 1 & 0 \\ 0 & 1 \end{bmatrix} \qquad 0 = \begin{bmatrix} 0 \\ 0 \end{bmatrix} \qquad R = [.2 \quad .3] \qquad Q = [.5]$$

If $I = [1]$ is the 1×1 identity matrix, then $I - Q$ is also a 1×1 matrix; $F = (I - Q)^{-1}$ is simply the multiplicative inverse of the single entry in $I - Q$. So

$$F = ([1] - [.5])^{-1} = [.5]^{-1} = [2]$$

$$FR = [2][.2 \quad .3] = [.4 \quad .6]$$

and the limiting matrix is

$$\overline{P} = \begin{matrix} & \begin{matrix} A & B & C \end{matrix} \\ \begin{matrix} A \\ B \\ C \end{matrix} & \begin{bmatrix} 1 & 0 & 0 \\ 0 & 1 & 0 \\ .4 & .6 & 0 \end{bmatrix} \end{matrix} \qquad \left[\begin{array}{c|c} I & 0 \\ \hline FR & 0 \end{array}\right]$$

(B) Since the successive state matrices are given by $S_k = S_0P^k$ (Theorem 1, Section 9.1) and P^k approaches $\overline{P}$, it follows that S_k approaches

$$S_0\overline{P} = [0 \quad 0 \quad 1] \begin{bmatrix} 1 & 0 & 0 \\ 0 & 1 & 0 \\ .4 & .6 & 0 \end{bmatrix} = [.4 \quad .6 \quad 0]$$

which agrees with the results in part (C) of Example 3.

(C) This time, the successive state matrices approach

$$S_0\overline{P} = [.5 \quad 0 \quad .5] \begin{bmatrix} 1 & 0 & 0 \\ 0 & 1 & 0 \\ .4 & .6 & 0 \end{bmatrix} = [.7 \quad .3 \quad 0]$$

which agrees with the results in part (D) of Example 3.

Matched Problem 4 Repeat Example 4 for the standard form P found in Matched Problem 3.

Recall that the limiting matrix for a regular Markov chain contains the long-run probabilities of going from any state to any other state. This is also true for the limiting matrix of an absorbing Markov chain. Let's compare the transition matrix P and its limiting matrix $\overline{P}$ from Example 4:

$$P = \begin{matrix} & A & B & C \\ A & 1 & 0 & 0 \\ B & 0 & 1 & 0 \\ C & .2 & .3 & .5 \end{matrix} \quad \text{approaches} \quad \overline{P} = \begin{matrix} & A & B & C \\ A & 1 & 0 & 0 \\ B & 0 & 1 & 0 \\ C & .4 & .6 & 0 \end{matrix}$$

The rows of P and $\overline{P}$ corresponding to the absorbing states A and B are identical. That is, if the probability of going from state A to state A is 1 at the beginning of the chain, then this probability will remain 1 for all trials in the chain and for the limiting matrix. The entries in the third row of $\overline{P}$ give the long-run probabilities of going from the nonabsorbing state C to states A, B, or C.

The fundamental matrix F provides some additional information about an absorbing chain. Recall from Example 4 that $F = [2]$. It can be shown that the entries in F determine the average number of trials that it takes to go from a given nonabsorbing state to an absorbing state. In the case of Example 4, the single entry 2 in F indicates that it will take an average of 2 years for a farmer to go from state C (owns a farm) to one of the absorbing states (sells the farm). Some will reach an absorbing state in 1 year, and some will take more than 2 years. But the average will be 2 years. These observations are summarized in Theorem 3, which we state without proof.

THEOREM 3 Properties of the Limiting Matrix $\overline{P}$

If P is a transition matrix in standard form for an absorbing Markov chain, F is the fundamental matrix, and $\overline{P}$ is the limiting matrix, then

(A) The entry in row i and column j of $\overline{P}$ is the long-run probability of going from state i to state j. For the nonabsorbing states, these probabilities are also the entries in the matrix FR used to form $\overline{P}$.

(B) The sum of the entries in each row of the fundamental matrix F is the average number of trials it will take to go from each nonabsorbing state to some absorbing state.

(Note that the rows of both F and FR correspond to the nonabsorbing states in the order given in the standard form P).

CONCEPTUAL INSIGHT

1. The zero matrix in the lower right corner of the limiting matrix $\overline{P}$ in Theorem 2 indicates that the long-run probability of going from any nonabsorbing state to any other nonabsorbing state is always 0. That is, in the long run, all elements in an absorbing Markov chain end up in one of the absorbing states.
2. If the transition matrix for an absorbing Markov chain is not in standard form, it is still possible to find a limiting matrix (see Problems 53 and 54, Exercises 9.3). However, it is customary to use standard form when investigating the limiting behavior of an absorbing chain.

Now that we have developed the necessary tools for analyzing the long-run behavior of an absorbing Markov chain, we apply these tools to an earlier application (see Example 4, Section 9.1).

EXAMPLE 5 **Student Retention** The following transition diagram is for part-time students enrolled in a university MBA program:

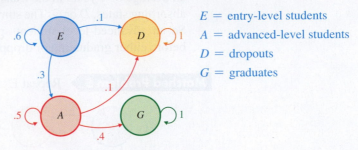

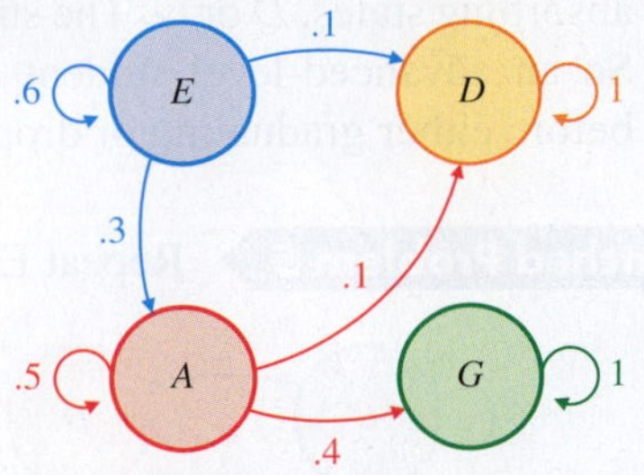

E = entry-level students
A = advanced-level students
D = dropouts
G = graduates

(A) In the long run, what percentage of entry-level students will graduate? What percentage of advanced-level students will not graduate?

(B) What is the average number of years that an entry-level student will remain in this program? An advanced-level student?

SOLUTION

(A) First, notice that this is an absorbing Markov chain with two absorbing states, state D and state G. A standard form for this absorbing chain is

$$P = \begin{array}{c} \\ D \\ G \\ E \\ A \end{array} \begin{array}{c} \begin{array}{cccc} D & G & E & A \end{array} \\ \left[\begin{array}{cc|cc} 1 & 0 & 0 & 0 \\ 0 & 1 & 0 & 0 \\ \hline .1 & 0 & .6 & .3 \\ .1 & .4 & 0 & .5 \end{array}\right] \end{array} \qquad \left[\begin{array}{c|c} I & 0 \\ \hline R & Q \end{array}\right]$$

The submatrices in this partition are

$$I = \begin{bmatrix} 1 & 0 \\ 0 & 1 \end{bmatrix} \qquad 0 = \begin{bmatrix} 0 & 0 \\ 0 & 0 \end{bmatrix} \qquad R = \begin{bmatrix} .1 & 0 \\ .1 & .4 \end{bmatrix} \qquad Q = \begin{bmatrix} .6 & .3 \\ 0 & .5 \end{bmatrix}$$

Therefore,

$$F = (I - Q)^{-1} = \left(\begin{bmatrix} 1 & 0 \\ 0 & 1 \end{bmatrix} - \begin{bmatrix} .6 & .3 \\ 0 & .5 \end{bmatrix} \right)^{-1}$$

$$= \begin{bmatrix} .4 & -.3 \\ 0 & .5 \end{bmatrix}^{-1}$$

$$= \begin{bmatrix} 2.5 & 1.5 \\ 0 & 2 \end{bmatrix}$$

Use row operations to find this matrix inverse.

and

$$FR = \begin{bmatrix} 2.5 & 1.5 \\ 0 & 2 \end{bmatrix} \begin{bmatrix} .1 & 0 \\ .1 & .4 \end{bmatrix} = \begin{bmatrix} .4 & .6 \\ .2 & .8 \end{bmatrix}$$

The limiting matrix is

$$\overline{P} = \begin{array}{c} \\ D \\ G \\ E \\ A \end{array} \begin{array}{c} \begin{array}{cccc} D & G & E & A \end{array} \\ \left[\begin{array}{cc|cc} 1 & 0 & 0 & 0 \\ 0 & 1 & 0 & 0 \\ \hline .4 & .6 & 0 & 0 \\ .2 & .8 & 0 & 0 \end{array}\right] \end{array} \qquad \left[\begin{array}{c|c} I & 0 \\ \hline FR & 0 \end{array}\right]$$

From this limiting form, we see that in the long run 60% of the entry-level students will graduate and 20% of the advanced-level students will not graduate.

(B) The sum of the first-row entries of the fundamental matrix F is $2.5 + 1.5 = 4$. According to Theorem 3, this indicates that an entry-level student will spend an average of 4 years in the transient states E and A before reaching one of the absorbing states, D or G. The sum of the second-row entries of F is $0 + 2 = 2$. So an advanced-level student spends an average of 2 years in the program before either graduating or dropping out.

Matched Problem 5 Repeat Example 5 for the following transition diagram:

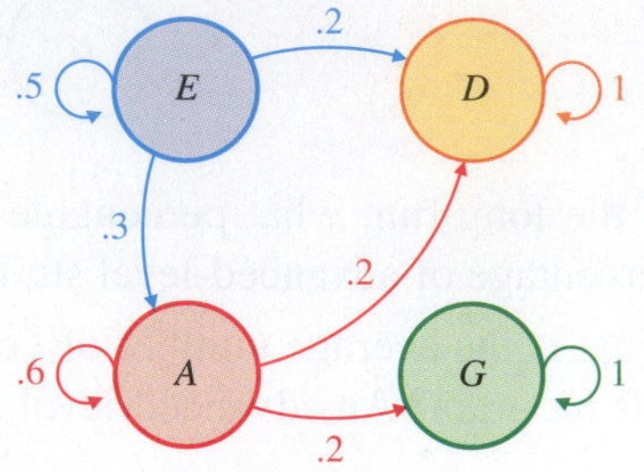

E = entry-level students
A = advanced-level students
D = dropouts
G = graduates

Graphing Calculator Approximations

Just as was the case for regular Markov chains, the limiting matrix $\overline{P}$ for an absorbing Markov chain with transition matrix P can be approximated by computing P^k on a graphing calculator for sufficiently large values of k. For example, computing P^{50} for the standard form P in Example 5 produces the following results:

$$P^{50} = \begin{bmatrix} 1 & 0 & 0 & 0 \\ 0 & 1 & 0 & 0 \\ .1 & 0 & .6 & .3 \\ .1 & .4 & 0 & .5 \end{bmatrix}^{50} = \begin{bmatrix} 1 & 0 & 0 & 0 \\ 0 & 1 & 0 & 0 \\ .4 & .6 & 0 & 0 \\ .2 & .8 & 0 & 0 \end{bmatrix} = \overline{P}$$

where we have replaced very small numbers displayed in scientific notation with 0 (see the Conceptual Insight on page 476, Section 9.2).

CAUTION Before you use P^k to approximate $\overline{P}$, be certain to determine that $\overline{P}$ exists. If you attempt to approximate a limiting matrix when none exists, the results can be misleading. For example, consider the transition matrix

$$P = \begin{bmatrix} 1 & 0 & 0 & 0 & 0 \\ .2 & .2 & 0 & .3 & .3 \\ 0 & 0 & 0 & .5 & .5 \\ 0 & 0 & 1 & 0 & 0 \\ 0 & 0 & 1 & 0 & 0 \end{bmatrix}$$

Computing P^{50} on a graphing calculator produces the following matrix:

$$P^{50} = \begin{bmatrix} 1 & 0 & 0 & 0 & 0 \\ .25 & 0 & .625 & .0625 & .0625 \\ 0 & 0 & 1 & 0 & 0 \\ 0 & 0 & 0 & .5 & .5 \\ 0 & 0 & 0 & .5 & .5 \end{bmatrix} \qquad (2)$$

It is tempting to stop at this point and conclude that the matrix in (2) must be a good approximation for $\overline{P}$. But to do so would be incorrect! If P^{50} approximates a limiting matrix $\overline{P}$, then P^{51} should also approximate the same matrix. However, computing P^{51} produces quite a different matrix:

$$P^{51} = \begin{bmatrix} 1 & 0 & 0 & 0 & 0 \\ .25 & 0 & .125 & .3125 & .3125 \\ 0 & 0 & 0 & .5 & .5 \\ 0 & 0 & 1 & 0 & 0 \\ 0 & 0 & 1 & 0 & 0 \end{bmatrix} \qquad (3)$$

Computing additional powers of P shows that the even powers of P approach matrix (2) while the odd powers approach matrix (3). The transition matrix P does not have a limiting matrix. ▲

A graphing calculator can be used to perform the matrix calculations necessary to find $\overline{P}$ exactly, as illustrated in Figure 4 for the transition matrix P from Example 5. The matrix names [I], [B], [C], and [F] denote I, R, Q, and F, respectively. This approach has the advantage of producing the fundamental matrix F, whose row sums provide additional information about the long-run behavior of the chain.

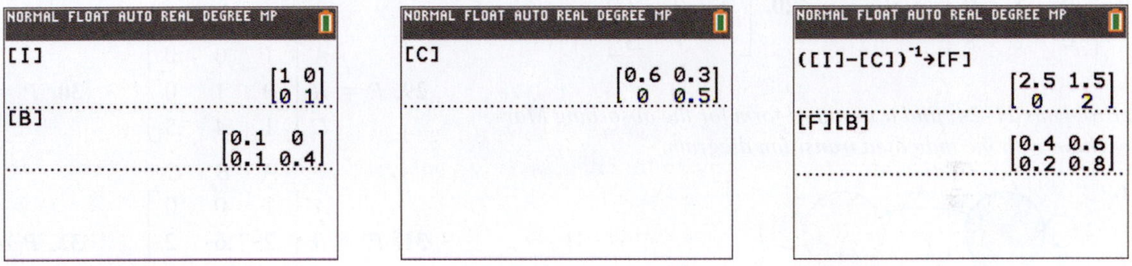

(A) Store I and R in the graphing calculator memory

(B) Store Q in the graphing calculator memory

(C) Compute F and FR

Figure 4 Matrix calculations

Exercises 9.3

A *In Problems 1–6, identify the absorbing states in the indicated transition matrix.*

1. $P = \begin{array}{c} \\ A \\ B \\ C \end{array}\!\!\begin{array}{c} \begin{array}{ccc} A & B & C \end{array} \\ \begin{bmatrix} .6 & .3 & .1 \\ 0 & 1 & 0 \\ 0 & 0 & 1 \end{bmatrix} \end{array}$

2. $P = \begin{array}{c} \\ A \\ B \\ C \end{array}\!\!\begin{array}{c} \begin{array}{ccc} A & B & C \end{array} \\ \begin{bmatrix} 0 & 1 & 0 \\ .3 & .2 & .5 \\ 0 & 0 & 1 \end{bmatrix} \end{array}$

3. $P = \begin{array}{c} \\ A \\ B \\ C \end{array}\!\!\begin{array}{c} \begin{array}{ccc} A & B & C \end{array} \\ \begin{bmatrix} 0 & 0 & 1 \\ 1 & 0 & 0 \\ 0 & 1 & 0 \end{bmatrix} \end{array}$

4. $P = \begin{array}{c} \\ A \\ B \\ C \end{array}\!\!\begin{array}{c} \begin{array}{ccc} A & B & C \end{array} \\ \begin{bmatrix} 1 & 0 & 0 \\ .3 & .4 & .3 \\ 0 & 0 & 1 \end{bmatrix} \end{array}$

5. $P = \begin{array}{c} \\ A \\ B \\ C \\ D \end{array}\!\!\begin{array}{c} \begin{array}{cccc} A & B & C & D \end{array} \\ \begin{bmatrix} 1 & 0 & 0 & 0 \\ 0 & 0 & 1 & 0 \\ .1 & .1 & .5 & .3 \\ 0 & 0 & 0 & 1 \end{bmatrix} \end{array}$

6. $P = \begin{array}{c} \\ A \\ B \\ C \\ D \end{array}\!\!\begin{array}{c} \begin{array}{cccc} A & B & C & D \end{array} \\ \begin{bmatrix} 0 & 1 & 0 & 0 \\ 1 & 0 & 0 & 0 \\ .1 & .2 & .3 & .4 \\ .7 & .1 & .1 & .1 \end{bmatrix} \end{array}$

In Problems 7–10, identify the absorbing states for each transition diagram, and determine whether or not the diagram represents an absorbing Markov chain.

7.

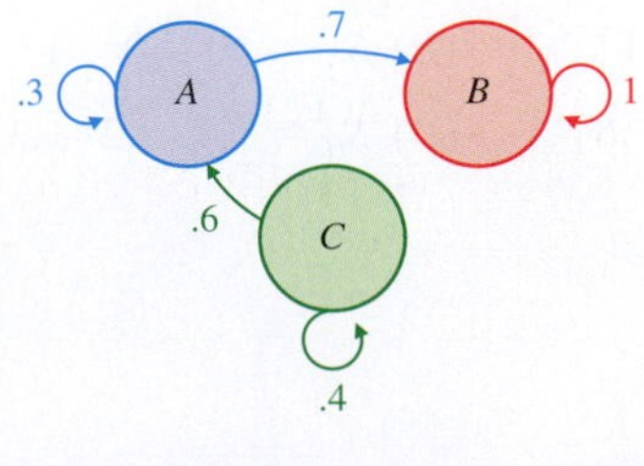

8.

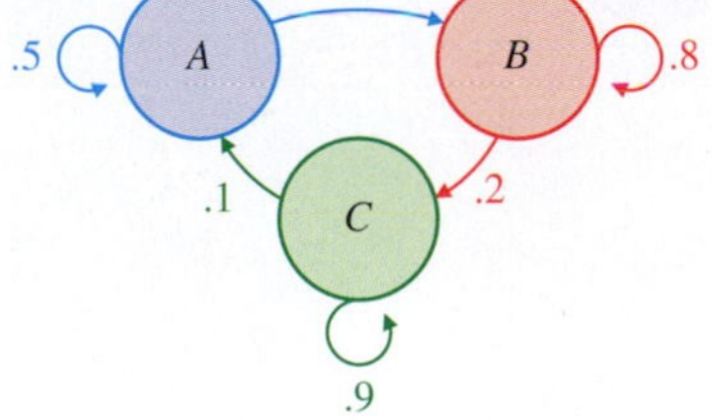

9.

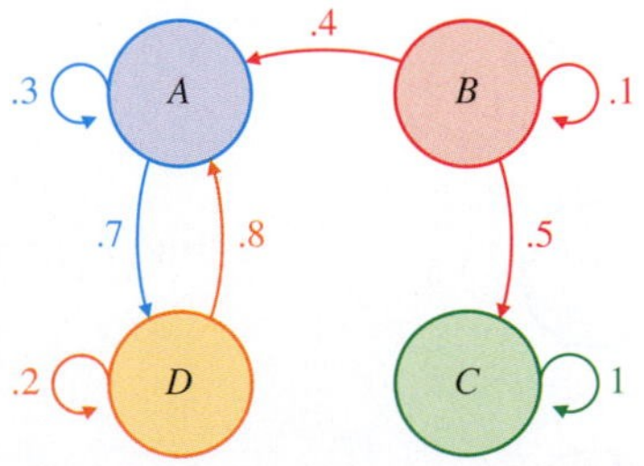

10.

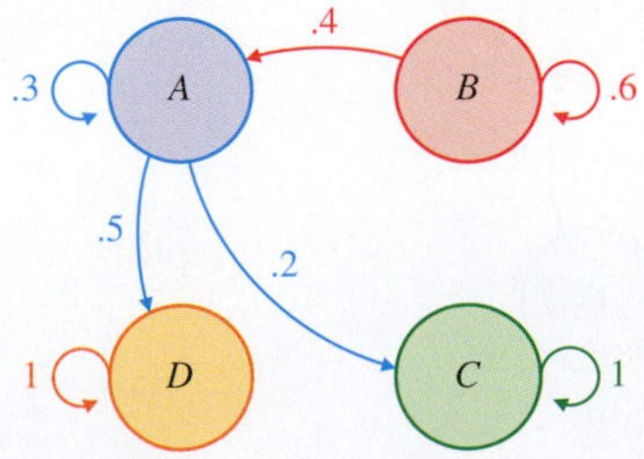

In Problems 11–20, could the given matrix be the transition matrix of an absorbing Markov chain?

11. $\begin{bmatrix} 0 & 1 \\ 1 & 0 \end{bmatrix}$ **12.** $\begin{bmatrix} 1 & 0 \\ 0 & 1 \end{bmatrix}$

13. $\begin{bmatrix} .3 & .7 \\ 0 & 1 \end{bmatrix}$ **14.** $\begin{bmatrix} .6 & .4 \\ 1 & 0 \end{bmatrix}$

15. $\begin{bmatrix} 1 & 0 & 0 \\ 0 & 1 & 0 \\ 0 & 0 & 1 \end{bmatrix}$ **16.** $\begin{bmatrix} 0 & 1 & 0 \\ 0 & 0 & 1 \\ 1 & 0 & 0 \end{bmatrix}$

17. $\begin{bmatrix} .9 & .1 & 0 \\ .1 & .9 & 0 \\ 0 & 0 & 1 \end{bmatrix}$ **18.** $\begin{bmatrix} .5 & .5 & 0 \\ .4 & .3 & .3 \\ 0 & 0 & 1 \end{bmatrix}$

19. $\begin{bmatrix} .9 & 0 & .1 \\ 0 & 1 & 0 \\ 0 & .2 & .8 \end{bmatrix}$ **20.** $\begin{bmatrix} 1 & 0 & 0 \\ 0 & 0 & 1 \\ 0 & .7 & .3 \end{bmatrix}$

B *In Problems 21–24, find a standard form for the absorbing Markov chain with the indicated transition diagram.*

21.

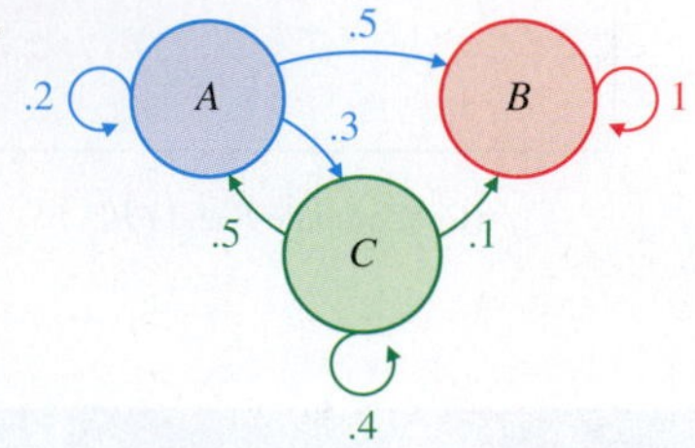

22.

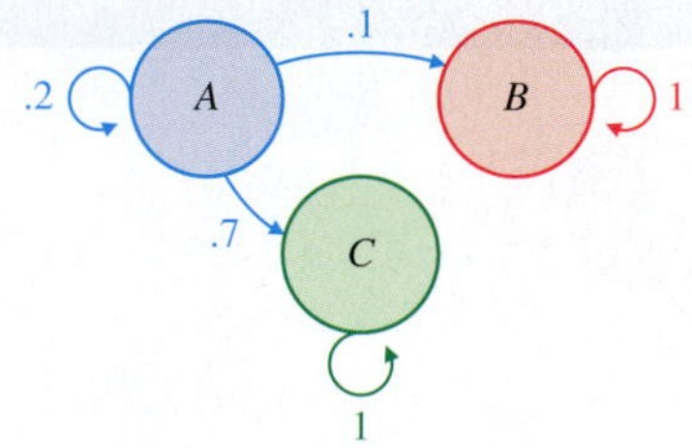

23.

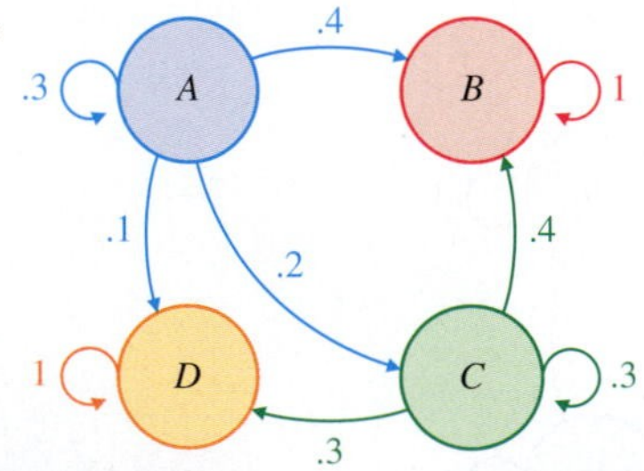

24.

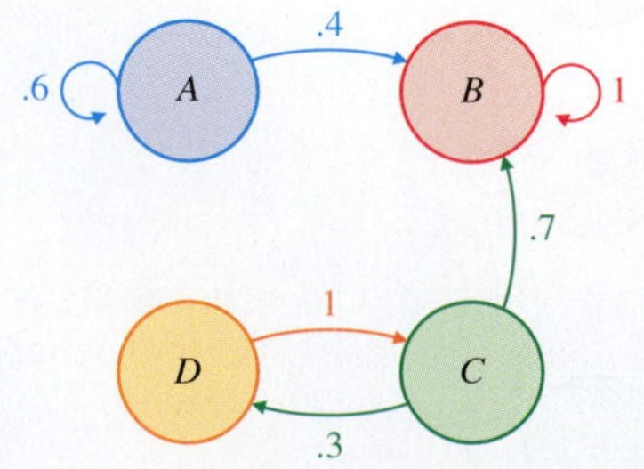

In Problems 25–28, find a standard form for the absorbing Markov chain with the indicated transition matrix.

25. $P = \begin{array}{c|ccc} & A & B & C \\ \hline A & .2 & .3 & .5 \\ B & 1 & 0 & 0 \\ C & 0 & 0 & 1 \end{array}$ **26.** $P = \begin{array}{c|ccc} & A & B & C \\ \hline A & 0 & 0 & 1 \\ B & 0 & 1 & 0 \\ C & .7 & .2 & .1 \end{array}$

27. $P = \begin{array}{c|cccc} & A & B & C & D \\ \hline A & .1 & .2 & .3 & .4 \\ B & 0 & 1 & 0 & 0 \\ C & .5 & .2 & .2 & .1 \\ D & 0 & 0 & 0 & 1 \end{array}$ **28.** $P = \begin{array}{c|cccc} & A & B & C & D \\ \hline A & 0 & .3 & .3 & .4 \\ B & 0 & 1 & 0 & 0 \\ C & 0 & 0 & 1 & 0 \\ D & .8 & .1 & .1 & 0 \end{array}$

In Problems 29–34, find the limiting matrix for the indicated standard form. Find the long-run probability of going from each nonabsorbing state to each absorbing state and the average number of trials needed to go from each nonabsorbing state to an absorbing state.

29. $P = \begin{array}{c|ccc} & A & B & C \\ \hline A & 1 & 0 & 0 \\ B & 0 & 1 & 0 \\ C & .1 & .4 & .5 \end{array}$ **30.** $P = \begin{array}{c|ccc} & A & B & C \\ \hline A & 1 & 0 & 0 \\ B & 0 & 1 & 0 \\ C & .3 & .2 & .5 \end{array}$

31. $P = \begin{array}{c|ccc} & A & B & C \\ \hline A & 1 & 0 & 0 \\ B & .2 & .6 & .2 \\ C & .4 & .2 & .4 \end{array}$ **32.** $P = \begin{array}{c|ccc} & A & B & C \\ \hline A & 1 & 0 & 0 \\ B & .1 & .6 & .3 \\ C & .2 & .2 & .6 \end{array}$

33. $P = \begin{array}{c|cccc} & A & B & C & D \\ \hline A & 1 & 0 & 0 & 0 \\ B & 0 & 1 & 0 & 0 \\ C & .1 & .2 & .6 & .1 \\ D & .2 & .2 & .3 & .3 \end{array}$ **34.** $P = \begin{array}{c|cccc} & A & B & C & D \\ \hline A & 1 & 0 & 0 & 0 \\ B & 0 & 1 & 0 & 0 \\ C & .1 & .1 & .7 & .1 \\ D & .3 & .1 & .4 & .2 \end{array}$

Problems 35–40 refer to the matrices in Problems 29–34. Use the limiting matrix $\overline{P}$ found for each transition matrix P in Problems 29–34 to determine the long-run behavior of the successive state matrices for the indicated initial-state matrices.

35. For matrix P from Problem 29 with
(A) $S_0 = [0 \quad 0 \quad 1]$ (B) $S_0 = [.2 \quad .5 \quad .3]$

36. For matrix P from Problem 30 with
(A) $S_0 = [0 \quad 0 \quad 1]$ (B) $S_0 = [.2 \quad .5 \quad .3]$

37. For matrix P from Problem 31 with
(A) $S_0 = [0 \quad 0 \quad 1]$ (B) $S_0 = [.2 \quad .5 \quad .3]$

38. For matrix P from Problem 32 with
(A) $S_0 = [0 \quad 0 \quad 1]$ (B) $S_0 = [.2 \quad .5 \quad .3]$

39. For matrix P from Problem 33 with
(A) $S_0 = [0 \quad 0 \quad 0 \quad 1]$ (B) $S_0 = [0 \quad 0 \quad 1 \quad 0]$
(C) $S_0 = [0 \quad 0 \quad .4 \quad .6]$ (D) $S_0 = [.1 \quad .2 \quad .3 \quad .4]$

40. For matrix P from Problem 34 with
(A) $S_0 = [0 \quad 0 \quad 0 \quad 1]$ (B) $S_0 = [0 \quad 0 \quad 1 \quad 0]$
(C) $S_0 = [0 \quad 0 \quad .4 \quad .6]$ (D) $S_0 = [.1 \quad .2 \quad .3 \quad .4]$

In Problems 41–48, discuss the validity of each statement. If the statement is always true, explain why. If not, give a counterexample.

41. If a Markov chain has an absorbing state, then it is an absorbing chain.

42. If a Markov chain has exactly two states and at least one absorbing state, then it is an absorbing chain.

43. If a Markov chain has exactly three states, one absorbing and two nonabsorbing, then it is an absorbing chain.

44. If a Markov chain has exactly three states, one nonabsorbing and two absorbing, then it is an absorbing chain.

45. If every state of a Markov chain is an absorbing state, then it is an absorbing chain.

46. If a Markov chain is absorbing, then it has a unique stationary matrix.

47. If a Markov chain is absorbing, then it is regular.

48. If a Markov chain is regular, then it is absorbing.

In Problems 49–52, use a graphing calculator to approximate the limiting matrix for the indicated standard form.

49. $$P = \begin{array}{c} \\ A \\ B \\ C \\ D \end{array}\begin{array}{c} \begin{array}{cccc} A & B & C & D \end{array} \\ \begin{bmatrix} 1 & 0 & 0 & 0 \\ 0 & 1 & 0 & 0 \\ .5 & .3 & .1 & .1 \\ .6 & .2 & .1 & .1 \end{bmatrix} \end{array}$$

50. $$P = \begin{array}{c} \\ A \\ B \\ C \\ D \end{array}\begin{array}{c} \begin{array}{cccc} A & B & C & D \end{array} \\ \begin{bmatrix} 1 & 0 & 0 & 0 \\ 0 & 1 & 0 & 0 \\ .1 & .1 & .5 & .3 \\ 0 & .2 & .3 & .5 \end{bmatrix} \end{array}$$

51. $$P = \begin{array}{c} \\ A \\ B \\ C \\ D \\ E \end{array}\begin{array}{c} \begin{array}{ccccc} A & B & C & D & E \end{array} \\ \begin{bmatrix} 1 & 0 & 0 & 0 & 0 \\ 0 & 1 & 0 & 0 & 0 \\ 0 & .4 & .5 & 0 & .1 \\ 0 & .4 & 0 & .3 & .3 \\ .4 & .4 & 0 & .2 & 0 \end{bmatrix} \end{array}$$

52. $$P = \begin{array}{c} \\ A \\ B \\ C \\ D \\ E \end{array}\begin{array}{c} \begin{array}{ccccc} A & B & C & D & E \end{array} \\ \begin{bmatrix} 1 & 0 & 0 & 0 & 0 \\ 0 & 1 & 0 & 0 & 0 \\ .5 & 0 & 0 & 0 & .5 \\ 0 & .4 & 0 & .2 & .4 \\ 0 & 0 & .1 & .7 & .2 \end{bmatrix} \end{array}$$

C 53. The following matrix P is a nonstandard transition matrix for an absorbing Markov chain:

$$P = \begin{array}{c} \\ A \\ B \\ C \\ D \end{array}\begin{array}{c} \begin{array}{cccc} A & B & C & D \end{array} \\ \begin{bmatrix} .2 & .2 & .6 & 0 \\ 0 & 1 & 0 & 0 \\ .5 & .1 & 0 & .4 \\ 0 & 0 & 0 & 1 \end{bmatrix} \end{array}$$

To find a limiting matrix for P, follow the steps outlined below.

Step 1 Using a transition diagram, rearrange the columns and rows of P to produce a standard form for this chain.

Step 2 Find the limiting matrix for this standard form.

Step 3 Using a transition diagram, reverse the process used in Step 1 to produce a limiting matrix for the original matrix P.

54. Repeat Problem 53 for

$$P = \begin{array}{c} \\ A \\ B \\ C \\ D \end{array}\begin{array}{c} \begin{array}{cccc} A & B & C & D \end{array} \\ \begin{bmatrix} 1 & 0 & 0 & 0 \\ .3 & .6 & 0 & .1 \\ .2 & .3 & .5 & 0 \\ 0 & 0 & 0 & 1 \end{bmatrix} \end{array}$$

55. Verify the results in Problem 53 by computing P^k on a graphing calculator for large values of k.

56. Verify the results in Problem 54 by computing P^k on a graphing calculator for large values of k.

57. Show that $S = [x \quad 1 - x \quad 0], 0 \le x \le 1$, is a stationary matrix for the transition matrix

$$P = \begin{array}{c} \\ A \\ B \\ C \end{array}\begin{array}{c} \begin{array}{ccc} A & B & C \end{array} \\ \begin{bmatrix} 1 & 0 & 0 \\ 0 & 1 & 0 \\ .1 & .5 & .4 \end{bmatrix} \end{array}$$

Discuss the generalization of this result to any absorbing Markov chain with two absorbing states and one nonabsorbing state.

58. Show that $S = [x \quad 1 - x \quad 0 \quad 0], 0 \le x \le 1$, is a stationary matrix for the transition matrix

$$P = \begin{array}{c} \\ A \\ B \\ C \\ D \end{array}\begin{array}{c} \begin{array}{cccc} A & B & C & D \end{array} \\ \begin{bmatrix} 1 & 0 & 0 & 0 \\ 0 & 1 & 0 & 0 \\ .1 & .2 & .3 & .4 \\ .6 & .2 & .1 & .1 \end{bmatrix} \end{array}$$

Discuss the generalization of this result to any absorbing Markov chain with two absorbing states and two nonabsorbing states.

59. An absorbing Markov chain has the following matrix P as a standard form:

$$P = \begin{array}{c} \\ A \\ B \\ C \\ D \end{array}\begin{array}{c} \begin{array}{cccc} A & B & C & D \end{array} \\ \begin{bmatrix} 1 & 0 & 0 & 0 \\ .2 & .3 & .1 & .4 \\ 0 & .5 & .3 & .2 \\ 0 & .1 & .6 & .3 \end{bmatrix} \end{array} \qquad \left[\begin{array}{c|c} I & 0 \\ \hline R & Q \end{array}\right]$$

Let w_k denote the maximum entry in Q^k. Note that $w_1 = .6$.

(A) Find w_2, w_4, w_8, w_{16}, and w_{32} to three decimal places.

(B) Describe Q^k when k is large.

60. Refer to the matrices P and Q of Problem 59. For k a positive integer, let $T_k = I + Q + Q^2 + \cdots + Q^k$.

(A) Explain why $T_{k+1} = T_kQ + I$.

(B) Using a graphing calculator and part (A) to quickly compute the matrices T_k, discover and describe the connection between $(I - Q)^{-1}$ and T_k when k is large.

Applications

61. **Loans.** A credit union classifies car loans into one of four categories: the loan has been paid in full (F), the account is in good standing (G) with all payments up to date, the account is in arrears (A) with one or more missing payments, or the account has been classified as a bad debt (B) and sold to a collection agency. Past records indicate that each month 10% of the accounts in good standing pay the loan in full, 80% remain in good standing, and 10% become in arrears. Furthermore, 10% of the accounts in arrears are paid in full, 40% become accounts in good standing, 40% remain in arrears, and 10% are classified as bad debts.

    (A) In the long run, what percentage of the accounts in arrears will pay their loan in full?

    (B) In the long run, what percentage of the accounts in good standing will become bad debts?

    (C) What is the average number of months that an account in arrears will remain in this system before it is either paid in full or classified as a bad debt?

62. **Employee training.** A chain of car muffler and brake repair shops maintains a training program for its mechanics. All new mechanics begin training in muffler repairs. Every 3 months, the performance of each mechanic is reviewed. Past records indicate that after each quarterly review, 30% of the muffler repair trainees are rated as qualified to repair mufflers and begin training in brake repairs, 20% are terminated for unsatisfactory performance, and the remainder continue as muffler repair trainees. Also, 30% of the brake repair trainees are rated as fully qualified mechanics requiring no further training, 10% are terminated for unsatisfactory performance, and the remainder continue as brake repair trainees.

    (A) In the long run, what percentage of muffler repair trainees will become fully qualified mechanics?

    (B) In the long run, what percentage of brake repair trainees will be terminated?

    (C) What is the average number of quarters that a muffler repair trainee will remain in the training program before being either terminated or promoted to fully qualified mechanic?

63. **Marketing.** Three electronics firms are aggressively marketing their graphing calculators to high school and college mathematics departments by offering volume discounts, complimentary display equipment, and assistance with curriculum development. Due to the amount of equipment involved and the necessary curriculum changes, once a department decides to use a particular calculator in their courses, they never switch to another brand or stop using calculators. Each year, 6% of the departments decide to use calculators from company A, 3% decide to use calculators from company B, 11% decide to use calculators from company C, and the remainder decide not to use any calculators in their courses.

    (A) In the long run, what is the market share of each company?

    (B) On average, how many years will it take a department to decide to use calculators from one of these companies in their courses?

64. **Pensions.** Once a year company employees are given the opportunity to join one of three pension plans: A, B, or C. Once an employee decides to join one of these plans, the employee cannot drop the plan or switch to another plan. Past records indicate that each year 4% of employees elect to join plan A, 14% elect to join plan B, 7% elect to join plan C, and the remainder do not join any plan.

    (A) In the long run, what percentage of the employees will elect to join plan A? Plan B? Plan C?

    (B) On average, how many years will it take an employee to decide to join a plan?

65. **Medicine.** After bypass surgery, patients are placed in an intensive care unit (ICU) until their condition stabilizes. Then they are transferred to a cardiac care ward (CCW), where they remain until they are released from the hospital. In a particular metropolitan area, a study of hospital records produced the following data: each day 2% of the patients in the ICU died, 52% were transferred to the CCW, and the remainder stayed in the ICU. Furthermore, each day 4% of the patients in the CCW developed complications and were returned to the ICU, 1% died while in the CCW, 22% were released from the hospital, and the remainder stayed in the CCW.

    (A) In the long run, what percentage of the patients in the ICU are released from the hospital?

    (B) In the long run, what percentage of the patients in the CCW die without ever being released from the hospital?

    (C) What is the average number of days that a patient in the ICU will stay in the hospital?

66. **Medicine.** The study discussed in Problem 65 also produced the following data for patients who underwent aortic valve replacements: each day 2% of the patients in the ICU died, 60% were transferred to the CCW, and the remainder stayed in the ICU. Furthermore, each day 5% of the patients in the CCW developed complications and were returned to the ICU, 1% died while in the CCW, 19% were released from the hospital, and the remainder stayed in the CCW.

    (A) In the long run, what percentage of the patients in the CCW are released from the hospital?

    (B) In the long run, what percentage of the patients in the ICU die without ever being released from the hospital?

    (C) What is the average number of days a patient in the CCW will stay in the hospital?

67. **Psychology.** A rat is placed in room F or room B of the maze shown in the figure. The rat wanders from room to room until it enters one of the rooms containing food, L or R.

Assume that the rat chooses an exit from a room at random and that once it enters a room with food it never leaves.

(A) What is the long-run probability that a rat placed in room B ends up in room R?

(B) What is the average number of exits that a rat placed in room B will choose until it finds food?

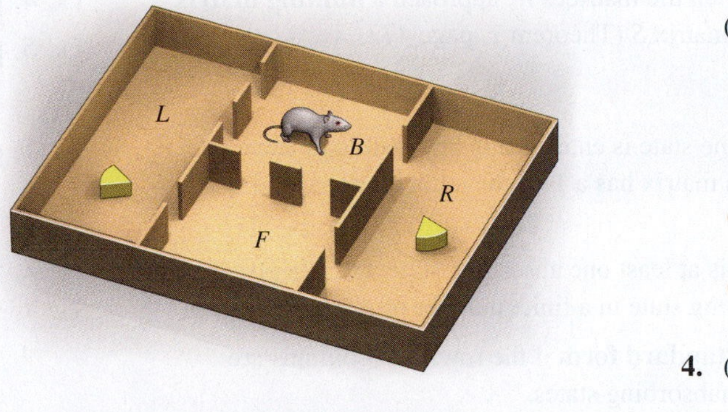

Figure for 67 and 68

68. Psychology. Repeat Problem 67 if the exit from room B to room R is blocked.

Answers to Matched Problems

1. (A) State B is absorbing.
   (B) State C is absorbing.
2. (A) Absorbing Markov chain
   (B) Not an absorbing Markov chain
3. (A)

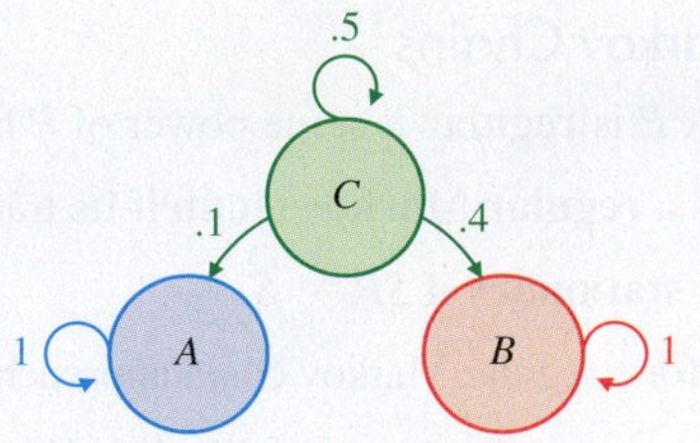

(B)
$$P = \begin{matrix} & \begin{matrix} A & B & C \end{matrix} \\ \begin{matrix} A \\ B \\ C \end{matrix} & \begin{bmatrix} 1 & 0 & 0 \\ 0 & 1 & 0 \\ .1 & .4 & .5 \end{bmatrix} \end{matrix}$$

(C) Company A will purchase 20% of the farms, and company B will purchase 80%.

(D) Company A will purchase 60% of the farms, and company B will purchase 40%.

4. (A)
$$\overline{P} = \begin{matrix} & \begin{matrix} A & B & C \end{matrix} \\ \begin{matrix} A \\ B \\ C \end{matrix} & \begin{bmatrix} 1 & 0 & 0 \\ 0 & 1 & 0 \\ .2 & .8 & 0 \end{bmatrix} \end{matrix}$$

(B) $[.2 \quad .8 \quad 0]$

(C) $[.6 \quad .4 \quad 0]$

5. (A) Thirty percent of entry-level students will graduate; 50% of advanced-level students will not graduate.

(B) An entry-level student will spend an average of 3.5 years in the program; an advanced-level student will spend an average of 2.5 years in the program.

Chapter 9 Summary and Review

Important Terms, Symbols, and Concepts

9.1 Properties of Markov Chains

EXAMPLES

- The progression of a system through a sequence of states is called a **stochastic process** if chance elements are involved in the transition from one state to the next.
- A **transition diagram** or **transition probability matrix** can be used to represent the probabilities of moving from one state to another. If those probabilities do not change with time, the stochastic process is called a **Markov chain**.
- If a Markov chain has n states, then the entry s_{ki} of the **kth-state matrix**

$$S_k = [s_{k1} \quad s_{k2} \quad \cdots \quad s_{kn}]$$

gives the probability of being in state i after the kth trial. The sum of the entries in S_k is 1. Ex. 1, p. 462

- The entry $p_{i,j}$ of the $n \times n$ **transition matrix** P gives the probability of moving from state i to state j on the next trial. The sum of the entries in each row of P is 1. Ex. 2, p. 463 Ex. 3, p. 464
- If S_0 is an initial-state matrix for a Markov chain, then $S_k = S_0P^k$ (Theorem 1, page 463). Ex. 4, p. 464

9.2 Regular Markov Chains

- A transition matrix P is **regular** if some power of P has only positive entries. Ex. 1, p. 471
- A Markov chain is a **regular Markov chain** if its transition matrix is regular.
- A state matrix S is **stationary** if $SP = S$. Ex. 2, p. 472
- The state matrices for a regular Markov chain approach a unique stationary matrix S (Theorem 1, page 472). Ex. 3, p. 473
- If P is the transition matrix for a regular Markov chain, then the matrices P^k approach a **limiting matrix** $\overline{P}$, where each row of $\overline{P}$ is equal to the unique stationary matrix S (Theorem 1, page 472). Ex. 4, p. 474 Ex. 5, p. 475

9.3 Absorbing Markov Chains

- A state in a Markov chain is an **absorbing state** if once the state is entered it is impossible to leave. A state is absorbing if and only if its row in the transition matrix has a 1 on the main diagonal and 0's elsewhere. Ex. 1, p. 481
- A Markov chain is an **absorbing Markov chain** if there is at least one absorbing state and it is possible to go from each nonabsorbing state to at least one absorbing state in a finite number of steps. Ex. 2, p. 482
- A transition matrix for an absorbing Markov chain is in **standard form** if the rows and columns are labeled so that all the absorbing states precede all the nonabsorbing states.
- If a standard form P for an absorbing Markov chain is partitioned as Ex. 3, p. 484 Ex. 4, p. 487

$$P = \left[\begin{array}{c|c} I & 0 \\ \hline R & Q \end{array}\right]$$

then P^k approaches a limiting matrix $\overline{P}$ as k increases, where

$$\overline{P} = \left[\begin{array}{c|c} I & 0 \\ \hline FR & 0 \end{array}\right]$$

The matrix $F = (I - Q)^{-1}$, where I is the identity matrix of the same size as Q, is called the **fundamental matrix** for P (Theorem 2, page 486). Ex. 5, p. 489

- The entry in row i and column j of $\overline{P}$ is the long-run probability of going from state i to state j. The sum of the entries in each row of F is the average number of trials that it will take to go from each nonabsorbing state to some absorbing state (Theorem 3, page 488).

Review Exercises

Work through all the problems in this chapter review and check your answers in the back of the book. Answers to all review problems are there along with section numbers in italics to indicate where each type of problem is discussed. Where weaknesses show up, review appropriate sections in the text.

A **1.** Given the transition matrix P and initial-state matrix S_0 shown below, find S_1 and S_2 and explain what each represents:

$$P = \begin{matrix} & \begin{matrix} A & B \end{matrix} \\ \begin{matrix} A \\ B \end{matrix} & \begin{bmatrix} .6 & .4 \\ .2 & .8 \end{bmatrix} \end{matrix} \qquad S_0 = [.3 \quad .7]$$

In Problems 2–6, P is a transition matrix for a Markov chain. Identify any absorbing states and classify the chain as regular, absorbing, or neither.

2. $P = \begin{matrix} & \begin{matrix} A & B \end{matrix} \\ \begin{matrix} A \\ B \end{matrix} & \begin{bmatrix} 1 & 0 \\ .7 & .3 \end{bmatrix} \end{matrix}$

3. $P = \begin{matrix} & \begin{matrix} A & B \end{matrix} \\ \begin{matrix} A \\ B \end{matrix} & \begin{bmatrix} 0 & 1 \\ .7 & .3 \end{bmatrix} \end{matrix}$

4. $P = \begin{matrix} & \begin{matrix} A & B \end{matrix} \\ \begin{matrix} A \\ B \end{matrix} & \begin{bmatrix} 0 & 1 \\ 1 & 0 \end{bmatrix} \end{matrix}$

5. $P = \begin{matrix} & \begin{matrix} A & B & C \end{matrix} \\ \begin{matrix} A \\ B \\ C \end{matrix} & \begin{bmatrix} .8 & 0 & .2 \\ 0 & 1 & 0 \\ 0 & 0 & 1 \end{bmatrix} \end{matrix}$

6. $P = \begin{matrix} & \begin{matrix} A & B & C & D \end{matrix} \\ \begin{matrix} A \\ B \\ C \\ D \end{matrix} & \begin{bmatrix} 1 & 0 & 0 & 0 \\ 0 & 1 & 0 & 0 \\ 0 & 0 & .3 & .7 \\ 0 & 0 & .6 & .4 \end{bmatrix} \end{matrix}$

In Problems 7–10, write a transition matrix for the transition diagram indicated, identify any absorbing states, and classify each Markov chain as regular, absorbing, or neither.

7.

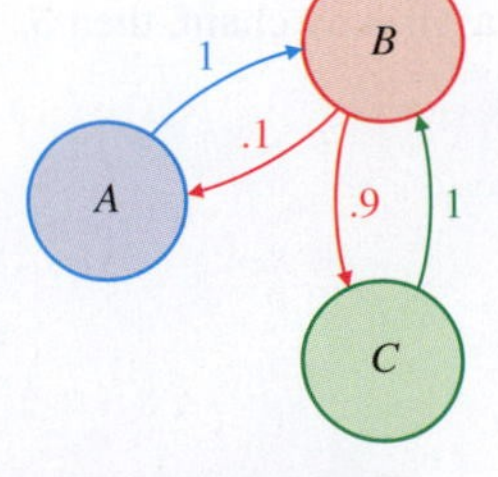

8.

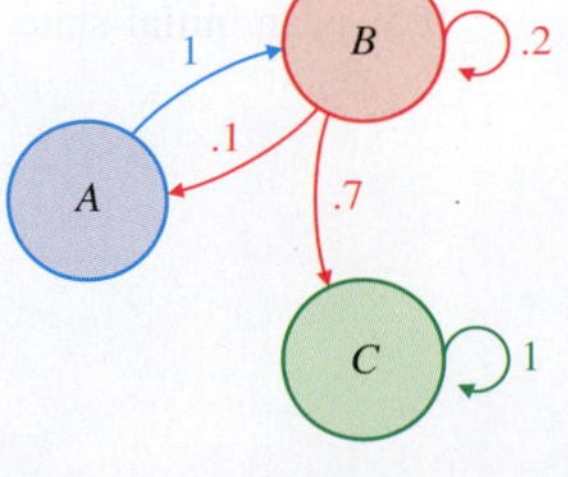

9.

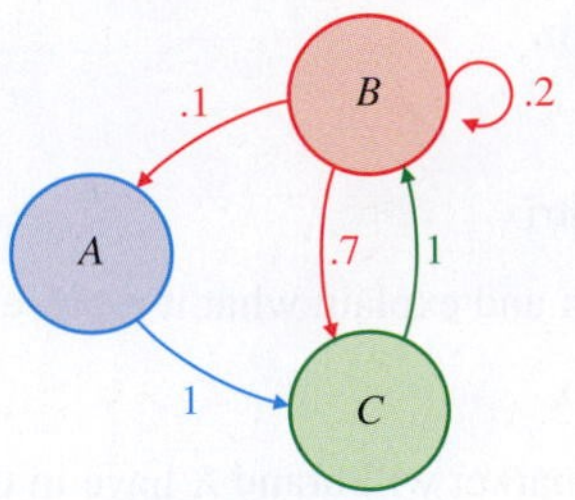

10.

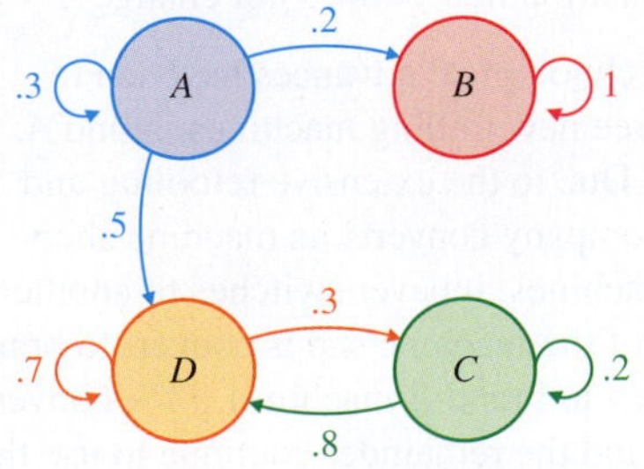

11. A Markov chain has three states, A, B, and C. The probability of going from state A to state B in one trial is .2, the probability of going from state A to state C in one trial is .5, the probability of going from state B to state A in one trial is .8, the probability of going from state B to state C in one trial is .2, the probability of going from state C to state A in one trial is .1, and the probability of going from state C to state B in one trial is .3. Draw a transition diagram and write a transition matrix for this chain.

12. Given the transition matrix

$$P = \begin{array}{c} \\ A \\ B \end{array}\begin{array}{c} \begin{array}{cc} A & B \end{array} \\ \begin{bmatrix} .4 & .6 \\ .9 & .1 \end{bmatrix} \end{array}$$

find the probability of

(A) Going from state A to state B in two trials

(B) Going from state B to state A in three trials

B *In Problems 13 and 14, solve the equation $SP = S$ to find the stationary matrix S and the limiting matrix $\overline{P}$.*

13. $P = \begin{array}{c} \\ A \\ B \end{array}\begin{array}{c} \begin{array}{cc} A & B \end{array} \\ \begin{bmatrix} .4 & .6 \\ .2 & .8 \end{bmatrix} \end{array}$

14. $P = \begin{array}{c} \\ A \\ B \\ C \end{array}\begin{array}{c} \begin{array}{ccc} A & B & C \end{array} \\ \begin{bmatrix} .4 & .6 & 0 \\ .5 & .3 & .2 \\ 0 & .8 & .2 \end{bmatrix} \end{array}$

In Problems 15 and 16, find the limiting matrix for the indicated standard form. Find the long-run probability of going from each nonabsorbing state to each absorbing state and the average number of trials needed to go from each nonabsorbing state to an absorbing state.

15. $P = \begin{array}{c} \\ A \\ B \\ C \end{array}\begin{array}{c} \begin{array}{ccc} A & B & C \end{array} \\ \begin{bmatrix} 1 & 0 & 0 \\ 0 & 1 & 0 \\ .3 & .1 & .6 \end{bmatrix} \end{array}$

16. $P = \begin{array}{c} \\ A \\ B \\ C \\ D \end{array}\begin{array}{c} \begin{array}{cccc} A & B & C & D \end{array} \\ \begin{bmatrix} 1 & 0 & 0 & 0 \\ 0 & 1 & 0 & 0 \\ .1 & .5 & .2 & .2 \\ .1 & .1 & .4 & .4 \end{bmatrix} \end{array}$

In Problems 17–20, use a graphing calculator to approximate the limiting matrix for the indicated transition matrix.

17. Matrix P from Problem 13

18. Matrix P from Problem 14

19. Matrix P from Problem 15

20. Matrix P from Problem 16

21. Find a standard form for the absorbing Markov chain with transition matrix

$$P = \begin{array}{c} \\ A \\ B \\ C \\ D \end{array}\begin{array}{c} \begin{array}{cccc} A & B & C & D \end{array} \\ \begin{bmatrix} .6 & .1 & .2 & .1 \\ 0 & 1 & 0 & 0 \\ .3 & .2 & .3 & .2 \\ 0 & 0 & 0 & 1 \end{bmatrix} \end{array}$$

In Problems 22 and 23, determine the long-run behavior of the successive state matrices for the indicated transition matrix and initial-state matrices.

22. $P = \begin{array}{c} \\ A \\ B \\ C \end{array}\begin{array}{c} \begin{array}{ccc} A & B & C \end{array} \\ \begin{bmatrix} 0 & 1 & 0 \\ 0 & 0 & 1 \\ .2 & .6 & .2 \end{bmatrix} \end{array}$

(A) $S_0 = [0 \quad 0 \quad 1]$

(B) $S_0 = [.5 \quad .3 \quad .2]$

23. $P = \begin{array}{c} \\ A \\ B \\ C \end{array}\begin{array}{c} \begin{array}{ccc} A & B & C \end{array} \\ \begin{bmatrix} 1 & 0 & 0 \\ 0 & 1 & 0 \\ .2 & .6 & .2 \end{bmatrix} \end{array}$

(A) $S_0 = [0 \quad 0 \quad 1]$

(B) $S_0 = [.5 \quad .3 \quad .2]$

24. Let P be a 2×2 transition matrix for a Markov chain. Can P be regular if two of its entries are 0? Explain.

25. Let P be a 3×3 transition matrix for a Markov chain. Can P be regular if three of its entries are 0? If four of its entries are 0? Explain.

C **26.** A red urn contains 2 red marbles, 1 blue marble, and 1 green marble. A blue urn contains 1 red marble, 3 blue marbles, and 1 green marble. A green urn contains 6 red marbles, 3 blue marbles, and 1 green marble. A marble is selected from an urn, the color is noted, and the marble is returned to the urn from which it was drawn. The next marble is drawn from the urn whose color is the same as the marble just drawn. This is a Markov process with three states: draw from the red urn, draw from the blue urn, or draw from the green urn.

(A) Draw a transition diagram for this process.

(B) Write the transition matrix P.

(C) Determine whether this chain is regular, absorbing, or neither.

(D) Find the limiting matrix $\overline{P}$, if it exists, and describe the long-run behavior of this process.

27. Repeat Problem 26 if the blue and green marbles are removed from the red urn.

28. Show that $S = [x \quad y \quad z \quad 0]$, where $0 \le x \le 1$, $0 \le y \le 1$, $0 \le z \le 1$, and $x + y + z = 1$, is a stationary matrix for the transition matrix

$$P = \begin{array}{c} \\ A \\ B \\ C \\ D \end{array}\begin{array}{c} \begin{array}{cccc} A & B & C & D \end{array} \\ \begin{bmatrix} 1 & 0 & 0 & 0 \\ 0 & 1 & 0 & 0 \\ 0 & 0 & 1 & 0 \\ .1 & .3 & .4 & .2 \end{bmatrix} \end{array}$$

Discuss the generalization of this result to any absorbing chain with three absorbing states and one nonabsorbing state.

29. Give an example of a transition matrix for a Markov chain that has no limiting matrix.

30. Give an example of a transition matrix for an absorbing Markov chain that has two different stationary matrices.

31. Give an example of a transition matrix for a regular Markov chain for which $[.3 \quad .1 \quad .6]$ is a stationary matrix.

32. Give an example of a transition matrix for an absorbing Markov chain for which $[.3 \quad .1 \quad .6]$ is a stationary matrix.

33. Explain why an absorbing Markov chain that has more than one state is not regular.

34. Explain why a regular Markov chain that has more than one state is not absorbing.

35. A Markov chain has transition matrix

$$P = \begin{bmatrix} .4 & .6 \\ .2 & .8 \end{bmatrix}$$

For $S = [.3 \quad .9]$, calculate SP. Is S a stationary matrix? Explain.

In Problems 36 and 37, use a graphing calculator to approximate the entries (to three decimal places) of the limiting matrix, if it exists, of the indicated transition matrix.

36. $$P = \begin{array}{c} \\ A \\ B \\ C \\ D \end{array}\begin{array}{c} \begin{array}{cccc} A & B & C & D \end{array} \\ \begin{bmatrix} .2 & .3 & .1 & .4 \\ 0 & 0 & 1 & 0 \\ 0 & .8 & 0 & .2 \\ 0 & 0 & 1 & 0 \end{bmatrix} \end{array}$$

37. $$P = \begin{array}{c} \\ A \\ B \\ C \\ D \end{array}\begin{array}{c} \begin{array}{cccc} A & B & C & D \end{array} \\ \begin{bmatrix} .1 & 0 & .3 & .6 \\ .2 & .4 & .1 & .3 \\ .3 & .5 & 0 & .2 \\ .9 & .1 & 0 & 0 \end{bmatrix} \end{array}$$

Applications

38. Product switching. A company's brand (X) has 20% of the market. A market research firm finds that if a person uses brand X, the probability is .7 that he or she will buy it next time. On the other hand, if a person does not use brand X (represented by X'), the probability is .5 that he or she will switch to brand X next time.

(A) Draw a transition diagram.

(B) Write a transition matrix.

(C) Write the initial-state matrix.

(D) Find the first-state matrix and explain what it represents.

(E) Find the stationary matrix.

(F) What percentage of the market will brand X have in the long run if the transition matrix does not change?

39. Marketing. Recent technological advances have led to the development of three new milling machines: brand A, brand B, and brand C. Due to the extensive retooling and startup costs, once a company converts its machine shop to one of these new machines, it never switches to another brand. Each year, 6% of the machine shops convert to brand A machines, 8% convert to brand B machines, 11% convert to brand C machines, and the remainder continue to use their old machines.

(A) In the long run, what is the market share of each brand?

(B) What is the average number of years that a company waits before converting to one of the new milling machines?

40. Internet. Table 1 gives the percentage of U.S. adults who at least occasionally used the Internet in the given year.

Table 1

Year	Percent
1995	14
2000	49
2005	68
2010	79

Source: Pew Internet & American Life Project Surveys

The following transition matrix P is proposed as a model for the data, where I represents the population of Internet users.

$$\begin{array}{c} \text{Current} \\ \text{year} \end{array} \begin{array}{c} \\ I \\ I' \end{array}\begin{array}{c} \text{Five years later} \\ \begin{array}{cc} I & I' \end{array} \\ \begin{bmatrix} .95 & .05 \\ .40 & .60 \end{bmatrix} \end{array} = P$$

(A) Let $S_0 = [.14 \quad .86]$ and find S_1, S_2, and S_3. Compute both matrices exactly and then round entries to two decimal places.

(B) Construct a new table comparing the results from part (A) with the data in Table 1.

(C) According to this transition matrix, what percentage of the adult U.S. population will use the Internet in the long run?

41. Employee training. In order to become a fellow of the Society of Actuaries, a person must pass a series of ten examinations. Passage of the first two preliminary exams is a prerequisite for employment as a trainee in the actuarial department of a large insurance company. Each year, 15% of the trainees complete the next three exams and become associates of the Society of Actuaries, 5% leave the company, never to return, and the remainder continue as trainees. Furthermore, each year, 17% of the associates complete the remaining five exams and become fellows of the Society of Actuaries, 3% leave the company, never to return, and the remainder continue as associates.

(A) In the long run, what percentage of the trainees will become fellows?

(B) In the long run, what percentage of the associates will leave the company?

(C) What is the average number of years that a trainee remains in this program before either becoming a fellow or being discharged?

42. Genetics. A given plant species has red, pink, or white flowers according to the genotypes RR, RW, and WW, respectively. If each of these genotypes is crossed with a red-flowering plant, the transition matrix is

Next generation

$$\text{This generation}\quad \begin{array}{c} \textit{Red} \\ \textit{Pink} \\ \textit{White} \end{array} \overset{\begin{array}{ccc} \textit{Red} & \textit{Pin} & \textit{White} \end{array}}{\begin{bmatrix} 1 & 0 & 0 \\ .5 & .5 & 0 \\ 0 & 1 & 0 \end{bmatrix}}$$

If each generation of the plant is crossed only with red plants to produce the next generation, show that eventually all the flowers produced by the plants will be red. (Find the limiting matrix.)

43. Smoking. Table 2 gives the percentage of U.S. adults who were smokers in the given year.

Table 2

Year	Percent
1985	30.1
1995	24.7
2005	20.9
2010	19.3

Source: American Lung Association

The following transition matrix P is proposed as a model for the data, where S represents the population of U.S. adult smokers.

Five years later

$$\text{Current year}\quad \begin{array}{c} S \\ S' \end{array} \overset{\begin{array}{cc} S & S' \end{array}}{\begin{bmatrix} .74 & .26 \\ .03 & .97 \end{bmatrix}} = P$$

(A) Let $S_0 = [.301 \quad .699]$, and find S_1, S_2, and S_3. Compute the matrices exactly and then round entries to three decimal places.

(B) Construct a new table comparing the results from part (A) with the data in Table 2.

(C) According to this transition matrix, what percentage of the adult U.S. population will be smokers in the long run?

10 Data Description and Probability Distributions

Introduction

In this chapter, we study various techniques for analyzing and displaying data. We use bar graphs, broken-line graphs, and pie graphs to present visual interpretations or comparisons of data. We use measures of central tendency (mean, median, and mode) and measures of dispersion (range, variance, and standard deviation) to describe and compare data sets.

Data collected from different sources—IQ scores, measurements of manufactured parts, healing times after surgery, for example—often exhibit surprising similarity. We might express such similarity by saying that all three data sets exhibit characteristics of a normal distribution. In Sections 10.4 and 10.5, we develop theoretical probability distributions (binomial distributions and normal distributions) that can be used as models of empirical data (see Problem 71 in Section 10.5).

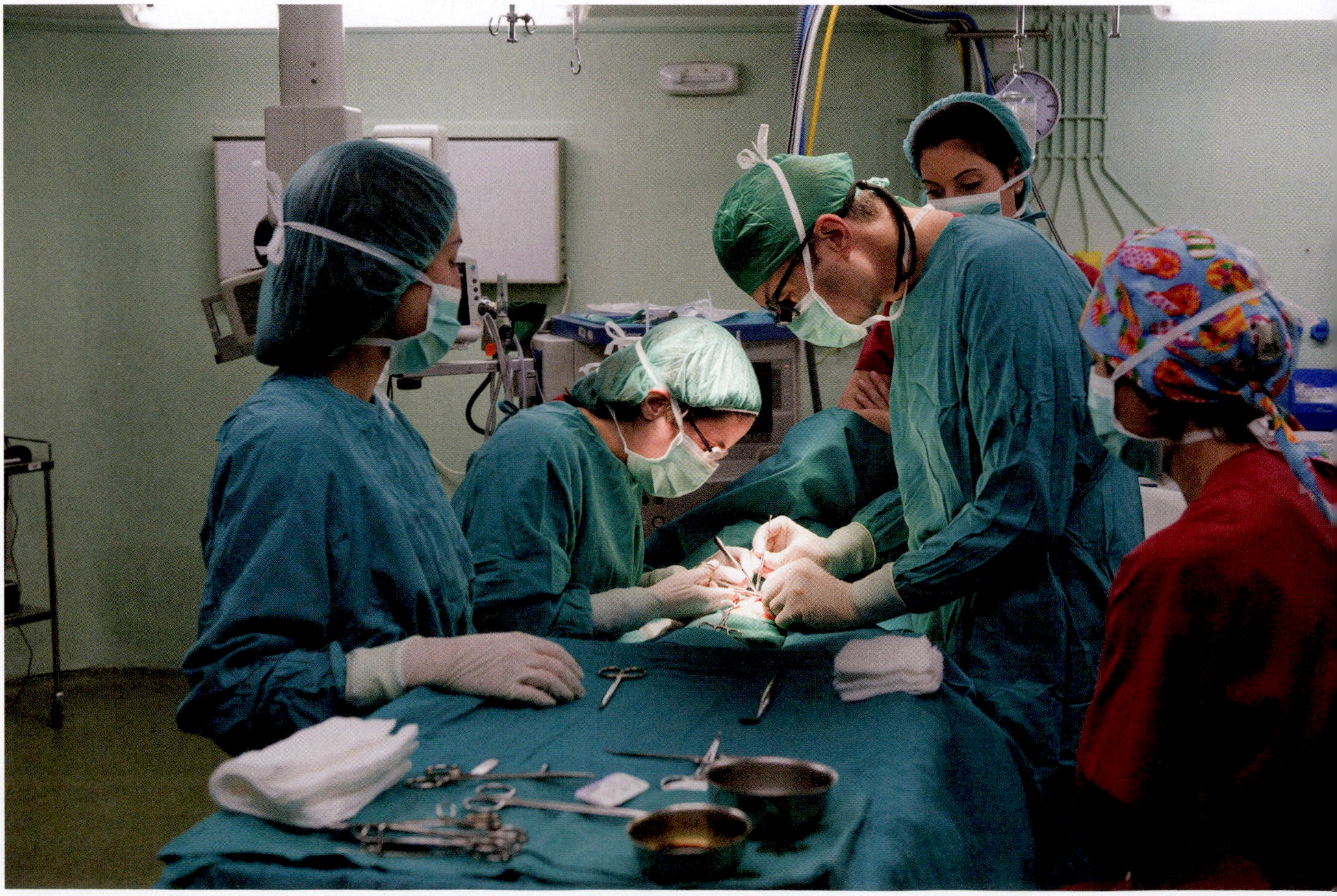

10.1 Graphing Data

- Bar Graphs, Broken-Line Graphs, and Pie Graphs
- Frequency Distributions
- Comments on Statistics
- Histograms
- Frequency Polygons and Cumulative Frequency Polygons

Websites, television, newspapers, magazines, books, and reports make substantial use of graphics to visually communicate complicated sets of data to the viewer. In this section, we look at bar graphs, broken-line graphs, and pie graphs and the techniques for producing them. It is important to remember that graphs are visual aids and should be prepared with care. The object is to provide the viewer with the maximum amount of information while minimizing the time and effort required to interpret the information in the graph.

Bar Graphs, Broken-Line Graphs, and Pie Graphs

Table 1 U.S. Public Debt

Year	Debt (billion $)
1975	533
1985	1,823
1995	4,974
2005	7,934
2015	18,151

Source: U.S. Treasury

Bar graphs are used widely because they are easy to construct and interpret. They are effective in presenting visual interpretations or comparisons of data. Consider Tables 1 and 2. Bar graphs are well suited to describe these two data sets. Vertical bars are usually used for time series—that is, data that changes over time, as in Table 1. The labels on the horizontal axis are then units of time (hours, days, years), as shown in Figure 1. Horizontal bars are generally used for data that changes by category, as in Table 2, because of the ease of labeling categories on the vertical axis of the bar graph (see Fig. 2). To increase clarity, space is left between the bars. Bar graphs for the data in Tables 1 and 2 are illustrated in Figures 1 and 2, respectively.

Table 2 Passenger Boardings at Busiest U.S. Airports, 2015

Airport	Boardings (million passengers)
Atlanta (ATL)	49.3
Chicago (ORD)	36.4
Los Angeles (LAX)	36.3
Dallas/Ft. Worth (DFW)	31.6
New York (JFK)	27.8

Source: Federal Aviation Administration

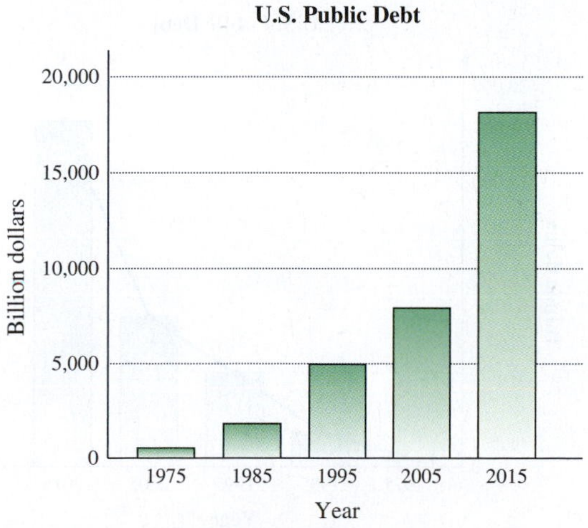

Figure 1 Vertical bar graph

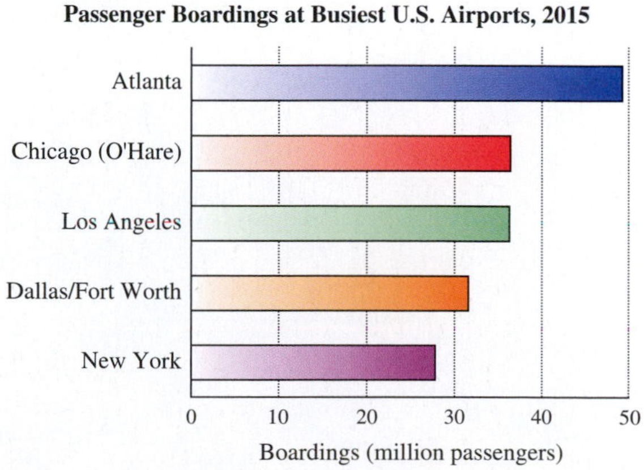

Figure 2 Horizontal bar graph

Two additional variations on bar graphs, the double bar graph and the divided bar graph, are illustrated in Figures 3 and 4, respectively.

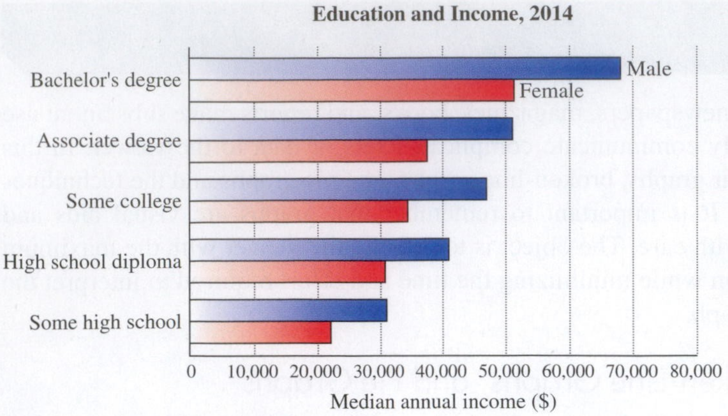

Figure 3 Double bar graph
Source: Institute of Education Sciences

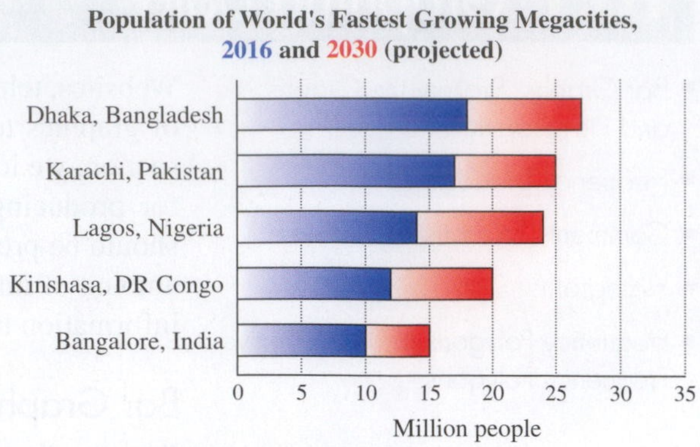

Figure 4 Divided bar graph
Source: UN Population Division

Explore and Discuss 1

(A) Using Figure 3, estimate the median annual income of a male with some college and a female who holds a bachelor's degree. Within which educational category is there the greatest difference between male and female income? The least difference?

(B) Using Figure 4, estimate the population of Kinshasa in the years 2016 and 2030. Which of the cities is projected to have the greatest increase in population from 2016 to 2030? The least increase? The greatest percentage increase? The least percentage increase?

A **broken-line graph** can be obtained from a vertical bar graph by joining the midpoints of the tops of consecutive bars with straight lines. For example, using Figure 1, we obtain the broken-line graph in Figure 5.

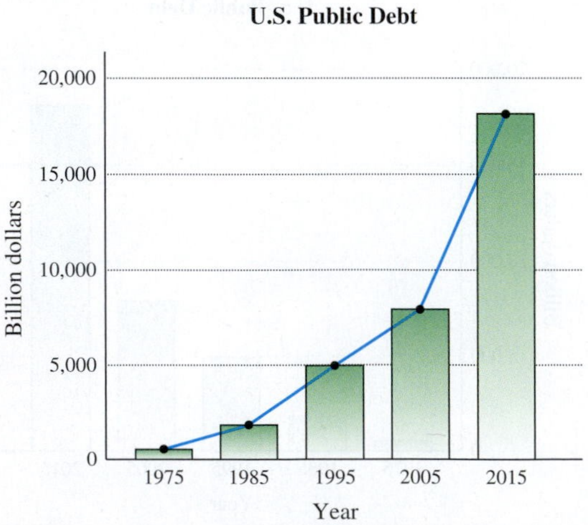

Figure 5 Broken-line graph

Broken-line graphs are particularly useful when we want to emphasize the change in one or more variables relative to time. Figures 6 and 7 illustrate two additional variations of broken-line graphs.

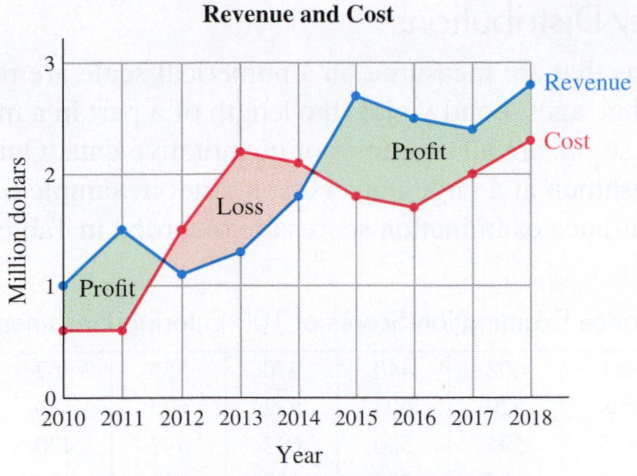

Figure 6 Broken-line graphs (variation)

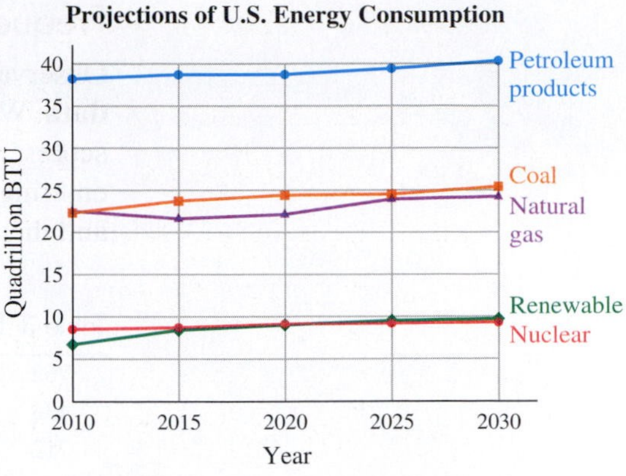

Figure 7 Broken-line graphs (variation)

Explore and Discuss 2

(A) Using Figure 6, estimate the revenue and costs in 2017. In which years is a profit realized? In which year is the greatest loss experienced?

(B) Using Figure 7, estimate the U.S. consumption of nuclear energy in 2030. Estimate the percentage of total consumption that will come from renewable energy in the year 2030.

A **pie graph** is generally used to show how a whole is divided among several categories. The amount in each category is expressed as a percentage, and then a circle is divided into segments (pieces of pie) proportional to the percentages of each category. The central angle of a segment is the percentage of 360° corresponding to the percentage of that category (see Fig. 8). In constructing pie graphs, we use relatively few categories, usually arrange the segments in ascending or descending order of size around the circle, and label each part.

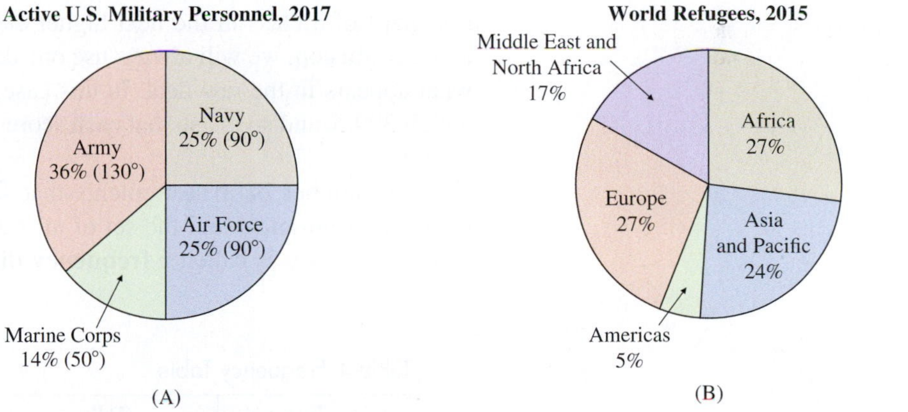

Figure 8 Pie graphs

Bar graphs, broken-line graphs, and pie graphs are easily constructed using a spreadsheet. After entering the data (see Fig. 9 for the data in Fig. 8A) and choosing the type of display (bar, broken-line, pie), the graph is drawn automatically. Various options for axes, gridlines, patterns, and text are available to improve the clarity of the visual display.

	A	B	C	D
1	Army	Navy	Air Force	Marine Corps
2	466,524	322,799	318,571	183,923

Figure 9

Frequency Distributions

Observations that are measured on a numerical scale are referred to as **quantitative data**. Weights, ages, bond yields, the length of a part in a manufacturing process, test scores, and so on, are all examples of quantitative data. Out of the total population of entering freshmen at a large university, a random sample of 100 students is selected, and their entrance examination scores are recorded in Table 3.

Table 3 Entrance Examination Scores of 100 Entering Freshmen

762	451	602	440	570	553	367	520	454	653
433	508	520	603	532	673	480	592	565	662
712	415	595	580	643	542	470	743	608	503
566	493	635	780	537	622	463	613	502	577
618	581	644	605	588	695	517	537	552	682
340	537	370	745	605	673	487	412	613	470
548	627	576	637	787	507	566	628	676	750
442	591	735	523	518	612	589	648	662	512
663	588	627	584	672	533	738	455	512	622
544	462	730	576	588	705	695	541	537	563

The mass of raw data in Table 3 certainly does not elicit much interest or exhibit much useful information. The data must be organized in some way so that it is comprehensible. This can be done by constructing a **frequency table**. We generally choose 5 to 20 **class intervals** of equal length to cover the data range—the more data, the greater the number of intervals. Then we tally the data relative to these intervals. The **data range** in Table 3 is $787 - 340 = 447$ (found by subtracting the smallest value in the data from the largest). If we choose 10 intervals, each of length 50, we will be able to cover all the scores. Table 4 shows the result of this tally.

At first, it might seem appropriate to start at 300 and form the class intervals: 300–350, 350–400, 400–450, and so on. But if we do this, where will we place 350 or 400? We could, of course, adopt a convention of placing scores falling on an upper boundary of a class in the next higher class (and some people do this); however, to avoid confusion, we will always use one decimal place more for class boundaries than what appears in the raw data. In this case, we chose the class intervals 299.5–349.5, 349.5–399.5, and so on, so that each score could be assigned to one and only one class interval.

The number of measurements that fall within a given class interval is called the **class frequency**, and the set of all such frequencies associated with their corresponding classes is called a **frequency distribution**. Table 4 represents a frequency

Table 4 Frequency Table

Class Interval	Tally	Frequency	Relative Frequency
299.5–349.5	|	1	.01
349.5–399.5	||	2	.02
399.5–449.5	𝍸	5	.05
449.5–499.5	𝍸 𝍸	10	.10
499.5–549.5	𝍸 𝍸 𝍸 𝍸 |	21	.21
549.5–599.5	𝍸 𝍸 𝍸 𝍸	20	.20
599.5–649.5	𝍸 𝍸 𝍸 ||||	19	.19
649.5–699.5	𝍸 𝍸 |	11	.11
699.5–749.5	𝍸 ||	7	.07
749.5–799.5	||||	4	.04
		100	1.00

distribution of the set of raw scores in Table 3. If we divide each frequency by the total number of items in the original data set (in our case 100), we obtain the **relative frequency** of the data falling in each class interval—that is, the percentage of the whole that falls in each class interval (see the last column in Table 4).

The relative frequencies also can be interpreted as probabilities associated with the experiment, "A score is drawn at random out of the 100 in the sample." An appropriate sample space for this experiment would be the set of simple outcomes

$$e_1 = \text{a score falls in the first class interval}$$
$$e_2 = \text{a score falls in the second class interval}$$
$$\vdots$$
$$e_{10} = \text{a score falls in the tenth class interval}$$

The set of relative frequencies is then referred to as the **probability distribution** for the sample space.

EXAMPLE 1 **Determining Probabilities from a Frequency Table** Referring to Table 4 and the probability distribution just described, determine the probability that

(A) A randomly drawn score is between 499.5 and 549.5.

(B) A randomly drawn score is between 449.5 and 649.5.

SOLUTION

(A) Since the relative frequency associated with the class interval 499.5–549.5 is .21, the probability that a randomly drawn score (from the sample of 100) will fall in this interval is .21.

(B) Since a score falling in the interval 449.5–649.5 is a compound event, we simply add the probabilities for the simple events whose union is this compound event. We add the probabilities corresponding to each class interval from 449.5 to 649.5 to obtain

$$.10 + .21 + .20 + .19 = .70$$

Matched Problem 1 Repeat Example 1 for the following intervals:

(A) 649.5–699.5 (B) 299.5–499.5

Comments on Statistics

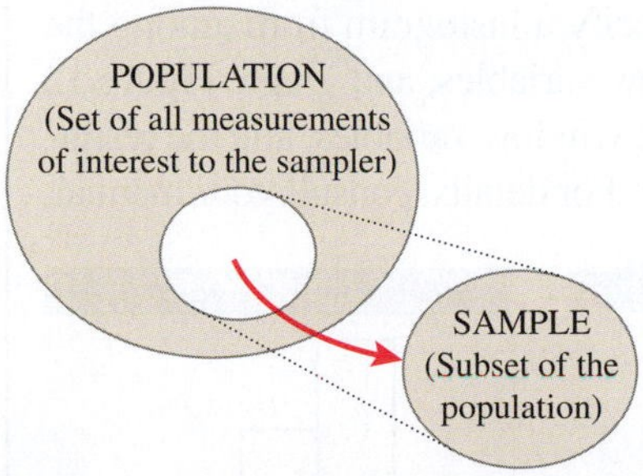

Figure 10 Inferential statistics: Based on information obtained from a sample, the goal of statistics is to make inferences about the population as a whole.

Now, of course, what we are really interested in is whether the probability distribution for the sample of 100 entrance examination scores has anything to do with the total population of entering freshmen at the university. This is a problem for the branch of mathematics called *statistics*, which deals with the process of making inferences about a total population based on random samples drawn from the population. Figure 10 schematically illustrates the *inferential statistical* process. We will not go too far into inferential statistics in this book since the subject is studied in detail in statistics courses but our work in probability provides a good foundation for this study.

Intuitively, in the entrance examination example, we would expect that the larger the sample size, the more closely the probability distribution for the sample will approximate that for the total population.

Histograms

A **histogram** is a special kind of vertical bar graph. In fact, if you rotate Table 4 counterclockwise 90°, the tally marks in the table take on the appearance of a bar graph. Histograms have no space between the bars; class boundaries are located on the horizontal axis, and frequencies are associated with the vertical axis.

Figure 11 is a histogram for the frequency distribution in Table 4. Note that we have included both frequencies and relative frequencies on the vertical scale. You can include either one or the other, or both, depending on what needs to be emphasized. The histogram is the most common graphical representation of frequency distributions.

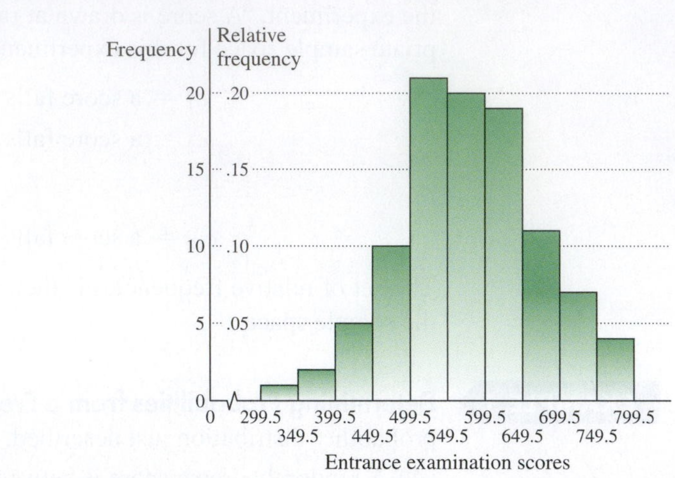

Figure 11 Histogram

EXAMPLE 2 **Constructing Histograms with a Graphing Calculator** Twenty vehicles were chosen at random upon arrival at a vehicle emissions inspection station, and the time elapsed (in minutes) from arrival to completion of the emissions test was recorded for each of the vehicles:

5	12	11	4	7	20	14	8	12	11
18	15	14	9	10	13	12	20	26	17

(A) Use a graphing calculator to draw a histogram of the data, choosing the five class intervals 2.5–7.5, 7.5–12.5, and so on.

(B) What is the probability that for a vehicle chosen at random from the sample, the time required at the inspection station will be less than 12.5 minutes? That it will exceed 22.5 minutes?

SOLUTION

(A) Various kinds of statistical plots can be drawn by most graphing calculators. To draw a histogram we enter the data as a list, specify a histogram from among the various statistical plotting options, set the window variables, and graph. Figure 12 shows the data entered as a list, the settings of the window variables, and the resulting histogram for a particular graphing calculator. For details, consult your manual.

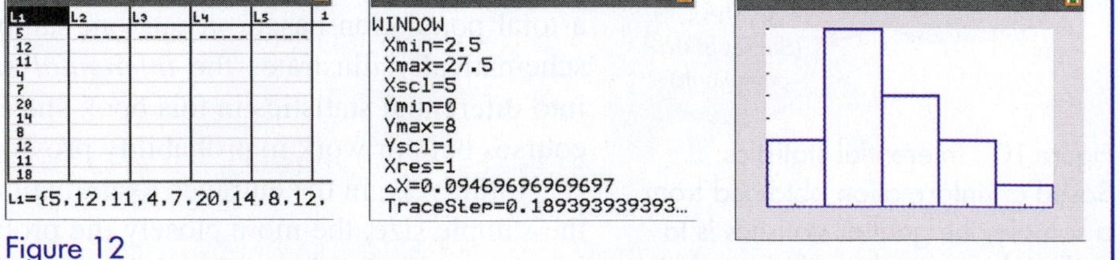

Figure 12

(B) From the histogram in Figure 12, we see that the first class has frequency 3 and the second has frequency 8. The upper boundary of the second class is 12.5, and the total number of data items is 20. Therefore, the probability that the time required will be less than 12.5 minutes is

$$\frac{3 + 8}{20} = \frac{11}{20} = .55$$

Similarly, since the frequency of the last class is 1, the probability that the time required will exceed 22.5 minutes is

$$\frac{1}{20} = .05$$

Matched Problem 2 The weights (in pounds) were recorded for 20 kindergarten children chosen at random:

51	46	37	39	48	42	41	44	57	36
47	44	41	50	45	46	34	39	42	44

(A) Use a graphing calculator to draw a histogram of the data, choosing the five class intervals 32.5–37.5, 37.5–42.5, and so on.

(B) What is the probability that a kindergarten child chosen at random from the sample weighs less than 42.5 pounds? More than 42.5 pounds?

Frequency Polygons and Cumulative Frequency Polygons

A **frequency polygon** is a broken-line graph where successive midpoints of the tops of the bars in a histogram are joined by straight lines. To draw a frequency polygon for a frequency distribution, you do not need to draw a histogram first; you can just locate the midpoints and join them with straight lines. Figure 13 is a frequency polygon for the frequency distribution in Table 4. If the amount of data becomes very large and we substantially increase the number of classes, then the frequency polygon will take on the appearance of a smooth curve called a **frequency curve**.

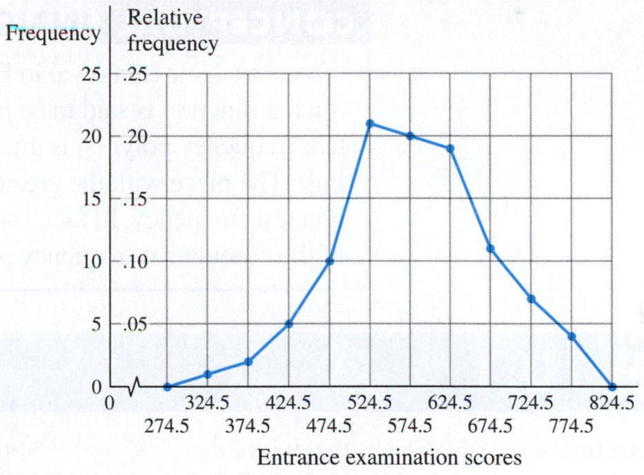

Figure 13 Frequency polygon

If we are interested in how many or what percentage of a total sample lies above or below a particular measurement, a **cumulative frequency table** and **polygon** are useful. Using the frequency distribution in Table 4, we accumulate the frequencies by starting with the first class and adding frequencies as we move down the column. The results are shown in Table 5.

Table 5 Cumulative Frequency Table

Class Interval	Frequency	Cumulative Frequency	Relative Cumulative Frequency
299.5–349.5	1	1	.01
349.5–399.5	2	3	.03
399.5–449.5	5	8	.08
449.5–499.5	10	18	.18
499.5–549.5	21	39	.39
549.5–599.5	20	59	.59
599.5–649.5	19	78	.78
649.5–699.5	11	89	.89
699.5–749.5	7	96	.96
749.5–799.5	4	100	1.00

To form a cumulative frequency polygon, or **ogive** as it is also called, the cumulative frequency is plotted over the upper boundary of the corresponding class. Figure 14 is the cumulative frequency polygon for the cumulative frequency table in Table 5. Notice that we can easily see that 78% of the students scored below 649.5, while only 18% scored below 499.5. We can conclude that the probability of a randomly selected score from the sample of 100 lying below 649.5 is .78 and above 649.5 is $1.00 - .78 = .22$.

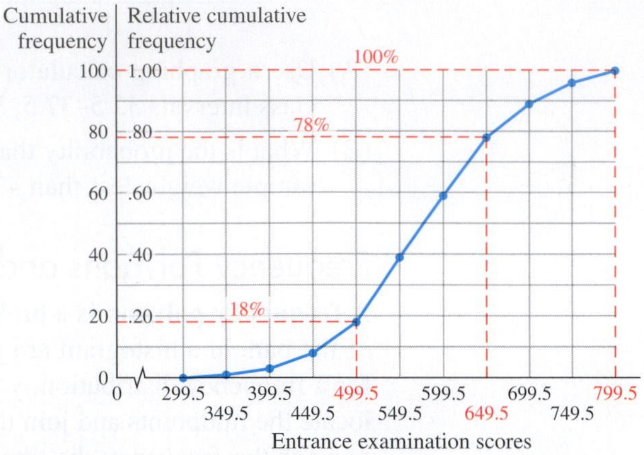

Figure 14 Cumulative frequency polygon (ogive)

CONCEPTUAL INSIGHT

Above each class interval in Figure 14, the cumulative frequency polygon is linear. Such a function is said to be *piecewise linear*. The slope of each piece of a cumulative frequency polygon is greater than or equal to zero (that is, the graph is never falling). The piece with the greatest slope corresponds to the class interval that has the greatest frequency. In fact, for any class interval, the frequency is equal to the slope of the cumulative frequency polygon multiplied by the width of the class interval.

Exercises 10.1

1. (A) Construct a frequency table and histogram for the following data set using a class interval width of 2, starting at 0.5.

6	7	2	7	9
6	4	7	6	6

(B) Construct a frequency table and histogram for the following data set using a class interval width of 2, starting at 0.5.

5	6	8	1	3
5	10	7	6	8

(C) How are the two histograms of parts (A) and (B) similar? How are the two data sets different?

2. (A) Construct a frequency table and histogram for the data set of part (A) of Problem 1 using a class interval width of 1, starting at 0.5.

(B) Construct a frequency table and histogram for the data set of part (B) of Problem 1 using a class interval width of 1, starting at 0.5.

(C) How are the histograms of parts (A) and (B) different?

3. The graphing calculator command shown in Figure A generated a set of 400 random integers from 2 to 24, stored as list L_1. The statistical plot in Figure B is a histogram of L_1, using a class interval width of 1, starting at 1.5.

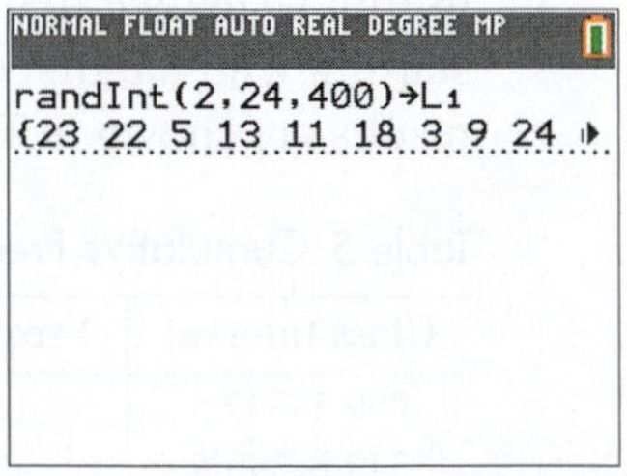

(A)

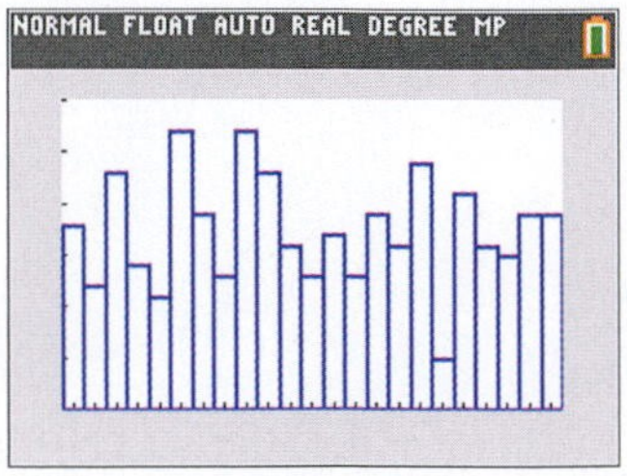

(B)

(A) Explain how the window variables can be changed to display a histogram of the same data set using a class interval width of 2, starting at 1.5. A width of 4, starting at 1.5.

(B) Describe the effect of increasing the class interval width on the shape of the histogram.

4. An experiment consists of rolling a pair of dodecahedral (12-sided) dice and recording their sum (the sides of each die are numbered from 1 to 12). The command shown in Figure A simulated 500 rolls of the dodecahedral dice. The statistical plot in Figure B is a histogram of the 500 sums using a class interval width of 1, starting at 1.5.
   (A) Explain how the window variables can be changed to display a histogram of the same data set using a class interval width of 2, starting at 1.5. A width of 3, starting at −0.5.
   (B) Describe the effect of increasing the class interval width on the shape of the histogram.

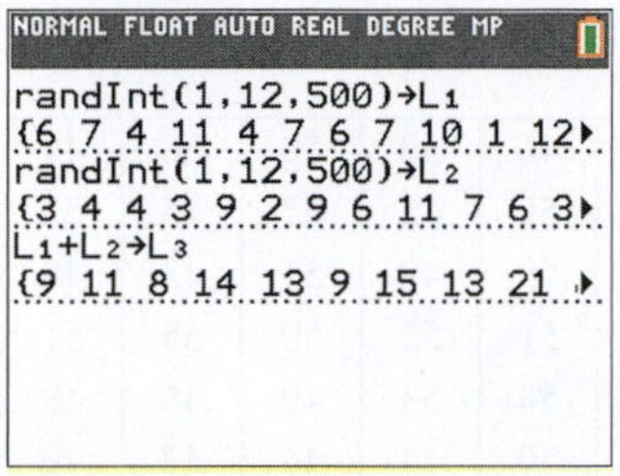

(A)

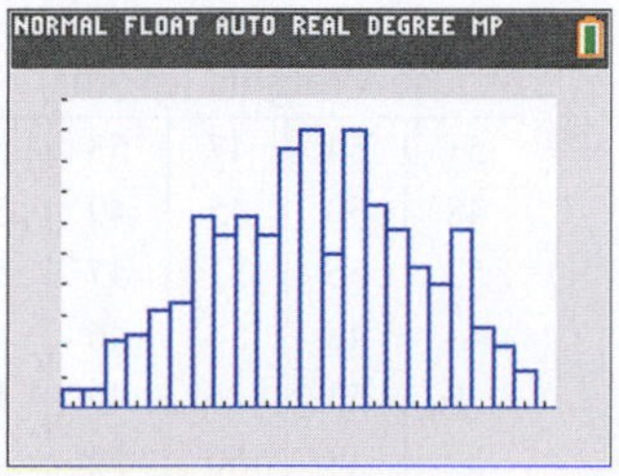

(B)

Applications

5. **Gross domestic product.** Graph the data in the following table using a bar graph. (*Source:* U.S. Bureau of Economic Analysis)

Gross Domestic Product (GDP)

Year	GDP (trillion $)
1975	5.49
1985	7.71
1995	10.28
2005	14.37
2015	16.49

6. **Corporation revenues.** Graph the data in the following table using a bar graph. (*Source: Fortune*)

Corporation Revenues, 2015

Corporation	Revenue (billion $)
Walmart	482
ExxonMobil	246
Apple	234
Berkshire Hathaway	211
McKesson	192
United Health	157

7. **Gold production.** Use the double bar graph on world gold production to determine the country that showed the greatest increase in gold production from 2010 to 2015. Which country showed the greatest percentage increase? The greatest percentage decrease? (*Source:* U.S. Geological Survey)

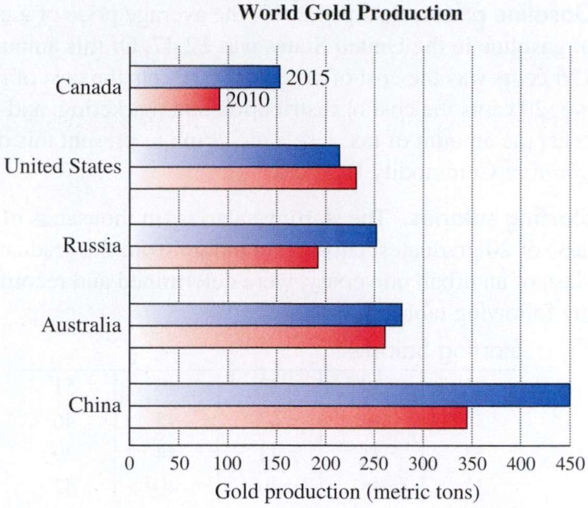

8. **Gasoline prices.** Graph the data in the following table using a divided bar graph. (*Source:* American Petroleum Institute)

State Gasoline Prices, April 2017

State	Price Before Tax ($ per gallon)	Tax
California	2.41	.57
Hawaii	2.42	.63
Michigan	1.90	.59
New York	1.83	.62
Texas	1.80	.38

9. **Postal service.** Graph the data in the following table using a broken line graph. (*Source:* U.S. Postal Service)

U.S. Postal Service Employees

Year	Number of Employees
1990	760,668
1995	753,384
2000	787,538
2005	704,716
2010	583,908
2015	491,863

10. **Postal service.** Refer to Problem 9. If the data were presented in a bar graph, would horizontal bars or vertical bars be used? Could the data be presented in a pie graph? Explain.

11. **Federal income.** Graph the data in the following table using a pie graph. (*Source:* U.S. Treasury)

Federal Income by Source, 2015

Source	Income (billion $)
Personal income tax	1,365
Social insurance taxes	943
Corporate income tax	292
Borrowing to cover deficit	390
Other	260

12. Gasoline prices. In April 2017, the average price of a gallon of gasoline in the United States was \$2.47. Of this amount, 126 cents was the cost of crude oil, 57 cents the cost of refining, 20 cents the cost of distribution and marketing, and 44 cents the amount of tax. Use a pie graph to present this data. (*Source:* Commodity HQ.com)

13. Starting salaries. The starting salaries (in thousands of dollars) of 20 graduates, chosen at random from the graduating class of an urban university, were determined and recorded in the following table:

Starting Salaries

44	39	37	49	51
38	42	47	45	46
33	41	43	44	39
37	45	39	40	42

(A) Construct a frequency and relative frequency table using a class interval width of 4 starting at 30.5.

(B) Construct a histogram.

(C) What is the probability that a graduate chosen from the sample will have a starting salary above \$42,500? Below \$38,500?

(D) Construct a histogram using a graphing calculator.

14. Commute times. Thirty-two people were chosen at random from among the employees of a large corporation. Their commute times (in hours) from home to work were recorded in the following table:

Commute Times

0.5	0.9	0.2	0.4	0.7	1.2	1.1	0.7
0.6	0.4	0.8	1.1	0.9	0.3	0.4	1.0
0.9	1.0	0.7	0.3	0.6	1.1	0.7	1.1
0.4	1.3	0.7	0.6	1.0	0.8	0.4	0.9

(A) Construct a frequency and relative frequency table using a class interval width of 0.2, starting at 0.15.

(B) Construct a histogram.

(C) What is the probability that a person chosen at random from the sample will have a commuting time of at least an hour? Of at most half an hour?

(D) Construct a histogram using a graphing calculator.

15. Common stocks. The following table shows price–earnings ratios of 100 common stocks chosen at random from the New York Stock Exchange.

Price–Earnings (PE) Ratios

7	11	6	6	10	6	31	28	13	19
6	18	9	7	5	5	9	8	10	6
10	3	4	6	7	9	9	19	7	9
17	33	17	12	7	5	7	10	7	9
18	17	4	6	11	13	7	6	10	7
7	9	8	15	16	11	10	7	5	14
12	10	6	7	7	13	10	5	6	4
10	6	7	11	19	17	6	9	6	5
6	13	4	7	6	12	9	14	9	7
18	5	12	8	8	8	13	9	13	15

(A) Construct a frequency and relative frequency table using a class interval of 5, starting at -0.5.

(B) Construct a histogram.

(C) Construct a frequency polygon.

(D) Construct a cumulative frequency and relative cumulative frequency table. What is the probability that a price–earnings ratio drawn at random from the sample will fall between 4.5 and 14.5?

(E) Construct a cumulative frequency polygon.

16. Mouse weights. One hundred healthy mice were weighed at the beginning of an experiment with the following results:

Mouse Weights (grams)

51	54	47	53	59	46	50	50	56	46
48	50	45	49	52	55	42	57	45	51
53	55	51	47	53	53	49	51	43	48
44	48	54	46	49	51	52	50	55	51
50	53	45	49	57	54	53	49	46	48
52	48	50	52	47	50	44	46	47	49
49	51	57	49	51	42	49	53	44	52
53	55	48	52	44	46	54	54	57	55
48	50	50	55	52	48	47	52	55	50
59	52	47	46	56	54	51	56	54	55

(A) Construct a frequency and relative frequency table using a class interval of 2, starting at 41.5.

(B) Construct a histogram.

(C) Construct a frequency polygon.

(D) Construct a cumulative frequency and relative cumulative frequency table. What is the probability of a mouse weight drawn at random from the sample lying between 45.5 and 53.5?

(E) Construct a cumulative frequency polygon.

17. Population growth. Graph the data in the following table using a broken-line graph.

Annual World Population Growth

Year	Growth (millions)
1900	11
1925	21
1950	46
1975	71
2000	80
2025	71

18. AIDS epidemic. One way to gauge the toll of the AIDS epidemic in Sub-Saharan Africa is to compare life expectancies with the figures that would have been projected in the absence of AIDS. Use the broken-line graphs shown to estimate the life expectancy of a child born in the year 2012. What would the life expectancy of the same child be in the absence of AIDS? For which years of birth is the life expectancy less than 50 years? If there were no AIDS epidemic, for which years of birth would the life expectancy be less than 50 years?

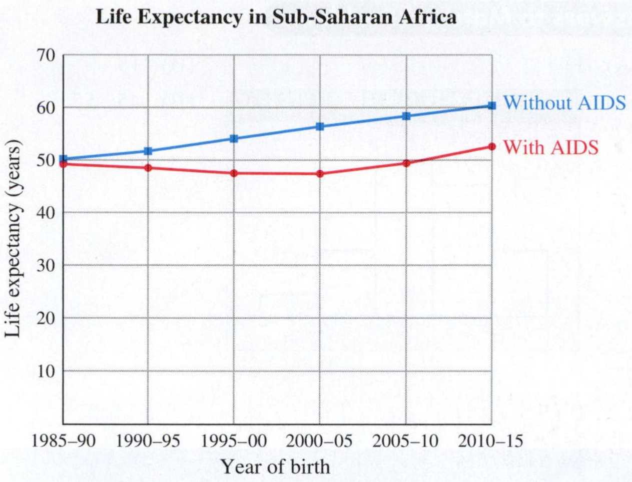

Figure for 18

19. **Nutrition.** Graph the data in the following table using a double bar graph.

Recommended Daily Allowances

Grams of	Males Age 15–18	Females Age 15–18
Carbohydrate	375	275
Protein	60	44
Fat	100	73

20. **Greenhouse gases.** The U.S. Environmental Protection Agency estimated that of all emissions of greenhouse gases by the United States in 2014, carbon dioxide accounted for 81%, methane for 11%, nitrous oxide for 6%, and fluorinated gases for 3%. Use a pie graph to present this data. Find the central angles of the graph.

21. **Nutrition.** Graph the nutritional information in the following table using a double bar graph.

Fast-Food Burgers: Nutritional Information

	Calories	Calories from Fat
2-oz burger, plain	268	108
2 addl. oz of beef	154	90
1 slice cheese	105	81
3 slices bacon	109	81
1 tbsp. mayonnaise	100	99

22. **Nutrition.** Refer to Problem 21. Suppose that you are trying to limit the fat in your diet to at most 30% of your calories, and your calories down to 2000 per day. Should you order the quarter-pound bacon cheeseburger with mayo for lunch? How would such a lunch affect your choice of breakfast and dinner? Discuss.

23. **Education.** For statistical studies, U.S. states are often grouped by region: Northeast, Midwest, South, and West. The 1965 total public school enrollment (in millions) in each region was 8.8, 11.8, 13.8, and 7.6, respectively. The projected 2020 enrollment was 7.8, 10.8, 20.4, and 13.6, respectively. Use two pie graphs to present this data and discuss any trends suggested by your graphs, (*Source:* U.S. Department of Education)

24. **Study abroad.** Would a pie graph be more effective or less effective than the bar graph shown in presenting information on the most popular destinations of U.S. college students who study abroad? Justify your answer. (*Source:* Institute of International Education)

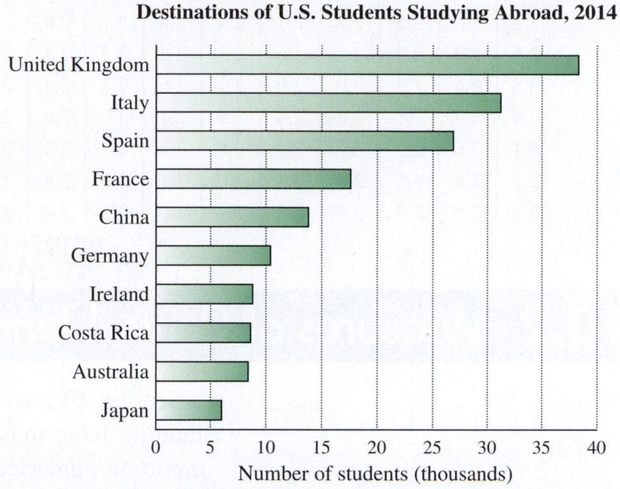

25. **Median age.** Use the broken-line graph shown to estimate the median age in 1900 and 2000. In which decades did the median age increase? In which did it decrease? Discuss the factors that may have contributed to the increases and decreases.

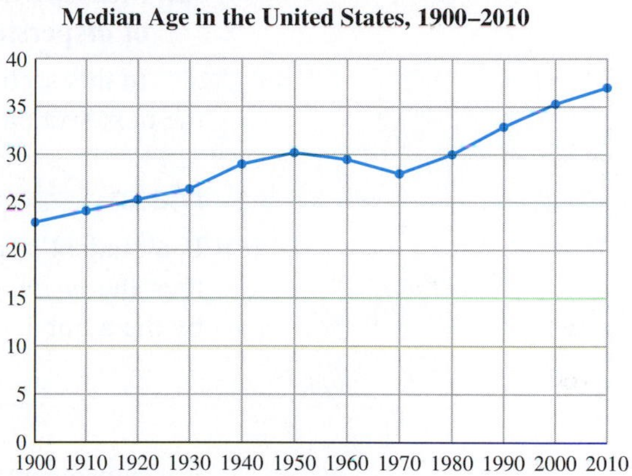

26. **State prisoners.** In 1980 in the United States, 6% of the inmates of state prisons were incarcerated for drug offenses, 30% for property crimes, 4% for public order offenses, and 60% for violent crimes; in 2014 the percentages were 16%, 19%, 12%, and 53%, respectively. Present the data using two pie graphs. Discuss factors that may account for the shift in percentages between 1980 and 2014. (*Source:* U.S. Department of Justice)

27. **Grade-point averages.** One hundred seniors were chosen at random from a graduating class at a university and their grade-point averages recorded:

(A) Construct a frequency and relative frequency table using a class interval of 0.2 starting at 1.95.

(B) Construct a histogram.

(C) Construct a frequency polygon.

(D) Construct a cumulative frequency and relative cumulative frequency table. What is the probability of a GPA drawn at random from the sample being over 2.95?

(E) Construct a cumulative frequency polygon.

Grade-Point Averages (GPA)

2.1	2.0	2.7	2.6	2.1	3.5	3.1	2.1	2.2	2.9
2.3	2.5	3.1	2.2	2.2	2.0	2.3	2.5	2.1	2.4
2.7	2.9	2.1	2.2	2.5	2.3	2.1	2.1	3.3	2.1
2.2	2.2	2.5	2.3	2.7	2.4	2.8	3.1	2.0	2.3
2.6	3.2	2.2	2.5	3.6	2.3	2.4	3.7	2.5	2.4
3.5	2.4	2.3	3.9	2.9	2.7	2.6	2.1	2.4	2.0
2.4	3.3	3.1	2.8	2.3	2.5	2.1	3.0	2.6	2.3
2.1	2.6	2.2	3.2	2.7	2.8	3.4	2.7	3.6	2.1
2.7	2.8	3.5	2.4	2.3	2.0	2.1	3.1	2.8	2.1
3.8	2.5	2.7	2.1	2.2	2.4	2.9	3.3	2.0	2.6

Answers to Matched Problems

1. (A) .11 (B) .18

2. (A)

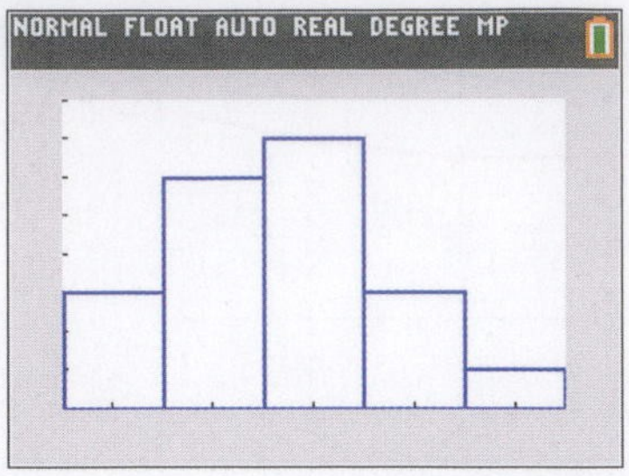

(B) .45; .55

10.2 Measures of Central Tendency

- Mean
- Median
- Mode

In Section 10.1 we saw that graphical techniques contributed substantially to understanding large masses of raw data. In this and the next section, we discuss several important numerical measures that are used to describe sets of data. These numerical descriptions are generally of two types:

1. Measures that indicate the approximate center of a distribution, called **measures of central tendency**; and
2. Measures that indicate the amount of scatter about a central point, called **measures of dispersion**.

In this section, we look at three widely used measures of central tendency, and in the next section we consider measures of dispersion.

Mean

If a student's scores on four exams are $x_1 = 82$, $x_2 = 75$, $x_3 = 86$, and $x_4 = 87$, then the *mean*, or *arithmetic average*, of the scores is the sum of the scores divided by the number of scores:

$$\frac{x_1 + x_2 + x_3 + x_4}{4} = \frac{82 + 75 + 86 + 87}{4} = \frac{330}{4} = 82.5$$

It is convenient, especially when there are many scores, to use a compact notation for the sum of the scores. We use the summation symbol Σ (the uppercase Greek letter sigma) to indicate that the scores should be added (see Appendix B.1). So

$$\sum_{i=1}^{4} x_i \quad \text{denotes} \quad x_1 + x_2 + x_3 + x_4$$

and the mean of the four exam scores is

$$\frac{\sum_{i=1}^{4} x_i}{4} = 82.5$$

The mean is a single number that, in a sense, represents the entire data set. It involves all the measurements in the set and it is easily computed. Because of these and other desirable properties, the mean is the most widely used measure of central tendency.

In statistics, we are concerned with both a sample mean and the mean of the corresponding population (the sample mean is often used as an estimator for the population mean), so it is important to use different symbols to represent these two means. It is customary to use a letter with an overbar, such as $\bar{x}$, to represent a sample mean and the Greek letter μ ("mu") to represent a population mean.

$$\bar{x} = \textbf{sample mean} \qquad \mu = \textbf{population mean}$$

DEFINITION Mean: Ungrouped Data

If $x_1, x_2, \ldots, x_n$ is a set of n measurements, then the **mean** of the set of measurements is given by

$$[\text{mean}] = \frac{\sum_{i=1}^{n} x_i}{n} = \frac{x_1 + x_2 + \cdots + x_n}{n} \quad (1)$$

where

$$\bar{x} = [\text{mean}] \text{ if data set is a sample}$$

$$\mu = [\text{mean}] \text{ if data set is the population}$$

EXAMPLE 1 **Finding the Mean** Find the mean for the sample measurements 3, 5, 1, 8, 6, 5, 4, and 6.

SOLUTION Solve using formula (1):

$$\bar{x} = \frac{\sum_{i=1}^{n} x_i}{n} = \frac{3 + 5 + 1 + 8 + 6 + 5 + 4 + 6}{8} = \frac{38}{8} = 4.75$$

Matched Problem 1 Find the mean for the sample measurements 3.2, 4.5, 2.8, 5.0, and 3.6.

Fifteen customers filled out a satisfaction survey, rating the service they received by checking a number from 1 to 5. Here are the ratings:

5, 3, 5, 4, 4, 3, 4, 5, 5, 5, 4, 2, 5, 5, 4

To find the sum, we could simply add the 15 numbers. But there is another option. Note that the score 5 appears seven times, 4 appears five times, 3 appears twice, 2 appears once, and 1 appears zero times. The sum of all 15 scores can therefore be obtained by multiplying each possible score x_i by its frequency f_i:

$$\sum_{i=1}^{5} x_i f_i = 5(7) + 4(5) + 3(2) + 2(1) + 1(0) = 63$$

The mean of the 15 scores is

$$\frac{\sum_{i=1}^{5} x_i f_i}{15} = \frac{63}{15} = 4.2$$

For grouped data, such as in Table 4 on page 504, we may not have the option of simply adding the scores; the original data set may be unavailable. Instead, we form a sum similar to that of the customer satisfaction survey, where each term is an x_i multiplied by its frequency f_i, as explained in the definition of the mean for grouped data.

DEFINITION Mean: Grouped Data

A data set of n measurements is grouped into k classes in a frequency table. If x_i is the midpoint of the ith class interval and f_i is the ith class frequency, then the **mean for the grouped data** is given by

$$[\text{mean}] = \frac{\sum_{i=1}^{k} x_i f_i}{n} = \frac{x_1 f_1 + x_2 f_2 + \cdots + x_k f_k}{n} \tag{2}$$

where

$$n = \sum_{i=1}^{k} f_i = \text{total number of measurements}$$

$$\bar{x} = [\text{mean}] \text{ if data set is a sample}$$

$$\mu = [\text{mean}] \text{ if data set is the population}$$

CAUTION Note that n is the total number of measurements in the entire data set—not the number of classes!

The mean computed by formula (2) is a **weighted average** of the midpoints of the class intervals. In general, this will be close to, but not exactly the same as, the mean computed by formula (1) for ungrouped data.

EXAMPLE 2 **Finding the Mean for Grouped Data** Find the mean for the sample data summarized in Table 4 on page 504.

SOLUTION In Table 1, we repeat part of Table 4, adding columns for the class midpoints x_i and the products $x_i f_i$.

Table 1 Entrance Examination Scores

Class Interval	Midpoint x_i	Frequency f_i	Product $x_i f_i$
299.5–349.5	324.5	1	324.5
349.5–399.5	374.5	2	749.0
399.5–449.5	424.5	5	2,122.5
449.5–499.5	474.5	10	4,745.0
499.5–549.5	524.5	21	11,014.5
549.5–599.5	574.5	20	11,490.0
599.5–649.5	624.5	19	11,865.5
649.5–699.5	674.5	11	7,419.5
699.5–749.5	724.5	7	5,071.5
749.5–799.5	774.5	4	3,098.0
		$n = \sum_{i=1}^{10} f_i = 100$	$\sum_{i=1}^{10} x_i f_i = 57{,}900.0$

The average entrance examination score for the sample of 100 entering freshmen is

$$\bar{x} = \frac{\sum_{i=1}^{k} x_i f_i}{n} = \frac{57{,}900}{100} = 579$$

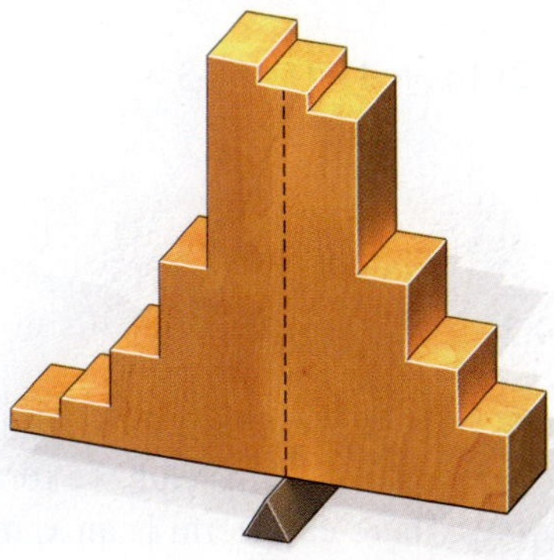

Figure 1 The balance point on the histogram is $\bar{x} = 579$.

If the histogram for the data in Table 1 (Fig. 11, Section 10.1) was drawn on a piece of wood of uniform thickness and the wood cut around the outside of the figure, then the resulting object would balance exactly at the mean $\bar{x} = 579$, as shown in Figure 1.

Matched Problem 2 Compute the mean for the grouped sample data listed in Table 2.

Table 2

Class Interval	Frequency
0.5–5.5	6
5.5–10.5	20
10.5–15.5	18
15.5–20.5	4

CONCEPTUAL INSIGHT

The mean for ungrouped data and the mean for grouped data can be interpreted as the expected values of appropriately chosen random variables (see Section 8.5).

Consider a set of n measurements $x_1, x_2, \ldots, x_n$ (ungrouped data). Let S be the sample space consisting of n simple events (the n measurements), each equally likely. Let X be the random variable that assigns the numerical value x_i to each simple event in S. Then each measurement x_i has probability $p_i = \frac{1}{n}$. The expected value of X is given by

$$\begin{aligned} E(X) &= x_1p_1 + x_2p_2 + \cdots + x_np_n \\ &= x_1 \cdot \frac{1}{n} + x_2 \cdot \frac{1}{n} + \cdots + x_n \cdot \frac{1}{n} \\ &= \frac{x_1 + x_2 + \cdots + x_n}{n} \\ &= [\text{mean}] \end{aligned}$$

Similarly, consider a set of n measurements grouped into k classes in a frequency table (grouped data). Let S' be the sample space consisting of n simple events (the n measurements), each equally likely. Let X' be the random variable that assigns the midpoint x_i of the ith class interval to the measurements that belong to that class interval. Then each midpoint x_i has probability $p_i = f_i \cdot \frac{1}{n}$, where f_i denotes the frequency of the ith class interval. The expected value of X' is given by

$$\begin{aligned} E(X') &= x_1p_1 + x_2p_2 + \cdots + x_kp_k \\ &= x_1\left(f_1 \cdot \frac{1}{n}\right) + x_2\left(f_2 \cdot \frac{1}{n}\right) + \cdots + x_k\left(f_k \cdot \frac{1}{n}\right) \\ &= \frac{x_1f_1 + x_2f_2 + \cdots + x_kf_k}{n} \\ &= [\text{mean}] \end{aligned}$$

Median

Occasionally, the mean can be misleading as a measure of central tendency. Suppose the annual salaries of seven people in a small company are \$34,000, \$40,000, \$56,000, \$36,000, \$36,000, \$156,000, and \$48,000. The mean salary is

$$\bar{x} = \frac{\sum_{i=1}^{n} x_i}{n} = \frac{\$406{,}000}{7} = \$58{,}000$$

Six of the seven salaries are below the average! The one large salary distorts the results.

A measure of central tendency that is not influenced by extreme values is the **median**. The following definition of median makes precise our intuitive notion of the "middle element" when a set of measurements is arranged in ascending or descending order. Some sets of measurements, for example, 5, 7, 8, 13, 21, have a middle element. Other sets, for example, 9, 10, 15, 20, 23, 24, have no middle element, or you might prefer to say that they have two middle elements. For any number between 15 and 20, half the measurements fall above the number and half fall below.

DEFINITION Median: Ungrouped Data

1. If the number of measurements in a set is odd, the **median** is the middle measurement when the measurements are arranged in ascending or descending order.
2. If the number of measurements in a set is even, the **median** is the mean of the two middle measurements when the measurements are arranged in ascending or descending order.

EXAMPLE 3 **Finding the Median** Find the median salary in the preceding list of seven salaries.

SOLUTION Arrange the salaries in increasing order and choose the middle one:

Salary	
\$34,000	
36,000	
36,000	
40,000	⟵ Median (\$40,000)
48,000	
56,000	⟵ Mean (\$58,000)
156,000	

In this case, the median is a better measure of central tendency than the mean.

Matched Problem 3 Add the salary \$100,000 to those in Example 3 and compute the median and mean for these eight salaries.

The median, as we have defined it, is easy to determine and is not influenced by extreme values. Our definition does have some minor handicaps, however. First, if the measurements we are analyzing were carried out in a laboratory and presented to us in a frequency table, we may not have access to the individual measurements. In that case we would not be able to compute the median using the above definition. Second, a set like 4, 4, 6, 7, 7, 7, 9 would have median 7 by our definition, but 7 does not possess the symmetry we expect of a "middle element" since there are three measurements below 7 but only one above.

To overcome these handicaps, we define a second concept, the *median for grouped data*. To guarantee that the median for grouped data exists and is unique, we assume that the frequency table for the grouped data has no classes of frequency 0.

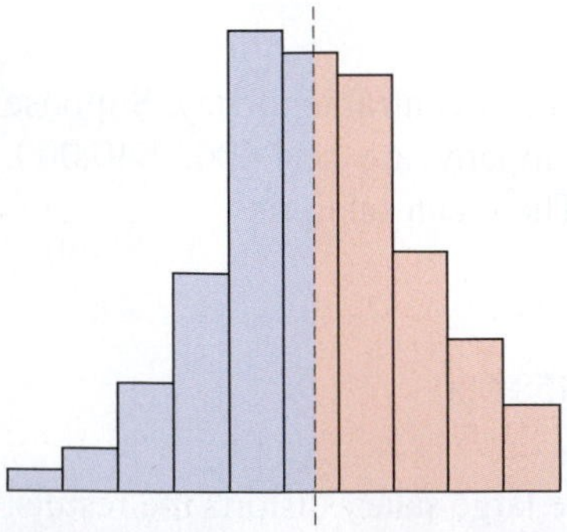

Figure 2 The area to the left of the median equals the area to the right.

DEFINITION Median: Grouped Data

The **median for grouped data** with no classes of frequency 0 is the number such that the histogram has the same area to the left of the median as to the right of the median (see Fig. 2).

EXAMPLE 4 **Finding the Median for Grouped Data** Compute the median for the grouped data in Table 3.

SOLUTION First we draw the histogram of the data (Fig. 3). The total area of the histogram is 15, which is just the sum of the frequencies, since all rectangles have a base of length 1. The area to the left of the median must be half the total area—that is, $\frac{15}{2} = 7.5$. Looking at Figure 3, we see that the median M lies between 6.5 and 7.5. Thus, the area to the left of M, which is the sum of the blue shaded areas in Figure 3, must be 7.5:

$$(1)(3) + (1)(1) + (1)(2) + (M - 6.5)(4) = 7.5$$

Solving for M gives $M = 6.875$. The median for the grouped data in Table 3 is 6.875.

Table 3

Class Interval	Frequency
3.5–4.5	3
4.5–5.5	1
5.5–6.5	2
6.5–7.5	4
7.5–8.5	3
8.5–9.5	2

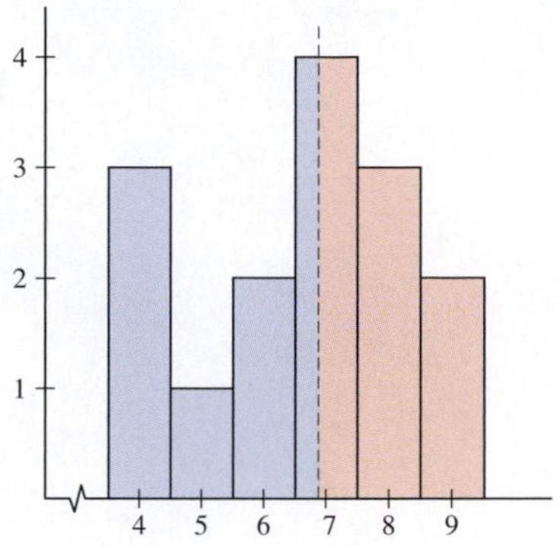

Figure 3

CAUTION The area of the leftmost rectangle in Figure 3 is $(1)(3) = 3$ because the base of the rectangle is 1 and the height is 3. The base is equal to the width of the class interval and the height is the frequency of that class. See Matched Problem 4. ▲

Matched Problem 4 Find the median for the grouped data in the following table:

Class Interval	Frequency
3.5–5.5	4
5.5–7.5	2
7.5–9.5	3
9.5–11.5	5
11.5–13.5	4
13.5–15.5	3

Mode

A third measure of central tendency is the *mode*.

DEFINITION Mode

The **mode** is the most frequently occurring measurement in a data set. There may be a unique mode, several modes, or, if no measurement occurs more than once, essentially no mode.

EXAMPLE 5 **Finding Mode, Median, and Mean**

Data Set	Mode	Median	Mean
(A) 4, 5, 5, 5, 6, 6, 7, 8, 12	5	6	6.44
(B) 1, 2, 3, 3, 3, 5, 6, 7, 7, 7, 23	3, 7	5	6.09
(C) 1, 3, 5, 6, 7, 9, 11, 15, 16	None	7	8.11

Data set (B) in Example 5 is referred to as **bimodal**, since there are two modes. Since no measurement in data set (C) occurs more than once, we say that it has no mode.

Matched Problem 5 Compute the mode(s), median, and mean for each data set:

(A) 2, 1, 2, 1, 1, 5, 1, 9, 4 (B) 2, 5, 1, 4, 9, 8, 7

(C) 8, 2, 6, 8, 3, 3, 1, 5, 1, 8, 3

The mode, median, and mean can be computed in various ways with the aid of a graphing calculator. In Figure 4A, the data set of Example 5B is entered as a list, and its median and mean are computed. The histogram in Figure 4B shows the two modes of the same data set.

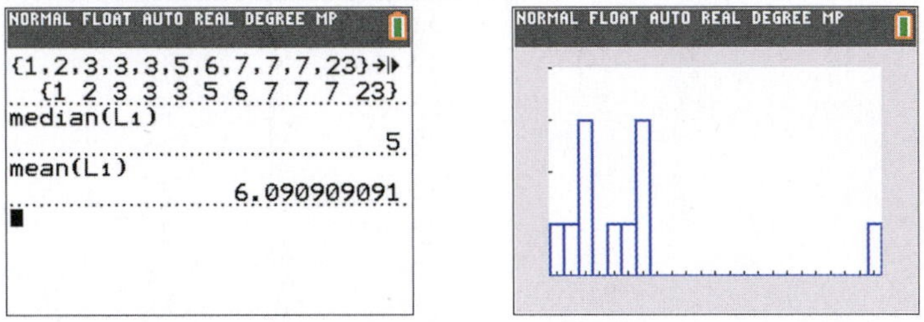

Figure 4 (A) (B)

As with the median, the mode is not influenced by extreme values. Suppose that in the data set of Example 5B, we replace 23 with 8. The modes remain 3 and 7 and the median is still 5; however, the mean changes to 4.73. The mode is most useful for large data sets because it emphasizes data concentration. For example, a clothing retailer would be interested in the mode of sizes due to customer demand of various items in a store.

The mode also can be used for qualitative attributes—that is, attributes that are not numerical. The mean and median are not suitable in these cases. For example, the mode can be used to give an indication of a favorite brand of ice cream or the worst movie of the year. Figure 5 shows the results of a random survey of 1,000 people on entree preferences when eating dinner out. According to this survey, we would say that the modal preference is beef. Note that the mode is the only measure of central tendency (location) that can be used for this type of data; mean and median make no sense.

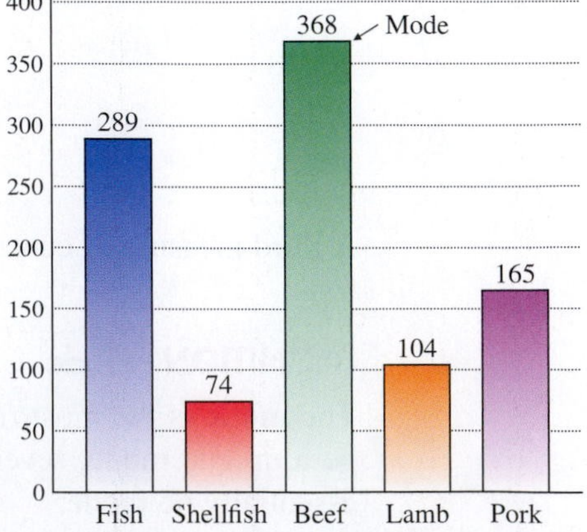

Figure 5 The modal preference for an entree is beef.

In actual practice, the mean is used the most, the median next, and the mode a distant third.

Explore and Discuss 1

For many sets of measurements the median lies between the mode and the mean. But this is not always so.

(A) In a class of seven students, the scores on an exam were 52, 89, 89, 92, 93, 96, 99. Show that the mean is less than the mode and that the mode is less than the median.

(B) Construct hypothetical sets of exam scores to show that all possible orders among the mean, median, and mode can occur.

Exercises 10.2

Skills Warm-up Exercises

W *In Problems 1–4, find the mean of the data set. (If necessary, review Section B.1).*

1. 5, 8, 6, 7, 9

2. 12, 15, 18, 21

3. −1, −2, 0, 2, 4

4. 85, 75, 65, 75

In Problems 5–8, find the indicated sum. (If necessary, review Section B.1).

5. $\sum_{i=1}^{6} x_i$ if $x_i = i + 3$ for $i = 1, 2, \ldots, 6$

6. $\sum_{i=1}^{5} x_i$ if $x_i = 100i$ for $i = 1, 2, \ldots, 5$

7. $\sum_{i=6}^{10} x_i f_i$ if $x_i = 2i$ and $f_i = 3i$ for $i = 6, 7, \ldots, 10$

8. $\sum_{i=5}^{8} x_i f_i$ if $x_i = i^2$ and $f_i = i$ for $i = 5, 6, 7, 8$

A *Find the mean, median, and mode for the sets of ungrouped data given in Problems 9 and 10.*

9. 1, 2, 2, 3, 3, 3, 3, 4, 4, 5

10. 1, 1, 1, 1, 2, 3, 4, 5, 5, 5

Find the mean, median, and/or mode, whichever are applicable, in Problems 11 and 12.

11.

Flavor	Number Preferring
Vanilla	139
Chocolate	376
Strawberry	89
Pistachio	105
Cherry	63
Almond mocha	228

12.

Car Color	Number Preferring
Red	1,324
White	3,084
Black	1,617
Blue	2,303
Brown	2,718
Gold	1,992

Find the mean for the sets of grouped data in Problems 13 and 14.

13.

Interval	Frequency
0.5–2.5	2
2.5–4.5	5
4.5–6.5	7
6.5–8.5	1

14.

Interval	Frequency
0.5–2.5	5
2.5–4.5	1
4.5–6.5	2
6.5–8.5	7

B **15.** Which single measure of central tendency—mean, median, or mode—would you say best describes the following set of measurements? Discuss the factors that justify your preference.

8.01	7.91	8.13	6.24	7.95
8.04	7.99	8.09	6.24	81.2

16. Which single measure of central tendency—mean, median, or mode—would you say best describes the following set of measurements? Discuss the factors that justify your preference.

47	51	80	91	85
69	91	95	81	60

17. A data set is formed by recording the results of 100 rolls of a fair die.

(A) What would you expect the mean of the data set to be? The median?

(B) Form such a data set by using a graphing calculator to simulate 100 rolls of a fair die, and find its mean and median.

18. A data set is formed by recording the sums on 200 rolls of a pair of fair dice.

(A) What would you expect the mean of the data set to be? The median?

(B) Form such a data set by using a graphing calculator to simulate 200 rolls of a pair of fair dice, and find the mean and median of the set.

C **19.** (A) Construct a set of four numbers that has mean 300, median 250, and mode 175.

(B) Let $m_1 > m_2 > m_3$. Devise and discuss a procedure for constructing a set of four numbers that has mean m_1, median m_2, and mode m_3.

20. (A) Construct a set of five numbers that has mean 200, median 150, and mode 50.

(B) Let $m_1 > m_2 > m_3$. Devise and discuss a procedure for constructing a set of five numbers that has mean m_1, median m_2, and mode m_3.

Applications

21. Price–earnings ratios. Find the mean, median, and mode for the data in the following table.

Price–Earnings Ratios for Eight Stocks in a Portfolio

5.3	10.1	18.7	35.5
12.9	8.4	16.2	10.1

22. Gasoline tax. Find the mean, median, and mode for the data in the following table. (*Source:* American Petroleum Institute.)

State Gasoline Tax, 2017

State	Tax (cents per gal.)
Wisconsin	32.9
New York	43.9
Connecticut	39.9
Nebraska	28.2
Kansas	24.0
Texas	20.0
California	38.1
Tennessee	21.4

23. Lightbulb lifetime. Find the mean and median for the data in the following table.

Life (Hours) of 50 Randomly Selected Lightbulbs

Interval	Frequency
799.5–899.5	3
899.5–999.5	10
999.5–1,099.5	24
1,099.5–1,199.5	12
1,199.5–1.299.5	1

24. Price–earnings ratios. Find the mean and median for the data in the following table.

Price–Earnings Ratios of 100 Randomly Chosen Stocks from the New York Stock Exchange

Interval	Frequency
−0.5–4.5	5
4.5–9.5	54
9.5–14.5	25
14.5–19.5	9
19.5–24.5	4
24.5–29.5	1
29.5–34.5	2

25. Student loan debt. Find the mean, median, and mode for the data in the following table that gives the percentages by state of students graduating from college in 2015 who had student loan debt. (*Source*: The Institute for College Access & Success)

College Graduates with Student Loan Debt

State	Percentage of 2015 College Graduates with Student Loan Debt
Arizona	56
Arkansas	57
Louisiana	51
New Mexico	58
Oklahoma	52
Texas	56

26. Tourism. Find the mean, median, and mode for the data in the following table. (*Source:* World Tourism Organization)

International Tourism Receipts, 2015

Country	Receipts (billion $)
United States	178.3
China	114.1
Spain	56.5
France	45.9
Thailand	44.6
United Kingdom	42.4
Italy	39.7
Germany	36.9

27. Mouse weights. Find the mean and median for the data in the following table.

Mouse Weights (grams)

Interval	Frequency
41.5–43.5	3
43.5–45.5	7
45.5–47.5	13
47.5–49.5	17
49.5–51.5	19
51.5–53.5	17
53.5–55.5	15
55.5–57.5	7
57.5–59.5	2

28. Blood cholesterol levels. Find the mean and median for the data in the following table.

Blood Cholesterol Levels (milligrams per deciliter)

Interval	Frequency
149.5–169.5	4
169.5–189.5	11
189.5–209.5	15
209.5–229.5	25
229.5–249.5	13
249.5–269.5	7
269.5–289.5	3
289.5–309.5	2

29. Immigration. Find the mean, median, and mode for the data in the following table. (*Source:* U.S. Census Bureau)

Top Ten Countries of Birth of U.S. Foreign-Born Population, 2015

Country	Number (thousands)
Mexico	11,643
China	2,677
India	2,390
Philippines	1,982
El Salvador	1,352
Vietnam	1,301
Cuba	1,211
Dominican Republic	1,063
Korea	1,060
Guatemala	928

30. Grade-point averages. Find the mean and median for the grouped data in the following table.

Graduating Class Grade-Point Averages

Interval	Frequency
1.95–2.15	21
2.15–2.35	19
2.35–2.55	17
2.55–2.75	14
2.75–2.95	9
2.95–3.15	6
3.15–3.35	5
3.35–3.55	4
3.55–3.75	3
3.75–3.95	2

31. Entrance examination scores. Compute the median for the grouped data of entrance examination scores given in Table 1 on page 514.

32. Presidents. Find the mean and median for the grouped data in the following table.

U.S. Presidents' Ages at Inauguration

Age	Number
39.5–44.5	2
44.5–49.5	7
49.5–54.5	12
54.5–59.5	13
59.5–64.5	7
64.5–69.5	2
69.5–74.5	2

Answers to Matched Problems

1. $\bar{x} = 3.82$
2. $\bar{x} \approx 10.1$
3. Median = \$44,000; mean = \$63,250
4. Median for grouped data = 10.1
5. Arrange each set of data in ascending order:

Data Set	Mode	Median	Mean
(A) 1, 1, 1, 1, 2, 2, 4, 5, 9	1	2	2.89
(B) 1, 2, 4, 5, 7, 8, 9	None	5	5.14
(C) 1, 1, 2, 3, 3, 3, 5, 6, 8, 8, 8	3, 8	3	4.36

10.3 Measures of Dispersion

- Range
- Standard Deviation: Ungrouped Data
- Standard Deviation: Grouped Data
- Significance of Standard Deviation

A measure of central tendency gives us a typical value that can be used to describe a whole set of data, but this measure does not tell us whether the data are tightly clustered or widely dispersed. We now consider two measures of variation—*range* and *standard deviation*—that will give some indication of data scatter.

Range

A measure of dispersion, or scatter, that is easy to compute and can be easily understood is the range. The **range for a set of ungrouped data** is the difference between the largest and the smallest values in the data set. The **range for a frequency distribution** is the difference between the upper boundary of the highest class and the lower boundary of the lowest class.

Consider the histograms in Figure 1. We see that the range adds only a little information about the amount of variation in a data set. The graphs clearly show that even though each data set has the same mean and range, all three sets differ in the amount of scatter, or variation, of the data relative to the mean. The data set in part (A) is tightly clustered about the mean; the data set in part (B) is dispersed away from the mean; and the data set in part (C) is uniformly distributed over its range.

Since the range depends only on the extreme values of the data, it does not give us any information about the dispersion of the data between these extremes. We need a measure of dispersion that will give us some idea of how the data are clustered or scattered relative to the mean. *Standard deviation* is such a measure.

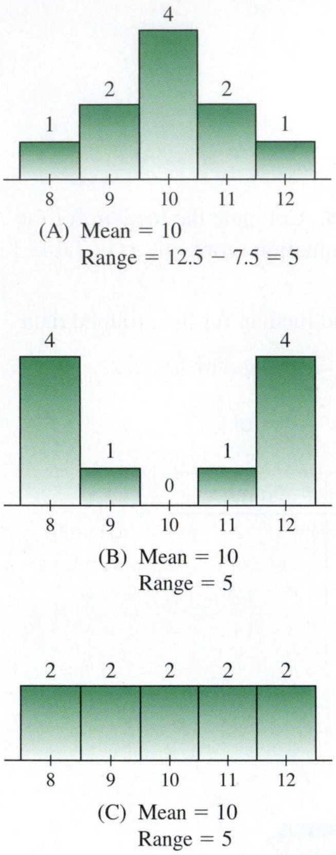

Figure 1

Standard Deviation: Ungrouped Data

We will develop the concepts of *variance* and *standard deviation*—both measures of variation—through a simple example. Suppose that a random sample of five stamped parts is selected from a manufacturing process, and these parts are found to have the following lengths (in centimeters):

$$5.2, 5.3, 5.2, 5.5, 5.3$$

Computing the sample mean, we obtain

$$\bar{x} = \frac{\sum_{i=1}^{n} x_i}{n} = \frac{5.2 + 5.3 + 5.2 + 5.5 + 5.3}{5}$$

$$= 5.3 \text{ centimeters}$$

Table 1

x_i	$(x_i - \bar{x})$
5.2	−0.1
5.3	0.0
5.2	−0.1
5.5	0.2
5.3	0.0

How much variation exists between the sample mean and all measurements in the sample? As a first attempt at measuring the variation, let us represent the **deviation** of a measurement from the mean by $(x_i - \bar{x})$. Table 1 lists all the deviations for this sample.

Using these deviations, what kind of formula can we find that will give us a single measure of variation? It appears that the average of the deviations might be a good measure. But look at what happens when we add the second column in Table 1. We get 0! It turns out that this will always happen for any data set. Now what? We could take the average of the absolute values of the deviations; however, this approach leads to problems relative to statistical inference. Instead, to get around the sign problem, we will take the average of the squares of the deviations and call this number the **variance** of the data set:

$$[\text{variance}] = \frac{\sum_{i=1}^{n} (x_i - \bar{x})^2}{n} \qquad (1)$$

Calculating the variance using the entries in Table 1, we have

$$[\text{variance}] = \frac{\sum_{i=1}^{5}(x_i - 5.3)^2}{5} = 0.012 \text{ square centimeter}$$

We still have a problem because the units in the variance are square centimeters instead of centimeters (the units of the original data set). To obtain the units of the original data set, we take the positive square root of the variance and call the result the **standard deviation** of the data set:

$$[\text{standard deviation}] = \sqrt{\frac{\sum_{i=1}^{n}(x_i - \bar{x})^2}{n}} = \sqrt{\frac{\sum_{i=1}^{5}(x_i - 5.3)^2}{5}} \qquad (2)$$

$$\approx 0.11 \text{ centimeter}$$

The *sample variance* is usually denoted by s^2 and the *population variance* by σ^2 (σ is the Greek lowercase letter "sigma"). The *sample standard deviation* is usually denoted by s and the *population standard deviation* by σ.

In inferential statistics, the sample variance s^2 is often used as an estimator for the population variance σ^2 and the sample standard deviation s for the population standard deviation σ. It can be shown that one can obtain better estimates of the population parameters in terms of the sample parameters (particularly when using small samples) if the divisor n is replaced by $n - 1$ when computing sample variances or sample standard deviations.

DEFINITION Variance: Ungrouped Data*

The **sample variance** s^2 of a set of n sample measurements $x_1, x_2, \ldots, x_n$ with mean $\bar{x}$ is given by

$$s^2 = \frac{\sum_{i=1}^{n}(x_i - \bar{x})^2}{n - 1} \qquad (3)$$

If $x_1, x_2, \ldots, x_n$ is the whole population with mean μ, then the **population variance** σ^2 is given by

$$\sigma^2 = \frac{\sum_{i=1}^{n}(x_i - \mu)^2}{n}$$

*In this section, we restrict our interest to the sample variance.

The standard deviation is just the positive square root of the variance.

DEFINITION Standard Deviation: Ungrouped Data†

The **sample standard deviation** s of a set of n sample measurements $x_1, x_2, \ldots, x_n$ with mean $\bar{x}$ is given by

$$s = \sqrt{\frac{\sum_{i=1}^{n}(x_i - \bar{x})^2}{n - 1}} \qquad (4)$$

If $x_1, x_2, \ldots, x_n$ is the whole population with mean μ, then the **population standard deviation** σ is given by

$$\sigma = \sqrt{\frac{\sum_{i=1}^{n}(x_i - \mu)^2}{n}}$$

†In this section, we restrict our interest to the sample standard deviation.

Computing the standard deviation for the original sample measurements (see Table 1), we now obtain

$$s = \sqrt{\frac{\sum_{i=1}^{5}(x_i - 5.3)^2}{5 - 1}} \approx 0.12 \text{ cm}$$

EXAMPLE 1 **Finding the Standard Deviation** Find the standard deviation for the sample measurements 1, 3, 5, 4, 3.

SOLUTION To find the standard deviation for the data set, we can utilize a table or use a calculator. Most will prefer the latter. Here is what we compute:

$$\bar{x} = \frac{1 + 3 + 5 + 4 + 3}{5} = 3.2$$

$$s = \sqrt{\frac{(1 - 3.2)^2 + (3 - 3.2)^2 + (5 - 3.2)^2 + (4 - 3.2)^2 + (3 - 3.2)^2}{5 - 1}}$$

$$\approx 1.48$$

Matched Problem 1 Find the standard deviation for the sample measurements 1.2, 1.4, 1.7, 1.3, 1.5.

Remark—Many graphing calculators can compute $\bar{x}$ and s directly after the sample measurements are entered—a helpful feature, especially when the sample is fairly large. This shortcut is illustrated in Figure 2 for a particular graphing calculator, where the data from Example 1 are entered as a list, and several different one-variable statistics are immediately calculated. Included among these statistics are the mean $\bar{x}$, the sample standard deviation s (denoted by Sx in Fig. 2B), the population standard deviation σ (denoted by σx), the number n of measurements, the smallest element of the data set (denoted by minX), the largest element of the data set (denoted by maxX), the median (denoted by Med), and several statistics we have not discussed.

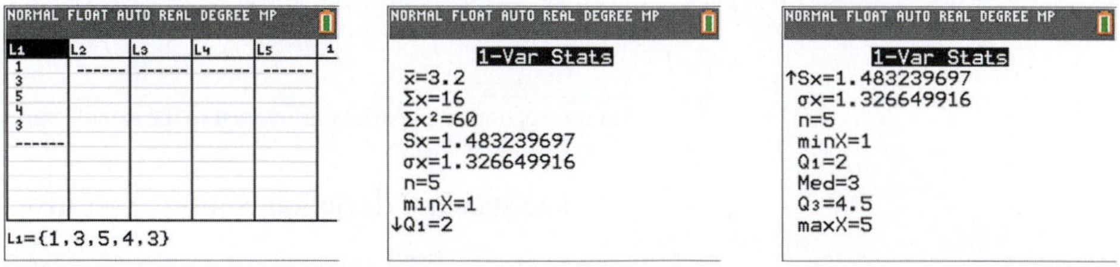

(A) Data (B) Statistics (C) Statistics (continued)

Figure 2

CONCEPTUAL INSIGHT

If the sample measurements in Example 1 are considered to constitute the whole population, then the population standard deviation σx is approximately equal to 1.33 [see Fig. 2(B)]. The computation of σx is the same as that of Example 1, except that the denominator $n - 1\ (= 5 - 1)$ under the radical sign is replaced by $n\ (= 5)$. Consequently, Sx $\approx$ 1.48 is greater than σx $\approx$ 1.33. Formulas (4) and (2) produce nearly the same results when the sample size n is large. The **law of large numbers** states that we can make a sample standard deviation s as close to the population standard deviation σ as we like by making the sample sufficiently large.

Explore and Discuss 1

(A) When is the sample standard deviation of a set of measurements equal to 0?

(B) Can the population standard deviation of a set of measurements ever be greater than the range? Explain why or why not.

Standard Deviation: Grouped Data

Formula (4) for sample standard deviation is extended to grouped sample data as described in the following box:

DEFINITION Standard Deviation: Grouped Data*

Suppose a data set of n sample measurements is grouped into k classes in a frequency table, where x_i is the midpoint and f_i is the frequency of the ith class interval. If $\bar{x}$ is the mean for the grouped data, then the **sample standard deviation** s for the grouped data is

$$s = \sqrt{\frac{\sum_{i=1}^{k}(x_i - \bar{x})^2 f_i}{n - 1}} \qquad (5)$$

where $n = \sum_{i=1}^{k} f_i =$ total number of measurements. If $x_1, x_2, \ldots, x_n$ is the whole population with mean μ, then the **population standard deviation σ** is given by

$$\sigma = \sqrt{\frac{\sum_{i=1}^{n}(x_i - \mu)^2 f_i}{n}}$$

*In this section, we restrict our interest to the sample standard deviation.

EXAMPLE 2 **Finding the Standard Deviation for Grouped Data** Find the standard deviation for each set of grouped sample data.

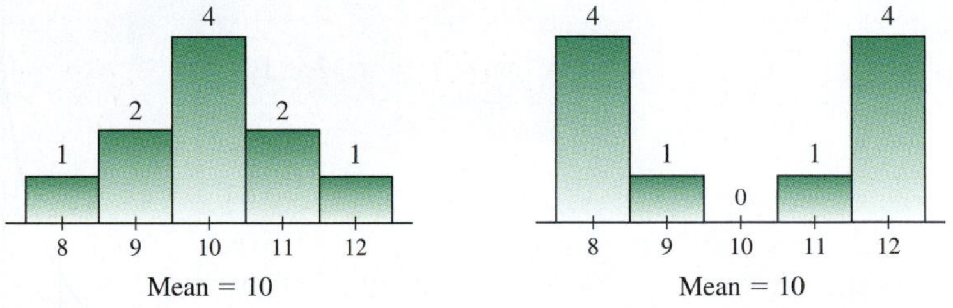

SOLUTION

(A) $$s = \sqrt{\frac{(8 - 10)^2(1) + (9 - 10)^2(2) + (10 - 10)^2(4) + (11 - 10)^2(2) + (12 - 10)^2(1)}{10 - 1}}$$

$$= \sqrt{\frac{12}{9}} \approx 1.15$$

(B) $$s = \sqrt{\frac{(8 - 10)^2(4) + (9 - 10)^2(1) + (10 - 10)^2(0) + (11 - 10)^2(1) + (12 - 10)^2(4)}{10 - 1}}$$

$$= \sqrt{\frac{34}{9}} \approx 1.94$$

Comparing the results of parts (A) and (B) in Example 2, we find that the larger standard deviation is associated with the data that deviate furthest from the mean.

Matched Problem 2 Find the standard deviation for the grouped sample data shown below.

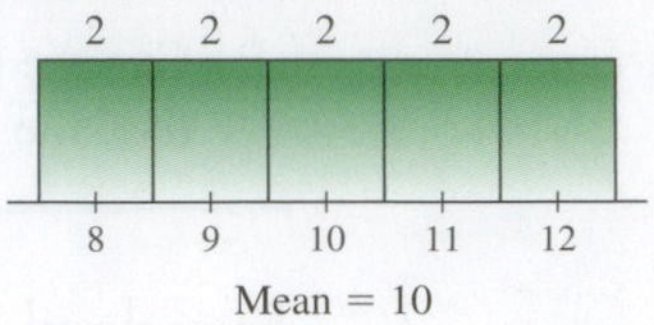

Remark—Figure 3 illustrates the shortcut computation of the mean and standard deviation on a graphing calculator when the data are grouped. The sample data of Example 2A are entered. List L_1 contains the midpoints of the class intervals, and list L_2 contains the corresponding frequencies. The mean, standard deviation, and other one-variable statistics are then calculated immediately.

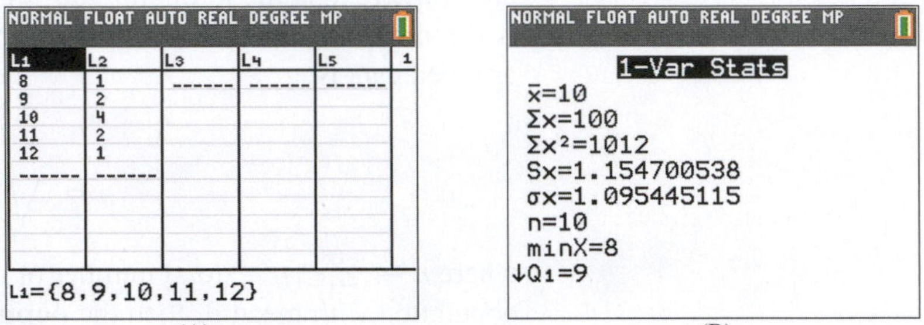

Figure 3 (A) (B)

Significance of Standard Deviation

The standard deviation can give us additional information about a frequency distribution of a set of raw data. Suppose we draw a smooth curve through the midpoints of the tops of the rectangles forming a histogram for a fairly large frequency distribution (see Fig. 4). If the resulting curve is approximately bell shaped, then it can be shown that approximately 68% of the data will lie in the interval from $\bar{x} - s$ to $\bar{x} + s$, about 95% of the data will lie in the interval from $\bar{x} - 2s$ to $\bar{x} + 2s$, and almost all the data will lie in the interval from $\bar{x} - 3s$ to $\bar{x} + 3s$. We will have much more to say about this in Section 10.5.

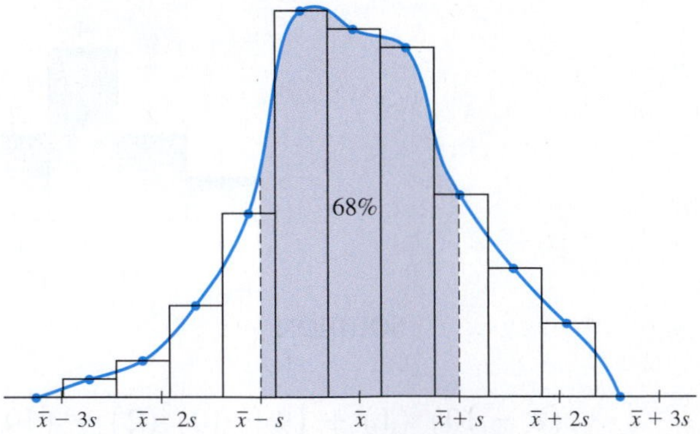

Figure 4

Exercises 10.3

Skills Warm-up Exercises

W *In Problems 1–8, find the indicated sum. (If necessary, review Section B.1).*

1. $\sum_{i=1}^{3} (x_i - 2)^2$ if $x_1 = -1, x_2 = 1, x_3 = 6$

2. $\sum_{i=1}^{5} (x_i - 10)^2$ if $x_1 = 8, x_2 = 9, x_3 = 10, x_4 = 11, x_5 = 12$

3. $\sum_{i=1}^{4} (x_i + 3)^2$ if $x_1 = -4, x_2 = -3, x_3 = -2, x_4 = -3$

4. $\sum_{i=1}^{3}(x_i + 15)^2$ if

$x_1 = -20, x_2 = -15, x_3 = -10$

5. $\sum_{i=1}^{3}(x_i - 4)^2 f_i$ if

$x_1 = 0, x_2 = 4, x_3 = 8, f_1 = 1, f_2 = 1, f_3 = 1$

6. $\sum_{i=1}^{3}(x_i - 4)^2 f_i$ if

$x_1 = 0, x_2 = 4, x_3 = 8, f_1 = 2, f_2 = 1, f_3 = 2$

7. $\sum_{i=1}^{5}(x_i - 15)^2 f_i$ if

$x_i = 5i, f_i = 2, \text{ for } i = 1, 2, \ldots, 5$

8. $\sum_{i=1}^{4}(x_i - 3.5)^2 f_i$ if

$x_i = i + 1, f_i = 3, \text{ for } i = 1, 2, 3, 4$

A

9. (A) Find the mean and standard deviation of the following set of ungrouped sample data.

4 2 3 5 3 1 6 4 2 3

(B) What proportion of the measurements lies within 1 standard deviation of the mean? Within 2 standard deviations? Within 3 standard deviations?

(C) Based on your answers to part (B), would you conjecture that the histogram is approximately bell shaped? Explain.

(D) To confirm your conjecture, construct a histogram with class interval width 1, starting at 0.5.

10. (A) Find the mean and standard deviation of the following set of ungrouped sample data.

3 5 1 2 1 5 4 5 1 3

(B) What proportion of the measurements lies within 1 standard deviation of the mean? Within 2 standard deviations? Within 3 standard deviations?

(C) Based on your answers to part (B), would you conjecture that the histogram is approximately bell shaped? Explain.

(D) To confirm your conjecture, construct a histogram with class interval width 1, starting at 0.5.

B

In Problems 11 and 12, find the standard deviation for each set of grouped sample data using formula (5) on page 525.

11.

Interval	Frequency
0.5–3.5	2
3.5–6.5	5
6.5–9.5	7
9.5–12.5	1

12.

Interval	Frequency
0.5–3.5	5
3.5–6.5	1
6.5–9.5	2
9.5–12.5	7

In Problems 13–18, discuss the validity of each statement. If the statement is always true, explain why. If not, give a counterexample.

13. The range for a set of sample measurements is less than or equal to the range of the whole population.

14. The range for a set of measurements is greater than or equal to 0.

15. The sample standard deviation is less than or equal to the sample variance.

16. Given a set of sample measurements that are not all equal, the sample standard deviation is a positive real number.

17. If x_1, x_2 is the whole population, then the population standard deviation is equal to one-half the distance from x_1 to x_2.

18. The sample variance of a set of measurements is always less than the population variance.

C

19. A data set is formed by recording the sums in 100 rolls of a pair of dice. A second data set is formed by recording the results of 100 draws of a ball from a box containing 11 balls numbered 2 through 12.

(A) Which of the two data sets would you expect to have the smaller standard deviation? Explain.

(B) To obtain evidence for your answer to part (A), use a graphing calculator to simulate both experiments and compute the standard deviations of each data set.

20. A data set is formed by recording the results of rolling a fair die 200 times. A second data set is formed by rolling a pair of dice 200 times, each time recording the minimum of the two numbers.

(A) Which of the two data sets would you expect to have the smaller standard deviation? Explain.

(B) To obtain evidence for your answer to part (A), use a graphing calculator to simulate both experiments and compute the standard deviations of each data set.

Applications

Find the mean and standard deviation for each of the sample data sets given in Problems 21–28.

Use the suggestions in the remarks following Examples 1 and 2 to perform some of the computations.

21. Earnings per share. The earnings per share (in dollars) for 12 companies selected at random from the list of *Fortune* 500 companies are:

2.35	1.42	8.05	6.71
3.11	2.56	0.72	4.17
5.33	7.74	3.88	6.21

22. Checkout times. The checkout times (in minutes) for 12 randomly selected customers at a large supermarket during the store's busiest time are:

4.6	8.5	6.1	7.8
10.9	9.3	11.4	5.8
9.7	8.8	6.7	13.2

23. Quality control. The lives (in hours of continuous use) of 100 randomly selected flashlight batteries are:

Interval	Frequency
6.95–7.45	2
7.45–7.95	10
7.95–8.45	23
8.45–8.95	30
8.95–9.45	21
9.45–9.95	13
9.95–10.45	1

24. Stock analysis. The price–earnings ratios of 100 randomly selected stocks from the New York Stock Exchange are:

Interval	Frequency
−0.5–4.5	5
4.5–9.5	54
9.5–14.5	25
14.5–19.5	13
19.5–24.5	0
24.5–29.5	1
29.5–34.5	2

25. Medicine. The reaction times (in minutes) of a drug given to a random sample of 12 patients are:

4.9	5.1	3.9	4.2
6.4	3.4	5.8	6.1
5.0	5.6	5.8	4.6

26. Nutrition: animals. The mouse weights (in grams) of a random sample of 100 mice involved in a nutrition experiment are:

Interval	Frequency
41.5–43.5	3
43.5–45.5	7
45.5–47.5	13
47.5–49.5	17
49.5–51.5	19
51.5–53.5	17
53.5–55.5	15
55.5–57.5	7
57.5–59.5	2

27. Reading scores. The grade-level reading scores from a reading test given to a random sample of 12 students in an urban high school graduating class are:

9	11	11	15
10	12	12	13
8	7	13	12

28. Grade-point average. The grade-point averages of a random sample of 100 students from a university's graduating class are:

Interval	Frequency
1.95–2.15	21
2.15–2.35	19
2.35–2.55	17
2.55–2.75	14
2.75–2.95	9
2.95–3.15	6
3.15–3.35	5
3.35–3.55	4
3.55–3.75	3
3.75–3.95	2

Answers to Matched Problems

1. $s \approx 0.19$

2. $s \approx 1.49$ (a value between those found in Example 2, as expected)

10.4 Bernoulli Trials and Binomial Distributions

- Bernoulli Trials
- Binomial Formula: Brief Review
- Binomial Distribution
- Application

In Section 10.1, we discussed frequency and relative frequency distributions, which were represented by tables and histograms. Frequency distributions and their corresponding probability distributions based on actual observations are *empirical* in nature. But there are many situations in which it is of interest to determine the kind of relative frequency distribution we might expect before any data have been collected. What we have in mind is a **theoretical**, or **hypothetical, probability distribution**—that is, a probability distribution based on assumptions and theory rather than actual observations or measurements. Theoretical probability distributions are used to approximate properties of real-world distributions, assuming that the theoretical and empirical distributions are closely matched.

Of the many interesting theoretical probability distributions, one in particular is the *binomial distribution.* The reason for the name "binomial distribution" is that the distribution is closely related to the binomial expansion of $(q + p)^n$, where n is a natural number. We start the discussion with a particular type of experiment called a *Bernoulli experiment,* or *trial.*

Bernoulli Trials

If we toss a coin, either a head occurs or it does not. If we roll a die, either a 3 shows or it fails to show. If you are vaccinated for smallpox, either you contract smallpox or you do not. What do all these situations have in common? All can be classified as experiments with two possible outcomes, each the complement of the other. An experiment for which there are only two possible outcomes, E or E', is called a **Bernoulli experiment**, or **trial**, named after Jacob Bernoulli (1654–1705), the Swiss scientist and mathematician who was one of the first to study the probability problems related to a two-outcome experiment.

In a Bernoulli experiment or trial, it is customary to refer to one of the two outcomes as a **success** S and to the other as a **failure** F. If we designate the probability of success by

Reminder

Uppercase "P" is an abbreviation for the word "probability." Lowercase "p" stands for a number between 0 and 1, inclusive.

$$P(S) = p$$

then the probability of failure is

$$P(F) = 1 - p = q \qquad \text{Note: } p + q = 1$$

EXAMPLE 1 **Probability of Success in a Bernoulli Trial** Suppose that we roll a fair die and ask for the probability of a 6 turning up. This can be viewed as a Bernoulli trial by identifying success with a 6 turning up and failure with any of the other numbers turning up. So,

$$p = \tfrac{1}{6} \qquad \text{and} \qquad q = 1 - \tfrac{1}{6} = \tfrac{5}{6}$$

Matched Problem 1 Find p and q for a single roll of a fair die, where success is a number divisible by 3 turning up.

Now, suppose that a Bernoulli trial is repeated a number of times. We might try to determine the probability of a given number of successes out of the given number of trials. For example, we might be interested in the probability of obtaining exactly three 5's in six rolls of a fair die or the probability that 8 people will not catch influenza out of the 10 who have been inoculated.

Suppose that a Bernoulli trial is repeated five times so that each trial is completely *independent* of any other, and p is the probability of success on each trial. Then the probability of the outcome $SSFFS$ would be

$$\begin{aligned} P(SSFFS) &= P(S)P(S)P(F)P(F)P(S) \\ &= ppqqp \\ &= p^3q^2 \end{aligned}$$

See Section 8.3.

In general, we define a *sequence of Bernoulli trials* as follows:

DEFINITION Bernoulli Trials

A sequence of experiments is called a **sequence of Bernoulli trials**, or a **binomial experiment**, if

1. Only two outcomes are possible in each trial.
2. The probability of success p for each trial is a constant (probability of failure is then $q = 1 - p$).
3. All trials are independent.

EXAMPLE 2 **Probability of an Outcome of a Binomial Experiment** If we roll a fair die five times and identify a success in a single roll with a 1 turning up, what is the probability of the sequence *SFFSS* occurring?

SOLUTION

$$p = \tfrac{1}{6} \qquad q = 1 - p = \tfrac{5}{6}$$

$$\begin{aligned} P(SFFSS) &= pqqpp \\ &= p^3q^2 \\ &= \left(\tfrac{1}{6}\right)^3\left(\tfrac{5}{6}\right)^2 \approx .003 \end{aligned}$$

Matched Problem 2 In Example 2, find the probability of the outcome *FSSSF*.

If we roll a fair die five times, what is the probability of obtaining exactly three 1's? Notice how this problem differs from Example 2. In that example we looked at only one way that three 1's can occur. Then in Matched Problem 2 we saw another way. So exactly three 1's may occur in the following two sequences (among others):

$$SFFSS \qquad FSSSF$$

We found that the probability of each sequence occurring is the same, namely,

$$\left(\tfrac{1}{6}\right)^3\left(\tfrac{5}{6}\right)^2$$

How many more sequences will produce exactly three 1's? To answer this question, think of the number of ways that the following five blank positions can be filled with three *S*'s and two *F*'s:

$$\begin{matrix} b_1 & b_2 & b_3 & b_4 & b_5 \\ \square & \square & \square & \square & \square \end{matrix}$$

A given sequence is determined, once the *S*'s are assigned. We are interested in the number of ways three blank positions can be selected for the *S*'s out of the five available blank positions b_1, b_2, b_3, b_4, and b_5. This problem should sound familiar—it is the problem of finding the number of combinations of 5 objects taken 3 at a time, that is, ${}_5C_3$. So, the number of different sequences of successes and failures that produce exactly three successes (exactly three 1's) is

$${}_5C_3 = \frac{5!}{3!2!} = 10$$

Since the probability of each sequence is the same,

$$p^3q^2 = \left(\tfrac{1}{6}\right)^3\left(\tfrac{5}{6}\right)^2$$

and there are 10 mutually exclusive sequences that produce exactly three 1's,

$$\begin{aligned} P(\text{exactly three successes}) &= {}_5C_3\left(\frac{1}{6}\right)^3\left(\frac{5}{6}\right)^2 \\ &= \frac{5!}{3!2!}\left(\frac{1}{6}\right)^3\left(\frac{5}{6}\right)^2 \\ &= (10)\left(\frac{1}{6}\right)^3\left(\frac{5}{6}\right)^2 \approx .032 \end{aligned}$$

Reasoning in essentially the same way, the following important theorem can be proved:

THEOREM 1 Probability of *x* Successes in *n* Bernoulli Trials

The probability of exactly x successes in n independent repeated Bernoulli trials, with the probability of success of each trial p (and of failure q), is

$$P(x \text{ successes}) = {}_nC_x p^x q^{n-x} \tag{1}$$

EXAMPLE 3 **Probability of x Successes in n Bernoulli Trials** If a fair die is rolled five times, what is the probability of rolling

(A) Exactly two 3's?

(B) At least two 3's?

SOLUTION

(A) Use formula (1) with $n = 5$, $x = 2$, and $p = \frac{1}{6}$:

$$P(x = 2) = {}_5C_2\left(\frac{1}{6}\right)^2\left(\frac{5}{6}\right)^3$$

$$= \frac{5!}{2!3!}\left(\frac{1}{6}\right)^2\left(\frac{5}{6}\right)^3 \approx .161$$

(B) Notice how this problem differs from part (A). Here we have

$$P(x \geq 2) = P(x = 2) + P(x = 3) + P(x = 4) + P(x = 5)$$

It is actually easier to compute the probability of the complement of this event, $P(x < 2)$, and use

$$P(x \geq 2) = 1 - P(x < 2)$$

where

$$P(x < 2) = P(x = 0) + P(x = 1)$$

We now compute $P(x = 0)$ and $P(x = 1)$:

$$P(x = 0) = {}_5C_0\left(\frac{1}{6}\right)^0\left(\frac{5}{6}\right)^5 = \left(\frac{5}{6}\right)^5 \approx .402$$

$$P(x = 1) = {}_5C_1\left(\frac{1}{6}\right)^1\left(\frac{5}{6}\right)^4 = \frac{5!}{1!4!}\left(\frac{1}{6}\right)^1\left(\frac{5}{6}\right)^4 \approx .402$$

Therefore,

$$P(x < 2) = .402 + .402 = .804$$

and

$$P(x \geq 2) = 1 - .804 = .196$$

Matched Problem 3 Using the same die experiment as in Example 3, what is the probability of rolling

(A) Exactly one 3? (B) At least one 3?

Binomial Formula: Brief Review

Before extending Bernoulli trials to binomial distributions, it is worthwhile to review the binomial formula, which is discussed in detail in Appendix B.3. To start, let us calculate directly the first five natural number powers of $(a + b)^n$:

$$(a + b)^1 = a + b$$

$$(a + b)^2 = a^2 + 2ab + b^2$$

$$(a + b)^3 = a^3 + 3a^2b + 3ab^2 + b^3$$

$$(a + b)^4 = a^4 + 4a^3b + 6a^2b^2 + 4ab^3 + b^4$$

$$(a + b)^5 = a^5 + 5a^4b + 10a^3b^2 + 10a^2b^3 + 5ab^4 + b^5$$

In general, it can be shown that a binomial expansion is given by the well-known **binomial formula:**

Binomial Formula

For n a natural number,

$$(a + b)^n = {}_nC_0a^n + {}_nC_1a^{n-1}b + {}_nC_2a^{n-2}b^2 + \cdots + {}_nC_nb^n$$

EXAMPLE 4 **Finding Binomial Expansions** Use the binomial formula to expand $(q + p)^3$.

SOLUTION

$$\begin{aligned}(q + p)^3 &= {}_3C_0q^3 + {}_3C_1q^2p + {}_3C_2qp^2 + {}_3C_3p^3 \\ &= q^3 + 3q^2p + 3qp^2 + p^3\end{aligned}$$

Matched Problem 4 Use the binomial formula to expand $(q + p)^4$.

Binomial Distribution

We now generalize the discussion of Bernoulli trials to *binomial distributions.* We start by considering a sequence of three Bernoulli trials. Let the random variable X_3 represent the number of successes in three trials, 0, 1, 2, or 3. We are interested in the probability distribution for this random variable.

Which outcomes of an experiment consisting of a sequence of three Bernoulli trials lead to the random variable values 0, 1, 2, and 3, and what are the probabilities associated with these values? Table 1 answers these questions.

Table 1

Simple Event	Probability of Simple Event	X_3 x successes in 3 trials	$P(X_3 = x)$
FFF	$qqq = q^3$	0	q^3
FFS	$qqp = q^2p$	1	$3q^2p$
FSF	$qpq = q^2p$		
SFF	$pqq = q^2p$		
FSS	$qpp = qp^2$	2	$3qp^2$
SFS	$pqp = qp^2$		
SSF	$ppq = qp^2$		
SSS	$ppp = p^3$	3	p^3

The terms in the last column of Table 1 are the terms in the binomial expansion of $(q + p)^3$, as we saw in Example 4. The last two columns in Table 1 provide a probability distribution for the random variable X_3. Note that both conditions for a probability distribution (see Section 8.5) are met:

1. $0 \le P(X_3 = x) \le 1, \quad x \in \{0, 1, 2, 3\}$
2. $1 = 1^3 = (q + p)^3$ Recall that $q + p = 1$.
$$\begin{aligned}&= {}_3C_0q^3 + {}_3C_1q^2p + {}_3C_2qp^2 + {}_3C_3p^3 \\ &= q^3 + 3q^2p + 3qp^2 + p^3 \\ &= P(X_3 = 0) + P(X_3 = 1) + P(X_3 = 2) + P(X_3 = 3)\end{aligned}$$

Reasoning in the same way for the general case, we see why the probability distribution of a random variable associated with the number of successes in a sequence of n Bernoulli trials is called a *binomial distribution*—the probability of each number is a term in the binomial expansion of $(q + p)^n$. For this reason, a sequence of Bernoulli trials is often referred to as a *binomial experiment.* In terms of a formula, which we already discussed from another point of view (see Theorem 1), we have

DEFINITION Binomial Distribution

$$\begin{aligned}P(X_n = x) &= P(x \text{ successes in } n \text{ trials}) \\ &= {}_nC_xp^xq^{n-x} \qquad x \in \{0, 1, 2, \ldots, n\}\end{aligned}$$

where p is the probability of success and q is the probability of failure on each trial.

Informally, we will write $P(x)$ in place of $P(X_n = x)$.

EXAMPLE 5 **Constructing Tables and Histograms for Binomial Distributions** Suppose a fair die is rolled three times and success on a single roll is considered to be rolling a number divisible by 3. For the binomial distribution,

(A) Write the probability function.

(B) Construct a table.

(C) Draw a histogram.

SOLUTION

(A) $p = \frac{1}{3}$ Since two numbers out of six are divisible by 3

$q = 1 - p = \frac{2}{3}$

$n = 3$

Therefore,

$$P(x) = P(x \text{ successes in 3 trials}) = {}_3C_x\left(\tfrac{1}{3}\right)^x\left(\tfrac{2}{3}\right)^{3-x}$$

(B)

x	$P(x)$
0	${}_3C_0\left(\frac{1}{3}\right)^0\left(\frac{2}{3}\right)^3 \approx .30$
1	${}_3C_1\left(\frac{1}{3}\right)^1\left(\frac{2}{3}\right)^2 \approx .44$
2	${}_3C_2\left(\frac{1}{3}\right)^2\left(\frac{2}{3}\right)^1 \approx .22$
3	${}_3C_3\left(\frac{1}{3}\right)^3\left(\frac{2}{3}\right)^0 \approx .04$
	1.00

(C)

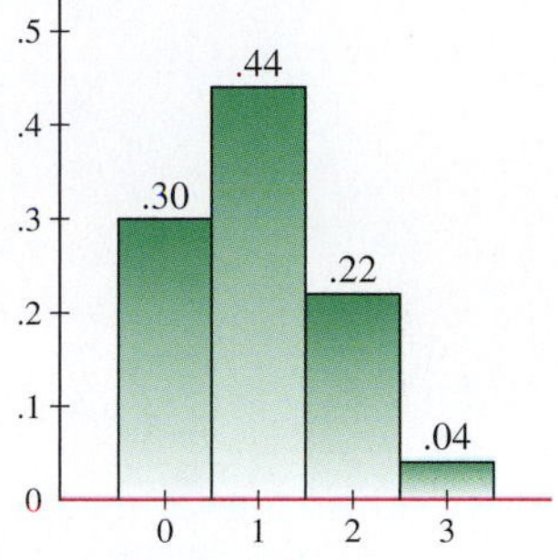

Figure 1

If we actually performed the binomial experiment in Example 5 a large number of times with a fair die, we would find that we would roll no number divisible by 3 in three rolls of a die about 30% of the time, one number divisible by 3 in three rolls about 44% of the time, two numbers divisible by 3 in three rolls about 22% of the time, and three numbers divisible by 3 in three rolls only 4% of the time. Note that the sum of all the probabilities is 1, as it should be.

The graphing calculator command in Figure 2A simulates 100 repetitions of the binomial experiment in Example 5. The number of successes on each trial is stored in list L_1. From Figure 2B, which shows a histogram of L_1, we note that the empirical probability of rolling one number divisible by 3 in three rolls is $\frac{48}{100} = 48\%$, close to the theoretical probability of 44%. The empirical probabilities of 0, 2, or 3 successes also would be close to the corresponding theoretical probabilities.

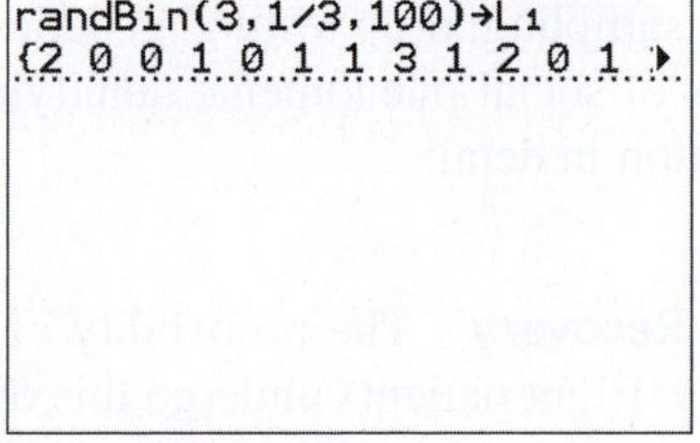

(A)

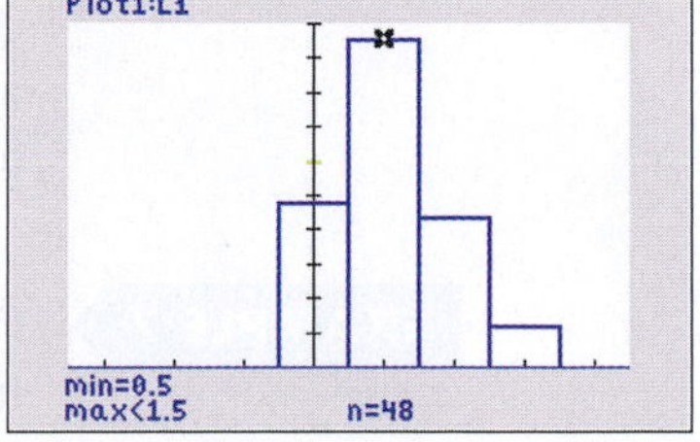

(B)

Figure 2

Matched Problem 5 Repeat Example 5, where the binomial experiment consists of two rolls of a die instead of three rolls.

Let X be a random variable with probability distribution

x_i	x_1	x_2	$\cdots$	x_n
p_i	p_1	p_2	$\cdots$	p_n

In Section 8.5 we defined the **expected value** of X to be

$$E(X) = x_1p_1 + x_2p_2 + \cdots + x_np_n$$

The expected value of X is also called the **mean** of the random variable X, often denoted by μ. The **standard deviation** of a random variable X having mean μ is defined by

$$\sigma = \sqrt{(x_1 - \mu)^2 \cdot p_1 + (x_2 - \mu)^2 \cdot p_2 + \cdots + (x_n - \mu)^2 \cdot p_n}$$

If a random variable has a binomial distribution, where n is the number of Bernoulli trials, p is the probability of success, and q the probability of failure, then the mean and standard deviation are given by the following formulas:

Mean: $\mu = np$

Standard deviation: $\sigma = \sqrt{npq}$

CONCEPTUAL INSIGHT

Let the random variable X_3 denote the number x of successes in a sequence of three Bernoulli trials. Then $x = 0, 1, 2,$ or 3. The expected value of X_3 is given by

$$\begin{aligned}
E(X_3) &= 0 \cdot P(0) + 1 \cdot P(1) + 2 \cdot P(2) + 3 \cdot P(3) && \text{See Table 1.}\\
&= 0 + 1 \cdot 3q^2p + 2 \cdot 3qp^2 + 3 \cdot p^3 && \text{Factor out } 3p.\\
&= 3p(q^2 + 2qp + p^2) && \text{Factor the perfect square.}\\
&= 3p(q + p)^2 && q + p = 1\\
&= 3p
\end{aligned}$$

This proves the formula $\mu = np$ in the case $n = 3$. Similar but more complicated computations can be used to justify the general formulas $\mu = np$ and $\sigma = \sqrt{npq}$ for the mean and standard deviation of random variables having binomial distributions.

EXAMPLE 6 **Computing the Mean and Standard Deviation of a Binomial Distribution** Compute the mean and standard deviation for the random variable in Example 5.

SOLUTION $n = 3 \quad p = \frac{1}{3} \quad q = 1 - \frac{1}{3} = \frac{2}{3}$

$$\mu = np = 3\left(\tfrac{1}{3}\right) = 1 \qquad \sigma = \sqrt{npq} = \sqrt{3\left(\tfrac{1}{3}\right)\left(\tfrac{2}{3}\right)} \approx .82$$

Matched Problem 6 Compute the mean and standard deviation for the random variable in Matched Problem 5.

Application

Binomial experiments are associated with a wide variety of practical problems: industrial sampling, drug testing, genetics, epidemics, medical diagnosis, opinion polls, analysis of social phenomena, qualifying tests, and so on. We will now consider one application in detail.

EXAMPLE 7 **Patient Recovery** The probability of recovering after a particular type of operation is .5. Eight patients undergo this operation. For the binomial distribution,

(A) Write the probability function.

(B) Construct a table.

(C) Construct a histogram.

(D) Find the mean and standard deviation.

SOLUTION

(A) Letting a recovery be a success, we have

$$p = .5 \qquad q = 1 - p = .5 \qquad n = 8$$

Hence,

$$P(x) = P(\text{exactly } x \text{ successes in 8 trials}) = {}_8C_x(.5)^x(.5)^{8-x} = {}_8C_x(.5)^8$$

(B)

x	$P(x)$
0	${}_8C_0(.5)^8 \approx .004$
1	${}_8C_1(.5)^8 \approx .031$
2	${}_8C_2(.5)^8 \approx .109$
3	${}_8C_3(.5)^8 \approx .219$
4	${}_8C_4(.5)^8 \approx .273$
5	${}_8C_5(.5)^8 \approx .219$
6	${}_8C_6(.5)^8 \approx .109$
7	${}_8C_7(.5)^8 \approx .031$
8	${}_8C_8(.5)^8 \approx .004$
	$.999 \approx 1$

The discrepancy in the sum is due to round-off errors.

(C)

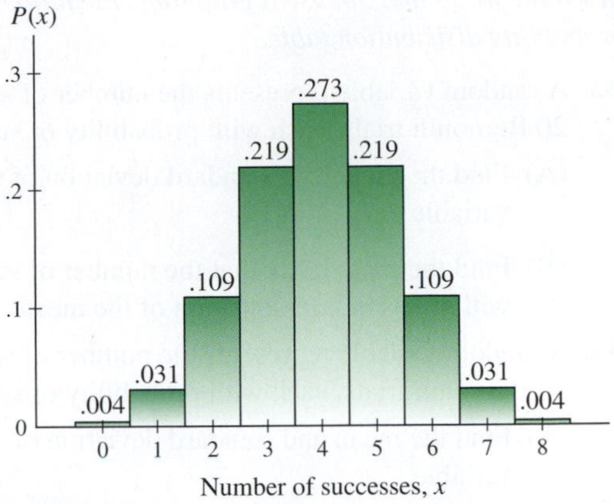

(D) $\mu = np = 8(.5) = 4 \qquad \sigma = \sqrt{npq} = \sqrt{8(.5)(.5)} \approx 1.41$

Matched Problem 7 Repeat Example 7 for four patients.

Exercises 10.4

A *Evaluate ${}_nC_x p^x q^{n-x}$ for the values of n, x, and p given in Problems 1–6.*

1. $n = 5, x = 1, p = \frac{1}{2}$

2. $n = 5, x = 2, p = \frac{1}{2}$

3. $n = 6, x = 3, p = .4$

4. $n = 6, x = 6, p = .4$

5. $n = 4, x = 3, p = \frac{2}{3}$

6. $n = 4, x = 3, p = \frac{1}{3}$

In Problems 7–12, a fair coin is tossed four times. What is the probability of obtaining

7. A head on the first toss and tails on each of the other tosses?

8. Exactly one head?

9. At least three tails?

10. Tails on each of the first three tosses?

11. No heads?

12. Four heads?

In Problems 13–18, construct a histogram for the binomial distribution $P(x) = {}_nC_x p^x q^{n-x}$, and compute the mean and standard deviation if

13. $n = 3, p = \frac{1}{4}$

14. $n = 3, p = \frac{3}{4}$

15. $n = 4, p = \frac{1}{3}$

16. $n = 5, p = \frac{1}{3}$

17. $n = 5, p = 0$

18. $n = 4, p = 1$

B *In Problems 19–24, round answers to four decimal places. A fair die is rolled four times. Find the probability of obtaining*

19. A 6, 6, 5, and 5 in that order.

20. Two 6's and two 5's in any order.

21. Exactly two 6's.

22. Exactly three 6's.

23. No 6's.

24. At least two 6's.

25. If a baseball player has a batting average of .350, what is the probability that the player will get the following number of hits in the next four times at bat?

(A) Exactly 2 hits (B) At least 2 hits

26. If a true–false test with 10 questions is given, what is the probability of scoring

(A) Exactly 70% just by guessing?

(B) 70% or better just by guessing?

27. A multiple-choice test consists of 10 questions, each with choices A, B, C, D, E (exactly one choice is correct). Which is more likely if you simply guess at each question: all your answers are wrong, or at least half are right? Explain.

28. If 60% of the electorate supports the mayor, what is the probability that in a random sample of 10 voters, fewer than half support her?

Construct a histogram for each of the binomial distributions in Problems 29–32. Compute the mean and standard deviation for each distribution.

29. $P(x) = {}_6C_x(.4)^x(.6)^{6-x}$ **30.** $P(x) = {}_6C_x(.6)^x(.4)^{6-x}$

31. $P(x) = {}_8C_x(.3)^x(.7)^{8-x}$ **32.** $P(x) = {}_8C_x(.7)^x(.3)^{8-x}$

In Problems 33 and 34, use a graphing calculator to construct a probability distribution table.

33. A random variable represents the number of successes in 20 Bernoulli trials, each with probability of success $p = .85$.

(A) Find the mean and standard deviation of the random variable.

(B) Find the probability that the number of successes lies within 1 standard deviation of the mean.

34. A random variable represents the number of successes in 20 Bernoulli trials, each with probability of success $p = .45$.

(A) Find the mean and standard deviation of the random variable.

(B) Find the probability that the number of successes lies within 1 standard deviation of the mean.

C *In Problems 35 and 36, a coin is loaded so that the probability of a head occurring on a single toss is $\frac{3}{4}$. In five tosses of the coin, what is the probability of getting*

35. All heads or all tails?

36. Exactly 2 heads or exactly 2 tails?

37. Toss a coin three times or toss three coins simultaneously, and record the number of heads. Repeat the binomial experiment 100 times and compare your relative frequency distribution with the theoretical probability distribution.

38. Roll a die three times or roll three dice simultaneously, and record the number of 5's that occur. Repeat the binomial experiment 100 times and compare your relative frequency distribution with the theoretical probability distribution.

39. Find conditions on p that guarantee the histogram for a binomial distribution is symmetrical about $x = n/2$. Justify your answer.

40. Consider two binomial distributions for 1,000 repeated Bernoulli trials—the first for trials with $p = .15$, and the second for trials with $p = .85$. How are the histograms for the two distributions related? Explain.

41. A random variable represents the number of heads in ten tosses of a coin.

(A) Find the mean and standard deviation of the random variable.

(B) Use a graphing calculator to simulate 200 repetitions of the binomial experiment, and compare the mean and standard deviation of the numbers of heads from the simulation to the answers for part (A).

42. A random variable represents the number of times a sum of 7 or 11 comes up in ten rolls of a pair of dice.

(A) Find the mean and standard deviation of the random variable.

(B) Use a graphing calculator to simulate 100 repetitions of the binomial experiment, and compare the mean and standard deviation of the numbers of 7's or 11's from the simulation to the answers for part (A).

Applications

43. Management training. Each year a company selects a number of employees for a management training program at a university. On average, 70% of those sent complete the program. Out of 7 people sent, what is the probability that

(A) Exactly 5 complete the program?

(B) 5 or more complete the program?

44. Employee turnover. If the probability of a new employee in a fast-food chain still being with the company at the end of 1 year is .6, what is the probability that out of 8 newly hired people,

(A) 5 will still be with the company after 1 year?

(B) 5 or more will still be with the company after 1 year?

45. Quality control. A manufacturing process produces, on average, 6 defective items out of 100. To control quality, each day a sample of 10 completed items is selected at random and inspected. If the sample produces more than 2 defective items, then the whole day's output is inspected, and the manufacturing process is reviewed. What is the probability of this happening, assuming that the process is still producing 6% defective items?

46. Guarantees. A manufacturing process produces, on average, 3% defective items. The company ships 10 items in each box and wants to guarantee no more than 1 defective item per box. If this guarantee applies to each box, what is the probability that the box will fail to meet the guarantee?

47. Quality control. A manufacturing process produces, on average, 5 defective items out of 100. To control quality, each day a random sample of 6 completed items is selected and inspected. If success on a single trial (inspection of 1 item) is finding the item defective, then the inspection of each of 6 items in the sample constitutes a binomial experiment. For the binomial distribution,

(A) Write the probability function.

(B) Construct a table.

(C) Draw a histogram.

(D) Compute the mean and standard deviation.

48. Management training. Each year a company selects 5 employees for a management training program at a university. On average, 40% of those sent complete the course in the top 10% of their class. If we consider an employee finishing in the top 10% of the class a success in a binomial experiment, then for the 5 employees entering the program, there exists a binomial distribution involving $P(x$ successes out of 5). For the binomial distribution,

(A) Write the probability function.

(B) Construct a table.

(C) Draw a histogram.

(D) Compute the mean and standard deviation.

49. Medical diagnosis. A tuberculosis patient is given a chest x-ray. Four tuberculosis x-ray specialists examine each x-ray independently. If each specialist can detect tuberculosis 80% of the time when it is present, what is the probability that at least 1 of the specialists will detect tuberculosis in this patient?

50. Harmful drug side effects. A pharmaceutical laboratory claims that a drug causes serious side effects in 20 people out of 1,000, on average. To check this claim, a hospital administers the drug to 10 randomly chosen patients and finds that 3 suffer from serious side effects. If the laboratory's claims are correct, what is the probability that the hospital gets these results?

51. Genetics. The probability that brown-eyed parents, both with the recessive gene for blue eyes, will have a child with brown eyes is .75. If such parents have 5 children, what is the probability that they will have

(A) All blue-eyed children?

(B) Exactly 3 children with brown eyes?

(C) At least 3 children with brown eyes?

52. Gene mutations. The probability of gene mutation under a given level of radiation is 3×10^{-5}. What is the probability of at least 1 gene mutation if 10^5 genes are exposed to this level of radiation?

53. Epidemics. If the probability of a person contracting influenza on exposure is .6, consider the binomial distribution for a family of 6 that has been exposed. For this distribution,

(A) Write the probability function.

(B) Construct a table.

(C) Draw a histogram.

(D) Compute the mean and standard deviation.

54. Drug side effects. The probability that a given drug will produce a serious side effect in a person using the drug is .02. In the binomial distribution for 450 people using the drug, what are the mean and standard deviation?

55. Testing. A multiple-choice test is given with 5 choices (only one is correct) for each of 10 questions. What is the probability of passing the test with a grade of 70% or better just by guessing?

56. Opinion polls. An opinion poll based on a small sample can be unrepresentative of the population. To see why, assume that 40% of the electorate favors a certain candidate. If a random sample of 7 is asked their preference, what is the probability that a majority will favor this candidate?

57. Testing. A multiple-choice test is given with 5 choices (only one is correct) for each of 5 questions. Answering each of the 5 questions by guessing constitutes a binomial experiment with an associated binomial distribution. For this distribution,

(A) Write the probability function.

(B) Construct a table.

(C) Draw a histogram.

(D) Compute the mean and standard deviation.

58. Sociology. The probability that a marriage will end in divorce within 10 years is .35. What are the mean and standard deviation for the binomial distribution involving 1,000 marriages?

59. Sociology. If the probability is .55 that a marriage will end in divorce within 20 years, what is the probability that out of 6 couples just married, in the next 20 years

(A) None will be divorced?

(B) All will be divorced?

(C) Exactly 2 will be divorced?

(D) At least 2 will be divorced?

Answers to Matched Problems

1. $p = \frac{1}{3}, q = \frac{2}{3}$ **2.** $p^3q^2 = \left(\frac{1}{6}\right)^3\left(\frac{5}{6}\right)^2 \approx .003$

3. (A) .402

(B) $1 - P(x = 0) = 1 - .402 = .598$

4. ${}_4C_0q^4 + {}_4C_1q^3p + {}_4C_2q^2p^2 + {}_4C_3qp^3 + {}_4C_4p^4 =$

$q_4 + 4q^3p + 6q^2p^2 + 4qp^3 + p^4$

5. (A) $P(x) = P(x \text{ successes in 2 trials})$

${}_2C_x\left(\frac{1}{3}\right)^x\left(\frac{2}{3}\right)^{2-x}, x \in \{0, 1, 2\}$

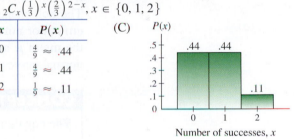

(B)

x	$P(x)$
0	$\frac{4}{9} \approx .44$
1	$\frac{4}{9} \approx .44$
2	$\frac{1}{9} \approx .11$

(C)

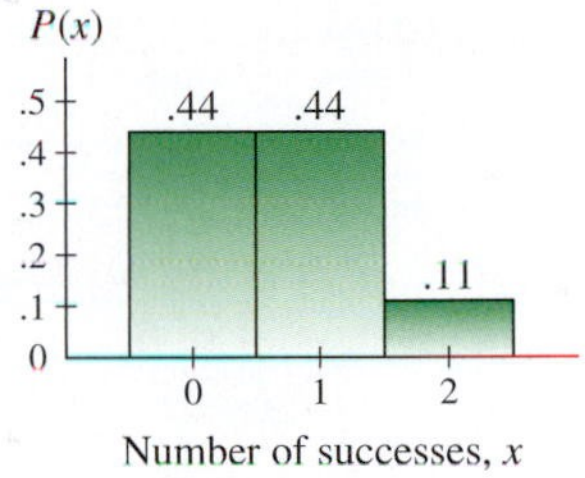

6. $\mu \approx .67; \sigma \approx .67$

7. (A) $P(x) = P(\text{exactly } x \text{ successes in 4 trials}) = {}_4C_x(.5)^4$

(B)

x	$P(x)$
0	.06
1	.25
2	.38
3	.25
4	.06
	1.00

(C)

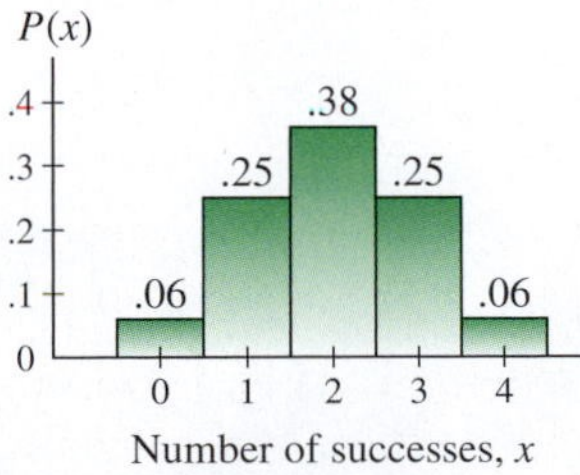

(D) $\mu = 2; \sigma = 1$

10.5 Normal Distributions

- Normal Distribution
- Areas under Normal Curves
- Approximating a Binomial Distribution with a Normal Distribution

Normal Distribution

If we take the histogram for a binomial distribution, such as the one we drew for Example 7, Section 10.4 ($n = 8, p = .5$), and join the midpoints of the tops of the bars with a smooth curve, we obtain the bell-shaped curve in Figure 1.

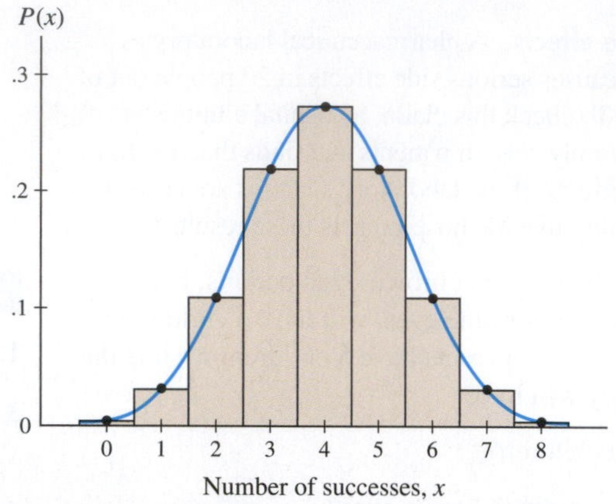

Figure 1 Binomial distribution and bell-shaped curve

The mathematical foundation for this type of curve was established by Abraham De Moivre (1667–1754), Pierre Laplace (1749–1827), and Carl Gauss (1777–1855). The bell-shaped curves studied by these famous mathematicians are called **normal curves** or **normal probability distributions**, and their equations are determined by the **mean** μ and **standard deviation** σ of the distribution. Figure 2 illustrates three normal curves with different means and standard deviations.

CONCEPTUAL INSIGHT

The equation for a normal curve is fairly complicated:

$$f(x) = \frac{1}{\sigma\sqrt{2\pi}} e^{-(x-\mu)^2/2\sigma^2}$$

where $\pi \approx 3.1416$ and $e \approx 2.7183$. Given the values of μ and σ, however, the function is completely specified, and we could plot points or use a graphing calculator to produce its graph. Substituting $x + h$ for x produces an equation of the same form but with a different value of μ. Therefore, in the terminology of Section 2.2, any horizontal translation of a normal curve is another normal curve.

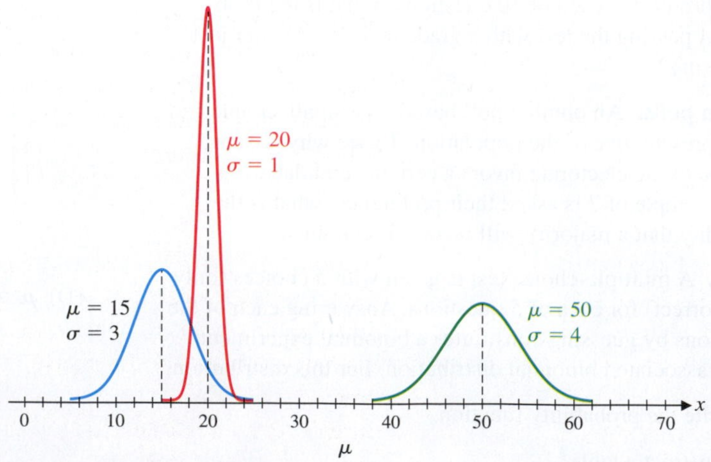

Figure 2 Normal probability distributions

Until now we have dealt with **discrete random variables**, that is, random variables that assume a finite or "countably infinite" number of values (we have dealt only with the finite case). Random variables associated with normal distributions are *continuous* in nature; they assume all values over an interval on a real number line. These are called **continuous random variables**. Random variables associated with people's heights, lightbulb lifetimes, or the lengths of time between breakdowns of a copy machine are continuous. The following is a list of some of the important properties of normal curves (normal probability distributions of a continuous random variable):

Properties of Normal Curves

1. Normal curves are bell shaped and symmetrical with respect to a vertical line.
2. The mean is the real number at the point where the axis of symmetry intersects the horizontal axis.
3. The shape of a normal curve is completely determined by its mean and a positive real number called the standard deviation. A small standard deviation indicates a tight clustering about the mean and a tall, narrow curve; a large standard deviation indicates a large deviation from the mean and a broad, flat curve (see Fig. 2).
4. Irrespective of the shape, the area between the curve and the x axis is always 1.
5. Irrespective of the shape, 68.3% of the area will lie within an interval of 1 standard deviation on either side of the mean, 95.4% within 2 standard deviations on either side, and 99.7% within 3 standard deviations on either side (see Fig. 3).

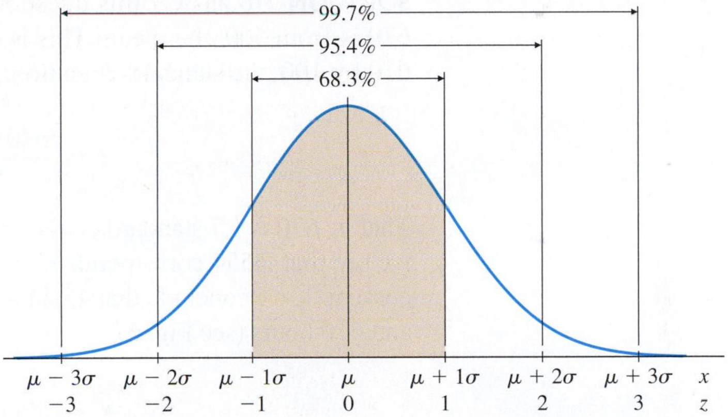

Figure 3 Normal curve areas

The normal probability distribution is the most important of all theoretical distributions, and it is a useful tool for solving practical problems. Not only does a normal curve provide a good approximation for a binomial distribution for large n, but it also approximates many other relative frequency distributions. For example, normal curves often provide good approximations for the relative frequency distributions for heights and weights of people, measurements of manufactured parts, scores on IQ tests, college entrance examinations, civil service tests, and measurements of errors in laboratory experiments.

Areas under Normal Curves

To use normal curves in practical problems, we must be able to determine areas under different parts of a normal curve. Remarkably, the area under a normal curve between a mean μ and a given number of standard deviations to the right (or left) of μ is the same, regardless of the shape of the normal curve. For example, the area under the normal curve with $\mu = 3$, $\sigma = 5$ from $\mu = 3$ to $\mu + 1.5\sigma = 10.5$ is equal to the area under the normal curve with $\mu = 15$, $\sigma = 2$ from $\mu = 15$ to $\mu + 1.5\sigma = 18$ (see Fig. 4, noting that the shaded regions have the same areas, or equivalently, the same numbers of pixels). Therefore, such areas for any normal curve can be easily determined from the areas for the **standard normal curve**, that is, the normal curve with mean 0 and standard deviation 1. In fact, if z represents the number of standard deviations that a measurement x is from a mean μ, then the area under a normal curve from μ to $\mu + z\sigma$ equals the area under the standard normal curve from 0 to z (see Fig. 5). Appendix C lists those areas for the standard normal curve.

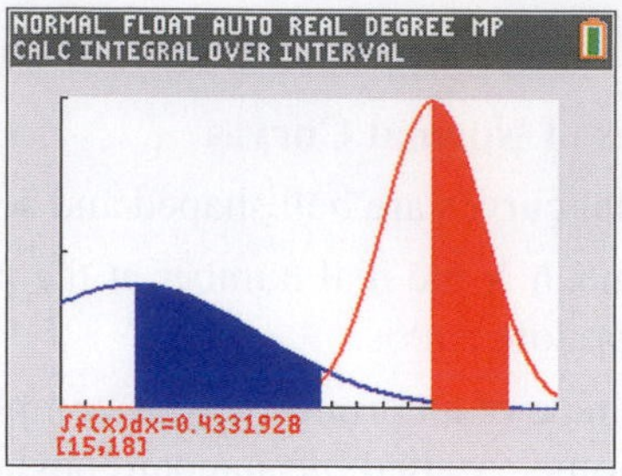

Figure 4

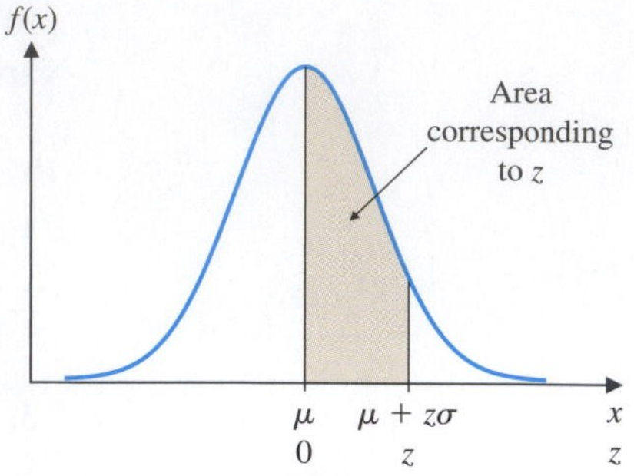

Figure 5 Areas and z values

EXAMPLE 1 **Finding Probabilities for a Normal Distribution** A manufacturing process produces lightbulbs with life expectancies that are normally distributed with a mean of 500 hours and a standard deviation of 100 hours. What percentage of the lightbulbs can be expected to last between 500 and 670 hours?

SOLUTION To answer this question, we determine how many standard deviations 670 is from 500, the mean. This is done by dividing the distance between 500 and 670 by 100, the standard deviation. Thus,

$$z = \frac{670 - 500}{100} = \frac{170}{100} = 1.70$$

That is, 670 is 1.7 standard deviations from 500, the mean. Referring to Appendix C, we see that .4554 corresponds to $z = 1.70$. And since the total area under a normal curve is 1, we conclude that 45.54% of the lightbulbs produced will last between 500 and 670 hours (see Fig. 6).

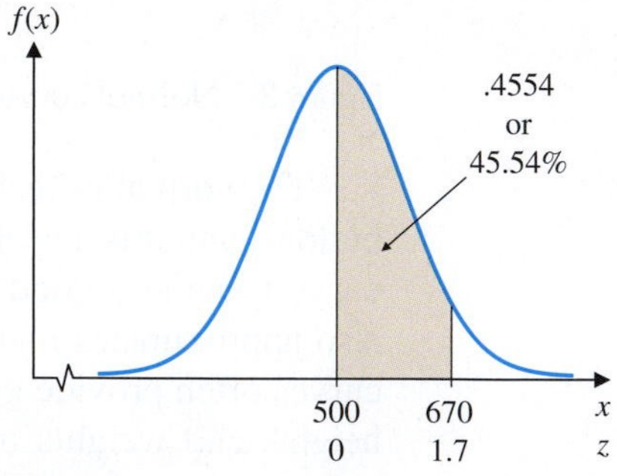

Figure 6 Lightbulb life expectancy: positive z

Matched Problem 1 What percentage of the lightbulbs in Example 1 can be expected to last between 500 and 750 hours?

In general, to find how many standard deviations that a measurement x is from a mean μ, first determine the distance between x and μ and then divide by σ:

$$z = \frac{\textbf{distance between } x \textbf{ and } \mu}{\textbf{standard deviation}} = \frac{x - \mu}{\sigma}$$

EXAMPLE 2 **Finding Probabilities for a Normal Distribution** From all lightbulbs produced (see Example 1), what is the probability that a lightbulb chosen at random lasts between 380 and 500 hours?

SOLUTION To answer this, we first find z:

$$z = \frac{x - \mu}{\sigma} = \frac{380 - 500}{100} = -1.20$$

It is usually a good idea to draw a rough sketch of a normal curve and insert relevant data (see Fig. 7).

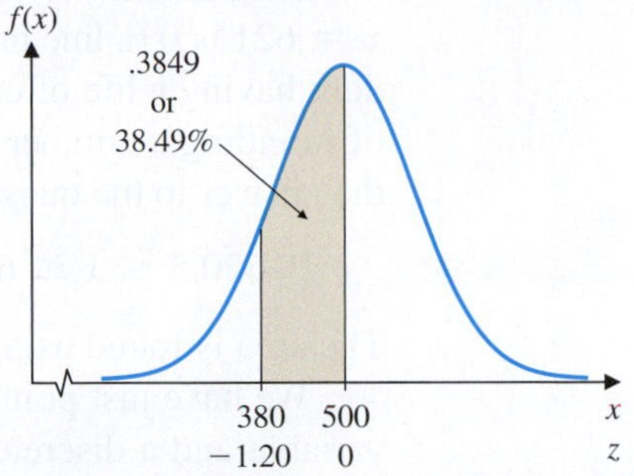

Figure 7 Lightbulb life expectancy: negative z

Appendix C does not include negative values for z, but because normal curves are symmetrical with respect to a vertical line through the mean, we simply use the absolute value (positive value) of z for the table. So the area corresponding to $z = -1.20$ is the same as the area corresponding to $z = 1.20$, which is .3849. And since the area under the whole normal curve is 1, we conclude that the probability of a lightbulb chosen at random lasting between 380 and 500 hours is .3849.

Matched Problem 2 Refer to Example 1. What is the probability that a lightbulb chosen at random lasts between 400 and 500 hours?

The first graphing calculator command in Figure 8A simulates the life expectancies of 100 lightbulbs by generating 100 random numbers from the normal distribution with $\mu = 500$, $\sigma = 100$ of Example 1. The numbers are stored in list L_1. Note from Figure 8A that the mean and standard deviation of L_1 are close to the mean and standard deviation of the normal distribution. From Figure 8B, which shows a histogram of L_1, we note that the empirical probability that a lightbulb lasts between 380 and 500 hours is

$$\frac{11 + 18 + 13}{100} = .42$$

which is close to the theoretical probability of .3849 computed in Example 2.

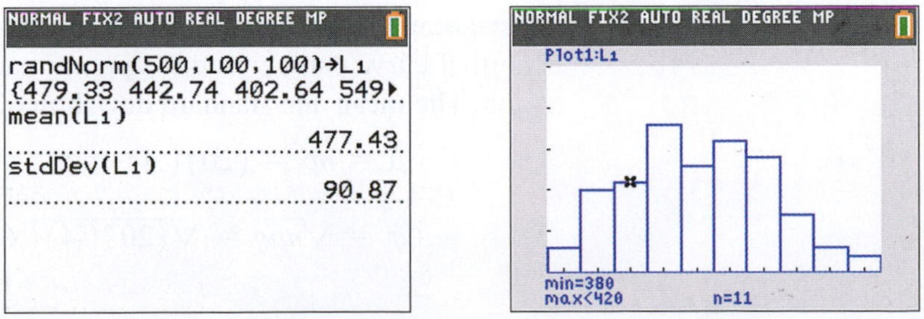

Figure 8

Several important properties of a continuous random variable with normal distribution are listed below:

Properties of a Normal Probability Distribution

1. $P(a \le x \le b) =$ area under the normal curve from a to b
2. $P(-\infty < x < \infty) = 1 =$ total area under the normal curve
3. $P(x = c) = 0$

In Example 2, what is the probability of a lightbulb chosen at random having a life of exactly 621 hours? The area above 621 and below the normal curve at $x = 621$ is 0 (a line has no width). Thus, the probability of a lightbulb chosen at random having a life of exactly 621 hours is 0. However, if the number 621 is the result of rounding a number between 620.5 and 621.5 (which is most likely the case), then the answer to the question is

$$P(620.5 \le x \le 621.5) = \text{area under the normal curve from 620.5 to 621.5}$$

The area is found using the procedures outlined in Example 2.

We have just pointed out an important distinction between a continuous random variable and a discrete random variable: For a probability distribution of a continuous random variable, the probability of x assuming a single value is always 0. On the other hand, for a probability distribution of a discrete random variable, the probability of x assuming a particular value from the set of permissible values is usually a positive number between 0 and 1.

Approximating a Binomial Distribution with a Normal Distribution

You no doubt discovered from problems in Exercises 10.4 that when a binomial random variable assumes a large number of values (that is, when n is large), the use of the probability distribution formula

$$P(x \text{ successes in } n \text{ trials}) = {}_nC_x p^x q^{n-x}$$

becomes tedious. It would be helpful if there was an easily computed approximation of this distribution for large n. Such a distribution is found in the form of an appropriately selected normal distribution.

To clarify ideas and relationships, let us consider an example of a normal distribution approximation of a binomial distribution with a relatively small value of n. Then we will consider an example with a large value of n.

EXAMPLE 3 **Market Research** A credit card company claims that their card is used by 40% of the people buying gasoline in a particular city. A random sample of 20 gasoline purchasers is made. If the company's claim is correct, what is the probability that

(A) From 6 to 12 people in the sample use the card?

(B) Fewer than 4 people in the sample use the card?

SOLUTION We begin by drawing a normal curve with the same mean and standard deviation as the binomial distribution (Fig. 9). A histogram superimposed on this normal curve can be used to approximate the histogram for the binomial distribution. The mean and standard deviation of the binomial distribution are

$$\mu = np = (20)(.4) = 8 \qquad n = \text{sample size}$$

$$\sigma = \sqrt{npq} = \sqrt{(20)(.4)(.6)} \approx 2.19 \qquad p = .4 \text{ (from the 40\% claim)}$$

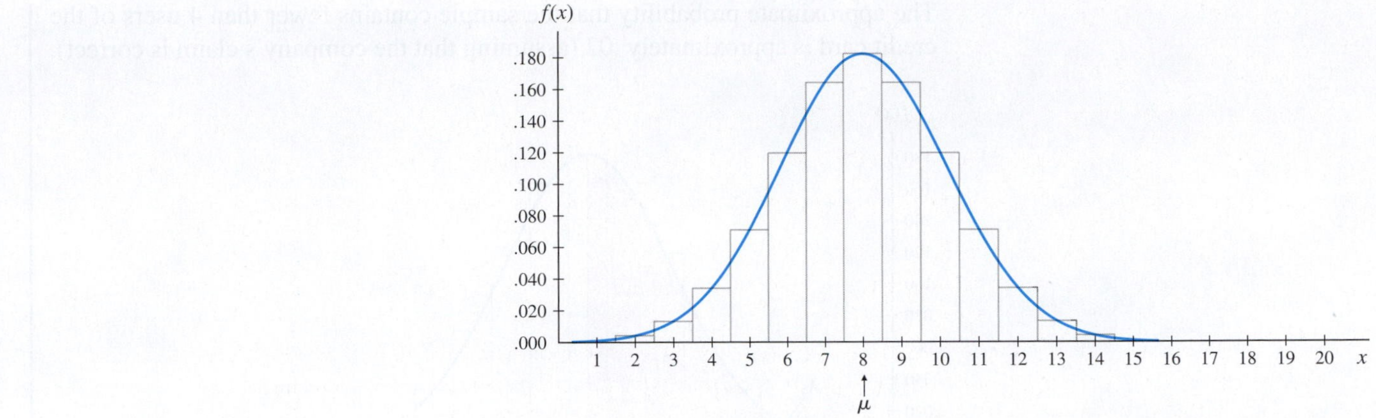

Figure 9

(A) To approximate the probability that 6 to 12 people in the sample use the credit card, we find the area under the normal curve from 5.5 to 12.5. We use 5.5 rather than 6 because the rectangle in the histogram corresponding to 6 extends from 5.5 to 6.5. Reasoning in the same way, we use 12.5 instead of 12. To use Appendix C, we split the area into two parts: A_1 to the left of the mean and A_2 to the right of the mean. The sketch in Figure 10 is helpful. Areas A_1 and A_2 are found as follows:

$$z_1 = \frac{x_1 - \mu}{\sigma} = \frac{5.5 - 8}{2.19} \approx -1.14 \qquad A_1 = .3729$$

$$z_2 = \frac{x_2 - \mu}{\sigma} = \frac{12.5 - 8}{2.19} \approx 2.05 \qquad A_2 = .4798$$

$$\text{Total area} = A_1 + A_2 = .8527$$

The approximate probability that the sample will contain between 6 and 12 users of the credit card is .85 (assuming that the firm's claim is correct).

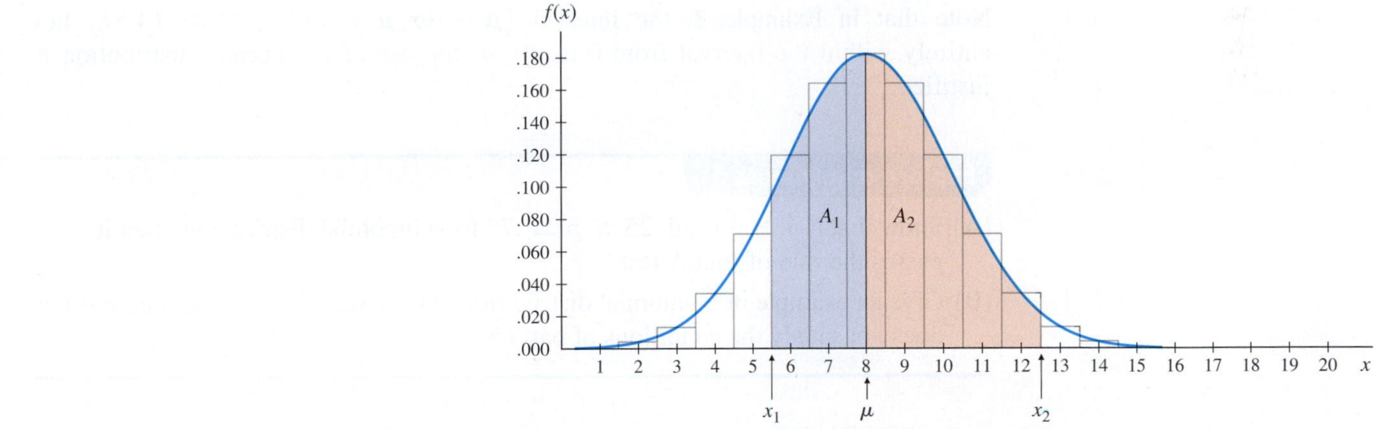

Figure 10

(B) To use the normal curve to approximate the probability that the sample contains fewer than 4 users of the credit card, we must find the area A_1 under the normal curve to the left of 3.5. The sketch in Figure 11 is useful. Since the total area under either half of the normal curve is .5, we first use Appendix C to find the area A_2 under the normal curve from 3.5 to the mean 8, and then subtract A_2 from .5:

$$z = \frac{x - \mu}{\sigma} = \frac{3.5 - 8}{2.19} \approx -2.05 \qquad A_2 = .4798$$

$$A_1 = .5 - A_2 = .5 - .4798 = .0202$$

The approximate probability that the sample contains fewer than 4 users of the credit card is approximately .02 (assuming that the company's claim is correct).

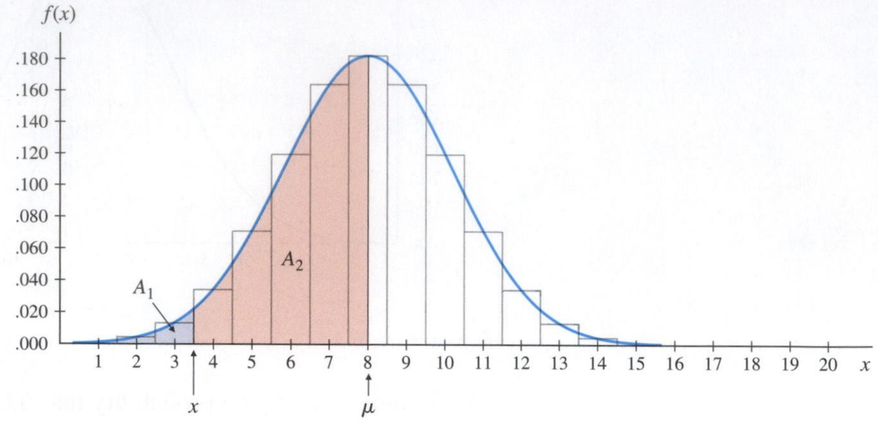

Figure 11

Matched Problem 3 In Example 3, use the normal curve to approximate the probability that in the sample there are

(A) From 5 to 9 users of the credit card.

(B) More than 10 users of the card.

You no doubt are wondering how large n should be before a normal distribution provides an adequate approximation for a binomial distribution. Without getting too involved, the following rule of thumb provides a good test:

Use a normal distribution to approximate a binomial distribution only if the interval $[\mu - 3\sigma, \mu + 3\sigma]$ lies entirely in the interval from 0 to n.

Note that in Example 3, the interval $[\mu - 3\sigma, \mu + 3\sigma] = [1.43, 14.57]$ lies entirely within the interval from 0 to 20; so the use of the normal distribution is justified.

Explore and Discuss 1

(A) Show that if $n \geq 30$ and $.25 \leq p \leq .75$ for a binomial distribution, then it passes the rule-of-thumb test.

(B) Give an example of a binomial distribution that passes the rule-of-thumb test but does not satisfy the conditions of part (A).

EXAMPLE 4 **Quality Control** A company manufactures 50,000 ballpoint pens each day. The manufacturing process produces 50 defective pens per 1,000, on average. A random sample of 400 pens is selected from each day's production and tested. What is the probability that the sample contains

(A) At least 14 and no more than 25 defective pens?

(B) 33 or more defective pens?

SOLUTION Is it appropriate to use a normal distribution to approximate this binomial distribution? The answer is yes, since the rule-of-thumb test passes with ease:

$$\mu = np = 400(.05) = 20 \qquad p = \frac{50}{1{,}000} = .05$$

$$\sigma = \sqrt{npq} = \sqrt{400(.05)(.95)} \approx 4.36$$

$$[\mu - 3\sigma, \mu + 3\sigma] = [6.92, 33.08]$$

This interval is well within the interval from 0 to 400.

(A) To find the approximate probability of the number of defective pens in a sample being at least 14 and not more than 25, we find the area under the normal curve from 13.5 to 25.5. To use Appendix C, we split the area into an area to the left of the mean and an area to the right of the mean, as shown in Figure 12.

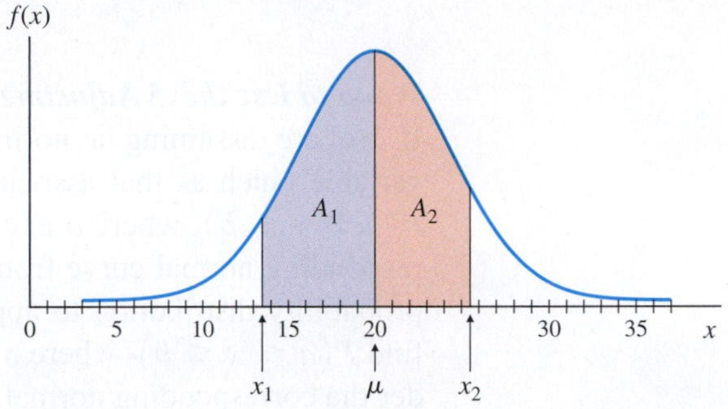

Figure 12

$$z_1 = \frac{x_1 - \mu}{\sigma} = \frac{13.5 - 20}{4.36} \approx -1.49 \qquad A_1 = .4319$$

$$z_2 = \frac{x_2 - \mu}{\sigma} = \frac{25.5 - 20}{4.36} \approx 1.26 \qquad A_2 = .3962$$

$$\text{Total area} = A_1 + A_2 = .8281$$

The approximate probability of the number of defective pens in the sample being at least 14 and not more than 25 is .83.

(B) Since the total area under a normal curve from the mean on is .5, we find the area A_1 (see Fig. 13) from Appendix C and subtract it from .5 to obtain A_2.

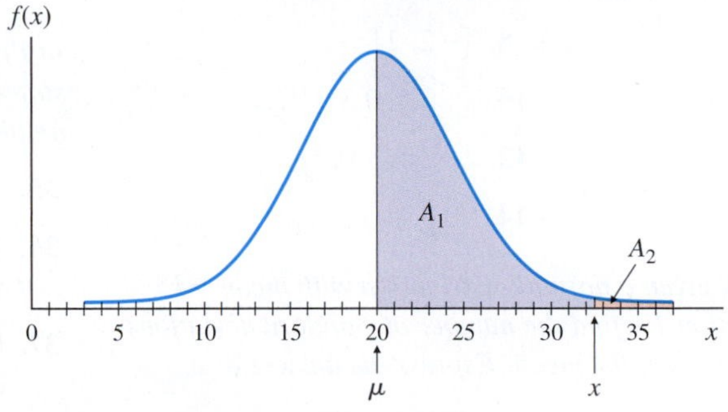

Figure 13

$$z = \frac{x - \mu}{\sigma} = \frac{32.5 - 20}{4.36} \approx 2.87 \qquad A_1 = .4979$$

$$A_2 = .5 - A_1 = .5 - .4979 = .0021 \approx .002$$

The approximate probability of finding 33 or more defective pens in the sample is .002. If a random sample of 400 included more than 33 defective pens, then the management could conclude that either a rare event has happened and the manufacturing process is still producing only 50 defective pens per 1,000, on average, or that something is wrong with the manufacturing process and it is producing more than 50 defective pens per 1,000, on average. The company might have a policy of checking the manufacturing process whenever 33 or more defective pens are found in a sample rather than believing a rare event has happened and that the manufacturing process is still running smoothly.

Matched Problem 4 Suppose in Example 4 that the manufacturing process produces 40 defective pens per 1,000, on average. What is the approximate probability that in the sample of 400 pens there are

(A) At least 10 and no more than 20 defective pens?

(B) 27 or more defective pens?

When to Use the .5 Adjustment

If we are assuming a normal probability distribution for a continuous random variable (such as that associated with heights or weights of people), then we find $P(a \le x \le b)$, where a and b are real numbers, by finding the area under the corresponding normal curve from a to b (see Example 2). However, if we use a normal probability distribution to approximate a binomial probability distribution, then we find $P(a \le x \le b)$, where a and b are nonnegative integers, by finding the area under the corresponding normal curve from $a - .5$ to $b + .5$ (see Examples 3 and 4).

Exercises 10.5

A *In Problems 1–6, use Appendix C to find the area under the standard normal curve from 0 to the indicated measurement.*

1. 2.00 **2.** 3.30

3. 1.24 **4.** 1.08

5. −2.75 **6.** −0.92

In Problems 7–14, use Appendix C to find the area under the standard normal curve and above the given interval on the horizontal axis.

7. $[-1, 1]$ **8.** $[-2, 2]$

9. $[-0.4, 0.7]$ **10.** $[-0.5, 0.3]$

11. $[0.2, 1.8]$ **12.** $[-2.1, -0.9]$

13. $(-\infty, 0.1]$ **14.** $[0.6, \infty)$

In Problems 15–20, given a normal distribution with mean −15 and standard deviation 10, find the number of standard deviations each measurement is from the mean. Express the answer as a positive number.

15. −22 **16.** 6.4

17. −1.8 **18.** −13.5

19. 10.9 **20.** −48.6

In Problems 21–26, given a normal distribution with mean 25 and standard deviation 5, find the area under the normal curve from the mean to the indicated measurement.

21. 27.2 **22.** 36.1

23. 12.8 **24.** 18.7

25. 28.3 **26.** 23.9

B *In Problems 27–34, consider the normal distribution with mean 60 and standard deviation 12. Find the area under the normal curve and above the given interval on the horizontal axis.*

27. $[48, 60]$ **28.** $[60, 84]$

29. $[57, 63]$ **30.** $[54, 66]$

31. $(-\infty, 54]$ **32.** $[63, \infty)$

33. $[51, \infty)$ **34.** $(-\infty, 78]$

In Problems 35–40, discuss the validity of each statement. If the statement is always true, explain why. If not, give a counterexample.

35. The mean of a normal distribution is a positive real number.

36. The standard deviation of a normal distribution is a positive real number.

37. If two normal distributions have the same mean and standard deviation, then they have the same shape.

38. If two normal distributions have the same mean, then they have the same standard deviation.

39. The area under a normal distribution and above the horizontal axis is equal to 1.

40. In a normal distribution, the probability is 0 that a score lies more than 3 standard deviations away from the mean.

In Problems 41–48, use the rule-of-thumb test to check whether a normal distribution (with the same mean and standard deviation as the binomial distribution) is a suitable approximation for the binomial distribution with

41. $n = 15, p = .7$ **42.** $n = 12, p = .6$

43. $n = 15, p = .4$ **44.** $n = 20, p = .6$

45. $n = 100, p = .05$ **46.** $n = 200, p = .03$

47. $n = 500, p = .05$ **48.** $n = 400, p = .08$

49. The probability of success in a Bernoulli trial is $p = .1$. Explain how to determine the number of repeated trials necessary to obtain a binomial distribution that passes the rule-of-thumb test for using a normal distribution as a suitable approximation.

50. For a binomial distribution with $n = 100$, explain how to determine the smallest and largest values of p that pass the rule-of-thumb test for using a normal distribution as a suitable approximation.

C *A binomial experiment consists of 500 trials. The probability of success for each trial is .4. What is the probability of obtaining the number of successes indicated in Problems 51–58? Approximate these probabilities to two decimal places using a normal curve. (This binomial experiment easily passes the rule-of-thumb test, as you can check. When computing the probabilities, adjust the intervals as in Examples 3 and 4.)*

51. 185–220 **52.** 190–205

53. 210–220 **54.** 175–185

55. 225 or more **56.** 212 or more

57. 175 or less **58.** 188 or less

To graph Problems 59–62, use a graphing calculator and refer to the normal probability distribution function with mean μ and standard deviation σ:

$$f(x) = \frac{1}{\sigma\sqrt{2\pi}} e^{-(x-\mu)^2/2\sigma^2} \qquad (1)$$

59. Graph equation (1) with $\sigma = 5$ and

(A) $\mu = 10$ (B) $\mu = 15$ (C) $\mu = 20$

Graph all three in the same viewing window with Xmin $= -10$, Xmax $= 40$, Ymin $= 0$, and Ymax $= 0.1$.

60. Graph equation (1) with $\sigma = 5$ and

(A) $\mu = 8$ (B) $\mu = 12$ (C) $\mu = 16$

Graph all three in the same viewing window with Xmin $= -10$, Xmax $= 30$, Ymin $= 0$, and Ymax $= 0.1$.

61. Graph equation (1) with $\mu = 20$ and

(A) $\sigma = 2$ (B) $\sigma = 4$

Graph both in the same viewing window with Xmin $= 0$, Xmax $= 40$, Ymin $= 0$, and Ymax $= 0.2$.

62. Graph equation (1) with $\mu = 18$ and

(A) $\sigma = 3$ (B) $\sigma = 6$

Graph both in the same viewing window with Xmin $= 0$, Xmax $= 40$, Ymin $= 0$, and Ymax $= 0.2$.

63. (A) If 120 scores are chosen from a normal distribution with mean 75 and standard deviation 8, how many scores x would be expected to satisfy $67 \le x \le 83$?

(B) Use a graphing calculator to generate 120 scores from the normal distribution with mean 75 and standard deviation 8. Determine the number of scores x such that $67 \le x \le 83$, and compare your results with the answer to part (A).

64. (A) If 250 scores are chosen from a normal distribution with mean 100 and standard deviation 10, how many scores x would be expected to be greater than 110?

(B) Use a graphing calculator to generate 250 scores from the normal distribution with mean 100 and standard deviation 10. Determine the number of scores greater than 110, and compare your results with the answer to part (A).

Applications

65. Sales. Salespeople for a solar technology company have average annual sales of \$200,000, with a standard deviation of \$20,000. What percentage of the salespeople would be expected to make annual sales of \$240,000 or more? Assume a normal distribution.

66. Guarantees. The average lifetime for a car battery is 170 weeks, with a standard deviation of 10 weeks. If the company guarantees the battery for 3 years, what percentage of the batteries sold would be expected to be returned before the end of the warranty period? Assume a normal distribution.

67. Quality control. A manufacturing process produces a critical part of average length 100 millimeters, with a standard deviation of 2 millimeters. All parts deviating by more than 5 millimeters from the mean must be rejected. What percentage of the parts must be rejected, on the average? Assume a normal distribution.

68. Quality control. An automated manufacturing process produces a component with an average width of 7.55 centimeters, with a standard deviation of 0.02 centimeter. All components deviating by more than 0.05 centimeter from the mean must be rejected. What percentage of the parts must be rejected, on the average? Assume a normal distribution.

69. Marketing claims. A company claims that 60% of the households in a given community use its product. A competitor surveys the community, using a random sample of 40 households, and finds only 15 households out of the 40 in the sample use the product. If the company's claim is correct, what is the probability of 15 or fewer households using the product in a sample of 40? Conclusion? Approximate a binomial distribution with a normal distribution.

70. Labor relations. A union representative claims 60% of the union membership will vote in favor of a particular settlement. A random sample of 100 members is polled, and out of these, 47 favor the settlement. What is the approximate probability of 47 or fewer in a sample of 100 favoring the settlement when 60% of all the membership favor the settlement? Conclusion? Approximate a binomial distribution with a normal distribution.

71. Medicine. The average healing time of a certain type of incision is 240 hours, with standard deviation of 20 hours. What percentage of the people having this incision would heal in 8 days or less? Assume a normal distribution.

72. Agriculture. The average height of a hay crop is 38 inches, with a standard deviation of 1.5 inches. What percentage of the crop will be 40 inches or more? Assume a normal distribution.

73. Genetics. In a family with 2 children, the probability that both children are girls is approximately .25. In a random sample of 1,000 families with 2 children, what is the approximate probability that 220 or fewer will have 2 girls? Approximate a binomial distribution with a normal distribution.

74. Genetics. In Problem 73, what is the approximate probability of the number of families with 2 girls in the sample being at least 225 and not more than 275? Approximate a binomial distribution with a normal distribution.

75. Testing. Scholastic Aptitude Tests (SATs) are scaled so that the mean score is 500 and the standard deviation is 100. What percentage of students taking this test should score 700 or more? Assume a normal distribution.

76. Politics. Candidate Harkins claims that she will receive 52% of the vote for governor. Her opponent, Mankey, finds that 470 out of a random sample of 1,000 registered voters favor Harkins. If Harkins's claim is correct, what is the probability that only 470 or fewer will favor her in a random sample of 1,000? Conclusion? Approximate a binomial distribution with a normal distribution.

77. Grading on a curve. An instructor grades on a curve by assuming that grades on a test are normally distributed. If the average grade is 70 and the standard deviation is 8, find the test scores for each grade interval if the instructor assigns grades as follows: 10% A's, 20% B's, 40% C's, 20% D's, and 10% F's.

78. Psychology. A test devised to measure aggressive–passive personalities was standardized on a large group of people. The scores were normally distributed with a mean of 50 and a standard deviation of 10. If we designate the highest 10% as aggressive, the next 20% as moderately aggressive, the middle 40% as average, the next 20% as moderately passive, and the lowest 10% as passive, what ranges of scores will be covered by these five designations?

Answers to Matched Problems

1. 49.38%
2. .3413
3. (A) .70 (B) .13
4. (A) .83 (B) .004

Chapter 10 Summary and Review

Important Terms, Symbols, and Concepts

10.1 Graphing data

EXAMPLES

- **Bar graphs, broken-line graphs,** and **pie graphs** are used to present visual interpretations or comparisons of data. Large sets of **quantitative data** can be organized in a **frequency table**, generally constructed by choosing 5 to 20 **class intervals** of equal length to cover the **data range**. The number of measurements that fall in a given class interval is called the **class frequency**, and the set of all such frequencies associated with their respective classes is called a **frequency distribution**. The **relative frequency** of a class is its frequency divided by the total number of items in the data set. — Ex. 1, p. 505
- A **histogram** is a vertical bar graph used to represent a frequency distribution. A **frequency polygon** is a broken-line graph obtained by joining successive midpoints of the tops of the bars in a histogram. A **cumulative frequency polygon**, or **ogive**, is obtained by plotting the cumulative frequency over the upper boundary of the corresponding class. — Ex. 2, p. 506

10.2 Measures of Central Tendency

- The **mean** of a set of quantitative data is the sum of all the measurements in the set divided by the total number of measurements in the set. — Ex. 1, p. 513
- The **mean for data grouped into classes** is a weighted average of the midpoints of the class intervals. — Ex. 2, p. 514
- When a data set has n measurements and these measurements are arranged in ascending or descending order, the **median** is the middle measurement when n is odd and the mean of the two middle measurements when n is even. — Ex. 3, p. 516
- The **median for grouped data** with no classes of frequency 0 is the number such that the histogram has the same area to the left of the median as to the right of the median. — Ex. 4, p. 517
- The **mode** is the most frequently occurring measurement in a data set. — Ex. 5, p. 518

10.3 Measures of Dispersion

- The **variance** and the **standard deviation** of a set of ungrouped measurements indicate how the data is dispersed relative to the mean. The same can be said about a set of grouped data. Ex. 1, p. 524 Ex. 2, p. 525

10.4 Bernoulli Trials and Binomial Distributions

- A sequence of experiments is called a **sequence of Bernoulli trials**, or a **binomial experiment**, if
  1. Only two outcomes are possible in each trial. Ex. 1, p. 529
  2. The probability of success p is the same for each trial. Ex. 2, p. 530
  3. All trials are independent. Ex. 3, p. 531
- If the random variable X_n represents the number of successes in n Bernoulli trials, then the probability distribution of X_n is the **binomial distribution** given by $P(X_n = x) = {}_nC_x p^x q^{n-x}, x = 0, 1, \ldots, n$. Ex. 5, p. 533
- The **mean** and **standard deviation** of a binomial distribution are given by the formulas $\mu = np$ and $\sigma = \sqrt{npq}$, respectively. Ex. 6, p. 534

10.5 Normal Distributions

- **Normal curves** are bell-shaped continuous curves that approximate the relative frequency distributions of many different types of measurements.
- The probability that a normally distributed measurement lies between a and b, denoted $P(a \le x \le b)$, is equal to the area under the normal curve from a to b. Ex. 1, p. 540 Ex. 2, p. 541
- To approximate a binomial distribution that is associated with a sequence of n Bernoulli trials, each having probability of success p, use a normal distribution with $\mu = np$ and $\sigma = \sqrt{npq}$. Ex. 3, p. 542 Ex. 4, p. 544

Review Exercises

Work through all the problems in this chapter review and check your answers in the back of the book. Answers to all review problems are there along with section numbers in italics to indicate where each type of problem is discussed. Where weaknesses show up, review appropriate sections in the text.

A

1. Use a bar graph and a broken-line graph to graph the data on voter turnout, as a percentage of the population eligible to vote, in U.S. presidential elections. (*Source:* U.S. Elections Project)

Voter Turnout in U.S. Presidential Elections

Year	Percentage of Eligible Voters
1996	53
2000	55
2004	61
2008	62
2012	59
2016	60

2. Use a pie graph to graph the data on educational attainment in the U.S. population of adults 25 years of age or older. (*Source:* U.S. Census Bureau, 2015 American Community Survey)

Educational Attainment in the United States

Attainment	Percentage (25 Years or Older)
Less than high school diploma	13
High school diploma	28
Some college or associate's degree	29
Bachelor's degree	19
Graduate or professional degree	12

3. (A) Draw a histogram for the binomial distribution

$$P(x) = {}_3C_x(.4)^x(.6)^{3-x}$$

(B) What are the mean and standard deviation?

4. For the set of sample measurements 1, 1, 2, 2, 2, 3, 3, 4, 4, 5, find the

(A) Mean (B) Median

(C) Mode (D) Standard deviation

5. If a normal distribution has a mean of 100 and a standard deviation of 10, then

(A) How many standard deviations is 118 from the mean?

(B) What is the area under the normal curve between the mean and 118?

B

6. Given the sample of 25 quiz scores listed in the following table from a class of 500 students:

(A) Construct a frequency table using a class interval of width 2, starting at 9.5.

(B) Construct a histogram.

(C) Construct a frequency polygon.

(D) Construct a cumulative frequency and relative cumulative frequency table.

(E) Construct a cumulative frequency polygon.

Quiz Scores

14	13	16	15	17
19	15	14	17	15
15	13	12	14	14
12	14	13	11	15
16	14	16	17	14

7. For the set of grouped sample data given in the table,

(A) Find the mean.

(B) Find the standard deviation.

(C) Find the median.

Interval	Frequency
0.5–3.5	1
3.5–6.5	5
6.5–9.5	7
9.5–12.5	2

8. (A) Construct a histogram for the binomial distribution

$$P(x) = {}_6C_x(.5)^x(.5)^{6-x}$$

(B) What are the mean and standard deviation?

9. What are the mean and standard deviation for a binomial distribution with $p = .6$ and $n = 1{,}000$?

In Problems 10 and 11, discuss the validity of each statement. If the statement is always true, explain why. If not, give a counterexample.

10. (A) If the data set $x_1, x_2, \ldots, x_n$ has mean $\bar{x}$, then the data set $x_1 + 5, x_2 + 5, \ldots, x_n + 5$ has mean $\bar{x} + 5$.

(B) If the data set $x_1, x_2, \ldots, x_n$ has standard deviation s, then the data set $x_1 + 5, x_2 + 5, \ldots, x_n + 5$ has standard deviation $s + 5$.

11. (A) If X represents a binomial random variable with mean μ, then $P(X \ge \mu) = .5$.

(B) If X represents a normal random variable with mean μ, then $P(X \ge \mu) = .5$.

(C) The area of a histogram of a binomial distribution is equal to the area above the x axis and below a normal curve.

12. If the probability of success in a single trial of a binomial experiment with 1,000 trials is .6, what is the probability of obtaining at least 550 and no more than 650 successes in 1,000 trials? (*Hint*: Approximate with a normal distribution.)

13. Given a normal distribution with mean 50 and standard deviation 6, find the area under the normal curve:

(A) Between 41 and 62

(B) From 59 on

14. A data set is formed by recording the sums of 100 rolls of a pair of dice. A second data set is formed by again rolling a pair of dice 100 times but recording the product, not the sum, of the two numbers.

(A) Which of the two data sets would you expect to have the smaller standard deviation? Explain.

(B) To obtain evidence for your answer to part (A), use a graphing calculator to simulate both experiments, and compute the standard deviations of each of the two data sets.

C **15.** For the sample quiz scores in Problem 6, find the mean and standard deviation using the data

(A) Without grouping.

(B) Grouped, with class interval of width 2, starting at 9.5.

16. A fair die is rolled five times. What is the probability of rolling

(A) Exactly three 6's?

(B) At least three 6's?

17. Two dice are rolled three times. What is the probability of getting a sum of 7 at least once?

18. Ten students take an exam worth 100 points.

(A) Construct a hypothetical set of exam scores for the ten students in which both the median and the mode are 30 points higher than the mean.

(B) Could the median and mode both be 50 points higher than the mean? Explain.

19. In the last presidential election, 39% of a city's registered voters actually cast ballots.

(A) In a random sample of 20 registered voters from that city, what is the probability that exactly 8 voted in the last presidential election?

(B) Verify by the rule-of-thumb test that the normal distribution with mean 7.8 and standard deviation 2.18 is a good approximation of the binomial distribution with $n = 20$ and $p = .39$.

(C) For the normal distribution of part (B), $P(x = 8) = 0$. Explain the discrepancy between this result and your answer from part (A).

20. A random variable represents the number of wins in a 12-game season for a football team that has a probability of .9 of winning any of its games.

(A) Find the mean and standard deviation of the random variable.

(B) Find the probability that the team wins each of its 12 games.

(C) Use a graphing calculator to simulate 100 repetitions of the binomial experiment associated with the random variable, and compare the empirical probability of a perfect season with the answer to part (B).

Applications

21. Retail sales. The daily number of bad checks received by a large department store in a random sample of 10 days out of the past year was 15, 12, 17, 5, 5, 8, 13, 5, 16, and 4. Find the

(A) Mean

(B) Median

(C) Mode

(D) Standard deviation

22. Preference survey. Find the mean, median, and/or mode, whichever are applicable, for the following employee cafeteria service survey:

Drink Ordered with Meal	Number
Coffee	435
Tea	137
Milk	298
Soft drink	522
Milk shake	392

23. Plant safety. The weekly record of reported accidents in a large auto assembly plant in a random sample of 35 weeks from the past 10 years is listed below:

34	33	36	35	37	31	37
39	34	35	37	35	32	35
33	35	32	34	32	32	39
34	31	35	33	31	38	34
36	34	37	34	36	39	34

(A) Construct a frequency and relative frequency table using class intervals of width 2, starting at 29.5.

(B) Construct a histogram and frequency polygon.

(C) Find the mean and standard deviation for the grouped data.

24. Personnel screening. The scores on a screening test for new technicians are normally distributed with mean 100 and standard deviation 10. Find the approximate percentage of applicants taking the test who score

(A) Between 92 and 108

(B) 115 or higher

25. Market research. A newspaper publisher claims that 70% of the people in a community read their newspaper. Doubting the assertion, a competitor randomly surveys 200 people in the community. Based on the publisher's claim (and assuming a binomial distribution),

(A) Compute the mean and standard deviation.

(B) Determine whether the rule-of-thumb test warrants the use of a normal distribution to approximate this binomial distribution.

(C) Calculate the approximate probability of finding at least 130 and no more than 155 readers in the sample.

(D) Determine the approximate probability of finding 125 or fewer readers in the sample.

(E) Use a graphing calculator to graph the relevant normal distribution.

26. Health care. A small town has three doctors on call for emergency service. The probability that any one doctor will be available when called is .90. What is the probability that at least one doctor will be available for an emergency call?

Appendix

A Basic Algebra Review

Appendix A reviews some important basic algebra concepts usually studied in earlier courses. The material may be studied systematically before beginning the rest of the book or reviewed as needed.

A.1 Real Numbers

- Set of Real Numbers
- Real Number Line
- Basic Real Number Properties
- Further Properties
- Fraction Properties

The rules for manipulating and reasoning with symbols in algebra depend, in large measure, on properties of the real numbers. In this section we look at some of the important properties of this number system. To make our discussions here and elsewhere in the book clearer and more precise, we occasionally make use of simple *set* concepts and notation.

Set of Real Numbers

Informally, a **real number** is any number that has a decimal representation. The decimal representation may be terminating or repeating or neither. The decimal representation 4.713 516 94 is terminating (the space after every third decimal place is used to help keep track of the number of decimal places). The decimal representation $5.254\ 7\overline{47}$ is repeating (the overbar indicates that the block "47" repeats indefinitely). The decimal representation 3.141 592 653 . . . of the number π, the ratio of the circumference to the diameter of a circle, is neither terminating nor repeating. Table 1 describes the set of real numbers and some of its important subsets. Figure 1 illustrates how these sets of numbers are related.

Table 1 Set of Real Numbers

Symbol	Name	Description	Examples
N	Natural numbers	Counting numbers (also called positive integers)	$1, 2, 3, \ldots$
Z	Integers	Natural numbers, their negatives, and 0	$\ldots, -2, -1, 0, 1, 2, \ldots$
Q	Rational numbers	Numbers that can be represented as a/b, where a and b are integers and $b \neq 0$; decimal representations are repeating or terminating	$-4, 0, 1, 25, \frac{-3}{5}, \frac{2}{3}, 3.67, -0.33\overline{3}, 5.272\ 7\overline{27}$
I	Irrational numbers	Numbers that can be represented as nonrepeating and nonterminating decimal numbers	$\sqrt{2}, \pi, \sqrt[3]{7}, 1.414\ 213 \ldots, 2.718\ 281\ 82 \ldots$
R	Real numbers	Rational and irrational numbers	

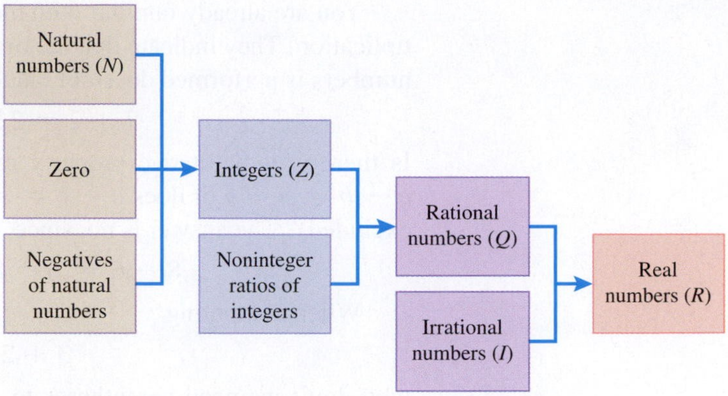

Figure 1 Real numbers and important subsets

The set of integers contains all the natural numbers and something else—their negatives and 0. The set of rational numbers contains all the integers and something else—noninteger ratios of integers. And the set of real numbers contains all the rational numbers and something else—irrational numbers.

Real Number Line

A one-to-one correspondence exists between the set of real numbers and the set of points on a line. That is, each real number corresponds to exactly one point, and each point corresponds to exactly one real number. A line with a real number associated with each point, and vice versa, as shown in Figure 2, is called a **real number line**, or simply a **real line**. Each number associated with a point is called the coordinate of the point.

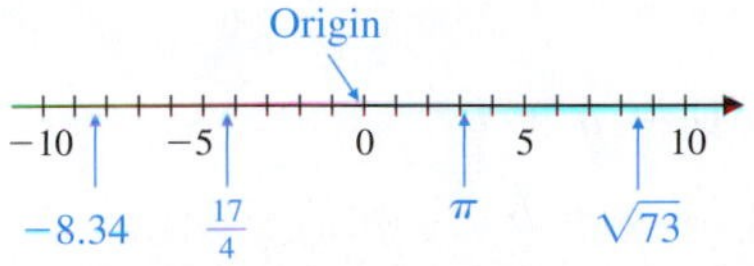

Figure 2 Real number line

The point with coordinate 0 is called the **origin**. The arrow on the right end of the line indicates a positive direction. The coordinates of all points to the right of the origin are called **positive real numbers**, and those to the left of the origin are called **negative real numbers**. The real number 0 is neither positive nor negative.

Basic Real Number Properties

We now take a look at some of the basic properties of the real number system that enable us to convert algebraic expressions into *equivalent forms*.

SUMMARY Basic Properties of the Set of Real Numbers

Let a, b, and c be arbitrary elements in the set of real numbers R.

Addition Properties

Associative: $(a + b) + c = a + (b + c)$

Commutative: $a + b = b + a$

Identity: 0 is the additive identity; that is, $0 + a = a + 0 = a$ for all a in R, and 0 is the only element in R with this property.

Inverse: For each a in R, $-a$, is its unique additive inverse; that is, $a + (-a) = (-a) + a = 0$ and $-a$ is the only element in R relative to a with this property.

Multiplication Properties

Associative: $(ab)c = a(bc)$

Commutative: $ab = ba$

Identity: 1 is the multiplicative identity; that is, $(1)a = a(1) = a$ for all a in R, and 1 is the only element in R with this property.

Inverse: For each a in R, $a \neq 0$, $1/a$ is its unique multiplicative inverse; that is, $a(1/a) = (1/a)a = 1$, and $1/a$ is the only element in R relative to a with this property.

Distributive Properties

$$a(b + c) = ab + ac \quad (a + b)c = ac + bc$$

You are already familiar with the **commutative properties** for addition and multiplication. They indicate that the order in which the addition or multiplication of two numbers is performed does not matter. For example,

$$7 + 2 = 2 + 7 \quad \text{and} \quad 3 \cdot 5 = 5 \cdot 3$$

Is there a commutative property relative to subtraction or division? That is, does $a - b = b - a$ or does $a \div b = b \div a$ for all real numbers a and b (division by 0 excluded)? The answer is no, since, for example,

$$8 - 6 \neq 6 - 8 \quad \text{and} \quad 10 \div 5 \neq 5 \div 10$$

When computing

$$3 + 2 + 6 \quad \text{or} \quad 3 \cdot 2 \cdot 6$$

why don't we need parentheses to indicate which two numbers are to be added or multiplied first? The answer is to be found in the **associative properties**. These properties allow us to write

$$(3 + 2) + 6 = 3 + (2 + 6) \quad \text{and} \quad (3 \cdot 2) \cdot 6 = 3 \cdot (2 \cdot 6)$$

so it does not matter how we group numbers relative to either operation. Is there an associative property for subtraction or division? The answer is no, since, for example,

$$(12 - 6) - 2 \neq 12 - (6 - 2) \quad \text{and} \quad (12 \div 6) \div 2 \neq 12 \div (6 \div 2)$$

Evaluate each side of each equation to see why.

What number added to a given number will give that number back again? What number times a given number will give that number back again? The answers are 0 and 1, respectively. Because of this, 0 and 1 are called the **identity elements** for the real numbers. Hence, for any real numbers a and b,

$$0 + 5 = 5 \quad \text{and} \quad (a + b) + 0 = a + b$$

$$1 \cdot 4 = 4 \quad \text{and} \quad (a + b) \cdot 1 = a + b$$

We now consider **inverses**. For each real number a, there is a unique real number $-a$ such that $a + (-a) = 0$. The number $-a$ is called the **additive inverse** of a, or the **negative** of a. For example, the additive inverse (or negative) of 7 is -7, since $7 + (-7) = 0$. The additive inverse (or negative) of -7 is $-(-7) = 7$, since $-7 + [-(-7)] = 0$.

CONCEPTUAL INSIGHT

Do not confuse negation with the sign of a number. If a is a real number, $-a$ is the negative of a and may be positive or negative. Specifically, if a is negative, then $-a$ is positive and if a is positive, then $-a$ is negative.

For each nonzero real number a, there is a unique real number $1/a$ such that $a(1/a) = 1$. The number $1/a$ is called the **multiplicative inverse** of a, or the **reciprocal** of a. For example, the multiplicative inverse (or reciprocal) of 4 is $\frac{1}{4}$, since $4\left(\frac{1}{4}\right) = 1$. (Also note that 4 is the multiplicative inverse of $\frac{1}{4}$.) The number 0 has no multiplicative inverse.

We now turn to the **distributive properties**, which involve both multiplication and addition. Consider the following two computations:

$$5(3 + 4) = 5 \cdot 7 = 35 \qquad 5 \cdot 3 + 5 \cdot 4 = 15 + 20 = 35$$

Thus,

$$5(3 + 4) = 5 \cdot 3 + 5 \cdot 4$$

and we say that multiplication by 5 *distributes* over the sum $(3 + 4)$. In general, **multiplication distributes over addition** in the real number system. Two more illustrations are

$$9(m + n) = 9m + 9n \qquad (7 + 2)u = 7u + 2u$$

EXAMPLE 1 **Real Number Properties** State the real number property that justifies the indicated statement.

Statement	Property Illustrated
(A) $x(y + z) = (y + z)x$	Commutative $(\cdot)$
(B) $5(2y) = (5 \cdot 2)y$	Associative $(\cdot)$
(C) $2 + (y + 7) = 2 + (7 + y)$	Commutative $(+)$
(D) $4z + 6z = (4 + 6)z$	Distributive
(E) If $m + n = 0$, then $n = -m$.	Inverse $(+)$

MATCHED PROBLEM 1 State the real number property that justifies the indicated statement.

(A) $8 + (3 + y) = (8 + 3) + y$

(B) $(x + y) + z = z + (x + y)$

(C) $(a + b)(x + y) = a(x + y) + b(x + y)$

(D) $5xy + 0 = 5xy$

(E) If $xy = 1, x \neq 0$, then $y = 1/x$.

Further Properties

Subtraction and *division* can be defined in terms of addition and multiplication, respectively:

DEFINITION Subtraction and Division

For all real numbers a and b,

Subtraction: $a - b = a + (-b)$ $\quad 7 - (-5) = 7 + [-(-5)] = 7 + 5 = 12$

Division: $a \div b = a\left(\frac{1}{b}\right), b \neq 0$ $\quad 9 \div 4 = 9\left(\frac{1}{4}\right) = \frac{9}{4}$

To subtract b from a, add the negative (the additive inverse) of b to a. To divide a by b, multiply a by the reciprocal (the multiplicative inverse) of b. Note that division by 0 is not defined, since 0 does not have a reciprocal. **0 can never be used as a divisor!**

The following properties of negatives can be proved using the preceding assumed properties and definitions.

THEOREM 1 Negative Properties

For all real numbers a and b,

1. $-(-a) = a$
2. $(-a)b = -(ab) = a(-b) = -ab$
3. $(-a)(-b) = ab$
4. $(-1)a = -a$
5. $\frac{-a}{b} = -\frac{a}{b} = \frac{a}{-b}, b \neq 0$
6. $\frac{-a}{-b} = -\frac{-a}{b} = -\frac{a}{-b} = \frac{a}{b}, b \neq 0$

We now state two important properties involving 0.

THEOREM 2 Zero Properties

For all real numbers a and b,

1. $a \cdot 0 = 0$ $\quad 0 \cdot 0 = 0 \quad (-35)(0) = 0$
2. $ab = 0$ if and only if $a = 0$ or $b = 0$
   If $(3x + 2)(x - 7) = 0$, then either $3x + 2 = 0$ or $x - 7 = 0$.

Fraction Properties

Recall that the quotient $a \div b(b \neq 0)$ written in the form a/b is called a **fraction**. The quantity a is called the **numerator**, and the quantity b is called the **denominator**.

THEOREM 3 Fraction Properties

For all real numbers $a, b, c, d,$ and k (division by 0 excluded):

1. $\frac{a}{b} = \frac{c}{d}$ if and only if $ad = bc$ $\qquad \frac{4}{6} = \frac{6}{9}$ since $4 \cdot 9 = 6 \cdot 6$

2. $\frac{ka}{kb} = \frac{a}{b}$ $\qquad \frac{7 \cdot 3}{7 \cdot 5} = \frac{3}{5}$

3. $\frac{a}{b} \cdot \frac{c}{d} = \frac{ac}{bd}$ $\qquad \frac{3}{5} \cdot \frac{7}{8} = \frac{3 \cdot 7}{5 \cdot 8}$

4. $\frac{a}{b} \div \frac{c}{d} = \frac{a}{b} \cdot \frac{d}{c}$ $\qquad \frac{2}{3} \div \frac{5}{7} = \frac{2}{3} \cdot \frac{7}{5}$

5. $\frac{a}{b} + \frac{c}{b} = \frac{a + c}{b}$ $\qquad \frac{3}{6} + \frac{5}{6} = \frac{3 + 5}{6}$

6. $\frac{a}{b} - \frac{c}{b} = \frac{a - c}{b}$ $\qquad \frac{7}{8} - \frac{3}{8} = \frac{7 - 3}{8}$

7. $\frac{a}{b} + \frac{c}{d} = \frac{ad + bc}{bd}$ $\qquad \frac{2}{3} + \frac{3}{5} = \frac{2 \cdot 5 + 3 \cdot 3}{3 \cdot 5}$

A fraction is a quotient, not just a pair of numbers. So if a and b are real numbers with $b \neq 0$, then $\frac{a}{b}$ corresponds to a point on the real number line. For example, $\frac{17}{2}$ corresponds to the point halfway between $\frac{16}{2} = 8$ and $\frac{18}{2} = 9$. Similarly, $-\frac{21}{5}$ corresponds to the point that is $\frac{1}{5}$ unit to the left of -4.

EXAMPLE 2 **Estimation** Round $\frac{22}{7} + \frac{18}{19}$ to the nearest integer.

SOLUTION Note that a calculator is not required: $\frac{22}{7}$ is a little greater than 3, and $\frac{18}{19}$ is a little less than 1. Therefore the sum, rounded to the nearest integer, is 4.

MATCHED PROBLEM 2 Round $\frac{6}{93}$ to the nearest integer.

Fractions with denominator 100 are called **percentages**. They are used so often that they have their own notation:

$$\frac{3}{100} = 3\% \qquad \frac{7.5}{100} = 7.5\% \qquad \frac{110}{100} = 110\%$$

So 3% is equivalent to 0.03, 7.5% is equivalent to 0.075, and so on.

EXAMPLE 3 **State Sales Tax** Find the sales tax that is owed on a purchase of \$947.69 if the tax rate is 6.5%.

SOLUTION $6.5\%(\$947.69) = 0.065(947.69) = \$61.60$

MATCHED PROBLEM 3 You intend to give a 20% tip, rounded to the nearest dollar, on a restaurant bill of \$78.47. How much is the tip?

Exercises A.1

All variables represent real numbers.

A *In Problems 1–6, replace each question mark with an appropriate expression that will illustrate the use of the indicated real number property.*

1. Commutative property $(\cdot)$: $uv = ?$
2. Commutative property $(+)$: $x + 7 = ?$
3. Associative property $(+)$: $3 + (7 + y) = ?$
4. Associative property $(\cdot)$: $x(yz) = ?$
5. Identity property $(\cdot)$: $1(u + v) = ?$
6. Identity property $(+)$: $0 + 9m = ?$

In Problems 7–26, indicate true (T) or false (F).

7. $5(8m) = (5 \cdot 8)m$

8. $a + cb = a + bc$

9. $5x + 7x = (5 + 7)x$

10. $uv(w + x) = uvw + uvx$

11. $-2(-a)(2x - y) = 2a(-4x + y)$

12. $8 \div (-5) = 8\left(\frac{1}{-5}\right)$

13. $(x + 3) + 2x = 2x + (x + 3)$

14. $\frac{x}{3y} \div \frac{5y}{x} = \frac{15y^2}{x^2}$

15. $\frac{2x}{-(x + 3)} = -\frac{2x}{x + 3}$

16. $-\frac{2x}{-(x - 3)} = \frac{2x}{x - 3}$

17. $(-3)\left(\frac{1}{-3}\right) = 1$

18. $(-0.5) + (0.5) = 0$

19. $-x^2y^2 = (-1)x^2y^2$

20. $[-(x + 2)](-x) = (x + 2)x$

21. $\frac{a}{b} + \frac{c}{d} = \frac{a + c}{b + d}$

22. $\frac{k}{k + b} = \frac{1}{1 + b}$

23. $(x + 8)(x + 6) = (x + 8)x + (x + 8)6$

24. $u(u - 2v) + v(u - 2v) = (u + v)(u - 2v)$

25. If $(x - 2)(2x + 3) = 0$, then either $x - 2 = 0$ or $2x + 3 = 0$.

26. If either $x - 2 = 0$ or $2x + 3 = 0$, then $(x - 2)(2x + 3) = 0$.

B **27.** If $uv = 1$, does either u or v have to be 1? Explain.

28. If $uv = 0$, does either u or v have to be 0? Explain.

29. Indicate whether the following are true (T) or false (F):

(A) All integers are natural numbers.

(B) All rational numbers are real numbers.

(C) All natural numbers are rational numbers.

30. Indicate whether the following are true (T) or false (F):

(A) All natural numbers are integers.

(B) All real numbers are irrational.

(C) All rational numbers are real numbers.

31. Give an example of a real number that is not a rational number.

32. Give an example of a rational number that is not an integer.

33. Given the sets of numbers N (natural numbers), Z (integers), Q (rational numbers), and R (real numbers), indicate to which set(s) each of the following numbers belongs:

(A) 8 (B) $\sqrt{2}$ (C) -1.414 (D) $\frac{-5}{2}$

34. Given the sets of numbers N, Z, Q, and R (see Problem 33), indicate to which set(s) each of the following numbers belongs:

(A) -3 (B) 3.14 (C) π (D) $\frac{2}{3}$

35. Indicate true (T) or false (F), and for each false statement find real number replacements for a, b, and c that will provide a counterexample. For all real numbers a, b, and c,

(A) $a(b - c) = ab - c$

(B) $(a - b) - c = a - (b - c)$

(C) $a(bc) = (ab)c$

(D) $(a \div b) \div c = a \div (b \div c)$

36. Indicate true (T) or false (F), and for each false statement find real number replacements for a and b that will provide a counterexample. For all real numbers a and b,

(A) $a + b = b + a$

(B) $a - b = b - a$

(C) $ab = ba$

(D) $a \div b = b \div a$

C **37.** If $c = 0.151515\ldots$, then $100c = 15.1515\ldots$ and

$$100c - c = 15.1515\ldots - 0.151515\ldots$$
$$99c = 15$$
$$c = \frac{15}{99} = \frac{5}{33}$$

Proceeding similarly, convert the repeating decimal $0.090909\ldots$ into a fraction. (All repeating decimals are rational numbers, and all rational numbers have repeating decimal representations.)

38. Repeat Problem 37 for $0.181818\ldots$.

Use a calculator to express each number in Problems 39 and 40 as a decimal to the capacity of your calculator. Observe the repeating decimal representation of the rational numbers and the nonrepeating decimal representation of the irrational numbers.

39. (A) $\frac{13}{6}$ (B) $\sqrt{21}$ (C) $\frac{7}{16}$ (D) $\frac{29}{111}$

40. (A) $\frac{8}{9}$ (B) $\frac{3}{11}$ (C) $\sqrt{5}$ (D) $\frac{11}{8}$

In Problems 41–44, without using a calculator, round to the nearest integer.

41. (A) $\frac{43}{13}$ (B) $\frac{37}{19}$

42. (A) $\frac{9}{17}$ (B) $-\frac{12}{25}$

43. (A) $\frac{7}{8} + \frac{11}{12}$ (B) $\frac{55}{9} - \frac{7}{55}$

44. (A) $\frac{5}{6} - \frac{18}{19}$ (B) $\frac{13}{5} + \frac{44}{21}$

Applications

45. Sales tax. Find the tax owed on a purchase of \$182.39 if the state sales tax rate is 9%. (Round to the nearest cent).

46. Sales tax. If you paid \$29.86 in tax on a purchase of \$533.19, what was the sales tax rate? (Write as a percentage, rounded to one decimal place).

47. Gasoline prices. If the price per gallon of gas jumped from \$4.25 to \$4.37, what was the percentage increase? (Round to one decimal place).

48. Gasoline prices. The price of gas increased 4% in one week. If the price last week was \$4.30 per gallon, what is the price now? (Round to the nearest cent).

Answers to Matched Problems

1. (A) Associative (+) (B) Commutative (+)
(C) Distributive (D) Identity (+)
(E) Inverse (·)

2. 0 **3.** \$16

A.2 Operations on Polynomials

- Natural Number Exponents
- Polynomials
- Combining Like Terms
- Addition and Subtraction
- Multiplication
- Combined Operations

This section covers basic operations on *polynomials*. Our discussion starts with a brief review of natural number exponents. Integer and rational exponents and their properties will be discussed in detail in subsequent sections. (Natural numbers, integers, and rational numbers are important parts of the real number system; see Table 1 and Figure 1 in Appendix A.1.)

Natural Number Exponents

We define a **natural number exponent** as follows:

DEFINITION Natural Number Exponent

For n a natural number and b any real number,

$$b^n = b \cdot b \cdot \cdots \cdot b \qquad n \text{ factors of } b$$

$$3^5 = 3 \cdot 3 \cdot 3 \cdot 3 \cdot 3 \qquad 5 \text{ factors of } 3$$

where n is called the exponent and b is called the **base**.

Along with this definition, we state the **first property of exponents**:

THEOREM 1 First Property of Exponents

For any natural numbers m and n, and any real number b:

$$b^m b^n = b^{m+n} \qquad (2t^4)(5t^3) = 2 \cdot 5t^{4+3} = 10t^7$$

Polynomials

Algebraic expressions are formed by using constants and variables and the algebraic operations of addition, subtraction, multiplication, division, raising to powers, and taking roots. Special types of algebraic expressions are called *polynomials*. A **polynomial in one variable** x is constructed by adding or subtracting constants and terms of the form ax^n, where a is a real number and n is a natural number. A **polynomial in two variables** x and y is constructed by adding and subtracting constants and terms of the form ax^my^n, where a is a real number and m and n are natural numbers. Polynomials in three and more variables are defined in a similar manner.

Polynomials		Not Polynomials	
8	0	$\frac{1}{x}$	$\frac{x-y}{x^2+y^2}$
$3x^3 - 6x + 7$	$6x + 3$		
$2x^2 - 7xy - 8y^2$	$9y^3 + 4y^2 - y + 4$	$\sqrt{x^3 - 2x}$	$2x^{-2} - 3x^{-1}$
$2x - 3y + 2$	$u^5 - 3u^3v^2 + 2uv^4 - v^4$		

Polynomial forms are encountered frequently in mathematics. For the efficient study of polynomials, it is useful to classify them according to their *degree*. If a term in a polynomial has only one variable as a factor, then the **degree of the term** is the power of the variable. If two or more variables are present in a term as factors, then the **degree of the term** is the sum of the powers of the variables. The **degree of a polynomial** is the degree of the nonzero term with the highest degree in the polynomial. Any nonzero constant is defined to be a **polynomial of degree 0**. The number 0 is also a polynomial but is not assigned a degree.

EXAMPLE 1 **Degree**

(A) The degree of the first term in $5x^3 + \sqrt{3}x - \frac{1}{2}$ is 3, the degree of the second term is 1, the degree of the third term is 0, and the degree of the whole polynomial is 3 (the same as the degree of the term with the highest degree).

(B) The degree of the first term in $8u^3v^2 - \sqrt{7}uv^2$ is 5, the degree of the second term is 3, and the degree of the whole polynomial is 5.

Matched Problem 1

(A) Given the polynomial $6x^5 + 7x^3 - 2$, what is the degree of the first term? The second term? The third term? The whole polynomial?

(B) Given the polynomial $2u^4v^2 - 5uv^3$, what is the degree of the first term? The second term? The whole polynomial?

In addition to classifying polynomials by degree, we also call a single-term polynomial a **monomial**, a two-term polynomial a **binomial**, and a three-term polynomial a **trinomial**.

Combining Like Terms

The concept of *coefficient* plays a central role in the process of combining *like terms*. A constant in a term of a polynomial, including the sign that precedes it, is called the **numerical coefficient**, or simply, the **coefficient**, of the term. If a constant does not appear, or only a + sign appears, the coefficient is understood to be 1. If only a − sign appears, the coefficient is understood to be −1. Given the polynomial

$$5x^4 - x^3 - 3x^2 + x - 7 \quad = 5x^4 + (-1)x^3 + (-3)x^2 + 1x + (-7)$$

the coefficient of the first term is 5, the coefficient of the second term is −1, the coefficient of the third term is −3, the coefficient of the fourth term is 1, and the coefficient of the fifth term is −7.

The following distributive properties are fundamental to the process of combining *like terms*.

THEOREM 2 Distributive Properties of Real Numbers

1. $a(b + c) = (b + c)a = ab + ac$
2. $a(b - c) = (b - c)a = ab - ac$
3. $a(b + c + \cdots + f) = ab + ac + \cdots + af$

Two terms in a polynomial are called **like terms** if they have exactly the same variable factors to the same powers. The numerical coefficients may or may not be the same. Since constant terms involve no variables, all constant terms are like terms. If a polynomial contains two or more like terms, these terms can be combined into a single term by making use of distributive properties. The following example illustrates the reasoning behind the process:

$$\begin{aligned} 3x^2y - 5xy^2 + x^2y - 2x^2y &= 3x^2y + x^2y - 2x^2y - 5xy^2 \\ &= (3x^2y + 1x^2y - 2x^2y) - 5xy^2 \\ &= (3 + 1 - 2)x^2y - 5xy^2 \\ &= 2x^2y - 5xy^2 \end{aligned}$$

Note the use of distributive properties.

Free use is made of the real number properties discussed in Appendix A.1.

How can we simplify expressions such as $4(x - 2y) - 3(2x - 7y)$? We clear the expression of parentheses using distributive properties, and combine like terms:

$$\begin{aligned} 4(x - 2y) - 3(2x - 7y) &= 4x - 8y - 6x + 21y \\ &= -2x + 13y \end{aligned}$$

EXAMPLE 2 **Removing Parentheses** Remove parentheses and simplify:

(A) $2(3x^2 - 2x + 5) + (x^2 + 3x - 7) = 2(3x^2 - 2x + 5) + 1(x^2 + 3x - 7)$

$$\begin{aligned} &= 6x^2 - 4x + 10 + x^2 + 3x - 7 \\ &= 7x^2 - x + 3 \end{aligned}$$

(B) $(x^3 - 2x - 6) - (2x^3 - x^2 + 2x - 3)$

$$\begin{aligned} &= 1(x^3 - 2x - 6) + (-1)(2x^3 - x^2 + 2x - 3) \\ &= x^3 - 2x - 6 - 2x^3 + x^2 - 2x + 3 \\ &= -x^3 + x^2 - 4x - 3 \end{aligned}$$

Be careful with the sign here

(C) $[3x^2 - (2x + 1)] - (x^2 - 1) = [3x^2 - 2x - 1] - (x^2 - 1)$

$$\begin{aligned} &= 3x^2 - 2x - 1 - x^2 + 1 \\ &= 2x^2 - 2x \end{aligned}$$

MATCHED PROBLEM 2 Remove parentheses and simplify:

(A) $3(u^2 - 2v^2) + (u^2 + 5v^2)$

(B) $(m^3 - 3m^2 + m - 1) - (2m^3 - m + 3)$

(C) $(x^3 - 2) - [2x^3 - (3x + 4)]$

Addition and Subtraction

Addition and subtraction of polynomials can be thought of in terms of removing parentheses and combining like terms, as illustrated in Example 2. Horizontal and vertical arrangements are illustrated in the next two examples. You should be able to work either way, letting the situation dictate your choice.

EXAMPLE 3 **Adding Polynomials** Add horizontally and vertically:

$$x^4 - 3x^3 + x^2, \quad -x^3 - 2x^2 + 3x, \quad \text{and} \quad 3x^2 - 4x - 5$$

SOLUTION Add horizontally:

$$\begin{aligned} &(x^4 - 3x^3 + x^2) + (-x^3 - 2x^2 + 3x) + (3x^2 - 4x - 5) \\ &= x^4 - 3x^3 + x^2 - x^3 - 2x^2 + 3x + 3x^2 - 4x - 5 \\ &= x^4 - 4x^3 + 2x^2 - x - 5 \end{aligned}$$

Or vertically, by lining up like terms and adding their coefficients:

$$\begin{array}{r} x^4 - 3x^3 + x^2 \phantom{{}+3x-5} \\ - x^3 - 2x^2 + 3x \phantom{{}-5} \\ 3x^2 - 4x - 5 \\ \hline x^4 - 4x^3 + 2x^2 - x - 5 \end{array}$$

MATCHED PROBLEM 3 Add horizontally and vertically:

$$3x^4 - 2x^3 - 4x^2, \quad x^3 - 2x^2 - 5x, \quad \text{and} \quad x^2 + 7x - 2$$

EXAMPLE 4 **Subtracting Polynomials** Subtract $4x^2 - 3x + 5$ from $x^2 - 8$, both horizontally and vertically.

SOLUTION

$$\begin{aligned}(x^2 - 8) - (4x^2 - 3x + 5) &= x^2 - 8 - 4x^2 + 3x - 5 \\ &= -3x^2 + 3x - 13\end{aligned}$$

or

$$\begin{array}{rrr} x^2 & & -\ 8 \\ -4x^2 & +\ 3x & -\ 5 \\ \hline -3x^2 & +\ 3x & -\ 13 \end{array}$$

← Change signs and add.

MATCHED PROBLEM 4 Subtract $2x^2 - 5x + 4$ from $5x^2 - 6$, both horizontally and vertically.

Multiplication

Multiplication of algebraic expressions involves the extensive use of distributive properties for real numbers, as well as other real number properties.

EXAMPLE 5 **Multiplying Polynomials** Multiply: $(2x - 3)(3x^2 - 2x + 3)$

SOLUTION

$$\begin{aligned}(2x - 3)(3x^2 - 2x + 3) &= 2x(3x^2 - 2x + 3) - 3(3x^2 - 2x + 3) \\ &= 6x^3 - 4x^2 + 6x - 9x^2 + 6x - 9 \\ &= 6x^3 - 13x^2 + 12x - 9\end{aligned}$$

Or, using a vertical arrangement,

$$\begin{array}{rrrr} & 3x^2 & -\ 2x & +\ 3 \\ & & 2x & -\ 3 \\ \hline 6x^3 & -\ 4x^2 & +\ 6x & \\ & -\ 9x^2 & +\ 6x & -\ 9 \\ \hline 6x^3 & -\ 13x^2 & +\ 12x & -\ 9 \end{array}$$

MATCHED PROBLEM 5 Multiply: $(2x - 3)(2x^2 + 3x - 2)$

Thus, to multiply two polynomials, multiply each term of one by each term of the other, and combine like terms.

Products of binomial factors occur frequently, so it is useful to develop procedures that will enable us to write down their products by inspection. To find the product $(2x - 1)(3x + 2)$ we proceed as follows:

$$\begin{aligned}(2x - 1)(3x + 2) &= 6x^2 + 4x - 3x - 2 \\ &= 6x^2 + x - 2\end{aligned}$$

The inner and outer products are like terms, so combine into a single term.

To speed the process, we do the step in the dashed box mentally.

Products of certain binomial factors occur so frequently that it is useful to learn formulas for their products. The following formulas are easily verified by multiplying the factors on the left.

THEOREM 3 Special Products

1. $(a - b)(a + b) = a^2 - b^2$
2. $(a + b)^2 = a^2 + 2ab + b^2$
3. $(a - b)^2 = a^2 - 2ab + b^2$

EXAMPLE 6 **Special Products** Multiply mentally, where possible.

(A) $(2x - 3y)(5x + 2y)$ (B) $(3a - 2b)(3a + 2b)$

(C) $(5x - 3)^2$ (D) $(m + 2n)^3$

SOLUTION

(A) $(2x - 3y)(5x + 2y) = 10x^2 + 4xy - 15xy - 6y^2$

$= 10x^2 - 11xy - 6y^2$

(B) $(3a - 2b)(3a + 2b) = (3a)^2 - (2b)^2$

$= 9a^2 - 4b^2$

(C) $(5x - 3)^2 = (5x)^2 - 2(5x)(3) + 3^2$

$= 25x^2 - 30x + 9$

(D) $(m + 2n)^3 = (m + 2n)^2(m + 2n)$

$= (m^2 + 4mn + 4n^2)(m + 2n)$

$= m^2(m + 2n) + 4mn(m + 2n) + 4n^2(m + 2n)$

$= m^3 + 2m^2n + 4m^2n + 8mn^2 + 4mn^2 + 8n^3$

$= m^3 + 6m^2n + 12mn^2 + 8n^3$

MATCHED PROBLEM 6 Multiply mentally, where possible.

(A) $(4u - 3v)(2u + v)$ (B) $(2xy + 3)(2xy - 3)$

(C) $(m + 4n)(m - 4n)$ (D) $(2u - 3v)^2$

(E) $(2x - y)^3$

Combined Operations

We complete this section by considering several examples that use all the operations just discussed. Note that in simplifying, we usually remove grouping symbols starting from the inside. That is, we remove parentheses ( ) first, then brackets [ ], and finally braces { }, if present. Also, we observe the following order of operations.

DEFINITION Order of Operations

Multiplication and division precede addition and subtraction, and taking powers precedes multiplication and division.

$2 \cdot 3 + 4 = 6 + 4 = 10, \quad \text{not} \quad 2 \cdot 7 = 14$

$\frac{10^2}{2} = \frac{100}{2} = 50, \quad \text{not} \quad 5^2 = 25$

EXAMPLE 7 **Combined Operations** Perform the indicated operations and simplify:

(A) $3x - \{5 - 3[x - x(3 - x)]\} = 3x - \{5 - 3[x - 3x + x^2]\}$

$= 3x - \{5 - 3x + 9x - 3x^2\}$

$= 3x - 5 + 3x - 9x + 3x^2$

$= 3x^2 - 3x - 5$

(B) $(x - 2y)(2x + 3y) - (2x + y)^2 = 2x^2 - xy - 6y^2 - (4x^2 + 4xy + y^2)$

$= 2x^2 - xy - 6y^2 - 4x^2 - 4xy - y^2$

$= -2x^2 - 5xy - 7y^2$

MATCHED PROBLEM 7 Perform the indicated operations and simplify:

(A) $2t - \{7 - 2[t - t(4 + t)]\}$

(B) $(u - 3v)^2 - (2u - v)(2u + v)$

Exercises A.2

A *Problems 1–8 refer to the following polynomials:*

(A) $2x - 3$ (B) $2x^2 - x + 2$ (C) $x^3 + 2x^2 - x + 3$

1. What is the degree of (C)?

2. What is the degree of (A)?

3. Add (B) and (C).

4. Add (A) and (B).

5. Subtract (B) from (C).

6. Subtract (A) from (B).

7. Multiply (B) and (C).

8. Multiply (A) and (C).

In Problems 9–30, perform the indicated operations and simplify.

9. $2(u - 1) - (3u + 2) - 2(2u - 3)$

10. $2(x - 1) + 3(2x - 3) - (4x - 5)$

11. $4a - 2a[5 - 3(a + 2)]$

12. $2y - 3y[4 - 2(y - 1)]$

13. $(a + b)(a - b)$

14. $(m - n)(m + n)$

15. $(3x - 5)(2x + 1)$

16. $(4t - 3)(t - 2)$

17. $(2x - 3y)(x + 2y)$

18. $(3x + 2y)(x - 3y)$

19. $(3y + 2)(3y - 2)$

20. $(2m - 7)(2m + 7)$

21. $-(2x - 3)^2$

22. $-(5 - 3x)^2$

23. $(4m + 3n)(4m - 3n)$

24. $(3x - 2y)(3x + 2y)$

25. $(3u + 4v)^2$

26. $(4x - y)^2$

27. $(a - b)(a^2 + ab + b^2)$

28. $(a + b)(a^2 - ab + b^2)$

29. $[(x - y) + 3z][(x - y) - 3z]$

30. $[a - (2b - c)][a + (2b - c)]$

B *In Problems 31–44, perform the indicated operations and simplify.*

31. $m - \{m - [m - (m - 1)]\}$

32. $2x - 3\{x + 2[x - (x + 5)] + 1\}$

33. $(x^2 - 2xy + y^2)(x^2 + 2xy + y^2)$

34. $(3x - 2y)^2(2x + 5y)$

35. $(5a - 2b)^2 - (2b + 5a)^2$

36. $(2x - 1)^2 - (3x + 2)(3x - 2)$

37. $(m - 2)^2 - (m - 2)(m + 2)$

38. $(x - 3)(x + 3) - (x - 3)^2$

39. $(x - 2y)(2x + y) - (x + 2y)(2x - y)$

40. $(3m + n)(m - 3n) - (m + 3n)(3m - n)$

41. $(u + v)^3$

42. $(x - y)^3$

43. $(x - 2y)^3$

44. $(2m - n)^3$

45. Subtract the sum of the last two polynomials from the sum of the first two: $2x^2 - 4xy + y^2$, $3xy - y^2$, $x^2 - 2xy - y^2$, $-x^2 + 3xy - 2y^2$

46. Subtract the sum of the first two polynomials from the sum of the last two: $3m^2 - 2m + 5$, $4m^2 - m$, $3m^2 - 3m - 2$, $m^3 + m^2 + 2$

C *In Problems 47–50, perform the indicated operations and simplify.*

47. $[(2x - 1)^2 - x(3x + 1)]^2$

48. $[5x(3x + 1) - 5(2x - 1)^2]^2$

49. $2\{(x - 3)(x^2 - 2x + 1) - x[3 - x(x - 2)]\}$

50. $-3x\{x[x - x(2 - x)] - (x + 2)(x^2 - 3)\}$

51. If you are given two polynomials, one of degree m and the other of degree n, where m is greater than n, what is the degree of their product?

52. What is the degree of the sum of the two polynomials in Problem 51?

53. How does the answer to Problem 51 change if the two polynomials can have the same degree?

54. How does the answer to Problem 52 change if the two polynomials can have the same degree?

55. Show by example that, in general, $(a + b)^2 \neq a^2 + b^2$. Discuss possible conditions on a and b that would make this a valid equation.

56. Show by example that, in general, $(a - b)^2 \neq a^2 - b^2$. Discuss possible conditions on a and b that would make this a valid equation.

Applications

57. Investment. You have \$10,000 to invest, part at 9% and the rest at 12%. If x is the amount invested at 9%, write an algebraic expression that represents the total annual income from both investments. Simplify the expression.

58. Investment. A person has \$100,000 to invest. If \$$x$ are invested in a money market account yielding 7% and twice that amount in certificates of deposit yielding 9%, and if the rest is invested in high-grade bonds yielding 11%, write an algebraic expression that represents the total annual income from all three investments. Simplify the expression.

59. Gross receipts. Four thousand tickets are to be sold for a musical show. If x tickets are to be sold for \$20 each and three times that number for \$30 each, and if the rest are sold for \$50 each, write an algebraic expression that represents the gross receipts from ticket sales, assuming all tickets are sold. Simplify the expression.

60. Gross receipts. Six thousand tickets are to be sold for a concert, some for \$20 each and the rest for \$35 each. If x is the number of \$20 tickets sold, write an algebraic expression that represents the gross receipts from ticket sales, assuming all tickets are sold. Simplify the expression.

61. Nutrition. Food mix A contains 2% fat, and food mix B contains 6% fat. A 10-kilogram diet mix of foods A and B is formed. If x kilograms of food A are used, write an algebraic expression that represents the total number of kilograms of fat in the final food mix. Simplify the expression.

62. Nutrition. Each ounce of food M contains 8 units of calcium, and each ounce of food N contains 5 units of calcium. A 160-ounce diet mix is formed using foods M and N. If x is the number of ounces of food M used, write an algebraic expression that represents the total number of units of calcium in the diet mix. Simplify the expression.

Answers to Matched Problems

1. (A) 5, 3, 0, 5 (B) 6, 4, 6
2. (A) $4u^2 - v^2$ (B) $-m^3 - 3m^2 + 2m - 4$ (C) $-x^3 + 3x + 2$
3. $3x^4 - x^3 - 5x^2 + 2x - 2$
4. $3x^2 + 5x - 10$
5. $4x^3 - 13x + 6$
6. (A) $8u^2 - 2uv - 3v^2$ (B) $4x^2y^2 - 9$ (C) $m^2 - 16n^2$ (D) $4u^2 - 12uv + 9v^2$ (E) $8x^3 - 12x^2y + 6xy^2 - y^3$
7. (A) $-2t^2 - 4t - 7$ (B) $-3u^2 - 6uv + 10v^2$

A.3 Factoring Polynomials

- Common Factors
- Factoring by Grouping
- Factoring Second-Degree Polynomials
- Special Factoring Formulas
- Combined Factoring Techniques

A positive integer is **written in factored form** if it is written as the product of two or more positive integers; for example, $120 = 10 \cdot 12$. A positive integer is **factored completely** if each factor is prime; for example, $120 = 2 \cdot 2 \cdot 2 \cdot 3 \cdot 5$. (Recall that an integer $p > 1$ is **prime** if p cannot be factored as the product of two smaller positive integers. So the first ten primes are 2, 3, 5, 7, 11, 13, 17, 19, 23, and 29). A **tree diagram** is a helpful way to visualize a factorization (Fig 1).

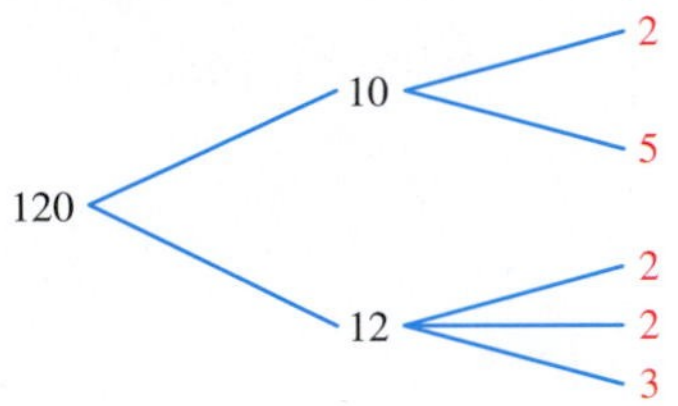

Figure 1

A polynomial is **written in factored form** if it is written as the product of two or more polynomials. The following polynomials are written in factored form:

$$4x^2y - 6xy^2 = 2xy(2x - 3y) \qquad 2x^3 - 8x = 2x(x - 2)(x + 2)$$
$$x^2 - x - 6 = (x - 3)(x + 2) \qquad 5m^2 + 20 = 5(m^2 + 4)$$

Unless stated to the contrary, we will limit our discussion of factoring polynomials to polynomials with integer coefficients.

A polynomial with integer coefficients is said to be **factored completely** if each factor cannot be expressed as the product of two or more polynomials with integer coefficients, other than itself or 1. All the polynomials above, as we will see by the conclusion of this section, are factored completely.

Writing polynomials in completely factored form is often a difficult task. But accomplishing it can lead to the simplification of certain algebraic expressions and to the solution of certain types of equations and inequalities. The distributive properties for real numbers are central to the factoring process.

Common Factors

Generally, a first step in any factoring procedure is to factor out all factors common to all terms.

EXAMPLE 1 **Common Factors** Factor out all factors common to all terms.

(A) $3x^3y - 6x^2y^2 - 3xy^3$

(B) $3y(2y + 5) + 2(2y + 5)$

SOLUTION

(A) $3x^3y - 6x^2y^2 - 3xy^3 = (3xy)x^2 - (3xy)2xy - (3xy)y^2$
$= 3xy(x^2 - 2xy - y^2)$

(B) $3y(2y + 5) + 2(2y + 5) = 3y(2y + 5) + 2(2y + 5)$
$= (3y + 2)(2y + 5)$

MATCHED PROBLEM 1 Factor out all factors common to all terms.

(A) $2x^3y - 8x^2y^2 - 6xy^3$ (B) $2x(3x - 2) - 7(3x - 2)$

Factoring by Grouping

Occasionally, polynomials can be factored by grouping terms in such a way that we obtain results that look like Example 1B. We can then complete the factoring following the steps used in that example. This process will prove useful in the next subsection, where an efficient method is developed for factoring a second-degree polynomial as the product of two first-degree polynomials, if such factors exist.

EXAMPLE 2 **Factoring by Grouping** Factor by grouping.

(A) $3x^2 - 3x - x + 1$

(B) $4x^2 - 2xy - 6xy + 3y^2$

(C) $y^2 + xz + xy + yz$

SOLUTION

(A) $3x^2 - 3x - x + 1$ Group the first two and the last two terms.
$= (3x^2 - 3x) - (x - 1)$ Factor out any common factors from each group. The common factor $(x - 1)$ can be taken out, and the factoring is complete.
$= 3x(x - 1) - (x - 1)$
$= (x - 1)(3x - 1)$

(B) $4x^2 - 2xy - 6xy + 3y^2 = (4x^2 - 2xy) - (6xy - 3y^2)$
$= 2x(2x - y) - 3y(2x - y)$
$= (2x - y)(2x - 3y)$

(C) If, as in parts (A) and (B), we group the first two terms and the last two terms of $y^2 + xz + xy + yz$, no common factor can be taken out of each group to complete the factoring. However, if the two middle terms are reversed, we can proceed as before:

$$\begin{aligned} y^2 + xz + xy + yz &= y^2 + xy + xz + yz \\ &= (y^2 + xy) + (xz + yz) \\ &= y(y + x) + z(x + y) \\ &= y(x + y) + z(x + y) \\ &= (x + y)(y + z) \end{aligned}$$

MATCHED PROBLEM 2 Factor by grouping.

(A) $6x^2 + 2x + 9x + 3$

(B) $2u^2 + 6uv - 3uv - 9v^2$

(C) $ac + bd + bc + ad$

Factoring Second-Degree Polynomials

We now turn our attention to factoring second-degree polynomials of the form

$$2x^2 - 5x - 3 \quad \text{and} \quad 2x^2 + 3xy - 2y^2$$

into the product of two first-degree polynomials with integer coefficients. Since many second-degree polynomials with integer coefficients cannot be factored in this way, it would be useful to know ahead of time that the factors we are seeking actually exist. The factoring approach we use, involving the *ac test*, determines at the beginning whether first-degree factors with integer coefficients do exist. Then, if they exist, the test provides a simple method for finding them.

THEOREM 1 *ac* Test for Factorability

If in polynomials of the form

$$ax^2 + bx + c \quad \text{or} \quad ax^2 + bxy + cy^2 \tag{1}$$

the product ac has two integer factors p and q whose sum is the coefficient b of the middle term; that is, if integers p and q exist so that

$$pq = ac \quad \text{and} \quad p + q = b \tag{2}$$

then the polynomials have first-degree factors with integer coefficients. If no integers p and q exist that satisfy equations (2), then the polynomials in equations (1) will not have first-degree factors with integer coefficients.

If integers p and q exist that satisfy equations (2) in the ac test, the factoring always can be completed as follows: Using $b = p + q$, split the middle terms in equations (1) to obtain

$$ax^2 + bx + c = ax^2 + px + qx + c$$

$$ax^2 + bxy + cy^2 = ax^2 + pxy + qxy + cy^2$$

Complete the factoring by grouping the first two terms and the last two terms as in Example 2. This process always works, and it does not matter if the two middle terms on the right are interchanged.

Several examples should make the process clear. After a little practice, you will perform many of the steps mentally and will find the process fast and efficient.

EXAMPLE 3 **Factoring Second-Degree Polynomials** Factor, if possible, using integer coefficients.

(A) $4x^2 - 4x - 3$ (B) $2x^2 - 3x - 4$ (C) $6x^2 - 25xy + 4y^2$

SOLUTION

(A) $4x^2 - 4x - 3$

Step 1 Use the ac test to test for factorability. Comparing $4x^2 - 4x - 3$ with $ax^2 + bx + c$, we see that $a = 4$, $b = -4$, and $c = -3$. Multiply a and c to obtain

$$ac = (4)(-3) = -12$$

List all pairs of integers whose product is -12, as shown in the margin. These are called **factor pairs** of -12. Then try to find a factor pair that sums to $b = -4$, the coefficient of the middle term in $4x^2 - 4x - 3$. (In practice, this part of Step 1 is often done mentally and can be done rather quickly.) Notice that the factor pair 2 and -6 sums to -4. By the ac test, $4x^2 - 4x - 3$ has first-degree factors with integer coefficients.

All factor pairs of $-12 = ac$

pq
$(1)(-12)$
$(-1)(12)$
$(2)(-6)$
$(-2)(6)$
$(3)(-4)$
$(-3)(4)$

Step 2 Split the middle term, using $b = p + q$, and complete the factoring by grouping. Using $-4 = 2 + (-6)$, we split the middle term in $4x^2 - 4x - 3$ and complete the factoring by grouping:

$$\begin{aligned} 4x^2 - 4x - 3 &= 4x^2 + 2x - 6x - 3 \\ &= (4x^2 + 2x) - (6x + 3) \\ &= 2x(2x + 1) - 3(2x + 1) \\ &= (2x + 1)(2x - 3) \end{aligned}$$

The result can be checked by multiplying the two factors to obtain the original polynomial.

(B) $2x^2 - 3x - 4$

Step 1 Use the *ac* test to test for factorability:

$$ac = (2)(-4) = -8$$

pq
$(-1)(8)$
$(1)(-8)$
$(-2)(4)$
$(2)(-4)$

All factor pairs of $-8 = ac$

Does -8 have a factor pair whose sum is -3? None of the factor pairs listed in the margin sums to $-3 = b$, the coefficient of the middle term in $2x^2 - 3x - 4$. According to the *ac* test, we can conclude that $2x^2 - 3x - 4$ does not have first-degree factors with integer coefficients, and we say that the polynomial is **not factorable**.

(C) $6x^2 - 25xy + 4y^2$

Step 1 Use the *ac* test to test for factorability:

$$ac = (6)(4) = 24$$

Mentally checking through the factor pairs of 24, keeping in mind that their sum must be $-25 = b$, we see that if $p = -1$ and $q = -24$, then

$$pq = (-1)(-24) = 24 = ac$$

and

$$p + q = (-1) + (-24) = -25 = b$$

So the polynomial is factorable.

Step 2 Split the middle term, using $b = p + q$, and complete the factoring by grouping. Using $-25 = (-1) + (-24)$, we split the middle term in $6x^2 - 25xy + 4y^2$ and complete the factoring by grouping:

$$\begin{aligned} 6x^2 - 25xy + 4y^2 &= 6x^2 - xy - 24xy + 4y^2 \\ &= (6x^2 - xy) - (24xy - 4y^2) \\ &= x(6x - y) - 4y(6x - y) \\ &= (6x - y)(x - 4y) \end{aligned}$$

The check is left to the reader.

MATCHED PROBLEM 3 Factor, if possible, using integer coefficients.

(A) $2x^2 + 11x - 6$
(B) $4x^2 + 11x - 6$
(C) $6x^2 + 5xy - 4y^2$

Special Factoring Formulas

The factoring formulas listed in the following box will enable us to factor certain polynomial forms that occur frequently. These formulas can be established by multiplying the factors on the right.

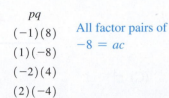

THEOREM 2 Special Factoring Formulas

Perfect square:	1. $u^2 + 2uv + v^2 = (u + v)^2$
Perfect square:	2. $u^2 - 2uv + v^2 = (u - v)^2$
Difference of squares:	3. $u^2 - v^2 = (u - v)(u + v)$
Difference of cubes:	4. $u^3 - v^3 = (u - v)(u^2 + uv + v^2)$
Sum of cubes:	5. $u^3 + v^3 = (u + v)(u^2 - uv + v^2)$

CAUTION Notice that $u^2 + v^2$ is not included in the list of special factoring formulas. In fact,

$$u^2 + v^2 \neq (au + bv)(cu + dv)$$

for any choice of real number coefficients a, b, c, and d.

EXAMPLE 4 **Factoring** Factor completely.

(A) $4m^2 - 12mn + 9n^2$ (B) $x^2 - 16y^2$ (C) $z^3 - 1$
(D) $m^3 + n^3$ (E) $a^2 - 4(b + 2)^2$

SOLUTION

(A) $4m^2 - 12mn + 9n^2 = (2m - 3n)^2$

(B) $x^2 - 16y^2 = x^2 - (4y)^2 = (x - 4y)(x + 4y)$

(C) $z^3 - 1 = (z - 1)(z^2 + z + 1)$ Use the *ac* test to verify that $z^2 + z + 1$ cannot be factored.

(D) $m^3 + n^3 = (m + n)(m^2 - mn + n^2)$ Use the *ac* test to verify that $m^2 - mn + n^2$ cannot be factored.

(E) $a^2 - 4(b + 2)^2 = [a - 2(b + 2)][a + 2(b + 2)]$

MATCHED PROBLEM 4 Factor completely:

(A) $x^2 + 6xy + 9y^2$ (B) $9x^2 - 4y^2$ (C) $8m^3 - 1$
(D) $x^3 + y^3z^3$ (E) $9(m - 3)^2 - 4n^2$

Combined Factoring Techniques

We complete this section by considering several factoring problems that involve combinations of the preceding techniques.

PROCEDURE Factoring Polynomials

Step 1 Take out any factors common to all terms.
Step 2 Use any of the special formulas listed in Theorem 2 that are applicable.
Step 3 Apply the *ac* test to any remaining second-degree polynomial factors.

Note: It may be necessary to perform some of these steps more than once. Furthermore, the order of applying these steps can vary.

EXAMPLE 5 **Combined Factoring Techniques** Factor completely.

(A) $3x^3 - 48x$ (B) $3u^4 - 3u^3v - 9u^2v^2$
(C) $3m^2 - 24mn^3$ (D) $3x^4 - 5x^2 + 2$

SOLUTION

(A) $3x^3 - 48x = 3x(x^2 - 16) = 3x(x - 4)(x + 4)$

(B) $3u^4 - 3u^3v - 9u^2v^2 = 3u^2(u^2 - uv - 3v^2)$

(C) $3m^4 - 24mn^3 = 3m(m^3 - 8n^3) = 3m(m - 2n)(m^2 + 2mn + 4n^2)$
(D) $3x^4 - 5x^2 + 2 = (3x^2 - 2)(x^2 - 1) = (3x^2 - 2)(x - 1)(x + 1)$

MATCHED PROBLEM 5 Factor completely.

(A) $18x^3 - 8x$
(B) $4m^3n - 2m^2n^2 + 2mn^3$
(C) $2t^4 - 16t$
(D) $2y^4 - 5y^2 - 12$

Exercises A.3

A *In Problems 1–8, factor out all factors common to all terms.*

1. $6m^4 - 9m^3 - 3m^2$
2. $6x^4 - 8x^3 - 2x^2$
3. $8u^3v - 6u^2v^2 + 4uv^3$
4. $10x^3y + 20x^2y^2 - 15xy^3$
5. $7m(2m - 3) + 5(2m - 3)$
6. $5x(x + 1) - 3(x + 1)$
7. $4ab(2c + d) - (2c + d)$
8. $12a(b - 2c) - 15b(b - 2c)$

In Problems 9–18, factor by grouping.

9. $2x^2 - x + 4x - 2$
10. $x^2 - 3x + 2x - 6$
11. $3y^2 - 3y + 2y - 2$
12. $2x^2 - x + 6x - 3$
13. $2x^2 + 8x - x - 4$
14. $6x^2 + 9x - 2x - 3$
15. $wy - wz + xy - xz$
16. $ac + ad + bc + bd$
17. $am - 3bm + 2na - 6bn$
18. $ab + 6 + 2a + 3b$

B *In Problems 19–56, factor completely. If a polynomial cannot be factored, say so.*

19. $3y^2 - y - 2$
20. $2x^2 + 5x - 3$
21. $u^2 - 2uv - 15v^2$
22. $x^2 - 4xy - 12y^2$
23. $m^2 - 6m - 3$
24. $x^2 + x - 4$
25. $w^2x^2 - y^2$
26. $25m^2 - 16n^2$
27. $9m^2 - 6mn + n^2$
28. $x^2 + 10xy + 25y^2$
29. $y^2 + 16$
30. $u^2 + 81$
31. $4z^2 - 28z + 48$
32. $6x^2 + 48x + 72$
33. $2x^4 - 24x^3 + 40x^2$
34. $2y^3 - 22y^2 + 48y$
35. $4xy^2 - 12xy + 9x$
36. $16x^2y - 8xy + y$
37. $6m^2 - mn - 12n^2$
38. $6s^2 + 7st - 3t^2$
39. $4u^3v - uv^3$
40. $x^3y - 9xy^3$
41. $2x^3 - 2x^2 + 8x$
42. $3m^3 - 6m^2 + 15m$
43. $8x^3 - 27y^3$
44. $5x^3 + 40y^3$
45. $x^4y + 8xy$
46. $8a^3 - 1$

C

47. $(x + 2)^2 - 9y^2$
48. $(a - b)^2 - 4(c - d)^2$
49. $5u^2 + 4uv - 2v^2$
50. $3x^2 - 2xy - 4y^2$
51. $6(x - y)^2 + 23(x - y) - 4$
52. $4(A + B)^2 - 5(A + B) - 6$
53. $y^4 - 3y^2 - 4$
54. $m^4 - n^4$
55. $15y(x - y)^3 + 12x(x - y)^2$
56. $15x^2(3x - 1)^4 + 60x^3(3x - 1)^3$

In Problems 57–60, discuss the validity of each statement. If the statement is true, explain why. If not, give a counterexample.

57. If n is a positive integer greater than 1, then $u^n - v^n$ can be factored.

58. If m and n are positive integers and $m \neq n$, then $u^m - v^n$ is not factorable.

59. If n is a positive integer greater than 1, then $u^n + v^n$ can be factored.

60. If k is a positive integer, then $u^{2k+1} + v^{2k+1}$ can be factored.

Answers to Matched Problems

1. (A) $2xy(x^2 - 4xy - 3y^2)$ (B) $(2x - 7)(3x - 2)$

2. (A) $(3x + 1)(2x + 3)$ (B) $(u + 3v)(2u - 3v)$
(C) $(a + b)(c + d)$

3. (A) $(2x - 1)(x + 6)$ (B) Not factorable
(C) $(3x + 4y)(2x - y)$

4. (A) $(x + 3y)^2$ (B) $(3x - 2y)(3x + 2y)$
(C) $(2m - 1)(4m^2 + 2m + 1)$
(D) $(x + yz)(x^2 - xyz + y^2z^2)$
(E) $[3(m - 3) - 2n][3(m - 3) + 2n]$

5. (A) $2x(3x - 2)(3x + 2)$ (B) $2mn(2m^2 - mn + n^2)$
(C) $2t(t - 2)(t^2 + 2t + 4)$
(D) $(2y^2 + 3)(y - 2)(y + 2)$

A.4 Operations on Rational Expressions

- Reducing to Lowest Terms
- Multiplication and Division
- Addition and Subtraction
- Compound Fractions

We now turn our attention to fractional forms. A quotient of two algebraic expressions (division by 0 excluded) is called a **fractional expression**. If both the numerator and the denominator are polynomials, the fractional expression is called a **rational expression**. Some examples of rational expressions are

$$\frac{1}{x^3 + 2x} \qquad \frac{5}{x} \qquad \frac{x + 7}{3x^2 - 5x + 1} \qquad \frac{x^2 - 2x + 4}{1}$$

In this section, we discuss basic operations on rational expressions. Since variables represent real numbers in the rational expressions we will consider, the properties of real number fractions summarized in Appendix A.1 will play a central role.

AGREEMENT Variable Restriction

Even though not always explicitly stated, we always assume that variables are restricted so that division by 0 is excluded.

For example, given the rational expression

$$\frac{2x + 5}{x(x + 2)(x - 3)}$$

the variable x is understood to be restricted from being 0, -2, or 3, since these values would cause the denominator to be 0.

Reducing to Lowest Terms

Central to the process of reducing rational expressions to *lowest terms* is the *fundamental property of fractions*, which we restate here for convenient reference:

THEOREM 1 Fundamental Property of Fractions

If a, b, and k are real numbers with $b, k \neq 0$, then

$$\frac{ka}{kb} = \frac{a}{b} \qquad \frac{5 \cdot 2}{5 \cdot 7} = \frac{2}{7} \qquad \frac{x(x + 4)}{2(x + 4)} = \frac{x}{2}, \quad x \neq -4$$

Using this property from left to right to eliminate all common factors from the numerator and the denominator of a given fraction is referred to as **reducing a fraction to lowest terms**. We are actually dividing the numerator and denominator by the same nonzero common factor.

Using the property from right to left—that is, multiplying the numerator and denominator by the same nonzero factor—is referred to as **raising a fraction to higher terms**. We will use the property in both directions in the material that follows.

EXAMPLE 1 **Reducing to Lowest Terms** Reduce each fraction to lowest terms.

(A) $$\frac{1 \cdot 2 \cdot 3 \cdot 4}{1 \cdot 2 \cdot 3 \cdot 4 \cdot 5 \cdot 6} = \frac{\cancel{1} \cdot \cancel{2} \cdot \cancel{3} \cdot \cancel{4}}{\cancel{1} \cdot \cancel{2} \cdot \cancel{3} \cdot \cancel{4} \cdot 5 \cdot 6} = \frac{1}{5 \cdot 6} = \frac{1}{30}$$

(B) $$\frac{2 \cdot 4 \cdot 6 \cdot 8}{1 \cdot 2 \cdot 3 \cdot 4} = \frac{\overset{2}{\cancel{2}} \cdot \overset{2}{\cancel{4}} \cdot \overset{2}{\cancel{6}} \cdot \overset{2}{\cancel{8}}}{\cancel{1} \cdot \cancel{2} \cdot \cancel{3} \cdot \cancel{4}} = 2 \cdot 2 \cdot 2 \cdot 2 = 16$$

MATCHED PROBLEM 1 Reduce each fraction to lowest terms.

(A) $\frac{1 \cdot 2 \cdot 3 \cdot 4 \cdot 5}{1 \cdot 2 \cdot 1 \cdot 2 \cdot 3}$ (B) $\frac{1 \cdot 4 \cdot 9 \cdot 16}{1 \cdot 2 \cdot 3 \cdot 4}$

CONCEPTUAL INSIGHT

Using Theorem 1 to divide the numerator and denominator of a fraction by a common factor is often referred to as **canceling**. This operation can be denoted by drawing a slanted line through each common factor and writing any remaining factors above or below the common factor. Canceling is often incorrectly applied to individual terms in the numerator or denominator, instead of to common factors. For example,

$$\frac{14-5}{2} = \frac{9}{2}$$ Theorem 1 does not apply. There are no common factors in the numerator.

$$\frac{14-5}{2} \neq \frac{\overset{7}{\cancel{14}}-5}{\underset{1}{\cancel{2}}} = 2$$ Incorrect use of Theorem 1. To cancel 2 in the denominator, 2 must be a factor of each term in the numerator.

EXAMPLE 2 **Reducing to Lowest Terms** Reduce each rational expression to lowest terms.

(A) $$\frac{6x^2+x-1}{2x^2-x-1} = \frac{(2x+1)(3x-1)}{(2x+1)(x-1)}$$ Factor numerator and denominator completely.

$$= \frac{3x-1}{x-1}$$ Divide numerator and denominator by the common factor $(2x+1)$.

(B) $$\frac{x^4-8x}{3x^3-2x^2-8x} = \frac{x(x-2)(x^2+2x+4)}{x(x-2)(3x+4)}$$

$$= \frac{x^2+2x+4}{3x+4}$$

MATCHED PROBLEM 2 Reduce each rational expression to lowest terms.

(A) $\frac{x^2-6x+9}{x^2-9}$ (B) $\frac{x^3-1}{x^2-1}$

Multiplication and Division

Since we are restricting variable replacements to real numbers, multiplication and division of rational expressions follow the rules for multiplying and dividing real number fractions summarized in Appendix A.1.

THEOREM 2 Multiplication and Division

If a, b, c, and d are real numbers, then

1. $\frac{a}{b}\cdot\frac{c}{d} = \frac{ac}{bd}$, $b, d \neq 0$ $\quad \frac{3}{5}\cdot\frac{x}{x+5} = \frac{3x}{5(x+5)}$
2. $\frac{a}{b}\div\frac{c}{d} = \frac{a}{b}\cdot\frac{d}{c}$, $b, c, d \neq 0$ $\quad \frac{3}{5}\div\frac{x}{x+5} = \frac{3}{5}\cdot\frac{x+5}{x}$

EXAMPLE 3 **Multiplication and Division** Perform the indicated operations and reduce to lowest terms.

(A) $$\frac{10x^3y}{3xy+9y}\cdot\frac{x^2-9}{4x^2-12x}$$ Factor numerators and denominators. Then divide any numerator and any denominator with a like common factor.

$$= \frac{\overset{5x^2}{\cancel{10x^3y}}}{\underset{3\cdot 1}{3\cancel{y}\cancel{(x+3)}}}\cdot\frac{\overset{1\cdot 1}{\cancel{(x-3)}\cancel{(x+3)}}}{\underset{2\cdot 1}{\cancel{4x}\cancel{(x-3)}}}$$

$$= \frac{5x^2}{6}$$

$$\text{(B)}\ \frac{4-2x}{4} \div (x-2) = \frac{\overset{1}{\cancel{2}}(2-x)}{\underset{2}{\cancel{4}}} \cdot \frac{1}{x-2} \qquad x-2 = \frac{x-2}{1}$$

$$= \frac{2-x}{2(x-2)} = \frac{\overset{-1}{\cancel{-(x-2)}}}{2\underset{1}{\cancel{(x-2)}}} \qquad b-a = -(a-b)\text{, a useful change in some problems}$$

$$= -\frac{1}{2}$$

MATCHED PROBLEM 3 Perform the indicated operations and reduce to lowest terms.

(A) $\dfrac{12x^2y^3}{2xy^2 + 6xy} \cdot \dfrac{y^2 + 6y + 9}{3y^3 + 9y^2}$ (B) $(4 - x) \div \dfrac{x^2 - 16}{5}$

Addition and Subtraction

Again, because we are restricting variable replacements to real numbers, addition and subtraction of rational expressions follow the rules for adding and subtracting real number fractions.

THEOREM 3 Addition and Subtraction

For a, b, and c real numbers,

1. $\dfrac{a}{b} + \dfrac{c}{b} = \dfrac{a+c}{b}, \quad b \neq 0 \qquad \dfrac{x}{x+5} + \dfrac{8}{x+5} = \dfrac{x+8}{x+5}$
2. $\dfrac{a}{b} - \dfrac{c}{b} = \dfrac{a-c}{b}, \quad b \neq 0 \qquad \dfrac{x}{3x^2y^2} - \dfrac{x+7}{3x^2y^2} = \dfrac{x-(x+7)}{3x^2y^2}$

We add rational expressions with the same denominators by adding or subtracting their numerators and placing the result over the common denominator. If the denominators are not the same, we raise the fractions to higher terms, using the fundamental property of fractions to obtain common denominators, and then proceed as described.

Even though any common denominator will do, our work will be simplified if the *least common denominator (LCD)* is used. Often, the LCD is obvious, but if it is not, the steps in the next box describe how to find it.

PROCEDURE Least Common Denominator

The least common denominator (LCD) of two or more rational expressions is found as follows:

1. Factor each denominator completely, including integer factors.
2. Identify each different factor from all the denominators.
3. Form a product using each different factor to the highest power that occurs in any one denominator. This product is the LCD.

EXAMPLE 4 **Addition and Subtraction** Combine into a single fraction and reduce to lowest terms.

(A) $\dfrac{3}{10} + \dfrac{5}{6} - \dfrac{11}{45}$ (B) $\dfrac{4}{9x} - \dfrac{5x}{6y^2} + 1$ (C) $\dfrac{1}{x-1} - \dfrac{1}{x} - \dfrac{2}{x^2-1}$

SOLUTION

(A) To find the LCD, factor each denominator completely:

$$\left.\begin{aligned}10 &= 2\cdot 5\\ 6 &= 2\cdot 3\\ 45 &= 3^2\cdot 5\end{aligned}\right\}\quad LCD = 2\cdot 3^2\cdot 5 = 90$$

Now use the fundamental property of fractions to make each denominator 90:

$$\frac{3}{10} + \frac{5}{6} - \frac{11}{45} = \frac{9\cdot 3}{9\cdot 10} + \frac{15\cdot 5}{15\cdot 6} - \frac{2\cdot 11}{2\cdot 45}$$

$$= \frac{27}{90} + \frac{75}{90} - \frac{22}{90}$$

$$= \frac{27 + 75 - 22}{90} = \frac{80}{90} = \frac{8}{9}$$

(B) $$\left.\begin{aligned}9x &= 3^2x\\ 6y^2 &= 2\cdot 3y^2\end{aligned}\right\}\quad LCD = 2\cdot 3^2xy^2 = 18xy^2$$

$$\frac{4}{9x} - \frac{5x}{6y^2} + 1 = \frac{2y^2\cdot 4}{2y^2\cdot 9x} - \frac{3x\cdot 5x}{3x\cdot 6y^2} + \frac{18xy^2}{18xy^2}$$

$$= \frac{8y^2 - 15x^2 + 18xy^2}{18xy^2}$$

(C) $$\frac{1}{x-1} - \frac{1}{x} - \frac{2}{x^2-1}$$

$$= \frac{1}{x-1} - \frac{1}{x} - \frac{2}{(x-1)(x+1)} \qquad LCD = x(x-1)(x+1)$$

$$= \frac{x(x+1) - (x-1)(x+1) - 2x}{x(x-1)(x+1)}$$

$$= \frac{x^2 + x - x^2 + 1 - 2x}{x(x-1)(x+1)}$$

$$= \frac{1 - x}{x(x-1)(x+1)}$$

$$= \frac{\overset{-1}{-\cancel{(x-1)}}}{x\underset{1}{\cancel{(x-1)}}(x+1)} = \frac{-1}{x(x+1)}$$

MATCHED PROBLEM 4 Combine into a single fraction and reduce to lowest terms.

(A) $\dfrac{5}{28} - \dfrac{1}{10} + \dfrac{6}{35}$ (B) $\dfrac{1}{4x^2} - \dfrac{2x+1}{3x^3} + \dfrac{3}{12x}$

(C) $\dfrac{2}{x^2 - 4x + 4} + \dfrac{1}{x} - \dfrac{1}{x-2}$

Compound Fractions

A fractional expression with fractions in its numerator, denominator, or both is called a **compound fraction**. It is often necessary to represent a compound fraction as a **simple fraction**—that is (in all cases we will consider), as the quotient of two polynomials. The process does not involve any new concepts. It is a matter of applying old concepts and processes in the correct sequence.

EXAMPLE 5 **Simplifying Compound Fractions** Express as a simple fraction reduced to lowest terms:

(A) $\dfrac{\dfrac{1}{5+h} - \dfrac{1}{5}}{h}$ (B) $\dfrac{\dfrac{y}{x^2} - \dfrac{x}{y^2}}{\dfrac{y}{x} - \dfrac{x}{y}}$

SOLUTION We will simplify the expressions in parts (A) and (B) using two different methods—each is suited to the particular type of problem.

(A) We simplify this expression by combining the numerator into a single fraction and using division of rational forms.

$$\frac{\frac{1}{5+h} - \frac{1}{5}}{h} = \left[\frac{1}{5+h} - \frac{1}{5}\right] \div \frac{h}{1}$$
$$= \frac{5-5-h}{5(5+h)} \cdot \frac{1}{h}$$
$$= \frac{-h}{5(5+h)h} = \frac{-1}{5(5+h)}$$

(B) The method used here makes effective use of the fundamental property of fractions in the form

$$\frac{a}{b} = \frac{ka}{kb} \qquad b, k \neq 0$$

Multiply the numerator and denominator by the LCD of all fractions in the numerator and denominator—in this case, x^2y^2:

$$\frac{x^2y^2\left(\frac{y}{x^2} - \frac{x}{y^2}\right)}{x^2y^2\left(\frac{y}{x} - \frac{x}{y}\right)} = \frac{x^2y^2\frac{y}{x^2} - x^2y^2\frac{x}{y^2}}{x^2y^2\frac{y}{x} - x^2y^2\frac{x}{y}} = \frac{y^3 - x^3}{xy^3 - x^3y}$$
$$= \frac{\overset{1}{\cancel{(y-x)}}(y^2 + xy + x^2)}{xy\underset{1}{\cancel{(y-x)}}(y+x)}$$
$$= \frac{y^2 + xy + x^2}{xy(y+x)} \quad \text{or} \quad \frac{x^2 + xy + y^2}{xy(x+y)}$$

MATCHED PROBLEM 5 Express as a simple fraction reduced to lowest terms:

(A) $\dfrac{\dfrac{1}{2+h} - \dfrac{1}{2}}{h}$ (B) $\dfrac{\dfrac{a}{b} - \dfrac{b}{a}}{\dfrac{a}{b} + 2 + \dfrac{b}{a}}$

Exercises A.4

A *In Problems 1–22, perform the indicated operations and reduce answers to lowest terms.*

1. $\dfrac{5 \cdot 9 \cdot 13}{3 \cdot 5 \cdot 7}$

2. $\dfrac{10 \cdot 9 \cdot 8}{3 \cdot 2 \cdot 1}$

3. $\dfrac{12 \cdot 11 \cdot 10 \cdot 9}{4 \cdot 3 \cdot 2 \cdot 1}$

4. $\dfrac{15 \cdot 10 \cdot 5}{20 \cdot 15 \cdot 10}$

5. $\dfrac{d^5}{3a} \div \left(\dfrac{d^2}{6a^2} \cdot \dfrac{a}{4d^3}\right)$

6. $\left(\dfrac{d^5}{3a} \div \dfrac{d^2}{6a^2}\right) \cdot \dfrac{a}{4d^3}$

7. $\dfrac{x^2}{12} + \dfrac{x}{18} - \dfrac{1}{30}$

8. $\dfrac{2y}{18} - \dfrac{-1}{28} - \dfrac{y}{42}$

9. $\dfrac{4m-3}{18m^3} + \dfrac{3}{4m} - \dfrac{2m-1}{6m^2}$

10. $\dfrac{3x+8}{4x^2} - \dfrac{2x-1}{x^3} - \dfrac{5}{8x}$

11. $\dfrac{x^2-9}{x^2-3x} \div (x^2 - x - 12)$

12. $\dfrac{2x^2+7x+3}{4x^2-1} \div (x+3)$

13. $\dfrac{2}{x} - \dfrac{1}{x-3}$

14. $\dfrac{5}{m-2} - \dfrac{3}{2m+1}$

15. $\dfrac{2}{(x+1)^2} - \dfrac{5}{x^2-x-2}$

16. $\dfrac{3}{x^2-5x+6} - \dfrac{5}{(x-2)^2}$

17. $\dfrac{x+1}{x-1} - 1$

18. $m - 3 - \dfrac{m-1}{m-2}$

19. $\dfrac{3}{a-1} - \dfrac{2}{1-a}$

20. $\dfrac{5}{x-3} - \dfrac{2}{3-x}$

21. $\dfrac{2x}{x^2-16} - \dfrac{x-4}{x^2+4x}$

22. $\dfrac{m+2}{m^2-2m} - \dfrac{m}{m^2-4}$

B *In Problems 23–34, perform the indicated operations and reduce answers to lowest terms. Represent any compound fractions as simple fractions reduced to lowest terms.*

23. $\dfrac{x^2}{x^2+2x+1} + \dfrac{x-1}{3x+3} - \dfrac{1}{6}$

24. $\dfrac{y}{y^2-y-2} - \dfrac{1}{y^2+5y-14} - \dfrac{2}{y^2+8y+7}$

25. $\dfrac{1-\dfrac{x}{y}}{2-\dfrac{y}{x}}$

26. $\dfrac{2}{5-\dfrac{3}{4x+1}}$

27. $\dfrac{c+2}{5c-5} - \dfrac{c-2}{3c-3} + \dfrac{c}{1-c}$

28. $\dfrac{x+7}{ax-bx} + \dfrac{y+9}{by-ay}$

29. $\dfrac{1+\dfrac{3}{x}}{x-\dfrac{9}{x}}$

30. $\dfrac{1-\dfrac{y^2}{x^2}}{1-\dfrac{y}{x}}$

31. $\dfrac{\dfrac{1}{2(x+h)} - \dfrac{1}{2x}}{h}$

32. $\dfrac{\dfrac{1}{x+h} - \dfrac{1}{x}}{h}$

33. $\dfrac{\dfrac{x}{y} - 2 + \dfrac{y}{x}}{\dfrac{x}{y} - \dfrac{y}{x}}$

34. $\dfrac{1+\dfrac{2}{x}-\dfrac{15}{x^2}}{1+\dfrac{4}{x}-\dfrac{5}{x^2}}$

In Problems 35–42, imagine that the indicated "solutions" were given to you by a student whom you were tutoring in this class.

(A) Is the solution correct? If the solution is incorrect, explain what is wrong and how it can be corrected.

(B) Show a correct solution for each incorrect solution.

35. $\dfrac{x^2+4x+3}{x+3} = \dfrac{x^2+4x}{x} = x+4$

36. $\dfrac{x^2-3x-4}{x-4} = \dfrac{x^2-3x}{x} = x-3$

37. $\dfrac{(x+h)^2 - x^2}{h} = (x+1)^2 - x^2 = 2x+1$

38. $\dfrac{(x+h)^3 - x^3}{h} = (x+1)^3 - x^3 = 3x^2+3x+1$

39. $\dfrac{x^2-3x}{x^2-2x-3} + x - 3 = \dfrac{x^2-3x+x-3}{x^2-2x-3} = 1$

40. $\dfrac{2}{x-1} - \dfrac{x+3}{x^2-1} = \dfrac{2x+2-x-3}{x^2-1} = \dfrac{1}{x+1}$

41. $\dfrac{2x^2}{x^2-4} - \dfrac{x}{x-2} = \dfrac{2x^2-x^2-2x}{x^2-4} = \dfrac{x}{x+2}$

42. $x + \dfrac{x-2}{x^2-3x+2} = \dfrac{x+x-2}{x^2-3x+2} = \dfrac{2}{x-2}$

C *Represent the compound fractions in Problems 43–46 as simple fractions reduced to lowest terms.*

43. $\dfrac{\dfrac{1}{3(x+h)^2} - \dfrac{1}{3x^2}}{h}$

44. $\dfrac{\dfrac{1}{(x+h)^2} - \dfrac{1}{x^2}}{h}$

45. $x - \dfrac{2}{1-\dfrac{1}{x}}$

46. $2 - \dfrac{1}{1-\dfrac{2}{a+2}}$

Answers to Matched Problems

1. (A) 10 (B) 24
2. (A) $\dfrac{x-3}{x+3}$ (B) $\dfrac{x^2+x+1}{x+1}$
3. (A) $2x$ (B) $\dfrac{-5}{x+4}$
4. (A) $\dfrac{1}{4}$ (B) $\dfrac{3x^2-5x-4}{12x^3}$ (C) $\dfrac{4}{x(x-2)^2}$
5. (A) $\dfrac{-1}{2(2+h)}$ (B) $\dfrac{a-b}{a+b}$

A.5 Integer Exponents and Scientific Notation

- Integer Exponents
- Scientific Notation

We now review basic operations on integer exponents and scientific notation.

Integer Exponents

DEFINITION Integer Exponents

For n an integer and a a real number:

1. For n a positive integer,

$$a^n = a \cdot a \cdot \cdots \cdot a \quad n \text{ factors of } a \quad 5^4 = 5 \cdot 5 \cdot 5 \cdot 5$$

2. For $n = 0$,

$$a^0 = 1 \quad a \neq 0 \quad 12^0 = 1$$

0^0 is not defined.

3. For n a negative integer,

$$a^n = \frac{1}{a^{-n}} \quad a \neq 0 \quad a^{-3} = \frac{1}{a^{-(-3)}} = \frac{1}{a^3}$$

[If n is negative, then $(-n)$ is positive.]

Note: It can be shown that for *all* integers n,

$$a^{-n} = \frac{1}{a^n} \quad \text{and} \quad a^n = \frac{1}{a^{-n}} \quad a \neq 0 \quad a^5 = \frac{1}{a^{-5}}, \quad a^{-5} = \frac{1}{a^5}$$

The following properties are very useful in working with integer exponents.

THEOREM 1 Exponent Properties

For n and m integers and a and b real numbers,

1. $a^m a^n = a^{m+n}$ $\qquad a^8 a^{-3} = a^{8+(-3)} = a^5$
2. $(a^n)^m = a^{mn}$ $\qquad (a^{-2})^3 = a^{3(-2)} = a^{-6}$
3. $(ab)^m = a^m b^m$ $\qquad (ab)^{-2} = a^{-2}b^{-2}$
4. $\left(\frac{a}{b}\right)^m = \frac{a^m}{b^m} \quad b \neq 0$ $\qquad \left(\frac{a}{b}\right)^5 = \frac{a^5}{b^5}$
5. $\frac{a^m}{a^n} = a^{m-n} = \frac{1}{a^{n-m}} \quad a \neq 0$ $\qquad \frac{a^{-3}}{a^7} = \frac{1}{a^{7-(-3)}} = \frac{1}{a^{10}}$

Exponents are frequently encountered in algebraic applications. You should sharpen your skills in using exponents by reviewing the preceding basic definitions and properties and the examples that follow.

EXAMPLE 1 **Simplifying Exponent Forms** Simplify, and express the answers using positive exponents only.

(A) $(2x^3)(3x^5) = 2 \cdot 3x^{3+5} = 6x^8$

(B) $x^5 x^{-9} = x^{-4} = \frac{1}{x^4}$

(C) $\frac{x^5}{x^7} = x^{5-7} = x^{-2} = \frac{1}{x^2}$ or $\frac{x^5}{x^7} = \frac{1}{x^{7-5}} = \frac{1}{x^2}$

(D) $\dfrac{x^{-3}}{y^{-4}} = \dfrac{y^4}{x^3}$

(E) $(u^{-3}v^2)^{-2} = (u^{-3})^{-2}(v^2)^{-2} = u^6v^{-4} = \dfrac{u^6}{v^4}$

(F) $\left(\dfrac{y^{-5}}{y^{-2}}\right)^{-2} = \dfrac{(y^{-5})^{-2}}{(y^{-2})^{-2}} = \dfrac{y^{10}}{y^4} = y^6$

(G) $\dfrac{4m^{-3}n^{-5}}{6m^{-4}n^3} = \dfrac{2m^{-3-(-4)}}{3n^{3-(-5)}} = \dfrac{2m}{3n^8}$

MATCHED PROBLEM 1 Simplify, and express the answers using positive exponents only.

(A) $(3y^4)(2y^3)$ (B) m^2m^{-6} (C) $(u^3v^{-2})^{-2}$

(D) $\left(\dfrac{y^{-6}}{y^{-2}}\right)^{-1}$ (E) $\dfrac{8x^{-2}y^{-4}}{6x^{-5}y^2}$

EXAMPLE 2 **Converting to a Simple Fraction** Write $\dfrac{1-x}{x^{-1}-1}$ as a simple fraction with positive exponents.

SOLUTION First note that

$$\frac{1-x}{x^{-1}-1} \neq \frac{x(1-x)}{-1} \qquad \text{A common error}$$

The original expression is a compound fraction, and we proceed to simplify it as follows:

$$\frac{1-x}{x^{-1}-1} = \frac{1-x}{\frac{1}{x}-1} \qquad \text{Multiply numerator and denominator by } x \text{ to clear internal fractions.}$$

$$= \frac{x(1-x)}{x\left(\frac{1}{x}-1\right)}$$

$$= \frac{x(1-x)}{1-x} = x$$

MATCHED PROBLEM 2 Write $\dfrac{1+x^{-1}}{1-x^{-2}}$ as a simple fraction with positive exponents.

Scientific Notation

In the real world, one often encounters very large and very small numbers. For example,

- The public debt in the United States in 2016, to the nearest billion dollars, was

  $19,573,000,000,000

- The world population in the year 2025, to the nearest million, is projected to be

  7,947,000,000

- The sound intensity of a normal conversation is

  0.000 000 000 316 watt per square centimeter

It is generally troublesome to write and work with numbers of this type in standard decimal form. The first and last example cannot even be entered into many calculators as they are written. But with exponents defined for all integers, we can now express any finite decimal form as the product of a number between 1 and 10 and an integer power of 10, that is, in the form

$$a \times 10^n \qquad 1 \le a < 10, \quad a \text{ in decimal form}, \quad n \text{ an integer}$$

A number expressed in this form is said to be in **scientific notation**. The following are some examples of numbers in standard decimal notation and in scientific notation:

Decimal and Scientific Notation	
$7 = 7 \times 10^0$	$0.5 = 5 \times 10^{-1}$
$67 = 6.7 \times 10$	$0.45 = 4.5 \times 10^{-1}$
$580 = 5.8 \times 10^2$	$0.0032 = 3.2 \times 10^{-3}$
$43{,}000 = 4.3 \times 10^4$	$0.000\,045 = 4.5 \times 10^{-5}$
$73{,}400{,}000 = 7.34 \times 10^7$	$0.000\,000\,391 = 3.91 \times 10^{-7}$

Note that the power of 10 used corresponds to the number of places we move the decimal to form a number between 1 and 10. The power is positive if the decimal is moved to the left and negative if it is moved to the right. Positive exponents are associated with numbers greater than or equal to 10, negative exponents are associated with positive numbers less than 1, and a zero exponent is associated with a number that is 1 or greater but less than 10.

EXAMPLE 3 **Scientific Notation**

(A) Write each number in scientific notation:

$$7{,}320{,}000 \quad \text{and} \quad 0.000\,000\,54$$

(B) Write each number in standard decimal form:

$$4.32 \times 10^6 \quad \text{and} \quad 4.32 \times 10^{-5}$$

SOLUTION

(A) $7{,}320{,}000 = 7.320\,000. \times 10^6 = 7.32 \times 10^6$

6 places left
Positive exponent

$0.000\,000\,54 = 0.000\,000\,5.4 \times 10^{-7} = 5.4 \times 10^{-7}$

7 places right
Negative exponent

(B) $4.32 \times 10^6 = 4{,}320{,}000.$ (6 places right; Positive exponent 6) $4.32 \times 10^{-5} = \dfrac{4.32}{10^5} = 0.000\,043\,2$ (Negative exponent −5; 5 places left)

Matched Problem 3

(A) Write each number in scientific notation: 47,100; 2,443,000,000; 1.45

(B) Write each number in standard decimal form: 3.07×10^8; 5.98×10^{-6}

Exercises A.5

A *In Problems 1–14, simplify and express answers using positive exponents only. Variables are restricted to avoid division by 0.*

1. $2x^{-9}$ **2.** $3y^{-5}$ **3.** $\dfrac{3}{2w^{-7}}$

4. $\dfrac{5}{4x^{-9}}$ **5.** $2x^{-8}x^{5}$ **6.** $3c^{-9}c^{4}$

7. $\dfrac{w^{-8}}{w^{-3}}$ **8.** $\dfrac{m^{-11}}{m^{-5}}$ **9.** $(2a^{-3})^2$

10. $7d^{-4}d^{4}$ **11.** $(a^{-3})^2$ **12.** $(5b^{-2})^2$

13. $(2x^4)^{-3}$ **14.** $(a^{-3}b^4)^{-3}$

In Problems 15–20, write each number in scientific notation.

15. 82,300,000,000 **16.** 5,380,000

17. 0.783 **18.** 0.019

19. 0.000 034 **20.** 0.000 000 007 832

In Problems 21–28, write each number in standard decimal notation.

21. 4×10^4 **22.** 9×10^6

23. 7×10^{-3} **24.** 2×10^{-5}

25. 6.171×10^7 **26.** 3.044×10^3

27. 8.08×10^{-4} **28.** 1.13×10^{-2}

B *In Problems 29–38, simplify and express answers using positive exponents only. Assume that variables are nonzero.*

29. $(22 + 31)^0$ **30.** $(2x^3y^4)^0$

31. $\dfrac{10^{-3} \cdot 10^4}{10^{-11} \cdot 10^{-2}}$ **32.** $\dfrac{10^{-17} \cdot 10^{-5}}{10^{-3} \cdot 10^{-14}}$

33. $(5x^2y^{-3})^{-2}$ **34.** $(2m^{-3}n^2)^{-3}$

35. $\left(\dfrac{-5}{2x^3}\right)^{-2}$ **36.** $\left(\dfrac{2a}{3b^2}\right)^{-3}$

37. $\dfrac{8x^{-3}y^{-1}}{6x^2y^{-4}}$ **38.** $\dfrac{9m^{-4}n^3}{12m^{-1}n^{-1}}$

In Problems 39–42, write each expression in the form $ax^p + bx^q$ or $ax^p + bx^q + cx^r$, where a, b, and c are real numbers and p, q, and r are integers. For example,

$$\frac{2x^4 - 3x^2 + 1}{2x^3} = \boxed{\frac{2x^4}{2x^3} - \frac{3x^2}{2x^3} + \frac{1}{2x^3}} = x - \frac{3}{2}x^{-1} + \frac{1}{2}x^{-3}$$

39. $\dfrac{7x^5 - x^2}{4x^5}$ **40.** $\dfrac{5x^3 - 2}{3x^2}$

41. $\dfrac{5x^4 - 3x^2 + 8}{2x^2}$ **42.** $\dfrac{2x^3 - 3x^2 + x}{2x^2}$

Write each expression in Problems 43–46 with positive exponents only, and as a single fraction reduced to lowest terms.

43. $\dfrac{3x^2(x-1)^2 - 2x^3(x-1)}{(x-1)^4}$

44. $\dfrac{5x^4(x+3)^2 - 2x^5(x+3)}{(x+3)^4}$

45. $2x^{-2}(x-1) - 2x^{-3}(x-1)^2$

46. $2x(x+3)^{-1} - x^2(x+3)^{-2}$

In Problems 47–50, convert each number to scientific notation and simplify. Express the answer in both scientific notation and in standard decimal form.

47. $\dfrac{9{,}600{,}000{,}000}{(1{,}600{,}000)(0.000\,000\,25)}$

48. $\dfrac{(60{,}000)(0.000\,003)}{(0.0004)(1{,}500{,}000)}$

49. $\dfrac{(1{,}250{,}000)(0.000\,38)}{0.0152}$

50. $\dfrac{(0.000\,000\,82)(230{,}000)}{(625{,}000)(0.0082)}$

51. What is the result of entering 2^{3^2} on a calculator?

52. Refer to Problem 51. What is the difference between $2^{(3^2)}$ and $(2^3)^2$? Which agrees with the value of 2^{3^2} obtained with a calculator?

53. If $n = 0$, then property 1 in Theorem 1 implies that $a^m a^0 = a^{m+0} = a^m$. Explain how this helps motivate the definition of a^0.

54. If $m = -n$, then property 1 in Theorem 1 implies that $a^{-n}a^n = a^0 = 1$. Explain how this helps motivate the definition of a^{-n}.

C *Write the fractions in Problems 55–58 as simple fractions reduced to lowest terms.*

55. $\dfrac{u + v}{u^{-1} + v^{-1}}$ **56.** $\dfrac{x^{-2} - y^{-2}}{x^{-1} + y^{-1}}$

57. $\dfrac{b^{-2} - c^{-2}}{b^{-3} - c^{-3}}$ **58.** $\dfrac{xy^{-2} - yx^{-2}}{y^{-1} - x^{-1}}$

Applications

Problems 59 and 60 refer to Table 1.

Table 1 U.S. Public Debt, Interest on Debt, and Population

Year	Public Debt ($)	Interest on Debt ($)	Population
2000	5,674,000,000,000	362,000,000,000	281,000,000
2016	19,573,000,000,000	433,000,000,000	323,000,000

59. Public debt. Carry out the following computations using scientific notation, and write final answers in standard decimal form.

(A) What was the per capita debt in 2016 (to the nearest dollar)?

(B) What was the per capita interest paid on the debt in 2016 (to the nearest dollar)?

(C) What was the percentage interest paid on the debt in 2016 (to two decimal places)?

60. Public debt. Carry out the following computations using scientific notation, and write final answers in standard decimal form.

(A) What was the per capita debt in 2000 (to the nearest dollar)?

(B) What was the per capita interest paid on the debt in 2000 (to the nearest dollar)?

(C) What was the percentage interest paid on the debt in 2000 (to two decimal places)?

Air pollution. *Air quality standards establish maximum amounts of pollutants considered acceptable in the air. The amounts are frequently given in parts per million (ppm). A standard of 30 ppm also can be expressed as follows:*

$$30 \text{ ppm} = \frac{30}{1{,}000{,}000} = \frac{3 \times 10}{10^6}$$

$$= 3 \times 10^{-5} = 0.000\,03 = 0.003\%$$

In Problems 61 and 62, express the given standard:

(A) In scientific notation

(B) In standard decimal notation

(C) As a percent

61. 9 ppm, the standard for carbon monoxide, when averaged over a period of 8 hours

62. 0.03 ppm, the standard for sulfur oxides, when averaged over a year

63. Crime. In 2015, the United States had a violent crime rate of 373 per 100,000 people and a population of 320 million people. How many violent crimes occurred that year? Compute the answer using scientific notation and convert the answer to standard decimal form (to the nearest thousand).

64. Population density. The United States had a 2016 population of 323 million people and a land area of 3,539,000 square miles. What was the population density? Compute the answer using scientific notation and convert the answer to standard decimal form (to one decimal place).

Answers to Matched Problems

1. (A) $6y^7$ (B) $\frac{1}{m^4}$ (C) $\frac{v^4}{u^6}$

(D) y^4 (E) $\frac{4x^3}{3y^6}$

2. $\frac{x}{x-1}$

3. (A) 4.7×10^4; 2.443×10^9; 1.45×10^0

(B) 307,000,000; 0.000 005 98

A.6 Rational Exponents and Radicals

- *n*th Roots of Real Numbers
- Rational Exponents and Radicals
- Properties of Radicals

Square roots may now be generalized to *nth roots*, and the meaning of exponent may be generalized to include all rational numbers.

*n*th Roots of Real Numbers

Consider a square of side r with area 36 square inches. We can write

$$r^2 = 36$$

and conclude that side r is a number whose square is 36. We say that r is a **square root** of b if $r^2 = b$. Similarly, we say that r is a **cube root** of b if $r^3 = b$. And, in general,

DEFINITION ***n*th Root**

For any natural number n,

$$r \text{ is an } \boldsymbol{n}\textbf{th root} \text{ of } b \text{ if } r^n = b$$

So 4 is a square root of 16, since $4^2 = 16$; -2 is a cube root of -8, since $(-2)^3 = -8$. Since $(-4)^2 = 16$, we see that -4 is also a square root of 16. It can be shown that any positive number has two real square roots; two real 4th roots; and, in general, two real nth roots if n is even. Negative numbers have no real square roots; no real 4th roots; and, in general, no real nth roots if n is even. The reason is that no real number raised to an even power can be negative. For odd roots, the situation is simpler. Every real number has exactly one real cube root; one real 5th root; and, in general, one real nth root if n is odd.

Additional roots can be considered in the *complex number system*. In this book, we restrict our interest to *real roots of real numbers*, and *root* will always be interpreted to mean "real root."

Rational Exponents and Radicals

We now turn to the question of what symbols to use to represent *n*th roots. For *n* a natural number greater than 1, we use

$$b^{1/n} \quad \text{or} \quad \sqrt[n]{b}$$

to represent a **real *n*th root of *b***. The exponent form is motivated by the fact that $(b^{1/n})^n = b$ if exponent laws are to continue to hold for rational exponents. The other form is called an ***n*th root radical**. In the expression below, the symbol $\sqrt{\ }$ is called a **radical**, *n* is the **index** of the radical, and *b* is the **radicand**:

Index → $\sqrt[n]{b}$ ← Radical

Radicand

When the index is 2, it is usually omitted. That is, when dealing with square roots, we simply use $\sqrt{b}$ rather than $\sqrt[2]{b}$. If there are two real *n*th roots, both $b^{1/n}$ and $\sqrt[n]{b}$ denote the positive root, called the **principal *n*th root**.

EXAMPLE 1 **Finding *n*th Roots** Evaluate each of the following:

(A) $4^{1/2}$ and $\sqrt{4}$ (B) $-4^{1/2}$ and $-\sqrt{4}$ (C) $(-4)^{1/2}$ and $\sqrt{-4}$

(D) $8^{1/3}$ and $\sqrt[3]{8}$ (E) $(-8)^{1/3}$ and $\sqrt[3]{-8}$ (F) $-8^{1/3}$ and $-\sqrt[3]{8}$

SOLUTION

(A) $4^{1/2} = \sqrt{4} = 2 \quad (\sqrt{4} \neq \pm 2)$ (B) $-4^{1/2} = -\sqrt{4} = -2$

(C) $(-4)^{1/2}$ and $\sqrt{-4}$ are not real numbers

(D) $8^{1/3} = \sqrt[3]{8} = 2$ (E) $(-8)^{1/3} = \sqrt[3]{-8} = -2$

(F) $-8^{1/3} = -\sqrt[3]{8} = -2$

MATCHED PROBLEM 1 Evaluate each of the following:

(A) $16^{1/2}$ (B) $-\sqrt{16}$ (C) $\sqrt[3]{-27}$ (D) $(-9)^{1/2}$ (E) $(\sqrt[4]{81})^3$

CAUTION The symbol $\sqrt{4}$ represents the single number 2, not ± 2. Do not confuse $\sqrt{4}$ with the solutions of the equation $x^2 = 4$, which are usually written in the form $x = \pm\sqrt{4} = \pm 2$. ▲

We now define b^r for any rational number $r = m/n$.

DEFINITION Rational Exponents

If *m* and *n* are natural numbers without common prime factors, *b* is a real number, and *b* is nonnegative when *n* is even, then

$$b^{m/n} = \begin{cases} (b^{1/n})^m = (\sqrt[n]{b})^m & 8^{2/3} = (8^{1/3})^2 = (\sqrt[3]{8})^2 = 2^2 = 4 \\ (b^m)^{1/n} = \sqrt[n]{b^m} & 8^{2/3} = (8^2)^{1/3} = \sqrt[3]{8^2} = \sqrt[3]{64} = 4 \end{cases}$$

and

$$b^{-m/n} = \frac{1}{b^{m/n}} \quad b \neq 0 \qquad 8^{-2/3} = \frac{1}{8^{2/3}} = \frac{1}{4}$$

Note that the two definitions of $b^{m/n}$ are equivalent under the indicated restrictions on *m*, *n*, and *b*.

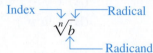

CONCEPTUAL INSIGHT

All the properties for integer exponents listed in Theorem 1 in Section A.5 also hold for rational exponents, provided that b is nonnegative when n is even. This restriction on b is necessary to avoid nonreal results. For example,

$$(-4)^{3/2} = \sqrt{(-4)^3} = \sqrt{-64} \quad \text{Not a real number}$$

To avoid nonreal results, all variables in the remainder of this discussion represent positive real numbers.

EXAMPLE 2 **From Rational Exponent Form to Radical Form and Vice Versa** Change rational exponent form to radical form.

(A) $x^{1/7} = \sqrt[7]{x}$

(B) $(3u^2v^3)^{3/5} = \sqrt[5]{(3u^2v^3)^3}$ or $(\sqrt[5]{3u^2v^3})^3$ The first is usually preferred.

(C) $y^{-2/3} = \dfrac{1}{y^{2/3}} = \dfrac{1}{\sqrt[3]{y^2}}$ or $\sqrt[3]{y^{-2}}$ or $\sqrt[3]{\dfrac{1}{y^2}}$

Change radical form to rational exponent form.

(D) $\sqrt[5]{6} = 6^{1/5}$ (E) $-\sqrt[3]{x^2} = -x^{2/3}$

(F) $\sqrt{x^2 + y^2} = (x^2 + y^2)^{1/2}$ Note that $(x^2 + y^2)^{1/2} \neq x + y$. Why?

MATCHED PROBLEM 2 Convert to radical form.

(A) $u^{1/5}$ (B) $(6x^2y^5)^{2/9}$ (C) $(3xy)^{-3/5}$

Convert to rational exponent form.

(D) $\sqrt[4]{9u}$ (E) $-\sqrt[7]{(2x)^4}$ (F) $\sqrt[3]{x^3 + y^3}$

EXAMPLE 3 **Working with Rational Exponents** Simplify each and express answers using positive exponents only. If rational exponents appear in final answers, convert to radical form.

(A) $(3x^{1/3})(2x^{1/2}) = 6x^{1/3+1/2} = 6x^{5/6} = 6\sqrt[6]{x^5}$

(B) $(-8)^{5/3} = [(-8)^{1/3}]^5 = (-2)^5 = -32$

(C) $(2x^{1/3}y^{-2/3})^3 = 8xy^{-2} = \dfrac{8x}{y^2}$

(D) $\left(\dfrac{4x^{1/3}}{x^{1/2}}\right)^{1/2} = \dfrac{4^{1/2}x^{1/6}}{x^{1/4}} = \dfrac{2}{x^{1/4-1/6}} = \dfrac{2}{x^{1/12}} = \dfrac{2}{\sqrt[12]{x}}$

MATCHED PROBLEM 3 Simplify each and express answers using positive exponents only. If rational exponents appear in final answers, convert to radical form.

(A) $9^{3/2}$ (B) $(-27)^{4/3}$ (C) $(5y^{1/4})(2y^{1/3})$

(D) $(2x^{-3/4}y^{1/4})^4$ (E) $\left(\dfrac{8x^{1/2}}{x^{2/3}}\right)^{1/3}$

EXAMPLE 4 **Working with Rational Exponents** Multiply, and express answers using positive exponents only.

(A) $3y^{2/3}(2y^{1/3} - y^2)$ (B) $(2u^{1/2} + v^{1/2})(u^{1/2} - 3v^{1/2})$

SOLUTION

(A) $3y^{2/3}(2y^{1/3} - y^2) = 6y^{2/3+1/3} - 3y^{2/3+2}$

$= 6y - 3y^{8/3}$

(B) $(2u^{1/2} + v^{1/2})(u^{1/2} - 3v^{1/2}) = 2u - 5u^{1/2}v^{1/2} - 3v$

MATCHED PROBLEM 4 Multiply, and express answers using positive exponents only.

(A) $2c^{1/4}(5c^3 - c^{3/4})$ (B) $(7x^{1/2} - y^{1/2})(2x^{1/2} + 3y^{1/2})$

EXAMPLE 5 **Working with Rational Exponents** Write the following expression in the form $ax^p + bx^q$, where a and b are real numbers and p and q are rational numbers:

$$\frac{2\sqrt{x} - 3\sqrt[3]{x^2}}{2\sqrt[3]{x}}$$

SOLUTION

$$\frac{2\sqrt{x} - 3\sqrt[3]{x^2}}{2\sqrt[3]{x}} = \frac{2x^{1/2} - 3x^{2/3}}{2x^{1/3}}$$ Change to rational exponent form.

$$= \frac{2x^{1/2}}{2x^{1/3}} - \frac{3x^{2/3}}{2x^{1/3}}$$ Separate into two fractions.

$$= x^{1/6} - 1.5x^{1/3}$$

MATCHED PROBLEM 5 Write the following expression in the form $ax^p + bx^q$, where a and b are real numbers and p and q are rational numbers:

$$\frac{5\sqrt[3]{x} - 4\sqrt{x}}{2\sqrt{x^3}}$$

Properties of Radicals

Changing or simplifying radical expressions is aided by several properties of radicals that follow directly from the properties of exponents considered earlier.

THEOREM 1 Properties of Radicals

If n is a natural number greater than or equal to 2, and if x and y are positive real numbers, then

1. $\sqrt[n]{x^n} = x$ $\quad\sqrt[3]{x^3} = x$
2. $\sqrt[n]{xy} = \sqrt[n]{x}\sqrt[n]{y}$ $\quad\sqrt[5]{xy} = \sqrt[5]{x}\,\sqrt[5]{y}$
3. $\sqrt[n]{\dfrac{x}{y}} = \dfrac{\sqrt[n]{x}}{\sqrt[n]{y}}$ $\quad\sqrt[4]{\dfrac{x}{y}} = \dfrac{\sqrt[4]{x}}{\sqrt[4]{y}}$

EXAMPLE 6 **Applying Properties of Radicals** Simplify using properties of radicals.

(A) $\sqrt[4]{(3x^4y^3)^4}$ (B) $\sqrt[4]{8}\,\sqrt[4]{2}$ (C) $\sqrt[3]{\dfrac{xy}{27}}$

SOLUTION

(A) $\sqrt[4]{(3x^4y^3)^4} = 3x^4y^3$ Property 1

(B) $\sqrt[4]{8}\sqrt[4]{2} = \sqrt[4]{16} = \sqrt[4]{2^4} = 2$ Properties 2 and 1

(C) $\sqrt[3]{\dfrac{xy}{27}} = \dfrac{\sqrt[3]{xy}}{\sqrt[3]{27}} = \dfrac{\sqrt[3]{xy}}{3}$ or $\dfrac{1}{3}\sqrt[3]{xy}$ Properties 3 and 1

MATCHED PROBLEM 6 Simplify using properties of radicals.

(A) $\sqrt[7]{(x^3 + y^3)^7}$ (B) $\sqrt[3]{8y^3}$ (C) $\dfrac{\sqrt[3]{16x^4y}}{\sqrt[3]{2xy}}$

What is the best form for a radical expression? There are many answers, depending on what use we wish to make of the expression. In deriving certain formulas, it is sometimes useful to clear either a denominator or a numerator of radicals. The process is referred to as **rationalizing** the denominator or numerator. Examples 7 and 8 illustrate the rationalizing process.

EXAMPLE 7 **Rationalizing Denominators** Rationalize each denominator.

(A) $\dfrac{6x}{\sqrt{2x}}$ (B) $\dfrac{6}{\sqrt{7}-\sqrt{5}}$ (C) $\dfrac{x-4}{\sqrt{x}+2}$

SOLUTION

(A) $\dfrac{6x}{\sqrt{2x}} = \dfrac{6x}{\sqrt{2x}} \cdot \dfrac{\sqrt{2x}}{\sqrt{2x}} = \dfrac{6x\sqrt{2x}}{2x} = 3\sqrt{2x}$

(B) $\dfrac{6}{\sqrt{7}-\sqrt{5}} = \dfrac{6}{\sqrt{7}-\sqrt{5}} \cdot \dfrac{\sqrt{7}+\sqrt{5}}{\sqrt{7}+\sqrt{5}}$

$= \dfrac{6(\sqrt{7}+\sqrt{5})}{2} = 3(\sqrt{7}+\sqrt{5})$

(C) $\dfrac{x-4}{\sqrt{x}+2} = \dfrac{x-4}{\sqrt{x}+2} \cdot \dfrac{\sqrt{x}-2}{\sqrt{x}-2}$

$= \dfrac{(x-4)(\sqrt{x}-2)}{x-4} = \sqrt{x}-2$

MATCHED PROBLEM 7 Rationalize each denominator.

(A) $\dfrac{12ab^2}{\sqrt{3ab}}$ (B) $\dfrac{9}{\sqrt{6}+\sqrt{3}}$ (C) $\dfrac{x^2-y^2}{\sqrt{x}-\sqrt{y}}$

EXAMPLE 8 **Rationalizing Numerators** Rationalize each numerator.

(A) $\dfrac{\sqrt{2}}{2\sqrt{3}}$ (B) $\dfrac{3+\sqrt{m}}{9-m}$ (C) $\dfrac{\sqrt{2+h}-\sqrt{2}}{h}$

SOLUTION

(A) $\dfrac{\sqrt{2}}{2\sqrt{3}} = \dfrac{\sqrt{2}}{2\sqrt{3}} \cdot \dfrac{\sqrt{2}}{\sqrt{2}} = \dfrac{2}{2\sqrt{6}} = \dfrac{1}{\sqrt{6}}$

(B) $\dfrac{3+\sqrt{m}}{9-m} = \dfrac{3+\sqrt{m}}{9-m} \cdot \dfrac{3-\sqrt{m}}{3-\sqrt{m}} = \dfrac{9-m}{(9-m)(3-\sqrt{m})} = \dfrac{1}{3-\sqrt{m}}$

(C) $\dfrac{\sqrt{2+h}-\sqrt{2}}{h} = \dfrac{\sqrt{2+h}-\sqrt{2}}{h} \cdot \dfrac{\sqrt{2+h}+\sqrt{2}}{\sqrt{2+h}+\sqrt{2}}$

$= \dfrac{h}{h(\sqrt{2+h}+\sqrt{2})} = \dfrac{1}{\sqrt{2+h}+\sqrt{2}}$

MATCHED PROBLEM 8 Rationalize each numerator.

(A) $\dfrac{\sqrt{3}}{3\sqrt{2}}$ (B) $\dfrac{2-\sqrt{n}}{4-n}$ (C) $\dfrac{\sqrt{3+h}-\sqrt{3}}{h}$

Exercises A.6

A *Change each expression in Problems 1–6 to radical form. Do not simplify.*

1. $6x^{3/5}$ **2.** $7y^{2/5}$ **3.** $(32x^2y^3)^{3/5}$

4. $(7x^2y)^{5/7}$ **5.** $(x^2+y^2)^{1/2}$ **6.** $x^{1/2}+y^{1/2}$

Change each expression in Problems 7–12 to rational exponent form. Do not simplify.

7. $5\sqrt[4]{x^3}$ **8.** $7m\sqrt[5]{n^2}$ **9.** $\sqrt[5]{(2x^2y)^3}$

10. $\sqrt[7]{(8x^4y)^3}$ **11.** $\sqrt[3]{x}+\sqrt[3]{y}$ **12.** $\sqrt[3]{x^2+y^3}$

In Problems 13–24, find rational number representations for each, if they exist.

13. $25^{1/2}$ **14.** $64^{1/3}$ **15.** $16^{3/2}$

16. $16^{3/4}$ **17.** $-49^{1/2}$ **18.** $(-49)^{1/2}$

19. $-64^{2/3}$ **20.** $(-64)^{2/3}$ **21.** $\left(\frac{4}{25}\right)^{3/2}$

22. $\left(\frac{8}{27}\right)^{2/3}$ **23.** $9^{-3/2}$ **24.** $8^{-2/3}$

In Problems 25–34, simplify each expression and write answers using positive exponents only. All variables represent positive real numbers.

25. $x^{4/5}x^{-2/5}$ **26.** $y^{-3/7}y^{4/7}$ **27.** $\frac{m^{2/3}}{m^{-1/3}}$

28. $\frac{x^{1/4}}{x^{3/4}}$ **29.** $(8x^3y^{-6})^{1/3}$ **30.** $(4u^{-2}v^4)^{1/2}$

31. $\left(\frac{4x^{-2}}{y^4}\right)^{-1/2}$ **32.** $\left(\frac{w^4}{9x^{-2}}\right)^{-1/2}$

33. $\frac{(8x)^{-1/3}}{12x^{1/4}}$ **34.** $\frac{6a^{3/4}}{15a^{-1/3}}$

Simplify each expression in Problems 35–40 using properties of radicals. All variables represent positive real numbers.

35. $\sqrt[5]{(2x+3)^5}$ **36.** $\sqrt[3]{(7+2y)^3}$

37. $\sqrt{6x}\sqrt{15x^3}\sqrt{30x^7}$ **38.** $\sqrt[5]{16a^4}\sqrt[5]{4a^2}\sqrt[5]{8a^3}$

39. $\frac{\sqrt{6x}\sqrt{10}}{\sqrt{15x}}$ **40.** $\frac{\sqrt{8}\sqrt{12y}}{\sqrt{6y}}$

B *In Problems 41–48, multiply, and express answers using positive exponents only.*

41. $3x^{3/4}(4x^{1/4} - 2x^8)$

42. $2m^{1/3}(3m^{2/3} - m^6)$

43. $(3u^{1/2} - v^{1/2})(u^{1/2} - 4v^{1/2})$

44. $(a^{1/2} + 2b^{1/2})(a^{1/2} - 3b^{1/2})$

45. $(6m^{1/2} + n^{-1/2})(6m - n^{-1/2})$

46. $(2x - 3y^{1/3})(2x^{1/3} + 1)$

47. $(3x^{1/2} - y^{1/2})^2$

48. $(x^{1/2} + 2y^{1/2})^2$

Write each expression in Problems 49–54 in the form $ax^p + bx^q$, where a and b are real numbers and p and q are rational numbers.

49. $\frac{\sqrt[3]{x^2} + 2}{2\sqrt[3]{x}}$ **50.** $\frac{12\sqrt{x} - 3}{4\sqrt{x}}$ **51.** $\frac{2\sqrt[4]{x^3} + \sqrt[3]{x}}{3x}$

52. $\frac{3\sqrt[3]{x^2} + \sqrt{x}}{5x}$ **53.** $\frac{2\sqrt[3]{x} - \sqrt{x}}{4\sqrt{x}}$ **54.** $\frac{x^2 - 4\sqrt{x}}{2\sqrt[3]{x}}$

Rationalize the denominators in Problems 55–60.

55. $\frac{12mn^2}{\sqrt{3mn}}$ **56.** $\frac{14x^2}{\sqrt{7x}}$ **57.** $\frac{2(x+3)}{\sqrt{x} - 2}$

58. $\frac{3(x+1)}{\sqrt{x+4}}$ **59.** $\frac{7(x-y)^2}{\sqrt{x} - \sqrt{y}}$ **60.** $\frac{3a - 3b}{\sqrt{a} + \sqrt{b}}$

Rationalize the numerators in Problems 61–66.

61. $\frac{\sqrt{5xy}}{5x^2y^2}$ **62.** $\frac{\sqrt{3mn}}{3mn}$

63. $\frac{\sqrt{x+h} - \sqrt{x}}{h}$ **64.** $\frac{\sqrt{2(a+h)} - \sqrt{2a}}{h}$

65. $\frac{\sqrt{t} - \sqrt{x}}{t^2 - x^2}$ **66.** $\frac{\sqrt{x} - \sqrt{y}}{\sqrt{x} + \sqrt{y}}$

Problems 67–70 illustrate common errors involving rational exponents. In each case, find numerical examples that show that the left side is not always equal to the right side.

67. $(x + y)^{1/2} \neq x^{1/2} + y^{1/2}$ **68.** $(x^3 + y^3)^{1/3} \neq x + y$

69. $(x + y)^{1/3} \neq \frac{1}{(x+y)^3}$ **70.** $(x + y)^{-1/2} \neq \frac{1}{(x+y)^2}$

C *In Problems 71–82, discuss the validity of each statement. If the statement is true, explain why. If not, give a counterexample.*

71. $\sqrt{x^2} = x$ for all real numbers x

72. $\sqrt{x^2} = |x|$ for all real numbers x

73. $\sqrt[3]{x^3} = |x|$ for all real numbers x

74. $\sqrt[3]{x^3} = x$ for all real numbers x

75. If $r < 0$, then r has no cube roots.

76. If $r < 0$, then r has no square roots.

77. If $r > 0$, then r has two square roots.

78. If $r > 0$, then r has three cube roots.

79. The fourth roots of 100 are $\sqrt{10}$ and $-\sqrt{10}$.

80. The square roots of $2\sqrt{6} - 5$ are $\sqrt{3} - \sqrt{2}$ and $\sqrt{2} - \sqrt{3}$.

81. $\sqrt{355 - 60\sqrt{35}} = 5\sqrt{7} - 6\sqrt{5}$

82. $\sqrt[3]{7 - 5\sqrt{2}} = 1 - \sqrt{2}$

In Problems 83–88, simplify by writing each expression as a simple or single fraction reduced to lowest terms and without negative exponents.

83. $-\frac{1}{2}(x - 2)(x + 3)^{-3/2} + (x + 3)^{-1/2}$

84. $2(x - 2)^{-1/2} - \frac{1}{2}(2x + 3)(x - 2)^{-3/2}$

85. $\frac{(x-1)^{1/2} - x\left(\frac{1}{2}\right)(x-1)^{-1/2}}{x - 1}$

86. $\frac{(2x-1)^{1/2} - (x+2)\left(\frac{1}{2}\right)(2x-1)^{-1/2}(2)}{2x - 1}$

87. $\frac{(x+2)^{2/3} - x\left(\frac{2}{3}\right)(x+2)^{-1/3}}{(x+2)^{4/3}}$

88. $\frac{2(3x-1)^{1/3} - (2x+1)\left(\frac{1}{3}\right)(3x-1)^{-2/3}(3)}{(3x-1)^{2/3}}$

In Problems 89–94, evaluate using a calculator. (Refer to the instruction book for your calculator to see how exponential forms are evaluated.)

89. $22^{3/2}$

90. $15^{5/4}$

91. $827^{-3/8}$

92. $103^{-3/4}$

93. $37.09^{7/3}$

94. $2.876^{8/5}$

In Problems 95 and 96, evaluate each expression on a calculator and determine which pairs have the same value. Verify these results algebraically.

95. (A) $\sqrt{3} + \sqrt{5}$ (B) $\sqrt{2 + \sqrt{3}} + \sqrt{2 - \sqrt{3}}$

(C) $1 + \sqrt{3}$ (D) $\sqrt[3]{10 + 6\sqrt{3}}$

(E) $\sqrt{8 + \sqrt{60}}$ (F) $\sqrt{6}$

96. (A) $2\sqrt[3]{2} + \sqrt{5}$ (B) $\sqrt{8}$

(C) $\sqrt{3} + \sqrt{7}$ (D) $\sqrt{3 + \sqrt{8}} + \sqrt{3 - \sqrt{8}}$

(E) $\sqrt{10 + \sqrt{84}}$ (F) $1 + \sqrt{5}$

Answers to Matched Problems

1. (A) 4 (B) -4
(C) -3 (D) Not a real number
(E) 27

2. (A) $\sqrt[5]{u}$ (B) $\sqrt[9]{(6x^2y^5)^2}$ or $(\sqrt[9]{(6x^2y^5)})^2$
(C) $1/\sqrt[5]{(3xy)^3}$ (D) $(9u)^{1/4}$
(E) $-(2x)^{4/7}$ (F) $(x^3 + y^3)^{1/3}$ (not $x + y$)

3. (A) 27 (B) 81
(C) $10y^{7/12} = 10\sqrt[12]{y^7}$ (D) $16y/x^3$
(E) $2/x^{1/18} = 2/\sqrt[18]{x}$

4. (A) $10c^{13/4} - 2c$ (B) $14x + 19x^{1/2}y^{1/2} - 3y$

5. $2.5x^{-7/6} - 2x^{-1}$

6. (A) $x^3 + y^3$ (B) $2y$ (C) $2x$

7. (A) $4b\sqrt{3ab}$ (B) $3(\sqrt{6} - \sqrt{3})$
(C) $(x + y)(\sqrt{x} + \sqrt{y})$

8. (A) $\dfrac{1}{\sqrt{6}}$ (B) $\dfrac{1}{2 + \sqrt{n}}$ (C) $\dfrac{1}{\sqrt{3 + h} + \sqrt{3}}$

A.7 Quadratic Equations

- Solution by Square Root
- Solution by Factoring
- Quadratic Formula
- Quadratic Formula and Factoring
- Other Polynomial Equations
- Application: Supply and Demand

In this section we consider equations involving second-degree polynomials.

DEFINITION Quadratic Equation

A **quadratic equation** in one variable is any equation that can be written in the form

$$ax^2 + bx + c = 0 \qquad a \neq 0 \qquad \text{Standard form}$$

where x is a variable and a, b, and c are constants.

The equations

$$5x^2 - 3x + 7 = 0 \qquad \text{and} \qquad 18 = 32t^2 - 12t$$

are both quadratic equations, since they are either in the standard form or can be transformed into this form.

We restrict our review to finding real solutions to quadratic equations.

Solution by Square Root

The easiest type of quadratic equation to solve is the special form where the first-degree term is missing:

$$ax^2 + c = 0 \qquad a \neq 0$$

The method of solution of this special form makes direct use of the square-root property:

THEOREM 1 Square-Root Property

If $a^2 = b$, then $a = \pm\sqrt{b}$.

EXAMPLE 1 **Square-Root Method** Use the square-root property to solve each equation.

(A) $x^2 - 7 = 0$ (B) $2x^2 - 10 = 0$

(C) $3x^2 + 27 = 0$ (D) $(x - 8)^2 = 9$

SOLUTION

(A) $x^2 - 7 = 0$

$x^2 = 7$ What real number squared is 7?

$x = \pm\sqrt{7}$ Short for $\sqrt{7}$ or $-\sqrt{7}$

(B) $2x^2 - 10 = 0$

$2x^2 = 10$

$x^2 = 5$ What real number squared is 5?

$x = \pm\sqrt{5}$

(C) $3x^2 + 27 = 0$

$3x^2 = -27$

$x^2 = -9$ What real number squared is -9?

No real solution, since no real number squared is negative.

(D) $(x - 8)^2 = 9$

$x - 8 = \pm\sqrt{9}$

$x - 8 = \pm 3$

$x = 8 \pm 3 = 5$ or 11

MATCHED PROBLEM 1 Use the square-root property to solve each equation.

(A) $x^2 - 6 = 0$ (B) $3x^2 - 12 = 0$

(C) $x^2 + 4 = 0$ (D) $(x + 5)^2 = 1$

Solution by Factoring

If the left side of a quadratic equation when written in standard form can be factored, the equation can be solved very quickly. The method of solution by factoring rests on a basic property of real numbers, first mentioned in Section A.1.

CONCEPTUAL INSIGHT

Theorem 2 in Section A.1 states that if a and b are real numbers, then $ab = 0$ if and only if $a = 0$ or $b = 0$. To see that this property is useful for solving quadratic equations, consider the following:

$$x^2 - 4x + 3 = 0 \quad (1)$$
$$(x - 1)(x - 3) = 0$$
$$x - 1 = 0 \quad \text{or} \quad x - 3 = 0$$
$$x = 1 \quad \text{or} \quad x = 3$$

You should check these solutions in equation (1).

If one side of the equation is not 0, then this method cannot be used. For example, consider

$$x^2 - 4x + 3 = 8 \quad (2)$$
$$(x - 1)(x - 3) = 8$$
$$x - 1 \neq 8 \quad \text{or} \quad x - 3 \neq 8$$
$$x = 9 \quad \text{or} \quad x = 11$$

$ab = 8$ does not imply that $a = 8$ or $b = 8$.

Verify that neither $x = 9$ nor $x = 11$ is a solution for equation (2).

EXAMPLE 2 **Factoring Method** Solve by factoring using integer coefficients, if possible.

(A) $3x^2 - 6x - 24 = 0$ (B) $3y^2 = 2y$ (C) $x^2 - 2x - 1 = 0$

SOLUTION

(A) $3x^2 - 6x - 24 = 0$ Divide both sides by 3, since 3 is a factor of each coefficient.

$x^2 - 2x - 8 = 0$ Factor the left side, if possible.

$(x - 4)(x + 2) = 0$

$x - 4 = 0$ or $x + 2 = 0$

$x = 4$ or $x = -2$

(B) $3y^2 = 2y$

$3y^2 - 2y = 0$ We lose the solution $y = 0$ if both sides are divided by y ($3y^2 = 2y$ and $3y = 2$ are not equivalent).

$y(3y - 2) = 0$

$y = 0$ or $3y - 2 = 0$

$3y = 2$

$y = \frac{2}{3}$

(C) $x^2 - 2x - 1 = 0$

This equation cannot be factored using integer coefficients. We will solve this type of equation by another method, considered below.

MATCHED PROBLEM 2 Solve by factoring using integer coefficients, if possible.

(A) $2x^2 + 4x - 30 = 0$ (B) $2x^2 = 3x$ (C) $2x^2 - 8x + 3 = 0$

Note that an equation such as $x^2 = 25$ can be solved by either the square-root or the factoring method, and the results are the same (as they should be). Solve this equation both ways and compare.

Also, note that the factoring method can be extended to higher-degree polynomial equations. Consider the following:

$$x^3 - x = 0$$
$$x(x^2 - 1) = 0$$
$$x(x - 1)(x + 1) = 0$$
$$x = 0 \quad \text{or} \quad x - 1 = 0 \quad \text{or} \quad x + 1 = 0$$
$$\text{Solution: } x = 0, 1, -1$$

Check these solutions in the original equation.

The factoring and square-root methods are fast and easy to use when they apply. However, there are quadratic equations that look simple but cannot be solved by either method. For example, as was noted in Example 2C, the polynomial in

$$x^2 - 2x - 1 = 0$$

cannot be factored using integer coefficients. This brings us to the well-known and widely used *quadratic formula*.

Quadratic Formula

There is a method called *completing the square* that will work for all quadratic equations. After briefly reviewing this method, we will use it to develop the quadratic formula, which can be used to solve any quadratic equation.

The method of **completing the square** is based on the process of transforming a quadratic equation in standard form,

$$ax^2 + bx + c = 0$$

into the form

$$(x + A)^2 = B$$

where A and B are constants. Then, this last equation can be solved easily (if it has a real solution) by the square-root method discussed above.

Consider the equation from Example 2C:

$$x^2 - 2x - 1 = 0 \qquad (3)$$

Since the left side does not factor using integer coefficients, we add 1 to each side to remove the constant term from the left side:

$$x^2 - 2x = 1 \qquad (4)$$

Now we try to find a number that we can add to each side to make the left side a square of a first-degree polynomial. Note the following square of a binomial:

$$(x + m)^2 = x^2 + 2mx + m^2$$

We see that the third term on the right is the square of one-half the coefficient of x in the second term on the right. To complete the square in equation (4), we add the square of one-half the coefficient of x, $(-\frac{2}{2})^2 = 1$, to each side. (This rule works only when the coefficient of x^2 is 1, that is, $a = 1$.) Thus,

$$x^2 - 2x + 1 = 1 + 1$$

The left side is the square of $x - 1$, and we write

$$(x - 1)^2 = 2$$

What number squared is 2?

$$x - 1 = \pm\sqrt{2}$$

$$x = 1 \pm \sqrt{2}$$

And equation (3) is solved!

Let us try the method on the general quadratic equation

$$ax^2 + bx + c = 0 \qquad a \neq 0 \qquad (5)$$

and solve it once and for all for x in terms of the coefficients a, b, and c. We start by multiplying both sides of equation (5) by $1/a$ to obtain

$$x^2 + \frac{b}{a}x + \frac{c}{a} = 0$$

Add $-c/a$ to both sides:

$$x^2 + \frac{b}{a}x = -\frac{c}{a}$$

Now we complete the square on the left side by adding the square of one-half the coefficient of x, that is, $(b/2a)^2 = b^2/4a^2$ to each side:

$$x^2 + \frac{b}{a}x + \frac{b^2}{4a^2} = \frac{b^2}{4a^2} - \frac{c}{a}$$

Writing the left side as a square and combining the right side into a single fraction, we obtain

$$\left(x + \frac{b}{2a}\right)^2 = \frac{b^2 - 4ac}{4a^2}$$

Now we solve by the square-root method:

$$x + \frac{b}{2a} = \pm\sqrt{\frac{b^2 - 4ac}{4a^2}}$$

$$x = -\frac{b}{2a} \pm \frac{\sqrt{b^2 - 4ac}}{2a} \qquad \text{Since } \pm\sqrt{4a^2} = \pm 2a \text{ for any real number } a$$

When this is written as a single fraction, it becomes the **quadratic formula**:

Quadratic Formula

If $ax^2 + bx + c = 0, a \neq 0$, then

$$x = \frac{-b \pm \sqrt{b^2 - 4ac}}{2a}$$

This formula is generally used to solve quadratic equations when the square-root or factoring methods do not work. The quantity $b^2 - 4ac$ under the radical is called the **discriminant**, and it gives us the useful information about solutions listed in Table 1.

Table 1

$b^2 - 4ac$	$ax^2 + bx + c = 0$
Positive	Two real solutions
Zero	One real solution
Negative	No real solutions

EXAMPLE 3 **Quadratic Formula Method** Solve $x^2 - 2x - 1 = 0$ using the quadratic formula.

SOLUTION

$x^2 - 2x - 1 = 0$

$$x = \frac{-b \pm \sqrt{b^2 - 4ac}}{2a} \qquad a = 1, b = -2, c = -1$$

$$= \frac{-(-2) \pm \sqrt{(-2)^2 - 4(1)(-1)}}{2(1)}$$

$$= \frac{2 \pm \sqrt{8}}{2} = \frac{2 \pm 2\sqrt{2}}{2} = 1 \pm \sqrt{2} \approx -0.414 \quad \text{or} \quad 2.414$$

CHECK

$x^2 - 2x - 1 = 0$

When $x = 1 + \sqrt{2}$,

$$(1 + \sqrt{2})^2 - 2(1 + \sqrt{2}) - 1 = 1 + 2\sqrt{2} + 2 - 2 - 2\sqrt{2} - 1 = 0$$

When $x = 1 - \sqrt{2}$,

$$(1 - \sqrt{2})^2 - 2(1 - \sqrt{2}) - 1 = 1 - 2\sqrt{2} + 2 - 2 + 2\sqrt{2} - 1 = 0$$

MATCHED PROBLEM 3 Solve $2x^2 - 4x - 3 = 0$ using the quadratic formula.

If we try to solve $x^2 - 6x + 11 = 0$ using the quadratic formula, we obtain

$$x = \frac{6 \pm \sqrt{-8}}{2}$$

which is not a real number. (Why?)

Quadratic Formula and Factoring

As in Section A.3, we restrict our interest in factoring to polynomials with integer coefficients. If a polynomial cannot be factored as a product of lower-degree polynomials with integer coefficients, we say that the polynomial is **not factorable in the integers**.

How can you factor the quadratic polynomial $x^2 - 13x - 2{,}310$? We start by solving the corresponding quadratic equation using the quadratic formula:

$$x^2 - 13x - 2{,}310 = 0$$

$$x = \frac{-(-13) \pm \sqrt{(-13)^3 - 4(1)(-2{,}310)}}{2}$$

$$x = \frac{-(-13) \pm \sqrt{9{,}409}}{2}$$

$$= \frac{13 \pm 97}{2} = 55 \quad \text{or} \quad -42$$

Now we write

$$x^2 - 13x - 2{,}310 = [x - 55][x - (-42)] = (x - 55)(x + 42)$$

Multiplying the two factors on the right produces the second-degree polynomial on the left.

What is behind this procedure? The following two theorems justify and generalize the process:

THEOREM 2 Factorability Theorem

A second-degree polynomial, $ax^2 + bx + c$, with integer coefficients can be expressed as the product of two first-degree polynomials with integer coefficients if and only if $\sqrt{b^2 - 4ac}$ is an integer.

THEOREM 3 Factor Theorem

If r_1 and r_2 are solutions to the second-degree equation $ax^2 + bx + c = 0$, then

$$ax^2 + bx + c = a(x - r_1)(x - r_2)$$

EXAMPLE 4 **Factoring with the Aid of the Discriminant** Factor, if possible, using integer coefficients.

(A) $4x^2 - 65x + 264$ (B) $2x^2 - 33x - 306$

SOLUTION (A) $4x^2 - 65x + 264$

Step 1 Test for factorability:

$$\sqrt{b^2 - 4ac} = \sqrt{(-65)^2 - 4(4)(264)} = 1$$

Since the result is an integer, the polynomial has first-degree factors with integer coefficients.

Step 2 Factor, using the factor theorem. Find the solutions to the corresponding quadratic equation using the quadratic formula:

$$4x^2 - 65x + 264 = 0$$

From step 1

$$x = \frac{-(-65) \pm 1}{2 \cdot 4} = \frac{33}{4} \quad \text{or} \quad 8$$

Thus,

$$4x^2 - 65x + 264 = 4\left(x - \frac{33}{4}\right)(x - 8)$$

$$= (4x - 33)(x - 8)$$

(B) $2x^2 - 33x - 306$

Step 1 Test for factorability:

$$\sqrt{b^2 - 4ac} = \sqrt{(-33)^2 - 4(2)(-306)} = \sqrt{3{,}537}$$

Since $\sqrt{3{,}537}$ is not an integer, the polynomial is not factorable in the integers.

MATCHED PROBLEM 4 Factor, if possible, using integer coefficients.

(A) $3x^2 - 28x - 464$ (B) $9x^2 + 320x - 144$

Other Polynomial Equations

There are formulas that are analogous to the quadratic formula, but considerably more complicated, that can be used to solve any cubic (degree 3) or quartic (degree 4) polynomial equation. It can be shown that no such general formula exists for solving quintic (degree 5) or polynomial equations of degree greater than five. Certain polynomial equations, however, can be solved easily by taking roots.

EXAMPLE 5 **Solving a Quartic Equation** Find all real solutions to $6x^4 - 486 = 0$.

SOLUTION

$$\begin{aligned} 6x^4 - 486 &= 0 && \text{Add 486 to both sides} \\ 6x^4 &= 486 && \text{Divide both sides by 6} \\ x^4 &= 81 && \text{Take the 4th root of both sides} \\ x &= \pm 3 \end{aligned}$$

MATCHED PROBLEM 5 Find all real solutions to $6x^5 + 192 = 0$.

Application: Supply and Demand

Supply-and-demand analysis is a very important part of business and economics. In general, producers are willing to supply more of an item as the price of an item increases and less of an item as the price decreases. Similarly, buyers are willing to buy less of an item as the price increases, and more of an item as the price decreases. We have a dynamic situation where the price, supply, and demand fluctuate until a price is reached at which the supply is equal to the demand. In economic theory, this point is called the **equilibrium point**. If the price increases from this point, the supply will increase and the demand will decrease; if the price decreases from this point, the supply will decrease and the demand will increase.

EXAMPLE 6 **Supply and Demand** At a large summer beach resort, the weekly supply-and-demand equations for folding beach chairs are

$$p = \frac{x}{140} + \frac{3}{4} \quad \text{Supply equation}$$

$$p = \frac{5{,}670}{x} \quad \text{Demand equation}$$

The supply equation indicates that the supplier is willing to sell x units at a price of p dollars per unit. The demand equation indicates that consumers are willing to buy x units at a price of p dollars per unit. How many units are required for supply to equal demand? At what price will supply equal demand?

SOLUTION Set the right side of the supply equation equal to the right side of the demand equation and solve for x.

$$\frac{x}{140} + \frac{3}{4} = \frac{5{,}670}{x} \quad \text{Multiply by } 140x\text{, the LCD.}$$

$$x^2 + 105x = 793{,}800 \quad \text{Write in standard form.}$$

$$x^2 + 105x - 793{,}800 = 0 \quad \text{Use the quadratic formula.}$$

$$x = \frac{-105 \pm \sqrt{105^2 - 4(1)(-793{,}800)}}{2}$$

$$x = 840 \text{ units}$$

The negative root is discarded since a negative number of units cannot be produced or sold. Substitute $x = 840$ back into either the supply equation or the demand equation to find the equilibrium price (we use the demand equation).

$$p = \frac{5{,}670}{x} = \frac{5{,}670}{840} = \$6.75$$

At a price of \$6.75 the supplier is willing to supply 840 chairs and consumers are willing to buy 840 chairs during a week.

MATCHED PROBLEM 6 Repeat Example 6 if near the end of summer, the supply-and-demand equations are

$$p = \frac{x}{80} - \frac{1}{20} \quad \text{Supply equation}$$

$$p = \frac{1{,}264}{x} \quad \text{Demand equation}$$

Exercises A.7

Find only real solutions in the problems below. If there are no real solutions, say so.

A *Solve Problems 1–4 by the square-root method.*

1. $2x^2 - 22 = 0$

2. $3m^2 - 21 = 0$

3. $(3x - 1)^2 = 25$

4. $(2x + 1)^2 = 16$

Solve Problems 5–8 by factoring.

5. $2u^2 - 8u - 24 = 0$

6. $3x^2 - 18x + 15 = 0$

7. $x^2 = 2x$

8. $n^2 = 3n$

Solve Problems 9–12 by using the quadratic formula.

9. $x^2 - 6x - 3 = 0$

10. $m^2 + 8m + 3 = 0$

11. $3u^2 + 12u + 6 = 0$

12. $2x^2 - 20x - 6 = 0$

B *Solve Problems 13–30 by using any method.*

13. $\frac{2x^2}{3} = 5x$

14. $x^2 = -\frac{3}{4}x$

15. $4u^2 - 9 = 0$

16. $9y^2 - 25 = 0$

17. $8x^2 + 20x = 12$

18. $9x^2 - 6 = 15x$

19. $x^2 = 1 - x$

20. $m^2 = 1 - 3m$

21. $2x^2 = 6x - 3$

22. $2x^2 = 4x - 1$

23. $y^2 - 4y = -8$

24. $x^2 - 2x = -3$

25. $(2x + 3)^2 = 11$

26. $(5x - 2)^2 = 7$

27. $\frac{3}{p} = p$

28. $x - \frac{7}{x} = 0$

29. $2 - \frac{2}{m^2} = \frac{3}{m}$

30. $2 + \frac{5}{u} = \frac{3}{u^2}$

In Problems 31–38, factor, if possible, as the product of two first-degree polynomials with integer coefficients. Use the quadratic formula and the factor theorem.

31. $x^2 + 40x - 84$

32. $x^2 - 28x - 128$

33. $x^2 - 32x + 144$

34. $x^2 + 52x + 208$

35. $2x^2 + 15x - 108$

36. $3x^2 - 32x - 140$

37. $4x^2 + 241x - 434$

38. $6x^2 - 427x - 360$

C **39.** Solve $A = P(1 + r)^2$ for r in terms of A and P; that is, isolate r on the left side of the equation (with coefficient 1) and end up with an algebraic expression on the right side involving A and P but not r. Write the answer using positive square roots only.

40. Solve $x^2 + 3mx - 3n = 0$ for x in terms of m and n.

41. Consider the quadratic equation

$$x^2 + 4x + c = 0$$

where c is a real number. Discuss the relationship between the values of c and the three types of roots listed in Table 1 on page 590.

42. Consider the quadratic equation

$$x^2 - 2x + c = 0$$

where c is a real number. Discuss the relationship between the values of c and the three types of roots listed in Table 1 on page 590.

In Problems 43–48, find all real solutions.

43. $x^3 + 8 = 0$

44. $x^3 - 8 = 0$

45. $5x^4 - 500 = 0$

46. $2x^3 + 250 = 0$

47. $x^4 - 8x^2 + 15 = 0$

48. $x^4 - 12x^2 + 32 = 0$

Applications

49. Supply and demand. A company wholesales shampoo in a particular city. Their marketing research department established the following weekly supply-and-demand equations:

$$p = \frac{x}{450} + \frac{1}{2} \quad \text{Supply equation}$$

$$p = \frac{6{,}300}{x} \quad \text{Demand equation}$$

How many units are required for supply to equal demand? At what price per bottle will supply equal demand?

50. Supply and demand. An importer sells an automatic camera to outlets in a large city. During the summer, the weekly supply-and-demand equations are

$$p = \frac{x}{6} + 9 \quad \text{Supply equation}$$

$$p = \frac{24{,}840}{x} \quad \text{Demand equation}$$

How many units are required for supply to equal demand? At what price will supply equal demand?

51. Interest rate. If P dollars are invested at $100r$ percent compounded annually, at the end of 2 years it will grow to $A = P(1 + r)^2$. At what interest rate will \$484 grow to \$625 in 2 years? (*Note:* If $A = 625$ and $P = 484$, find r.)

52. Interest rate. Using the formula in Problem 51, determine the interest rate that will make \$1,000 grow to \$1,210 in 2 years.

53. Ecology. To measure the velocity v (in feet per second) of a stream, we position a hollow L-shaped tube with one end under the water pointing upstream and the other end pointing straight up a couple of feet out of the water. The water will then be pushed up the tube a certain distance h (in feet) above the surface of the stream. Physicists have shown that $v^2 = 64h$. Approximately how fast is a stream flowing if $h = 1$ foot? If $h = 0.5$ foot?

54. Safety research. It is of considerable importance to know the least number of feet d in which a car can be stopped, including reaction time of the driver, at various speeds v (in miles per hour). Safety research has produced the formula $d = 0.044v^2 + 1.1v$. If it took a car 550 feet to stop, estimate the car's speed at the moment the stopping process was started.

Answers to Matched Problems

1. (A) $\pm\sqrt{6}$ (B) ± 2
(C) No real solution (D) $-6, -4$

2. (A) $-5, 3$ (B) $0, \frac{3}{2}$
(C) Cannot be factored using integer coefficients

3. $(2 \pm \sqrt{10})/2$

4. (A) Cannot be factored using integer coefficients
(B) $(9x - 4)(x + 36)$

5. -2

6. 320 chairs at \$3.95 each

Appendix

C Area under the Standard Normal Curve

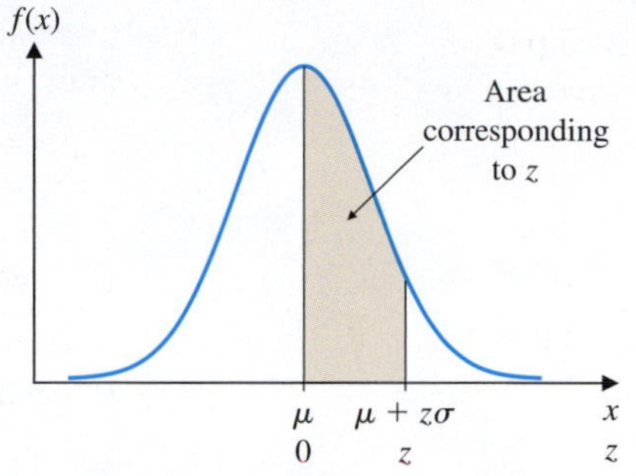

Area under the Standard Normal Curve

(Table Entries Represent the Area under the Standard Normal Curve from 0 to z, $z \geq 0$)

z	.00	.01	.02	.03	.04	.05	.06	.07	.08	.09
0.0	0.0000	0.0040	0.0080	0.0120	0.0160	0.0199	0.0239	0.0279	0.0319	0.0359
0.1	0.0398	0.0438	0.0478	0.0517	0.0557	0.0596	0.0636	0.0675	0.0714	0.0753
0.2	0.0793	0.0832	0.0871	0.0910	0.0948	0.0987	0.1026	0.1064	0.1103	0.1141
0.3	0.1179	0.1217	0.1255	0.1293	0.1331	0.1368	0.1406	0.1443	0.1480	0.1517
0.4	0.1554	0.1591	0.1628	0.1664	0.1700	0.1736	0.1772	0.1808	0.1844	0.1879
0.5	0.1915	0.1950	0.1985	0.2019	0.2054	0.2088	0.2123	0.2157	0.2190	0.2224
0.6	0.2257	0.2291	0.2324	0.2357	0.2389	0.2422	0.2454	0.2486	0.2517	0.2549
0.7	0.2580	0.2611	0.2642	0.2673	0.2704	0.2734	0.2764	0.2794	0.2823	0.2852
0.8	0.2881	0.2910	0.2939	0.2967	0.2995	0.3023	0.3051	0.3078	0.3106	0.3133
0.9	0.3159	0.3186	0.3212	0.3238	0.3264	0.3289	0.3315	0.3340	0.3365	0.3389
1.0	0.3413	0.3438	0.3461	0.3485	0.3508	0.3531	0.3554	0.3577	0.3599	0.3621
1.1	0.3643	0.3665	0.3686	0.3708	0.3729	0.3749	0.3770	0.3790	0.3810	0.3830
1.2	0.3849	0.3869	0.3888	0.3907	0.3925	0.3944	0.3962	0.3980	0.3997	0.4015
1.3	0.4032	0.4049	0.4066	0.4082	0.4099	0.4115	0.4131	0.4147	0.4162	0.4177
1.4	0.4192	0.4207	0.4222	0.4236	0.4251	0.4265	0.4279	0.4292	0.4306	0.4319
1.5	0.4332	0.4345	0.4357	0.4370	0.4382	0.4394	0.4406	0.4418	0.4429	0.4441
1.6	0.4452	0.4463	0.4474	0.4484	0.4495	0.4505	0.4515	0.4525	0.4535	0.4545
1.7	0.4554	0.4564	0.4573	0.4582	0.4591	0.4599	0.4608	0.4616	0.4625	0.4633
1.8	0.4641	0.4649	0.4656	0.4664	0.4671	0.4678	0.4686	0.4693	0.4699	0.4706
1.9	0.4713	0.4719	0.4726	0.4732	0.4738	0.4744	0.4750	0.4756	0.4761	0.4767
2.0	0.4772	0.4778	0.4783	0.4788	0.4793	0.4798	0.4803	0.4808	0.4812	0.4817
2.1	0.4821	0.4826	0.4830	0.4834	0.4838	0.4842	0.4846	0.4850	0.4854	0.4857
2.2	0.4861	0.4864	0.4868	0.4871	0.4875	0.4878	0.4881	0.4884	0.4887	0.4890
2.3	0.4893	0.4896	0.4898	0.4901	0.4904	0.4906	0.4909	0.4911	0.4913	0.4916
2.4	0.4918	0.4920	0.4922	0.4925	0.4927	0.4929	0.4931	0.4932	0.4934	0.4936
2.5	0.4938	0.4940	0.4941	0.4943	0.4945	0.4946	0.4948	0.4949	0.4951	0.4952
2.6	0.4953	0.4955	0.4956	0.4957	0.4959	0.4960	0.4961	0.4962	0.4963	0.4964
2.7	0.4965	0.4966	0.4967	0.4968	0.4969	0.4970	0.4971	0.4972	0.4973	0.4974
2.8	0.4974	0.4975	0.4976	0.4977	0.4977	0.4978	0.4979	0.4979	0.4980	0.4981
2.9	0.4981	0.4982	0.4982	0.4983	0.4984	0.4984	0.4985	0.4985	0.4986	0.4986
3.0	0.4987	0.4987	0.4987	0.4988	0.4988	0.4989	0.4989	0.4989	0.4990	0.4990
3.1	0.4990	0.4991	0.4991	0.4991	0.4992	0.4992	0.4992	0.4992	0.4993	0.4993
3.2	0.4993	0.4993	0.4994	0.4994	0.4994	0.4994	0.4994	0.4995	0.4995	0.4995
3.3	0.4995	0.4995	0.4995	0.4996	0.4996	0.4996	0.4996	0.4996	0.4996	0.4997
3.4	0.4997	0.4997	0.4997	0.4997	0.4997	0.4997	0.4997	0.4997	0.4997	0.4998
3.5	0.4998	0.4998	0.4998	0.4998	0.4998	0.4998	0.4998	0.4998	0.4998	0.4998
3.6	0.4998	0.4998	0.4999	0.4999	0.4999	0.4999	0.4999	0.4999	0.4999	0.4999
3.7	0.4999	0.4999	0.4999	0.4999	0.4999	0.4999	0.4999	0.4999	0.4999	0.4999
3.8	0.4999	0.4999	0.4999	0.4999	0.4999	0.4999	0.4999	0.4999	0.4999	0.4999
3.9	0.5000	0.5000	0.5000	0.5000	0.5000	0.5000	0.5000	0.5000	0.5000	0.5000

ANSWERS

Diagnostic Prerequisite Test

Section references are provided in parentheses following each answer to guide students to the specific content in the book where they can find help or remediation.

1. (A) $(y + z)x$ (B) $(2 + x) + y$ (C) $2x + 3x$ *(A.1)* **2.** $x^3 - 3x^2 + 4x + 8$ *(A.2)* **3.** $x^3 + 3x^2 - 2x + 12$ *(A.2)* **4.** $-3x^5 + 2x^3 - 24x^2 + 16$ *(A.2)* **5.** (A) 1 (B) 1 (C) 2 (D) 3 *(A.2)* **6.** (A) 3 (B) 1 (C) -3 (D) 1 *(A.2)* **7.** $14x^2 - 30x$ *(A.2)* **8.** $6x^2 - 5xy - 4y^2$ *(A.2)* **9.** $(x + 2)(x + 5)$ *(A.3)* **10.** $x(x + 3)(x - 5)$ *(A.3)* **11.** $7/20$ *(A.1)* **12.** 0.875 *(A.1)* **13.** (A) 4.065×10^{12} (B) 7.3×10^{-3} *(A.5)* **14.** (A) 255,000,000 (B) 0.000 406 *(A.5)* **15.** (A) T (B) F *(A.1)* **16.** 0 and -3 are two examples of infinitely many. *(A.1)* **17.** $6x^5y^{15}$ *(A.5)* **18.** $3u^4/v^2$ *(A.5)* **19.** 6×10^2 *(A.5)* **20.** x^6/y^4 *(A.5)* **21.** $u^{7/3}$ *(A.6)* **22.** $3a^2/b$ *(A.6)* **23.** $\frac{5}{9}$ *(A.5)* **24.** $x + 2x^{1/2}y^{1/2} + y$ *(A.6)* **25.** $\dfrac{a^2 + b^2}{ab}$ *(A.4)* **26.** $\dfrac{a^2 - c^2}{abc}$ *(A.4)* **27.** $\dfrac{y^5}{x}$ *(A.4)* **28.** $\dfrac{1}{xy^2}$ *(A.4)* **29.** $\dfrac{-1}{7(7 + h)}$ *(A.4)* **30.** $\dfrac{xy}{y - x}$ *(A.6)* **31.** (A) Subtraction (B) Commutative $(+)$ (C) Distributive (D) Associative $(\cdot)$ (E) Negatives (F) Identity $(+)$ *(A.1)* **32.** (A) 6 (B) 0 *(A.1)* **33.** $4x = x - 4$; $x = -4/3$ *(1.1)* **34.** $-15/7$ *(1-2)* **35.** $(4/7, 0)$ *(1.2)* **36.** $(0, -4)$ *(1-2)* **37.** $x = 0, 5$ *(A.7)* **38.** $x = \pm\sqrt{7}$ *(A.7)* **39.** $x = -4, 5$ *(A.7)* **40.** $x = 1, \frac{1}{6}$ *(A.7)*

Chapter 1

Exercises 1.1

1. $x = 5$ **3.** $x = 2$ **5.** $x = -19$ **7.** $4 \le x < 13$ **9.** $-2 < x < 7$ **11.** $x \le 4$ **13.** $(-8, 2]$ **15.** $(-\infty, 9)$ **17.** $(-7, -5]$ **19.** $x = -\dfrac{3}{2}$ **21.** $y < -\dfrac{15}{2}$ **23.** $u = -\dfrac{3}{4}$ **25.** $x = 10$ **27.** $y \ge 3$ **29.** $x = 36$ **31.** $m < \dfrac{36}{7}$ **33.** $3 \le x < 7$ or $[3, 7)$

35. $-20 \le C \le 20$ or $[-20, 20]$

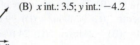

37. $y = \dfrac{3}{4}x - 3$ **39.** $y = -(A/B)x + (C/B) = (-Ax + C)/B$ **41.** $C = \dfrac{5}{9}(F - 32)$ **43.** $-2 < x \le 1$ or $(-2, 1]$

45. Negative **47.** 4,500 \$35 tickets and 5,000 \$55 tickets **49.** Fund A: \$180,000; Fund B: \$320,000 **51.** \$15,405 **53.** (A) \$420 (B) \$55 **55.** 34 rounds **57.** \$32,000 **59.** 5,851 books **61.** (B) 6,180 books (C) At least \$11.50 **63.** 5,000 **65.** 12.6 yr

Exercises 1.2

1. (D) **3.** (C) **5.**

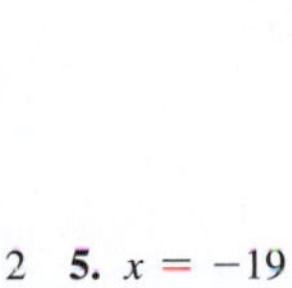

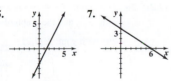

7.

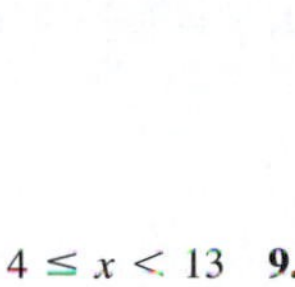

9. Slope = 5; y int. = -7 **11.** Slope = $-\dfrac{5}{2}$; y int. = -9 **13.** Slope = $\dfrac{1}{4}$; y int. = $\dfrac{2}{3}$ **15.** Slope = 2; x int. = -5 **17.** Slope = 8; x int. = 5 **19.** Slope = $\dfrac{6}{7}$; x int. = -7 **21.** $y = 2x + 1$ **23.** $y = -\dfrac{1}{3}x + 6$ **25.** x int.: $\dfrac{1}{2}$; y int.: 1; $y = -2x + 1$ **27.** x int.: -3; y int.: 1; $y = \dfrac{x}{3} + 1$ **29.**

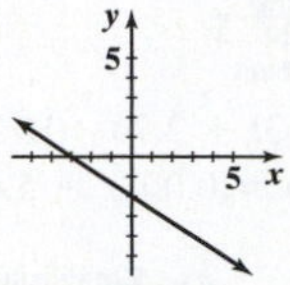

31.

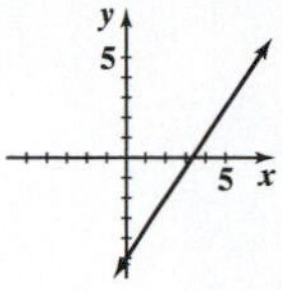

33.

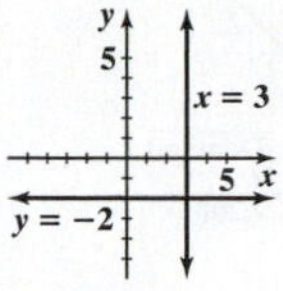

35. -4 **37.** $-\dfrac{3}{5}$ **39.** 2 **41.**

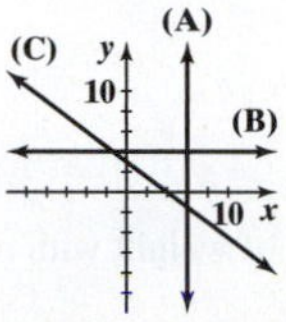

43.

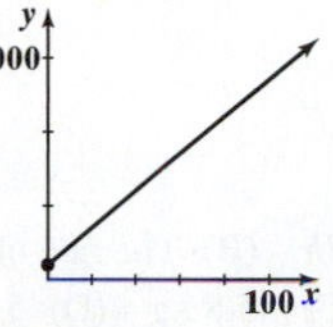

45. (A)

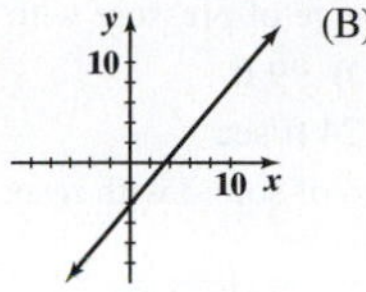

(B) x int.: 3.5; y int.: -4.2

(C)

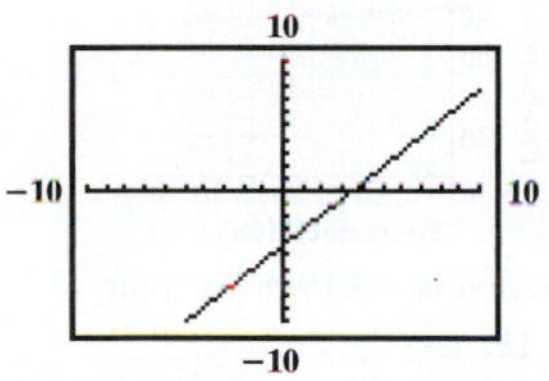

(D) x int.: 3.5; y int.: -4.2

47. $x = 4, y = -3$ **49.** $x = -1.5, y = -3.5$ **51.** $y = 5x - 15$ **53.** $y = -2x + 7$ **55.** $y = \dfrac{1}{3}x - \dfrac{20}{3}$ **57.** $y = -3.2x + 30.86$ **59.** (A) $m = \dfrac{2}{3}$ (B) $-2x + 3y = 11$ (C) $y = \dfrac{2}{3}x + \dfrac{11}{3}$ **61.** (A) $m = -\dfrac{5}{4}$ (B) $5x + 4y = -14$ (C) $y = -\dfrac{5}{4}x - \dfrac{7}{2}$ **63.** (A) Not defined (B) $x = 5$ (C) None **65.** (A) $m = 0$ (B) $y = 5$ (C) $y = 5$ **67.** The graphs have the same y int., (0, 2). **69.** $C = 124 + 0.12x$; 1,050 donuts **71.** (A) $C = 75x + 1,647$

(B)

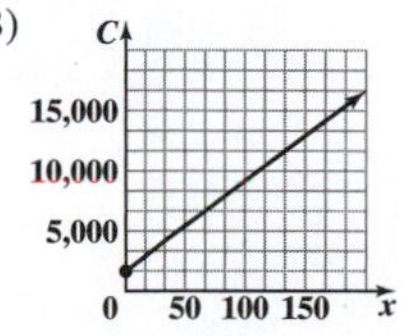

(C) The y int., \$1,647, is the fixed cost and the slope, \$75, is the cost per club.

73. (A) $R = 1.4C - 7$ (B) \$137 **75.** (A) $V = -7,500t + 157,000$ (B) \$112,000 (C) During the 12th year (D)

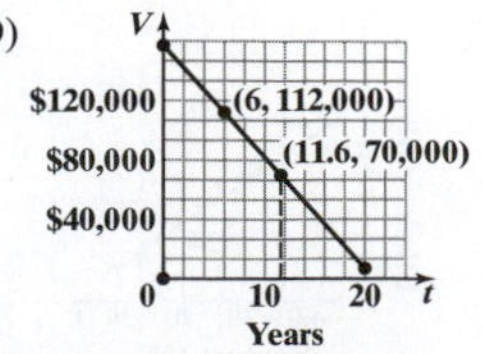

77. (A) $T = -1.84x + 212$ (B) 205.56°F (C) 6,522 ft

(D)

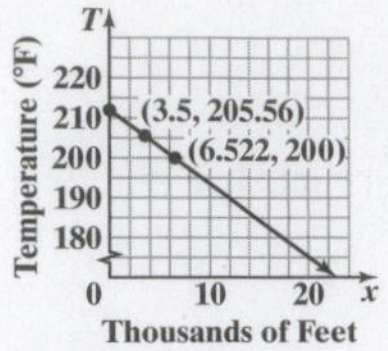

79. (A) $T = 70 - 3.6A$ (B) 10,000 ft

81. (A) $N = -0.0063t + 2.76$ (B) 2.45 persons **83.** (A) $f = -0.49t + 21$ (B) 2028 **85.** (A) $p = 0.001x + 5.4$ (B) $p = -0.001x + 13$ (C) (3,800, 9.2) (D)

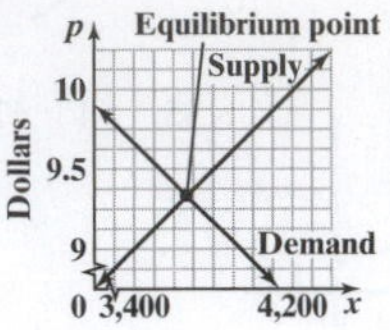

87. (A) $s = \frac{2}{5}w$ (B) 8 in. (C) 9 lb

Exercises 1.3

1. (A) $w = 49 + 1.7h$ (B) The rate of change of weight with respect to height is 1.7 kg/in. (C) 55.8 kg (D) 5′6.5″

3. (A) $P = 0.445d + 14.7$ (B) The rate of change of pressure with respect to depth is 0.445lb/in.^{2} per ft. (C) 37 lb/in.^{2} (D) 99 ft

5. (A) $a = 2{,}880 - 24t$ (B) -24 ft/sec (C) 24 ft/sec

7. $s = 0.6t + 331$; the rate of change of the speed of sound with respect to temperature is 0.6 m/s per °C. **9.** (A)

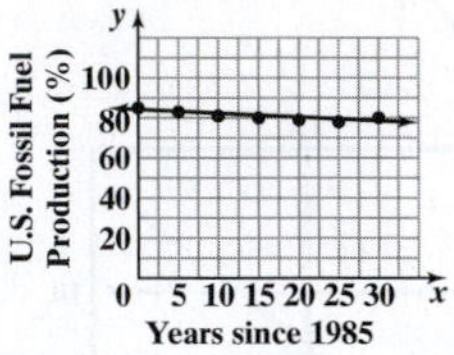

(B) The rate of change of fossil fuel production is -0.19% per year. (C) 76% of total production (D) 2058 **11.** (A)

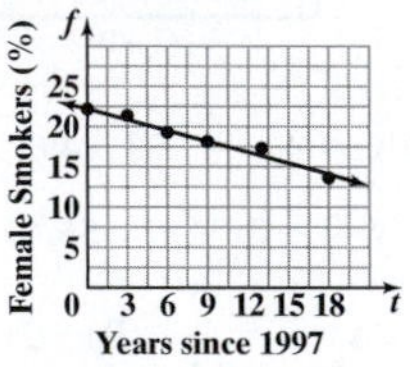

(B) 2025 **13.** (A)

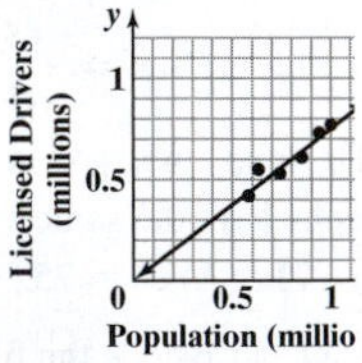

(B) 1,050,000 (C) 1,359,000

15. (A)

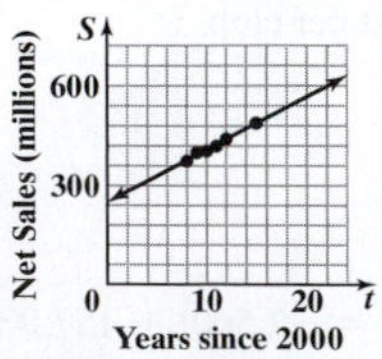

(B) \$662 billion

17. (A)

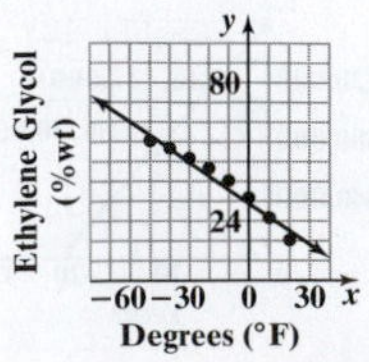

(B) 2°F (C) 22.75%

19. (A) The rate of change of height with respect to Dbh is 1.37 ft/in. (B) Height increases by approximately 1.37 ft. (C) 18 ft (D) 20 in.

21. (A) Undergraduate male enrollment is increasing at a rate of 87,000 students per year; undergraduate female enrollment is increasing at a rate of 140,000 students per year. (B) Male: 8.6 million; female: 11.5 million (C) 2026 **23.** $y = 0.061x + 50.703$; 54.67°F **25.** Men: $y = -0.070x + 49.058$; women: $y = -0.085x + 54.858$; yes **27.** Supply: $y = 0.2x + 0.87$; demand: $y = -0.15x + 3.5$; equilibrium price = \$2.37

Chapter 1 Review Exercises

1. $x = 2.8$ *(1.1)* **2.** $x = 2$ *(1.1)* **3.** $y = 1.8 - 0.4x$ *(1.1)*

4. $x = \frac{4}{3}y + \frac{7}{3}$ *(1.1)* **5.** $y < \frac{13}{4}$ or $\left(-\infty, \frac{13}{4}\right)$

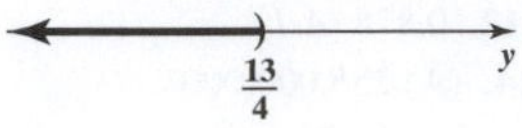

(1.1) **6.** $1 \le x < 3$ or $[1, 3)$

(1.1) **7.** $x \ge \frac{9}{2}$ or $\left[\frac{9}{2}, \infty\right)$

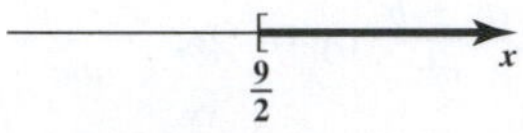

(1.1) **8.** *(1.2)*

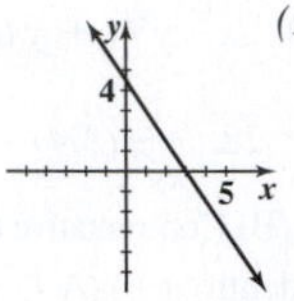

9. $2x + 3y = 12$ *(1.2)* **10.** x int. = 9; y int. = -6; slope = $\frac{2}{3}$ *(1.2)*

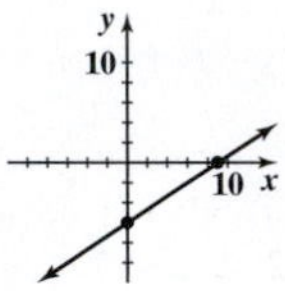

11. $y = -\frac{2}{3}x + 6$ *(1.2)*

12. Vert. line: $x = -6$; hor. line: $y = 5$ *(1.2)* **13.** (A) $y = -\frac{2}{3}x$ (B) $y = 3$ *(1.2)* **14.** (A) $3x + 2y = 1$ (B) $y = 5$ (C) $x = -2$ *(1.2)*

15. $x = \frac{25}{2}$ *(1.1)* **16.** $u = 36$ *(1.1)* **17.** $x = \frac{30}{11}$ *(1.1)* **18.** $x = 21$ *(1.1)*

19. $x = 4$ *(1.1)* **20.** $x < 4$ or $(-\infty, 4)$ *(1.1)*

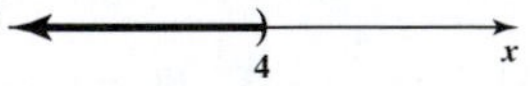

21. $x \ge 1$ or $[1, \infty)$ *(1.1)*

22. $x < -\frac{143}{17}$ or $\left(-\infty, -\frac{143}{17}\right)$

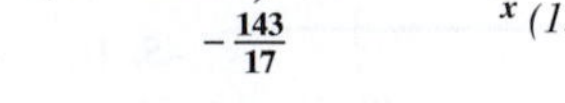

(1.1)

23. $1 < x \le 4$ or $(1, 4]$

(1.1)

24. $\frac{3}{8} \le x \le \frac{7}{8}$ or $\left[\frac{3}{8}, \frac{7}{8}\right]$

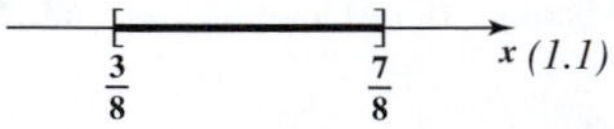

(1.1)

25.

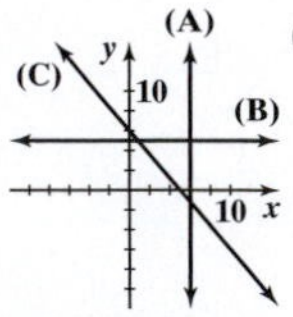

(1.2) **26.** The graph of $x = -3$ is a vert. line with x int. -3, and the graph of $y = 2$ is a hor. line with y int. 2. *(1.2)*

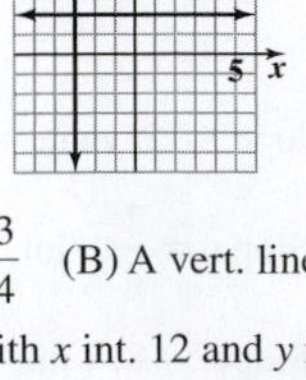

27. (A) An oblique line through the origin with slope $-\frac{3}{4}$ (B) A vert. line with x int. $-\frac{4}{3}$ (C) The x axis (D) An oblique line with x int. 12 and y int. 9 *(1.2)* **28.** $\frac{2A - bh}{h}$ *(1.1)* **29.** $\frac{S - P}{St}$ *(1.1)* **30.** $a < 0$ and b any real number *(1.1)* **31.** Less than *(1.1)* **32.** The graphs appear to be perpendicular to each other. (It can be shown that if the slopes of two slant lines are the negative reciprocals of each other, then the two lines are perpendicular.) *(1.2)*

33. \$75,000 *(1.1)* **34.** 9,375 DVDs *(1.1)* **35.** (A) $m = 132 - 0.6x$ (B) $M = 187 - 0.85x$ (C) Between 120 and 170 beats per minute (D) Between 102 and 144.5 beats per minute *(1.3)*
36. (A) $V = 224{,}000 - 15{,}500t$ (B) \$38,000 *(1.2)* **37.** (A) $R = 1.6C$ (B) \$192 (C) \$110 (D) The slope is 1.6. This is the rate of change of retail price with respect to cost. *(1.2)* **38.** \$400; \$800 *(1.1)* **39.** Demand: $p = 5.24 - 0.00125x$; 1,560 bottles *(1.2)* **40.** (A)

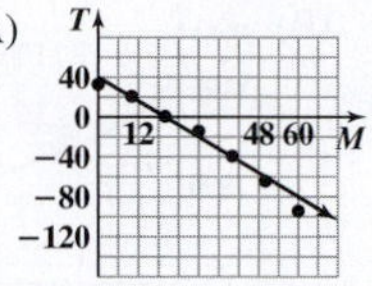

(B) $-30°F$ (C) 45% *(1.3)* **41.** (A) The dropout rate is decreasing at a rate of 0.308 percentage points per year. (B)

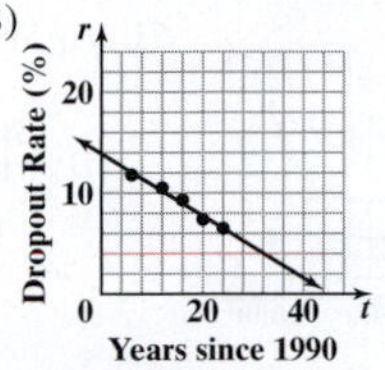

(C) 2026 *(1.3)*

42. (A) The CPI is increasing at a rate of 4.295 units per year. (B) 276.62 *(1.3)*
43. (A) The rate of change of tree height with respect to Dbh is 0.74. (B) Tree height increases by about 0.74 ft. (C) 21 ft (D) 16 in. *(1.3)*

Chapter 2

Exercises 2.1

1.
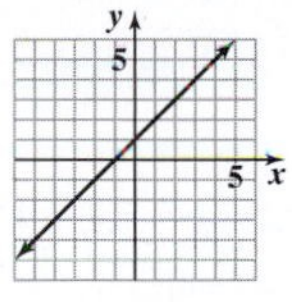

3.
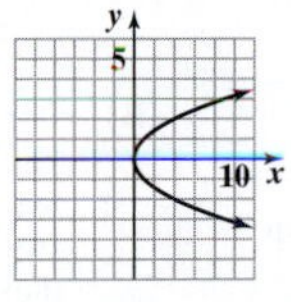

5.
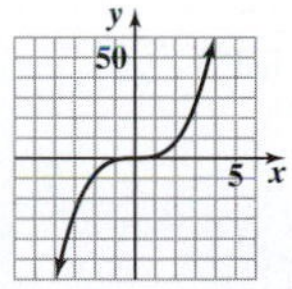

7.
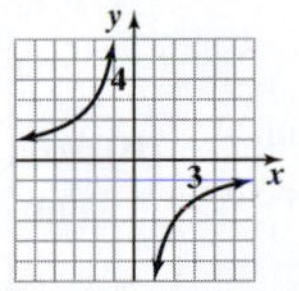

9. A function **11.** Not a function **13.** A function **15.** A function **17.** Not a function **19.** A function **21.** Linear **23.** Neither **25.** Linear **27.** Constant

29.
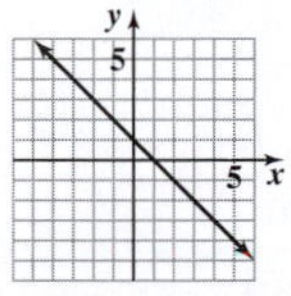

31.
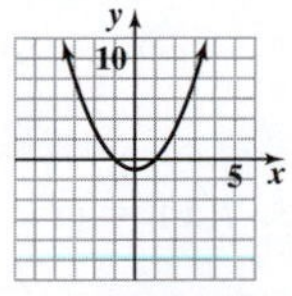

33.
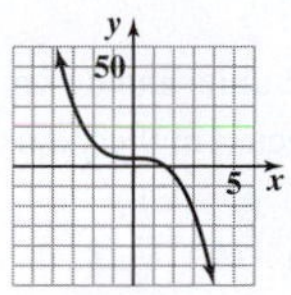

35.
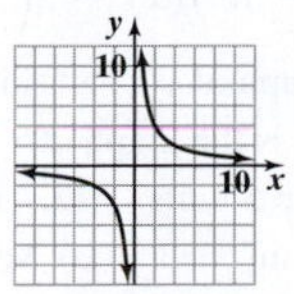

37.
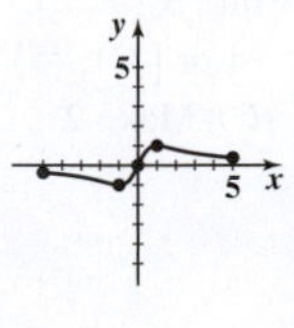

39. $y = 0$ **41.** $y = 4$

43. $x = -5$ **45.** $x = -6$ **47.** All real numbers **49.** All real numbers except -4 **51.** $x \le 7$ **53.** Yes; all real numbers **55.** No; for example, when $x = 0$, $y = \pm 2$ **57.** Yes; all real numbers except 0 **59.** No; when $x = 1$, $y = \pm 1$ **61.** $25x^2 - 4$ **63.** $x^2 + 4x$ **65.** $x^4 - 4$ **67.** $x - 4$ **69.** $h^2 - 4$ **71.** $4h + h^2$ **73.** $4h + h^2$ **75.** (A) $4x + 4h - 3$ (B) $4h$ (C) 4 **77.** (A) $4x^2 + 8xh + 4h^2 - 7x - 7h + 6$ (B) $8xh + 4h^2 - 7h$ (C) $8x + 4h - 7$ **79.** (A) $20x + 20h - x^2 - 2xh - h^2$ (B) $20h - 2xh - h^2$ (C) $20 - 2x - h$

81. $P(w) = 2w + \dfrac{50}{w}, w > 0$ **83.** $A(l) = l(50 - l), 0 < l < 50$

85. (A)
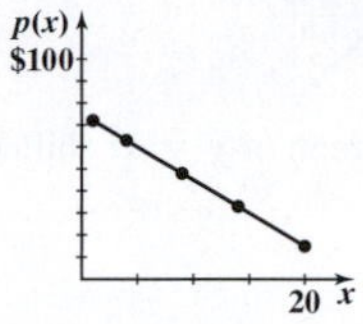

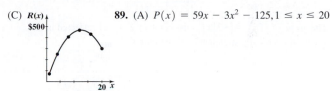

(B) \$54; \$42

87. (A) $R(x) = (75 - 3x)x, 1 \le x \le 20$ (B)

x	R(x)
1	72
4	252
8	408
12	468
16	432
20	300

(C)
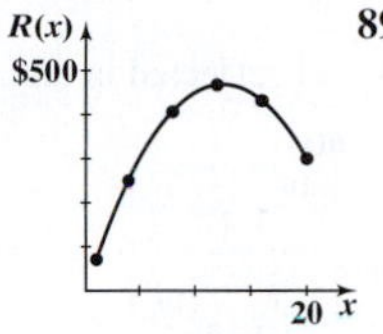

89. (A) $P(x) = 59x - 3x^2 - 125, 1 \le x \le 20$

(B)

x	P(x)
1	−69
4	63
8	155
12	151
16	51
20	−145

(C)
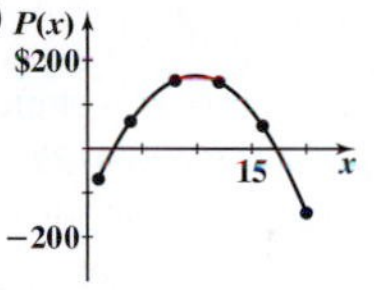

91. $v = \dfrac{75 - w}{15 + w}$; 1.9032 cm/sec

Exercises 2.2

1. Domain: all real numbers; range: $[-4, \infty)$ **3.** Domain: all real numbers; range: all real numbers **5.** Domain: $[0, \infty)$; range: $(-\infty, 8]$ **7.** Domain: all real numbers; range: all real numbers **9.** Domain: all real numbers; range: $[9, \infty)$ **11.**

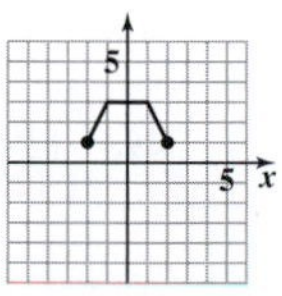

13.
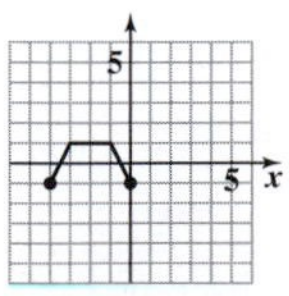

15.
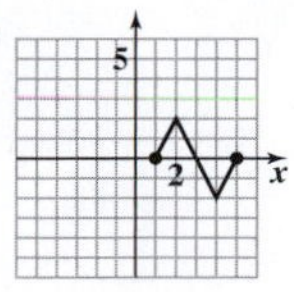

17.
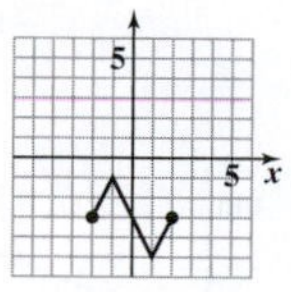

19.
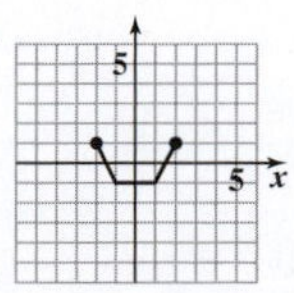

21.
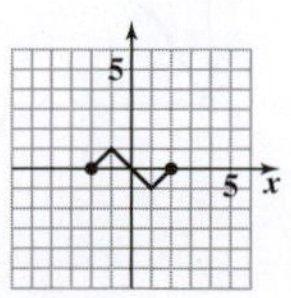

23. 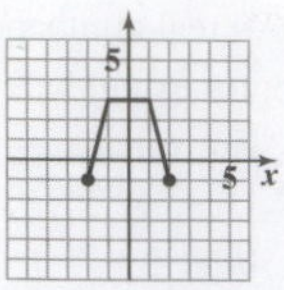

25. 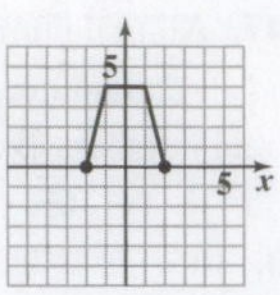

27. The graph of $g(x) = -|x + 3|$ is the graph of $y = |x|$ reflected in the x axis and shifted 3 units to the left.

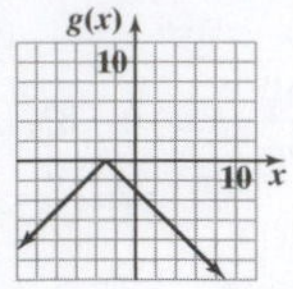

29. The graph of $f(x) = (x - 4)^2 - 3$ is the graph of $y = x^2$ shifted 4 units to the right and 3 units down.

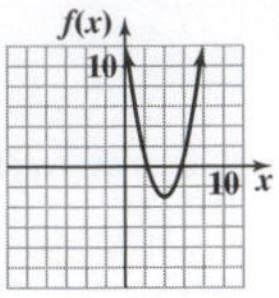

31. The graph of $f(x) = 7 - \sqrt{x}$ is the graph of $y = \sqrt{x}$ reflected in the x axis and shifted 7 units up.

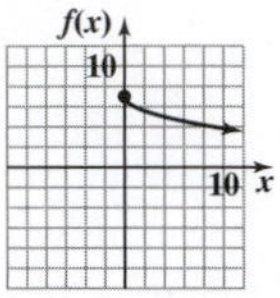

33. The graph of $h(x) = -3|x|$ is the graph of $y = |x|$ reflected in the x axis and vertically stretched by a factor of 3.

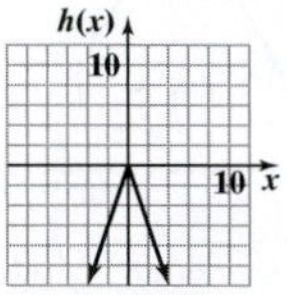

35. The graph of the basic function $y = x^2$ is shifted 2 units to the left and 3 units down. Equation: $y = (x + 2)^2 - 3$. **37.** The graph of the basic function $y = x^2$ is reflected in the x axis and shifted 3 units to the right and 2 units up. Equation: $y = 2 - (x - 3)^2$. **39.** The graph of the basic function $y = \sqrt{x}$ is reflected in the x axis and shifted 4 units up. Equation: $y = 4 - \sqrt{x}$. **41.** The graph of the basic function $y = x^3$ is shifted 2 units to the left and 1 unit down. Equation: $y = (x + 2)^3 - 1$.

43. $g(x) = \sqrt{x - 2} - 3$

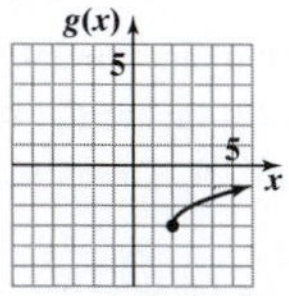

45. $g(x) = -|x + 3|$

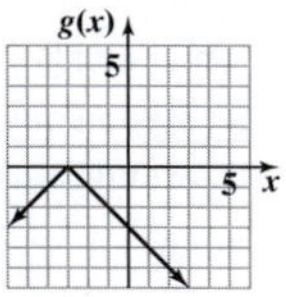

47. $g(x) = -(x - 2)^3 - 1$

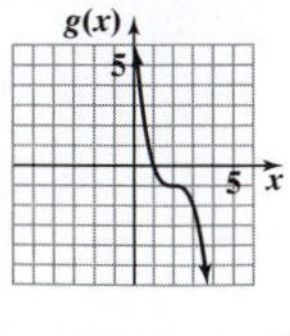

49. 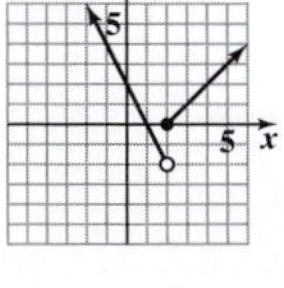

51. 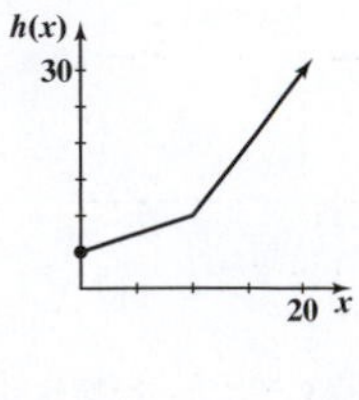

53. 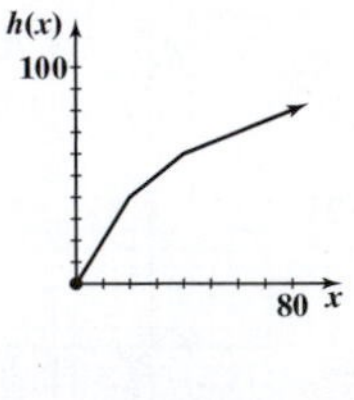

55. The graph of the basic function $y = |x|$ is reflected in the x axis and vertically shrunk by a factor of 0.5. Equation: $y = -0.5|x|$. **57.** The graph of the basic function $y = x^2$ is reflected in the x axis and vertically stretched by a factor of 2. Equation: $y = -2x^2$. **59.** The graph of the basic function $y = \sqrt[3]{x}$ is reflected in the x axis and vertically stretched by a factor of 3. Equation: $y = -3\sqrt[3]{x}$. **61.** Reversing the order does not change the result. **63.** Reversing the order can change the result. **65.** Reversing the order can change the result. **67.** (A) The graph of the basic function $y = \sqrt{x}$ is reflected in the x axis, vertically expanded by a factor of 4, and shifted up 115 units. (B)

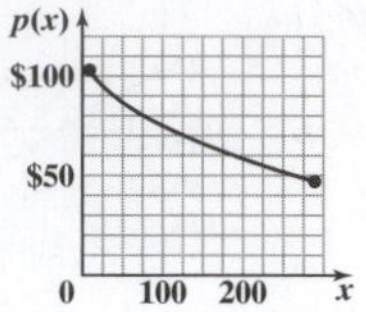

69. (A) The graph of the basic function $y = x^3$ is vertically contracted by a factor of 0.000 48 and shifted right 500 units and up 60,000 units.

(B)

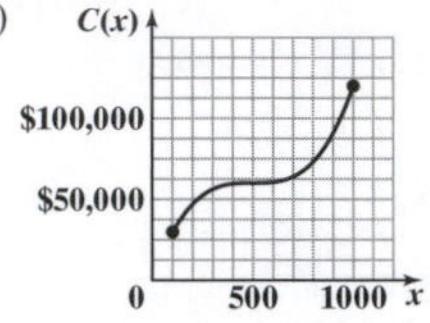

71. (A) $S(x) = \begin{cases} 8.5 + 0.065x & \text{if } 0 \le x \le 700 \\ -9 + 0.09x & \text{if } x > 700 \end{cases}$

(B)

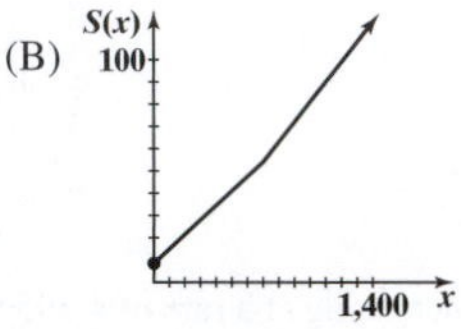

73. (A) $T(x) = \begin{cases} 0.02x & \text{if } 0 < x \le 25{,}000 \\ 0.04x - 500 & \text{if } 25{,}000 < x \le 100{,}000 \\ 0.06x - 2{,}500 & \text{if } x > 100{,}000 \end{cases}$

(B)

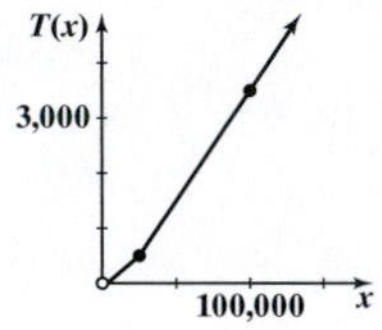

(C) \$1,700; \$4,100

75. (A) The graph of the basic function $y = x$ is vertically stretched by a factor of 5.5 and shifted down 220 units.

(B)

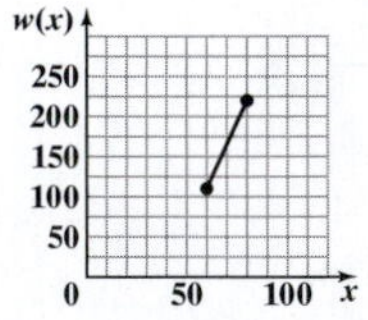

77. (A) The graph of the basic function $y = \sqrt{x}$ is vertically stretched by a factor of 7.08.

(B)

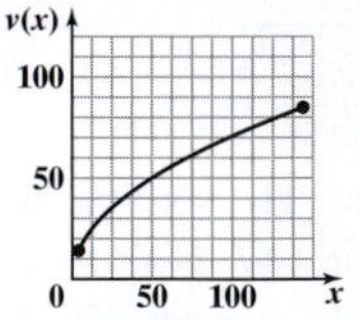

Exercises 2.3

1. $f(x) = (x - 5)^2 - 25$ **3.** $f(x) = (x + 10)^2 - 50$

5. $f(x) = -2(x - 1)^2 - 3$ **7.** $f(x) = 2\left(x + \frac{1}{2}\right)^2 + \frac{1}{2}$

9. The graph of $f(x)$ is the graph of $y = x^2$ shifted right 2 units and down 1 unit. **11.** The graph of $m(x)$ is the graph of $y = x^2$ reflected in the x axis, then shifted right 3 units and up 5 units. **13.** (A) m (B) g (C) f (D) n **15.** (A) x int.: 1, 3; y int.: -3 (B) Vertex: (2, 1) (C) Max.: 1 (D) Range: $y \le 1$ or $(-\infty, 1]$ **17.** (A) x int.: -3, -1; y int.: 3 (B) Vertex: $(-2, -1)$ (C) Min.: -1 (D) Range: $y \ge -1$ or $[-1, \infty)$ **19.** (A) x int.: $3 \pm \sqrt{2}$; y int.: -7 (B) Vertex: $(3, 2)$ (C) Max.: 2 (D) Range: $y \le 2$ or $(-\infty, 2]$

21. (A) x int.: $-1 \pm \sqrt{2}$; y int.: -1 (B) Vertex: $(-1, -2)$ (C) Min.: -2 (D) Range: $y \geq -2$ or $[-2, \infty)$ **23.** $y = -[x - (-2)]^2 + 5$ or $y = -(x + 2)^2 + 5$ **25.** $y = (x - 1)^2 - 3$ **27.** Vertex form: $(x - 4)^2 - 4$ (A) x int.: 2, 6; y int.: 12 (B) Vertex: $(4, -4)$ (C) Min.: -4 (D) Range: $y \geq -4$ or $[-4, \infty)$ **29.** Vertex form: $-4(x - 2)^2 + 1$ (A) x int.: 1.5, 2.5; y int.: -15 (B) Vertex: $(2, 1)$ (C) Max.: 1 (D) Range: $y \leq 1$ or $(-\infty, 1]$ **31.** Vertex form: $0.5(x - 2)^2 + 3$ (A) x int.: none; y int.: 5 (B) Vertex: $(2, 3)$ (C) Min.: 3 (D) Range: $y \geq 3$ or $[3, \infty)$ **33.** (A) $-4.87, 8.21$ (B) $-3.44, 6.78$ (C) No solution **35.** 651.0417 **37.** $g(x) = 0.25(x - 3)^2 - 9.25$ (A) x int.: $-3.08, 9.08$; y int.: -7 (B) Vertex: $(3, -9.25)$ (C) Min.: -9.25 (D) Range: $y \geq -9.25$ or $[-9.25, \infty)$ **39.** $f(x) = -0.12(x - 4)^2 + 3.12$ (A) x int.: $-1.1, 9.1$; y int.: 1.2 (B) Vertex: $(4, 3.12)$ (C) Max.: 3.12 (D) Range: $y \leq 3.12$ or $(-\infty, 3.12]$ **41.** $(-\infty, -5) \cup (3, \infty)$ **43.** $[-3, 2]$ **45.** $x = -5.37, 0.37$ **47.** $-1.37 < x < 2.16$ **49.** $x \leq -0.74$ or $x \geq 4.19$ **51.** Axis: $x = 2$; vertex: (2, 4); range: $y \geq 4$ or $[4, \infty)$; no x int. **53.** (A)

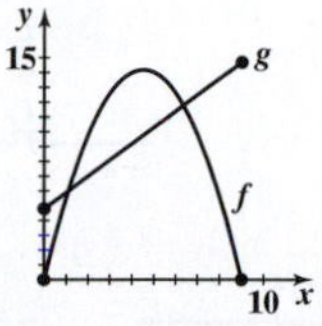

(B) 1.64, 7.61 (C) $1.64 < x < 7.61$ (D) $0 \leq x < 1.64$ or $7.61 < x \leq 10$

55. (A)

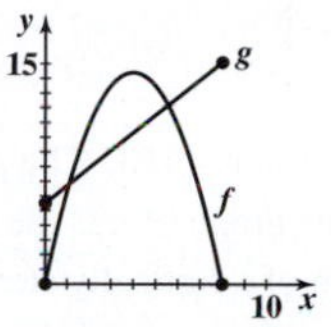

(B) 1.10, 5.57 (C) $1.10 < x < 5.57$ (D) $0 \leq x < 1.10$ or $5.57 < x \leq 8$

65. (A)

x	28	30	32	34	36
Mileage	45	52	55	51	47
$f(x)$	45.3	51.8	54.2	52.4	46.5

(B)

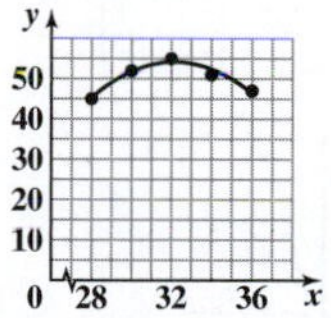

(C) $f(31) = 53.50$ thousand miles; (D) $f(35) = 49.95$ thousand miles;

69. (A)

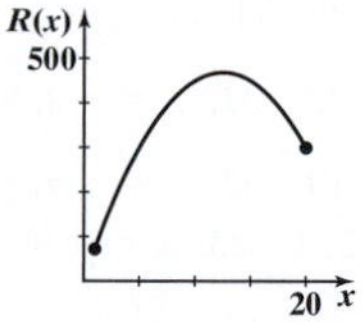

(B) 12.5 (12,500,000 chips); \$468,750,000 (C) \$37.50

71. (A)

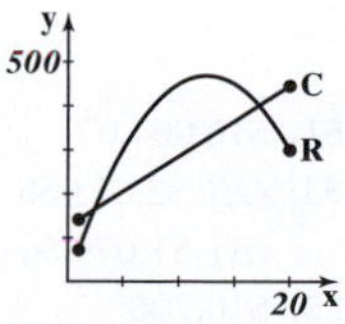

(B) 2,415,000 chips and 17,251,000 chips (C) Loss: $1 \leq x < 2.415$ or $17.251 < x \leq 20$; profit: $2.415 < x < 17.251$

73. (A) $P(x) = 59x - 3x^2 - 125$

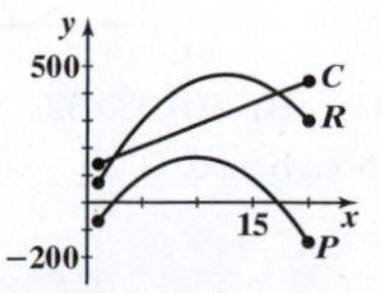

(C) Intercepts and break-even points: 2,415,000 chips and 17,251,000 chips
(D) Maximum profit is \$165,083,000 at a production level of 9,833,000 chips. This is much smaller than the maximum revenue of \$468,750,000.

75. $x = 0.14$ cm **77.** 10.6 mph

Exercises 2.4

1. (A) 1 (B) -3 (C) 21 **3.** (A) 2 (B) $-5, -4$ (C) 20 **5.** (A) 6 (B) None (C) 15 **7.** (A) 5 (B) $0, -6$ (C) 0 **9.** (A) 11 (B) $-5, -2, 5$ (C) $-12{,}800$ **11.** (A) 4 (B) Negative **13.** (A) 5 (B) Negative **15.** (A) 1 (B) Negative **17.** (A) 6 (B) Positive **19.** 10 **21.** 1 **23.** (A) x int.: -2; y int.: -1 (B) Domain: all real numbers except 2 (C) Vertical asymptote: $x = 2$; horizontal asymptote: $y = 1$ (D)

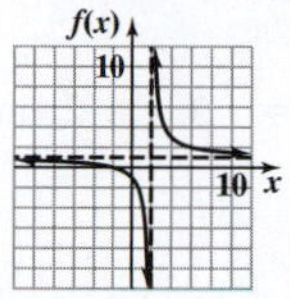

25. (A) x int.: 0; y int.: 0 (B) Domain: all real numbers except -2 (C) Vertical asymptote: $x = -2$; horizontal asymptote: $y = 3$ (D)

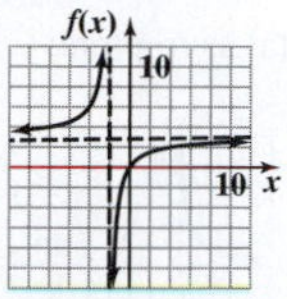

27. (A) x int.: 2; y int.: -1 (B) Domain: all real numbers except 4 (C) Vertical asymptote: $x = 4$; horizontal asymptote: $y = -2$ (D)

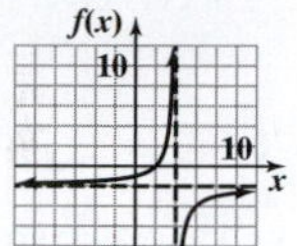

29. (A)

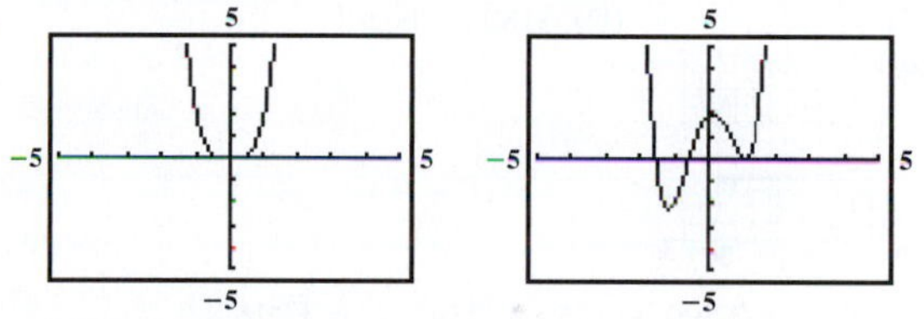

(B)

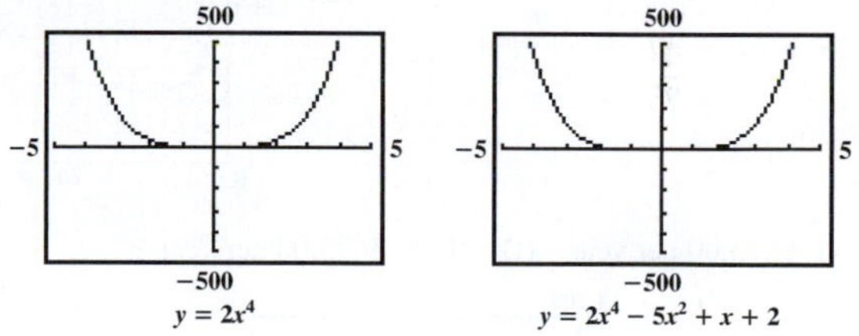

$y = 2x^4$ $y = 2x^4 - 5x^2 + x + 2$

31. (A)

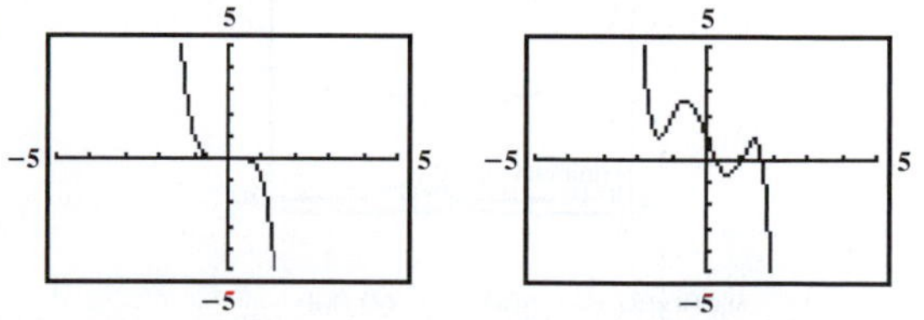

(B)

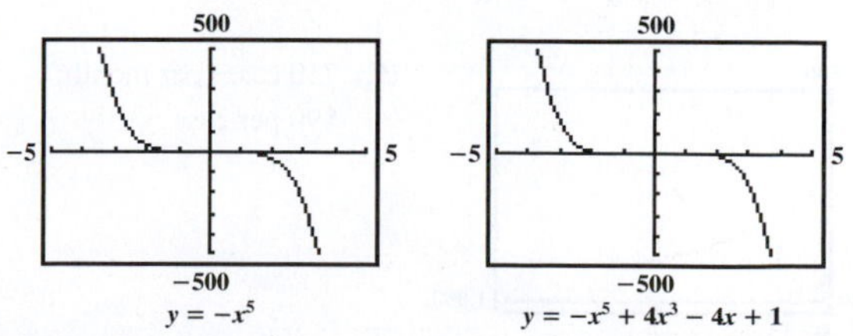

$y = -x^5$ $y = -x^5 + 4x^3 - 4x + 1$

33. $y = \frac{5}{6}$ **35.** $y = \frac{1}{4}$ **37.** $y = 0$ **39.** None **41.** $x = -1, x = 1, x = -3, x = 3$ **43.** $x = 5$ **45.** $x = -6, x = 6$ **47.** (A) x int.: 0; y int.: 0 (B) Vertical asymptotes: $x = -2, x = 3$; horizontal asymptote: $y = 2$

(C) 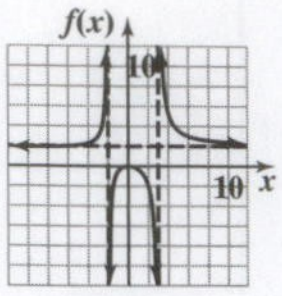

(D) 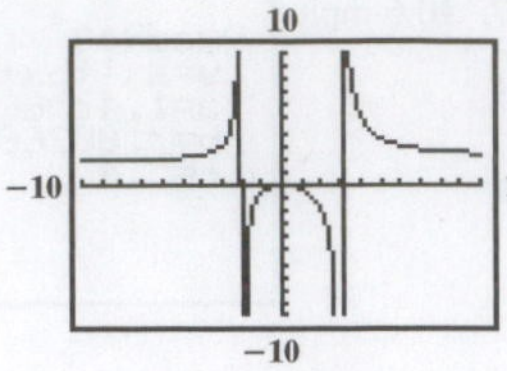

49. (A) x int.: $\pm\sqrt{3}$; y int.: $-\dfrac{2}{3}$ (B) Vertical asymptotes: $x = -3$, $x = 3$; horizontal asymptote: $y = -2$

(C) 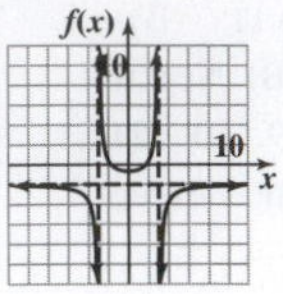

(D) 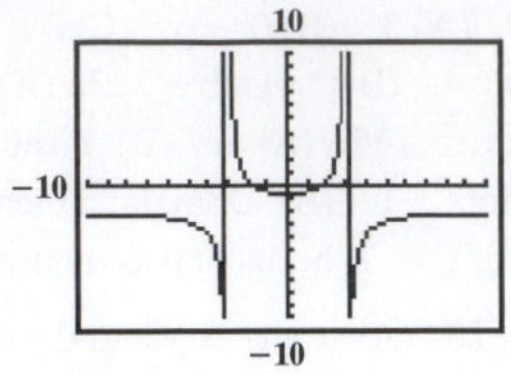

51. (A) x int.: 6; y int.: −4 (B) Vertical asymptotes: $x = -3$, $x = 2$; horizontal asymptote: $y = 0$

(C) 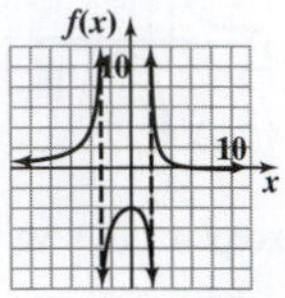

(D) 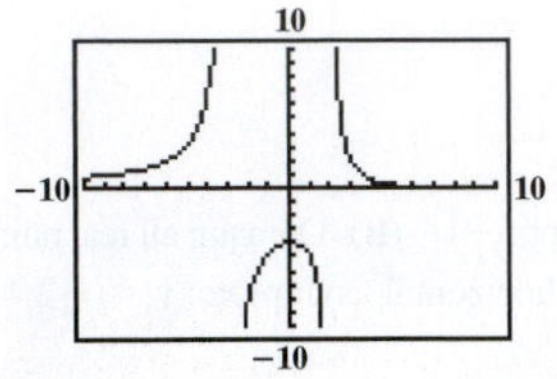

53. $f(x) = x^2 - x - 2$ **55.** $f(x) = 4x - x^3$

57. (A) $C(x) = 180x + 200$ (B) $\overline{C}(x) = \dfrac{180x + 200}{x}$

(C) 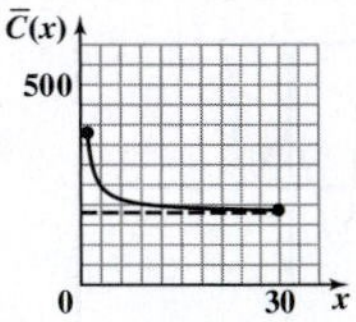

(D) $180 per board

59. (A) $\overline{C}(n) = \dfrac{2{,}500 + 175n + 25n^2}{n}$ (B) 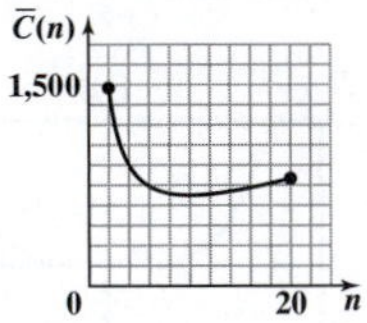

(C) 10 yr; $675.00 per year (D) 10 yr; $675.00 per year

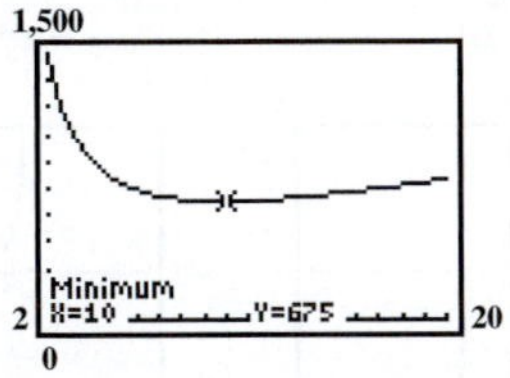

61. (A) $\overline{C}(x) = \dfrac{0.00048(x - 500)^3 + 60{,}000}{x}$

(B) 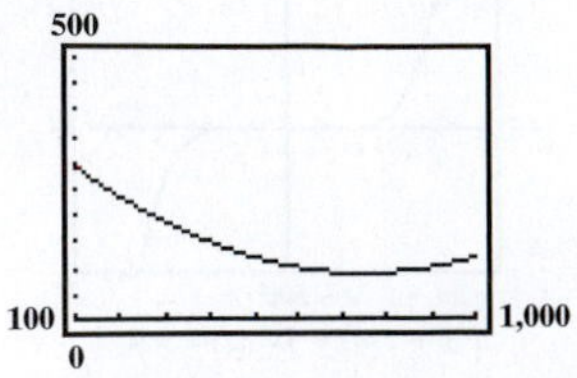

(C) 750 cases per month; $90 per case

63. (A) 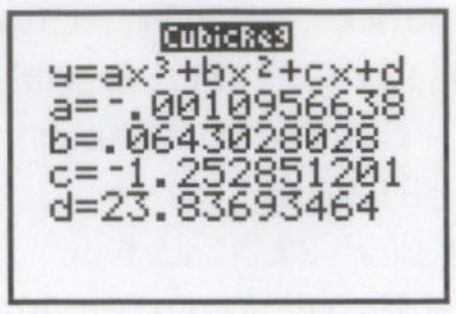

(B) 1.7 lb **65.** (A) 0.06 cm/sec

(B) 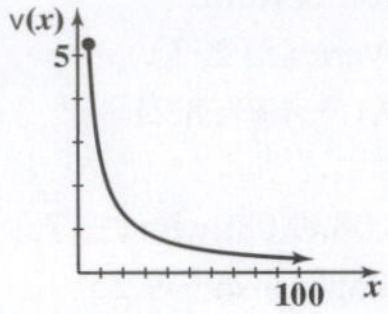

67. (A) 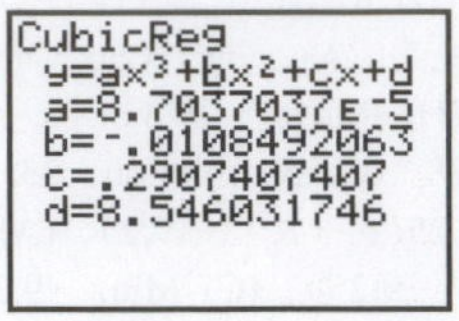

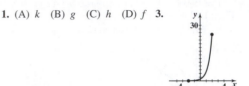

(B) 5.5

Exercises 2.5

1. (A) k (B) g (C) h (D) f **3.** 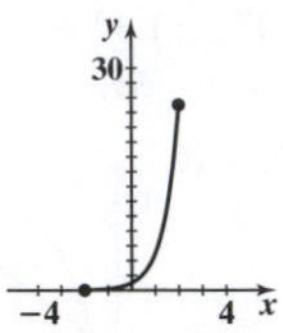

5. 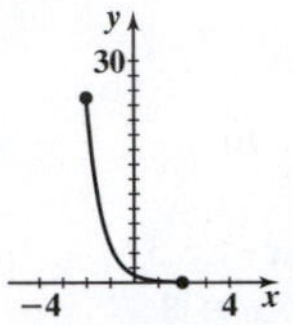

7. 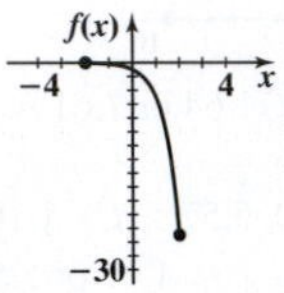

9. 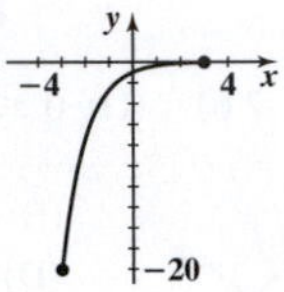

11. The graph of g is the graph of f reflected in the x axis. **13.** The graph of g is the graph of f shifted 1 unit to the left. **15.** The graph of g is the graph of f shifted 1 unit up. **17.** The graph of g is the graph of f vertically stretched by a factor of 2 and shifted to the left 2 units.

19. (A) 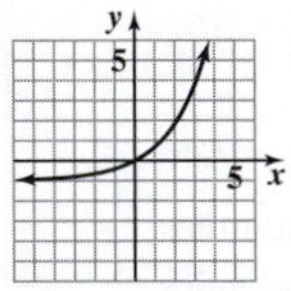

(B) 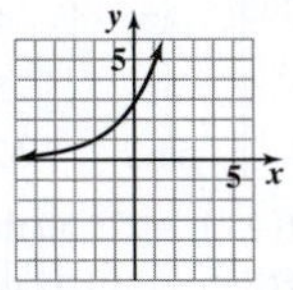

(C) 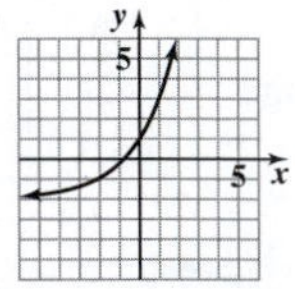

(D) 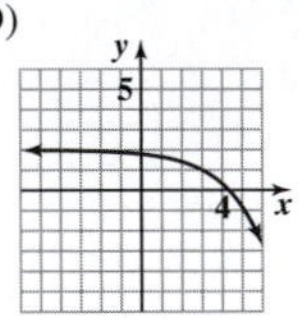

21. 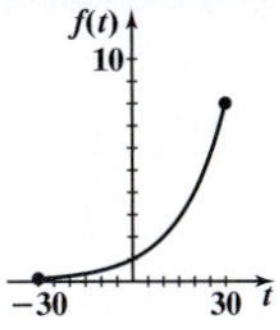

23. 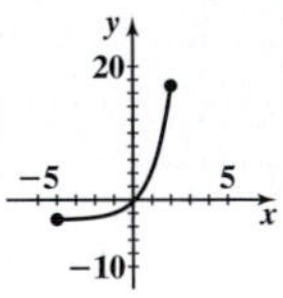

25. 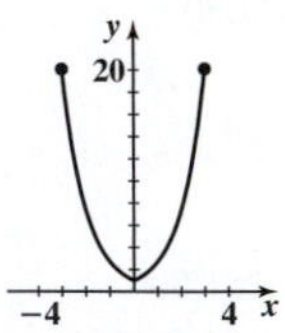

27. $a = 1, -1$ **29.** $x = 32$ **31.** $x = -2, 5$ **33.** $x = -9$ **35.** $x = 3, 19$ **37.** $x = -4, -3$ **39.** $x = -7$ **41.** $x = -2, 2$ **43.** $x = 1/4$ **45.** No solution

47. 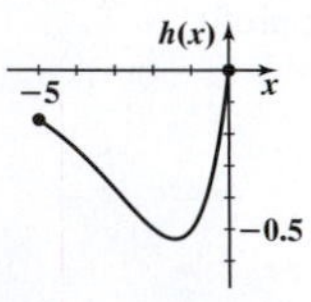

49. 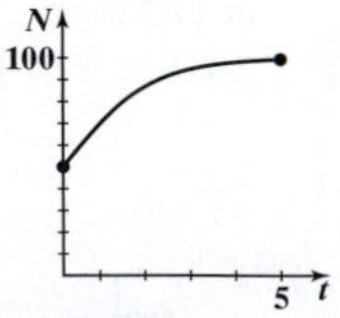

51. $16,064.07 **53.** (A) $2,633.56 (B) $7,079.54 **55.** $10,706

57. (A) $10,095.41 (B) $10,080.32 (C) $10,085.27 **59.** N approaches 2 as t increases without bound.

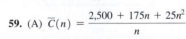

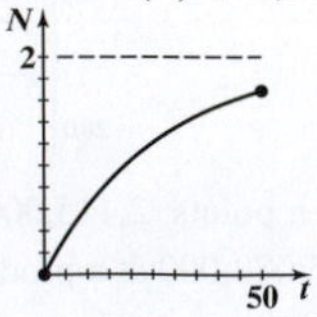

61. (A) 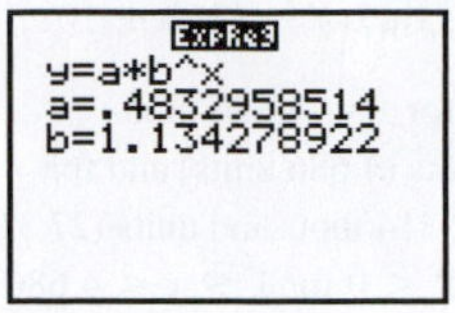

(B) 9.94 billion **63.** (A) 10% (B) 1% **65.** (A) $P = 12e^{0.0402x}$ (B) 17.9 million **67.** (A) $P = 127e^{-0.0016x}$ (B) 124 million

Exercises 2.6

1. $27 = 3^3$ **3.** $10^0 = 1$ **5.** $8 = 4^{3/2}$ **7.** $\log_7 49 = 2$ **9.** $\log_4 8 = \frac{3}{2}$ **11.** $\log_b A = u$ **13.** 6 **15.** −5 **17.** 7 **19.** −3 **21.** Not defined **23.** $\log_b P - \log_b Q$ **25.** $5 \log_b L$ **27.** q^p **29.** $x = 48$ **31.** $b = 4$ **33.** $y = -3$ **35.** $b = 1/3$ **37.** $x = 8$ **39.** False **41.** True **43.** True **45.** False **47.** $x = 2$ **49.** $x = 8$ **51.** $x = 7$ **53.** No solution

55. 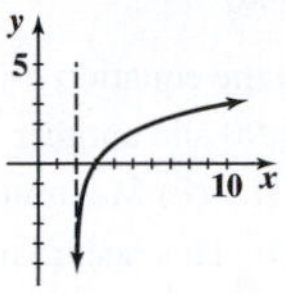

57. The graph of $y = \log_2(x - 2)$ is the graph of $y = \log_2 x$ shifted to the right 2 units. **59.** Domain: $(-1, \infty)$; range: all real numbers **61.** (A) 3.547 43 (B) −2.160 32 (C) 5.626 29 (D) −3.197 04 **63.** (A) 13.4431 (B) 0.0089 (C) 16.0595 (D) 0.1514 **65.** 1.0792 **67.** 1.4595 **69.** 18.3559 **71.** Increasing: $(0, \infty)$

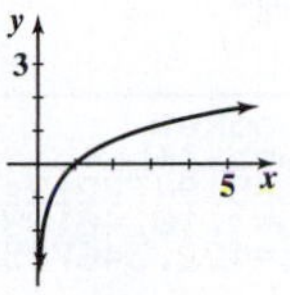

73. Decreasing: $(0, 1]$ Increasing: $[1, \infty)$

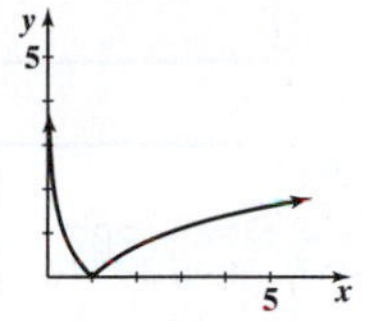

75. Increasing: $(-2, \infty)$

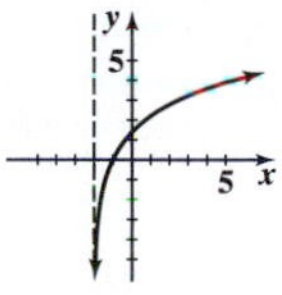

77. Increasing: $(0, \infty)$

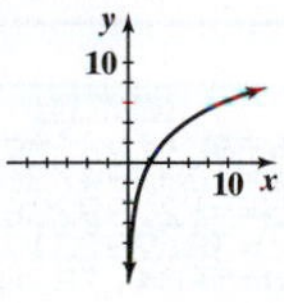

79. Because $b^0 = 1$ for any permissible base $b (b > 0, b \neq 1)$. **81.** $x > \sqrt{x} > \ln x$ for $1 < x \leq 16$ **83.** 4 yr **85.** 9.87 yr; 9.80 yr **87.** 7.51 yr **89.** (A) 5,373

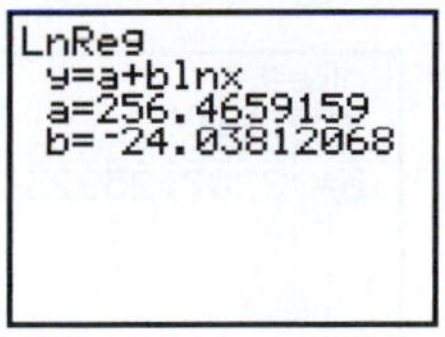

(B) 7,220

93. 168 bushels/acre

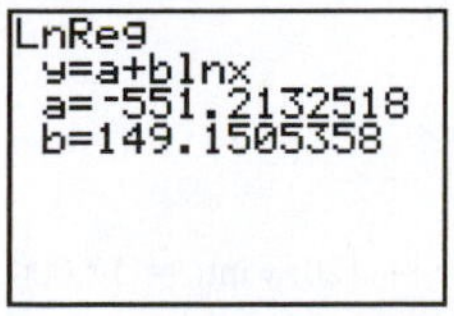

95. 912 yr

Chapter 2 Review Exercises

1. 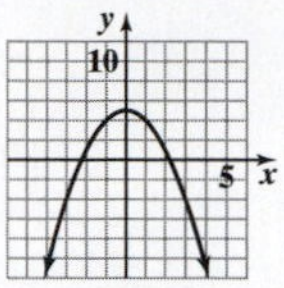 (2.1) **2.** 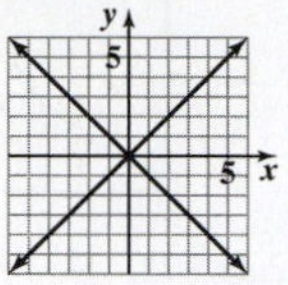 (2.1) **3.** 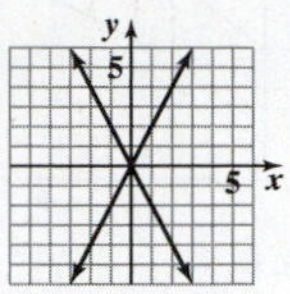 (2.1)

4. (A) Not a function (B) A function (C) A function (D) Not a function *(2.1)* **5.** (A) −2 (B) −8 (C) 0 (D) Not defined *(2.1)* **6.** $v = \ln u$ *(2.6)* **7.** $y = \log x$ *(2.6)* **8.** $M = e^N$ *(2.6)* **9.** $u = 10^v$ *(2.6)* **10.** $x = 9$ *(2.6)* **11.** $x = 6$ *(2.6)* **12.** $x = 4$ *(2.6)* **13.** $x = 2.157$ *(2.6)* **14.** $x = 13.128$ *(2.6)* **15.** $x = 1{,}273.503$ *(2.6)* **16.** $x = 0.318$ *(2.6)* **17.** (A) $y = 4$ (B) $x = 0$ (C) $y = 1$ (D) $x = -1$ or 1 (E) $y = -2$ (F) $x = -5$ or 5 *(2.1)*

18. (A) 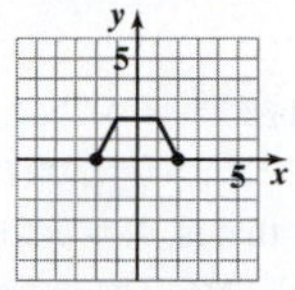 (B) 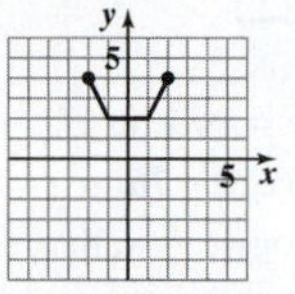 (C) 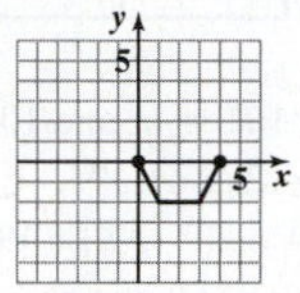

(D) 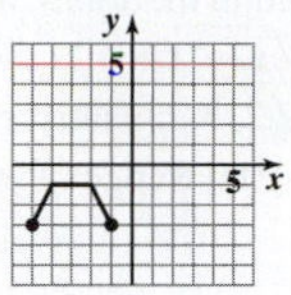 *(2.2)*

19. $f(x) = -(x - 2)^2 + 4$. The graph of $f(x)$ is the graph of $y = x^2$ reflected in the x axis, then shifted right 2 units and up 4 units. *(2.2)* **20.** (A) g (B) m (C) n (D) f *(2.2, 2.3)*

21. (A) x intercepts: −4, 0; y intercept: 0 (B) Vertex: $(-2, -4)$ (C) Minimum: −4 (D) Range: $y \geq -4$ or $[-4, \infty)$ *(2.3)* **22.** Quadratic *(2.3)* **23.** Linear *(2.1)* **24.** None *(2.1, 2.3)* **25.** Constant *(2.1)* **26.** $x = 8$ *(2.6)* **27.** $x = 3$ *(2.6)* **28.** $x = 3$ *(2.5)* **29.** $x = -1, 3$ *(2.5)* **30.** $x = 0, \frac{3}{2}$ *(2.5)* **31.** $x = -2$ *(2.6)* **32.** $x = \frac{1}{2}$ *(2.6)* **33.** $x = 27$ *(2.6)* **34.** $x = 13.3113$ *(2.6)* **35.** $x = 158.7552$ *(2.6)* **36.** $x = 0.0097$ *(2.6)* **37.** $x = 1.4359$ *(2.6)* **38.** $x = 1.4650$ *(2.6)* **39.** $x = 92.1034$ *(2.6)* **40.** $x = 9.0065$ *(2.6)* **41.** $x = 2.1081$ *(2.6)* **42.** (A) All real numbers except $x = -2$ and 3 (B) $x < 5$ *(2.1)* **43.** Vertex form: $4\left(x + \frac{1}{2}\right)^2 - 4$; x intercepts: $-\frac{3}{2}$ and $\frac{1}{2}$; y intercept: −3; vertex: $(-\frac{1}{2}, -4)$; minimum: −4; range: $y \geq -4$ or $[-4, \infty)$ *(2.3)* **44.** $(-1.54, -0.79)$; $(0.69, 0.99)$ *(2.5, 2.6)*

45. 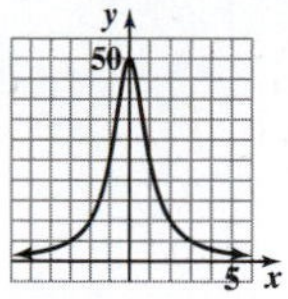 *(2.1)* **46.** 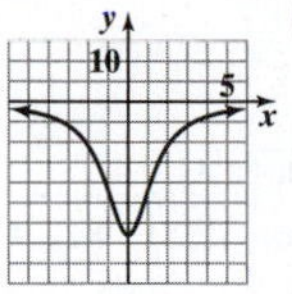 *(2.1)* **47.** 6 *(2.1)*

48. −19 *(2.1)* **49.** $10x - 4$ *(2.1)* **50.** $21 - 5x$ *(2.1)* **51.** (A) −1 (B) $-1 - 2h$ (C) $-2h$ (D) −2 *(2.1)* **52.** (A) $a^2 - 3a + 1$ (B) $a^2 + 2ah + h^2 - 3a - 3h + 1$ (C) $2ah + h^2 - 3h$ (D) $2a + h - 3$ *(2.1)* **53.** The graph of function m is the graph of $y = |x|$ reflected in the x axis and shifted to the right 4 units. *(2.2)* **54.** The graph of function g is the graph of $y = x^3$ vertically contracted by a factor of 0.3 and shifted up 3 units. *(2.2)* **55.** The graph of $y = x^2$ is vertically expanded by a factor of 2, reflected in the x axis, and shifted to the left 3 units. Equation: $y = -2(x + 3)^2$. *(2.2)* **56.** $f(x) = 2\sqrt{x + 3} - 1$ *(2.2)*

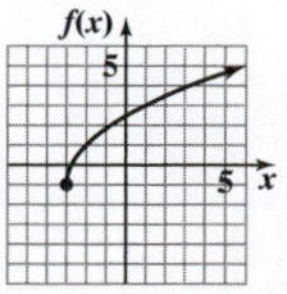

57. $y = 0$ *(2.4)* **58.** $y = \frac{3}{4}$ *(2.4)* **59.** None *(2.4)* **60.** $x = -10, x = 10$ *(2.4)* **61.** $x = -2$ *(2.4)* **62.** True *(2.3)* **63.** False *(2.3)* **64.** False *(2.3)* **65.** True *(2.4)* **66.** True *(2.5)* **67.** True *(2.3)*

68. 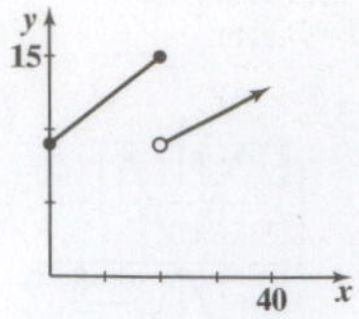

(2.2) **69.** 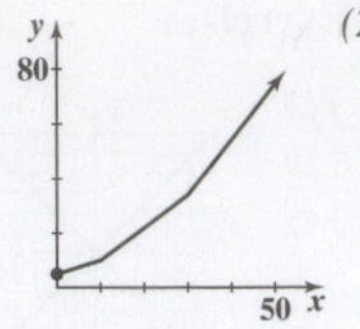

(2.2)

70. $y = -(x-4)^2 + 3$ *(2.2, 2.3)* **71.** $f(x) = -0.4(x-4)^2 + 7.6$
(A) x intercepts: $-0.4, 8.4$; y intercept: 1.2 (B) Vertex: $(4.0, 7.6)$
(C) Maximum: 7.6 (D) Range: $y \le 7.6$ or $(-\infty, 7.6]$ *(2.3)*

72. 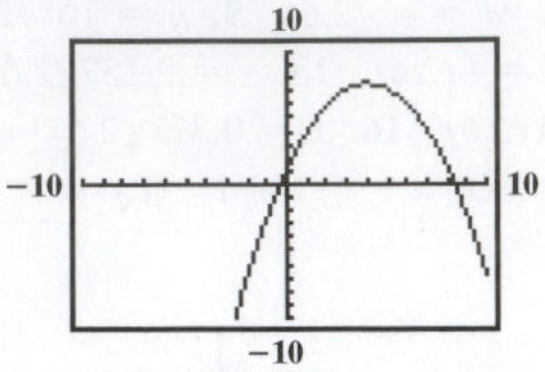

(A) x intercepts: $-.4, 8.4$; y intercept: 1.2 (B) Vertex: $(4.0, 7.6)$ (C) Maximum: 7.6 (D) Range: $y \le 7.6$ or $(-\infty, 7.6]$ *(2.3)*

73. $\log 10^{\pi} = \pi$ and $10^{\log\sqrt{2}} = \sqrt{2}$; $\ln e^{\pi} = \pi$ and $e^{\ln\sqrt{2}} = \sqrt{2}$ *(2.6)*
74. $x = 2$ *(2.6)* **75.** $x = 2$ *(2.6)* **76.** $x = 1$ *(2.6)* **77.** $x = 300$ *(2.6)*
78. $y = ce^{-5t}$ *(2.6)* **79.** If $\log_1 x = y$, then $1^y = x$; that is, $1 = x$ for all positive real numbers x, which is not possible. *(2.6)* **80.** The graph of $y = \sqrt[3]{x}$ is vertically expanded by a factor of 2, reflected in the x axis, and shifted 1 unit left and 1 unit down. Equation: $y = -2\sqrt[3]{x+1} - 1$. *(2.2)*
81. $G(x) = 0.3(x+2)^2 - 8.1$ (A) x intercepts: $-7.2, 3.2$; y intercept: -6.9 (B) Vertex: $(-2, -8.1)$ (C) Minimum: -8.1 (D) Range: $y \ge -8.1$ or $[-8.1, \infty)$ *(2.3)*

82. 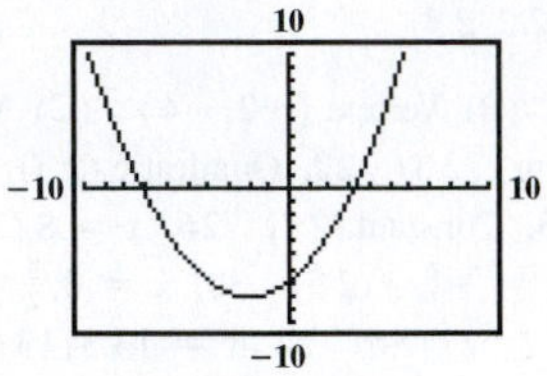

(A) x intercepts: $-7.2, 3.2$; y intercept: -6.9 (B) Vertex: $(-2, -8.1)$ (C) Minimum: -8.1 (D) Range: $y \ge -8.1$ or $[-8.1, \infty)$ *(2.3)*

83. (A) $S(x) = \begin{cases} 3 & \text{if } 0 \le x \le 20 \\ 0.057x + 1.86 & \text{if } 20 < x \le 200 \\ 0.0346x + 6.34 & \text{if } 200 < x \le 1{,}000 \\ 0.0217x + 19.24 & \text{if } x > 1{,}000 \end{cases}$

(B) 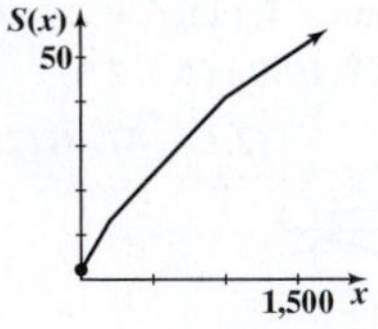

(2.2) **84.** \$5,321.95 *(2.5)*
85. \$5,269.51 *(2.5)*

86. 201 months ($\approx$16.7 years) *(2.5)* **87.** 9.38 yr *(2.5)*

88. (A) 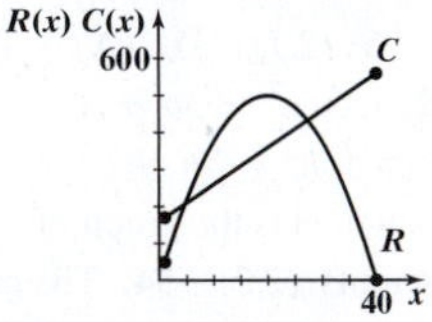

(B) $R = C$ for $x = 4.686$ thousand units (4,686 units) and for $x = 27.314$ thousand units (27,314 units); $R < C$ for $1 \le x < 4.686$ or $27.314 < x \le 40$; $R > C$ for $4.686 < x < 27.314$. (C) Maximum revenue is 500 thousand dollars (\$500,000). This occurs at an output of 20 thousand units (20,000 units). At this output, the wholesale price is $p(20) = \$25$. *(2.3)*

89. (A) $P(x) = R(x) - C(x) = x(50 - 1.25x) - (160 + 10x)$
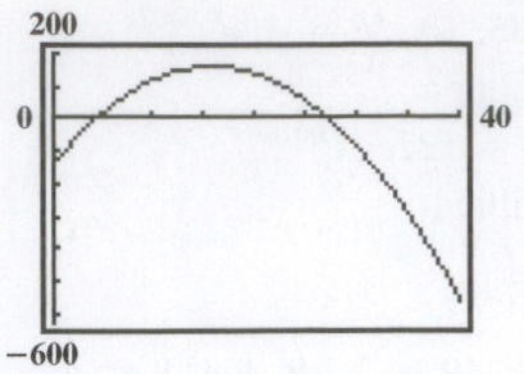

(B) $P = 0$ for $x = 4.686$ thousand units (4,686 units) and for $x = 27.314$ thousand units (27,314 units); $P < 0$ for $1 \le x < 4.686$ or $27.314 < x \le 40$; $p > 0$ for $4.686 < x < 27.314$. (C) Maximum profit is 160 thousand dollars (\$160,000). This occurs at an output of 16 thousand units (16,000 units). At this output, the wholesale price is $p(16) = \$30$. *(2.3)*

90. (A) $A(x) = -\frac{3}{2}x^2 + 420x$ (B) Domain: $0 \le x \le 280$
(C) 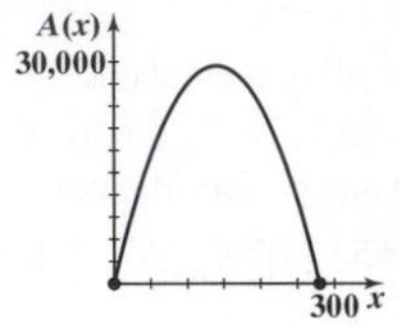

(D) There are two solutions to the equation $A(x) = 25{,}000$, one near 90 and another near 190. (E) 86 ft; 194 ft (F) Maximum combined area is 29,400 ft^2. This occurs for $x = 140$ ft and $y = 105$ ft. *(2.3)*

91. (A) 2,833 sets 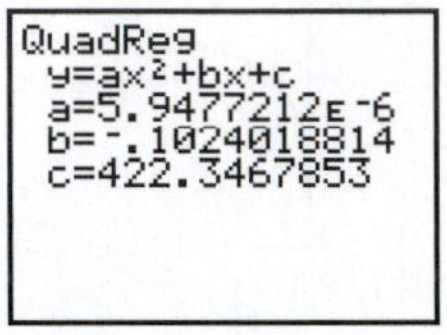

(B) 4,836 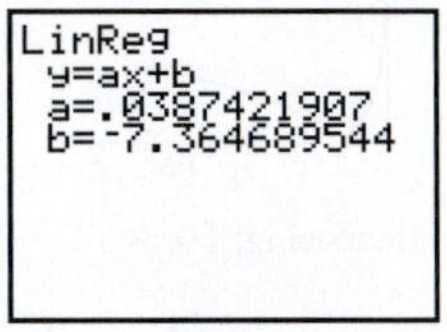

(C) Equilibrium price: \$131.59; equilibrium quantity: 3,587 cookware sets *(2.3)*

92. (A) 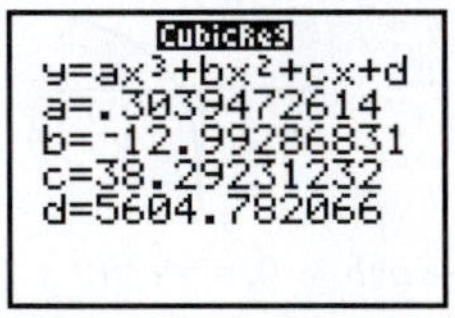

(B) 4976 *(2.4)*

93. (A) $N = 2^{2t}$ or $N = 4^t$ (B) 15 days *(2.5)* **94.** $k = 0.009\,42$; 489 ft *(2.6)*

95. (A) 6,134,000 *(2.6)* 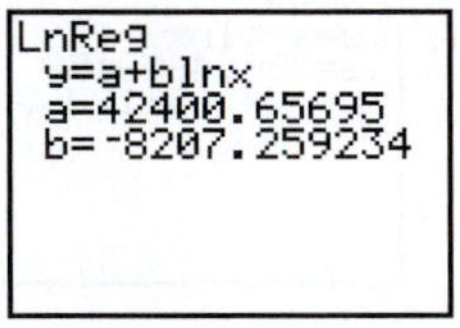

96. 23.1 yr *(2.5)* **97.** (A) \$1,319 billion (B) 2031 *(2.5)* 

Chapter 3

Exercises 3.1

1. \$25.42 **3.** 64% **5.** Slope = 120; y int. = 12,000 **7.** Slope = 50; y int. = 2,000 **9.** 0.062 **11.** 13.7% **13.** 0.0025 **15.** 0.84% **17.** $\frac{1}{2}$ yr **19.** $\frac{1}{3}$ yr **21.** $\frac{5}{4}$ yr **23.** $\frac{10}{13}$ yr **25.** \$42 **27.** \$1,800 **29.** 0.12 or 12% **31.** $\frac{1}{2}$ yr **33.** \$4,612.50 **35.** \$875 **37.** 0.36 or 36% **39.** 1 yr

41. $r = I/(Pt)$ **43.** $P = A/(1 + rt)$ **45.** $t = \frac{A - P}{Pr}$ **47.** The graphs are linear, all with y intercept \$1,000; their slopes are 40, 80, and 120, respectively. **49.** \$45 **51.** \$30 **53.** \$7,647.20 **55.** 8.1% **57.** 18% **59.** \$1,604.40; 20.88% **61.** 4.298% **63.** \$992.38 **65.** \$24.31 **67.** \$27 **69.** 5.396% **71.** 6.986% **73.** 12.085% **75.** 109.895% **77.** 87.158% **79.** \$118.94 **81.** \$1,445.89 **83.** \$7.27 **85.** \$824.85 **87.** 630% **89.** 771%

Exercises 3.2

1. $P = 950$ **3.** $x = 17$ **5.** $i = 0.5$ **7.** $n = 4$ **9.** \$5,983.40 **11.** \$4,245.07 **13.** \$3,125.79 **15.** \$2,958.11 **17.** 2.5 yr **19.** 6.79% **21.** 1.65% **23.** 0.46% **25.** 0.02% **27.** 2.43% **29.** 3.46% **31.** 6.36% **33.** 8.76% **35.** 2.92% **37.** (A) \$126.25; \$26.25 (B) \$126.90; \$26.90 (C) \$127.05; \$27.05 **39.** (A) \$5,524.71 (B) \$6,104.48 **41.** \$12,175.69 **43.** All three graphs are increasing, curve upward, and have the same y intercept; the greater the interest rate, the greater the increase. The amounts at the end of 8 years are \$1,376.40, \$1,892.46, and \$2,599.27, respectively.

45.

Period	Interest	Amount
0		\$1,000.00
1	\$97.50	\$1,097.50
2	\$107.01	\$1,204.51
3	\$117.44	\$1,321.95
4	\$128.89	\$1,450.84
5	\$141.46	\$1,592.29
6	\$155.25	\$1,747.54

47. (A) \$7,440.94 (B) \$5,536.76 **49.** (A) \$19,084.49 (B) \$11,121.45 **51.** (A) 3.97% (B) 2.32% **53.** (A) 5.28% (B) 5.27% **55.** $11\frac{2}{3}$ yr

57. 3.75 yr **59.** $n \approx 12$ **61.** (A) $7\frac{1}{4}$ yr (B) 6 yr **63.** (A) 7.7 yr (B) 6.3 yr **65.** \$65,068.44 **67.** \$282,222.44 **69.** \$19.78 per ft^2 per mo **71.** (A) In 2026, 250 years after the signing, it would be worth \$175,814.55. (B) If interest were compounded monthly, daily, or continuously, it would be worth \$179,119.92, \$180,748.53, or \$180,804.24, respectively.

(C)

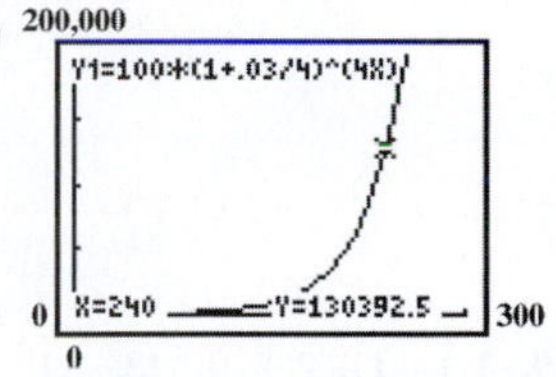

73. 9.66% **75.** 2 yr, 10 mo **77.** \$163,295.21 **79.** 3,615 days; 8.453 yr

81.

Years	Exact Rate	Rule of 72
6	12.2	12.0
7	10.4	10.3
8	9.1	9.0
9	8.0	8.0
10	7.2	7.2
11	6.5	6.5
12	5.9	6.0

83. 14 quarters **85.** To maximize earnings, choose 10% simple interest for investments lasting fewer than 11 years and 7% compound interest otherwise. **87.** 3.33%

89. 7.02% **91.** \$15,843.80 **93.** 4.53% **95.** 13.44% **97.** 17.62%

Exercises 3.3

1. 1,023 **3.** 3,000 **5.** 71,744,530 **7.** $i = 0.02$; $n = 80$ **9.** $i = 0.0375$; $n = 24$ **11.** $i = 0.0075$; $n = 48$ **13.** $i = 0.0595$; $n = 12$ **15.** $FV = \$13,435.19$ **17.** $PMT = \$310.62$ **19.** $n = 17$ **21.** $i = 0.09$

25. $n = \dfrac{\ln\left(1 + i\dfrac{FV}{PMT}\right)}{\ln(1 + i)}$ **27.** Value: \$84,895.40; interest: \$24,895.40

29. \$20,931.01 **31.** \$667.43 **33.** \$763.39

35.

Period	Amount	Interest	Balance
1	\$1,000.00	\$0.00	\$1,000.00
2	\$1,000.00	\$83.20	\$2,083.20
3	\$1,000.00	\$173.32	\$3,256.52
4	\$1,000.00	\$270.94	\$4,527.46
5	\$1,000.00	\$376.69	\$5,904.15

37. First year: \$33.56; second year: \$109.64; third year: \$190.41 **39.** \$111,050.77 **41.** \$1,308.75 **43.** (A) 1.540% (B) \$202.12 **45.** 33 months **47.** 7.77% **49.** 0.75% **51.** After 11 quarterly

Exercises 3.4

1. 511/256 **3.** 33,333,333/1,000,000 **5.** 171/256 **7.** $i = 0.006$; $n = 48$ **9.** $i = 0.02475$; $n = 40$ **11.** $i = 0.02525$; $n = 32$ **13.** $i = 0.0548$; $n = 9$ **15.** $PV = \$3,458.41$ **17.** $PMT = \$586.01$ **19.** $n = 29$ **21.** $i = 0.029$ **27.** \$35,693.18 **29.** \$11,241.81; \$1,358.19 **31.** \$69.58; \$839.84 **33.** 31 months **35.** 71 months **37.** For 0% financing, the monthly payments should be \$242.85, not \$299. If a loan of \$17,485 is amortized in 72 payments of \$299, the rate is 7.11% compounded monthly. **39.** The monthly payments with 0% financing are \$455. If you take the rebate, the monthly payments are \$434.24. You should choose the rebate. **41.** \$314.72; \$17,319.68

43.

Payment Number	Payment	Interest	Unpaid Balance Reduction	Unpaid Balance
0				\$5,000.00
1	\$706.29	\$140.00	\$566.29	4,433.71
2	706.29	124.14	582.15	3,851.56
3	706.29	107.84	598.45	3,253.11
4	706.29	91.09	615.20	2,637.91
5	706.29	73.86	632.43	2,005.48
6	706.29	56.15	650.14	1,355.34
7	706.29	37.95	668.34	687.00
8	706.24	19.24	687.00	0.00
Totals	\$5,650.27	\$650.27	\$5,000.00	

45. First year: \$466.05; second year: \$294.93; third year: \$107.82 **47.** \$97,929.78; \$116,070.22 **49.** \$143.85/mo; \$904.80 **51.** Monthly payment: \$908.99 (A) \$125,862 (B) \$81,507 (C) \$46,905 **53.** (A) Monthly payment: \$1,015.68; interest: \$114,763 (B) 197 months; interest saved: \$23,499 **55.** (A) 157 (B) 243 (C) The withdrawals continue forever. **57.** (A) Monthly withdrawals: \$1,229.66; total interest: \$185,338.80 (B) Monthly deposits: \$162.65 **59.** \$65,584 **61.** \$34,692 **63.** All three graphs are decreasing, curve downward, and have the same x intercept; the unpaid balances are always in the ratio 2:3:4. The monthly payments are \$402.31, \$603.47, and \$804.62, with total interest amounting to \$94,831.60, \$142,249.20, and \$189,663.20, respectively. **65.** 14.45% **67.** 10.21%

Chapter 3 Review Exercises

1. $A = \$104.50$ *(3.1)* **2.** $P = \$800$ *(3.1)* **3.** $t = 0.75$ yr, or 9 mo *(3.1)* **4.** $r = 6\%$ *(3.1)* **5.** $A = \$1,393.68$ *(3.2)* **6.** $P = \$3,193.50$ *(3.2)* **7.** $A = \$5,824.92$ *(3.2)* **8.** $P = \$22,612.86$ *(3.2)* **9.** $FV = \$69,770.03$ *(3.3)* **10.** $PMT = \$115.00$ *(3.3)* **11.** $PV = \$33,944.27$ *(3.4)* **12.** $PMT = \$166.07$ *(3.4)* **13.** $n \approx 16$ *(3.2)* **14.** $n \approx 41$ *(3.3)* **15.** \$3,350.00; \$350.00 *(3.1)* **16.** \$19,654 *(3.2)* **17.** \$12,944.67 *(3.2)*

18. (A)

Period	Interest	Amount
0		\$400.00
1	\$21.60	\$421.60
2	\$22.77	\$444.37
3	\$24.00	\$468.36
4	\$25.29	\$493.65

(3.2)

(B)

Period	Interest	Payment	Balance
1		$100.00	$100.00
2	$5.40	$100.00	$205.40
3	$11.09	$100.00	$316.49
4	$17.09	$100.00	$433.58

(3.3)

19. To maximize earnings, choose 13% simple interest for investments lasting less than 9 years and 9% compound interest for investments lasting 9 years or more. *(3.2)* **20.** $164,402 *(3.2)* **21.** 7.83% *(3.2)* **22.** 9% compounded quarterly, since its effective rate is 9.31%, while the effective rate of 9.25% compounded annually is 9.25% *(3.2)* **23.** $25,861.65; $6,661.65 *(3.3)* **24.** 288% *(3.1)* **25.** $1,725.56 *(3.1)* **26.** $29,354 *(3.2)* **27.** $18,021 *(3.2)* **28.** 15% *(3.1)* **29.** The monthly payments with 0% financing are $450. If you take the rebate, the monthly payments are $426.66. You should choose the rebate. *(3.4)* **30.** (A) 6.43% (B) 6.45% *(3.2)* **31.** 9 quarters or 2 yr, 3 mo *(3.2)* **32.** 139 mo; 93 mo *(3.2)* **33.** (A) $571,499 (B) $1,973,277 *(3.3)* **34.** 10.45% *(3.2)* **35.** (A) 174% (B) 65.71% *(3.1)* **36.** $725.89 *(3.3)* **37.** (A) $140,945.57 (B) $789.65 (C) $136,828 *(3.3, 3.4)* **38.** $102.99; $943.52 *(3.4)* **39.** $576.48 *(3.3)* **40.** 3,374 days; 10 yr *(3.2)* **41.** $175.28; $2,516.80 *(3.4)* **42.** $13,418.78 *(3.2)* **43.** 5 yr, 10 mo *(3.3)* **44.** 18 yr *(3.4)* **45.** 28.8% *(3.1)*

46.

Payment Number	Payment	Interest	Unpaid Balance Reduction	Unpaid Balance
0				$1,000.00
1	$265.82	$25.00	$240.82	759.18
2	265.82	18.98	246.84	512.34
3	265.82	12.81	253.01	259.33
4	265.81	6.48	259.33	0.00
Totals	$1,063.27	$63.27	$1,000.00	

47. 28 months *(3.3)* **48.** $55,347.48; $185,830.24 *(3.3)* **49.** 2.47% *(3.2)* **50.** 6.33% *(3.1)* **51.** 44 deposits *(3.3)* **52.** (A) $1,189.52 (B) $72,963.07 (C) $7,237.31 *(3.4)* **53.** The certificate would be worth $53,394.30 when the 360th payment is made. By reducing the principal, the loan would be paid off in 252 months. If the monthly payment were then invested at 7% compounded monthly, it would be worth $67,234.20 at the time of the 360th payment. *(3.2, 3.3, 3.4)* **54.** The lower rate would save $12,247.20 in interest payments. *(3.4)* **55.** $3,807.59 *(3.2)* **56.** 5.79% *(3.2)* **57.** $4,844.96 *(3.1)* **58.** $6,697.11 *(3.4)* **59.** 7.24% *(3.2)* **60.** (A) $398,807 (B) $374,204 *(3.3)* **61.** $15,577.64 *(3.2)* **62.** (A) 30 yr: $569.26; 15 yr: $749.82 (B) 30 yr: $69,707.99; 15 yr: $37,260.74 *(3.4)* **63.** $20,516 *(3.4)* **64.** 33.52% *(3.4)* **65.** (A) 10.74% (B) 15 yr: 40 yr *(3.3)*

Chapter 4

Exercises 4.1

1. $(0, 7)$ **3.** $(24, 0)$ **5.** $(5, -18)$ **7.** $y - 7 = -6(x - 2)$ **9.** (B); no solution **11.** (A); $x = -3, y = 1$ **13.** $x = 2, y = 4$ **15.** No solution (parallel lines) **17.** $x = 4, y = 5$ **19.** $x = 1, y = 4$ **21.** $u = 2, v = -3$ **23.** $m = 8, n = 6$ **25.** $x = 1, y = 1$ **27.** No solution (inconsistent) **29.** Infinitely many solutions (dependent) **31.** $m = \frac{1}{2}, n = \frac{11}{10}$ **33.** $x = -1, y = 2$ **35.** $x = 7, y = 3$ **37.** $x = \frac{4}{5}, y = -\frac{2}{3}$ **39.** $x = 0, y = 0$ **41.** $x = 14, y = 5$ **43.** Price tends to come down. **49.** $(1.125, 0.125)$ **51.** No solution (parallel lines) **53.** $(4.176, -1.235)$ **55.** $(-3.310, -2.241)$

57.

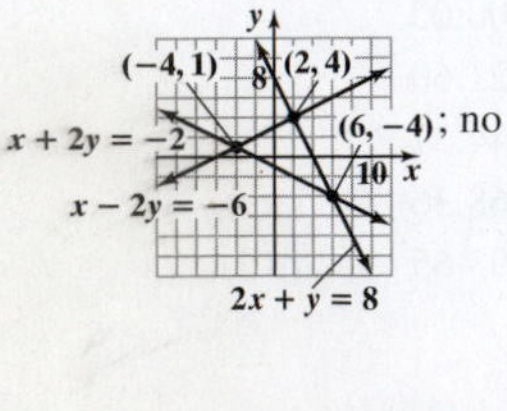

; no

59.

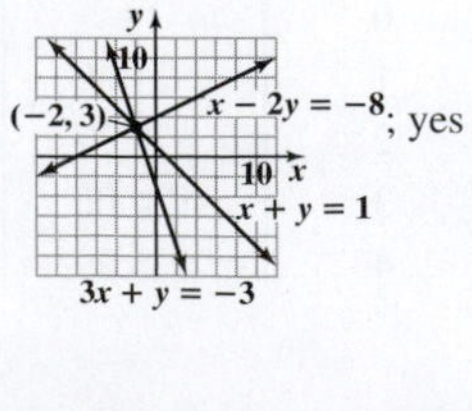

; yes

61.

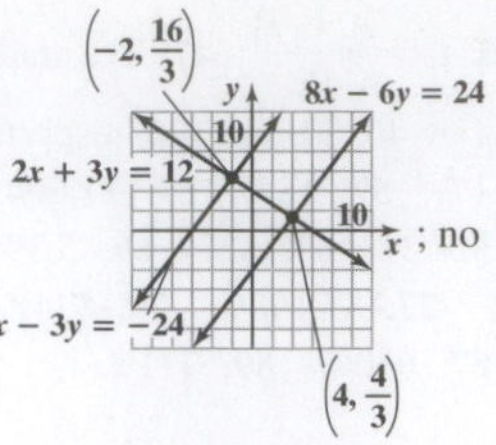

; no

63. (A) $(20, -24)$ (B) $(-4, 6)$ (C) No solution **65.** (A) Supply: 143 T-shirts; demand: 647 T-shirts (B) Supply: 857 T-shirts; demand: 353 T-shirts (C) Equilibrium price $= \$6.50$; equilibrium quantity $= 500$ T-shirts

(D)

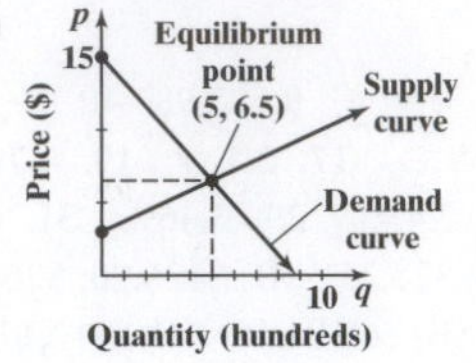

67. (A) $p = 1.5x + 1.95$ (B) $p = -1.5x + 7.8$ (C) Equilibrium price: $4.875; equilibrium quantity: 1.95 billion bushels

(D)

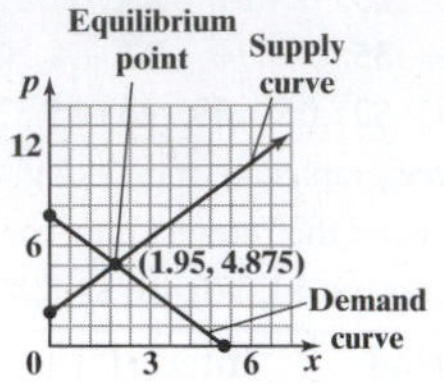

69. (A) 120 mowers

(B)

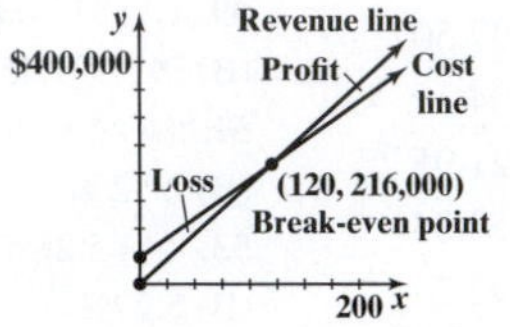

71. (A) $C = 24{,}000 + 7.45x$; $R = 19.95x$ (B) 1,920

(C)

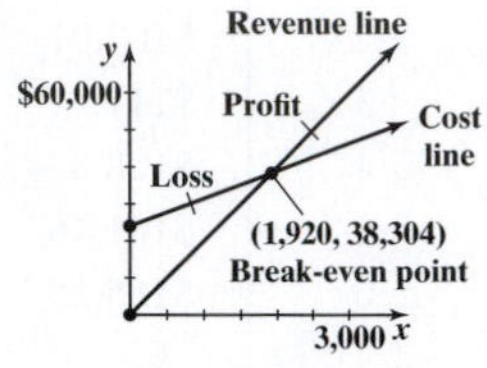

73. Base price $= \$17.95$; surcharge $= \$2.45/\text{lb}$ **75.** 5,720 lb robust blend; 6,160 lb mild blend **77.** Mix *A*: 80 g; mix *B*: 60 g **79.** Operate the Mexico plant for 75 hours and the Taiwan plant for 50 hours. **81.** (A) $a = 196$, $b = -16$ (B) 196 ft (C) 3.5 sec **83.** 40 sec, 24 sec, 120 mi

85. (A)

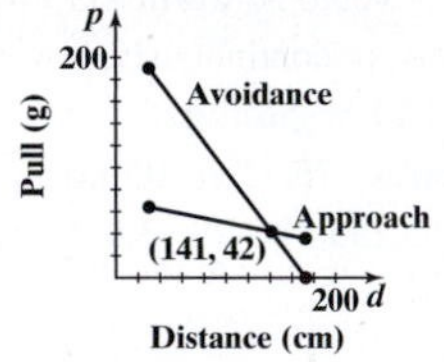

(B) $d = 141$ cm (approx.) (C) Vacillate

Exercises 4.2

1. 6; 3 **3.** 3×3; 2×1 **5.** *D* **7.** *B* **9.** 2, 1 **11.** 2, 8, 0 **13.** -1

15. $\begin{bmatrix} 3 & 5 \\ 2 & -4 \end{bmatrix}$; $\left[\begin{array}{cc|c} 3 & 5 & 8 \\ 2 & -4 & -7 \end{array}\right]$ **17.** $\begin{bmatrix} 1 & 4 \\ 6 & 0 \end{bmatrix}$; $\left[\begin{array}{cc|c} 1 & 4 & 15 \\ 6 & 0 & 18 \end{array}\right]$

19. $2x_1 + 5x_2 = 7$, $x_1 + 4x_2 = 9$ **21.** $4x_1 = -10$, $8x_2 = 40$

23. $\left[\begin{array}{cc|c} 1 & -3 & 5 \\ 2 & -4 & 6 \end{array}\right]$ **25.** $\left[\begin{array}{cc|c} 2 & -4 & 6 \\ 2 & -6 & 10 \end{array}\right]$ **27.** $\left[\begin{array}{cc|c} 3 & -7 & 11 \\ 1 & -3 & 5 \end{array}\right]$

29. $\left[\begin{array}{cc|c} -1 & 2 & -3 \\ 1 & -3 & 5 \end{array}\right]$ **31.** $\left[\begin{array}{cc|c} 1 & -1 & 1 \\ 1 & -3 & 5 \end{array}\right]$ **33.** $\left[\begin{array}{cc|c} 2 & -4 & 6 \\ -1 & 1 & -1 \end{array}\right]$

35. $\frac{1}{3}R_2 \rightarrow R_2$ **37.** $6R_1 + R_2 \rightarrow R_2$ **39.** $\frac{1}{3}R_2 + R_1 \rightarrow R_1$ **41.** $R_1 \leftrightarrow R_2$

43. $\{(6, 6)\}$

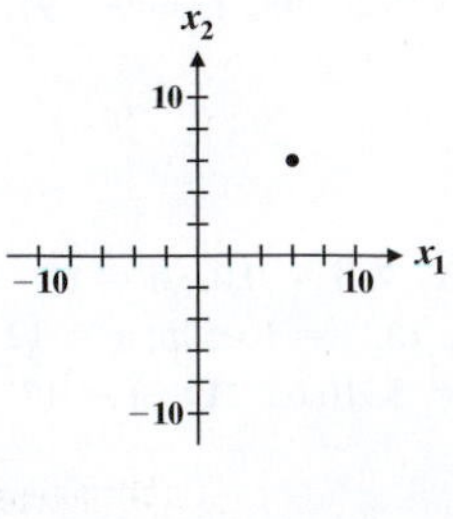

45. $\{(\frac{2}{3}t - 1, t) \mid t \text{ is any real number}\}$

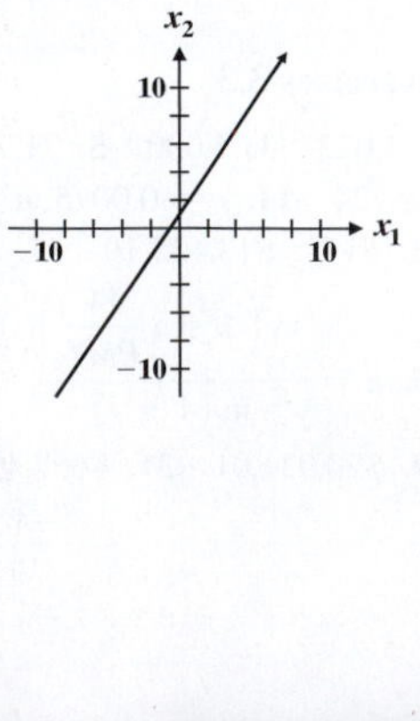

47. $x_1 = 3, x_2 = 2$; each pair of lines has the same intersection point.

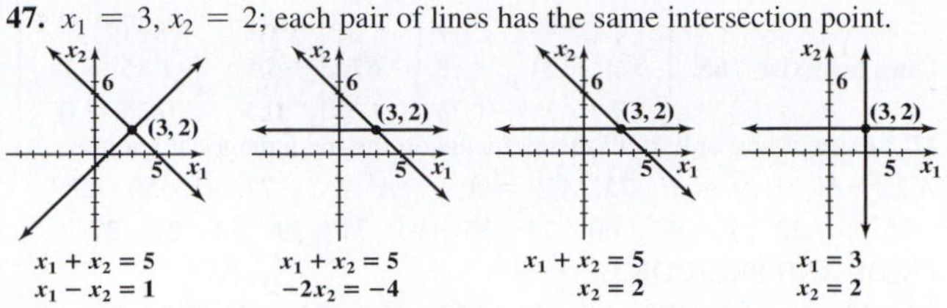

49. $x_1 = -4, x_2 = 6$ **51.** No solution **53.** $x_1 = 2t + 15, x_2 = t$ for any real number t **55.** $x_1 = 3, x_2 = 1$ **57.** $x_1 = 2, x_2 = 1$ **59.** $x_1 = 2, x_2 = 4$ **61.** No solution **63.** $x_1 = 1, x_2 = 4$ **65.** Infinitely many solutions: $x_2 = s$, $x_1 = 2s - 3$ for any real number s **67.** Infinitely many solutions; $x_2 = s$, $x_1 = \frac{1}{2}s + \frac{1}{2}$ for any real number s **69.** $x_1 = -1, x_2 = 3$ **71.** No solution **73.** Infinitely many solutions: $x_2 = t, x_1 = \frac{3}{2}t + 2$ for any real number t **75.** $x_1 = 2, x_2 = -1$ **77.** $x_1 = 2, x_2 = -1$ **79.** $x_1 = 1.1, x_2 = 0.3$ **81.** $x_1 = -23.125, x_2 = 7.8125$ **83.** $x_1 = 3.225, x_2 = -6.9375$

Exercises 4.3

1. $\left[\begin{array}{ccc|c} 1 & 2 & 3 & 12 \\ 1 & 7 & -5 & 15 \end{array}\right]$ **3.** $\left[\begin{array}{ccc|c} 1 & 0 & 6 & 2 \\ 0 & 1 & -1 & 5 \\ 1 & 3 & 0 & 7 \end{array}\right]$ **5.** $x_1 - 3x_2 = 4$, $3x_1 + 2x_2 = 5$, $-x_1 + 6x_2 = 3$

7. $5x_1 - 2x_2 + 8x_4 = 4$ **9.** Reduced form **11.** Not reduced form; $R_2 \leftrightarrow R_3$ or $R_2 + R_1 \rightarrow R_1$ **13.** Not reduced form; $\frac{1}{3}R_1 \rightarrow R_1$ **15.** Not reduced form; $-5R_2 + R_3 \rightarrow R_3$ **17.** Not reduced form; $-\frac{1}{2}R_2 \rightarrow R_2$ **19.** $x_1 = -2$, $x_2 = 3, x_3 = 0$ **21.** $x_1 = 2t + 3, x_2 = -t - 5, x_3 = t$ for any real number t **23.** No solution **25.** $x_1 = 3t + 5, x_2 = -2t - 7, x_3 = t$ for any real number t **27.** $x_1 = 2s + 3t - 5, x_2 = s, x_3 = -3t + 2, x_4 = t$ for any real numbers s and t **29.** 19 **31.** 21, 25, 27 **33.** False **35.** True **37.** False

39. $\left[\begin{array}{cc|c} 1 & 0 & -7 \\ 0 & 1 & 3 \end{array}\right]$ **41.** $\left[\begin{array}{ccc|c} 1 & 0 & -1 & 23 \\ 0 & 1 & 2 & -7 \end{array}\right]$ **43.** $\left[\begin{array}{ccc|c} 1 & 0 & 0 & -5 \\ 0 & 1 & 0 & 4 \\ 0 & 0 & 1 & -2 \end{array}\right]$

45. $\left[\begin{array}{ccc|c} 1 & 0 & 2 & 1 \\ 0 & 1 & -2 & -1 \\ 0 & 0 & 0 & 0 \end{array}\right]$ **47.** $x_1 = -2, x_2 = 3, x_3 = 1$ **49.** $x_1 = 0$, $x_2 = -2, x_3 = 2$ **51.** $x_1 = 2t + 3, x_2 = t - 2$, $x_3 = t$ for any real number t **53.** $x_1 = 1, x_2 = 2$ **55.** No solution **57.** $x_1 = t - 1, x_2 = 2t + 2, x_3 = t$ for any real number t **59.** $x_1 = -2s + t + 1, x_2 = s, x_3 = t$ for any real numbers s and t **61.** No solution **63.** (A) Dependent system with two parameters and an infinite number of solutions (B) Dependent system with one parameter and an infinite number of solutions (C) Independent system with a unique solution (D) Impossible **65.** $x_1 = 2s - 3t + 3, x_2 = s + 2t + 2, x_3 = s, x_4 = t$ for s and t any real numbers **67.** $x_1 = -0.5, x_2 = 0.2, x_3 = 0.3, x_4 = -0.4$ **69.** $x_1 = 2s - 1.5t + 1, x_2 = s, x_3 = -t + 1.5, x_4 = 0.5t - 0.5, x_5 = t$ for any real numbers s and t **71.** $a = 2, b = -4, c = -7$

73. (A) x_1 = no. of one-person boats
x_2 = no. of two-person boats
x_3 = no. of four-person boats
$0.5x_1 + x_2 + 1.5x_3 = 380$
$0.6x_1 + 0.9x_2 + 1.2x_3 = 330$
$0.2x_1 + 0.3x_2 + 0.5x_3 = 120$
20 one-person boats, 220 two-person boats, and 100 four-person boats

(B) $0.5x_1 + x_2 + 1.5x_3 = 380$
$0.6x_1 + 0.9x_2 + 1.2x_3 = 330$
$(t - 80)$ one-person boats, $(420 - 2t)$ two-person boats, and t four-person boats, where t is an integer satisfying $80 \le t \le 210$

(C) $0.5x_1 + x_2 = 380$
$0.6x_1 + 0.9x_2 = 330$
$0.2x_1 + 0.3x_2 = 120$
There is no production schedule that will use all the labor-hours in all departments.

75. x_1 = no. of 8,000-gal tank cars
x_2 = no. of 16,000-gal tank cars
x_3 = no. of 24,000-gal tank cars
$x_1 + x_2 + x_3 = 24$
$8{,}000x_1 + 16{,}000x_2 + 24{,}000x_3 = 520{,}000$
$(t - 17)$ 8,000-gal tank cars, $(41 - 2t)$ 16,000-gal tank cars, and t 24,000-gal tank cars, where t = 17, 18, 19, or 20

77. The minimum monthly cost is $24,100 when 7 16,000-gallon and 17 24,000-gallon tank cars are leased.

79. x_1 = federal income tax
x_2 = state income tax
x_3 = local income tax
$x_1 + 0.5x_2 + 0.5x_3 = 3{,}825{,}000$
$0.2x_1 + x_2 + 0.2x_3 = 1{,}530{,}000$
$0.1x_1 + 0.1x_2 + x_3 = 765{,}000$
Tax liability is 57.65%.

81. x_1 = taxable income of company A
x_2 = taxable income of company B
x_3 = taxable income of company C
x_4 = taxable income of company D
$x_1 - 0.08x_2 - 0.03x_3 - 0.07x_4 = 2.272$
$-0.12x_1 + x_2 - 0.11x_3 - 0.13x_4 = 2.106$
$-0.11x_1 - 0.09x_2 + x_3 - 0.08x_4 = 2.736$
$-0.06x_1 - 0.02x_2 - 0.14x_3 + x_4 = 3.168$
Taxable incomes are $2,927,000 for company A, $3,372,000 for company B, $3,675,000 for company C, and $3,926,000 for company D.

83. (A) x_1 = no. of ounces of food A
x_2 = no. of ounces of food B
x_3 = no. of ounces of food C
$30x_1 + 10x_2 + 20x_3 = 340$
$10x_1 + 10x_2 + 20x_3 = 180$
$10x_1 + 30x_2 + 20x_3 = 220$
8 oz of food A, 2 oz of food B, and 4 oz of food C

(B) $30x_1 + 10x_2 = 340$
$10x_1 + 10x_2 = 180$
$10x_1 + 30x_2 = 220$
There is no combination that will meet all the requirements.

(C) $30x_1 + 10x_2 + 20x_3 = 340$
$10x_1 + 10x_2 + 20x_3 = 180$
8 oz of food A, $(10 - 2t)$ oz of food B, and t oz of food C where $0 \le t \le 5$

85. x_1 = no. of barrels of mix A
x_2 = no. of barrels of mix B
x_3 = no. of barrels of mix C
x_4 = no. of barrels of mix D
$30x_1 + 30x_2 + 30x_3 + 60x_4 = 900$
$50x_1 + 75x_2 + 25x_3 + 25x_4 = 750$
$30x_1 + 20x_2 + 20x_3 + 50x_4 = 700$
$(10 - t)$ barrels of mix A, $(t - 5)$ barrels of mix B, $(25 - 2t)$ barrels of mix C, and t barrels of mix D, where t is an integer satisfying $5 \le t \le 10$

87. 0 barrels of mix A, 5 barrels of mix B, 5 barrels of mix C, and 10 barrels of mix D **89.** $y = 0.01x^2 + x + 75$, 450 million **91.** $y = 0.004x^2 + 0.06x + 77.6$; 1995–2000: 79.4 years; 2000–2005: 80.4 years

93.

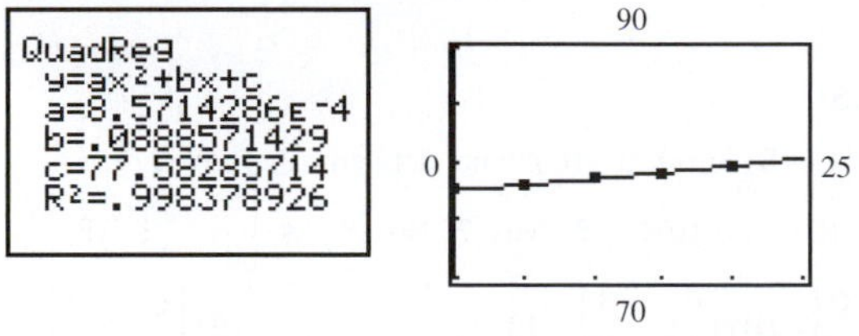

95. x_1 = no. of hours for company A
x_2 = no. of hours for company B
$30x_1 + 20x_2 = 600$
$10x_1 + 20x_2 = 400$
Company A: 10 hr; company B: 15 hr

97. (A) 6th St. and Washington Ave.: $x_1 + x_2 = 1{,}200$; 6th St. and Lincoln Ave.: $x_2 + x_3 = 1{,}000$; 5th St. and Lincoln Ave.: $x_3 + x_4 = 1{,}300$ (B) $x_1 = 1{,}500 - t, x_2 = t - 300$, $x_3 = 1{,}300 - t$, and $x_4 = t$, where $300 \le t \le 1{,}300$
(C) 1,300; 300 (D) Washington Ave.: 500; 6th St.: 700; Lincoln Ave.: 300

Exercises 4.4

1. $[4 \quad 15]$ **3.** $\begin{bmatrix} 2 & 4 \\ -2 & 6 \end{bmatrix}$ **5.** Not defined **7.** $[21 \quad -35 \quad 63 \quad 28]$

9. $\begin{bmatrix} 5 \\ -3 \end{bmatrix}$ **11.** $\begin{bmatrix} 2 & 4 \\ 1 & -5 \end{bmatrix}$ **13.** $\begin{bmatrix} 1 & -5 \\ -2 & -4 \end{bmatrix}$ **15.** $\begin{bmatrix} 5 & 10 \\ 15 & 20 \end{bmatrix}$ **17.** $\begin{bmatrix} 5 & 10 \\ 15 & 20 \end{bmatrix}$

19. $\begin{bmatrix} 7 & 9 \\ 0 & 0 \end{bmatrix}$ **21.** $\begin{bmatrix} 0 & 3 \\ 0 & 7 \end{bmatrix}$ **23.** $[-23]$ **25.** $\begin{bmatrix} -15 & 10 \\ 12 & -8 \end{bmatrix}$ **27.** $[1]$

29. $\begin{bmatrix} -2 & 0 & 2 \\ -1 & 0 & 1 \\ -3 & 0 & 3 \end{bmatrix}$ **31.** $\begin{bmatrix} -12 & 12 & 18 \\ 20 & -18 & -6 \end{bmatrix}$ **33.** Not defined

35. $\begin{bmatrix} 11 & 2 \\ 4 & 27 \end{bmatrix}$ **37.** $\begin{bmatrix} 6 & 4 \\ 0 & -3 \end{bmatrix}$ **39.** $\begin{bmatrix} -1.3 & -0.7 \\ -0.2 & -0.5 \\ 0.1 & 1.1 \end{bmatrix}$ **41.** $\begin{bmatrix} -66 & 69 & 39 \\ 92 & -18 & -36 \end{bmatrix}$

43. Not defined **45.** $\begin{bmatrix} -18 & 48 \\ 54 & -34 \end{bmatrix}$ **47.** $\begin{bmatrix} -26 & -15 & -25 \\ -4 & -18 & 4 \\ 2 & 43 & -19 \end{bmatrix}$

49. $AB = \begin{bmatrix} 0 & 0 \\ 0 & 0 \end{bmatrix}, BA = \begin{bmatrix} a^2+ab & a^2+ab \\ -a^2-ab & -a^2-ab \end{bmatrix}$ **51.** $\begin{bmatrix} 0 & 0 \\ 0 & 0 \end{bmatrix}$

53. B^n approaches $\begin{bmatrix} 0.25 & 0.75 \\ 0.25 & 0.75 \end{bmatrix}$; AB^n approaches $[0.25 \quad 0.75]$

55. $a = -1, b = 1, c = 3, d = -5$ **57.** $a = 3, b = 4, c = 1, d = 2$

59. False **61.** True **63.** (A) True (B) True (C) True

65.

Guitar	Banjo	
\$51.50	\$40.50	Materials
\$87.00	\$120.00	Labor

67.

	Basic car	Air	AM/FM radio	Cruise control
Model A	\$2,937	\$459	\$200	\$118
Model B	\$2,864	\$201	\$88	\$52
Model C	\$2,171	\$417	\$177	\$101

69. (A) \$19.84 (B) \$38.19 (C) MN gives the labor costs at each plant.

(D)

$MN =$	*MA*	*VA*	
	\$19.84	\$16.90	One-person boat
	\$31.49	\$26.81	Two-person boat
	\$44.87	\$38.19	Four-person boat

71. (A) 70 g (B) 30 g (C) MN gives the amount (in grams) of protein, carbohydrate, and fat in 20 oz of each mix.

(D)

$MN =$	Mix X	Mix Y	Mix Z	
	70	60	50	Protein
	380	360	340	Carbohydrate
	50	40	30	Fat

73. (A) \$9,950 (B) \$16,400

(C)

$NM =$	Cost per town	
	\$9,950	Berkeley
	\$16,400	Oakland

The entries are the total cost per town.

Exercises 4.5

1. (A) -4; $1/4$ (B) 3; $-1/3$ (C) 0; not defined **3.** (A) $-2/3$; $3/2$ (B) $1/7$; -7 (C) -1.6; 0.625 **5.** No **7.** No **9.** (A) $\begin{bmatrix} 2 & -3 \\ 0 & 0 \end{bmatrix}$ (B) $\begin{bmatrix} 2 & 0 \\ 4 & 0 \end{bmatrix}$

11. (A) $\begin{bmatrix} 0 & 0 \\ 4 & 5 \end{bmatrix}$ (B) $\begin{bmatrix} 0 & -3 \\ 0 & 5 \end{bmatrix}$ **13.** (A) $\begin{bmatrix} 2 & -3 \\ 4 & 5 \end{bmatrix}$ (B) $\begin{bmatrix} 2 & -3 \\ 4 & 5 \end{bmatrix}$

15. $\begin{bmatrix} -2 & 1 & 3 \\ 2 & 4 & -2 \\ 5 & 1 & 0 \end{bmatrix}$ **17.** $\begin{bmatrix} -2 & 1 & 3 \\ 2 & 4 & -2 \\ 5 & 1 & 0 \end{bmatrix}$ **19.** Yes **21.** No **23.** Yes **25.** No **27.** Yes

39. $\begin{bmatrix} -1 & 0 \\ -3 & 1 \end{bmatrix}$ **41.** $\begin{bmatrix} 3 & -2 \\ -1 & 1 \end{bmatrix}$ **43.** $\begin{bmatrix} 7 & -3 \\ -2 & 1 \end{bmatrix}$

45. $\begin{bmatrix} -5 & -12 & 3 \\ -2 & -4 & 1 \\ 2 & 5 & -1 \end{bmatrix}$ **47.** $\begin{bmatrix} 6 & -2 & -1 \\ -5 & 2 & 1 \\ -3 & 1 & 1 \end{bmatrix}$ **49.** $\begin{bmatrix} -2 & -3 \\ 3 & 4 \end{bmatrix}$

51. Does not exist **53.** $\begin{bmatrix} 1.5 & -0.5 \\ -2 & 1 \end{bmatrix}$ **55.** $\begin{bmatrix} 0.35 & 0.01 \\ 0.05 & 0.03 \end{bmatrix}$

57. $\begin{bmatrix} \frac{1}{3} & 0 \\ 0 & \frac{1}{3} \end{bmatrix}$ **59.** $\begin{bmatrix} \frac{1}{2} & 0 \\ 0 & 2 \end{bmatrix}$ **61.** $\begin{bmatrix} 1 & 2 & 2 \\ -2 & -3 & -4 \\ -1 & -2 & -1 \end{bmatrix}$

63. Does not exist **65.** $\begin{bmatrix} -3 & -2 & 1.5 \\ 4 & 3 & -2 \\ 3 & 2 & -1.25 \end{bmatrix}$ **67.** $\begin{bmatrix} -1.75 & -0.375 & 0.5 \\ -5.5 & -1.25 & 1 \\ 0.5 & 0.25 & 0 \end{bmatrix}$

71. M^{-1} exists if and only if all the elements on the main diagonal are nonzero.

73. $A^{-1} = A$; $A^2 = I$ **75.** $A^{-1} = A$; $A^2 = I$ **77.** 41 50 28 35 37 55 22 31 47 60 24 36 49 71 39 54 21 22

79. PRIDE AND PREJUDICE

81. 37 47 10 58 103 67 47 123 121 75 53 142 58 68 23 91 90 74 38 117 83 59 39 103 113 97 45 147 76 57 38 95 **83.** RAWHIDE TO WALTER REED **85.** 30 28 58 15 38 13 19 26 12 30 39 56 48 43 40 9 30 29 12 33

87. DOUBLE DOUBLE TOIL AND TROUBLE

Exercises 4.6

1. $-3/5$ **3.** $-7/4$ **5.** $9/8$ **7.** $1/10$ **9.** $3x_1 + x_2 = 5$ $\quad 2x_1 - x_2 = -4$

11. $\begin{aligned} -3x_1 + x_2 \phantom{{}+ x_3} &= 3 \\ 2x_1 \phantom{{}+ x_2} + x_3 &= -4 \\ -x_1 + 3x_2 - 2x_3 &= 2 \end{aligned}$ **13.** $\begin{bmatrix} 3 & -4 \\ 2 & 1 \end{bmatrix}\begin{bmatrix} x_1 \\ x_2 \end{bmatrix} = \begin{bmatrix} 1 \\ 5 \end{bmatrix}$

15. $\begin{bmatrix} 1 & -3 & 2 \\ -2 & 3 & 0 \\ 1 & 1 & 4 \end{bmatrix}\begin{bmatrix} x_1 \\ x_2 \\ x_3 \end{bmatrix} = \begin{bmatrix} -3 \\ 1 \\ -2 \end{bmatrix}$ **17.** $x_1 = -8, x_2 = 2$ **19.** $x_1 = 0, x_2 = 4$ **21.** $x_1 = 3, x_2 = -2$

23. $x_1 = 11, x_2 = 4$ **25.** $x_1 = 3, x_2 = 2$ **27.** No solution

29. $x_1 = 2, x_2 = 5$ **31.** (A) $x_1 = -3, x_2 = 2$ (B) $x_1 = -1, x_2 = 2$ (C) $x_1 = -8, x_2 = 3$ **33.** (A) $x_1 = 17, x_2 = -5$ (B) $x_1 = 7, x_2 = -2$ (C) $x_1 = 24, x_2 = -7$ **35.** (A) $x_1 = 1, x_2 = 0, x_3 = 0$ (B) $x_1 = -7, x_2 = -2, x_3 = 3$ (C) $x_1 = 17, x_2 = 5, x_3 = -7$

37. (A) $x_1 = 8, x_2 = -6, x_3 = -2$ (B) $x_1 = -6, x_2 = 6, x_3 = 2$ (C) $x_1 = 20, x_2 = -16, x_3 = -10$ **39.** $X = A^{-1}B$ **41.** $X = BA^{-1}$

43. $X = A^{-1}BA$ **45.** $x_1 = 2t + 2.5, x_2 = t$, for any real number t

47. No solution **49.** $x_1 = 13t + 3, x_2 = 8t + 1, x_3 = t$, for any real number t

51. $X = (A - B)^{-1}C$ **53.** $X = (A + I)^{-1}C$

55. $X = (A + B)^{-1}(C + D)$ **57.** $x_1 = 3, x_2 = 8$

59. $x_1 = 10.2, x_2 = 4.4, x_3 = 7.3$

61. $x_1 = 3.1, x_2 = 4.3, x_3 = -2.7, x_4 = 8$

63. $x_1 =$ no. of \$25 tickets

$x_2 =$ no. of \$35 tickets

$x_1 + x_2 = 10{,}000$ seats

$25x_1 + 35x_2 = k$ Return

(A) Concert 1: 7,500 \$25 tickets, 2,500 \$35 tickets

Concert 2: 5,000 \$25 tickets, 5,000 \$35 tickets

Concert 3: 2,500 \$25 tickets, 7,500 \$35 tickets

(B) No (C) $250{,}000 + 10t$, $0 \le t \le 10{,}000$

65. $x_1 =$ no. of hours plant *A* operates

$x_2 =$ no. of hours plant *B* operates

$10x_1 + 8x_2 = k_1$ no. of car frames produced

$5x_1 + 8x_2 = k_2$ no. of truck frames produced

Order 1: 280 hr at plant *A* and 25 hr at plant *B*

Order 2: 160 hr at plant *A* and 150 hr at plant *B*

Order 3: 80 hr at plant *A* and 225 hr at plant *B*

67. $x_1 =$ president's bonus

$x_2 =$ executive vice-president's bonus

$x_3 =$ associate vice-president's bonus

$x_4 =$ assistant vice-president's bonus

$$\begin{aligned} x_1 + 0.03x_2 + 0.03x_3 + 0.03x_4 &= 60{,}000 \\ 0.025x_1 + x_2 + 0.025x_3 + 0.025x_4 &= 50{,}000 \\ 0.02x_1 + 0.02x_2 + x_3 + 0.02x_4 &= 40{,}000 \\ 0.015x_1 + 0.015x_2 + 0.015x_3 + x_4 &= 30{,}000 \end{aligned}$$

President: \$56,600; executive vice-president: \$47,000; associate vice-president: \$37,400; assistant vice-president: \$27,900

69. (A) $x_1 =$ no. of ounces of mix *A*

$x_2 =$ no. of ounces of mix *B*

$0.20x_1 + 0.14x_2 = k_1$ Protein

$0.04x_1 + 0.03x_2 = k_2$ Fat
Diet 1: 50 oz mix *A* and 500 oz mix *B*
Diet 2: 450 oz mix *A* and 0 oz mix *B*
Diet 3: 150 oz mix *A* and 500 oz mix *B* (B) No

Exercises 4.7

1. −3 **3.** 100 **5.** 4 **7.** 8 **9.** 40¢ from *A*; 20¢ from *E*
11. $\begin{bmatrix} 0.6 & -0.2 \\ -0.2 & 0.9 \end{bmatrix}; \begin{bmatrix} 1.8 & 0.4 \\ 0.4 & 1.2 \end{bmatrix}$ **13.** $X = \begin{bmatrix} x_1 \\ x_2 \end{bmatrix} = \begin{bmatrix} 16.4 \\ 9.2 \end{bmatrix}$
15. 30¢ from *A*; 10¢ from *B*; 20¢ from *E* **17.** $\begin{bmatrix} 0.7 & -0.2 & -0.2 \\ -0.1 & 0.9 & -0.1 \\ -0.2 & -0.1 & 0.9 \end{bmatrix}$
19. Agriculture: $18 billion; building: $15.6 billion; energy: $22.4 billion
21. $\begin{bmatrix} 1.4 & 0.4 \\ 0.6 & 1.6 \end{bmatrix}; \begin{bmatrix} 24 \\ 46 \end{bmatrix}$ **23.** $I - M$ is singular; *X* does not exist.
25. $\begin{bmatrix} 1.58 & 0.24 & 0.58 \\ 0.4 & 1.2 & 0.4 \\ 0.22 & 0.16 & 1.22 \end{bmatrix}; \begin{bmatrix} 38.6 \\ 18 \\ 17.4 \end{bmatrix}$ **27.** (A) Agriculture: $80 million; manufacturing: $64 million. (B) The final demand for agriculture increases to $54 million and the final demand for manufacturing decreases to $38 million.
29. $\begin{bmatrix} 0.25 & 0.1 \\ 0.25 & 0.3 \end{bmatrix}$
31. The total output of the energy sector should be 75% of the total output of the mining sector.
33. Each element should be between 0 and 1, inclusive.
35. Coal: $28 billion; steel: $26 billion
37. Agriculture: $148 million; tourism: $146 million
39. Agriculture: $40.1 billion; manufacturing: $29.4 billion; energy: $34.4 billion
41. Year 1: agriculture: $65 billion; energy: $83 billion; labor: $71 billion; manufacturing: $88 billion
Year 2: agriculture: $81 billion; energy: $97 billion; labor: $83 billion; manufacturing: $99 billion
Year 3: agriculture: $117 billion; energy: $124 billion; labor: $106 billion; manufacturing: $120 billion

Chapter 4 Review Exercises

1. $x = 4, y = 4$ *(4.1)* **2.** $x = 4, y = 4$ *(4.1)* **3.** (A) Not in reduced form; $R_1 \leftrightarrow R_2$ (B) Not in reduced form; $\frac{1}{3}R_2 \leftrightarrow R_2$ (C) Reduced form (D) Not in reduced form; $(-1)R_2 + R_1 \rightarrow R_1$ *(4.3)* **4.** (A) 2×5, 3×2 (B) $a_{24} = 3$, $a_{15} = 2$, $b_{31} = -1$, $b_{22} = 4$ (C) *AB* is not defined; *BA* is defined *(4.2, 4.4)* **5.** (A) $x_1 = 8, x_2 = 2$ (B) $x_1 = -15.5, x_2 = 23.5$ *(4.6)* **6.** $\begin{bmatrix} 3 & 3 \\ 4 & 2 \end{bmatrix}$ *(4.4)*
7. Not defined *(4.4)* **8.** $\begin{bmatrix} -3 & 0 \\ 1 & -1 \end{bmatrix}$ *(4.4)* **9.** $\begin{bmatrix} 4 & 3 \\ 7 & 4 \end{bmatrix}$ *(4.4)* **10.** Not defined *(4.4)*
11. $\begin{bmatrix} 5 \\ 5 \end{bmatrix}$ *(4.4)* **12.** $\begin{bmatrix} 2 & 3 \\ 4 & 6 \end{bmatrix}$ *(4.4)* **13.** $[8]$ *(4.4)* **14.** Not defined *(4.4)*
15. $\begin{bmatrix} -2 & 3 \\ 3 & -4 \end{bmatrix}$ *(4.5)* **16.** $x_1 = 9, x_2 = -11$ *(4.1)* **17.** $x_1 = 9, x_2 = -11$ *(4.2)*
18. $x_1 = 9, x_2 = -11$; $x_1 = 16, x_2 = -19$; $x_1 = -2, x_2 = 4$ *(4.6)*
19. Not defined *(4.4)* **20.** $\begin{bmatrix} 10 & -8 \\ 4 & 6 \end{bmatrix}$ *(4.4)* **21.** $\begin{bmatrix} -2 & 8 \\ 8 & 6 \end{bmatrix}$ *(4.4)*
22. $\begin{bmatrix} -2 & -1 & -3 \\ 4 & 2 & 6 \\ 6 & 3 & 9 \end{bmatrix}$ *(4.4)* **23.** $[9]$ *(4.4)* **24.** $\begin{bmatrix} 10 & -5 & 1 \\ -1 & -4 & -5 \\ 1 & -7 & -2 \end{bmatrix}$ *(4.4)*
25. $\begin{bmatrix} -\frac{5}{2} & 2 & -\frac{1}{2} \\ 1 & -1 & 1 \\ \frac{1}{2} & 0 & -\frac{1}{2} \end{bmatrix}$ *(4.5)* **26.** (A) $x_1 = 2, x_2 = 1, x_3 = -1$ (B) $x_1 = -5t - 12, x_2 = 3t + 7, x_3 = t$ for *t* any real number (C) $x_1 = -2t + 5$, $x_2 = t + 3$, $x_3 = t$ for *t* any real number *(4.3)*
27. $x_1 = 2, x_2 = 1, x_3 = -1$; $x_1 = 1, x_2 = -2, x_3 = 1$; $x_1 = -1$, $x_2 = 2, x_3 = -2$ *(4.6)* **28.** The system has an infinite no. of solutions for $k = 3$ and a unique solution for any other value of *k*. *(4.3)*
29. $(I - M)^{-1} = \begin{bmatrix} 1.4 & 0.3 \\ 0.8 & 1.6 \end{bmatrix}; X = \begin{bmatrix} 48 \\ 56 \end{bmatrix}$ *(4.7)* **30.** $\begin{bmatrix} 0.3 & 0.4 \\ 0.15 & 0.2 \end{bmatrix}$ *(4.7)*
31. $I - M$ is singular; *X* does not exist. *(4.5)* **32.** $x = 3.46, y = 1.69$ *(4.1)*

33. $\begin{bmatrix} -0.9 & -0.1 & 5 \\ 0.8 & 0.2 & -4 \\ 0.1 & -0.1 & 0 \end{bmatrix}$ *(4.5)* **34.** $x_1 = 1{,}400, x_2 = 3{,}200, x_3 = 2{,}400$ *(4.6)*
35. $x_1 = 1{,}400, x_2 = 3{,}200, x_3 = 2{,}400$ *(4.3)*
36. $(I\text{-}M)^{-1} = \begin{bmatrix} 1.3 & 0.4 & 0.7 \\ 0.2 & 1.6 & 0.3 \\ 0.1 & 0.8 & 1.4 \end{bmatrix}; X = \begin{bmatrix} 81 \\ 49 \\ 62 \end{bmatrix}$ *(4.7)*
37. (A) Unique solution (B) Either no solution or an infinite no. of solutions *(4.6)*
38. (A) Unique solution (B) No solution (C) Infinite no. of solutions *(4.3)*
39. (B) is the only correct solution. *(4.6)* **40.** (A) $C = 243{,}000 + 22.45x$; $R = 59.95x$ (B) $x = 6{,}480$ machines; $R = C = \$388{,}476$ (C) Profit occurs if $x > 6{,}480$; loss occurs if $x < 6{,}480$. *(4.1)*

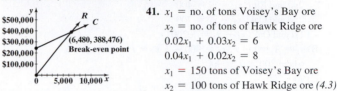

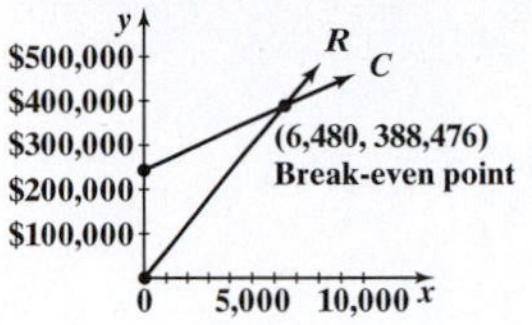

41. x_1 = no. of tons Voisey's Bay ore
x_2 = no. of tons of Hawk Ridge ore
$0.02x_1 + 0.03x_2 = 6$
$0.04x_1 + 0.02x_2 = 8$
$x_1 = 150$ tons of Voisey's Bay ore
$x_2 = 100$ tons of Hawk Ridge ore *(4.3)*

42. (A) $\begin{bmatrix} x_1 \\ x_2 \end{bmatrix} = \begin{bmatrix} -25 & 37.5 \\ 50 & -25 \end{bmatrix}\begin{bmatrix} 6 \\ 8 \end{bmatrix} = \begin{bmatrix} 150 \\ 100 \end{bmatrix}$
$x_1 = 150$ tons of Voisey's Bay ore
$x_2 = 100$ tons of Hawk Ridge ore
(B) $\begin{bmatrix} x_1 \\ x_2 \end{bmatrix} = \begin{bmatrix} -25 & 37.5 \\ 50 & -25 \end{bmatrix}\begin{bmatrix} 7.5 \\ 7 \end{bmatrix} = \begin{bmatrix} 75 \\ 200 \end{bmatrix}$
$x_1 = 75$ tons of Voisey's Bay ore
$x_2 = 200$ tons of Hawk Ridge ore *(4.6)*

43. (A) x_1 = no. of 3,000-ft³ hoppers
x_2 = no. of 4,500-ft³ hoppers
x_3 = no. of 6,000-ft³ hoppers
$x_1 + x_2 + x_3 = 20$
$3{,}000x_1 + 4{,}500x_2 + 6{,}000x_3 = 108{,}000$
$x_1 = (t - 12)$ 3,000-ft³ hoppers
$x_2 = (32 - 2t)$ 4,500-ft³ hoppers
$x_3 = t$ 6,000-ft³ hoppers
where t = 12, 13, 14, 15, or 16
(B) The minimum monthly cost is $5,700 when 8 4,500-ft³ and 12 6,000-ft³ hoppers are leased. *(4.3)*

44. (A) Elements in *MN* give the cost of materials for each alloy from each supplier.
(B)

	Supplier A	Supplier B	
$MN =$	$7,620	$7,530	Alloy 1
	$13,880	$13,930	Alloy 2

(C) Supplier A Supplier B
$[1 \; 1]MN = [\$21{,}500 \quad \$21{,}460]$
Total material costs *(4.4)*

45. (A) $6.35 (B) Elements in *MN* give the total labor costs for each calculator at each plant.
(C)

	CA	TX	
$MN =$	$3.65	$3.00	Model A
	$6.35	$5.20	Model B

(4.4)

46. x_1 = amount invested at 5% x_2 = amount invested at 10%
$x_1 + x_2 = 5{,}000$ $0.05x_1 + 0.1x_2 = 400$ $2,000 at 5%, $3,000 at 10% *(4.3)*
47. $2,000 at 5% and $3,000 at 10% *(4.6)* **48.** No to both. The annual yield must be between $250 and $500 inclusive. *(4.6)*
49. x_1 = no. of $8 tickets
x_2 = no. of $12 tickets
x_3 = no. of $20 tickets
$x_1 + x_2 + x_3 = 25{,}000$
$8x_1 + 12x_2 + 20x_3 = k_1$ Return required
$x_1 - x_3 = 0$
Concert 1: 5000 $8 tickets, 15,000 $12 tickets, and 5,000 $20 tickets
Concert 2: 7,500 $8 tickets, 10,000 $12 tickets, and 7,500 $20 tickets
Concert 3: 10,000 $8 tickets, 5,000 $12 tickets, and 10,000 $20 tickets *(4.6)*

50. $x_1 + x_2 + x_3 = 25{,}000$
$8x_1 + 12x_2 + 20x_3 = k_1$ Return required
Concert 1: $(2t - 5{,}000)$ \$8 tickets, $(30{,}000 - 3t)$ \$12 tickets, and t \$20 tickets, where t is an integer satisfying $2{,}500 \le t \le 10{,}000$
Concert 2: $(2t - 7{,}500)$ \$8 tickets, $(32{,}500 - 3t)$ \$12 tickets, and t \$20 tickets, where t is an integer satisfying $3{,}750 \le t \le 10{,}833$
Concert 3: $(2t - 10{,}000)$ \$8 tickets, $(35{,}000 - 3t)$ \$12 tickets, and t \$20 tickets, where t is an integer satisfying $5{,}000 \le t \le 11{,}666$ *(4.3)*
51. (A) Agriculture: \$80 billion; fabrication: \$60 billion
(B) Agriculture: \$135 billion; fabrication: \$145 billion *(4.7)*
52. BEWARE THE IDES OF MARCH *(4.5)*
53. (A) 1st and Elm: $x_1 + x_4 = 1{,}300$
2nd and Elm: $x_1 - x_2 = 400$
2nd and Oak: $x_2 + x_3 = 700$
1st and Oak: $x_3 - x_4 = -200$
(B) $x_1 = 1{,}300 - t$, $x_2 = 900 - t$, $x_3 = t - 200$, $x_4 = t$, where $200 \le t \le 900$
(C) 900; 200
(D) Elm St.: 800; 2nd St.: 400; Oak St.: 300 *(4.3)*

Chapter 5

Exercises 5.1

1. No **3.** Yes **5.** No **7.** No **9.**

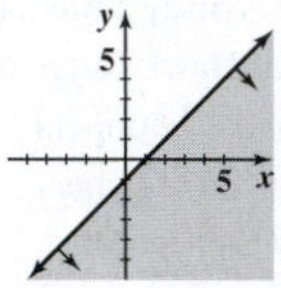

11.

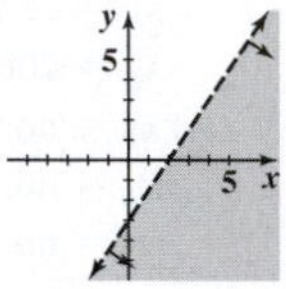

13.

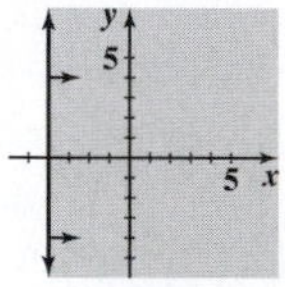

15.

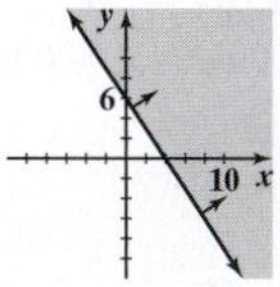

17.

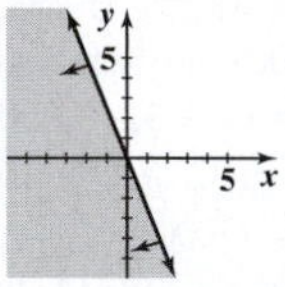

19. (A)

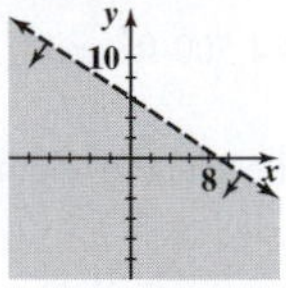

(B)

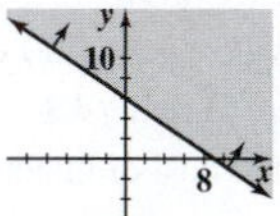

21. (A)

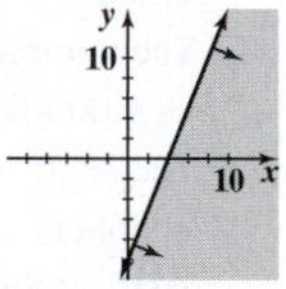

(B)

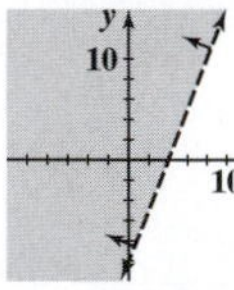

23. Let a = no. of applicants; $a < 10$
25. Let h = no. of hours of practice per day; $h \ge 2.5$ **27.** Let p = monthly take-home pay; $p > \$3{,}000$ **29.** Let t = tax rate; $t < 40\%$
31. Let e = enrollment; $e \le 30$
33. $2x + 3y = -6$; $2x + 3y \ge -6$ **35.** $y = 3$; $y < 3$ **37.** $4x - 5y = 0$; $4x - 5y \ge 0$ **39.** Let x = enrollment in finite mathematics; let y = enrollment in calculus; $x + y < 300$ **41.** Let x = revenue; y = cost; $x \le y - 20{,}000$ **43.** Let x = no. of grams of saturated fat; let y = no. of grams of unsaturated fat; $x > 3y$
45.

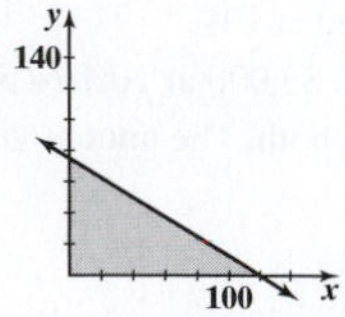

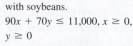

47.

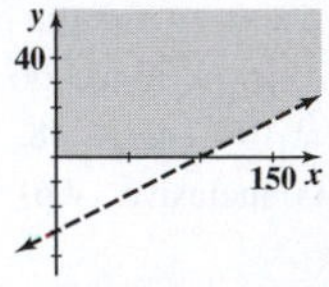

49.

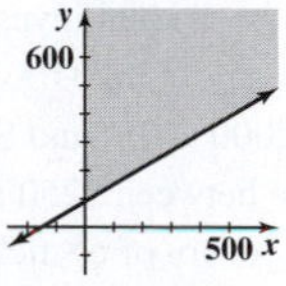

51.

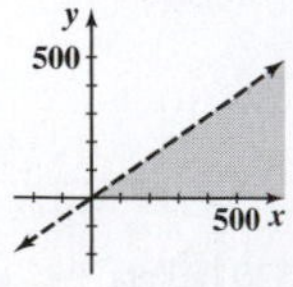

53. The solution set is the empty set and has no graph.
55. Let x = no. of acres planted with corn.
Let y = no. of acres planted with soybeans.
$90x + 70y \le 11{,}000$, $x \ge 0$, $y \ge 0$

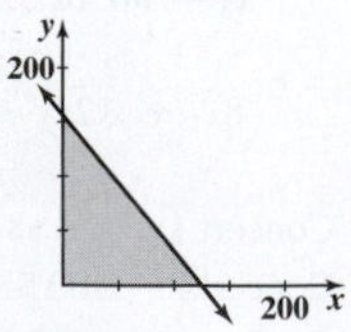

57. Let x = no. of lbs of brand A.
Let y = no. of lbs of brand B.
(A) $0.26x + 0.16y \ge 120$, $x \ge 0$, $y \ge 0$ (B) $0.03x + 0.08y \le 28$, $x \ge 0$, $y \ge 0$

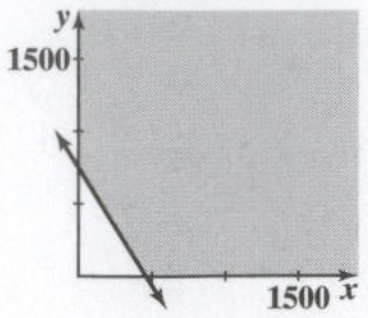

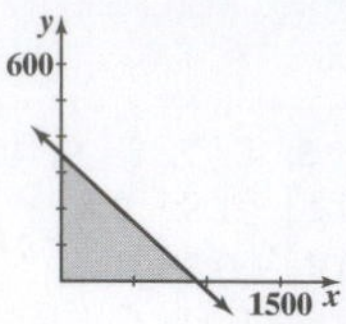

59. Let x = no. of lbs of the standard blend.
Let y = no. of lbs of the deluxe blend.
$0.3x + 0.09y \ge 0.20(x + y)$, $x \ge 0$, $y \ge 0$

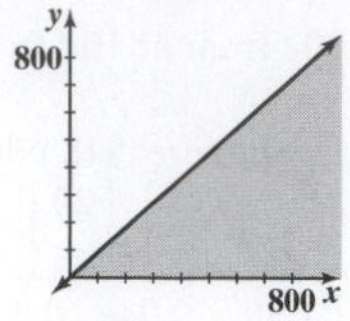

61. Let x = no. of weeks Plant A is operated.
Let y = no. of weeks Plant B is operated.
$10x + 8y \ge 400$, $x \ge 0$, $y \ge 0$

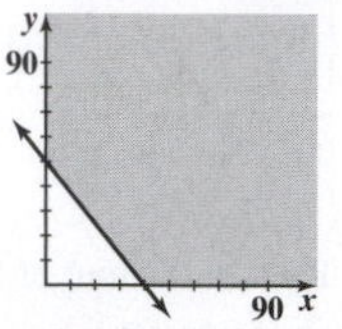

63. Let x = no. of radio spots.
Let y = no. of television spots.
$200x + 800y \le 10{,}000$, $x \ge 0$, $y \ge 0$

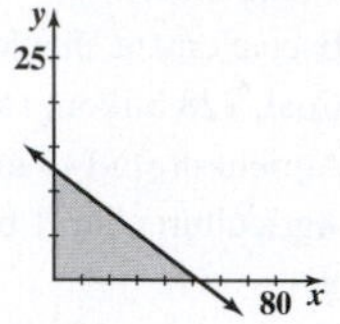

65. Let x = no. of regular mattresses cut per day.
Let y = no. of king mattresses cut per day
$5x + 6y \le 3{,}000$, $x \ge 0$, $y \ge 0$

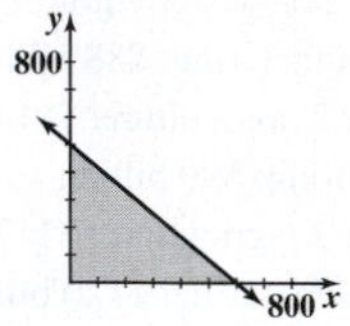

Exercises 5.2

1. Yes **3.** No **5.** No **7.** Yes **9.** IV **11.** I **13.**

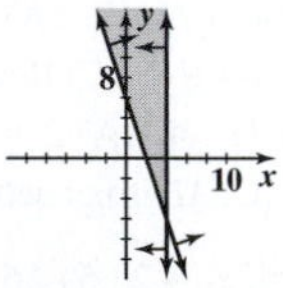

15.

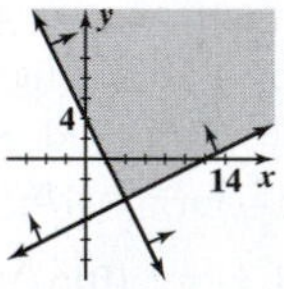

17. IV; (8, 0), (18, 0), (6, 4) **19.** I; (0, 16), (6, 4), (18, 0)

21. Bounded **23.** Unbounded **25.** Bounded **27.** Unbounded
29. Bounded

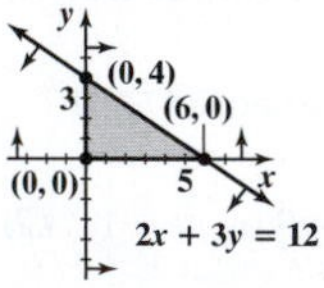

31. Bounded

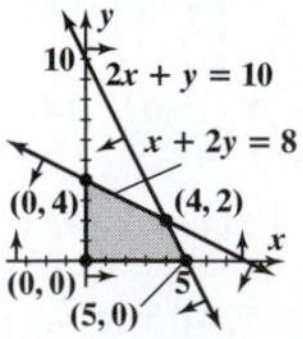

33. Unbounded

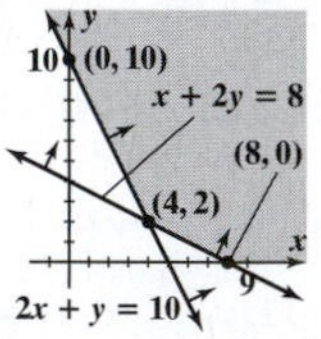

35. Bounded

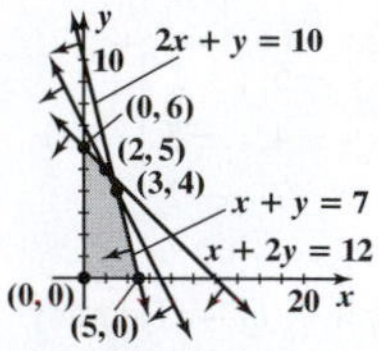

37. Unbounded

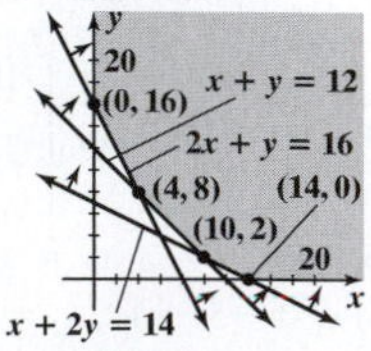

39. Bounded

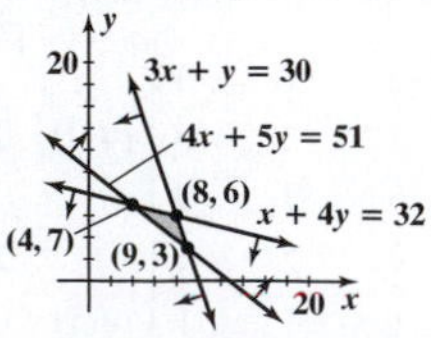

41. Empty

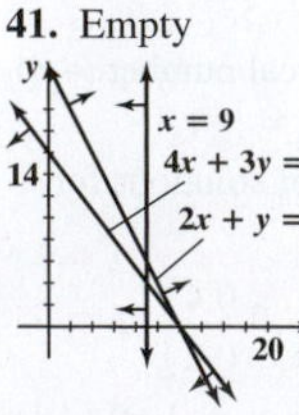

43. Unbounded

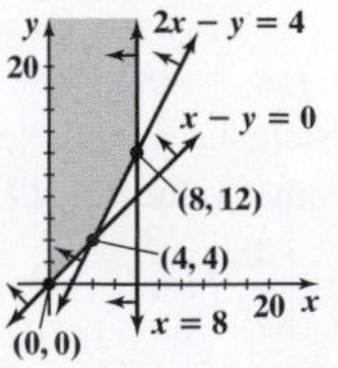

45. Bounded

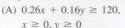

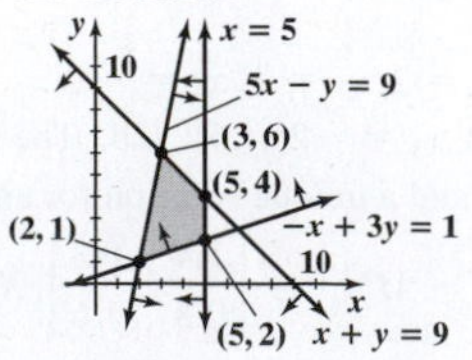

47. Bounded

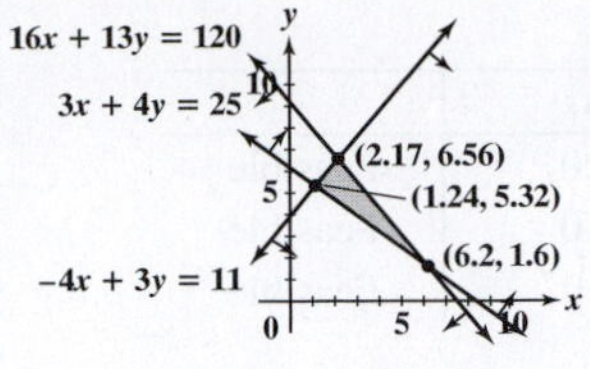

49. (A) $3x + 4y = 36$ and $3x + 2y = 30$ intersect at $(8, 3)$;
$3x + 4y = 36$ and $x = 0$ intersect at $(0, 9)$;
$3x + 4y = 36$ and $y = 0$ intersect at $(12, 0)$;
$3x + 2y = 30$ and $x = 0$ intersect at $(0, 15)$;
$3x + 2y = 30$ and $y = 0$ intersect at $(10, 0)$;
$x = 0$ and $y = 0$ intersect at $(0, 0)$
(B) $(8, 3)$, $(0, 9)$, $(10, 0)$, $(0, 0)$

51. $6x + 4y \le 108$
$x + y \le 24$
$x \ge 0$
$y \ge 0$

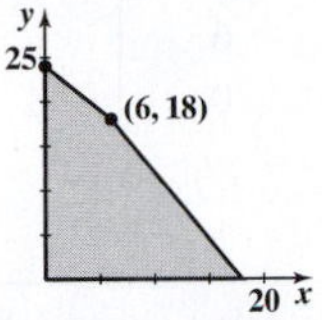

53. (A) All production schedules in the feasible region that are on the graph of $50x + 60y = 1{,}100$ will result in a profit of \$1,100. (B) There are many possible choices. For example, producing 5 trick skis and 15 slalom skis will produce a profit of \$1,150. All the production schedules in the feasible region that are on the graph of $50x + 60y = 1{,}150$ will result in a profit of \$1,150.

55. $20x + 10y \ge 460$
$30x + 30y \ge 960$
$5x + 10y \ge 220$
$x \ge 0$
$y \ge 0$

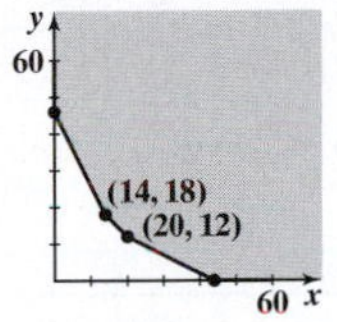

57. $10x + 20y \le 800$
$20x + 10y \le 640$
$x \ge 0$
$y \ge 0$

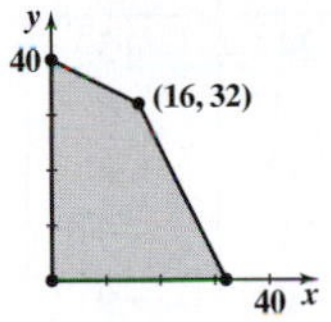

Exercises 5.3

1. Max $Q = 154$; Min $Q = 0$ **3.** Max $Q = 120$; Min $Q = -60$ **5.** Max $Q = 0$; Min $Q = -32$ **7.** Max $Q = 40$; Min $Q = -48$ **9.** Max $P = 16$ at $x = 7$ and $y = 9$ **11.** Max $P = 84$ at $x = 7$ and $y = 9$, at $x = 0$ and $y = 12$, and at every point on the line segment joining the preceding two points. **13.** Min $C = 32$ at $x = 0$ and $y = 8$ **15.** Min $C = 36$ at $x = 4$ and $y = 3$ **17.** Max $P = 240$ at $x = 24$ and $y = 0$ **19.** Min $C = 90$ at $x = 0$ and $y = 10$ **21.** Max $P = 30$ at $x = 4$ and $y = 2$ **23.** Min $z = 14$ at $x = 4$ and $y = 2$; no max **25.** Max $P = 260$ at $x = 2$ and $y = 5$ **27.** Min $z = 140$ at $x = 14$ and $y = 0$; no max **29.** Min $P = 20$ at $x = 0$ and $y = 2$; Max $P = 150$ at $x = 5$ and $y = 0$ **31.** Feasible region empty; no optimal solutions **33.** Min $P = 140$ at $x = 3$ and $y = 8$; Max $P = 260$ at $x = 8$ and $y = 10$, at $x = 12$ and $y = 2$, or at any point on the line segment from $(8, 10)$ to $(12, 2)$ **35.** Max $P = 26{,}000$ at $x = 400$ and $y = 600$ **37.** Max $P = 5{,}507$ at $x = 6.62$ and $y = 4.25$ **39.** Max $z = 2$ at $x = 4$ and $y = 2$; min z does not exist **41.** $5 < a < 30$ **43.** $a > 30$ **45.** $a = 30$ **47.** $a = 0$

49. (A) Let: x = no. of trick skis
y = no. of slalom skis
Maximize $P = 40x + 30y$
subject to $6x + 4y \le 108$
$x + y \le 24$
$x \ge 0, y \ge 0$
Max profit = \$780 when 6 trick skis and 18 slalom skis are produced.

(B) Max profit decreases to \$720 when 18 trick skis and no slalom skis are produced.

(C) Max profit increases to \$1,080 when no trick skis and 24 slalom skis are produced.

51. (A) Let x = no. of days to operate plant A
y = no. of days to operate plant B
Maximize $C = 1000x + 900y$
subject to $20x + 25y \ge 200$
$60x + 50y \ge 500$
$x \ge 0, y \ge 0$
Plant A: 5 days; Plant B: 4 days; min cost \$8,600
(B) Plant A: 10 days; Plant B: 0 days; min cost \$6,000
(C) Plant A: 0 days; Plant B: 10 days; min cost \$8,000

53. Let x = no. of buses
y = no. of vans
Maximize $C = 1{,}200x + 100y$
subject to $40x + 8y \ge 400$
$3x + y \le 36$
$x \ge 0, y \ge 0$
7 buses, 15 vans; min cost \$9,900

55. Let x = amount invested in the CD
y = amount invested in the mutual fund
Maximize $P = 0.05x + 0.09y$
subject to $x + y \le 60{,}000$
$y \ge 10{,}000$
$x \ge 2y$
$x, y \ge 0$
\$40,000 in the CD and \$20,000 in the mutual fund; max return is \$3,800

57. (A) Let x = no. of gallons produced by the old process
y = no. of gallons produced by the new process
Maximize $P = 60x + 20y$
subject to $20x + 5y \le 16{,}000$
$40x + 20y \le 30{,}000$
$x \ge 0, y \ge 0$
Max P = \$450 when 750 gal are produced using the old process exclusively.
(B) Max P = \$380 when 400 gal are produced using the old process and 700 gal are produced using the new process.
(C) Max P = \$288 when 1,440 gal are produced using the new process exclusively.

59. (A) Let x = no. of bags of brand A
y = no. of bags of brand B
Maximize $N = 8x + 3y$
subject to $4x + 4y \ge 1{,}000$
$2x + y \le 400$
$x \ge 0, y \ge 0$
150 bags brand A, 100 bags brand B;
Max nitrogen = 1,500 lb
(B) 0 bags brand A, 250 bags brand B;
Min nitrogen = 750 lb

61. Let x = no. of cubic yards of mix A
y = no. of cubic yards of mix B
Minimize $C = 30x + 35y$
subject to $20x + 10y \ge 460$
$30x + 30y \ge 960$
$5x + 10y \ge 220$
$x \ge 0, y \ge 0$
20 yd^3 A, 12 yd^3 B; \$1,020

63. Let x = no. of mice used
y = no. of rats used
Maximize $P = x + y$
subject to $10x + 20y \le 800$
$20x + 10y \le 640$
$x \ge 0, y \ge 0$
48; 16 mice, 32 rats

Chapter 5 Review Exercises

1. *(5.1)*

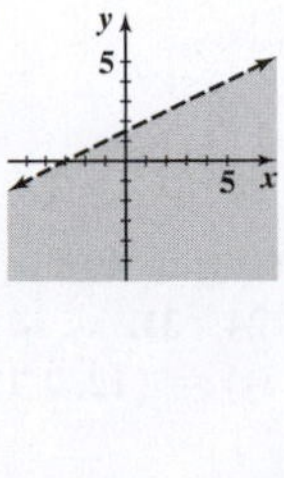

2. *(5.1)*

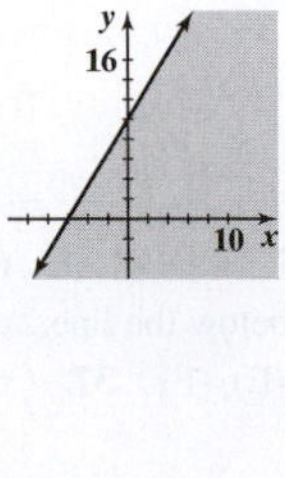

3. Bounded *(5.2)*

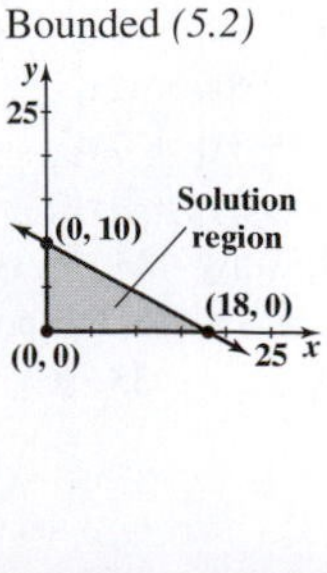

4. Unbounded *(5.2)* **5.** Bounded *(5.2)* **6.** Unbounded *(5.2)*

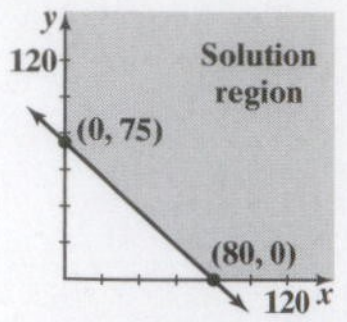

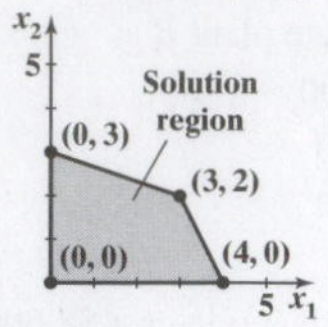

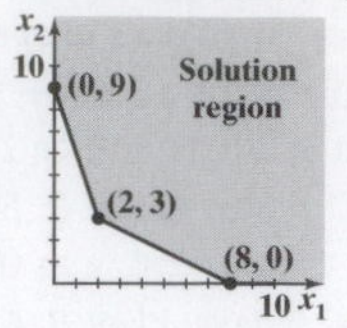

7. $2x - 3y = 12$; $2x - 3y \le 12$ *(5.1)* **8.** $4x + y = 8$; $4x + y \ge 8$ *(5.1)* **9.** Max $P = 24$ at $x = 0$ and $y = 4$ *(5.3)* **10.** Min $C = 40$ at $x = 0$ and $y = 20$ *(5.3)* **11.** Max $P = 26$ at $x = 2$ and $y = 5$ *(5.3)* **12.** Min $C = 51$ at $x = 3$ and $y = 9$ *(5.3)* **13.** Max $P = 36$ at $x = 8$ and $y = 6$ *(5.3)*

14. Let x = no. of calculator boards. *(5.1)*
y = no. of toaster boards

(A) $4x + 3y \le 300$, $x \ge 0$, $y \ge 0$ (B) $2x + y \le 120$, $x \ge 0$, $y \ge 0$

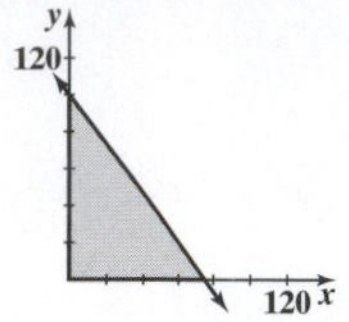

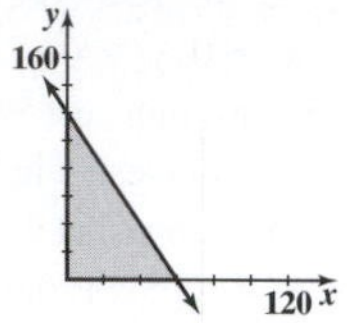

15. (A) Let x = no. of regular sails
y = no. of competition sails

Maximize $P = 100x + 200y$
subject to $2x + 3y \le 150$
$4x + 10y \le 380$
$x, y \ge 0$

Max P = \$8,500 when 45 regular and 20 competition sails are produced.

(B) Max profit increases to \$9,880 when 38 competition and no regular sails are produced.

(C) Max profit decreases to \$7,500 when no competition and 75 regular sails are produced. *(5.3)*

16. (A) Let x = no. of grams of mix A
y = no. of grams of mix B

Minimize $C = 0.04x + 0.09y$
subject to $2x + 5y \ge 850$
$2x + 4y \ge 800$
$4x + 5y \ge 1{,}150$
$x, y \ge 0$

Min C = \$16.50 when 300 g mix A and 50 g mix B are used

(B) The minimum cost decreases to \$13.00 when 100 g mix A and 150 g mix B are used

(C) The minimum cost increases to \$17.00 when 425 g mix A and no mix B are used *(5.3)*

Chapter 6

Exercises 6.1

1. 56 **3.** 55 **5.** 21 **7.** 70 **9.** $(x_1, x_2, s_1, s_2) = (0, 2, 0, 2)$

11. $(x_1, x_2, s_1, s_2) = (8, 0, -6, 0)$

13.
$$\begin{aligned} 2x_1 + 3x_2 + s_1 \quad\quad &= 9 \\ 6x_1 + 7x_2 \quad + s_2 &= 13 \end{aligned}$$

15.
$$\begin{aligned} 12x_1 - 14x_2 + s_1 \quad\quad\quad &= 55 \\ 19x_1 + 5x_2 \quad + s_2 \quad &= 40 \\ -8x_1 + 11x_2 \quad\quad + s_3 &= 64 \end{aligned}$$

17. $6x_1 + 5x_2 + s_1 = 18$

19.
$$\begin{aligned} 4x_1 - 3x_2 + s_1 \quad\quad\quad\quad &= 12 \\ 5x_1 + 2x_2 \quad + s_2 \quad\quad &= 25 \\ -3x_1 + 7x_2 \quad\quad + s_3 \quad &= 32 \\ 2x_1 + x_2 \quad\quad\quad + s_4 &= 9 \end{aligned}$$

21. s_1, s_2 **23.** x_1, s_2 **25.** (A), (B), (E), (F) **27.** Max $P = 40$ at $x_1 = 0$, $x_2 = 8$ **29.** The points below the line $2x_1 + 3x_2 = 24$ **31.** x_2, s_1, s_3 **33.** x_2, s_3 **35.** (C), (D), (E), (F) **37.** $(x_1, x_2, s_1, s_2, s_3) = (12, 0, 12, 6, 0)$

39. $(x_1, x_2, s_1, s_2, s_3) = (8, 16, 0, -2, 0)$

41. $4x_1 + 5x_2 + s_1 = 20$

x_1	x_2	s_1	
0	0	20	Feasible
0	4	0	Feasible
5	0	0	Feasible

43.
$$\begin{aligned} x_1 + x_2 + s_1 \quad &= 6 \\ x_1 + 4x_2 \quad + s_2 &= 12 \end{aligned}$$

x_1	x_2	s_1	s_2	
0	0	6	12	Feasible
0	6	0	−12	Not feasible
0	3	3	0	Feasible
6	0	0	6	Feasible
12	0	−6	0	Not feasible
4	2	0	0	Feasible

45.
$$\begin{aligned} 2x_1 + 5x_2 + s_1 \quad &= 20 \\ x_1 + 2x_2 \quad + s_2 &= 9 \end{aligned}$$

x_1	x_2	s_1	s_2	
0	0	20	9	Feasible
0	4	0	1	Feasible
0	9/2	−5/2	0	Not feasible
10	0	0	−1	Not feasible
9	0	2	0	Feasible
5	2	0	0	Feasible

47.
$$\begin{aligned} x_1 + 2x_2 + s_1 \quad\quad &= 24 \\ x_1 + x_2 \quad + s_2 \quad &= 15 \\ 2x_1 + x_2 \quad\quad + s_3 &= 24 \end{aligned}$$

x_1	x_2	s_1	s_2	s_3	
0	0	24	15	24	Feasible
0	12	0	3	12	Feasible
0	15	−6	0	9	Not feasible
0	24	−24	−9	0	Not feasible
24	0	0	−9	−24	Not feasible
15	0	9	0	−6	Not feasible
12	0	12	3	0	Feasible
6	9	0	0	3	Feasible
8	8	0	−1	0	Not feasible
9	6	3	0	0	Feasible

49. Corner points: $(0, 0)$, $(0, 4)$, $(5, 0)$

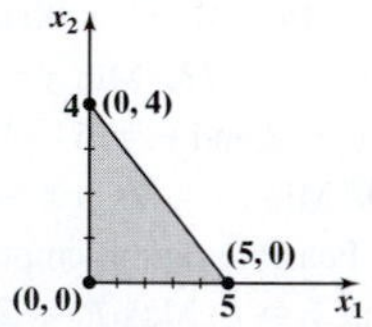

51. Corner points: $(0, 0)$, $(0, 3)$, $(6, 0)$, $(4, 2)$

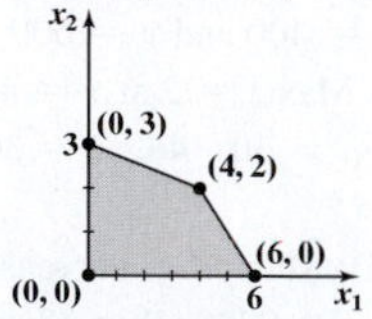

53. Corner points: $(0, 0)$, $(0, 4)$, $(9, 0)$, $(5, 2)$

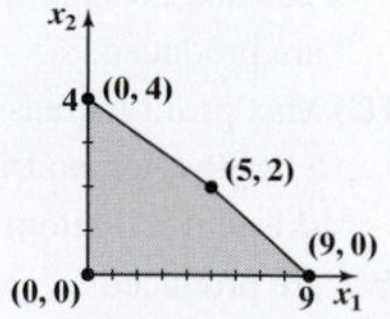

59. Max $P = 50$ at $x_1 = 5, x_2 = 0$ **61.** Max $P = 100$ at $x_1 = 4, x_2 = 2$
63. Max $P = 225$ at $x_1 = 9, x_2 = 0$ **65.** Max $P = 540$ at $x_1 = 6, x_2 = 9$
67. Every point with coordinates $(0, x_2)$, where $x_2 \geq 0$, satisfies the problem constraint. So the feasible region is unbounded. **69.** Every point with coordinates (x_1, x_2), where $x_1 = x_2 \geq 0$, satisfies both problem constraints. So the feasible region is unbounded. **71.** If (x_1, x_2) is a point in the feasible region, then $0 \leq x_1 \leq 100$ and $0 \leq x_2 \leq 50 + 2x_1 \leq 50 + 2(100) = 250$, so the feasible region is bounded; Max $P = 8,250$ at $x_1 = 100, x_2 = 250$.
73. $_{10}C_4 = 210$ **75.** $_{72}C_{30} \approx 1.64 \times 10^{20}$

Exercises 6.2

1. (A) Basic: x_2, s_1, P; nonbasic: x_1, s_2 (B) $x_1 = 0, x_2 = 12, s_1 = 15$, $s_2 = 0, P = 50$ (C) Additional pivot required **3.** (A) Basic: x_2, x_3, s_3, P; nonbasic: x_1, s_1, s_2 (B) $x_1 = 0, x_2 = 15, x_3 = 5, s_1 = 0, s_2 = 0$, $s_3 = 12, P = 45$ (C) No optimal solution

5.

Enter ↓

$$\begin{array}{c} \\ \text{Exit} \to s_1 \\ s_2 \\ P \end{array} \left[\begin{array}{ccccc|c} x_1 & x_2 & s_1 & s_2 & P & \\ ① & 4 & 1 & 0 & 0 & 4 \\ 3 & 5 & 0 & 1 & 0 & 24 \\ \hline -8 & -5 & 0 & 0 & 1 & 0 \end{array} \right]$$

$$\begin{array}{c} x_1 \\ \sim s_2 \\ P \end{array} \left[\begin{array}{ccccc|c} 1 & 4 & 1 & 0 & 0 & 4 \\ 0 & -7 & -3 & 1 & 0 & 12 \\ \hline 0 & 27 & 8 & 0 & 1 & 32 \end{array} \right]$$

7.

Enter ↓

$$\begin{array}{c} \\ \text{Exit} \to x_2 \\ s_2 \\ s_3 \\ P \end{array} \left[\begin{array}{cccccc|c} x_1 & x_2 & s_1 & s_2 & s_3 & P & \\ ② & 1 & 1 & 0 & 0 & 0 & 4 \\ 3 & 0 & 1 & 1 & 0 & 0 & 8 \\ 0 & 0 & 2 & 0 & 1 & 0 & 2 \\ \hline -4 & 0 & -3 & 0 & 0 & 1 & 5 \end{array} \right]$$

$$\begin{array}{c} x_1 \\ \sim s_2 \\ s_3 \\ P \end{array} \left[\begin{array}{cccccc|c} 1 & \frac{1}{2} & \frac{1}{2} & 0 & 0 & 0 & 2 \\ 0 & -\frac{3}{2} & -\frac{1}{2} & 1 & 0 & 0 & 2 \\ 0 & 0 & 2 & 0 & 1 & 0 & 2 \\ \hline 0 & 2 & -1 & 0 & 0 & 1 & 13 \end{array} \right]$$

9. (A)
$$\begin{aligned} 2x_1 + x_2 + s_1 \quad\quad &= 10 \\ x_1 + 3x_2 \quad\; + s_2 &= 10 \\ -15x_1 - 10x_2 \quad\quad\; + P &= 0 \\ x_1, x_2, s_1, s_2 &\geq 0 \end{aligned}$$

(B)

Enter ↓

$$\begin{array}{c} \\ \text{Exit} \to s_1 \\ s_2 \\ P \end{array} \left[\begin{array}{ccccc|c} x_1 & x_2 & s_1 & s_2 & P & \\ ② & 1 & 1 & 0 & 0 & 10 \\ 1 & 3 & 0 & 1 & 0 & 10 \\ \hline -15 & -10 & 0 & 0 & 1 & 0 \end{array} \right]$$

(C) Max $P = 80$ at $x_1 = 4$ and $x_2 = 2$

11. (A)
$$\begin{aligned} 2x_1 + x_2 + s_1 \quad\quad &= 10 \\ x_1 + 3x_2 \quad\; + s_2 &= 10 \\ -30x_1 - x_2 \quad\quad\; + P &= 0 \\ x_1, x_2, s_1, s_2 &\geq 0 \end{aligned}$$

(B)

Enter ↓

$$\begin{array}{c} \\ \text{Exit} \to s_1 \\ s_2 \\ P \end{array} \left[\begin{array}{ccccc|c} x_1 & x_2 & s_1 & s_2 & P & \\ ② & 1 & 1 & 0 & 0 & 10 \\ 1 & 3 & 0 & 1 & 0 & 10 \\ \hline -30 & -1 & 0 & 0 & 1 & 0 \end{array} \right]$$

(C) Max $P = 150$ at $x_1 = 5$ and $x_2 = 0$

13. Max $P = 260$ at $x_1 = 2$ and $x_2 = 5$ **15.** No optimal solution exists.
17. Max $P = 7$ at $x_1 = 3$ and $x_2 = 5$ **19.** Max $P = 90$ at $x_1 = 6$ and $x_2 = 0$ **21.** No optimal solution exists. **23.** Max $P = 58$ at $x_1 = 12, x_2 = 0$, and $x_3 = 2$ **25.** Max $P = 17$ at $x_1 = 4, x_2 = 3$ and $x_3 = 0$

27. Max $P = 22$ at $x_1 = 1, x_2 = 6$, and $x_3 = 0$
29. Max $P = 26,000$ at $x_1 = 400$ and $x_2 = 600$
31. Max $P = 450$ at $x_1 = 0, x_2 = 180$, and $x_2 = 30$
33. Max $P = 88$ at $x_1 = 24$ and $x_2 = 8$

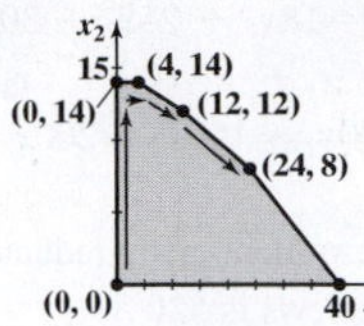

35. No solution **37.** Choosing either col. produces the same optimal solution: max $P = 13$ at $x_1 = 13$ and $x_2 = 10$ **39.** Choosing col. 1: max $P = 60$ at $x_1 = 12, x_2 = 8$, and $x_3 = 0$. Choosing col. 2: max $P = 60$ at $x_1 = 0, x_2 = 20$, and $x_3 = 0$.
41. Let $x_1 = $ no. of A components
$x_2 = $ no. of B components
$x_3 = $ no. of C components
Maximize $P = 7x_1 + 8x_2 + 10x_3$
subject to $2x_1 + 3x_2 + 2x_3 \leq 1,000$
$x_1 + x_2 + 2x_3 \leq 800$
$x_1, x_2, x_3 \geq 0$
200 A components, 0 B components, and 300 C components; max profit is $4,400
43. Let $x_1 = $ amount invested in government bonds
$x_2 = $ amount invested in mutual funds
$x_3 = $ amount invested in money markey funds
Maximize $P = 0.08x_1 + 0.13x_2 + 0.15x_3$
subject to $x_1 + x_2 + x_3 \leq 100,000$
$-x_1 + x_2 + x_3 \leq 0$
$x_1, x_2, x_2 \geq 0$
$50,000 in government bonds, $0 in mutual funds, and $50,000 in money market funds; max return is $11,500
45. Let $x_1 = $ no. of ads placed in daytime shows
$x_2 = $ no. of ads placed in prime-time shows
$x_3 = $ no. of ads placed in late-night shows
Maximize $P = 14,000x_1 + 24,000x_2 + 18,000x_3$
subject to $x_1 + x_2 + x_3 \leq 15$
$1,000x_1 + 2,000x_2 + 1,500x_3 \leq 20,000$
$x_1, x_2, x_3 \geq 0$
10 daytime ads, 5 prime-time ads, and 0 late-night ads; max no. of potential customers is 260,000
47. Let $x_1 = $ no. of colonial houses
$x_2 = $ no. of split-level houses
$x_3 = $ no. of ranch houses
Maximize $P = 20,000x_1 + 18,000x_2 + 24,000x_3$
subject to $\frac{1}{2}x_1 + \frac{1}{2}x_2 + x_3 \leq 30$
$60,000x_1 + 60,000x_2 + 80,000x_3 \leq 3,200,000$
$4,000x_1 + 3,000x_3 + 4,000x_3 \leq 180,000$
$x_1, x_2, x_3 \geq 0$
20 colonial, 20 split-level, and 10 ranch houses; max profit is $1,000,000
49. The model is the same as the model for Problem 43 except that $P = 17,000x_1 + 18,000x_2 + 24,000x_3$.
0 colonial, 40 split-level, and 10 ranch houses; max profit is $960,000; 20,000 labor-hours are not used
51. The model is the same as the model for Problem 43 except that $P = 25,000x_1 + 18,000x_2 + 24,000x_3$.
45 colonial, 0 split-level, and 0 ranch houses; max profit is $1,125,000; 7.5 acres of land and $500,000 of capital are not used
53. Let $x_1 = $ no. of grams of food A Maximize $P = 3x_1 + 4x_2 + 5x_3$
$x_2 = $ no. of grams of food B subject to $x_1 + 3x_2 + 2x_3 \leq 30$
$x_3 = $ no. of grams of food C $2x_1 + x_2 + 2x_3 \leq 24$
$x_1, x_2, x_3 \geq 0$

0 g food A, 3 g food B, and 10.5 g food C; max protein is 64.5 units

55. Let x_1 = no. of undergraduate students
x_2 = no. of graduate students
x_3 = no. of faculty members

$$\begin{aligned}\text{Maximize } & P = 18x_1 + 25x_2 + 30x_3\\ \text{subject to } & x_1 + x_2 + x_3 \le 20\\ & 100x_1 + 150x_2 + 200x_3 \le 3{,}200\\ & x_1, x_2, x_3 \ge 0\end{aligned}$$

0 undergraduate students, 16 graduate students, and 4 faculty members; max no. of interviews is 520

Exercises 6.3

1. $\begin{bmatrix} -5 & 2 \\ 3 & 7 \end{bmatrix}$ **3.** $\begin{bmatrix} 1 & 6 & 9 \\ 7 & 3 & 4 \\ 8 & 5 & 2 \end{bmatrix}$ **5.** $\begin{bmatrix} 2 & 5 \\ 1 & 2 \\ -6 & 0 \\ 0 & 1 \\ -1 & 3 \end{bmatrix}$ **7.** $\begin{bmatrix} 1 & 0 & 8 & 4 \\ 2 & 2 & 0 & -1 \\ -1 & -7 & 1 & 3 \end{bmatrix}$

9. (A) $\begin{aligned}\text{Maximize } & P = 4y_1 + 5y_2\\ \text{subject to } & y_1 + 2y_2 \le 8\\ & 3y_1 + y_2 \le 9\\ & y_1, y_2 \ge 0\end{aligned}$

(B) $\begin{aligned} y_1 + 2y_2 + x_1 \quad\quad\quad &= 8\\ 3y_1 + y_2 \quad + x_2 \quad &= 9\\ -4y_1 - 5y_2 \quad\quad + P &= 0\end{aligned}$

(C)
$$\begin{array}{ccccc|c} y_1 & y_2 & x_1 & x_2 & P & \\ 1 & 2 & 1 & 0 & 0 & 8 \\ 3 & 1 & 0 & 1 & 0 & 9 \\ \hline -4 & -5 & 0 & 0 & 1 & 0 \end{array}$$

11. (A) Max $P = 121$ at $y_1 = 3$ and $y_2 = 5$ (B) Min $C = 121$ at $x_1 = 1$ and $x_2 = 2$

13. (A) $\begin{aligned}\text{Maximize } & P = 13y_1 + 12y_2\\ \text{subject to } & 4y_1 + 3y_2 \le 9\\ & y_1 + y_2 \le 2\\ & y_1, y_2 \ge 0\end{aligned}$

(B) Min $C = 26$ at $x_1 = 0$ and $x_2 = 13$

15. (A) $\begin{aligned}\text{Maximize } & P = 15y_1 + 8y_2\\ \text{subject to } & 2y_1 + y_2 \le 7\\ & 3y_1 + 2y_2 \le 12\\ & y_1, y_2 \ge 0\end{aligned}$

(B) Min $C = 54$ at $x = 6$ and $x_2 = 1$

17. (A) $\begin{aligned}\text{Maximize } & P = 8y_1 + 4y_2\\ \text{subject to } & 2y_1 - 2y_2 \le 11\\ & y_1 + 3y_2 \le 4\\ & y_1, y_2 \ge 0\end{aligned}$

(B) Min $C = 32$ at $x_1 = 0$ and $x_2 = 8$

19. (A) $\begin{aligned}\text{Maximize } & P = 6y_1 + 4y_2\\ \text{subject to } & -3y_1 + y_2 \le 7\\ & y_1 - 2y_2 \le 9\\ & y_1, y_2 \ge 0\end{aligned}$

(B) No optimal solution exists.

21. Min $C = 24$ at $x_1 = 8$ and $x_2 = 0$ **23.** Min $C = 20$ at $x_1 = 0$ and $x_2 = 4$ **25.** Min $C = 140$ at $x_1 = 14$ and $x_2 = 0$ **27.** Min $C = 44$ at $x_1 = 6$ and $x_2 = 2$ **29.** Min $C = 100$ at $x_1 = 0$ and $x_2 = 4$ **31.** Min $C = 24$ at $x_1 = 0$, $x_2 = 0$, and $x_3 = 2$ **33.** Min $C = 43$ at $x_1 = 0$, $x_2 = 1$, and $x_3 = 3$ **35.** No optimal solution exists. **37.** 2 variables and 4 problem constraints **39.** 2 constraints and any no. of variables

41. (A) $\begin{aligned}\text{Maximize } & P = 7y_1 + 10y_2\\ \text{subject to } & 5y_1 + 4y_2 \le 4\\ & 2y_1 + 6y_2 \le -1\\ & y_1, y_2 \ge 0\end{aligned}$

(B) No

43. (A) $\begin{aligned}\text{Minimize } & C = 3x_1 + x_2 + 5x_3\\ \text{subject to } & -2x_1 + 6x_2 + x_3 \ge -10\\ & -5x_1 + x_2 + 4x_3 \ge 15\\ & x_1, x_2, x_3 \ge 0\end{aligned}$

(B) $\begin{aligned}\text{Maximize } & P = -10y_1 + 15y_2\\ \text{subject to } & -2y_1 - 5y_2 \le 3\\ & 6y_1 + y_2 \le 1\\ & y_1 + 4y_2 \le 5\\ & y_1, y_2 \ge 0\end{aligned}$

(C) Yes

45. Min $C = 44$ at $x_1 = 0$, $x_2 = 3$, and $x_3 = 5$

47. Min $C = 166$ at $x_1 = 0$, $x_2 = 12$, $x_3 = 20$, and $x_4 = 3$

49. Let x_1 = no. of hours the Cedarburg plant is operated
x_2 = no. of hours the Grafton plant is operated
x_3 = no. of hours the West Bend plant is operated

$$\begin{aligned}\text{Minimize } & C = 70x_1 + 75x_2 + 90x_3\\ \text{subject to } & 20x_1 + 10x_2 + 20x_3 \ge 300\\ & 10x_1 + 20x_2 + 20x_3 \ge 200\\ & x_1, x_2, x_3 \ge 0\end{aligned}$$

Cedarburg plant 10 hr per day, West Bend plant 5 hr per day, Grafton plant not used; min cost is \$1,150

51. The model is the same as the model for Problem 49 except that the second constraint (deluxe ice cream) is $10x_1 + 20x_2 + 20x_3 \ge 300$. West Bend Plant 15 hr per day, Cedarburg and Grafton plants not used; min cost is \$1,350

53. The model is the same as the model for Problem 49 except that the second constraint (deluxe ice cream) is $10x_1 + 20x_2 + 20x_3 \ge 400$. Grafton plant 10 hr per day, West Bend plant 10 hr per day, Cedarburg plant not used; min cost is \$1,650

55. Let x_1 = no. of ounces of food L
x_2 = no. of ounces of food M
x_3 = no. of ounces of food N

$$\begin{aligned}\text{Minimize } & C = 20x_1 + 24x_2 + 18x_3\\ \text{subject to } & 20x_1 + 10x_2 + 10x_3 \ge 300\\ & 10x_1 + 10x_2 + 10x_3 \ge 200\\ & 10x_1 + 15x_2 + 10x_3 \ge 240\\ & x_1, x_2, x_3 \ge 0\end{aligned}$$

10 oz L, 8 oz M, 2 oz N; min cholesterol intake is 428 units

57. Let x_1 = no. of students bused from North Division to Central
x_2 = no. of students bused from North Division to Washington
x_3 = no. of students bused from South Division to Central
x_4 = no. of students bused from South Division to Washington

$$\begin{aligned}\text{Minimize } & C = 5x_1 + 2x_2 + 3x_3 + 4x_4\\ \text{subject to } & x_1 + x_2 \ge 300\\ & x_3 + x_4 \ge 500\\ & x_1 + x_3 \le 400\\ & x_2 + x_4 \le 500\\ & x_1, x_2, x_3, x_4 \ge 0\end{aligned}$$

300 students bused from North Division to Washington, 400 from South Division to Central, and 100 from South Division to Washington; min cost is \$2,200

Exercises 6.4

1. (A) $\begin{aligned}\text{Maximize } & P = 5x_1 + 2x_2 - Ma_1\\ \text{subject to } & x_1 + 2x_2 + s_1 \quad\quad = 12\\ & x_1 + x_2 \quad - s_2 + a_1 = 4\\ & x_1, x_2, s_1, s_2, a_1 \ge 0\end{aligned}$

(B)
$$\begin{array}{cccccc|c} x_1 & x_2 & s_1 & s_2 & a_1 & P & \\ 1 & 2 & 1 & 0 & 0 & 0 & 12 \\ 1 & 1 & 0 & -1 & 1 & 0 & 4 \\ \hline -M-5 & -M-2 & 0 & M & 0 & 1 & -4M \end{array}$$

(C) $x_1 = 12$, $x_2 = 0$, $s_1 = 0$, $s_2 = 8$, $a_1 = 0$, $P = 60$
(D) Max $P = 60$ at $x_1 = 12$ and $x_2 = 0$

3. (A) Maximize $P = 3x_1 + 5x_2 - Ma_1$

$$\begin{array}{ll} \text{subject to} & 2x_1 + x_2 + s_1 \qquad\quad = 8 \\ & x_1 + x_2 \qquad\quad + a_1 = 6 \\ & x_1, x_2, s_1, a_1 \ge 0 \end{array}$$

(B)

$$\begin{array}{c} \begin{array}{cccccc} x_1 & x_2 & s_1 & a_1 & P & \end{array} \\ \left[\begin{array}{ccccc|c} 2 & 1 & 1 & 0 & 0 & 8 \\ 1 & 1 & 0 & 1 & 0 & 6 \\ \hline -M-3 & -M-5 & 0 & 0 & 1 & -6M \end{array}\right] \end{array}$$

(C) $x_1 = 0, x_2 = 6, s_1 = 2, a_1 = 0, P = 30$

(D) Max $P = 30$ at $x_1 = 0$ and $x_2 = 6$

5. (A) Maximize $P = 4x_1 + 3x_2 - Ma_1$

$$\begin{array}{ll} \text{subject to} & -x_1 + 2x_2 + s_1 \qquad\qquad = 2 \\ & x_1 + x_2 \qquad - s_2 + a_1 = 4 \\ & x_1, x_2, s_1, s_2, a_1 \ge 0 \end{array}$$

(B)

$$\begin{array}{c} \begin{array}{ccccccc} x_1 & x_2 & s_1 & s_2 & a_1 & P & \end{array} \\ \left[\begin{array}{cccccc|c} -1 & 2 & 1 & 0 & 0 & 0 & 2 \\ 1 & 1 & 0 & -1 & 1 & 0 & 4 \\ \hline -M-4 & -M-3 & 0 & M & 0 & 1 & -4M \end{array}\right] \end{array}$$

(C) No optimal solution exists.

(D) No optimal solution exists.

7. (A) Maximize $P = 5x_1 + 10x_2 - Ma_1$

$$\begin{array}{ll} \text{subject to} & x_1 + x_2 + s_1 \qquad\qquad = 3 \\ & 2x_1 + 3x_2 \qquad - s_2 + a_1 = 12 \\ & x_1, x_2, s_1, s_2, a_1 \ge 0 \end{array}$$

(B)

$$\begin{array}{c} \begin{array}{ccccccc} x_1 & x_2 & s_1 & s_2 & a_1 & P & \end{array} \\ \left[\begin{array}{cccccc|c} 1 & 1 & 1 & 0 & 0 & 0 & 3 \\ 2 & 3 & 0 & -1 & 1 & 0 & 12 \\ \hline -2M-5 & -3M-10 & 0 & M & 0 & 1 & -12M \end{array}\right] \end{array}$$

(C) $x_1 = 0, x_2 = 3, s_1 = 0, s_2 = 0, a_1 = 3, P = -3M + 30$

(D) No optimal solution exists.

9. Min $P = 1$ at $x_1 = 3$ and $x_2 = 5$; max $P = 16$ at $x_1 = 8$ and $x_2 = 0$ **11.** Max $P = 44$ at $x_1 = 2$ and $x_2 = 8$ **13.** No optimal solution exists. **15.** Min $C = -9$ at $x_1 = 0$, $x_2 = \frac{7}{4}$, and $x_3 = \frac{3}{4}$ **17.** Max $P = 32$ at $x_1 = 0$, $x_2 = 4$, and $x_3 = 2$ **19.** Max $P = 65$ at $x_1 = \frac{35}{2}$, $x_2 = 0$, and $x_3 = \frac{15}{2}$ **21.** Max $P = 120$ at $x_1 = 20$, $x_2 = 0$, and $x_3 = 20$ **23.** Problem 5: unbounded feasible region:

Problem 7: empty feasible region:

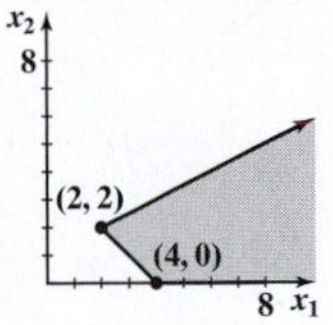

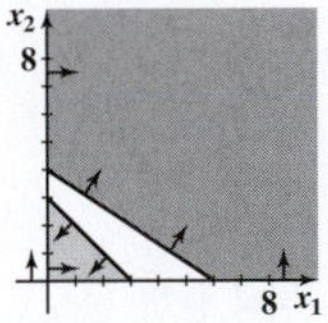

25. Min $C = -30$ at $x_1 = 0$, $x_2 = \frac{3}{4}$, and $x_3 = 0$ **27.** Max $P = 17$ at $x_1 = \frac{49}{5}$, $x_2 = 0$, and $x_3 = \frac{22}{5}$ **29.** Min $C = \frac{135}{2}$ at $x_1 = \frac{15}{4}$, $x_2 = \frac{3}{4}$, and $x_3 = 0$ **31.** Max $P = 372$ at $x_1 = 28$, $x_2 = 4$, and $x_3 = 0$

33. Let x_1 = no. of ads placed in the *Sentinel*
x_2 = no. of ads placed in the *Journal*
x_3 = no. of ads placed in the *Tribune*

$$\begin{array}{ll} \text{Minimize} & C = 200x_1 + 200x_2 + 100x_3 \\ \text{subject to} & x_1 + x_2 + x_3 \le 10 \\ & 2{,}000x_1 + 500x_2 + 1{,}500x_3 \ge 16{,}000 \\ & x_1, x_2, x_3 \ge 0 \end{array}$$

2 ads in the *Sentinel*, 0 ads in the *Journal*, 8 ads in the *Tribune;* min cost is \$1,200

35. Let x_1 = no. of bottles of brand A
x_2 = no. of bottles of brand B
x_3 = no. of bottles of brand C

$$\begin{array}{ll} \text{Minimize} & C = 0.6x_1 + 0.4x_2 + 0.9x_3 \\ \text{subject to} & 10x_1 + 10x_2 + 20x_3 \ge 100 \\ & 2x_1 + 3x_2 + 4x_3 \le 24 \\ & x_1, x_2, x_3 \ge 0 \end{array}$$

0 bottles of A, 4 bottles of B, 3 bottles of C; min cost is \$4.30

37. Let x_1 = no. of cubic yards of mix A
x_2 = no. of cubic yards of mix B
x_3 = no. of cubic yards of mix C

$$\begin{array}{ll} \text{Maximize} & P = 12x_1 + 16x_2 + 8x_3 \\ \text{subject to} & 12x_1 + 8x_2 + 16x_3 \le 700 \\ & 16x_1 + 8x_2 + 16x_3 \ge 800 \\ & x_1, x_2, x_3 \ge 0 \end{array}$$

25 yd^3 A, 50 yd^3 B, 0 yd^3 C; max is 1,100 lb

39. Let x_1 = no. of car frames produced at the Milwaukee plant
x_2 = no. of truck frames produced at the Milwaukee plant
x_3 = no. of car frames produced at the Racine plant
x_4 = no. of truck frames produced at the Racine plant

$$\begin{array}{ll} \text{Maximize} & P = 50x_1 + 70x_2 + 50x_3 + 70x_4 \\ \text{subject to} & x_1 + x_3 \le 250 \\ & x_2 + x_4 \le 350 \\ & x_1 + x_2 \le 300 \\ & x_3 + x_4 \le 200 \\ & 150x_1 + 200x_2 \le 50{,}000 \\ & 135x_3 + 180x_4 \le 35{,}000 \\ & x_1, x_2, x_3, x_4 \ge 0 \end{array}$$

41. Let x_1 = no. of barrels of A used in regular gasoline
x_2 = no. of barrels of A used in premium gasoline
x_3 = no. of barrels of B used in regular gasoline
x_4 = no. of barrels of B used in premium gasoline
x_5 = no. of barrels of C used in regular gasoline
x_6 = no. of barrels of C used in premium gasoline

$$\begin{array}{ll} \text{Maximize} & P = 10x_1 + 18x_2 + 8x_3 + 16x_4 + 4x_5 + 12x_6 \\ \text{subject to} & x_1 + x_2 \le 40{,}000 \\ & x_3 + x_4 \le 25{,}000 \\ & x_5 + x_6 \le 15{,}000 \\ & x_1 + x_3 + x_5 \ge 30{,}000 \\ & x_2 + x_4 + x_6 \ge 25{,}000 \\ & -5x_1 + 5x_3 + 15x_5 \ge 0 \\ & -15x_2 - 5x_4 + 5x_6 \ge 0 \\ & x_1, x_2, x_3, x_4, x_5, x_6 \ge 0 \end{array}$$

43. Let x_1 = percentage invested in high-tech funds
x_2 = percentage invested in global funds
x_3 = percentage invested in corporate bonds
x_4 = percentage invested in municipal bonds
x_5 = percentage invested in CDs

$$\begin{array}{ll} \text{Maximize} & P = 0.11x_1 + 0.1x_2 + 0.09x_3 + 0.08x_4 + 0.05x_5 \\ \text{subject to} & x_1 + x_2 + x_3 + x_4 + x_5 = 1 \\ & 2.7x_1 + 1.8x_2 + 1.2x_3 + 0.5x_4 \le 1.8 \\ & x_5 \ge 0.2 \\ & x_1, x_2, x_3, x_4, x_5 \ge 0 \end{array}$$

45. Let x_1 = no. of ounces of food L
x_2 = no. of ounces of food M
x_3 = no. of ounces of food N

$$\begin{array}{ll} \text{Minimize} & C = 0.4x_1 + 0.6x_2 + 0.8x_3 \\ \text{subject to} & 30x_1 + 10x_2 + 30x_3 \ge 400 \\ & 10x_1 + 10x_2 + 10x_3 \ge 200 \\ & 10x_1 + 30x_2 + 20x_3 \ge 300 \\ & 8x_1 + 4x_2 + 6x_3 \le 150 \\ & 60x_1 + 40x_2 + 50x_3 \le 900 \\ & x_1, x_2, x_3 \ge 0 \end{array}$$

47. Let x_1 = no. of students from town A enrolled in school I
x_2 = no. of students from town A enrolled in school II
x_3 = no. of students from town B enrolled in school I
x_4 = no. of students from town B enrolled in school II
x_5 = no. of students from town C enrolled in school I
x_6 = no. of students from town C enrolled in school II

$$
\begin{array}{llll}
\text{Minimize} & C = 4x_1 + 8x_2 + 6x_3 + 4x_4 + 3x_5 + 9x_6 \\
\text{subject to} & x_1 + x_2 & = 500 \\
& x_3 + x_4 & = 1{,}200 \\
& x_5 + x_6 & = 1{,}800 \\
& x_1 + x_3 + x_5 & \le 2{,}000 \\
& x_2 + x_4 + x_6 & \le 2{,}000 \\
& x_1 + x_3 + x_5 & \ge 1{,}400 \\
& x_2 + x_4 + x_6 & \ge 1{,}400 \\
& x_1 & \le 300 \\
& x_2 & \le 300 \\
& x_3 & \le 720 \\
& x_4 & \le 720 \\
& x_5 & \le 1{,}080 \\
& x_6 & \le 1{,}080 \\
& x_1, x_2, x_3, x_4, x_5, x_6 \ge 0
\end{array}
$$

Chapter 6 Review Exercises

1. x_2, s_2 *(6.1)* **2.** x_2, s_1 *(6.1)* **3.** $x_1 = 14, x_2 = 0, s_1 = 4, s_2 = 0$ *(6.1)* **4.** $x_1 = 6, x_2 = 4, s_1 = 0, s_2 = 0$ *(6.1)* **5.** (A), (B), (E), and (F) *(6.1)* **6.** The points above the line $2x_1 + 5x_2 = 32$ *(6.1)* **7.** Max $P = 700$ at $x_1 = 14$ and $x_2 = 0$ *(6.1)* **8.** 84 *(6.1)*

9.
$$
\begin{aligned}
2x_1 + x_2 + s_1 \quad &= 8 \\
x_1 + 2x_2 \quad + s_2 &= 10
\end{aligned}
$$
(6.1)

10. 2 basic and 2 nonbasic variables *(6.1)*

11.

x_1	x_2	s_1	s_2	Feasible?
0	0	8	10	Yes
0	8	0	−6	No
0	5	3	0	Yes
4	0	0	6	Yes
10	0	−12	0	No
2	4	0	0	Yes

(6.1)

12. Enter ↓ (column x_1)

$$
\begin{array}{r}
\\ \text{Exit} \rightarrow s_1 \\ s_2 \\ P
\end{array}
\begin{array}{ccccc|c}
x_1 & x_2 & s_1 & s_2 & P & \\
\textcircled{2} & 1 & 1 & 0 & 0 & 8 \\
1 & 2 & 0 & 1 & 0 & 10 \\
\hline
-6 & -2 & 0 & 0 & 1 & 0
\end{array}
$$
(6.2)

13. Max $P = 24$ at $x_1 = 4$ and $x_2 = 0$ *(6.2)*

14. Basic variables: x_2, s_2, s_3, P; nonbasic variables: x_1, x_3, s_1

Enter ↓ (column x_1)

$$
\begin{array}{r}
\\ x_2 \\ s_2 \\ \text{Exit} \rightarrow s_3 \\ P
\end{array}
\begin{array}{ccccccc|c}
x_1 & x_2 & x_3 & s_1 & s_2 & s_3 & P & \\
2 & 1 & 3 & -1 & 0 & 0 & 0 & 20 \\
3 & 0 & 4 & 1 & 1 & 0 & 0 & 30 \\
\textcircled{2} & 0 & 5 & 2 & 0 & 1 & 0 & 10 \\
\hline
-8 & 0 & -5 & 3 & 0 & 0 & 1 & 50
\end{array}
$$

$$
\sim
\begin{array}{r}
\\ x_2 \\ s_2 \\ x_1 \\ P
\end{array}
\begin{array}{ccccccc|c}
x_1 & x_2 & x_3 & s_1 & s_2 & s_3 & P & \\
0 & 1 & -2 & -3 & 0 & -1 & 0 & 10 \\
0 & 0 & -\frac{7}{2} & -2 & 1 & -\frac{3}{2} & 0 & 15 \\
1 & 0 & \frac{5}{2} & 1 & 0 & \frac{1}{2} & 0 & 5 \\
\hline
0 & 0 & 15 & 11 & 0 & 4 & 1 & 90
\end{array}
$$
(6.2)

15. (A) $x_1 = 0, x_2 = 2, s_1 = 0, s_2 = 5, P = 12$; additional pivoting required (B) $x_1 = 0, x_2 = 0, s_1 = 0, s_2 = 7, P = 22$; no optimal solution exists (C) $x_1 = 6, x_2 = 0, s_1 = 15, s_2 = 0, P = 10$; optimal solution *(6.2)*

16.
$$
\begin{array}{ll}
\text{Maximize} & P = 15y_1 + 20y_2 \\
\text{subject to} & y_1 + 2y_2 \le 5 \\
& 3y_1 + y_2 \le 2 \\
& y_1, y_2 \ge 0
\end{array}
$$
(6.3)

17.
$$
\begin{aligned}
y_1 + 2y_2 + x_1 \quad &= 5 \\
3y_1 + y_2 \quad + x_2 &= 2 \\
-15y_1 - 20y_2 \quad + P &= 0
\end{aligned}
$$
(6.3)

18.
$$
\begin{array}{cccccc|c}
y_1 & y_2 & x_1 & x_2 & P & \\
1 & 2 & 1 & 0 & 0 & 5 \\
3 & 1 & 0 & 1 & 0 & 2 \\
\hline
-15 & -20 & 0 & 0 & 1 & 0
\end{array}
$$
(6.3)

19. Max $P = 40$ at $y_1 = 0$ and $y_2 = 2$ *(6.2)* **20.** Min $C = 40$ at $x_1 = 0$ and $x_2 = 20$ *(6.3)* **21.** Max $P = 26$ at $x_1 = 2$ and $x_2 = 5$ *(6.2)*

22.
$$
\begin{array}{ll}
\text{Maximize} & P = 10y_1 + 15y_2 + 3y_3 \\
\text{subject to} & y_1 + y_2 \le 3 \\
& y_1 + 2y_2 + y_3 \le 8 \\
& y_1, y_2, y_3 \ge 0
\end{array}
$$
(6.3)

23. Min $C = 51$ at $x_1 = 9$ and $x_2 = 3$ *(6.3)* **24.** No optimal solution exists. *(6.2)* **25.** Max $P = 23$ at $x_1 = 4, x_2 = 1$, and $x_3 = 0$ *(6.2)* **26.** Max $P = 84$ at $x_1 = 0, x_2 = 12$, and $x_3 = 0$ *(6.1)* **27.** Two pivot operations *(6.2)*

28. (A) Modified problem:

$$
\begin{array}{ll}
\text{Maximize} & P = x_1 + 3x_2 - Ma_1 \\
\text{subject to} & x_1 + x_2 - s_1 + a_1 = 6 \\
& x_1 + 2x_2 + s_2 = 8 \\
& x_1, x_2, s_1, s_2, a_1 \ge 0
\end{array}
$$

(B) Preliminary simplex tableau:

$$
\begin{array}{cccccc|c}
x_1 & x_2 & s_1 & a_1 & s_2 & P & \\
1 & 1 & -1 & 1 & 0 & 0 & 6 \\
1 & 2 & 0 & 0 & 1 & 0 & 8 \\
\hline
-1 & -3 & 0 & M & 0 & 1 & 0
\end{array}
$$

Initial simplex tableau:

$$
\begin{array}{cccccc|c}
x_1 & x_2 & s_1 & a_1 & s_2 & P & \\
1 & 1 & -1 & 1 & 0 & 0 & 6 \\
1 & 2 & 0 & 0 & 1 & 0 & 8 \\
\hline
-M-1 & -M-3 & M & 0 & 0 & 1 & -6M
\end{array}
$$

(C) $x_1 = 4, x_2 = 2, s_1 = 0, a_1 = 0, s_2 = 0, P = 10$ (D) Since $a_1 = 0$, the optimal solution to the original problem is Max $P = 10$ at $x_1 = 4$ and $x_2 = 2$ *(6.4)*

29. (A) Modified problem:

$$
\begin{array}{ll}
\text{Maximize} & P = x_1 + x_2 - Ma_1 \\
\text{subject to} & x_1 + x_2 - s_1 + a_1 = 5 \\
& x_1 + 2x_2 + s_2 = 4 \\
& x_1, x_2, s_1, s_2, a_1 \ge 0
\end{array}
$$

(B) Preliminary simplex tableau:

$$
\begin{array}{cccccc|c}
x_1 & x_2 & s_1 & a_1 & s_2 & P & \\
1 & 1 & -1 & 1 & 0 & 0 & 5 \\
1 & 2 & 0 & 0 & 1 & 0 & 4 \\
\hline
-1 & -1 & 0 & M & 0 & 1 & 0
\end{array}
$$

Initial simplex tableau:

$$
\begin{array}{cccccc|c}
x_1 & x_2 & s_1 & a_1 & s_2 & P & \\
1 & 1 & -1 & 1 & 0 & 0 & 5 \\
1 & 2 & 0 & 0 & 1 & 0 & 4 \\
\hline
-M-1 & -M-1 & M & 0 & 0 & 1 & -5M
\end{array}
$$

(C) $x_1 = 4, x_2 = 0, s_1 = 0, s_2 = 0, a_1 = 1, P = -M + 4$
(D) Since $a_1 \ne 0$, the original problem has no optimal solution. *(6.4)*

30.
$$
\begin{array}{ll}
\text{Maximize} & P = 2x_1 + 3x_2 + x_3 - Ma_1 - Ma_2 \\
\text{subject to} & x_1 - 3x_2 + x_3 + s_1 = 7 \\
& x_1 + x_2 - 2x_3 - s_2 + a_1 = 2 \\
& 3x_1 + 2x_2 - x_3 + a_2 = 4 \\
& x_1, x_2, s_1, s_2, a_1, a_2 \ge 0
\end{array}
$$
(6.4)

31. The basic simplex method with slack variables solves standard maximization problems involving $\le$ constraints with nonnegative constants on the right side. *(6.2)* **32.** The dual problem method solves minimization problems with positive coefficients in the objective function. *(6.3)* **33.** The big M method solves any linear programming problem. *(6.4)*

34. Max $P = 36$ at $x_1 = 6$, $x_2 = 8$ *(6.2)*

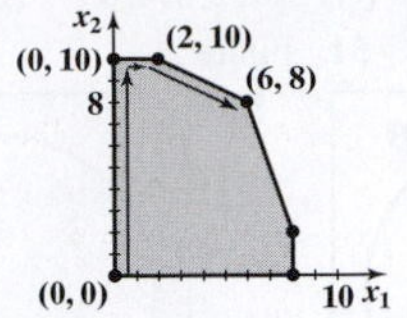

35. Min $C = 15$ at $x_1 = 3$ and $x_2 = 3$ *(6.3)* **36.** Min $C = 15$ at $x_1 = 3$ and $x_2 = 3$ *(6.4)* **37.** Min $C = 9{,}960$ at $x_1 = 0$, $x_2 = 240$, $x_3 = 400$, and $x_4 = 60$ *(6.3)*

38. (A) Let x_1 = amount invested in oil stock
x_2 = amount invested in steel stock
x_3 = amount invested in government bonds

$$\begin{aligned} \text{Maximize} \quad & P = 0.12x_1 + 0.09x_2 + 0.05x_3 \\ \text{subject to} \quad & x_1 + x_2 + x_3 \le 150{,}000 \\ & x_1 \le 50{,}000 \\ & x_1 + x_2 - x_3 \le 25{,}000 \\ & x_1, x_2, x_3 \ge 0 \end{aligned}$$

Max return is \$12,500 when \$50,000 is invested in oil stock, \$37,500 is invested in steel stock, and \$62,500 in government bonds. (B) Max return is \$13,625 when \$87,500 is invested in steel stock and \$62,500 in government bonds. *(6.2)*

39. (A) Let x_1 = no. of regular chairs
x_2 = no. of rocking chairs
x_3 = no. of chaise lounges

$$\begin{aligned} \text{Maximize} \quad & P = 17x_1 + 24x_2 + 31x_3 \\ \text{subject to} \quad & x_1 + 2x_2 + 3x_3 \le 2{,}500 \\ & 2x_1 + 2x_2 + 4x_3 \le 3{,}000 \\ & 3x_1 + 3x_2 + 2x_3 \le 3{,}500 \\ & x_1, x_2, x_3 \ge 0 \end{aligned}$$

Max P = \$30,000 when 250 regular chairs, 750 rocking chairs, and 250 chaise lounges are produced. (B) Maximum profit increases to \$32,750 when 1,000 regular chairs, 0 rocking chairs, and 250 chaise lounges are produced. (C) Maximum profit decreases to \$28,750 when 125 regular chairs, 625 rocking chairs, and 375 chaise lounges are produced. *(6.2)*

40. Let x_1 = no. of motors shipped from factory *A* to plant *X*
x_2 = no. of motors shipped from factory *A* to plant *Y*
x_3 = no. of motors shipped from factory *B* to plant *X*
x_4 = no. of motors shipped from factory *B* to plant *Y*

$$\begin{aligned} \text{Minimize} \quad & C = 5x_1 + 8x_2 + 9x_3 + 7x_4 \\ \text{subject to} \quad & x_1 + x_2 \le 1{,}500 \\ & +x_3 + x_4 \le 1{,}000 \\ & x_1 + x_3 \ge 900 \\ & x_2 + x_4 \ge 1{,}200 \\ & x_1, x_2, x_3, x_4 \ge 0 \end{aligned}$$

Min C = \$13,100 when 900 motors are shiped from factory *A* to plant *X*, 200 motors are shipped from factory *A* to plant *Y*, and 1,000 motors are shipped from factory *B* to plant *Y* *(6.3)*

41. Let x_1 = no. of lbs of long-grain rice used in brand *A*
x_2 = no. of lbs of long-grain rice used in brand *B*
x_3 = no. of lbs of wild rice used in brand *A*
x_4 = no. of lbs of wild rice used in brand *B*

$$\begin{aligned} \text{Maximize} \quad & P = 0.8x_1 + 0.5x_2 - 1.9x_3 - 2.2x_4 \\ \text{subject to} \quad & 0.1x_1 - 0.9x_3 \le 0 \\ & 0.05x_2 - 0.95x_4 \le 0 \\ & x_1 + x_2 \le 8{,}000 \\ & x_3 + x_4 \le 500 \\ & x_1, x_2, x_3, x_4 \ge 0 \end{aligned}$$

Max profit is \$3,350 when 1,350 lb long-grain rice and 150 lb wild rice are used to produce 1,500 lb of brand *A*, and 6,650 lb long-grain rice and 350 lb wild rice are used to produce 7,000 lb of brand *B*. *(6.2)*

Chapter 7

Exercises 7.1

1. 1, 2, 4, 5, 10, 20 **3.** 11, 22, 33, 44, 55 **5.** 23, 29 **9.** 91 is not odd; false **11.** 91 is prime and 91 is odd; false **13.** If 91 is odd, then 91 is prime; false **15.** If the moon is a cube, then rain is wet; true **17.** The moon is a cube or rain is wet; true **19.** If rain is not wet, then the moon is not a cube; true **21.** Disjunction; true **23.** Conditional; false **25.** Negation; false **27.** Conjunction; true **29.** Converse: If triangle *ABC* is equiangular, then triangle *ABC* is equilateral. Contrapositive: If triangle *ABC* is not equiangular, then triangle *ABC* is not equilateral. **31.** Converse: If $f(x)$ is an increasing function, then $f(x)$ is a linear function with positive slope. Contrapositive: If $f(x)$ is not an increasing function, then $f(x)$ is not a linear function with positive slope. **33.** Converse: If *n* is an integer that is a multiple of 2 and a multiple of 4, then *n* is an integer that is a multiple of 8. Contrapositive: If *n* is an integer that is not a multiple of 2 or not a multiple of 4, then *n* is an integer that is not a multiple of 8.

35. Contingency

p	q	$\neg p \wedge q$
T	T	F
T	F	F
F	T	T
F	F	F

37. Contingency

p	q	$\neg p \rightarrow q$
T	T	T
T	F	T
F	T	T
F	F	F

39. Contingency

p	q	$q \wedge (p \vee q)$
T	T	T
T	F	F
F	T	T
F	F	F

41. Tautology

p	q	$p \vee (p \rightarrow q)$
T	T	T
T	F	T
F	T	T
F	F	T

43. Contingency

p	q	$p \rightarrow (p \wedge q)$
T	T	T
T	F	F
F	T	T
F	F	T

45. Contingency

p	q	$(p \rightarrow q) \rightarrow \neg p$
T	T	F
T	F	T
F	T	T
F	F	T

47. Contingency

p	q	$\neg p \rightarrow (p \vee q)$
T	T	T
T	F	T
F	T	T
F	F	F

49. Contingency

p	q	$q \to (\neg p \wedge q)$
T	T	F
T	F	T
F	T	T
F	F	T

51. Contradiction

p	q	$(\neg p \wedge q) \wedge (q \to p)$
T	T	F
T	F	F
F	T	F
F	F	F

53.

p	q	$p \vee q$
T	T	T
T	F	T
F	T	T
F	F	F

55.

p	q	$\neg p \wedge q$	$p \vee q$
T	T	F	T
T	F	F	T
F	T	T	T
F	F	F	F

57.

p	q	$\neg p \to (q \wedge \neg q)$
T	T	T
T	F	T
F	T	F
F	F	F

59.

p	q	$\neg p \to (p \vee q)$	$p \vee q$
T	T	T	T
T	F	T	T
F	T	T	T
F	F	F	F

61.

p	q	$q \wedge (p \vee q)$	$q \vee (p \wedge q)$
T	T	T	T
T	F	F	F
F	T	T	T
F	F	F	F

63.

p	q	$p \vee (p \to q)$	$p \to (p \vee q)$
T	T	T	T
T	F	T	T
F	T	T	T
F	F	T	T

65. $p \to \neg q \equiv \neg p \vee \neg q$ *By (4)*
$\equiv \neg(p \wedge q)$ *By (6)*
67. $\neg(p \to q) \equiv \neg(\neg p \vee q)$ *By (4)*
$\equiv \neg(\neg p) \wedge \neg q$ *By (6)*
$\equiv p \wedge \neg q$ *By (1)*
69. $q \wedge \neg p$ **71.** Contingency **73.** No

Exercises 7.2

1. No **3.** No **5.** No **7.** T **9.** T **11.** T **13.** F **15.** $\{2,3\}$ **17.** $\{1,2,3,4\}$ **19.** $\{1,4,7,10,13\}$ **21.** $\varnothing$ **23.** $\{5,-5\}$ **25.** $\{-3\}$ **27.** $\{1,3,5,7,9\}$ **29.** $A' = \{1,5\}$ **31.** 100 **33.** 60 **35.** 81 **37.** 37 **39.** 18 **41.** 19 **43.** 100 **45.** (A) $\{1,2,3,4,6\}$ (B) $\{1,2,3,4,6\}$ **47.** $\{1,2,3,4,6\}$ **49.** Infinite **51.** Finite

53.
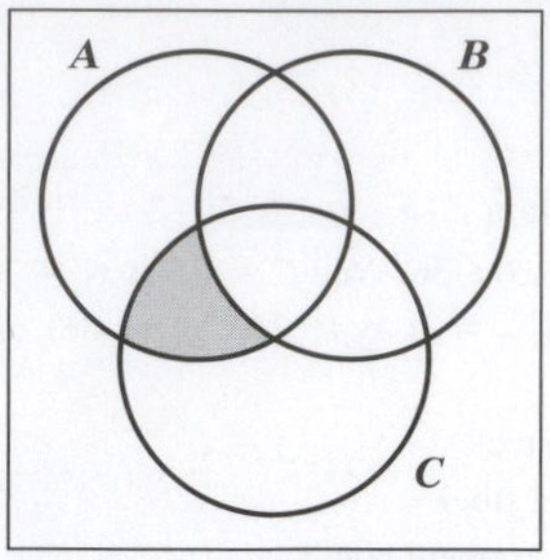

55.
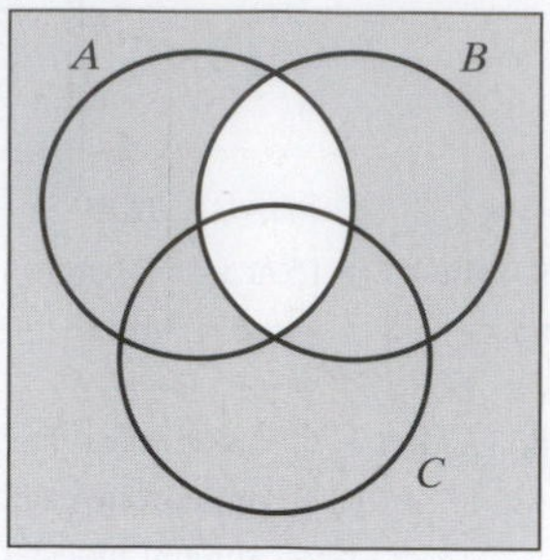

57.
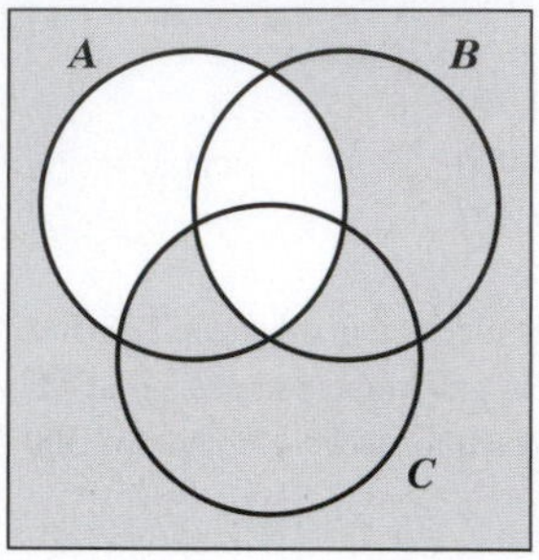

59. Disjoint **61.** Not disjoint **63.** True **65.** False **67.** False **69.** False **71.** False **73.** (A) 2 (B) 4 (C) 8 (D) 16 **75.** 85 **77.** 33 **79.** 14 **81.** 66 **83.** 54 **85.** 106 **87.** 0 **89.** 6 **91.** A+, AB+ **93.** A−, A+, B+, AB−, AB+, O+ **95.** O+, O− **97.** B−, B+

Exercises 7.3

1. 13 **3.** 13 **5.** 9 **7.** (A) 4 ways: (B) $2 \cdot 2 = 4$ ways

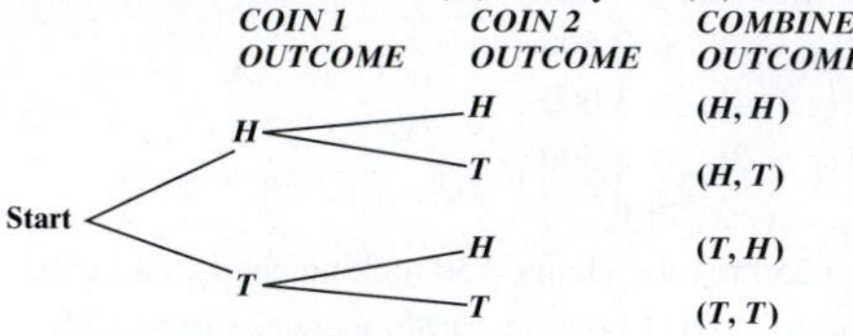

9. (A) 12 combined outcomes:

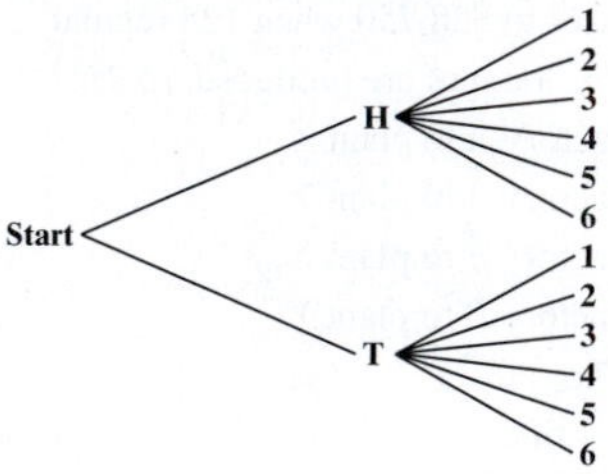

(B) $2 \cdot 6 = 12$ combined outcomes

11. (A) 9 (B) 18 **13.** 60; 125; 80 **15.** (A) 8 (B) 144
17. $n(A \cap B') = 50, n(A \cap B) = 50, n(A' \cap B) = 40, n(A' \cap B') = 60$
19. $n(A \cap B') = 5, n(A \cap B) = 30, n(A' \cap B) = 55, n(A' \cap B') = 10$
21. $n(A \cap B') = 160, n(A \cap B) = 30, n(A' \cap B) = 50, n(A' \cap B') = 60$
23. $n(A \cap B') = 30, n(A \cap B) = 30, n(A' \cap B) = 10, n(A' \cap B') = 10$

25.

	A	A'	Totals
B	30	60	90
B'	40	70	110
Totals	70	130	200

27.

	A	A'	Totals
B	20	35	55
B'	25	20	45
Totals	45	55	100

29.

	A	A'	Totals
B	58	8	66
B'	17	7	24
Totals	75	15	90

31.

	A	A'	Totals
B	0	145	145
B'	110	45	155
Totals	110	190	300

33. (A) True (B) False **35.** $5 \cdot 3 \cdot 4 \cdot 2 = 120$ **37.** $52^5 = 380{,}204{,}032$ **39.** $10 \cdot 9 \cdot 8 \cdot 7 \cdot 6 = 30{,}240$; $10 \cdot 10 \cdot 10 \cdot 10 \cdot 10 = 100{,}000$; $10 \cdot 9 \cdot 9 \cdot 9 \cdot 9 = 65{,}610$ **41.** $26 \cdot 26 \cdot 26 \cdot 10 \cdot 10 \cdot 10 = 17{,}576{,}000$; $26 \cdot 25 \cdot 24 \cdot 10 \cdot 9 \cdot 8 = 11{,}232{,}000$ **43.** No, the same 8 combined choices are available either way. **45.** z **47.** x and z **49.** 14 **51.** 14 **53.** (A) 6 combined outcomes: (B) $3 \cdot 2 = 6$

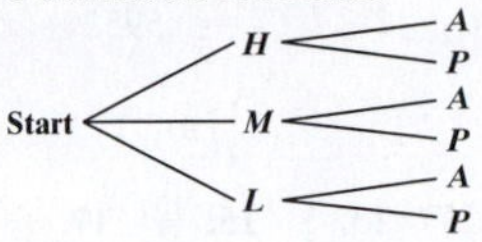

55. 12 **57.** (A) 1,010 (B) 190 (C) 270 **59.** 1,570 **61.** (A) 102,000 (B) 689,000 (C) 1,470,000 (D) 1,372,000 **63.** (A) 12 classifications: (B) $2 \cdot 2 \cdot 3 = 12$ **65.** 2,905

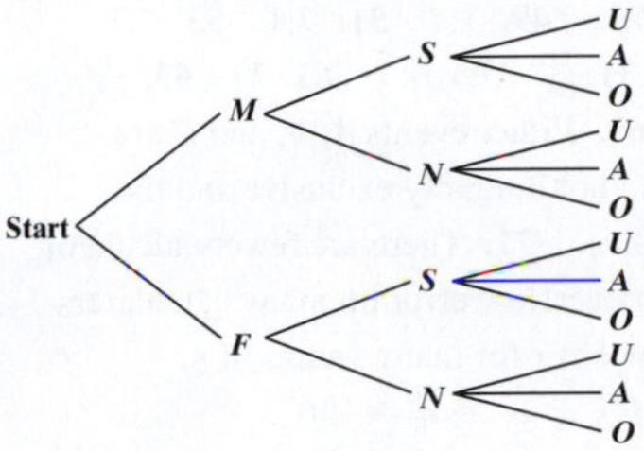

Exercises 7.4

1. 220 **3.** 252 **5.** 9,900 **7.** 40,320 **9.** 5,040 **11.** 720 **13.** 990 **15.** 70 **17.** 249,500 **19.** 1,287 **21.** 13,366,080 **23.** 0.1114 **25.** 0.2215 **27.** $n(n-1)$ **29.** $n(n+1)/2$ **31.** Permutation **33.** Combination **35.** Neither **37.** ${}_{10}P_3 = 10 \cdot 9 \cdot 8 = 720$ **39.** ${}_7C_3 = 35$; ${}_7P_3 = 210$ **41.** ${}_{26}C_6 = 230{,}230$ **43.** ${}_{12}C_5 = 792$ **45.** ${}_{48}C_3 = 17{,}296$ **47.** $13^4 = 28{,}561$ **49.** 3,744 **51.** ${}_8C_3 \cdot {}_{10}C_4 \cdot {}_7C_2 = 246{,}960$ **53.** The numbers are the same read up or down, since ${}_nC_r = {}_nC_{n-r}$. **55.** True **57.** False **59.** True **61.** (A) ${}_8C_2 = 28$ (B) ${}_8C_3 = 56$ (C) ${}_8C_4 = 70$ **63.** ${}_6P_4 = 360$ **65.** (A) ${}_8P_5 = 6{,}720$ (B) ${}_8C_5 = 56$ (C) $2 \cdot {}_6C_4 = 30$ **67.** $n(K \cap H') = 120$, $n(K \cap H) = 72$, $n(K' \cap H) = 435$, $n(K' \cap H') = 699$ **69.** ${}_{24}C_{12} = 2{,}704{,}156$ **71.** (A) ${}_{24}C_3 = 2{,}024$ (B) ${}_{19}C_3 = 969$ **73.** (A) ${}_{30}C_{10} = 30{,}045{,}015$ (B) ${}_8C_2 \cdot {}_{12}C_5 \cdot {}_{10}C_3 = 2{,}661{,}120$ **75.** (A) ${}_6C_3 \cdot {}_5C_2 = 200$ (B) ${}_6C_4 \cdot {}_5C_1 = 75$ (C) ${}_6C_5 = 6$ (D) ${}_{11}C_5 = 462$ (E) ${}_6C_4 \cdot {}_5C_1 + {}_6C_5 = 81$ **77.** 336; 512 **79.** ${}_4P_2 = 12$

Chapter 7 Review Exercises

1. 3^4 is not less than 4^3; true *(7.1)* **2.** 2^3 is less than 3^2 or 3^4 is less than 4^3; true *(7.1)* **3.** 2^3 is less than 3^2 and 3^4 is less than 4^3; false *(7.1)* **4.** If 2^3 is less than 3^2, then 3^4 is less than 4^3; false *(7.1)* **5.** If 3^4 is less than 4^3, then 2^3 is less than 3^2; true *(7.1)* **6.** If 3^4 is not less than 4^3, then 2^3 is not less than 3^2; false *(7.1)* **7.** T *(7.2)* **8.** F *(7.2)* **9.** T *(7.2)* **10.** F *(7.2)* **11.** Conditional; true *(7.1)* **12.** Disjunction; false *(7.1)* **13.** Conjunction; false *(7.1)* **14.** Negation; true *(7.1)* **15.** Converse: If the square matrix A does not have an inverse, then the square matrix A has a row of zeros. Contrapositive: If the square matrix A has an inverse, then the square matrix A does not have a row of zeros. *(7.1)* **16.** Converse: If the square matrix A has an inverse, then the square matrix A is an identity matrix. Contrapositive: If the square matrix A does not have an inverse, then the square matrix A is not an identity matrix. *(7.1)* **17.** $\{1, 2, 3, 4, 5\}$ *(7.2)* **18.** $\{2, 3, 4\}$ *(7.2)*

19. $\varnothing$ *(7.2)* **20.** (A) 12 combined outcomes: (B) $6 \cdot 2 = 12$ *(7.3)*

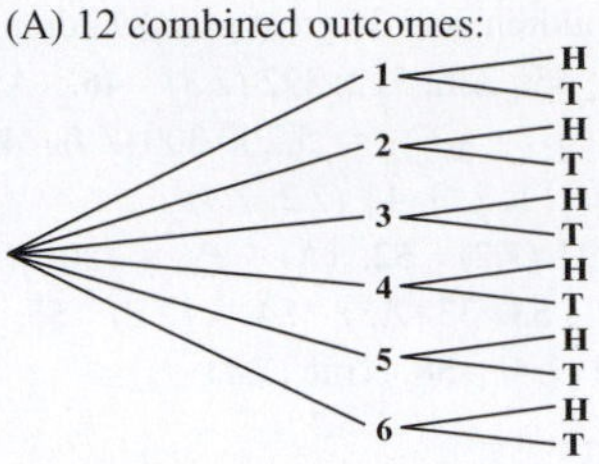

21. (A) 65 (B) 75 (C) 150 (D) 85 *(7.3)* **22.** (A) 35 (B) 105 (C) 115 (D) 45 *(7.3)* **23.** 24 *(7.4)* **24.** 360,360 *(7.4)* **25.** 3,003 *(7.4)* **26.** 56 *(7.4)* **27.** 6720 *(7.4)* **28.** 9,295 *(7.4)* **29.** $6 \cdot 5 \cdot 4 \cdot 3 \cdot 2 \cdot 1 = 720$ *(7.3)* **30.** ${}_6P_6 = 6! = 720$ *(7.4)*

31. Contingency *(7.1)*

p	q	$(p \to q) \wedge (q \to p)$
T	T	T
T	F	F
F	T	F
F	F	T

32. Contingency *(7.1)*

p	q	$p \vee (q \to p)$
T	T	T
T	F	T
F	T	F
F	F	T

33. Contradiction *(7.1)*

p	q	$(p \vee \neg p) \to (q \wedge \neg q)$
T	T	F
T	F	F
F	T	F
F	F	F

34. Contingency *(7.1)*

p	q	$\neg q \wedge (p \to q)$
T	T	F
T	F	F
F	T	F
F	F	T

35. Tautology *(7.1)*

p	q	$\neg p \to (p \to q)$
T	T	T
T	F	T
F	T	T
F	F	T

36. Contingency *(7.1)*

p	q	$\neg(p \vee \neg q)$
T	T	F
T	F	F
F	T	T
F	F	F

37. Infinite *(7.2)* **38.** Finite *(7.2)* **39.** Infinite *(7.2)* **40.** Infinite *(7.2)* **41.** Disjoint *(7.2)* **42.** Not disjoint *(7.2)*

43. *(7.2)*

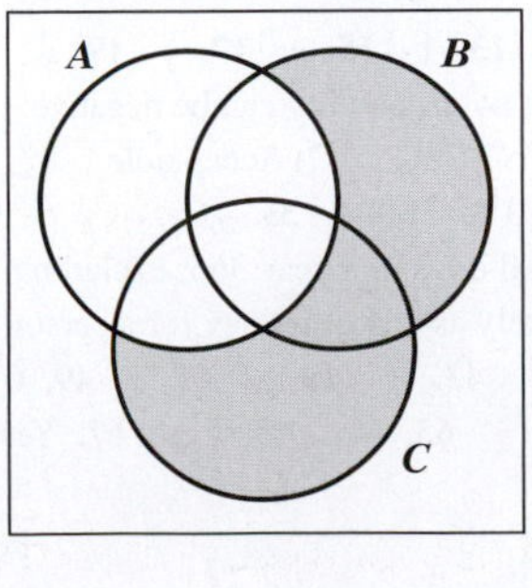

44. 5 children, 15 grandchildren, and 30 great grandchildren, for a total of 50 descendants *(7.3, 7.4)* **45.** 336; 512; 392 *(7.3)* **46.** (A) $_6P_3 = 120$ (B) $_5C_2 = 10$ *(7.4)* **47.** $_{25}C_{12} = {}_{25}C_{13} = 5{,}200{,}300$ *(7.4)* **48.** $N_1 \cdot N_2 \cdot N_3$ *(7.3)* **49.** $\{-1, 0, 1\}$ *(7.2)* **50.** $\{1, 2, 3, 4\}$ *(7.2, 7.4)* **51.** $\{1, 4, 9, 16, 25, 36, 49\}$ *(7.2)* **52.** (A) $_{10}P_3 = 720$ (B) $_6P_3 = 120$ (C) $_{10}C_3 = 120$ *(7.3, 7.4)* **53.** 33 *(7.3)* **54.** x *(7.3)* **55.** x, z, and w *(7.3)* **56.** True *(7.4)* **57.** False *(7.4)* **58.** True *(7.4)*

59. *(7.1)*

p	q	$p \wedge q$
T	T	T
T	F	F
F	T	F
F	F	F

60. *(7.1)*

p	q	$p \to q$
T	T	T
T	F	F
F	T	T
F	F	T

61. *(7.1)*

p	q	$\neg p \to (q \wedge \neg q)$
T	T	T
T	F	T
F	T	F
F	F	F

62. *(7.1)*

p	q	$p \vee q$	$\neg p \to q$
T	T	T	T
T	F	T	T
F	T	T	T
F	F	F	F

63. *(7.1)*

p	q	$p \wedge (p \to q)$
T	T	T
T	F	F
F	T	F
F	F	F

64. *(7.1)*

p	q	$\neg(p \wedge \neg q)$	$p \to q$
T	T	T	T
T	F	F	F
F	T	T	T
F	F	T	T

65. $2^5 = 32$; 6 *(7.3)* **66.** Yes, it is both if $r = 0$ or $r = 1$. *(7.4)* **67.** 120 *(7.3)* **68.** (A) 610 (B) 390 (C) 270 *(7.3)* **69.** $_{40}C_6 = 3{,}838{,}380$ *(7.4)* **70.** (A) $67! \approx 3.647 \times 10^{94}$ *(7.4)*

Chapter 8

Exercises 8.1

1. E **3.** F **5.** F **7.** $\frac{1}{3}$ **9.** $\frac{4}{15}$ **11.** 0 **13.** 1 **15.** $\frac{1}{4}$ **17.** $\frac{1}{2}$ **19.** $\frac{1}{52}$ **21.** $\frac{4}{13}$ **23.** 0 **25.** $\frac{1}{2}$ **27.** (A) Reject; no probability can be negative (B) Reject; $P(J) + P(G) + P(P) + P(S) \neq 1$ (C) Acceptable **29.** $P(J) + P(P) = .56$ **31.** $\frac{1}{8}$ **33.** $1/10^4 = .0001$ **35.** $_{26}C_5/_{52}C_5 \approx .025$ **37.** $_{12}C_5/_{52}C_5 \approx .000\,305$ **39.** $S = \{$All days in a year, 365, excluding leap year$\}$; $\frac{1}{365}$, assuming each day is as likely as any other day for a person to be born **41.** $1/_5P_5 = 1/5! = .008\,33$ **43.** $\frac{5}{36}$ **45.** $\frac{1}{6}$ **47.** $\frac{7}{9}$ **49.** 0 **51.** $\frac{1}{3}$ **53.** $\frac{2}{9}$ **55.** $\frac{2}{3}$ **57.** $\frac{1}{4}$ **59.** $\frac{1}{4}$ **61.** $\frac{3}{4}$ **63.** No **65.** Yes **67.** Yes

69. (A) Yes (B) Yes, because we would expect, on average, 20 heads in 40 flips; $P(\text{H}) = \frac{37}{40} = .925$; $P(\text{T}) = \frac{3}{40} = .075$ **71.** $\frac{1}{16}$ **73.** $\frac{3}{16}$ **75.** $\frac{3}{16}$ **77.** $\frac{1}{2}$ **79.** $_{26}C_5/_{52}C_5 \approx .0253$ **81.** $_{12}C_2 \cdot {}_{40}C_4/_{52}C_6 \approx .2963$ **83.** $_{48}C_4/_{52}C_4 \approx .7187$ **85.** $_{13}C_2 \cdot {}_{13}C_2 \cdot {}_{26}C_3/_{52}C_7 \approx .1182$ **87.** (A) $\frac{8}{50} = .16$ (B) $\frac{1}{6} \approx .167$ (C) Answer depends on results of simulation. **89.** (A) Represent the outcomes H and T by 1 and 2, respectively, and select 500 random integers from the integers 1 and 2. (B) Answer depends on results of simulation. (C) Each is $\frac{1}{2} = .5$ **91.** (A) $_{12}P_4 \approx .000\,084$ (B) $1/12^4 \approx .000\,048$ **93.** (A) $_6C_3 \cdot {}_5C_2/_{11}C_5 \approx .433$ (B) $_6C_4 \cdot {}_5C_1/_{11}C_5 \approx .162$ (C) $_6C_5/_{11}C_5 \approx .013$ (D) $(_6C_4 \cdot {}_5C_1 + {}_6C_5)/_{11}C_5 \approx .175$ **95.** (A) $1/_8P_3 \approx .0030$ (B) $1/8^3 \approx .0020$ **97.** (A) $_5C_3/_9C_3 \approx .1190$ (B) $(_5C_3 + {}_5C_2 \cdot {}_4C_1)/_9C_3 \approx .5952$

Exercises 8.2

1. $\frac{1}{3}$ **3.** $\frac{7}{24}$ **5.** $\frac{2}{7}$ **7.** .38 **9.** .88 **11.** .27 **13.** $\frac{1}{4}$ **15.** $\frac{10}{13}$ **17.** $\frac{3}{52}$ **19.** $\frac{11}{26}$ **21.** $\frac{5}{26}$ **23.** $\frac{43}{52}$ **25.** .76 **27.** .52 **29.** .6 **31.** .36 **33.** .49 **35.** $\frac{5}{18}$ **37.** $\frac{11}{36}$ **39.** (A) $\frac{3}{5}; \frac{5}{3}$ (B) $\frac{1}{3}; \frac{3}{1}$ (C) $\frac{2}{3}; \frac{3}{2}$ (D) $\frac{11}{9}; \frac{9}{11}$ **41.** (A) $\frac{3}{11}$ (B) $\frac{11}{18}$ (C) $\frac{4}{5}$ or .8 (D) .49 **43.** False **45.** False **47.** True **49.** 1:1 **51.** 7:1 **53.** 2:1 **55.** 1:2 **57.** (A) $\frac{1}{8}$ (B) \$8 **59.** (A) .31; $\frac{31}{69}$ (B) .6; $\frac{3}{2}$ **61.** $\frac{11}{26}$ **63.** $\frac{7}{13}$ **65.** .78 **67.** $\frac{16}{100} + \frac{12}{100} - \frac{4}{100} = .24$ **69.** Either events A, B, and C are mutually exclusive, or events A and B are not mutually exclusive and the other pairs of events are mutually exclusive. **71.** There are fewer calculator steps, and, in addition, 365! produces an overflow error on many calculators, while $_{365}P_n$ does not produce an overflow error for many values of n. **73.** $P(E) = 1 - \frac{12!}{(12-n)!12^n}$ **77.** (A) $\frac{11}{50} + \frac{7}{50} = \frac{18}{50} = .36$ (B) $\frac{6}{36} + \frac{5}{36} = \frac{11}{36} \approx .306$ (C) Answer depends on results of simulation. **79.** (A) $P(C \cup L) = P(C) + P(L) - P(C \cap L) = .45 + .75 - .35 = .85$ (B) $P(C' \cap L') = .15$ **81.** (A) $P(M_1 \cup A) = P(M_1) + P(A) - P(M_1 \cap A) = .2 + .3 - .05 = .45$ (B) $P[(M_2 \cap A') \cup (M_3 \cap A')] = P(M_2 \cap A') + P(M_3 \cap A') = .2 + .35 = .55$ **83.** .83 **85.** $P(A''S) = \frac{50}{1{,}000} = .05$ **87.** (A) $P(U \cup N) = .22$; $\frac{11}{39}$ (B) $P[(D \cap A) \cup (R \cap A)] = .3$; $\frac{7}{3}$

Exercises 8.3

1. One answer is:

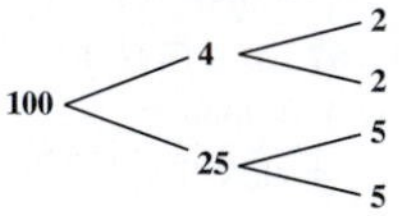

3. One answer is:

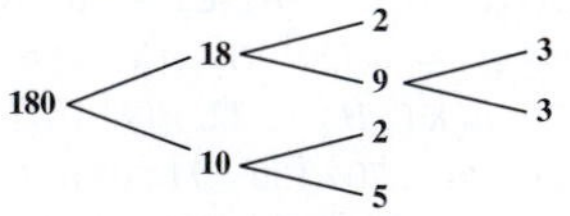

5. One answer is:

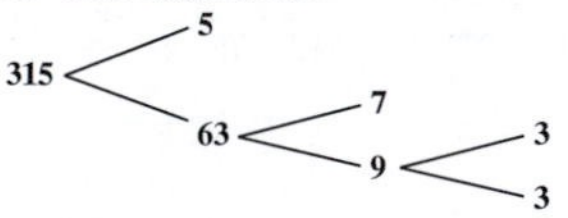

7. 1/13 **9.** 1/4 **11.** 1 **13.** 1/2 **15.** 2/9 **17.** 2/5 **19.** 6/25 **21.** 2/5 **23.** .10 **25.** .03 **27.** .3 **29.** .1 **31.** .35 **33.** 0 **35.** Dependent **37.** Independent **39.** Dependent

41. Dependent **43.** (A) $\frac{1}{2}$ (B) $2(\frac{1}{2})^8 \approx .00781$ **45.** (A) $\frac{1}{4}$ (B) Dependent **47.** (A) .24 (B) .34 **49.** (A) Independent and not mutually exclusive (B) Dependent and mutually exclusive. **51.** $(\frac{1}{2})(\frac{1}{2}) = \frac{1}{4}; \frac{1}{2} + \frac{1}{2} - \frac{1}{4} = \frac{3}{4}$ **53.** (A) $(\frac{1}{4})(\frac{13}{51}) \approx .0637$ (B) $(\frac{1}{4})(\frac{1}{4}) = .0625$ **55.** (A) $\frac{3}{13}$ (B) Independent **57.** (A) Dependent (B) Independent

59.

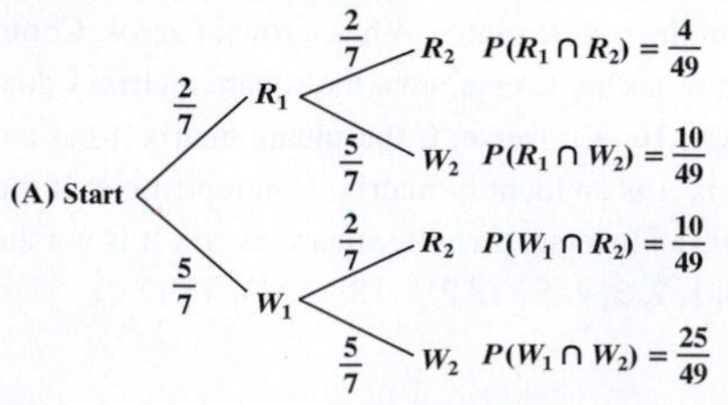

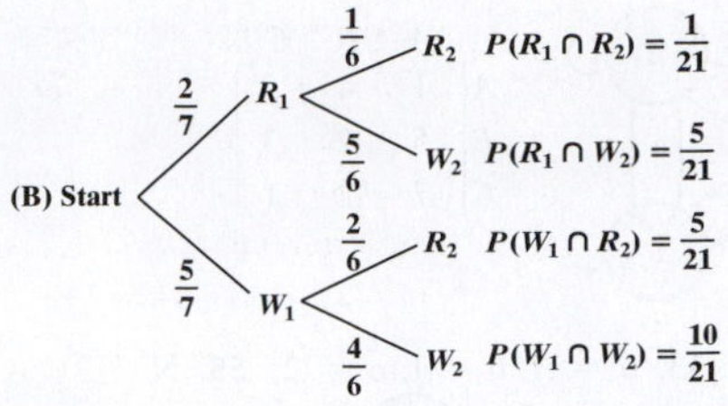

61. (A) $\frac{24}{49}$ (B) $\frac{11}{21}$ **63.** False **65.** True **67.** False **69.** True **71.** $\frac{5}{18}$
73. (A) .167 (B) .25 (C) .25

81. (A)

	H	S	Totals
Y	.400	.200	.600
N	.150	.250	.400
Totals	.550	.450	1.000

(B) $P(Y) = .6$; $P(Y|H) = \frac{.400}{.550} \approx .727$ (C) $P(S) = .450$; $P(S|Y) = \frac{.200}{.600} \approx .333$
83. (A) .167 (B) .25 (C) .25

85. (A)

	C	C'	Totals
R	.06	.44	.50
R'	.02	.48	.50
Totals	.08	.92	1.00

(B) Dependent (C) $P(C|R) = .12$ and $P(C) = .08$; since $P(C|R) > P(C)$, the red dye should be banned.

87. (A) $P(A|F) = \frac{.130}{.520} = .250$; $P(A|F') = \frac{.120}{.480} = .250$ (B) $P(A) = .250$ (C) No; no **89.** .33

Exercises 8.4

1. 2/5 **3.** 3/2 **5.** 9/10 **7.** $(.6)(.7) = .42$ **9.** $(.6)(.7)+(.4)(.2) = .50$ **11.** .84 **13.** .417 **15.** .375 **17.** .10 **19.** .15 **21.** .75 **23.** $\frac{1}{3}$ **25.** $\frac{1}{3}$

27. $\frac{1}{5}$ **29.**

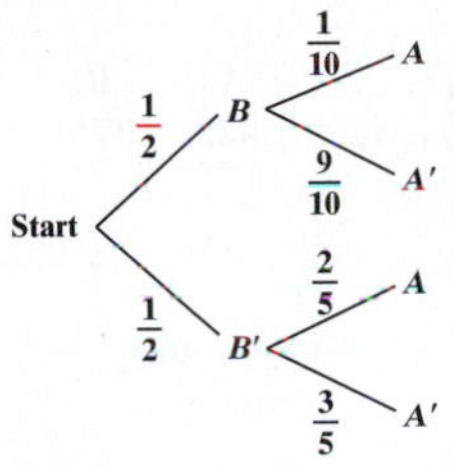

31. .25 **33.** .333 **35.** .50 **37.** .745 **41.** (A) True (B) True **43.** .235 **45.** $\frac{101}{204} \approx .4951$ **47.** $\frac{50}{101} \approx .4950$ **49.** $\frac{51}{101} \approx .5050$ **53.** .913; .226 **55.** .091; .545; .364 **57.** .667; .000 412 **59.** .231; .036 **61.** .941; .0588

Exercises 8.5

1. 71 **3.** 75 **5.** 142 **7.** $E(X) = -0.1$ **9.** \$43.75 **11.** \$0.148 **13.** \$25

15. Probability distribution:

x_i	0	1	2
p_i	$\frac{1}{4}$	$\frac{1}{2}$	$\frac{1}{4}$

$E(X) = 1$

17. Payoff table:

x_i	\$1	–\$1
p_i	$\frac{1}{2}$	$\frac{1}{2}$

$E(X) = 0$; game is fair

19. Payoff table:

x_i	–\$3	–\$2	–\$1	\$0	\$1	\$2
p_i	$\frac{1}{6}$	$\frac{1}{6}$	$\frac{1}{6}$	$\frac{1}{6}$	$\frac{1}{6}$	$\frac{1}{6}$

$E(X) = -50¢$; game is not fair

21. –\$0.50 **23.** –\$0.035; \$0.035 **25.** \$40. Let x = amount you should lose if a 6 turns up. Set up a payoff table; then set the expected value of the game equal to zero and solve for x. **27.** Win \$1 **29.** –\$0.154 **31.** \$2.75 **33.** A_2; \$210

35. Payoff table:

x_i	\$35	–\$1
p_i	$\frac{1}{38}$	$\frac{37}{38}$

$E(X) = -5.26¢$

37. .002

39. Payoff table:

x_i	\$499	\$99	\$19	\$4	–\$1
p_i	.0002	.0006	.001	.004	.9942

$E(X) = -80¢$

41. (A)

x_i	0	1	2
p_i	$\frac{7}{15}$	$\frac{7}{15}$	$\frac{1}{15}$

(B) .60

43. (A)

x_i	–\$5	\$195	\$395	\$595
p_i	.985	.0149	.000 059 9	.000 000 06

(B) $E(X) \approx -\$2$

45. (A) –\$56 (B) The value per game is $\frac{-\$56}{200} = -\$0.28$, compared with an expected value of $-\$0.0526$. (C) The simulated gain or loss depends on the results of the simulation; the expected loss is \$26.32. **47.** \$36.27

49. Payoff table:

x_i	\$4,850	–\$150
p_i	.01	.99

$E(X) = -\$100$

51. Site A, with $E(X) = \$3.6$ million **53.** 1.54 **55.** –\$1.338

Chapter 8 Review Exercises

1. ${}_{13}C_5/{}_{52}C_5 \approx .0005$ *(8.1)* **2.** $1/{}_{15}P_2 \approx .0048$ *(8.1)* **3.** $1/{}_{10}P_3 \approx .0014$; $1/{}_{10}C_3 \approx .0083$ *(8.1)* **4.** .05 *(8.1)*

5. Payoff table:

x_i	–\$2	–\$1	\$0	\$1	\$2
p_i	$\frac{1}{5}$	$\frac{1}{5}$	$\frac{1}{5}$	$\frac{1}{5}$	$\frac{1}{5}$

$E(X) = 0$; game is fair *(8.5)*

6. (A) .7 (B) .6 *(8.2)* **7.** $P(R \cup G) = .8$; odds for $R \cup G$ are 8 to 2 *(8.2)* **8.** $\frac{5}{11} \approx .455$ *(8.2)* **9.** .27 *(8.3)* **10.** .20 *(8.3)* **11.** .02 *(8.3)* **12.** .03 *(8.3)* **13.** .15 *(8.3)* **14.** .1304 *(8.3)* **15.** .1 *(8.3)* **16.** No, since $P(T|Z) \neq P(T)$ *(8.3)* **17.** Yes, since $P(S \cap X) = P(S)P(X)$ *(8.3)* **18.** .4 *(8.3)* **19.** .2 *(8.3)* **20.** .3 *(8.3)* **21.** .08 *(8.3)* **22.** .18 *(8.3)* **23.** .26 *(8.3)* **24.** .31 *(8.4)* **25.** .43 *(8.4)* **26.** (A) $\frac{5}{16}$ (B) $\frac{1}{4}$ (C) As the sample in part (A) increases in size, approximate empirical probabilities should approach the theoretical probabilities. *(8.1)* **27.** No *(8.1)* **28.** Yes *(8.1)*

29. Payoff table:

x_i	\$5	–\$4	\$2
p_i	.25	.5	.25

$E(X) = -25¢$; game is not fair *(8.5)*

30. (A) $\frac{1}{3}$ (B) $\frac{2}{9}$ *(8.3)* **31.** (A) $\frac{2}{13}$; 2 to 11; (B) $\frac{4}{13}$; 4 to 9; (C) $\frac{12}{13}$; 12 to 1 *(8.2)* **32.** (A) 1 to 8 (B) \$8 *(8.2)*
33. (A) $P(2\text{ heads}) = .21$, $P(1\text{ head}) = .48$, $P(0\text{ heads}) = .31$ (B) $P(2\text{ heads}) = .25$, $P(1\text{ head}) = .50$, $P(0\text{ heads}) = .25$ (C) 2 heads, 250; 1 head, 500; 0 heads, 250 *(8.1, 8.5)*
34. $\frac{1}{2}$; since the coin has no memory, the 10th toss is independent of the preceding 9 tosses. *(8.3)*

35. (A)

x_i	2	3	4	5	6	7	8	9	10	11	12
p_i	$\frac{1}{36}$	$\frac{2}{36}$	$\frac{3}{36}$	$\frac{4}{36}$	$\frac{5}{36}$	$\frac{6}{36}$	$\frac{5}{36}$	$\frac{4}{36}$	$\frac{3}{36}$	$\frac{2}{36}$	$\frac{1}{36}$

(B) $E(X) = 7$ *(8.5)*

36. $A = \{(1,3),(2,2),(3,1),(2,6),(3,5),(4,4),(5,3),(6,2),(6,6)\}$; $B = \{(1,5),(2,4),(3,3),(4,2),(5,1),(6,6)\}$;
$P(A) = \frac{1}{4}$; $P(B) = \frac{1}{6}$; $P(A \cap B) = \frac{1}{36}$; $P(A \cup B) = \frac{7}{18}$ *(8.2)*

37. (1) Probability of an event cannot be negative; (2) sum of probabilities of simple events must be 1; (3) probability of an event cannot be greater than 1. *(8.1)*

38.

	A	A'	Totals
B	15	30	45
B'	35	20	55
Totals	50	50	100

(8.2)

39. (A) .6 (B) $\frac{5}{6}$ *(8.3)* **40.** (A) $\frac{1}{13}$ (B) Independent *(8.3)* **41.** (A) $\frac{6}{25}$ (B) $\frac{3}{10}$ *(8.3)* **42.** Part (B) *(8.3)* **43.** (A) 1.2 (B) 1.2 *(8.5)* **44.** (A) $\frac{3}{5}$ (B) $\frac{1}{3}$ (C) $\frac{5}{8}$ (D) $\frac{9}{14}$ *(8.3, 8.4)* **45.** No *(8.3)* **46.** (A) ${}_{13}C_5/{}_{52}C_5$ (B) ${}_{13}C_3 \cdot {}_{13}C_2/{}_{52}C_5$ *(8.1)* **47.** ${}_8C_2/{}_{10}C_4 = \frac{2}{15}$ *(8.1)* **48.** Events S and H are mutually exclusive. Hence, $P(S \cap H) = 0$, while $P(S) \neq 0$ and $P(H) \neq 0$. Therefore, $P(S \cap H) \neq P(S)P(H)$, which implies S and H are dependent. *(8.3)* **49.** (A) $\frac{8}{50} = .16$ (B) $\frac{9}{36} = .25$ (C) The empirical probability depends on the results of the simulation; the theoretical probability is $\frac{5}{36} \approx .139$. *(8.1)* **50.** The empirical probability depends on the results of the simulation; the theoretical probability is $\frac{2}{52} \approx .038$. *(8.3)* **51.** False *(8.2)* **52.** True *(8.2)* **53.** False *(8.3)* **54.** False *(8.2)* **55.** True *(8.3)* **56.** False *(8.2)* **57.** (A) .350 (B) $\frac{3}{8} = .375$ (C) 375 *(8.1)* **58.** $-.0172$; .0172; no *(8.5)* **59.** $1 - {}_7C_3/{}_{10}C_3 = \frac{17}{24}$ *(8.1)* **60.** $\frac{12}{51} \approx .235$ *(8.3)* **61.** $\frac{12}{51} \approx .235$ *(8.3)*
62. (A)

x_i	2	3	4	5	6
p_i	$\frac{9}{36}$	$\frac{12}{36}$	$\frac{10}{36}$	$\frac{4}{36}$	$\frac{1}{36}$

(B) $E(X) = \frac{10}{3}$ *(8.5)*
63. $E(X) \approx -\$0.167$; no; $\$(10/3) \approx \$3.33$ *(8.5)* **64.** (A) $\frac{1}{4}$; 1 to 3 (B) \$3 *(8.2, 8.4)* **65.** $1 - 10!/(5!10^5) \approx .70$ *(8.2)* **66.** $P(A|B) = P(B|A)$ if and only if $P(A) = P(B)$ or $P(A \cap B) = 0$. *(8.3)* **67.** (A) .8 (B) .2 (C) .5 *(8.2)* **68.** $P(A \cap P) = P(A)P(P|A) = .34$ *(8.3)* **69.** (A) $P(A) = .290$; $P(B) = .290$; $P(A \cap B) = .100$; $P(A|B) = .345$; $P(B|A) = .345$ (B) No, since $P(A \cap B) \neq P(A)P(B)$. (C) $P(C) = .880$; $P(D) = .120$; $P(C \cap D) = 0$; $P(C|D) = 0$; $P(D|C) = 0$ (D) Yes, since $C \cap D = \varnothing$; dependent, since $P(C \cap D) = 0$ and $P(C)P(D) \neq 0$ *(8.3)* **70.** Plan A: $E(X) = \$7.6$ million; plan B: $E(X) = \$7.8$ million; plan B *(8.5)*
71. Payoff table:

x_i	\$1,830	$-\$170$
p_i	.08	.92

$E(X) = -\$10$ *(8.5)*
72. $1 - ({}_{10}C_4/{}_{12}C_4) \approx .576$ *(8.2)*
73. (A)

x_i	0	1	2
p_i	$\frac{12}{22}$	$\frac{9}{22}$	$\frac{1}{22}$

(B) $E(X) = \frac{1}{2}$ *(8.5)* **74.** .955 *(8.4)* **75.** $\frac{6}{7} \approx .857$ *(8.4)*

Chapter 9

Exercises 9.1

1. $\begin{bmatrix} 16 \\ 14 \end{bmatrix}$ **3.** Not defined **5.** $[14 \quad 17]$ **7.** Not defined
9. $S_1 = [.1 \quad .9]$; $S_2 = [.16 \quad .84]$ **11.** $S_1 = [.46 \quad .54]$; $S_2 = [.376 \quad .624]$ **13.** $S_1 = [.25 \quad .75]$; $S_2 = [.25 \quad .75]$ **15.** $S_1 = [.5 \quad .5]$; $S_2 = [.65 \quad .35]$ **17.** $S_1 = [.71 \quad .29]$; $S_2 = [.587 \quad .413]$ **19.** $S_1 = [.65 \quad .35]$; $S_2 = [.605 \quad .395]$ **21.** $S_1 = [.7 \quad .2 \quad .1]$; $S_2 = [.33 \quad .35 \quad .32]$ **23.** $S_1 = [.35 \quad .35 \quad .3]$; $S_2 = [.465 \quad .3 \quad .235]$ **25.** $S_1 = [.53 \quad .28 \quad .19]$; $S_2 = [.397 \quad .325 \quad .278]$ **27.** $S_1 = [.2 \quad .3 \quad .5]$; $S_2 = [.27 \quad .4 \quad .33]$ **29.** $S_1 = [.28 \quad .44 \quad .28]$; $S_2 = [.212 \quad .492 \quad .296]$ **31.** $S_1 = [.24 \quad .37 \quad .39]$; $S_2 = [.241 \quad .446 \quad .313]$

33. 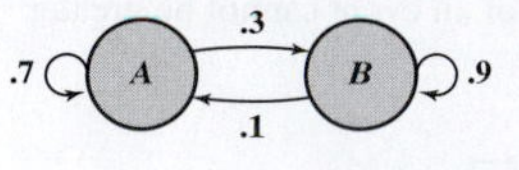

35. $P = \begin{bmatrix} .2 & .3 & .5 \\ .1 & .8 & .1 \\ .4 & .2 & .4 \end{bmatrix}$ **37.** Yes **39.** No

41. No **43.** Yes **45.** 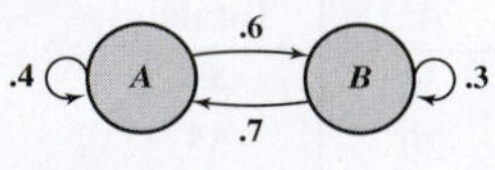

$\begin{array}{c|cc} & A & B \\ \hline A & .4 & .6 \\ B & .7 & .3 \end{array}$

47. No **49.** 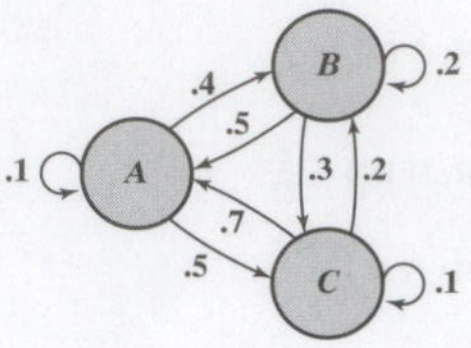

$\begin{array}{c|ccc} & A & B & C \\ \hline A & .1 & .4 & .5 \\ B & .5 & .2 & .3 \\ C & .7 & .2 & .1 \end{array}$

51. $a = .5, b = .6, c = .7$ **53.** $a = .7, b = 1, c = .2$ **55.** No

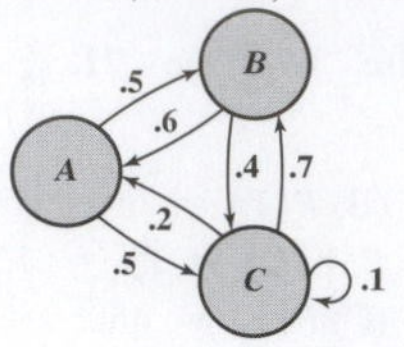

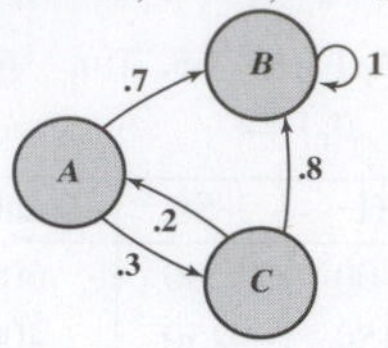

57. 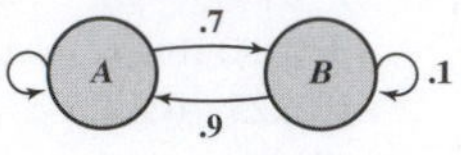

$\begin{array}{c|cc} & A & B \\ \hline A & .3 & .7 \\ B & .9 & .1 \end{array}$

59. 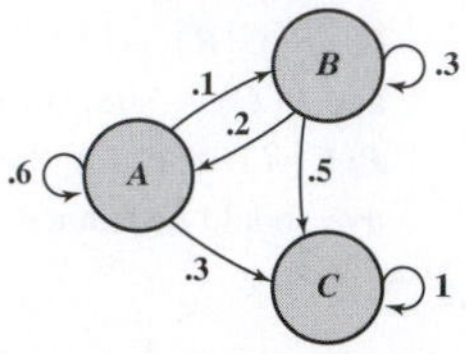

$\begin{array}{c|ccc} & A & B & C \\ \hline A & .6 & .1 & .3 \\ B & .2 & .3 & .5 \\ C & 0 & 0 & 1 \end{array}$ **61.** .35 **63.** .212

65. $S_2 = \overset{A}{[.43} \quad \overset{B}{.35} \quad \overset{C}{.22]}$; the probabilities of going from state A to states A, B, and C in two trials

67. $S_3 = \overset{A}{[.212} \quad \overset{B}{.298} \quad \overset{C}{.49]}$; the probabilities of going from state C to states A, B, and C in three trials

69. $n = 9$ **71.** $P^4 = \begin{array}{c|cc} & A & B \\ \hline A & .4375 & .5625 \\ B & .375 & .625 \end{array}$; $S_4 = \overset{A}{[.425} \quad \overset{B}{.575]}$

73. $P^4 = \begin{array}{c|ccc} & A & B & C \\ \hline A & .36 & .16 & .48 \\ B & .6 & 0 & .4 \\ C & .4 & .24 & .36 \end{array}$; $S_4 = \overset{A}{[.452} \quad \overset{B}{.152} \quad \overset{C}{.396]}$

77. (A) $\begin{array}{c|cccc} & A & B & C & D \\ \hline A & .0154 & .3534 & .0153 & .6159 \\ B & 0 & 1 & 0 & 0 \\ C & .0102 & .2962 & .0103 & .6833 \\ D & 0 & 0 & 0 & 1 \end{array}$ (B) .6159 (C) .2962 (D) 0

81. (A) $[.25 \quad .75]$ (B) $[.25 \quad .75]$ (C) $[.25 \quad .75]$ (D) $[.25 \quad .75]$ (E) The state matrices appear to approach the same matrix, $S = [.25 \quad .75]$, regardless of the values in the initial-state matrix.

83. $Q = \begin{bmatrix} .25 & .75 \\ .25 & .75 \end{bmatrix}$; the rows of Q are the same as the matrix S from Problem 81 **85.** (A) R = rain, R' = no rain

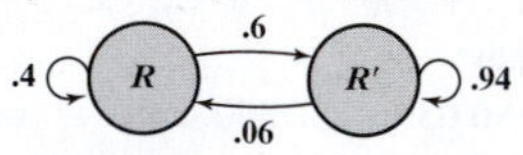

(B) $\begin{array}{c|cc} & R & R' \\ \hline R & .4 & .6 \\ R' & .06 & .94 \end{array}$ (C) Saturday: .196; Sunday: .12664

87. (A) 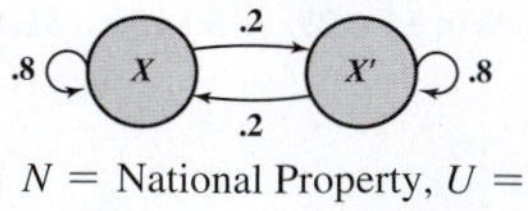

(B) $\begin{array}{c|cc} & X & X' \\ \hline X & .8 & .2 \\ X' & .2 & .8 \end{array}$ (C) 32%; 39.2%

89. (A) N = National Property, U = United Family, O = other companies

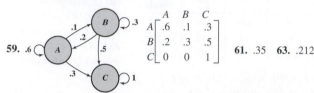

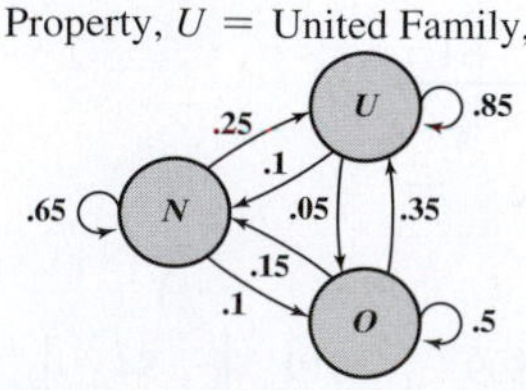

(B) $\begin{array}{c|ccc} & N & U & O \\ \hline N & .65 & .25 & .1 \\ U & .1 & .85 & .05 \\ C & .15 & .35 & .5 \end{array}$ (C) 38.5%; 32% (D) 45%; 53.65%

91. (A) B = beginning agent, I = intermediate agent, T = terminated agent, Q = qualified agent

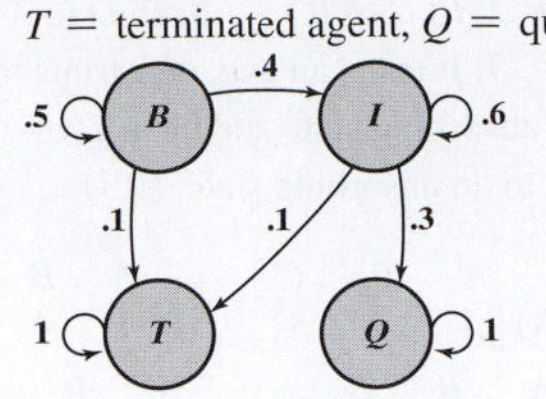

(B) $\begin{array}{c} \\ B \\ I \\ T \\ Q \end{array}\begin{array}{c} \begin{array}{cccc} B & I & T & Q \end{array} \\ \begin{bmatrix} .5 & .4 & .1 & 0 \\ 0 & .6 & .1 & .3 \\ 0 & 0 & 1 & 0 \\ 0 & 0 & 0 & 1 \end{bmatrix} \end{array}$ (C) .12; .3612

93. (A) $\begin{array}{c} \\ HMO \\ PPO \\ FFS \end{array}\begin{array}{c} \begin{array}{ccc} HMO & PPO & FFS \end{array} \\ \begin{bmatrix} .8 & .15 & .05 \\ .2 & .7 & .1 \\ .25 & .3 & .45 \end{bmatrix} \end{array}$ (B) HMO: 34.75%; PPO: 37%; FFS: 28.25% (C) HMO: 42.2625%; PPO: 39.5875%; FFS: 18.15%

95. (A) $\begin{array}{c} \\ H \\ R \end{array}\begin{array}{c} \begin{array}{cc} H & R \end{array} \\ \begin{bmatrix} .847 & .153 \\ .174 & .826 \end{bmatrix} \end{array}$ (B) 45.6% (C) 49.8%

Exercises 9.2

1. $\begin{bmatrix} 1 & 0 \\ 0 & 0 \end{bmatrix}$ **3.** $\begin{bmatrix} 1 & 0 \\ 0 & 1 \end{bmatrix}$ **5.** $\begin{bmatrix} 1 & 0 & 0 \\ 0 & 0 & 0 \\ 0 & 0 & 1 \end{bmatrix}$ **7.** $\begin{bmatrix} 0 & 0 & 0 \\ 0 & 0 & 0 \\ 0 & 0 & 0 \end{bmatrix}$ **9.** Yes

11. No **13.** No **15.** Yes **17.** No **19.** Yes **21.** No

23. $S = [.4 \quad .6]$; $\overline{P} = \begin{bmatrix} .4 & .6 \\ .4 & .6 \end{bmatrix}$ **25.** $S = [.375 \quad .625]$; $\overline{P} = \begin{bmatrix} .375 & .625 \\ .375 & .625 \end{bmatrix}$

27. $S = [.3 \quad .5 \quad .2]$; $\overline{P} = \begin{bmatrix} .3 & .5 & .2 \\ .3 & .5 & .2 \\ .3 & .5 & .2 \end{bmatrix}$

29. $S = [.6 \quad .24 \quad .16]$; $\overline{P} = \begin{bmatrix} .6 & .24 & .16 \\ .6 & .24 & .16 \\ .6 & .24 & .16 \end{bmatrix}$ **31.** $SP = [.2 \quad .5]$; no, the sum of the entries in S is not 1.

33. $SP = [0 \quad 0]$; no, the sum of the entries in S is not 1 **35.** False **37.** True **39.** False **41.** $S = [.3553 \quad .6447]$ **43.** $S = [.3636 \quad .4091 \quad .2273]$

45. (A)

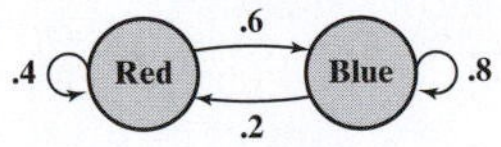

(B) $\begin{array}{c} \\ Red \\ Blue \end{array}\begin{array}{c} \begin{array}{cc} Red & Blue \end{array} \\ \begin{bmatrix} .4 & .6 \\ .2 & .8 \end{bmatrix} \end{array}$

(C) $[.25 \quad .75]$; in the long run, the red urn will be selected 25% of the time and the blue urn 75% of the time. **47.** (A) The state matrices alternate between $[.2 \quad .8]$ and $[.8 \quad .2]$; so they do not approach any one matrix. (B) The state matrices are all equal to S_0, so S_0 is a stationary matrix. (C) The powers of P alternate between P and I (the 2×2 identity); so they do not approach a limiting matrix. (D) Parts (B) and (C) of Theorem 1 are not valid for this matrix. Since P is not regular, this is not a contradiction. **49.** (A) Since P is not regular, it may have more than one stationary matrix. (B) $[.5 \quad 0 \quad .5]$ is another stationary matrix. (C) P has an infinite number of stationary matrices.

51. $\overline{P} = \begin{bmatrix} 1 & 0 & 0 \\ .25 & 0 & .75 \\ 0 & 0 & 1 \end{bmatrix}$; each row of $\overline{P}$ is a stationary matrix for P.

53. (A) .39; .3; .284; .277 (B) Each entry of the second column of P^{k+1} is the product of a row of P and the second column of P^k. Each entry of the latter is $\le M_k$, so the product is $\le M_k$. **55.** 72.5%

57. (A) $S_1 = [.512 \quad .488]$; $S_2 = [.568 \quad .432]$; $S_3 = [.609 \quad .391]$; $S_4 = [.639 \quad .361]$

(B)

Year	Data (%)	Model (%)
1970	43.3	43.3
1980	51.5	51.2
1990	57.5	56.8
2000	59.8	60.9
2010	58.5	63.9

(C) 71.4%

59. GTT: 25%; NCJ: 25%; Dash: 50% **61.** Poor: 20%; satisfactory: 40%; preferred: 40% **63.** 51% **65.** Stationary matrix = $[.25 \quad .50 \quad .25]$

67. (A) $[.25 \quad .75]$ (B) 42.5%; 51.25% (C) 60% rapid transit; 40% automobile

69. (A) $S_1 = [.334 \quad .666]$; $S_2 = [.343 \quad .657]$; $S_3 = [.347 \quad .653]$; $S_4 = [.349 \quad .651]$

(B)

Year	Data (%)	Model (%)
1970	30.9	30.9
1980	33.3	33.4
1990	34.4	34.3
2000	35.6	34.7
2010	37.1	34.9

(C) 35%

Exercises 9.3

1. B, C **3.** No absorbing states **5.** A, D **7.** B is an absorbing state; absorbing chain **9.** C is an absorbing state; not an absorbing chain **11.** No **13.** Yes

15. Yes **17.** No **19.** Yes **21.** $\begin{array}{c} \\ B \\ A \\ C \end{array}\begin{array}{c} \begin{array}{ccc} B & A & C \end{array} \\ \begin{bmatrix} 1 & 0 & 0 \\ .5 & .2 & .3 \\ .1 & .5 & .4 \end{bmatrix} \end{array}$ **23.** $\begin{array}{c} \\ B \\ D \\ A \\ C \end{array}\begin{array}{c} \begin{array}{cccc} B & D & A & C \end{array} \\ \begin{bmatrix} 1 & 0 & 0 & 0 \\ 0 & 1 & 0 & 0 \\ .4 & .1 & .3 & .2 \\ .4 & .3 & 0 & .3 \end{bmatrix} \end{array}$

25. $\begin{array}{c} \\ C \\ A \\ B \end{array}\begin{array}{c} \begin{array}{ccc} C & A & B \end{array} \\ \begin{bmatrix} 1 & 0 & 0 \\ .5 & .2 & .3 \\ 0 & 1 & 0 \end{bmatrix} \end{array}$ **27.** $\begin{array}{c} \\ B \\ D \\ A \\ C \end{array}\begin{array}{c} \begin{array}{cccc} B & D & A & C \end{array} \\ \begin{bmatrix} 1 & 0 & 0 & 0 \\ 0 & 1 & 0 & 0 \\ .2 & .4 & .1 & .3 \\ .2 & .1 & .5 & .2 \end{bmatrix} \end{array}$

29. $\overline{P} = \begin{array}{c} \\ A \\ B \\ C \end{array}\begin{array}{c} \begin{array}{ccc} A & B & C \end{array} \\ \begin{bmatrix} 1 & 0 & 0 \\ 0 & 1 & 0 \\ .2 & .8 & 0 \end{bmatrix} \end{array}$; $P(C \text{ to } A) = .2$; $P(C \text{ to } B) = .8$. It will take an avg. of 2 trials to go from C to either A or B.

31. $\overline{P} = \begin{array}{c} \\ A \\ B \\ C \end{array}\begin{array}{c} \begin{array}{ccc} A & B & C \end{array} \\ \begin{bmatrix} 1 & 0 & 0 \\ 1 & 0 & 0 \\ 1 & 0 & 0 \end{bmatrix} \end{array}$; $P(B \text{ to } A) = 1$; $P(C \text{ to } A) = 1$. It will take an avg. of 4 trials to go from B to A, and an avg. of 3 trials to go from C to A.

33. $\overline{P} = \begin{array}{c} \\ A \\ B \\ C \\ D \end{array}\begin{array}{c} \begin{array}{cccc} A & B & C & D \end{array} \\ \begin{bmatrix} 1 & 0 & 0 & 0 \\ 0 & 1 & 0 & 0 \\ .36 & .64 & 0 & 0 \\ .44 & .56 & 0 & 0 \end{bmatrix} \end{array}$; $P(C \text{ to } A) = .36$; $P(C \text{ to } B) = .64$; $P(D \text{ to } A) = .44$; $P(D \text{ to } B) = .56$.

It will take an avg. of 3.2 trials to go from C to either A or B, and an avg. of 2.8 trials to go from D to either A or B. **35.** (A) $[.2 \quad .8 \quad 0]$ (B) $[.26 \quad .74 \quad 0]$ **37.** (A) $[1 \quad 0 \quad 0]$ (B) $[1 \quad 0 \quad 0]$ **39.** (A) $[.44 \quad .56 \quad 0 \quad 0]$ (B) $[.36 \quad .64 \quad 0 \quad 0]$ (C) $[.408 \quad .592 \quad 0 \quad 0]$ (D) $[.384 \quad .616 \quad 0 \quad 0]$ **41.** False

43. False **45.** True **47.** False **49.** $\begin{array}{c} \\ A \\ B \\ C \\ D \end{array}\begin{array}{c} \begin{array}{cccc} A & B & C & D \end{array} \\ \begin{bmatrix} 1 & 0 & 0 & 0 \\ 0 & 1 & 0 & 0 \\ .6375 & .3625 & 0 & 0 \\ .7375 & .2625 & 0 & 0 \end{bmatrix} \end{array}$

51. $\begin{array}{c} \\ A \\ B \\ C \\ D \\ E \end{array}\begin{array}{c} \begin{array}{ccccc} A & B & C & D & E \end{array} \\ \begin{bmatrix} 1 & 0 & 0 & 0 & 0 \\ 0 & 1 & 0 & 0 & 0 \\ .0875 & .9125 & 0 & 0 & 0 \\ .1875 & .8125 & 0 & 0 & 0 \\ .4375 & .5625 & 0 & 0 & 0 \end{bmatrix} \end{array}$

53. $\begin{array}{cc} & \begin{matrix} A & \;B & \;C & \;D \end{matrix} \\ \begin{matrix} A \\ B \\ C \\ D \end{matrix} & \begin{bmatrix} 0 & .52 & 0 & .48 \\ 0 & 1 & 0 & 0 \\ 0 & .36 & 0 & .64 \\ 0 & 0 & 0 & 1 \end{bmatrix} \end{array}$

59. (A) .370; .297; .227; .132; .045 (B) For large k, all entries of Q^k are close to 0. **61.** (A) 75% (B) 12.5% (C) 7.5 months **63.** (A) Company A: 30%; company B: 15%; company C: 55% (B) 5 yr **65.** (A) 91.52% (B) 4.96% (C) 6.32 days **67.** (A) .375 (B) 1.75 exits

Chapter 9 Review Exercises

1. $S_1 = \begin{matrix} A & B \\ [.32 & .68] \end{matrix}$; $S_2 = \begin{matrix} A & B \\ [.328 & .672] \end{matrix}$. The probability of being in state A after one trial is .32 and after two trials is .328; the probability of being in state B after one trial is .68 and after two trials is .672. *(9.1)*
2. State A is absorbing; chain is absorbing. *(9.2, 9.3)* **3.** No absorbing states; chain is regular. *(9.2, 9.3)* **4.** No absorbing states; chain is neither. *(9.2, 9.3)* **5.** States B and C are absorbing; chain is absorbing. *(9.2, 9.3)*
6. States A and B are absorbing; chain is neither. *(9.2, 9.3)*

7. $\begin{array}{cc} & \begin{matrix} A & B & C \end{matrix} \\ \begin{matrix} A \\ B \\ C \end{matrix} & \begin{bmatrix} 0 & 1 & 0 \\ .1 & 0 & .9 \\ 0 & 1 & 0 \end{bmatrix} \end{array}$; No absorbing states; chain is neither. *(9.1, 9.2, 9.3)*

8. $\begin{array}{cc} & \begin{matrix} A & B & C \end{matrix} \\ \begin{matrix} A \\ B \\ C \end{matrix} & \begin{bmatrix} 0 & 1 & 0 \\ .1 & .2 & .7 \\ 0 & 0 & 1 \end{bmatrix} \end{array}$; C is absorbing; chain is absorbing. *(9.1, 9.2, 9.3)*

9. $\begin{array}{cc} & \begin{matrix} A & B & C \end{matrix} \\ \begin{matrix} A \\ B \\ C \end{matrix} & \begin{bmatrix} 0 & 0 & 1 \\ .1 & .2 & .7 \\ 0 & 1 & 0 \end{bmatrix} \end{array}$; No absorbing states; chain is regular. *(9.1, 9.2, 9.3)*

10. $\begin{array}{cc} & \begin{matrix} A & B & C & D \end{matrix} \\ \begin{matrix} A \\ B \\ C \\ D \end{matrix} & \begin{bmatrix} .3 & .2 & 0 & .5 \\ 0 & 1 & 0 & 0 \\ 0 & 0 & .2 & .8 \\ 0 & 0 & .3 & .7 \end{bmatrix} \end{array}$; B is absorbing; chain is neither. *(9.1, 9.2, 9.3)*

11. 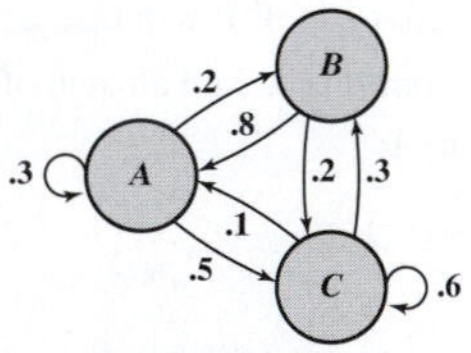

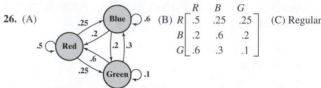

$\begin{array}{cc} & \begin{matrix} A & B & C \end{matrix} \\ \begin{matrix} A \\ B \\ C \end{matrix} & \begin{bmatrix} .3 & .2 & .5 \\ .8 & 0 & .2 \\ .1 & .3 & .6 \end{bmatrix} \end{array}$ *(9.1)* **12.** (A) .3 (B) .675 *(9.1)*

13. $S = \begin{matrix} A & B \\ [.25 & .75] \end{matrix}$; $\overline{P} = \begin{array}{cc} & \begin{matrix} A & B \end{matrix} \\ \begin{matrix} A \\ B \end{matrix} & \begin{bmatrix} .25 & .75 \\ .25 & .75 \end{bmatrix} \end{array}$ *(9.2)*

14. $S = \begin{matrix} A & B & C \\ [.4 & .48 & .12] \end{matrix}$; $\overline{P} = \begin{array}{cc} & \begin{matrix} A & B & C \end{matrix} \\ \begin{matrix} A \\ B \\ C \end{matrix} & \begin{bmatrix} .4 & .48 & .12 \\ .4 & .48 & .12 \\ .4 & .48 & .12 \end{bmatrix} \end{array}$ *(9.2)*

15. $\begin{array}{cc} & \begin{matrix} A & B & C \end{matrix} \\ \begin{matrix} A \\ B \\ C \end{matrix} & \begin{bmatrix} 1 & 0 & 0 \\ 0 & 1 & 0 \\ .75 & .25 & 0 \end{bmatrix} \end{array}$; $P(C \text{ to } A) = .75$; $P(C \text{ to } B) = .25$. It takes an average of 2.5 trials to go from C to an absorbing state. *(9.3)*

16. $\begin{array}{cc} & \begin{matrix} A & B & C & D \end{matrix} \\ \begin{matrix} A \\ B \\ C \\ D \end{matrix} & \begin{bmatrix} 1 & 0 & 0 & 0 \\ 0 & 1 & 0 & 0 \\ .2 & .8 & 0 & 0 \\ .3 & .7 & 0 & 0 \end{bmatrix} \end{array}$; $P(C \text{ to } A) = .2$; $P(C \text{ to } B) = .8$; $P(D \text{ to } A) = .3$; $P(D \text{ to } B) = .7$. It takes an avg. of 2 trials to go from C to an absorbing state and an avg. of 3 trials to go from D to an absorbing state. *(9.3)*

21. $\begin{array}{cc} & \begin{matrix} B & D & A & C \end{matrix} \\ \begin{matrix} B \\ D \\ A \\ C \end{matrix} & \begin{bmatrix} 1 & 0 & 0 & 0 \\ 0 & 1 & 0 & 0 \\ .1 & .1 & .6 & .2 \\ .2 & .2 & .3 & .3 \end{bmatrix} \end{array}$ *(9.3)* **22.** (A) $\begin{matrix} A & B & C \\ [.1 & .4 & .5] \end{matrix}$ (B) $\begin{matrix} A & B & C \\ [.1 & .4 & .5] \end{matrix}$ *(9.3)* **23.** (A) $\begin{matrix} A & B & C \\ [.25 & .75 & 0] \end{matrix}$ (B) $\begin{matrix} A & B & C \\ [.55 & .45 & 0] \end{matrix}$ *(9.3)*

24. No. Each row of P would contain a 0 and a 1, but none of the four matrices with this property is regular. *(9.2)*

25. Yes; for example, $P = \begin{bmatrix} 0 & 0 & 1 \\ 0 & 0 & 1 \\ .2 & .3 & .5 \end{bmatrix}$ is regular. *(9.2)*

26. (A) 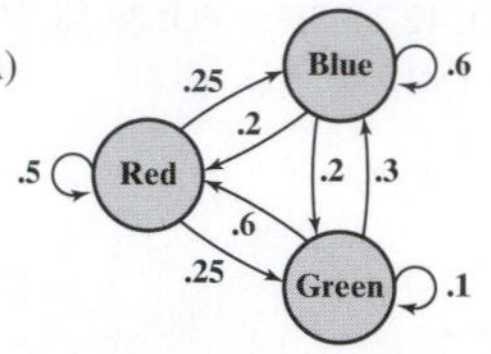

(B) $\begin{array}{cc} & \begin{matrix} R & B & G \end{matrix} \\ \begin{matrix} R \\ B \\ G \end{matrix} & \begin{bmatrix} .5 & .25 & .25 \\ .2 & .6 & .2 \\ .6 & .3 & .1 \end{bmatrix} \end{array}$ (C) Regular

(D) $\begin{array}{cc} & \begin{matrix} R & B & G \end{matrix} \\ \begin{matrix} R \\ B \\ G \end{matrix} & \begin{bmatrix} .4 & .4 & .2 \\ .4 & .4 & .2 \\ .4 & .4 & .2 \end{bmatrix} \end{array}$ In the long run, the red urn will be selected 40% of the time, the blue urn 40% of the time, and the green urn 20% of the time. *(9.2)*

27. (A) 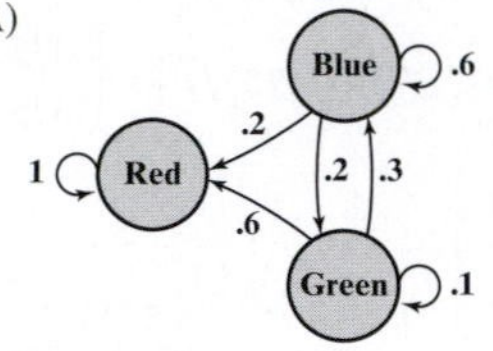

(B) $\begin{array}{cc} & \begin{matrix} R & B & G \end{matrix} \\ \begin{matrix} R \\ B \\ G \end{matrix} & \begin{bmatrix} 1 & 0 & 0 \\ .2 & .6 & .2 \\ .6 & .3 & .1 \end{bmatrix} \end{array}$ (C) Absorbing

(D) $\begin{array}{cc} & \begin{matrix} R & B & G \end{matrix} \\ \begin{matrix} R \\ B \\ G \end{matrix} & \begin{bmatrix} 1 & 0 & 0 \\ 1 & 0 & 0 \\ 1 & 0 & 0 \end{bmatrix} \end{array}$ Once the red urn is selected, the blue and green urns will never be selected again. It will take an avg. of 3.67 trials to reach the red urn from the blue urn and an avg. of 2.33 trials to reach the red urn from the green urn. *(9.3)*

29. $\begin{bmatrix} 0 & 1 \\ 1 & 0 \end{bmatrix}$ is one example *(9.2)* **30.** $\begin{bmatrix} 1 & 0 \\ 0 & 1 \end{bmatrix}$ is one example *(9.3)*

31. $\begin{bmatrix} .3 & .1 & .6 \\ .3 & .1 & .6 \\ .3 & .1 & .6 \end{bmatrix}$ is one example *(9.2)*

32. $\begin{bmatrix} 1 & 0 & 0 \\ 0 & 1 & 0 \\ 0 & 0 & 1 \end{bmatrix}$ is one example *(9.3)*

33. If P is the transition matrix of an absorbing Markov chain with more than one state, then P has a row with 1 on the main diagonal and 0's elsewhere. Every power of P has that same row, so no power of P has all positive entries, and the Markov chain is not regular. *(9.2, 9.3)* **34.** If P is the transition matrix of a regular Markov chain, then some power of P has all positive entries. This is impossible for an absorbing Markov chain with more than one state (see Problem 33), so the Markov chain is not absorbing. *(9.2, 9.3)*

35. $SP = [.3 \quad .9]$; no, the sum of the entries in S is not 1. *(9.1)*

36. No limiting matrix *(9.2, 9.3)* **37.** $P = \begin{array}{cc} & \begin{matrix} A & B & C & D \end{matrix} \\ \begin{matrix} A \\ B \\ C \\ D \end{matrix} & \begin{bmatrix} .392 & .163 & .134 & .311 \\ .392 & .163 & .134 & .311 \\ .392 & .163 & .134 & .311 \\ .392 & .163 & .134 & .311 \end{bmatrix} \end{array}$ *(9.2)*

38. (A)
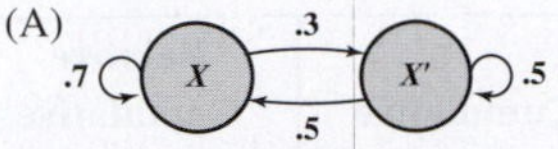

(B) $\begin{matrix} & X & X' \\ X & .7 & .3 \\ X' & .5 & .5 \end{matrix}$ (C) $\begin{matrix} X & X' \\ [.2 & .8] \end{matrix}$

(D) $\begin{matrix} X & X' \\ [.54 & .46] \end{matrix}$; 54% of the consumers will purchase brand X on the next purchase.

(E) $\begin{matrix} X & X' \\ [.625 & .375] \end{matrix}$ (F) 62.5% *(9.2)* **39.** (A) Brand *A*: 24%; brand *B*: 32%; brand *C*: 44% (B) 4 yr *(9.3)*

40. (A) $S_1 = [.48 \quad .52]$; $S_2 = [.66 \quad .34]$; $S_3 = [.76 \quad .24]$

(B)

Year	Data (%)	Model (%)
1995	14	14
2000	49	48
2005	68	66
2010	79	76

(C) 89% *(9.2)*

41. (A) 63.75% (B) 15% (C) 8.75 yr *(9.3)*

42. $\bar{P} = \begin{matrix} & Red & Pink & White \\ Red & 1 & 0 & 0 \\ Pink & 1 & 0 & 0 \\ White & 1 & 0 & 0 \end{matrix}$ *(9.3)*

43. (A) $S_1 = [.244 \quad .756]$; $S_2 = [.203 \quad .797]$; $S_3 = [.174 \quad .826]$

(B)

Year	Data (%)	Model (%)
1985	30.1	30.1
1995	24.7	24.4
2005	20.9	20.3
2010	19.3	17.4

(C) 10.3% *(9.2)*

Chapter 10

Exercises 10.1

1. (A) and (B)

Class Interval	Tally	Frequency	Relative Frequency
0.5–2.5	|	1	.1
2.5–4.5	|	1	.1
4.5–6.5	||||	4	.4
6.5–8.5	|||	3	.3
8.5–10.5	|	1	.1

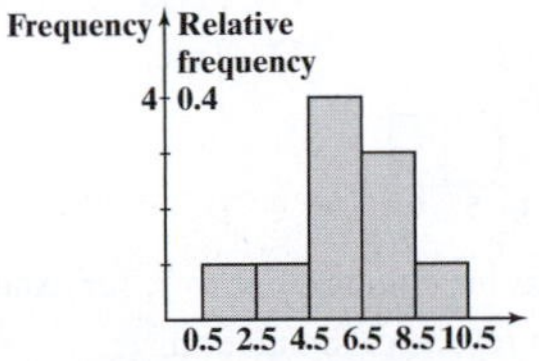

(C) The frequency tables and histograms are identical, but the data set in part (B) is more spread out than that of part (A).

3. (A) Let *X*min = 1.5, *X*max = 25.5, change *X*scl from 1 to 2, and multiply *Y*max and *Y*scl by 2; change *X*scl from 1 to 4, and multiply *Y*max and *Y*scl by 4.
(B) The shape becomes more symmetrical and more rectangular.

5.
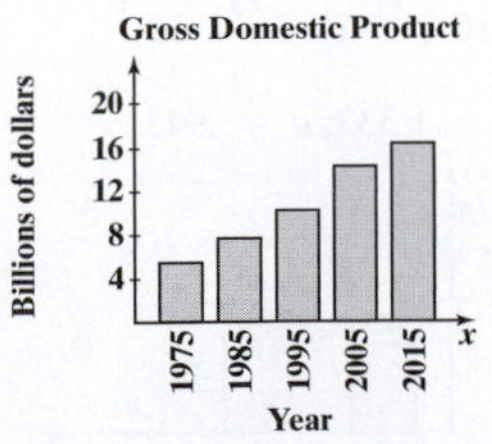

7. China; Canada; United States

9.
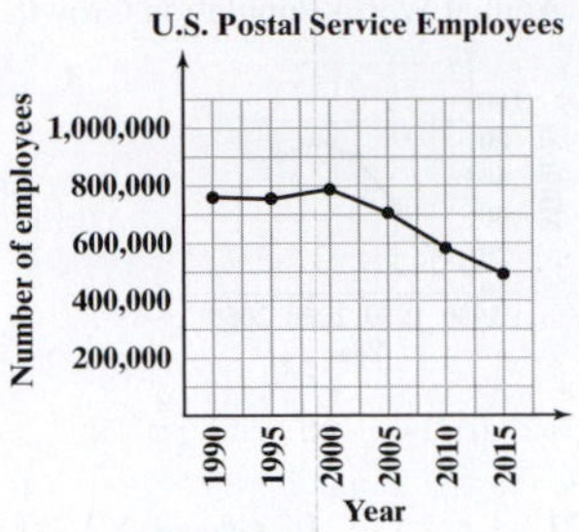

11.
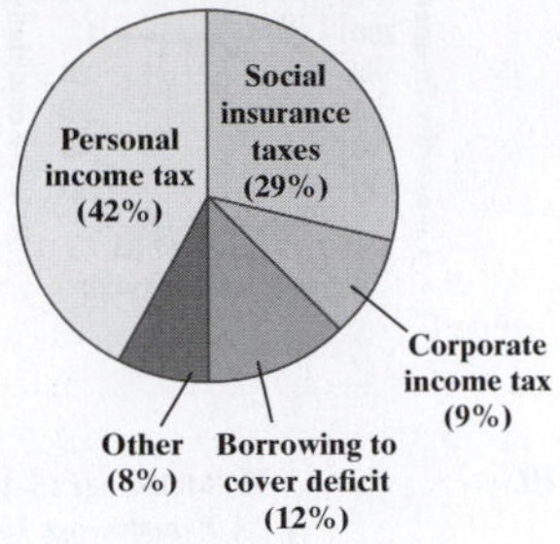

13. (A)

Class Interval	Tally	Frequency	Relative Frequency
30.5–34.5	|	1	.05
34.5–38.5	|||	3	.15
38.5–42.5	~~||||~~ ||	7	.35
42.5–46.5	~~||||~~ |	6	.30
46.5–50.5	||	2	.10
50.5–54.5	|	1	.05

(B)
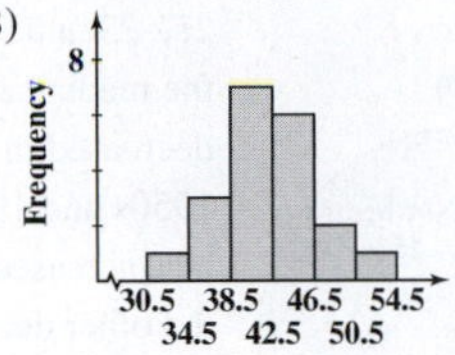

(C) .45; .2

(D)
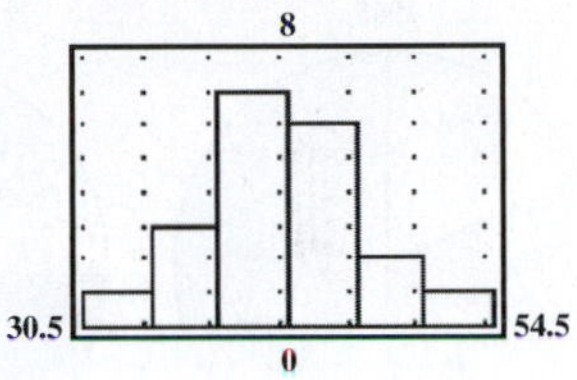

15. (A)

Class Interval	Frequency	Relative Frequency
−0.5–4.5	5	.05
4.5–9.5	54	.54
9.5–14.5	25	.25
14.5–19.5	13	.13
19.5–24.5	0	.00
24.5–29.5	1	.01
29.5–34.5	2	.02
	100	1.00

(B)
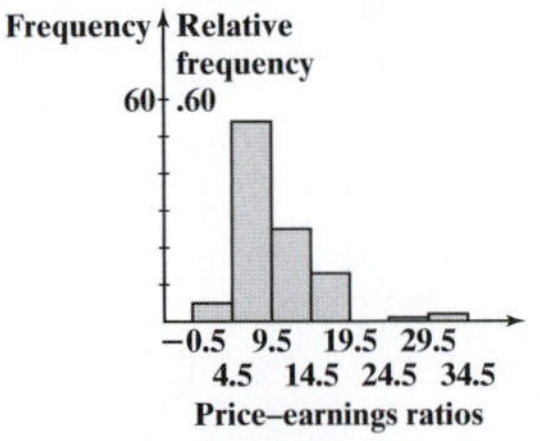

(C)
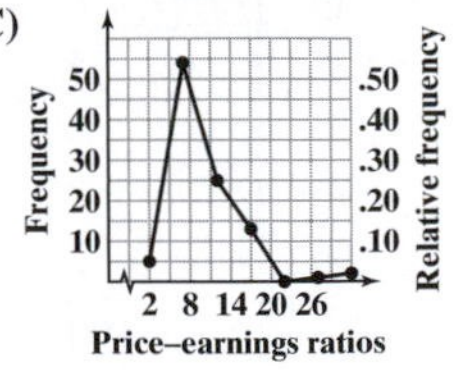

(D)

Class Interval	Frequency	Cumulative Frequency	Relative Cumulative Frequency
−0.5–4.5	5	5	.05
4.5–9.5	54	59	.59
9.5–14.5	25	84	.84
14.5–19.5	13	97	.97
19.5–24.5	0	97	.97
24.5–29.5	1	98	.98
29.5–34.5	2	100	1.00

P(PE ratio between 4.5 and 14.5) = .79

(E)

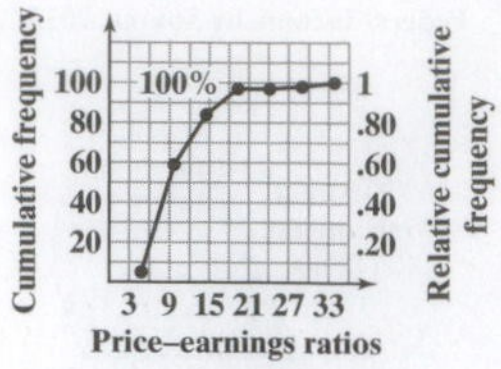

17. Annual World Population Growth

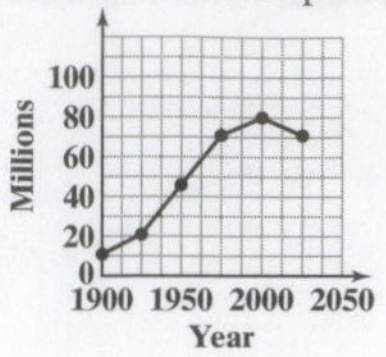

19.

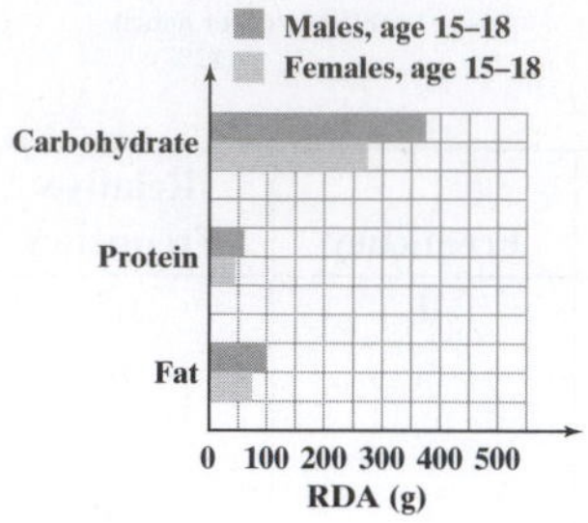

21.

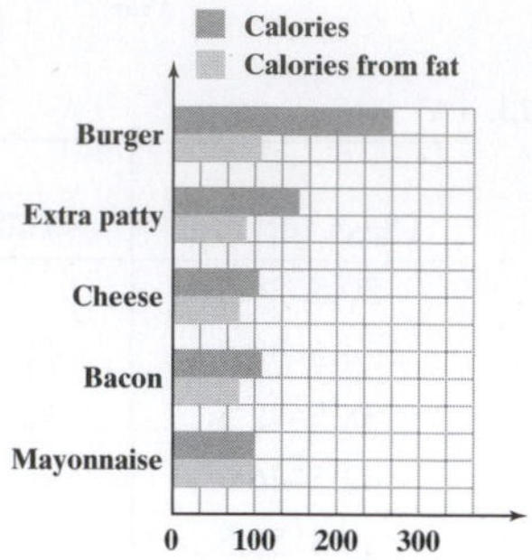

23. Total Public School Enrollment by Region

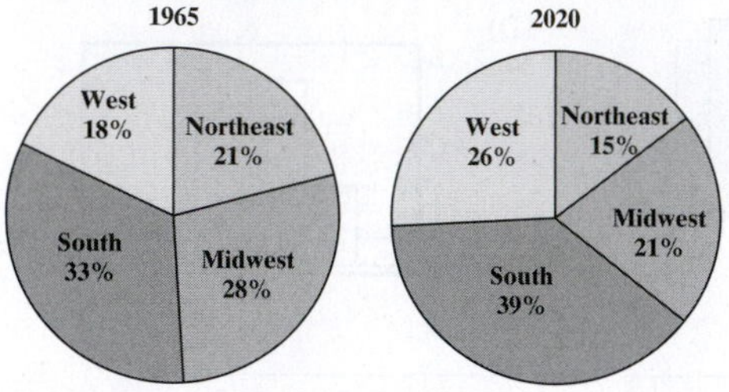

25. 23 and 35; the median age decreased in the 1950s and 1960s but increased in the other decades.

27. (A)

Class Interval	Frequency	Relative Frequency
1.95–2.15	21	.21
2.15–2.35	19	.19
2.35–2.55	17	.17
2.55–2.75	14	.14
2.75–2.95	9	.09
2.95–3.15	6	.06
3.15–3.35	5	.05
3.35–3.55	4	.04
3.55–3.75	3	.03
3.75–3.95	2	.02
	100	1.00

(B)

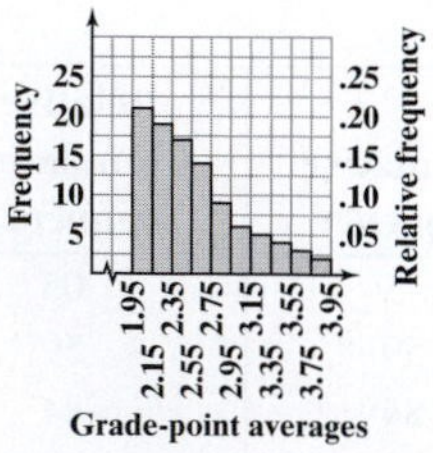

(C)

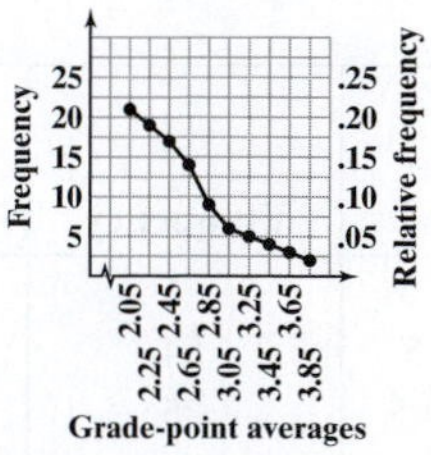

(D)

Class Interval	Frequency	Cumulative Frequency	Relative Cumulative Frequency
1.95–2.15	21	21	.21
2.15–2.35	19	40	.40
2.35–2.55	17	57	.57
2.55–2.75	14	71	.71
2.75–2.95	9	80	.80
2.95–3.15	6	86	.86
3.15–3.35	5	91	.91
3.35–3.55	4	95	.95
3.55–3.75	3	98	.98
3.75–3.95	2	100	1.00

$P(\text{GPA} > 2.95) = .2$

(E)

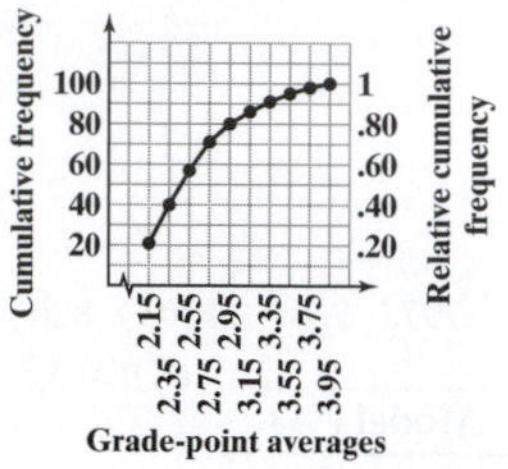

Exercises 10.2

1. 7 **3.** 0.6 **5.** 39 **7.** 1,980 **9.** Mean = 3; median = 3; mode = 3 **11.** Modal preference is chocolate. **13.** Mean ≈ 4.4 **15.** The median **17.** (A) Close to 3.5; close to 3.5 (B) Answer depends on results of simulation. **19.** (A) 175, 175, 325, 525 (B) Let the four numbers be u, v, w, x, where u and v are both equal to m_3. Choose w so that the mean of w and m_3 is m_2; then choose x so that the mean of u, v, w, and x is m_1.
21. Mean ≈ 14.7; median = 11.5; mode = 10.1
23. Mean = 1,045.5 hr; median = 1,049.5 hr
25. Mean = 55; median = 56; mode = 56
27. Mean = 50.5 g; median = 50.55 g
29. Mean = 2,560,700; median = 1,326,500; no mode **31.** Median = 577

Exercises 10.3

1. 26 **3.** 2 **5.** 32 **7.** 500 **9.** (A) $\bar{x} = 3.3$; $s \approx 1.494$ (B) 70%; 100%; 100% (C) Yes (D)

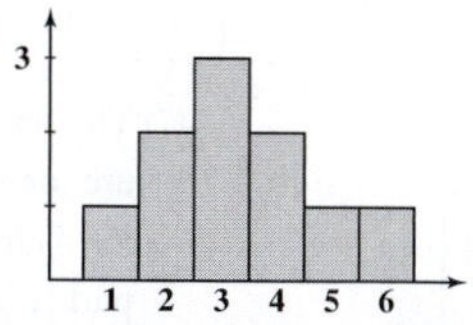

11. 2.5 **13.** True **15.** False **17.** True

19. (A) The first data set. It is more likely that the sum is close to 7, for example, than to 2 or 12. (B) Answer depends on results of simulation. **21.** $\bar{x} \approx \$4.35$; $s \approx \$2.45$ **23.** $\bar{x} \approx 8.7$ hr; $s \approx 0.6$ hr **25.** $\bar{x} \approx 5.1$ min; $s \approx 0.9$ min **27.** $\bar{x} \approx 11.1$; $s \approx 2.3$

Exercises 10.4

1. $\frac{5}{32} \approx .156$ **3.** .276 **5.** $\frac{32}{81} \approx .395$ **7.** $\frac{1}{16}$ **9.** $\frac{5}{16}$ **11.** $\frac{1}{16}$

13. $\mu = .75$; $\sigma = .75$

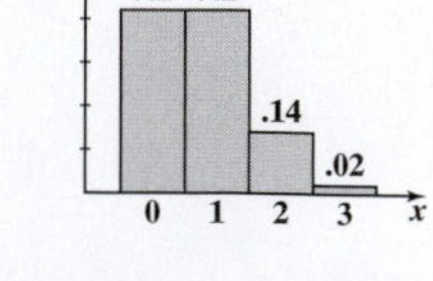

15. $\mu = 1.333$; $\sigma = .943$

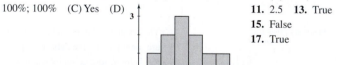

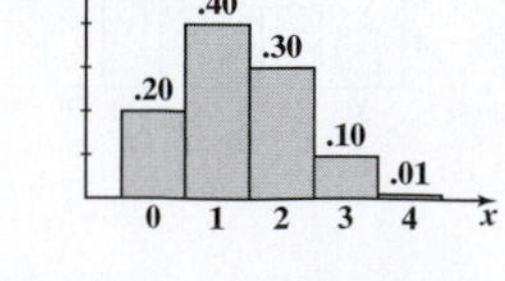

17. $\mu = 0; \sigma = 0$

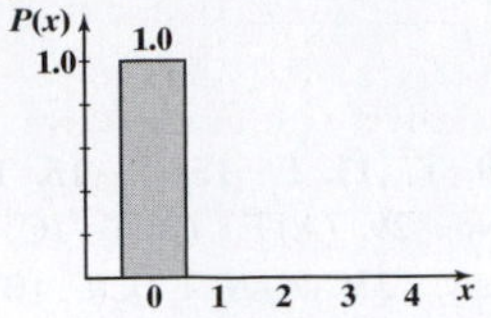

19. .0008 **21.** .1157 **23.** .4823 **25.** (A) .311 (B) .437 **27.** It is more likely that all answers are wrong (.107) than that at least half are right (.033).

29. $\mu = 2.4; \sigma = 1.2$

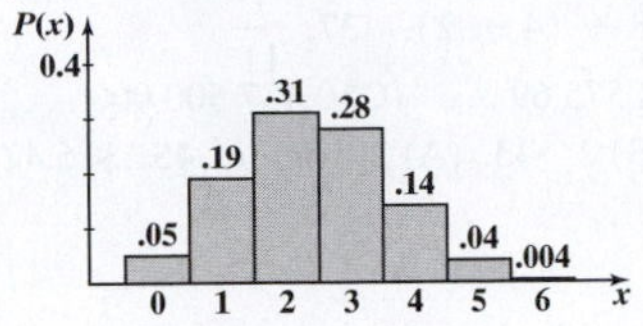

31. $\mu = 2.4; \sigma = 1.3$

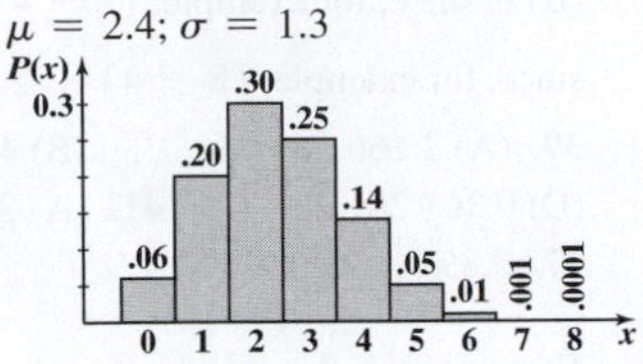

33. (A) $\mu = 17; \sigma = 1.597$ (B) .654 **35.** .238

37. The theoretical probability distribution is given by $P(x) = {}_3C_x(.5)^x(.5)^{3-x} = {}_3C_x(.5)^3$

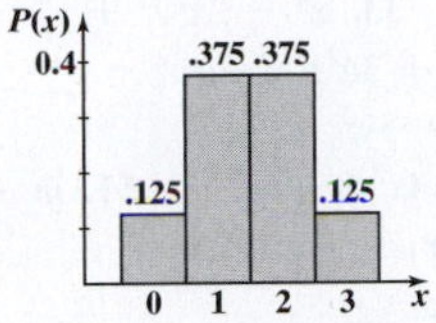

Frequency of Heads in 100 Tosses of Three Coins

Number of Heads	Theoretical Frequency	Actual Frequency
0	12.5	(List your experimental results here)
1	37.5	
2	37.5	
3	12.5	

39. $p = .5$ **41.** (A) $\mu = 5; \sigma = 1.581$ (B) Answer depends on results of simulation. **43.** (A) .318 (B) .647 **45.** .0188

47. (A) $P(x) = {}_6C_x(.05)^x(.95)^{6-x}$

(B)

x	$P(x)$
0	.735
1	.232
2	.031
3	.002
4	.000
5	.000
6	.000

(C)

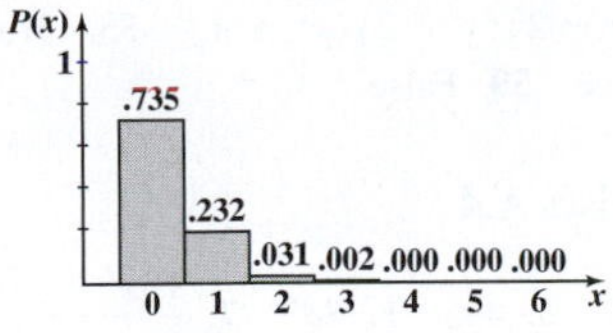

(D) $\mu = .30; \sigma = .53$ **49.** .998

51. (A) .001 (B) .264 (C) .896 **53.** (A) $P(x) = {}_6C_x(.6)^x(.4)^{6-x}$

(B)

x	$P(x)$
0	.004
1	.037
2	.138
3	.276
4	.311
5	.187
6	.047

(C)

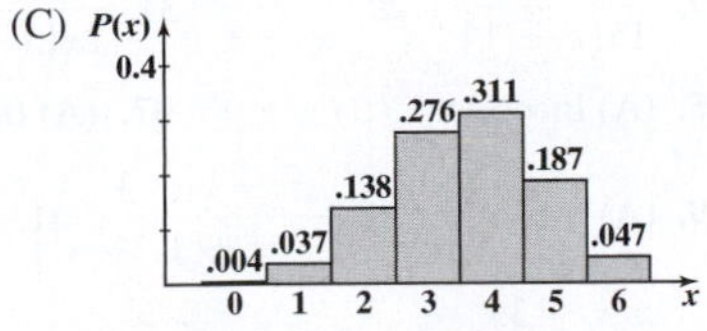

(D) $\mu = 3.6; \sigma = 1.2$ **55.** .000 864

57. (A) $P(x) = {}_5C_x(.2)^x(.8)^{5-x}$

(B)

x	$P(x)$
0	.328
1	.410
2	.205
3	.051
4	.006
5	.000

(C)

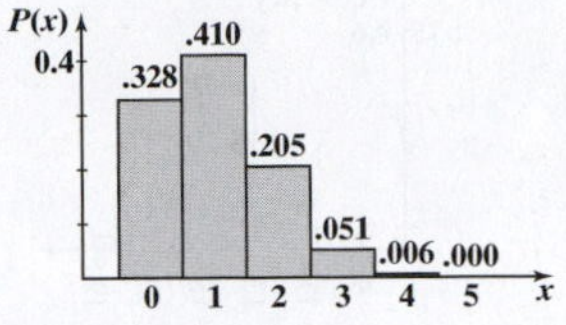

(D) $\mu = 1; \sigma = .89$ **59.** (A) .0083 (B) .0277 (C) .1861 (D) .9308

Exercises 10.5

1. .4772 **3.** .3925 **5.** .4970 **7.** .6826 **9.** .4134 **11.** .3848 **13.** .5398 **15.** .7 **17.** 1.32 **19.** 2.59 **21.** .1700 **23.** .4927 **25.** .2454 **27.** .3413 **29.** .1974 **31.** .3085 **33.** .7734 **35.** False **37.** True **39.** True **41.** No **43.** Yes **45.** No **47.** Yes **49.** Solve the inequality $np - 3\sqrt{npq} \geq 0$ to obtain $n \geq 81$. **51.** .89 **53.** .16 **55.** .01 **57.** .01

59.

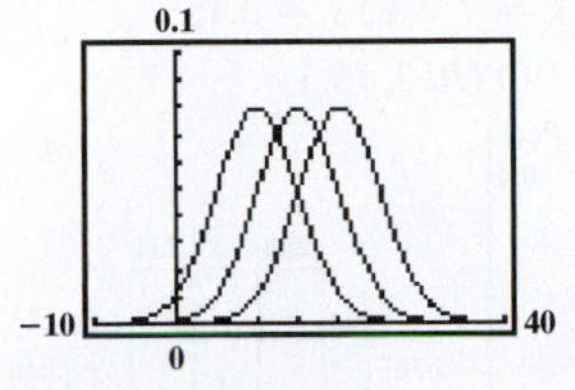

61.

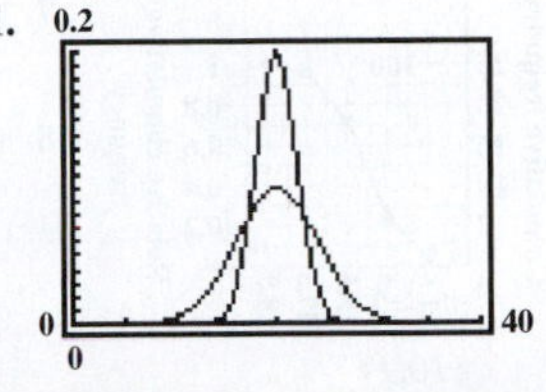

63. (A) Approx. 82 (B) Answer depends on results of simulation. **65.** 2.28% **67.** 1.24% **69.** .0031; either a rare event has happened or the company's claim is false. **71.** 0.82% **73.** .0158 **75.** 2.28% **77.** A's, 80.2 or greater; B's, 74.2–80.2; C's, 65.8–74.2; D's, 59.8–65.8; F's, 59.8 or lower

Chapter 10 Review Exercises

1. *(10.1)*

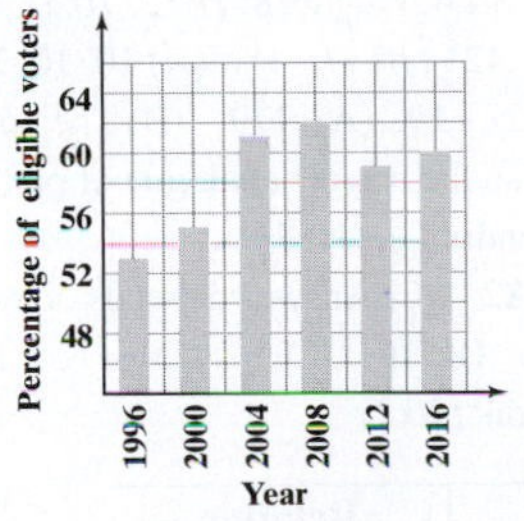

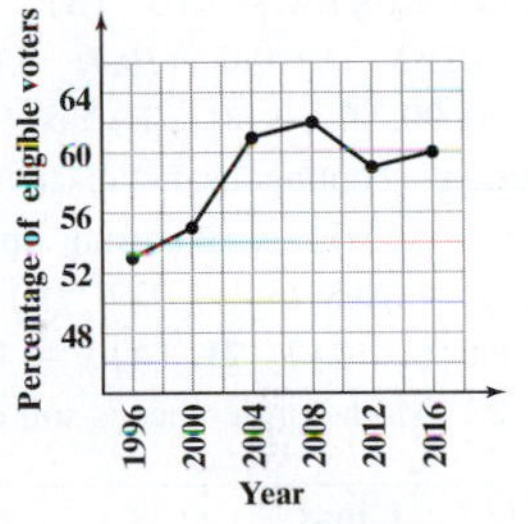

2. *(10.1)*

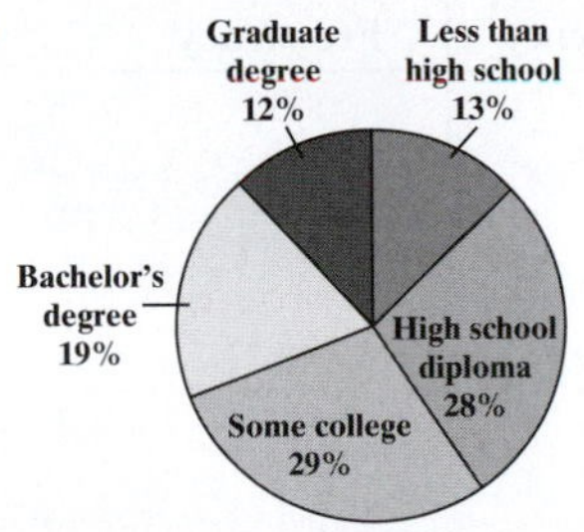

3. (A)

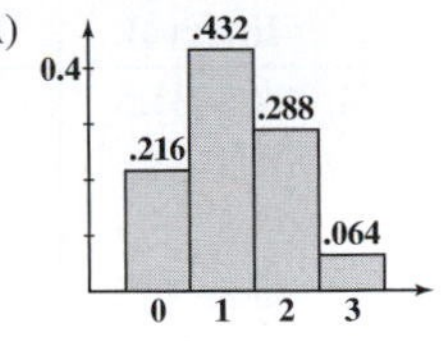

(B) $\mu = 1.2; \sigma = .85$ *(10.4)* **4.** (A) $\bar{x} = 2.7$ (B) 2.5 (C) 2 (D) $s = 1.34$ *(10.2, 10.3)* **5.** (A) 1.8 (B) .4641 *(10.5)*

6. (A)

Class Interval	Frequency	Relative Frequency
9.5–11.5	1	.04
11.5–13.5	5	.20
13.5–15.5	12	.48
15.5–17.5	6	.24
17.5–19.5	1	.04
	25	1.00

(B)

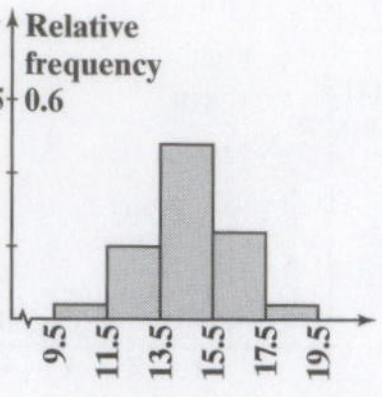

(C)

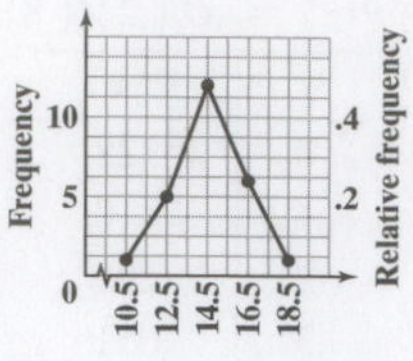

(D)

Class Interval	Frequency	Cumulative Frequency	Relative Cumulative Frequency
9.5–11.5	1	1	.04
11.5–13.5	5	6	.24
13.5–15.5	12	18	.72
15.5–17.5	6	24	.96
17.5–19.5	1	25	1.00

(E)

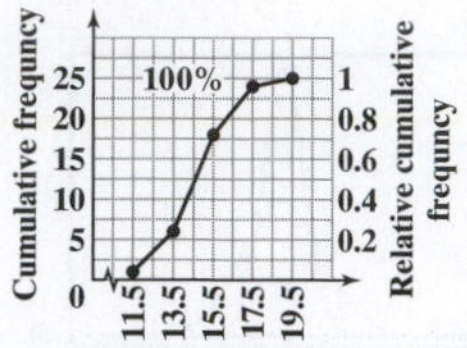

(10.1)

7. (A) $\bar{x} = 7$ (B) $s = 2.45$ (C) 7.14 *(10.2, 10.3)*

8. (A)

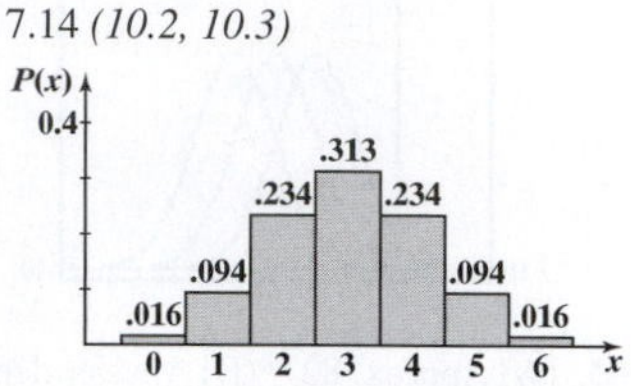

(B) $\mu = 3$; $\sigma = 1.22$ *(10.4)* **9.** $\mu = 600$; $\sigma = 15.49$ *(10.4)* **10.** (A) True (B) False *(10.2, 10.3)* **11.** (A) False (B) True (C) True *(10.4, 10.5)* **12.** .999 *(10.5)* **13.** (A) .9104 (B) .0668 *(10.5)* **14.** (A) The first data set. Sums range from 2 to 12, but products range from 1 to 36. (B) Answer depend on results of simulation. *(10.3)* **15.** (A) $\bar{x} = 14.6$; $s = 1.83$ (B) $\bar{x} = 14.6$; $s = 1.78$; *(10.2, 10.3)* **16.** (A) .0322 (B) .0355 *(10.4)* **17.** .421 *(10.4)* **18.** (A) 10, 10, 20, 20, 90, 90, 90, 90, 90, 90 (B) No *(10.2)* **19.** (A) .179 (B) The normal distribution is continuous, not discrete, so the correct analogue of part (A) is $P(7.5 \le x \le 8.5) \approx .18$ (using Appendix C). *(10.5)* **20.** (A) $\mu = 10.8$; $\sigma = 1.039$ (B) .282 (C) Answer depends on results of simulation. *(10.4)* **21.** (A) $\bar{x} = 10$ (B) 10 (C) 5 (D) $s = 5.14$ *(10.2, 10.3)* **22.** Modal preference is soft drink *(10.2)*

23. (A)

Class Interval	Frequency	Relative Frequency
29.5–31.5	3	.086
31.5–33.5	7	.2
33.5–35.5	14	.4
35.5–37.5	7	.2
37.5–39.5	4	.114
	35	1.00

(B)

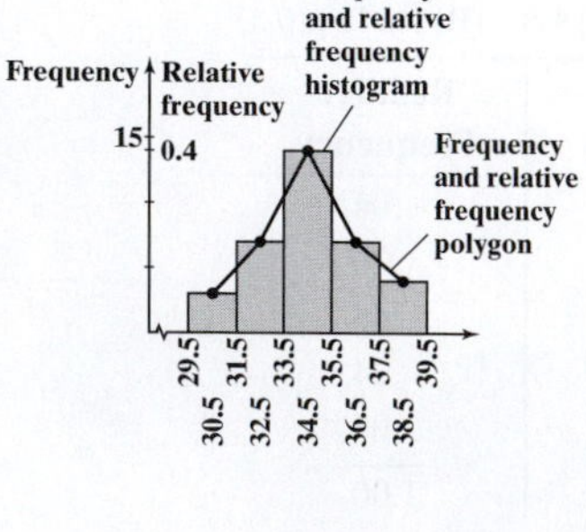

(C) $\bar{x} = 34.61$; $s = 2.22$ *(10.1, 10.2, 10.3)*

24. (A) 57.62% (B) 6.68% *(10.5)*

25. (A) $\mu = 140$; $\sigma = 6.48$ (B) Yes (C) .939 (D) .0125 (E)

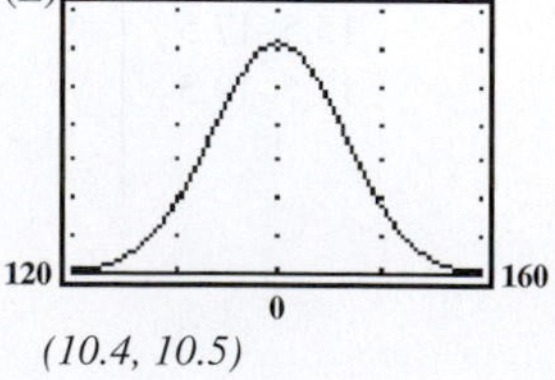

(10.4, 10.5)

26. .999 *(10.4)*

Appendix A

Exercises A.1

1. vu **3.** $(3 + 7) + y$ **5.** $u + v$ **7.** T **9.** T **11.** F **13.** T **15.** T **17.** T **19.** T **21.** F **23.** T **25.** T **27.** No **29.** (A) F (B) T (C) T **31.** $\sqrt{2}$ and π are two examples of infinitely many. **33.** (A) N, Z, Q, R (B) R (C) Q, R (D) Q, R **35.** (A) F, since, for example, $2(3 - 1) \ne 2 \cdot 3 - 1$ (B) F, since, for example, $(8 - 4) - 2 \ne 8 - (4 - 2)$ (C) T (D) F, since, for example, $(8 \div 4) \div 2 \ne 8 \div (4 \div 2)$. **37.** $\frac{1}{11}$ **39.** (A) 2.166 666 666 . . . (B) 4.582 575 69 . . . (C) 0.437 500 000 . . . (D) 0.261 261 261 . . . **41.** (A) 3 (B) 2 **43.** (A) 2 (B) 6 **45.** $16.42 **47.** 2.8%

Exercises A.2

1. 3 **3.** $x^3 + 4x^2 - 2x + 5$ **5.** $x^3 + 1$ **7.** $2x^5 + 3x^4 - 2x^3 + 11x^2 - 5x + 6$ **9.** $-5u + 2$ **11.** $6a^2 + 6a$ **13.** $a^2 - b^2$ **15.** $6x^2 - 7x - 5$ **17.** $2x^2 + xy - 6y^2$ **19.** $9y^2 - 4$ **21.** $-4x^2 + 12x - 9$ **23.** $16m^2 - 9n^2$ **25.** $9u^2 + 24uv + 16v^2$ **27.** $a^3 - b^3$ **29.** $x^2 - 2xy + y^2 - 9z^2$ **31.** 1 **33.** $x^4 - 2x^2y^2 + y^4$ **35.** $-40ab$ **37.** $-4m + 8$ **39.** $-6xy$ **41.** $u^3 + 3u^2v + 3uv^2 + v^3$ **43.** $x^3 - 6x^2y + 12xy^2 - 8y^3$ **45.** $2x^2 - 2xy + 3y^2$ **47.** $x^4 - 10x^3 + 27x^2 - 10x + 1$ **49.** $4x^3 - 14x^2 + 8x - 6$ **51.** $m + n$ **53.** No change **55.** $(1 + 1)^2 \ne 1^2 + 1^2$; either a or b must be 0 **57.** $0.09x + 0.12(10{,}000 - x) = 1{,}200 - 0.03x$ **59.** $20x + 30(3x) + 50(4{,}000 - x - 3x) = 200{,}000 - 90x$ **61.** $0.02x + 0.06(10 - x) = 0.6 - 0.04x$

Exercises A.3

1. $3m^2(2m^2 - 3m - 1)$ **3.** $2uv(4u^2 - 3uv + 2v^2)$ **5.** $(7m + 5)(2m - 3)$ **7.** $(4ab - 1)(2c + d)$ **9.** $(2x - 1)(x + 2)$ **11.** $(y - 1)(3y + 2)$ **13.** $(x + 4)(2x - 1)$ **15.** $(w + x)(y - z)$ **17.** $(a - 3b)(m + 2n)$ **19.** $(3y + 2)(y - 1)$ **21.** $(u - 5v)(u + 3v)$ **23.** Not factorable **25.** $(wx - y)(wx + y)$ **27.** $(3m - n)^2$ **29.** Not factorable **31.** $4(z - 3)(z - 4)$ **33.** $2x^2(x - 2)(x - 10)$ **35.** $x(2y - 3)^2$ **37.** $(2m - 3n)(3m + 4n)$ **39.** $uv(2u - v)(2u + v)$ **41.** $2x(x^2 - x + 4)$ **43.** $(2x - 3y)(4x^2 + 6xy + 9y^2)$ **45.** $xy(x + 2)(x^2 - 2x + 4)$ **47.** $[(x + 2) - 3y][(x + 2) + 3y]$ **49.** Not factorable **51.** $(6x - 6y - 1)(x - y + 4)$ **53.** $(y - 2)(y + 2)(y^2 + 1)$ **55.** $3(x - y)^2(5xy - 5y^2 + 4x)$ **57.** True **59.** False

Exercises A.4

1. 39/7 **3.** 495 **5.** $8d^6$ **7.** $\frac{15x^2 + 10x - 6}{180}$ **9.** $\frac{15m^2 + 14m - 6}{36m^3}$ **11.** $\frac{1}{x(x - 4)}$ **13.** $\frac{x - 6}{x(x - 3)}$ **15.** $\frac{-3x - 9}{(x - 2)(x + 1)^2}$ **17.** $\frac{2}{x - 1}$ **19.** $\frac{5}{a - 1}$ **21.** $\frac{x^2 + 8x - 16}{x(x - 4)(x + 4)}$ **23.** $\frac{7x^2 - 2x - 3}{6(x + 1)^2}$ **25.** $\frac{x(y - x)}{y(2x - y)}$ **27.** $\frac{-17c + 16}{15(c - 1)}$ **29.** $\frac{1}{x - 3}$ **31.** $\frac{-1}{2x(x + h)}$ **33.** $\frac{x - y}{x + y}$ **35.** (A) Incorrect (B) $x + 1$ **37.** (A) Incorrect (B) $2x + h$ **39.** (A) Incorrect (B) $\frac{x^2 - x - 3}{x + 1}$ **41.** (A) Correct **43.** $\frac{-2x - h}{3(x + h)^2x^2}$ **45.** $\frac{x(x - 3)}{x - 1}$

Exercises A.5

1. $2/x^9$ **3.** $3w^7/2$ **5.** $2/x^3$ **7.** $1/w^5$ **9.** $4/a^6$ **11.** $1/a^6$ **13.** $1/8x^{12}$ **15.** 8.23×10^{10} **17.** 7.83×10^{-1} **19.** 3.4×10^{-5} **21.** 40,000 **23.** 0.007 **25.** 61,710,000 **27.** 0.000 808 **29.** 1 **31.** 10^{14} **33.** $y^6/25x^4$

35. $4x^6/25$ **37.** $4y^3/3x^5$ **39.** $\dfrac{7}{4} - \dfrac{1}{4}x^{-3}$ **41.** $\dfrac{5}{2}x^2 - \dfrac{3}{2} + 4x^{-2}$

43. $\dfrac{x^2(x-3)}{(x-1)^3}$ **45.** $\dfrac{2(x-1)}{x^3}$ **47.** 2.4×10^{10}; 24,000,000,000

49. 3.125×10^4; 31,250 **51.** 64 **55.** uv **57.** $\dfrac{bc(c+b)}{c^2 + bc + b^2}$

59. (A) \$60,598 (B) \$1,341 (C) 2.21% **61.** (A) 9×10^{-6} (B) 0.000 009
(C) 0.0009% **63.** 1,194,000

Exercises A.6

1. $6\sqrt[5]{x^3}$ **3.** $\sqrt[5]{(32x^2y^3)^3}$ **5.** $\sqrt{x^2+y^2}$ (not $x+y$) **7.** $5x^{3/4}$

9. $(2x^2y)^{3/5}$ **11.** $x^{1/3} + y^{1/3}$ **13.** 5 **15.** 64 **17.** -7 **19.** -16

21. $\dfrac{8}{125}$ **23.** $\dfrac{1}{27}$ **25.** $x^{2/5}$ **27.** m **29.** $2x/y^2$ **31.** $xy^2/2$

33. $1/(24x^{7/12})$ **35.** $2x+3$ **37.** $30x^5\sqrt{3x}$ **39.** 2 **41.** $12x - 6x^{35/4}$

43. $3u - 13u^{1/2}v^{1/2} + 4v$ **45.** $36m^{3/2} - \dfrac{6m^{1/2}}{n^{1/2}} + \dfrac{6m}{n^{1/2}} - \dfrac{1}{n}$

47. $9x - 6x^{1/2}y^{1/2} + y$ **49.** $\dfrac{1}{2}x^{1/3} + x^{-1/3}$ **51.** $\dfrac{2}{3}x^{-1/4} + \dfrac{1}{3}x^{-2/3}$

53. $\dfrac{1}{2}x^{-1/6} - \dfrac{1}{4}$ **55.** $4n\sqrt{3mn}$ **57.** $\dfrac{2(x+3)\sqrt{x-2}}{x-2}$

59. $7(x-y)(\sqrt{x} + \sqrt{y})$ **61.** $\dfrac{1}{xy\sqrt{5xy}}$ **63.** $\dfrac{1}{\sqrt{x+h} + \sqrt{x}}$

65. $\dfrac{1}{(t+x)(\sqrt{t} + \sqrt{x})}$ **67.** $x = y = 1$ is one of many choices.

69. $x = y = 1$ is one of many choices. **71.** False **73.** False **75.** False

77. True **79.** True **81.** False **83.** $\dfrac{x+8}{2(x+3)^{3/2}}$ **85.** $\dfrac{x-2}{2(x-1)^{3/2}}$

87. $\dfrac{x+6}{3(x+2)^{5/3}}$ **89.** 103.2 **91.** 0.0805 **93.** 4,588

95. (A) and (E); (B) and (F); (C) and (D)

Exercises A.7

1. $\pm\sqrt{11}$ **3.** $-\dfrac{4}{3}, 2$ **5.** $-2, 6$ **7.** $0, 2$ **9.** $3 \pm 2\sqrt{3}$ **11.** $-2 \pm \sqrt{2}$

13. $0, \dfrac{15}{2}$ **15.** $\pm\dfrac{3}{2}$ **17.** $\dfrac{1}{2}, -3$ **19.** $(-1 \pm \sqrt{5})/2$ **21.** $(3 \pm \sqrt{3})/2$

23. No real solution **25.** $(-3 \pm \sqrt{11})/2$ **27.** $\pm\sqrt{3}$ **29.** $-\dfrac{1}{2}, 2$

31. $(x-2)(x+42)$ **33.** Not factorable in the integers

35. $(2x-9)(x+12)$ **37.** $(4x-7)(x+62)$ **39.** $r = \sqrt{A/P} - 1$

41. If $c < 4$, there are two distinct real roots; if $c = 4$, there is one real double
root; and if $c > 4$, there are no real roots. **43.** -2 **45.** $\pm\sqrt{10}$

47. $\pm\sqrt{3}, \pm\sqrt{5}$ **49.** 1,575 bottles at \$4 each **51.** 13.64%

53. 8 ft/sec; $4\sqrt{2}$ or 5.66 ft/sec

INDEX

NOTE: Page numbers preceded by G refer to online Chapter 11: Games and Decisions (`goo.gl/6VBjkQ`). *Page numbers preceded by A refer to online Appendix B: Special Topics* (`goo.gl/mjbXrG`).

D

E

G

H

I

INDEX OF APPLICATIONS

NOTE: Page numbers preceded by G refer to online Chapter 11: Games and Decisions (goo.gl/6VBjkQ). *Page numbers preceded by A refer to online Appendix B: Special Topics* (goo.gl/mjbXrG).

Business and Economics

Life Sciences

Social Sciences

A Library of Elementary Functions

Basic Functions

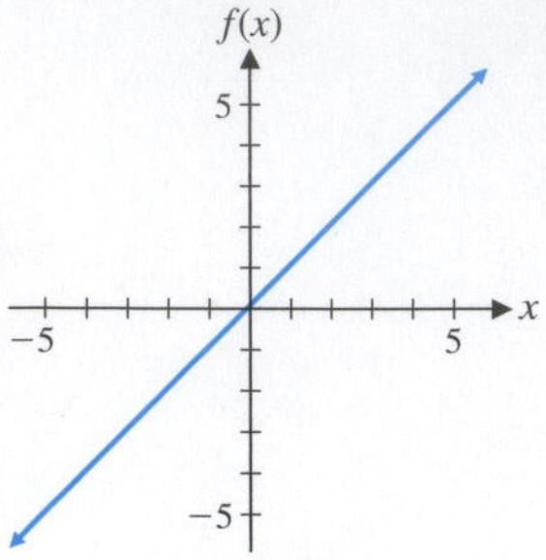

Identity function
$f(x) = x$

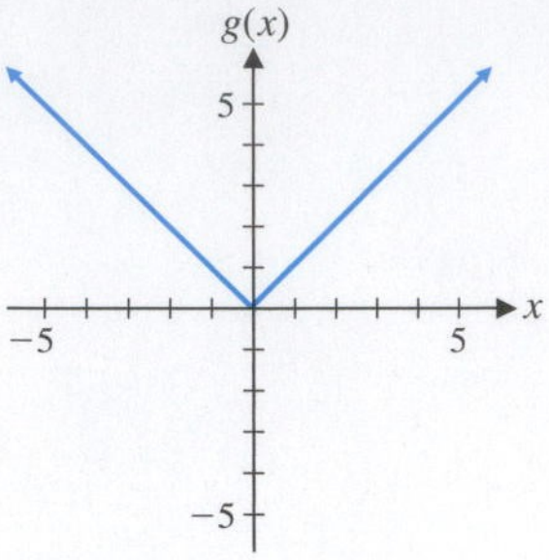

Absolute value function
$g(x) = |x|$

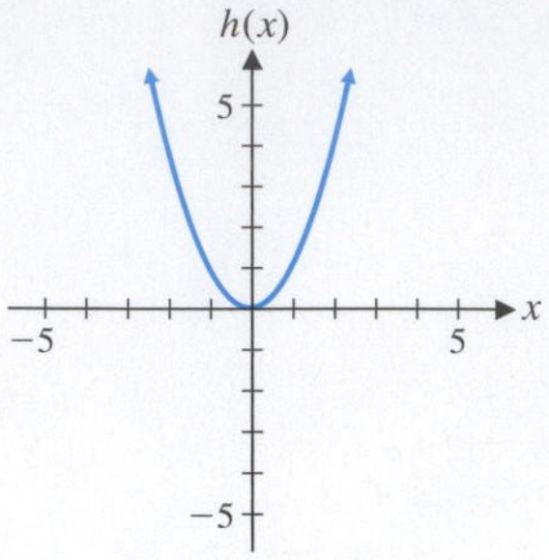

Square function
$h(x) = x^2$

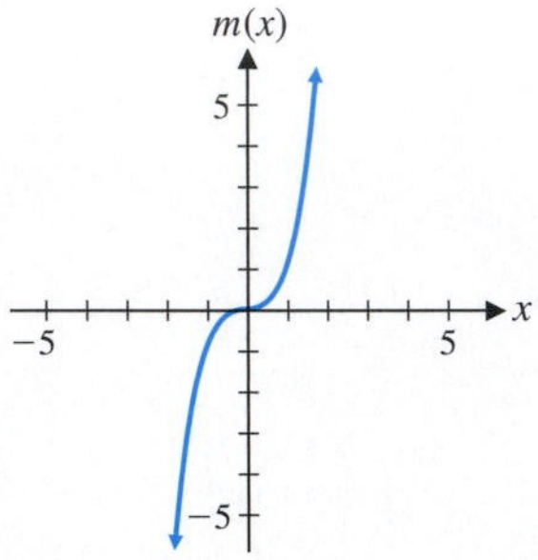

Cube function
$m(x) = x^3$

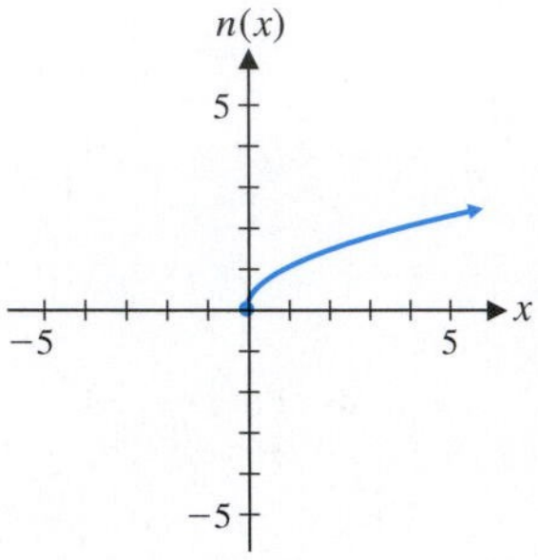

Square root function
$n(x) = \sqrt{x}$

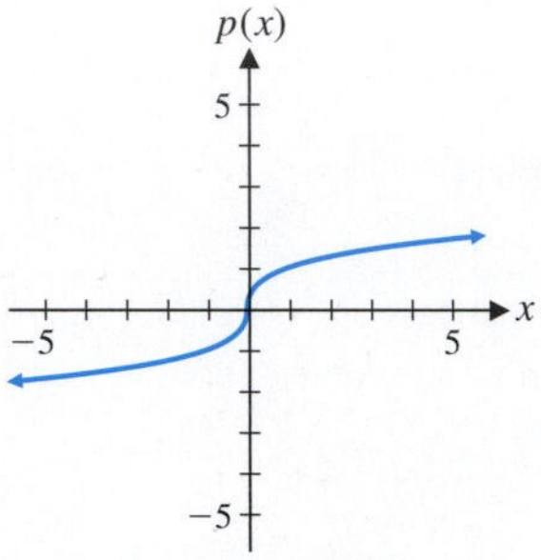

Cube root function
$p(x) = \sqrt[3]{x}$

Linear and Constant Functions

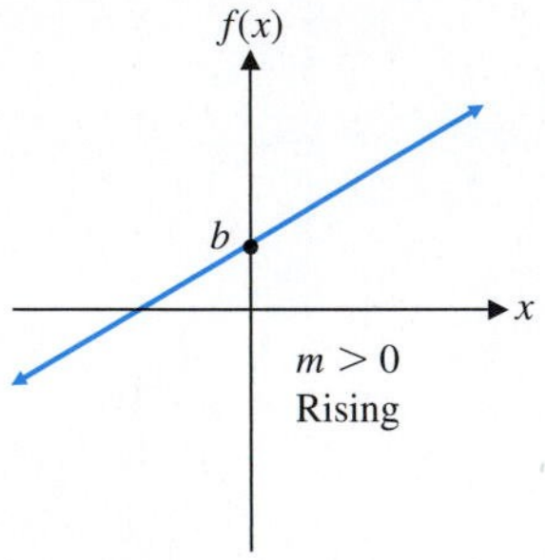

Linear function
$f(x) = mx + b$

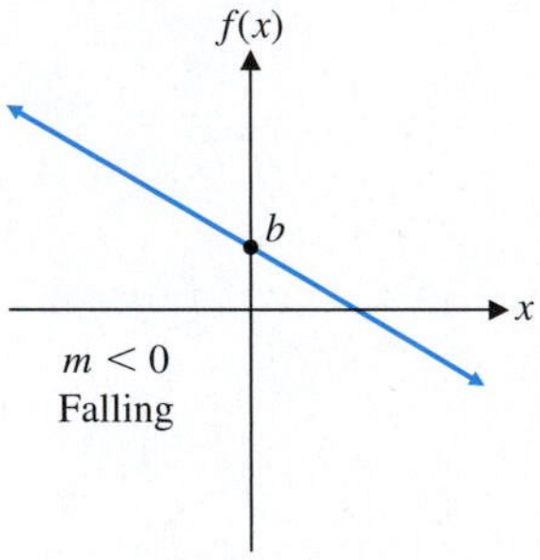

Linear function
$f(x) = mx + b$

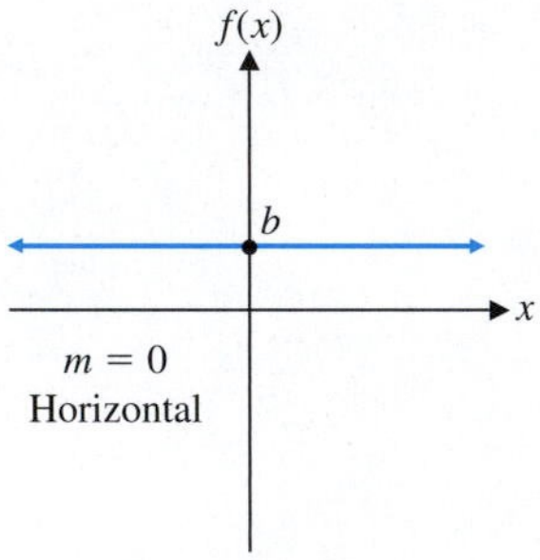

Constant function
$f(x) = b$

Quadratic Functions

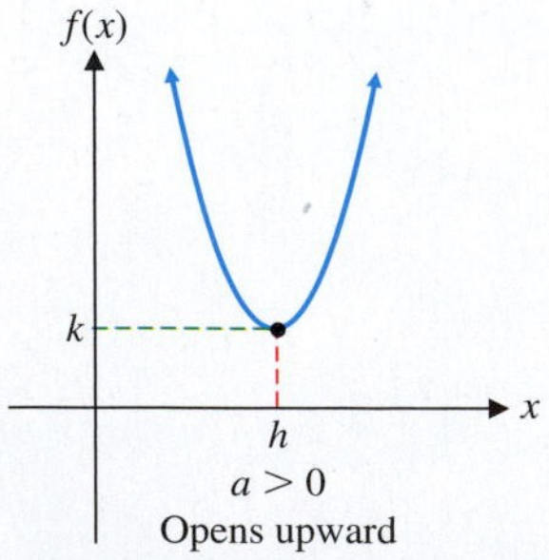

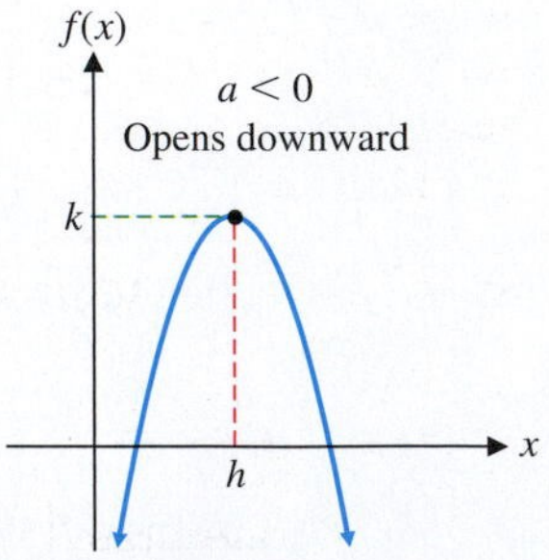

$f(x) = ax^2 + bx + c = a(x - h)^2 + k$

Exponential and Logarithmic Functions

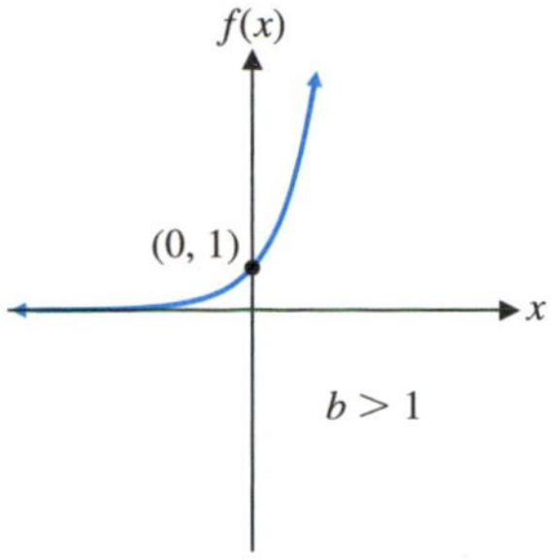

Exponential function
$f(x) = b^x$

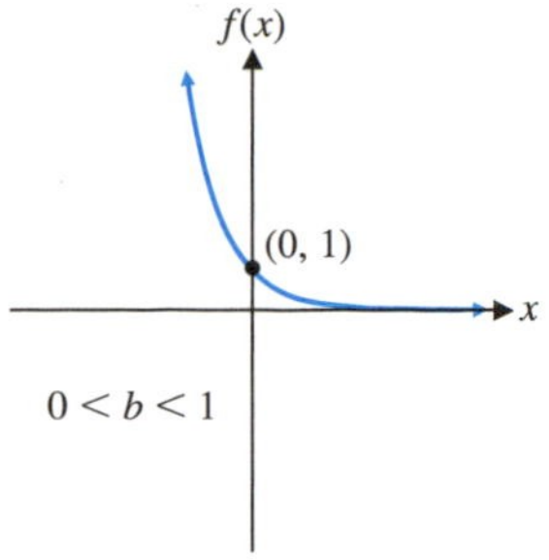

Exponential function
$f(x) = b^x$

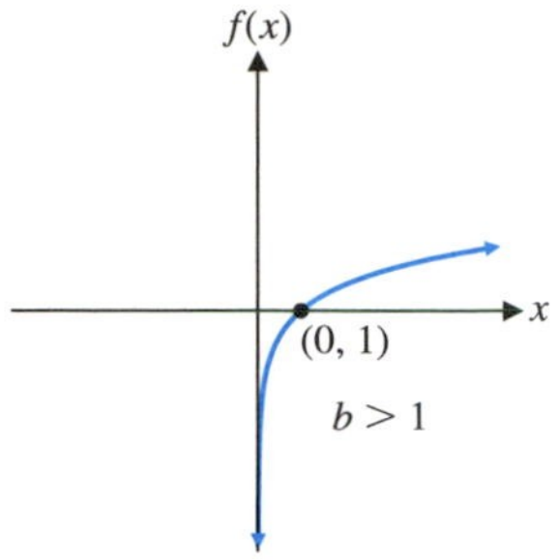

Logarithmic function
$f(x) = \log_b x$

Representative Polynomial Functions (degree > 2)

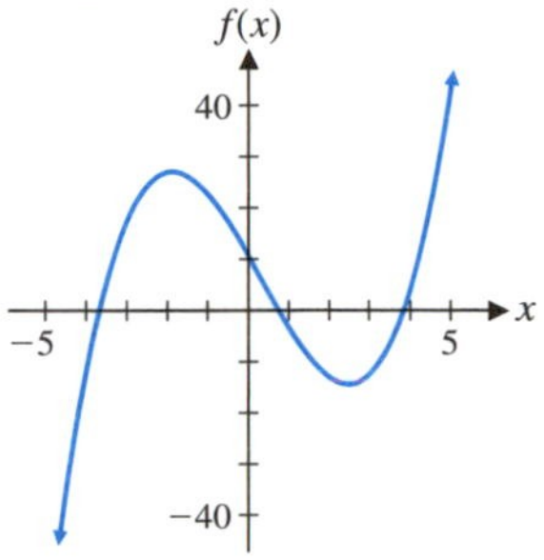

Third-degree polynomial
$f(x) = x^3 - x^2 - 14x + 11$

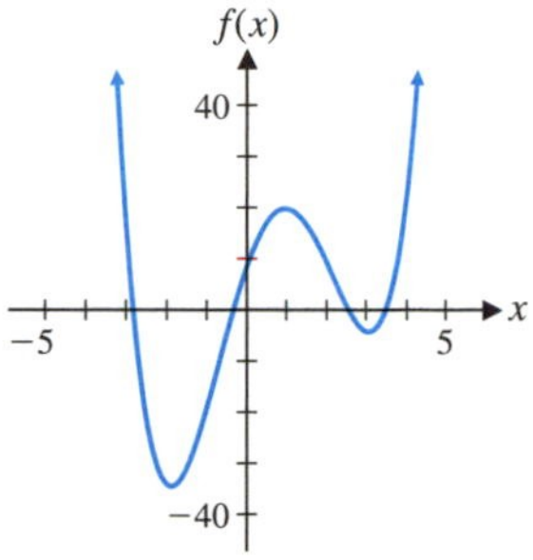

Fourth-degree polynomial
$f(x) = x^4 - 3x^3 - 9x^2 + 23x + 8$

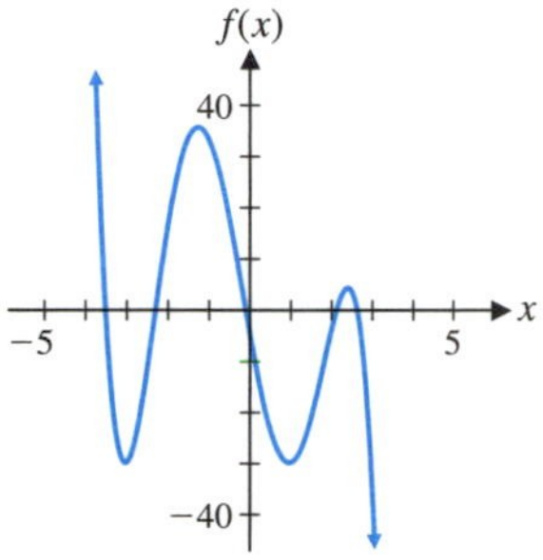

Fifth-degree polynomial
$f(x) = -x^5 - x^4 + 14x^3 + 6x^2 - 45x - 3$

Representative Rational Functions

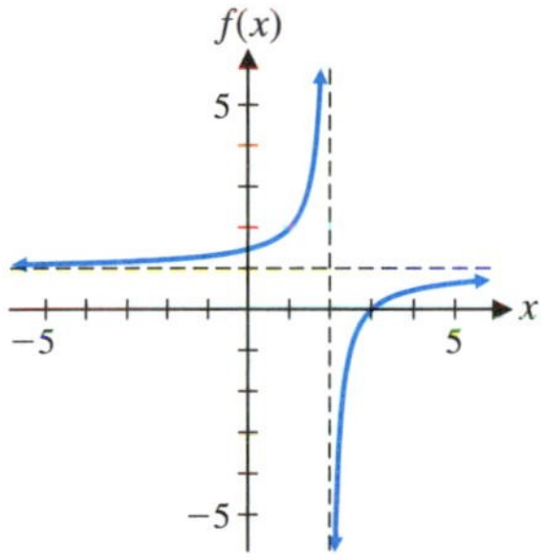

$f(x) = \dfrac{x - 3}{x - 2}$

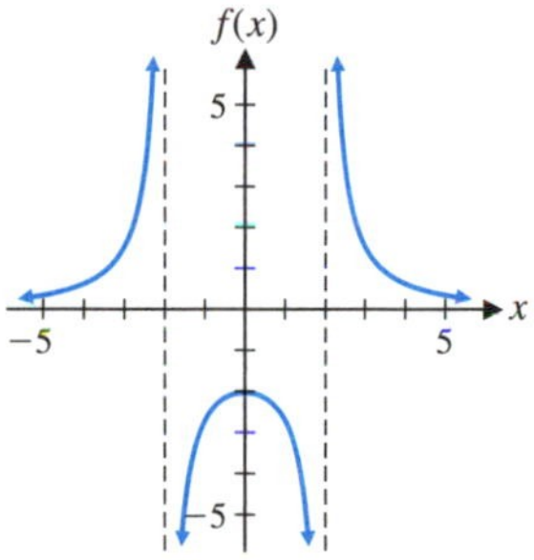

$f(x) = \dfrac{8}{x^2 - 4}$

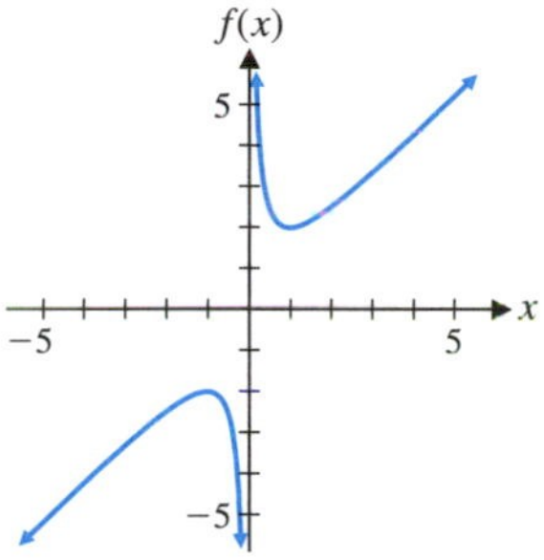

$f(x) = x + \dfrac{1}{x}$

Graph Transformations

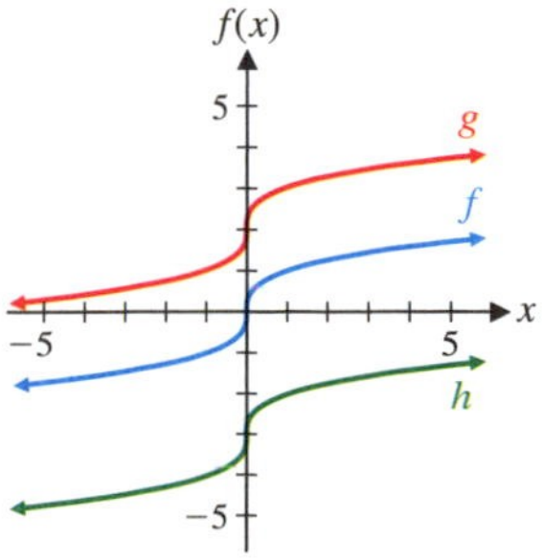

Vertical shift
$g(x) = f(x) + 2$
$h(x) = f(x) - 3$

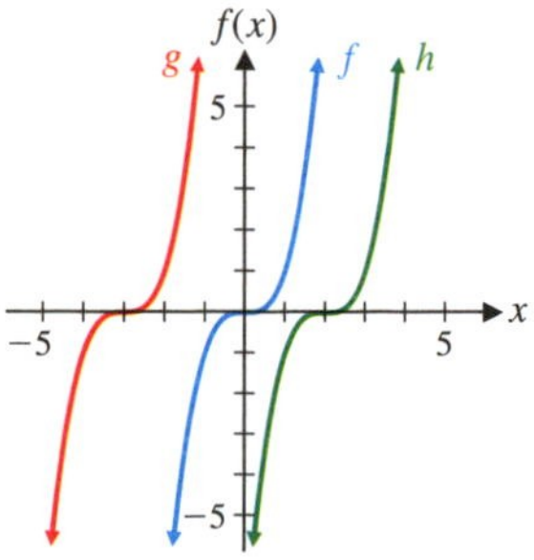

Horizontal shift
$g(x) = f(x + 3)$
$h(x) = f(x - 2)$

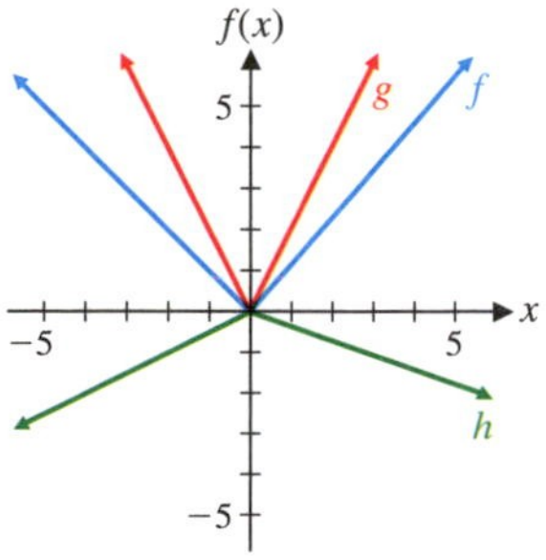

Stretch, shrink and reflection
$g(x) = 2f(x)$
$h(x) = -0.5f(x)$

Finite Math Reference

Mathematics of Finance

Simple Interest

$I = Prt \qquad A = P(1 + rt)$

(interest I on principal P at annual rate r for t years; amount $A = P + I$)

Compound Interest

$$A = P(1 + i)^n = P\left(1 + \frac{r}{m}\right)^{mt}$$

(amount A on principal P at annual rate r, compounded m times per year for t years; interest rate per period is $i = \frac{r}{m}$; number of periods is $n = mt$)

Continuous Compound Interest

$A = Pe^{rt}$ (amount A on principal P at annual rate r, compounded continuously for t years)

Annual Percentage Yield

$APY = \left(1 + \frac{r}{m}\right)^m - 1$ (APY, or effective rate, of annual rate r compounded m times per year)

$APY = e^r - 1$ (APY, or effective rate, of annual rate r compounded continuously)

Future Value of an Ordinary Annuity

$$FV = PMT\frac{(1 + i)^n - 1}{i}$$

(future value FV of n equal periodic payments PMT, made at the end of each period, at rate i per period)

Present Value of an Ordinary Annuity

$$PV = PMT\frac{1 - (1 + i)^{-n}}{i}$$

(present value PV of n equal periodic payments PMT, made at the end of each period, at rate i per period)

Logic

p	q	Negation (not p) $\neg p$	Disjunction (p or q) $p \vee q$	Conjunction (p and q) $p \wedge q$	Conditional (if p then q) $p \rightarrow q$
T	T	F	T	T	T
T	F	F	T	F	F
F	T	T	T	F	T
F	F	T	F	F	T

$q \rightarrow p$ is the converse of $p \rightarrow q$

$\neg q \rightarrow \neg p$ is the contrapositive of $p \rightarrow q$